Precalculus Demystified

Demystified Series

Advanced Statistics Demystified
Algebra Demystified
Anatomy Demystified
Astronomy Demystified
Biology Demystified
Business Statistics Demystified
Calculus Demystified
Chemistry Demystified
College Algebra Demystified
Differential Equations Demystified
Digital Electronics Demystified
Earth Science Demystified
Electronics Demystified
Everyday Math Demystified
Geometry Demystified
Math Word Problems Demystified
Microbiology Demystified
Physics Demystified
Physiology Demystified
Pre-Algebra Demystified
Precalculus Demystified
Project Management Demystified
Robotics Demystified
Statistics Demystified
Trigonometry Demystified

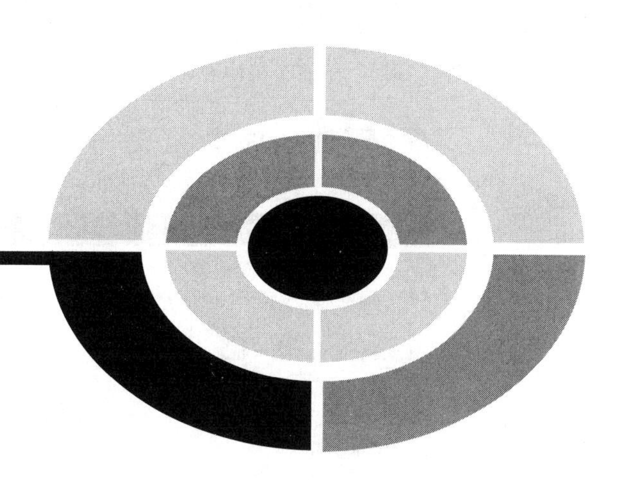

Precalculus Demystified

RHONDA HUETTENMUELLER

McGRAW-HILL

New York Chicago San Francisco Lisbon London
Madrid Mexico City Milan New Delhi San Juan
Seoul Singapore Sydney Toronto

For Ed Landesman

The McGraw·Hill Companies

Library of Congress Cataloging-in-Publication Data

Huettenmueller, Rhonda.
Precalculus demystified / Rhonda Huettenmueller.
p. cm.
Includes index.
ISBN 0-07-143927-7
1. Calculus. I. Title.

QA303.2.H84 2005
512—dc22

2004065569

7 8 9 0 DOC/DOC 0 1 0 9 8 7

ISBN 0-07-143927-7

The sponsoring editor for this book was Judy Bass and the production supervisor was Pamela A. Pelton. It was set in Times Roman by Keyword Publishing Services Ltd. The art director for the cover was Margaret Webster-Shapiro. Cover design by Handel Low.

Printed and bound by RR Donnelley.

CONTENTS

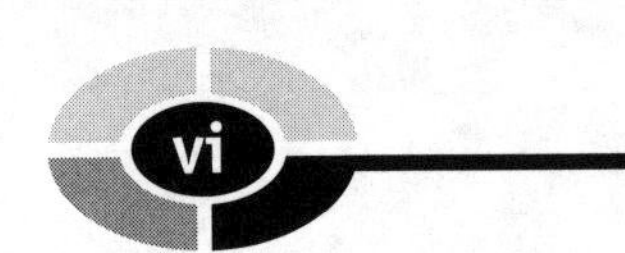

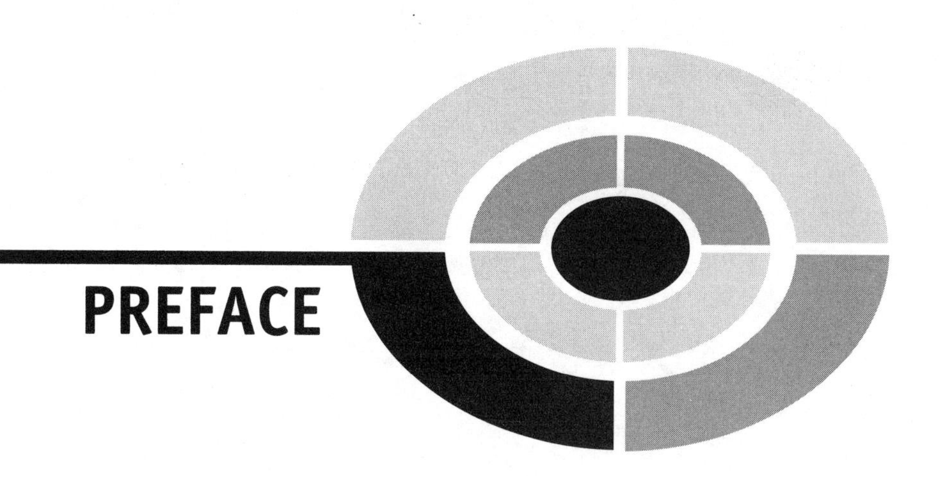

PREFACE

The goal of this book is to give you the skills and knowledge necessary to succeed in calculus. Much of the difficulty calculus students face is with algebra. They have to solve equations, find equations of lines, study graphs, solve word problems, and rewrite expressions—all of these require a solid background in algebra. You will get experience with all this and more in this book. Not only will you learn about the basic functions in this book, you also will strengthen your algebra skills because all of the examples and most of the solutions are given with a lot of detail. Enough steps are given in the problems to make the reasoning easy to follow.

The basic functions covered in this book are linear, polynomial, and rational functions, as well as exponential, logarithmic, and trigonometric functions. Because understanding the slope of a line is crucial to making sense of calculus, the interpretation of a line's slope is given extra attention. Other calculus topics introduced in this book are Newton's Quotient, the average rate of change, increasing/decreasing intervals of a function, and optimizing functions. Your experience with these ideas will help you when you learn calculus.

Concepts are presented in clear, simple language, followed by detailed examples. To make sure you understand the material, each section ends with a set of practice problems. Each chapter ends with a multiple-choice test, and there is a final exam at the end of the book. You will get the most from this book if you work steadily from the beginning to the end. Because much of the material is sequential, you should review the ideas in the previous section. Study for each end-of-chapter test as if it really were a test, and take it without looking at examples and without using notes. This will let you know what you have learned and where, if anywhere, you need to spend more time.

Good luck.

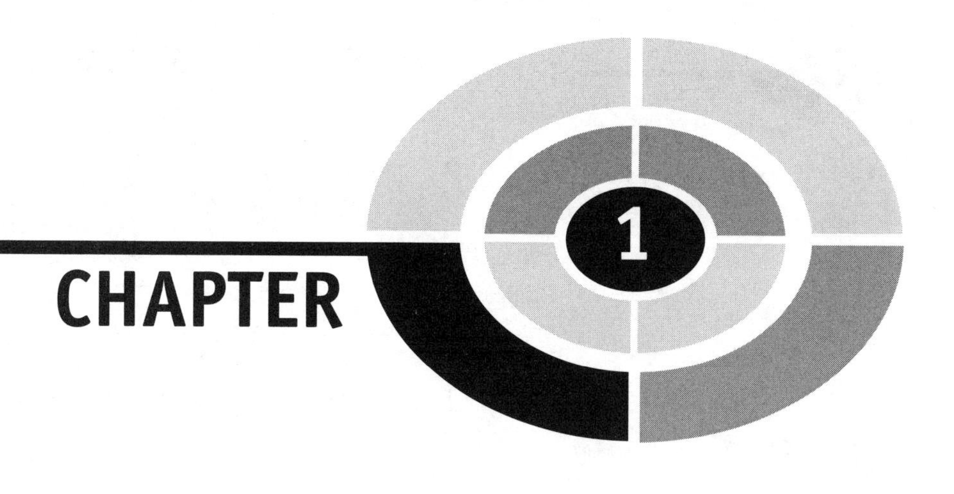

The Slope and Equation of a Line

The slope of a line and the meaning of the slope are important in calculus. In fact, the slope formula is the basis for differential calculus. The slope of a line measures its tilt. The sign of the slope tells us if the line tilts up (if the slope is positive) or tilts down (if the slope is negative). The larger the number, the steeper the slope.

We can put any two points on the line, (x_1, y_1) and (x_2, y_2), in the slope formula to find the slope of the line.

$$m = \frac{y_2 - y_1}{x_2 - x_1}$$

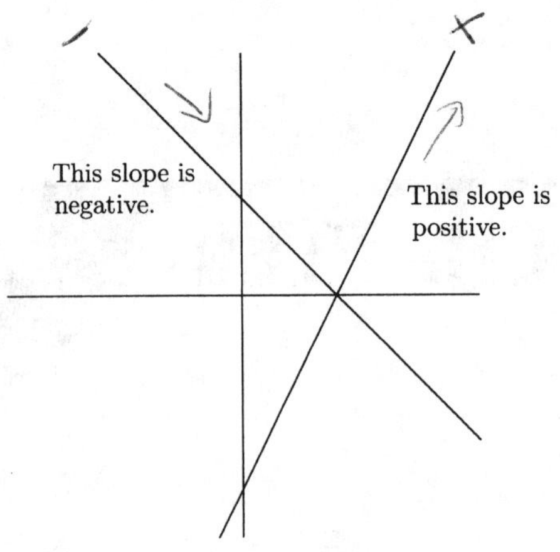

Fig. 1.1.

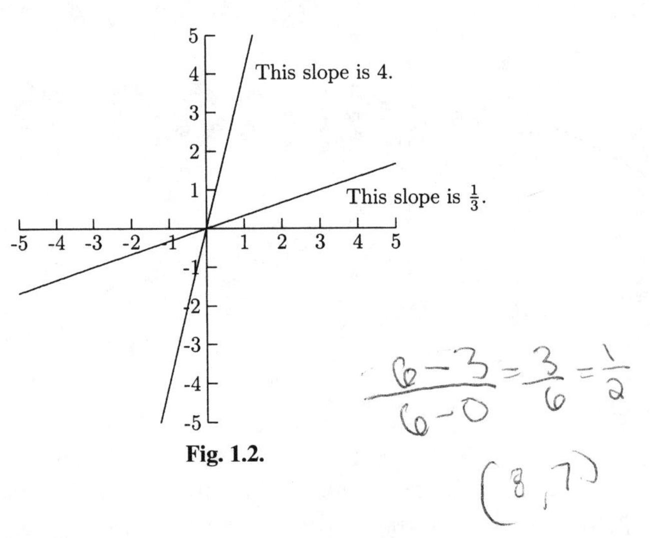

Fig. 1.2.

For example, (0, 3), (−2, 2), (6, 6), and $(-1, \frac{5}{2})$ are all points on the same line. We can pick any pair of points to compute the slope.

$$m = \frac{2-3}{-2-0} = \frac{-1}{-2} = \frac{1}{2} \qquad m = \frac{\frac{5}{2}-2}{-1-(-2)} = \frac{\frac{1}{2}}{1} = \frac{1}{2}$$

$$m = \frac{3-6}{0-6} = \frac{-3}{-6} = \frac{1}{2}$$

A slope of $\frac{1}{2}$ means that if we increase the x-value by 2, then we need to increase the y-value by 1 to get another point on the line. For example, knowing that (0, 3) is on the line means that we know $(0+2, 3+1) = (2, 4)$ is also on the line.

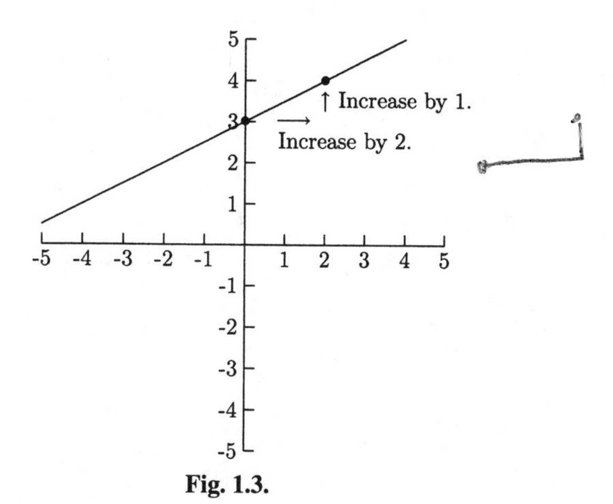

Fig. 1.3.

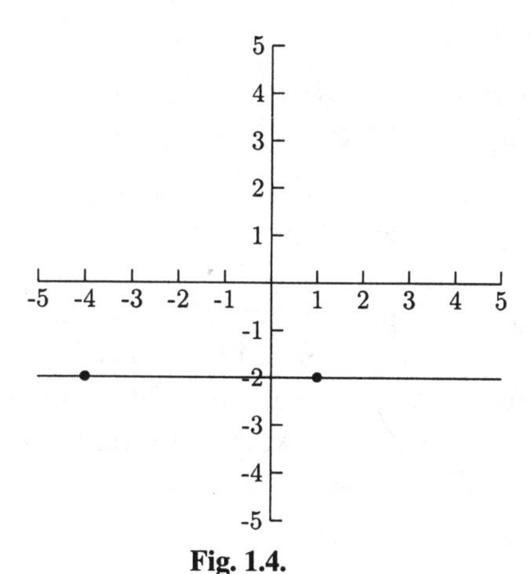

Fig. 1.4.

As we can see from Figure 1.4, $(-4, -2)$ and $(1, -2)$ are two points on a horizontal line. We will put these points in the slope formula.

$$m = \frac{-2 - (-2)}{1 - (-4)} = \frac{0}{5} = 0$$

The slope of every horizontal line is 0. The y-values on a horizontal line do not change but the x-values do.

What happens to the slope formula for a vertical line?

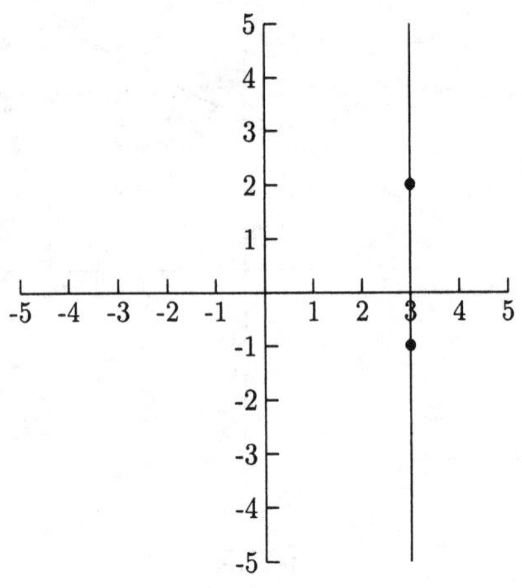

Fig. 1.5.

The points (3, 2) and (3, −1) are on the vertical line in Figure 1.5. Let's see what happens when we put them in the slope formula.

$$m = \frac{-1-2}{3-3} = \frac{-3}{0}$$

This is not a number so the slope of a vertical line does not exist (we also say that it is undefined). The x-values on a vertical line do not change but the y-values do.

Any line is the graph of a linear equation. The equation of a horizontal line is $y = a$ (where a is the y-value of every point on the line). Some examples of horizontal lines are $y = 4$, $y = 1$, and $y = -5$.

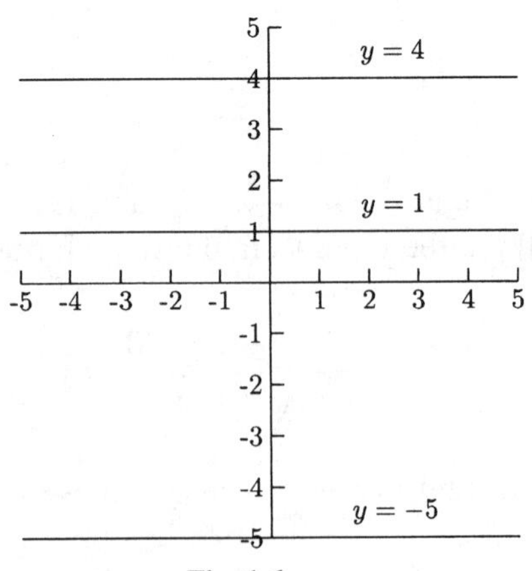

Fig. 1.6.

The equation of a vertical line is $x = a$ (where a is the x-value of every point on the line). Some examples are $x = -3$, $x = 2$, and $x = 4$.

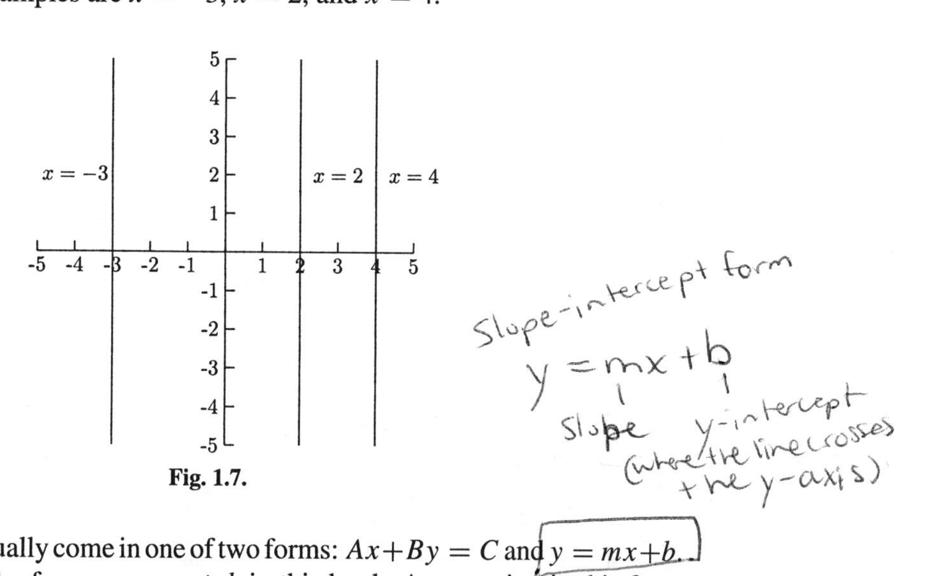

Fig. 1.7.

Other equations usually come in one of two forms: $Ax+By = C$ and $y = mx+b$. We will usually use the form $y = mx + b$ in this book. An equation in this form gives us two important pieces of information. The first is m, the slope. The second is b, the y-intercept (where the line crosses the y-axis). For this reason, this form is called the *slope-intercept* form. In the line $y = \frac{2}{3}x + 4$, the slope of the line is $\frac{2}{3}$ and the y-intercept is (0, 4), or simply, 4.

We can find an equation of a line by knowing its slope and any point on the line. There are two common methods for finding this equation. One is to put m, x, and y (x and y are the coordinates of the point we know) in $y = mx + b$ and use algebra to find b. The other is to put these same numbers in the *point-slope* form of the line, $y - y_1 = m(x - x_1)$. We will use both methods in the next example.

EXAMPLES

- Find an equation of the line with slope $-\frac{3}{4}$ containing the point (8, −2).

  We will let $m = -\frac{3}{4}$, $x = 8$, and $y = -2$ in $y = mx + b$ to find b.

$$-2 = -\frac{3}{4}(8) + b$$

$$4 = b$$

  The line is $y = -\frac{3}{4}x + 4$.

Now we will let $m = -\frac{3}{4}$, $x_1 = 8$ and $y_1 = -2$ in $y - y_1 = m(x - x_1)$.

$$y - (-2) = -\frac{3}{4}(x - 8)$$

$$y + 2 = -\frac{3}{4}x + 6$$

$$y = -\frac{3}{4}x + 4$$

- Find an equation of the line with slope 4, containing the point (0, 3).
  We know the slope is 4 and we know the y-intercept is 3 (because (0, 3) is on the line), so we can write the equation without having to do any work: $y = 4x + 3$.
- Find an equation of the horizontal line that contains the point (5, −6).
  Because the y-values are the same on a horizontal line, we know that this equation is $y = -6$. We can still find the equation algebraically using the fact that $m = 0$, $x = 5$ and $y = -6$. Then $y = mx + b$ becomes $-6 = 0(5) + b$. From here we can see that $b = -6$, so $y = 0x - 6$, or simply, $y = -6$.
- Find an equation of the vertical line containing the point (10, −1).
  Because the x-values are the same on a vertical line, we know that the equation is $x = 10$. We cannot find this equation algebraically because m does not exist.

We can find an equation of a line if we know any two points on the line. First we need to use the slope formula to find m. Then we will pick one of the points to put into $y = mx + b$.

EXAMPLES

Find an equation of the line containing the given points.

- (−2, 3) and (10, 15)

$$m = \frac{15 - 3}{10 - (-2)} = 1$$

We will use $x = -2$ and $y = 3$ in $y = mx + b$ to find b.

$$3 = 1(-2) + b$$

$$5 = b$$

The equation is $y = 1x + 5$, or simply $y = x + 5$.

- $(\frac{1}{2}, -1)$ and $(4, 3)$

$$m = \frac{3-(-1)}{4-\frac{1}{2}} = \frac{4}{\frac{7}{2}} = 4 \div \frac{7}{2} = 4 \cdot \frac{2}{7} = \frac{8}{7}$$

  Using $x = 4$ and $y = 3$ in $y = mx + b$, we have

$$3 = \frac{8}{7}(4) + b$$

$$-\frac{11}{7} = b.$$

  The equation is $y = \frac{8}{7}x - \frac{11}{7}$.
- $(0, 1)$ and $(12, 1)$
  The y-values are the same, making this a horizontal line. The equation is $y = 1$.

If a graph is clear enough, we can find two points on the line or even its slope. If fact, if the slope and y-intercept are easy enough to see on the graph, we know right away what the equation is.

EXAMPLES

-

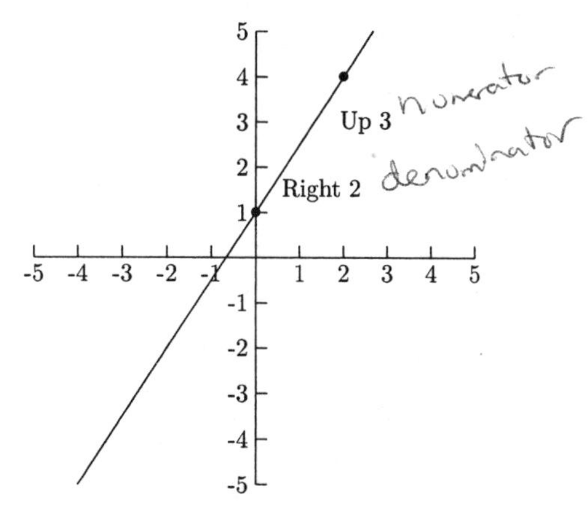

Fig. 1.8.

The line in Figure 1.8 crosses the y-axis at 1, so $b = 1$. From this point, we can go right 2 and up 3 to reach the point $(2, 4)$ on the line. "Right 2" means that the denominator of the slope is 2. "Up 3" means that the numerator of the slope is 3. The slope is $\frac{3}{2}$, so the equation of the line is $y = \frac{3}{2}x + 1$.

•

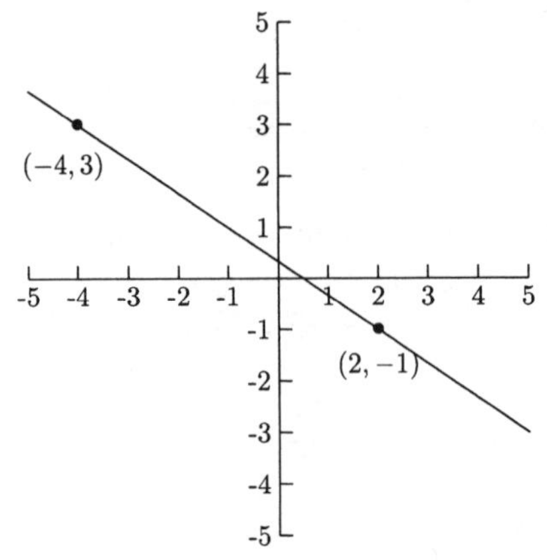

Fig. 1.9.

The y-intercept is not easy to determine, but we do have two points. We can either find the slope by using the slope formula, or visually (as we did above). We can find the slope visually by asking how we can go from $(-4, 3)$ to $(2, -1)$: Down 4 (making the numerator of the slope -4) and right 6 (making the denominator 6). If we use the slope formula, we have

$$m = \frac{-1-3}{2-(-4)} = \frac{-4}{6} = -\frac{2}{3}.$$

Using $x = 2$ and $y = -1$ in $y = mx + b$, we have $-1 = -\frac{2}{3}(2) + b$. From this, we have $b = \frac{1}{3}$. The equation is $y = -\frac{2}{3}x + \frac{1}{3}$.

•

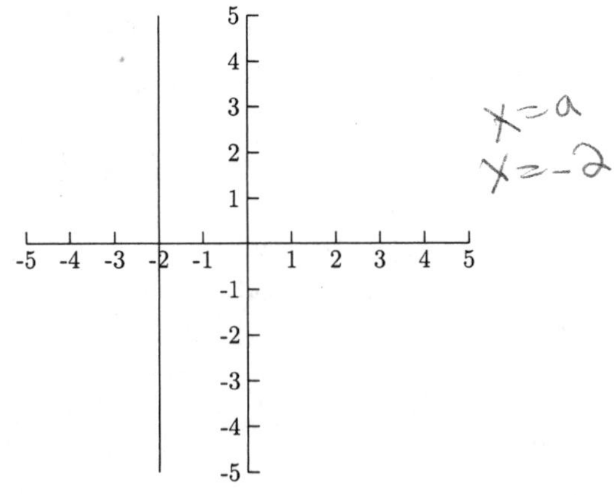

Fig. 1.10.

The line in Figure 1.10 is vertical, so it has the form $x = a$. All of the x-values are -2, so the equation is $x = -2$.

When an equation for a line is in the form $Ax + By = C$, we can find the slope by solving the equation for y. This will put the equation in the form $y = mx + b$.

EXAMPLE

- Find the slope of the line $6x - 2y = 3$.

$$6x - 2y = 3$$
$$-2y = -6x + 3$$
$$y = 3x - \frac{3}{2}$$

The slope is 3 (or $\frac{3}{1}$).

Two lines are parallel if their slopes are equal (or if both lines are vertical).

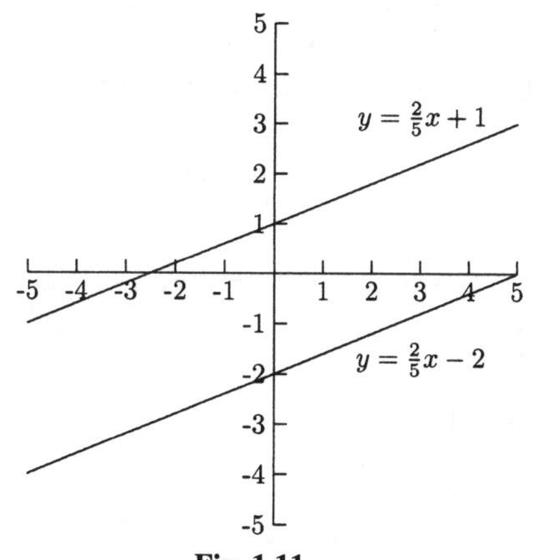

Fig. 1.11.

Two lines are perpendicular if their slopes are *negative reciprocals* of each other (or if one line is horizontal and the other is vertical). Two numbers are negative reciprocals of each other if one is positive and the other is negative and inverting one gets the other (if we ignore the sign).

EXAMPLES

- $\frac{5}{6}$ and $-\frac{6}{5}$ are negative reciprocals
- $-\frac{3}{4}$ and $\frac{4}{3}$ are negative reciprocals

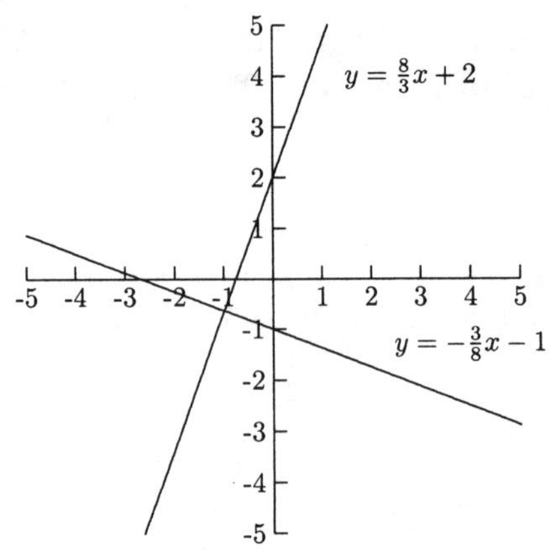

Fig. 1.12.

- -2 and $\frac{1}{2}$ are negative reciprocals
- 1 and -1 are negative reciprocals

We can decide whether two lines are parallel or perpendicular or neither by putting them in the form $y = mx + b$ and comparing their slopes.

EXAMPLES

Determine whether the lines are parallel or perpendicular or neither.

- $4x - 3y = -15$ and $4x - 3y = 6$

$$4x - 3y = -15 \qquad\qquad 4x - 3y = 6$$

$$-3y = -4x - 15 \qquad\qquad -3y = -4x + 6$$

$$y = \frac{4}{3}x + 5 \qquad\qquad y = \frac{4}{3}x - 2$$

The lines have the same slope, so they are parallel.

- $3x - 5y = 20$ and $5x - 3y = -15$

$$3x - 5y = 20 \qquad\qquad 5x - 3y = -15$$

$$-5y = -3x + 20 \qquad\qquad -3y = -5x - 15$$

$$y = \frac{3}{5}x - 4 \qquad\qquad y = \frac{5}{3}x + 5$$

The slopes are reciprocals of each other but not *negative* reciprocals, so they are not perpendicular. They are not parallel, either.

- $x - y = 2$ and $x + y = -8$

$$x - y = 2 \qquad\qquad x + y = -8$$

$$y = x - 2 \qquad\qquad y = -x - 8$$

  The slope of the first line is 1 and the second is -1. Because 1 and -1 are negative reciprocals, these lines are perpendicular.
- $y = 10$ and $x = 3$
  The line $y = 10$ is horizontal, and the line $x = 3$ is vertical. They are perpendicular.

Sometimes we need to find an equation of a line when we know only a point on the line and an equation of another line that is either parallel or perpendicular to it. We need to find the slope of the line whose equation we have and use this to find the equation of the line we are looking for.

EXAMPLES

- Find an equation of the line containing the point $(-4, 5)$ that is parallel to the line $y = 2x + 1$.
  The slope of $y = 2x + 1$ is 2. This is the same as the line we want, so we will let $x = -4$, $y = 5$, and $m = 2$ in $y = mx + b$. We get $5 = 2(-4) + b$, so $b = 13$. The equation of the line we want is $y = 2x + 13$.
- Find an equation of the line with x-intercept 4 that is perpendicular to $x - 3y = 12$.
  The x-intercept is 4 means that the point $(4, 0)$ is on the line. The slope of the line we want will be the negative reciprocal of the slope of the line $x - 3y = 12$. We will find the slope of $x - 3y = 12$ by solving for y.

$$x - 3y = 12$$

$$y = \frac{1}{3}x - 4$$

  The slope we want is -3, which is the negative reciprocal of $\frac{1}{3}$. When we let $x = 4$, $y = 0$, and $m = -3$ in $y = mx + b$, we have $0 = -3(4) + b$, which gives us $b = 12$. The line is $y = -3x + 12$.
- Find an equation of the line containing the point $(3, -8)$, perpendicular to the line $y = 9$.
  The line $y = 9$ is horizontal, so the line we want is vertical. The vertical line passing through $(3, -8)$ is $x = 3$.

PRACTICE

When asked to find an equation for a line, put your answer in the form $y = mx + b$ unless the line is horizontal ($y = a$) or vertical ($x = a$).

1. Find the slope of the line containing the points (4, 12) and (−6, 1).
2. Find the slope of the line with x-intercept 5 and y-intercept −3.
3. Find an equation of the line containing the point $(-10, 4)$ with slope $-\frac{2}{5}$.
4. Find an equation of the line with y-intercept −5 and slope 2.
5. Find an equation of the line in Figure 1.13.

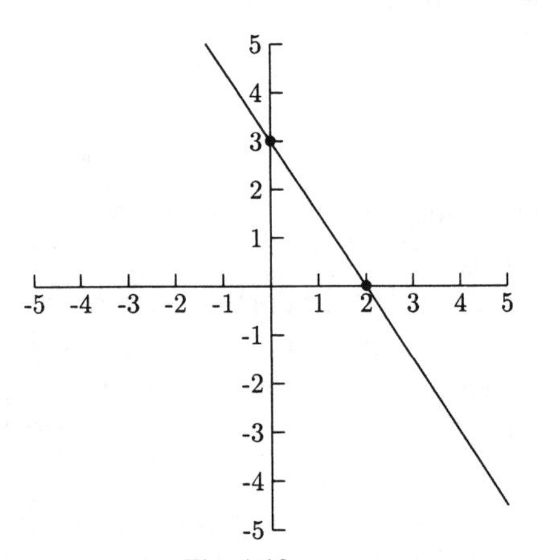

Fig. 1.13.

6. Find an equation of the line containing the points $(\frac{3}{4}, 1)$ and $(-2, -1)$.
7. Determine whether the lines $3x - 7y = 28$ and $7x + 3y = 3$ are parallel or perpendicular or neither.
8. Find an equation of the line containing (2, 3) and perpendicular to the line $x - y = 5$.
9. Find an equation of the line parallel to the line $x = 6$ containing the point $(-3, 2)$.
10. Determine whether the lines $2x - 3y = 1$ and $-4x + 6y = 5$ are parallel or perpendicular or neither.

SOLUTIONS

1. $m = \frac{1-12}{-6-4} = \frac{-11}{-10} = \frac{11}{10}$

2. The x-intercept is 5 and the y-intercept is -3 mean that the points (5, 0) and (0, -3) are on the line.

$$m = \frac{-3-0}{0-5} = \frac{-3}{-5} = \frac{3}{5}$$

3. Put $x = -10$, $y = 4$, and $m = -\frac{2}{5}$ in $y = mx + b$ to find b.

$$4 = -\frac{2}{5}(-10) + b$$

$$0 = b$$

The equation is $y = -\frac{2}{5}x + 0$, or simply $y = -\frac{2}{5}x$.

4. $m = 2$, $b = -5$, so the line is $y = 2x - 5$.

5. From the graph, we can see that the y-intercept is 3. We can use the indicated points (0, 3) and (2, 0) to find the slope in two ways. One way is to put these numbers in the slope formula.

$$m = \frac{0-3}{2-0} = -\frac{3}{2}$$

The other way is to move from (0, 3) to (2, 0) by going down 3 (so the numerator of the slope is -3) and right 2 (so the denominator is 2). Either way, we have the slope $-\frac{3}{2}$. Because the y-intercept is 3, the equation is $y = -\frac{3}{2}x + 3$.

6. $m = \frac{-1-1}{-2-\frac{3}{4}} = \frac{-2}{-\frac{11}{4}} = -2 \div -\frac{11}{4} = -2 \cdot -\frac{4}{11} = \frac{8}{11}$

We will use $x = -2$ and $y = -1$ in $y = mx + b$.

$$-1 = \frac{8}{11}(-2) + b$$

$$\frac{5}{11} = b$$

The equation is $y = \frac{8}{11}x + \frac{5}{11}$.

7. We will solve for y in each equation so that we can compare their slopes.

$$3x - 7y = 28 \qquad\qquad 7x + 3y = 3$$

$$y = \frac{3}{7}x - 4 \qquad\qquad y = -\frac{7}{3}x + 1$$

The slopes are negative reciprocals of each other, so these lines are perpendicular.

8. Once we have found the slope for the line $x - y = 5$, we will use its negative reciprocal as the slope of the line we want.

$$x - y = 5$$

$$y = x - 5$$

The slope of this line is 1. The negative reciprocal of 1 is -1. We will use $x = 2,\ y = 3$, and $m = -1$ in $y = mx + b$.

$$3 = -1(2) + b$$

$$5 = b$$

The equation is $y = -1x + 5$, or simply $y = -x + 5$.

9. The line $x = 6$ is vertical, so the line we want is also vertical. The vertical line that goes through $(-3, 2)$, is $x = -3$.

10. We will solve for y in each equation and compare their slopes.

$$2x - 3y = 1 \qquad\qquad -4x + 6y = 5$$

$$y = \frac{2}{3}x - \frac{1}{3} \qquad\qquad y = \frac{2}{3}x + \frac{5}{6}$$

The slopes are the same, so these lines are parallel.

Applications of Lines and Slopes

We can use the slope of a line to decide whether points in the plane form certain shapes. Here, we will use the slope to decide whether or not three points form a right triangle and whether or not four points form a parallelogram. After we plot the points, we can decide which points to put into the slope formula.

EXAMPLES

- Show that $(-1, 2)$, $(4, -3)$, and $(5, 0)$ are the vertices of a right triangle.

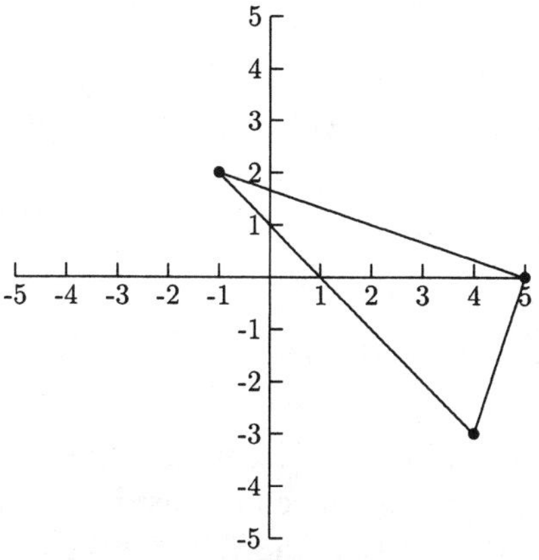

Fig. 1.14.

From the graph in Figure 1.14, we can see that the line segment between $(5, 0)$ and $(-1, 2)$ should be perpendicular to the line segment between $(5, 0)$ and $(4, -3)$. Once we have found the slopes of these line segments, we will see that they are negative reciprocals.

$$m = \frac{2-0}{-1-5} = -\frac{1}{3} \qquad\qquad m = \frac{-3-0}{4-5} = 3$$

- Show that $(-3, 1)$, $(3, -5)$, $(4, -1)$, and $(-2, 5)$ are the vertices of a parallelogram.

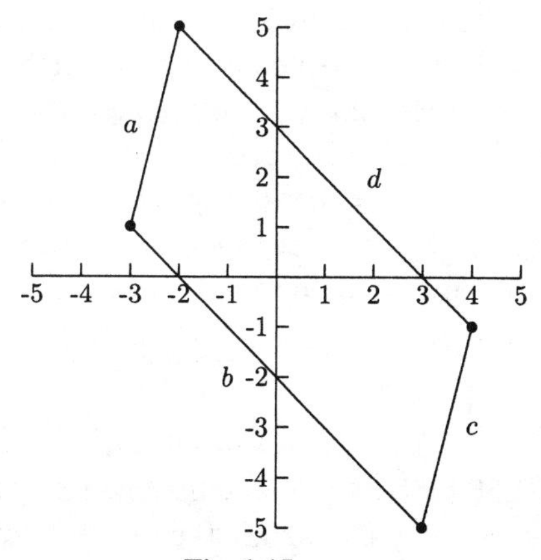

Fig. 1.15.

From the graph in Figure 1.15, we see that we need to show that line segments a and c are parallel and that line segments b and d are parallel.

The slope for segment a is $m = \dfrac{5-1}{-2-(-3)} = 4$,

and the slope for segment c is $m = \dfrac{-1-(-5)}{4-3} = 4$.

The slope for segment b is $m = \dfrac{-5-1}{3-(-3)} = -1$,

and the slope for segment d is $m = \dfrac{-1-5}{4-(-2)} = -1$.

There are many applications of linear equations to business and science. Suppose the property tax rate for a school district is \$1.50 per \$100 valuation. This is a linear relationship between the value of the property and the amount of tax on the property. The slope of the line in this relationship is

$$\frac{\text{Tax change}}{\text{Value change}} = \frac{\$1.50}{\$100}.$$

As the value of property increases by \$100, the tax increases by \$1.50. Two variables are linearly related if a fixed increase of one variable causes a fixed increase or decrease in the other variable. These changes are proportional. For example, if a property increases in value by \$50, then its tax would increase by \$0.75.

We can find an equation (also called a model) that describes the relationship between two variables if we are given two points or one point and the slope. As in most word problems, we will need to find the information in the statement of the problem, it is seldom spelled out for us. One of the first things we need to do is to decide which quantity will be represented by x and which by y. Sometimes it does not matter. In the problems that follow, it will matter. If we are instructed to "give variable 1 in terms of variable 2," then variable 1 will be y and variable 2 will be x. This is because in the equation $y = mx + b$, y is given in terms of x. For example, if we are asked to give the property tax in terms of property value, then y would represent the property tax and x would represent the property value.

EXAMPLES

- A family paid \$52.50 for water in January when they used 15,000 gallons and \$77.50 in May when they used 25,000 gallons. Find an equation that gives the amount of the water bill in terms of the number of gallons of water used.

Because we need to find the cost in terms of water used, we will let y represent the cost and x, the amount of water used. Our ordered pairs will be (water, cost): (15,000, 52.50) and (25,000, 77.50). Now we can compute the slope.

$$m = \frac{77.50 - 52.50}{25{,}000 - 15{,}000} = 0.0025$$

We will use $x = 15{,}000$, $y = 52.50$, and $m = 0.0025$ in $y = mx + b$ to find b.

$$52.50 = 0.0025(15{,}000) + b$$

$$15 = b$$

The equation is $y = 0.0025x + 15$. With this equation, the family can predict its water bill by putting the amount of water used in the equation. For example, 32,000 gallons would cost $0.0025(32{,}000) + 15 = \$95$.

- A bakery sells a special bread. It costs \$6000 to produce 10,000 loaves of bread per day and \$5900 to produce 9500 loaves. Find an equation that gives the daily costs in terms of the number of loaves of bread produced.
  Because we want the cost in terms of the number of loaves produced, we will let y represent the daily cost and x, the number of loaves produced. Our points will be of the form (number of loaves, daily cost): (10,000, 6000) and (9500, 5900).

$$m = \frac{5900 - 6000}{9500 - 10{,}000} = \frac{1}{5}$$

We will use $x = 10{,}000$, $y = 6000$, and $m = \frac{1}{5}$ in $y = mx + b$.

$$6000 = \frac{1}{5}(10{,}000) + b$$

$$4000 = b$$

The equation is $y = \frac{1}{5}x + 4000$.

The slope, and sometimes the y-intercept, have important meanings in applied problems. In the first example, the household water bill was computed using $y = 0.0025x + 15$. The slope means that each gallon costs \$0.0025 (or 0.25 cents). As the number of gallons increases by 1, the cost increases by \$0.0025. The y-intercept is the cost when 0 gallons are used. This additional monthly charge is \$15. The slope in the bakery problem means that five loaves of bread costs \$1 to produce (or each loaf costs \$0.20). The y-intercept tells us the bakery's daily fixed costs are \$4000. Fixed costs are costs that the bakery must pay regardless of the number of loaves produced. Fixed costs might include rent, equipment payments, insurance, taxes, etc.

In the following examples, information about the slope will be given and a point will be given or implied.

- The dosage of medication given to an adult cow is 500 mg plus 9 mg per pound. Find an equation that gives the amount of medication (in mg) per pound of weight.
  We will use 500 mg as the y-intercept. The slope is

$$\frac{\text{increase in medication}}{\text{increase in weight}} = \frac{9}{1}.$$

  The equation is $y = 9x + 500$, where x is in pounds and y is in milligrams.
- At the surface of the ocean, a certain object has 1500 pounds of pressure on it. For every foot below the surface, the pressure on the object increases about 43 pounds. Find an equation that gives the pressure (in pounds) on the object in terms of its depth (in feet) in the ocean.
  At 0 feet, the pressure on the object is 1500 lbs, so the y-intercept is 1500. The slope is

$$\frac{\text{increase in pressure}}{\text{increase in depth}} = \frac{43}{1} = 43.$$

  This makes the equation $y = 43x + 1500$, where x is the depth in feet and y is the pressure in pounds.
- A pancake mix requires $\frac{3}{4}$ cup of water for each cup of mix. Find an equation that gives the amount of water needed in terms of the amount of pancake mix.
  Although no point is directly given, we can assume that (0, 0) is a point on the line because when there is no mix, no water is needed. The slope is

$$\frac{\text{increase in water}}{\text{increase in mix}} = \frac{3/4}{1} = \frac{3}{4}.$$

  The equation is $y = \frac{3}{4}x + 0$, or simply $y = \frac{3}{4}x$.

PRACTICE

1. Show that the points $(-5, 1)$, $(2, 0)$, and $(-2, -3)$ are the vertices of a right triangle.

2. Show that the points $(-2,-3)$, $(3,6)$, $(-5,2)$, and $(6,1)$ are the vertices of a parallelogram.
3. A sales representative earns a monthly base salary plus a commission on sales. Her pay this month will be \$2000 on sales of \$10,000. Last month, her pay was \$2720 on sales of \$16,000. Find an equation that gives her monthly pay in terms of her sales level.
4. The temperature scales Fahrenheit and Celsius are linearly related. Water freezes at 0°C and 32°F. Water boils at 212°F and 100°C. Find an equation that gives degrees Celsius in terms of degrees Fahrenheit.
5. A sales manager believes that each \$100 spent on television advertising results in an increase of 45 units sold. If sales were 8250 units sold when \$3600 was spent on television advertising, find an equation that gives the sales level in terms of the amount spent on advertising.

SOLUTIONS

1.

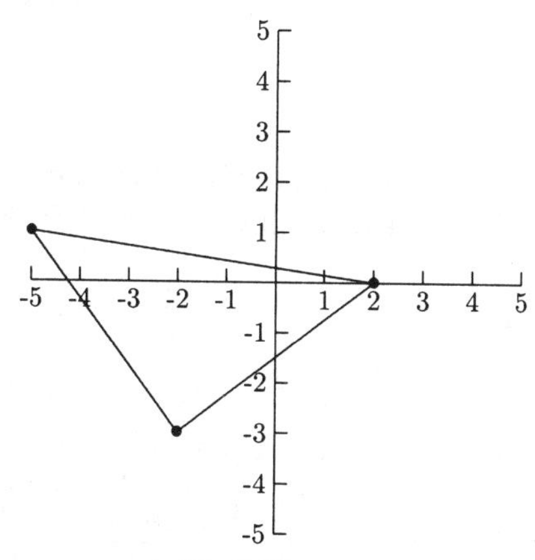

Fig. 1.16.

We will show that the slope of the line segment between $(-5, 1)$ and $(-2,-3)$ is the negative reciprocal of the slope of the line segment between $(-2,-3)$ and $(2, 0)$. This will show that the angle at $(-2,-3)$ is a right angle.

$$m = \frac{-3-1}{-2-(-5)} = -\frac{4}{3} \qquad m = \frac{0-(-3)}{2-(-2)} = \frac{3}{4}$$

2.

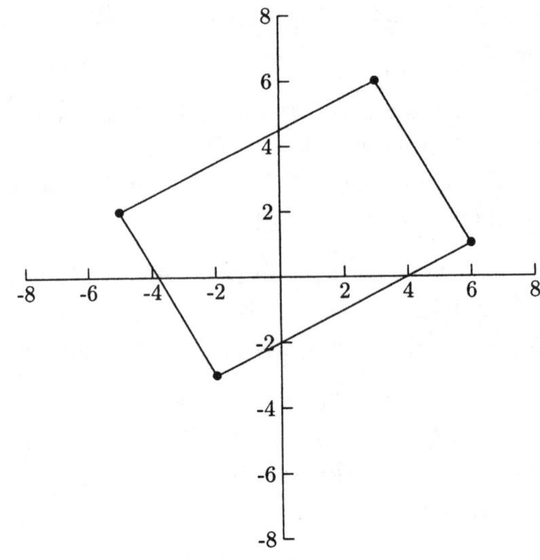

Fig. 1.17.

We will show that the slope of the line segment between $(-5, 2)$ and $(-2, -3)$ is equal to the slope of the line segment between $(3, 6)$ and $(6, 1)$.

$$m = \frac{-3 - 2}{-2 - (-5)} = -\frac{5}{3} \qquad m = \frac{1 - 6}{6 - 3} = -\frac{5}{3}$$

Now we will show that the slope of the line segment between $(-5, 2)$ and $(3, 6)$ is equal to the slope of the line segment between $(-2, -3)$ and $(6, 1)$.

$$m = \frac{6 - 2}{3 - (-5)} = \frac{1}{2} \qquad m = \frac{1 - (-3)}{6 - (-2)} = \frac{1}{2}$$

3. Because we want pay in terms of sales, y will represent pay, and x will represent monthly sales. The points are (10,000, 2000) and (16,000, 2720).

$$m = \frac{2720 - 2000}{16{,}000 - 10{,}000} = \frac{3}{25}$$

(This means that for every \$25 in sales, the representative earns \$3.) We will use $x = 10{,}000$, $y = 2000$, and $m = \frac{3}{25}$ in $y = mx + b$.

$$2000 = \frac{3}{25}(10{,}000) + b$$

$$800 = b$$

The equation is $y = \frac{3}{25}x + 800$. (The y-intercept is 800 means that her monthly base pay is \$800.)

4. The points are (degrees Fahrenheit, degrees Celcius): $(32, 0)$ and $(212, 100)$.

$$m = \frac{100 - 0}{212 - 32} = \frac{5}{9}$$

(This means that a 9°F increase in temperature corresponds to an increase of 5°C.) We will use $F = 32$, $C = 0$, and $m = \frac{5}{9}$ in $C = mF + b$.

$$0 = \frac{5}{9}(32) + b$$
$$-\frac{160}{9} = b$$

The equation is $C = \frac{5}{9}F - \frac{160}{9}$. (The y-intercept is $-\frac{160}{9}$ means that the temperature 0°F corresponds to $-\frac{160}{9}$°C.)

5. The points are (amount spent on advertising, number of units sold). The slope is

$$\frac{\text{increase in sales}}{\text{increase in advertising spending}} = \frac{45}{100} = \frac{9}{20},$$

and our point is (3600, 8250).

$$8250 = \frac{9}{20}(3600) + b$$
$$6630 = b$$

The equation is $y = \frac{9}{20}x + 6630$. (The slope means that every \$20 spent on television advertising results in an extra 9 units sold. The y-intercept is 6630 means that if nothing is spent on television advertising, 6630 units would be sold.)

CHAPTER 1 REVIEW

1. Find the slope of the line containing the points (3, 1) and (4, −2).
   (a) $\frac{1}{3}$ (b) -3 (c) $-\frac{1}{3}$ (d) 3

2. Are the lines $2x + y = 4$ and $2x - 4y = 5$ parallel, perpendicular, or neither?
   (a) Parallel (b) Perpendicular
   (c) Neither (d) Cannot be determined

3. Are the lines $x = 4$ and $y = -4$ parallel, perpendicular, or neither?
   (a) Parallel (b) Perpendicular
   (c) Neither (d) Cannot be determined

4. What is the equation of the line containing the points $(0, -1)$ and $(5, 1)$?
   (a) $y = -1$ (b) $y = \frac{5}{2}x - 1$
   (c) $y = -4x - 1$ (d) $y = \frac{2}{5}x - 1$

5. Find an equation of the line containing the point $(-1, -5)$ and parallel to the line $y = 2x - 4$.
   (a) $y = 2x - 3$ (b) $y = 2x - 5$
   (c) $y = 2x - 1$ (d) $y = 2x + 4$

6. Find an equation of the line containing the point $(3, 3)$ and perpendicular to the line $y = 2x + 5$.
   (a) $y = -\frac{1}{2}x + \frac{9}{2}$ (b) $y = \frac{1}{2}x + 3$
   (c) $y = \frac{1}{2}x + \frac{3}{2}$ (d) $y = -\frac{1}{2}x + 3$

7. Find an equation of the line in Figure 1.18.
   (a) $y = -\frac{1}{2}x + 4$ (b) $y = \frac{1}{2}x + 4$
   (c) $y = -2x + 4$ (d) $y = 2x + 4$

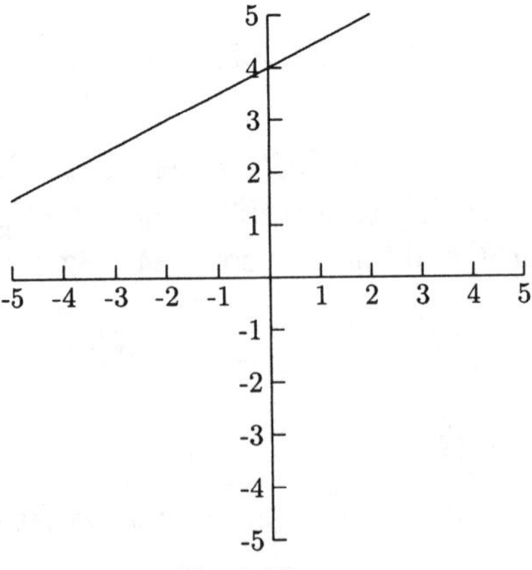

Fig. 1.18.

8. Find an equation of the horizontal line that goes through the point (4, 9).
   (a) $x = 4$ (b) $y = 9$
   (c) Cannot be determined

9. Are the points $(-5, -1)$, $(1, 4)$, and $(6, -2)$ the vertices of a right triangle?
   (a) Yes (b) No
   (c) Cannot be determined

10. A government agency leases a photocopier for a fixed monthly charge plus a charge for each photocopy. In one month, the bill was $350 for 4000 copies. In the following month, the bill was $375 for 5000 copies. Find the monthly bill in terms of the number of copies used.
    (a) $y = 1.267x - 4718$ (b) $y = 40x - 10{,}000$
    (c) $y = 0.789x - 3570$ (d) $y = 0.025x + 250$

SOLUTIONS

1. B	2. B	3. B	4. D	5. A
6. A	7. B	8. B	9. A	10. D

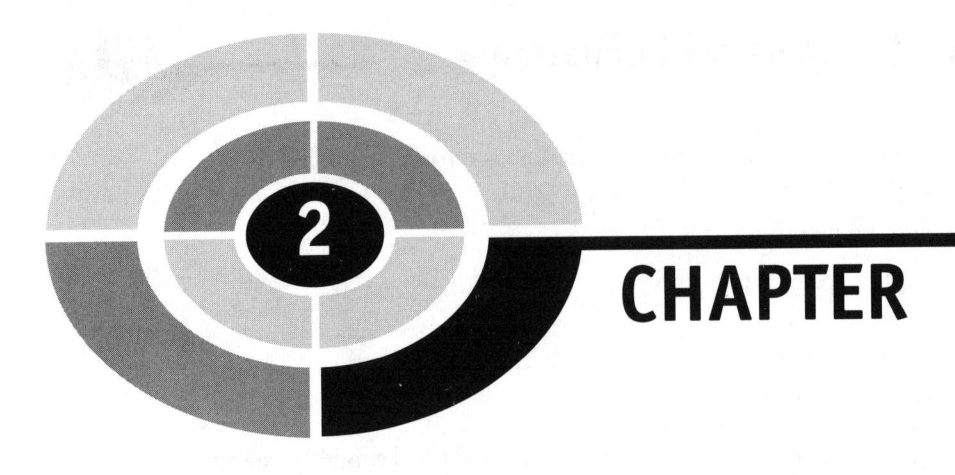

Introduction to Functions

A *relation* between two sets A and B is a collection of ordered pairs, where the first coordinate comes from A and the second comes from B. For example, if $A = \{1, 2, 3, 4\}$ and $B = \{a, b, c\}$, one relation is the three pairs $\{(1, c), (1, a), (3, a)\}$. A *function* on sets A and B is a special kind of relation where *every* element of A is paired with *exactly one* element from B. The relation above fails to be a function in two ways. Not every element of A is paired with an element from B, 1 and 3 are used but not 2 and 4. Also, the element 1 is used *twice*, not *once*. There are no such restrictions on B; that is, elements from B can be paired with elements from A many times or not at all. For example, $\{(1, a), (2, a), (3, b), (4, b)\}$ is a function from A to B.

Functions exist all around us. If a worker is paid by the hour, his weekly pay is a function of how many hours he worked. For any number of hours worked, there is exactly one pay amount that corresponds to that time. If A is the set of all triangles and B is the set of real numbers, then we have a function that pairs each triangle with exactly one real number that is its area. We will be concerned with functions from real numbers to real numbers. A will either be all of the real numbers or will be some part of the real numbers, and B will be the real numbers.

A linear function is one of the most basic kinds of functions. These functions have the form $f(x) = mx + b$. The only difference between $f(x) = mx + b$ and $y = mx + b$ is that y is replaced by $f(x)$. Very often $f(x)$ and y will be the same. The letter f is the name of the function. Other common names of functions are g and h. The notation $f(x)$ is pronounced "f of x" or "f at x."

Evaluating a function at a quantity means to substitute the quantity for x (or whatever the variable is). For example, evaluating the function $f(x) = 2x - 5$ at 6 means to substitute 6 for x.

$$f(6) = 2(6) - 5 = 7$$

We might also say $f(6) = 7$. The quantity inside the parentheses is x and the quantity on the right of the equal sign is y. One advantage to this notation is that we have both the x- and y-values without having to say anything about x and y. Functions that have no variables in them are called *constant functions.* All y-values for these functions are the same.

EXAMPLES

- Find $f(-2)$, $f(0)$, and $f(6)$ for $f(x) = \sqrt{x+3}$.
  We need to substitute -2, 0, and 6 for x in the function.

$$f(-2) = \sqrt{-2+3} = \sqrt{1} = 1$$

$$f(0) = \sqrt{0+3} = \sqrt{3}$$

$$f(6) = \sqrt{6+3} = \sqrt{9} = 3$$

- Find $f(-8)$, $f(\pi)$, and $f(10)$ for $f(x) = 16$.
  $f(x) = 16$ is a constant function, so the y-value is 16 no matter what quantity is in the parentheses.

$$f(-8) = 16 \qquad f(\pi) = 16 \qquad f(10) = 16$$

A *piecewise* function is a function with two or more formulas for computing y. The formula to use depends on where x is. There will be an interval for x written next to each formula for y.

-

$$f(x) = \begin{cases} x - 1 & \text{if } x \leq -2 \\ 2x & \text{if } -2 < x < 2 \\ x^2 & \text{if } x \geq 2 \end{cases}$$

In this example, there are three formulas for y: $y = x - 1$, $y = 2x$, and $y = x^2$, and three intervals for x: $x \leq -2$, $-2 < x < 2$, and $x \geq 2$. When evaluating this function, we need to decide to which interval x belongs. Then we will use the corresponding formula for y.

EXAMPLES

- Find $f(5)$, $f(-3)$, and $f(0)$ for the function above.
  For $f(5)$, does $x = 5$ belong to $x \leq -2$, $-2 < x < 2$, or $x \geq 2$? Because $5 \geq 2$, we will use $y = x^2$, the formula written next to $x \geq 2$.

$$f(5) = 5^2 = 25$$

  For $f(-3)$, does $x = -3$ belong to $x \leq -2$, $-2 < x < 2$, or $x \geq 2$? Because $-3 \leq -2$, we will use $y = x - 1$, the formula written next to $x \leq -2$.

$$f(-3) = -3 - 1 = -4$$

  For $f(0)$, does $x = 0$ belong to $x \leq -2$, $-2 < x < 2$, or $x \geq 2$? Because $-2 < 0 < 2$, we will use $y = 2x$, the formula written next to $-2 < x < 2$.

$$f(0) = 2(0) = 0$$

- Find $f(3)$, $f(1)$, and $f(-4)$ for

$$f(x) = \begin{cases} -x & \text{if } x \leq 1 \\ 5 & \text{if } x > 1 \end{cases}$$

$$f(3) = 5 \qquad \text{because } 3 > 1$$

$$f(1) = -1 \qquad \text{because } 1 \leq 1$$

$$f(-4) = -(-4) = 4 \qquad \text{because } -4 \leq 1$$

Piecewise functions come up in daily life. For example, suppose a company pays the regular hourly wage for someone who works up to eight hours but time and a half for someone who works more than eight hours but no more than ten hours and double time for more than ten hours. Then a worker whose regular hourly pay is \$10 has the daily pay function below.

$$p(h) = \begin{cases} 10h & \text{if } 0 \leq h \leq 8 \\ 15(h-8) + 80 & \text{if } 8 < h \leq 10 \\ 20(h-10) + 110 & \text{if } 10 < h \leq 24 \end{cases}$$

Below is an example of a piecewise function taken from an Internal Revenue Service (IRS) publication. The y-value is the amount of personal income tax for a single person. The x-value is the amount of taxable income.

$$f(x) = \begin{cases} 4316 & \text{if } 30{,}000 \le x < 30{,}050 \\ 4329 & \text{if } 30{,}050 \le x < 30{,}100 \\ 4341 & \text{if } 30{,}100 \le x < 30{,}150 \\ 4354 & \text{if } 30{,}150 \le x < 30{,}200 \end{cases}$$

A single person whose taxable income was \$30,120 would pay \$4341. (Source: 2003, 1040 Forms and Instructions)

PRACTICE

1. Find $f(-1)$ and $f(0)$ for $f(x) = 3x^2 + 2x - 1$.
2. Evaluate $f(x) = \frac{1}{x+1}$ at $x = -3$, $x = 1$, and $x = \frac{1}{2}$.
3. Evaluate $g(x) = \sqrt{x - 6}$ at $x = 6$, $x = 8$, and $x = 10$.
4. Find $f(5)$, $f(3)$, $f(2)$, $f(0)$, and $f(-1)$.

$$f(x) = \begin{cases} x^2 + x & \text{if } x \le -1 \\ 10 & \text{if } -1 < x \le 2 \\ -6x & \text{if } x > 2 \end{cases}$$

5. The function below gives the personal income tax for a single person for the 2003 year. If a single person had a taxable income of \$63,575, what is her tax?

$$f(x) = \begin{cases} 12{,}666 & \text{if } 63{,}400 \le x < 63{,}450 \\ 12{,}679 & \text{if } 63{,}450 \le x < 63{,}500 \\ 12{,}691 & \text{if } 63{,}500 \le x < 63{,}550 \\ 12{,}704 & \text{if } 63{,}550 \le x < 63{,}600 \end{cases}$$

SOLUTIONS

1.
$$f(-1) = 3(-1)^2 + 2(-1) - 1 = 3 - 2 - 1 = 0$$
$$f(0) = 3(0)^2 + 2(0) - 1 = -1$$

2.
$$f(-3) = \frac{1}{-3+1} = \frac{1}{-2} \text{ or } -\frac{1}{2}$$
$$f(1) = \frac{1}{1+1} = \frac{1}{2}$$
$$f\left(\frac{1}{2}\right) = \frac{1}{\frac{1}{2}+1} = \frac{1}{\frac{1}{2}+\frac{2}{2}} = \frac{1}{\frac{3}{2}} = 1 \div \frac{3}{2} = 1 \cdot \frac{2}{3} = \frac{2}{3}$$

3.
$$g(6) = \sqrt{6-6} = \sqrt{0} = 0$$
$$g(8) = \sqrt{8-6} = \sqrt{2}$$
$$g(10) = \sqrt{10-6} = \sqrt{4} = 2$$

4.
$$f(5) = -6(5) = -30 \qquad f(3) = -6(3) = -18$$
$$f(2) = 10 \qquad f(0) = 10$$
$$f(-1) = (-1)^2 + (-1) = 0$$

5. The tax is \$12,704 because $63{,}550 \le 63{,}575 < 63{,}600$.

Functions can be evaluated at quantities that are not numbers, but the idea is the same—substitute the quantity in the parentheses for x and simplify.

EXAMPLES

- Evaluate $f(a+3)$, $f(a^2)$, $f(u-v)$, and $f(a+h)$ for $f(x) = 8x+5$. We will let $x = a+3$, $x = a^2$, $x = u-v$, and $x = a+h$ in the function.

$$f(a+3) = 8(a+3)+5 = 8a+24+5 = 8a+29$$
$$f(a^2) = 8(a^2)+5 = 8a^2+5$$
$$f(u-v) = 8(u-v)+5 = 8u-8v+5$$
$$f(a+h) = 8(a+h)+5 = 8a+8h+5$$

- Evaluate $f(10a)$, $f(-a)$, $f(a+h)$, and $f(x+1)$ for $f(x)=x^2+3x-4$.

$$f(10a)=(10a)^2+3(10a)-4=10^2a^2+30a-4=100a^2+30a-4$$

$$f(-a)=(-a)^2+3(-a)-4=a^2-3a-4$$

Remember, $(-a)^2=(-a)(-a)=a^2$, not $-a^2$.

$$f(a+h)=(a+h)^2+3(a+h)-4=(a+h)(a+h)+3(a+h)-4$$

$$=a^2+2ah+h^2+3a+3h-4$$

$$f(x+1)=(x+1)^2+3(x+1)-4=(x+1)(x+1)+3(x+1)-4$$

$$=x^2+2x+1+3x+3-4=x^2+5x$$

- Find $f(a-12)$, $f(a^2+1)$, $f(a+h)$, and $f(x+3)$ for $f(x)=-4$. This is a constant function, so the y-value is -4 no matter what is in the parentheses.

$$f(a-12)=-4 \qquad f(a^2+1)=-4$$

$$f(a+h)=-4 \qquad f(x+3)=-4$$

- Find $f(2u+v)$, $f(\frac{1}{u})$, and $f(2x)$ for

$$f(x)=\frac{x+1}{x+2}.$$

$$f(2u+v)=\frac{2u+v+1}{2u+v+2}$$

$$f\left(\frac{1}{u}\right)=\frac{\frac{1}{u}+1}{\frac{1}{u}+2}=\frac{\frac{1}{u}+\frac{u}{u}\cdot 1}{\frac{1}{u}+\frac{u}{u}\cdot 2}$$

$$=\frac{\frac{1}{u}+\frac{u}{u}}{\frac{1}{u}+\frac{2u}{u}}=\frac{\frac{1+u}{u}}{\frac{1+2u}{u}}$$

$$=\frac{1+u}{u}\div\frac{1+2u}{u}=\frac{1+u}{u}\cdot\frac{u}{1+2u}=\frac{1+u}{1+2u}$$

$$f(2x)=\frac{2x+1}{2x+2}$$

Very early in an introductory calculus course, students use function evaluation to evaluate an important formula called *Newton's Quotient*.

$$\frac{f(a+h)-f(a)}{h}$$

When evaluating Newton's Quotient, we will be given a function such as $f(x) = x^2 + 3$. We need to find $f(a+h)$ and $f(a)$. Once we have these two quantities, we will put them into the quotient and simplify. Simplifying the quotient is usually the messiest part. For $f(x) = x^2 + 3$, we have $f(a+h) = (a+h)^2 + 3 = (a+h)(a+h) + 3 = a^2 + 2ah + h^2 + 3$, and $f(a) = a^2 + 3$. We will substitute $a^2 + 2ah + h^2 + 3$ for $f(a+h)$ and $a^2 + 3$ for $f(a)$.

$$\frac{f(a+h)-f(a)}{h} = \frac{\overbrace{a^2+2ah+h^2+3}^{f(a+h)} - \overbrace{(a^2+3)}^{f(a)}}{h}$$

Now we need to simplify this fraction.

$$\frac{a^2+2ah+h^2+3-(a^2+3)}{h} = \frac{a^2+2ah+h^2+3-a^2-3}{h}$$

$$= \frac{2ah+h^2}{h} \qquad \text{Factor } h.$$

$$= \frac{h(2a+h)}{h} = 2a+h$$

EXAMPLES

Evaluate Newton's Quotient for the given functions.

- $f(x) = 3x^2$

$$f(a+h) = 3(a+h)^2 = 3(a+h)(a+h) = 3(a^2+2ah+h^2)$$

$$= 3a^2 + 6ah + 3h^2$$

$$f(a) = 3a^2$$

$$\frac{f(a+h)-f(a)}{h} = \frac{3a^2+6ah+3h^2-3a^2}{h} = \frac{6ah+3h^2}{h}$$

$$= \frac{h(6a+3h)}{h} = 6a+3h$$

- $f(x) = x^2 - 2x + 5$

$$f(a+h) = (a+h)^2 - 2(a+h) + 5 = (a+h)(a+h) - 2(a+h) + 5$$

$$= a^2 + 2ah + h^2 - 2a - 2h + 5$$

$$f(a) = a^2 - 2a + 5$$

$$\frac{f(a+h) - f(a)}{h} = \frac{a^2 + 2ah + h^2 - 2a - 2h + 5 - (a^2 - 2a + 5)}{h}$$

$$= \frac{a^2 + 2ah + h^2 - 2a - 2h + 5 - a^2 + 2a - 5}{h}$$

$$= \frac{2ah + h^2 - 2h}{h} = \frac{h(2a + h - 2)}{h}$$

$$= 2a + h - 2$$

- $f(x) = \frac{1}{x}$

$$f(a+h) = \frac{1}{a+h} \quad \text{and} \quad f(a) = \frac{1}{a}$$

$$\frac{f(a+h) - f(a)}{h} = \frac{\frac{1}{a+h} - \frac{1}{a}}{h}$$

$$= \frac{\frac{1}{a+h} \cdot \frac{a}{a} - \frac{1}{a} \cdot \frac{a+h}{a+h}}{h}$$

$$= \frac{\frac{a}{a(a+h)} - \frac{a+h}{a(a+h)}}{h}$$

$$= \frac{\frac{a-(a+h)}{a(a+h)}}{h} = \frac{\frac{a-a-h}{a(a+h)}}{h}$$

$$= \frac{\frac{-h}{a(a+h)}}{h} = \frac{-h}{a(a+h)} \div h$$

$$= \frac{-h}{a(a+h)} \cdot \frac{1}{h} = \frac{-1}{a(a+h)}$$

Do not worry—you will not spend a lot of time evaluating Newton's Quotient in calculus, there are formulas that do most of the work. What is Newton's Quotient, anyway? It is nothing more than the slope formula where $x_1 = a,\ y_1 = f(a),\ x_2 = a + h$, and $y_2 = f(a + h)$.

$$m = \frac{y_2 - y_1}{x_2 - x_1} = \frac{f(a+h) - f(a)}{a + h - a} = \frac{f(a+h) - f(a)}{h}$$

PRACTICE

1. Evaluate $f(u + 1),\ f(u^3),\ f(a + h)$, and $f(2x - 1)$ for $f(x) = 7x - 4$.
2. Find $f(-a),\ f(2a),\ f(a + h)$, and $f(x + 5)$ for $f(x) = 2x^2 - x + 3$.
3. Find $f(u + v),\ f(u^2),\ f(\frac{1}{u})$, and $f(x^2 + 3)$ for

$$f(x) = \frac{10x + 1}{3x + 2}$$

4. Evaluate Newton's Quotient for $f(x) = 3x^2 + 2x - 1$.
5. Evaluate Newton's Quotient for $f(x) = \frac{15}{2x-3}$.

SOLUTIONS

1.

$$f(u + 1) = 7(u + 1) - 4 = 7u + 7 - 4 = 7u + 3$$

$$f(u^3) = 7(u^3) - 4 = 7u^3 - 4$$

$$f(a + h) = 7(a + h) - 4 = 7a + 7h - 4$$

$$f(2x - 1) = 7(2x - 1) - 4 = 14x - 7 - 4 = 14x - 11$$

2.

$$f(-a) = 2(-a)^2 - (-a) + 3 = 2a^2 + a + 3$$

$$f(2a) = 2(2a)^2 - 2a + 3 = 2(4a^2) - 2a + 3 = 8a^2 - 2a + 3$$

$$f(a + h) = 2(a + h)^2 - (a + h) + 3 = 2(a + h)(a + h) - (a + h) + 3$$

$$= 2(a^2 + 2ah + h^2) - a - h + 3 = 2a^2 + 4ah + 2h^2 - a - h + 3$$

$$f(x + 5) = 2(x + 5)^2 - (x + 5) + 3 = 2(x + 5)(x + 5) - (x + 5) + 3$$

$$= 2(x^2 + 10x + 25) - x - 5 + 3 = 2x^2 + 19x + 48$$

3.
$$f(u+v)=\frac{10(u+v)+1}{3(u+v)+2}=\frac{10u+10v+1}{3u+3v+2}$$

$$f(u^2)=\frac{10u^2+1}{3u^2+2}$$

$$f\left(\frac{1}{u}\right)=\frac{10\left(\frac{1}{u}\right)+1}{3\left(\frac{1}{u}\right)+2}$$

$$=\frac{\frac{10}{u}+1}{\frac{3}{u}+2}=\frac{\frac{10}{u}+1\cdot\frac{u}{u}}{\frac{3}{u}+2\cdot\frac{u}{u}}=\frac{\frac{10}{u}+\frac{u}{u}}{\frac{3}{u}+\frac{2u}{u}}$$

$$=\frac{\frac{10+u}{u}}{\frac{3+2u}{u}}=\frac{10+u}{u}\div\frac{3+2u}{u}$$

$$=\frac{10+u}{u}\cdot\frac{u}{3+2u}=\frac{10+u}{3+2u}$$

$$f(x^2+3)=\frac{10(x^2+3)+1}{3(x^2+3)+2}$$

$$=\frac{10x^2+31}{3x^2+11}$$

4. $f(a+h)=3(a+h)^2+2(a+h)-1=3(a+h)(a+h)+2(a+h)-1$

$$=3(a^2+2ah+h^2)+2a+2h-1$$

$$=3a^2+6ah+3h^2+2a+2h-1$$

$$f(a)=3a^2+2a-1$$

$$\frac{f(a+h)-f(a)}{h}=\frac{3a^2+6ah+3h^2+2a+2h-1-(3a^2+2a-1)}{h}$$

$$=\frac{3a^2+6ah+3h^2+2a+2h-1-3a^2-2a+1}{h}$$

$$=\frac{6ah+3h^2+2h}{h}$$

$$=\frac{h(6a+3h+2)}{h}=6a+3h+2$$

5. $f(a+h) = \dfrac{15}{2(a+h)-3} = \dfrac{15}{2a+2h-3}$ and $f(a) = \dfrac{15}{2a-3}$

$$\frac{f(a+h)-f(a)}{h} = \frac{\frac{15}{2a+2h-3} - \frac{15}{2a-3}}{h}$$

$$= \frac{\frac{15}{2a+2h-3} \cdot \frac{2a-3}{2a-3} - \frac{15}{2a-3} \cdot \frac{2a+2h-3}{2a+2h-3}}{h}$$

$$= \frac{\frac{15(2a-3)-15(2a+2h-3)}{(2a+2h-3)(2a-3)}}{h} = \frac{\frac{30a-45-30a-30h+45}{(2a+2h-3)(2a-3)}}{h}$$

$$= \frac{\frac{-30h}{(2a+2h-3)(2a-3)}}{h} = \frac{-30h}{(2a+2h-3)(2a-3)} \div h$$

$$= \frac{-30h}{(2a+2h-3)(2a-3)} \cdot \frac{1}{h} = \frac{-30}{(2a+2h-3)(2a-3)}$$

Domain and Range

The *domain* of a function from set A to set B is all of set A. The *range* is either all or part of set B. In our example at the beginning of the chapter, we had $A = \{1, 2, 3, 4\}$, $B = \{a, b, c\}$ and our function was $\{(1, a), (2, a), (3, b), (4, b)\}$. The domain of this function is $\{1, 2, 3, 4\}$, and the range is all of the elements from B that were paired with elements from A. These were $\{a, b\}$.

For the functions in this book, the domain will consist of all the real numbers we are allowed to use for x. The range will be all of the y-values. In this chapter, we will find the domain algebraically. In Chapter 3, we will find both the domain and range from graphs of functions.

Very often, we find the domain of a function by thinking about what we cannot do. For now the things we cannot do are limited to division by zero and taking even roots of negative numbers. If a function has x in a denominator, set the denominator equal to zero and solve for x. The domain will *not* include the solution(s) to this equation (assuming the equation has a solution). If a function has x under an even root sign, set the quantity under the sign greater than or equal to zero to find the domain. Later when we learn about logarithm functions and functions from trigonometry, we will have other things we cannot do. The domain and range are usually given in interval notation. There is a review of interval notation in the Appendix.

EXAMPLES

- $f(x) = \dfrac{x^2 - 4}{x + 3}$

  We cannot let $x + 3$ to be zero, so we cannot let $x = -3$. The domain is $x \neq -3$, or $(-\infty, -3) \cup (-3, \infty)$.

- $f(x) = \dfrac{1}{x^3 + 2x^2 - x - 2}$

  We will use factoring by grouping to factor the denominator. (There is a review of factoring by grouping in the Appendix.)

$$x^3 + 2x^2 - x - 2 = 0$$
$$x^2(x + 2) - 1(x + 2) = 0$$
$$(x + 2)(x^2 - 1) = 0$$
$$(x + 2)(x - 1)(x + 1) = 0$$

$$x + 2 = 0 \qquad x - 1 = 0 \qquad x + 1 = 0$$
$$x = -2 \qquad x = 1 \qquad x = -1$$

The domain is all real numbers except 1, -1, and -2. The domain is shaded on the number line in Figure 2.1.

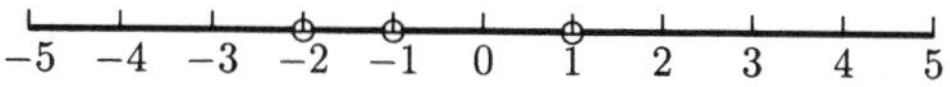

–5 –4 –3 –2 –1 0 1 2 3 4 5

Fig. 2.1.

The domain is $(-\infty, -2) \cup (-2, -1) \cup (-1, 1) \cup (1, \infty)$.

- $g(x) = \dfrac{x + 5}{x^2 + 1}$

  Because $x^2 + 1 = 0$ has no real number solution, we can let x equal any real number. The domain is all real numbers, or $(-\infty, \infty)$.

- $f(x) = \sqrt{x - 8}$

  Because we can only take the square root of nonnegative numbers, $x - 8$ must be nonnegative. We represent "$x - 8$ must be nonnegative" as "$x - 8 \geq 0$." Solving $x - 8 \geq 0$, we get $x \geq 8$. The domain is $x \geq 8$, or $[8, \infty)$.

- $f(x) = \sqrt{x^2 - x - 2}$

  (The Appendix has a review on solving nonlinear inequalities.) We need to solve $x^2 - x - 2 \geq 0$. Factoring $x^2 - x - 2$, we have $(x - 2)(x + 1)$.

$$x - 2 = 0 \qquad\qquad x + 1 = 0$$

$$x = 2 \qquad\qquad x = -1$$

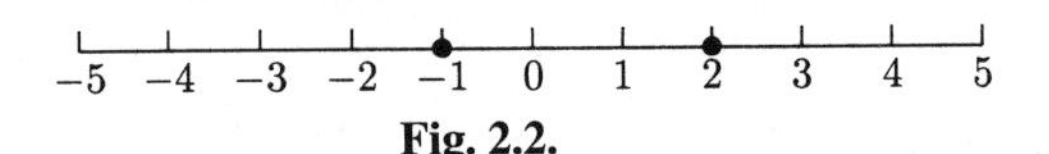

Fig. 2.2.

We will use $x = -2$ for the number to the left of -1, $x = 0$ for the number between -1 and 2, and $x = 3$ for the number to the right of 2 in $x^2 - x - 2 \geq 0$ to see which of these numbers makes it true.

Is $(-2)^2 - (-2) - 2 \geq 0$? Yes. Put "True" to the left of -1.
Is $0^2 - 0 - 2 \geq 0$? No. Put "False" between -1 and 2.
Is $3^2 - 3 - 2 \geq 0$? Yes. Put "True" to the right of 2.

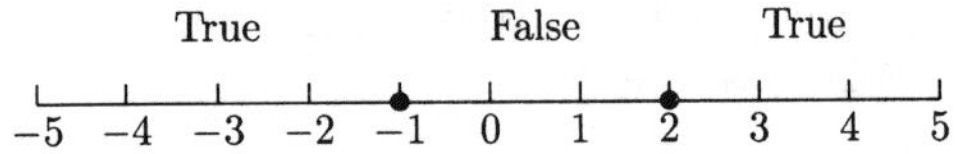

True False True

-5 -4 -3 -2 -1 0 1 2 3 4 5

Fig. 2.3.

The inequality is true for $x \leq -1$ and $x \geq 2$, so the domain is $(-\infty, -1] \cup [2, \infty)$.

- $f(x) = \sqrt[4]{x^2 + 5}$

  Because $x^2 + 5$ is always positive, we can let x be any real number. The domain is $(-\infty, \infty)$.

- $g(x) = \sqrt[3]{x + 7}$

We can take the cube root of any number, so the domain is all real numbers, or $(-\infty, \infty)$.

- $f(x) = x^4 - x^2 + 1$

There is no x in a denominator and no x under an even root sign, so the domain is all real numbers, or $(-\infty, \infty)$.
There are some functions that have x in a denominator and under an even root. At times, it will be useful to shade a number line to keep track of the domain.

- $f(x) = \dfrac{x^2 + x - 3}{\sqrt{4 - x}}$

We cannot let $\sqrt{4 - x}$ be zero, and we cannot let $4 - x$ be negative. These restrictions mean that we must have $4 - x > 0$ (instead of $4 - x \geq 0$). The domain is $4 > x$ (or $x < 4$), which is the interval $(-\infty, 4)$.

- $h(x) = \dfrac{15 - x}{x^2 + 3x - 4} + \sqrt{x + 10}$

For $\sqrt{x + 10}$ we need $x + 10 \geq 0$, or $x \geq -10$.

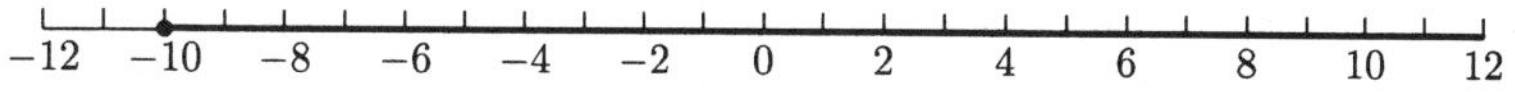

−12 −10 −8 −6 −4 −2 0 2 4 6 8 10 12

Fig. 2.4.

We also need for $x^2 + 3x - 4$ not to be zero.

$$x^2 + 3x - 4 = 0$$

$$(x + 4)(x - 1) = 0$$

$$x + 4 = 0 \qquad\qquad x - 1 = 0$$

$$x = -4 \qquad\qquad x = 1$$

We cannot let $x = -4$ and $x = 1$, so we will remove these numbers from $x \geq -10$. The domain is $[-10, -4) \cup (-4, 1) \cup (1, \infty)$.

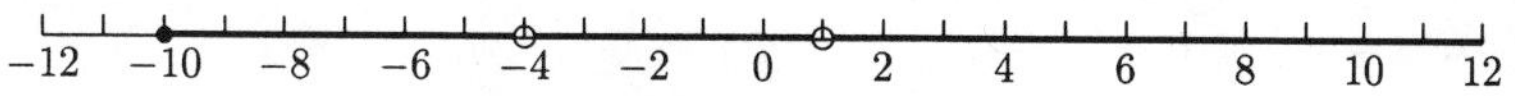

−12 −10 −8 −6 −4 −2 0 2 4 6 8 10 12

Fig. 2.5.

PRACTICE

For Problems 2–11, give the domain in interval notation.

1. A function consists of the ordered pairs $\{(h, 5), (z, 3), (i, 12)\}$. List the elements in the domain.
2. $f(x) = \dfrac{2x+3}{x-8}$
3. $f(x) = \dfrac{-1}{x^2-2x}$
4. $f(x) = \dfrac{x-3}{x^2+10}$
5. $g(x) = \sqrt[3]{6-x}$
6. $h(x) = \sqrt{x+3}$
7. $f(x) = \sqrt{4-x^2}$
8. $f(x) = \sqrt{3x^2+5}$
9. $f(x) = \dfrac{1}{\sqrt{x-9}}$
10. $f(x) = 4x^3 - 2x + 5$
11. $f(x) = \dfrac{\sqrt{x+5}}{x^2+2x-8}$

SOLUTIONS

1. The domain consists of the first coordinate of the ordered pairs—h, z, and i.
2. We cannot let $x - 8 = 0$, so we cannot let $x = 8$. The domain is $x \neq 8$, or $(-\infty, 8) \cup (8, \infty)$.
3. We cannot let $x^2 - 2x = x(x-2) = 0$, so we cannot let $x = 0$ or $x = 2$. The domain is all real numbers except 0 and 2, or $(-\infty, 0) \cup (0, 2) \cup (2, \infty)$.
4. Because $x^2 + 10 = 0$ has no real number solution, the domain is all real numbers, or $(-\infty, \infty)$.
5. We can take the cube root of any number, so the domain is all real numbers, or $(-\infty, \infty)$.
6. We must have $x + 3 \geq 0$, or $x \geq -3$. The domain is $[-3, \infty)$.

7. We need to solve $4 - x^2 = (2 - x)(2 + x) \geq 0$.

False
True
False
−5 −4 −3 −2 −1 0 1 2 3 4 5

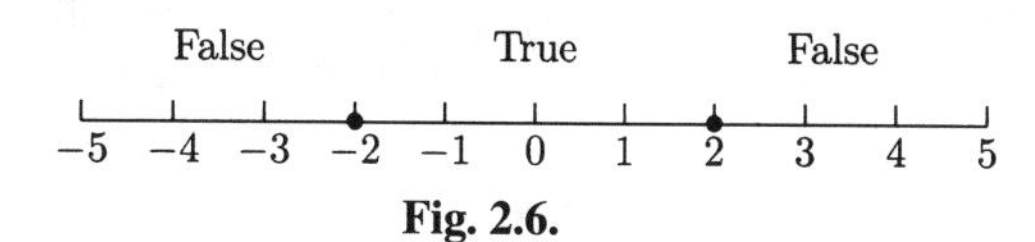

Fig. 2.6.

The domain is $[-2, 2]$.

8. Because $3x^2 + 5 \geq 0$ is true for all real numbers, the domain is $(-\infty, \infty)$.
9. We need $x - 9 > 0$. The domain is $x > 9$, or $(9, \infty)$.
10. The domain is all real numbers, or $(-\infty, \infty)$.
11. From $x + 5 \geq 0$, we have $x \geq -5$.

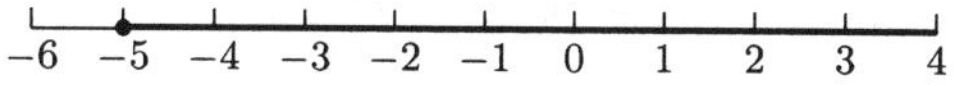

−6 −5 −4 −3 −2 −1 0 1 2 3 4

Fig. 2.7.

Now we need to solve $x^2 + 2x - 8 = (x + 4)(x - 2) = 0$.

$$x + 4 = 0 \qquad\qquad x - 2 = 0$$

$$x = -4 \qquad\qquad x = 2$$

Now we need to remove −4 and 2 from $x \geq -5$. The domain is $[-5, -4) \cup (-4, 2) \cup (2, \infty)$.

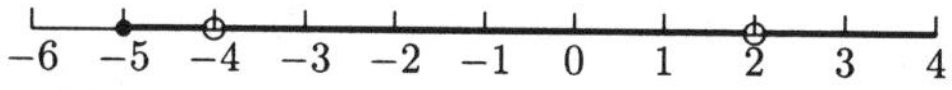

−6 −5 −4 −3 −2 −1 0 1 2 3 4

Fig. 2.8.

At times the domain of a function will matter when we are solving an applied problem. For example, suppose there is a 10″ × 18″ piece of cardboard that will be made into an open-topped box. After cutting a square x by x inches from each corner, the sides will be folded up to form the box.

18 − 2x
x
x
10 − 2x

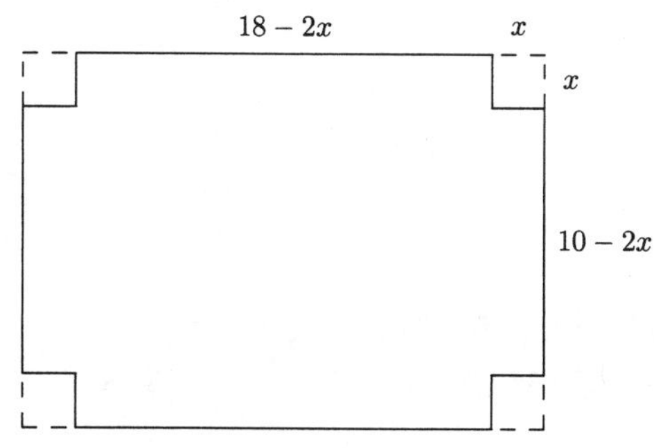

Fig. 2.9.

The volume of the box is a function of x, $V(x) = x(18-2x)(10-2x)$. What is the domain of this function? We obviously cannot cut a negative number of inches from each corner. If we cut 0 inches from each corner, we do not have a box, so x must be positive. Finally, the box is only 10 inches wide, so we can cut up to five inches from each corner. These facts make the domain $0 < x < 5$. Maximizing the volume of this box is a typical problem in a first semester of calculus. The solutions to the mathematical problem are $\frac{14\pm\sqrt{61}}{3}$ (approximately 2.0635, and 7.27008). Only one of these numbers is in the domain of the applied function, so only one of these numbers is the solution.

CHAPTER 2 REVIEW

1. Evaluate $f(x) = 4 - 2x^2$ at $x = 3$.
   (a) -14 (b) -12 (c) -10 (d) -8

2. Evaluate $f(-1)$ for

$$f(x) = \begin{cases} 5 & \text{if } x < 0 \\ x+3 & \text{if } x \geq 0 \end{cases}$$

   (a) -1 (b) 5 (c) 2 (d) 2, 5

3. Evaluate $f(u^2+v)$ for $f(x) = 4x+6$.
   (a) $(u^2+v)(4x+6)$ (b) $4u^2+v+6$
   (c) $4v^2x+6$ (d) $4u^2+4v+6$

4. What is the domain for $f(x) = \sqrt{x^2+1}$?
   (a) $(-\infty, \infty)$ (b) $(-\infty, -1] \cup [1, \infty)$
   (c) $(-\infty, -1) \cup (1, \infty)$ (d) $[-1, 1]$

5. Evaluate $\frac{f(a+h)-f(a)}{h}$ for $f(x) = x^2+3$.
   (a) $2a+h^2$ (b) $2a+h^2+3$
   (c) $2a+h$ (d) $2a+h+3$

6. What is the domain for $f(x) = \sqrt{x-5}$?
   (a) $(-\infty, 5) \cup (5, \infty)$ (b) $[5, \infty)$
   (c) $(-\infty, 5]$ (d) $(-\infty, -5) \cup (5, \infty)$

7. What is the domain for $f(x) = \frac{1}{x^2-9}$?

(a) $(-\infty, 9) \cup (9, \infty)$ (b) $(-\infty, 3) \cup (3, \infty)$

(c) $[3, \infty)$ (d) $(-\infty, -3) \cup (-3, 3) \cup (3, \infty)$

8. What is the domain for the function $\{(a, 6), (b, 6), (d, 9)\}$?

(a) $\{a, b, d\}$ (b) $\{6, 9\}$

(c) $\{a, b, d, 6, 9\}$ (d) $\{a, b, d, 9\}$

9. What is the domain for

$$f(x) = \frac{x-5}{\sqrt{x-4}}?$$

(a) $[4, 5) \cup (5, \infty)$ (b) $(-\infty, 4) \cup (4, \infty)$

(c) $[4, \infty)$ (d) $(4, \infty)$

10. What is the domain for

$$f(x) = \frac{\sqrt{x-4}}{x-5}?$$

(a) $[4, 5) \cup (5, \infty)$ (b) $(-\infty, 4) \cup (4, \infty)$

(c) $[4, \infty)$ (d) $(4, \infty)$

SOLUTIONS

1. A	2. B	3. D	4. A	5. C
6. B	7. D	8. A	9. D	10. A

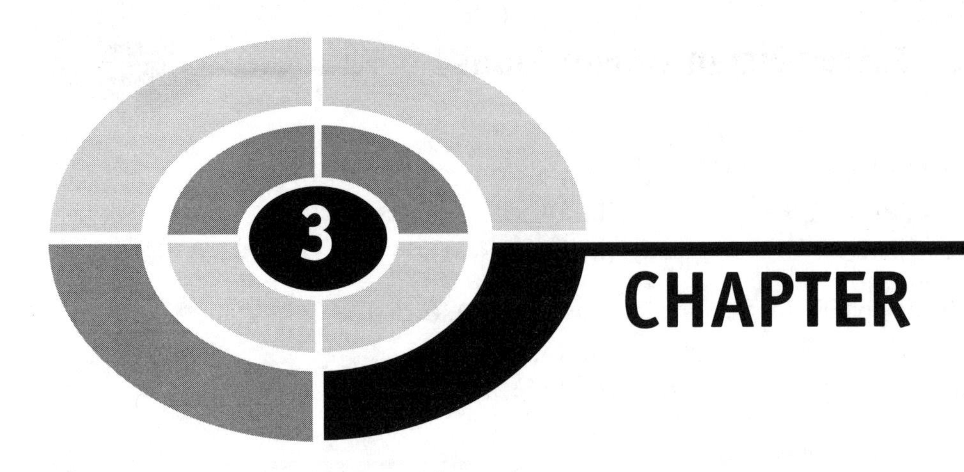

Functions and Their Graphs

The graph of a function can give us a great deal of information about the function. In this chapter we will use the graph of a function to evaluate the function, find the x- and y-intercepts (if any), the domain and range, and determine where the function is increasing or decreasing (an important idea in calculus).

To say that $f(-3) = 1$ means that the point $(-3, 1)$ is on the graph of $f(x)$. If $(5, 4)$ is a point on the graph of $f(x)$, then $f(5) = 4$.

EXAMPLE

- The graph in Figure 3.1 is the graph of $f(x) = x^3 - x^2 - 4x + 4$. Find $f(-1)$, $f(0)$, $f(3)$, and $f(-2)$.

  The point $(-1, 6)$ is on the graph means that $f(-1) = 6$.
  The point $(0, 4)$ is on the graph means that $f(0) = 4$.

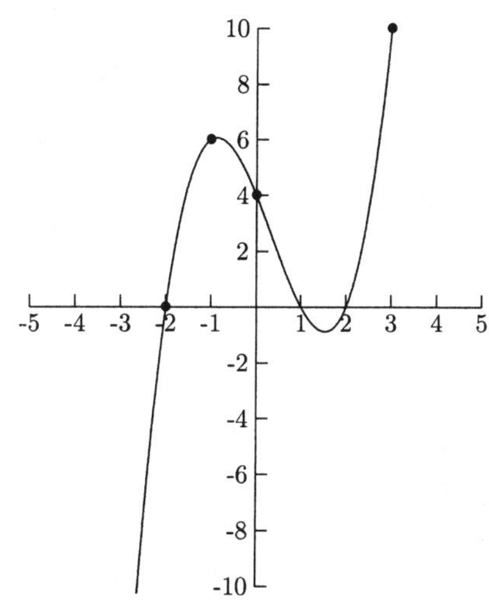

Fig. 3.1.

The point (3, 10) is on the graph means that $f(3) = 10$.
The point $(-2, 0)$ is on the graph means that $f(-2) = 0$.

The graph also shows the *intercepts* of the graph. Remember that an x-intercept is a point where the graph touches the x-axis, and the y-intercept is a point where the graph touches the y-axis. We can tell that the y-intercept for the graph in Figure 3.1 is 4 (or (0, 4)) and the x-intercepts are -2, 1, and 2 (or $(-2, 0)$, $(1, 0)$ and $(2, 0)$).

An equation "gives y as a function of x" means that for every x-value, there is a unique y-value. From this fact we can look at a graph of an equation to decide if the equation gives y as a function of x. If an x-value has more than one y-value in the equation, then there will be more than one point on the graph that has the same x-coordinate. A line through points that have the same x-coordinate is vertical. This is the idea behind the *Vertical Line Test.* The graph of an equation passes the Vertical Line Test if every vertical line touches the graph at one point or not at all. If so, then the equation is a function.

The graph of $y^2 = x$ is shown in Figure 3.2. The vertical line $x = 4$ touches the graph in two places, (4, 2) and $(4, -2)$, so y is not a function of x in the equation $y^2 = x$.

The domain of a function consists of all possible x-values. We can find the domain of a function by looking at its graph. The graph's extension horizontally shows the function's domain. The range of a function consists of all possible y-values. The graph's vertical extention shows the function's range.

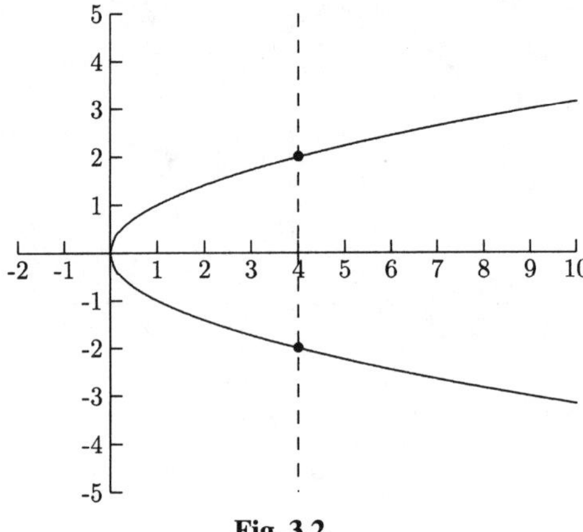

Fig. 3.2.

EXAMPLES

Give the domain and range in interval notation.

- 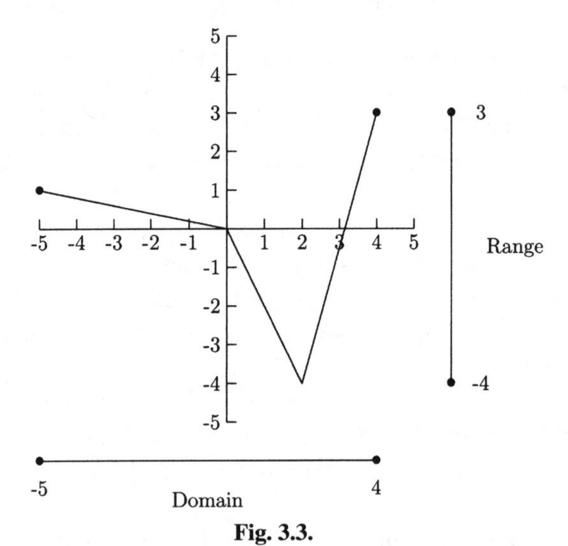

Fig. 3.3.

The graph extends horizontally from $x = -5$ to $x = 4$. Because there are closed dots on these endpoints (instead of open dots), $x = -5$ and $x = 4$ are part of the domain, too. The domain is $[-5, 4]$. The graph extends vertically from $y = -4$ to $y = 3$. The range is $[-4, 3]$.

- 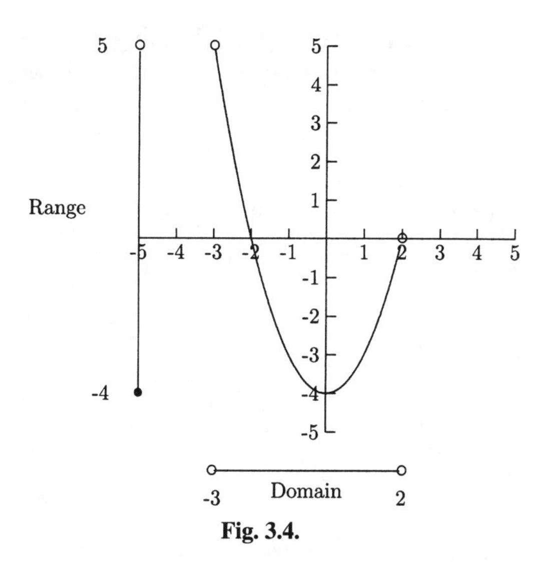

Fig. 3.4.

The graph extends horizontally from $x = -3$ to $x = 2$. Because open dots are used on $(-3, 5)$ and $(2, 0)$, these points are not on the graph, so $x = -3$ and $x = 2$ are not part of the domain. The domain is $(-3, 2)$. The graph extends vertically from $y = -4$ and $y = 5$. The range is $[-4, 5)$. We need to use a bracket around -4 because $(0, -4)$ *is* a point on the graph, and a parenthesis around 5 because the point $(-3, 5)$ is not a point on the graph.

- 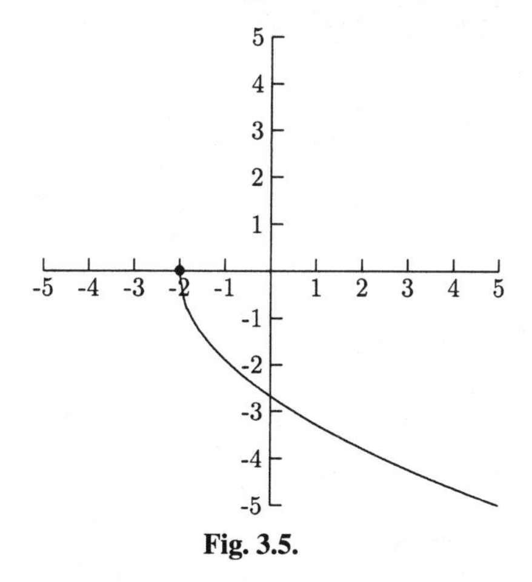
Fig. 3.5.

The graph extends horizontally from $x = -2$ on the left and vertically from below $y = 0$. The domain is $[-2, \infty)$, and the range is $(-\infty, 0]$.

A function is increasing on an interval if moving toward the right in the interval means the graph is going up. A function is decreasing on an interval if moving toward the right in the interval means the graph is going down. The function whose graph is in Figure 3.6 is increasing from $x = -3$ to $x = 0$ as well as to the right of $x = 2$. It is decreasing to the left of $x = -3$ and between $x = 0$ and $x = 2$. Using interval notation, we say the function is increasing on the intervals $(-3, 0)$ and $(2, \infty)$ and decreasing on the intervals $(-\infty, -3)$ and $(0, 2)$. For reasons covered in calculus, parentheses are used for the interval notation.

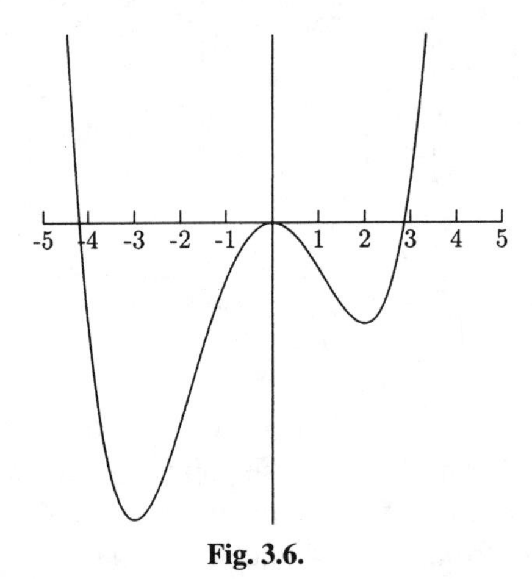

Fig. 3.6.

A function is constant on an interval if the y-values do not change. This part of the graph will be part of a horizontal line.

EXAMPLES

Determine the intervals on which the functions are increasing, decreasing or constant.

- 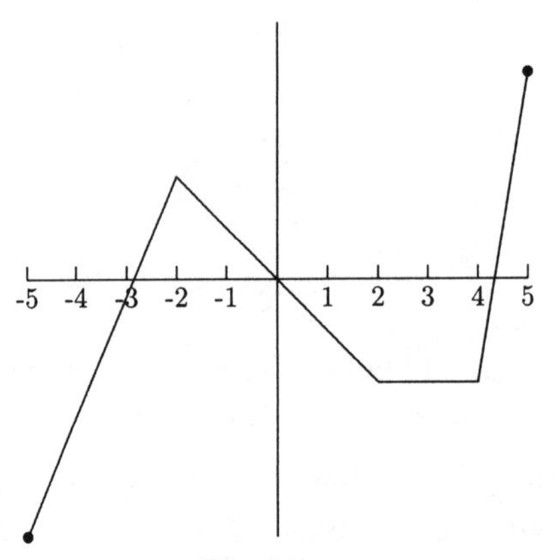

Fig. 3.7.

This function is increasing on $(-5, -2)$ and $(4, 5)$. It is decreasing on $(-2, 2)$ and constant on $(2, 4)$.

•

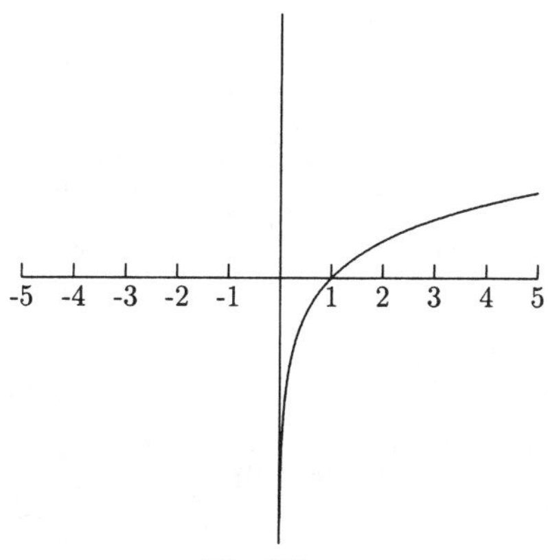

Fig. 3.8.

The function is increasing on all of its domain, $(0, \infty)$.

PRACTICE

1. Is the graph in Figure 3.9 the graph of a function?

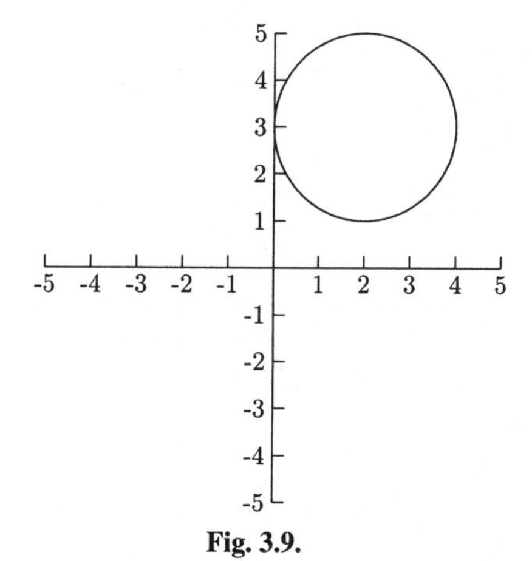

Fig. 3.9.

2. Refer to the graph of $f(x)$ in Figure 3.10.

   (a) What is $f(-3)$?
   (b) What is $f(5)$?
   (c) What is the domain?

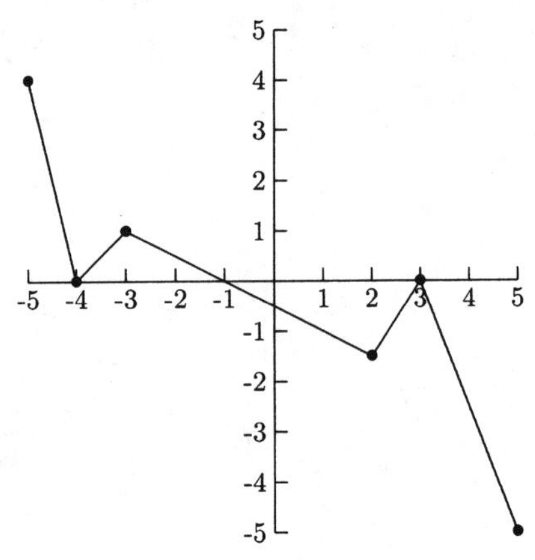

Fig. 3.10.

(d) What is the range?
(e) What are the x-intercepts?
(f) What is the y-intercept?
(g) What is/are the increasing interval(s)?
(h) What is/are the decreasing interval(s)?

3. Refer to the graph of $f(x)$ in Figure 3.11.

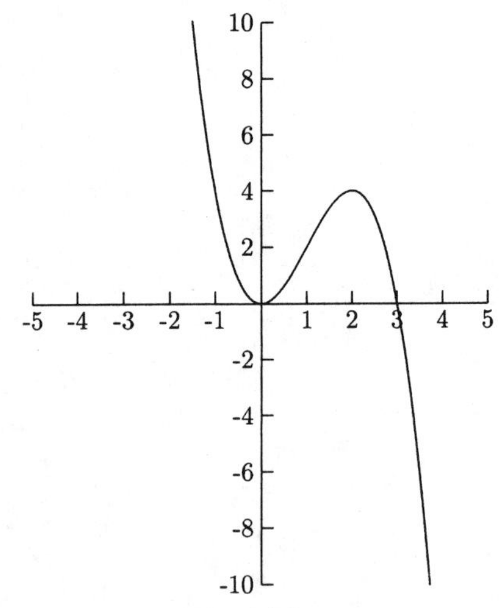

Fig. 3.11.

(a) What is $f(2)$? $f(1)$?
(b) What are the x-intercepts? What is the y-intercept?

(c) What is the domain? Range?

(d) What is the increasing interval? What are the decreasing intervals?

SOLUTIONS

1. No. The graph fails the Vertical Line Test.
2. (a) $f(-3) = 1$ because $(-3, 1)$ is a point on the graph.
   (b) $f(5) = -5$ because $(5, -5)$ is a point on the graph.
   (c) The domain is $[-5, 5]$.
   (d) The range is $[-5, 4]$.
   (e) The x-intercepts are -4, -1, and 3.
   (f) The y-intercept is $-\frac{1}{2}$.
   (g) The increasing intervals are $(-4, -3)$ and $(2, 3)$.
   (h) The decreasing intervals are $(-5, -4)$, $(-3, 2)$ and $(3, 5)$.
3. (a) $f(2) = 4$ because $(2, 4)$ is a point on the graph. $f(1) = 2$ because $(1, 2)$ is a point on the graph.
   (b) The x-intercepts are 0 and 3. The y-intercept is 0.
   (c) The domain and range are each all real numbers, $(-\infty, \infty)$.
   (d) The increasing interval is $(0, 2)$, and the decreasing intervals are $(-\infty, 0)$ and $(2, \infty)$.

Graphs are useful tools to present a lot of information in a small space. Being able to read a graph and draw conclusions from it are important in many subjects in addition to mathematics. In the example below, we will practice drawing conclusions based on information given in the graph in Figure 3.12. This graph shows the daily balance of a checking account for about two weeks. No more than one transaction (a deposit or a check written) is made in one day. For example, the balance at the end of the second day is \$350 and \$300 at the end of the third day, so a \$50 check was written on the third day.

1. On what day was a check for \$200 written?
   On the 12th day when the balance dropped from \$150 to −\$50.
2. What is the largest deposit?
   The largest increase was \$200, on the 8th day when the balance increased from \$200 to \$400.
3. What is the largest check written?
   The largest check was written on the tenth day when the balance dropped from \$400 to \$150.

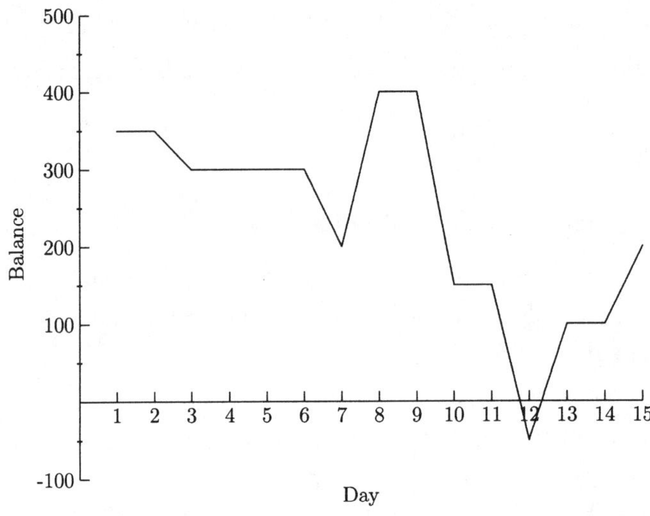

Fig. 3.12.

4. When was the account overdrawn?
   The balance was negative on the 12th day.

Average Rate of Change

Calculus deals with the rate of change. A familiar example of a rate of change is speed (or more accurately, velocity). Velocity is the rate of change of distance per unit of time. A car traveling in city traffic will generally have a lower rate of change of distance per hour than a car traveling on an interstate freeway. A glass of water placed in a refrigerator will have a lower rate of temperature change than a glass of water placed in a freezer. In calculus, you will study instantaneous rates of change of functions at different values of x. We will study the *average rate of change* in this book. As you will see in the following examples, the average rate of change can hide a lot of variation.

EXAMPLES

- Suppose $1000 was invested in company stock of some manufacturing company. The value of the investment at the beginning of each year is given in Table 3.1.

    1. How much did the stock increase per year on average from the beginning of Year 3 to the beginning of Year 6?

Table 3.1

Year	Value (in dollars)	Change from the previous year
1	1000	New investment of $1000
2	1205	Gain of $205
3	1162	Loss of $43
4	1025	Loss of $137
5	1190	Gain of $165
6	1252	Gain of $62
7	1434	Gain of $182
8	1621	Gain of $187
9	2015	Gain of $394
10	2845	Gain of $830

For this three-year period the investment increased in value from $1162 to $1252. The average rate of change is

$$\frac{1252 - 1162}{6 - 3} = \frac{90}{3} = 30 \text{ per year.}$$

2. What was the average annual loss from the beginning of Year 2 to the beginning of Year 5?
   The average rate of change during this three-year period is

$$\frac{1190 - 1205}{5 - 2} = \frac{-15}{3} = -5 \text{ per year.}$$

   The negative symbol means that this change is a loss, not a gain.
3. What was the average annual increase over the full period?
   The average increase in the investment over the full nine years is

$$\frac{2845 - 1000}{10 - 1} = \frac{1845}{9} = 205 \text{ per year.}$$

- Find the average rate of change between $(-3, 9)$ and $(-1, 3)$ and between $(1, 1.5)$ and $(3, 1.125)$ for the function whose graph is given in Figure 3.13.
  The average rate of change of a function between two points on the graph is the slope of the line containing the two points. For the points $(-3, 9)$ and $(-1, 3)$, $x_1 = -3$, $y_1 = 9$ and $x_2 = -1$ and $y_2 = 3$.

$$\text{Average rate of change} = \frac{y_2 - y_1}{x_2 - x_1} = \frac{3 - 9}{-1 - (-3)} = \frac{-6}{2} = \frac{-3}{1} = -3$$

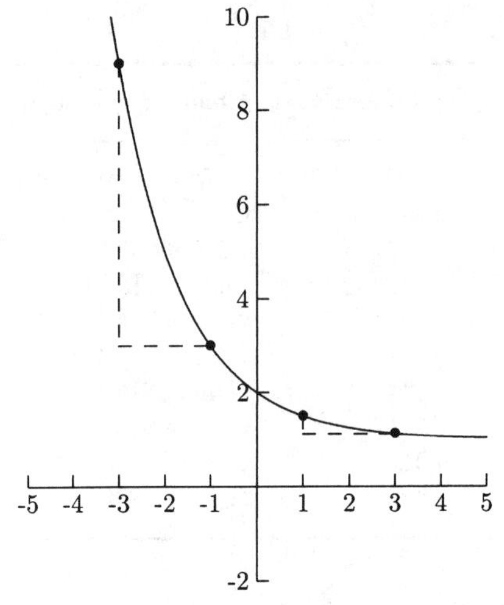

Fig. 3.13.

Between $x = -3$ and $x = -1$, the y-values of this function *decrease* by 3 as x increases by 1, on average.
For the points (1, 1.5) and (3, 1.125) $x_1 = 1$, $y_1 = 1.5$ and $x_2 = 3$, $y_2 = 1.125$.

$$\text{Average rate of change} = \frac{y_2 - y_1}{x_2 - x_1} = \frac{1.125 - 1.5}{3 - 1} = \frac{-0.375}{2} = -0.1875$$

Between $x = 1$ and $x = 3$, the y-values of this function decrease on average by 0.1875 as x increases by 1.

- Find the average rate of change of $f(x) = -3x^2 + 10$ between $x = -1$ and $x = 2$.
  Once we have found the y-values by putting these x-values into the function, we will find the slope of the line containing these two points.

$$y_1 = f(x_1) = f(-1) = -3(-1)^2 + 10 = 7$$

$$y_2 = f(x_2) = f(2) = -3(2)^2 + 10 = -2$$

$$\text{Average rate of change} = \frac{y_2 - y_1}{x_2 - x_1} = \frac{-2 - 7}{2 - (-1)} = \frac{-9}{3} = \frac{-3}{1} = -3$$

Between $x = -1$ and $x = 2$, this function decreases on average by 3 as x increases by 1.

PRACTICE

1. A sales representative's pay is based on his sales. Table 3.2 shows his salary during one year.

Table 3.2

Month	Pay
January (1)	2100
February (2)	2000
March (3)	2400
April (4)	2700
May (5)	2500
June (6)	3000
July (7)	3500
August (8)	3600
September (9)	2500
October (10)	2000
November (11)	2000
December (12)	2100

How much did his monthly pay change on average between January and July? Between July and December? Between October and December?

2. Find the average rate of change between the indicated points of the function whose graph is given in Figure 3.14.

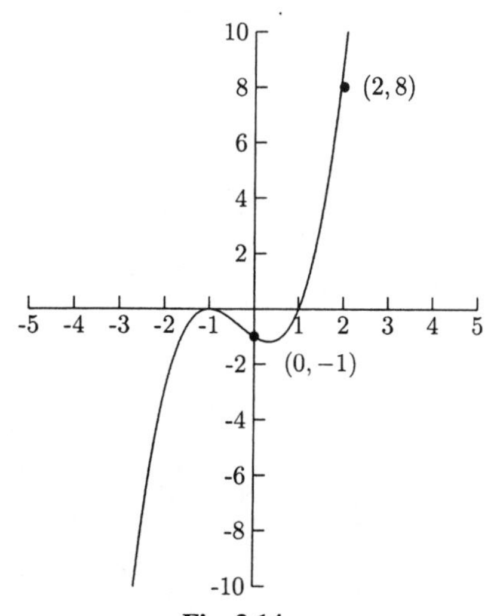

Fig. 3.14.

3. Find the average rate of change for $f(x) = 2 - x^3$ between $x = -2$ and $x = 1$.
4. Find the average rate of change for $f(x) = 6x - 3$ between $x = -5$ and $x = 3$ and between $x = 0$ and $x = 8$.

SOLUTIONS

1. The average monthly increase between January and July is the slope of the line containing the points (1,2100) and (7,3500).

$$\frac{3500 - 2100}{7 - 1} \approx 233$$

The average monthly decrease between July and December is the slope of the line containing the points (7,3500) and (12,2100).

$$\frac{2100 - 3500}{12 - 7} = -280$$

The average monthly increase from October to December is the slope of the line containing the points (10,2000) and (12,2100).

$$\frac{2100 - 2000}{12 - 10} = 50$$

2. $x_1 = 0,\ y_1 = -1$ and $x_2 = 2,\ y_2 = 8$

$$\text{Average rate of change} = \frac{8 - (-1)}{2 - 0} = \frac{9}{2}$$

3.

$$y_1 = f(x_1) = f(-2) = 2 - (-2)^3 = 10$$

$$y_2 = f(x_2) = f(1) = 2 - (1)^3 = 1$$

$$\text{Average rate of change} = \frac{1 - 10}{1 - (-2)} = -3$$

4. For $x_1 = -5$ and $x_2 = 3$—

$$y_1 = f(x_1) = f(-5) = 6(-5) - 3 = -33$$

$$y_2 = f(x_2) = f(3) = 6(3) - 3 = 15$$

$$\text{Average rate of change} = \frac{15 - (-33)}{3 - (-5)} = 6$$

For $x_1 = 0$ and $x_2 = 8$—

$$y_1 = f(x_1) = f(0) = 6(0) - 3 = -3$$

$$y_2 = f(x_2) = f(8) = 6(8) - 3 = 45$$

$$\text{Average rate of change} = \frac{45 - (-3)}{8 - 0} = 6$$

The average rate of change between *any* two points on a linear function is the slope.

Newton's Quotient gives the average rate of change of $f(x)$ between $x_1 = a$ and $x_2 = a + h$.

$$y_1 = f(x_1) = f(a) \qquad y_2 = f(x_2) = f(a + h)$$

$$\text{Average rate of change} = \frac{y_2 - y_1}{x_2 - x_1} = \frac{f(a + h) - f(a)}{a + h - a} = \frac{f(a + h) - f(a)}{a}$$

Even and Odd Functions

A graph is *symmetric* if one half looks like the other half. We might also say that one half of the graph is a reflection of the other.

When a graph has symmetry, we usually say that it is symmetric with respect to a line or a point. The graph in Figure 3.15 is symmetric with respect to the x-axis because the half of the graph above the x-axis is a reflection of the half below the x-axis. The graph in Figure 3.16 is symmetric with respect to the y-axis.

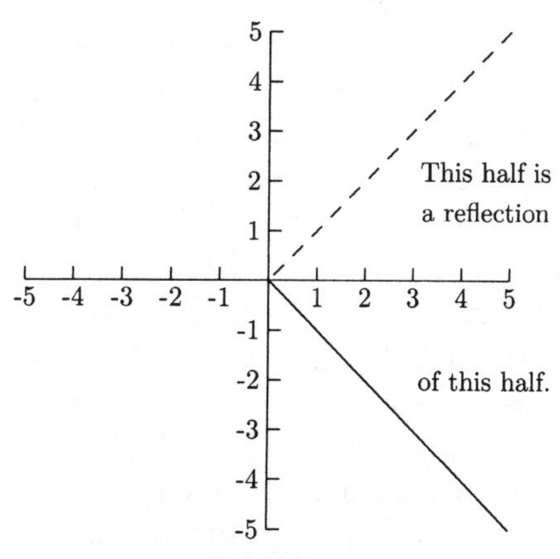

Fig. 3.15.

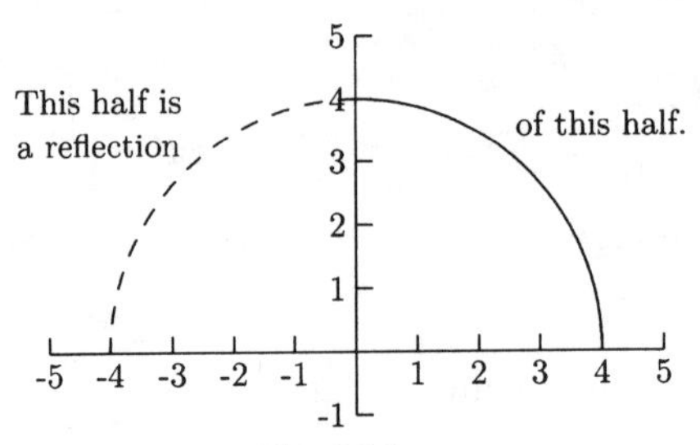

Fig. 3.16.

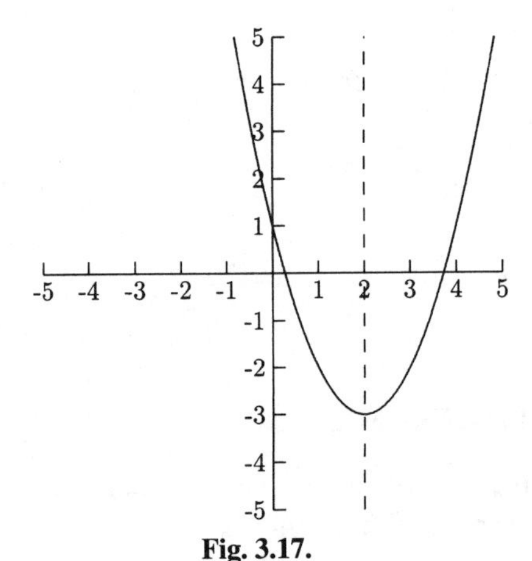

Fig. 3.17.

The graph in Figure 3.17 is symmetric with respect to the vertical line $x = 2$.

One type of symmetry that is a little harder to see is *origin symmetry*. A graph has origin symmetry if folding the graph along the x-axis then again along the y-axis would have one part of the graph coincide with the other part. The graphs in Figures 3.18 and 3.19 have origin symmetry.

Knowing in advance whether or not the graph of a function is symmetric can make sketching the graph less work. We can use algebra to decide if the graph of a function has y-axis symmetry or origin symmetry. Except for the function $f(x) = 0$, the graph of a function will not have x-axis symmetry because x-axis symmetry would cause a graph to fail the Vertical Line Test.

For the graph of a function to be symmetric with respect to the y-axis, a point on the left side of the y-axis will have a mirror image on the right side of the graph.

The graph of a function with y-axis symmetry has the property that (x, y) is on the graph means that $(-x, y)$ is also on the graph. The functional notation for this idea is $f(x) = f(-x)$. "$f(x) = f(-x)$" says that the y value for x $(f(x))$

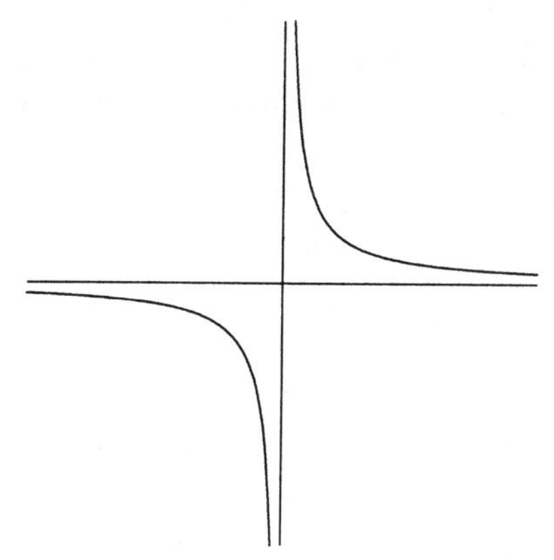

Fig. 3.18.

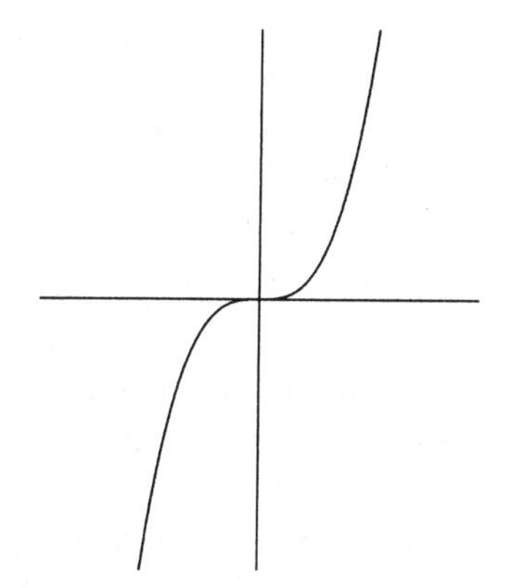

Fig. 3.19.

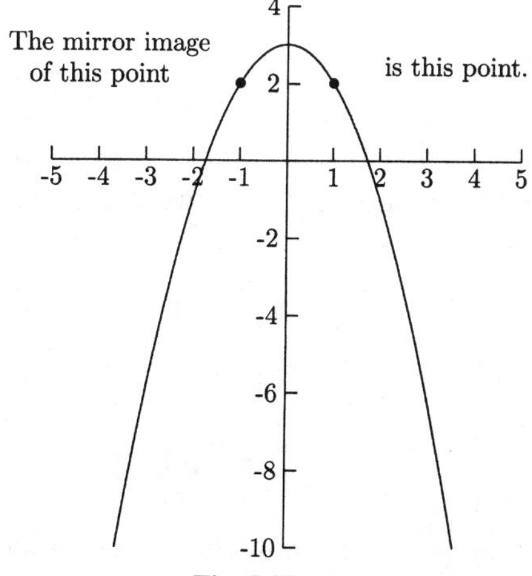

Fig. 3.20.

is the same as the y-value for $-x$ ($f(-x)$). If evaluating a function at $-x$ does not change the equation, then its graph will have y-axis symmetry. Such functions are called *even functions*.

For a function whose graph is symmetric with respect to the origin, the mirror image of (x, y) is $(-x, -y)$.

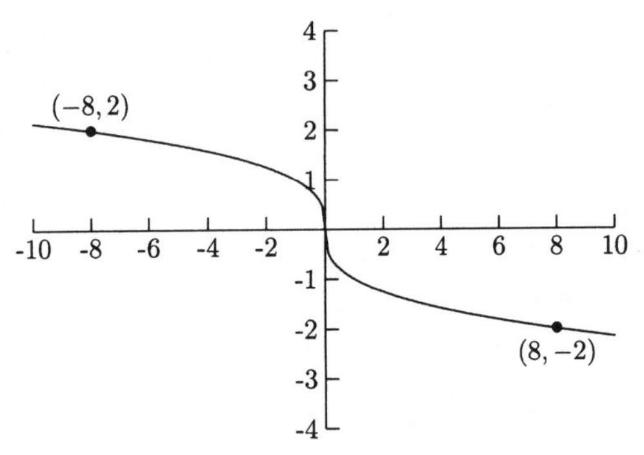

Fig. 3.21.

The functional notation for this idea is $f(-x) = -f(x)$. "$f(-x) = -f(x)$" says that the y-value for $-x$ ($f(-x)$) is the opposite of the y-value for x ($-f(x)$). If evaluating a function at $-x$ changes the equation to its negative, then the graph of the function will be symmetric with respect to the origin. These functions are called *odd functions*.

In order to work the following problems, we will need the following facts.

$$a(-x)^{\text{even power}} = ax^{\text{even power}} \quad \text{and} \quad a(-x)^{\text{odd power}} = -ax^{\text{odd power}}$$

EXAMPLES

Determine if the given function is even (its graph is symmetric with respect to the y-axis), odd (its graph is symmetric with respect to the origin), or neither.

- $f(x) = x^2 - 2$
  Does evaluating $f(x)$ at $-x$ change the function? If so, is $f(-x) = -(x^2 - 2) = -f(x)$?

$$f(-x) = (-x)^2 - 2 = x^2 - 2$$

Evaluating $f(x)$ at $-x$ does not change the function, so the function is even.

- $f(x) = x^3 + 5x$
  Does evaluating $f(x)$ at $-x$ change the function? If so, is $f(-x) = -(x^3 + 5x) = -f(x)$?

$$f(-x) = (-x)^3 + 5(-x) = -x^3 - 5x = -(x^3 + 5x) = -f(x)$$

  Evaluating $f(x)$ at $-x$ gives us $-f(x)$, so the function is odd.

- $f(x) = \dfrac{x}{x+1}$
  Does evaluating $f(x)$ at $-x$ change the function? If so, is $f(-x) = -\frac{x}{x+1} = -f(x)$?

$$f(-x) = \frac{-x}{-x+1}$$

  Because $f(-x)$ is not the same as $f(x)$ nor the same as $-f(x)$, the function is neither even nor odd.

PRACTICE

For 1–4, determine whether or not the graph has symmetry. If it does, determine the kind of symmetry it has. For 5–8, determine if the functions are even, odd, or neither.

1.

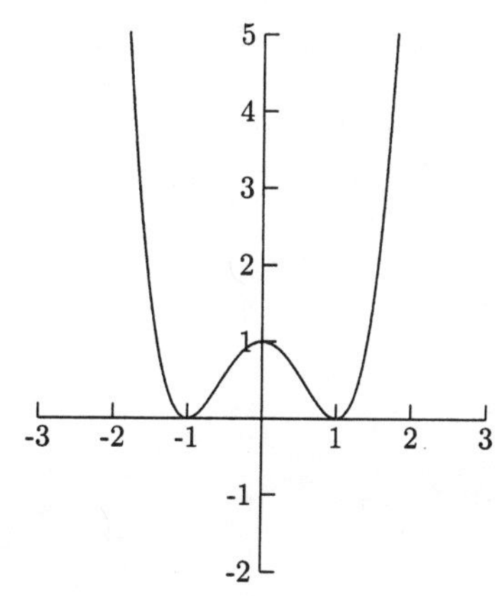

Fig. 3.22.

2.

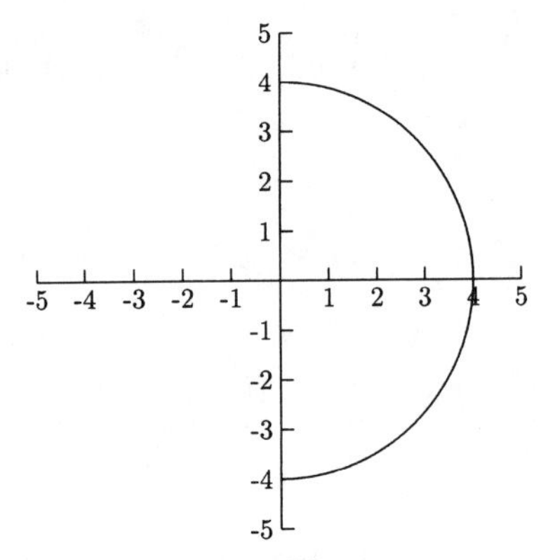

Fig. 3.23.

3.

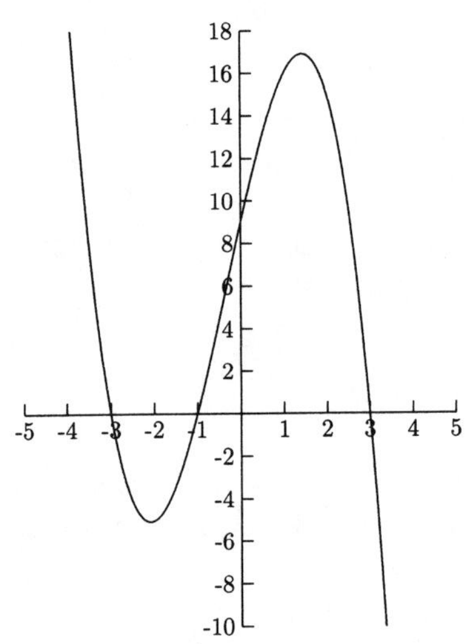

Fig. 3.24.

4.

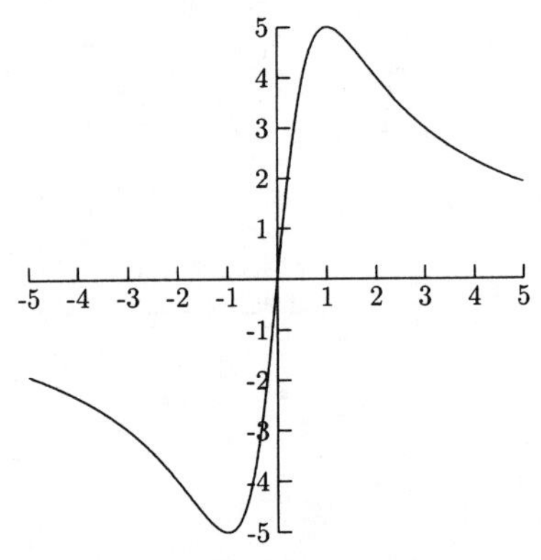

Fig. 3.25.

5. $f(x) = x^3 + 6$
6. $f(x) = 3x^2 - 2$
7. $f(x) = \dfrac{x^2 - 3}{x^3 + 2x}$
8. $g(x) = \sqrt[3]{x}$

SOLUTIONS

1. This graph has y-axis symmetry.
2. This graph has x-axis symmetry.
3. This graph does not have symmetry.
4. This graph has origin symmetry.
5. $f(-x) = (-x)^3 + 6 = -x^3 + 6$

   $f(-x) \neq f(x)$ and $f(-x) \neq -f(x)$, making $f(x)$ neither even nor odd.
6. $f(-x) = 3(-x)^2 - 2 = 3x^2 - 2$

   $f(-x) = f(x)$, making $f(x)$ even.
7. $$f(-x) = \frac{(-x)^2 - 3}{(-x)^3 + 2(-x)} = \frac{x^2 - 3}{-x^3 - 2x} = \frac{x^2 - 3}{-(x^3 + 2x)}$$

   $$= -\frac{x^2 - 3}{x^3 + 2x} = -f(x)$$

   $f(-x) = -f(x)$, making $f(x)$ odd.
8. $g(-x) = \sqrt[3]{-x} = -\sqrt[3]{x} = -g(x)$

   $g(-x) = -g(x)$, making $g(x)$ odd.

CHAPTER 3 REVIEW

Problems 1–2 refer to the graph in Figure 3.26.

1. Find $f(1)$.

   (a) -1 (b) -2 (c) 1 (d) 2

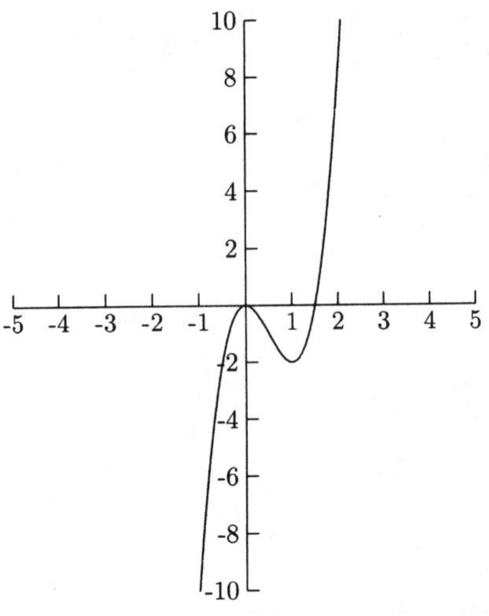

Fig. 3.26.

2. Where is the function decreasing?
(a) $(-\infty, 0) \cup (1, \infty)$ (b) $(0, -2)$ (c) $(0, 1)$ (d) $(1, \infty)$

Problems 3–6 refer to the graph of $f(x)$ in Figure 3.27.

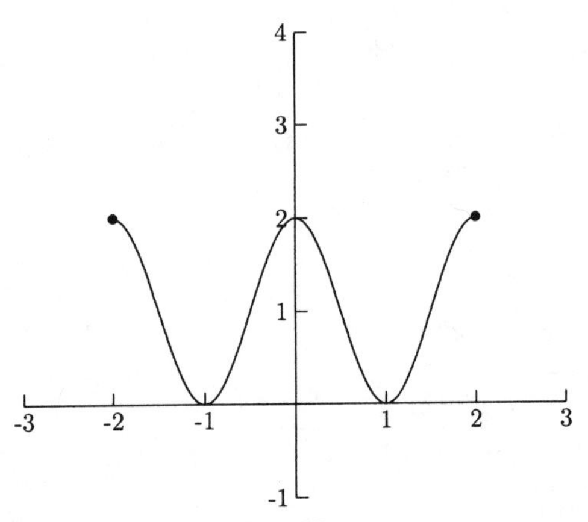

Fig. 3.27.

3. What is the domain?
(a) $[0, 2]$ (b) $[2, 0]$ (c) $[-2, 2]$ (d) $[-2, 0]$

4. What is the range?
(a) $[0, 2]$ (b) $[2, 0]$ (c) $[-2, 2]$ (d) $[-2, 0]$

5. What is the average rate of change of the function between $x = -2$ and $x = 1$?
   (a) $-\frac{1}{2}$ (b) $-\frac{2}{3}$ (c) $-\frac{3}{4}$ (d) -1

6. Is the graph in Figure 3.27 symmetric?
   (a) Yes, with respect to the x-axis. (c) Yes, with respect to the origin.
   (b) Yes, with respect to the y-axis. (d) No.

7. Find the average rate of change for $f(x) = \frac{1}{x+1}$ between $x = 0$ and $x = 2$.
   (a) -3 (b) $-\frac{1}{3}$ (c) 3 (d) $\frac{1}{3}$

8. Is the function $f(x) = 3x^2 + 5$ even, odd, or neither?
   (a) Even
   (b) Odd
   (c) Neither
   (d) Cannot be determined without the graph

9. Is the function $f(x) = 3x^3 + 5$ even, odd, or neither?
   (a) Even
   (b) Odd
   (c) Neither
   (d) Cannot be determined without the graph

10. Is the function $f(x) = 4x^2/(x^3 + x)$ even, odd, or neither?
   (a) Even
   (b) Odd
   (c) Neither
   (d) Cannot be determined without the graph

SOLUTIONS

1. B	2. C	3. C	4. A	5. B
6. B	7. B	8. A	9. C	10. B

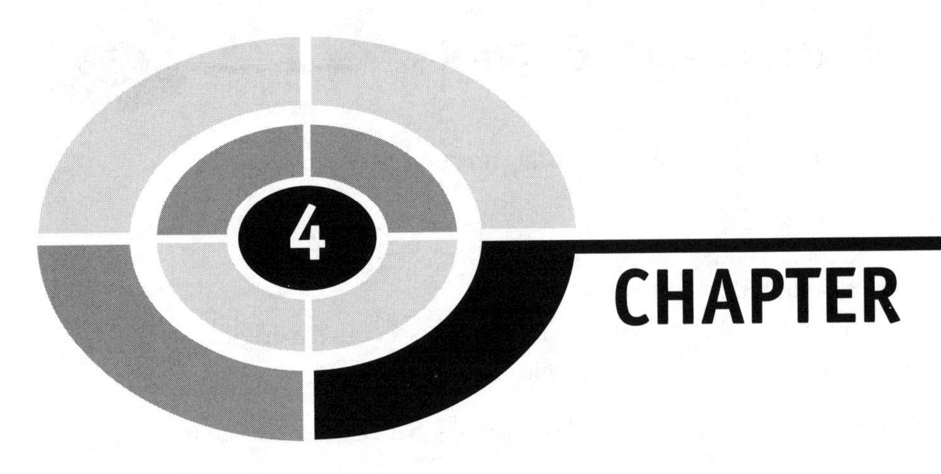

Combinations of Functions and Inverse Functions

Most of the functions studied in calculus are some combination of only a few families of functions, most of the combinations are arithmetic. We can add two functions, $f + g(x)$, subtract them, $f - g(x)$, multiply them, $fg(x)$, and divide them $\frac{f}{g}(x)$. The domain of $f + g(x)$, $f - g(x)$, and $fg(x)$, is the intersection of the domain of $f(x)$ and $g(x)$. In other words, their domain is where the domain of $f(x)$ overlaps the domain of $g(x)$. The domain of $\frac{f}{g}(x)$ is the same, except we need to remove any x that makes $g(x) = 0$.

EXAMPLES

Find $f + g(x)$, $f - g(x)$, $fg(x)$, and $\frac{f}{g}(x)$ and their domain.

- $f(x) = x^2 - 2x + 5$ and $g(x) = 6x - 10$

$$f + g(x) = f(x) + g(x) = (x^2 - 2x + 5) + (6x - 10) = x^2 + 4x - 5$$

$$f - g(x) = f(x) - g(x) = (x^2 - 2x + 5) - (6x - 10) = x^2 - 8x + 15$$

$$fg(x) = f(x)g(x) = (x^2 - 2x + 5)(6x - 10) = 6x^3 - 10x^2 - 12x^2 + 20x + 30x - 50$$

$$= 6x^3 - 22x^2 + 50x - 50$$

$$\frac{f}{g}(x) = \frac{f(x)}{g(x)} = \frac{x^2 - 2x + 5}{6x - 10}$$

The domain of $f + g(x)$, $f - g(x)$, and $fg(x)$ is $(-\infty, \infty)$. The domain of $\frac{f}{g}(x)$ is $x \neq \frac{5}{3}$ (from $6x - 10 = 0$), or $(-\infty, \frac{5}{3}) \cup (\frac{5}{3}, \infty)$.

- $f(x) = x - 3$ and $g(x) = \sqrt{x + 2}$

$$f + g(x) = x - 3 + \sqrt{x + 2} \qquad f - g(x) = x - 3 - \sqrt{x + 2}$$

$$fg(x) = (x - 3)\sqrt{x + 2} \qquad \frac{f}{g}(x) = \frac{x - 3}{\sqrt{x + 2}}$$

The domain for $f + g(x)$, $f - g(x)$, and $fg(x)$ is $[-2, \infty)$ (from $x + 2 \geq 0$). The domain for $\frac{f}{g}(x)$ is $(-2, \infty)$ because we need $\sqrt{x + 2} \neq 0$.

An important combination of two functions is *function composition*. This involves evaluating one function at the other. The notation for composing f with g is $f \circ g(x)$. By definition, $f \circ g(x) = f(g(x))$, this means that we substitute $g(x)$ for x in $f(x)$.

EXAMPLES

Find $f \circ g(x)$ and $g \circ f(x)$.

- $f(x) = x^2 + 1$ and $g(x) = 3x + 2$

$$f \circ g(x) = f(g(x))$$
$$= f(3x + 2) \qquad \text{Replace } g(x) \text{ with } 3x + 2.$$
$$= (3x + 2)^2 + 1 \qquad \text{Substitute } 3x + 2 \text{ for } x \text{ in } f(x).$$
$$= (3x + 2)(3x + 2) + 1 = 9x^2 + 12x + 5$$

$$g \circ f(x) = g(f(x))$$
$$= g(x^2 + 1) \qquad \text{Replace } f(x) \text{ with } x^2 + 1.$$
$$= 3(x^2 + 1) + 2 \qquad \text{Substitute } x^2 + 1 \text{ for } x \text{ in } g(x).$$
$$= 3x^2 + 3 + 2 = 3x^2 + 5$$

- $f(x) = \sqrt{5x - 2}$ and $g(x) = x^2$

$$f \circ g(x) = f(g(x))$$
$$= f(x^2) \qquad \text{Replace } g(x) \text{ with } x^2.$$
$$= \sqrt{5x^2 - 2} \qquad \text{Substitute } x^2 \text{ for } x \text{ in } f(x).$$

$$g \circ f(x) = g(f(x))$$
$$= g(\sqrt{5x - 2}) \qquad \text{Replace } f(x) \text{ with } \sqrt{5x - 2}.$$
$$= (\sqrt{5x - 2})^2 \qquad \text{Substitute } \sqrt{5x - 2} \text{ for } x \text{ in } g(x).$$
$$= 5x - 2$$

- $f(x) = \dfrac{1}{x + 1}$ and $g(x) = \dfrac{2x - 1}{x + 3}$

$$f \circ g(x) = f(g(x)) = f\left(\frac{2x - 1}{x + 3}\right)$$
$$= \frac{1}{\frac{2x-1}{x+3} + 1} = \frac{1}{\frac{2x-1}{x+3} + 1 \cdot \frac{x+3}{x+3}}$$
$$= \frac{1}{\frac{2x-1+x+3}{x+3}} = \frac{1}{\frac{3x+2}{x+3}}$$

$$= 1 \div \frac{3x+2}{x+3} = 1 \cdot \frac{x+3}{3x+2}$$

$$= \frac{x+3}{3x+2}$$

$$g \circ f(x) = g(f(x)) = g\left(\frac{1}{x+1}\right)$$

$$= \frac{2(\frac{1}{x+1}) - 1}{\frac{1}{x+1} + 3} = \frac{\frac{2}{x+1} - 1 \cdot \frac{x+1}{x+1}}{\frac{1}{x+1} + 3 \cdot \frac{x+1}{x+1}}$$

$$= \frac{\frac{2-(x+1)}{x+1}}{\frac{1+3(x+1)}{x+1}} = \frac{\frac{-x+1}{x+1}}{\frac{3x+4}{x+1}}$$

$$= \frac{-x+1}{x+1} \div \frac{3x+4}{x+1} = \frac{-x+1}{x+1} \cdot \frac{x+1}{3x+4}$$

$$= \frac{-x+1}{3x+4}$$

At times, we only need to find $f \circ g(x)$ for a particular value of x. The y-value for $g(x)$ becomes the x-value for $f(x)$.

EXAMPLE

- Find $f \circ g(-1)$, $f \circ g(0)$, and $g \circ f(1)$ for $f(x) = 4x + 3$ and $g(x) = 2 - x^2$.

$f \circ g(-1) = f(g(-1))$	Compute $g(-1)$.
$= f(1)$	$g(-1) = 2 - (-1)^2 = 1$
$= 4(1) + 3 = 7$	Evaluate $f(x)$ at $x = 1$.

$f \circ g(0) = f(g(0))$	Compute $g(0)$.
$= f(2)$	$g(0) = 2 - 0^2 = 2$
$= 4(2) + 3 = 11$	Evaluate $f(x)$ at $x = 2$.

$$\begin{aligned} g \circ f(1) &= g(f(1)) && \text{Compute } f(1). \\ &= g(7) && f(1) = 4(1) + 3 = 7 \\ &= 2 - 7^2 = -47 && \text{Evaluate } g(x) \text{ at } x = 7. \end{aligned}$$

We can compose two functions at a single x-value by looking at the graphs of the individual functions. To find $f \circ g(a)$, we will look at the graph of $g(x)$ to find the point whose x-coordinate is a. The y-coordinate of this point will be $g(a)$. Then we will look at the graph of $f(x)$ to find the point whose x-coordinate is $g(a)$. The y-coordinate of this point will be $f(g(a)) = f \circ g(a)$.

EXAMPLE

Refer to Figure 4.1. The solid graph is the graph of $f(x)$, and the dashed graph is the graph of $g(x)$.

- Find $f \circ g(-1)$, $f \circ g(3)$, $f \circ g(5)$, and $g \circ f(0)$.

$$\begin{aligned} f \circ g(-1) &= f(g(-1)) && \text{Look for } x = -1 \text{ on } g(x). \\ &= f(-2) && (-1, -2) \text{ is on the graph of } g(x), \text{ so } g(-1) = -2. \\ &= 0 && (-2, 0) \text{ is on the graph of } f(x), \text{ so } f(-2) = 0. \end{aligned}$$

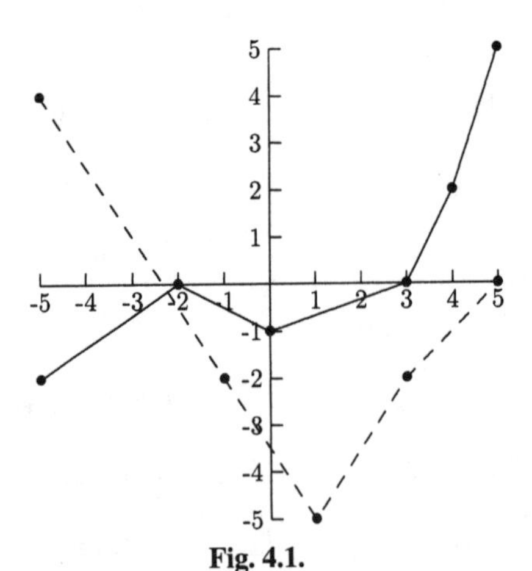

Fig. 4.1.

$f \circ g(3) = f(g(3))$ Look for $x = 3$ on $g(x)$.

$= f(-2)$ $(3, -2)$ is on the graph of $g(x)$, so $g(3) = -2$.

$= 0$ $(-2, 0)$ is on the graph of $f(x)$, so $f(-2) = 0$.

$f \circ g(5) = f(g(5))$ Look for $x = 5$ on $g(x)$.

$= f(0)$ $(5, 0)$ is on the graph of $g(x)$, so $g(5) = 0$.

$= -1$ $(0, -1)$ is on the graph of $f(x)$, so $f(0) = -1$.

$g \circ f(0) = g(f(0))$ Look for $x = 0$ on $f(x)$.

$= g(-1)$ $(0, -1)$ is on the graph of $f(x)$, so $f(0) = -1$.

$= -2$ $(-1, -2)$ is on the graph of $g(x)$, so $g(-1) = -2$.

Unfortunately, finding the domain for the composition of two functions is not straightforward. The definition for the domain of $f \circ g(x)$ is the set of all real numbers x such that $g(x)$ is in the domain of $f(x)$. When finding the domain for $f \circ g(x)$, begin with the domain with $g(x)$. Then remove any x-value whose y-value is not in the domain for $f(x)$. For example if $f(x) = \frac{1}{x}$ $g(x) = x + 3$, the y-values for $g(x)$ are $x + 3$. We need for $x + 3$ to be nonzero for $f \circ g(x) = \frac{1}{x+3}$.

EXAMPLES

Find the domain for $f \circ g(x)$.

- $f(x) = \frac{1}{x^2}$ and $g(x) = \sqrt{2x - 6}$

  The domain for $g(x)$ is $x \geq 3$ (from $2x - 6 \geq 0$). Are there any x-values in $[3, \infty)$ we cannot put into $\frac{1}{(\sqrt{2x-6})^2}$? We cannot allow $(\sqrt{2x - 6})^2$ to be zero, so we cannot allow $x = 3$. The domain for $f \circ g(x)$ is $(3, \infty)$.

- $f(x) = \frac{1}{x}$ and $g(x) = \frac{x-1}{x+1}$

  The domain for $g(x)$ is $x \neq -1$. Are there any x-values we need to remove from $x \neq -1$? We need to find any real numbers that are not in the domain for

  $$f \circ g(x) = f(g(x)) = f\left(\frac{x-1}{x+1}\right) = \frac{1}{\frac{x-1}{x+1}}$$

  The denominator of this fraction is $\frac{x-1}{x+1}$, so we cannot allow $\frac{x-1}{x+1}$ to be zero. A fraction equals zero only when the numerator is zero, so we cannot allow $x - 1$ to be zero. We must remove $x = 1$ from the domain of $g(x)$.

The domain of $f \circ g(x)$ is $x \neq -1, 1$, or $(-\infty, -1) \cup (-1, 1) \cup (1, \infty)$. This function simplifies to $f \circ g(x) = \frac{x+1}{x-1}$, which hides the fact that we cannot let $x = -1$.

Any number of functions can be composed together. Functions can even be composed with themselves. When composing three or more functions together, we will work from the right to the left, performing one composition at a time.

EXAMPLES

Find $f \circ f(x)$ and $f \circ g \circ h(x)$.

- $f(x) = x^3$, $g(x) = 2x - 5$, and $h(x) = x^2 + 1$.

$$f \circ f(x) = f(f(x)) = f(x^3) = (x^3)^3 = x^9$$

For $f \circ g \circ h(x)$, we will begin with $g \circ h(x) = g(h(x)) = g(x^2 + 1) = 2(x^2 + 1) - 5 = 2x^2 - 3$. Now we need to evaluate $f(x)$ at $2x^2 - 3$.

$$\begin{aligned} f \circ g \circ h(x) &= f(g(h(x)) \\ &= f(2x^2 - 3) = (2x^2 - 3)^3 \end{aligned}$$

- $f(x) = 3x + 7$, $g(x) = |x - 2|$, and $h(x) = x^4 - 5$

$$f \circ f(x) = f(f(x)) = f(3x + 7) = 3(3x + 7) + 7 = 9x + 28$$

$$f \circ g \circ h(x) = f \circ g(h(x))$$

$$g(h(x)) = g(x^4 - 5) = |(x^4 - 5) - 2| = |x^4 - 7|$$

$$\begin{aligned} f \circ g(h(x)) &= f(g(h(x))) = f(|x^4 - 7|) \\ &= 3|x^4 - 7| + 7 \end{aligned}$$

In order for calculus students to use some formulas, they need to recognize complicated functions as a combination of simpler functions. Sums, differences, products, and quotients are easy to see, but some compositions of functions are less obvious.

EXAMPLES

Find functions $f(x)$ and $g(x)$ so that $h(x) = f \circ g(x)$.

- $h(x) = \sqrt{x + 16}$

Although there are many possibilities for $f(x)$ and $g(x)$, there is usually one pair of functions that is obvious. Usually we want $g(x)$ to be the computation that is done first and $f(x)$, the computation to be done last. Here, when computing the y-value for $h(x)$, we would calculate $x + 16$. This will be $g(x)$. The last calculation will be to take the square root. This will be $f(x)$. If we let $f(x) = \sqrt{x}$ and $g(x) = x + 16$, we have $f \circ g(x) = f(g(x)) = f(x + 16) = \sqrt{x + 16} = h(x)$.

- $h(x) = \frac{2}{x^2+1}$
  When computing a y-value for $h(x)$, we would first find $x^2 + 1$. This will be $g(x)$. This number will be the denominator of a fraction whose numerator is 2. This will be $f(x)$, a fraction whose numerator is 2 and whose denominator is x. If $f(x) = \frac{2}{x}$ and $g(x) = x^2 + 1$,

$$f \circ g(x) = f(g(x)) = f(x^2 + 1) = \frac{2}{x^2 + 1} = h(x).$$

PRACTICE

1. $f(x) = 3x^2 + x$ and $g(x) = x - 4$
   (a) Find $f + g(x)$, $f - g(x)$, $fg(x)$, and $\frac{f}{g}(x)$.
   (b) What is the domain for $\frac{f}{g}(x)$?
   (c) Find $f \circ g(x)$ and $g \circ f(x)$.
   (d) What is the domain for $f \circ g(x)$?
   (e) Find $f \circ g(1)$ and $g \circ f(0)$.
   (f) Find $f \circ f(x)$.
2. Find $f \circ g(x)$, $g \circ f(x)$, and the domain for $f \circ g(x)$.

$$f(x) = \frac{2x - 3}{x + 4} \quad \text{and} \quad g(x) = \frac{x}{x - 1}$$

3. Refer to the graphs in Figure 4.2. The solid graph is the graph of $f(x)$, and the dashed graph is the graph of $g(x)$. Find $f \circ g(1)$, $f \circ g(4)$, and $g \circ f(-2)$.
4. Find $f \circ g \circ h(x)$ for $f(x) = \frac{1}{x+3}$, $g(x) = 4x + 9$, and $h(x) = 5x^2 - 1$.
5. Find functions $f(x)$ and $g(x)$ so that $h(x) = f \circ g(x)$, where $h(x) = (x - 5)^3 + 2$.

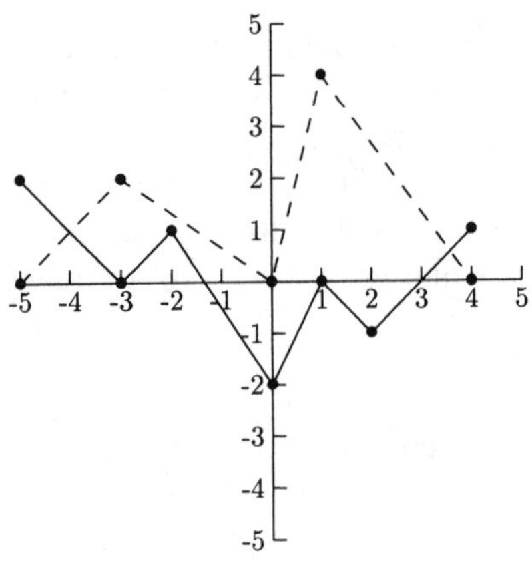

Fig. 4.2.

SOLUTIONS

1. (a)

$$f + g(x) = (3x^2 + x) + (x - 4) = 3x^2 + 2x - 4$$

$$f - g(x) = (3x^2 + x) - (x - 4) = 3x^2 + 4$$

$$fg(x) = (3x^2 + x)(x - 4) = 3x^3 - 11x^2 - 4x$$

$$\frac{f}{g}(x) = \frac{3x^2 + x}{x - 4}$$

(b) The domain is $x \neq 4$, (from $x - 4 = 0$), or $(-\infty, 4) \cup (4, \infty)$.

(c)

$$f \circ g(x) = f(g(x)) = f(x - 4) = 3(x - 4)^2 + (x - 4)$$

$$= 3(x - 4)(x - 4) + x - 4 = 3x^2 - 23x + 44$$

$$g \circ f(x) = g(f(x)) = g(3x^2 + x) = 3x^2 + x - 4$$

(d) The domain for $g(x)$ is all real numbers. We can let x be any real number for $f(x)$, so we do not need to remove anything from the domain of $g(x)$. The domain of $f \circ g(x)$ is all real numbers, or $(-\infty, \infty)$.

(e)

$$
\begin{aligned}
f \circ g(1) &= f(g(1)) \\
&= f(-3) && g(1) = 1 - 4 = -3 \\
&= 24 && f(-3) = 3(-3)^2 + (-3) = 24 \\
g \circ f(0) &= g(f(0)) \\
&= g(0) && f(0) = 3(0)^2 + 0 = 0 \\
&= -4 && g(0) = 0 - 4 = -4
\end{aligned}
$$

(f)

$$
\begin{aligned}
f \circ f(x) &= f(f(x)) = f(3x^2 + x) = 3(3x^2 + x)^2 + (3x^2 + x) \\
&= 3(3x^2 + x)(3x^2 + x) + 3x^2 + x = 27x^4 + 18x^3 + 6x^2 + x
\end{aligned}
$$

2.

$$
\begin{aligned}
f \circ g(x) &= f(g(x)) = f\left(\frac{x}{x-1}\right) \\
&= \frac{2\left(\frac{x}{x-1}\right) - 3}{\frac{x}{x-1} + 4} = \frac{\frac{2x}{x-1} - 3 \cdot \frac{x-1}{x-1}}{\frac{x}{x-1} + 4 \cdot \frac{x-1}{x-1}} \\
&= \frac{\frac{2x - 3(x-1)}{x-1}}{\frac{x + 4(x-1)}{x-1}} = \frac{\frac{-x+3}{x-1}}{\frac{5x-4}{x-1}} \\
&= \frac{-x+3}{x-1} \div \frac{5x-4}{x-1} = \frac{-x+3}{x-1} \cdot \frac{x-1}{5x-4} \\
&= \frac{-x+3}{5x-4}
\end{aligned}
$$

$$
\begin{aligned}
g \circ f(x) &= g(f(x)) = g\left(\frac{2x-3}{x+4}\right) \\
&= \frac{\frac{2x-3}{x+4}}{\frac{2x-3}{x+4} - 1} = \frac{\frac{2x-3}{x+4}}{\frac{2x-3}{x+4} - 1 \cdot \frac{x+4}{x+4}} \\
&= \frac{\frac{2x-3}{x+4}}{\frac{2x-3-(x+4)}{x+4}} = \frac{\frac{2x-3}{x+4}}{\frac{x-7}{x+4}}
\end{aligned}
$$

$$= \frac{2x-3}{x+4} \div \frac{x-7}{x+4} = \frac{2x-3}{x+4} \cdot \frac{x+4}{x-7}$$

$$= \frac{2x-3}{x-7}$$

The domain of $g(x)$ is $x \neq 1$. Now we need to see if there is anything we need to remove from $x \neq 1$. Before simplifying $f \circ g(x)$, we have

$$\frac{2\left(\frac{x}{x-1}\right) - 3}{\frac{x}{x-1} + 4}.$$

The denominator of this fraction cannot be zero, so we must have $\frac{x}{x-1} + 4 \neq 0$.

$$\frac{x}{x-1} + 4 = 0$$

$$(x-1)\left(\frac{x}{x-1} + 4\right) = (x-1)0$$

$$x + 4(x-1) = 0$$

$$5x - 4 = 0$$

$$x = \frac{4}{5}$$

The domain is $x \neq 1, \frac{4}{5}$, or $(-\infty, \frac{4}{5}) \cup (\frac{4}{5}, 1) \cup (1, \infty)$.

While it seems that $x = -4$ might not be allowed in the domain of $f \circ g(x)$, $x = -4$ is in the domain.

$$f \circ g(-4) = f(g(-4))$$

$$= f\left(\frac{4}{5}\right) \qquad g(-4) = \frac{-4}{-4-1} = \frac{4}{5}$$

$$= -\frac{7}{24} \qquad f\left(\frac{4}{5}\right) = \frac{2(4/5) - 3}{4/5 + 4} = -\frac{7}{24}$$

3.

$f \circ g(1) = f(g(1))$	Look for $x = 1$ on $g(x)$.
$= f(4)$	$(1, 4)$ is on the graph of $g(x)$, so $g(1) = 4$.
$= 1$	$(4, 1)$ is on the graph of $f(x)$, so $f(4) = 1$.

$$f \circ g(4) = f(g(4)) \qquad \text{Look for } x = 4 \text{ on } g(x).$$
$$= f(0) \qquad (4, 0) \text{ is on the graph of } g(x), \text{ so } g(4) = 0.$$
$$= -2 \qquad (0, -2) \text{ is on the graph of } f(x), \text{ so } f(0) = -2.$$

$$g \circ f(-2) = g(f(-2)) \qquad \text{Look for } x = -2 \text{ on the graph of } f(x).$$
$$= g(1) \qquad (-2, 1) \text{ is on the graph of } f(x), \text{ so } f(-2) = 1.$$
$$= 4 \qquad (1, 4) \text{ is on the graph of } g(x), \text{ so } g(1) = 4.$$

4.

$$f \circ g \circ h(x) = f \circ g(h(x)) = f(g(h(x)))$$
$$g(h(x)) = g(5x^2 - 1) = 4(5x^2 - 1) + 9 = 20x^2 + 5$$
$$f(g(h(x))) = f(20x^2 + 5) = \frac{1}{(20x^2 + 5) + 3} = \frac{1}{20x^2 + 8}$$

5. One possibility is $g(x) = x - 5$ and $f(x) = x^3 + 2$.

$$f \circ g(x) = f(g(x)) = f(x - 5) = (x - 5)^3 + 2 = h(x)$$

Inverse Functions

We can apply many operations on functions that we can apply to real numbers—adding, multiplying, raising to powers, etc. These operations can have identities and functions have inverses in the same way they do with real numbers. The additive identity for function addition is $i(x) = 0$. Each function has an additive inverse, $-f(x)$ is the additive inverse for $f(x)$. The multiplicative identity for function multiplication is $i(x) = 1$, and the multiplicative inverse for $f(x)$ is $\frac{1}{f(x)}$.

If we look at function composition as an operation on functions, then we can ask whether or not there is an identity for this operation and whether or not functions have inverses for this operation. There is an identity for this operation, $i(x) = x$. For any function $f(x)$, $f \circ i(x) = f(i(x)) = f(x)$. *Some* functions have inverses. Later we will see which functions have inverses and how to find inverses. The notation for the inverse function of $f(x)$ is $f^{-1}(x)$. This is different from $(f(x))^{-1}$, which is the multiplicative inverse for $f(x)$. For now, we will be given two functions that are said to be inverses of each other. We will use function composition to verify that they are.

EXAMPLES

Verify that $f(x)$ and $g(x)$ are inverses.

- $f(x) = 2x + 5$ and $g(x) = \frac{1}{2}x - \frac{5}{2}$
  We will show that $f \circ g(x) = x$ and $g \circ f(x) = x$.

$$f \circ g(x) = f(g(x)) = f\left(\frac{1}{2}x - \frac{5}{2}\right)$$

$$= 2\left(\frac{1}{2}x - \frac{5}{2}\right) + 5 = x - 5 + 5 = x$$

$$g \circ f(x) = g(f(x)) = g(2x+5) = \frac{1}{2}(2x+5) - \frac{5}{2}$$

$$= x + \frac{5}{2} - \frac{5}{2} = x$$

- $f(x) = 5x^3 - 6$ and $g(x) = \sqrt[3]{\frac{x+6}{5}}$

$$f \circ g(x) = f(g(x)) = f\left(\sqrt[3]{\frac{x+6}{5}}\right) = 5\left(\sqrt[3]{\frac{x+6}{5}}\right)^3 - 6$$

$$= 5\left(\frac{x+6}{5}\right) - 6 = x + 6 - 6 = x$$

$$g \circ f(x) = g(f(x)) = g(5x^3 - 6) = \sqrt[3]{\frac{5x^3 - 6 + 6}{5}}$$

$$= \sqrt[3]{\frac{5x^3}{5}} = \sqrt[3]{x^3} = x$$

- $f(x) = \dfrac{2x-1}{x+4}$ and $g(x) = \dfrac{4x+1}{2-x}$

$$f \circ g(x) = f(g(x)) = f\left(\frac{4x+1}{2-x}\right) = \frac{2\left(\frac{4x+1}{2-x}\right) - 1}{\frac{4x+1}{2-x} + 4}$$

$$= \frac{\frac{2(4x+1)}{2-x} - 1 \cdot \frac{2-x}{2-x}}{\frac{4x+1}{2-x} + 4 \cdot \frac{2-x}{2-x}}$$

$$= \frac{\frac{8x+2-(2-x)}{2-x}}{\frac{4x+1+4(2-x)}{2-x}} = \frac{\frac{9x}{2-x}}{\frac{9}{2-x}}$$

$$= \frac{9x}{2-x} \div \frac{9}{2-x} = \frac{9x}{2-x} \cdot \frac{2-x}{9} = x$$

$$g \circ f(x) = g(f(x)) = g\left(\frac{2x-1}{x+4}\right) = \frac{4\left(\frac{2x-1}{x+4}\right)+1}{2-\frac{2x-1}{x+4}}$$

$$= \frac{\frac{4(2x-1)}{x+4} + 1 \cdot \frac{x+4}{x+4}}{2 \cdot \frac{x+4}{x+4} - \frac{2x-1}{x+4}} = \frac{\frac{8x-4+x+4}{x+4}}{\frac{2(x+4)-(2x-1)}{x+4}}$$

$$= \frac{\frac{9x}{x+4}}{\frac{9}{x+4}} = \frac{9x}{x+4} \div \frac{9}{x+4}$$

$$= \frac{9x}{x+4} \cdot \frac{x+4}{9} = x$$

If we think of a function as a collection of points on a graph, or ordered pairs, then the only thing that makes $f(x)$ different from $f^{-1}(x)$ is that their x-coordinates and y-coordinates are reversed. For example, if $(3, -1)$ is a point on the graph of $f(x)$, then $(-1, 3)$ is a point on the graph of $f^{-1}(x)$.

EXAMPLE

The graph of a function $f(x)$ is given in Figure 4.3. Use the graph of $f(x)$ to sketch the graph of $f^{-1}(x)$.

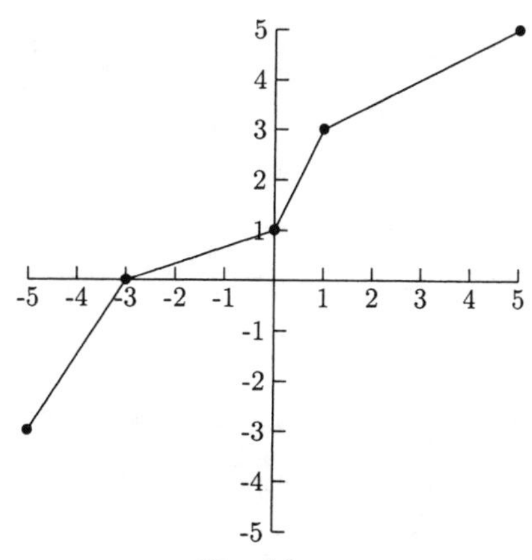

Fig. 4.3.

We will make a table of values for $f(x)$ and switch the x and y columns for $f^{-1}(x)$.

Table 4.1

x	$y = f(x)$
-5	-3
-3	0
0	1
1	3
5	5

To get the table for $f^{-1}(x)$, we will switch the x- and y-values.

Table 4.2

x	$y = f^{-1}(x)$
-3	-5
0	-3
1	0
3	1
5	5

The solid graph is the graph of $f(x)$, and the dashed graph is the graph of $f^{-1}(x)$.

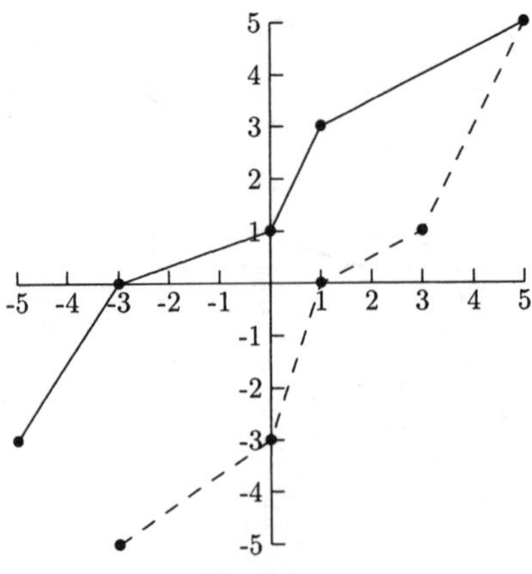

Fig. 4.4.

If $f(x)$ is a function that has an inverse, then the graph of $f^{-1}(x)$ is a reflection of the graph of $f(x)$ across the line $y = x$.

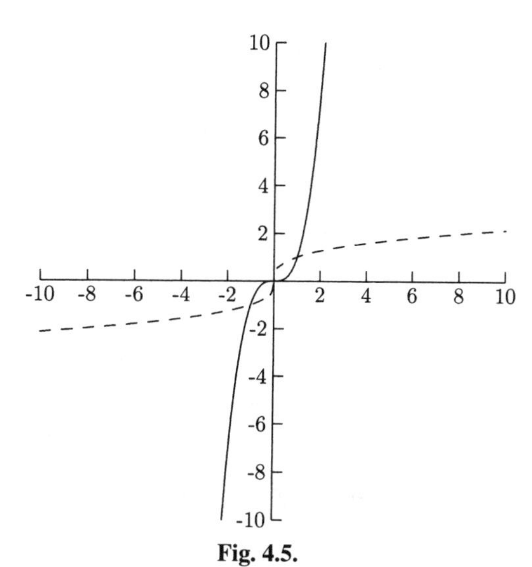

Fig. 4.5.

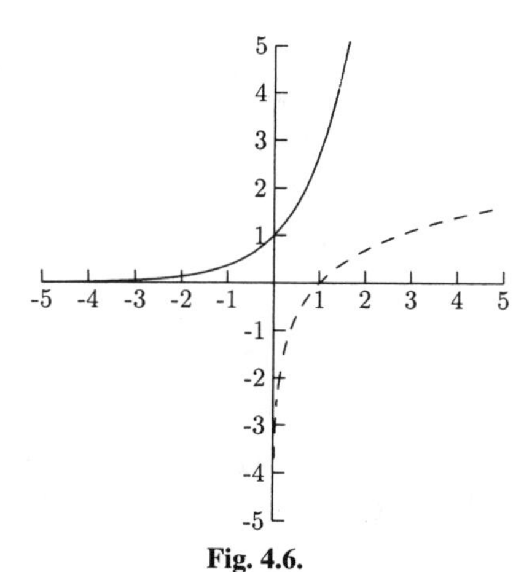

Fig. 4.6.

A function has an inverse if its graph passes the Horizontal Line Test—if any horizontal line touches the graph in more than one place, then the function will not have an inverse. Functions whose graphs pass the Horizontal Line Test are called *one-to-one* functions. For a one-to-one function, every x will be paired with exactly one y *and* every y will be paired with exactly one x.

EXAMPLE

- The graph of $f(x)$ is given in Figure 4.7. Is $f(x)$ one to one?

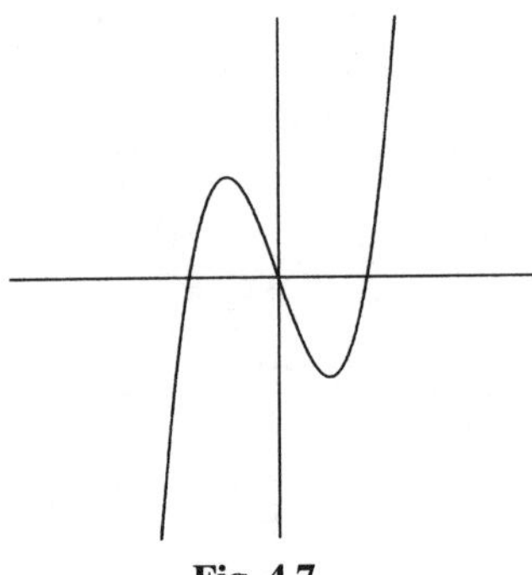

Fig. 4.7.

This graph fails the Horizontal Line Test, so $f(x)$ is not one to one.

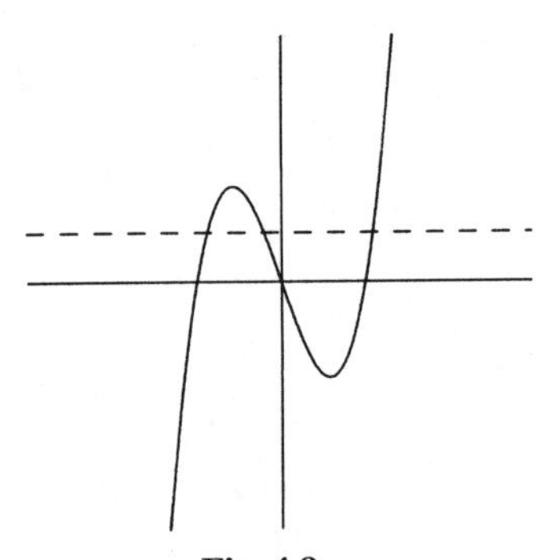

Fig. 4.8.

For functions that are not one to one, we can restrict the domain to force the function to be one to one. The function whose graph is in Figure 4.9, $f(x) = x^2 - 3$, is not one to one. If we restrict the domain to $x \geq 0$, then the new function is one to one.

Finding the inverse function is not hard, but it can be a little tedious. The steps below show the process of algebraically switching x and y.

1. Replace $f(x)$ with x, and replace x with y.
2. Solve this equation for y.
3. Replace y with $f^{-1}(x)$.

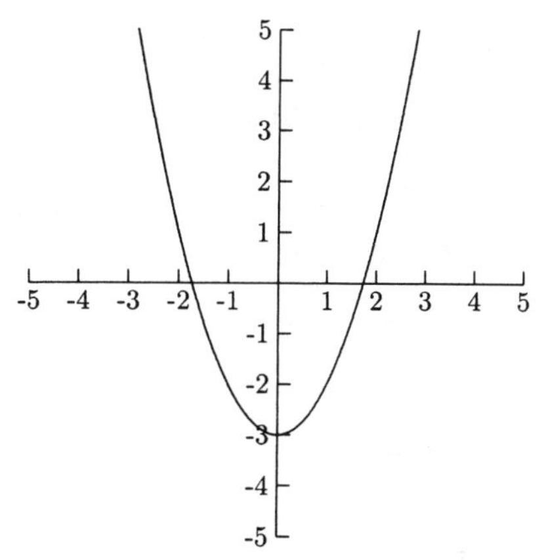

Fig. 4.9.

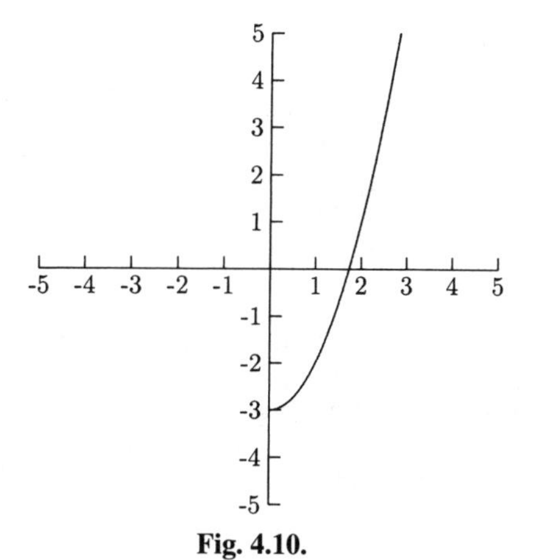

Fig. 4.10.

EXAMPLES

Find $f^{-1}(x)$.

- $f(x) = 6x + 14$

$$x = 6y + 14 \qquad \text{Step 1}$$

$$x - 14 = 6y \qquad \text{Step 2}$$

$$\frac{x - 14}{6} = y$$

$$f^{-1}(x) = \frac{x - 14}{6} \qquad \text{Step 3}$$

- $f(x) = 9(x-4)^5$

$$x = 9(y-4)^5 \qquad \text{Step 1}$$

$$\frac{x}{9} = (y-4)^5 \qquad \text{Step 2}$$

$$\sqrt[5]{\frac{x}{9}} = y - 4$$

$$\sqrt[5]{\frac{x}{9}} + 4 = y$$

$$f^{-1}(x) = \sqrt[5]{\frac{x}{9}} + 4 \qquad \text{Step 3}$$

- $f(x) = \frac{1-x}{2-x}$

$$x = \frac{1-y}{2-y} \qquad \text{Step 1}$$

$$x(2-y) = 1 - y \qquad \text{Step 2}$$

$$2x - xy = 1 - y$$

$$2x - 1 = xy - y \qquad y \text{ terms on one side, non-}y \text{ terms on other side}$$

$$2x - 1 = y(x-1) \qquad \text{Factor } y$$

$$\frac{2x-1}{x-1} = y$$

$$f^{-1}(x) = \frac{2x-1}{x-1} \qquad \text{Step 3}$$

PRACTICE

1. Show that $f(x) = \frac{1}{2}x + 7$ and $g(x) = 2x - 14$ are inverses.
2. Show that $f(x) = \sqrt[3]{x-8}$ and $g(x) = x^3 + 8$ are inverses.
3. Show that $f(x) = \frac{x+2}{x-3}$ and $g(x) = \frac{3x+2}{x-1}$ are inverses.
4. Use the graph of $f(x)$ in Figure 4.11 to sketch the graph of $f^{-1}(x)$.
5. Find $f^{-1}(x)$ for $f(x) = 5x + 12$.
6. Find $g^{-1}(x)$ for $g(x) = \sqrt[3]{2x} - 1$.
7. Find $f^{-1}(x)$ for $f(x) = \frac{2x-3}{6x+1}$

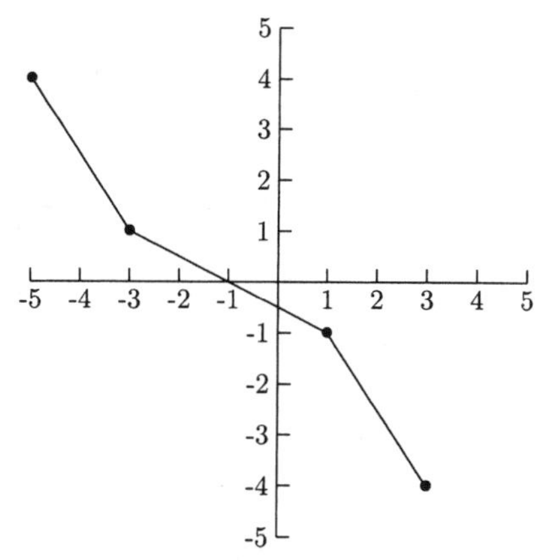

Fig. 4.11.

SOLUTIONS

1.

$$f \circ g(x) = f(g(x)) = f(2x - 14) = \frac{1}{2}(2x - 14) + 7 = x - 7 + 7 = x$$

$$g \circ f(x) = g(f(x)) = g\left(\frac{1}{2}x + 7\right) = 2\left(\frac{1}{2}x + 7\right) - 14 = x + 14 - 14 = x$$

2.

$$f \circ g(x) = f(g(x)) = f(x^3 + 8) = \sqrt[3]{(x^3 + 8) - 8} = \sqrt[3]{x^3} = x$$

$$g \circ f(x) = g(f(x)) = g(\sqrt[3]{x - 8}) = (\sqrt[3]{x - 8})^3 + 8 = x - 8 + 8 = x$$

3.

$$f \circ g(x) = f(g(x)) = f\left(\frac{3x+2}{x-1}\right) = \frac{\frac{3x+2}{x-1} + 2}{\frac{3x+2}{x-1} - 3}$$

$$= \frac{\frac{3x+2}{x-1} + 2 \cdot \frac{x-1}{x-1}}{\frac{3x+2}{x-1} - 3 \cdot \frac{x-1}{x-1}} = \frac{\frac{3x+2+2(x-1)}{x-1}}{\frac{3x+2-3(x-1)}{x-1}}$$

$$= \frac{\frac{5x}{x-1}}{\frac{5}{x-1}} = \frac{5x}{x-1} \div \frac{5}{x-1}$$

$$= \frac{5x}{x-1} \cdot \frac{x-1}{5} = x$$

$$g \circ f(x) = g(f(x)) = g\left(\frac{x+2}{x-3}\right) = \frac{3\left(\frac{x+2}{x-3}\right)+2}{\frac{x+2}{x-3}-1}$$

$$= \frac{\frac{3(x+2)}{x-3}+2\cdot\frac{x-3}{x-3}}{\frac{x+2}{x-3}-1\cdot\frac{x-3}{x-3}} = \frac{\frac{3x+6+2(x-3)}{x-3}}{\frac{x+2-(x-3)}{x-3}}$$

$$= \frac{\frac{5x}{x-3}}{\frac{5}{x-3}} = \frac{5x}{x-3} \div \frac{5}{x-3}$$

$$= \frac{5x}{x-3} \cdot \frac{x-3}{5} = x$$

4. The solid graph in Figure 4.12 is the graph of $f(x)$, and the dashed graph is the graph of $f^{-1}(x)$.

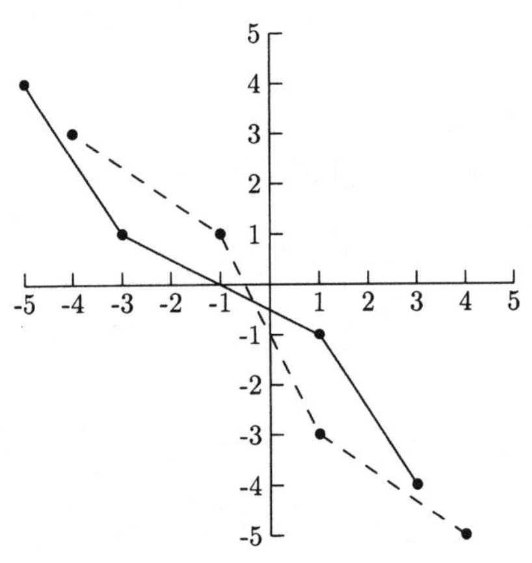

Fig. 4.12.

5.

$$x = 5y + 12$$

$$x - 12 = 5y$$

$$\frac{x-12}{5} = y \text{ so, } f^{-1}(x) = \frac{x-12}{5}$$

6.

$$x = \sqrt[3]{2y} - 1$$

$$x + 1 = \sqrt[3]{2y}$$

$$(x+1)^3 = 2y$$

$$\frac{(x+1)^3}{2} = y \text{ so } g^{-1}(x) = \frac{(x+1)^3}{2}$$

7.

$$x = \frac{2y-3}{6y+1}$$

$$x(6y+1) = 2y-3$$

$$6xy + x = 2y - 3$$

$$x + 3 = 2y - 6xy$$

$$x + 3 = y(2-6x)$$

$$\frac{x+3}{2-6x} = y \text{ so } f^{-1}(x) = \frac{x+3}{2-6x}$$

CHAPTER 4 REVIEW

Problems 1–5 refer to $f(x) = \frac{1}{x-3}$ and $g(x) = 2x + 4$.

1. Find the domain for $f + g(x)$.
   (a) $(3, \infty)$ (b) $(-\infty, 3) \cup (3, \infty)$
   (c) $(-\infty, 3] \cup [3, \infty)$ (d) $[3, \infty)$
2. Find $f \circ g(x)$.
   (a) $\frac{1}{2x+1}$ (b) $x - 3$ (c) $\frac{2}{x-3} + 4$ (d) $\frac{2x+4}{x-3}$
3. Find $g \circ f(x)$.
   (a) $\frac{1}{2x+1}$ (b) $x - 3$ (c) $\frac{2}{x-3} + 4$ (d) $\frac{2x+4}{x-3}$
4. Find $f \circ g(4)$
   (a) 12 (b) $\frac{1}{9}$ (c) 6 (d) 48
5. Find $f^{-1}(x)$.
   (a) $x - 3$ (b) $\frac{3x+1}{x+1}$ (c) $\frac{3x+1}{x}$ (d) $\frac{3x+1}{x-3}$
6. The graph of $f(x)$ is given in Figure 4.13. Does $f(x)$ have an inverse?
   (a) Yes (b) No (c) Cannot be determined

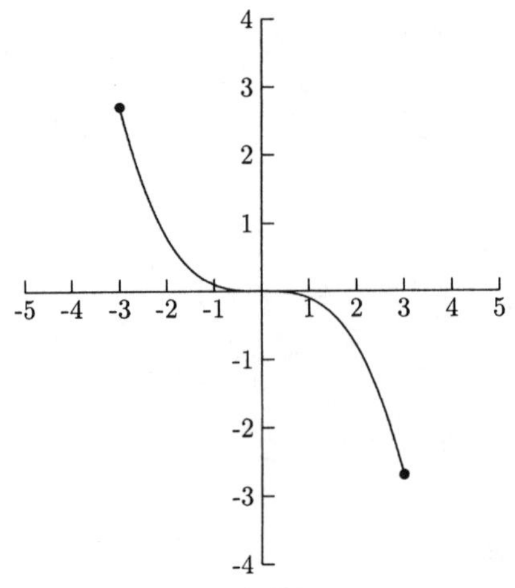

Fig. 4.13.

7. Are $f(x) = \frac{1}{2}x + 3$ and $g(x) = 2x - 3$ inverses?

   (a) Yes (b) No (c) Cannot be determined

8. What is the domain for $f \circ g(x)$ where

$$f(x) = \frac{1}{x} \quad \text{and} \quad g(x) = \frac{x-2}{x+2}?$$

   (a) $(-\infty, -2) \cup (-2, 0) \cup (0, 2) \cup (2, \infty)$
   (b) $(-\infty, -2) \cup (-2, 2) \cup (2, \infty)$

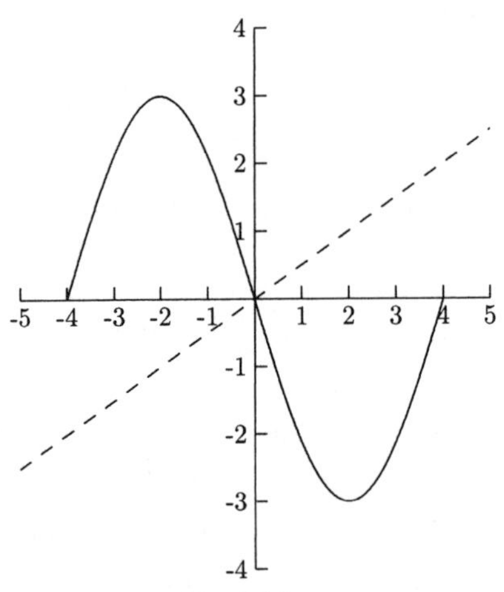

Fig. 4.14.

(c) $(-\infty, 0) \cup (0, 2) \cup (2, \infty)$
(d) $(-\infty, -2) \cup (-2, 0) \cup (0, \infty)$

9. The solid graph in Figure 4.14 is the graph of $f(x)$. The dashed graph is the graph of $g(x)$. Find $f \circ g(4)$.
(a) -2 (b) -3 (c) 2 (d) 3

SOLUTIONS

1. B	2. A	3. C	4. B	5. C
6. A	7. B	8. B	9. B	

Translations and Special Functions

Calculus students work with only a few families of functions—absolute value, nth root, cubic, quadratic, polynomial, rational, exponential, logarithmic, and trigonometric functions. Two or more of these functions might be combined arithmetically or by using function composition. In this chapter, we will look at the absolute value function (whose graph is in Figure 5.1), the square root function (whose graph is in Figure 5.2), and the cubic function (whose graph is in Figure 5.3).

We will also look at how these functions are affected by some simple changes. Knowing the effects certain changes have on a function will make sketching its graph by hand much easier. This understanding will also help you to use a graphing calculator. One of the simplest changes to a function is to add a number. This change will cause the graph to shift vertically or horizontally.

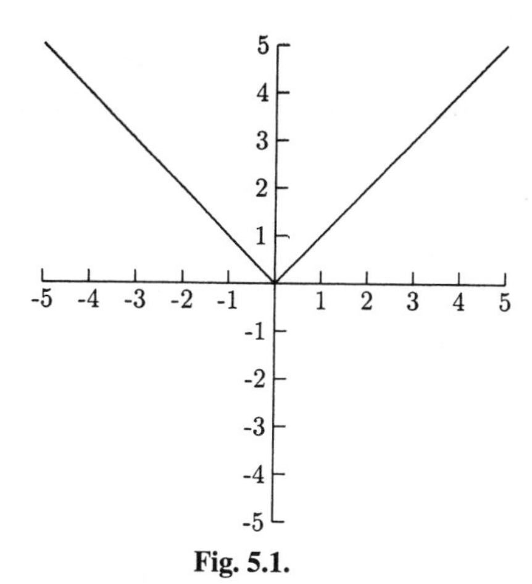

Fig. 5.1.

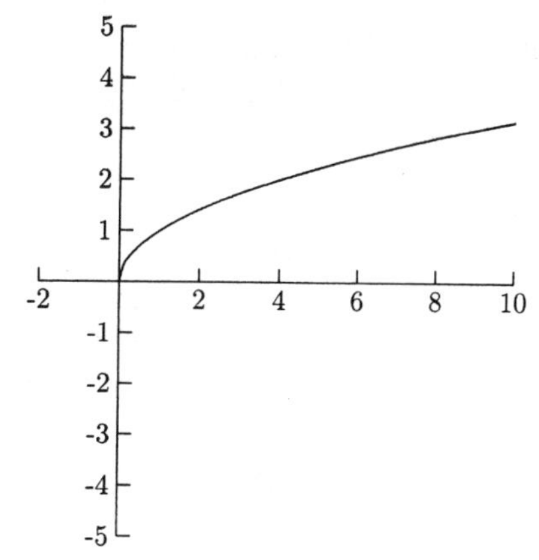

Fig. 5.2.

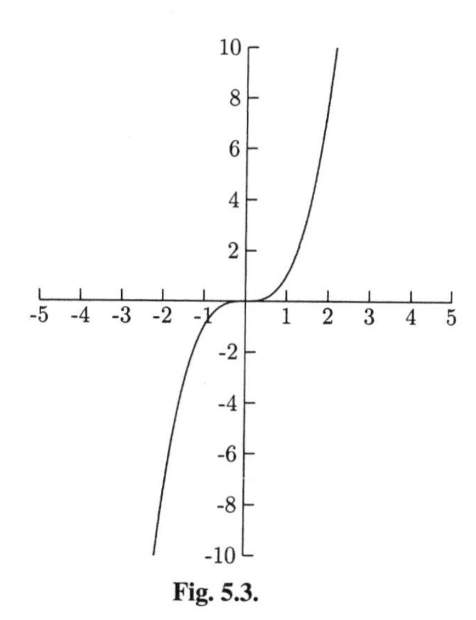

Fig. 5.3.

What effect does adding 1 to a function have on its graph? It depends on where we put "+1." Adding 1 to x will shift the graph *left* one unit. Adding 1 to y will shift the graph *up* one unit.

- $y = |x+1|$, 1 is added to x, shifting the graph to the left 1 unit. See Figure 5.4.
- $y = |x| + 1$, 1 is added to y (which is $|x|$) shifting the graph up 1 unit. See Figure 5.5.

For the graphs in this chapter, the solid graph will be the graph of the original function, and the dashed graph will be the graph of the transformed function.

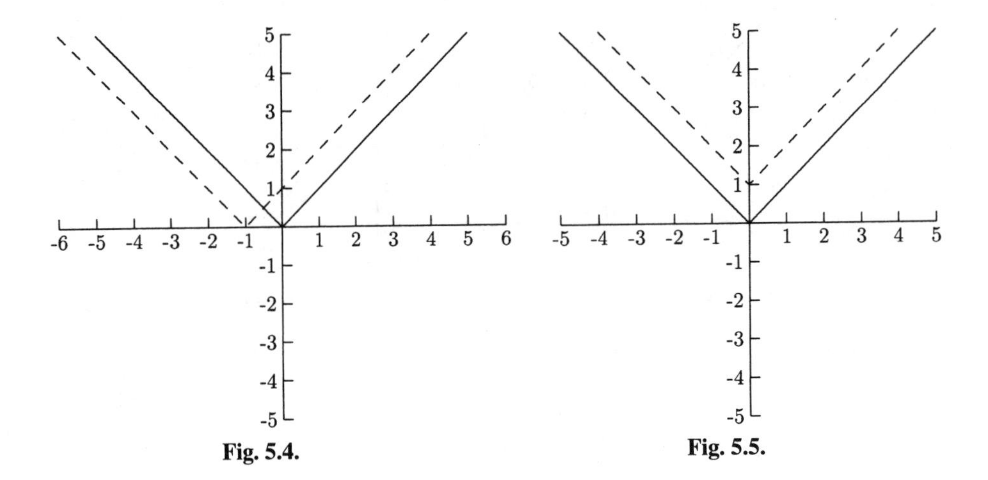

Fig. 5.4. **Fig. 5.5.**

- $y = \sqrt{x+2}$, 2 is added to x, shifting the graph to the left 2 units. See Figure 5.6.
- $y = \sqrt{x} + 2$, 2 is added to y (which is $\sqrt{x}$) shifting the graph up 2 units. See Figure 5.7.

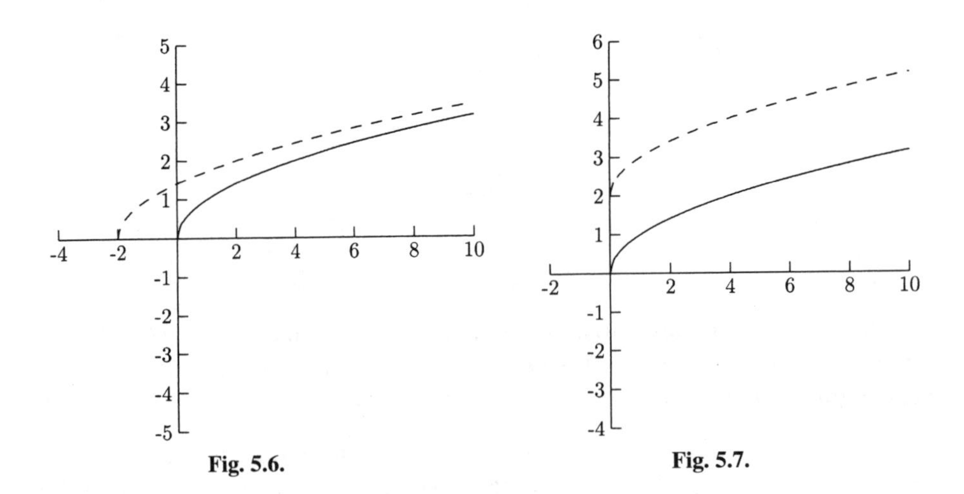

Fig. 5.6. **Fig. 5.7.**

Subtracting a number from x will shift the graph to the right while subtracting a number from y will shift the graph down.

- $y = (x - 1)^3$, 1 is subtracted from x, shifting the graph to the right 1 unit. See Figure 5.8.
- $y = x^3 - 1$, 1 is subtracted from y (which is x^3), shifting the graph down 1 unit. See Figure 5.9.

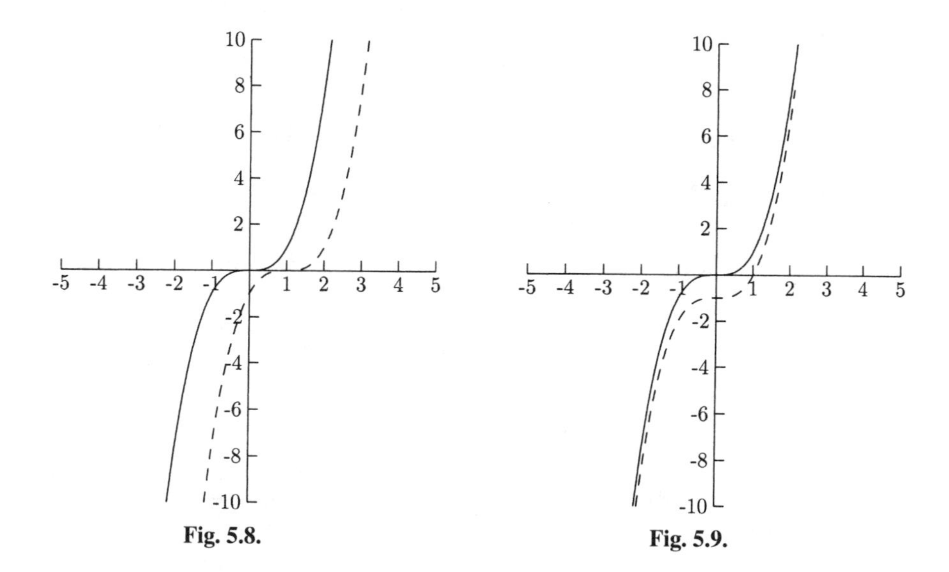

Fig. 5.8.

Fig. 5.9.

Multiplying the x-values or y-values by a number changes the graph, usually by stretching or compressing it. Multiplying the x-values or y-values by -1 will reverse the graph. If a is a number larger than 1 ($a > 1$), then multiplying x by a will horizontally compress the graph, but multiplying y by a will vertically stretch the graph. If a is positive but less than 1 ($0 < a < 1$), then multiplying x by a will horizontally stretch the graph, but multiplying y by a will vertically compress the graph.

- $y = \sqrt{2x}$, the graph is horizontally compressed. See Figure 5.10.
- $y = 2\sqrt{x}$, the graph is vertically stretched. See Figure 5.11.
- $y = \left(\frac{1}{2}x\right)^3$, the graph is horizontally stretched. See Figure 5.12.
- $y = \frac{1}{2}x^3$, the graph is vertically compressed. See Figure 5.13.

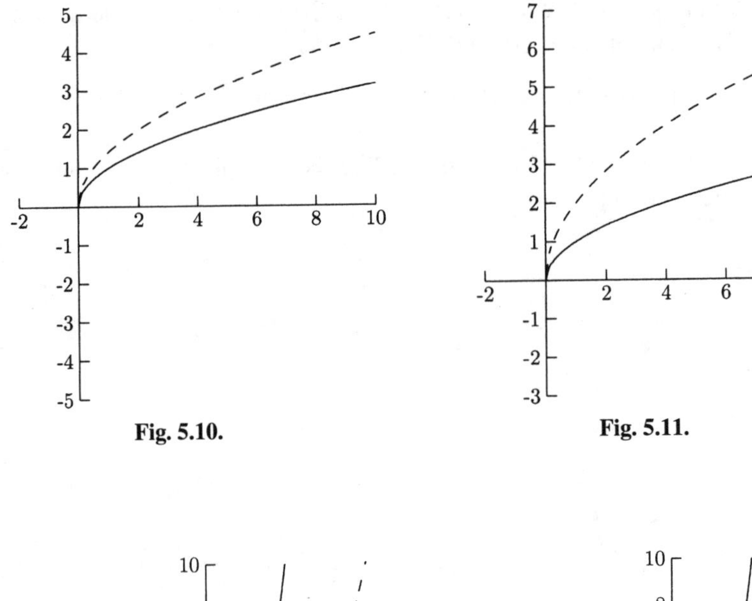

Fig. 5.10. Fig. 5.11.

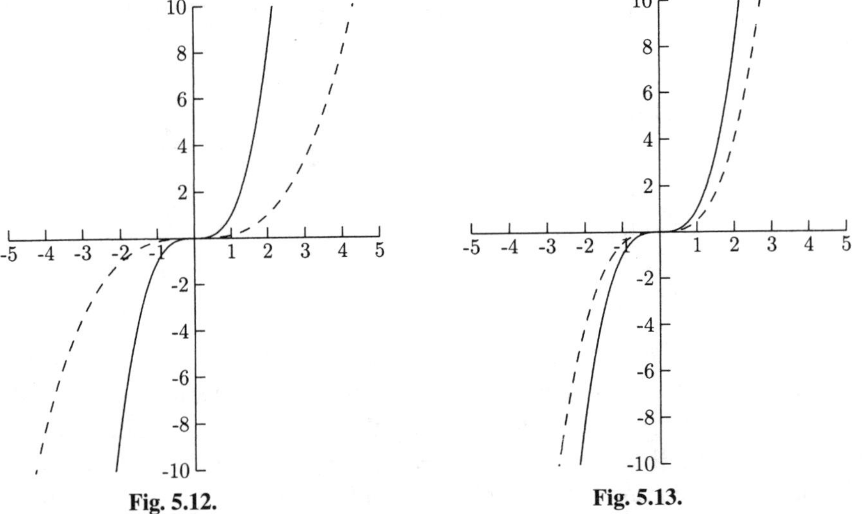

Fig. 5.12. Fig. 5.13.

For many functions, but not all, vertical compression is the same as horizontal stretching, and vertical stretching is the same as horizontal compression.

Multiplying the x-values by -1 will reverse the graph horizontally. This is called *reflecting the graph across the y-axis*. Multiplying the y-values by -1 will reverse the graph vertically. This is called *reflecting the graph across the x-axis*.

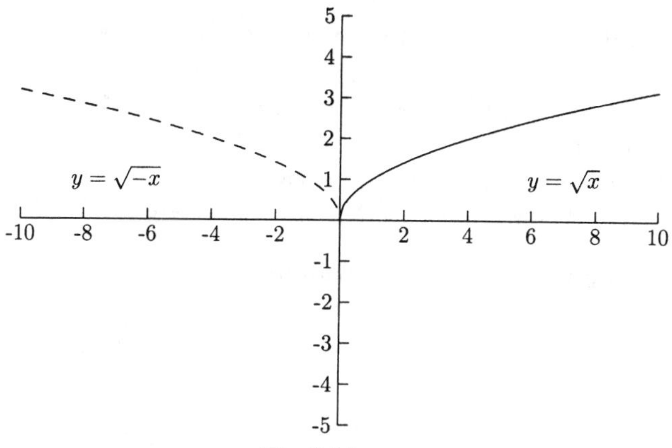

Fig. 5.14.

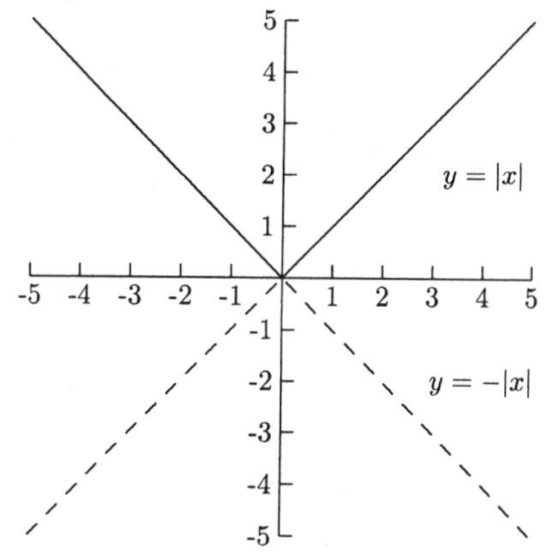

Fig. 5.15.

When a function is even, reflecting the graph across the y-axis does not change the graph. When a function is odd, reflecting the graph across the y-axis is the same as reflecting it across the x-axis.

We can use function notation to summarize these transformations.

$$y = af(x + h) + k$$

- If h is positive, the graph is shifted to the left h units.
- If h is negative, the graph is shifted to the right h units.
- If k is positive, the graph is shifted up k units.
- If k is negative, the graph is shifted down k units.

- If $a > 1$, the graph is vertically stretched. The larger a is, the greater the stretch.
- If $0 < a < 1$, the graph is vertically compressed. The closer to 0 a is, the greater the compression.
- The graph of $-f(x)$ is reflected across the x-axis.
- The graph of $f(-x)$ is reflected across the y-axis.

The graphs below are various transformations of the graph of $y = |x|$.

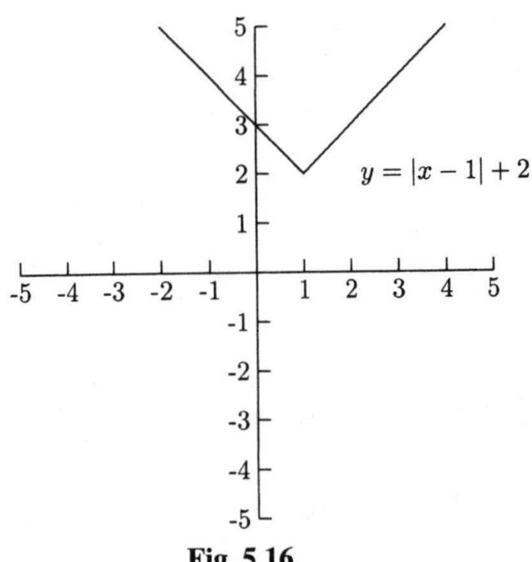

Fig. 5.16.

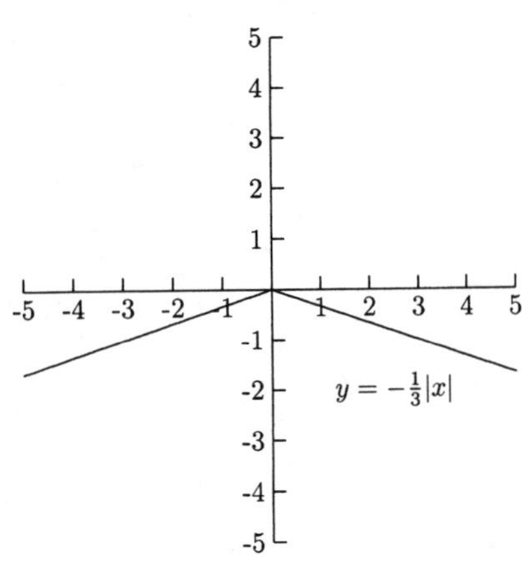

Fig. 5.17.

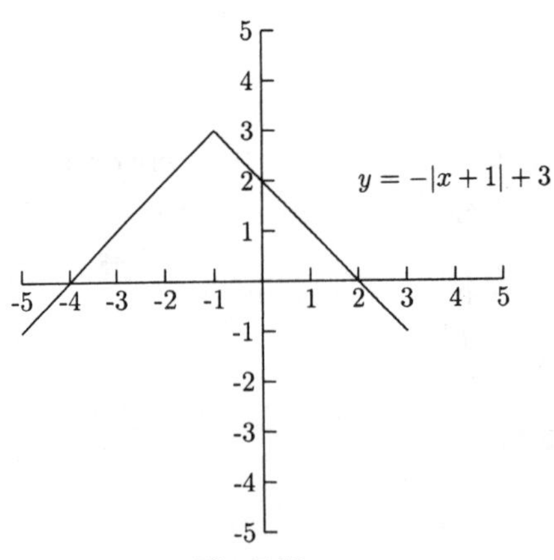

Fig. 5.18.

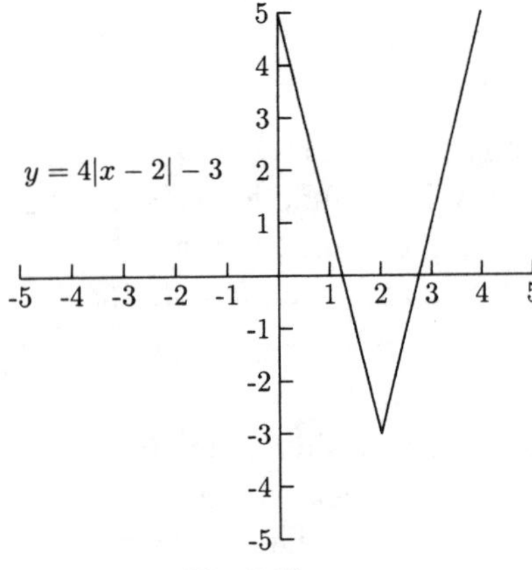

Fig. 5.19.

EXAMPLES

The graph of $y = f(x)$ is given in Figure 5.20. Sketch the transformations. We will sketch the graph by moving the points $(-4, 5)$, $(-1, -1)$, $(1, 3)$, and $(4, 0)$.

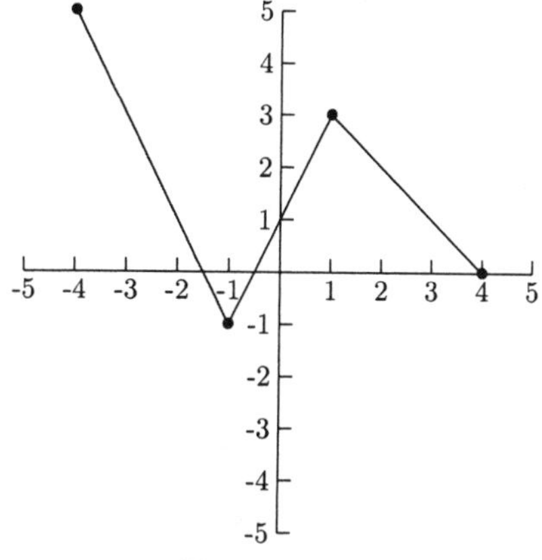

Fig. 5.20.

- $y = f(x+1) - 3$

Table 5.1

Original point	Left 1 $x - 1$	Down 3 $y - 3$	Plot this point
$(-4, 5)$	$-4 - 1 = -5$	$5 - 3 = 2$	$(-5, 2)$
$(-1, -1)$	$-1 - 1 = -2$	$-1 - 3 = -4$	$(-2, -4)$
$(1, 3)$	$1 - 1 = 0$	$3 - 3 = 0$	$(0, 0)$
$(4, 0)$	$4 - 1 = 3$	$0 - 3 = -3$	$(3, -3)$

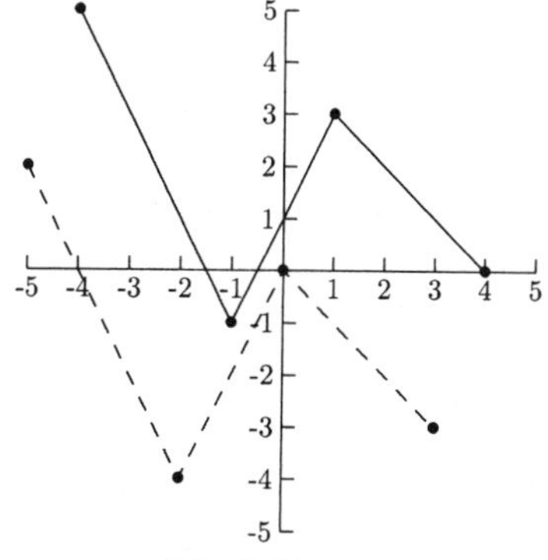

Fig. 5.21.

- $y = -f(x)$

Table 5.2

Original point	x does not change x	Opposite of y $-y$	Plot this point
$(-4, 5)$	-4	-5	$(-4, -5)$
$(-1, -1)$	-1	$-(-1) = 1$	$(-1, 1)$
$(1, 3)$	1	-3	$(1, -3)$
$(4, 0)$	4	$-0 = 0$	$(4, 0)$

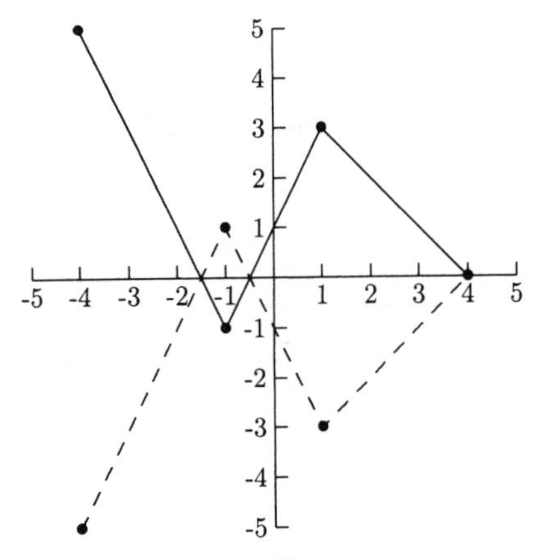

Fig. 5.22.

- $y = 2f(x - 3)$

Table 5.3

Original point	Right 3 $x+3$	Stretched $2y$	Plot this point
$(-4, 5)$	$-4 + 3 = -1$	$2(5) = 10$	$(-1, 10)$
$(-1, -1)$	$-1 + 3 = 2$	$2(-1) = -2$	$(2, -2)$
$(1, 3)$	$1 + 3 = 4$	$2(3) = 6$	$(4, 6)$
$(4, 0)$	$4 + 3 = 7$	$2(0) = 0$	$(7, 0)$

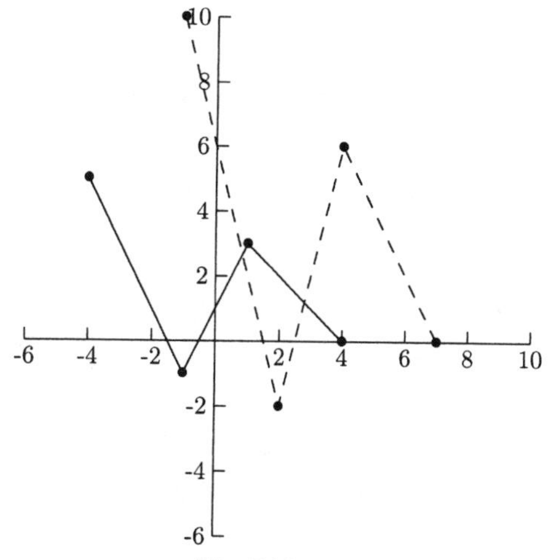

Fig. 5.23.

- $y = \frac{1}{2}f(-x) + 2$

Table 5.4

Original point	Opposite of x $-x$	Compressed and up 2 $\frac{1}{2}y + 2$	Plot this point
$(-4, 5)$	$-(-4) = 4$	$\frac{1}{2}(5) + 2 = \frac{9}{2}$	$(4, \frac{9}{2})$
$(-1, -1)$	$-(-1) = 1$	$\frac{1}{2}(-1) + 2 = \frac{3}{2}$	$(1, \frac{3}{2})$
$(1, 3)$	-1	$\frac{1}{2}(3) + 2 = \frac{7}{2}$	$(-1, \frac{7}{2})$
$(4, 0)$	-4	$\frac{1}{2}(0) + 2 = 2$	$(-4, 2)$

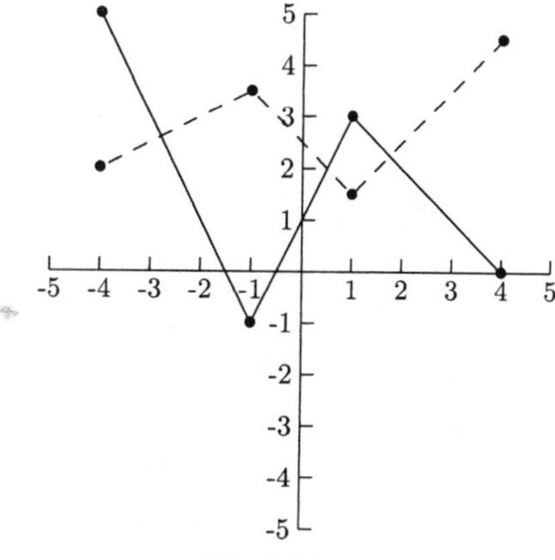

Fig. 5.24.

PRACTICE

For 1–4, match the graph with its function. Some functions will be left over.

1.

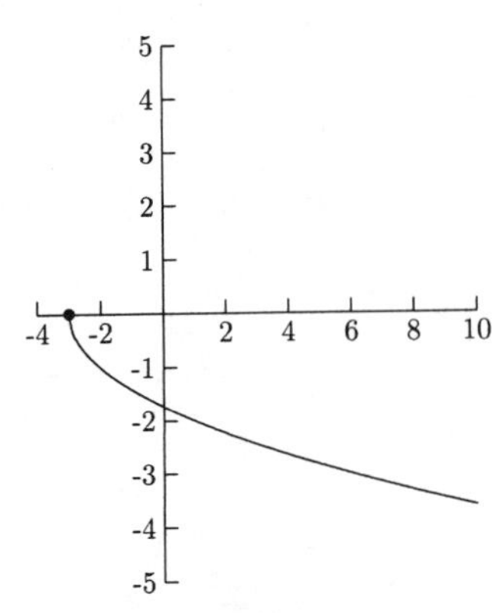

Fig. 5.25.

2.

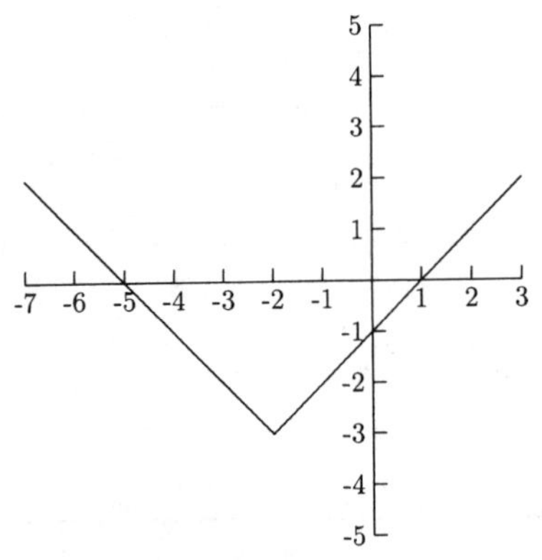

Fig. 5.26.

3.

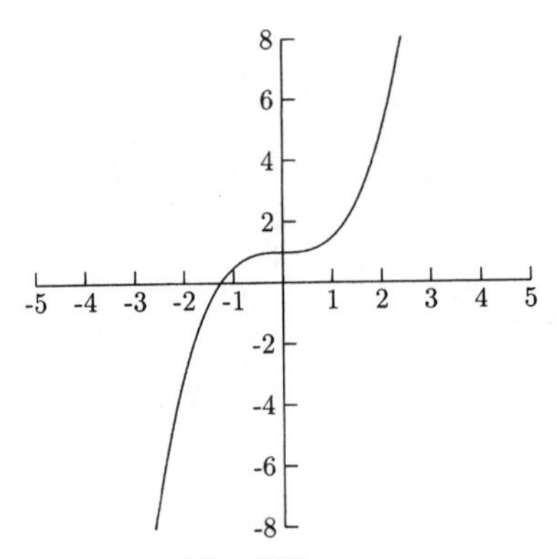

Fig. 5.27.

4.

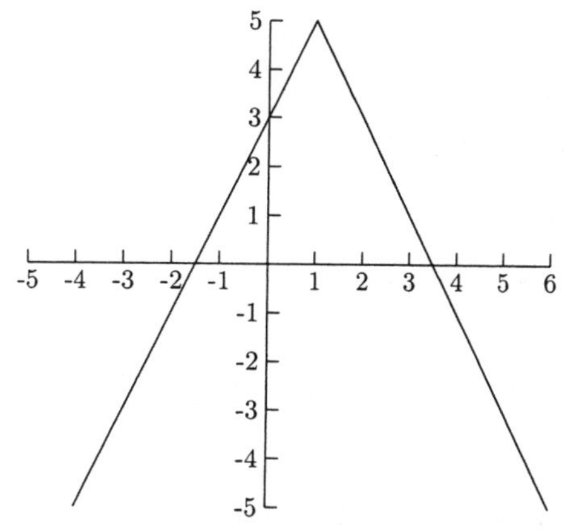

Fig. 5.28.

$f(x) = -2|x - 1| + 5$ $\quad f(x) = -\sqrt{x+3}$ $\quad f(x) = \sqrt{3-x}$

$f(x) = -\frac{1}{2}x^3 + 1$ $\quad f(x) = |x + 2| - 3$ $\quad f(x) = \frac{1}{2}x^3 + 1$

For Problems 5–8, use the statements below to describe the transformations on $f(x)$. Some of the statements will be used more than once, and others will not be used.

(A) shifts the graph to the left.
(B) shifts the graph to the right.
(C) shifts the graph up.
(D) shifts the graph down.
(E) reflects the graph across the y-axis.
(F) reflects the graph across the x-axis.
(G) vertically compresses the graph.
(H) vertically stretches the graph.
(I) reflects the graph across the x-axis and vertically compresses the graph.
(J) reflects the graph across the y-axis and vertically stretches the graph.

5. For the function $f(-x) + 3$,
   (a) What does "+3" do?
   (b) What does the negative sign on x do?
6. For the function $3f(x - 1) - 4$,
   (a) What does "3" do?

(b) What does "-1" do?
(c) What does "-4" do?

7. For the function $-\frac{1}{2}f(x+3)+1$,

(a) What does "$-\frac{1}{2}$" do?
(b) What does "$+3$" do?
(c) What does "$+1$" do?

8. For the function $\frac{1}{3}f(-x)-1$,

(a) What does "$\frac{1}{3}$" do?
(b) What does the negative sign on x do?
(c) What does "-1" do?

Refer to the graph of $f(x)$ in Figure 5.29 for Problems 9–10.

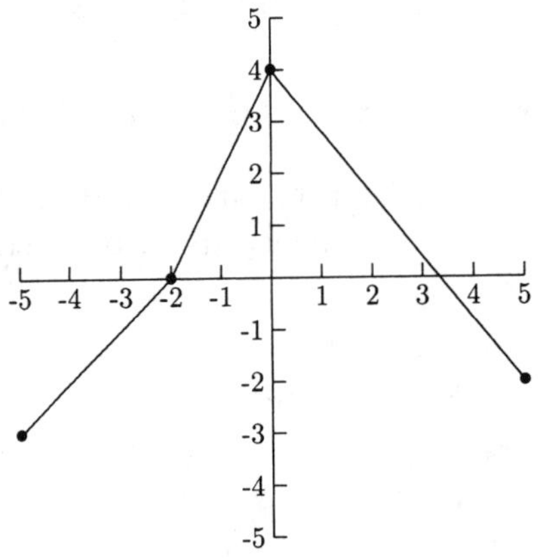

Fig. 5.29.

9. Sketch the graph of $f(-x)-1$.
10. Sketch the graph of $-\frac{1}{2}f(x+3)+1$.

SOLUTIONS

1. $f(x)=-\sqrt{x+3}$
2. $f(x)=|x+2|-3$
3. $f(x)=\frac{1}{2}x^3+1$
4. $f(x)=-2|x-1|+5$

5. C, E
6. H, B, D
7. I, A, C
8. G, E, D
9.

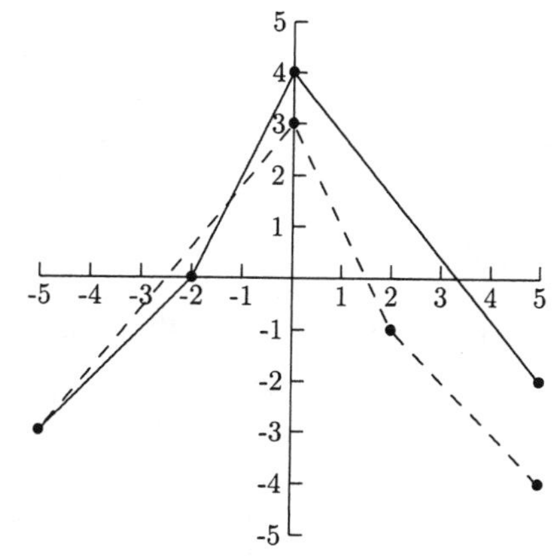

Fig. 5.30.

Table 5.5

Original point	Opposite of x $-x$	Down 1 $y-1$	Plot this point
$(-5, -3)$	$-(-5) = 5$	$-3 - 1 = -4$	$(5, -4)$
$(-2, 0)$	$-(-2) = 2$	$0 - 1 = -1$	$(2, -1)$
$(0, 4)$	$-0 = 0$	$4 - 1 = 3$	$(0, 3)$
$(5, -2)$	-5	$-2 - 1 = -3$	$(-5, -3)$

10.

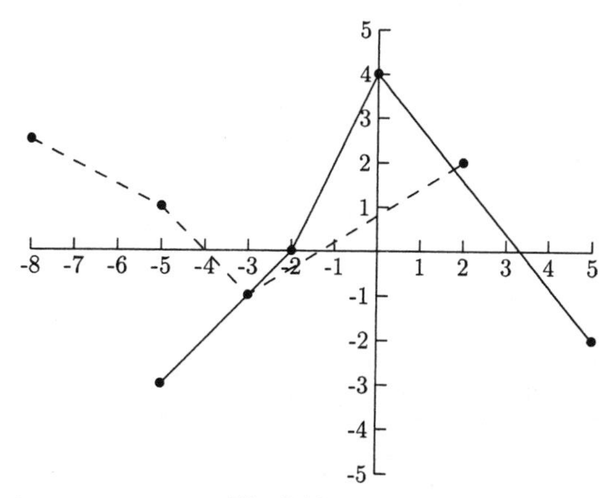

Fig. 5.31.

Table 5.6

Original point	Left 3 $x-3$	Opposite of y, compressed, up 1 $-\frac{1}{2}y+1$	Plot this point
$(-5,-3)$	$-5-3=-8$	$-\frac{1}{2}(-3)+1=\frac{5}{2}$	$(-8,\frac{5}{2})$
$(-2,0)$	$-2-3=-5$	$-\frac{1}{2}(0)+1=1$	$(-5,1)$
$(0,4)$	$0-3=-3$	$-\frac{1}{2}(4)+1=-1$	$(-3,-1)$
$(5,-2)$	$5-3=2$	$-\frac{1}{2}(-2)+1=2$	$(2,2)$

CHAPTER 5 REVIEW

Match the graphs in Figures 5.32–5.34 with the functions in Problems 1–3.

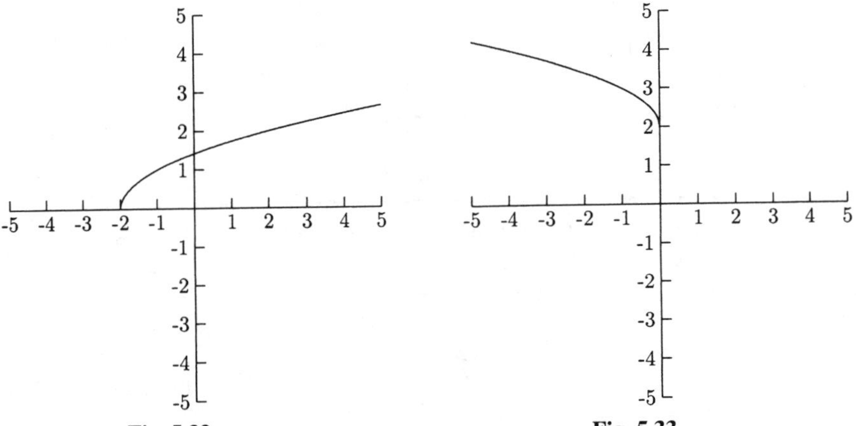

Fig. 5.32.

Fig. 5.33.

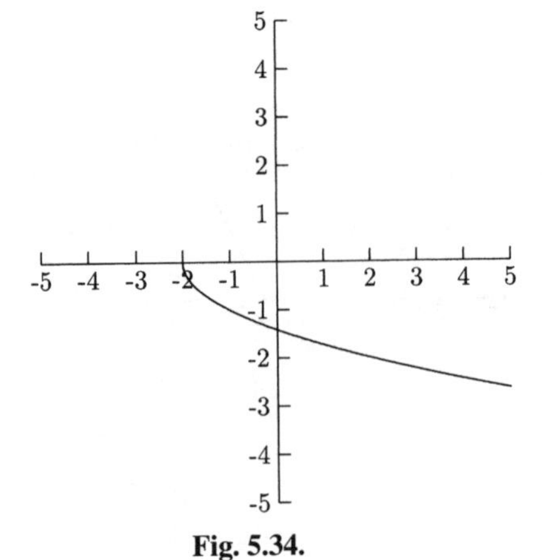

Fig. 5.34.

1. $f(x) = \sqrt{x+2}$
2. $f(x) = -\sqrt{x+2}$
3. $f(x) = \sqrt{-x} + 2$
4. The graph of $f(x) = (x+1) + 2$ is the graph of $f(x)$
   (a) shifted to the left one unit and down two units.
   (b) shifted to the left one unit and up two units.
   (c) shifted to the right one unit and down two units.
   (d) shifted to the right one unit and up two units.

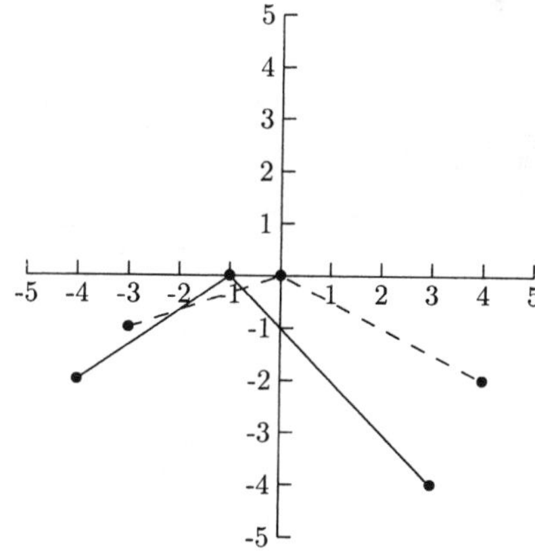

Fig. 5.35.

5. The solid graph in Figure 5.35 is the graph of $f(x)$, and the dashed graph is the graph of a transformation. What is the transformation?
   (a) $\frac{1}{2}f(x-1)$ (b) $\frac{1}{2}f(x+1)$
   (c) $f(x-1) + \frac{1}{2}$ (d) $f(x+1) + \frac{1}{2}$

SOLUTIONS

1. Figure 5.32 2. Figure 5.34 3. Figure 5.33 4. B 5. A

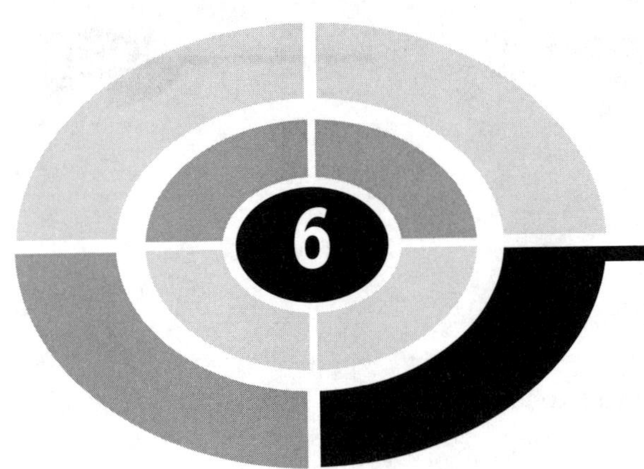

CHAPTER

Quadratic Functions

The graph of every quadratic function, $f(x) = ax^2 + bx + c$, is a transformation of the graph of $y = x^2$. (See Figure 6.1.)

When the function is written in the form $f(x) = a(x-h)^2 + k$, we have a pretty good idea of what its graph looks like: h will cause the graph to shift horizontally, and k will cause it to shift vertically. The point $(0, 0)$ on $y = x^2$ has shifted to (h, k). This point is the *vertex*. On a parabola that opens up (when a is positive), the vertex is the lowest point on the graph. The vertex is the highest point on a parabola that opens down (when a is negative).

We need to know the vertex when sketching a parabola. Once we have the vertex, we will find two points to its left and two points to its right. We should choose points in such a way that shows the curvature around the vertex and how fast the ends are going up or down. It does not matter which points we choose, but a good rule of thumb is to find $h-2a$, $h-a$, $h+a$, and $h+2a$. Because a parabola is symmetric

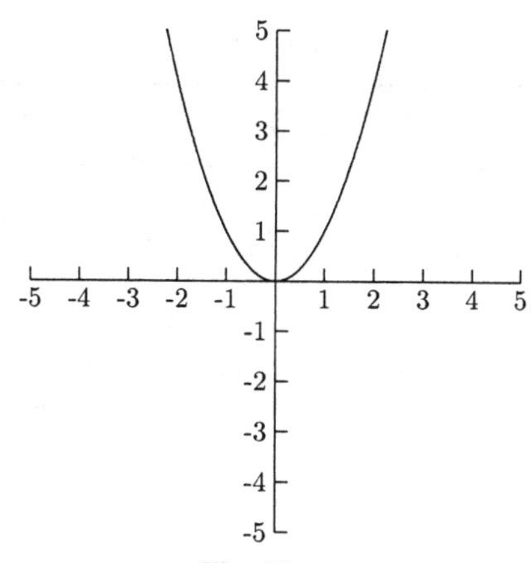

Fig. 6.1.

about the line $x = h$ (the vertical line that goes through the vertex), the y-values for $h - a$ and $h + a$ will be the same and the y-values for $h - 2a$ and $h + 2a$ will be the same, too. We will also find the intercepts.

EXAMPLES

Sketch the graph for the following quadratic functions. Find the y-intercept and the x-intercepts, if any.

- $f(x) = (x - 1)^2 - 4$

  $a = 1, h = 1, k = -4$ The parabola opens up and the vertex is $(1, -4)$. For the y-intercept, let $x = 0$ in the function. The y-intercept is $(0-1)^2-4 = -3$. For the x-intercepts, let $y = 0$ and solve for x.

$$(x - 1)^2 - 4 = 0$$

$$(x - 1)^2 = 4 \qquad \text{Take the square root of each side.}$$

$$x - 1 = \pm 2$$

$$x = 1 \pm 2 = 1 + 2, 1 - 2 = 3, -1$$

The x-intercepts are 3 and -1.

Table 6.1

	x	y	Plot this point
$h-2a$	$1-2(1)=-1$	$(-1-1)^2-4=0$	$(-1,0)$
$h-a$	$1-1=0$	$(0-1)^2-4=-3$	$(0,-3)$
h	1	-4	$(1,-4)$
$h+a$	$1+1=2$	$(2-1)^2-4=-3$	$(2,-3)$
$h+2a$	$1+2(1)=3$	$(3-1)^2-4=0$	$(3,0)$

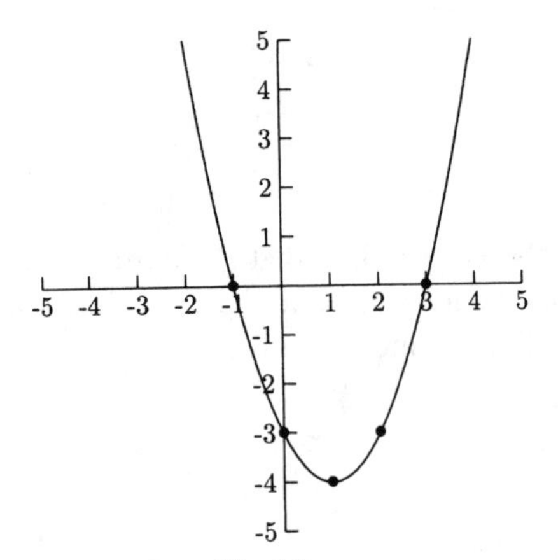

Fig. 6.2.

- $g(x)=-2(x+1)^2+18$

  $a=-2$, $h=-1$, $k=18$ The parabola opens down, and the vertex is $(-1,18)$.

  $$y=-2(0+1)^2+18 \qquad -2(x+1)^2+18=0$$

  $$y=16 \qquad -2(x+1)^2=-18$$

  $$(x+1)^2=9$$

  $$x+1=\pm 3$$

  $$x=-1\pm 3=-1-3,$$

  $$-1+3=-4,2$$

  The y-intercept is 16 and the x-intercepts are -4 and 2.

Table 6.2

	x	y	Plot this point
$h-2a$	$-1-2(-2)=3$	$-2(3+1)^2+18=-14$	$(3,-14)$
$h-a$	$-1-(-2)=1$	$-2(1+1)^2+18=10$	$(1,10)$
h	-1	18	$(-1,18)$
$h+a$	$-1+(-2)=-3$	$-2(-3+1)^2+18=10$	$(-3,10)$
$h+2a$	$-1+2(-2)=-5$	$-2(-5+1)^2+18=-14$	$(-5,-14)$

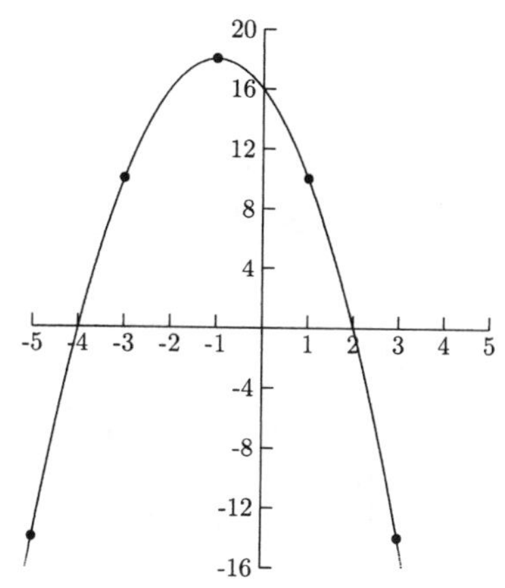

Fig. 6.3.

- $f(x)=\frac{1}{2}(x+1)^2+2$
  $a=\frac{1}{2}$, $h=-1$, $k=2$ The parabola opens up, and the vertex is $(-1,2)$. Because the parabola opens up ($a=\frac{1}{2}$ is positive) and the vertex is above the x-axis ($k=2$ is positive), there will be no x-intercept. If we were to solve the equation $\frac{1}{2}(x+1)^2+2=0$, we would not get a real number solution. The y-intercept is $y=\frac{1}{2}(0+1)^2+2=2\frac{1}{2}$.

Table 6.3

	x	y	Plot this point
$h-2a$	$-1-2(\frac{1}{2})=-2$	$\frac{1}{2}(-2+1)^2+2=2\frac{1}{2}$	$(-2,2\frac{1}{2})$
$h-a$	$-1-\frac{1}{2}=-1\frac{1}{2}$	$\frac{1}{2}(-1\frac{1}{2}+1)^2+2=2\frac{1}{8}$	$(-1\frac{1}{2},2\frac{1}{8})$
h	-1	2	$(-1,2)$
$h+a$	$-1+\frac{1}{2}=-\frac{1}{2}$	$\frac{1}{2}(-\frac{1}{2}+1)^2+2=2\frac{1}{8}$	$(-\frac{1}{2},2\frac{1}{8})$
$h+2a$	$-1+2(\frac{1}{2})=0$	$\frac{1}{2}(0+1)^2+2=2\frac{1}{2}$	$(0,2\frac{1}{2})$

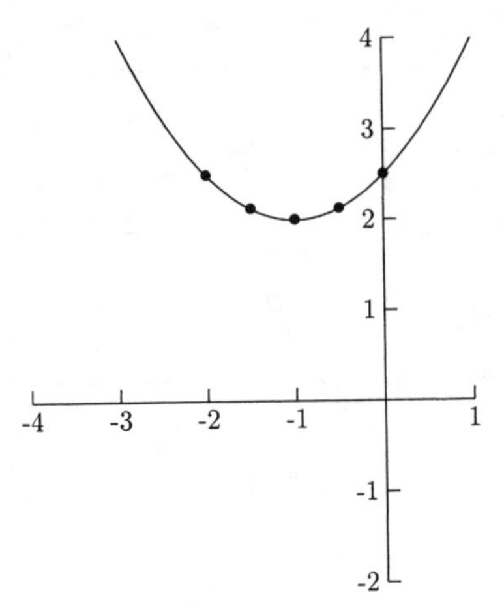

Fig. 6.4.

By knowing the vertex and one other point on the graph, we can find an equation for the quadratic function. Once we know the vertex, we have h and k in $y = a(x - h)^2 + k$. By using another point (x, y), we can find a.

EXAMPLE

- The vertex for a quadratic function is $(-3, 4)$, and the y-intercept is -10. Find an equation for this function.
  Let $h = -3$, $k = 4$ in $y = a(x - h)^2 + k$ to get $y = a(x + 3)^2 + 4$. Saying that the y-intercept is -10 is another way of saying $(0, -10)$ is a point on the graph. We can let $x = 0$ and $y = -10$ in $y = a(x + 3)^2 + 4$ to find a.

$$-10 = a(0 + 3)^2 + 4$$

$$-14 = 9a$$

$$-\frac{14}{9} = a$$

One equation for this function is $y = -\frac{14}{9}(x + 3)^2 + 4$.

Quadratic equations are not normally written in the convenient form $f(x) = a(x - h)^2 + k$. We can complete the square on $f(x) = ax^2 + bx + c$ to find (h, k). Begin by completing the square on the x^2 and x terms.

EXAMPLES

Find the vertex.

- $y = x^2 - 6x - 2$

$$y = x^2 - 6x - 2$$

$$y = \left[x^2 - 6x + \left(\frac{6}{2}\right)^2\right] - 2 + ?$$

We need to balance putting $+(6/2)^2 = 9$ in the parentheses by adding -9 to -2

$$y = (x^2 - 6x + 9) - 2 - 9$$

$$y = (x - 3)^2 - 11 \qquad \text{The vertex is } (3, -11).$$

- $f(x) = 4x^2 + 8x + 1$

We will begin by factoring $a = 4$ from $4x^2 + 8x$. Then we will complete the square on the x^2 and x terms.

$$f(x) = 4x^2 + 8x + 1$$

$$f(x) = 4(x^2 + 2x) + 1$$

$$f(x) = 4(x^2 + 2x + 1) + 1 + ?$$

By putting $+1$ in the parentheses, we are adding $4(1) = 4$. We need to balance this by adding -4 to 1.

$$f(x) = 4(x^2 + 2x + 1) + 1 + (-4)$$

$$f(x) = 4(x + 1)^2 - 3 \qquad \text{The vertex is } (-1, -3).$$

When factoring an unusual quantity from two or more terms, it is not obvious what terms go in the parentheses. We can find the terms that go in the parentheses by writing the terms to be factored as numerators of fractions and the number to be factored as the denominator. The terms that go in the parentheses are the simplified fractions.

- $f(x) = -3x^2 + 9x + \frac{1}{4}$

  We need to factor $a = -3$ from $-3x^2 + 9x$.

$$\frac{-3x^2}{-3} + \frac{9x}{-3} = x^2 - 3x$$

$$f(x) = -3x^2 + 9x + \frac{1}{4}$$

$$f(x) = -3(x^2 - 3x) + \frac{1}{4}$$

$$f(x) = -3\left(x^2 - 3x + \frac{9}{4}\right) + \frac{1}{4} + ? \qquad \left(\frac{3}{2}\right)^2 = \frac{9}{4}$$

By putting $+\frac{9}{4}$ in the parentheses, we are adding $-3\left(\frac{9}{4}\right) = -\frac{27}{4}$. We need to balance this by adding $\frac{27}{4}$ to $\frac{1}{4}$

$$f(x) = -3\left(x^2 - 3x + \frac{9}{4}\right) + \frac{1}{4} + \frac{27}{4} = -3\left(x^2 - 3x + \frac{9}{4}\right) + \frac{28}{4}$$

$$f(x) = -3\left(x - \frac{3}{2}\right)^2 + 7 \qquad \text{The vertex is } \left(\frac{3}{2}, 7\right)$$

- $g(x) = \frac{2}{3}x^2 + x - 2$

  Factoring $a = \frac{2}{3}$ from $\frac{2}{3}x^2 + x$, we have

$$\frac{(2/3)x^2}{2/3} + \frac{x}{2/3} = x^2 + x \div \frac{2}{3} = x^2 + x \cdot \frac{3}{2} = x^2 + \frac{3}{2}x.$$

$$g(x) = \frac{2}{3}\left(x^2 + \frac{3}{2}x\right) - 2$$

$$= \frac{2}{3}\left(x^2 + \frac{3}{2}x + \frac{9}{16}\right) - 2 + ? \qquad \left(\frac{1}{2} \cdot \frac{3}{2}\right)^2 = \frac{9}{16}$$

By adding $\frac{9}{16}$ in the parentheses, we are adding $\frac{2}{3} \cdot \frac{9}{16} = \frac{3}{8}$. We need to balance this by adding $-3/8$ to -2.

$$g(x) = \frac{2}{3}\left(x^2 + \frac{3}{2}x + \frac{9}{16}\right) - 2 - \frac{3}{8}$$

$$g(x) = \frac{2}{3}\left(x + \frac{3}{4}\right)^2 - \frac{19}{8} \qquad \text{The vertex is } \left(-\frac{3}{4}, -\frac{19}{8}\right)$$

One advantage to the form $f(x) = ax^2 + bx + c$ is that it is usually easier to use to find the intercepts. We can use factoring and the quadratic formula when it is in this form. Also, c is the y-intercept. Because a is the same number in both forms, we can tell whether the parabola opens up or down when the equation is in either form. It can be tedious to complete the square on $f(x) = ax^2 + bx + c$ to find the vertex. Fortunately, there is a shortcut.

$$h = \frac{-b}{2a} \quad \text{and} \quad k = f\left(\frac{-b}{2a}\right)$$

This shortcut comes from completing the square to rewrite $f(x) = ax^2 + bx + c$ as $f(x) = a(x - h)^2 + k$.

$$\begin{aligned} f(x) &= ax^2 + bx + c \\ &= a\left(x^2 + \frac{b}{a}x\right) + c \\ &= a\left(x^2 + \frac{b}{a}x + \left(\frac{b}{2a}\right)^2\right) + c - a \cdot \left(\frac{b}{2a}\right)^2 \\ &= a\left(x + \frac{b}{2a}\right)^2 + c - \frac{b^2}{4a} \qquad \text{The vertex is } \left(\frac{-b}{2a}, c - \frac{b^2}{4a}\right) \end{aligned}$$

It is easier to find k by evaluating the function at $x = \frac{-b}{2a}$ than by using this formula.

EXAMPLE

Use the shortcut to find the vertex.

- $f(x) = -3x^2 + 9x + 4$

$$h = \frac{-b}{2a} = \frac{-9}{2(-3)} = \frac{3}{2} \text{ and } k = f\left(\frac{3}{2}\right) = -3\left(\frac{3}{2}\right)^2 + 9\left(\frac{3}{2}\right) + 4 = \frac{43}{4}$$

The vertex is $(\frac{3}{2}, \frac{43}{4})$.

An important topic in calculus is optimizing functions; that is, finding a maximum and/or minimum value for the function. Precalculus students can use algebra to optimize quadratic functions. A quadratic function has a minimum value (if its graph opens up) or a maximum value (if its graph opens down). If a is positive, then k is the minimum functional value. If a is negative, then k is the maximum functional value. These values occur at $x = h$.

EXAMPLES

Find the minimum or maximum functional value and where it occurs.

- $f(x) = -(x-3)^2 + 25$

  The parabola opens down because $a = -1$ is negative. This means that $k = 25$ is the maximum functional value. It occurs at $x = 3$.

- $y = 0.01x^2 - 6x + 2000$

$$h = \frac{-b}{2a} = \frac{-(-6)}{2(0.01)} = 300 \text{ and } k = 0.01(300)^2 - 6(300) + 2000 = 1100$$

  $a = 0.01$ is positive, so $k = 1100$ is the minimum functional value. The minimum occurs at $x = 300$.

PRACTICE

For Problems 1–3, sketch the graph and identify the vertex and intercepts.

1. $y = -(x-1)^2 + 4$
2. $f(x) = \frac{2}{3}(x+1)^2 + 2$
3. $y = -\frac{1}{2}x^2 - x + 12$
4. Rewrite $f(x) = -\frac{3}{5}x^2 - 6x - 11$ in the form $f(x) = a(x-h)^2 + k$, using completing the square.
5. Find the maximum or minimum functional value for $g(x) = -0.002x^2 + 5x + 150$.

6. Find an equation for the quadratic function whose vertex is $(2, 5)$ and whose graph contains the point $(-8, 15)$.

SOLUTIONS

1. The vertex is $(1, 4)$. The y-intercept is $-(0-1)^2+4=3$.

$$-(x-1)^2+4=0$$
$$-(x-1)^2=-4$$
$$(x-1)^2=4$$
$$(x-1)=\pm 2$$
$$x=1\pm 2=1+2,\ 1-2=3,\ -1$$

The x-intercepts are 3 and -1.

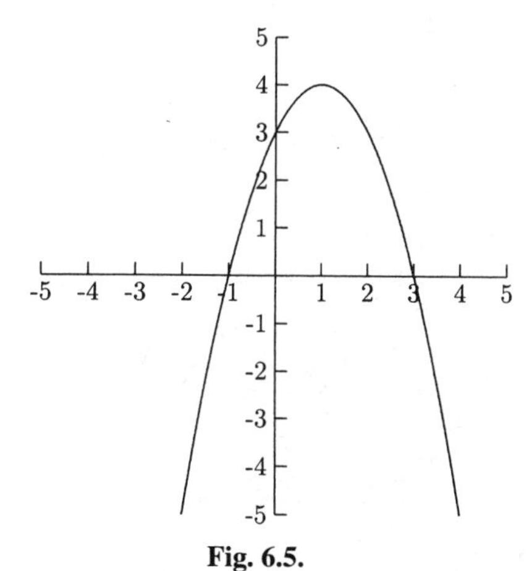

Fig. 6.5.

2. The vertex is $(-1, 2)$. The y-intercept is $\frac{2}{3}(0+1)+2=\frac{8}{3}$. There are two ways we can tell that there are no x-intercepts. The parabola opens up and the vertex is above the x-axis, so the parabola is always above the x-axis. Also, the equation $\frac{2}{3}(x+1)^2+2=0$ has no real number solution.

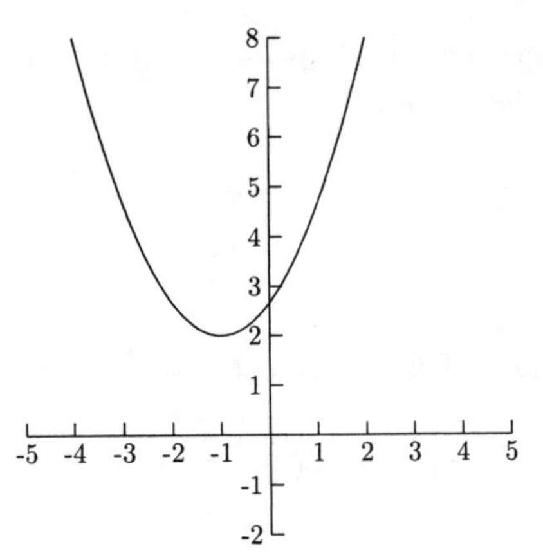

Fig. 6.6.

3. The vertex is $(-1, \frac{25}{2})$.

$$h = \frac{-b}{2a} = \frac{-(-1)}{2 \cdot \frac{-1}{2}} = -1 \text{ and } k = -\frac{1}{2}(-1)^2 - (-1) + 12 = \frac{25}{2}$$

The y-intercept is $-\frac{1}{2}(0)^2 - 0 + 12 = 12$.

$$0 = -\frac{1}{2}x^2 - x + 12$$

$$-2(0) = -2\left(-\frac{1}{2}x^2 - x + 12\right)$$

$$0 = x^2 + 2x - 24$$

$$0 = (x+6)(x-4)$$

$$x + 6 = 0 \qquad\qquad x - 4 = 0$$

$$x = -6 \qquad\qquad x = 4$$

The x-intercepts are -6 and 4.

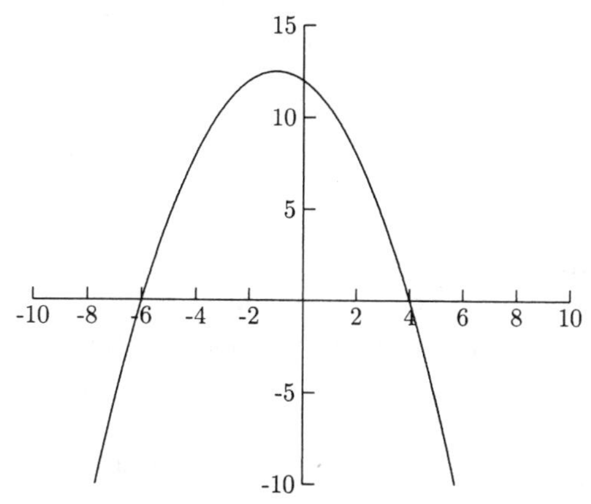

Fig. 6.7.

4. $f(x) = -\frac{3}{5}x^2 - 6x - 11$

$$f(x) = -\frac{3}{5}(x^2 + 10x) - 11 \qquad \frac{-6x}{-3/5} = -6x \div -\frac{3}{5} = -6x \cdot -\frac{5}{3} = 10x$$

$$f(x) = -\frac{3}{5}(x^2 + 10x + 25) - 11 + 15$$

$$f(x) = -\frac{3}{5}(x + 5)^2 + 4$$

5. This function has a maximum value because $a = -0.002$ is negative. The answer is k.

$$h = \frac{-b}{2a} = \frac{-5}{2(-0.002)} = 1250 \text{ and}$$

$$k = g(1250) = -0.002(1250)^2 + 5(1250) + 150 = 3275$$

The maximum functional value is 3275.

6. $h = 2$, $k = 5$ which makes $y = a(x - h)^2 + k$ become $y = a(x - 2)^2 + 5$. We can find a by letting $x = -8$ and $y = 15$.

$$y = a(x - 2)^2 + 5$$

$$15 = a(-8 - 2)^2 + 5$$

$$10 = a(-10)^2$$

$$10 = 100a$$

$$0.1 = a$$

The equation is $y = 0.1(x - 2)^2 + 5$.

These techniques to maximize/minimize quadratic functions can be applied to problems outside of mathematics. We can maximize the enclosed area, minimize the surface area of a box, maximize revenue, and optimize many other problems. In the first group of problems, the functions to be optimized will be given. In the second, we will have to find the functions based on the information given in the problem. The answers to every problem below will be one or both coordinates of the vertex.

EXAMPLES

- The weekly profit function for a product is given by $P(x) = -0.0001x^2 + 3x - 12{,}500$, where x is the number of units produced per week, and $P(x)$ is the profit (in dollars). What is the maximum weekly profit? How many units should be produced for this profit?
  The profit function is a quadratic function which has a maximum value. What information does the vertex give us? h is the number of units needed to maximize the weekly profit, and k is the maximum weekly profit.

  $$h = \frac{-b}{2a} = \frac{-3}{2(-0.0001)} = 15{,}000 \text{ and}$$

  $$k = -0.0001(15{,}000)^2 + 3(15{,}000) - 12{,}500 = 10{,}000$$

  Maximize the weekly profit by producing 15,000 units. The maximum weekly profit is \$10,000.
- The number of units of a product sold depends on the amount of money spent on advertising. If $y = -26x^2 + 2600x + 10{,}000$ gives the number of units sold after x thousands of dollars is spent on advertising, find the amount spent on advertising that results in the most sales.
  h will give us the amount to spend on advertising in order to maximize sales, and k will tell us the maximum sales level. We only need to find h.

  $$h = \frac{-b}{2a} = \frac{-2600}{2(-26)} = 50$$

  \$50 thousand should be spent on advertising to maximize sales.

The height of an object propelled upward (neglecting air resistance) is given by the quadratic function $s(t) = -16t^2 + v_0t + s_0$, where s is the height in feet, and

t is the number of seconds after the initial thrust. The initial velocity (in feet per second) of the object is v_0, and s_0 is the initial height (in feet) of the object. For example, if an object is tossed up at the rate of 10 feet per second, then $v_0 = 10$. If an object is propelled upward from a height of 50 feet, then $s_0 = 50$. If an object is dropped, its initial velocity is 0, so $v_0 = 0$.

EXAMPLES

- An object is tossed upward with an initial velocity of 15 feet per second from a height of four feet. What is the object's maximum height? How long does it take the object to reach its maximum height?
  Because the initial velocity is 15 feet per second, $v_0 = 15$, and the initial height is four feet, so $s_0 = 4$. The function that gives the height of the object (in feet) after t seconds is $s(t) = -16t^2 + 15t + 4$.
  $$h = \frac{-b}{2a} = \frac{-15}{2(-16)} = 0.46875 \text{ and}$$
  $$k = -16(0.46875)^2 + 15(0.46875) + 4 = 7.515625$$
  The object reaches its maximum height of 7.515625 feet after 0.46875 seconds.
- A projectile is fired from the ground with an initial velocity of 120 miles per hour. What is the projectile's maximum height? How long does it take to reach its maximum height?
  Because the projectile is being fired from the ground, its initial height is 0, so $s_0 = 0$. The initial velocity is given as 120 miles per hour—we need to convert this to feet per second. There are 5280 feet per mile, so 120 miles is 120(5280) = 633,600 feet. There are 60(60) = 3600 seconds per hour.
  $$\frac{120 \text{ miles}}{1 \text{ hour}} = \frac{633{,}600 \text{ feet}}{3600 \text{ seconds}} = 176 \text{ feet per second}$$
  Now we have the function: $s(t) = -16t^2 + 176t + 0 = -16t^2 + 176t$.
  $$h = \frac{-b}{2a} = \frac{-176}{2(-16)} = 5.5 \text{ and } k = -16(5.5)^2 + 176(5.5) = 484$$
  The projectile reaches its maximum height of 484 feet after 5.5 seconds.

Another problem involving the maximum vertical height is one where we know the horizontal distance traveled instead of the time it has traveled. The x-coordinates describe the object's horizontal distance, and the y-coordinates describe its height. Here we will find the maximum height and how far it traveled horizontally to reach the maximum height.

EXAMPLE

- A ball is thrown across a field. Its path can be described by the equation $y = -0.002x^2 + 0.2x + 5$, where x is the horizontal distance (in feet) and y is the height (in feet). See Figure 6.8. What is the ball's maximum height? How far had it traveled horizontally to reach its maximum height?

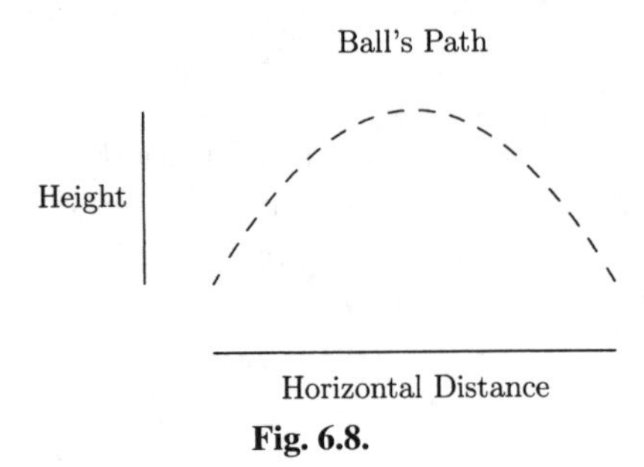

Fig. 6.8.

k will answer the first question, and h will answer the second.

$$h = \frac{-b}{2a} = \frac{-0.2}{2(-0.002)} = 50 \text{ and } k = -0.002(50)^2 + 0.2(50) + 5 = 10$$

The ball reached a maximum height of 10 feet when it traveled 50 feet horizontally.

The revenue of a product or service can depend on its price in two ways. An increase in the price means that more revenue per unit is earned but fewer units are sold. A decrease in the price means that less revenue is earned per unit but more units are sold. Quadratic functions model some of these relationships. In the next problems, a current price and sales level are given. We will be told how a price increase or decrease affects the sales level. We will let x represent the number of price increases/decreases. Suppose every \$10 decrease in the price results in an increase of five customers. Then the revenue function is (old price $- 10x$)(old sales level $+ 5x$). If every \$50 increase in the price results in a loss of one customer, then the revenue function is (old price $+ 50x$)(old sales level $- 1x$). These functions are quadratic functions which have a maximum value. The vertex tells us the maximum revenue and how many times to decrease/increase the price to get the maximum revenue.

EXAMPLES

- A management firm has determined that 60 apartments in a complex can be rented if the monthly rent is \$900, and that for each \$50 increase in the rent, three tenants are lost with little chance of being replaced. What rent should be charged to maximize revenue? What is the maximum revenue?

Let x represent the number of \$50 increases in the rent. This means if the rent is raised \$50, $x = 1$, if the rent is increased \$100, $x = 2$, and if the rent is increased \$150, $x = 3$. The rent function is $900 + 50x$. The number of tenants depends on the number of \$50 increases in the rent. So, if the rent is raised \$50, there will be $60 - 3(1)$ tenants; if the rent is raised \$100, the there will be $60 - 3(2)$ tenants; and if the rent is raised \$150, there will be $60 - 3(3)$ tenants. If the rent is raised \$50x, there will be $60 - 3x$ tenants. The revenue function is

$$R = (900 + 50x)(60 - 3x) = -150x^2 + 300x + 54{,}000.$$

$$h = \frac{-b}{2a} = \frac{-300}{2(-150)} = 1 \text{ and } k = -150(1)^2 + 300(1) + 54{,}000 = 54{,}150$$

The maximum revenue is \$54,150. Maximize revenue by charging $900 + 50(1) = \$950$ per month for rent.

- A cinema multiplex averages 2500 tickets sold on a Saturday when ticket prices are \$8. Concession revenue averages \$1.50 per ticket sold. A research firm has determined that for each \$0.50 increase in the ticket price, 100 fewer tickets will be sold. What is the maximum revenue (including concession revenue) and what ticket price maximizes the revenue?
  Let x represent the number of \$0.50 increases in the price. The ticket price is $8 + 0.50x$. The average number of tickets sold is $2500 - 100x$. The average ticket revenue is $(8.00 + 0.50x)(2500 - 100x)$. The average concession revenue is $1.50(2500 - 100x)$. The total revenue is

$$R = (8.00 + 0.50x)(2500 - 100x) + 1.50(2500 - 100x)$$
$$= -50x^2 + 300x + 23{,}750.$$

$$h = \frac{-b}{2a} = \frac{-300}{2(-50)} = 3 \text{ and } k = -50(3)^2 + 300(3) + 23{,}750 = 24{,}200$$

To maximize revenue, the ticket price should be $8.00 + 0.50(3) = \$9.50$, and the maximum revenue is \$24,200.

- The manager of a performing arts company offers a group discount price of \$45 per person for groups of 20 or more and will drop the price by \$1.50 per person for each additional person. What is the maximum revenue? What size group will maximize the revenue?

Because the price does not change until more than 20 people are in the group, we will let x represent the additional people in the group. What is the price per person if the group size is more than 20? If one extra person is in the group, the price is $45 - 1(1.50)$. If there are two extra people, the price is $45 - 2(1.50)$; and if there are three extra people, the price is $45 - 3(1.50)$. So, if there are x additional people, the price is $45 - 1.50x$. The revenue is

$$R = (20 + x)(45 - 1.50x) = -1.50x^2 + 15x + 900.$$

$$h = \frac{-b}{2a} = \frac{-15}{2(-1.50)} = 5 \text{ and } k = -1.50(5)^2 + 15(5) + 900 = 937.50$$

The group size that maximizes revenue is $20 + 5 = 25$. The maximum revenue is \$937.50.

Optimizing geometric figures are common calculus and precalculus problems. In many of these problems, there are more than two variables. We will be given enough information in the problem to eliminate one of the variables. For example, if we want the area of a rectangle, the formula is $A = LW$. If we know the perimeter is 20, then we can use the equation $2L + 2W = 20$ to solve for either L or W and substitute this quantity in the area function, reducing the equation from three to two variables. The new area function will be quadratic.

EXAMPLES

- A parks department has 1200 meters of fencing available to enclose two adjacent playing fields. (See Figure 6.9.) What dimensions will maximize the enclosed area? What is the maximum enclosed area?

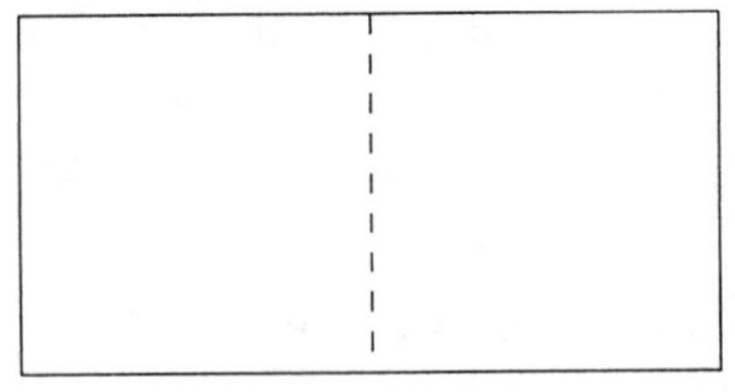

Fig. 6.9.

The total enclosed area is $A = LW$. Because there is 1200 meters of fencing available, we must have $L + W + W + W + L = 1200$ (see Figure 6.10).

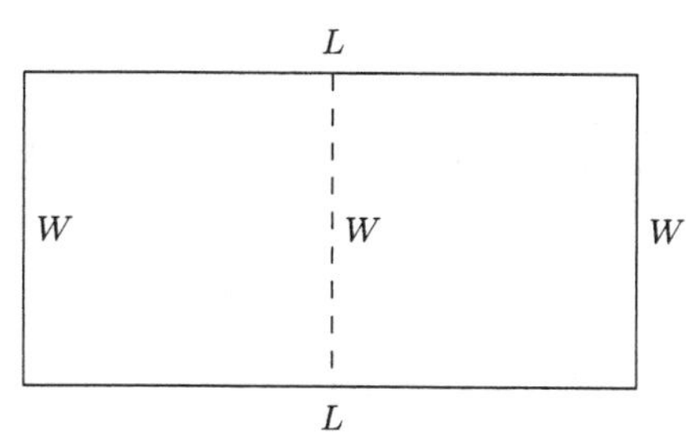

Fig. 6.10.

We can solve for L or W in this equation and substitute it in $A = LW$, reducing the equation to two variables. We will solve for L in $2L + 3W = 1200$.

$$2L + 3W = 1200$$

$$L = \frac{1200 - 3W}{2}$$

Now $A = LW$ becomes $A = \frac{1200-3W}{2} \cdot W = -\frac{3}{2}W^2 + 600W$. This function has a maximum value.

$$h = \frac{-b}{2a} = \frac{-600}{2(-3/2)} = 200 \text{ and } k = -\frac{3}{2}(200)^2 + 600(200) = 60{,}000$$

The width that maximizes the enclosed area is 200 meters, the length is $\frac{1200-3(200)}{2} = 300$ meters. The maximum enclosed area is 60,000 square meters.

Another common fencing problem is one where only three sides of a rectangular area needs to be fenced. The fourth side is some other boundary like a stream or the side of a building. We will call two sides W and the third side L. Then "$2W + L =$ amount of fencing" allows us to solve for L and substitute "$L =$ amount of fencing $-2W$" in $A = LW$ to reduce the area formula to two variables.

EXAMPLE

- A farmer has 1000 feet of fencing materials available to fence a rectangular pasture next to a river. If the side along the river does not need to be fenced, what dimensions will maximize the enclosed area? What is the maximum enclosed area?

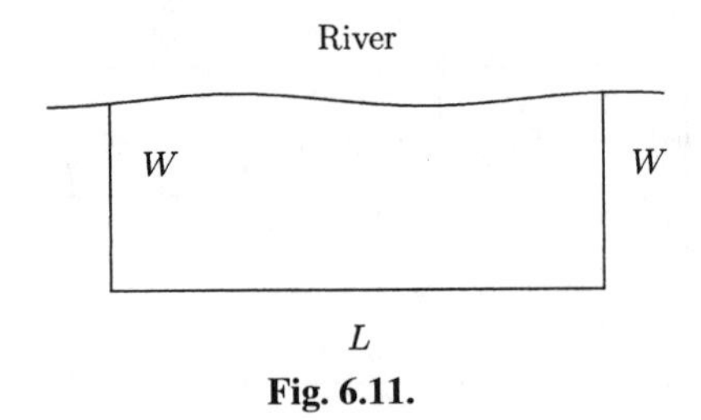

Fig. 6.11.

Using the fact that $2W + L = 1000$, we can solve for L and substitute this quantity in the area formula $A = LW$.

$$2W + L = 1000$$
$$L = 1000 - 2W$$
$$A = LW$$
$$A = (1000 - 2W)W = -2W^2 + 1000W$$

This quadratic function has a maximum value.

$$h = \frac{-b}{2a} = \frac{-1000}{2(-2)} = 250 \text{ and } k = -2(250)^2 + 1000(250) = 125{,}000$$

Maximize the enclosed area by letting $W = 250$ feet and $L = 1000 - 2(250) = 500$ feet. The maximum enclosed area is 125,000 square feet.

In the last problems, we will maximize the area of a figure but will have to work a little harder to find the area function to maximize.

EXAMPLES

- A window is to be constructed in the shape of a rectangle surmounted by a semicircle (see Figure 6.12). The perimeter of the window needs to be 18 feet. What dimensions will admit the greatest amount of light?

  The dimensions that will admit the greatest amount of light are the same that will maximize the area of the window. The area of the window is the rectangular area plus the area of the semicircle. The area of the rectangular region is LW. Because the width of the window is the diameter (or twice the

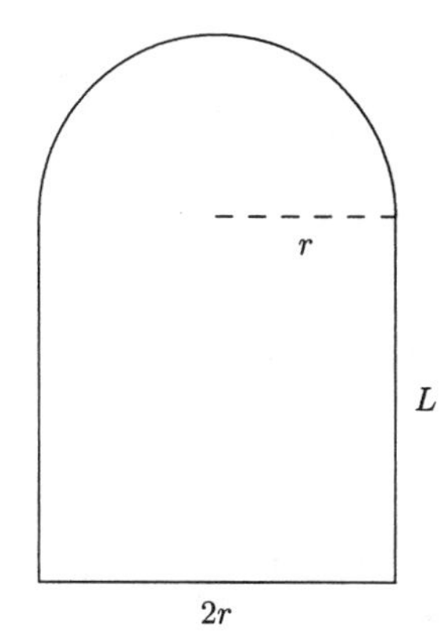

Fig. 6.12.

radius) of the semicircle, we can rewrite the area as $L(2r) = 2rL$. The area of the semicircle is half of the area of a circle with radius r, or $\frac{1}{2}\pi r^2$. The total area of the window is

$$A = 2rL + \frac{1}{2}\pi r^2.$$

Now we will use the fact that the perimeter is 18 feet to help us replace L with an expression using r. The perimeter is made up of the two sides $(2L)$ and the bottom of the rectangle $(2r)$ and the length around the semicircle. The length around the outside of the semicircle is half of the circumference of a circle with radius r, $\frac{1}{2}(2\pi r) = \pi r$. The total perimeter is $P = 2L + 2r + \pi r$. This is equal to 18. We will solve the equation $2L + 2r + \pi r = 18$ for L.

$$2L + 2r + \pi r = 18$$

$$2L = 18 - 2r - \pi r$$

$$L = \frac{18 - 2r - \pi r}{2} = 9 - r - \frac{1}{2}\pi r$$

Now we will substitute $9 - r - \frac{1}{2}\pi r$ for L in the area formula.

$$A = 2rL + \frac{1}{2}\pi r^2$$

$$A = 2r(9 - r - \frac{1}{2}\pi r) + \frac{1}{2}\pi r^2$$

$$A = 18r - 2r^2 - \pi r^2 + \frac{1}{2}\pi r^2 = 18r - 2r^2 - \frac{1}{2}\pi r^2 = 18r - \left(2 + \frac{1}{2}\pi\right)r^2$$

$$A = -\left(2 + \frac{1}{2}\pi\right)r^2 + 18r$$

This quadratic function has a maximum value.

$$h = \frac{-b}{2a} = \frac{-18}{2\left[-\left(2 + \frac{1}{2}\pi\right)\right]} = \frac{18}{4 + \pi} \approx 2.52$$

Maximize the amount of light admitted in the window by letting the radius of the semicircle be about 2.52 feet, and the length about $9 - 2.52 - \frac{\pi}{2}(2.52) \approx$ 2.52 feet.

- A track is to be constructed so that it is shaped like Figure 6.13, a rectangle with a semicircle at each end. If the inside perimeter of the track is to be $\frac{1}{4}$ mile, what is the maximum area of the rectangle?

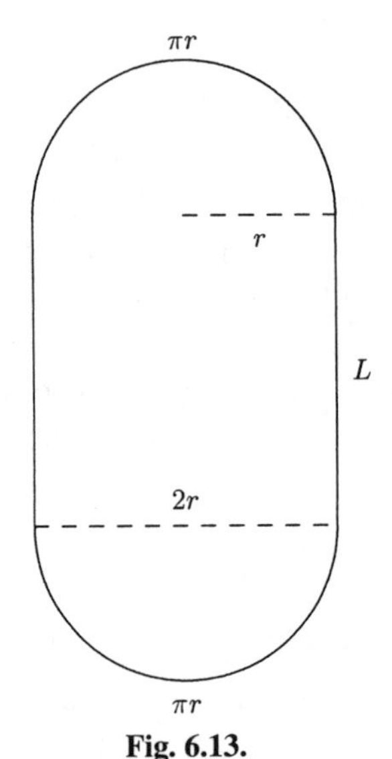

Fig. 6.13.

The length of the rectangle is L. Its width is the diameter of the semicircles (or twice their radius). The area formula for the rectangle is $A = LW = L(2r) = 2rL$. The perimeter of the figure is the two sides of the rectangle ($2L$) plus the length around each semicircle (πr). The total perimeter is $2L + 2\pi r$. Although we could work with the dimensions in miles, it will be easier to convert 1/4 mile to feet. There are 5280/4 = 1320 feet in 1/4 mile.

We will solve $2L + 2\pi r = 1320$ for L. Solving for r works, too.

$$2L + 2\pi r = 1320$$

$$2L = 1320 - 2\pi r$$

$$L = \frac{1320 - 2\pi r}{2} = 660 - \pi r$$

$$A = 2rL$$

$$A = 2r(660 - \pi r) = -2\pi r^2 + 1320r$$

The area function has a maximum value.

$$h = \frac{-b}{2a} = \frac{-1320}{2(-2\pi)} = \frac{330}{\pi}$$

$$k = -2\pi \left(\frac{330}{\pi}\right)^2 + 1320\left(\frac{330}{\pi}\right)$$

$$= \frac{217{,}800}{\pi} \approx 69{,}328$$

The maximum area of the rectangular region is about 69,328 square feet.

- A rectangle is to be constructed so that it is bounded below by the x-axis, on the left by the y-axis, and above by the line $y = -2x + 12$. (See Figure 6.14). What is the maximum area of the rectangle?

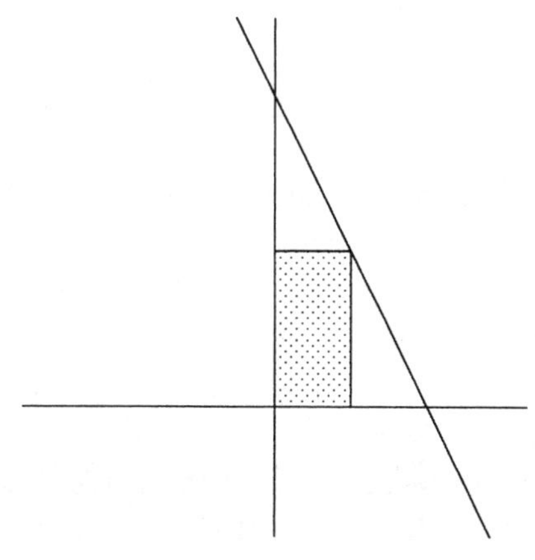

Fig. 6.14.

The coordinates of the corners will help us to see how we can find the length and width of the rectangle.

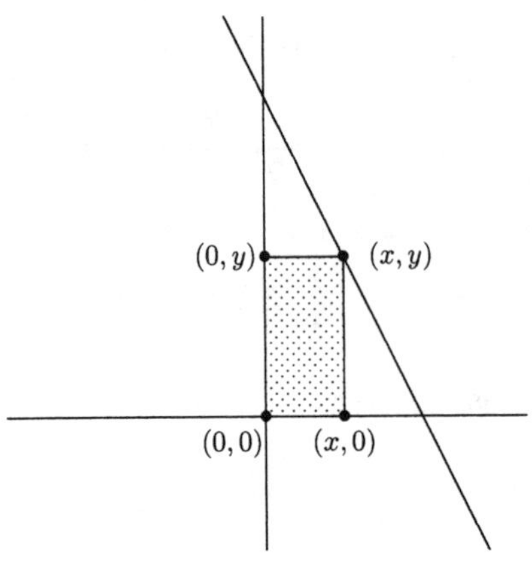

Fig. 6.15.

The height of the rectangle is y and the width is x. This makes the area $A = xy$. We need to eliminate x or y. Because $y = -2x + 12$, we can substitute $-2x + 12$ for y in $A = xy$ to make it the quadratic function $A = xy = x(-2x + 12) = -2x^2 + 12x$.

$$h = \frac{-b}{2a} = \frac{-12}{2(-2)} = 3 \text{ and } k = -2(3)^2 + 12(3) = 18$$

The maximum area is 18 square units.

PRACTICE

1. The average cost of a product can be approximated by the function $C(x) = 0.00025x^2 - 0.25x + 70.5$, where x is the number of units produced and $C(x)$ is the average cost in dollars. What level of production will minimize the average cost?

2. A frog jumps from a rock to the shore of a pond. Its path is given by the equation $y = -\frac{5}{72}x^2 + \frac{5}{3}x$, where x is the horizontal distance in inches, and y is the height in inches. What is the frog's maximum height? How far had it traveled horizontally when it reached its maximum height?

3. A projectile is fired upward from a ten-foot platform. The projectile's initial velocity is 108 miles per hour. What is the projectile's maximum height? When will it reach its maximum height?

4. Attendance at home games for a college basketball team averages 1000 and the ticket price is \$12. Concession sales average \$2 per person. A student survey reveals that for every \$0.25 decrease in the ticket price, 25 more students will attend the home games. What ticket price will maximize revenue? What is the maximum revenue?

5. A school has 1600 feet of fencing available to enclose three playing fields (see Figure 6.16). What dimensions will maximize the enclosed area?

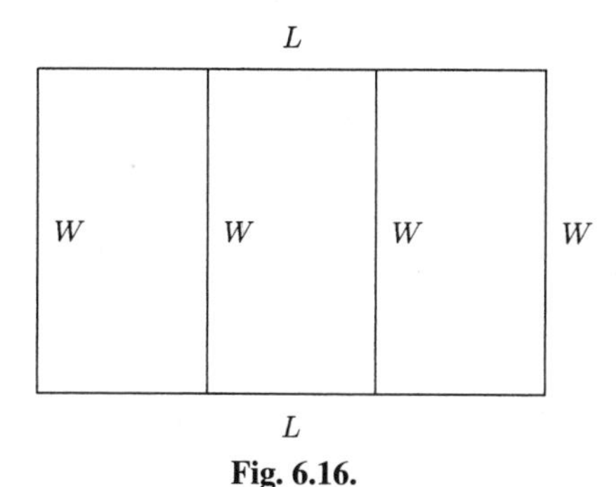

Fig. 6.16.

6. The manager of a large warehouse wants to enclose an area behind the building. He has 900 feet of fencing available. What dimensions will maximize the enclosed area? What is the maximum area?

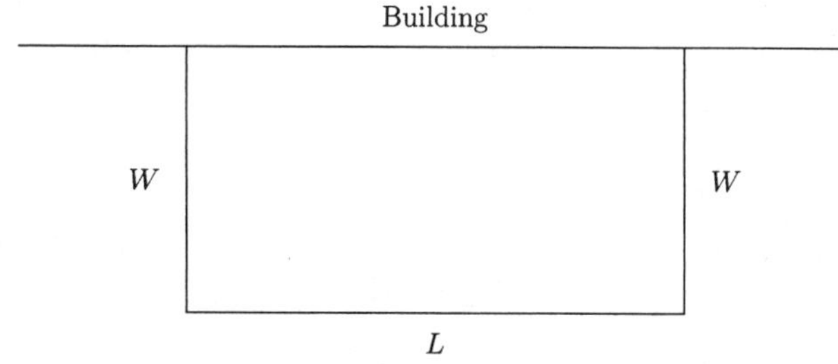

Fig. 6.17.

7. A swimming pool is to be constructed in the shape of a rectangle with a semicircle at one end (see Figure 6.12). If the perimeter is to be 120 feet, what dimensions will maximize the area? What is the maximum area?

8. A rectangle is to be constructed so that it is bounded by the x-axis, the y-axis, and the line $y = -3x + 4$ (see Figure 6.18). What is the maximum area of the rectangle?

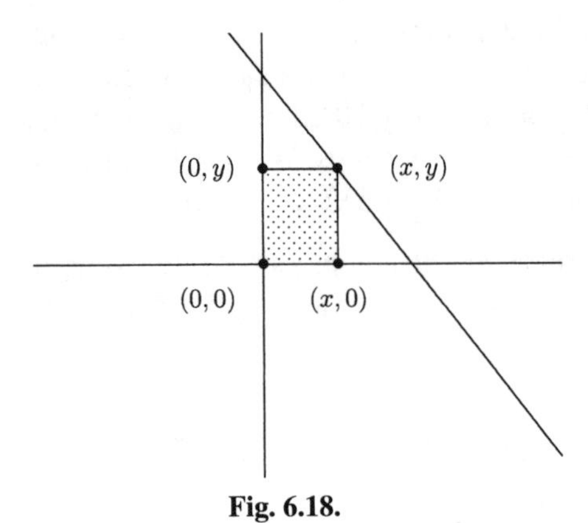

Fig. 6.18.

SOLUTIONS

1. We only need to find h.

$$h = \frac{-b}{2a} = \frac{-(-0.25)}{2(0.00025)} = 500$$

Minimize the average cost by producing 500 units.

2. k answers the first question, h answers the second.

$$h = \frac{-b}{2a} = \frac{-5/3}{2(-5/72)} = \frac{5/3}{5/36} = \frac{5}{3} \div \frac{5}{36} = \frac{5}{3} \cdot \frac{36}{5} = 12$$

$$k = -\frac{5}{72}(12)^2 + \frac{5}{3}(12) = 10$$

The frog reached a maximum height of 10 inches and had traveled 12 inches horizontally when it reached its maximum height.

3. The formula $s(t) = -16t^2 + v_0 t + s_0$ is in feet and seconds, so we need to convert 108 miles per hour to feet per second. There are 5280 feet in a mile and $60(60) = 3600$ seconds in an hour.

$$\frac{108 \text{ miles}}{1 \text{ hour}} = \frac{108(5280) \text{ feet}}{3600 \text{ seconds}} = 158.4 \text{ feet per second}$$

Replacing v_0 with 158.4 and s_0 with 10, we have the function giving the height of the projectile after t seconds, $s(t) = -16t^2 + 158.4t + 10$.

$$h = \frac{-b}{2a} = \frac{-158.4}{2(-16)} = 4.95 \text{ and}$$

$$k = -16(4.95)^2 + 158.4(4.95) + 10 = 402.04$$

The projectile reaches a maximum height of 402.04 feet after 4.95 seconds.

4. We will let x represent the number of \$0.25 decreases in the ticket price. The ticket price is $12 - 0.25x$ and the average number attending the games is $1000 + 25x$. Ticket revenue is $(12 - 0.25x)(1000 + 25x)$. Revenue from concession sales is $2(1000 + 25x)$. Total revenue is

$$R = (12 - 0.25x)(1000 + 25x) + 2(1000 + 25x)$$

$$= -6.25x^2 + 100x + 14{,}000$$

$$h = \frac{-b}{2a} = \frac{-100}{2(-6.25)} = 8 \text{ and}$$

$$k = -6.25(8)^2 + 100(8) + 14{,}000 = 14{,}400$$

The ticket price that will maximize revenue is $12 - 0.25(8) = \$10$ and the maximum revenue is \$14,400.

5. The total area is $A = LW$. Because there is 1600 feet of fencing available, $2L + 4W = 1600$. Solving this equation for L, we have $L = 800 - 2W$. Substitute $800 - 2W$ for L in $A = LW$.

$$A = LW$$

$$= (800 - 2W)W = -2W^2 + 800W$$

$$h = \frac{-b}{2a} = \frac{-800}{2(-2)} = 200$$

Maximize the enclosed area by letting the width be 200 feet and the length be $800 - 2(200) = 400$ feet.

6. The enclosed area is $A = LW$. Because 900 feet of fencing is available, $2W + L = 900$. Solving this for L, we have $L = 900 - 2W$. We will substitute $900 - 2W$ for L in $A = LW$.

$$A = LW$$

$$= (900 - 2W)W = -2W^2 + 900W$$

$$h = \frac{-b}{2a} = \frac{-900}{2(-2)} = 225 \text{ and } k = -2(225)^2 + 900(225) = 101{,}250$$

Maximize the enclosed area by letting the width be 225 feet and the length $900 - 2(225) = 450$ feet. The maximum enclosed area is 101,250 square feet.

7. The area of the rectangle is $2rL$ (the width is twice the radius of the semicircle). The area of the semicircle is half the area of a circle with radius r, $\frac{1}{2}\pi r^2$. The total area of the pool is $A = 2rL + \frac{1}{2}\pi r^2$. After finding an equation for the perimeter, we will solve the equation for L and substitute this for L in $A = 2rL + \frac{1}{2}\pi r^2$. The perimeter of the rectangular part is $L + 2r + L = 2r + 2L$. The length around the semicircle is half the circumference of a circle with radius r, $\frac{1}{2}(2\pi r) = \pi r$. The total length around the pool is $2L + 2r + \pi r$ which equals 120 feet.

$$2L + 2r + \pi r = 120$$

$$2L = 120 - 2r - \pi r$$

$$L = \frac{120 - 2r - \pi r}{2} = 60 - r - \frac{1}{2}\pi r$$

$$A = 2rL + \frac{1}{2}\pi r^2$$

$$= 2r\left(60 - r - \frac{1}{2}\pi r\right) + \frac{1}{2}\pi r^2 \quad \text{Substitute } 60 - r - \frac{1}{2}\pi r \text{ for } L.$$

$$= 120r - 2r^2 - \pi r^2 + \frac{1}{2}\pi r^2$$

$$= -2r^2 - \frac{1}{2}\pi r^2 + 120r$$

$$= \left(-2 - \frac{1}{2}\pi\right) r^2 + 120r$$

$$h = \frac{-b}{2a} = \frac{-120}{2(-2 - \frac{1}{2}\pi)} = \frac{-120}{-4 - \pi} = \frac{-120}{-(4 + \pi)}$$

$$= \frac{120}{4 + \pi} \approx 16.8$$

$$k = \left(-2 - \frac{1}{2}\pi\right)\left(\frac{120}{4+\pi}\right)^2 + 120\left(\frac{120}{4+\pi}\right)$$
$$= \frac{7200}{4+\pi} \approx 1008.2$$

Maximize the area by letting the radius of the semicircle be about 16.8 feet and the length of the rectangle about $60 - 16.8 - \frac{1}{2}\pi(16.8) \approx 16.8$ feet. The maximum area is about 1008.2 square feet.

8. The area is $A = LW$. The length of the rectangle is y (the distance from (0, 0) and (0, y)). The width is x (the distance from (0, 0) to (x, 0)). The area is now $A = xy$. Because $y = -3x + 4$, we can substitute $-3x + 4$ for y in $A = xy$.

$$A = xy$$
$$A = x(-3x + 4) = -3x^2 + 4x$$

$$h = \frac{-4}{2(-3)} = \frac{2}{3} \text{ and } k = -3\left(\frac{2}{3}\right)^2 + 4\left(\frac{2}{3}\right) = \frac{4}{3}$$

The maximum area is $\frac{4}{3}$ square units.

CHAPTER 6 REVIEW

1. What is the vertex for $f(x) = -2(x-1)^2 + 4$?
   (a) (1, 4) (b) (−1, 4) (c) (−2, 4) (d) (2, 4)

2. Complete the square on $y = x^2 - 6x + 10$ to write it in the form $y = a(x-h)^2 + k$.
   (a) $y = (x-3)^2 + 9$ (b) $y = (x-3)^2 + 19$
   (c) $y = (x-3)^2 - 9$ (d) $y = (x-3)^2 + 1$

3. Complete the square on $y = 3x^2 - x + 1$ to write it in the form $y = a(x-h)^2 + k$.
   (a) $y = 3(x - \frac{1}{6})^2 + \frac{13}{12}$ (b) $y = 3(x - \frac{1}{6})^2 + \frac{11}{12}$
   (c) $y = 3(x - \frac{1}{6})^2 + \frac{35}{36}$ (d) $y = 3(x - \frac{1}{6})^2 + \frac{37}{36}$

4. What are the x- and y-intercepts for $f(x) = 2x^2 + x - 6$?
   (a) The y-intercept is -6, and the x-intercepts are $-\frac{3}{2}$ and 2.
   (b) The y-intercept is -6, and the x-intercepts are $\frac{3}{2}$ and 2.
   (c) The y-intercept is -6, and the x-intercepts are $\frac{3}{2}$ and -2.
   (d) The y-intercept is -6, and the x-intercepts are $-\frac{3}{2}$ and -2.

5. What is the vertex for $f(x) = -0.02x^2 + 3x - 10$?
   (a) $(75, -122.5)$ (b) $(-75, -347.5)$
   (c) $(75, 102.5)$ (d) $(75, 5615)$

6. Find the maximum or minimum functional value for $f(x) = 6(x-25)^2 + 100$.
   (a) The maximum functional value is 25.
   (b) The maximum functional value is 100.
   (c) The minimum functional value is 25.
   (d) The minimum functional value is 100.

7. Find the quadratic function with vertex $(4, -2)$ and with the point $(5, -\frac{5}{3})$ on its graph.
   (a) $f(x) = \frac{1}{3}(x-4)^2 - 2$ (b) $f(x) = \frac{1}{27}(x+4)^2 - 2$
   (c) $f(x) = (x-4)^2 + 2$ (d) $f(x) = -\frac{1}{81}(x+4)^2 + 2$

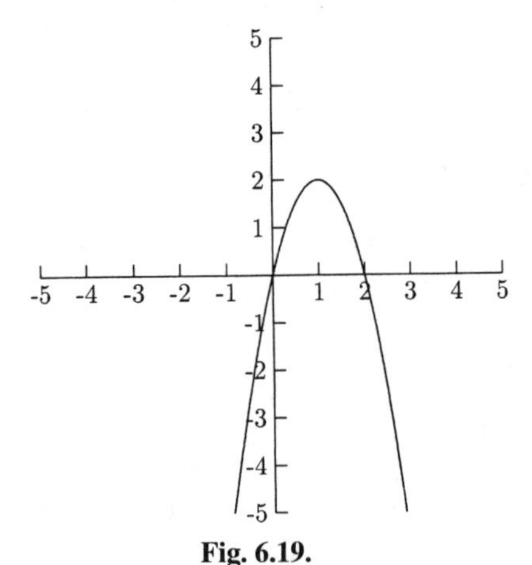

Fig. 6.19.

8. What is the function whose graph is in Figure 6.19?
   (a) $f(x) = -2(x+1)^2 + 2$ (b) $f(x) = -2(x-1)^2 + 2$
   (c) $f(x) = 2(x+1)^2 + 2$ (d) $f(x) = 2(x-1)^2 + 2$

9. A hot dog vendor at a local fair averages 140 hot dogs per day when the price is \$3. If for every \$0.25 increase in the price, 10 fewer hot dogs are sold on average, what price maximizes the revenue?
   (a) \$3.00 (b) \$3.25 (c) \$3.50 (d) \$3.75

10. A warehouse manager wants to fence a rectangular area behind his warehouse. He has 120 meters of fencing available. If the side against the building does not need to be fenced, what is the maximum enclosed area?
   (a) 1500 square meters (b) 1700 square meters
   (c) 1600 square meters (d) 1800 square meters

SOLUTIONS

1. A	2. D	3. B	4. C	5. C
6. D	7. A	8. B	9. B	10. D

Polynomial Functions

A polynomial function is a function in the form $f(x) = a_n x^n + a_{n-1}x^{n-1} + \cdots + a_1 x + a_0$, where each a_i is a real number and the powers on x are whole numbers. There is no x under a root sign and no x in a denominator. The number a_i is called a *coefficient*. For example, in the polynomial function $f(x) = -2x^3 + 5x^2 - 4x + 8$, the coefficients are -2, 5, -4, and 8. The *constant* term (the term with no variable) is 8. The powers on x are 3, 2, and 1. The *degree* of the polynomial (and polynomial function) is the highest power on x. In this example, the degree is 3. Quadratic functions are degree 2. Linear functions of the form $f(x) = mx + b$ (if $m \neq 0$) are degree 1. Constant functions of the form $f(x) = b$ are degree 0 (this is because $x^0 = 1$, making $f(x) = bx^0$).

The *leading term* of a polynomial (and polynomial function) is the term having x to the highest power. Usually, but not always, the leading term is written first. The *leading coefficient* is the coefficient on the leading term. In our example, the leading term is $-2x^3$, and the leading coefficient is -2. By looking at the leading term only, we can tell roughly what the graph looks like. The graph of any polynomial will either go up on both ends, go down on both ends, or go up on one end and down on the other. This is called the *end behavior* of the graph. The figures below illustrate the end behavior of polynomial functions. The shape of the dashed part of the graph depends on the individual function.

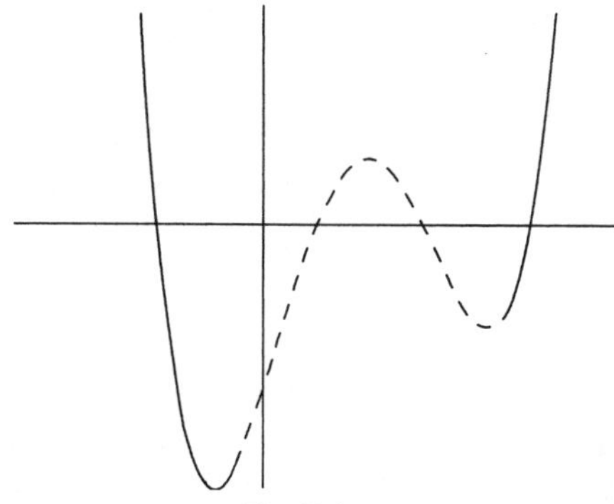

Fig. 7.1.

This graph goes up on both ends.

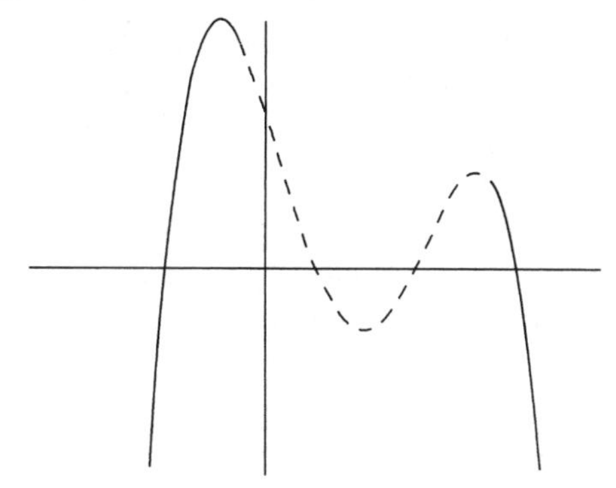

Fig. 7.2.

This graph goes down on both ends.

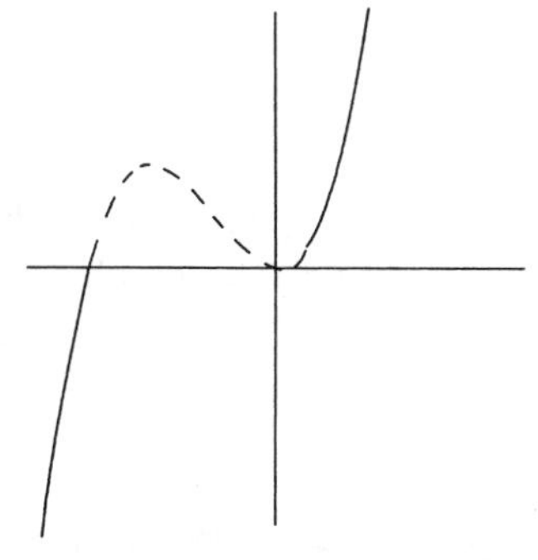

Fig. 7.3.

This graph goes down on the left and up on the right.

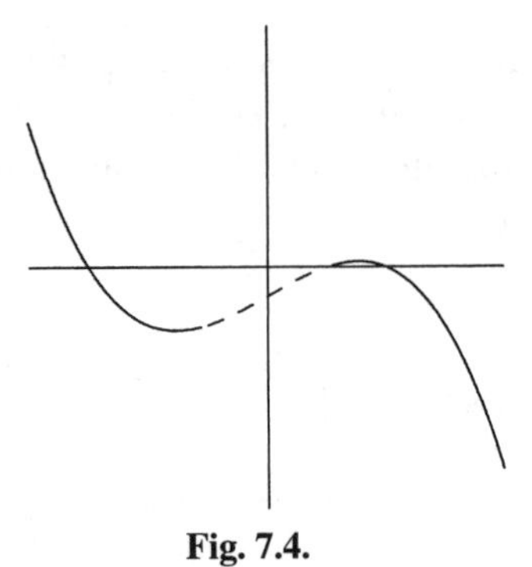

Fig. 7.4.

This graph goes up on the left and down on the right.

If the degree of the polynomial is an even number, the graph will look like the graph in Figure 7.1 or in Figure 7.2. If the leading coefficient is a positive number, the graph will look like the graph in Figure 7.1. If the leading coefficient is a negative number, the graph will look like the graph in Figure 7.2. If the degree of the polynomial is an odd number, the graph will look like the one in Figure 7.3 or in Figure 7.4. If the leading coefficient is a positive number, the graph will look like the graph in Figure 7.3. If the leading coefficient is a negative number, the graph will look like the graph in Figure 7.4.

How can one term in a polynomial function give us this information? For polynomial functions, the leading term dominates all of the other terms. For x-values large enough (both large positive numbers and large negative numbers), the other terms don't contribute much to the size of the y-values.

EXAMPLES

Match the graph of the given function with one of the graphs in Figures 7.1–7.4.

- $f(x) = 4x^5 + 6x^3 - 2x^2 + 8x + 11$

  We only need to look at the leading term, $4x^5$. The degree, 5, is odd, and the leading coefficient, 4, is positive. The graph of this function looks like the one in Figure 7.3.

- $P(x) = 5 + 2x - 6x^2$

  The leading term is $-6x^2$. The degree, 2, is even, and the leading coefficient, -6, is negative. The graph of this function looks like the one in Figure 7.2.

- $h(x) = -2x^3 + 4x^2 - 7x + 9$

  The leading term is $-2x^3$. The degree, 3, is odd, and the leading coefficient, -2, is negative. The graph of this function looks like the one in Figure 7.4.

- $g(x) = x^4 + 4x^3 - 8x^2 + 3x - 5$

  The leading term is x^4. The degree, 4, is even, and the leading coefficient, 1, is positive. The graph of this function looks like the one in Figure 7.1.

Finding the x-intercepts (if any) for the graph of a polynomial function is very important. The x-intercept of any graph is where the graph touches the x-axis. This happens when the y-coordinate of the point is 0. We found the x-intercepts for some quadratic functions by factoring and setting each factor equal to zero. This is how we will find the x-intercepts for polynomial functions. It is not always easy to do. In fact, some polynomials are so hard to factor that the best we can do is approximate the x-intercepts (using graphing calculators or calculus). This will not be the case for the polynomials in this book, however. Every polynomial here will be factorable using techniques covered later.

Because an x-intercept for $f(x) = a_n x^n + a_{n-1}x^{n-1} + \cdots + a_1 x + a_0$ is a solution to the equation $0 = a_n x^n + a_{n-1}x^{n-1} + \cdots + a_1 x + a_0$, x-intercepts are also called *zeros* of the polynomial. All of the following statements have the same meaning for a polynomial. Let c be a real number, and let $P(x)$ be a polynomial function.

1. c is an x-intercept of the graph of $P(x)$.
2. c is a zero for $P(x)$.
3. $x - c$ is a factor of $P(x)$.

EXAMPLES

- $x - 1$ is a factor means that 1 is an x-intercept and a zero.
- $x + 5$ is a factor means that -5 is an x-intercept and a zero.
- x is a factor means that 0 is an x-intercept and a zero.
- 3 is a zero means that $x - 3$ is a factor and 3 is an x-intercept.

We can find the zeros of a function (or at least the approximate zeros) by looking at its graph.

The x-intercepts of the graph in Figure 7.5 are 2 and -2. Now we know that $x - 2$ and $x + 2$ (which is $x - (-2)$) are factors of the polynomial.

The graph of the polynomial function in Figure 7.6 has x-intercepts of -1, 1, and 2. This means that $x - 1$, $x - 2$, and $x + 1$ (as $x - (-1)$) are factors of the polynomial.

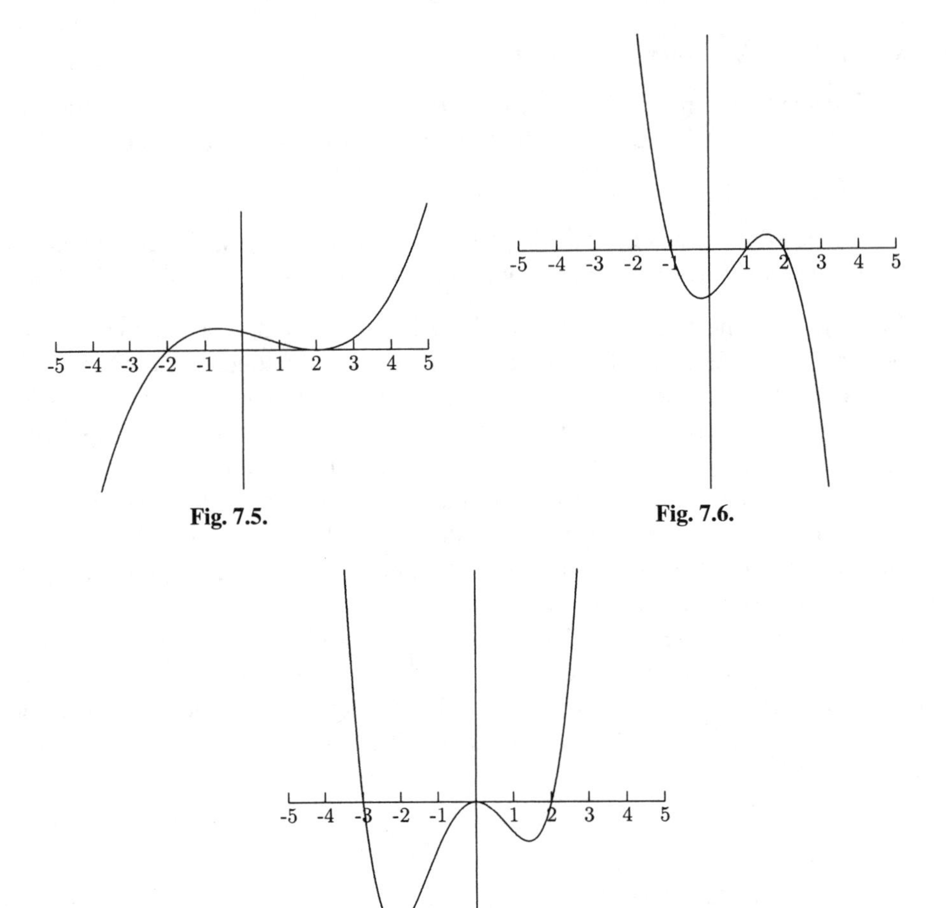

Fig. 7.5.

Fig. 7.6.

Fig. 7.7.

The x-intercepts for the graph in Figure 7.7 are -3, 0, and 2, making $x+3$, x (as $x-0$), and $x-2$ factors of the polynomial.

Now that we know about the end behavior of the graphs of polynomial functions and the relationship between x-intercepts and factors, we can look at a polynomial and have a pretty good idea of what its graph looks like. In the next set of examples, we will match the graphs from the previous section with their polynomial functions.

EXAMPLES

Match the functions with the graphs in Figures 7.5–7.7.

- $f(x) = \frac{1}{10}x^2(x+3)(x-2) = \frac{1}{10}x^4 + \frac{1}{10}x^3 - \frac{3}{5}x^2$

Because $f(x)$ is a polynomial whose degree is even and whose leading coefficient is positive, we will look for a graph that goes up on the left and up on the right. Because the factors are x^2, $x+3$, and $x-2$, we will also look for a graph with x-intercepts 0, -3, and 2. The graph in Figure 7.7 satisfies these conditions.

- $g(x) = -\frac{1}{2}(x-1)(x-2)(x+1) = -\frac{1}{2}x^3 + x^2 + \frac{1}{2}x - 1$

  Because $g(x)$ is a polynomial whose degree is odd and whose leading coefficient is negative, we will look for a graph that goes up on the left and down on the right. The factors are $x-1$, $x-2$, and $x+1$, we will also look for a graph with 1, 2, and -1 as x-intercepts. The graph in Figure 7.6 satisfies these conditions.

- $P(x) = \frac{1}{10}(x-2)^2(x+2) = \frac{1}{10}x^3 - \frac{1}{5}x^2 - \frac{2}{5}x + \frac{4}{5}$

  Because $P(x)$ is a polynomial whose degree is odd and whose leading coefficient is positive, we will look for a graph that goes down on the left and up on the right. The x-intercepts are 2 and -2. The graph in Figure 7.5 satisfies these conditions.

Sketching Graphs of Polynomials

To sketch the graph of most polynomial functions accurately, we need to use calculus (don't let that scare you—the calculus part is easier than the algebra part!) We can still get a pretty good graph using algebra alone. The general method is to plot x-intercepts (if there are any), a point to the left of the smallest x-intercept, a point between any two x-intercepts, and a point to the right of the largest x-intercept. Because y-intercepts are easy to find, it wouldn't hurt to plot these, too.

EXAMPLES

- $f(x) = -(2x-1)(x+2)(x-3)$

  The x-intercepts are -2, 3, and $\frac{1}{2}$ (from $2x - 1 = 0$). In addition to the x-intercepts, we will plot the points for $x = -2.5$ (to the left of $x = -2$), $x = -1$ (between $x = -2$ and $x = \frac{1}{2}$), $x = 2$ (between $x = \frac{1}{2}$ and $x = 3$), and $x = 3.5$ (to the right of $x = 3$).

Table 7.1

x	$f(x)$
-2.5	16.5
-2	0
-1	-12
0	-6
$\frac{1}{2}$	0
2	12
3	0
3.5	-16.5

The reason we used $x = -2.5$ instead of $x = -3$ and $x = 3.5$ instead of $x = 4$ is that their y-values were too large for our graph.

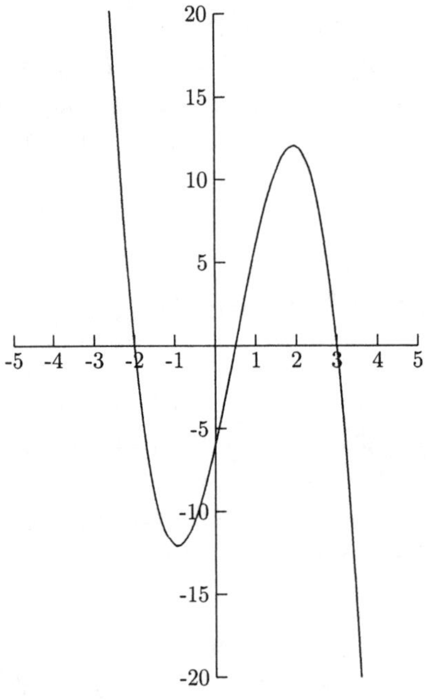

Fig. 7.8.

PRACTICE

Match the graph of the given function with one of the graphs in Figures 7.1–7.4.

1. $f(x) = -8x^3 + 4x^2 - 9x + 3$

2. $f(x) = 4x^5 + 10x^4 - 3x^3 + x^2$
3. $P(x) = -x^2 + x - 6$
4. $g(x) = 1 + x + x^2 + x^3$

Identify the x-intercepts and factors for the polynomial function whose graphs are given.

5.

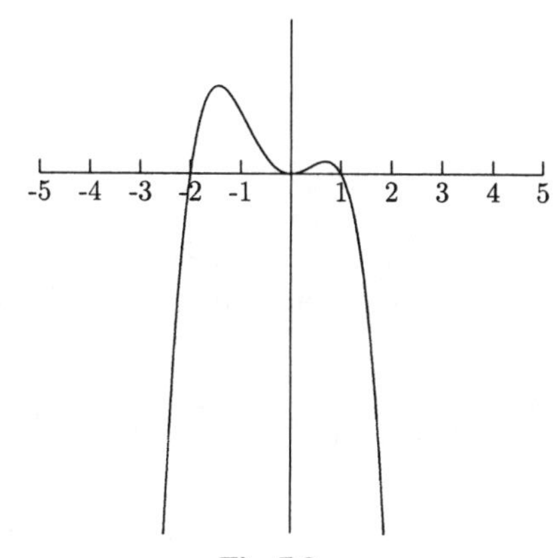

Fig. 7.9.

6.

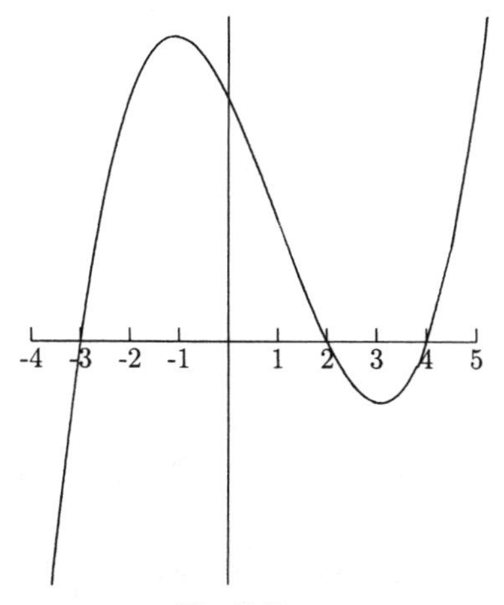

Fig. 7.10.

7.

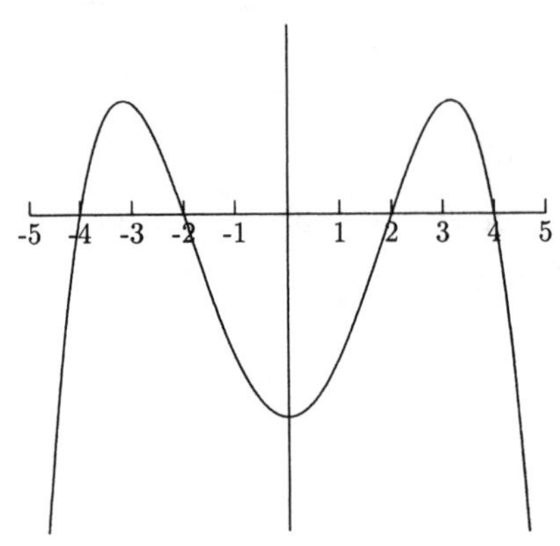

Fig. 7.11.

Match the polynomial function with one of the graphs in Figures 7.9 through 7.11.

8. $f(x) = -\frac{1}{8}(x+4)(x+2)(x-2)(x-4) = -\frac{1}{8}x^4 + \frac{5}{2}x^2 - 8$

9. $P(x) = -\frac{1}{2}x^2(x+2)(x-1) = -\frac{1}{2}x^4 - \frac{1}{2}x^3 + x^2$

10. $R(x) = \frac{1}{2}(x+3)(x-2)(x-4) = \frac{1}{2}x^3 - \frac{3}{2}x^2 - 5x + 12$

11. Sketch the graph of $f(x) = \frac{1}{2}x(x-2)(x+2)$.

12. Sketch the graph of $h(x) = -\frac{1}{10}(x+4)(x+1)(x-2)(x-3)$.

SOLUTIONS

1. Figure 7.4
2. Figure 7.3
3. Figure 7.2
4. Figure 7.3
5. The x-intercepts are -2, 0, and 1, so $x+2$, x, and $x-1$ are factors of the polynomial.
6. The x-intercepts are -3, 2, and 4, so $x+3$, $x-2$, and $x-4$ are factors of the polynomial.

7. The x-intercepts are -4, -2, 2 and 4, so $x + 4$, $x + 2$, $x - 2$ and $x - 4$ are factors of the polynomial.
8. Figure 7.11
9. Figure 7.9
10. Figure 7.10
11.

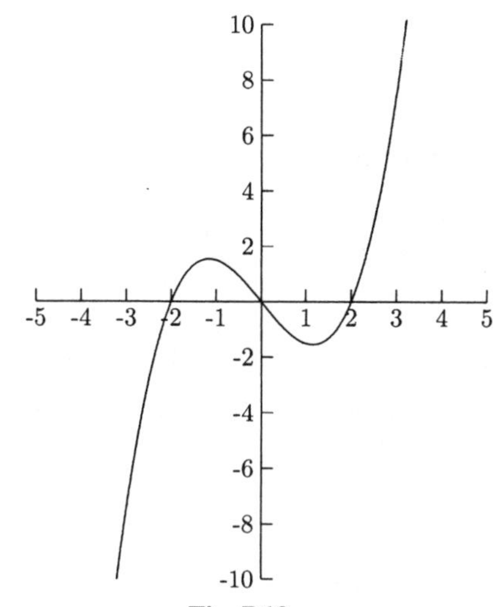

Fig. 7.12.

12.

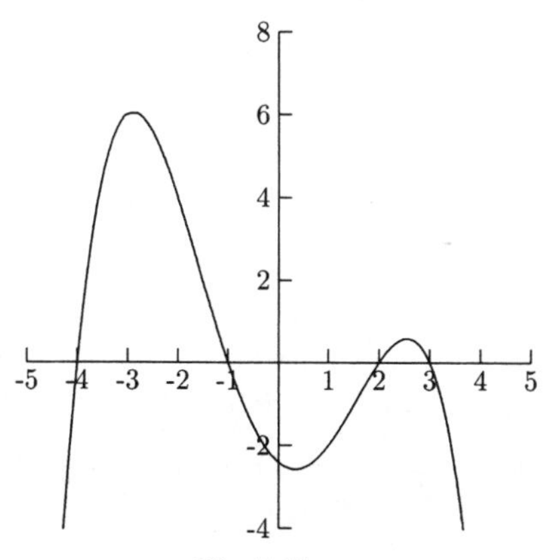

Fig. 7.13.

Polynomial Division

Polynomials can be divided in much the same way as whole numbers. When we take the quotient of two whole numbers (where the divisor is not zero), we get a quotient and a remainder. The same happens when we take the quotient of two polynomials. Polynomial division is useful when factoring polynomials.

Polynomial division problems usually come in one of two forms.

$$\frac{\text{dividend polynomial}}{\text{divisor polynomial}} \text{ or dividend polynomial} \div \text{divisor polynomial}$$

According to the division algorithm for polynomials, for any polynomials $f(x)$ and $g(x)$ (with $g(x)$ not the zero function)

$$\frac{f(x)}{g(x)} = q(x) + \frac{r(x)}{g(x)},$$

where $q(x)$ is the quotient (which might be 0) and $r(x)$ is the remainder, which has degree *strictly* less than the degree of $g(x)$. Multiplying by $g(x)$ to clear the fraction, we also get $f(x) = g(x)q(x) + r(x)$. First we will perform polynomial division using long division.

$$\begin{array}{r|l} & q(x) \\ \hline g(x) & f(x) \\ & \\ \hline & r(x) \end{array}$$

EXAMPLES

Find the quotient and remainder using long division.

- $\dfrac{4x^2 + 3x - 5}{x + 2}$

  $x + 2\,\overline{)\,4x^2 + 3x - 5}$

  We will begin by dividing the leading term of the dividend by the leading term of the divisor. For the first step in this example, we will divide $4x^2$ by x. You might see right away that $4x^2 \div x$ is $4x$. If not, write $4x^2 \div x$ as a fraction then reduce: $\frac{4x^2}{x} = 4x$. This will be the first term of the quotient.

  $$\begin{array}{r|l} & 4x \\ \hline x + 2 & 4x^2 + 3x - 5 \end{array}$$

Multiply $4x$ by the divisor: $4x(x+2) = 4x^2 + 8x$. Subtract this from the first two terms of the dividend. Be careful to subtract all of $4x^2 + 8x$, not just $4x^2$.

$$\begin{array}{r|l} & \quad 4x \\ \hline x+2 & \quad 4x^2 + 3x - 5 \\ & -(4x^2 + 8x) \\ \hline & \qquad\quad -5x \end{array}$$

Bring down the next term.

$$\begin{array}{r|l} & \quad 4x \\ \hline x+2 & \quad 4x^2 + 3x - 5 \\ & -(4x^2 + 8x) \\ \hline & \qquad\quad -5x - 5 \end{array}$$

Start the process again with $-5x \div x = -5$. The next term in the quotient is -5. Multiply $x+2$ by -5: $-5(x+2) = -5x - 10$. Subtract this from $-5x - 5$.

$$\begin{array}{r|l} & \quad 4x - 5 \\ \hline x+2 & \quad 4x^2 + 3x - 5 \\ & -(4x^2 + 8x) \\ \hline & \qquad\quad -5x - 5 \\ & \qquad -(-5x - 10) \\ \hline & \qquad\qquad\qquad 5 \end{array}$$

We are done because $5 \div x = \frac{5}{x}$ cannot be a term in a polynomial. The remainder is 5 and the quotient is $4x - 5$.

- $x^2 + 2x - 3 \overline{\big) 3x^4 + 5x^3 - 4x^2 + 7x - 1}$

Divide $3x^4$ by x^2 to get the first term of the quotient: $\frac{3x^4}{x^2} = 3x^2$. Multiply $x^2 + 2x - 3$ by $3x^2$: $3x^2(x^2 + 2x - 3) = 3x^4 + 6x^3 - 9x^2$. Subtract this from the first three terms in the dividend.

$$\begin{array}{r|l} & \quad 3x^2 \\ \hline x^2+2x-3 & \quad 3x^4 + 5x^3 - 4x^2 + 7x - 1 \\ & -(3x^4 + 6x^3 - 9x^2) \\ \hline & \qquad\quad -x^3 + 5x^2 \end{array}$$

Divide $-x^3$ by x^2 to get the second term in the quotient: $\frac{-x^3}{x^2} = -x$. Multiply $x^2 + 2x - 3$ by $-x$: $-x(x^2 + 2x - 3) = -x^3 - 2x^2 + 3x$. Subtract this from $-x^3 + 5x^2 + 7x$.

$$\begin{array}{r|l} & 3x^2 - x \\ \hline x^2 + 2x - 3 & 3x^4 + 5x^3 - 4x^2 + 7x - 1 \\ & -(3x^4 + 6x^3 - 9x^2) \\ \hline & \quad -x^3 + 5x^2 + 7x \\ & \quad -(-x^3 - 2x^2 + 3x) \\ \hline & \qquad 7x^2 + 4x \end{array}$$

Divide $7x^2$ by x^2 to get the third term in the quotient: $\frac{7x^2}{x^2} = 7$. Multiply $x^2 + 2x - 3$ by 7: $7(x^2 + 2x - 3) = 7x^2 + 14x - 21$. Subtract this from $7x^2 + 4x - 1$.

$$\begin{array}{r|l} & 3x^2 - x + 7 \\ \hline x^2 + 2x - 3 & 3x^4 + 5x^3 - 4x^2 + 7x - 1 \\ & -(3x^4 + 6x^3 - 9x^2) \\ \hline & \quad -x^3 + 5x^2 + 7x \\ & \quad -(-x^3 - 2x^2 + 3x) \\ \hline & \qquad 7x^2 + 4x - 1 \\ & \qquad -(7x^2 + 14x - 21) \\ \hline & \qquad\quad - 10x + 20 \end{array}$$

Because $\frac{-10x}{x^2}$ cannot be a term in a polynomial, we are done. The quotient is $3x^2 - x + 7$, and the remainder is $-10x + 20$.

It is important that every power of x, from the highest power to the constant term, be represented in the polynomial. Although it is possible to perform long division without all powers represented, it is very easy to make an error. Also, it is not possible to perform synthetic division (later in this chapter) without a coefficient for *every* term. If a power of x is not written, we need to rewrite the polynomial (either the dividend, divisor, or both) using a coefficient of 0 on the missing powers. For example, we would write $x^3 - 1$ as $x^3 + 0x^2 + 0x - 1$.

EXAMPLE

- $(x^3 - 8) \div (x + 1)$

  Rewrite as $(x^3 + 0x^2 + 0x - 8) \div (x + 1)$

$$\begin{array}{r|l} & x^2 - x + 1 \\ \hline x + 1 & x^3 + 0x^2 + 0x - 8 \\ & -(x^3 + x^2) \\ \hline & \quad -x^2 + 0x \\ & \quad -(-x^2 - x) \\ \hline & \qquad x - 8 \\ & \qquad -(x + 1) \\ \hline & \qquad\quad -9 \end{array}$$

  The quotient is $x^2 - x + 1$, and the remainder is -9.

Polynomial division is a little trickier when the leading coefficient of the divisor is not 1. The terms of the quotient are harder to find and are likely to be fractions.

EXAMPLES

Find the quotient and remainder using long division.

- $\dfrac{x^2 - x + 2}{2x - 1}$

  Find the first term in the quotient by dividing the first term of the dividend by the first term in the divisor:

$$\frac{x^2}{2x} = \frac{x}{2} = \frac{1}{2}x.$$

$$\begin{array}{r|l} & \tfrac{1}{2}x \\ \hline 2x - 1 & x^2 - x + 2 \\ & -(x^2 - \tfrac{1}{2}x) \\ \hline & -\tfrac{1}{2}x + 2 \end{array}$$

  The second term in the quotient is

$$\frac{-\frac{1}{2}x}{2x} = \frac{-\frac{1}{2}}{2} = -\frac{1}{2} \div 2 = -\frac{1}{2} \cdot \frac{1}{2} = -\frac{1}{4}.$$

  Multiply $2x - 1$ by $-\frac{1}{4}$: $-\frac{1}{4}(2x - 1) = -\frac{1}{2}x + \frac{1}{4}$.

$$\begin{array}{r|l} & \tfrac{1}{2}x - \tfrac{1}{4} \\ \hline 2x - 1 & x^2 - x + 2 \\ & -(x^2 - \tfrac{1}{2}x) \\ \hline & -\tfrac{1}{2}x + 2 \\ & -(-\tfrac{1}{2}x + \tfrac{1}{4}) \\ \hline & \tfrac{7}{4} \end{array}$$

  The quotient is $\frac{1}{2}x - \frac{1}{4}$, and the remainder is $\frac{7}{4}$.

- $(4x^2 + 5x - 6) \div \left(\dfrac{2}{3}x - 1\right)$

  Find the first term in the quotient by dividing the leading term in the quotient by the first term in the divisor.

$$\frac{4x^2}{\frac{2}{3}x} = \frac{4x}{\frac{2}{3}} = 4x \div \frac{2}{3} = 4x \cdot \frac{3}{2} = 6x$$

$$6x\left(\frac{2}{3}x - 1\right) = 4x^2 - 6x$$

$$\begin{array}{r|l} & \quad 6x \\ \hline \frac{2}{3}x - 1 & \quad 4x^2 + 5x - 6 \\ & -(4x^2 - 6x) \\ \hline & \qquad 11x - 6 \end{array}$$

$$\frac{11x}{\frac{2}{3}x} = \frac{11}{\frac{2}{3}} = 11 \div \frac{2}{3} = 11 \cdot \frac{3}{2} = \frac{33}{2}$$

$$\frac{33}{2}\left(\frac{2}{3}x - 1\right) = 11x - \frac{33}{2}$$

$$\begin{array}{r|l} & 6x + \frac{33}{2} \\ \hline \frac{2}{3}x - 1 & \quad 4x^2 + 5x - 6 \\ & -(4x^2 - 6x) \\ \hline & \qquad 11x - 6 \\ & \quad -(11x - \frac{33}{2}) \\ \hline & \qquad\qquad \frac{21}{2} \end{array}$$

The quotient is $6x + \frac{33}{2}$, and the remainder is $\frac{21}{2}$.

Synthetic division of polynomials is much easier than long division. It only works when the divisor is of a certain form, though. Here, we will use synthetic division when the divisor is of the form "x − number" or "x + number."

For a problem of the form

$$\frac{a_n x^n + a_{n-1}x^{n-1} + \cdots + a_1 x + a_0}{x - c} \text{ or}$$

$$(a_n x^n + a_{n-1}x^{n-1} + \cdots + a_1 x + a_0) \div (x - c),$$

write

$$c \,\underline{|\, a_n \quad a_{n-1} \quad \ldots \quad a_1 \quad a_0}$$

Every power of x must be represented.

In synthetic division, the tedious work in long division is reduced to a few steps.

EXAMPLES

Find the quotient and remainder using synthetic division.

- $$\frac{4x^3 - 5x^2 + x - 8}{x - 2}$$

$$\begin{array}{r|rrrr} 2 & 4 & -5 & 1 & -8 \end{array}$$

Bring down the first coefficient.

$$\begin{array}{r|rrrr} 2 & 4 & -5 & 1 & -8 \\ & & & & \\ \hline & 4 & & & \end{array}$$

Multiply this coefficient by 2 (the c) and put the product under -5, the next coefficient.

$$\begin{array}{r|rrrr} 2 & 4 & -5 & 1 & -8 \\ & & 8 & & \\ \hline & 4 & & & \end{array}$$

Add -5 and 8. Put the sum under 8.

$$\begin{array}{r|rrrr} 2 & 4 & -5 & 1 & -8 \\ & & 8 & & \\ \hline & 4 & 3 & & \end{array}$$

Multiply 3 by 2 and put the product under 1, the next coefficient.

$$\begin{array}{r|rrrr} 2 & 4 & -5 & 1 & -8 \\ & & 8 & 6 & \\ \hline & 4 & 3 & & \end{array}$$

Add 1 and 6. Put the sum under 6.

$$\begin{array}{r|rrrr} 2 & 4 & -5 & 1 & -8 \\ & & 8 & 6 & \\ \hline & 4 & 3 & 7 & \end{array}$$

Multiply 7 by 2. Put the product under -8, the last coefficient.

$$\begin{array}{r|rrrr} 2 & 4 & -5 & 1 & -8 \\ & & 8 & 6 & 14 \\ \hline & 4 & 3 & 7 & \end{array}$$

Add -8 and 14. Put the sum under 14. This is the last step.

$$\begin{array}{r|rrrr} 2 & 4 & -5 & 1 & -8 \\ & & 8 & 6 & 14 \\ \hline & 4 & 3 & 7 & 6 \end{array}$$

The numbers on the last row are the coefficients of the quotient and the remainder. The remainder is a constant (which is a term of degree 0), and the degree of the quotient is exactly one less degree than the degree of the dividend. In this example, the degree of the dividend is 3, so the degree of the quotient is 2. The last number on the bottom row is the remainder. The numbers before it are the coefficients of the quotient, in order from the highest degree to the lowest. The remainder in this example is 6. The coefficients of the quotient are 4, 3, and 7. The quotient is $4x^2 + 3x + 7$.

- $(3x^4 - x^2 + 2x + 9) \div (x + 5)$
  Because $x + 5 = x - (-5)$, $c = -5$.

$$\begin{array}{r|rrrrr} -5 & 3 & 0 & -1 & 2 & 9 \end{array}$$

Bring down 3, the first coefficient. Multiply it by -5. Put $3(-5) = -15$ under 0.

$$\begin{array}{r|rrrrr} -5 & 3 & 0 & -1 & 2 & 9 \\ & & -15 & & & \\ \hline & 3 & & & & \end{array}$$

Add $0 + (-15) = -15$. Multiply -15 by -5 and put $(-15)(-5) = 75$ under -1.

$$\begin{array}{r|rrrrr} -5 & 3 & 0 & -1 & 2 & 9 \\ & & -15 & 75 & & \\ \hline & 3 & -15 & & & \end{array}$$

Add -1 and 75. Multiply $-1 + 75 = 74$ by -5 and put $(74)(-5) = -370$ under 2.

$$\begin{array}{r|rrrrr} -5 & 3 & 0 & -1 & 2 & 9 \\ & & -15 & 75 & -370 & \\ \hline & 3 & -15 & 74 & & \end{array}$$

Add 2 to -370. Multiply $2 + (-370) = -368$ by -5 and put $(-368)(-5) = 1840$ under 9.

$$\begin{array}{r|rrrrr} -5 & 3 & 0 & -1 & 2 & 9 \\ & & -15 & 75 & -370 & 1840 \\ \hline & 3 & -15 & 74 & -368 & \end{array}$$

Add 9 to 1840. Put $9 + 1840 = 1849$ under 1840.

$$\begin{array}{r|rrrrr} -5 & 3 & 0 & -1 & 2 & 9 \\ & & -15 & 75 & -370 & 1840 \\ \hline & 3 & -15 & 74 & -368 & 1849 \end{array}$$

The dividend has degree 4, so the quotient has degree 3. The quotient is $3x^3 - 15x^2 + 74x - 368$ and the remainder is 1849.

When dividing a polynomial $f(x)$ by $x - c$, the remainder tells us two things. If we get a remainder of 0, then both the divisor, $(x - c)$, and quotient are factors of $f(x)$. Another fact we get from the remainder is that $f(c) =$ remainder.

$$f(x) = (x - c)q(x) + \text{remainder}$$

$$f(c) = (c - c)q(c) + \text{remainder} \qquad \text{Evaluate } f(x) \text{ at } x = c.$$

$$f(c) = 0q(c) + \text{remainder}$$

$$f(c) = \text{remainder}$$

The fact that $f(c)$ is the remainder is called the *Remainder Theorem.* It is useful when trying to evaluate complicated polynomials. We can also use this fact to check our work with synthetic division and long division (providing the divisor is $x - c$).

- $(x^3 - 6x^2 + 4x - 5) \div (x - 3)$
  By the Remainder Theorem, we should get the remainder to be $3^3 - 6(3^2) + 4(3) - 5 = -20$.

$$\begin{array}{r|rrrr} 3 & 1 & -6 & 4 & -5 \\ & & 3 & -9 & -15 \\ \hline & 1 & -3 & -5 & -20 \end{array}$$

EXAMPLE

Use synthetic division and the Remainder Theorem to evaluate $f(c)$.

- $f(x) = 14x^3 - 16x^2 + 10x + 8$; $c = 1$.
  We will first perform synthetic division with $x - c = x - 1$.

$$\begin{array}{r|rrrr} 1 & 14 & -16 & 10 & 8 \\ & & 14 & -2 & 8 \\ \hline & 14 & -2 & 8 & 16 \end{array}$$

The remainder is 16, so $f(1) = 16$.

Now we will use synthetic division and the Remainder Theorem to factor polynomials. Suppose $x = c$ is a zero for a polynomial $f(x)$. Let us see what happens when we divide $f(x)$ by $x - c$.

$$f(x) = (x - c)q(x) + r(x)$$

Because $x = c$ is a zero, the remainder is 0, so $f(x) = (x - c)q(x) + 0$, which means $f(x) = (x - c)q(x)$. The next step in completely factoring $f(x)$ is factoring $q(x)$, if necessary.

EXAMPLES

Completely factor the polynomials.

- $f(x) = x^3 - 4x^2 - 7x + 10$, $c = 1$ is a zero.
  We will use the fact that $c = 1$ is a zero to get started. We will use synthetic division to divide $f(x)$ by $x - 1$.

$$\begin{array}{r|rrrr} 1 & 1 & -4 & -7 & 10 \\ & & 1 & -3 & -10 \\ \hline & 1 & -3 & -10 & 0 \end{array}$$

  The quotient is $x^2 - 3x - 10$. We now have $f(x)$ partially factored.

$$\begin{aligned} f(x) &= x^3 - 4x^2 - 7x + 10 \\ &= (x-1)(x^2 - 3x - 10) \end{aligned}$$

  Because the quotient is quadratic, we can factor it directly or by using the quadratic formula.

$$x^2 - 3x - 10 = (x-5)(x+2)$$

  Now we have the complete factorization of $f(x)$:

$$\begin{aligned} f(x) &= x^3 - 4x^2 - 7x + 10 \\ &= (x-1)(x-5)(x+2). \end{aligned}$$

- $R(x) = x^3 - 2x + 1$, $c = 1$ is a zero.

$$\begin{array}{r|rrrr} 1 & 1 & 0 & -2 & 1 \\ & & 1 & 1 & -1 \\ \hline & 1 & 1 & -1 & 0 \end{array}$$

$$R(x) = x^3 - 2x + 1 = (x-1)(x^2 + x - 1)$$

  We will use the quadratic formula to find the two zeros of $x^2 + x - 1$.

$$x = \frac{-1 \pm \sqrt{1^2 - 4(1)(-1)}}{2(1)}$$

$$\frac{-1 \pm \sqrt{5}}{2} = \frac{-1 + \sqrt{5}}{2}, \frac{-1 - \sqrt{5}}{2}$$

The factors for these zeros are $x - \frac{-1+\sqrt{5}}{2}$ and $x - \frac{-1-\sqrt{5}}{2}$.

$$R(x) = (x-1)\left(x - \frac{-1+\sqrt{5}}{2}\right)\left(x - \frac{-1-\sqrt{5}}{2}\right)$$

PRACTICE

For Problems 1–4 use long division to find the quotient and remainder. For Problems 5 and 6, use synthetic division

1. $(6x^3 - 2x^2 + 5x - 1) \div (x^2 + 3x + 2)$
2. $(x^3 - x^2 + 2x + 5) \div (3x - 4)$
3. $\dfrac{3x^3 - x^2 + 4x + 2}{-\frac{1}{2}x^2 + 1}$
4. $\dfrac{x^3 - 1}{x - 1}$
5. $\dfrac{x^3 + 2x^2 + x - 8}{x + 3}$
6. $(x^3 + 8) \div (x + 2)$
7. Use synthetic division and the Remainder Theorem to evaluate $f(c)$.

   $f(x) = 6x^4 - 8x^3 + x^2 + 2x - 5;\ c = -2$
8. Completely factor the polynomial. $f(x) = x^3 + 2x^2 - x - 2;\ c = 1$ is a zero.
9. Completely factor the polynomial. $P(x) = x^3 - 5x^2 + 5x + 3;\ c = 3$ is a zero.

SOLUTIONS

1.

$$\begin{array}{r|l} & \quad 6x - 20 \\ \hline x^2+3x+2 & \;6x^3 - 2x^2 + 5x - 1 \\ & -(6x^3 + 18x^2 + 12x) \\ \hline & \qquad -20x^2 - 7x - 1 \\ & \quad -(-20x^2 - 60x - 40) \\ \hline & \qquad\qquad 53x + 39 \end{array}$$

The quotient is $6x - 20$, and the remainder is $53x + 39$.

2.

$$\begin{array}{r|l} & \frac{1}{3}x^2 + \frac{1}{9}x \\ \hline 3x-4 & x^3 - x^2 + 2x + 5 \\ & -(x^3 - \frac{4}{3}x^2) \\ \hline & \frac{1}{3}x^2 + 2x \\ & -(\frac{1}{3}x^2 - \frac{4}{9}x) \\ \hline & \frac{22}{9}x + 5 \end{array}$$

$$\frac{\frac{22}{9}x}{3x} = \frac{\frac{22}{9}}{3} = \frac{22}{9} \cdot \frac{1}{3} = \frac{22}{27}$$

$$\frac{22}{27}(3x-4) = \frac{22}{9}x - \frac{88}{27}$$

$$\begin{array}{r|l} & \frac{1}{3}x^2 + \frac{1}{9}x + \frac{22}{27} \\ \hline 3x-4 & x^3 - x^2 + 2x + 5 \\ & -(x^3 - \frac{4}{3}x^2) \\ \hline & \frac{1}{3}x^2 + 2x \\ & -(\frac{1}{3}x^2 - \frac{4}{9}x) \\ \hline & \frac{22}{9}x + 5 \\ & -(\frac{22}{9}x - \frac{88}{27}) \\ \hline & \frac{223}{27} \end{array}$$

The quotient is $\frac{1}{3}x^2 + \frac{1}{9}x + \frac{22}{27}$, and the remainder is $\frac{223}{27}$.

3. $$\frac{3x^3}{-\frac{1}{2}x^2} = \frac{3x}{-\frac{1}{2}} = 3x \div -\frac{1}{2} = 3x \cdot (-2) = -6x$$

$$-6x(-\tfrac{1}{2}x^2 + 0x + 1) = 3x^3 + 0x^2 - 6x$$

$$\begin{array}{r|l} & -6x \\ \hline -\frac{1}{2}x^2 + 0x + 1 & 3x^3 - x^2 + 4x + 2 \\ & -(3x^3 - 0x^2 - 6x) \\ \hline & -x^2 + 10x + 2 \end{array}$$

$$\frac{-x^2}{-\frac{1}{2}x^2} = \frac{1}{\frac{1}{2}} = 1 \div \frac{1}{2} = 1 \cdot 2 = 2$$

$$2\left(-\frac{1}{2}x^2 + 0x + 1\right) = -x^2 + 0x + 2$$

$$
\begin{array}{r|l}
 & -6x + 2 \\
\cline{2-2}
-\frac{1}{2}x^2 + 0x + 1 & 3x^3 - x^2 + 4x + 2 \\
 & -(3x^3 - 0x^2 - 6x) \\
\cline{2-2}
 & \quad -x^2 + 10x + 2 \\
 & \quad -(-x^2 + 0x + 2) \\
\cline{2-2}
 & \qquad 10x + 0
\end{array}
$$

The quotient is $-6x + 2$, and the remainder is $10x$.

4.

$$
\begin{array}{r|l}
 & x^2 + x + 1 \\
\cline{2-2}
x - 1 & x^3 + 0x^2 + 0x - 1 \\
 & -(x^3 - x^2) \\
\cline{2-2}
 & \quad x^2 + 0x \\
 & \quad -(x^2 - x) \\
\cline{2-2}
 & \qquad x - 1 \\
 & \qquad -(x - 1) \\
\cline{2-2}
 & \qquad\quad 0
\end{array}
$$

The quotient is $x^2 + x + 1$, and the remainder is 0.

5.

$$
\begin{array}{r|rrrr}
-3 & 1 & 2 & 1 & -8 \\
 & & -3 & 3 & -12 \\
\hline
 & 1 & -1 & 4 & -20
\end{array}
$$

The quotient is $x^2 - x + 4$, and the remainder is -20.

6.

$$
\begin{array}{r|rrrr}
-2 & 1 & 0 & 0 & 8 \\
 & & -2 & 4 & -8 \\
\hline
 & 1 & -2 & 4 & 0
\end{array}
$$

The quotient is $x^2 - 2x + 4$, and the remainder is 0.

7.

$$
\begin{array}{r|rrrrr}
-2 & 6 & -8 & 1 & 2 & -5 \\
 & & -12 & 40 & -82 & 160 \\
\hline
 & 6 & -20 & 41 & -80 & 155
\end{array}
$$

The remainder is 155, so $f(-2) = 155$.

8.

$$
\begin{array}{r|rrrr}
1 & 1 & 2 & -1 & -2 \\
 & & 1 & 3 & 2 \\
\hline
 & 1 & 3 & 2 & 0
\end{array}
$$

$$
\begin{aligned}
f(x) &= (x - 1)(x^2 + 3x + 2) \\
&= (x - 1)(x + 1)(x + 2)
\end{aligned}
$$

9.

$$\begin{array}{r|rrrr} 3 & 1 & -5 & 5 & 3 \\ & & 3 & -6 & -3 \\ \hline & 1 & -2 & -1 & 0 \end{array}$$

$$P(x) = (x-3)(x^2 - 2x - 1)$$

In order to factor $x^2 - 2x - 1$, we must first find its zeros.

$$x = \frac{-(-2) \pm \sqrt{(-2)^2 - 4(1)(-1)}}{2(1)}$$

$$= \frac{2 \pm \sqrt{8}}{2} = \frac{2 \pm 2\sqrt{2}}{2}$$

$$= \frac{2(1 \pm \sqrt{2})}{2} = 1 \pm \sqrt{2}$$

$$= 1 + \sqrt{2}, 1 - \sqrt{2}$$

Because $x = 1 + \sqrt{2}$ is a zero, $x - (1 + \sqrt{2}) = x - 1 - \sqrt{2}$ is a factor.
Because $x = 1 - \sqrt{2}$ is a zero, $x - (1 - \sqrt{2}) = x - 1 + \sqrt{2}$ is a factor.

$$P(x) = (x-3)(x - 1 - \sqrt{2})(x - 1 + \sqrt{2})$$

In the above examples and practice problems, a zero was given to help us get started with factoring. Usually, we have to find a starting point ourselves. The *Rational Zero Theorem* gives us a place to start. The Rational Zero Theorem says that if a polynomial function $f(x)$, with integer coefficients, has a rational number p/q as a zero, then p is a divisor of the constant term and q is a divisor of the leading coefficient. Not all polynomials have rational zeros, but most of those in precalculus courses do.

The Rational Zero Theorem is used to create a list of candidates for zeros. These candidates are rational numbers whose numerators divide the polynomial's constant term and whose denominators divide its leading coefficient. Once we have this list, we will try each number in the list to see which, if any, are zeros. Once we have found a zero, we can begin to factor the polynomial.

EXAMPLES

List the possible rational zeros.

- $f(x) = 4x^3 + 6x^2 - 2x + 9$

The numerators in our list will be the divisors of 9: 1, 3, and 9 as well as their negatives, -1, -3, and -9. The denominators will be the divisors of 4: 1, 2, and 4. The list of possible rational zeros is—

$$\frac{1}{1}, \frac{3}{1}, \frac{9}{1}, -\frac{1}{1}, -\frac{3}{1}, -\frac{9}{1}, \frac{1}{2}, \frac{3}{2}, \frac{9}{2},$$
$$-\frac{1}{2}, -\frac{3}{2}, -\frac{9}{2}, \frac{1}{4}, \frac{3}{4}, \frac{9}{4}, -\frac{1}{4}, -\frac{3}{4}, \text{ and } -\frac{9}{4}.$$

This list could be written with a little less effort as ± 1, ± 3, ± 9, $\pm\frac{1}{2}$, $\pm\frac{3}{2}$, $\pm\frac{9}{2}$, $\pm\frac{1}{4}$, $\pm\frac{3}{4}$, $\pm\frac{9}{4}$.

We only need to list the numerators with negative numbers and not the denominators. The reason is that no new numbers are added to the list, only duplicates of numbers already there. For example, $\frac{-1}{2}$ and $\frac{1}{-2}$ are the same number.

- $g(x) = 6x^4 - 5x^3 + 2x - 8$

  The possible numerators are the divisors of 8: ± 1, ± 2, ± 4, and ± 8. The possible denominators are the divisors of 6: 1, 2, 3, and 6. The list of possible rational zeros is—

$$\pm 1, \pm 2, \pm 4, \pm 8, \pm\frac{1}{2}, \pm\frac{2}{2}, \pm\frac{4}{2}, \pm\frac{8}{2}, \pm\frac{1}{3}, \pm\frac{2}{3}, \pm\frac{4}{3}, \pm\frac{8}{3}, \pm\frac{1}{6}, \pm\frac{2}{6},$$
$$\pm\frac{4}{6}, \pm\frac{8}{6}.$$

  There are several duplicates on this list. There will be duplicates when the constant term and leading coefficient have common factors. The duplicates don't really hurt anything, but they could waste time when checking the list for zeros.

Now that we have a starting place, we can factor many polynomials. Here is the strategy. First we will see if the polynomial can be factored directly. If not, we need to list the possible rational zeros. Then we will try the numbers in this list, one at a time, until we find a zero. Once we have found a zero, we will use polynomial division (long division or synthetic division) to find the quotient. Next, we will factor the quotient. If the quotient is a quadratic factor, we will either factor it directly or use the quadratic formula to find its zeros. If the quotient is a polynomial of degree 3 or higher, we will need to start over to factor the quotient. Eventually, the quotient will be a quadratic factor.

EXAMPLES

Completely factor each polynomial.

- $f(x) = 3x^4 - 2x^3 - 7x^2 - 2x$
  First we will factor x from each term: $f(x) = x(3x^3 - 2x^2 - 7x - 2)$. The possible rational zeros for $3x^3 - 2x^2 - 7x - 2$ are $\pm 1,\ \pm 2,\ \pm\frac{1}{3},\ \pm\frac{2}{3}$.

$$3(1)^3 - 2(1)^2 - 7(1) - 2 \neq 0$$

$$3(-1)^3 - 2(-1)^2 - 7(-1) - 2 = 0$$

We will use synthetic division to find the quotient for $(3x^3 - 2x^2 - 7x - 2) \div (x + 1)$.

$$\begin{array}{r|rrrr} -1 & 3 & -2 & -7 & -2 \\ & & -3 & 5 & 2 \\ \hline & 3 & -5 & -2 & 0 \end{array}$$

The quotient is $3x^2 - 5x - 2$ which factors into $(3x + 1)(x - 2)$.

$$\begin{aligned} f(x) &= 3x^4 - 2x^3 - 7x^2 - 2x \\ &= x(3x^3 - 2x^2 - 7x - 2) \\ &= x(x + 1)(3x^2 - 5x - 2) \\ &= x(x + 1)(3x + 1)(x - 2) \end{aligned}$$

- $h(x) = 3x^3 + 4x^2 - 18x + 5$
  The possible rational zeros are $\pm 1,\ \pm 5,\ \pm\frac{1}{3}$, and $\pm\frac{5}{3}$.

$$h(1) = 3(1^3) + 4(1^2) - 18(1) + 5 \neq 0$$

$$h(-1) = 3(-1)^3 + 4(-1)^2 - 18(-1) + 5 \neq 0$$

$$h(5) = 3(5^3) + 4(5^2) - 18(5) + 5 \neq 0$$

Continuing in this way, we see that $h(-5) \neq 0,\ h(\frac{1}{3}) \neq 0,\ h(-\frac{1}{3}) \neq 0$ and $h(\frac{5}{3}) = 0$.

$$\begin{array}{r|rrrr} \frac{5}{3} & 3 & 4 & -18 & 5 \\ & & 5 & 15 & -5 \\ \hline & 3 & 9 & -3 & 0 \end{array}$$

$$h(x) = \left(x - \frac{5}{3}\right)(3x^2 + 9x - 3)$$

$$= \left(x - \frac{5}{3}\right)(3)(x^2 + 3x - 1) = \left[3\left(x - \frac{5}{3}\right)\right](x^2 + 3x - 1)$$

$$= (3x - 5)(x^2 + 3x - 1)$$

We will find the zeros of $x^2 + 3x - 1$ using the quadratic formula.

$$x = \frac{-3 \pm \sqrt{3^2 - 4(1)(-1)}}{2(1)}$$

$$= \frac{-3 \pm \sqrt{13}}{2} = \frac{-3 + \sqrt{13}}{2}, \quad \frac{-3 - \sqrt{13}}{2}$$

$$h(x) = (3x - 5)\left(x - \frac{-3 + \sqrt{13}}{2}\right)\left(x - \frac{-3 - \sqrt{13}}{2}\right)$$

For a polynomial such as $f(x) = 5x^3 + 20x^2 - 9x - 36$, the list of possible rational zeros is quite long—36! There are ways of getting around having to test every one of them. The fastest way is to use a graphing calculator to sketch the graph of $y = 5x^3 + 20x^2 - 9x - 36$. There appears to be an x-intercept at $x = -4$ (remember that x-intercepts are zeros.)

$$\begin{array}{r|rrrr} -4 & 5 & 20 & -9 & -36 \\ & & -20 & 0 & 36 \\ \hline & 5 & 0 & -9 & 0 \end{array}$$

$f(x) = (x + 4)(5x^2 - 9)$ We will solve $5x^2 - 9 = 0$ to find the other zeros.

$$5x^2 - 9 = 0$$

$$5x^2 = 9$$

$$x^2 = \frac{9}{5}$$

$$x = \pm\sqrt{\frac{9}{5}} = \pm\frac{3}{\sqrt{5}}$$

$$= \pm\frac{3}{\sqrt{5}} \cdot \frac{\sqrt{5}}{\sqrt{5}}$$

$$= \pm\frac{3\sqrt{5}}{5} = \frac{3\sqrt{5}}{5}, \quad -\frac{3\sqrt{5}}{5}$$

$$f(x) = (x+4)\left(x - \frac{3\sqrt{5}}{5}\right)\left(x + \frac{3\sqrt{5}}{5}\right)$$

There are also a couple of algebra facts that can help eliminate some of the possible rational zeros. The first we will learn is *Descartes' Rule of Signs*. The second is the *Upper and Lower Bounds Theorem*. Descartes' Rule of Signs counts the number of positive zeros and negative zeros. For instance, according to the rule $f(x) = x^3 + x^2 + 4x + 6$ has no positive zeros at all. This shrinks the list of possible rational zeros from ± 1, ± 2, ± 3, and ± 6 to $-1, -2, -3$, and -6. Another advantage of the sign test is that if we know that there are two positive zeros and we have found one of them, then we *know* that there is exactly one more.

The Upper and Lower Bounds Theorem gives us an idea of how large (in both the positive and negative directions) the zeros can be. For example, we can use the Upper and Lower Bounds Theorem to show that all of the zeros for $f(x) = 5x^3 + 20x^2 - 9x - 36$ are between -5 and 5. This shrinks the list of possible rational zeros from ± 1, ± 2, ± 3, ± 4, ± 6, ± 9, ± 12, ± 18, ± 36, $\pm\frac{1}{5}$, $\pm\frac{2}{5}$, $\pm\frac{3}{5}$, $\pm\frac{4}{5}$, $\pm\frac{6}{5}$, $\pm\frac{9}{5}$ $\pm\frac{12}{5}$, $\pm\frac{18}{5}$, and $\pm\frac{36}{5}$ to ± 1, ± 2, ± 3, ± 4, $\pm\frac{1}{5}$, $\pm\frac{2}{5}$, $\pm\frac{3}{5}$, $\pm\frac{4}{5}$, $\pm\frac{6}{5}$, $\pm\frac{9}{5}$, $\pm\frac{12}{5}$, and $\pm\frac{18}{5}$.

Descartes' Rule of Signs counts the number of positive zeros and the number of negative zeros by counting sign changes. The maximum number of positive zeros for a polynomial function is the number of sign changes in $f(x) = a_n x^n + a_{n-1}x^{n-1} + \cdots + a_1 x + a_0$. The possible number of positive zeros is the number of sign changes minus an even whole number. For example, if there are 5 sign changes, there are 5 or 3 or 1 positive zeros. If there are 6 sign changes, there are 6 or 4 or 2 or 0 positive zeros. The polynomial function $f(x) = 3x^4 - 2x^3 + 7x^2 + 5x - 8$ has 3 sign changes: from 3 to -2, from -2 to 7, and from 5 to -8. There are either 3 or 1 positive zeros. The maximum number of negative zeros is the number of sign changes in the polynomial $f(-x)$. The possible number of negative zeros is the number of sign changes in $f(-x)$ minus an even whole number.

EXAMPLES

Use Descartes' Rule of Signs to count the possible number of positive zeros and negative zeros for the polynomial functions.

- $f(x) = 5x^3 - 6x^2 - 10x + 4$
  There are 2 sign changes: from 5 to -6 and from -10 to 4. This means that there are either 2 or 0 positive zeros. Before we count the possible number of negative zeros, remember from earlier in the book that for a number a, $a(-x)^{\text{even power}} = ax^{\text{even power}}$ and $a(-x)^{\text{odd power}} = -ax^{\text{odd power}}$.

$$\begin{aligned} f(-x) &= 5(-x)^3 - 6(-x)^2 - 10(-x) + 4 \\ &= -5x^3 - 6x^2 + 10x + 4 \end{aligned}$$

  There is 1 sign change, from -6 to 10, so there is exactly 1 negative zero.
- $P(x) = x^5 + x^3 + x + 4$
  There are no sign changes, so there are no positive zeros.

$$\begin{aligned} P(-x) &= (-x)^5 + (-x)^3 + (-x) + 4 \\ &= -x^5 - x^3 - x + 4 \end{aligned}$$

  There is 1 sign change, so there is exactly 1 negative zero.

The Upper and Lower Bounds Theorem helps us to find a range of x-values that will contain all real zeros. It does *not* tell us what these bounds are. We make a guess as to what these bounds are then check them. For a negative number $x = a$, the statement "a is a lower bound for the real zeros" means that there is no number to the left of $x = a$ on the x-axis that is a zero. For a positive number $x = b$, the statement "b is an upper bound for the real zeros" means that there is no number to the right of $x = b$ on the x-axis that is a zero. In other words, all of the x-intercepts are between a and b.

To determine whether a negative number $x = a$ is a lower bound for a polynomial, we need to use synthetic division. If the numbers in the bottom row alternate between nonpositive and nonnegative numbers, then $x = a$ is a lower bound for the negative zeros. A "nonpositive" number is 0 or negative, and a "nonnegative" number is 0 or positive.

To determine whether a positive number $x = b$ is an upper bound for the positive zeros, again we need to use synthetic division. If the numbers on the bottom row are all nonnegative, then $x = b$ is an upper bound on the positive zeros.

EXAMPLES

Show that the given values for a and b are lower, and upper bounds, respectively, for the following polynomials.

- $f(x) = x^4 + x^3 - 16x^2 - 4x + 48$; $a = -5$ and $b = 5$

−5	1	1	−16	−4	48
		−5	20	−20	120
	1	−4	4	−24	168

The bottom row alternates between positive and negative numbers, so $a = -5$ is a lower bound for the negative zeros of $f(x)$.

5	1	1	−16	−4	48
		5	30	70	330
	1	6	14	66	378

The entries on the bottom row are all positive, so $b = 5$ is an upper bound for the positive zeros of $f(x)$. All of the real zeros for $f(x)$ are between $x = -5$ and $x = 5$.

If 0 appears on the bottom row when testing for an upper bound, we can consider 0 to be positive. If 0 appears in the bottom row when testing for a lower bound, we can consider 0 to be negative if the previous entry is positive and positive if the previous entry is negative. In other words, consider 0 to be the opposite sign as the previous entry.

- $P(x) = 4x^4 + 20x^3 + 7x^2 + 3x - 6$ with $a = -5$

−5	4	20	7	3	−6
		−20	0	−35	160
	4	0	7	−32	154

Because 0 follows a positive number, we will consider 0 to be negative. This makes the bottom row alternate between positive and negative entries, so $a = -5$ is a lower bound for the negative zeros of $P(x)$.

The Upper and Lower Bounds Theorem has some limitations. For instance, it does not tell us *how* to find upper and lower bounds for the zeros of a polynomial. For any polynomial, there are infinitely many upper and lower bounds. For instance, if $x = 5$ is an upper bound, then any number larger than 5 is also an upper bound. For many polynomials, a starting place is the quotient of the constant term and the leading coefficient and its negative: $\pm\frac{\text{constant term}}{\text{leading coefficient}}$. First show that these are bounds for the zeros, then work your way inward. For example, if $f(x) = 2x^3 - 7x^2 + x + 50$, let $a = -\frac{50}{2} = -25$ and $b = \frac{50}{2} = 25$. Then, let a and b get closer together, say $a = -10$ and $b = 10$.

PRACTICE

1. List the candidates for rational zeros. Do not try to find the zeros. $f(x) = 3x^4 + 8x^3 - 11x^2 + 3x + 4$

2. List the candidates for rational zeros. Do not try to find the zeros. $P(x) = 6x^4 - 24$
3. Completely factor $h(x) = 2x^3 + 5x^2 - 23x + 10$.
4. Completely factor $P(x) = 7x^3 + 26x^2 - 15x + 2$.
5. Use Descartes' Rule of Signs to count the possible number of positive zeros and the possible number of negative zeros of $f(x) = 2x^4 - 6x^3 - x^2 + 4x - 8$.
6. Use Descartes' Rule of Signs to count the possible number of positive zeros and the possible number of negative zeros of $f(x) = -x^3 - x^2 + x + 1$.
7. Show that the given values for a and b are lower and upper, respectively, bounds for the zeros of $f(x) = x^3 - 6x^2 + x + 5$; $a = -3,\ b = 7$.
8. Show that the given values for a and b are lower and upper, respectively, bounds for the zeros of $f(x) = x^4 - x^2 - 2$; $a = -2,\ b = 2$.
9. Sketch the graph for $g(x) = x^3 - x^2 - 17x - 15$.

SOLUTIONS

1. Possible numerators: $\pm 1,\ \pm 2,\ \pm 4$

   Possible denominators: 1 and 3

   Possible rational zeros: $\pm 1,\ \pm 2,\ \pm 4,\ \pm\frac{1}{3},\ \pm\frac{2}{3},\ \pm\frac{4}{3}$

2. Possible numerators: $\pm 1,\ \pm 2,\ \pm 3,\ \pm 4,\ \pm 6,\ \pm 8,\ \pm 12,\ \pm 24$

   Possible denominators: $1,\ 2,\ 3,\ 6$

   Possible rational zeros (with duplicates omitted): $\pm 1,\ \pm 2,\ \pm 3,\ \pm 4,\ \pm 6,\ \pm 8,\ \pm 12,\ \pm 24,\ \pm\frac{1}{2},\ \pm\frac{3}{2},\ \pm\frac{1}{3},\ \pm\frac{2}{3},\ \pm\frac{4}{3},\ \pm\frac{8}{3},\ \pm\frac{1}{6}$

3. The possible rational zeros are $\pm 1,\ \pm 2,\ \pm 5,\ \pm 10,\ \pm\frac{1}{2}$, and $\pm\frac{5}{2}$. Because $h(2) = 0$, $x = 2$ is a zero of $h(x)$.

$$\begin{array}{r|rrrr} 2 & 2 & 5 & -23 & 10 \\ & & 4 & 18 & -10 \\ \hline & 2 & 9 & -5 & 0 \end{array}$$

$$h(x) = (x - 2)(2x^2 + 9x - 5)$$

$$h(x) = (x - 2)(2x - 1)(x + 5)$$

4. The possible rational zeros are $\pm 1,\ \pm 2,\ \pm\frac{1}{7}$, and $\pm\frac{2}{7}$. Because $P(\frac{2}{7}) = 0$, $x = \frac{2}{7}$ is a zero for $P(x)$.

$$\begin{array}{r|rrrr} \frac{2}{7} & 7 & 26 & -15 & 2 \\ & & 2 & 8 & -2 \\ \hline & 7 & 28 & -7 & 0 \end{array}$$

$$\begin{aligned} P(x) &= \left(x - \frac{2}{7}\right)(7x^2 + 28x - 7) \\ &= \left(x - \frac{2}{7}\right)(7)(x^2 + 4x - 1) = \left[7\left(x - \frac{2}{7}\right)\right](x^2 + 4x - 1) \\ &= (7x - 2)(x^2 + 4x - 1) \end{aligned}$$

We will use the quadratic formula to find the zeros for $x^2 + 4x - 1$.

$$\begin{aligned} x &= \frac{-4 \pm \sqrt{4^2 - 4(1)(-1)}}{2(1)} = \frac{-4 \pm \sqrt{20}}{2} \\ &= \frac{-4 \pm 2\sqrt{5}}{2} = \frac{2(-2 \pm \sqrt{5})}{2} \\ &= -2 \pm \sqrt{5} = -2 + \sqrt{5},\ -2 - \sqrt{5} \end{aligned}$$

$$\begin{aligned} x^2 + 4x - 1 &= (x - (-2 + \sqrt{5}))(x - (-2 - \sqrt{5})) \\ &= (x + 2 - \sqrt{5})(x + 2 + \sqrt{5}) \end{aligned}$$

$$P(x) = (7x - 2)(x + 2 - \sqrt{5})(x + 2 + \sqrt{5})$$

5. There are 3 sign changes in $f(x)$, so there are 3 or 1 positive zeros.

$$\begin{aligned} f(-x) &= 2(-x)^4 - 6(-x)^3 - (-x)^2 + 4(-x) - 8 \\ &= 2x^4 + 6x^3 - x^2 - 4x - 8 \end{aligned}$$

   There is 1 sign change in $f(-x)$, so there is exactly 1 negative zero.

6. There is 1 sign change in $f(x)$, so there is exactly 1 positive zero.

$$\begin{aligned} f(-x) &= -(-x)^3 - (-x)^2 + (-x) + 1 \\ &= x^3 - x^2 - x + 1 \end{aligned}$$

   There are 2 sign changes in $f(-x)$, so there are 2 or 0 negative zeros.

7.

$$\begin{array}{r|rrrr} -3 & 1 & -6 & 1 & 5 \\ & & -3 & 27 & -84 \\ \hline & 1 & -9 & 28 & -79 \end{array}$$

The entries on the bottom row alternate between positive and negative (or nonnegative and nonpositive), so $a = -3$ is a lower bound for the zeros of $f(x)$.

$$\begin{array}{r|rrrr} 7 & 1 & -6 & 1 & 5 \\ & & 7 & 7 & 56 \\ \hline & 1 & 1 & 8 & 61 \end{array}$$

The entries on the bottom are positive (nonnegative), so $b = 7$ is an upper bound for the positive zeros of $f(x)$.

8.

$$\begin{array}{r|rrrrr} -2 & 1 & 0 & -1 & 0 & -2 \\ & & -2 & 4 & -6 & 12 \\ \hline & 1 & -2 & 3 & -6 & 10 \end{array}$$

The entries on the bottom row alternate between positive and negative, so $a = -2$ is a lower bound for the negative zeros of $f(x)$.

$$\begin{array}{r|rrrrr} 2 & 1 & 0 & -1 & 0 & -2 \\ & & 2 & 4 & 6 & 12 \\ \hline & 1 & 2 & 3 & 6 & 10 \end{array}$$

The entries on the bottom row are all positive, so $b = 2$ is an upper bound for the positive zeros of $f(x)$.

9. The x-intercepts are -3, -1, and 5. We will plot points for $x = -3.5$, $x = -2$, $x = 0$, $x = 3$, and $x = 5.5$.

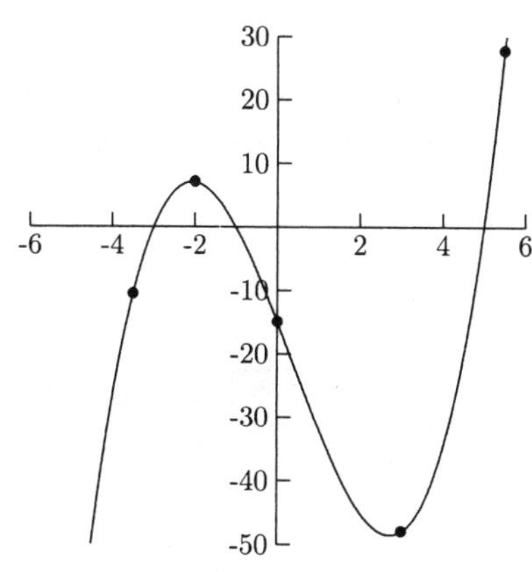

Fig. 7.14.

Complex Numbers

Until now, zeros of polynomials have been real numbers. The next topic involves *complex* zeros. These zeros come from even roots of negative numbers like $\sqrt{-1}$. Before working with complex zeros of polynomials, we will first learn some complex number arithmetic. Complex numbers are normally written in the form $a+bi$, where a and b are real numbers and $i=\sqrt{-1}$. A number such as $4+\sqrt{-9}$ would be written as $4+3i$ because $\sqrt{-9}=\sqrt{9}\sqrt{-1}=3i$. Real numbers are complex numbers where $b=0$.

EXAMPLES

Write the complex numbers in the form $a+bi$, where a and b are real numbers.

- $\sqrt{-64}=\sqrt{64}\sqrt{-1}=8i$
- $\sqrt{-27}=\sqrt{27}\sqrt{-1}=\sqrt{27}\,i=\sqrt{9\cdot 3}\,i=\sqrt{9}\sqrt{3}\,i$
  $=3\sqrt{3}\,i$ Be careful, $\sqrt{3i}\neq\sqrt{3}\,i$.
- $6+\sqrt{-8}=6+\sqrt{8}\,i=6+\sqrt{4\cdot 2}\,i=6+\sqrt{4}\sqrt{2}\,i=6+2\sqrt{2}\,i$

Adding complex numbers is a matter of adding like terms. Add the real parts, a and c, and the imaginary parts, b and d.

$$(a+bi)+(c+di)=(a+c)+(b+d)i$$

Subtract two complex numbers by distributing the minus sign in the parentheses then adding the like terms.

$$a+bi-(c+di)=a+bi-c-di=(a-c)+(b-d)i$$

EXAMPLES

Perform the arithmetic. Write the sum or difference in the form $a+bi$, where a and b are real numbers.

- $(3-5i)+(4+8i)=(3+4)+(-5+8)i=7+3i$
- $2i-6+9i=-6+11i$
- $7-\sqrt{-18}+3+5\sqrt{-2}=7-\sqrt{18}\,i+3+5\sqrt{2}\,i$
  $=7-\sqrt{9\cdot 2}\,i+3+5\sqrt{2}\,i=7-3\sqrt{2}\,i+3+5\sqrt{2}\,i$
  $=10+2\sqrt{2}\,i$
- $11-3i-(7+6i)=11-3i-7-6i=4-9i$

- $7+\sqrt{-8}-(1-\sqrt{-18})=7+\sqrt{8}\,i-1+\sqrt{18}\,i$
$=7+2\sqrt{2}\,i-1+3\sqrt{2}\,i=6+5\sqrt{2}\,i$

Multiplying complex numbers is not as straightforward as adding and subtracting them. First we will take the product of two purely imaginary numbers (numbers whose real parts are 0). Remember that $i=\sqrt{-1}$, which makes $i^2=-1$. In most complex number multiplication problems, we will have a term with i^2. Replace i^2 with -1. Multiply two complex numbers in the form $a+bi$ using the FOIL method, substituting -1 for i^2 and combining like terms.

EXAMPLES

Write the product in the form $a+bi$, where a and b are real numbers.

- $(5i)(6i)=30i^2=30(-1)=-30$
- $(2i)(-9i)=-18i^2=-18(-1)=18$
- $(\sqrt{-6})(\sqrt{-9})=(\sqrt{6}\,i)(\sqrt{9}\,i)=(\sqrt{6})(3)i^2=3\sqrt{6}(-1)=-3\sqrt{6}$
- $(4+2i)(5+3i)=20+12i+10i+6i^2=20+22i+6(-1)=14+22i$
- $(8-2i)(8+2i)=64+16i-16i-4i^2=64-4(-1)=68$

The complex numbers $a+bi$ and $a-bi$ are called *complex conjugates*. The only difference between a complex number and its conjugate is the sign between the real part and the imaginary part. The product of any complex number and its conjugate is a real number.

$$\begin{aligned}(a+bi)(a-bi)&=a^2-abi+abi-b^2i^2\\&=a^2-b^2(-1)\\&=a^2+b^2\end{aligned}$$

EXAMPLES

- The complex conjugate of $3+2i$ is $3-2i$.
- The complex conjugate of $-7-i$ is $-7+i$.
- The complex conjugate of $10i$ is $-10i$.
- $(7-2i)(7+2i)$. Here, $a=7$ and $b=2$, so $a^2=49$ and $b^2=4$, making $(7-2i)(7+2i)=49+4=53$.
- $(1-i)(1+i)$. Here $a=1$ and $b=1$, so $a^2=1$ and $b^2=1$, making $(1-i)(1+i)=1+1=2$.

Dividing two complex numbers can be a little complicated. These problems are normally written in fraction form. If the denominator is purely imaginary, we can simply multiply the fraction by $\frac{i}{i}$ and simplify.

EXAMPLES

Perform the division. Write the quotient in the form $a + bi$, where a and b are real numbers.

- $$\frac{2+3i}{i} = \frac{2+3i}{i} \cdot \frac{i}{i} = \frac{(2+3i)i}{i^2}$$

$$= \frac{2i+3i^2}{i^2} = \frac{2i+3(-1)}{-1}$$

$$= \frac{-3+2i}{-1} = -(-3+2i)$$

$$= 3 - 2i$$

- $$\frac{4+5i}{2i} = \frac{4+5i}{2i} \cdot \frac{i}{i} = \frac{4i+5i^2}{2i^2}$$

$$= \frac{4i+5(-1)}{2(-1)}$$

$$= \frac{4i-5}{-2} = \frac{-(4i-5)}{2} = \frac{-(-5+4i)}{2}$$

$$= \frac{5-4i}{2} = \frac{5}{2} - 2i$$

When the divisor (denominator) is in the form $a + bi$, multiplying the fraction by $\frac{i}{i}$ will not work.

$$\frac{2-5i}{3+6i} \cdot \frac{i}{i} = \frac{2i-5i^2}{3i+6i^2} = \frac{5+2i}{-6+3i}$$

What *does* work is to multiply the fraction by the denominator's conjugate over itself. This works because the product of any complex number and its conjugate is a real number. We will use the FOIL method in the numerator (if necessary) and the fact that $(a+bi)(a-bi) = a^2 + b^2$ in the denominator.

EXAMPLES

Write the quotient in the form $a + bi$, where a and b are real numbers.

- $$\frac{2+7i}{6+i} = \frac{2+7i}{6+i} \cdot \frac{6-i}{6-i} = \frac{12-2i+42i-7i^2}{6^2+1^2}$$
  $$= \frac{12+40i-7(-1)}{37} = \frac{12+40i+7}{37}$$
  $$= \frac{19+40i}{37} = \frac{19}{37} + \frac{40}{37}i$$

- $$\frac{4-9i}{5-2i} = \frac{4-9i}{5-2i} \cdot \frac{5+2i}{5+2i} = \frac{20+8i-45i-18i^2}{5^2+2^2}$$
  $$= \frac{20-37i-18(-1)}{25+4} = \frac{20-37i+18}{29}$$
  $$= \frac{38-37i}{29} = \frac{38}{29} - \frac{37}{29}i$$

There are reasons to write complex numbers in the form $a + bi$. One is that complex numbers are plotted in the plane (real numbers are plotted on the number line), where the x-axis becomes the *real* axis and the y-axis becomes the *imaginary* axis. The number $3 - 4i$ is plotted in Figure 7.15.

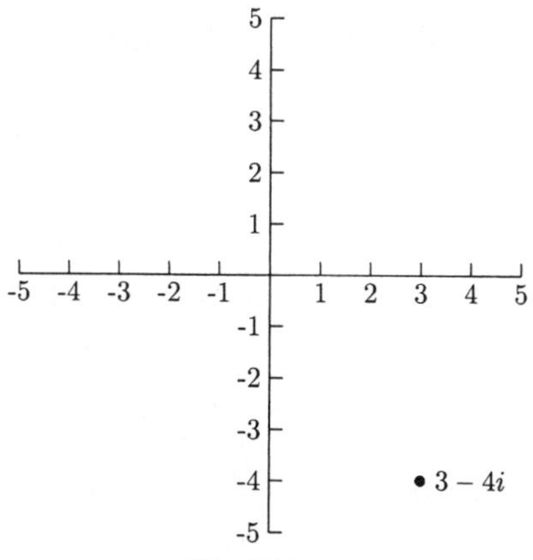

Fig. 7.15.

PRACTICE

For Problems 1–3, write the complex number in the form $a + bi$, where a and b are real numbers.

1. $\sqrt{-25}$
2. $\sqrt{-24}$
3. $14 - \sqrt{-36}$

For Problems 4–15, perform the arithmetic. Write answers in the form $a + bi$, where a and b are real numbers.

4. $18 - 4i + (-15) + 2i$
5. $5 + i + 5 - i$
6. $7 + i + 12 + i$
7. $-5 + \sqrt{-12} + 7 + 4\sqrt{-12}$
8. $\sqrt{-48} - (-1 - \sqrt{-75})$
9. $(2i)(10i)$
10. $(4\sqrt{-25})(2\sqrt{-25})$
11. $\sqrt{-6} \cdot \sqrt{-15}$
12. $(15 + 3i)(-2 + i)$
13. $(3 + 2i)(3 - 2i)$
14. $(8 - 10i)(8 + 10i)$
15. $(1 - 9i)(1 + 9i)$

For Problems 16–18, identify the complex conjugate.

16. $15 + 7i$
17. $-3 + i$
18. $-9i$

For Problems 19–21, write the quotient in the form $a + bi$, where a and b are real numbers.

19. $\frac{4-9i}{-3i}$
20. $\frac{4+2i}{1-3i}$
21. $\frac{6+4i}{6-4i}$

SOLUTIONS

1. $\sqrt{-25} = \sqrt{25}\, i = 5i$
2. $\sqrt{-24} = \sqrt{24}\, i = \sqrt{4 \cdot 6}\, i = 2\sqrt{6}\, i$
3. $14 - \sqrt{-36} = 14 - \sqrt{36}\, i = 14 - 6i$
4. $18 - 4i + (-15) + 2i = 3 - 2i$
5. $5 + i + 5 - i = 10 + 0i = 10$
6. $7 + i + 12 + i = 19 + 2i$
7. $$\begin{aligned} -5 + \sqrt{-12} + 7 + 4\sqrt{-12} &= -5 + \sqrt{12}\, i + 7 + 4\sqrt{12}\, i \\ &= -5 + \sqrt{4 \cdot 3}\, i + 7 + 4\sqrt{4 \cdot 3}\, i \\ &= -5 + 2\sqrt{3}\, i + 7 + 4 \cdot 2\sqrt{3}\, i \\ &= -5 + 2\sqrt{3}\, i + 7 + 8\sqrt{3}\, i \\ &= 2 + 10\sqrt{3}\, i \end{aligned}$$
8. $$\begin{aligned} \sqrt{-48} - (-1 - \sqrt{-75}) &= \sqrt{48}\, i + 1 + \sqrt{75}\, i \\ &= \sqrt{16 \cdot 3}\, i + 1 + \sqrt{25 \cdot 3}\, i \\ &= 4\sqrt{3}\, i + 1 + 5\sqrt{3}\, i = 1 + 9\sqrt{3}\, i \end{aligned}$$
9. $(2i)(10i) = 20i^2 = 20(-1) = -20$
10. $(4\sqrt{-25})(2\sqrt{-25}) = 4(5i)[2(5i)] = 200i^2 = 200(-1) = -200$
11. $\sqrt{-6} \cdot \sqrt{-15} = \sqrt{6}\, i \cdot \sqrt{15}\, i = \sqrt{6 \cdot 15}\, i^2 = \sqrt{90}\, i^2 = 3\sqrt{10}(-1)$ $= -3\sqrt{10}$
12. $(15+3i)(-2+i) = -30+15i-6i+3i^2 = -30+9i+3(-1) = -33+9i$
13. $(3+2i)(3-2i) = 9-6i+6i-4i^2 = 9-4(-1) = 13$ (or $3^2+2^2 = 13$)
14. $(8 - 10i)(8 + 10i) = 64 + 80i - 80i - 100i^2 = 64 - 100(-1) = 164$ (or $8^2 + 10^2 = 164$)
15. $(1-9i)(1+9i) = 1+9i-9i-81i^2 = 1-81(-1) = 82$ (or $1^2+9^2 = 82$)
16. The complex conjugate of $15 + 7i$ is $15 - 7i$.
17. The complex conjugate of $-3 + i$ is $-3 - i$.
18. The complex conjugate of $-9i$ is $9i$.
19. $$\begin{aligned} \frac{4-9i}{-3i} &= \frac{4-9i}{-3i} \cdot \frac{i}{i} = \frac{4i - 9i^2}{-3i^2} \\ &= \frac{4i - 9(-1)}{-3(-1)} = \frac{9+4i}{3} = 3 + \frac{4}{3}i \end{aligned}$$

20. $$\frac{4+2i}{1-3i} = \frac{4+2i}{1-3i} \cdot \frac{1+3i}{1+3i} = \frac{4+12i+2i+6i^2}{1^2+3^2}$$
$$= \frac{4+14i+6(-1)}{10} = \frac{-2+14i}{10} = -\frac{1}{5} + \frac{7}{5}i$$

21. $$\frac{6+4i}{6-4i} = \frac{6+4i}{6-4i} \cdot \frac{6+4i}{6+4i} = \frac{36+24i+24i+16i^2}{6^2+4^2}$$
$$= \frac{36+48i+16(-1)}{36+16} = \frac{20+48i}{52} = \frac{5}{13} + \frac{12}{13}i$$

Complex Solutions to Quadratic Equations

Every quadratic equation has a solution, real or complex. The real solutions for a quadratic equation are the x-intercepts, for the graph of the quadratic function.

The graph for $f(x) = x^2 + 1$ has no x-intercepts.

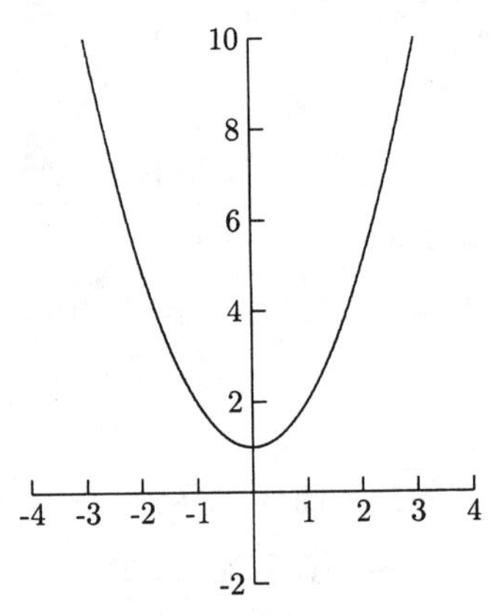

Fig. 7.16.

The equation $x^2 + 1 = 0$ does have two complex solutions.

$$x^2 + 1 = 0$$
$$x^2 = -1$$
$$x = \pm\sqrt{-1}$$
$$= \pm i$$

EXAMPLE

Solve the equation and write the solutions in the form $a + bi$, where a and b are real numbers.

- $3x^2 + 8x + 14 = 0$

$$
\begin{aligned}
x &= \frac{-8 \pm \sqrt{8^2 - 4(3)(14)}}{2(3)} = \frac{-8 \pm \sqrt{-104}}{6} \\
&= \frac{-8 \pm 2\sqrt{26}\,i}{6} = \frac{2(-4 \pm \sqrt{26}\,i)}{6} \\
&= \frac{-4 \pm \sqrt{26}\,i}{3} = -\frac{4}{3} \pm \frac{\sqrt{26}}{3} i \\
&= -\frac{4}{3} + \frac{\sqrt{26}}{3} i, \quad -\frac{4}{3} - \frac{\sqrt{26}}{3} i
\end{aligned}
$$

In this problem, the complex solutions to the quadratic equation came in conjugate pairs. This always happens when the solutions are complex numbers. A quadratic expression that has complex zeros is called *irreducible* (over the reals) because it cannot be factored using real numbers. For example, the polynomial function $f(x) = x^4 - 1$ can be factored using real numbers as $(x^2 - 1)(x^2 + 1) = (x - 1)(x + 1)(x^2 + 1)$. The factor $x^2 + 1$ is irreducible because it is factored as $(x - i)(x + i)$.

We can tell which quadratic factors are irreducible without having to use the quadratic formula. We only need part of the quadratic formula, $b^2 - 4ac$. When this number is negative, the quadratic factor has two complex zeros, $\frac{-b \pm \sqrt{\text{negative number}}}{2a}$. When this number is positive, there are two real number solutions, $\frac{-b \pm \sqrt{\text{positive number}}}{2a}$. When this number is zero, there is one real zero, $\frac{-b \pm \sqrt{0}}{2a} = \frac{-b}{2a}$. For this reason, $b^2 - 4ac$ is called the *discriminant*.

The graphs of some polynomials having irreducible quadratic factors need extra points plotted to get a more accurate graph. The graph in Figure 7.17 shows the graph of $f(x) = x^4 - 3x^2 - 4$ using our usual method—plotting the x-intercepts, a point to the left of the smallest x-intercept, a point between each consecutive pair of x-intercepts, and a point to the right of the largest x-intercept.

See what happens to the graph when we plot the points for $x = 1$ and $x = -1$.

The graph of $f(x) = (x - 2)(x^2 + 6x + 10)$ is sketched in Figure 7.19. The graphs we have sketched have several vertices between x-intercepts. When this happens, we need calculus to find them.

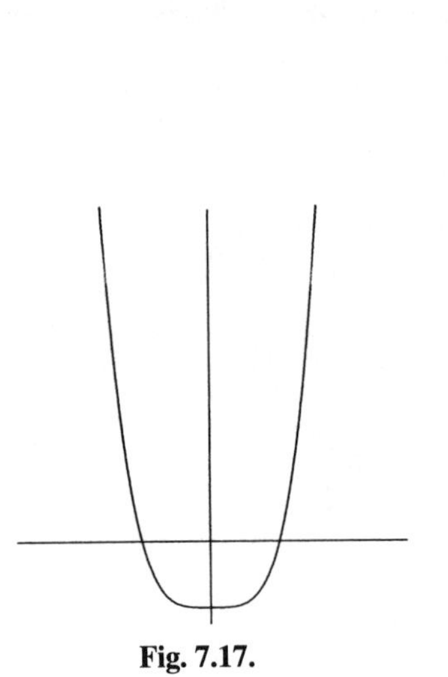

Fig. 7.17.

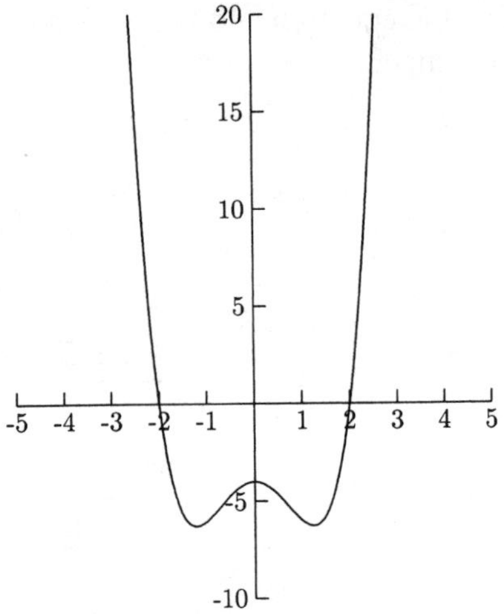

Fig. 7.18.

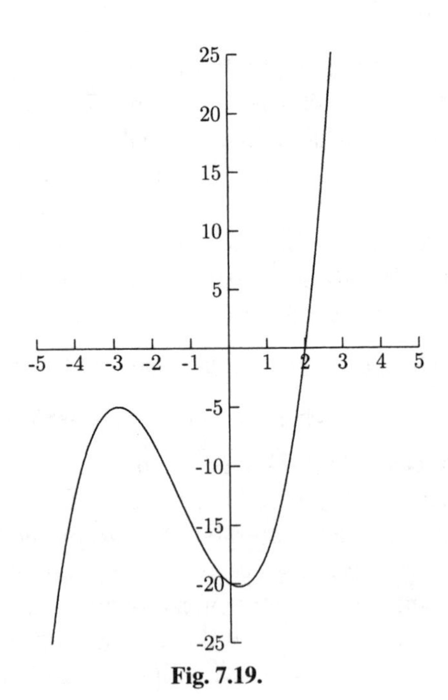

Fig. 7.19.

The Fundamental Theorem of Algebra

By the *Fundamental Theorem of Algebra*, every polynomial of degree n has exactly n zeros (some might be counted more than once). Because $x = c$ is a zero implies $x - c$ is a factor, every polynomial can be completely factored in the form $a(x - c_n)(x - c_{n-1})\ldots(x - c_1)$, where a is a real number and c_i is real or complex. Factors in the form $x - c$ are called *linear factors*. Factors such as $2x + 1$ can be written in the form $x - c$ by factoring 2: $2(x + \frac{1}{2})$ or $2(x - (-\frac{1}{2}))$.

To completely factor a polynomial, we usually need to first find its zeros. At times, we will use the Rational Zero Theorem, polynomial division, and the quadratic formula.

EXAMPLES

Find all zeros, real and complex.

- $h(x) = x^4 - 16$

$$x^4 - 16 = (x^2 - 4)(x^2 + 4) = (x - 2)(x + 2)(x^2 + 4)$$

The real zeros are 2 and -2. We will find the complex zeros by solving $x^2 + 4 = 0$.

$$x^2 + 4 = 0$$

$$x^2 = -4$$

$$x = \pm\sqrt{-4} = \pm 2i$$

The complex zeros are $\pm 2i$.

- $P(x) = x^4 + 6x^3 + 9x^2 - 6x - 10$

The possible rational zeros are ± 1, ± 2, ± 5, and ± 10. $P(1) = 0$.

$$\begin{array}{r|rrrrr} 1 & 1 & 6 & 9 & -6 & -10 \\ & & 1 & 7 & 16 & 10 \\ \hline & 1 & 7 & 16 & 10 & 0 \end{array}$$

$P(x) = (x - 1)(x^3 + 7x^2 + 16x + 10)$

Because $x^3 + 7x^2 + 16x + 10$ has no sign changes, there are no positive zeros; $x = -1$ is a zero for $x^3 + 7x^2 + 16x + 10$.

$$\begin{array}{r|rrrr} -1 & 1 & 7 & 16 & 10 \\ & & -1 & -6 & -10 \\ \hline & 1 & 6 & 10 & 0 \end{array}$$

$P(x) = (x - 1)(x + 1)(x^2 + 6x + 10)$

Solve $x^2 + 6x + 10 = 0$ to find the complex zeros.

$$x = \frac{-6 \pm \sqrt{6^2 - 4(1)(10)}}{2(1)} = \frac{-6 \pm \sqrt{-4}}{2}$$

$$= \frac{-6 \pm 2i}{2} = \frac{2(-3 \pm i)}{2} = -3 \pm i$$

The zeros are $\pm 1, -3 \pm i$.

If we know a complex number is a zero for a polynomial, we automatically know another zero—the complex conjugate is also a zero. This gives us a quadratic factor for the polynomial. Once we have this computed, we can use long division to find the quotient, which will be another factor of the polynomial. Each time we factor a polynomial, we are closer to finding its zeros.

EXAMPLES

Find all zeros, real and complex.

- $f(x) = 3x^4 + x^3 + 17x^2 + 4x + 20$ and $x = 2i$ is a zero.
  Because $x = 2i$ is a zero, its conjugate, $-2i$, is another zero. This tells us that two factors are $x - 2i$ and $x + 2i$.

  $$(x - 2i)(x + 2i) = x^2 + 2ix - 2ix - 4i^2 = x^2 - 4(-1) = x^2 + 4$$

  We will divide $f(x)$ by $x^2 + 4 = x^2 + 0x + 4$.

  $$\begin{array}{r|l} & 3x^2 + x + 5 \\ \hline x^2 + 0x + 4 & 3x^4 + x^3 + 17x^2 + 4x + 20 \\ & -(3x^4 + 0x^3 + 12x^2) \\ \hline & x^3 + 5x^2 + 4x \\ & -(x^3 + 0x^2 + 4x) \\ \hline & 5x^2 + 0x + 20 \\ & -(5x^2 + 0x + 20) \\ \hline & 0 \end{array}$$

  $f(x) = (x^2 + 4)(3x^2 + x + 5)$
  Solving $3x^2 + x + 5 = 0$, we get the solutions

  $$x = \frac{-1 \pm \sqrt{1^2 - 4(3)(5)}}{2(3)} = \frac{-1 \pm \sqrt{-59}}{6} = \frac{-1 \pm \sqrt{59}\,i}{6}.$$

  The zeros are $\pm 2i$, $\frac{-1 \pm \sqrt{59}\,i}{6}$.

- $h(x) = 2x^3 - 7x^2 + 170x - 246$, $x = 1 + 9i$ is a zero.
  Because $x = 1 + 9i$ is a zero, we know that $x = 1 - 9i$ is also a zero. We also know that $x - (1 + 9i) = x - 1 - 9i$ and $x - (1 - 9i) = x - 1 + 9i$ are factors. We will multiply these two factors.

$$(x - 1 - 9i)(x - 1 + 9i) = x^2 - x + 9ix - x + 1 - 9i - 9ix + 9i - 81i^2$$

$$= x^2 - 2x + 1 - 81(-1) = x^2 - 2x + 82$$

$$\begin{array}{r|llll}
 & 2x- & 3 & & \\ \hline
x^2 - 2x + 82 & 2x^3- & 7x^2+ & 170x- & 246 \\
 & -(2x^3- & 4x^2+ & 164x) & \\ \hline
 & & -3x^2+ & 6x- & 246 \\
 & & -(-3x^2+ & 6x- & 246) \\ \hline
 & & & & 0
\end{array}$$

  $h(x) = (2x - 3)(x^2 - 2x + 82)$
  The zeros are $1 \pm 9i$ and $\frac{3}{2}$ (from $2x - 3 = 0$).

A consequence of the Fundamental Theorem of Algebra is that a polynomial of degree n will have n zeros, though not necessarily n different zeros. For example, the polynomial $f(x) = (x - 2)^3 = (x - 2)(x - 2)(x - 2)$ has $x = 2$ as a zero three times. The number of times an x-value is a zero is called its *multiplicity*. In the above example, $x = 2$ is a zero with multiplicity 3.

EXAMPLE

- $f(x) = x^4(x + 3)^2(x - 6)$
  $x = 0$ is a zero with multiplicity 4 (We can think of x^4 as $(x - 0)^4$.)
  $x = -3$ is a zero with multiplicity 2
  $x = 6$ is a zero with multiplicity 1

Now, instead of finding the zeros for a given polynomial, we will find a polynomial with the given zeros. Because we will know the zeros, we will know the factors. Once we know the factors of a polynomial, we pretty much know the polynomial.

EXAMPLES

Find a polynomial with integer coefficients having the given degree and zeros.

- Degree 3 with zeros 1, 2, and 5
  Because $x = 1$ is a zero, $x - 1$ is a factor. Because $x = 2$ is a zero, $x - 2$ is a factor. And because $x = 5$ is a zero, $x - 5$ is a factor. Such a polynomial

will be of the form $a(x-1)(x-2)(x-5)$, where a is some nonzero number. We will want to choose a so that the coefficients are integers.

$$\begin{aligned}a(x-1)(x-2)(x-5) &= a(x-1)[(x-2)(x-5)]\\ &= a(x-1)(x^2-7x+10)\\ &= a(x^3-7x^2+10x-x^2+7x-10)\\ &= a(x^3-8x^2+17x-10)\end{aligned}$$

Because the coefficients are already integers, we can let $a = 1$. One polynomial of degree three having integer coefficients and 1, 2, and 5 as zeros is $x^3-8x^2+17x-10$.

- Degree 4 with zeros $-3, 2-5i$, with -3 a zero of multiplicity 2
  Because -3 is a zero of multiplicity 2, $(x+3)^2 = x^2+6x+9$ is a factor. Because $2-5i$ is a zero, $2+5i$ is another zero. Another factor of the polynomial is

$$\begin{aligned}(x-(2-5i))(x-(2+5i)) &= (x-2+5i)(x-2-5i)\\ &= x^2-2x-5ix-2x+4+10i+5ix\\ &\quad -10i-25i^2\\ &= x^2-4x+4-25(-1) = x^2-4x+29.\end{aligned}$$

The polynomial has the form $a(x^2+6x+9)(x^2-4x+29)$, where a is any real number that makes all coefficients integers.

$$\begin{aligned}a(x^2+6x+9)(x^2-4x+29) &= a(x^4-4x^3+29x^2+6x^3-24x^2\\ &\quad +174x+9x^2-36x+261)\\ &= a(x^4+2x^3+14x^2+138x+261)\end{aligned}$$

Because the coefficients are already integers, we can let $a = 1$. One polynomial that satisfies the given conditions is $x^4+2x^3+14x^2+138x+261$.

In the previous problems, there were infinitely many answers because a could be any integer. In the following problem, there will be exactly one polynomial that satisfies the given conditions. This means that a will likely be a number other than 1.

- Degree 3 with zeros -1, -2, and 4, where the coefficient for x is -20

$$\begin{aligned} a(x+1)(x+2)(x-4) &= a(x+1)[(x+2)(x-4)] \\ &= a(x+1)(x^2-2x-8) \\ &= a(x^3-2x^2-8x+x^2-2x-8) \\ &= a(x^3-x^2-10x-8) \\ &= ax^3-ax^2-10ax-8a \end{aligned}$$

Because we need the coefficient of x to be -20, we need $-10ax = -20x$, so we need $a = 2$ (from $-10a = -20$). The polynomial that satisfies the conditions is $2x^3 - 2x^2 - 20x - 16$.

PRACTICE

For Problems 1–6 solve the equations and write complex solutions in the form $a + bi$, where a and b are real numbers.

1. $9x^2 + 4 = 0$
2. $6x^2 + 8x + 9 = 0$
3. $x^4 - 81 = 0$
4. $x^3 + 13x - 34 = 0$
5. $x^4 - x^3 + 8x^2 - 9x - 9 = 0$; $x = -3i$ is a solution.
6. $x^3 - 5x^2 + 7x + 13 = 0$; $x = 3 - 2i$ is a solution.

For Problems 7–10 find a polynomial with integer coefficients having the given conditions.

7. Degree 3 with zeros 0, -4, and 6
8. Degree 4 with zeros -1 and $6 - 7i$, where $x = -1$ has multiplicity 2.
9. Degree 3, zeros 4, and ± 1, with leading coefficient 3
10. Degree 4 with zeros i and $4i$, with constant term -16
11. State each zero and its multiplicity for $f(x) = x^2(x+4)(x+9)^6(x-5)^3$

SOLUTIONS

1. $9x^2 + 4 = 0$

$$9x^2 = -4$$

$$x^2 = -\frac{4}{9}$$

$$x = \pm\sqrt{-\frac{4}{9}} = \pm\frac{2}{3}i = \frac{2}{3}i,\ -\frac{2}{3}i$$

2. $x = \dfrac{-8 \pm \sqrt{8^2 - 4(6)(9)}}{2(6)}$

$$= \frac{-8 \pm \sqrt{-152}}{12} = \frac{-8 \pm 2\sqrt{38}\,i}{12}$$

$$= \frac{2(-4 \pm \sqrt{38}\,i)}{12} = \frac{-4 \pm \sqrt{38}\,i}{6}$$

$$= -\frac{4}{6} \pm \frac{\sqrt{38}}{6}i = -\frac{2}{3} \pm \frac{\sqrt{38}}{6}i$$

$$= -\frac{2}{3} + \frac{\sqrt{38}}{6}i,\ -\frac{2}{3} - \frac{\sqrt{38}}{6}i$$

3. $x^4 - 81 = (x^2 - 9)(x^2 + 9) = (x - 3)(x + 3)(x^2 + 9)$

$$x^2 + 9 = 0$$

$$x^2 = -9$$

$$x = \pm\sqrt{-9} = \pm 3i$$

The solutions are ± 3, $\pm 3i$.

4. $x = 2$ is a solution, so $x - 2$ is a factor of $x^3 + 13x - 34$. Using synthetic division, we can find the quotient, which will be another factor.

$$\begin{array}{r|rrrr} 2 & 1 & 0 & 13 & -34 \\ & & 2 & 4 & 34 \\ \hline & 1 & 2 & 17 & 0 \end{array}$$

The quotient is $x^2 + 2x + 17$. We will find the other solutions by solving $x^2 + 2x + 17 = 0$.

$$x = \frac{-2 \pm \sqrt{2^2 - 4(1)(17)}}{2(1)} = \frac{-2 \pm \sqrt{-64}}{2} = \frac{-2 \pm 8i}{2}$$

$$= \frac{2(-1 \pm 4i)}{2} = -1 \pm 4i$$

The solutions are 2 and $-1 \pm 4i$.

5. $x = -3i$ is a solution, so $x = 3i$ is a solution, also. One factor of $x^4 - x^3 + 8x^2 - 9x - 9$ is $(x - 3i)(x + 3i) = x^2 + 9 = x^2 + 0x + 9$.

$$\begin{array}{r|rrrrr}
\multicolumn{1}{r}{} & x^2- & x- & 1 & & \\ \cline{2-6}
x^2+0x+9 & x^4- & x^3+ & 8x^2 & -9x & -9 \\
 & -(x^4+ & 0x^3+ & 9x^2) & & \\ \cline{2-5}
 & & -x^3- & x^2 & -9x & \\
 & & -(-x^3+ & 0x^2- & 9x) & \\ \cline{3-6}
 & & & -x^2 & +0x & -9 \\
 & & & -(-x^2+ & 0x & -9) \\ \cline{4-6}
 & & & & & 0
\end{array}$$

Solve $x^2 - x - 1 = 0$.

$$x = \frac{-(-1) \pm \sqrt{(-1)^2 - 4(1)(-1)}}{2(1)} = \frac{1 \pm \sqrt{5}}{2}$$

The solutions are $\pm 3i$, $\frac{1 \pm \sqrt{5}}{2}$.

6. $x = 3 - 2i$ is a solution, so $x = 3 + 2i$ is also a solution. One factor of $x^3 - 5x^2 + 7x + 13$ is

$$\begin{aligned}
(x-(3-2i))(x-(3+2i)) &= (x-3+2i)(x-3-2i) \\
&= x^2-3x-2ix-3x+9+6i+2ix-6i-4i^2 \\
&= x^2-6x+9-4(-1) = x^2-6x+13.
\end{aligned}$$

$$\begin{array}{r|rrrr}
\multicolumn{1}{r}{} & x+ & 1 & & \\ \cline{2-5}
x^2-6x+13 & x^3- & 5x^2+ & 7x+ & 13 \\
 & -(x^3- & 6x^2+ & 13x) & \\ \cline{2-5}
 & & x^2- & 6x+ & 13 \\
 & & -(x^2- & 6x+ & 13) \\ \cline{3-5}
 & & & & 0
\end{array}$$

The solutions are $3 \pm 2i$ and -1.

7. One polynomial with integer coefficients, with degree 3 and zeros 0, -4 and 6 is

$$x(x + 4)(x - 6) = x(x^2 - 2x - 24) = x^3 - 2x^2 - 24x.$$

8. One polynomial with integer coefficients, with degree 4 and zeros -1, $6-7i$, where $x=-1$ has multiplicity 2 is

$$(x+1)^2(x-(6-7i))(x-(6+7i)) = (x+1)^2(x-6+7i)(x-6-7i)$$
$$= [(x+1)(x+1)][x^2-6x-7ix-6x+36+42i+7ix-42i-49i^2]$$
$$= (x^2+2x+1)(x^2-12x+85)$$
$$= x^4-12x^3+85x^2+2x^3-24x^2+170x+x^2-12x+85$$
$$= x^4-10x^3+62x^2+158x+85.$$

9. The factors are $x-4$, $x-1$, and $x+1$.

$$a(x-4)(x-1)(x+1) = a(x-4)[(x-1)(x+1)] = a(x-4)(x^2-1)$$
$$= a[(x-4)(x^2-1)] = a(x^3-4x^2-x+4)$$
$$= ax^3-4ax^2-ax+4a$$

We want the leading coefficient to be 3, so $a=3$. The polynomial that satisfies the conditions is $3x^3-12x^2-3x+12$.

10. The factors are $x+i$, $x-i$, $x-4i$, and $x+4i$.

$$a(x+i)(x-i)(x-4i)(x+4i) = a[(x+i)(x-i)][(x-4i)(x+4i)]$$
$$= a(x^2+1)(x^2+16) = a(x^4+17x^2+16)$$
$$= ax^4+17ax^2+16a$$

We want $16a=-16$, so $a=-1$. The polynomial that satisfies the conditions is $-x^4-17x^2-16$.

11. $x=0$ is a zero with multiplicity 2.
$x=-4$ is a zero with multiplicity 1.
$x=-9$ is a zero with multiplicity 6.
$x=5$ is a zero with multiplicity 3.

CHAPTER 7 REVIEW

1. What are the x-intercepts of $f(x)=x^2(x+3)(x-2)$?
(a) -3 and 2 (b) 3 and -2 (c) 0, -3, and 2 (d) 0, 3, and -2

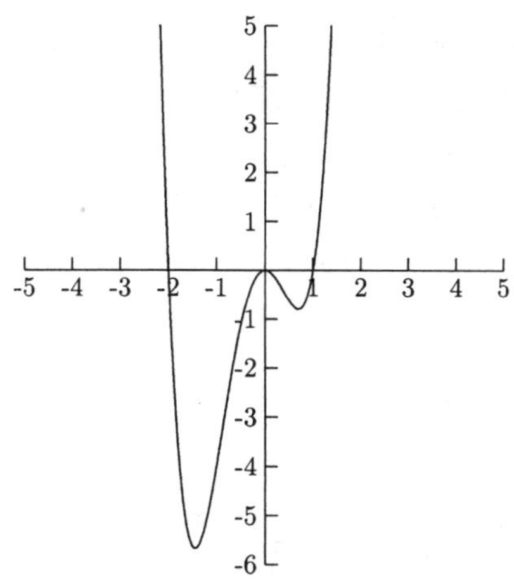

Fig. 7.20.

2. The graph in Figure 7.20 is the graph of which function?
   (a) $f(x) = 2x^2(x-1)(x+2) = 2x^4 + 2x^3 - 4x^2$
   (b) $f(x) = -2x^2(x-1)(x+2) = -2x^4 - 2x^3 + 4x^2$
   (c) $f(x) = 2x(x-1)(x+2) = 2x^3 + 2x^2 - 4x$
   (d) $f(x) = -2x(x-1)(x+2) = -2x^3 - 2x^2 + 4x$

3. What is the quotient and remainder for

$$\frac{x^3+1}{x^2+x+2}?$$

   (a) The quotient is $x - 1$, and the remainder is $-3x - 3$.
   (b) The quotient is $x - 1$, and the remainder is $-x + 3$.
   (c) The quotient is $x + 1$, and the remainder is $x + 3$.
   (d) The quotient is $x + 1$, and the remainder is $3x + 3$.

4. Use synthetic division to find the quotient and remainder for $(2x^3 - x^2 + 2x + 4) \div (x - 3)$.
   (a) The quotient is $2x^2 + x + 5$, and the remainder is 19.
   (b) The quotient is $2x^2 + 5x + 7$, and the remainder is 29.
   (c) The quotient is $2x^2 + 5x + 17$, and the remainder is 55.
   (d) The quotient is $2x^2 + x + 3$, and the remainder is 7.

5. What is the quotient for $(x^4 + x^3 - 3x + 5) \div (-2x^2 + x - 6)$?
   (a) The quotient is $-\frac{1}{2}x^2 - \frac{1}{4}x - \frac{11}{8}$.
   (b) The quotient is $-\frac{1}{2}x^2 + \frac{1}{4}x + \frac{13}{8}$.
   (c) The quotient is $-\frac{1}{2}x^2 - \frac{3}{4}x - \frac{15}{8}$.
   (d) The quotient is $-\frac{1}{2}x^2 - \frac{3}{4}x + \frac{9}{8}$.

6. Completely factor $P(x) = 4x^3 + 4x^2 - x - 1$.
   (a) $(x-1)^2(4x+1)$ (b) $(x+1)(2x-1)(2x+1)$
   (c) $(x+1)^2(4x-1)$ (d) $(x-1)(2x-1)(2x+1)$

7. Find all solutions for $x^2 + 2x + 4 = 0$.
   (a) $-1 \pm \sqrt{3}\,i$ (b) $1 \pm \sqrt{3}\,i$ (c) $1 \pm \sqrt{5}$ (d) $-1 \pm \sqrt{5}$

8. What is the quotient for $\frac{1-i}{2+3i}$?
   (a) $\frac{2}{13}$ (b) $\frac{5}{13} + \frac{1}{13}i$ (c) $\frac{5}{13} - \frac{1}{13}i$ (d) $-\frac{1}{13} - \frac{5}{13}i$

9. According to the Rational Zero Theorem, which is *NOT* a possible rational zero for $f(x) = 4x^5 - 6x^3 + 2x^2 - 6x - 9$?
   (a) -4 (b) $\frac{3}{2}$ (c) 3 (d) -9

10. According to Descartes' Rule of Signs, how many positive zeros does $f(x) = 4x^5 - 6x^3 + 2x^2 - 9$ have?
    (a) 3 (b) 2 or 0 (c) 2 (d) 3 or 1

11. Find all zeros for $f(x) = x^3 - 6x^2 + 13x - 10$.
    (a) $-2,\ 2 \pm i$ (b) $2,\ 2 \pm i$ (c) $2,\ 1 \pm 2i$ (d) $-2,\ 1 \pm 2i$

SOLUTIONS

1. C 2. A 3. B 4. C 5. D
6. B 7. A 8. D 9. A 10. D 11. B

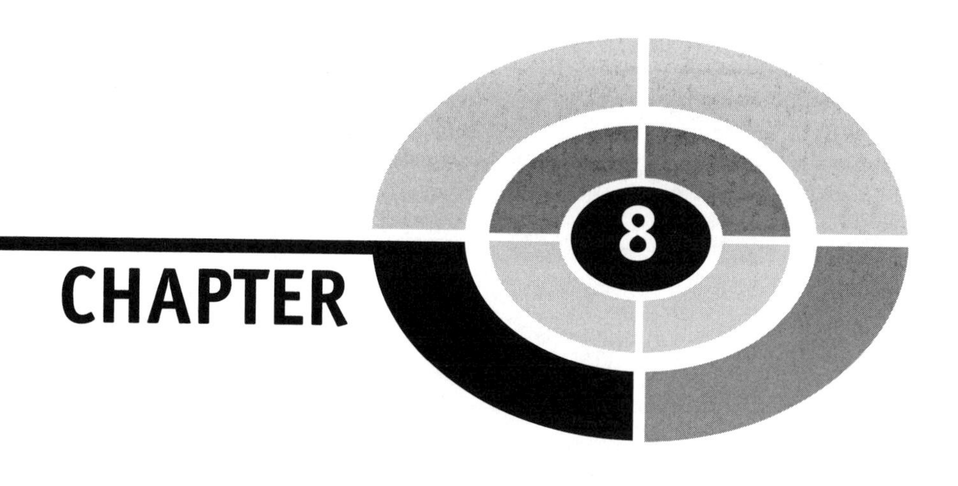

Rational Functions

A *rational function* is a function that can be written as one polynomial divided by another.

$$f(x) = \frac{P(x)}{Q(x)} = \frac{a_n x^n + a_{n-1}x^{n-1} + \cdots + a_1 x + a_0}{b_m x^m + b_{m-1}x^{m-1} + \cdots + b_1 x + b_0}$$

Polynomial functions are a special kind of rational function whose denominator function is $Q(x) = 1$. While the graph of every polynomial function has exactly one y-intercept, the graph of a rational function might not have a y-intercept. If it has a y-intercept, it can be found by setting x equal to zero.If it has any x-intercepts, they can be found by setting the numerator equal to zero.

The graphs of rational functions often come in pieces. For every x-value that causes a zero in the denominator, there will be a break in the graph. If the function is reduced to lowest terms (the numerator and denominator have no common factors), then there will be a *vertical asymptote* at these breaks. The graph rises (or falls) very fast near these asymptotes. The graph in Figure 8.1 is the graph of $f(x) = \frac{1}{x-1}$. It has a vertical asymptote at the line $x = 1$ because $x = 1$ causes a zero in the denominator.

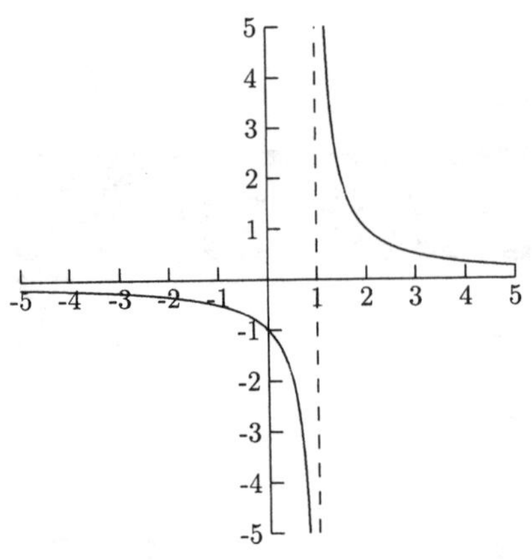

Fig. 8.1.

A vertical asymptote shows that the y-values get large when the x-values get close to a zero in the denominator. To see this, we will evaluate $f(x) = \frac{1}{x-1}$ at $x = 0.99$ and $x = 1.01$, two x-values close to a zero in the denominator.

$$f(0.99) = \frac{1}{0.99 - 1} = -100 \qquad \text{and} \qquad f(1.01) = \frac{1}{1.01 - 1} = 100$$

The graph flattens out horizontally near a *horizontal asymptote*. The graph in Figure 8.1 has the x-axis as its horizontal asymptote. A horizontal asymptote shows that as x gets very large, the y-values get very close to a fixed number. In the function $f(x) = \frac{1}{x-1}$, there is a horizontal asymptote at $y = 0$ (the x-axis). This means that as x gets large, the y-values get close to 0.

$$f(100) = \frac{1}{100 - 1} = \frac{1}{99} \approx 0.010101 \text{ and}$$

$$f(-100) = \frac{1}{-100 - 1} = -\frac{1}{101} \approx -0.009901$$

Vertical asymptotes are easy to find—set the denominator equal to zero and solve for x. Whether or not a graph has a horizontal asymptote depends on the degree of the numerator and of the denominator.

- If the degree of the numerator is larger than the degree of the denominator, there is no horizontal asymptote.
- If the degree of the denominator is larger than the degree of the numerator, there is a horizontal asymptote at $y = 0$, which is the x-axis.

- If the degree of the numerator equals the degree of the denominator, there is a horizontal asymptote at $y = \frac{a_n}{b_m}$, where a_n is the leading coefficient of the numerator and b_m is the leading coefficient of the denominator.

EXAMPLES

Find the intercepts, vertical asymptotes, and horizontal asymptotes.

- $f(x) = \dfrac{x^2 - 16}{3x + 1}$

  Solving $3x + 1 = 0$ we get $x = -\frac{1}{3}$. The vertical line $x = -\frac{1}{3}$ is the vertical asymptote for this graph. There is no horizontal asymptote because the degree of the numerator, 2, is more than the degree of the denominator, 1. The x-intercepts are ± 4 (from $x^2 - 16 = 0$) and the y-intercept is

$$\frac{0^2 - 16}{3(0) + 1} = -16.$$

- $g(x) = \dfrac{15}{x^2 - 4x - 5}$

  When we solve $x^2 - 4x - 5 = 0$, we get the solutions $x = 5, -1$. This graph has two vertical asymptotes, the vertical lines $x = 5$ and $x = -1$. The x-axis is the horizontal asymptote because the degree of the numerator, 0, is less than the degree of the denominator, 2. (A reminder, the degree of a constant term is 0, $15 = 15x^0$.) There is no x-intercept because the numerator of this fraction is always 15, it is never 0. The y-intercept is

$$\frac{15}{0^2 - 4(0) - 5} = -3.$$

- $f(x) = \dfrac{3x^2}{x^2 + 2}$

  Because $x^2 + 2 = 0$ has no real solutions, this graph has no vertical asymptote. There is a horizontal asymptote at $y = \frac{3}{1} = 3$ because the degree of the numerator and denominator is the same. The x-intercept is 0 (from $3x^2 = 0$). The y-intercept is

$$\frac{3(0)^2}{0^2 + 2} = \frac{0}{2} = 0.$$

The reason we can find the horizontal asymptotes so easily is that for large values of x, only the leading terms in the numerator and denominator really matter. The examples below will show an algebraic reason for the rules above. For any fixed number c any positive power on x,

$$\frac{c}{x^{\text{power}}}$$

is almost 0 for large values of x. For example, in $\frac{-10}{x^2}$, if we let x be any large number, the fraction will be close to 0.

$$\frac{-10}{(100)^2} = -0.001$$

The larger x is, the closer $\frac{-10}{x^2}$ is to 0.

EXAMPLES

- $f(x) = \dfrac{3x^3 + 5x^2 + x - 6}{2x^4 + 8x^2 - 1}$

  From above, we know that the x-axis, or the horizontal line $y = 0$, is a horizontal asymptote. Here is why. Because the highest power on x is 4, we will multiply the fraction by $\frac{1/x^4}{1/x^4}$, which reduces to 1, so we are not changing the fraction.

$$\frac{3x^3 + 5x^2 + x - 6}{2x^4 + 8x^2 - 1} \cdot \frac{\frac{1}{x^4}}{\frac{1}{x^4}} = \frac{\frac{3x^3}{x^4} + \frac{5x^2}{x^4} + \frac{x}{x^4} - \frac{6}{x^4}}{\frac{2x^4}{x^4} + \frac{8x^2}{x^4} - \frac{1}{x^4}} = \frac{\frac{3}{x} + \frac{5}{x^2} + \frac{1}{x^3} - \frac{6}{x^4}}{2 + \frac{8}{x^2} - \frac{1}{x^4}}$$

  For large values of x, $3/x$, $5/x^2$, $1/x^3$, $6/x^4$, $8/x^2$, and $1/x^4$ are very close to zero, so for large values of x,

$$\frac{\frac{3}{x} + \frac{5}{x^2} + \frac{1}{x^3} - \frac{6}{x^4}}{2 + \frac{8}{x^2} - \frac{1}{x^4}} \text{ is close to } \frac{0 + 0 + 0 - 0}{2 + 0 - 0} = \frac{0}{2} = 0.$$

- $g(x) = \dfrac{4x^3 + 8x^2 - 5x + 3}{9x^3 - x^2 + 8x - 2}$

  The degree of the numerator equals the degree of the denominator, so the graph of this function has a horizontal asymptote at the line $y = 4/9$. Here is why. Because the largest power on x is 3, we will multiply the fraction by $\frac{1/x^3}{1/x^3}$.

$$\frac{4x^3 + 8x^2 - 5x + 3}{9x^3 - x^2 - 8x - 2} \cdot \frac{\frac{1}{x^3}}{\frac{1}{x^3}} = \frac{\frac{4x^3}{x^3} + \frac{8x^2}{x^3} - \frac{5x}{x^3} + \frac{3}{x^3}}{\frac{9x^3}{x^3} - \frac{x^2}{x^3} - \frac{8x}{x^3} - \frac{2}{x^3}} = \frac{4 + \frac{8}{x} - \frac{5}{x^2} + \frac{3}{x^3}}{9 - \frac{1}{x} - \frac{8}{x^2} - \frac{2}{x^3}}$$

For large values of x, $\dfrac{4 + \frac{8}{x} - \frac{5}{x^2} + \frac{3}{x^3}}{9 - \frac{1}{x} - \frac{8}{x^2} - \frac{2}{x^3}}$ is close to $\dfrac{4 + 0 - 0 + 0}{9 - 0 - 0 - 0} = \dfrac{4}{9}$.

These steps are not necessary to find the horizontal asymptotes, only the three rules earlier in the chapter.

PRACTICE

Find the intercepts, vertical asymptotes, and horizontal asymptotes.

1. $f(x) = \dfrac{x + 2}{2x + 3}$
2. $g(x) = \dfrac{-3x}{x^2 + x - 20}$
3. $h(x) = \dfrac{x^2 - 1}{x^2 + 1}$
4. $R(x) = \dfrac{9x^2 - 1}{8x + 3}$
5. $f(x) = \dfrac{x^3 + 1}{x^2 + 4}$
6. $f(x) = \dfrac{2}{x^2}$

SOLUTIONS

1. The vertical asymptote is $x = -\frac{3}{2}$, from $2x + 3 = 0$. The horizontal asymptote is $y = \frac{1}{2}$ because the numerator and denominator have the same degree. The x-intercept is -2, from $x + 2 = 0$. The y-intercept is

$$\frac{0 + 2}{2(0) + 3} = \frac{2}{3}.$$

2. The vertical asymptotes are $x = -5$ and $x = 4$, from $x^2 + x - 20 = 0$. The horizontal asymptote is $y = 0$ because the denominator has the higher degree. The x-intercept is 0, from $-3x = 0$. The y-intercept is

$$\frac{-3(0)}{0^2 + 0 - 20} = \frac{0}{-20} = 0.$$

3. There is no vertical asymptote because $x^2 + 1 = 0$ has no real solution. The horizontal asymptote is $y = 1/1 = 1$ because the numerator and denominator have the same degree. The x-intercepts are ± 1, from $x^2 - 1 = 0$. The y-intercept is

$$\frac{0^2 - 1}{0^2 + 1} = \frac{-1}{1} = -1.$$

4. The vertical asymptote is $x = -\frac{3}{8}$, from $8x + 3 = 0$. There is no horizontal asymptote because the numerator has the higher degree. The x-intercepts are $\pm\frac{1}{3}$, from $9x^2 - 1 = 0$. The y-intercept is

$$\frac{9(0)^2 - 1}{8(0) + 3} = \frac{-1}{3}.$$

5. There is no vertical asymptote because $x^2+4 = 0$ has no real solution. There is no horizontal asymptote because the numerator has the higher degree. The x-intercept is -1, from $x^3 + 1 = 0$. The y-intercept is

$$\frac{0^3 + 1}{0^2 + 4} = \frac{1}{4}.$$

6. The vertical asymptote is $x = 0$, from $x^2 = 0$. The horizontal asymptote $y = 0$ because the denominator has the higher degree. There is no x-intercept because the numerator is 2, never 0. There is no y-intercept because $2/0^2$ is not defined.

When sketching the graph of a rational function, we use dashed lines for the asymptotes . We will sketch the graphs of rational functions in much the same way we sketched the graphs of polynomial functions. In addition to the points we plot for polynomial functions, we need to plot points to illustrate the asymptotic behavior of the graph. To show how a graph behaves near a vertical asymptote, we need to plot a point to its left and to its right. To show how a graph behaves near a horizontal asymptote, we need to plot points with large enough x-values, both positive and negative, to show how the graph flattens out. When a graph has both horizontal and vertical asymptotes, we will also plot a couple of mid-sized x-values.

EXAMPLES

Sketch the graph of the rational function.

- $f(x) = \dfrac{2x + 1}{x - 4}$

The x-intercept is $-\frac{1}{2}$, the y-intercept is $-\frac{1}{4}$. The vertical asymptote is $x = 4$, and the horizontal asymptote is $y = 2$. We will use dashed lines for the asymptotes and plot the points for $x = 3$, $x = 5$, $x = -10$, and $x = 10$ to show how the graph behaves near the asymptotes.

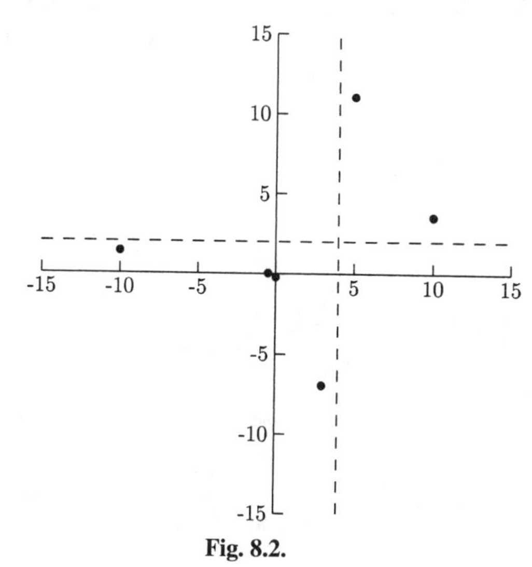

Fig. 8.2.

It is not obvious what the graph looks like so we will plot a point for $x = 7$. Then we will draw a smooth curve between the points.

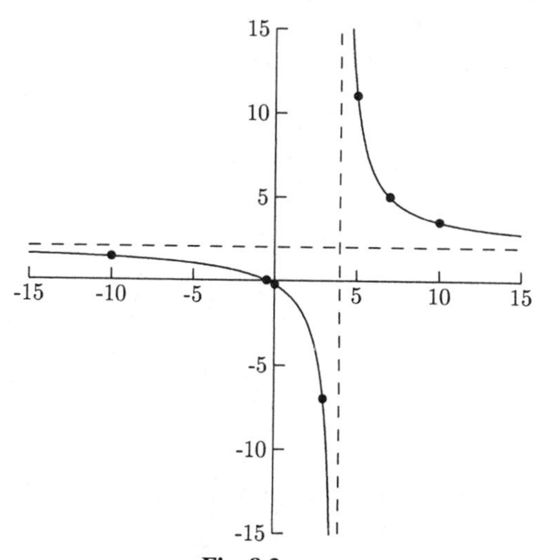

Fig. 8.3.

- $g(x) = \dfrac{1}{x^2+1}$

  There is no vertical asymptote because $x^2 + 1 = 0$ has no real solution. The x-axis is the horizontal asymptote. This graph has no x-intercept. The y-intercept is 1. We will use $x = 5, -5$ to show the graph's horizontal asymptotic behavior. The function is even, so the left half is a reflection of the right half. We will plot points for $x = 1, 2$. The y-values for $x = -1, -2$ will be the same.

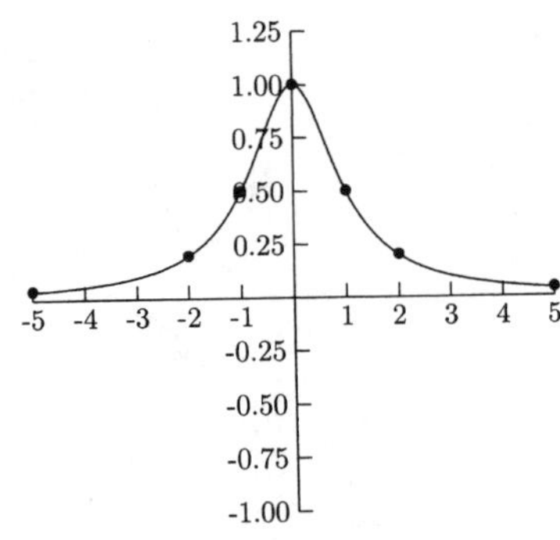

Fig. 8.4.

- $R(x) = \dfrac{x^2+1}{x^2-1}$

  The vertical asymptotes are $x = -1$ and $x = 1$. The horizontal asymptote is $y = 1$. There is no x-intercept, and the y-intercept is -1. We will use $x = 5, -5$ for the horizontal asymptote and $x = -0.9, 0.9, -1.1, 1.1$ for the vertical asymptotes. To get a better idea of what the graph looks like, we will need to plot other points. We will use $x = 2$ and $x = -2$.

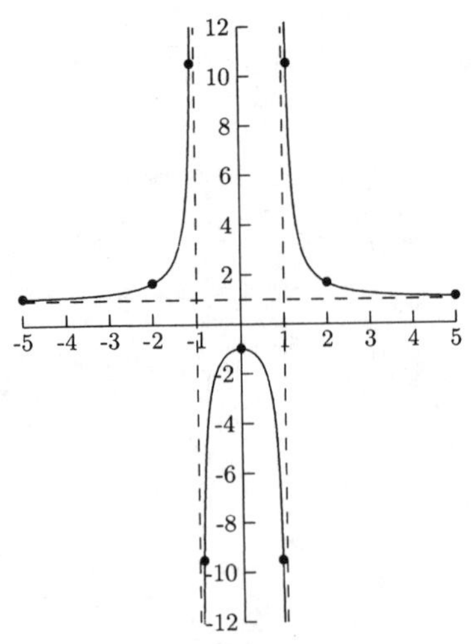

Fig. 8.5.

If the degree of the numerator is exactly one more than the degree of the denominator, then the graph has a *slant asymptote*. We can find the equation of a slant asymptote (a line whose slope is a nonzero number) by performing polynomial division. The equation for the slant asymptote is $y =$ quotient.

EXAMPLES

Find an equation for the slant asymptote.

- $f(x) = \dfrac{4x^2 + 3x - 5}{x + 2}$

  When we divide $4x^2 + 3x - 5$ by $x + 2$, we get a quotient of $4x - 5$. The slant asymptote is the line $y = 4x - 5$.

$$
\begin{array}{r|l}
 & \quad 4x \; - 5 \\
\hline
x + 2 & 4x^2 + 3x - 5 \\
 & -(4x^2 + 8x) \\
\hline
 & \quad -5x - 5 \\
 & \quad -(-5x - 10) \\
\hline
 & \qquad\qquad 5
\end{array}
$$

- $f(x) = \dfrac{x^3 + 2x^2 - 1}{x^2 + x + 2}$

$$
\begin{array}{r|l}
 & \quad x \; + 1 \\
\hline
x^2 + x + 2 & x^3 + 2x^2 + 0x - 1 \\
 & -(x^3 + x^2 + 2x) \\
\hline
 & \qquad x^2 - 2x - 1 \\
 & \quad -(x^2 + x + 2) \\
\hline
 & \qquad -3x - 3
\end{array}
$$

The slant asymptote is $y = x + 1$.

When sketching the graph of a rational function that has a slant asymptote, we can show the behavior of the graph near the slant asymptote by plotting points for larger x-values. We can tell if an x-value is large enough by checking its y-values in both the line and rational function. If they are fairly close, then the x-value is large enough.

EXAMPLES

Sketch the graph of rational function.

- $f(x) = \dfrac{x^2 + x - 6}{x + 2}$

The x-intercepts are -3 and 2. The y-intercept is -3. The vertical asymptote is $x = -2$.

$$\begin{array}{r|rrr} -2 & 1 & 1 & -6 \\ & & -2 & 2 \\ \hline & 1 & -1 & -4 \end{array}$$

The quotient is $x - 1$, so the slant asymptote is $y = x - 1$. We will use $x = 10$ and $x = -10$ to show the graph's behavior near the slant asymptote. We will also plot points for $x = -1$ and $x = -2.5$ for the vertical asymptote.

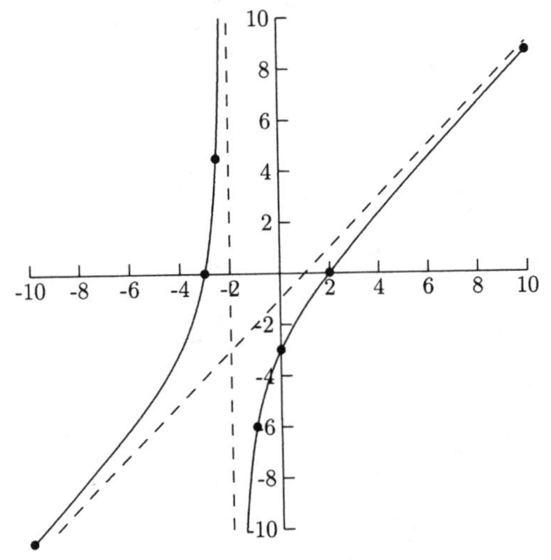

Fig. 8.6.

- $h(x) = \dfrac{x^3}{x^2 - 1}$

  The x-intercept is 0, the y-intercept is 0, too. The vertical asymptotes are $x = -1$ and $x = 1$.

$$\begin{array}{r|l} & \quad x \\ \hline x^2 + 0x - 1 & x^3 + \quad 0x^2 + 0x + 0 \\ & -(x^3 + 0x^2 \; - x) \\ \hline & \qquad\qquad x \end{array}$$

The quotient is x, so the slant asymptote is $y = x$. We will plot points for $x = -5$ and $x = 5$ to show the graph's behavior near the slant asymptote, $x = -1.1$, 1.1, -0.9, 0.9 for the vertical asymptotes, and $x = -2$, 2 for in-between points.

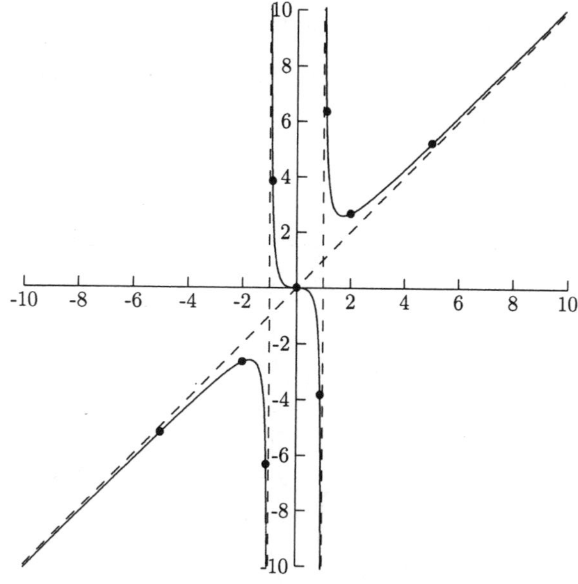

Fig. 8.7.

PRACTICE

Find the asymptotes and intercepts and sketch the graph.

1. $f(x) = \dfrac{1}{x+2}$

2. $g(x) = \dfrac{x}{x^2 - 1}$

3. $h(x) = \dfrac{2x-4}{x+2}$

4. Hint: Rewrite as one fraction.

$$f(x) = \frac{1}{x} + \frac{1}{x-2}$$

5. $f(x) = \dfrac{x^2 + x - 12}{x - 2}$

SOLUTIONS

1. The asymptotes are $x = -2$ and $y = 0$ (the x-axis). There is no x-intercept. The y-intercept is $\frac{1}{2}$.

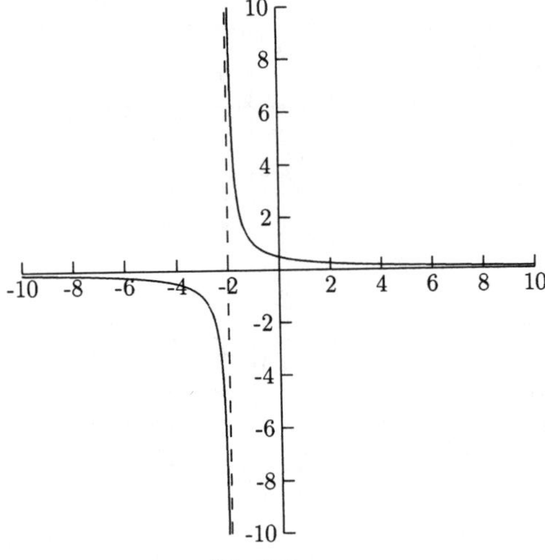

Fig. 8.8.

2. The asymptotes are $x = -1$, $x = 1$, and $y = 0$. The x-intercept and y-intercept is 0.

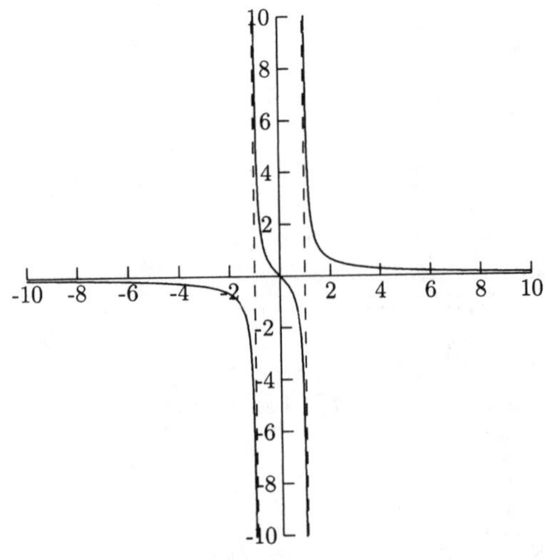

Fig. 8.9.

3. The asymptotes are $x = -2$ and $y = 2$. The x-intercept is 2, and the y-intercept is -2.

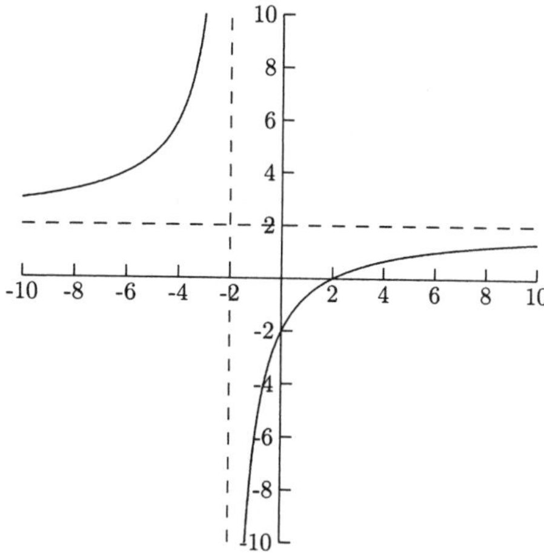

Fig. 8.10.

4. $$f(x) = \frac{1}{x} + \frac{1}{x-2} = \frac{1}{x} \cdot \frac{x-2}{x-2} + \frac{1}{x-2} \cdot \frac{x}{x} = \frac{x-2+x}{x(x-2)}$$

$$= \frac{2x-2}{x(x-2)} = \frac{2x-2}{x^2-2x}$$

The asymptotes are $x = 0$, $x = 2$, and $y = 0$. The x-intercept is 1, and there is no y-intercept.

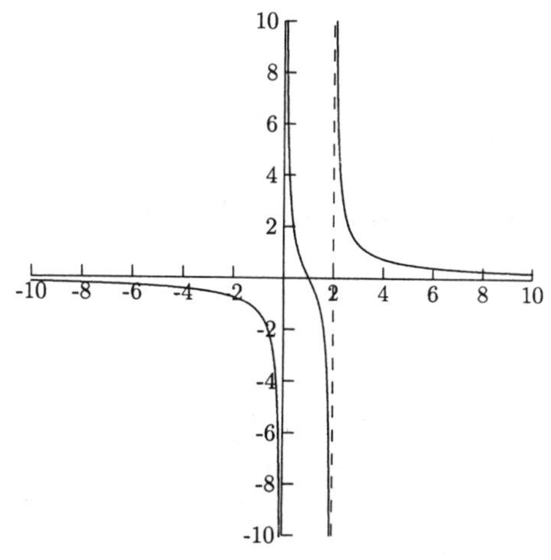

Fig. 8.11.

5. The vertical asymptote is $x = 2$. The x-intercepts are -4 and 3. The y-intercept is 6. We can use synthetic division to perform polynomial division.

$$\begin{array}{r|rrr} 2 & 1 & 1 & -12 \\ & & 2 & 6 \\ \hline & 1 & 3 & -6 \end{array}$$

The quotient is $x + 3$, so the slant asymptote is $y = x + 3$.

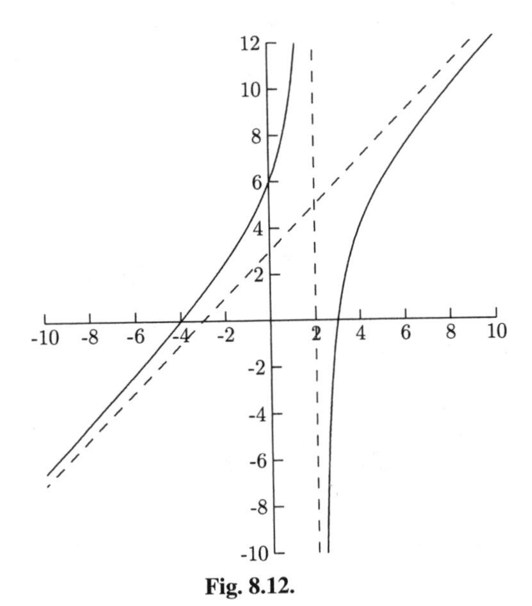

Fig. 8.12.

CHAPTER 8 REVIEW

1. What is the horizontal asymptote for the graph of

$$f(x) = \frac{2x^4 + 6x - 7}{5x^3 - 8x + 2}?$$

(a) $y = 0$ (b) $y = \frac{2}{5}$ (c) There is no horizontal asymptote.

(d) Cannot be determined without the graph.

2. What is the horizontal asymptote for the graph of

$$f(x) = \frac{2x^3 + 6x - 7}{5x^3 - 8x + 2}?$$

(a) $y = 0$ (b) $y = \frac{2}{5}$ (c) There is no horizontal asymptote.
(d) Cannot be determined without the graph.

3. What is the horizontal asymptote for the graph of

$$f(x) = \frac{2x^2 + 6x - 7}{5x^3 - 8x + 2}?$$

(a) $y = 0$ (b) $y = \frac{2}{5}$ (c) There is no horizontal asymptote.
(d) Cannot be determined without the graph.

4. What is/are the vertical asymptote(s) for the graph of

$$f(x) = \frac{x - 3}{x^2 + x - 2} = \frac{x - 3}{(x + 2)(x - 1)}?$$

(a) $x = 3$ (b) $x = -2$ and $x = 1$
(c) $x = 3, x = -2$, and $x = 1$ (d) There are no vertical asymptotes.

5. What are the intercepts for the graph of

$$f(x) = \frac{x^2 + 1}{x - 4}?$$

(a) There are no x-intercepts, and the y-intercept is $-\frac{1}{4}$
(b) The x-intercepts are ± 1, and the y-intercept is $-\frac{1}{4}$
(c) The x-intercepts are ± 1, and there is no y-intercept.
(d) There are no intercepts.

6. What is the slant asymptote for the graph of

$$f(x) = \frac{2x^2 + x - 1}{x + 2}?$$

(a) $y = 2x + 5$ (b) $y = 2x - 3$ (c) $y = 5$
(d) There is no slant asymptote.

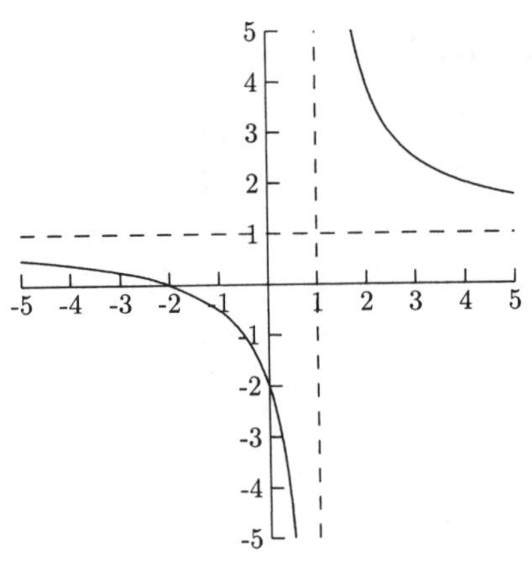

Fig. 8.13.

7. The graph in Figure 8.13 is the graph of which rational function?

   (a) $r(x) = \dfrac{x-2}{x+1}$

   (b) $q(x) = \dfrac{x-2}{x-1}$

   (c) $f(x) = \dfrac{x+2}{x-1}$

   (d) $g(x) = \dfrac{x+2}{x+1}$

SOLUTIONS

1. C	2. B	3. A	4. B	5. A
6. B	7. C			

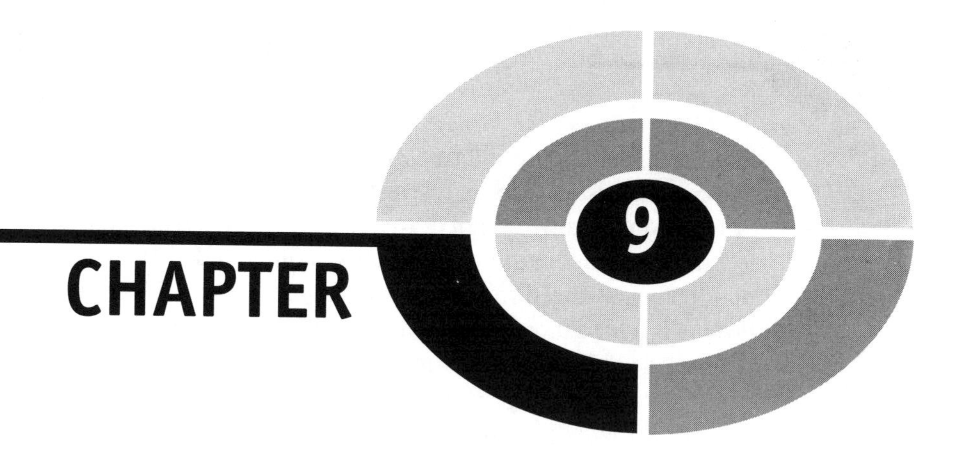

Exponents and Logarithms

Compound Growth

A quantity (such as a population, amount of money, or radiation level) changes *exponentially* if the growth or loss is a fixed percentage over a period of time. To see how this works, we will see how the value of an account grows over four years if \$100 is deposited and earns 5% interest, compounded annually. *Compounded annually* means that the interest earned in the previous year earns interest.

After one year, \$100 has grown to $100 + 0.05(100) = 100 + 5 = \$105$. In the second year, the original \$100 earns 5% plus the \$5 earns 5% interest: $105 + (105)(0.05) = \$110.25$. Now this amount earns interest in the third year: $110.25 + (110.25)(0.05) = \$115.76$. Finally, this amount earns interest in the fourth year: $115.76 + (115.76)(0.05) = \$121.55$. If interest is not compounded, that is, the interest does not earn interest, the account would only be worth \$120. The extra \$1.55 is interest earned on interest.

Compound growth is not dramatic over the short run but it is over time. If \$100 is left in an account earning 5% interest, compounded annually, for 20 years instead of four years, the difference between the compound growth and noncompound growth is a little more interesting. After 20 years, the compound amount is \$265.33 compared to \$200 for simple interest (noncompound growth). A graph of the growth of each type over 40 years is given in Figure 9.1. The line is the growth for simple (noncompounded) interest, and the curve is the growth with compound interest.

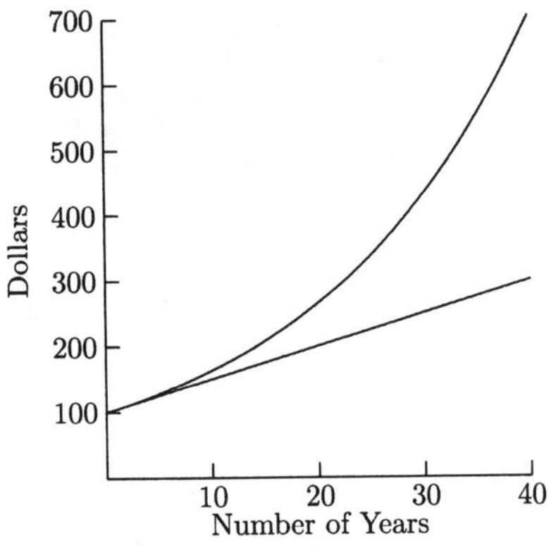

Fig. 9.1.

We can use a formula to compute the value of an account earning compounded interest. If P dollars is invested for t years, earning r interest rate, then it will grow to A dollars, where $A = P(1 + r)^t$.

EXAMPLES

Find the compound amount.

- \$5000, after three years, earning 6% interest, compounded annually
  We will use the formula $A = P(1 + r)^t$. $P = 5000$, $r = 0.06$, and $t = 3$. We want to know A, the compound amount.

$$A = 5000(1 + 0.06)^3 = 5000(1.06)^3 = 5000(1.191016)$$
$$= 5955.08$$

The compound amount is \$5955.08.

- \$10,000 after eight years, $7\frac{1}{4}\%$ interest, compounded annually

$$A = 10{,}000(1 + 0.0725)^8 = 10{,}000(1.0725)^8 \approx 10{,}000(1.7505656)$$
$$\approx 17{,}505.66$$

The compound amount is \$17,505.66

Many investments pay more often than once a year, some paying interest daily. Instead of using the annual interest rate, we need to use the interest rate per period, and instead of using the number of years, we need to use the number of periods. If there are n compounding periods per year, then the interest rate per period is $\frac{r}{n}$ and the total number of periods is nt. The compound amount formula becomes

$$A = P\left(1 + \frac{r}{n}\right)^{nt}.$$

EXAMPLES

Find the compound amount.

- \$5000, after three years, earning 6% annual interest
  (a) compounded semiannually
  (b) compounded monthly

  For (a), interest compounded semiannually means that it is compounded twice each year, so $n = 2$.

$$A = 5000\left(1 + \frac{0.06}{2}\right)^{2(3)} = 5000(1.03)^6 \approx 5000(1.194052) \approx 5970.26$$

  The compound amount is \$5970.26.
  For (b), interest compounded monthly means that it is compounded 12 times each year, so $n = 12$.

$$A = 5000\left(1 + \frac{0.06}{12}\right)^{12(3)} = 5000(1.005)^{36} \approx 5000(1.19668) \approx 5983.40$$

  The compound amount is \$5983.40.
- \$10,000, after eight years, earning $7\frac{1}{4}\%$ annual interest, compounded weekly
  Interest that is paid weekly is paid 52 times each year, so $n = 52$.

$$A = 10{,}000\left(1 + \frac{0.0725}{52}\right)^{52(8)} \approx 10{,}000(1.001394231)^{416}$$
$$\approx 10{,}000(1.785317) \approx 17{,}853.17$$

  The compound amount is \$17,853.17.

The more often interest is compounded per year, the more interest is earned. \$1000 earning 8% annual interest, compounded annually, is worth \$1080 after one year. If interest is compounded quarterly, it is worth \$1082.43 after one year. And if interest is compounded daily, it is worth \$1083.28 after one year. What if interest is compounded each hour? Each second? It turns out that the most this investment could be worth (at 8% interest) is \$1083.29, when interest is compounded each and every instant of time. Each instant of time, a tiny amount of interest is earned. This is called *continuous* compounding. The formula for the compound amount for interest compounded continuously is $A = Pe^{rt}$, where A, P, r, and t are the same quantities as before. The letter e stands for a constant called Euler's number. It is approximately 2.718281828. You probably have an e or e^x key on your calculator. Although e is irrational, it can be approximated by numbers of the form

$$\left(1 + \frac{1}{m}\right)^m,$$

where m is a large rational number. The larger m is, the better the approximation for e. If we make the substitution $m = \frac{n}{r}$ and use some algebra, we can see how $(1 + \frac{r}{n})^{nt}$ is very close to e^{rt}, for large values of n. If interest is compounded every minute, n would be 525,600, a rather large number!

EXAMPLE

- Find the compound amount of \$5000 after eight years, earning 12% annual interest, compounded continuously.

$$A = 5000e^{0.12(8)} = 5000e^{0.96} \approx 5000(2.611696) \approx 13{,}058.48$$

The compound amount is \$13,058.48.

The compound growth formula for continuously compounded interest is used for other growth and decay problems. The general exponential growth model is $n(t) = n_0e^{rt}$, where $n(t)$ replaces A and n_0 replaces P. Their meanings are the same—$n(t)$ is still the compound growth, and n_0 is still the beginning amount. The variable t represents time in this formula; although, time will not always be measured in years. The growth rate and t need to have the same unit of measure. If the growth rate is in days, then t needs to be in days. If the growth rate is in hours, then t needs to be in hours, and so on. If the "population" is getting smaller, then the formula is $n(t) = n_0e^{-rt}$.

EXAMPLES

- The population of a city is estimated to be growing at the rate of 10% per year. In 2000, its population was 160,000. Estimate its population in the year 2005.
  The year 2000 corresponds to $t = 0$, so the year 2005 corresponds to $t = 5$; n_0, the population in year $t = 0$, is 160,000. The population is growing at the rate of 10% per year, so $r = 0.10$. The formula $n(t) = n_0e^{rt}$ becomes $n(t) = 160{,}000e^{0.10t}$. We want to find $n(t)$ for $t = 5$.

$$n(5) = 160{,}000e^{0.10(5)} \approx 263{,}795$$

  The city's population is expected to be 264,000 in the year 2005 (estimates and projections are normally rounded off).
- A county is losing population at the rate of 0.7% per year. If the population in 2001 is 1,000,000, what is it expected to be in the year 2008?
  $n_0 = 1{,}000{,}000$, $t = 0$ is the year 2001, $t = 7$ is the year 2008, and $r = 0.007$. Because the county is losing population, we will use the decay model: $n(t) = n_0e^{-rt}$. The model for this county's population is $n(t) = 1{,}000{,}000e^{-.007t}$. We want to find $n(t)$ for $t = 7$.

$$n(7) = 1{,}000{,}000e^{-.007(7)} \approx 952{,}181$$

  The population is expected to be 952,000 in the year 2008.
- In an experiment, a culture of bacteria grew at the rate of 35% per hour. If 1000 bacteria were present at 10:00, how many were present at 10:45?

$$n_0 = 1000, \; r = 0.35, \; t \text{ is the number of hours after 10:00}$$

  The growth model becomes $n(t) = 1000e^{0.35t}$. We want to find $n(t)$ for 45 minutes, or $t = 0.75$ hours.

$$n(0.75) = 1000e^{0.35(0.75)} = 1000e^{0.2625} \approx 1300$$

  At 10:45, there were approximately 1300 bacteria present in the culture.

Present Value

Suppose a couple wants to give their newborn grandson a gift of \$50,000 on his 20th birthday. They can earn $7\frac{1}{2}\%$ interest, compounded annually. How much should they deposit now so that it grows to \$50,000 in 20 years? To answer this question, we will use the formula $A = P(1+r)^t$, where we know that $A = 50{,}000$ but are looking for P.

$$50{,}000 = P(1+0.075)^{20}$$

$$= P(1.075)^{20}$$

$$\frac{50{,}000}{(1.075)^{20}} = P$$

The couple should deposit \$11,770.66 now so that the investment grows to \$50,000 in 20 years.

We say that \$11,770.66 is the *present value* of \$50,000 due in 20 years, earning $7\frac{1}{2}\%$ interest, compounded annually. The present value formula is $P = A(1+r)^{-t}$, for interest compounded annually, and $P = A(1+\frac{r}{n})^{-nt}$, for interest compounded n times per year.

EXAMPLE

- Find the present value of \$20,000 due in $8\frac{1}{2}$ years, earning 6% annual interest, compounded monthly.

$$P = 20{,}000\left(1+\frac{0.06}{12}\right)^{-12(8.5)} = 20{,}000(1.005)^{-102} \approx 12{,}025.18$$

The present value is \$12,025.18.

PRACTICE

For Problems 1–7 find the compound amount.

1. \$800, after ten years, $6\frac{1}{2}\%$ interest, compounded annually
2. \$1200 after six years, $9\frac{1}{2}\%$ interest, compounded annually
3. A 20-year-old college student opens a retirement account with \$2000. If her account pays $8\frac{1}{4}\%$ interest, compounded annually, how much will be in the account when she reaches age 65?

4. \$800, after ten years, earning $6\frac{1}{4}\%$ annual interest

   (a) compounded quarterly

   (b) compounded weekly

5. \$9000, after five years, earning $6\frac{3}{4}\%$ annual interest, compounded daily (assume 365 days per year).

6. \$800, after 10 years, earning $6\frac{1}{2}\%$ annual interest, compounded continuously.

7. \$9000, after 5 years, earning $6\frac{3}{4}\%$ annual interest, compounded continuously.

8. The population of a city in the year 2002 is 2,000,000 and is expected to grow 1.5% per year. Estimate the city's population for the year 2012.

9. A construction company estimates that a piece of equipment is worth \$150,000 when new. If it loses value continuously at the annual rate of 10%, what would its value be in 10 years?

10. Under certain conditions a culture of bacteria grow at the rate of about 200% per hour. If 8000 bacteria are present in a dish, how many will be in the dish after 30 minutes?

11. Find the present value of \$9000 due in five years, earning 7% annual interest, compounded annually.

12. Find the present value of \$50,000 due in 10 years, earning 4% annual interest, compounded quarterly.

13. Find the present value of \$125,000 due in $4\frac{1}{2}$ years, earning $6\frac{1}{2}\%$ annual interest, compounded weekly.

SOLUTIONS

1. $A = 800(1 + 0.065)^{10} = 800(1.065)^{10} \approx 800(1.877137) \approx 1501.71$

   The compound amount is \$1501.71.

2. $A = 1200(1 + 0.095)^{6} = 1200(1.095)^{6} \approx 1200(1.72379) \approx 2068.55$

   The compound amount is \$2068.55.

3. $A = 2000(1 + 0.0825)^{45} = 2000(1.0825)^{45} \approx 2000(35.420585) \approx 70{,}841.17$

   The account will be worth \$70,841.17.

4. (a) $n = 4$

$$A = 800\left(1 + \frac{0.0625}{4}\right)^{4(10)} = 800(1.015625)^{40} \approx 800(1.85924)$$

$$\approx 1487.39$$

The compound amount is \$1487.39.

(b) $n = 52$

$$A = 800\left(1 + \frac{0.0625}{52}\right)^{52(10)} = 800(1.00120192)^{520}$$

$$\approx 800(1.86754) \approx 1494.04$$

The compound amount is \$1494.04.

5. $n = 365$

$$A = 9000\left(1 + \frac{0.0675}{365}\right)^{365(5)} \approx 9000(1.000184932)^{1825}$$

$$\approx 9000(1.4013959) \approx 12{,}612.56$$

The compound amount is \$12,612.56.

6. $A = 800e^{0.065(10)} = 800e^{0.65} \approx 800(1.915540829) \approx 1532.43$

The compound amount is \$1532.43.

7. $A = 9000e^{0.0675(5)} = 9000e^{0.3375} \approx 9000(1.401439608) \approx 12{,}612.96$

The compound amount is \$12,612.96.

8. $n_0 = 2{,}000{,}000$, $r = 0.015$ The growth formula is $n(t) = 2{,}000{,}000e^{0.015t}$ and we want to find $n(t)$ when $t = 10$.

$$n(10) = 2{,}000{,}000e^{0.015(10)} \approx 2{,}323{,}668$$

The population in the year 2012 is expected to be about 2.3 million.

9. $n_0 = 150{,}000$, $r = 0.10$ We will use the decay formula because value is being lost. The formula is $n(t) = 150{,}000e^{-0.10t}$. We want to find $n(t)$ when $t = 10$.

$$n(10) = 150{,}000e^{-0.10(10)} \approx 55{,}181.92$$

The equipment will be worth about \$55,000 after 10 years.

10. $n_0 = 8000$, $r = 2$ The growth formula is $n(t) = 8000e^{2t}$. We want to find $n(t)$ when $t = 0.5$.

$$n(0.5) = 8000e^{2(0.5)} \approx 21{,}746$$

About 21,700 bacteria will be present after 30 minutes.

11. $P = 9000(1.07)^{-5} \approx 6416.88$

The present value is \$6416.88.

12. $P = 50{,}000\left(1 + \frac{0.04}{4}\right)^{-4(10)} = 50{,}000(1.01)^{-40} \approx 33{,}582.66$

The present value is \$33,582.66.

13. $P = 125{,}000\left(1 + \frac{0.065}{52}\right)^{-52(4.5)} = 125{,}000(1.00125)^{-234} \approx 93{,}316.45$

The present value is \$93,316.45.

Graphs of Exponential Functions

A basic exponential function is of the form $f(x) = a^x$, where a is any positive number except 1. The graph of $f(x) = a^x$ comes in two shapes depending whether $0 < a < 1$ (a is positive but smaller than 1) or $a > 1$. Figure 9.2 is the graph of $f(x) = (\frac{1}{2})^x$ and Figure 9.3 is the graph of $f(x) = 2^x$.

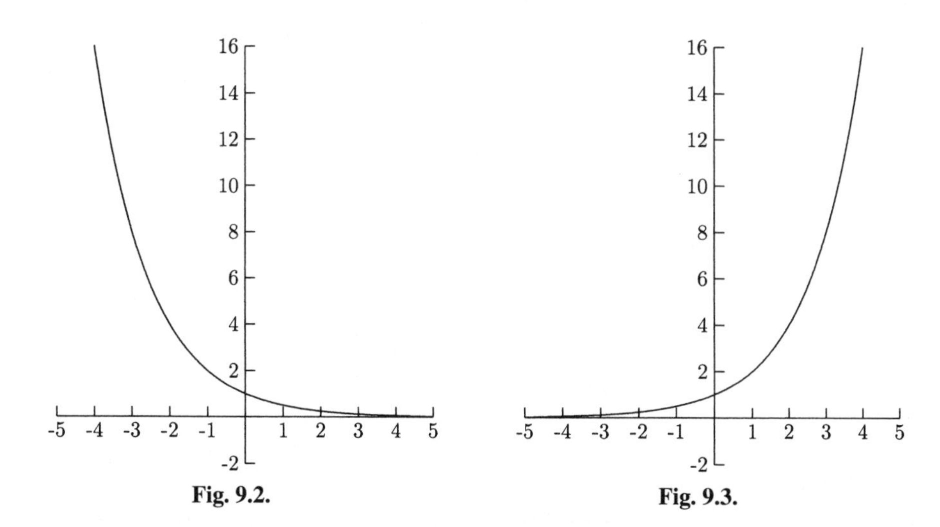

Fig. 9.2. Fig. 9.3.

Sketch the graph of $f(x) = a^x$ by plotting points for $x = -3$, $x = -2$, $x = -1$, $x = 0$, $x = 1$, $x = 2$, and $x = 3$. If a is too large or too small, points for $x = -3$

and $x = 3$ might be too awkward to graph because their y-values are too large or too close to 0. Before we begin sketching graphs, we will review the following exponent properties.

$$a^{-n} = \frac{1}{a^n} \qquad \left(\frac{1}{a}\right)^{-n} = a^n$$

EXAMPLES

Sketch the graphs.

- $f(x) = 2.5^x$
  We will begin with $x = -3, -2, -1, 0, 1, 2,$ and 3 in a table of values.

Table 9.1

x	$f(x)$
-3	$0.064\ (2.5^{-3} = \frac{1}{2.5^3})$
-2	$0.16\ (2.5^{-2} = \frac{1}{2.5^2})$
-1	$0.40\ (2.5^{-1} = \frac{1}{2.5})$
0	1
1	2.5
2	6.25
3	15.625

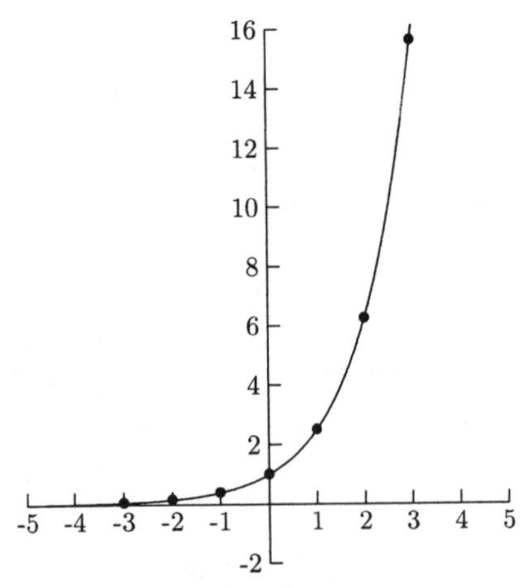

Fig. 9.4.

- $g(x) = (\frac{1}{3})^x$

Table 9.2

x	$f(x)$
-3	$27\ ((\frac{1}{3})^{-3} = 3^3)$
-2	$9\ ((\frac{1}{3})^{-2} = 3^2)$
-1	$3\ ((\frac{1}{3})^{-1} = 3^1)$
0	1
1	0.33
2	0.11
3	0.037

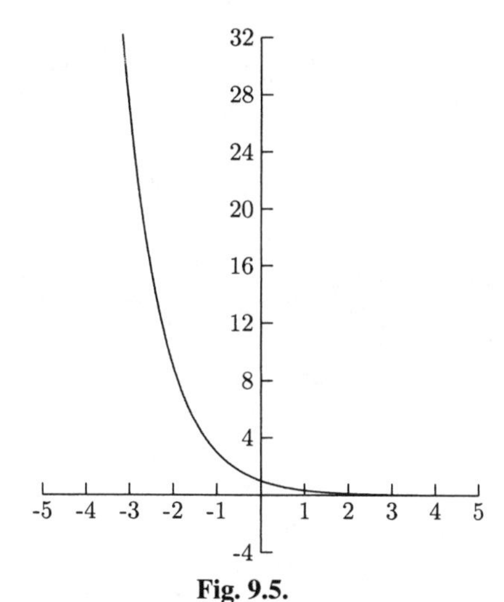

Fig. 9.5.

PRACTICE

Sketch the graphs.

1. $f(x) = (\frac{3}{2})^x$
2. $g(x) = (\frac{2}{3})^x$
3. $h(x) = e^x$ (Use the e or e^x key on your calculator.)

SOLUTIONS

1.

Table 9.3

x	$f(x)$
-3	$0.30\ ((\frac{3}{2})^{-3} = (\frac{2}{3})^3 = \frac{8}{27})$
-2	$0.44\ ((\frac{3}{2})^{-2} = (\frac{2}{3})^2 = \frac{4}{9})$
-1	$0.67\ ((\frac{3}{2})^{-1} = \frac{2}{3})$
0	1
1	1.5
2	2.25
3	3.375

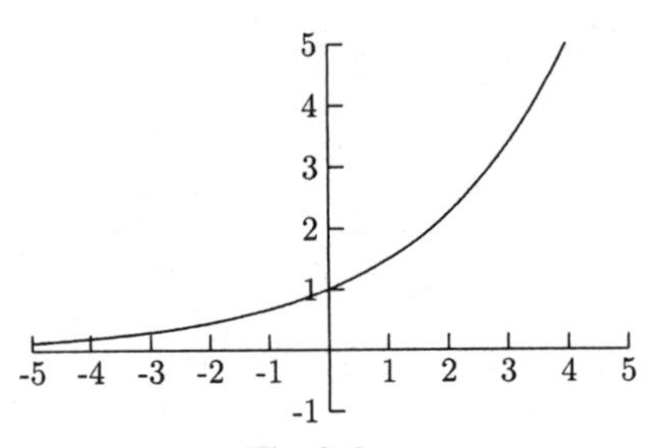

Fig. 9.6.

2.

Table 9.4

x	$f(x)$
-3	$3.375\ ((\frac{2}{3})^{-3} = (\frac{3}{2})^3)$
-2	$2.25\ ((\frac{2}{3})^{-2} = (\frac{3}{2})^2)$
-1	$1.5\ ((\frac{2}{3})^{-1} = \frac{3}{2})$
0	1
1	0.67
2	0.44
3	0.30

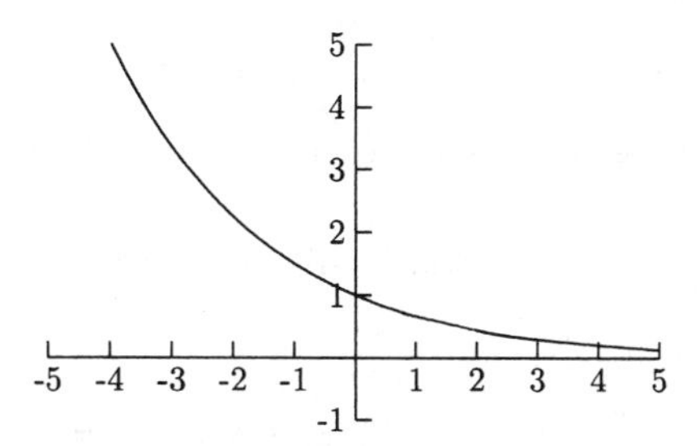

Fig. 9.7.

3.

Table 9.5

x	$f(x)$
-3	0.05
-2	0.14
-1	0.37
0	1
1	2.72
2	7.39
3	20.09

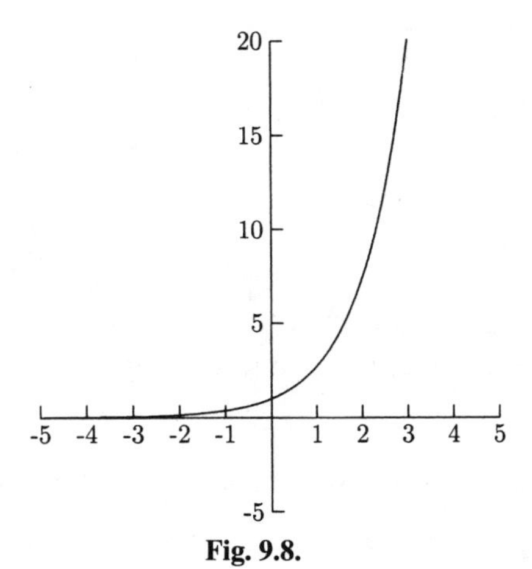

Fig. 9.8.

Transformations of the graphs of exponential functions behave in the same way as transformations of other functions.

EXAMPLES

- The graph of $f(x) = -2^x$ is the graph of $y = 2^x$ reflected about the x-axis (flipped upside down).
- The graph of $g(x) = 2^{-x}$ is the graph of $y = 2^x$ reflected about the y-axis (flipped sideways).
- The graph of $h(x) = 2^{x+1}$ is the graph of $y = 2^x$ shifted to the left 1 unit.
- The graph of $f(x) = -3 + 2^x$ is the graph of $y = 2^x$ shifted down 3 units.

Logarithms

A common question for investors is, "How long will it take for my investment to double?" If \$1000 is invested so that it earns 8% interest, compounded annually, how long will it take to grow to \$2000? To answer the question using the compound growth formula, we need to solve for t in the equation $2000 = 1000(1.08)^t$. We will divide both sides of the equation by 1000 to get $2 = (1.08)^t$. Now what? It does not make sense to "take the tth root" of both sides. We need to use logarithms. In mathematical terms, the logarithm and exponent functions are inverses. Logarithms (or *logs*) are very useful in solving many science and business problems.

The logarithmic equation $\log_a x = y$ is another way of writing the exponential equation $a^y = x$. Verbally, we say, "log base a of x is (or equals) y." For "$\log_a x$, we say," (the) log base a of x.

EXAMPLES

Rewrite the logarithmic equation as an exponential equation.

- $\log_3 9 = 2$

  The base of the logarithm is the base of the exponent, so 3 will be raised to a power. The number that is equal to the log is the power, so the power on 3 is 2.

  $$\log_3 9 = 2 \text{ rewritten as an exponent is } 3^2 = 9$$

- $\log_2 \frac{1}{8} = -3$

  The base is 2 and the power is -3.

  $$2^{-3} = \frac{1}{8}$$

- $\log_9 3 = \frac{1}{2}$

  The base is 9 and the power is $\frac{1}{2}$.

$$9^{\frac{1}{2}} = 3$$

Now we will work in the other direction, rewriting exponential equations as logarithmic equations. The equation $4^3 = 64$ written as a logarithmic equation is $\log_4 64 = 3$.

EXAMPLES

- $3^4 = 81$

  The base of the logarithm is 3, and we are taking the log of 81. The equation rewritten as a logarithmic equation is $\log_3 81 = 4$

- $a^3 = 4$

  The base is a, and we are taking the log of 4. The equation rewritten as a logarithmic equation is $\log_a 4 = 3$.

- $8^{2/3} = 4$

  The base is 8, and we are taking the log of 4. The equation rewritten as a logarithmic equation is $\log_8 4 = \frac{2}{3}$.

PRACTICE

For Problems 1–5, rewrite the logarithmic equations as exponential equations. For Problems 6–12 rewrite the exponential equations as logarithmic equations.

1. $\log_4 16 = 2$
2. $\log_{100} 10 = \frac{1}{2}$
3. $\log_e 2 = 0.6931$
4. $\log_{(x+1)} 9 = 2$
5. $\log_7 \frac{1}{49} = -2$
6. $5^2 = 25$
7. $4^0 = 1$
8. $7^{-1} = \frac{1}{7}$

9. $125^{1/3} = 5$
10. $10^{-4} = 0.0001$
11. $e^{1/2} = 1.6487$
12. $8^x = 5$

SOLUTIONS

1. $\log_4 16 = 2$ rewritten as an exponential equation is $4^2 = 16$
2. $\log_{100} 10 = \frac{1}{2}$ rewritten as an exponential equation is $100^{\frac{1}{2}} = 10$
3. $\log_e 2 = 0.6931$ rewritten as an exponential equation is $e^{0.6931} = 2$
4. $\log_{(x+1)} 9 = 2$ rewritten as an exponential equation is $(x+1)^2 = 9$
5. $\log_7 \frac{1}{49} = -2$ rewritten as an exponential equation is $7^{-2} = \frac{1}{49}$
6. $5^2 = 25$ rewritten as a logarithmic equation is $\log_5 25 = 2$
7. $4^0 = 1$ rewritten as a logarithmic equation is $\log_4 1 = 0$
8. $7^{-1} = \frac{1}{7}$ rewritten as a logarithmic equation is $\log_7 \frac{1}{7} = -1$
9. $125^{1/3} = 5$ rewritten as a logarithmic equation is $\log_{125} 5 = \frac{1}{3}$
10. $10^{-4} = 0.0001$ rewritten as a logarithmic equation is $\log_{10} 0.0001 = -4$
11. $e^{1/2} = 1.6487$ rewritten as a logarithmic equation is $\log_e 1.6487 = \frac{1}{2}$
12. $8^x = 5$ rewritten as a logarithmic equation is $\log_8 5 = x$

The first two logarithm properties we will learn are the cancelation properties. They come directly from rewriting one form of an equation in the other form.

$$\log_a a^x = x \text{ and } a^{log_a x} = x$$

When the bases of the exponent and logarithm are the same, they cancel. Let us see why these properties are true. What would the expression $\log_a a^x$ be? We will rewrite the equation "$\log_a a^x = ?$" as an exponential equation: $a^? = a^x$. Now we can see that "?" is x. This is why $\log_a a^x = x$. What would $a^{\log_a x}$ be? We will rewrite "$a^{\log_a x} = ?$" as a logarithmic equation: $\log_a ? = \log_a x$, so "?" is x, and $a^{\log_a x} = x$.

EXAMPLES

- $5^{\log_5 2}$

  The bases of the logarithm and exponent are both 5, so $5^{\log_5 2}$ simplifies to 2.

$$10^{\log_{10} 8} = 8 \qquad 4^{\log_4 x} = x \qquad e^{\log_e 6} = 6$$
$$29^{\log_{29} 1} = 1 \quad \log_m m^r = r \quad \log_7 7^{ab} = ab$$

  Sometimes we need to use exponent properties before using the property $\log_a a^x = x$.

$$\sqrt[n]{a^m} = a^{\frac{m}{n}} \text{ and } \frac{1}{a^m} = a^{-m}$$

EXAMPLES

- $\log_9 3 = \log_9 \sqrt{9} = \log_9 9^{1/2} = \dfrac{1}{2}$
- $\log_7 \dfrac{1}{49} = \log_7 \dfrac{1}{7^2} = \log_7 7^{-2} = -2$
- $\log_{10} \sqrt[4]{10} = \log_{10} 10^{1/4} = \dfrac{1}{4}$
- $\log_{10} \sqrt[5]{100} = \log_{10} \sqrt[5]{10^2} = \log_{10} 10^{2/5} = \dfrac{2}{5}$

Two types of logarithms occur frequently enough to have their own notation. They are $\log_e$ and $\log_{10}$. The notation for $\log_e$ is "ln" (pronounced "ell-in") and is called the *natural log*. The notation for $\log_{10}$ is "log" (no base is written) and is called the *common log*. The cancelation properties for these special logarithms are

$$\ln e^x = x \qquad e^{\ln x} = x \qquad \text{and} \qquad \log 10^x = x \qquad 10^{\log x} = x.$$

EXAMPLES

- $e^4 = x - 1$ rewritten as a log equation is $\ln(x - 1) = 4$
- $10^x = 6$ rewritten as a log equation is $\log 6 = x$
- $\ln 2x = 25$ rewritten as an exponent equation is $e^{25} = 2x$
- $\log(2x - 9) = 4$ rewritten as an exponent equation is $10^4 = 2x - 9$

  - $\ln e^{15} = 15$
  - $e^{\ln 14} = 14$
  - $\ln e^{-4} = -4$
  - $10^{\log 5} = 5$
  - $\log 10^{1/2} = \frac{1}{2}$
  - $\log 10^{-4} = -4$

PRACTICE

1. Rewrite as a logarithm: $e^{3x} = 4$
2. Rewrite as a logarithm: $10^{x-1} = 15$
3. Rewrite as an exponent: $\ln 6 = x + 1$
4. Rewrite as an exponent: $\log 5x = 3$

   Use logarithm properties to simplify the expression.

5. $9^{\log_9 3}$
6. $10^{\log_{10} 14}$
7. $5^{\log_5 x}$
8. $\log_{15} 15^2$
9. $\log_{10} 10^{-8}$
10. $\log_e e^x$
11. $\log_7 \sqrt{7}$
12. $\log_5 \frac{1}{5}$
13. $\log_3 \frac{1}{\sqrt{3}}$
14. $\log_4 \frac{1}{16}$
15. $\log_{25} \frac{1}{5}$
16. $\log_8 \frac{1}{2}$
17. $\log_{10} \sqrt{1000}$
18. $\ln e^5$
19. $\log 10^{\sqrt{x}}$
20. $10^{\log 9}$
21. $e^{\ln 6}$
22. $\log 10^{3x-1}$
23. $\ln e^{x+1}$

SOLUTIONS

1. $\ln 4 = 3x$
2. $\log 15 = x - 1$

3. $e^{x+1} = 6$
4. $10^3 = 5x$
5. $9^{\log_9 3} = 3$
6. $10^{\log_{10} 14} = 14$
7. $5^{\log_5 x} = x$
8. $\log_{15} 15^2 = 2$
9. $\log_{10} 10^{-8} = -8$
10. $\log_e e^x = x$
11. $\log_7 \sqrt{7} = \log_7 7^{1/2} = \frac{1}{2}$
12. $\log_5 \frac{1}{5} = \log_5 5^{-1} = -1$
13. $\log_3 \frac{1}{\sqrt{3}} = \log_3 \frac{1}{3^{1/2}} = \log_3 3^{-1/2} = -\frac{1}{2}$
14. $\log_4 \frac{1}{16} = \log_4 \frac{1}{4^2} = \log_4 4^{-2} = -2$
15. $\log_{25} \frac{1}{5} = \log_{25} \frac{1}{\sqrt{25}} = \log_{25} \frac{1}{25^{\frac{1}{2}}} = \log_{25} 25^{-1/2} = -\frac{1}{2}$
16. $2 = \sqrt[3]{8}$

$$\log_8 \frac{1}{2} = \log_8 \frac{1}{\sqrt[3]{8}} = \log_8 \frac{1}{8^{\frac{1}{3}}} = \log_8 8^{-1/3} = -\frac{1}{3}$$

17. $1000 = 10^3$, so $\log_{10} \sqrt{1000} = \log_{10} \sqrt{10^3} = \log_{10} 10^{3/2} = 3/2$
18. $\ln e^5 = 5$
19. $\log 10^{\sqrt{x}} = \sqrt{x}$
20. $10^{\log 9} = 9$
21. $e^{\ln 6} = 6$
22. $\log 10^{3x-1} = 3x - 1$
23. $\ln e^{x+1} = x + 1$

Exponent and Logarithm Equations (Part I)

Equations with exponents and logarithms come in many forms. Sometimes more than one strategy will work to solve them. We will first solve equations of the form "log = number" and "log = log." We will solve an equation of the form "log = number" by rewriting the equation as an exponential equation.

EXAMPLES

Solve the equation for x.

- $\log_3(x+1) = 4$

  Rewrite the equation as an exponential equation.

$$\begin{aligned} \log_3(x+1) &= 4 \\ 3^4 &= x+1 \\ 81 &= x+1 \\ 80 &= x \end{aligned}$$

- $\log_2(3x-4) = 5$

$$\begin{aligned} 2^5 &= 3x-4 \\ 32 &= 3x-4 \\ 12 &= x \end{aligned}$$

The logarithms cancel for equations in the form "log = log" as long as the bases are the same. For example, the solution to the equation $\log_8 x = \log_8 10$ is $x = 10$. The cancelation law $a^{\log_a x} = x$ makes this work.

$$\begin{aligned} \log_8 x &= \log_8 10 \\ 8^{\log_8 x} &= 8^{\log_8 10} \\ x &= 10 \quad \text{(By the cancelation law)} \end{aligned}$$

EXAMPLES

Solve for x.

- $\log_6(x + 1) = \log_6 2x$

$$\log_6(x + 1) = \log_6 2x$$
$$x + 1 = 2x \qquad \text{The logs cancel.}$$
$$1 = x$$

- $\log 4 = \log(x - 1)$

$$\log 4 = \log(x - 1)$$
$$4 = x - 1 \qquad \text{The logs cancel.}$$
$$5 = x$$

PRACTICE

Solve for x.

1. $\log_7(2x + 1) = 2$
2. $\log_4(x + 6) = 2$
3. $\log 5x = 1$
4. $\log_2(8x - 1) = 4$
5. $\log_3(4x - 1) = \log_3 2$
6. $\log_2(3 - x) = \log_2 17$
7. $\ln 15x = \ln(x + 4)$
8. $\log \frac{x}{x-1} = \log \frac{1}{2}$

SOLUTIONS

1. $\log_7(2x + 1) = 2$

$$7^2 = 2x + 1$$
$$24 = x$$

2. $\log_4(x+6) = 2$

$$4^2 = x + 6$$
$$10 = x$$

3. $\log 5x = 1$

$$10^1 = 5x$$
$$2 = x$$

4. $\log_2(8x - 1) = 4$

$$2^4 = 8x - 1$$
$$\frac{17}{8} = x$$

5. $\log_3(4x - 1) = \log_3 2$

$$4x - 1 = 2$$
$$x = \frac{3}{4}$$

6. $\log_2(3 - x) = \log_2 17$

$$3 - x = 17$$
$$x = -14$$

7. $\ln 15x = \ln(x + 4)$

$$15x = x + 4$$
$$x = \frac{4}{14} = \frac{2}{7}$$

8. $\log \frac{x}{x-1} = \log \frac{1}{2}$

$$\frac{x}{x-1} = \frac{1}{2} \quad \text{Cross-multiply.}$$
$$2x = x - 1$$
$$x = -1$$

We need to use calculators to find approximate solutions for exponential equations whose base is e or 10. We will rewrite the exponential equation as a logarithmic equation, solve for x, and then use a calculator to get an approximate solution.

EXAMPLES

Solve for x. Give solutions accurate to four decimal places.

- $e^{2x} = 3$

$$e^{2x} = 3 \quad \text{Rewrite as a logarithmic equation.}$$
$$2x = \ln 3$$
$$x = \frac{\ln 3}{2}$$
$$x \approx \frac{1.0986}{2} \approx 0.5493$$

- $10^{x+1} = 9$

$$10^{x+1} = 9 \quad \text{Rewrite as a logarithmic equation.}$$
$$x + 1 = \log 9$$
$$x = -1 + \log 9$$
$$x \approx -1 + 0.9542 \approx -0.0458$$

- $2500 = 1000e^{x-4}$

$$2500 = 1000e^{x-4} \quad \text{Divide both sides by 1000 before rewriting the equation.}$$
$$e^{x-4} = 2.5 \quad \text{Rewrite as a logarithmic equation.}$$
$$x - 4 = \ln 2.5$$
$$x = 4 + \ln 2.5 \approx 4 + 0.9163 \approx 4.9163$$

PRACTICE

Solve for x. Give your solutions accurate to four decimal places.

1. $10^{3x} = 7$
2. $e^{2x+5} = 15$

3. $5000 = 2500e^{4x}$
4. $32 = 8 \cdot 10^{6x-4}$
5. $200 = 400e^{-0.06x}$

SOLUTIONS

1. $10^{3x} = 7$

$$3x = \log 7$$
$$x = \frac{\log 7}{3} \approx \frac{0.8451}{3} \approx 0.2817$$

2. $e^{2x+5} = 15$

$$2x + 5 = \ln 15$$
$$2x = -5 + \ln 15$$
$$x = \frac{-5 + \ln 15}{2} \approx \frac{-5 + 2.7081}{2} \approx -1.1460$$

3. $5000 = 2500e^{4x}$

$$\frac{5000}{2500} = e^{4x}$$
$$4x = \ln\left(\frac{5000}{2500}\right)$$
$$4x = \ln 2$$
$$x = \frac{\ln 2}{4} \approx \frac{0.6931}{4} \approx 0.1733$$

4. $32 = 8 \cdot 10^{6x-4}$ Divide both sides by 8.

$$4 = 10^{6x-4}$$
$$6x - 4 = \log 4$$
$$6x = 4 + \log 4$$
$$x = \frac{4 + \log 4}{6} \approx \frac{4 + 0.6021}{6} \approx 0.767$$

5. $200 = 400e^{-0.06x}$

$$\frac{1}{2} = e^{-0.06x}$$

$$-0.06x = \ln\left(\frac{1}{2}\right)$$

$$x = \frac{\ln(\frac{1}{2})}{-0.06} \approx \frac{-0.69315}{-0.06} \approx 11.5525$$

The logarithm function $f(x) = \log_a x$ is the inverse of $g(x) = a^x$. The graph of $f(x)$ is the graph of $g(x)$ with the x- and y-values reversed. To sketch the graph by hand, we will rewrite the logarithm function as an exponent equation and graph the exponent equation.

EXAMPLES

Sketch the graph of the logarithmic functions.

- $y = \log_2 x$

  Rewrite the equation in exponential form, $x = 2^y$, and let the exponent, y, be the numbers $-3, -2, -1, 0, 1, 2$, and 3.

Table 9.6

x	y
$\frac{1}{8}$	-3
$\frac{1}{4}$	-2
$\frac{1}{2}$	-1
1	0
2	1
4	2
8	3

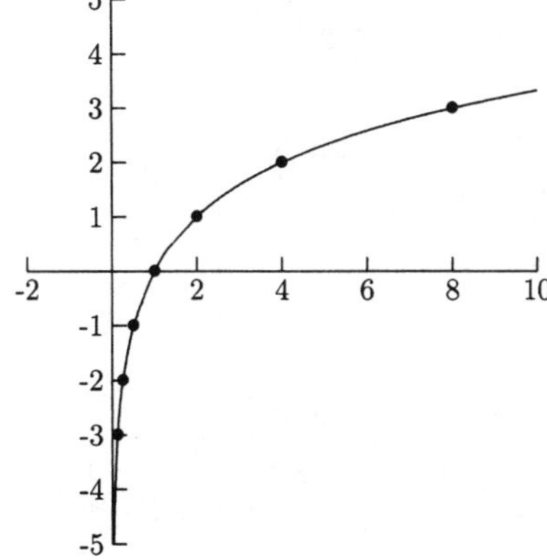

Fig. 9.9.

- $y = \ln x$

  Rewritten as an exponent equation, this is $x = e^y$. Let $y = -3, -2, -1, 0,$ 1, 2, and 3.

Table 9.7

x	y
0.05	−3
0.14	−2
0.37	−1
1	0
2.72	1
7.39	2
20.09	3

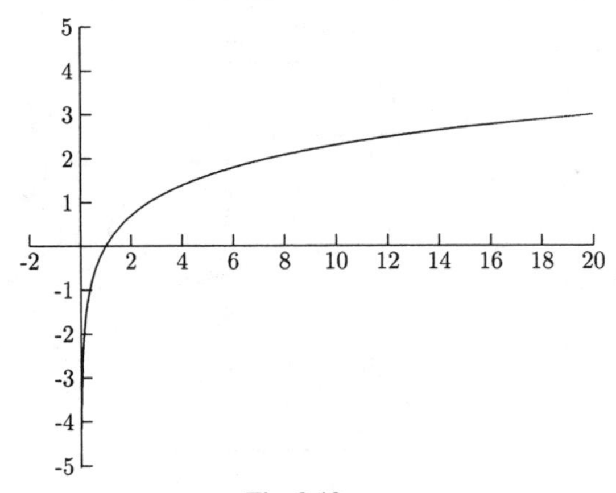

Fig. 9.10.

As you can see by these graphs, the domain of the function $f(x) = \log_a x$ is all positive real numbers, $(0, \infty)$.

PRACTICE

Sketch the graph of the logarithmic function.

1. $y = \log_{1.5} x$
2. $y = \log_3 x$

SOLUTIONS

1.

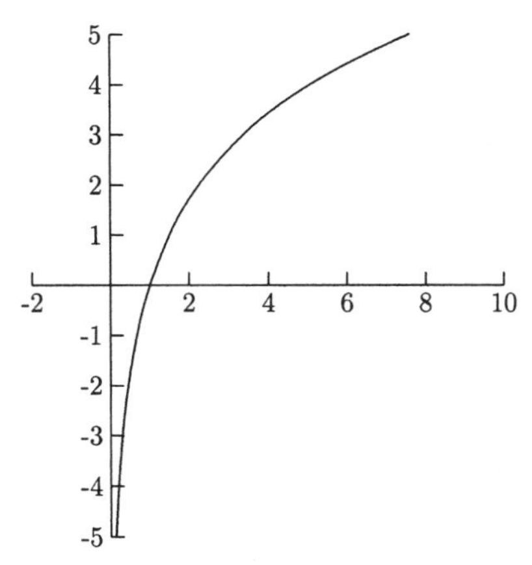

Fig. 9.11.

2.

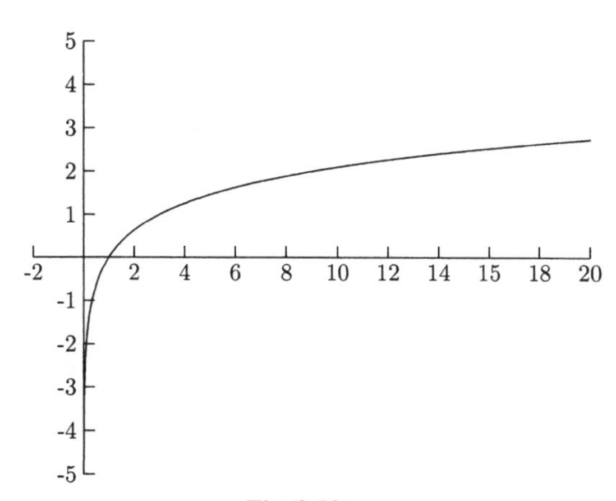

Fig. 9.12.

As long as a is larger than 1, all graphs for $f(x) = \log_a x$ look pretty much the same. The larger a is, the flatter the graph is to the right of $x = 1$. Knowing this and knowing how to graph transformations, we have a good idea of the graphs of many logarithmic functions.

- The graph of $f(x) = \log_2(x - 2)$ is the graph of $y = \log_2 x$ shifted to the right 2 units.
- The graph of $f(x) = -5 + \log_3 x$ is the graph of $y = \log_3 x$ shifted down 5 units.

- $f(x) = \frac{1}{3}\log x$ is the graph of $y = \log x$ flattened vertically by a factor of one-third.

The domain of $f(x) = \log_a x$ is all positive numbers. This means that we cannot take the log of 0 or the log of a negative number. The reason is that a is a positive number. A positive number raised to *any* power is always another positive number.

EXAMPLES

Find the domain. Give your answers in interval notation.

- $f(x) = \log_5(2 - x)$

  Because we are taking the log of $2 - x$, $2 - x$ needs to be positive.

$$2 - x > 0$$
$$-x > -2$$
$$x < 2$$

  The domain is $(-\infty, 2)$.

- $f(x) = \log(x^2 - x - 2)$

$$x^2 - x - 2 > 0$$
$$(x - 2)(x + 1) > 0$$

  Put $x = 2$ and $x = -1$ on the number line and test to see where $(x - 2)(x + 1) > 0$ is true.

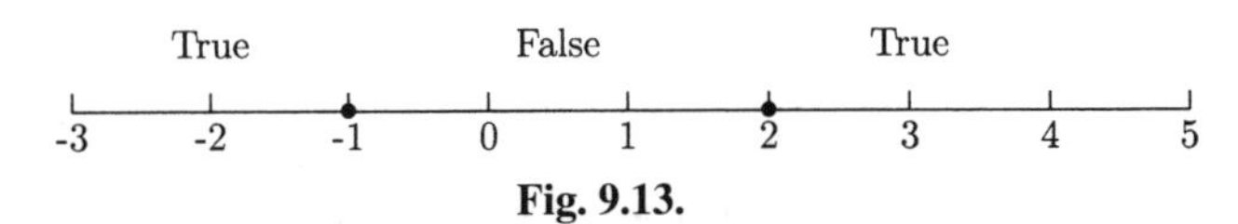

Fig. 9.13.

  We want the "True" intervals, so the domain is $(-\infty, -1) \cup (2, \infty)$.

- $g(x) = \ln(x^2 + 1)$

  Because $x^2 + 1$ is always positive, the domain is all real numbers, $(-\infty, \infty)$.

PRACTICE

Find the domain. Give your answers in interval notation.

1. $f(x) = \ln(10 - 2x)$

2. $h(x) = \log(x^2 - 4)$
3. $f(x) = \log(x^2 + 4)$

SOLUTIONS

1. Solve $10 - 2x > 0$. The domain is $x < 5$, $(-\infty, 5)$.
2. Solve $x^2 - 4 > 0$

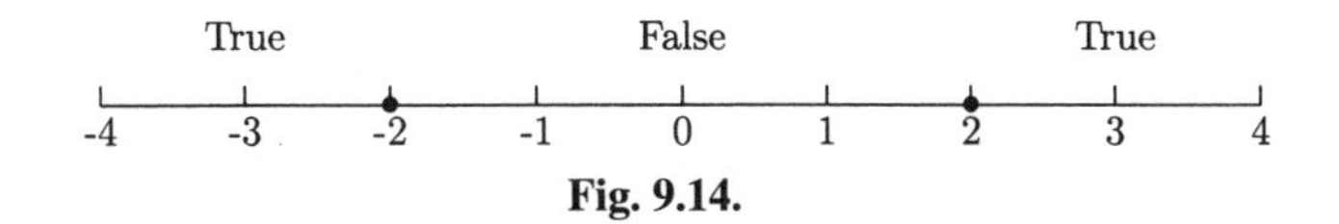

Fig. 9.14.

The domain is $(-\infty, -2) \cup (2, \infty)$.

3. Because $x^2 + 4 > 0$ is always positive, the domain is all real numbers, $(-\infty, \infty)$.

Exponent and Logarithmic Equations (Part II)

For some logarithmic equations, a solution might be extraneous solution. That is, such a solution is a solution to the rewritten equations but not to the original equations. Some solutions to the rewritten equations will cause logarithms of 0 or of negative numbers. We can check them in the original equation to see which solutions are true solutions.

EXAMPLES

Solve for x.

- $\log_2(x^2 + 3x - 10) = 3$

  We will rewrite this as an exponent equation: $2^3 = x^2 + 3x - 10$ and solve for x.

$$x^2 + 3x - 10 = 8$$

$$x^2 + 3x - 18 = 0$$

$$(x + 6)(x - 3) = 0$$

The solutions are $x = -6$ and $x = 3$. We will check them in the original equation.

$$\log_2((-6)^2 + 3(-6) - 10) = 3? \qquad \log_2(3^2 + 3(3) - 10) = 3?$$

$$\log_2 8 = 3 \text{ True} \qquad \log_2 8 = 3 \text{ True}$$

The solutions to the original equation are $x = -6$ and $x = 3$.

- $\log_5(x^2 + 5x - 4) = \log_5(x + 1)$

  The logs cancel leaving $x^2 + 5x - 4 = x + 1$.

$$x^2 + 5x - 4 = x + 1$$

$$x^2 + 4x - 5 = 0$$

$$(x + 5)(x - 1) = 0$$

The solutions are $x = -5$ and $x = 1$. We cannot allow $x = -5$ as a solution because $\log_5(-5 + 1)$ is not defined. We need to check $x = 1$.

$$\log_5(1^2 + 5(1) - 4) = \log_5(1 + 1) \quad \text{is true}$$

The solution is $x = 1$.

PRACTICE

Solve for x.

1. $\ln(x^2 + x - 20) = \ln(3x + 4)$
2. $\log_4(2x^2 - 3x + 59) = 3$

SOLUTIONS

1. $\ln(x^2 + x - 20) = \ln(3x + 4)$

$$x^2 + x - 20 = 3x + 4$$

$$x^2 - 2x - 24 = 0$$

$$(x - 6)(x + 4) = 0$$

The solutions are $x = 6$ and $x = -4$. Because $\ln[3(-4) + 4]$ is not defined, we only need to check $x = 6$.

$$\ln(6^2 + 6 - 20) = \ln[3(6) + 4] \quad \text{is true.}$$

The only solution is $x = 6$.

2. $\log_4(2x^2 - 3x + 59) = 3$

$$2x^2 - 3x + 59 = 4^3 \qquad (4^3 = 64)$$

$$2x^2 - 3x - 5 = 0$$

$$(2x - 5)(x + 1) = 0$$

We need to check the solutions $x = \frac{5}{2}$ and $x = -1$.

$$\log_4\left[2\left(\frac{5}{2}\right)^2 - 3\left(\frac{5}{2}\right) + 59\right] = 3? \qquad \log_4[2(-1)^2 - 3(-1) + 59] = 3?$$

$$\log_4 64 = 3 \text{ is true} \qquad \log_4 64 = 3 \text{ is true}$$

The solutions are $x = \frac{5}{2}$ and $x = -1$.

Three More Important Logarithm Properties

The following three logarithm properties come directly from the exponent properties $a^m \cdot a^n = a^{m+n}$, $\frac{a^m}{a^n} = a^{m-n}$, and $a^{mn} = (a^m)^n$.

1. $\log_b mn = \log_b m + \log_b n$
2. $\log_b \dfrac{m}{n} = \log_b m - \log_b n$
3. $\log_b m^t = t \log_b m$

We will see why Property 1 works. Let $x = \log_b m$ and $y = \log_b n$. Rewriting these equations as exponential equations, we get $b^x = m$ and $b^y = n$. Multiplying m and n, we have $mn = b^x \cdot b^y = b^{x+y}$. Rewriting the equation $mn = b^{x+y}$ as a logarithmic equation, we get $\log_b mn = x + y$. Because $x = \log_b m$ and $y = \log_b n$, $\log_b mn = x + y$ becomes $\log_b mn = \log_b m + \log_b n$.

EXAMPLES

Use Property 1 to rewrite the logarithms.

- $\log_4 7x = \log_4 7 + \log_4 x$
- $\ln 15t = \ln 15 + \ln t$
- $\log_6 19t^2 = \log_6 19 + \log_6 t^2$
- $\log 100y^4 = \log 10^2 + \log y^4 = 2 + \log y^4$

- $\log_9 3 + \log_9 27 = \log_9 3(27) = \log_9 81 = 2$
- $\ln x + \ln \sqrt{y} = \ln x\sqrt{y}$

Use Property 2 to rewrite the logarithms.

- $\log\left(\frac{x}{4}\right) = \log x - \log 4$
- $\ln\left(\frac{5}{x}\right) = \ln 5 - \ln x$
- $\log_{15} 3 - \log_{15} 2 = \log_{15}\left(\frac{3}{2}\right)$
- $\ln 16 - \ln t = \ln \frac{16}{t}$
- $\log_4\left(\frac{4}{3}\right) = \log_4 4 - \log_4 3 = 1 - \log 3$

The exponent property $\sqrt[n]{a^m} = a^{m/n}$ allows us to apply the third logarithm property to roots as well as to powers. The third logarithm property is especially useful in science and business applications.

EXAMPLES

Use Property 3 to rewrite the logarithms.

- $\log_4 3^x = x \log_4 3$
- $\frac{1}{3}\ln t = \ln t^{1/3}$
- $\log_6 \sqrt{2x} = \log_6 (2x)^{1/2} = \frac{1}{2}\log_6 2x$
- $\ln \sqrt[4]{t^3} = \ln t^{3/4} = \frac{3}{4}\ln t$
- $\log x^2 = 2\log x$
- $-3\log 8 = \log 8^{-3}$

PRACTICE

Use Property 1 to rewrite the logarithms in Problems 1–6.

1. $\ln 59t$
2. $\log 0.10y$
3. $\log_{30} 148x^2$

4. $\log_6 3 + \log_6 12$
5. $\log_5 9 + \log_5 10$
6. $\log 5 + \log 20$

Use Property 2 to rewrite the logarithms in Problems 7–12.

7. $\log_4 \frac{10}{9x}$
8. $\log_2 \frac{7}{8}$
9. $\ln \frac{t}{4}$
10. $\log \frac{100}{x^2}$
11. $\log_7 2 - \log_7 4$
12. $\log_8 x - \log_8 3$

Use Property 3 to rewrite the logarithms in Problems 13–20.

13. $\ln 5^x$
14. $\log_{12} \sqrt{3}$
15. $\log \sqrt{16x}$
16. $\log_5 6^{-t}$
17. $2 \log_8 3$
18. $(x + 6) \log_4 3$
19. $\log_{16} 10^{2x}$
20. $-2 \log_4 5$

SOLUTIONS

1. $\ln 59t = \ln 59 + \ln t$
2. $\log 0.10y = \log 0.10 + \log y = \log 10^{-1} + \log y = -1 + \log y$
3. $\log_{30} 148x^2 = \log_{30} 148 + \log_{30} x^2$
4. $\log_6 3 + \log_6 12 = \log_6(3 \cdot 12) = \log_6 36 = \log_6 6^2 = 2$
5. $\log_5 9 + \log_5 10 = \log_5(9 \cdot 10) = \log_5 90$
6. $\log 5 + \log 20 = \log(5 \cdot 20) = \log 100 = \log 10^2 = 2$
7. $\log_4 \dfrac{10}{9x} = \log_4 10 - \log_4 9x$

8. $\log_2 \frac{7}{8} = \log_2 7 - \log_2 8 = \log_2 7 - \log_2 2^3 = (\log_2 7) - 3$
9. $\ln \frac{t}{4} = \ln t - \ln 4$
10. $\log \frac{100}{x^2} = \log 100 - \log x^2 = \log 10^2 - \log x^2 = 2 - \log x^2$
11. $\log_7 2 - \log_7 4 = \log_7 \frac{2}{4} = \log_7 \frac{1}{2}$
12. $\log_8 x - \log_8 3 = \log_8 \frac{x}{3}$
13. $\ln 5^x = x \ln 5$
14. $\log_{12} \sqrt{3} = \log_{12} 3^{1/2} = \frac{1}{2} \log_{12} 3$
15. $\log \sqrt{16x} = \log(16x)^{1/2} = \frac{1}{2} \log 16x$
16. $\log_5 6^{-t} = -t \log_5 6$
17. $2 \log_8 3 = \log_8 3^2 = \log_8 9$
18. $(x + 6) \log_4 3 = \log_4 3^{x+6}$
19. $\log_{16} 10^{2x} = 2x \log_{16} 10$
20. $-2 \log_4 5 = \log_4 5^{-2} = \log_4 \frac{1}{5^2} = \log_4 \frac{1}{25}$

Sometimes we will need to use several logarithm properties to rewrite more complicated logarithms. The hardest part of this is to use the properties in the correct order. For example, which property should be used first on $\log \frac{x}{y^3}$? Do we first use the third property or the second property? We will use the second property first. For the expression $\log(\frac{x}{y})^3$, we would use the third property first.

Going in the other direction, we need to use all three properties in the expression $\log_2 9 - \log_2 x + 3 \log_2 y$. We need to use the second property to combine the first two terms.

$$\log_2 9 - \log_2 x + 3 \log_2 y = \log_2 \frac{9}{x} + 3 \log_2 y$$

We cannot use the first property on $\log_2 \frac{9}{x} + 3 \log_2 y$ until we have used the third property to move the 3.

$$\log_2 \frac{9}{x} + 3 \log_2 y = \log_2 \frac{9}{x} + \log_2 y^3 = \log_2 y^3 \frac{9}{x} = \log_2 \frac{9y^3}{x}$$

EXAMPLES

Rewrite as a single logarithm.

- $\log_2 3x - 4\log_2 y$

  We need use the third property to move the 4, then we can use the second property.

$$\log_2 3x - 4\log_2 y = \log_2 3x - \log_2 y^4 = \log_2 \frac{3x}{y^4}$$

- $3\log 4x + 2\log 3 - 2\log y$

$$\begin{aligned} 3\log 4x + 2\log 3 - 2\log y &= \log(4x)^3 + \log 3^2 - \log y^2 && \text{Property 3} \\ &= \log 4^3x^3 \cdot 3^2 - \log y^2 && \text{Property 1} \\ &= \log 576x^3 - \log y^2 = \log \frac{576x^3}{y^2} && \text{Property 2} \end{aligned}$$

- $t\ln 4 + \ln 5$

$$t\ln 4 + \ln 5 = \ln 4^t + \ln 5 = \ln(5 \cdot 4^t) \qquad (\text{not } \ln 20^t)$$

  Expand each logarithm.

- $\ln \dfrac{3\sqrt{x}}{y^2}$

$$\ln \frac{3\sqrt{x}}{y^2} = \ln 3(x^{1/2}) - \ln y^2 = \ln 3 + \ln x^{1/2} - \ln y^2 = \ln 3 + \frac{1}{2}\ln x - 2\ln y$$

- $\log_7 \dfrac{4}{10xy^2}$

$$\begin{aligned} \log_7 \frac{4}{10xy^2} &= \log_7 4 - \log_7 10xy^2 = \log_7 4 - (\log_7 10 + \log_7 x + \log_7 y^2) \\ &= \log_7 4 - (\log_7 10 + \log_7 x + 2\log_7 y) \text{ or} \\ &\quad \log_7 4 - \log_7 10 - \log_7 x - 2\log_7 y \end{aligned}$$

PRACTICE

For Problems 1–5, rewrite each as a single logarithm.

1. $2\log x + 3\log y$

2. $\log_6 2x - 2\log_6 3$
3. $3\ln t - \ln 4 + 2\ln 5$
4. $t\ln 6 + 2\ln 5$
5. $\frac{1}{2}\log x - 2\log 2y + 3\log z$

For Problems 6–10, expand each logarithm.

6. $\log \frac{4x}{y}$
7. $\ln \frac{6}{\sqrt{y}}$
8. $\log_4 \frac{10x}{\sqrt[3]{z}}$
9. $\ln \frac{\sqrt{4x}}{5y^2}$
10. $\log \sqrt{\frac{2y^3}{x}}$

SOLUTIONS

1. $2\log x + 3\log y = \log x^2 + \log y^3 = \log x^2y^3$

2. $$\begin{aligned}\log_6 2x - 2\log_6 3 &= \log_6 2x - \log_6 3^2\\ &= \log_6 2x - \log_6 9 = \log_6 \frac{2x}{9}\end{aligned}$$

3. $$\begin{aligned}3\ln t - \ln 4 + 2\ln 5 &= \ln t^3 - \ln 4 + \ln 5^2\\ &= \ln \frac{t^3}{4} + \ln 25\\ &= \ln 25\frac{t^3}{4} = \ln \frac{25t^3}{4}\end{aligned}$$

4. $t\ln 6 + 2\ln 5 = \ln 6^t + \ln 5^2 = \ln[25(6^t)]$

5. $$\begin{aligned}\frac{1}{2}\log x - 2\log 2y + 3\log z &= \log x^{1/2} - \log(2y)^2 + \log z^3\\ &= \log x^{1/2} - \log 2^2y^2 + \log z^3\\ &= \log x^{1/2} - \log 4y^2 + \log z^3\end{aligned}$$

$$= \log \frac{x^{1/2}}{4y^2} + \log z^3 = \log z^3 \frac{x^{1/2}}{4y^2}$$

$$= \log \frac{z^3 x^{1/2}}{4y^2} \text{ or } \log \frac{z^3 \sqrt{x}}{4y^2}$$

6. $\log \frac{4x}{y} = \log 4x - \log y = \log 4 + \log x - \log y$

7. $\ln \frac{6}{\sqrt{y}} = \ln 6 - \ln \sqrt{y} = \ln 6 - \ln y^{1/2} = \ln 6 - \frac{1}{2} \ln y$

8. $\log_4 \frac{10x}{\sqrt[3]{z}} = \log_4 10x - \log_4 \sqrt[3]{z} = \log_4 10x - \log_4 z^{1/3}$

$$= \log_4 10 + \log_4 x - \frac{1}{3} \log_4 z$$

9. $\ln \frac{\sqrt{4x}}{5y^2} = \ln \sqrt{4x} - \ln 5y^2 = \ln(4x)^{1/2} - \ln 5y^2$

$$= \frac{1}{2} \ln 4x - (\ln 5 + \ln y^2) = \frac{1}{2}(\ln 4 + \ln x) - (\ln 5 + 2 \ln y)$$

$$\text{or } \frac{1}{2} \ln 4 + \frac{1}{2} \ln x - \ln 5 - 2 \ln y$$

10. $\log \sqrt{\frac{2y^3}{x}} = \log \left(\frac{2y^3}{x}\right)^{1/2} = \frac{1}{2} \log \frac{2y^3}{x}$

$$= \frac{1}{2}(\log 2y^3 - \log x) = \frac{1}{2}(\log 2 + \log y^3 - \log x)$$

$$= \frac{1}{2}(\log 2 + 3 \log y - \log x) \text{ or } \frac{1}{2} \log 2 + \frac{3}{2} \log y - \frac{1}{2} \log x$$

More Logarithm Equations

With these logarithm properties we can solve more logarithm equations. We will use these properties to rewrite equations either in the form "log = log" or "log = number." When the equation is in the form "log = log," the logs cancel. When the equation is in the form "log = number," we will rewrite the equation as an exponential equation. Instead of checking solutions in the original equation, we only need to make sure that the original logarithms are defined for the solutions.

EXAMPLES

- $\log_2(x-5) + \log_2(x+2) = 3$

  We will use Property 1 to rewrite the equation in the form "log = number."

$$\log_2(x-5) + \log_2(x+2) = 3$$
$$\log_2(x-5)(x+2) = 3$$
$$(x-5)(x+2) = 2^3$$
$$x^2 - 3x - 10 = 8$$
$$x^2 - 3x - 18 = 0$$
$$(x-6)(x+3) = 0$$

  The solutions are $x = 6$ and $x = -3$. Because $\log_2(x+2)$ is not defined for $x = -3$, the only solution is $x = 6$.

- $2\log_5(x+1) - \log_5(x-3) = \log_5 25$

  We will use Property 3 followed by Property 2 to rewrite the equation in the form "log = log."

$$2\log_5(x+1) - \log_5(x-3) = \log_5 25$$
$$\log_5(x+1)^2 - \log_5(x-3) = \log_5 25$$
$$\log_5 \frac{(x+1)^2}{x-3} = \log_5 25$$
$$\frac{(x+1)^2}{x-3} = 25$$
$$(x+1)^2 = 25(x-3)$$
$$(x+1)(x+1) = 25x - 75$$
$$x^2 + 2x + 1 = 25x - 75$$
$$x^2 - 23x + 76 = 0$$
$$(x-4)(x-19) = 0$$

  Both $\log_5(x+1)$ and $\log_5(x-3)$ are defined for $x = 4$ and $x = 19$. The solutions are $x = 4$ and $x = 19$.

PRACTICE

1. $\log_3(2x+1) + \log_3(x+4) = 2$
2. $\ln(3x-4) + \ln(x+2) = \ln(2x+1) + \ln(x+2)$
3. $\log_2(5x+1) - \log_2(x-1) = 3$
4. $2\log_7(x+1) = 2$

SOLUTIONS

1. $\log_3(2x+1) + \log_3(x+4) = 2$ Use Property 1.

$\log_3(2x+1)(x+4) = 2$ Rewrite as an exponent equation.

$$(2x+1)(x+4) = 3^2$$
$$2x^2 + 9x + 4 = 9$$
$$2x^2 + 9x - 5 = 0$$
$$(2x-1)(x+5) = 0$$

Both $\log_3(2x+1)$ and $\log_3(x+5)$ are undefined for $x = -5$, so the only solution is $x = \frac{1}{2}$.

2. $\ln(3x-4) + \ln(x+2) = \ln(2x+1) + \ln(x+2)$ Use Property 1.

$\ln(3x-4)(x+2) = \ln(2x+1)(x+2)$ The logs cancel.

$$(3x-4)(x+2) = (2x+1)(x+2)$$
$$3x^2 + 2x - 8 = 2x^2 + 5x + 2$$
$$x^2 - 3x - 10 = 0$$
$$(x-5)(x+2) = 0$$

All of $\ln(3x-4)$, $\ln(x+2)$, and $\ln(2x+1)$ are not defined for $x = -2$, so the only solution is $x = 5$.

3. $\log_2(5x+1) - \log_2(x-1) = 3$ Use Property 2.

$\log_2 \dfrac{5x+1}{x-1} = 3$ Rewrite as an exponent.

$\dfrac{5x+1}{x-1} = 2^3 = 8$ Cross-multiply.

$$5x + 1 = 8(x-1)$$

$$5x + 1 = 8x - 8$$

$$x = 3$$

4. $2\log_7(x+1) = 2$ Use Property 3.

$\log_7(x+1)^2 = 2$ Rewrite as an exponent.

$$(x+1)^2 = 7^2$$

$$(x+1)(x+1) = 49$$

$$x^2 + 2x + 1 = 49$$

$$x^2 + 2x - 48 = 0$$

$$(x+8)(x-6) = 0$$

The only solution is $x = 6$ because $\log_7(x+1)$ is not defined at $x = -8$. We could have solved this problem in fewer steps if we had divided both sides by 2 in the first step, getting $\log_7(x+1) = 1$.

The domains for $f(x) = \log(x-1)(x+2)$ and $g(x) = \log(x-1) + \log(x+2)$ are not the same, which *seems* to contradict the first logarithm property. Neither $\log(x-1)$ nor $\log(x+2)$ is defined for $x = -3$ because $-3 - 1$ and $-3 + 2$ are negative. But $\log(x-1)(x+2)$ is defined for $x = -3$ because $(-3-1)(-3+2)$ is *positive*. The domain of $f(x)$ will include x-values for which both $(x-1)$ and $(x+2)$ are negative.

The Change of Base Formula

There are countless bases for logarithms but calculators usually have only two logarithms—log and ln. How can we use our calculators to approximate $\log_2 5$? We can use the change of base formula but first, let us use logarithm properties to find this number. Let $x = \log_2 5$. Then $2^x = 5$. Take the common log of each side.

$\log 2^x = \log 5$ Now use the third log property.

$x \log 2 = \log 5$ Divide both sides by the number log 2.

$$x = \frac{\log 5}{\log 2} \approx \frac{0.698970004}{0.301029996} \approx 2.321928095$$

This means that $2^{2.321928095}$ is very close to 5.

We just proved that $\log_2 5 = \frac{\log_{10} 5}{\log_{10} 2}$. Replace 2 with b, 5 with x, and 10 with a and we have the change of base formula.

$$\log_b x = \frac{\log_a x}{\log_a b}$$

This formula converts a logarithm with old base b to new base a. Usually, the new base is either e or 10.

EXAMPLE

- Evaluate $\log_7 15$. Give your solution accurate to four decimal places.

$$\log_7 15 = \frac{\log 15}{\log 7} \approx \frac{1.176091259}{0.84509804} \approx 1.3917$$

$$= \frac{\ln 15}{\ln 7} \approx \frac{2.708050201}{1.945910149} \approx 1.3917$$

The change of base formula can be used to solve equations like $4^{2x+1} = 8$ by rewriting the equation in logarithmic form and using the change of base formula. The equation becomes $\log_4 8 = 2x + 1$. Because $\log_4 8 = \frac{\ln 8}{\ln 4}$, the equation can be written as $2x + 1 = \frac{\ln 8}{\ln 4}$.

$$2x + 1 = \frac{\ln 8}{\ln 4}$$

$$2x = -1 + \frac{\ln 8}{\ln 4}$$

$$x = \frac{1}{2}\left(-1 + \frac{\ln 8}{\ln 4}\right) = \frac{1}{4}$$

EXAMPLE

- $8^x = \frac{1}{3}$

Rewriting this as a logarithm equation, we get $x = \log_8 \frac{1}{3}$. Now we can use the change of base formula.

$$x = \log_8 \frac{1}{3} = \frac{\ln \frac{1}{3}}{\ln 8} \approx -0.5283$$

PRACTICE

Evaluate the logarithms. Give your solution accurate to four decimal places.

1. $\log_6 25$
2. $\log_{20} 5$

   Solve for x. Give your solutions accurate to four decimal places.

3. $3^{x+2} = 12$
4. $15^{3x-2} = 10$
5. $24^{3x+5} = 9$

SOLUTIONS

1. $$\log_6 25 = \frac{\ln 25}{\ln 6} \approx \frac{3.218875825}{1.791759469} \approx 1.7965$$

$$= \frac{\log 25}{\log 6} \approx \frac{1.397940009}{0.7781525} \approx 1.7965$$

2. $$\log_{20} 5 = \frac{\ln 5}{\ln 20} \approx \frac{1.609437912}{2.995732274} \approx 0.5372$$

$$= \frac{\log 5}{\log 20} \approx \frac{0.698970004}{1.301029996} \approx 0.5372$$

3. Rewrite $3^{x+2} = 12$ as a logarithm equation: $x + 2 = \log_3 12$

$$x + 2 = \log_3 12 \qquad \text{Use the change of base formula.}$$

$$= \frac{\ln 12}{\ln 3}$$

$$x = -2 + \frac{\ln 12}{\ln 3} \approx 0.2619$$

4. Rewrite $15^{3x-2} = 10$ as a logarithm equation: $3x - 2 = \log_{15} 10$

$$3x - 2 = \log_{15} 10$$

$$= \frac{\ln 10}{\ln 15} \qquad \text{Use the change of base formula.}$$

$$3x = 2 + \frac{\ln 10}{\ln 15}$$

$$x = \frac{1}{3}\left(2 + \frac{\ln 10}{\ln 15}\right) \approx 0.9501$$

5. Rewrite $24^{3x+5} = 9$ as a logarithm equation: $3x + 5 = \log_{24} 9$.

$$3x + 5 = \log_{24} 9 \qquad \text{Use the change of base formula.}$$

$$= \frac{\ln 9}{\ln 24}$$

$$3x = -5 + \frac{\ln 9}{\ln 24}$$

$$x = \frac{1}{3}\left(-5 + \frac{\ln 9}{\ln 24}\right) \approx -1.4362$$

When both sides of an exponential equation have an exponent, we will use another method to solve for x. We will take either the natural log or the common log of each side and will use the third logarithm property to move the exponents in front of the logarithm. Once we have used the third logarithm property, we will perform the following steps to find x.

1. Distribute the logarithms.
2. Collect the x terms on one side of the equation and the non-x terms on the other side.
3. Factor x.
4. Divide both sides of the equation by x's coefficient (found in Step 3).

EXAMPLES

- $3^{2x} = 2^{x+1}$

  We will begin by taking the natural log of each side.

$$\ln 3^{2x} = \ln 2^{x+1} \qquad \text{Use the third log property.}$$

$$2x \ln 3 = (x+1)\ln 2$$

$$2x \ln 3 = x \ln 2 + \ln 2 \qquad \text{Distribute } \ln 2 \text{ over } (x+1).$$

  Now we want both terms with an x in them on one side of the equation and the term without x in it on the other side. This means that we will move $x \ln 2$ to the left side of the equation.

$$2x \ln 3 - x \ln 2 = \ln 2 \qquad \text{Factor } x \text{ on the left side.}$$

$$x(2\ln 3 - \ln 2) = \ln 2 \qquad \text{Divide each side by } 2\ln 3 - \ln 2.$$

$$x = \frac{\ln 2}{2\ln 3 - \ln 2} \qquad \text{We are finished here.}$$

$$x = \frac{\ln 2}{\ln \frac{9}{2}} \qquad \text{This is easier to calculate.}$$

$$x \approx 0.4608$$

- $10^{x+4} = 6^{3x-1}$

  Because one of the bases is 10, we will use common logarithms. This will simplify some of the steps. We will begin by taking the common log of both sides.

$$\log 10^{x+4} = \log 6^{3x-1} \qquad \text{The left side simplifies to } x+4.$$

$$x + 4 = \log 6^{3x-1} \qquad \text{Use the third log property.}$$

$$x + 4 = (3x-1)\log 6 \qquad \text{Distribute } \log 6 \text{ in } (3x-1).$$

$$x + 4 = 3x \log 6 - \log 6 \qquad \text{Collect } x \text{ terms on one side.}$$

$$x - 3x \log 6 = -4 - \log 6 \qquad \text{Factor } x \text{ on the left.}$$

$$x(1 - 3\log 6) = -4 - \log 6 \qquad \text{Divide both sides by } 1 - 3\log 6.$$

$$x = \frac{-4 - \log 6}{1 - 3\log 6} = \frac{-4 - \log 6}{1 - \log 216} \approx 3.5806$$

PRACTICE

Solve for x. Give your solutions accurate to four decimal places.

1. $4^x = 5^{x-1}$
2. $6^{2x} = 8^{3x-1}$
3. $10^{2-x} = 5^{x+3}$

SOLUTIONS

1. Take the natural log of each side of $4^x = 5^{x-1}$.

$$\ln 4^x = \ln 5^{x-1} \qquad \text{Use the third log property.}$$

$$x \ln 4 = (x-1)\ln 5$$

$$x \ln 4 = x \ln 5 - \ln 5 \qquad \text{This is Step 1.}$$

$$x \ln 4 - x \ln 5 = -\ln 5 \qquad \text{This is Step 2.}$$

$$x(\ln 4 - \ln 5) = -\ln 5 \qquad \text{This is Step 3.}$$

$$x = \frac{-\ln 5}{\ln 4 - \ln 5} \qquad \text{This is Step 4.}$$

$$\approx 7.2126$$

2. Take the natural log of each side of $6^{2x} = 8^{3x-1}$.

$$\ln 6^{2x} = \ln 8^{3x-1} \qquad \text{Use the third log property.}$$

$$2x \ln 6 = (3x-1)\ln 8$$

$$2x \ln 6 = 3x \ln 8 - \ln 8 \qquad \text{This is Step 1.}$$

$$2x \ln 6 - 3x \ln 8 = -\ln 8 \qquad \text{This is Step 2.}$$

$$x(2\ln 6 - 3\ln 8) = -\ln 8 \qquad \text{This is Step 3.}$$

$$x = \frac{-\ln 8}{2\ln 6 - 3\ln 8} \qquad \text{This is Step 4.}$$

$$\approx 0.7833$$

3. Take the common log of each side of $10^{2-x} = 5^{x+3}$. This lets us use the fact that $\log 10^{2-x} = 2 - x$.

$$\begin{aligned}
\log 10^{2-x} &= \log 5^{x+3} \\
2 - x &= (x+3)\log 5 \\
2 - x &= x\log 5 + 3\log 5 && \text{This is Step 1.} \\
-x - x\log 5 &= -2 + 3\log 5 && \text{This is Step 2.} \\
x(-1 - \log 5) &= -2 + 3\log 5 && \text{This is Step 3.} \\
x &= \frac{-2 + 3\log 5}{-1 - \log 5} && \text{This is Step 4.} \\
&\approx -0.0570
\end{aligned}$$

Applications of Logarithm and Exponential Equations

Now that we can solve exponential and logarithmic equations, we can solve many applied problems. We will need the compound growth formula for an investment earning interest rate r, compounded n times per year for t years, $A(t) = P(1+\frac{r}{n})^{nt}$ and the exponential growth formula for a population growing at the rate of r per year for t years, $n(t) = n_0 e^{rt}$. In the problems below, we will be looking for the time required for an investment to grow to a specified amount.

EXAMPLES

- How long will it take for \$1000 to grow to \$1500 if it earns 8% annual interest, compounded monthly?
  In the formula $A(t) = P(1+\frac{r}{n})^{nt}$ we know $A(t) = 1500$, $P = 1000$, $r = 0.08$, and $n = 12$. We do not know t.

$$1500 = 1000\left(1 + \frac{0.08}{12}\right)^{12t}$$

We will solve this equation for t and will round up to the nearest month.

$$1500 = 1000\left(1 + \frac{0.08}{12}\right)^{12t} \qquad \text{Divide both sides by 1000.}$$

$$1.5 = \left(1 + \frac{0.08}{12}\right)^{12t}$$

$$1.5 = 1.00667^{12t} \qquad \text{Take the natural log of both sides.}$$

$$\ln 1.5 = \ln 1.00667^{12t} \qquad \text{Use the third log property.}$$

$$\ln 1.5 = 12t \ln 1.00667 \qquad \text{Divide both sides by } 12 \ln 1.00667.$$

$$\frac{\ln 1.5}{12 \ln 1.00667} = t$$

$$t \approx 5.085$$

In five years and one month, the investment will grow to about \$1500.

- How long will it take an investment to double if it earns $6\frac{1}{2}\%$ annual interest, compounded daily?
  An investment of \$P doubles when it grows to \$2P, so let $A(t) = 2P$ in the compound growth formula.

$$2P = P\left(1 + \frac{0.065}{365}\right)^{365t} \qquad \text{Divide both sides by } P.$$

$$2 = \left(1 + \frac{0.065}{365}\right)^{365t}$$

$$2 = 1.000178^{365t} \qquad \text{Take the natural log of both sides.}$$

$$\ln 2 = \ln 1.000178^{365t} \qquad \text{Use the third log property.}$$

$$\ln 2 = 365t \ln 1.000178 \qquad \text{Divide both sides by } 365 \ln 1.000178.$$

$$\frac{\ln 2}{365 \ln 1.000178} = t$$

$$t \approx 10.66$$

In about 10 years, 8 months, the investment will double.

PRACTICE

Give your answers rounded up to the nearest compounding period.

1. How long will it take \$2000 to grow to \$40,000 if it earns 9% annual interest, compounded annually?

2. How long will it take for \$5000 to grow to \$7500 if it earns $6\frac{1}{2}\%$ annual interest, compounded weekly?
3. How long will it take an investment to double if it earns $6\frac{1}{4}\%$ annual interest, compounded quarterly?

SOLUTIONS

1. $$40{,}000 = 2000(1+0.09)^t$$

$$20 = 1.09^t$$

$$\ln 20 = \ln 1.09^t$$

$$\ln 20 = t \ln 1.09$$

$$\frac{\ln 20}{\ln 1.09} = t$$

$$34.76 \approx t$$

The \$2000 investment will grow to \$40,000 in 35 years.

2. $$7500 = 5000\left(1+\frac{0.065}{52}\right)^{52t}$$

$$1.5 = 1.00125^{52t}$$

$$\ln 1.5 = \ln 1.00125^{52t}$$

$$\ln 1.5 = 52t \ln 1.00125$$

$$\frac{\ln 1.5}{52 \ln 1.00125} = t$$

$$t \approx 6.24$$

In 6 years, 13 weeks ($0.24 \times 52 = 12.48$ rounds up to 13), the \$5000 investment will grow to \$7500.

3. $$2P = P\left(1+\frac{0.0625}{4}\right)^{4t}$$

$$2 = 1.015625^{4t}$$

$$\ln 2 = \ln 1.015625^{4t}$$

$$\ln 2 = 4t \ln 1.015625$$

$$\frac{\ln 2}{4 \ln 1.015625} = t$$

$$t \approx 11.18$$

In 11 years and 3 months (0.18 rounded up to the nearest quarter is 0.25, one quarter is 3 months), the investment will double.

This method works with population models where the population (either of people, animals, insects, bacteria, etc.) grows or decays at a certain percent every period. We will use the growth formula $n(t) = n_0 e^{rt}$. If the population is decreasing, we will use the decay formula, $n(t) = n_0 e^{-rt}$. Because we will be working with the base e, instead of taking the log of both sides, we will be rewriting the equations as log equations (this is equivalent to taking the natural log of both sides).

EXAMPLES

- A school district estimates that its student population will grow about 5% per year for the next 15 years. How long will it take the student population to grow from the current 8000 students to 12,000?
  We will solve for t in the equation $12{,}000 = 8000e^{0.05t}$.

$$12{,}000 = 8000e^{0.05t} \quad \text{Divide both sides by 8000.}$$

$$1.5 = e^{0.05t} \quad \text{Rewrite as a log.}$$

$$0.05t = \ln 1.5$$

$$t = \frac{\ln 1.5}{0.05} \approx 8.1$$

  The population is expected to reach 12,000 in about 8 years.
- The population of a certain city in the year 2004 is about 650,000. If it is losing 2% of its population each year, when will the population decline to 500,000?
  Because the population is declining, we will use the formula $n(t) = n_0 e^{-rt}$. Solve for t in the equation $500{,}000 = 650{,}000e^{-0.02t}$.

$$500{,}000 = 650{,}000e^{-0.02t}$$

$$\frac{10}{13} = e^{-0.02t} \quad \text{Rewrite as a log.}$$

$$-0.02t = \ln \frac{10}{13}$$

$$t = \frac{\ln \frac{10}{13}}{-0.02} \approx 13.1$$

The population is expected to drop to 500,000 around the year 2017.

- At 2:00 a culture contained 3000 bacteria. They are growing at the rate of 150% per hour. When will there be 5400 bacteria in the culture?
  A growth rate of 150% per hour means that $r = 1.5$ and that t is measured in hours.

$$5400 = 3000e^{1.5t}$$

$$1.8 = e^{1.5t}$$

$$1.5t = \ln 1.8$$

$$t = \frac{\ln 1.8}{1.5} \approx 0.39$$

At about 2:24 ($0.39 \times 60 = 23.4$ minutes) there will be 5400 bacteria in the culture.

PRACTICE

1. In 2003 a rural area had 1800 birds of a certain species. If the bird population is increasing at the rate of 15% per year, when will it reach 3000?
2. In 2002, the population of a certain city was 2 million. If the city's population is declining at the rate of 1.8% per year, when will it fall to 1.5 million?
3. At 9:00 a petrie dish contained 5000 bacteria. The bacteria population is growing at the rate of 160% per hour. When will the dish contain 20,000 bacteria?

SOLUTIONS

1. $3000 = 1800e^{0.15t}$

   $\frac{5}{3} = e^{0.15t}$

$$0.15t = \ln \frac{5}{3}$$

$$t = \frac{\ln \frac{5}{3}}{0.15} \approx 3.4$$

The bird population should reach 3000 in the year 2006.

2.
$$1.5 = 2e^{-0.018t}$$
$$0.75 = e^{-0.018t}$$
$$-0.018t = \ln 0.75$$
$$t = \frac{\ln 0.75}{-0.018} \approx 16$$

In the year 2018, the population will decline to 1.5 million.

3.
$$20{,}000 = 5000e^{1.6t}$$
$$4 = e^{1.6t}$$
$$1.6t = \ln 4$$
$$t = \tfrac{\ln 4}{1.6} \approx 0.87$$

At about 9:52 ($0.87 \times 60 = 52.2$ minutes), there will be 20,000 bacteria in the dish.

Finding the Growth Rate

We can find the growth rate of a population if we have reason to believe that it is growing exponentially and if we know the population level at two different times. We will use the first population level as n_0. Because we will know another population level, we have a value for $n(t)$ and for t. This means that the equation $n(t) = n_0 e^{rt}$ will have only one unknown, r. We can find r using natural logarithms in the same way we found t in the problems above.

EXAMPLES

- The population of a country is growing exponentially. In the year 2000, it was 10 million and in 2005, it was 12 million. What is the growth rate? In the year $t = 0$ (2000), the population was 10 million, so $n_0 = 10$. The growth formula becomes $n(t) = 10e^{rt}$. When $t = 5$ (the year 2005), the population is 12 million, so $n(t) = 12$. We will solve the equation $12 = 10e^{5r}$ for r.

$$12 = 10e^{5r}$$

$$1.2 = e^{5r}$$

$$5r = \ln 1.2$$

$$r = \frac{\ln 1.2}{5} \approx 0.036$$

The country's population is growing at the rate of 3.6% per year.

- Suppose a bacteria culture contains 2500 bacteria at 1:00 and at 1:30 there are 6000. What is the hourly growth rate?
  Because we are asked to find the hourly growth rate, t must be measured in hours and not minutes. Initially, at $t = 0$, the population is 2500, so $n_0 = 2500$. Half an hour later, the population is 6000, so $t = 0.5$ and $n(t) = 6000$. We will solve for r in the equation $6000 = 2500e^{0.5r}$.

$$6000 = 2500e^{0.5r}$$

$$2.4 = e^{0.5r}$$

$$0.5r = \ln 2.4$$

$$r = \frac{\ln 2.4}{0.5} \approx 1.75$$

The bacteria are increasing at the rate of 175% per hour.

- A certain species of fish is introduced in a large lake. Wildlife biologists expect the fish's population to double every four months for the first few years. What is the annual growth rate?
  If n_0 represents the fish's population when first put in the lake, then it will double to $2n_0$ after $t = 4$ months $= \frac{4}{12}$ years $= \frac{1}{3}$ years. The growth formula becomes $2n_0 = n_0e^{\frac{1}{3}r}$. This equation has two unknowns, n_0 and r, not one. But after we divide both sides of the equation by n_0, r becomes the only unknown.

$$2n_0 = n_0e^{\frac{1}{3}r}$$

$$2 = e^{\frac{1}{3}r}$$

$$\frac{1}{3}r = \ln 2$$

$$r = 3\ \ln 2 \approx 2.08$$

The fish population is expected to grow at the rate of 208% per year.

PRACTICE

1. The population of school children in a city grew from 125,000 to 200,000 in five years. Assuming exponential growth, find the annual growth rate for the number of school children.
2. A corporation that owns a chain of retail stores operated 500 stores in 2000 and 700 stores in 2003. Assuming that the number of stores is growing exponentially, what is its annual growth rate?
3. At 10:30, 1500 bacteria are present in a culture. At 11:00, 3500 are present. What is the hourly growth rate?

SOLUTIONS

1. $200{,}000 = 125{,}000e^{5r}$

$$1.6 = e^{5r}$$

$$5r = \ln 1.6$$

$$r = \frac{\ln 1.6}{5} \approx 0.094$$

The population of school children grew at the rate of 9.4% per year.

2. $700 = 500e^{3r}$

$$1.4 = e^{3r}$$

$$3r = \ln 1.4$$

$$r = \frac{\ln 1.4}{3} \approx 0.112$$

The number of stores is growing at the rate of 11.2% per year.

3. $3500 = 1500e^{0.5r}$

$$\frac{7}{3} = e^{0.5r}$$

$$0.5r = \ln \frac{7}{3}$$

$$r = \frac{\ln \frac{7}{3}}{0.5} \approx 1.69$$

The bacteria are increasing at the rate of 169% per hour.

Radioactive Decay

Some radioactive substances decay at the rate of nearly 100% per year and others at nearly 0% per year. For this reason, we use the *half-life* of a radioactive substance to describe how fast its radioactivity decays. For example, bismuth-210 has a half-life of 5 days. After 5 days, 16 grams of bismuth-210 decays to 8 grams of bismuth-210 (and 8 grams of another substance); after 10 days, 4 grams remain, and after 15 days, only 2 grams remains. We can use logarithms and the half-life to find the rate of decay. We will use the decay formula $n(t) = n_0 e^{-rt}$ in the following problems.

EXAMPLES

- Find the daily decay rate of bismuth-210.
  Because its half-life is 5 days, at $t = 5$, one-half of n_0 remains, so $n(t) = \frac{1}{2}n_0$.

$$\frac{1}{2}n_0 = n_0 e^{-5r} \qquad \text{Divide both sides by } n_0.$$

$$\frac{1}{2} = e^{-5r} \qquad \text{Rewrite as a log.}$$

$$-5r = \ln \frac{1}{2}$$

$$r = \frac{\ln \frac{1}{2}}{-5} \approx 0.1386$$

Bismuth-210 decays at the rate of 13.86% per day.

- The half-life of radium-226 is 1600 years. What is its annual decay rate?

$$\frac{1}{2}n_0 = n_0 e^{-1600r} \qquad \text{Divide both sides by } n_0.$$

$$\frac{1}{2} = e^{-1600r} \qquad \text{Rewrite as a log.}$$

$$-1600r = \ln\frac{1}{2}$$

$$r = \frac{\ln\frac{1}{2}}{-1600} \approx 0.000433$$

The decay rate for radium-226 is about 0.0433% per year.

In the same way we found the decay rate from the half-life, we can find the half-life from the decay rate. In the formula $\frac{1}{2}n_0 = n_0 e^{-rt}$, we know r and want to find t.

EXAMPLE

- Suppose a radioactive substance decays at the rate of 2.5% per hour. What is its half-life?

$$\frac{1}{2}n_0 = n_0 e^{-0.025t} \qquad \text{Divide both sides by } n_0.$$

$$\frac{1}{2} = e^{-0.025t} \qquad \text{Rewrite as a log.}$$

$$-0.025t = \ln\frac{1}{2}$$

$$t = \frac{\ln\frac{1}{2}}{-0.025} \approx 27.7$$

The half-life is 27.7 hours.

PRACTICE

1. Suppose a substance has a half-life of 45 days. Find its daily decay rate.

2. The half-life of lead-210 is 22.3 years. Find its annual decay rate.
3. Suppose the half-life for a substance is 1.5 seconds. What is its decay rate per second?
4. Suppose a radioactive substance decays at the rate of 0.1% per day. What is its half-life?
5. A radioactive substance decays at the rate of 0.02% per year. What is its half-life?

SOLUTIONS

1. $$\frac{1}{2}n_0 = n_0e^{-45r}$$

$$\frac{1}{2} = e^{-45r}$$

$$-45r = \ln\frac{1}{2}$$

$$r = \frac{\ln\frac{1}{2}}{-45} \approx 0.0154$$

The decay rate is 1.5% per day.

2. $$\frac{1}{2}n_0 = n_0e^{-22.3r}$$

$$\frac{1}{2} = e^{-22.3r}$$

$$-22.3r = \ln\frac{1}{2}$$

$$r = \frac{\ln\frac{1}{2}}{-22.3} \approx 0.0311$$

The decay rate is 3.1% per year.

3. $$\frac{1}{2}n_0 = n_0e^{-1.5r}$$

$$\frac{1}{2} = e^{-1.5r}$$

$$-1.5r = \ln \frac{1}{2}$$

$$r = \frac{\ln \frac{1}{2}}{-1.5} \approx 0.462$$

The substance decays at the rate of 46.2% per second.

4.
$$\frac{1}{2}n_0 = n_0 e^{-0.001t}$$

$$\frac{1}{2} = e^{-0.001t}$$

$$-0.001t = \ln \frac{1}{2}$$

$$t = \frac{\ln \frac{1}{2}}{-0.001} \approx 693.1$$

The half-life is 693 days.

5.
$$\frac{1}{2}n_0 = n_0 e^{-0.0002t}$$

$$\frac{1}{2} = e^{-0.0002t}$$

$$-0.0002t = \ln \frac{1}{2}$$

$$t = \frac{\ln \frac{1}{2}}{-0.0002} \approx 3466$$

The half-life is about 3466 years.

All living things have carbon-14 in them. Once they die, the carbon-14 is not replaced and begins to decay. The half-life of carbon-14 is approximately 5700 years. This information is used to find the age of many archeological finds. We will first find the annual decay rate for carbon-14 then will answer some typical carbon-14 dating questions.

$$\frac{1}{2}n_0 = n_0 e^{-5700r}$$

$$\frac{1}{2} = e^{-5700r}$$

$$-5700r = \ln \frac{1}{2}$$

$$r = \frac{\ln \frac{1}{2}}{-5700} \approx 0.000121605$$

Carbon-14 decays at the rate of 0.012% per year.

EXAMPLES

- How long will it take for 80% of the carbon-14 to decay in an animal after it has died?
  If 80% of the initial amount has decayed, then 20% remains, or $0.20n_0$.

$$0.20n_0 = n_0e^{-0.00012t}$$

$$0.20 = e^{-0.00012t}$$

$$-0.00012t = \ln 0.20$$

$$t = \frac{\ln 0.20}{-0.00012} \approx 13,412$$

  After about 13,400 years, 80% of the carbon-14 will have decayed.
- Suppose a bone is discovered and has 60% of its carbon-14. How old is the bone? 60% of its carbon-14 is $0.60n_0$.

$$0.60n_0 = n_0e^{-0.00012t}$$

$$0.60 = e^{-0.00012t}$$

$$-0.00012t = \ln 0.60$$

$$t = \frac{\ln 0.60}{-0.00012} \approx 4257$$

  The bone is about 4260 years old.

- Suppose an animal dies today. How much of its carbon-14 will remain after 250 years?

$$n(250) = n_0e^{-0.00012(250)} \approx 0.97n_0$$

  About 97% of its carbon-14 will remain after 250 years.

PRACTICE

1. Suppose a piece of wood from an archeological dig is being carbon-14 dated, and found to have 70% of its carbon-14 remaining. Estimate the age of the piece of wood.
2. How long would it take for an object to lose 25% of its carbon-14?
3. Suppose a tree fell 400 years ago. How much of its carbon-14 remains?

SOLUTIONS

1.

$$\begin{aligned} 0.70n_0 &= n_0e^{-0.00012t} \\ 0.70 &= e^{-0.00012t} \\ -0.00012t &= \ln 0.70 \\ t &= \frac{\ln 0.70}{-0.00012} \approx 2972 \end{aligned}$$

The wood is about 2970 years old.

2. An object has lost 25% of its carbon-14 when 75% of it remains.

$$\begin{aligned} 0.75n_0 &= n_0e^{-0.00012t} \\ 0.75 &= e^{-0.00012t} \\ -0.00012t &= \ln 0.75 \\ t &= \frac{\ln 0.75}{-0.00012} \approx 2397 \end{aligned}$$

After about 2400 years, an object will lose 25% of its carbon-14.

3.

$$n(400) = n_0e^{-0.00012(400)} \approx 0.953n_0$$

About 95% of its carbon-14 remains after 400 years.

CHAPTER 9 REVIEW

1. If $10,000 is invested earning 6% annual interest, compounded quarterly, what will it be worth after eight years?

   (a) $15,938.48 (b) $16,103.24 (c) $11,264.93 (d) $10,613.64

2. What is the present value of \$50,000 due in 10 years, earning 8% annual interest, compounded annually?
   (a) \$107,946.25 (b) \$19,277.16
   (c) \$23,159.67 (d) \$27,013.44

3. Rewrite $\log_a x = w$ as an exponential equation.
   (a) $a^w = x$ (b) $a^x = w$ (c) $x^a = w$ (d) $w^a = x$

4. Rewrite $7^m = n$ as a logarithmic equation.
   (a) $\log_n 7 = m$ (b) $\log_7 m = n$ (c) $\log_m 7 = n$ (d) $\log_7 n = m$

5. $e^{\ln 7} =$
   (a) $\ln 7$ (b) 7 (c) e^7 (d) $(\ln 7)e$

6. Rewrite as a single logarithm.

$$\ln x - 2\ln y + \ln z$$

   (a) $\ln \frac{xz}{y^2}$ (b) $\ln \left(\frac{xz}{y}\right)^2$ (c) $\ln \frac{xz}{2y}$ (d) $\frac{\ln xz}{2\ln y}$

7. Expand the logarithm.

$$\log_5 \sqrt[3]{\frac{ab^2}{c}}$$

   (a) $\sqrt[3]{\log_5 a + \log_5 b^2 - \log_5 c}$

   (b) $\frac{1}{3}\log_5 a + 2\log_5 b - \log_5 c$

   (c) $\sqrt[3]{\log_5 a} + \sqrt[3]{\log_5 b^2} - \sqrt[3]{\log_5 c}$

   (d) $\frac{1}{3}[\log_5 a + 2\log_5 b - \log_5 c]$

8. Solve for x: $\log_6(x-1) = 2$.
   (a) $x = 3$ (b) $x = 37$ (c) $x = 13$ (d) $x = 65$

9. Solve for x: $\log_4(x+1) + \log_4(x-1) = \log_4 8$.
   (a) $x = 4$ (b) $x = \pm 3$ (c) $x = 3$ (d) No solution

10. What is the domain for $f(x) = \log(x + 4)$?
    (a) $(-\infty, -4) \cup (-4, \infty)$
    (b) $(-\infty, -4)$
    (c) $(-4, \infty)$
    (d) $[-4, \infty)$

11. Solve for x: $3^{x+1} = 15$.
    (a) $x = -1 + \dfrac{\ln 3}{\ln 15}$
    (b) $x = -1 + \dfrac{\ln 15}{\ln 3}$
    (c) $x = 4$
    (d) No solution

12. How long will it take for an investment to double if it earns 10% annual interest, compounded quarterly?
    (a) About 4 years
    (b) About 5 years
    (c) About 6 years
    (d) About 7 years

13. The half-life of a substance is about 40 years. What is its annual decay rate?
    (a) About 1% (b) About 1.5% (c) About 1.7% (d) About 2.1%

SOLUTIONS

1. B	2. C	3. A	4. D	5. B	6. A	
7. D	8. B	9. C	10. C	11. B	12. D	13. C

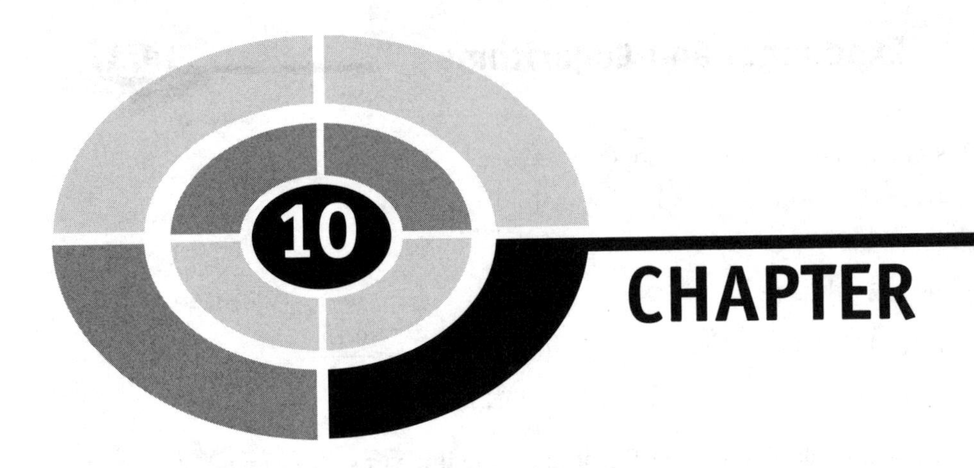

Systems of Equations and Inequalities

A system of equations is a collection of two or more equations whose graphs might or might not intersect (share a common point or points). If the graphs do intersect, then we say that the solution to the system is the point or points where the graphs intersect. For example, the solution to the system

$$\begin{cases} x + y = 4 \\ 3x - y = 0 \end{cases}$$

is $(1, 3)$ because the point $(1, 3)$ is on both graphs. See Figure 10.1.

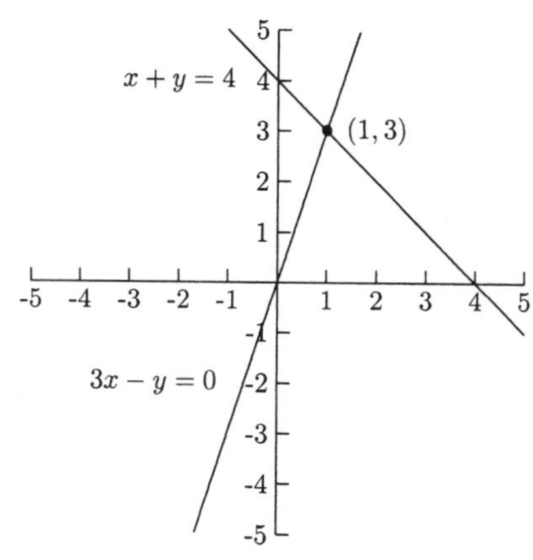

Fig. 10.1.

We say that $(1, 3)$ *satisfies* the system because if we let $x = 1$ and $y = 3$ in each equation, they will both be true.

$1 + 3 = 4$ This is a true statement.

$3(1) - 3 = 0$ This is a true statement.

There are several methods for solving systems of equations. One of them is by sketching the graphs and seeing where, if anywhere, the graphs intersect. Even with a graphing calculator, though, these solutions might only be approximations. When the equations are lines, *matrices* can be used. Graphing calculators are useful for these, too. We will use two algebraic methods in this chapter and two matrix methods in the next. One of the algebraic methods is *substitution* and the other is *elimination by addition*. Both methods will work with many kinds of systems of equations, but we will start out with systems of linear equations.

Substitution works by solving for one variable in one equation and making a substitution in the other equation. Usually, it does not matter which variable we use or which equation we begin with, but some choices are easier than others.

EXAMPLES

Solve the systems of equations. Put your solutions in the form of a point, (x, y).

- $$\begin{cases} x + y = 5 \\ -2x + y = -1 \end{cases}$$

We have four places to start.

1. Solve for x in the first equation: $x = 5 - y$

2. Solve for y in the first equation: $y = 5 - x$
3. Solve for x in the second equation: $x = \frac{1}{2} + \frac{1}{2}y$
4. Solve for y in the second equation: $y = 2x - 1$

The third option looks like it would be the most trouble, so we will use one of the others. We will use the first option. Because $x = 5 - y$ came from the *first* equation, we will substitute $5 - y$ for x in the *second* equation. Then $-2x + y = -1$ becomes $-2(5 - y) + y = -1$. Now we have one equation with one variable.

$$-2(5 - y) + y = -1$$

$$-10 + 2y + y = -1$$

$$3y = 9$$

$$y = 3$$

Now that we know $y = 3$, we could use any of the equations above to find x. We know that $x = 5 - y$, so we will use this.

$$x = 5 - 3 = 2$$

The solution is $x = 2$ and $y = 3$ or the point (2, 3). It is a good idea to check the solution.

$$2 + 3 = 5 \qquad \text{This is true.}$$

$$-2(2) + 3 = -1 \qquad \text{This is true.}$$

- $\begin{cases} 4x - y = 12 & \text{A} \\ 3x + y = 2 & \text{B} \end{cases}$

We will solve for y in equation B: $y = 2 - 3x$. Next we will substitute $2 - 3x$ for y in equation A and solve for x.

$$4x - y = 12$$

$$4x - (2 - 3x) = 12$$

$$4x - 2 + 3x = 12$$

$$7x = 14$$

$$x = 2$$

Now that we know $x = 2$, we will put $x = 2$ in one of the above equations. We will use $y = 2 - 3x$: $y = 2 - 3(2) = -4$. The solution is $x = 2$, $y = -4$, or

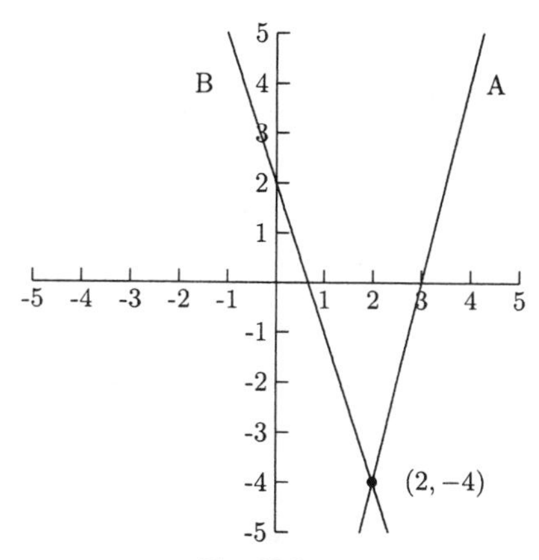

Fig. 10.2.

$(2, -4)$. The graphs in Figure 10.2 verify that the solution $(2, -4)$ is on both lines.

- $$\begin{cases} y = 4x + 1 & \text{A} \\ y = 3x + 2 & \text{B} \end{cases}$$

  Both equations are already solved for y, so all we need to do is to set them equal to each other.

$$A = B$$

$$4x + 1 = 3x + 2$$

$$x = 1$$

  Use either equation A or equation B to find y when $x = 1$. We will use A: $y = 4x + 1 = 4(1) + 1 = 5$. The solution is $x = 1$ and $y = 5$, or $(1, 5)$. We can see from the graphs in Figure 10.3 that $(1, 5)$ is the solution to the system.

Solving a system of equations by substitution can be messy when none of the coefficients is 1. Fortunately, there is another way. We can always *add* the two equations to eliminate one of the variables. Sometimes, though, we need to multiply one or both equations by a number to make it work.

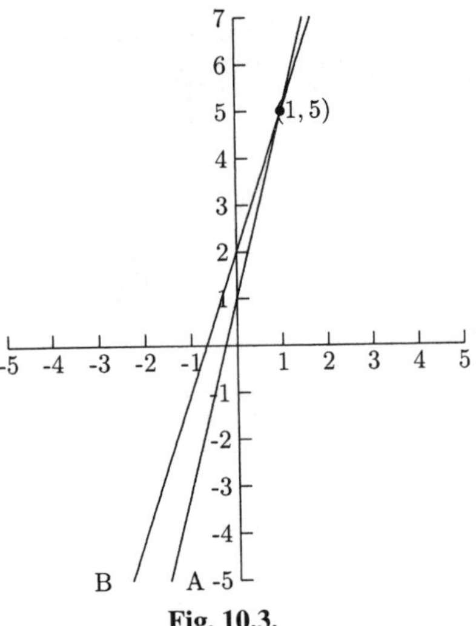

Fig. 10.3.

EXAMPLE

Solve the systems of equations. Put your solutions in the form of a point, (x, y).

- $\begin{cases} 2x - 3y = 16 & \text{A} \\ 5x + 3y = -2 & \text{B} \end{cases}$

  Add the equations by adding like terms. Because we will be adding $-3y$ to $3y$, the y-term will cancel, leaving one equation with only one variable.

$$\begin{aligned} 2x - 3y &= 16 \\ \underline{5x + 3y} &\underline{= -2} \\ 7x + 0y &= 14 \\ x &= 2 \end{aligned}$$

  We can put $x = 2$ into either A or B to find y. We will put $x = 2$ into A.

$$\begin{aligned} 2x - 3y &= 16 \\ 2(2) - 3y &= 16 \\ -3y &= 12 \\ y &= -4 \end{aligned}$$

  The solution is $(2, -4)$.

Sometimes we need to multiply one or both equations by some number or numbers so that one of the variables cancels. Multiplying both sides of *any* equation by a nonzero number never changes the solution.

EXAMPLES

- $\begin{cases} 3x + 6y = -12 & \text{A} \\ 2x + 6y = -14 & \text{B} \end{cases}$

  Because the coefficients on y are the same, we only need to make one of them negative. Multiply either A or B by -1, then add.

$$\begin{array}{rl} -3x - 6y = 12 & -\text{A} \\ \underline{2x + 6y = -14} & +\text{B} \\ -x = -2 & \\ x = 2 & \\ 3(2) + 6y = -12 & \text{Put } x = 2 \text{ in A} \\ y = -3 & \end{array}$$

  The solution is $(2, -3)$.

- $\begin{cases} 2x + 7y = 1 & \text{A} \\ 4x - 2y = 18 & \text{B} \end{cases}$

  Several options will work. We could multiply A by -2 so that we could add $-4x$ (in -2A) to $4x$ in B. We could multiply A by 2 and multiply B by 7 so that we could add $14y$ (in 2A) to $-14y$ (in 7B). We could also divide B by -2 so that we could add $2x$ (in A) to $-2x$ (in $-\frac{1}{2}$B). We will add $-2\text{A} + \text{B}$.

$$\begin{array}{rl} -4x - 14y = -2 & -2\text{A} \\ \underline{4x - 2y = 18} & +\text{B} \\ -16y = 16 & \\ y = -1 & \\ 2x + 7(-1) = 1 & \text{Put } y = -1 \text{ in A} \\ x = 4 & \end{array}$$

  The solution is $(4, -1)$.

Both equations in each of the following systems will need to be changed to eliminate one of the variables.

EXAMPLES

- $$\begin{cases} 8x - 5y = -2 & \text{A} \\ 3x + 2y = 7 & \text{B} \end{cases}$$

There are many options. Some are 3A − 8B, −3A + 8B, and 2A + 5B. We will compute 2A + 5B.

$$\begin{array}{rl} 16x - 10y = -4 & \text{2A} \\ \underline{15x + 10y = 35} & \text{+5B} \\ 31x = 31 & \\ x = 1 & \\ 8(1) - 5y = -2 & \text{Put } x = 1 \text{ in A} \\ y = 2 & \end{array}$$

The solution is $(1, 2)$.

- $$\begin{cases} \frac{2}{3}x - \frac{1}{4}y = \frac{25}{72} & \text{A} \\ \frac{1}{2}x + \frac{2}{5}y = -\frac{1}{30} & \text{B} \end{cases}$$

First, we will eliminate the fractions. The LCD for A is 72, and the LCD for B is 30.

$$\begin{array}{rl} 48x - 18y = 25 & \text{72A} \\ 15x + 12y = -1 & \text{30B} \end{array}$$

Now we will multiply the first equation by 2 and the second by 3.

$$\begin{array}{r} 96x - 36y = 50 \\ \underline{45x + 36y = -3} \\ 141x = 47 \\ x = \dfrac{47}{141} = \dfrac{1}{3} \\ 96\left(\dfrac{1}{3}\right) - 36y = 50 \\ y = -\dfrac{1}{2} \end{array}$$

The solution is $(\frac{1}{3}, -\frac{1}{2})$.

Applications of Systems of Equations

Systems of two linear equations can be used to solve many kinds of word problems. In these problems, two facts will be given about two variables. Each pair of facts can be represented by a linear equation. This gives us a system of two equations with two variables.

EXAMPLES

- A movie theater charges \$4 for each children's ticket and \$6.50 for each adult's ticket. One night 200 tickets were sold, amounting to \$1100 in ticket sales. How many of each type of ticket was sold?

  Let x represent the number of children's tickets sold and y, the number of adult tickets sold. One equation comes from the fact that a total of 200 adult and children's tickets were sold, giving us $x + y = 200$. The other equation comes from the fact that the ticket revenue was \$1100. The ticket revenue from children's tickets is $4x$, and the ticket revenue from adult tickets is $6.50y$. Their sum is 1100 giving us $4x + 6.50y = 1100$.

$$\begin{cases} 4x + 6.50y = 1100 & \text{A} \\ x + y = 200 & \text{B} \end{cases}$$

  We could use either substitution or addition to solve this system. Substitution is a little faster. We will solve for x in B.

$$x = 200 - y$$

$$4(200 - y) + 6.50y = 1100 \quad \text{Put } 200 - y \text{ into A}$$

$$800 - 4y + 6.50y = 1100$$

$$y = 120$$

$$x = 200 - y = 200 - 120 = 80$$

  Eighty children's tickets were sold, and 120 adult tickets were sold.

- A farmer had a soil test performed. He was told that his field needed 1080 pounds of Mineral A and 920 pounds of Mineral B. Two mixtures of fertilizers provide these minerals. Each bag of Brand I provides 25 pounds of Mineral A and 15 pounds of Mineral B. Brand II provides 20 pounds of Mineral A and 20 pounds of Mineral B. How many bags of each brand should he buy?

  Let x represent the number of bags of Brand I and y represent the number of bags of Brand II. Then the number of pounds of Mineral A he will get from Brand I is $25x$ and the number of pounds of Mineral B is $15x$. The number of pounds of Mineral A he will get from Brand II is $20y$ and the number of pounds of Mineral B is $20y$. He needs 1080 pounds of Mineral A, $25x$ pounds will come from Brand I and $20y$ will come from Brand II. This gives us the equation $25x + 20y = 1080$. He needs 920 pounds of Mineral B, $15x$ will come from Brand I and $20y$ will come from Brand II. This gives us the equation $15x + 20y = 920$.

$$\begin{cases} 25x + 20y = 1080 & \text{A} \\ 15x + 20y = 920 & \text{B} \end{cases}$$

We will compute A − B.

$$\begin{array}{rcll} 25x + 20y &=& 1080 & \text{A} \\ -15x - 20y &=& -920 & -\text{B} \\ \hline 10x &=& 160 & \\ x &=& 16 & \\ 25(16) + 20y &=& 1080 & \\ y &=& 34 & \end{array}$$

He needs 16 bags of Brand I and 34 bags of Brand II.

- A furniture manufacturer has some discontinued fabric and trim in stock. He can use them on sofas and chairs. There are 160 yards of fabric and 110 yards of trim. Each sofa takes 6 yards of fabric and 4.5 yards of trim. Each chair takes 4 yards of fabric and 2 yards of trim. How many sofas and chairs should be produced in order to use all the fabric and trim?
  Let x represent the number of sofas to be produced and y, the number of chairs. The manufacturer needs to use 160 yards of fabric, $6x$ will be used on sofas and $4y$ yards on chairs. This gives us the equation $6x + 4y = 160$. There are 110 yards of trim, $4.5x$ yards will be used on the sofas and $2y$ on the chairs. This gives us the equation $4.5x + 2y = 110$.

$$\begin{cases} 6x + 4y = 160 & \text{F} \\ 4.5x + 2y = 110 & \text{T} \end{cases}$$

We will compute F − 2T.

$$\begin{array}{rl} 6x + 4y = 160 & \text{F} \\ \underline{-9x - 4y = -220} & -2\text{T} \\ -3x = -60 & \\ x = 20 & \\ 6(20) + 4y = 160 & \\ y = 10 & \end{array}$$

The manufacturer needs to produce 20 sofas and 10 chairs.

PRACTICE

For Problems 1–9, solve the systems of equations. Put your solutions in the form of a point, (x, y).

1.

$$\begin{cases} 2x + 3y = 1 & \text{A} \\ x - 2y = -3 & \text{B} \end{cases}$$

2.

$$\begin{cases} x + y = 3 & \text{A} \\ x + 4y = 0 & \text{B} \end{cases}$$

3.

$$\begin{cases} -2x + 7y = 19 & \text{A} \\ 2x - 4y = -10 & \text{B} \end{cases}$$

4.

$$\begin{cases} 15x - y = 9 & \text{A} \\ 2x + y = 8 & \text{B} \end{cases}$$

5.

$$\begin{cases} -3x + 2y = 12 & \text{A} \\ 4x + 2y = -2 & \text{B} \end{cases}$$

6.

$$\begin{cases} 6x - 5y = 1 & \text{A} \\ 3x - 2y = 1 & \text{B} \end{cases}$$

7.

$$\begin{cases} 5x - 9y = -26 & \text{A} \\ 3x + 2y = 14 & \text{B} \end{cases}$$

8.

$$\begin{cases} 7x + 2y = 1 & \text{A} \\ 2x + 3y = -7 & \text{B} \end{cases}$$

9.

$$\begin{cases} \frac{3}{4}x + \frac{1}{5}y = \frac{23}{60} & \text{A} \\ \frac{1}{6}x - \frac{1}{4}y = -\frac{1}{9} & \text{B} \end{cases}$$

10. A grocery store sells two different brands of milk. The price for the name brand is $3.50 per gallon, and the price for the store's brand is $2.25 per gallon. On one Saturday, 4500 gallons of milk were sold for sales of $12,875. How many of each brand were sold?

11. A gardener wants to add 39 pounds of Nutrient A and 16 pounds of Nutrient B to her garden. Each bag of Brand X provides 3 pounds of Nutrient A and 2 pounds of Nutrient B. Each bag of Brand Y provides 4 pounds of Nutrient A and 1 pound of Nutrient B. How many bags of each brand should she buy?

12. A clothing manufacturer has 70 yards of a certain fabric and 156 buttons in stock. It manufacturers jackets and slacks that use this fabric and button. Each jacket requires $1\frac{1}{3}$ yards of fabric and 4 buttons. Each pair of slacks required $1\frac{3}{4}$ yards of fabric and 3 buttons. How many jackets and pairs of slacks should the manufacturer produce to use all the available fabric and buttons?

SOLUTIONS

1. Solve for x in B: $x = -3 + 2y$ and substitute this for x in A.

$$2x + 3y = 1$$
$$2(-3 + 2y) + 3y = 1$$
$$-6 + 4y + 3y = 1$$
$$7y = 7$$
$$y = 1 \qquad \text{Put } y = 1 \text{ in } x = -3 + 2y$$
$$x = -3 + 2(1) = -1$$

The solution is $(-1, 1)$.

2. Solve for x in B: $x = -4y$ and substitute this for x in A.

$$x + y = 3$$
$$-4y + y = 3$$
$$-3y = 3$$
$$y = -1 \qquad \text{Put } y = -1 \text{ in } x = -4y$$
$$x = -4(-1) = 4$$

The solution is $(4, -1)$.

3. We will add A + B.

$$\begin{array}{rl} -2x + 7y = 19 & \text{A} \\ \underline{2x - 4y = -10} & +\text{B} \\ 3y = 9 & \\ y = 3 & \\ -2x + 7(3) = 19 & \text{Put } y = 3 \text{ in A} \\ x = 1 & \end{array}$$

The solution is $(1, 3)$.

4.

$$\begin{array}{rl} 15x - y = 9 & \text{A} \\ \underline{2x + y = 8} & +\text{B} \\ 17x = 17 & \\ x = 1 & \\ 15(1) - y = 9 & \text{Put } x = 1 \text{ in A} \\ y = 6 & \end{array}$$

The solution is $(1, 6)$.

5. We will add $-\text{A} + \text{B}$.

$$\begin{array}{rl} 3x - 2y = -12 & -\text{A} \\ \underline{4x + 2y = \quad -2} & +\text{B} \\ 7x = -14 & \\ x = -2 & \\ -3(-2) + 2y = 12 & \text{Put } x = -2 \text{ in A} \\ y = 3 & \end{array}$$

The solution is $(-2, 3)$.

6. We will compute $\text{A} - 2\text{B}$.

$$\begin{array}{rl} 6x - 5y = 1 & \text{A} \\ \underline{-6x + 4y = -2} & -2\text{B} \\ -y = -1 & \\ y = 1 & \\ 6x - 5(1) = 1 & \text{Put } y = 1 \text{ in A} \\ x = 1 & \end{array}$$

The solution is $(1, 1)$.

7. We will compute 3A – 5B.

$$
\begin{aligned}
15x - 27y &= -78 \quad 3\text{A}\\
\underline{-15x - 10y} &\underline{= -70} \quad -5\text{B}\\
-37y &= -148\\
y &= 4\\
5x - 9(4) &= -26 \quad \text{Put } y = 4 \text{ in A}\\
x &= 2
\end{aligned}
$$

The solution is (2, 4).

8. We will compute 3A – 2B.

$$
\begin{aligned}
21x + 6y &= 3 \quad 3\text{A}\\
\underline{-4x - 6y} &\underline{= 14} \quad -2\text{B}\\
17x &= 17\\
x &= 1\\
7(1) + 2y &= 1 \quad \text{Put } x = 1 \text{ in A}\\
y &= -3
\end{aligned}
$$

The solution is (1, –3).

9. First clear the fractions.

$$
\begin{aligned}
45x + 12y &= 23 \quad 60\text{A}\\
6x - 9y &= -4 \quad 36\text{B}
\end{aligned}
$$

Add 3 times the first to 4 times the second.

$$
\begin{aligned}
135x + 36y &= 69\\
\underline{24x - 36y} &\underline{= -16}\\
159x &= 53\\
x &= \frac{53}{159} = \frac{1}{3}
\end{aligned}
$$

$$45\left(\frac{1}{3}\right) + 12y = 23$$

$$y = \frac{2}{3}$$

The solution is $(\frac{1}{3}, \frac{2}{3})$.

10. Let x represent the number of gallons of the name brand sold and y represent the number of gallons of the store brand sold. The total number of gallons sold is 4500, giving us $x + y = 4500$. Revenue from the name brand is $3.50x$ and is $2.25y$ for the store brand. Total revenue is \$12,875, giving us the equation $3.50x + 2.25y = 12{,}875$.

$$\begin{cases} x + y = 4{,}500 \\ 3.50x + 2.25y = 12{,}875 \end{cases}$$

We will use substitution.

$$x = 4500 - y$$

$$3.50(4500 - y) + 2.25y = 12{,}875$$

$$y = 2300$$

$$x = 4500 - y = 4500 - 2300 = 2200$$

The store sold 2200 gallons of the name brand and 2300 gallons of the store brand.

11. Let x represent the number of bags of Brand X and y, the number of bags of Brand Y. She will get $3x$ pounds of Nutrient A from x bags of Brand X and $4y$ pounds from y bags of Brand Y, so we need $3x + 4y = 39$. She will get $2x$ pounds of Nutrient B from x bags of Brand X and $1y$ pounds of Nutrient B from y bags of Brand Y, so we need $2x + y = 16$. We will use substitution.

$$y = 16 - 2x$$

$$3x + 4(16 - 2x) = 39$$

$$x = 5$$

$$y = 16 - 2x = 16 - 2(5) = 6$$

The gardener needs to buy 5 bags of Brand X and 6 bags of Brand Y.

12. Let x represent the number of jackets to be produced and y the number of pairs of slacks. To use 70 yards of fabric, we need $1\frac{1}{3}x + 1\frac{3}{4}y = 70$. To use 156 buttons, we need $4x + 3y = 156$.

$$1\frac{1}{3}x + 1\frac{3}{4}y = 70$$

$$\frac{4}{3}x + \frac{7}{4}y = 70 \quad \text{F}$$

$$4x + 3y = 156 \quad \text{B}$$

$$16x + 21y = 840 \quad 12\text{F}$$

$$\underline{-16x - 12y = -624} \quad -4\text{B}$$

$$9y = 216$$

$$y = 24$$

$$4x + 3(24) = 156$$

$$x = 21$$

The manufacturer should produce 21 jackets and 24 pairs of slacks.

Two lines in the plane either intersect in one point, are parallel, or are really the same line. Until now, our lines have intersected in one point. When solving a system of two linear equations that are parallel or are on the same line, both variables will cancel and we are left with a true statement such as "$3 = 3$" or a false statement such as "$5 = 1$." We will get a true statement when the two lines are the same and a false statement when they are parallel.

EXAMPLES

- $\begin{cases} 2x - 3y = 6 & \text{A} \\ -4x + 6y = 8 & \text{B} \end{cases}$

$$4x - 6y = 12 \quad 2\text{A}$$

$$\underline{-4x + 6y = 8} \quad +\text{B}$$

$$0 = 20$$

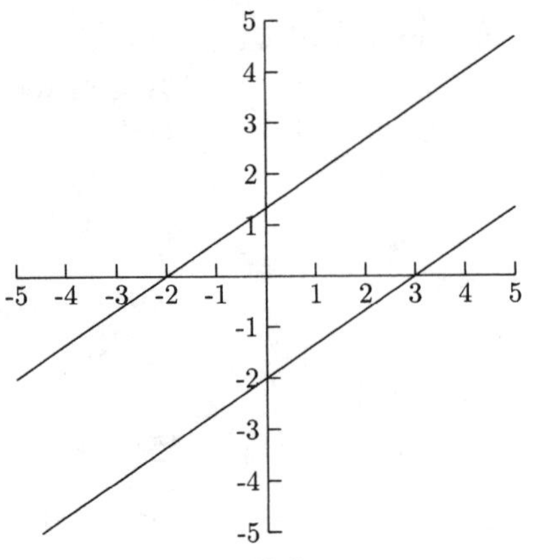

Fig. 10.4.

This is a false statement, so the lines are parallel. They are sketched in Figure 10.4

- $\begin{cases} y = \frac{2}{3}x - 1 \\ 2x - 3y = 3 \end{cases}$

We will use substitution.

$$2x - 3\left(\frac{2}{3}x - 1\right) = 3$$

$$2x - 2x + 3 = 3$$

$$0 = 0$$

Because $0 = 0$ is a true statement, these lines are the same.

When the system of equations is not a pair of lines, there could be no solutions, one solution, or more than one solution. The same methods used for pairs of lines will work with other kinds of systems.

EXAMPLES

- $\begin{cases} y = x^2 - 2x - 3 & \text{A} \\ 3x - y = 7 & \text{B} \end{cases}$

Elimination by addition would not work to eliminate x^2 because B has no x^2 term to cancel x^2 in A. Solving for x in B and substituting it in for x in A

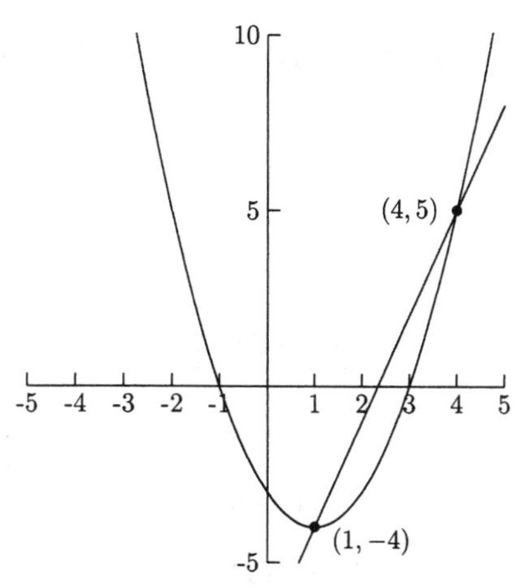

Fig. 10.5.

would work to eliminate x. Both addition and substitution will work to eliminate y. We will use addition to eliminate y.

$$y = x^2 - 2x - 3 \quad \text{A}$$

$$\underline{3x - y = 7} \qquad +\text{B}$$

$$3x = x^2 - 2x + 4$$

$$0 = x^2 - 5x + 4$$

$$0 = (x - 1)(x - 4)$$

The solutions occur when $x = 1$ or $x = 4$. We need to find two y-values. We will let $x = 1$ and $x = 4$ in A.

$$y = 1^2 - 2(1) - 3 = -4 \qquad (1, -4) \text{ is one solution.}$$

$$y = 4^2 - 2(4) - 3 = 5 \qquad (4, 5) \text{ is the other solution.}$$

We can see from the graphs in Figure 10.5 that these solutions are correct.

- $\begin{cases} x^2 + y^2 = 25 & \text{A} \\ y = -\frac{1}{3}x^2 + 7 & \text{B} \end{cases}$

We could solve for x^2 in A and substitute this in B. We cannot add the equations to eliminate y or y^2 because A does not have a y term to cancel y in B and B does not have a y^2 term to cancel y^2 in A. We will move $-\frac{1}{3}x^2$

to the left side of B and multiply B by -3. Then we can add this to A to eliminate x^2.

$$\frac{1}{3}x^2 + y = 7 \quad \text{B}$$

$$x^2 + y^2 = 25 \quad \text{A}$$

$$\underline{-x^2 - 3y = -21} \quad -3\text{B}$$

$$y^2 - 3y = 4$$

$$y^2 - 3y - 4 = 0$$

$$(y - 4)(y + 1) = 0$$

The solutions occur when $y = 4, -1$. Put $y = 4, -1$ in A to find their x-values.

$$x^2 + 4^2 = 25$$

$$x^2 = 9$$

$$x = \pm 3 \quad (-3, 4) \text{ and } (3, 4) \text{ are solutions.}$$

$$x^2 + (-1)^2 = 25$$

$$x^2 = 24$$

$$x = \pm\sqrt{24} = \pm 2\sqrt{6} \quad (2\sqrt{6}, -1) \text{ and } (-2\sqrt{6}, -1) \text{ are solutions.}$$

- $$\begin{cases} x^2 + y^2 = 4 & \text{A} \\ y = \frac{2}{x} & \text{B} \end{cases}$$

Addition will not work on this system but substitution will. We will substitute $y = \frac{2}{x}$ for y in A.

$$x^2 + \left(\frac{2}{x}\right)^2 = 4$$

$$x^2 + \frac{4}{x^2} = 4 \qquad \text{The LCD is } x^2$$

$$x^2\left(x^2 + \frac{4}{x^2}\right) = x^2(4)$$

$$x^4 + 4 = 4x^2$$

$$x^4 - 4x^2 + 4 = 0$$

$$(x^2 - 2)(x^2 - 2) = 0$$

$$x^2 = 2$$

$$x = \pm\sqrt{2}$$

We will put $x = \sqrt{2}$ and $x = -\sqrt{2}$ in $y = \frac{2}{x}$.

$$y = \frac{2}{\sqrt{2}} = \frac{2\sqrt{2}}{\sqrt{2}\sqrt{2}} = \frac{2\sqrt{2}}{2} = \sqrt{2}; \quad (\sqrt{2}, \sqrt{2}) \text{ is a solution.}$$

$$y = \frac{2}{-\sqrt{2}} = \frac{2\sqrt{2}}{-\sqrt{2}\sqrt{2}} = \frac{2\sqrt{2}}{-2} = -\sqrt{2}; \quad (-\sqrt{2}, -\sqrt{2}) \text{ is a solution.}$$

PRACTICE

Solve the systems of equations. Put your solutions in the form of a point, (x, y).

1.

$$\begin{cases} y = x^2 - 4 & \text{A} \\ x + y = 8 & \text{B} \end{cases}$$

2.

$$\begin{cases} x^2 + y^2 + 6x - 2y = -5 & \text{A} \\ y = -2x - 5 & \text{B} \end{cases}$$

3.

$$\begin{cases} x^2 - y^2 = 16 & \text{A} \\ x^2 + y^2 = 16 & \text{B} \end{cases}$$

4.

$$\begin{cases} 4x^2 + y^2 = 5 & \text{A} \\ y = \frac{1}{x} & \text{B} \end{cases}$$

SOLUTIONS

1.

$$y = x^2 - 4 \quad \text{A}$$
$$\underline{-x - y = -8} \quad -\text{B}$$
$$-x = x^2 - 12$$
$$0 = x^2 + x - 12 = (x + 4)(x - 3)$$

There are solutions for $x = -4$ and $x = 3$. Put these in A.

$$y = (-4)^2 - 4 = 12; \qquad (-4, 12) \text{ is a solution.}$$
$$y = 3^2 - 4 = 5; \qquad (3, 5) \text{ is a solution.}$$

2. Substitute $-2x - 5$ for y in A.

$$x^2 + (-2x - 5)^2 + 6x - 2(-2x - 5) = -5$$
$$x^2 + 4x^2 + 20x + 25 + 6x + 4x + 10 = -5$$
$$5x^2 + 30x + 40 = 0 \quad \text{Divide by 5}$$
$$x^2 + 6x + 8 = 0$$
$$(x + 4)(x + 2) = 0$$

There are solutions for $x = -4$ and $x = -2$. We will put these in B instead of A because there is less computation to do in B.

$$y = -2(-4) - 5 = 3; \qquad (-4, 3) \text{ is a solution.}$$
$$y = -2(-2) - 5 = -1; \qquad (-2, -1) \text{ is a solution.}$$

3.

$$x^2 - y^2 = 16 \quad \text{A}$$
$$\underline{x^2 + y^2 = 16} \quad +\text{B}$$
$$2x^2 = 32$$
$$x^2 = 16$$
$$x = \pm 4$$

Put $x = 4$ and $x = -4$ in A.

$$(-4)^2 - y^2 = 16 \qquad 4^2 - y^2 = 16$$

$$16 - y^2 = 16 \qquad 16 - y^2 = 16$$

$$y^2 = 0 \qquad y^2 = 0$$

$$y = 0 \qquad y = 0$$

The solutions are $(-4, 0)$ and $(4, 0)$.

4. Substitute $\frac{1}{x}$ for y in A.

$$4x^2 + \left(\frac{1}{x}\right)^2 = 5$$

$$x^2\left(4x^2 + \frac{1}{x^2}\right) = x^2(5)$$

$$4x^4 + 1 = 5x^2$$

$$4x^4 - 5x^2 + 1 = 0$$

$$(4x^2 - 1)(x^2 - 1) = 0$$

$$(2x - 1)(2x + 1)(x - 1)(x + 1) = 0$$

The solutions are $x = \pm\frac{1}{2}$ (from $2x - 1 = 0$ and $2x + 1 = 0$) and $x = \pm 1$. Put these in B.

$$y = \frac{1}{\frac{1}{2}} = 2; \qquad (\frac{1}{2}, 2) \text{ is a solution.}$$

$$y = \frac{1}{-\frac{1}{2}} = -2; \qquad (-\frac{1}{2}, -2) \text{ is a solution.}$$

$$y = \frac{1}{1} = 1; \qquad (1, 1) \text{ is a solution.}$$

$$y = \frac{1}{-1} = -1; \qquad (-1, -1) \text{ is a solution.}$$

Systems of Inequalities

The solution (if any) for a system of inequalities is usually a region in the plane. The solution to a polynomial inequality (the only kind in this book) is the region above or below the curve. We will begin with linear inequalities.

When sketching the graph for an inequality, we will use a solid graph for "$\leq$" and "$\geq$" inequalities, and a dashed graph for "$<$" and "$>$" inequalities. We can decide which side of the graph to shade by choosing *any* point not on the graph itself. We will put this point into the inequality. If it makes the inequality true, we will shade the side that has that point. If it makes the inequality false, we will shade the other side. *Every* point in the shaded region is a solution to the inequality.

EXAMPLES

- $2x + 3y \leq 6$

  We will sketch the line $2x + 3y = 6$, using a solid line because the inequality is "$\leq$."

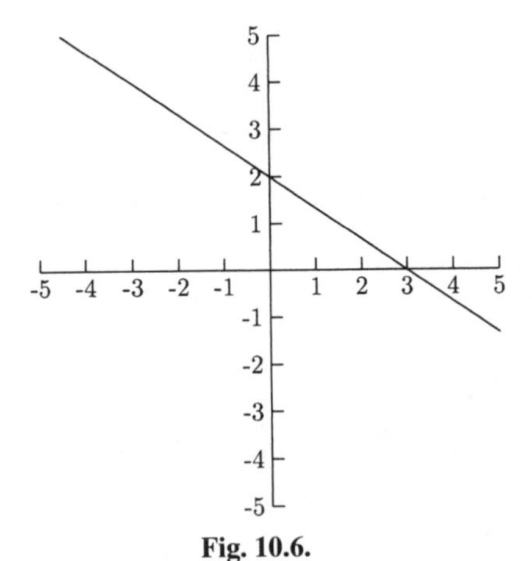

Fig. 10.6.

We will always use the origin, $(0, 0)$ in our inequalities unless the graph goes through the origin. Does $x = 0$ and $y = 0$ make $2x + 3y \leq 6$ true? $2(0) + 3(0) \leq 6$ is a true statement, so we will shade the side that has the origin.

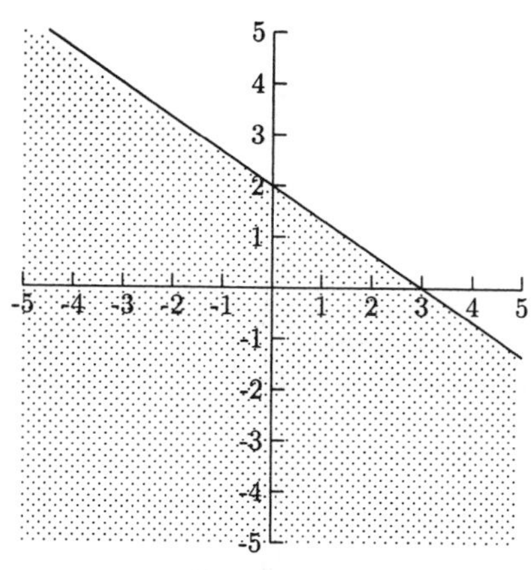

Fig. 10.7.

- $x - 2y > 4$

  We will sketch the line $x - 2y = 4$ using a dashed line because the inequality is ">."

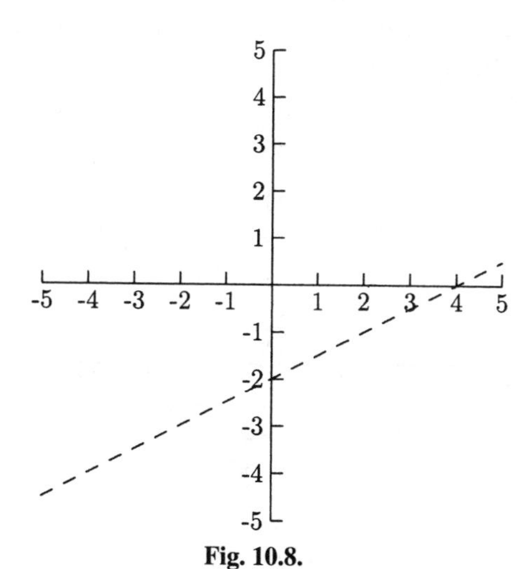

Fig. 10.8.

Now we need to decide which side of the line to shade. When we put (0, 0) in $x - 2y > 4$, we get the false statement $0 - 2(0) > 4$. We need to shade the side of the line that does *not* have the origin.

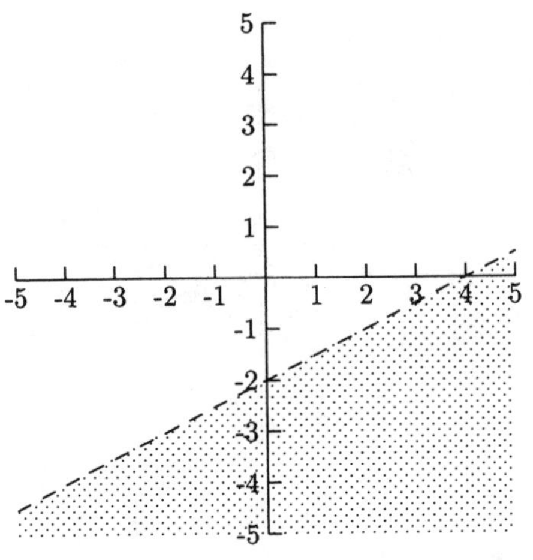

Fig. 10.9.

- $y < 3x$

  We use a dashed line to sketch the line $y = 3x$. Because the line goes through $(0, 0)$, we cannot use it to determine which side of the line to shade. This is because any point on the line makes the equality true. We want to know where the inequality is true. The point $(1, 0)$ is not on the line, so we can use it. $0 < 3(1)$ is true so we will shade the side of the line that has the point $(1, 0)$, which is the right side.

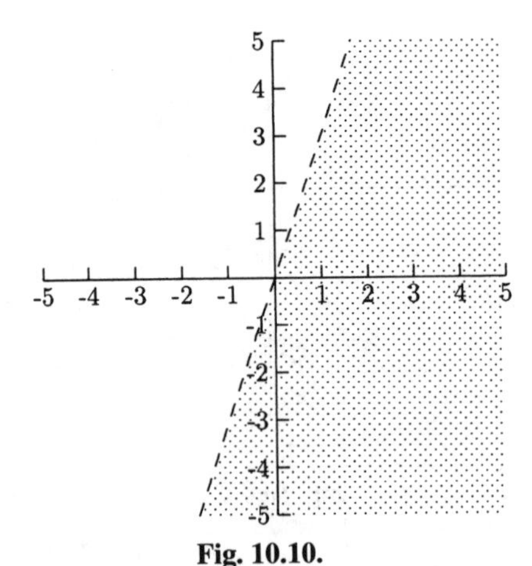

Fig. 10.10.

- $x \geq -3$

  The line $x = -3$ is a vertical line through $x = -3$. Because we want $x \geq -3$ we will shade to the right of the line.

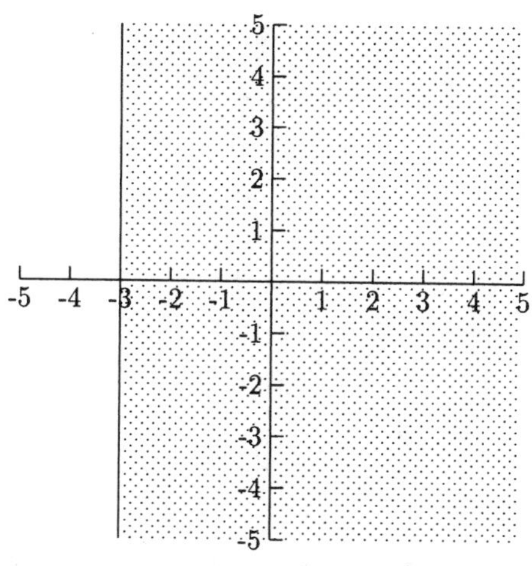

Fig. 10.11.

- $y < 2$

  The line $y = 2$ is a horizontal line at $y = 2$. Because we want $y < 2$, we will shade below the line.

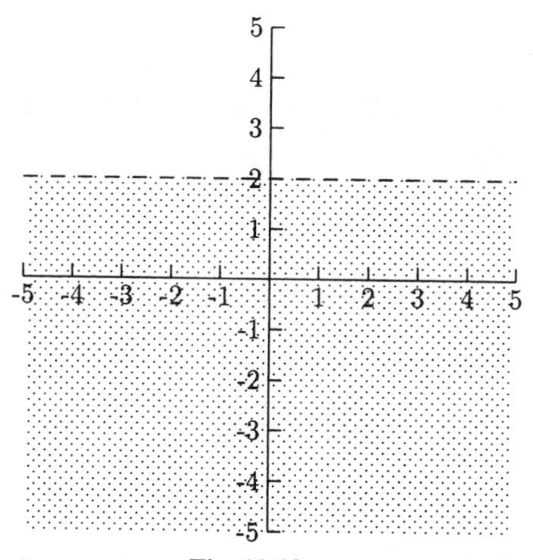

Fig. 10.12.

Graphing the solution region for nonlinear inequalities is done the same way—graph the inequality, using a solid graph for "≤" and "≥" inequalities and a dashed graph for "<" and ">" inequalities, then checking a point to see which side of the graph to shade.

EXAMPLES

- $y \leq x^2 - x - 2$

  The equality is $y = x^2 - x - 2 = (x - 2)(x + 1)$. The graph for this equation is a parabola.

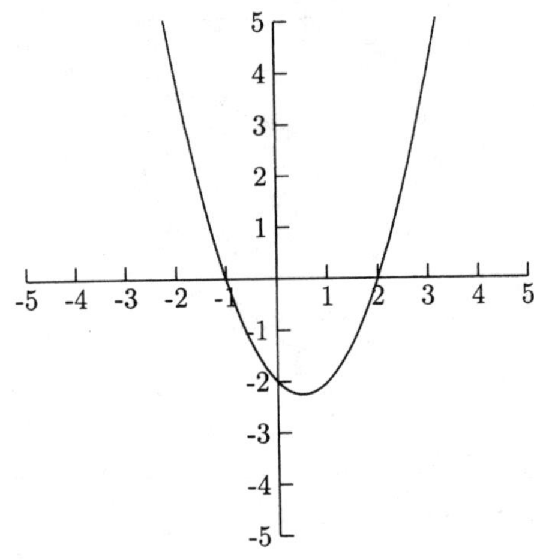

Fig. 10.13.

Because (0, 0) is not on the graph, we can use it to decide which side to shade; $0 \leq 0^2 - 0 - 2$ is false, so we shade below the graph, the side that does not contain (0, 0).

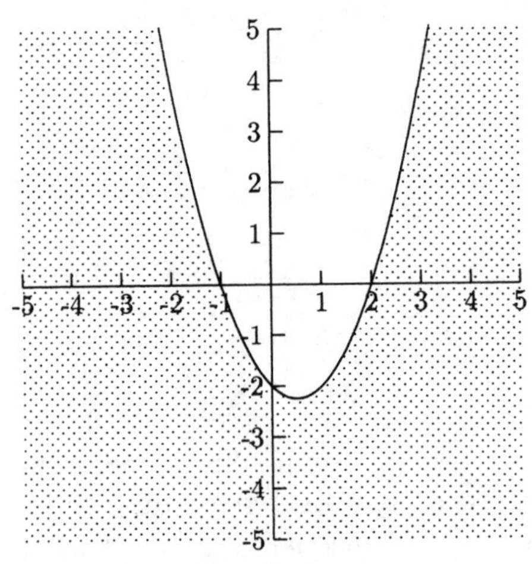

Fig. 10.14.

- $y > (x+2)(x-2)(x-4)$

  When we check (0, 0) in the inequality, we get the false statement $0 > (0+2)(0-2)(0-4)$. We will shade above the graph, the region that does not contain (0, 0).

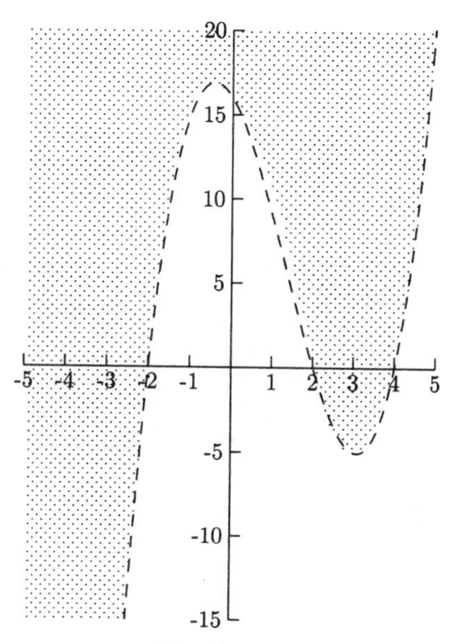

Fig. 10.15.

The solution (if there is one) to a system of two or more inequalities is the region that is part of each solution for the individual inequalities. For example, if we have a system of two inequalities and shade the solution to one inequality in blue and the other in yellow, then the solution to the system would be the region in green.

EXAMPLES

- $$\begin{cases} x - y < 3 \\ x + 2y > 1 \end{cases}$$

  Sketch the solution for each inequality. The solution to $x - y < 3$ is the region shaded vertically. The solution to $x + 2y > 1$ is the region shaded horizontally.

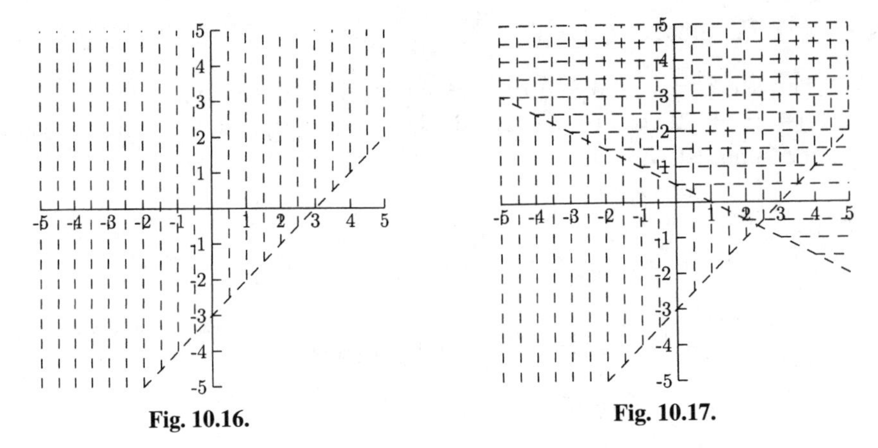

Fig. 10.16. Fig. 10.17.

The region that is in both solutions is above and between the lines.

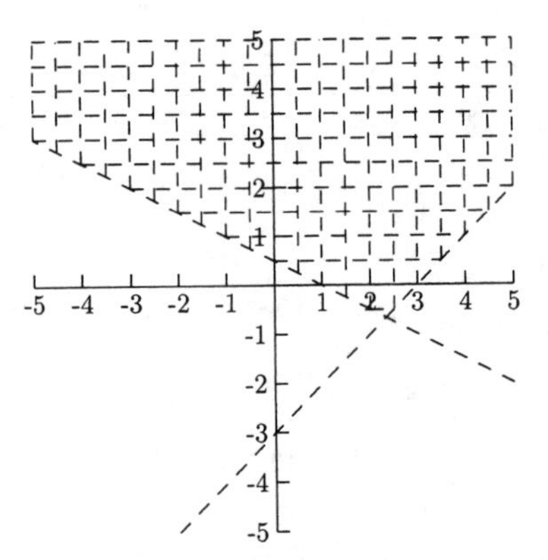

Fig. 10.18.

- $\begin{cases} y \leq 4 - x^2 \\ x - 7y \leq 4 \end{cases}$

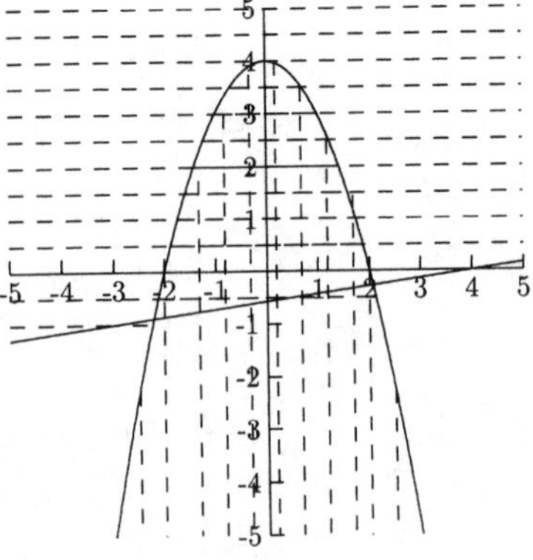

Fig. 10.19.

The solution to $y \leq 4 - x^2$ is the region shaded vertically. The solution to $x - 7y \leq 4$ is the region shaded horizontally. The region that is in both solutions is above the line and inside the parabola.

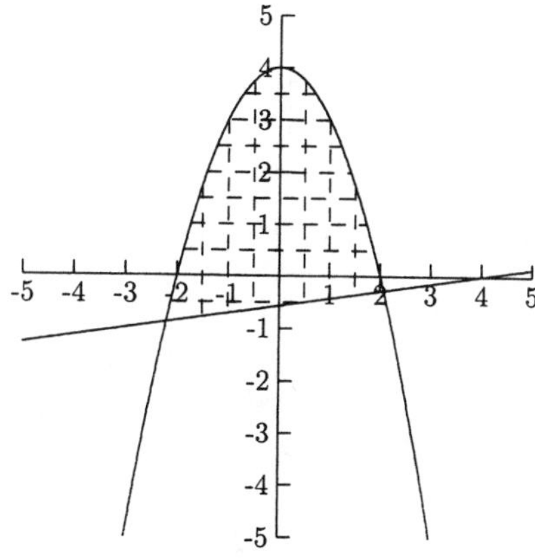

Fig. 10.20.

Because a solid graph indicates that points on the graph are also solutions, to be absolutely accurate, the correct solution uses dashed graphs for the part of the graphs that are not on the border of the shaded region.

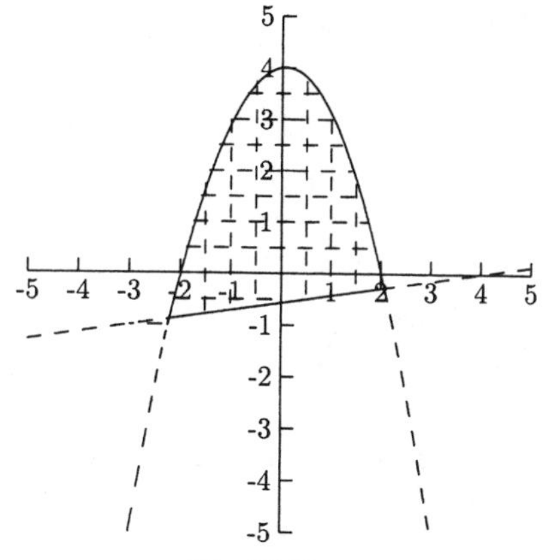

Fig. 10.21.

We will not quibble with this technicality here.

- $\begin{cases} 2x + y \leq 5 \\ x \geq 0 \\ y \geq 0 \end{cases}$

The inequalities $x \geq 0$ and $y \geq 0$ mean that we only need the top right corner of the graph. These inequalities are common in word problems.

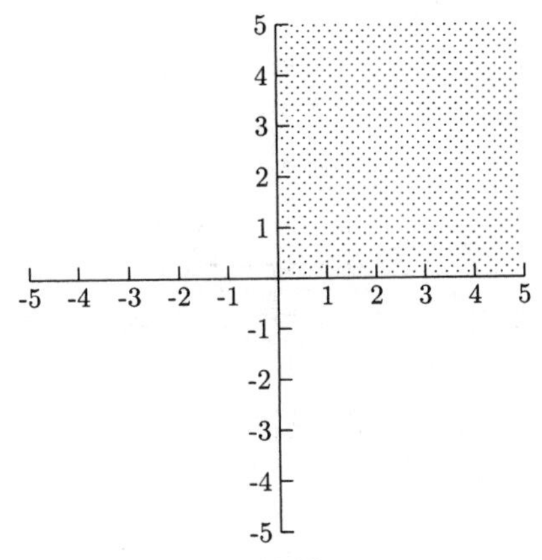

Fig. 10.22.

The solution to the system is the region in the top right corner of the graph below the line $2x + y = 5$.

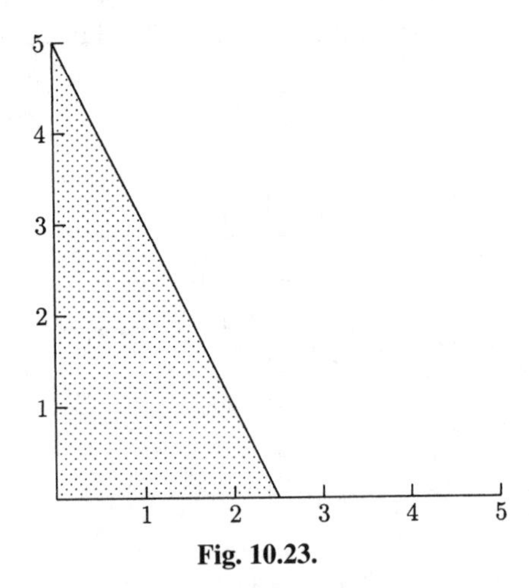

Fig. 10.23.

Some systems of inequalities have no solution. In the following example, the regions do not overlap, so there are no ordered pairs (points) that make both inequalities true.

- $\begin{cases} y \geq x^2 + 4 \\ x - y \geq 1 \end{cases}$

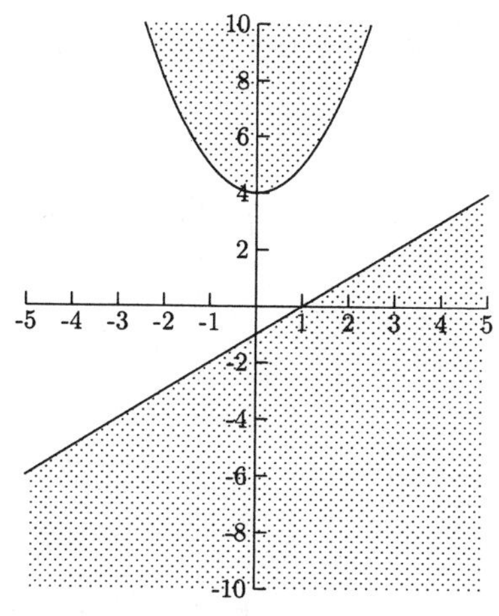

Fig. 10.24.

It is easy to lose track of the solution for a system of three or more inequalities. There are a couple of things you can do to make it easier. First, make sure the graph is large enough, using graph paper if possible. Second, shade the solution for each inequality in a different way, with different colors or shaded with horizontal, vertical, and slanted lines. The solution (if there is one) would be shaded all different ways. You could also shade one region at a time, erasing the part of the previous region that is not part of the inequality.

EXAMPLES

- $$\begin{cases} x + y \leq 4 \\ x \geq 1 \\ y \leq x \end{cases}$$

  First we will shade the solution for $x + y \leq 4$.

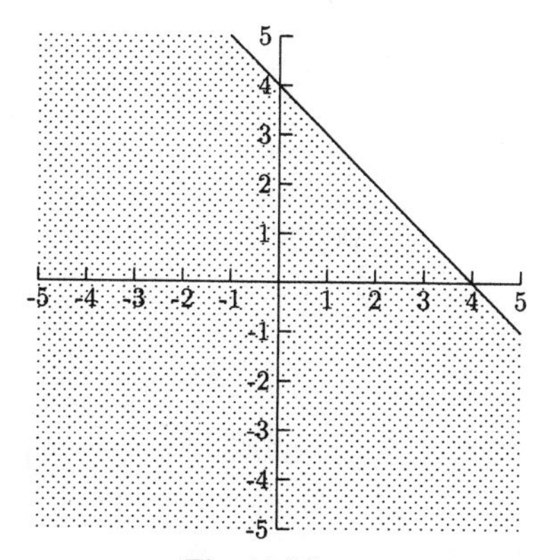

Fig. 10.25.

The region for $x \geq 1$ is the right of the line $x = 1$, so we will erase the region to the *left* of $x = 1$.

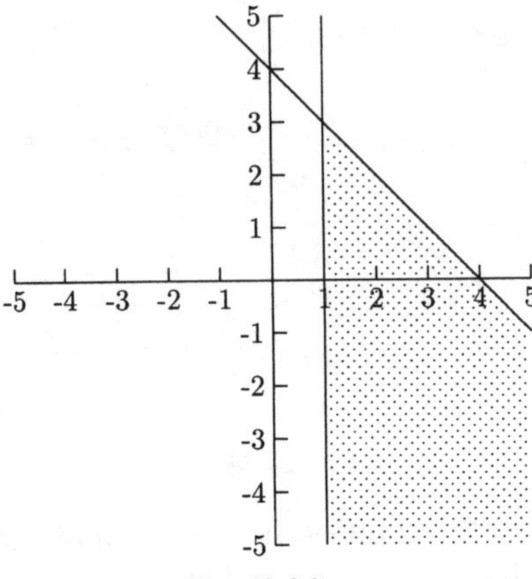

Fig. 10.26.

The solution to $y \leq x$ is the region below the line $y = x$, so we will erase the shading *above* the line $y = x$.
The shaded region in Figure 10.27 is the solution for the system.

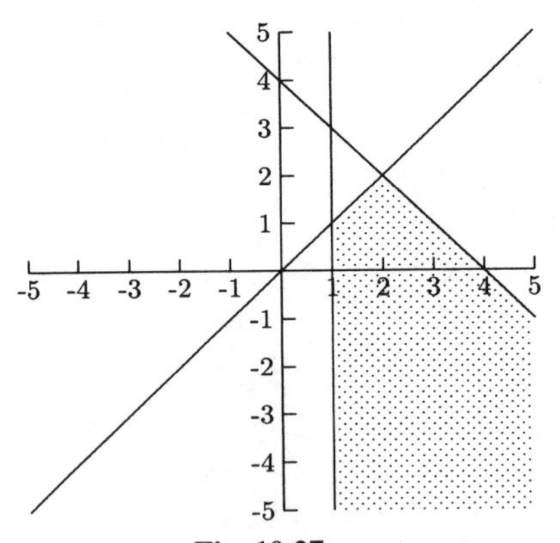

Fig. 10.27.

- $\begin{cases} y > x^2 - 16 \\ x < 2 \\ y < -5 \\ -x + y < -8 \end{cases}$

We will begin with $y > x^2 - 16$.

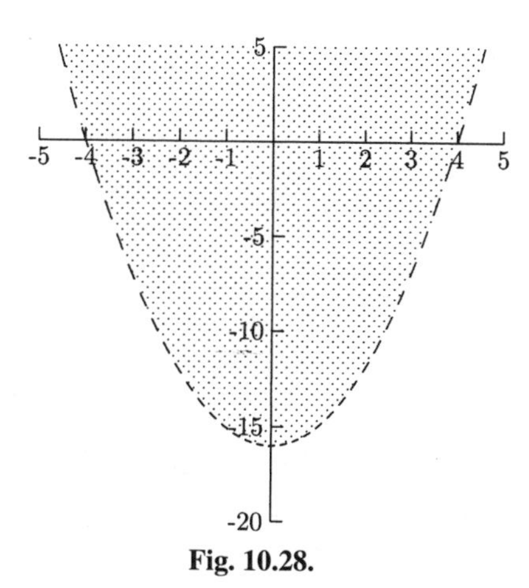

Fig. 10.28.

The solution to $x < 2$ is the region to the left of the line $x = 2$. We will erase the shading to the right of $x = 2$.

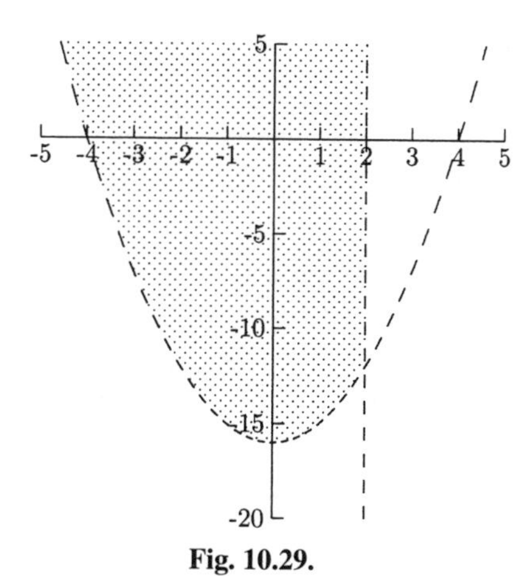

Fig. 10.29.

The solution to $y < -5$ is the region below the line $y = -5$. We will erase the shading above the line $y = -5$.

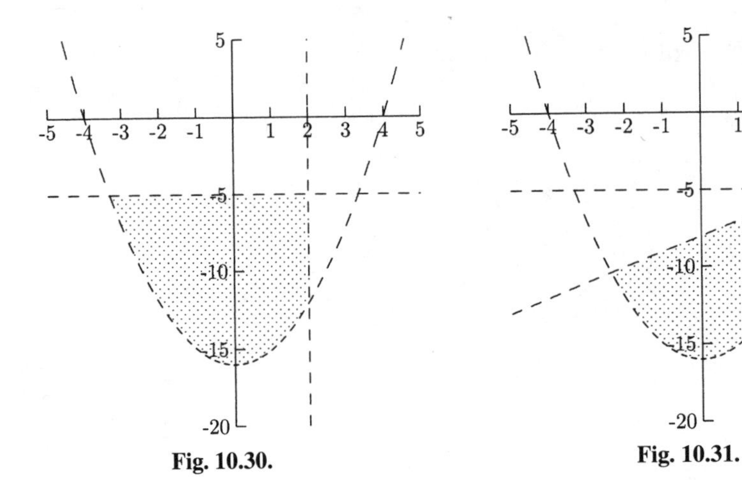

Fig. 10.30.

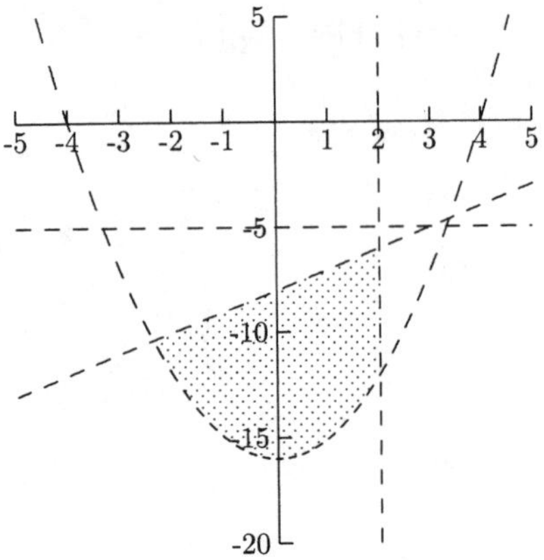

Fig. 10.31.

The solution to $-x + y < -8$ is the region below the line $-x + y = -8$, so we will erase the shading above the line. The solution to the system is in Figure 10.31.

PRACTICE

Graph the solution.

1. $2x - 4y < 4$
2. $x > 1$
3. $y \leq -1$
4. $y \leq x^2 - 4$
5. $y > x^3$
6. $y < |x|$
7. $y \geq (x - 3)(x + 1)(x + 3)$
8. $\begin{cases} 2x - y \leq 6 \\ x \geq 3 \end{cases}$
9. $\begin{cases} y > x^2 + 2x - 3 \\ x + y < 5 \end{cases}$
10. $\begin{cases} 2x + 3y \geq 6 \\ x \geq 0 \\ y \geq 0 \end{cases}$

11. $\begin{cases} 2x + y \geq 1 \\ -x + 2y \leq 4 \\ 5x - 3y \leq 15 \end{cases}$

SOLUTIONS

1.

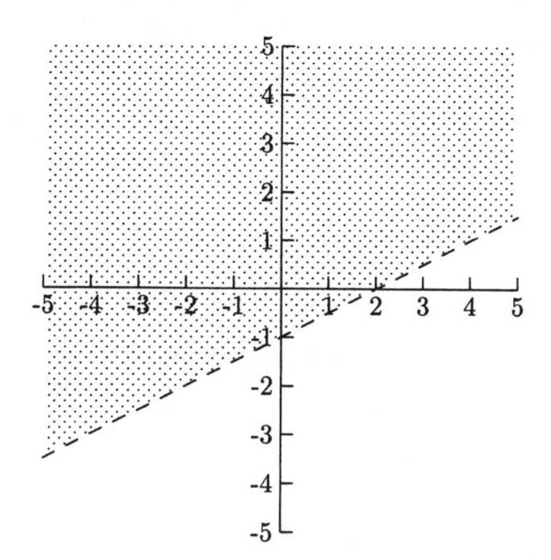

Fig. 10.32.

2.

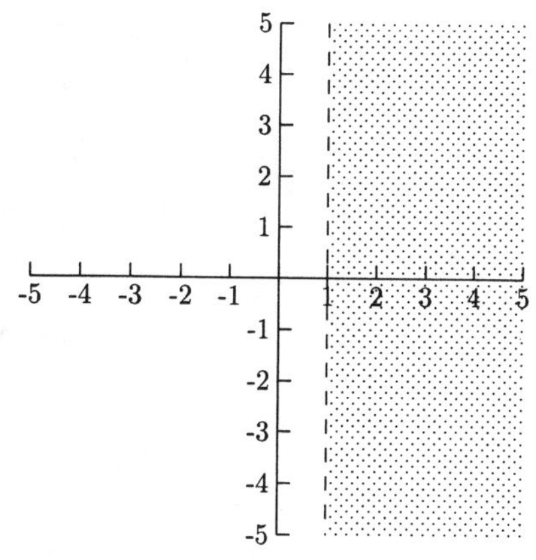

Fig. 10.33.

3.

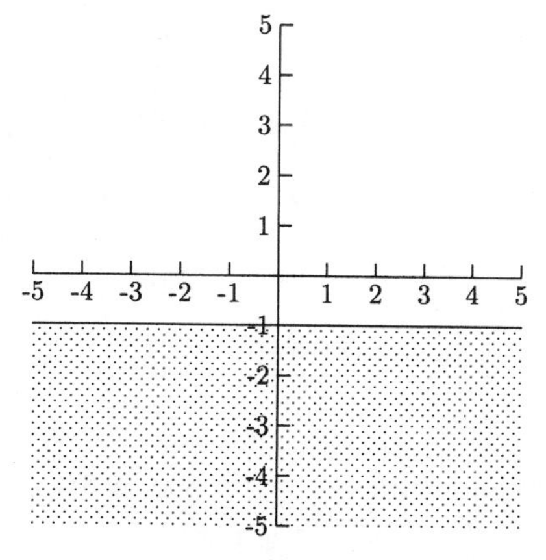

Fig. 10.34.

4.

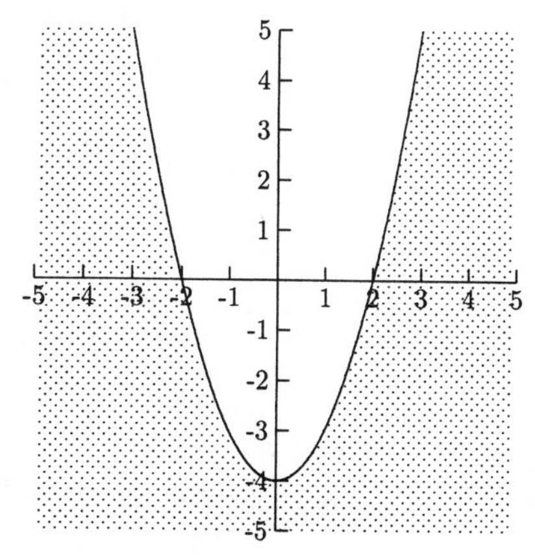

Fig. 10.35.

5.

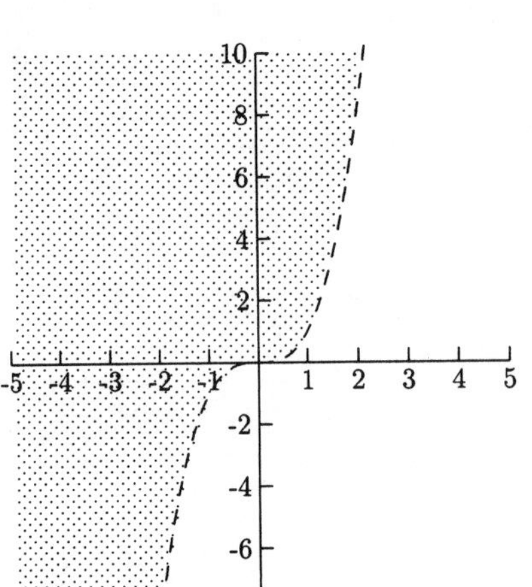

Fig. 10.36.

6.

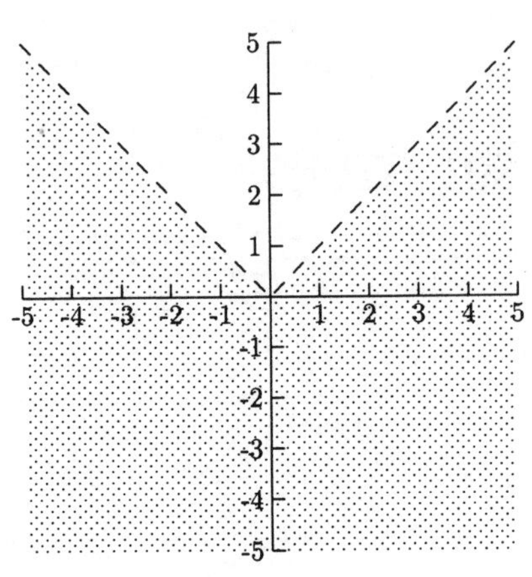

Fig. 10.37.

7.

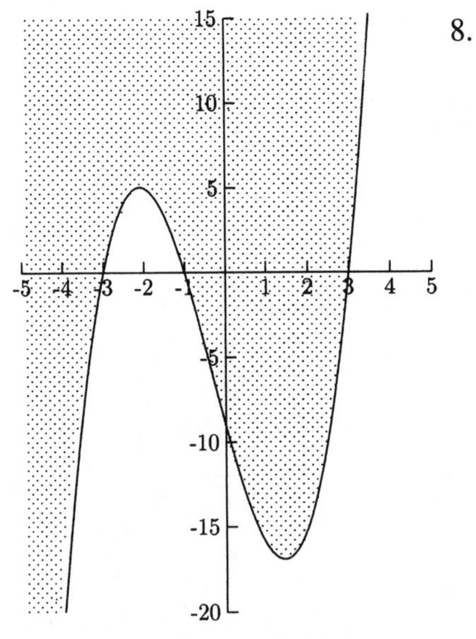

Fig. 10.38.

8.

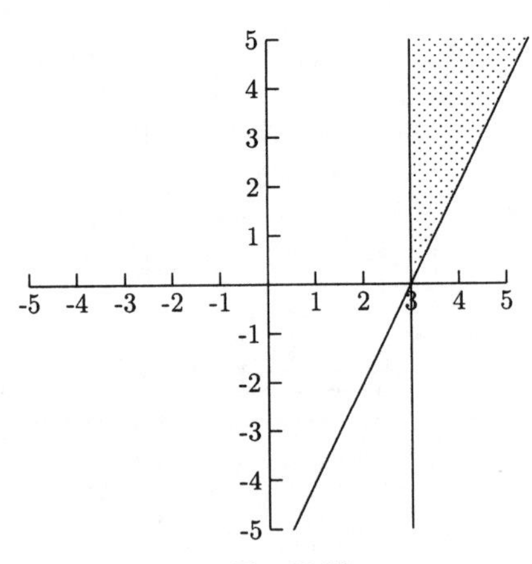

Fig. 10.39.

9.

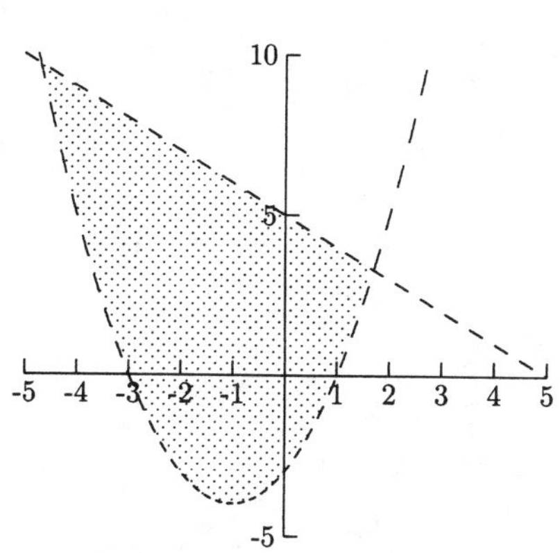

Fig. 10.40.

10.

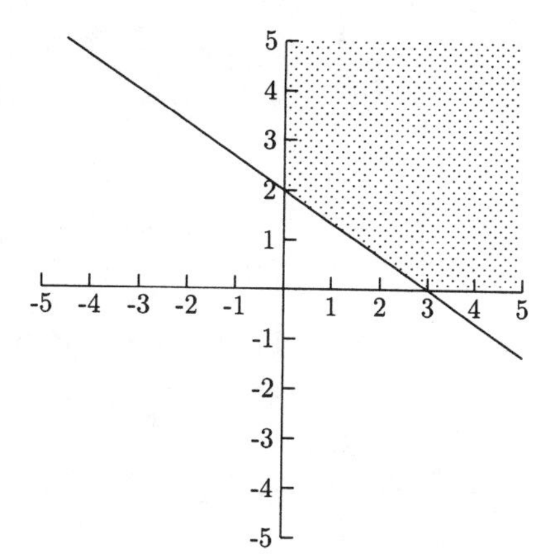

Fig. 10.41.

11.

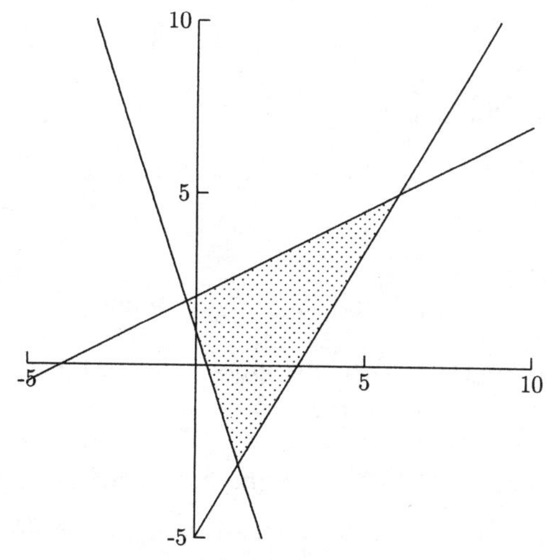

Fig. 10.42.

CHAPTER 10 REVIEW

In some of the following problems, you will be asked to find quantities such as $x + 2y$ for a system of equations. Solve the system and put the solution in the formula. For example, if the solution is $x = 3$ and $y = 5$, then $x + 2y$ becomes $3 + 2(5) = 13$.

1. Find $x + 2y$ for the system.

$$\begin{cases} 5x - 3y = 29 \\ 2x + 3y = -1 \end{cases}$$

(a) -2 (b) -1 (c) 1 (d) 2

2. Find $x + 2y$ for the system.

$$\begin{cases} y = 2x + 7 \\ y = -4x + 1 \end{cases}$$

(a) 8 (b) 9 (c) 10 (d) 11

3. Find $x + y$ for the system.

$$\begin{cases} 3x + 2y = 16 \\ 2x + 5y = 18 \end{cases}$$

(a) 4 (b) 5 (c) 6 (d) 7

4. Find $x + y$ for the system.

$$\begin{cases} y \quad = x^2 - 3x - 4 \\ x - y = -8 \end{cases}$$

(a) 2 and 14 (b) 3 and 12 (c) 4 and 20 (d) 5 and 15

5. The graph in Figure 10.43 is the graph of which inequality?
(a) $y > 2x + 2$ (b) $y \geq 2x + 2$ (c) $y \leq 2x + 2$ (d) $y < 2x + 2$

6. The graph in Figure 10.44 is the graph of which inequality?
(a) $y > x^2 - 2x + 1$ (b) $y \geq x^2 - 2x + 1$
(c) $y \leq x^2 - 2x + 1$ (d) $y < x^2 - 2x + 1$

7. The graph in Figure 10.45 is the graph for which system?

(a) $\begin{cases} y \leq -x^2 + 4x \\ y \leq x \end{cases}$ (b) $\begin{cases} y \leq -x^2 + 4x \\ y \geq x \end{cases}$

(c) $\begin{cases} y \geq -x^2 + 4x \\ y \geq x \end{cases}$ (d) $\begin{cases} y \geq -x^2 + 4x \\ y \leq x \end{cases}$

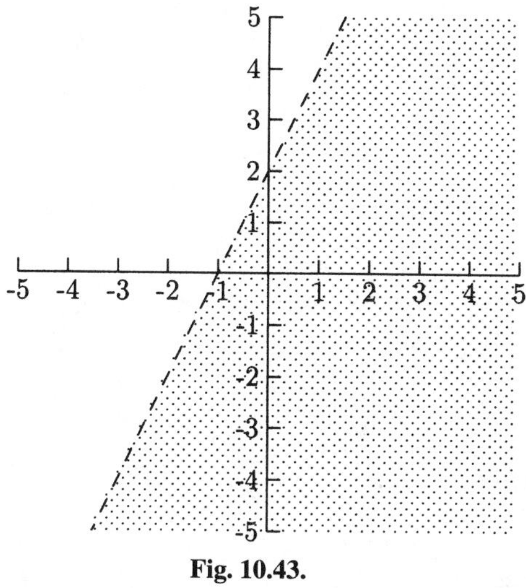

Fig. 10.43.

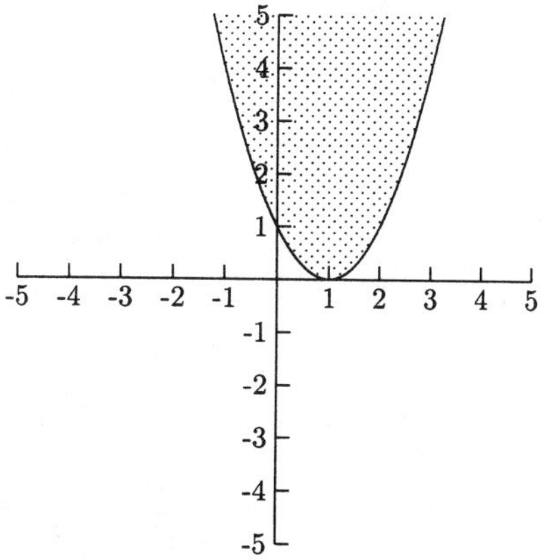

Fig. 10.44.

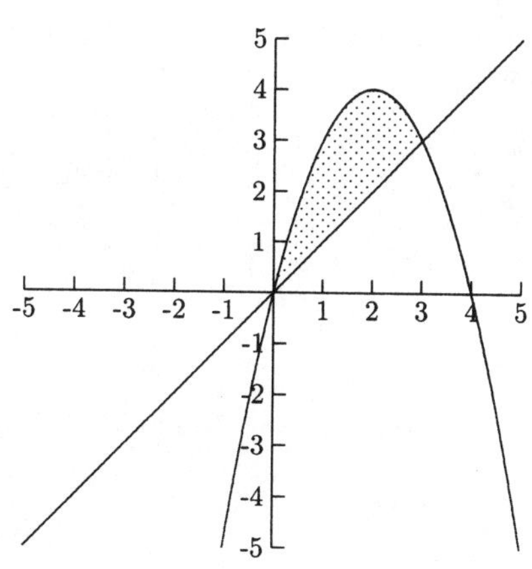

Fig. 10.45.

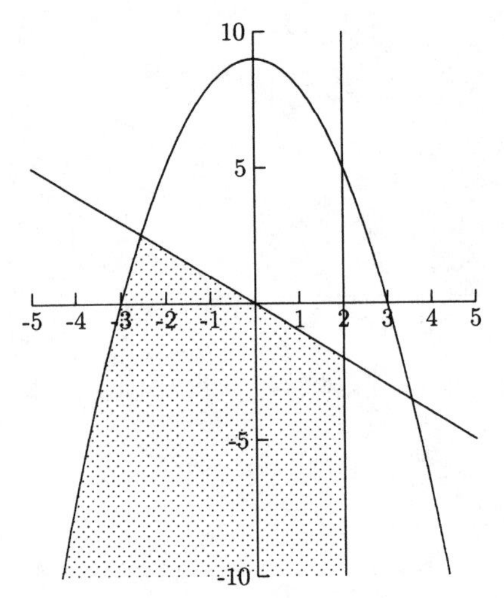

Fig. 10.46.

8. The graph in Figure 10.46 is the graph of which system?

(a) $\begin{cases} y \leq -x^2 + 9 \\ y \geq -x \\ x \geq 2 \end{cases}$

(b) $\begin{cases} y \leq -x^2 + 9 \\ y \geq -x \\ x \leq 2 \end{cases}$

(c) $\begin{cases} y \leq -x^2 + 9 \\ y \leq -x \\ x \geq 2 \end{cases}$

(d) $\begin{cases} y \leq -x^2 + 9 \\ y \leq -x \\ x \leq 2 \end{cases}$

SOLUTIONS

1. A 2. B 3. C 4. C 5. D 6. B 7. B 8 D

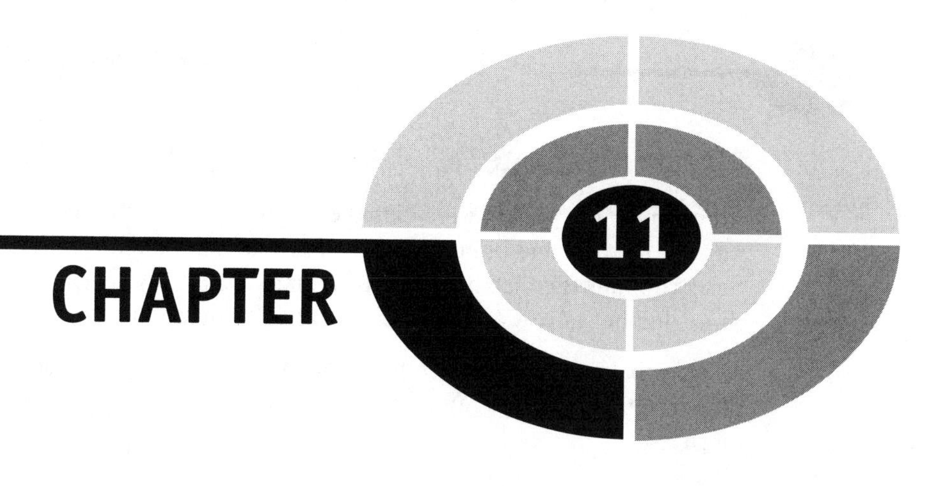

CHAPTER 11

Matrices

A matrix is an array of numbers or symbols made up of rows and columns. Matrices are used in science and business to represent several variables and relationships at once. For example, suppose there are three brands of fertilizers that provide different levels of three minerals that a gardener might need. The following matrix shows how much of each mineral is provided by each brand.

	Mineral A	Mineral B	Mineral C
Brand X	6	2	1
Brand Y	2	1	2
Brand Z	1	3	6

We will learn some matrix arithmetic as well as two matrix methods used to solve systems of linear equations. Most of the calculations are tedious. Fortunately graphing calculators and computer programs (including spreadsheets) can do most of them.

Matrix Arithmetic

The numbers in a matrix are called cells or entries. A matrix's size is given by the number of rows and columns it has. For example, a matrix that has two rows and three columns is called a 2×3 (pronounced "2 by 3") matrix. A matrix that has the same number of rows as columns is called a square matrix.

Two matrices need to be the same size before we can add them or find their difference. The sum of two or more matrices is the sum of their corresponding entries.

$$\begin{bmatrix} 2 & -1 \\ 3 & 4 \end{bmatrix} + \begin{bmatrix} 5 & 9 \\ 4 & 1 \end{bmatrix} = \begin{bmatrix} 2+5 & -1+9 \\ 3+4 & 4+1 \end{bmatrix} = \begin{bmatrix} 7 & 8 \\ 7 & 5 \end{bmatrix}$$

Subtract one matrix from another by subtracting their corresponding entries.

$$\begin{bmatrix} 2 & -1 \\ 3 & 4 \end{bmatrix} - \begin{bmatrix} 5 & 9 \\ 4 & 1 \end{bmatrix} = \begin{bmatrix} 2-5 & -1-9 \\ 3-4 & 4-1 \end{bmatrix} = \begin{bmatrix} -3 & -10 \\ -1 & 3 \end{bmatrix}$$

The *scalar product* of a matrix is a matrix whose entries are multiplied by a fixed number.

$$3\begin{bmatrix} 6 & -4 \\ 2 & 1 \\ 5 & 0 \end{bmatrix} = \begin{bmatrix} \mathbf{3}\cdot 6 & \mathbf{3}\cdot(-4) \\ \mathbf{3}\cdot 2 & \mathbf{3}\cdot 1 \\ \mathbf{3}\cdot 5 & \mathbf{3}\cdot 0 \end{bmatrix} = \begin{bmatrix} 18 & -12 \\ 6 & 3 \\ 15 & 0 \end{bmatrix}$$

It might seem that matrix multiplication is carried out the same way addition and subtraction are—multiply their corresponding entries. This operation is not very useful. The matrix multiplication operation that is useful requires more work. Two matrices do not need to be the same size, but the number of columns of the first matrix must be the same as the number of rows of the second matrix. This is because we get the entries of the product matrix by multiplying the rows of the first matrix by the columns of the second matrix. Here, we will multiply a 3×3 matrix by a 3×2 matrix.

$$\begin{bmatrix} A & B & C \\ D & E & F \\ G & H & I \end{bmatrix} \cdot \begin{bmatrix} K & L \\ M & N \\ O & P \end{bmatrix} = \begin{bmatrix} \text{Row 1} \times \text{Column 1} & \text{Row 1} \times \text{Column 2} \\ \text{Row 2} \times \text{Column 1} & \text{Row 2} \times \text{Column 2} \\ \text{Row 3} \times \text{Column 1} & \text{Row 3} \times \text{Column 2} \end{bmatrix}$$

Row 1 of the first matrix is $A\ B\ C$ and Column 1 of the second matrix is $\begin{matrix} K \\ M \\ O \end{matrix}$. The first entry on the product matrix is Row 1 × Column 1, which is this sum.

$$\begin{array}{cccc} & \text{Row 1} & & \text{Column 1} \\ & A & \times & K \\ & B & \times & M \\ + & C & \times & O \\ \hline \end{array}$$

$$\begin{bmatrix} \text{Row 1} \times \text{Column 1} & \text{Row 1} \times \text{Column 2} \\ \text{Row 2} \times \text{Column 1} & \text{Row 2} \times \text{Column 2} \\ \text{Row 3} \times \text{Column 1} & \text{Row 3} \times \text{Column 2} \end{bmatrix} = \begin{bmatrix} AK + BM + CO & AL + BN + CP \\ DK + EM + FO & DL + EN + FP \\ GK + HM + IO & GL + HN + IP \end{bmatrix}$$

EXAMPLES

- $\begin{bmatrix} 1 & -8 & 2 \\ 5 & 0 & -1 \\ 2 & 1 & 1 \end{bmatrix} \cdot \begin{bmatrix} 4 & -7 \\ -2 & 1 \\ 3 & 0 \end{bmatrix}$

$$= \begin{bmatrix} 1 \cdot 4 + (-8)(-2) + 2 \cdot 3 & 1(-7) + (-8)1 + 2 \cdot 0 \\ 5 \cdot 4 + 0(-2) + (-1)3 & 5(-7) + 0 \cdot 1 + (-1)0 \\ 2 \cdot 4 + 1(-2) + 1 \cdot 3 & 2(-7) + 1 \cdot 1 + 1 \cdot 0 \end{bmatrix} = \begin{bmatrix} 26 & -15 \\ 17 & -35 \\ 9 & -13 \end{bmatrix}$$

- $\begin{bmatrix} -6 & 2 \\ 7 & 1 \end{bmatrix} \cdot \begin{bmatrix} 4 & 1 & 0 \\ -3 & 5 & 2 \end{bmatrix}$

$$= \begin{bmatrix} -6 \cdot 4 + 2(-3) & -6 \cdot 1 + 2 \cdot 5 & -6 \cdot 0 + 2 \cdot 2 \\ 7 \cdot 4 + 1(-3) & 7 \cdot 1 + 1 \cdot 5 & 7 \cdot 0 + 1 \cdot 2 \end{bmatrix} = \begin{bmatrix} -30 & 4 & 4 \\ 25 & 12 & 2 \end{bmatrix}$$

An identity matrix is a square matrix with 1s on the main diagonal (from the upper left corner to the bottom right corner) and 0s everywhere else. The following are the 2×2 and 3×3 identity matrices.

$$\begin{bmatrix} 1 & 0 \\ 0 & 1 \end{bmatrix} \text{ and } \begin{bmatrix} 1 & 0 & 0 \\ 0 & 1 & 0 \\ 0 & 0 & 1 \end{bmatrix}$$

If we multiply any matrix by its corresponding identity matrix, we will get the original matrix back.

$$\begin{bmatrix} 1 & 0 \\ 0 & 1 \end{bmatrix} \cdot \begin{bmatrix} 3 & 6 & -2 \\ 2 & 1 & 5 \end{bmatrix} = \begin{bmatrix} 1 \cdot 3 + 0 \cdot 2 & 1 \cdot 6 + 0 \cdot 1 & 1(-2) + 0 \cdot 5 \\ 0 \cdot 3 + 1 \cdot 2 & 0 \cdot 6 + 1 \cdot 1 & 0(-2) + 1 \cdot 5 \end{bmatrix}$$

$$= \begin{bmatrix} 3 & 6 & -2 \\ 2 & 1 & 5 \end{bmatrix}$$

Matrix multiplication is not commutative. Reversing the order of the multiplication usually gets a different matrix, if the multiplication is even possible.

The matrix

$$\begin{bmatrix} 1 & -3 \\ 2 & 4 \end{bmatrix} \cdot \begin{bmatrix} 0 & 1 \\ 2 & -1 \end{bmatrix} = \begin{bmatrix} 1 \cdot 0 + (-3)2 & 1 \cdot 1 + (-3)(-1) \\ 2 \cdot 0 + 4 \cdot 2 & 2 \cdot 1 + 4(-1) \end{bmatrix} = \begin{bmatrix} -6 & 4 \\ 8 & -2 \end{bmatrix}$$

is not the same as

$$\begin{bmatrix} 0 & 1 \\ 2 & -1 \end{bmatrix} \cdot \begin{bmatrix} 1 & -3 \\ 2 & 4 \end{bmatrix} = \begin{bmatrix} 0 \cdot 1 + 1 \cdot 2 & 0(-3) + 1 \cdot 4 \\ 2 \cdot 1 + (-1)2 & 2(-3) + (-1)4 \end{bmatrix} = \begin{bmatrix} 2 & 4 \\ 0 & -10 \end{bmatrix}.$$

PRACTICE

Compute the following.

1. $\begin{bmatrix} 4 & 0 & -2 \\ 1 & 1 & 5 \end{bmatrix} - \begin{bmatrix} -3 & -2 & 2 \\ 6 & -4 & 3 \end{bmatrix}$

2. $5\begin{bmatrix} 3 & -6 \\ 2 & 4 \end{bmatrix}$

3. $\begin{bmatrix} 2 & -5 \\ 3 & 8 \end{bmatrix} \cdot \begin{bmatrix} 1 & 4 & -1 \\ 0 & -1 & 2 \end{bmatrix}$

4. $\begin{bmatrix} 1 & 0 & 3 \\ 2 & 1 & 0 \\ 3 & 1 & -2 \end{bmatrix} \cdot \begin{bmatrix} 4 & 2 & 1 \\ 1 & -3 & 1 \\ 3 & 6 & 2 \end{bmatrix}$

SOLUTIONS

1. $\begin{bmatrix} 4-(-3) & 0-(-2) & -2-2 \\ 1-6 & 1-(-4) & 5-3 \end{bmatrix} = \begin{bmatrix} 7 & 2 & -4 \\ -5 & 5 & 2 \end{bmatrix}$

2. $\begin{bmatrix} 5 \cdot 3 & 5 \cdot (-6) \\ 5 \cdot 2 & 5 \cdot 4 \end{bmatrix} = \begin{bmatrix} 15 & -30 \\ 10 & 20 \end{bmatrix}$

3. $\begin{bmatrix} 2 \cdot 1 + (-5)0 & 2 \cdot 4 + (-5)(-1) & 2(-1) + (-5)2 \\ 3 \cdot 1 + 8 \cdot 0 & 3 \cdot 4 + 8(-1) & 3(-1) + 8 \cdot 2 \end{bmatrix} = \begin{bmatrix} 2 & 13 & -12 \\ 3 & 4 & 13 \end{bmatrix}$

4. $\begin{bmatrix} 1 \cdot 4 + 0 \cdot 1 + 3 \cdot 3 & 1 \cdot 2 + 0(-3) + 3 \cdot 6 & 1 \cdot 1 + 0 \cdot 1 + 3 \cdot 2 \\ 2 \cdot 4 + 1 \cdot 1 + 0 \cdot 3 & 2 \cdot 2 + 1(-3) + 0 \cdot 6 & 2 \cdot 1 + 1 \cdot 1 + 0 \cdot 2 \\ 3 \cdot 4 + 1 \cdot 1 + (-2)3 & 3 \cdot 2 + 1(-3) + (-2)6 & 3 \cdot 1 + 1 \cdot 1 + (-2)2 \end{bmatrix}$

$$= \begin{bmatrix} 13 & 20 & 7 \\ 9 & 1 & 3 \\ 7 & -9 & 0 \end{bmatrix}$$

Row Operations and Inverses

We will use *row operations* to solve systems of equations and to find the multiplicative inverse of a matrix. These operations are similar to the elimination by addition method studied in Chapter 10. We will add two rows at a time (or some multiple of the rows) to make a particular entry 0. For example in the matrix $\begin{bmatrix} 1 & -3 & 2 \\ 4 & 1 & 6 \end{bmatrix}$ we might want to change the entry with a 4 in it to 0. To do so, we can multiply the first row (Row 1) by −4 and add it to the second row (Row 2).

$$-4 \text{ Row } 1 = -4(\,1 \quad -3 \quad 2\,) = -4 \quad 12 \quad -8$$

−4 Row 1	−4	12	−8
+Row 2	4	1	6
New Row 2	0	13	−2

The matrix changes to $\begin{bmatrix} 1 & -3 & 2 \\ 0 & 13 & -2 \end{bmatrix}$.

EXAMPLE

Using Row 2 and Row 3, change the entry with a 3 in it on the second row to 0.

$$\begin{bmatrix} 1 & 8 & 5 \\ -2 & 1 & 3 \\ 1 & 0 & 4 \end{bmatrix}$$

When adding the rows together, we need the last entry in each column to be opposites. If we multiply Row 2 by −4 and Row 3 by 3, we will be adding −4(3) to 3(4) to get 0. Multiplying Row 2 by 4 and Row 3 by −3 also works.

−4 Row 2	=	−4(−2)	−4(1)	−4(3)	=	8	−4	−12
+3 Row 3	=	3(1)	3(0)	3(4)	=	3	0	12
				New Row 2		11	−4	0

The new matrix is $\begin{bmatrix} 1 & 8 & 5 \\ 11 & -4 & 0 \\ 1 & 0 & 4 \end{bmatrix}$.

Our first use for row operations is to find the inverse of a matrix (if it has one). If we multiply a matrix by its inverse, we get the corresponding identity matrix. For example,

$$\begin{bmatrix} 1 & -2 \\ -1 & 4 \end{bmatrix} \cdot \begin{bmatrix} 2 & 1 \\ \frac{1}{2} & \frac{1}{2} \end{bmatrix} = \begin{bmatrix} 1 & 0 \\ 0 & 1 \end{bmatrix}.$$

To find the inverse of $\begin{bmatrix} A & B \\ C & D \end{bmatrix}$, we first need to write the *augmented* matrix. An augmented matrix for this method has the original matrix on the left and the identity matrix on the right.

$$\left[\begin{array}{cc|cc} A & B & 1 & 0 \\ C & D & 0 & 1 \end{array}\right]$$

We will use row operations to change the left half of the matrix to the 2×2 identity matrix. The inverse matrix will be the right half of the augmented matrix in Step 6.

Step 1 Use row operations to make the C entry a 0 for the new Row 2.
Step 2 Use row operations to make the B entry a 0 for the new Row 1.
Step 3 Write the next matrix.
Step 4 Divide Row 1 by the A entry.
Step 5 Divide Row 2 by the D entry.
Step 6 Write the new matrix. The inverse matrix will be the right half of this matrix.

EXAMPLE

- $\begin{bmatrix} 1 & -2 \\ -1 & 4 \end{bmatrix}$

The augmented matrix is $\left[\begin{array}{cc|cc} 1 & -2 & 1 & 0 \\ -1 & 4 & 0 & 1 \end{array}\right]$.

Step 1 We want to change -1, the C entry, to 0.

Row 1	1	−2	1	0
+Row 2	−1	4	0	1
New Row 2	0	2	1	1

Step 2 We want to change -2, the B entry, to 0.

2 Row 1	2	−4	2	0
+ Row 2	−1	4	0	1
New Row 1	1	0	2	1

Step 3 $\left[\begin{array}{cc|cc} 1 & 0 & 2 & 1 \\ 0 & 2 & 1 & 1 \end{array}\right]$

Step 4 This step is not necessary because dividing Row 1 by 1, the A entry, will not change any of its entries.

Step 5 Divide Row 2 by 2, the D entry. $\frac{1}{2}(0 \quad 2 \quad 1 \quad 1) = 0 \quad 1 \quad \frac{1}{2} \quad \frac{1}{2}$.

Step 6 $\begin{bmatrix} 1 & 0 & | & 2 & 1 \\ 0 & 1 & | & \frac{1}{2} & \frac{1}{2} \end{bmatrix}$

The inverse matrix is $\begin{bmatrix} 2 & 1 \\ \frac{1}{2} & \frac{1}{2} \end{bmatrix}$.

Finding the inverse of a 3×3 matrix takes a few more steps. Again, we will begin by writing the augmented matrix.

$$\begin{bmatrix} A & B & C \\ D & E & F \\ G & H & I \end{bmatrix} \longrightarrow \begin{bmatrix} A & B & C & | & 1 & 0 & 0 \\ D & E & F & | & 0 & 1 & 0 \\ G & H & I & | & 0 & 0 & 1 \end{bmatrix}$$

We will use row operations to turn the left half of the augmented matrix into the 3×3 identity matrix. There are many methods for getting from the first matrix to the last. The method outlined below will always work, assuming the matrix has an inverse.

Step 1 Use Row 1 and Row 2 to make the D entry to 0 for new Row 2.
Step 2 Use Row 1 and Row 3 to make the G entry to 0 for new Row 3.
Step 3 Write the next matrix. $\begin{bmatrix} \text{Old Row 1} \\ \text{New Row 2} \\ \text{New Row 3} \end{bmatrix}$
Step 4 Use Row 1 and Row 2 to make the B entry a 0 for new Row 1.
Step 5 Use Row 2 and Row 3 to make the H entry a 0 for new Row 3.
Step 6 Write the next matrix. $\begin{bmatrix} \text{New Row 1} \\ \text{Old Row 2} \\ \text{New Row 3} \end{bmatrix}$.
Step 7 Use Row 1 and Row 3 to make the C entry a 0 for new Row 1.
Step 8 Use Row 2 and Row 3 to make the F entry a 0 for new Row 2.
Step 9 Write the next matrix. $\begin{bmatrix} \text{New Row 1} \\ \text{New Row 2} \\ \text{Old Row 3} \end{bmatrix}$
Step 10 Divide Row 1 by A, Row 2 by E, and Row 3 by I. The inverse is the right half of the augmented matrix.

EXAMPLES

- Find the inverse matrix

$$\begin{bmatrix} 1 & 0 & -1 \\ 2 & 2 & 3 \\ 4 & -2 & 1 \end{bmatrix}$$

The augmented matrix is

$$\left[\begin{array}{ccc|ccc} 1 & 0 & -1 & 1 & 0 & 0 \\ 2 & 2 & 3 & 0 & 1 & 0 \\ 4 & -2 & 1 & 0 & 0 & 1 \end{array}\right].$$

Step 1 Use Row 1 and Row 2 to make the D entry a 0 by computing -2 Row 1 + Row 2.

$$\begin{array}{lcccccc} -2 \text{ Row } 1 & -2 & 0 & 2 & -2 & 0 & 0 \\ + \text{ Row } 2 & 2 & 2 & 3 & 0 & 1 & 0 \\ \hline \text{New Row } 2 & 0 & 2 & 5 & -2 & 1 & 0 \end{array}$$

Step 2 Use Row 1 and Row 3 to make the G entry a 0 by computing -4 Row 1 + Row 3.

$$\begin{array}{lcccccc} -4 \text{ Row } 1 & -4 & 0 & 4 & -4 & 0 & 0 \\ + \text{ Row } 3 & 4 & -2 & 1 & 0 & 0 & 1 \\ \hline \text{New Row } 3 & 0 & -2 & 5 & -4 & 0 & 1 \end{array}$$

Step 3

$$\left[\begin{array}{ccc|ccc} 1 & 0 & -1 & 1 & 0 & 0 \\ 0 & 2 & 5 & -2 & 1 & 0 \\ 0 & -2 & 5 & -4 & 0 & 1 \end{array}\right]$$

Step 4 This step is not necessary because the B entry is already 0. New Row 1 is old Row 1.

Step 5 Use Row 2 and Row 3 to make the H entry a 0 by computing Row 2 + Row 3.

$$\begin{array}{lcccccc} \text{Row } 2 & 0 & 2 & 5 & -2 & 1 & 0 \\ + \text{Row } 3 & 0 & -2 & 5 & -4 & 0 & 1 \\ \hline \text{New Row } 3 & 0 & 0 & 10 & -6 & 1 & 1 \end{array}$$

Step 6

$$\left[\begin{array}{ccc|ccc} 1 & 0 & -1 & 1 & 0 & 0 \\ 0 & 2 & 5 & -2 & 1 & 0 \\ 0 & 0 & 10 & -6 & 1 & 1 \end{array}\right]$$

Step 7 Use Row 1 and Row 3 to make the C entry a 0 by computing 10 Row 1 + Row 3.

$$\begin{array}{lcccccc} 10 \text{ Row } 1 & 10 & 0 & -10 & 10 & 0 & 0 \\ + \text{ Row } 3 & 0 & 0 & 10 & -6 & 1 & 1 \\ \hline \text{New Row } 1 & 10 & 0 & 0 & 4 & 1 & 1 \end{array}$$

Step 8 Use Row 2 and Row 3 to make the F entry a 0 by computing -2 Row 2 + Row 3.

-2 Row 2	0	-4	-10	4	-2	0
+ Row 3	0	0	10	-6	1	1
New Row 2	0	-4	0	-2	-1	1

Step 9

$$\left[\begin{array}{ccc|ccc} 10 & 0 & 0 & 4 & 1 & 1 \\ 0 & -4 & 0 & -2 & -1 & 1 \\ 0 & 0 & 10 & -6 & 1 & 1 \end{array}\right]$$

Step 10 Divide the Row 1 by 10, the Row 2 by -4, and Row 3 by 10 to get the next matrix.

$$\left[\begin{array}{ccc|ccc} 1 & 0 & 0 & \frac{2}{5} & \frac{1}{10} & \frac{1}{10} \\ 0 & 1 & 0 & \frac{1}{2} & \frac{1}{4} & -\frac{1}{4} \\ 0 & 0 & 1 & -\frac{3}{5} & \frac{1}{10} & \frac{1}{10} \end{array}\right]$$

The inverse matrix is

$$\begin{bmatrix} \frac{2}{5} & \frac{1}{10} & \frac{1}{10} \\ \frac{1}{2} & \frac{1}{4} & -\frac{1}{4} \\ -\frac{3}{5} & \frac{1}{10} & \frac{1}{10} \end{bmatrix}.$$

- Find the inverse matrix.

$$\begin{bmatrix} 6 & 0 & 2 \\ 1 & -1 & 0 \\ 0 & 1 & 1 \end{bmatrix}$$

The augmented matrix is

$$\left[\begin{array}{ccc|ccc} 6 & 0 & 2 & 1 & 0 & 0 \\ 1 & -1 & 0 & 0 & 1 & 0 \\ 0 & 1 & 1 & 0 & 0 & 1 \end{array}\right].$$

Step 1 Use Row 1 and Row 2 to make the 1 entry a 0.

Row 1	6	0	2	1	0	0
+(-6) Row 2	-6	6	0	0	-6	0
New Row 2	0	6	2	1	-6	0

Step 2 This step is not necessary because 0 is already in the G entry. New Row 3 is old Row 3.

Step 3

$$\begin{bmatrix} 6 & 0 & 2 & | & 1 & 0 & 0 \\ 0 & 6 & 2 & | & 1 & -6 & 0 \\ 0 & 1 & 1 & | & 0 & 0 & 1 \end{bmatrix}$$

Step 4 This step is not necessary because the B entry is already 0. New Row 1 is old Row 1.

Step 5 Use Row 2 and Row 3 to make the 1 entry a 0.

Row 2	0	6	2	1	−6	0
+(−6) Row 3	0	−6	−6	0	0	−6
New Row 3	0	0	−4	1	−6	−6

Step 6

$$\begin{bmatrix} 6 & 0 & 2 & | & 1 & 0 & 0 \\ 0 & 6 & 2 & | & 1 & -6 & 0 \\ 0 & 0 & -4 & | & 1 & -6 & -6 \end{bmatrix}$$

Step 7 Use Row 1 and Row 3 to make 2, the C entry, a 0.

2 Row 1	12	0	4	2	0	0
+ Row 3	0	0	−4	1	−6	−6
New Row 1	12	0	0	3	−6	−6

Step 8 Use Row 2 and Row 3 to make 2, the F entry, a 0.

2 Row 2	0	12	4	2	−12	0
+ Row 3	0	0	−4	1	−6	−6
New Row 2	0	12	0	3	−18	−6

Step 9

$$\begin{bmatrix} 12 & 0 & 0 & | & 3 & -6 & -6 \\ 0 & 12 & 0 & | & 3 & -18 & -6 \\ 0 & 0 & -4 & | & 1 & -6 & -6 \end{bmatrix}$$

Step 10 Divide Row 1 and Row 2 by 12 and Row 3 by −4.

$$\begin{bmatrix} 1 & 0 & 0 & | & \frac{1}{4} & -\frac{1}{2} & -\frac{1}{2} \\ 0 & 1 & 0 & | & \frac{1}{4} & -\frac{3}{2} & -\frac{1}{2} \\ 0 & 0 & 1 & | & -\frac{1}{4} & \frac{3}{2} & \frac{3}{2} \end{bmatrix}$$

The inverse matrix is

$$\begin{bmatrix} \frac{1}{4} & -\frac{1}{2} & -\frac{1}{2} \\ \frac{1}{4} & -\frac{3}{2} & -\frac{1}{2} \\ -\frac{1}{4} & \frac{3}{2} & \frac{3}{2} \end{bmatrix}.$$

PRACTICE

Find the inverse matrix.

1. $\begin{bmatrix} 1 & -1 \\ 2 & 3 \end{bmatrix}$

2. $\begin{bmatrix} -3 & 5 & 1 \\ 1 & 1 & -2 \\ 2 & -1 & 6 \end{bmatrix}$

SOLUTIONS

1. $\left[\begin{array}{cc|cc} 1 & -1 & 1 & 0 \\ 2 & 3 & 0 & 1 \end{array}\right]$

Step 1

−2 Row 1	−2	2	−2	0
+ Row 2	2	3	0	1
New Row 2	0	5	−2	1

Step 2

3 Row 1	3	−3	3	0
+Row 2	2	3	0	1
New Row 1	5	0	3	1

Step 3

$$\left[\begin{array}{cc|cc} 5 & 0 & 3 & 1 \\ 0 & 5 & -2 & 1 \end{array}\right]$$

Step 4 Divide Row 1 by 5.
Step 5 Divide Row 2 by 5.
Step 6

$$\left[\begin{array}{cc|cc} 1 & 0 & \frac{3}{5} & \frac{1}{5} \\ 0 & 1 & -\frac{2}{5} & \frac{1}{5} \end{array}\right]$$

The inverse matrix is

$$\begin{bmatrix} \frac{3}{5} & \frac{1}{5} \\ -\frac{2}{5} & \frac{1}{5} \end{bmatrix}.$$

2. $$\left[\begin{array}{rrr|rrr} -3 & 5 & 1 & 1 & 0 & 0 \\ 1 & 1 & -2 & 0 & 1 & 0 \\ 2 & -1 & 6 & 0 & 0 & 1 \end{array}\right]$$

Step 1

Row 1	−3	5	1	1	0	0
+3 Row 2	3	3	−6	0	3	0
New Row 2	0	8	−5	1	3	0

Step 2

2 Row 1	−6	10	2	2	0	0
+3 Row 3	6	−3	18	0	0	3
New Row 3	0	7	20	2	0	3

Step 3

$$\left[\begin{array}{rrr|rrr} -3 & 5 & 1 & 1 & 0 & 0 \\ 0 & 8 & -5 & 1 & 3 & 0 \\ 0 & 7 & 20 & 2 & 0 & 3 \end{array}\right]$$

Step 4

8 Row 1	−24	40	8	8	0	0
+(−5)Row 2	0	−40	25	−5	−15	0
New Row 1	−24	0	33	3	−15	0

Step 5

−7 Row 2	0	−56	35	−7	−21	0
+8 Row 3	0	56	160	16	0	24
New Row 3	0	0	195	9	−21	24

Step 6

$$\left[\begin{array}{rrr|rrr} -24 & 0 & 33 & 3 & -15 & 0 \\ 0 & 8 & -5 & 1 & 3 & 0 \\ 0 & 0 & 195 & 9 & -21 & 24 \end{array}\right]$$

To make the numbers smaller, replace Row 1 with $\frac{1}{3}$ Row 1 and Row 3 by $\frac{1}{3}$Row 3.

$$\left[\begin{array}{rrr|rrr} -8 & 0 & 11 & 1 & -5 & 0 \\ 0 & 8 & -5 & 1 & 3 & 0 \\ 0 & 0 & 65 & 3 & -7 & 8 \end{array}\right]$$

Step 7

65 Row 1	−520	0	715	65	−325	0
+(−11)Row 3	0	0	−715	−33	77	−88
New Row 1	−520	0	0	32	−248	−88

Step 8

13 Row 2	0	104	−65	13	39	0
+ Row 3	0	0	65	3	−7	8
New Row 2	0	104	0	16	32	8

Step 9

$$\left[\begin{array}{ccc|ccc} -520 & 0 & 0 & 32 & -248 & -88 \\ 0 & 104 & 0 & 16 & 32 & 8 \\ 0 & 0 & 65 & 3 & -7 & 8 \end{array}\right].$$

Step 10 Divide Row 1 by −520, Row 2 by 104, and Row 3 by 65.

$$\left[\begin{array}{ccc|ccc} 1 & 0 & 0 & -\frac{4}{65} & \frac{31}{65} & \frac{11}{65} \\ 0 & 1 & 0 & \frac{2}{13} & \frac{4}{13} & \frac{1}{13} \\ 0 & 0 & 1 & \frac{3}{65} & -\frac{7}{65} & \frac{8}{65} \end{array}\right]$$

The inverse matrix is

$$\begin{bmatrix} -\frac{4}{65} & \frac{31}{65} & \frac{11}{65} \\ \frac{2}{13} & \frac{4}{13} & \frac{1}{13} \\ \frac{3}{65} & -\frac{7}{65} & \frac{8}{65} \end{bmatrix}.$$

Matrices and Systems of Equations

There are three ways we can use matrices to solve a system of linear equations. Two of them will be discussed in this book. Solving systems using these methods will be very much like finding inverses. We will begin with 2×2 systems (two equations and two variables) and an augmented matrix of the form $\left[\begin{array}{cc|c} A & B & E \\ C & D & F \end{array}\right]$. A, B, C, and D are the coefficients of x and y in the equations and E and F are the constant terms. We will use the same steps above to change this matrix to one of the form $\left[\begin{array}{cc|c} 1 & 0 & \textit{number} \\ 0 & 1 & \textit{number} \end{array}\right]$. The numbers in the last column will be the solution.

EXAMPLE

- $$\begin{cases} 2x - 3y = 17 \\ -x + y = -7 \end{cases}$$

  The coefficients 2, −3, −1, and 1 are the entries in the left side of the matrix. The constant terms 17 and −7 are the entries on the right side of the matrix. The augmented matrix is $\left[\begin{array}{cc|c} 2 & -3 & 17 \\ -1 & 1 & -7 \end{array}\right]$.

Step 1 We want -1, the C entry, to be 0.

Row 1	2	-3	17
$+2$ Row 2	-2	2	-14
New Row 2	0	-1	3

Step 2 We want -3, the B entry, to be 0.

Row 1	2	-3	17
$+3$ Row 2	-3	3	-21
New Row 1	-1	0	-4

Step 3

$$\left[\begin{array}{rr|r} -1 & 0 & -4 \\ 0 & -1 & 3 \end{array}\right]$$

This row represents the equation $-1x + 0y = -4$
This row represents the equation $0x + (-1)y = 3$

Step 4 Divide Row 1 by -1.
Step 5 Divide Row 2 by -1.
Step 6

$$\left[\begin{array}{rr|r} 1 & 0 & 4 \\ 0 & 1 & -3 \end{array}\right]$$

This row represents the equation $1x + 0y = 4$.
This row represents the equation $0x + 1y = -3$.

The solution is $x = 4$ and $y = -3$.
Begin solving a 3×3 system (three equations and three variables) by writing the augmented matrix $\left[\begin{array}{ccc|c} A & B & C & J \\ D & E & F & K \\ G & H & I & L \end{array}\right]$. Using the same steps we used to find the inverse of a matrix, we want to change this matrix to one of the form $\left[\begin{array}{ccc|c} 1 & 0 & 0 & number \\ 0 & 1 & 0 & number \\ 0 & 0 & 1 & number \end{array}\right]$. The numbers in the fourth column will be the solution.

EXAMPLE

- $$\begin{cases} x \qquad + 3z = 3 \\ -x + y - z \ = 5 \\ 2x + y \qquad = -2 \end{cases}$$

The augmented matrix is $\left[\begin{array}{rrr|r} 1 & 0 & 3 & 3 \\ -1 & 1 & -1 & 5 \\ 2 & 1 & 0 & -2 \end{array}\right]$.

Step 1 Use Row 1 and Row 2 to change the *D* entry to 0.

Row 1	1	0	3	3
+ Row 2	−1	1	−1	5
New Row 2	0	1	2	8

Step 2 Use Row 1 and Row 3 to change the *G* entry to 0.

−2 Row 1	−2	0	−6	−6
+ Row 3	2	1	0	−2
New Row 3	0	1	−6	−8

Step 3

$$\left[\begin{array}{ccc|c} 1 & 0 & 3 & 3 \\ 0 & 1 & 2 & 8 \\ 0 & 1 & -6 & -8 \end{array}\right]$$

Step 4 Because the *B* entry is already 0, this step is not necessary. New Row 1 is old Row 1.

Step 5 Use Row 2 and Row 3 to change the *H* entry to 0.

−Row 2	0	−1	−2	− 8
+ Row 3	0	1	−6	− 8
New Row 3	0	0	−8	−16

Step 6

$$\left[\begin{array}{ccc|c} 1 & 0 & 3 & 3 \\ 0 & 1 & 2 & 8 \\ 0 & 0 & -8 & -16 \end{array}\right]$$

Step 7 Use Row 1 and Row 3 to make the *C* entry a 0.

8 Row 1	8	0	24	24
+3 Row 3	0	0	−24	−48
New Row 1	8	0	0	−24

Step 8 Use Row 2 and Row 3 to make the *F* entry a 0.

4 Row 2	0	4	8	32
+ Row 3	0	0	−8	−16
New Row 2	0	4	0	16

Step 9

$$\left[\begin{array}{ccc|c} 8 & 0 & 0 & -24 \\ 0 & 4 & 0 & 16 \\ 0 & 0 & -8 & -16 \end{array}\right]$$

Step 10 Divide Row 1 by 8, Row 2 by 4, and Row 3 by −8.

$$\left[\begin{array}{ccc|c} 1 & 0 & 0 & -3 \\ 0 & 1 & 0 & 4 \\ 0 & 0 & 1 & 2 \end{array}\right]$$ The solution is $x = -3$, $y = 4$, and $z = 2$.

The second method we will use to solve systems of equations involves finding the inverse of a matrix and multiplying two matrices. We begin by creating the coefficient matrix and the constant matrix for the system.

$$\begin{cases} Ax + By = E \\ Cx + Dy = F \end{cases}$$

The coefficient matrix is $\begin{bmatrix} A & B \\ C & D \end{bmatrix}$ and the constant matrix is $\begin{bmatrix} E \\ F \end{bmatrix}$. We will find the inverse of the coefficient matrix and multiply the inverse by the constant matrix. The product matrix will consist of one column of two numbers. These two numbers will be the solution to the system.

EXAMPLE

- $$\begin{cases} -2x + y = -7 \\ x - 3y = 1 \end{cases}$$

The coefficient matrix and constant matrix are

$$\begin{bmatrix} -2 & 1 \\ 1 & -3 \end{bmatrix} \text{ and } \begin{bmatrix} -7 \\ 1 \end{bmatrix}.$$

$$\left[\begin{array}{cc|cc} -2 & 1 & 1 & 0 \\ 1 & -3 & 0 & 1 \end{array}\right]$$

$$\begin{array}{lrrrr} \text{Row 1} & -2 & 1 & 1 & 0 \\ +2\text{ Row 2} & 2 & -6 & 0 & 2 \\ \hline \text{New Row 2} & 0 & -5 & 1 & 2 \end{array} \quad \text{and} \quad \begin{array}{lrrrr} 3\text{ Row 1} & -6 & 3 & 3 & 0 \\ +\text{ Row 2} & 1 & -3 & 0 & 1 \\ \hline \text{New Row 1} & -5 & 0 & 3 & 1 \end{array}$$

The next matrix is $\left[\begin{array}{cc|cc} -5 & 0 & 3 & 1 \\ 0 & -5 & 1 & 2 \end{array}\right]$. We need to divide Row 1 and Row 2 by −5.

$$\left[\begin{array}{cc|cc} 1 & 0 & -\frac{3}{5} & -\frac{1}{5} \\ 0 & 1 & -\frac{1}{5} & -\frac{2}{5} \end{array}\right]$$ The inverse matrix is $\begin{bmatrix} -\frac{3}{5} & -\frac{1}{5} \\ -\frac{1}{5} & -\frac{2}{5} \end{bmatrix}$.

Multiply the inverse matrix by the coefficient matrix.

$$\begin{bmatrix} -\frac{3}{5} & -\frac{1}{5} \\ -\frac{1}{5} & -\frac{2}{5} \end{bmatrix} \cdot \begin{bmatrix} -7 \\ 1 \end{bmatrix} = \begin{bmatrix} -\frac{3}{5} \cdot (-7) + \left(-\frac{1}{5}\right) \cdot 1 \\ -\frac{1}{5} \cdot (-7) + \left(-\frac{2}{5}\right) \cdot 1 \end{bmatrix} = \begin{bmatrix} 4 \\ 1 \end{bmatrix}$$

The solution is $x = 4$ and $y = 1$.
The strategy is the same for a 3×3 system of equations.

$$\begin{cases} Ax + By + Cz = J \\ Dx + Ey + Fz = K \\ Gx + Hy + Iz = L \end{cases}$$

The coefficient matrix and the constant matrix are

$$\begin{bmatrix} A & B & C \\ D & E & F \\ G & H & I \end{bmatrix} \text{ and } \begin{bmatrix} J \\ K \\ L \end{bmatrix}.$$

We will find the inverse matrix of the coefficient matrix and multiply it by the constant matrix.

EXAMPLE

- $$\begin{cases} -3x + 2y + z = 3 \\ 2x + y - z = 5 \\ -y + 2z = -3 \end{cases}$$

The coefficient matrix and constant matrix are

$$\begin{bmatrix} -3 & 2 & 1 \\ 2 & 1 & -1 \\ 0 & -1 & 2 \end{bmatrix} \text{ and } \begin{bmatrix} 3 \\ 5 \\ -3 \end{bmatrix}.$$

$$\left[\begin{array}{ccc|ccc} -3 & 2 & 1 & 1 & 0 & 0 \\ 2 & 1 & -1 & 0 & 1 & 0 \\ 0 & -1 & 2 & 0 & 0 & 1 \end{array}\right]$$

2 Row 1	−6	4	2	2	0	0
+3 Row 2	6	3	−3	0	3	0
New Row 2	0	7	−1	2	3	0

New Row 3 is old Row 3. The next matrix is $\left[\begin{array}{rrr|rrr} -3 & 2 & 1 & 1 & 0 & 0 \\ 0 & 7 & -1 & 2 & 3 & 0 \\ 0 & -1 & 2 & 0 & 0 & 1 \end{array}\right]$.

7 Row 1	−21	14	7	7	0	0
+(−2) Row 2	0	−14	2	−4	−6	0
New Row 1	−21	0	9	3	−6	0

Row 2	0	7	−1	2	3	0
+7 Row 3	0	−7	14	0	0	7
New Row 3	0	0	13	2	3	7

The next matrix is $\left[\begin{array}{rrr|rrr} -21 & 0 & 9 & 3 & -6 & 0 \\ 0 & 7 & -1 & 2 & 3 & 0 \\ 0 & 0 & 13 & 2 & 3 & 7 \end{array}\right]$.

13 Row 1	−273	0	117	39	−78	0
+(−9) Row 3	0	0	−117	−18	−27	−63
New Row 1	−273	0	0	21	−105	−63

13 Row 2	0	91	−13	26	39	0
+ Row 3	0	0	13	2	3	7
New Row 2	0	91	0	28	42	7

The next matrix is $\left[\begin{array}{rrr|rrr} -273 & 0 & 0 & 21 & -105 & -63 \\ 0 & 91 & 0 & 28 & 42 & 7 \\ 0 & 0 & 13 & 2 & 3 & 7 \end{array}\right]$.

Divide Row 1 by −273, Row 2 by 91, and Row 3 by 13.

$$\left[\begin{array}{rrr|rrr} 1 & 0 & 0 & -\frac{1}{13} & \frac{5}{13} & \frac{3}{13} \\ 0 & 1 & 0 & \frac{4}{13} & \frac{6}{13} & \frac{1}{13} \\ 0 & 0 & 1 & \frac{2}{13} & \frac{3}{13} & \frac{7}{13} \end{array}\right]$$

Multiply the inverse matrix by the constant matrix.

$$\begin{bmatrix} -\frac{1}{13} & \frac{5}{13} & \frac{3}{13} \\ \frac{4}{13} & \frac{6}{13} & \frac{1}{13} \\ \frac{2}{13} & \frac{3}{13} & \frac{7}{13} \end{bmatrix} \cdot \begin{bmatrix} 3 \\ 5 \\ -3 \end{bmatrix} = \begin{bmatrix} 3\left(-\frac{1}{13}\right) + 5\left(\frac{5}{13}\right) + (-3)\frac{3}{13} \\ 3\left(\frac{4}{13}\right) + 5\left(\frac{6}{13}\right) + (-3)\left(\frac{1}{13}\right) \\ 3\left(\frac{2}{13}\right) + 5\left(\frac{3}{13}\right) + (-3)\left(\frac{7}{13}\right) \end{bmatrix} = \begin{bmatrix} 1 \\ 3 \\ 0 \end{bmatrix}$$

The solution is $x = 1$, $y = 3$ and $z = 0$.

PRACTICE

1. Use the first method to solve the system.

$$\begin{cases} -5x + 2y + 3z = -8 \\ x + y - z = -5 \\ 2x + y + 3z = 23 \end{cases}$$

2. Use the first method to solve the system.

$$\begin{cases} 6x + 2z = -12 \\ x - y = -3 \\ y + z = 1 \end{cases}$$

3. Use the second method to solve the system.

$$\begin{cases} x + z = 6 \\ 3x - y + 2z = 17 \\ 6x + y - z = 5 \end{cases}$$

SOLUTIONS

1. The augmented matrix is $\left[\begin{array}{rrr|r} -5 & 2 & 3 & -8 \\ 1 & 1 & -1 & -5 \\ 2 & 1 & 3 & 23 \end{array}\right]$.

Row 1	−5	2	3	−8
+5Row 2	5	5	−5	−25
New Row 2	0	7	−2	−33

2 Row 1	−10	4	6	−16
+5 Row 3	10	5	15	115
New Row 3	0	9	21	99

The next matrix is $\left[\begin{array}{rrr|r} -5 & 2 & 3 & -8 \\ 0 & 7 & -2 & -33 \\ 0 & 9 & 21 & 99 \end{array}\right]$.

7 Row 1	−35	14	21	−56
+(−2) Row 2	0	−14	4	66
New Row 1	−35	0	25	10

9 Row 2	0	63	−18	−297
+(−7) Row 3	0	−63	−147	−693
New Row 3	0	0	−165	−990

The next matrix is $\left[\begin{array}{rrr|r} -35 & 0 & 25 & 10 \\ 0 & 7 & -2 & -33 \\ 0 & 0 & -165 & -990 \end{array}\right]$.

We can make the numbers in Row 1 and Row 3 smaller by dividing Row 1 by 5 and Row 3 by −165.

$$\left[\begin{array}{ccc|c} -7 & 0 & 5 & 2 \\ 0 & 7 & -2 & -33 \\ 0 & 0 & 1 & 6 \end{array}\right]$$

Row 1	−7	0	5	2
+(−5) Row 3	0	0	−5	−30
New Row 1	−7	0	0	−28

Row 2	0	7	−2	−33
+2 Row 3	0	0	2	12
New Row 2	0	7	0	−21

The next matrix is

$$\left[\begin{array}{ccc|c} -7 & 0 & 0 & -28 \\ 0 & 7 & 0 & -21 \\ 0 & 0 & 1 & 6 \end{array}\right].$$

Divide Row 1 by −7 and Row 2 by 7.

$$\left[\begin{array}{ccc|c} 1 & 0 & 0 & 4 \\ 0 & 1 & 0 & -3 \\ 0 & 0 & 1 & 6 \end{array}\right]$$

The solution is $x = 4$, $y = -3$, and $z = 6$.

2. The augmented matrix is $\left[\begin{array}{ccc|c} 6 & 0 & 2 & -12 \\ 1 & -1 & 0 & -3 \\ 0 & 1 & 1 & 1 \end{array}\right]$.

Row 1	6	0	2	−12
+(−6) Row 2	−6	6	0	18
New Row 2	0	6	2	6

New Row 3 is old Row 3. The next matrix is $\left[\begin{array}{ccc|c} 6 & 0 & 2 & -12 \\ 0 & 6 & 2 & 6 \\ 0 & 1 & 1 & 1 \end{array}\right]$.

Row 2	0	6	2	6
+(−6) Row 3	0	−6	−6	−6
New Row 3	0	0	−4	0

New Row 1 is old Row 1. The next matrix is $\left[\begin{array}{ccc|c} 6 & 0 & 2 & -12 \\ 0 & 6 & 2 & 6 \\ 0 & 0 & -4 & 0 \end{array}\right]$.

$$\begin{array}{r|rrrr} \text{Row 1} & 6 & 0 & 2 & -12 \\ +\frac{1}{2}\text{ Row 3} & 0 & 0 & -2 & 0 \\ \hline \text{New Row 1} & 6 & 0 & 0 & -12 \end{array} \qquad \begin{array}{r|rrrr} \text{New Row 2} & 0 & 6 & 2 & 6 \\ +\frac{1}{2}\text{ Row 3} & 0 & 0 & -2 & 0 \\ \hline \text{New Row 2} & 0 & 6 & 0 & 6 \end{array}$$

The next matrix is $\left[\begin{array}{ccc|c} 6 & 0 & 0 & -12 \\ 0 & 6 & 0 & 6 \\ 0 & 0 & -4 & 0 \end{array}\right]$. Divide Row 1 and Row 2 by 6 and Row 3 by -4.

$\left[\begin{array}{ccc|c} 1 & 0 & 0 & -2 \\ 0 & 1 & 0 & 1 \\ 0 & 0 & 1 & 0 \end{array}\right]$ The solution is $x = -2$, $y = 1$, and $z = 0$.

3. The augmented matrix is

$$\left[\begin{array}{ccc|ccc} 1 & 0 & 1 & 1 & 0 & 0 \\ 3 & -1 & 2 & 0 & 1 & 0 \\ 6 & 1 & -1 & 0 & 0 & 1 \end{array}\right].$$

$$\begin{array}{r|rrrrrr} -3\text{ Row 1} & -3 & 0 & -3 & -3 & 0 & 0 \\ +\text{ Row 2} & 3 & -1 & 2 & 0 & 1 & 0 \\ \hline \text{New Row 2} & 0 & -1 & -1 & -3 & 1 & 0 \end{array} \qquad \begin{array}{r|rrrrrr} -6\text{ Row 1} & -6 & 0 & -6 & -6 & 0 & 0 \\ +\text{ Row 3} & 6 & 1 & -1 & 0 & 0 & 1 \\ \hline \text{New Row 3} & 0 & 1 & -7 & -6 & 0 & 1 \end{array}$$

The next matrix is

$$\left[\begin{array}{ccc|ccc} 1 & 0 & 1 & 1 & 0 & 0 \\ 0 & -1 & -1 & -3 & 1 & 0 \\ 0 & 1 & -7 & -6 & 0 & 1 \end{array}\right].$$

$$\begin{array}{r|rrrrrr} \text{Row 2} & 0 & -1 & -1 & -3 & 1 & 0 \\ +\text{ Row 3} & 0 & 1 & -7 & -6 & 0 & 1 \\ \hline \text{New Row 3} & 0 & 0 & -8 & -9 & 1 & 1 \end{array}$$

New Row 1 is old Row 1. The next matrix is

$$\left[\begin{array}{ccc|ccc} 1 & 0 & 1 & 1 & 0 & 0 \\ 0 & -1 & -1 & -3 & 1 & 0 \\ 0 & 0 & -8 & -9 & 1 & 1 \end{array}\right].$$

$$\begin{array}{lrrrrrr} \text{8 Row 1} & 8 & 0 & 8 & 8 & 0 & 0 \\ \text{+ Row 3} & 0 & 0 & -8 & -9 & 1 & 1 \\ \hline \text{New Row 1} & 8 & 0 & 0 & -1 & 1 & 1 \end{array}$$

$$\begin{array}{lrrrrrr} -8\text{ Row 2} & 0 & 8 & 8 & 24 & -8 & 0 \\ \text{+ Row 3} & 0 & 0 & -8 & -9 & 1 & 1 \\ \hline \text{New Row 2} & 0 & 8 & 0 & 15 & -7 & 1 \end{array}$$

The next matrix is $\left[\begin{array}{rrr|rrr} 8 & 0 & 0 & -1 & 1 & 1 \\ 0 & 8 & 0 & 15 & -7 & 1 \\ 0 & 0 & -8 & -9 & 1 & 1 \end{array}\right]$.

Divide Row 1 and Row 2 by 8 and Row 3 by -8.

$$\left[\begin{array}{rrr|rrr} 1 & 0 & 0 & -\frac{1}{8} & \frac{1}{8} & \frac{1}{8} \\ 0 & 1 & 0 & \frac{15}{8} & -\frac{7}{8} & \frac{1}{8} \\ 0 & 0 & 1 & \frac{9}{8} & -\frac{1}{8} & -\frac{1}{8} \end{array}\right]$$

The inverse matrix is

$$\begin{bmatrix} -\frac{1}{8} & \frac{1}{8} & \frac{1}{8} \\ \frac{15}{8} & -\frac{7}{8} & \frac{1}{8} \\ \frac{9}{8} & -\frac{1}{8} & -\frac{1}{8} \end{bmatrix}.$$

Multiply the inverse matrix, by the coefficient matrix $\begin{bmatrix} 6 \\ 17 \\ 5 \end{bmatrix}$.

$$\begin{bmatrix} -\frac{1}{8} & \frac{1}{8} & \frac{1}{8} \\ \frac{15}{8} & -\frac{7}{8} & \frac{1}{8} \\ \frac{9}{8} & -\frac{1}{8} & -\frac{1}{8} \end{bmatrix} \cdot \begin{bmatrix} 6 \\ 17 \\ 5 \end{bmatrix} = \begin{bmatrix} 6\left(-\frac{1}{8}\right) + 17\left(\frac{1}{8}\right) + 5\left(\frac{1}{8}\right) \\ 6\left(\frac{15}{8}\right) + 17\left(-\frac{7}{8}\right) + 5\left(\frac{1}{8}\right) \\ 6\left(\frac{9}{8}\right) + 17\left(-\frac{1}{8}\right) + 5\left(-\frac{1}{8}\right) \end{bmatrix} = \begin{bmatrix} 2 \\ -3 \\ 4 \end{bmatrix}$$

The solution is $x = 2$, $y = -3$ and $z = 4$

The last computation we will learn is finding a matrix's *determinant*. Although we will not use the determinant here, it is used in vector mathematics courses, some theoretical algebra courses, and in algebra courses that cover Cramer's Rule (used to solve systems of linear equations). An interesting fact about determinants is that a square matrix has an inverse only when its determinant is a nonzero number.

The usual notation for a determinant is to enclose the matrix with two vertical bars instead of two brackets. The determinant for the matrix $\begin{bmatrix} A & B \\ C & D \end{bmatrix}$ is $\begin{vmatrix} A & B \\ C & D \end{vmatrix}$.

Finding the determinant for a 2×2 matrix is not hard.

$$\begin{vmatrix} A & B \\ C & D \end{vmatrix} = AD - BC$$

EXAMPLE

- $\begin{vmatrix} 4 & -3 \\ 5 & 2 \end{vmatrix} = 4(2) - (-3)(5) = 23$

  We find the determinant of larger matrices by breaking down the larger matrix into several 2×2 sub-matrices. For larger matrices, there are numerous formulas for computing their determinants. Some of them come from *expanding the matrix* along each row and along each column. This means that we will multiply the entries in a row or a column by the determinant of a smaller matrix. This smaller matrix comes from deleting the row and column an entry is in. When working with a 3×3 matrix, these sub-matrices will be 2×2 matrices.
  Suppose we want to expand the following matrix along the first row.

$$\begin{bmatrix} A & B & C \\ D & E & F \\ G & H & I \end{bmatrix}$$

We will multiply the A entry by the submatrix obtained by removing the first row ABC and the first column $\begin{matrix} A \\ D \\ G \end{matrix}$. This leaves us with the matrix $\begin{bmatrix} - & - & - \\ - & E & F \\ - & H & I \end{bmatrix}$. Our first calculation will be

$$A\begin{vmatrix} E & F \\ H & I \end{vmatrix} = A(EI - FH).$$

Similarly, when we use entry B, we will need to remove the first row $A\ B\ C$ and the second column $\begin{matrix} B \\ E \\ H \end{matrix}$. This leaves us with $\begin{vmatrix} D & F \\ G & I \end{vmatrix}$. There is a complication—the signs on the entries must alternate when we perform these expansions. For our matrix, the signs will alternate beginning with A not changing, but B and D changing.

$$\begin{matrix} A & -B & C \\ -D & E & -F \\ G & -H & J \end{matrix}$$

For our 3×3 matrix, the expansion along the first row looks like this.

$$A\begin{vmatrix} E & F \\ H & I \end{vmatrix} - B\begin{vmatrix} D & F \\ G & I \end{vmatrix} + C\begin{vmatrix} D & E \\ G & H \end{vmatrix} = A(EI - FH) - B(DI - FG) + C(DH - EG)$$

The expansion along the second column looks like this.

$$-B\begin{vmatrix} D & F \\ G & I \end{vmatrix} + E\begin{vmatrix} A & C \\ G & I \end{vmatrix} - H\begin{vmatrix} A & C \\ D & F \end{vmatrix} = -B(DI - FG) + E(AI - CG) - H(AF - CD)$$

EXAMPLE

- Find the determinant for $\begin{bmatrix} 4 & 1 & -3 \\ 2 & 0 & 4 \\ -2 & 2 & 1 \end{bmatrix}$.

We will use two calculations, along Row 2 and along Column 3. By Row 2 we have

$$-2\begin{vmatrix} 1 & -3 \\ 2 & 1 \end{vmatrix} + 0\begin{vmatrix} 4 & -3 \\ -2 & 1 \end{vmatrix} - 4\begin{vmatrix} 4 & 1 \\ -2 & 2 \end{vmatrix}$$

$$= -2(1 \cdot 1 - (-3)2) + 0(4 \cdot 1 - (-3)(-2)) - 4(4 \cdot 2 - 1(-2)) = -54.$$

By Column 3 we have

$$-3\begin{vmatrix} 2 & 0 \\ -2 & 2 \end{vmatrix} - 4\begin{vmatrix} 4 & 1 \\ -2 & 2 \end{vmatrix} + 1\begin{vmatrix} 4 & 1 \\ 2 & 0 \end{vmatrix}$$

$$= -3(2 \cdot 2 - 0(-2)) - 4(4 \cdot 2 - 1(-2)) + 1(4 \cdot 0 - 1 \cdot 2) = -54$$

The method is the same for larger matrices except that there are more levels of work.

$$\begin{bmatrix} A & B & C & D \\ E & F & G & H \\ I & J & K & L \\ M & N & O & P \end{bmatrix}$$

Expanding this matrix along Row 1 gives us

$$A\begin{vmatrix} F & G & H \\ J & K & L \\ N & O & P \end{vmatrix} - B\begin{vmatrix} E & G & H \\ I & K & L \\ M & O & P \end{vmatrix} + C\begin{vmatrix} E & F & H \\ I & J & L \\ M & N & P \end{vmatrix} - D\begin{vmatrix} E & F & G \\ I & J & K \\ M & N & O \end{vmatrix}.$$

Each of these four determinants must be computed using the previous method for a 3×3 matrix.

PRACTICE

1. $\begin{bmatrix} -8 & 1 & 3 \\ 2 & 5 & 0 \\ 6 & -4 & 2 \end{bmatrix}$

SOLUTION

1. Expanding this matrix along Row 2, we have

$$-2\begin{vmatrix} 1 & 3 \\ -4 & 2 \end{vmatrix} + 5\begin{vmatrix} -8 & 3 \\ 6 & 2 \end{vmatrix} - 0\begin{vmatrix} -8 & 1 \\ 6 & -4 \end{vmatrix}$$

$$= -2(1 \cdot 2 - 3(-4)) + 5(-8 \cdot 2 - 3 \cdot 6) - 0((-8)(-4) - 1 \cdot 6) = -198$$

CHAPTER 11 REVIEW

1. $\begin{bmatrix} 8 & 4 \\ -1 & -5 \end{bmatrix} - \begin{bmatrix} 3 & -1 \\ -2 & 6 \end{bmatrix} =$

   (a) $\begin{bmatrix} 5 & 3 \\ -3 & 1 \end{bmatrix}$ (b) $\begin{bmatrix} 11 & 3 \\ -3 & 1 \end{bmatrix}$

   (c) $\begin{bmatrix} 5 & 5 \\ -1 & 1 \end{bmatrix}$ (d) $\begin{bmatrix} 5 & 5 \\ 1 & -11 \end{bmatrix}$

2. $\begin{bmatrix} 7 & 4 \\ 1 & 1 \end{bmatrix} \cdot \begin{bmatrix} 0 & -2 \\ 3 & 1 \end{bmatrix} =$

   (a) $\begin{bmatrix} 0 & -8 \\ 3 & 1 \end{bmatrix}$ (b) $\begin{bmatrix} 12 & -10 \\ 3 & -1 \end{bmatrix}$

   (c) $\begin{bmatrix} 8 & -10 \\ 4 & 2 \end{bmatrix}$ (d) $\begin{bmatrix} 0 & -10 \\ 4 & -1 \end{bmatrix}$

3. $\begin{bmatrix} -1 & 4 \\ 0 & 2 \end{bmatrix} \cdot \begin{bmatrix} 1 & 1 & -1 \\ 2 & -1 & 3 \end{bmatrix} =$

(a) $\begin{bmatrix} 7 & 4 & -2 \\ 2 & 2 & -2 \end{bmatrix}$ (b) $\begin{bmatrix} 7 & 2 \\ 4 & 2 \\ -2 & -2 \end{bmatrix}$

(c) $\begin{bmatrix} 7 & -5 & 13 \\ 4 & -2 & 6 \end{bmatrix}$ (d) The product does not exist.

4. What is the determinant for $\begin{bmatrix} -8 & 1 \\ 4 & 5 \end{bmatrix}$?

(a) −36 (b) −37 (c) −44 (d) −27

5. What is the inverse for $\begin{bmatrix} 2 & -3 \\ -1 & 1 \end{bmatrix}$?

(a) $\begin{bmatrix} \frac{1}{2} & -\frac{1}{3} \\ -1 & 1 \end{bmatrix}$ (b) $\begin{bmatrix} -1 & -3 \\ -1 & -2 \end{bmatrix}$

(c) $\begin{bmatrix} \frac{1}{5} & -\frac{4}{5} \\ \frac{1}{2} & 2 \end{bmatrix}$ (d) $\begin{bmatrix} -1 & 1 \\ 2 & -3 \end{bmatrix}$

6. What is the determinant for $\begin{bmatrix} 6 & 0 & 2 \\ 1 & -1 & 0 \\ 0 & 1 & 1 \end{bmatrix}$?

(a) −4 (b) −5 (c) −6 (d) −7

7. What is the inverse for $\begin{bmatrix} 1 & 0 & 3 \\ -1 & 1 & -1 \\ 2 & 1 & 0 \end{bmatrix}$?

(a) $\begin{bmatrix} -\frac{1}{8} & -\frac{3}{8} & \frac{3}{8} \\ \frac{1}{4} & \frac{3}{4} & \frac{1}{4} \\ \frac{3}{8} & \frac{1}{8} & -\frac{1}{8} \end{bmatrix}$ (b) $\begin{bmatrix} \frac{1}{4} & \frac{3}{4} & \frac{1}{4} \\ \frac{3}{8} & \frac{1}{8} & -\frac{1}{8} \\ -\frac{1}{8} & -\frac{3}{8} & \frac{3}{8} \end{bmatrix}$

(c) $\begin{bmatrix} \frac{3}{8} & \frac{1}{8} & -\frac{1}{8} \\ -\frac{1}{8} & -\frac{3}{8} & \frac{3}{8} \\ \frac{1}{4} & \frac{3}{4} & \frac{1}{4} \end{bmatrix}$ (d) $\begin{bmatrix} \frac{1}{8} & \frac{3}{8} & -\frac{1}{8} \\ -\frac{3}{8} & -\frac{1}{8} & \frac{3}{8} \\ \frac{3}{4} & \frac{1}{4} & \frac{1}{4} \end{bmatrix}$

For Problems 8–10, use different matrix methods to solve the systems.

8. What is $x + y$ for the system?

$$\begin{cases} 5x - 8y = 29 \\ 2x + 2y = -4 \end{cases}$$

(a) -2 (b) -3 (c) -4 (d) -5

9. What is $x + y + z$ for the system?

$$\begin{cases} x + y + z = 1 \\ 2x - y + z = 3 \\ -x + y - 3z = -7 \end{cases}$$

(a) 0 (b) 1 (c) 2 (d) 3

10. What is $x + y + z$ for the system?

$$\begin{cases} 6x + \quad - z = -22 \\ x + y - z = -6 \\ y + z = 5 \end{cases}$$

(a) -1 (b) -2 (c) 1 (d) 2

SOLUTIONS

1. D	2. B	3. C	4. C	5. B
6. A	7. A	8. A	9. B	10. D

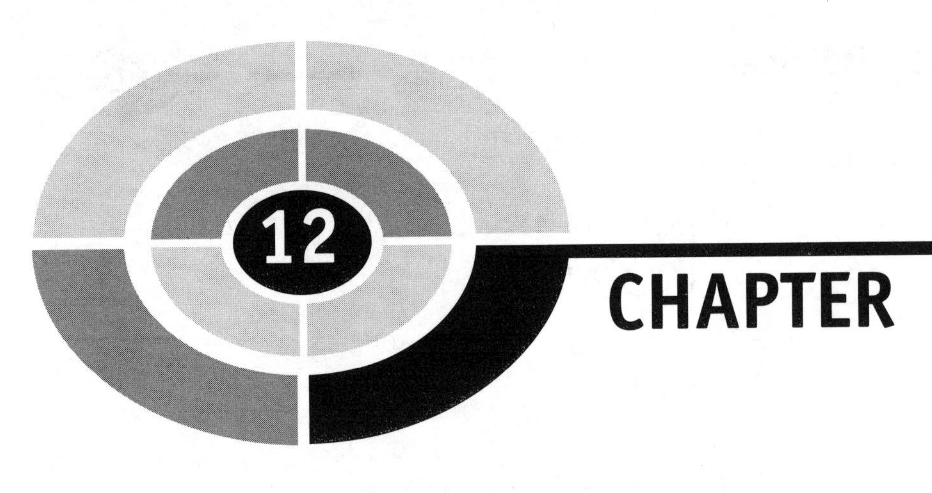

Conic Sections

A *conic section* is a shape obtained when a cone is sliced. The study of conic sections began over two thousand years ago and we use their properties today. Planets in our solar system move around the sun in elliptical orbits. The cross-section of many reflecting surfaces is in the shape of a parabola. In fact, all of the conic sections have useful reflecting properties. There are three conic sections—parabolas, ellipses (including circles), and hyperbolas.

Parabolas

In Chapter 6, we learned how to graph parabolas when their equations were in the form $y = a(x - h)^2 + k$ or $y = ax^2 + bx + c$. Now we will learn the formal definition for a parabola and another form for its equation.

DEFINITION: A parabola is the set of all points whose distance to a fixed point and a fixed line are the same.

The fixed point is the *focus*. The fixed line is the *directrix*. For example, the focus for the parabola $y = -\frac{1}{2}x^2 - 3x + 2$ is $(-3, 6)$, and the directrix is the horizontal line $y = 7$. The point $(0, 2)$ is on the parabola. Its distance from the line $y = 7$ is 5.

Its distance from the focus $(-3, 6)$ is also 5.

$$\sqrt{(-3-0)^2 + (6-2)^2} = \sqrt{25} = 5$$

The new form for a parabola that opens up or down is $(x-h)^2 = 4p(y-k)$. The vertex is still at (h, k), but p helps us to find the focus and the equation for the directrix. The focus is the point $(h, k+p)$, and the directrix is the horizontal line $y = k - p$. The form for the equation for a parabola that opens to the side is $(y-k)^2 = 4p(x-h)$. The focus for a parabola that opens to the right or to the left is the point $(h+p, k)$, and the directrix is the vertical line $x = h - p$. This information is summarized in Table 12.1 and in Figures 12.1 and 12.2.

Table 12.1

$(x-h)^2 = 4p(y-k)$	$(y-k)^2 = 4p(x-h)$
The vertex is (h, k).	The vertex is (h, k).
The parabola opens up if p is positive and down if p is negative.	The parabola opens to the right if p is positive and to the left if p is negative.
The focus is $(h, k+p)$.	The focus is $(h+p, k)$.
The directrix is $y = k - p$.	The directrix is $x = h - p$.
The axis of symmetry is $x = h$.	The axis of symmetry is $y = k$.

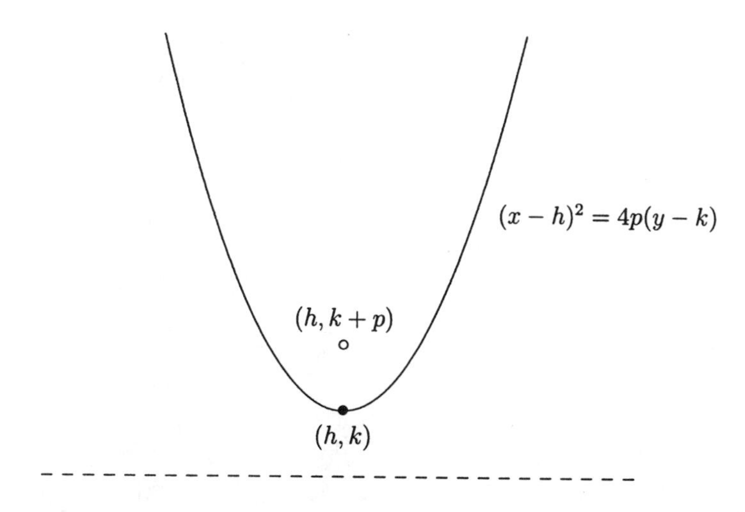

Fig. 12.1.

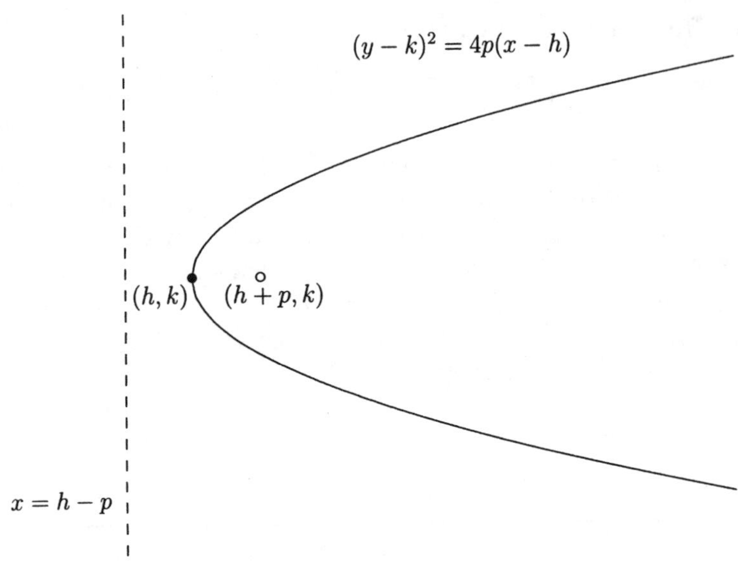

Fig. 12.2.

In the following examples, we will be asked to match the equation to its graph. The vertex for each parabola will be at $(0, 0)$. We can decide which graph goes to which equation either by finding the focus or the directrix in the equation and finding which graph has this focus or directrix.

EXAMPLES

Match the graphs in Figures 12.3 through 12.6 with their equations.

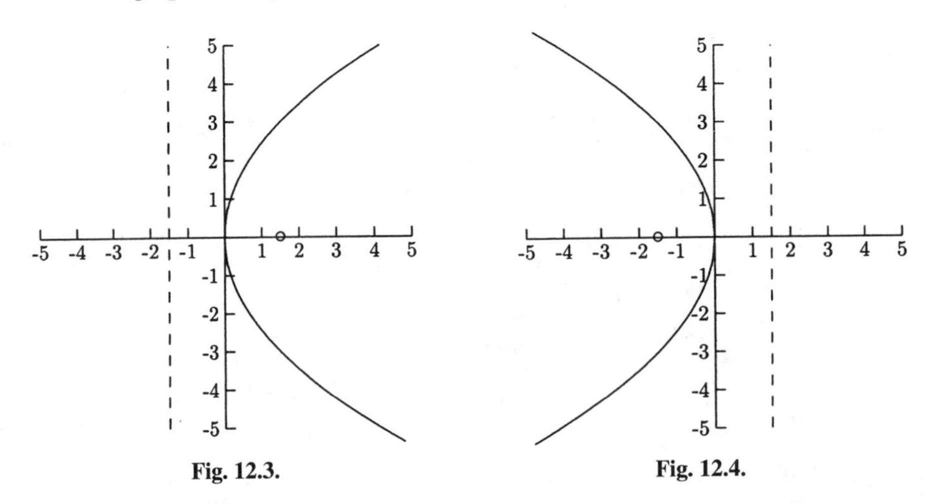

Fig. 12.3. **Fig. 12.4.**

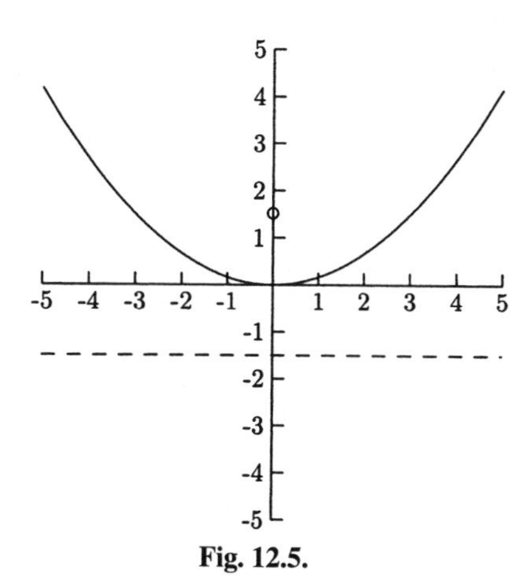

Fig. 12.5.

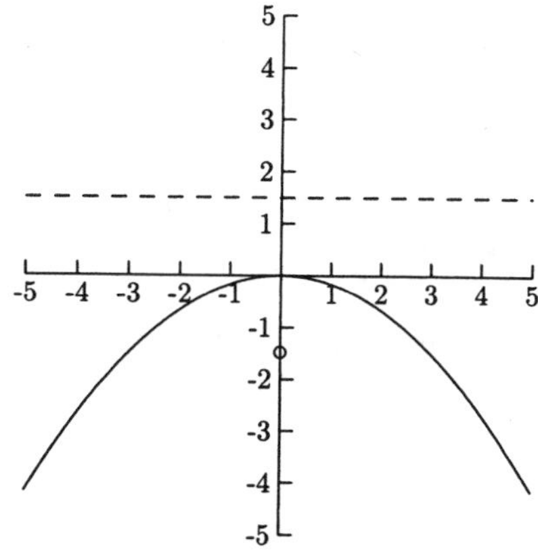

Fig. 12.6.

- $x^2 = 6y$

  The equation is in the form $(x - h)^2 = 4p(y - k)$, so the parabola will open up or down. We have $p = \frac{3}{2}$ (from $6 = 4p$). Now we know three things—that the parabola opens up (because p is positive), that the focus is $(h, k + p) = (0, 0 + \frac{3}{2}) = (0, \frac{3}{2})$, and the directrix is $y = -\frac{3}{2}$ (from $k - p = 0 - \frac{3}{2}$). The graph that behaves this way is in Figure 12.5.

- $y^2 = 6x$

  The equation is in the form $(y - k)^2 = 4p(x - h)$, so the parabola opens to the left or to the right. We have $p = \frac{3}{2}$ (from $6 = 4p$). Now we know that the parabola opens to the right, that the focus is $(h + p, k) = (0 + \frac{3}{2}, 0) = (\frac{3}{2}, 0)$, and that the directrix is $x = -\frac{3}{2}$ (from $h - p = 0 - \frac{3}{2}$). The graph for this equation is in Figure 12.3.

- $y^2 = -6x$

  The equation is in the form $(y - k)^2 = 4p(x - h)$, so the parabola opens to the left or to the right. We have $p = -\frac{3}{2}$ (from $-6 = 4p$). The parabola opens to the left, the focus is $(h + p, k) = (0 + -\frac{3}{2}, 0) = (-\frac{3}{2}, 0)$, and the directrix is $x = \frac{3}{2}$ (from $h - p = 0 - (-\frac{3}{2})$). The graph for this equation is in Figure 12.4.

- $x^2 = -6y$

  The equation is in the form $(x - h)^2 = 4p(y - k)$, so the parabola opens up or down. We have $p = -\frac{3}{2}$ (from $-6 = 4p$). The parabola opens down, the focus is $(h, k + p) = (0, 0 + (-\frac{3}{2})) = (0, -\frac{3}{2})$. The directrix is $y = \frac{3}{2}$ (from $k - p = 0 - (-\frac{3}{2})$). The graph for this equation is in Figure 12.6.

Using the information in Table 12.1, we can find the vertex, focus, directrix, and whether the parabola opens up, down, to the left, or to the right by looking at its equation.

EXAMPLES

Find the vertex, focus, and directrix. Determine if the parabola opens up, down, to the left, or to the right.

- $(x-3)^2 = 4(y-2)$

  This equation is in the form $(x-h)^2 = 4p(y-k)$. The vertex is (3, 2). Once we have found p, we can find the focus and directrix and how the parabola opens. $p = 1$ (from $4 = 4p$). The parabola opens up because p is positive; the focus is $(h, k+p) = (3, 2+1) = (3, 3)$; and the directrix is $y = 1$ (from $y = k - p = 2 - 1 = 1$).

- $(y+1)^2 = 8(x-3)$

  The equation is in the form $(y-k)^2 = 4p(x-h)$. The vertex is $(3, -1)$, $p = 2$ (from $8 = 4p$); the parabola opens to the right; the focus is $(h+p, k) = (3+2, -1) = (5, -1)$; and the directrix is $x = 1$ (from $x = h - p = 3 - 2 = 1$).

If we know any two of the vertex, focus, and directrix, we can find an equation of the parabola. From the information given, we first need to decide which form to use. Knowing the directrix is the fastest way to decide this. If the directrix is a horizontal line ($y =$ *number*), then the equation is $(x-h)^2 = 4p(y-k)$. If the directrix is a vertical line ($x =$ *number*), then the equation is $(y-k)^2 = 4p(x-h)$. If we do not have the directrix, we need to look at the coordinates of the vertex and focus. Either both the x-coordinates will be the same or both y-coordinates will be. If both x-coordinates are the same, the parabola opens up or down. We need to use the form $(x-h)^2 = 4p(y-k)$. If both y-coordinates are the same, the parabola opens to the side. We need to use the form $(y-k)^2 = 4p(x-h)$. Once we have decided which form to use, we might need to use algebra to find h, k, and p. For example, if we know the focus is $(2, -1)$ and the directrix is $x = 6$, then we know $h - p = 6$ and $h + p = 2$ and $k = -1$. The equations $h - p = 6$ and $h + p = 2$ form a system of equations.

$$\begin{aligned} h - p &= 6 \\ \underline{h + p} &\underline{= 2} \\ 2h &= 8 \end{aligned}$$

$$h = 4$$

$$4 - p = 6 \qquad \text{Let } h = 4 \text{ in } h - p = 6$$

$$p = -2$$

Now that we have all three numbers and the form, we are ready to write the equation: $(y + 1)^2 = -8(x - 4)$.

EXAMPLES

Find an equation for the parabola.

- The directrix is $y = 2$, and the vertex is $(3, 1)$.
Because the directrix is a horizontal line, the equation we want is $(x - h)^2 = 4p(y - k)$. The vertex is $(3, 1)$, giving us $h = 3$ and $k = 1$. From $y = k - p$ and $y = 2$, we have $1 - p = 2$, making $p = -1$. The equation is $(x - 3)^2 = -4(y - 1)$.
- The focus is $(4, -2)$, and the vertex is $(0, -2)$.
The y-coordinates are the same, so this parabola opens to the side, and the equation we need is $(y - k)^2 = 4p(x - h)$. The vertex is $(h, k) = (0, -2)$, giving us $h = 0$ and $k = -2$. The focus is $(h + p, k) = (4, -2)$. From this we have $h + p = 0 + p = 4$, making $p = 4$. The equation is $(y + 2)^2 = 16x$.
-

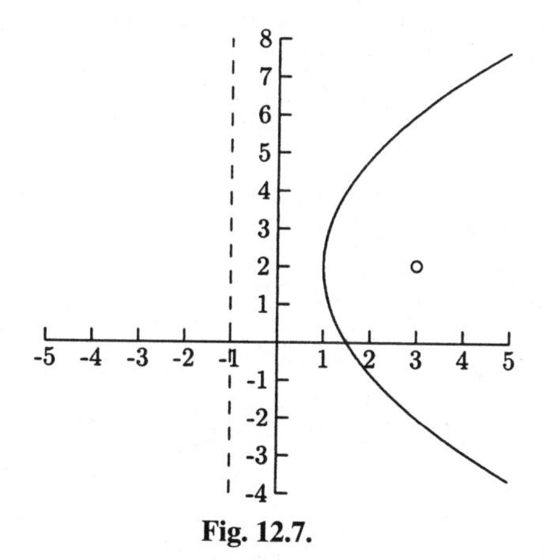

Fig. 12.7.

The directrix is the vertical line $x = -1$, and the focus is $(3, 2)$. Because the parabola opens to the right, the form we need is $(y - k)^2 = 4p(x - h)$.

From the focus we have $(h + p, k) = (3, 2)$, so $h + p = 3$ and $k = 2$. The directrix is $x = -1$ and $x = h - p$ so $h - p = -1$.

$$\begin{aligned} h - p &= -1 \\ \underline{h + p} &\underline{= 3} \\ 2h &= 2 \\ h &= 1 \\ 1 + p &= 3 \qquad \text{Let } h = 1 \text{ in } h + p = 3 \\ p &= 2 \end{aligned}$$

The equation is $(y - 2)^2 = 8(x - 1)$.

PRACTICE

1. Identify the vertex, focus, and directrix for $(y - 5)^2 = 10(x - 1)$.
2. Identify the vertex, focus, and directrix for $(x + 6)^2 = -\frac{1}{2}(y - 4)$.
3. Find an equation for the parabola that has directrix $y = -2$ and focus $(4, 10)$.

For Problems 4–6, match the equation with its graph in Figures 12.8–12.10.

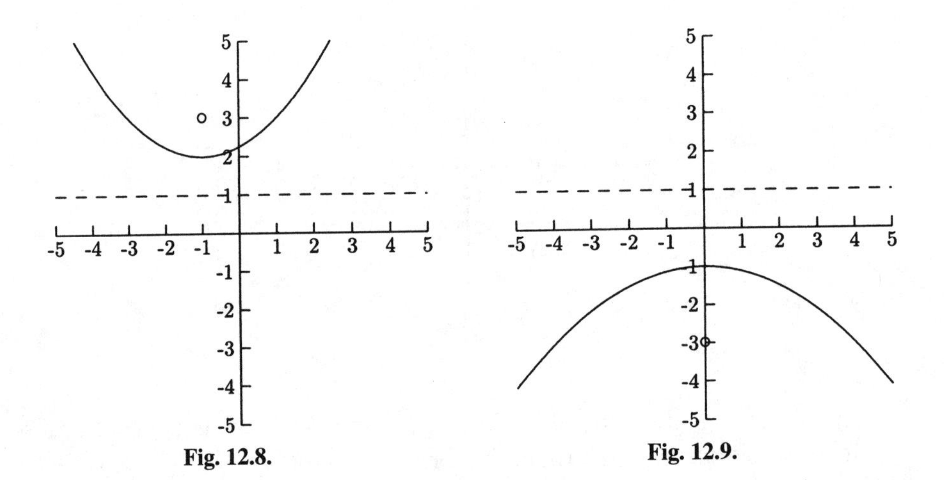

Fig. 12.8. Fig. 12.9.

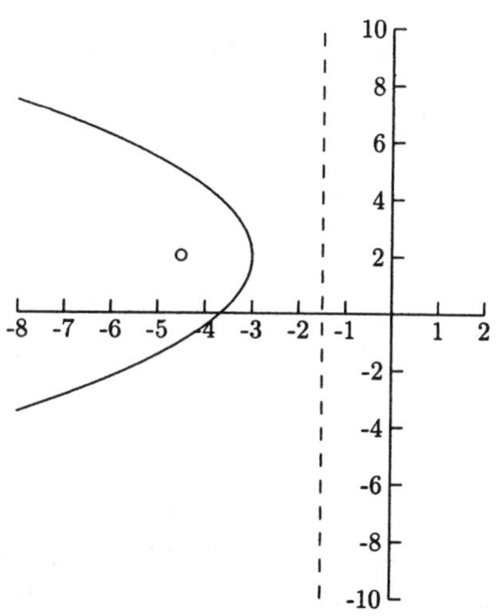

Fig. 12.10.

4. $x^2 = -8(y+1)$
5. $(x+1)^2 = 4(y-2)$
6. $(y-2)^2 = -6(x+3)$

SOLUTIONS

1. $h = 1$, $k = 5$, and $p = \frac{5}{2}$ (from $4p = 10$). The vertex is $(1, 5)$; the focus is $(h + p, k) = (1 + \frac{5}{2}, 5) = (\frac{7}{2}, 5)$ and the directrix is $x = -\frac{3}{2}$ (from $h - p = 1 - \frac{5}{2}$).
2. $h = -6$, $k = 4$, and $p = -\frac{1}{8}$ (from $4p = -\frac{1}{2}$). The vertex is $(-6, 4)$; the focus is $(h, k + p) = (-6, 4 + (-\frac{1}{8})) = (-6, \frac{31}{8})$; and the directrix is $y = \frac{33}{8}$ (from $k - p = 4 - (-\frac{1}{8})$).
3. We want to use the equation $(x-h)^2 = 4p(y-k)$. The focus is $(h, k+p)$, so $h = 4$ and $k + p = 10$. The directrix is $y = k - p$, so $k - p = -2$.

$$k + p = 10$$
$$\underline{k - p = -2}$$
$$2k = 8$$
$$k = 4$$

$$4 + p = 10 \qquad \text{Let } k = 4 \text{ in } k + p = 10$$

$$p = 6$$

The equation is $(x - 4)^2 = 24(y - 4)$.

4. Figure 12.9
5. Figure 12.8
6. Figure 12.10

Ellipses

Most ellipses look like flattened circles. Usually one diameter is longer than the other. In Figure 12.11, the horizontal diameter is longer than the vertical diameter. In Figure 12.12 the vertical diameter is longer than the horizontal diameter. The longer diameter is the *major axis*, and the shorter diameter is the *minor axis*. An ellipse has seven important points—the center, two endpoints of the major axis (the vertices), two endpoints of the minor axis, and two points along the major axis called the *foci* (plural for *focus*). When the equation of an ellipse is in the form

$$\frac{(x-h)^2}{a^2} + \frac{(y-k)^2}{b^2} = 1 \text{ or } \frac{(x-h)^2}{b^2} + \frac{(y-k)^2}{a^2} = 1,$$

we can find these points without much trouble.

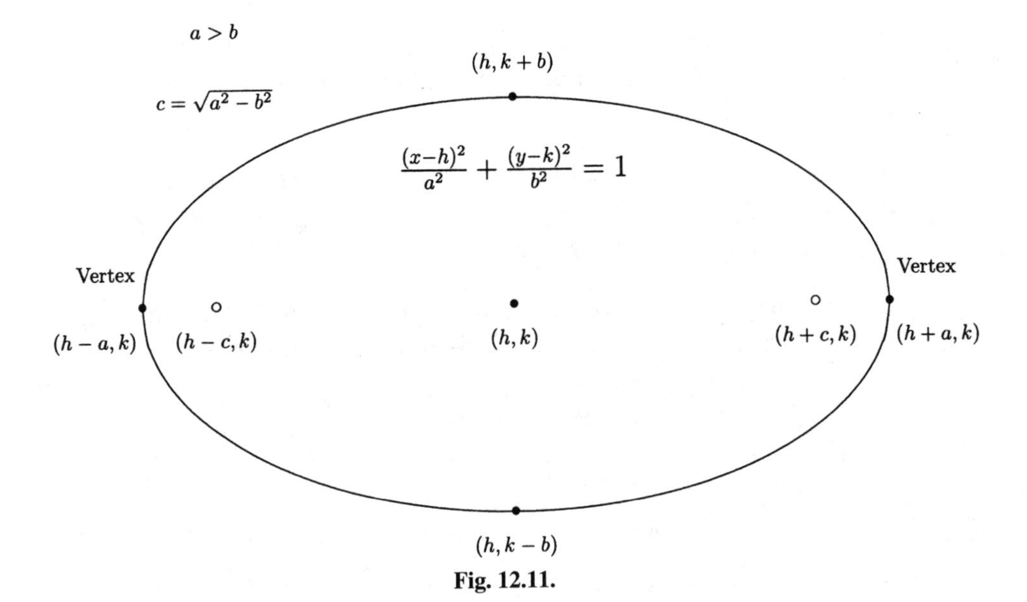

Fig. 12.11.

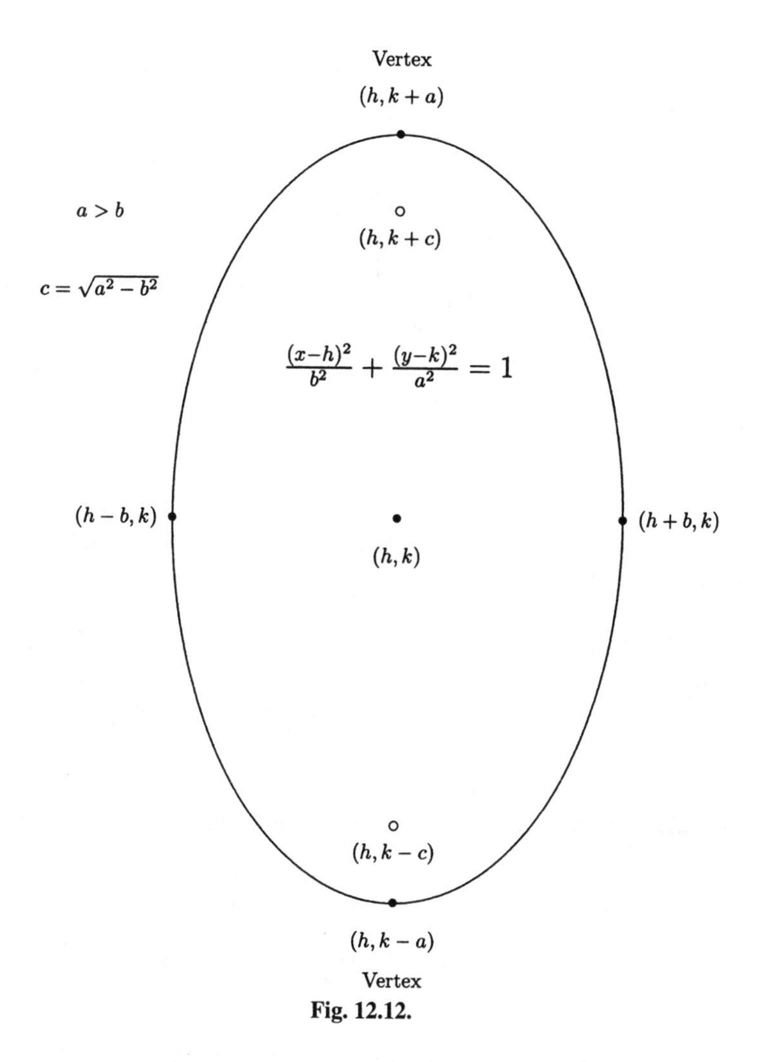

Fig. 12.12.

If all we want to do is to sketch the graph, all we really need to do is to plot the endpoints of the diameters and draw a rounded curve connecting them. For example, if we want to sketch the graph of $\frac{(x+1)^2}{4} + \frac{(y-1)^2}{9} = 1$, $a = 3$, $b = 2$, $h = -1$, and $k = 1$. According to Figure 12.12, the diameters have coordinates $(-1 - 2, 1) = (-3, 1)$, $(-1 + 2, 1) = (1, 1)$, $(-1, 1 + 3) = (-1, 4)$, and $(-1, 1 - 3) = (-1, -2)$. (See Figure 12.13.)

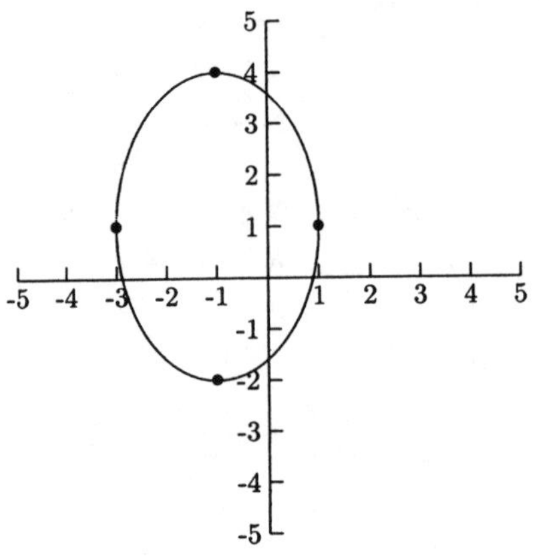

Fig. 12.13.

DEFINITION: An ellipse is the set of all points whose distances to two fixed points (the foci) is constant.

For example, the foci for $\frac{x^2}{25} + \frac{y^2}{9} = 1$ are $(-4, 0)$ and $(4, 0)$. If we take any point on this ellipse and calculate its distance to $(-4, 0)$ and to $(4, 0)$ and add these numbers, the sum will be 10. Two points on this ellipse are $(0, 3)$ and $(\frac{5}{3}, \sqrt{8})$.

$$\text{Distance from } (0, 3) \text{ to } (-4, 0) + \text{ Distance from } (0, 3) \text{ to } (4, 0)$$

$$= \sqrt{(-4-0)^2 + (0-3)^2} + \sqrt{(4-0)^2 + (0-3)^2}$$

$$= \sqrt{16+9} + \sqrt{16+9} = \sqrt{25} + \sqrt{25} = 10$$

$$\text{Distance from } (5/3, \sqrt{8}) \text{ to } (-4, 0) + \text{ Distance from } (5/3, \sqrt{8}) \text{ to } (4, 0)$$

$$= \sqrt{\left(-4 - \frac{5}{3}\right)^2 + (0 - \sqrt{8})^2} + \sqrt{\left(4 - \frac{5}{3}\right)^2 + (0 - \sqrt{8})^2}$$

$$= \sqrt{\frac{289}{9} + 8} + \sqrt{\frac{49}{9} + 8} = \sqrt{\frac{361}{9}} + \sqrt{\frac{121}{9}} = 10$$

In the next set of problems, we will be given an equation for an ellipse. From the equation, we can find h, k, a, and b. With these numbers and the information in Figures 12.11 or 12.12 we can find the center, foci, and vertices.

EXAMPLES

Find the center, foci, and vertices for the ellipse.

- $\dfrac{(x-3)^2}{16}+\dfrac{(y+5)^2}{25}=1$

  From the equation, we see that $h=3$, $k=-5$, a^2 and b^2 are 4^2 and 5^2, but which is a and which is b? a needs to be the larger number, so $a=5$ and $b=4$. This makes $c=\sqrt{a^2-b^2}=\sqrt{25-16}=3$. We need to use the information in Figure 12.12 because the larger denominator is under $(y-k)^2$.

  Center: $(h,k)=(3,-5)$

  Foci: $(h,k-c)=(3,-5-3)=(3,-8)$ and $(h,k+c)=(3,-5+3)=(3,-2)$

  Vertices: $(h,k-a)=(3,-5-5)=(3,-10)$ and $(h,k+a)=(3,-5+5)=(3,0)$

- $\dfrac{x^2}{16}+(y-2)^2=1$

  To make it easier to find h, k, a, and b, we will rewrite the equation.

$$\frac{(x-0)^2}{16}+\frac{(y-2)^2}{1}=1$$

  Now we can see that $h=0$, $k=2$, $a=4$, $b=1$, $c=\sqrt{a^2-b^2}=\sqrt{16-1}=\sqrt{15}$. Because the larger denominator is under $(x-0)^2$, we need to use the information in Figure 12.11.

  Center: $(h,k)=(0,2)$

  Foci: $(h-c,k)=(0-\sqrt{15},2)=(-\sqrt{15},2)$ and $(h+c,k)=(0+\sqrt{15},2)=(\sqrt{15},2)$

  Vertices: $(h-a,k)=(0-4,2)=(-4,2)$ and $(h+a,k)=(0+4,2)=(4,2)$

Now that we can find this important information from an equation of an ellipse, we are ready to match graphs of ellipses to their equations.

EXAMPLES

Match the equations with the graphs in Figures 12.14–12.16.

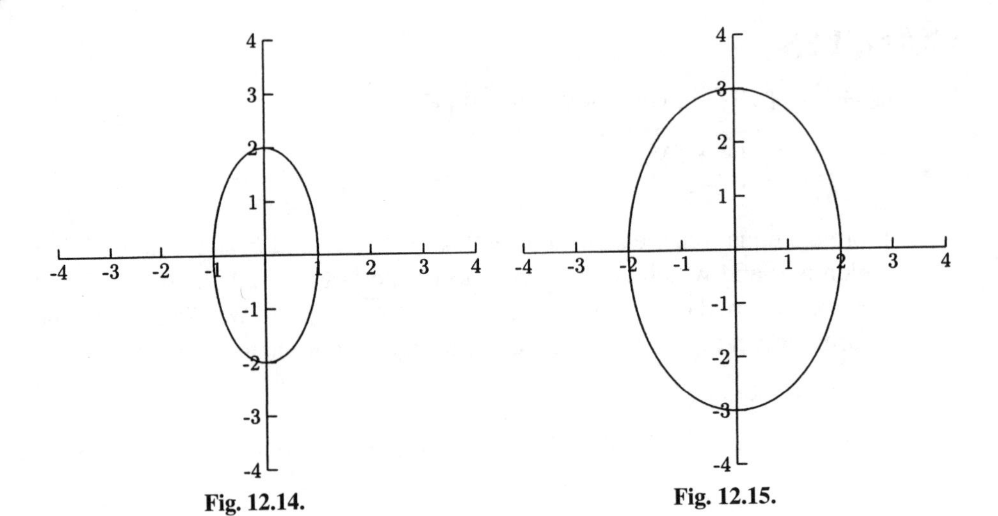

Fig. 12.14.

Fig. 12.15.

Fig. 12.16.

- $\dfrac{x^2}{4} + \dfrac{y^2}{9} = 1$

  The larger denominator is under y^2, so we need to use the information in Figure 12.12. Because $a = 3$, we need to look for a graph with vertices $(0, 3)$ and $(0, -3)$. This graph is in Figure 12.15.

- $\frac{x^2}{9} + \frac{y^2}{4} = 1$

  The larger denominator is under x^2, so we need to use the information in Figure 12.11. Because $a = 3$, the vertices are $(-3, 0)$ and $(3, 0)$. This graph is in Figure 12.16.

- $x^2 + \frac{y^2}{4} = 1$

  The larger denominator is under y^2, so we need to use the information in Figure 12.12. Because $a = 2$, the vertices are $(0, 2)$ and $(0, -2)$. This graph is in Figure 12.14.

With as little as three points, we can find an equation of an ellipse. Using the formulas in Figures 12.11 and 12.12 and some algebra, we can find h, k, a, and b.

EXAMPLES

Find an equation of the ellipse.

- The vertices are $(-4, 2)$ and $(6, 2)$, and $(1, 5)$ is a point on the graph. The y-coordinates are the same, so the major axis (the larger diameter) is horizontal, which means we need to use the information in Figure 12.11. The vertices are $(h - a, k)$ and $(h + a, k)$. This means that $h - a = -4$ and $h + a = 6$, and $k = 2$.

$$h - a = -4$$

$$\underline{h + a = 6}$$

$$2h = 2$$

$$h = 1$$

$$1 - a = -4 \qquad \text{Let } h = 1 \text{ in } h - a = -4$$

$$a = 5$$

  So far we know that

$$\frac{(x-1)^2}{25} + \frac{(y-2)^2}{b^2} = 1.$$

Because $(1, 5)$ is on the graph, $\frac{(1-1)^2}{25} + \frac{(5-2)^2}{b^2} = 1$. Solving this equation for b, we find that $b = 3$. The equation is

$$\frac{(x-1)^2}{25} + \frac{(y-2)^2}{9} = 1.$$

- The foci are $(-4, -10)$ and $(-4, 14)$ and $(-4, 15)$ is a vertex.
  The x-values of foci are the same, so the major axis is vertical. This tells us that we need to use the information in Figure 12.12.
  $(h, k-c) = (-4, -10)$ and $(h, k+c) = (-4, 14)$, so $h = -4$, $k-c = -10$ and $k + c = 14$.

$$k - c = -10$$
$$\underline{k + c = 14}$$
$$2k = 4$$
$$k = 2$$
$$2 - c = -10 \qquad \text{Let } k = 2 \text{ in } k - c = -10$$
$$c = 12$$

Because $(-4, 15)$ is a vertex, $k + a = 15$, so $2 + a = 15$ and $a = 13$. All we need to finish is to find b. Let $a = 13$ and $c = 12$ in $c = \sqrt{a^2 - b^2}$: $12 = \sqrt{13^2 - b^2}$. Solving this for b, we have $b = 5$. The equation is

$$\frac{(x+4)^2}{25} + \frac{(y-2)^2}{169} = 1.$$

The *eccentricity* of an ellipse is a number that measures how flat it is. The formula is $e = \frac{c}{a}$. This number ranges between 0 and 1. The closer to 1 the eccentricity of an ellipse is, the flatter it is. If $e = \frac{c}{a} = 0$, then the ellipse is a circle. In a circle, the center and foci are all the same point, and a and b are the same number. For example, $\frac{(x-5)^2}{9} + \frac{(y-4)^2}{9} = 1$ is a circle with center $(5, 4)$ and radius $\sqrt{9} = 3$. Usually we see equations of circles in the form $(x - h)^2 + (y - k)^2 = r^2$.

EXAMPLES

Find the ellipse's eccentricity.

- $\frac{x^2}{9} + \frac{y^2}{25} = 1$
  $a = 5, b = 3, c = \sqrt{25 - 9} = 4$ and $e = \frac{c}{a} = \frac{4}{5}$

- $\dfrac{(x+8)^2}{144} + \dfrac{(y+6)^2}{169} = 1$

  $a = 13$, $b = 12$, $c = \sqrt{169 - 144} = 5$, $e = \frac{c}{a} = \frac{5}{13}$ This ellipse is more rounded than the first because e is closer to 0.

PRACTICE

1. Identify the center, foci, vertices, and eccentricity for

$$\frac{x^2}{169} + \frac{(y-10)^2}{25} = 1.$$

2. Identify the center, foci, vertices, and eccentricity for

$$\frac{(x+9)^2}{20^2} + \frac{(y+2)^2}{29^2} = 1.$$

3. Identify the center and radius for the circle

$$\frac{(x+6)^2}{49} + \frac{(y-1)^2}{49} = 1.$$

For Problems 4–7, match the equation with the graph in Figures 12.17–12.20.

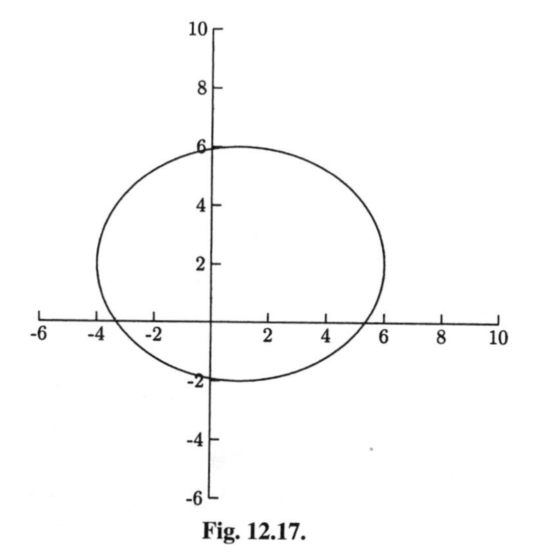

Fig. 12.17.

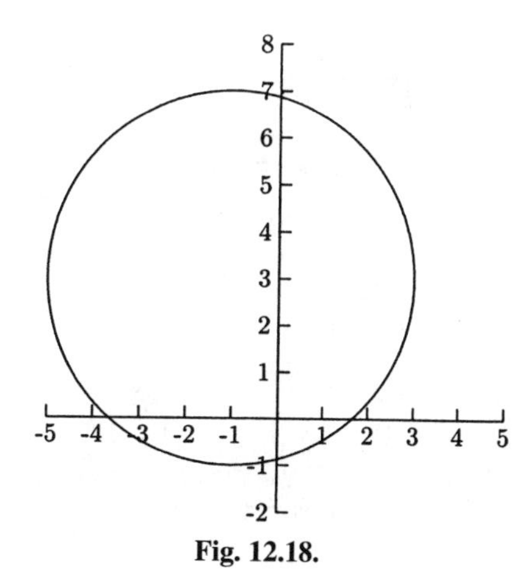

Fig. 12.18.

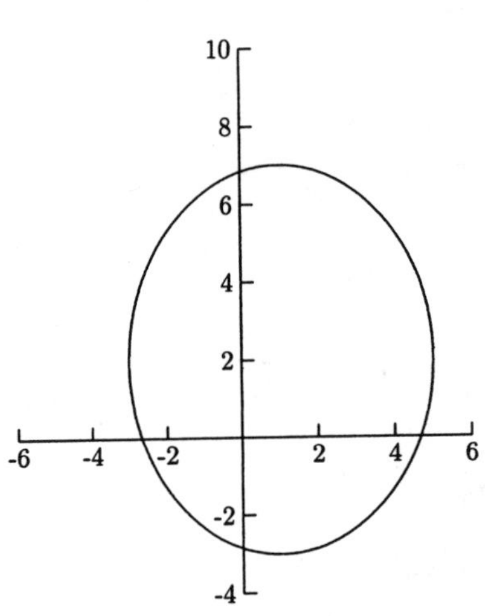

Fig. 12.19.

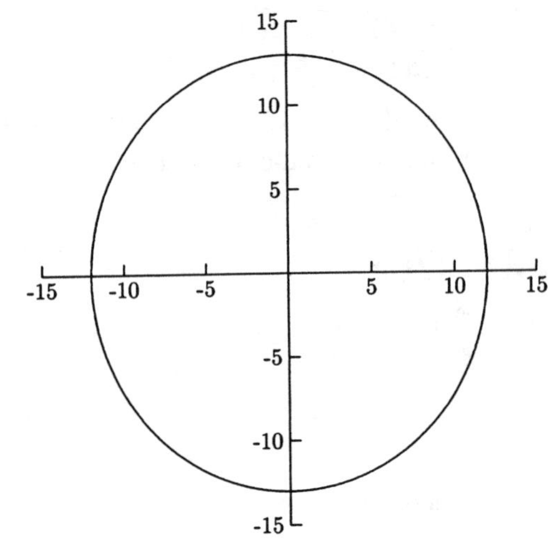

Fig. 12.20.

4. $\dfrac{(x-1)^2}{16}+\dfrac{(y-2)^2}{25}=1$

5. $\dfrac{x^2}{144}+\dfrac{y^2}{169}=1$

6. $\dfrac{(x+1)^2}{16}+\dfrac{(y-3)^2}{16}=1$

7. $\dfrac{(x-1)^2}{25}+\dfrac{(y-2)^2}{16}=1$

SOLUTIONS

1. $h=0, k=10, a=13, b=5, c=\sqrt{a^2-b^2}=\sqrt{169-25}=12$

   Center: (0, 10)

   Foci: $(h-c,k)=(0-12,10)=(-12,10)$ and $(h+c,k)=(0+12,10)=(12,10)$

   Vertices: $(h-a,k)=(0-13,10)=(-13,10)$ and $(h+a,k)=(0+13,10)=(13,10)$

   Eccentricity: $\dfrac{c}{a}=\dfrac{12}{13}$

2. $h = -9, k = -2, a = 29, b = 20, c = \sqrt{29^2 - 20^2} = 21$

   Center: $(-9, -2)$

   Foci: $(h, k-c) = (-9, -2-21) = (-9, -23)$ and $(h, k+c) = (-9, -2+21) = (-9, 19)$

   Vertices: $(h, k - a) = (-9, -2 - 29) = (-9, -31)$ and $(h, k + a) = (-9, -2 + 29) = (-9, 27)$

   Eccentricity: $\frac{c}{a} = \frac{21}{29}$

3. The center is $(-6, 1)$, and the radius is 7.
4. Figure 12.19
5. Figure 12.20
6. Figure 12.18
7. Figure 12.17

Hyperbolas

The last conic section is the hyperbola. Hyperbolas are formed when a slice is made through both parts of a double cone. The graph of a hyperbola comes in two pieces called *branches*. Like ellipses, hyperbolas have a center, two foci, and two vertices. Hyperbolas also have two slant asymptotes. The definition of a hyperbola involves the distance between points on the graph and two fixed points.

DEFINITION: A hyperbola is the set of all points such that the difference of the distance between a point and two fixed points (the foci) is constant.

For example, the foci for $\frac{x^2}{9} - \frac{y^2}{16} = 1$ are $(-5, 0)$ and $(5, 0)$. For any point on the hyperbola, the distance between this point and one focus minus the distance between the same point and the other focus is 6. Two points on the hyperbola are $(6, \sqrt{48})$ and $(12, \sqrt{240})$.

Distance from $(6, \sqrt{48})$ to $(-5, 0)$ – Distance from $(6, \sqrt{48})$ to $(5, 0)$

$$= \sqrt{(-5-6)^2 + (0-\sqrt{48})^2} - \sqrt{(5-6)^2 + (0-\sqrt{48})^2}$$

$$= \sqrt{121 + 48} - \sqrt{1 + 48} = 13 - 7 = 6$$

And

$$\text{Distance from } (12, \sqrt{240}) \text{ to } (-5, 0) - \text{Distance from } (12, \sqrt{240}) \text{ to } (5, 0)$$
$$= \sqrt{(-5-12)^2 + (0, -\sqrt{240})^2} - \sqrt{(5-12)^2 + (0-\sqrt{240})^2}$$
$$= \sqrt{289+240} - \sqrt{49+240} = 23 - 17 = 6$$

Equations of hyperbolas come in one of two forms.

$$\frac{(x-h)^2}{a^2} - \frac{(y-k)^2}{b^2} = 1 \quad \text{or} \quad \frac{(y-k)^2}{a^2} - \frac{(x-h)^2}{b^2} = 1$$

If the x^2 term is positive, one branch opens to the left and the other to the right. If the y^2 term is positive, one branch opens up and the other down. The formulas for these two forms are in Figures 12.21 and 12.22.

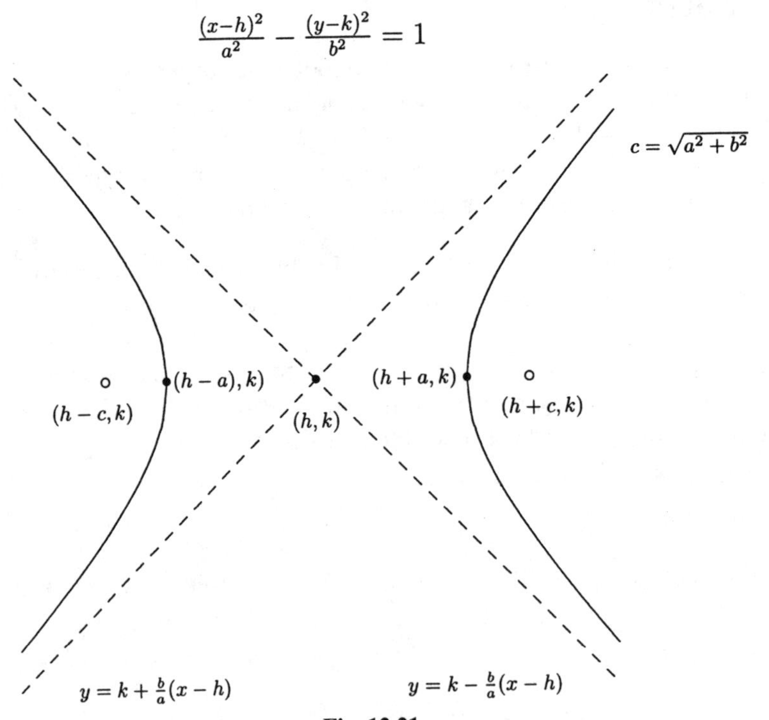

Fig. 12.21.

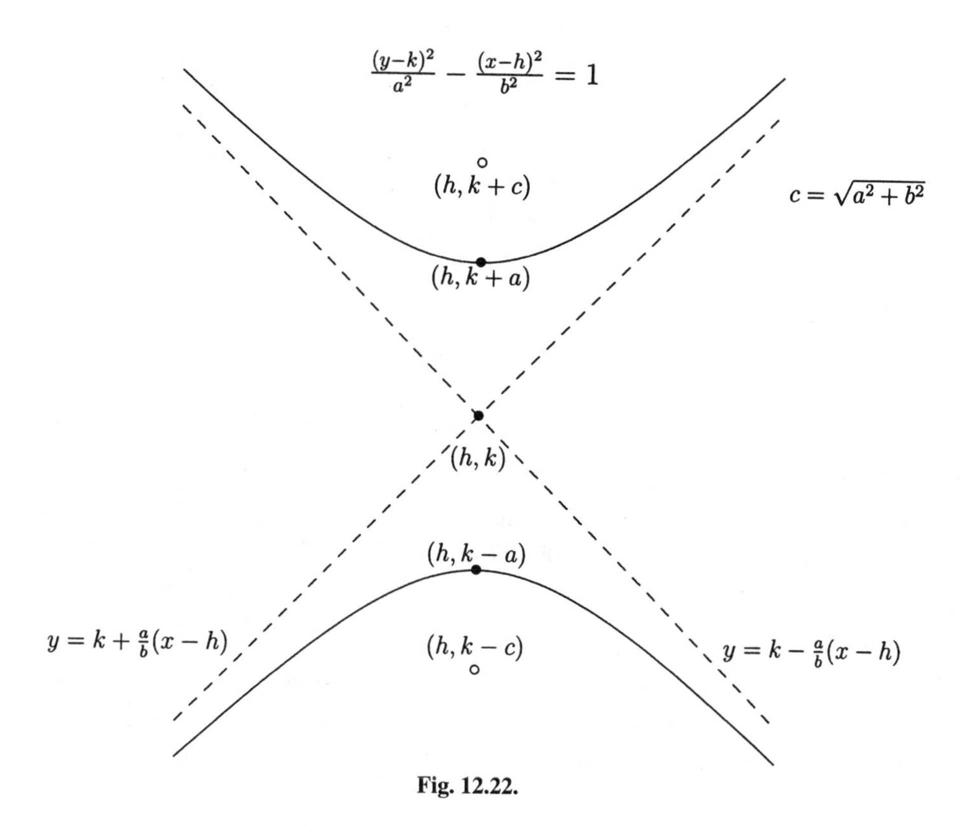

Fig. 12.22.

We can sketch a hyperbola by plotting the vertices and sketching the asymptotes, using dashed lines. We should also plot two points to the left and two points to the right of the vertices.

EXAMPLE

- Sketch the graph for $\frac{y^2}{4} - x^2 = 1$.

  Because y^2 is positive, we will use the information in Figure 12.22. The center is $(0, 0)$, $a = 2$, and $b = 1$. The vertices are $(h, k+a) = (0, 0+2) = (0, 2)$ and $(h, k-a) = (0, 0-2) = (0, -2)$. The asymptote formulas are $y = k - \frac{a}{b}(x-h)$ and $y = k + \frac{a}{b}(x-h)$. Using our numbers for h, k, a, and b, we have $y = -2x$ and $y = 2x$. We will use $x = 4$ and $x = -4$ for our extra points. If we let $x = 4$ or $x = -4$, we get two y-values, $\pm\sqrt{68}$. These give us four more points—$(4, \sqrt{68})$, $(4, -\sqrt{68})$, $(-4, \sqrt{68})$, and $(-4, -\sqrt{68})$. (see Figure 12.23.)

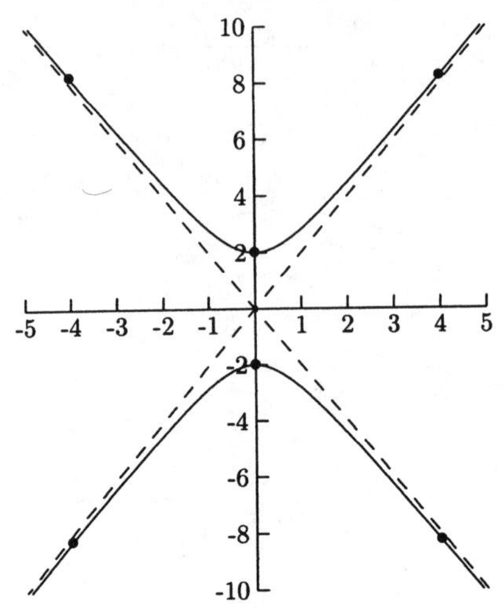

Fig. 12.23.

In the next examples, we will find the center, vertices, foci, and asymptotes for given hyperbolas. Once we have determined whether x^2 is positive or y^2 is positive, we can decide on which formulas to use, those in Figure 12.21 or Figure 12.22.

EXAMPLES

Find the center, vertices, foci, and asymptotes for the hyperbola.

- $$\frac{(x+7)^2}{36} - \frac{(y+4)^2}{64} = 1$$

  Because x^2 is positive, we will use the information in Figure 12.21.

  $h = -7, k = -4, a = 6, b = 8, c = \sqrt{36+64} = 10$

  Center: $(-7, -4)$

  Vertices: $(h-a, k) = (-7-6, -4) = (-13, -4)$ and $(h+a, k) = (-7+6, -4) = (-1, -4)$

  Foci: $(h-c, k) = (-7-10, -4) = (-17, -4)$ and $(h+c, k) = (-7+10, -4) = (3, -4)$

  Asymptotes: $y = k - \frac{b}{a}(x-h) = -4 - \frac{8}{6}(x+7) = -\frac{4}{3}x - \frac{40}{3}$ and $y = k + \frac{b}{a}(x-h) = -4 + \frac{8}{6}(x+7) = \frac{4}{3}x + \frac{16}{3}$

- $\dfrac{y^2}{144} - \dfrac{(x-1)^2}{25} = 1$

  Because y^2 is positive, we need to use the information in Figure 12.22.

  $h = 1, k = 0, a = 12, b = 5, c = \sqrt{144 + 25} = 13$

  Center: $(1, 0)$

  Vertices: $(h, k-a) = (1, 0-12) = (1, -12)$ and $(h, k+a) = (1, 0+12) = (1, 12)$

  Foci: $(h, k-c) = (1, 0-13) = (1, -13)$ and $(h, k+c) = (1, 0+13) = (1, 13)$

  Asymptotes: $y = k - \frac{a}{b}(x-h) = 0 - \frac{12}{5}(x-1) = -\frac{12}{5}x + \frac{12}{5}$ and $y = k + \frac{a}{b}(x-h) = 0 + \frac{12}{5}(x-1) = \frac{12}{5}x - \frac{12}{5}$

In the next problem, we will match equations of hyperbolas with their graphs. Being able to identify the vertices will not be enough. We will also need to use the equations of the asymptotes to find b (we will know a from the vertices). Because the center of each hyperbola will be at (0, 0), the asymptotes will either be $y = \frac{a}{b}x$ and $y = -\frac{a}{b}x$ or $y = \frac{b}{a}x$ and $y = -\frac{b}{a}x$.

EXAMPLES

Match the equation with its graph in Figures 12.24–12.27.

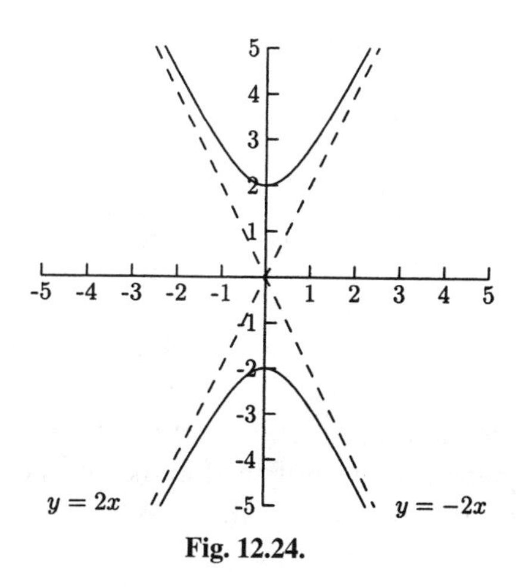

Fig. 12.24.

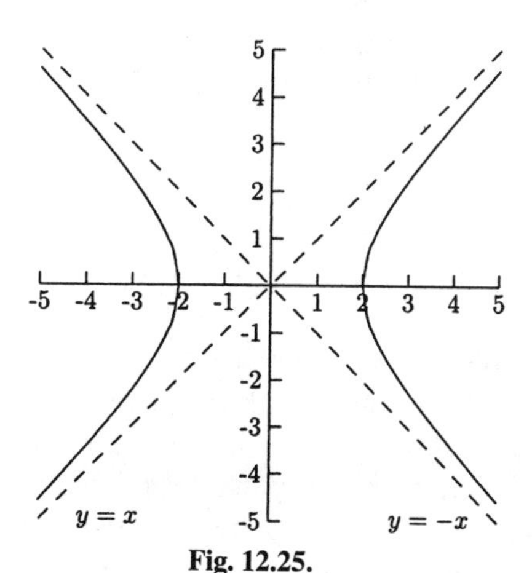

Fig. 12.25.

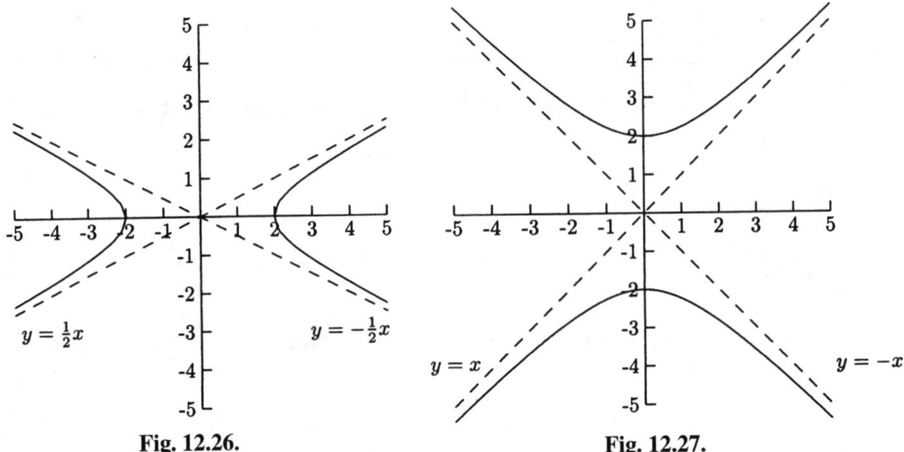

Fig. 12.26.

Fig. 12.27.

- $\dfrac{x^2}{4} - \dfrac{y^2}{4} = 1$

  The vertices are $(-2, 0)$ and $(2, 0)$. The slopes of the asymptotes are -1 and 1. The graph is in Figure 12.25.

- $\dfrac{x^2}{4} - y^2 = 1$

  The vertices are $(-2, 0)$ and $(2, 0)$. The slopes of the asymptotes are $-\frac{1}{2}$ and $\frac{1}{2}$. The graph is in Figure 12.26.

- $\dfrac{y^2}{4} - \dfrac{x^2}{4} = 1$

  The vertices are $(0, -2)$ and $(0, 2)$. The slopes of the asymptotes are -1 and 1. The graph is in Figure 12.27.

- $\dfrac{y^2}{4} - x^2 = 1$

  The vertices are $(0, -2)$ and $(0, 2)$. The slopes of the asymptotes are -2 and 2. The graph is in Figure 12.24.

We can find the equation for a hyperbola when we know some points or a point and the asymptotes. If we have the vertices and foci, then finding an equation for a hyperbola will be similar to finding an equation for an ellipse. If we are given the vertices and asymptotes or foci and asymptotes, we will need to use the slope of one of the asymptotes to find either a or b (we will know one but not the other from the vertices or foci). The first thing we need to decide is which formulas to use—those in Figures 12.21 or Figure 12.22. If the vertices or foci are on the same horizontal line (the y-coordinates are the same), we will use Figure 12.21. If they are on the same vertical line (the x-coordinates are the same), we will use Figure 12.22.

EXAMPLES

Find an equation for the hyperbola.

- The vertices are $(3, -1)$ and $(3, 7)$ and $y = \frac{4}{3}x - 1$ is an asymptote.
  The vertices are on the same vertical line, so we need to use the information in Figure 12.22. The vertices are $(h, k - a) = (3, -1)$ and $(h, k + a) = (3, 7)$. This gives us $h = 3$, $k - a = -1$ and $k + a = 7$.

$$k + a = 7$$

$$\underline{k - a = -1}$$

$$2k = 6$$

$$k = 3$$

$$3 + a = 7 \qquad \text{Let } k = 3 \text{ in } k + a = 7$$

$$a = 4$$

  The center is $(3, 3)$ and $a = 4$. Once we have b, we will be done. The slope of one of the asymptotes in Figure 12.22 is $\frac{a}{b}$, so we have $\frac{a}{b} = \frac{4}{b} = \frac{4}{3}$, so $b = 3$. The equation is

$$\frac{(y-3)^2}{16} - \frac{(x-3)^2}{9} = 1.$$

- The vertices are $(-8, 5)$ and $(4, 5)$, and the foci are $(-12, 5)$ and $(8, 5)$.
  The vertices and foci are on the same horizontal line, so we need to use the information in Figure 12.21. The vertices are $(h - a, k) = (-8, 5)$ and $(h + a, k) = (4, 5)$. Now we know $k = 5$ and we have the system $h - a = -8$ and $h + a = 4$.

$$h - a = -8$$

$$\underline{h + a = 4}$$

$$2h = -4$$

$$h = -2$$

$$-2 - a = -8 \qquad \text{Let } h = -2 \text{ in } h - a = -8$$

$$a = 6$$

A focus is $(h - c, k) = (-2 - c, 5) = (-12, 5)$, which gives us $-2 - c = -12$. Now that we see that $c = 10$, we can put this and $a = 6$ in $c = \sqrt{a^2 + b^2}$ to find b.

$$10 = \sqrt{36 + b^2}$$

$$100 = 36 + b^2$$

$$8 = b$$

The equation is

$$\frac{(x+2)^2}{36} - \frac{(y-5)^2}{64} = 1.$$

PRACTICE

1. Find the center, vertices, foci, and asymptotes for

$$\frac{y^2}{16} - \frac{(x-5)^2}{9} = 1.$$

2. Find the center, vertices, foci, and asymptotes for

$$\frac{(x+8)^2}{49} - \frac{(y+6)^2}{576} = 1.$$

3. Find an equation for the hyperbola having vertices $(-4, 2)$ and $(12, 2)$ and foci $(-6, 2)$ and $(14, 2)$.
4. Find an equation for the hyperbola having vertices $(-8, 0)$ and $(-4, 0)$ and with an asymptote $y = \frac{1}{2}x + 3$.

   In Problems 5–7, match the graphs in Figures 12.28–12.30 with their equations.

5. $(y - 1)^2 - (x - 1)^2 = 1$
6. $(x - 1)^2 - (y - 1)^2 = 1$
7. $\dfrac{(x-1)^2}{4} - (y - 1)^2 = 1$

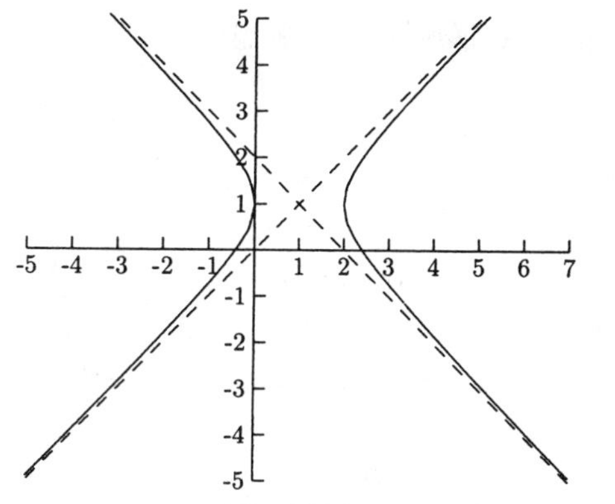

Fig. 12.28.

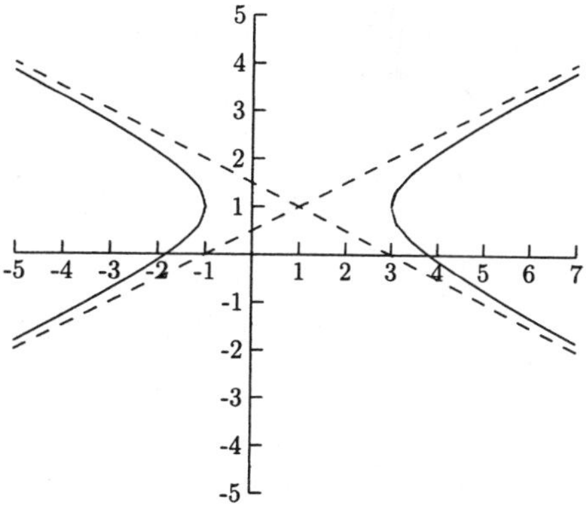

Fig. 12.29.

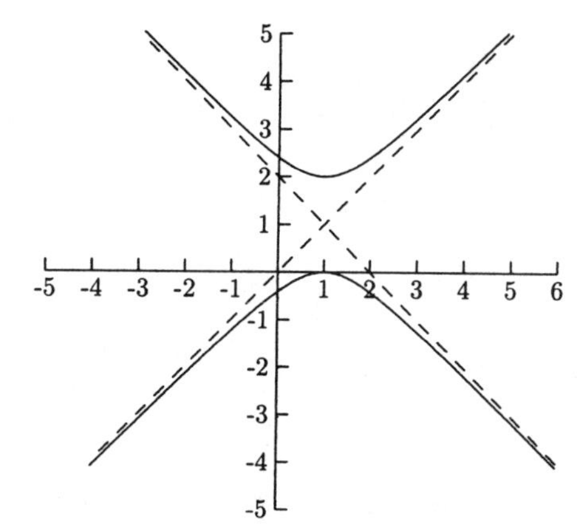

Fig. 12.30.

SOLUTIONS

1. $h = 5$, $k = 0$, $a = 4$, $b = 3$, and $c = \sqrt{16+9} = 5$

   Center: $(5, 0)$

   Vertices: $(h, k - a) = (5, 0 - 4) = (5, -4)$ and $(h, k + a) = (5, 0+4) = (5, 4)$

   Foci: $(h, k-c) = (5, 0-5) = (5, -5)$ and $(h, k+c) = (5, 0+5) = (5, 5)$

   Asymptotes: $y = k - \frac{a}{b}(x - h) = 0 - \frac{4}{3}(x - 5) = -\frac{4}{3}x + \frac{20}{3}$ and $y = k + \frac{a}{b}(x - h) = 0 + \frac{4}{3}(x - 5) = \frac{4}{3}x - \frac{20}{3}$

2. $h = -8, k = -6, a = 7, b = 24$, and $c = \sqrt{49 + 576} = 25$

   Center: $(-8, -6)$

   Vertices: $(h - a, k) = (-8 - 7, -6) = (-15, -6)$ and $(h + a, k) = (-8 + 7, -6) = (-1, -6)$

   Foci: $(h - c, k) = (-8 - 25, -6) = (-33, -6)$ and $(h + c, k) = (-8 + 25, -6) = (17, -6)$

   Asymptotes: $y = k - \frac{b}{a}(x - h) = -6 - \frac{24}{7}(x + 8) = -\frac{24}{7}x - \frac{234}{7}$ and

   $y = k + \frac{b}{a}(x - h) = -6 + \frac{24}{7}(x + 8) = \frac{24}{7}x + \frac{150}{7}$

3. The vertices are $(-4, 2)$ and $(12, 2)$, which gives us $k = 2$ and $(h - a, k) = (-4, 2)$ and $(h + a, k) = (12, 2)$.

$$h - a = -4$$
$$\underline{h + a = 12}$$
$$2h = 8$$
$$h = 4$$
$$4 - a = -4 \qquad \text{Let } h = 4 \text{ in } h - a = -4$$
$$a = 8$$

A focus is $(-6, 2)$, which gives us $(h - c, k) = (-6, 2)$ and $h - c = 4 - c = -6$. Solving $4 - c = -6$ gives us $c = 10$. We can find b by letting $a = 8$ and $c = 10$ in $c = \sqrt{a^2 + b^2}$.

$$c = \sqrt{a^2 + b^2}$$
$$10 = \sqrt{64 + b^2}$$
$$100 = 64 + b^2$$
$$6 = b$$

The equation is

$$\frac{(x - 4)^2}{64} - \frac{(y - 2)^2}{36} = 1.$$

4. $(h-a,k)=(-8,0)$ and $(h+a,k)=(-4,0)$, so $k=0$ and we have the following system.

$$\begin{aligned} h-a &= -8 \\ h+a &= -4 \\ \hline 2h &= -12 \\ h &= -6 \\ -6-a &= -8 \qquad \text{Let } h=-6 \text{ in } h-a=-8 \\ a &= 2. \end{aligned}$$

The slope of an asymptote is $\frac{1}{2}$, so $\frac{b}{a}=\frac{b}{2}=\frac{1}{2}$ and $b=1$. The equation is

$$\frac{(x+6)^2}{4}-y^2=1.$$

5. Figure 12.30
6. Figure 12.28
7. Figure 12.29

In order to use a graphing calculator to graph a conic section, the equation probably needs to be entered as two separate functions. For example, the graph of $y^2=x$ could be entered as $y=\sqrt{x}$ and $y=-\sqrt{x}$. To use a graphing calculator to graph a conic section that is not a function, solve for y. When taking the square root of both sides, we use a "$\pm$" symbol on one of the sides. It is this sign that gives us two separate equations.

EXAMPLES

Solve for y.

- $(y-1)^2+\dfrac{(x+3)^2}{9}=1$

$$(y-1)^2+\frac{(x+3)^2}{9}=1$$

$$(y-1)^2=1-\frac{(x+3)^2}{9}$$

$$y-1=\pm\sqrt{1-\frac{(x+3)^2}{9}}$$

$$y=1\pm\sqrt{1-\frac{(x+3)^2}{9}}$$

$$y=1+\sqrt{1-\frac{(x+3)^2}{9}},\quad 1-\sqrt{1-\frac{(x+3)^2}{9}}$$

- $\dfrac{x^2}{9}-\dfrac{(y+2)^2}{4}=1$

$$\frac{x^2}{9}-\frac{(y+2)^2}{4}=1$$

$$-\frac{(y+2)^2}{4}=1-\frac{x^2}{9}$$

$$\frac{(y+2)^2}{4}=-1+\frac{x^2}{9}$$

$$(y+2)^2=4\left(-1+\frac{x^2}{9}\right)$$

$$y+2=\pm\sqrt{4\left(-1+\frac{x^2}{9}\right)}$$

$$y=-2\pm\sqrt{4\left(-1+\frac{x^2}{9}\right)}$$

$$y=-2+\sqrt{4\left(-1+\frac{x^2}{9}\right)},\quad y=-2-\sqrt{4\left(-1+\frac{x^2}{9}\right)}$$

Equations of conic sections do not always come in the convenient forms we have been using. Sometimes they come in the general form $Ax^2+Bxy+Cy^2+Dx+Ey+F=0$. When A and C are equal (and $B=0$), the graph is a circle. If A and C are positive and not equal (and $B=0$), the graph is an ellipse. If A and C have different signs (and $B=0$), the graph is a hyperbola. If only one of A or C is nonzero (and $B=0$), the graph is a parabola. There are some conic sections whose entire graph is one point. These are called *degenerate conics*. We can rewrite an equation in the general form in the standard form (the form we have been using) by completing the square.

EXAMPLES

Rewrite the equation in standard form.

- $x^2 - 2x - 4y = 11$

  Because there is no y^2 term, the graph of this equation is a parabola that opens up or down. The standard equation is $(x-h)^2 = 4p(y-k)$. We need to have the x terms on one side of the equation and the other terms on the other side.

$$x^2 - 2x - 4y = 11$$

$$x^2 - 2x = 4y + 11$$

$$x^2 - 2x + \left(\frac{2}{2}\right)^2 = 4y + 11 + \left(\frac{2}{2}\right)^2$$

$$x^2 - 2x + 1 = 4y + 12$$

$$(x-1)^2 = 4(y+3)$$

- $-9x^2 + 16y^2 - 18x - 160y + 247 = 0$

  Because the signs on x^2 and y^2 are different, the graph of this equation is a hyperbola. The standard form for this equation is $\frac{(y-k)^2}{a^2} - \frac{(x-h)^2}{b^2} = 1$.

$$-9x^2 + 16y^2 - 18x - 160y + 247 = 0$$

$$16y^2 - 160y - 9x^2 - 18x = -247$$

$$16(y^2 - 10y) - 9(x^2 + 2x) = -247$$

$$16\left(y^2 - 10y + \left(\frac{10}{2}\right)^2\right) - 9\left(x^2 + 2x + \left(\frac{2}{2}\right)^2\right)$$

$$= -247 + 16\left(\frac{10}{2}\right)^2 - 9\left(\frac{2}{2}\right)^2$$

$$16(y-5)^2 - 9(x+1)^2 = 144$$

$$\frac{16(y-5)^2}{144} - \frac{9(x+1)^2}{144} = \frac{144}{144}$$

$$\frac{(y-5)^2}{9} - \frac{(x+1)^2}{16} = 1$$

PRACTICE

1. Solve for y

$$\frac{y^2}{4} - \frac{(x-3)^2}{9} = 1$$

2. Solve for y

$$\frac{(x+10)^2}{25} + \frac{(y+3)^2}{25} = 1$$

3. Rewrite the equation in standard form: $36x^2+9y^2-216x-72y+144=0$.

SOLUTIONS

1. $\frac{y^2}{4} - \frac{(x-3)^2}{9} = 1$

$$\frac{y^2}{4} = 1 + \frac{(x-3)^2}{9}$$

$$y^2 = 4\left(1 + \frac{(x-3)^2}{9}\right)$$

$$y = \pm\sqrt{4\left(1 + \frac{(x-3)^2}{9}\right)}$$

2. $\frac{(x+10)^2}{25} + \frac{(y+3)^2}{25} = 1$

$$\frac{(y+3)^2}{25} = 1 - \frac{(x+10)^2}{25}$$

$$(y+3)^2 = 25\left(1 - \frac{(x+10)^2}{25}\right)$$

$$y + 3 = \pm\sqrt{25\left(1 - \frac{(x+10)^2}{25}\right)}$$

$$y = -3 \pm \sqrt{25\left(1 - \frac{(x+10)^2}{25}\right)}$$

3. $$36x^2 + 9y^2 - 216x - 72y + 144 = 0$$

$$36x^2 - 216x + 9y^2 - 72y = -144$$

$$36(x^2 - 6x) + 9(y^2 - 8y) = -144$$

$$36(x^2 - 6x + 9) + 9(y^2 - 8y + 16) = -144 + 36(9) + 9(16)$$

$$36(x-3)^2 + 9(y-4)^2 = 324$$

$$\frac{36(x-3)^2}{324} + \frac{9(y-4)^2}{324} = \frac{324}{324}$$

$$\frac{(x-3)^2}{9} + \frac{(y-4)^2}{36} = 1$$

CHAPTER 12 REVIEW

1. What is the directrix for the parabola $(y+1)^2 = -6(x-3)$?

   (a) $x = \frac{3}{2}$ (b) $x = \frac{9}{2}$ (c) $y = -\frac{5}{2}$ (d) $y = \frac{1}{2}$

2. What is the focus for the parabola $(y+1)^2 = -6(x-3)$?

   (a) $(\frac{3}{2}, -1)$ (b) $(\frac{9}{2}, -1)$ (c) $(3, \frac{1}{2})$ (d) $(3, -\frac{5}{2})$

3. What are the vertices for the ellipse

$$\frac{(x-1)^2}{9} + \frac{(y-2)^2}{25} = 1?$$

   (a) (−2,2) and (4,2) (b) (−4,2) and (6,2)

   (c) (1,−3) and (1,7) (d) (1,−1) and (1,5)

4. What are the foci for the ellipse

$$\frac{(x-1)^2}{9} + \frac{(y-2)^2}{25} = 1?$$

(a) $(-3, 2)$ and $(5, 2)$
(b) $(1-\sqrt{34}, 2)$ and $(1+\sqrt{34}, 2)$
(c) $(1, -2)$ and $(1, 6)$
(d) $(1, 2-\sqrt{34})$ and $(1, 2+\sqrt{34})$

5. Which line is an asymptote for the hyperbola

$$(x-5)^2 - \frac{(y+1)^2}{4} = 1?$$

(a) $y = 2x - 11$
(b) $y = -2x - 9$
(c) $y = \frac{1}{2}x - \frac{7}{2}$
(d) $y = -\frac{1}{2}x + \frac{3}{2}$

6. Solve for y.

$$(x-4)^2 - \frac{(y-6)^2}{25} = 1$$

(a) $y = 6 \pm 5\sqrt{-1 + (x-4)^2}$
(b) $y = 6 \pm 5\sqrt{1 - (x-4)^2}$
(c) $y = -6 \pm 5\sqrt{1 + (x-4)^2}$
(d) $y = -6 \pm 5\sqrt{1 - (x-4)^2}$

7. What is the center and radius for the circle $(x+3)^2 + (y-4)^2 = 9$?

(a) The center is $(-3, 4)$, and the radius is 81.

(b) The center is $(-3, 4)$, and the radius is 3.

(c) The center is $(3, -4)$, and the radius is 81.

(d) The center is $(3, -4)$, and the radius is 3.

8. The graph in Figure 12.31 is the graph of which equation?

(a) $y^2 = 4x$ (b) $y^2 = -4x$ (c) $x^2 = -4y$ (d) $x^2 = 4y$

9. Find an equation of the ellipse with vertices $(8, -6)$ and $(8, 4)$ with a focus at $(8, 2)$.

(a) $\frac{(x-8)^2}{16} + \frac{(y+1)^2}{25} = 1$
(b) $\frac{(x-8)^2}{25} + \frac{(y+1)^2}{16} = 1$
(c) $\frac{(x-8)^2}{16} - \frac{(y+1)^2}{25} = 1$
(d) $\frac{(x-8)^2}{25} + \frac{(y+1)^2}{16} = 1$

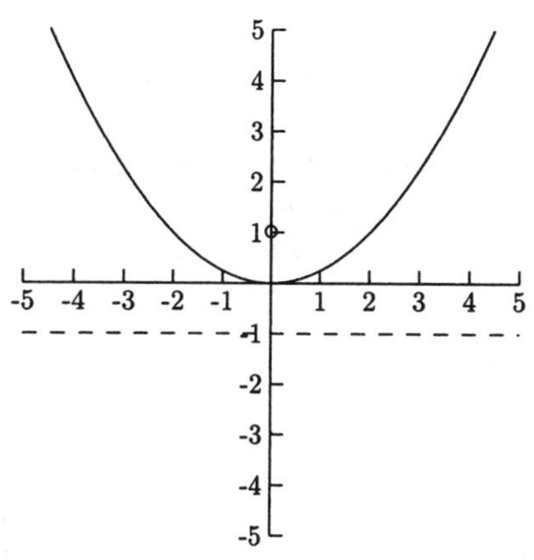

Fig. 12.31.

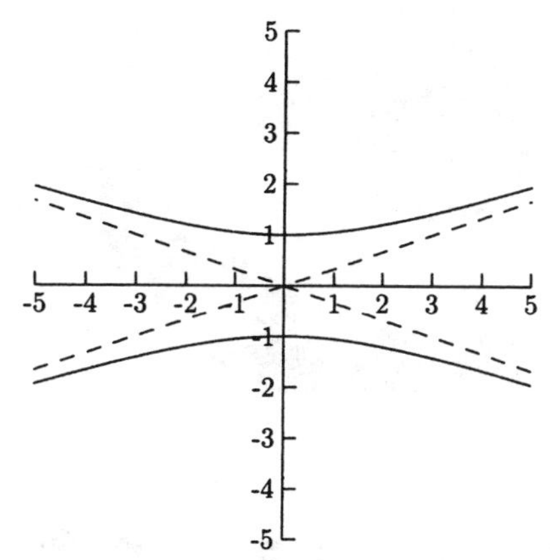

Fig. 12.32.

10. Which equation is the equation of a hyperbola with center $(1, 0)$ and with asymptote $y = 2x - 2$?

    (a) $y^2 - \dfrac{(x-1)^2}{4} = 1$ (b) $\dfrac{y^2}{4} + (x-1)^2 = 1$

    (c) $\dfrac{(x-1)^2}{4} - y^2 = 1$ (d) $(x-1)^2 - \dfrac{y^2}{4} = 1$

11. The graph in Figure 12.32 is the graph of which equation?

    (a) $x^2 - \dfrac{y^2}{9} = 1$ (b) $\dfrac{x^2}{9} - y^2 = 1$

    (c) $\dfrac{y^2}{9} - x^2 = 1$ (d) $y^2 - \dfrac{x^2}{9} = 1$

SOLUTIONS

1. B 2. A 3. C 4. C 5. A 6. A
7. B 8. D 9. A 10. D 11. D

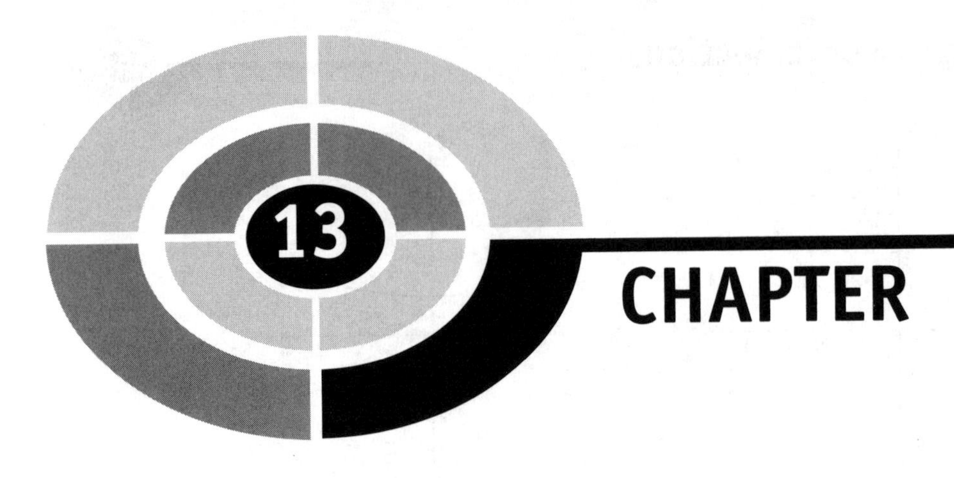

Trigonometry

Trigonometry has been used for over two thousand years to solve many real-world problems, among them surveying, navigating, and problems in engineering. Another important use is analytic—the trigonometric functions and their graphs are important in several mathematics courses. The *unit circle* is the basis of analytic trigonometry. The unit circle is the circle centered at the origin that has radius 1. See Figure 13.1.

Angles have two sides, the initial side and the terminal side. On the unit circle, the initial side is the positive part of the x-axis. The terminal side is the side that rotates. See Figure 13.2

A positive angle rotates counterclockwise, ↶. A negative angle rotates clockwise, ↷. Angles on the unit circle are often measured in *radians*. Radian measure is based on the circumference of the unit circle, $C = 2\pi r$. The radius is 1, so $2\pi r = 2\pi$. An angle that rotates all the way around the circle is 2π radians, half-way around is π radians, one-third the way is $\frac{1}{3}(2\pi) = \frac{2\pi}{3}$ radians, and so on. The relationship 2π radians = 360° gives us two equations.

$$\frac{\pi}{180} \text{ radians} = 1^\circ \quad \text{and} \quad \frac{180^\circ}{\pi} = 1 \text{ radian}$$

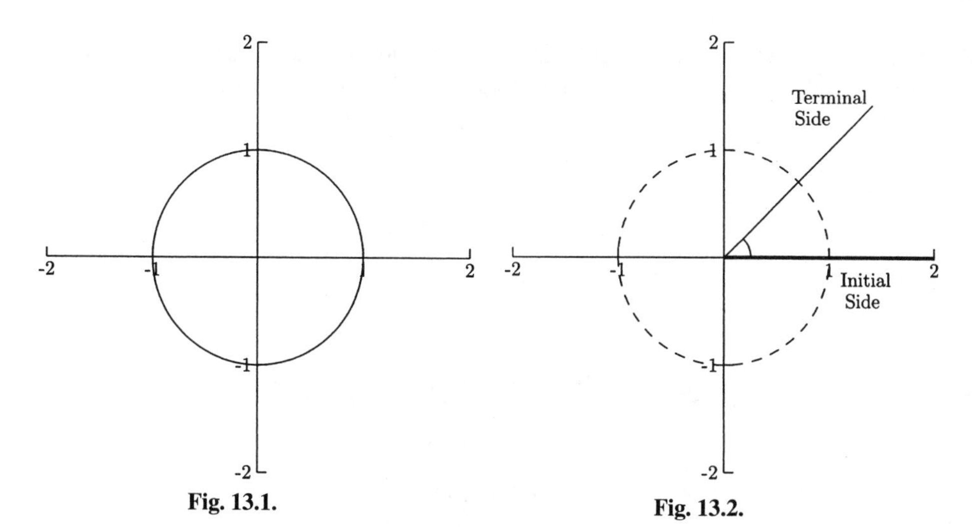

Fig. 13.1.

Fig. 13.2.

These equations help us to convert radian measure to degrees and degree measure to radians. We can convert radians to degrees by multiplying the angle by $180/\pi$. We can convert degrees to radians by multiplying the angle by $\pi/180$.

EXAMPLES

- Convert $4\pi/5$ radians to degree measure.
  Because we are going from radians to degrees, we will multiply the angle by $180/\pi$.

$$\frac{4\pi}{5} \cdot \frac{180}{\pi} = 144^\circ$$

- Convert $5\pi/6$ radians to degree measure.

$$\frac{5\pi}{6} \cdot \frac{180}{\pi} = 150^\circ$$

- Convert 48° to radian measure.
  Because we are going from degrees to radians, we will multiply the angle by $\pi/180$.

$$48 \cdot \frac{\pi}{180} = \frac{4\pi}{15} \text{ radians}$$

- Convert -72° to radian measure.

$$-72 \cdot \frac{\pi}{180} = -\frac{2\pi}{5}$$

Two angles are *coterminal* if their terminal sides are the same. For example, the terminal sides of the angles 300° and −60° are the same. See Figure 13.3.

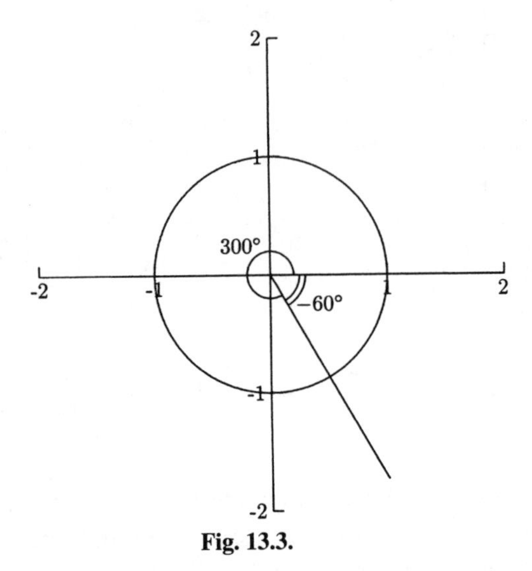

Fig. 13.3.

Two angles are coterminal if their difference is a multiple of 360° or 2π radians. In the example above, the difference of 300° and −60° is $300° - (-60°) = 360°$.

EXAMPLES

Determine whether or not the angles are coterminal.

- 18° and 738°

  Is the difference between 18° and 738° a multiple of 360? $738° - 18° = 720°$, $720° = 2 \cdot 360°$, so the angles are coterminal.
- −170° and 350°

  $350° - (-170°) = 350° + 170° = 520°$ and 520° is not a multiple of 360°, so the angles are not coterminal.
- $\pi/8$ radians and $-7\pi/8$ radians

  Is the difference of $\pi/8$ and $-7\pi/8$ a multiple of 2π?

$$\frac{\pi}{8} - \left(-\frac{7\pi}{8}\right) = \frac{8\pi}{8} = \pi \text{ radians}$$

Because π radians is not a multiple of 2π radians, the angles are not coterminal.

Every angle, θ (the Greek letter *theta*), has a *reference angle*, $\bar{\theta}$, associated with it. The reference angle is the smallest angle between the terminal side of the angle and the x-axis. A reference angle will be between 0 and $\pi/2$ radians, or 0° and 90°. The reference angle for all of the angles shown in Figures 13.4 through 13.7 is $\frac{\pi}{6}$.

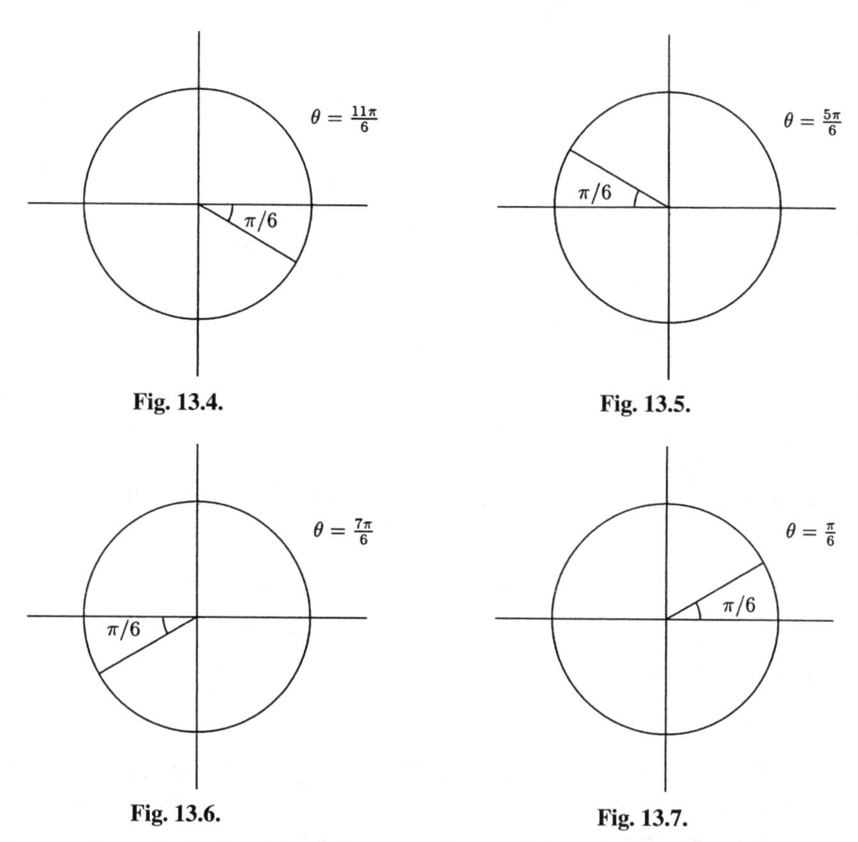

Fig. 13.4. **Fig. 13.5.** **Fig. 13.6.** **Fig. 13.7.**

The xy plane is divided into four quadrants. The trigonometric functions of angles in the different quadrants will have different signs. It is important to be familiar with the signs of the trigonometric functions in the different quadrants. One reason is that formulas have $\pm$ signs in them, and the sign of $+$ or $-$ depends on the quadrant in which the angle lies. Before we find reference angles, we will become familiar with the quadrants in the xy plane. (see Figure 13.8.)

EXAMPLES

Determine the quadrant in which the point lies.

- (5, −3)

  $x = 5$ is positive, and $y = -3$ is negative, the point is in Quadrant IV.

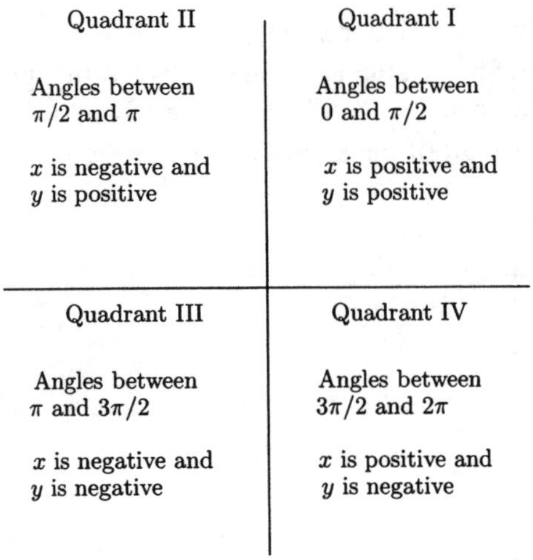

Quadrant II	Quadrant I
Angles between $\pi/2$ and π	Angles between 0 and $\pi/2$
x is negative and y is positive	x is positive and y is positive
Quadrant III	**Quadrant IV**
Angles between π and $3\pi/2$	Angles between $3\pi/2$ and 2π
x is negative and y is negative	x is positive and y is negative

Fig. 13.8.

- (4, 7)
  Both $x = 4$ and $y = 7$ are positive, the point is in Quadrant I.
- $(-1, -6)$
  Both $x = -1$ and $y = -6$ are negative, the point is in Quadrant III.
- $(-2, 10)$
  $x = -2$ is negative, $y = 10$ is positive, the point is in Quadrant II.

Below is an outline for finding the reference angle.

1. If the angle is not between 0 radians and 2π radians, find an angle between these two angles by adding or subtracting a multiple of 2π. Call this angle θ.
2. If θ is Quadrant I, θ is its own reference angle.
3. If θ is in Quadrant II, the reference angle is $\pi - \theta$.
4. If θ is in Quadrant III, the reference angle is $\theta - \pi$.
5. If θ is in Quadrant IV, the reference angle is $2\pi - \theta$.

EXAMPLES

Find the reference angle.

- $\theta = \dfrac{9\pi}{8}$
  This angle is in Quadrant III (bigger than π but smaller than $3\pi/2$), so $\bar{\theta} = 9\pi/8 - \pi = \pi/8$.

- $\theta = \dfrac{7\pi}{3}$

  This angle is not between 0 and 2π, so we need to add or subtract some multiple of 2π so that we do have an angle between 0 and 2π. The coterminal angle we need is $7\pi/3 - 2\pi = 7\pi/3 - 6\pi/3 = \pi/3$, $\pi/3$ is its own reference angle because it is in Quadrant I, so $\bar{\theta} = \pi/3$.

- $\theta = \dfrac{5\pi}{7}$

  This angle is in Quadrant II (between $\pi/2$ and π), so $\bar{\theta} = \pi - 5\pi/7 = 7\pi/7 - 5\pi/7 = 2\pi/7$.

- $\theta = -\dfrac{2\pi}{3}$

  This angle is not between 0 and 2π. It is coterminal with $2\pi + (-2\pi/3) = 6\pi/3 - 2\pi/3 = 4\pi/3$. The angles are in Quadrant III, so $\bar{\theta} = 4\pi/3 - \pi = 4\pi/3 - 3\pi/3 = \pi/3$.

There are six trigonometric functions, but four of them are written in terms of two of the main functions—sine and cosine. Although trigonometry was developed to solve problems involving triangles, there is a very close relationship between sine and cosine and the unit circle. Suppose an angle θ is given. The x-coordinate of the point on the unit circle for θ is cosine of the angle (written $\cos\theta$). The y-coordinate of the point is sine of the angle (written $\sin\theta$). For example, suppose the point determined by the angle θ is $(3/5, 4/5)$. Then $\cos\theta = 3/5$ and $\sin\theta = 4/5$. See Figure 13.9.

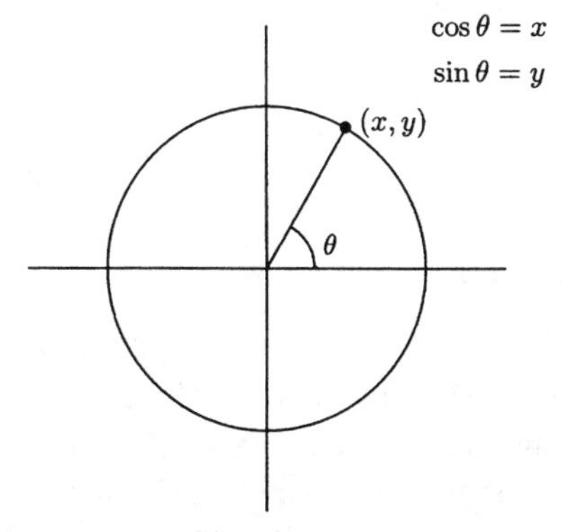

Fig. 13.9.

EXAMPLES

Find $\sin\theta$ and $\cos\theta$.

- 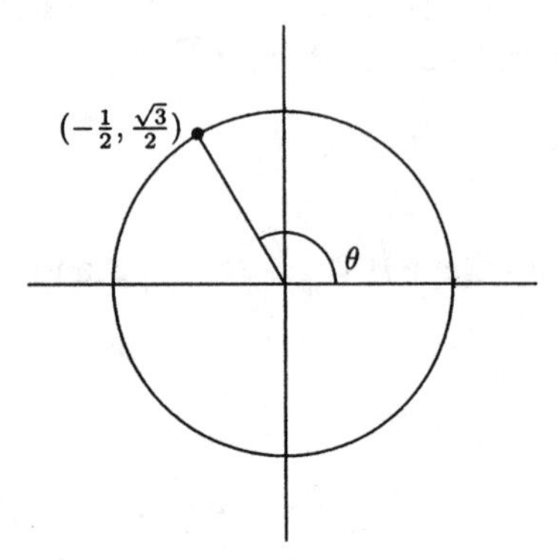

Fig. 13.10.

$\sin\theta = \sqrt{3}/2$ and $\cos\theta = -1/2$

- 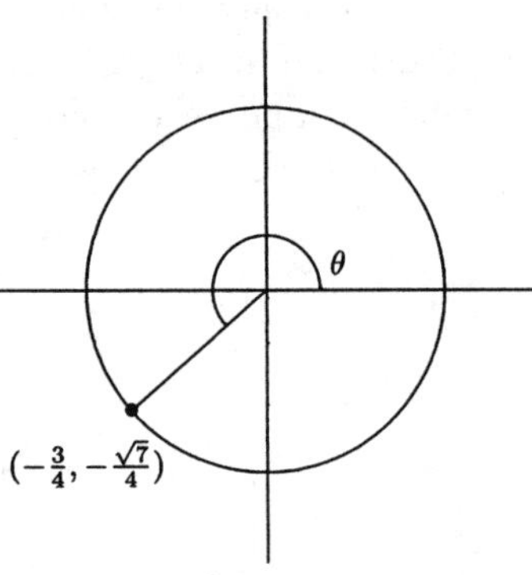

Fig. 13.11.

$\sin\theta = -\sqrt{7}/4$ and $\cos\theta = -3/4$

The equation for the unit circle is $x^2 + y^2 = 1$. For an angle θ, we can replace x with $\cos\theta$ and y with $\sin\theta$. This changes the equation to $\cos^2\theta + \sin^2\theta = 1$ ($\cos^2\theta$ means $(\cos\theta)^2$ and $\sin^2\theta$ means $(\sin\theta)^2$). This is an important equation. It allows us to find $\cos\theta$ if we know $\sin\theta$ and $\sin\theta$ if we know $\cos\theta$. Solving this equation for $\cos\theta$ gives us $\cos\theta = \pm\sqrt{1-\sin^2\theta}$. Solving it for $\sin\theta$ gives us $\sin\theta = \pm\sqrt{1-\cos^2\theta}$. For example, if we know $\sin\theta = 1/2$, we can find $\cos\theta$.

$$\cos\theta = \pm\sqrt{1-\sin^2\theta} = \pm\sqrt{1-\left(\frac{1}{2}\right)^2} = \pm\sqrt{\frac{3}{4}} = \pm\frac{\sqrt{3}}{2}$$

Is $\cos\theta = \sqrt{3}/2$ or $-\sqrt{3}/2$? We cannot answer this without knowing where θ is. If we know that θ is in Quadrants I or IV, then $\cos\theta = \sqrt{3}/2$ because cosine is positive in Quadrants I and IV. If we know that θ is in Quadrants II or III, then $\cos\theta = -\sqrt{3}/2$ because cosine is negative in Quadrants II and III.

EXAMPLES

Find $\sin\theta$ and $\cos\theta$.

- The terminal point for θ is $(-12/13, y)$, and θ is in Quadrant II.

  $\cos\theta = -12/13$

  Is $\sin\theta = \sqrt{1-\left(-\frac{12}{13}\right)^2}$ or $-\sqrt{1-\left(-\frac{12}{13}\right)^2}$?

  Because the y-values in Quadrant II are positive, $\sin\theta$ is positive.

$$\sin\theta = \sqrt{1-\left(-\frac{12}{13}\right)^2} = \sqrt{\frac{25}{169}} = \frac{5}{13}$$

- The terminal point for θ is $(x, -1/9)$, and θ is in Quadrant III.

  Both sine and cosine are negative in Quadrant III, so we will use the negative square root. Using $\sin\theta = -1/9$, we have

$$\cos\theta = -\sqrt{1-\left(-\frac{1}{9}\right)^2} = -\sqrt{\frac{80}{81}} = -\frac{4\sqrt{5}}{9}$$

The values for sine and cosine of the following angles should be memorized: $0, \pi/6, \pi/4, \pi/3$, and $\pi/2$. See Figure 13.12.

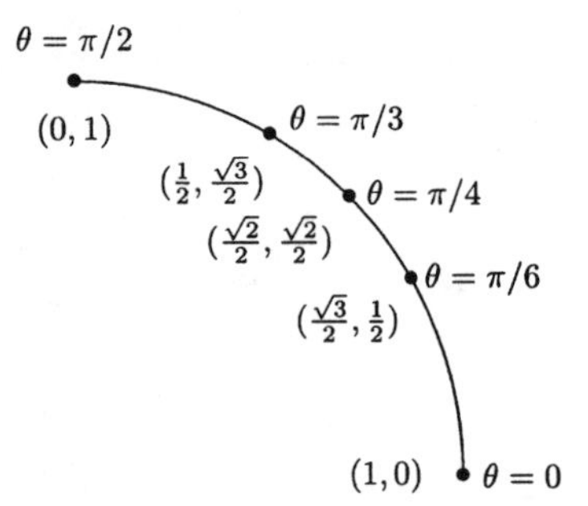

Fig. 13.12.

All of these angles are also reference angles in the other three quadrants. You should either memorize or be able to quickly compute them. The information is in the table below.

Table 13.1

	θ	$\cos\theta$	$\sin\theta$		θ	$\cos\theta$	$\sin\theta$
	0	1	0		π	-1	0
Quadrant I	$\frac{\pi}{6}$	$\frac{\sqrt{3}}{2}$	$\frac{1}{2}$	Quadrant III	$\pi + \frac{\pi}{6} = \frac{7\pi}{6}$	$-\frac{\sqrt{3}}{2}$	$-\frac{1}{2}$
Quadrant I	$\frac{\pi}{4}$	$\frac{\sqrt{2}}{2}$	$\frac{\sqrt{2}}{2}$	Quadrant III	$\pi + \frac{\pi}{4} = \frac{5\pi}{4}$	$-\frac{\sqrt{2}}{2}$	$-\frac{\sqrt{2}}{2}$
Quadrant I	$\frac{\pi}{3}$	$\frac{1}{2}$	$\frac{\sqrt{3}}{2}$	Quadrant III	$\pi + \frac{\pi}{3} = \frac{4\pi}{3}$	$-\frac{1}{2}$	$-\frac{\sqrt{3}}{2}$
	$\frac{\pi}{2}$	0	1		$\frac{3\pi}{2}$	0	-1
Quadrant II	$\pi - \frac{\pi}{3} = \frac{2\pi}{3}$	$-\frac{1}{2}$	$\frac{\sqrt{3}}{2}$	Quadrant IV	$2\pi - \frac{\pi}{3} = \frac{5\pi}{3}$	$\frac{1}{2}$	$-\frac{\sqrt{3}}{2}$
Quadrant II	$\pi - \frac{\pi}{4} = \frac{3\pi}{4}$	$-\frac{\sqrt{2}}{2}$	$\frac{\sqrt{2}}{2}$	Quadrant IV	$2\pi - \frac{\pi}{4} = \frac{7\pi}{4}$	$\frac{\sqrt{2}}{2}$	$-\frac{\sqrt{2}}{2}$
Quadrant II	$\pi - \frac{\pi}{6} = \frac{5\pi}{6}$	$-\frac{\sqrt{3}}{2}$	$\frac{1}{2}$	Quadrant IV	$2\pi - \frac{\pi}{6} = \frac{11\pi}{6}$	$\frac{\sqrt{3}}{2}$	$-\frac{1}{2}$

The other trigonometric functions are tangent (tan), cotangent (cot), secant (sec), and cosecant (csc). All of them can be written as a ratio with sine, cosine, or both.

$$\tan\theta = \frac{\sin\theta}{\cos\theta} = \frac{y}{x} \qquad \cot\theta = \frac{\cos\theta}{\sin\theta} = \frac{x}{y}$$

$$\sec\theta = \frac{1}{\cos\theta} = \frac{1}{x} \qquad \csc\theta = \frac{1}{\sin\theta} = \frac{1}{y}$$

Sine and cosine can be evaluated at any angle. This is not true for the other trigonometric functions. For example $\tan \pi/2 = \sin \pi/2 / \cos \pi/2$ and $\sec \pi/2 = 1/\cos \pi/2$ are not defined because $\cos \pi/2 = 0$. We can find all six trigonometric functions for an angle θ if we either know both coordinates of its terminal point or if we know one coordinate and the quadrant where θ lies.

Before we begin the next set of problems, we will review a shortcut that will save some computation for $\tan\theta$. A compound fraction of the form $(a/b)/(c/b)$ simplifies to a/c.

$$\frac{a/b}{c/b} = \frac{a}{b} \div \frac{c}{b} = \frac{a}{b} \cdot \frac{b}{c} = \frac{a}{c}$$

EXAMPLES

- $\dfrac{1/8}{5/8} = \dfrac{1}{5}$
- $\dfrac{4/7}{2/7} = \dfrac{4}{2} = 2$
- $\dfrac{-2/3}{1/3} = -\dfrac{2}{1} = -2$
- $\dfrac{1/9}{-5/9} = -\dfrac{1}{5}$

Find all six trigonometric functions for θ.

- The terminal point for θ is $(24/25, 7/25)$

$$\cos\theta = \frac{24}{25} \qquad \sin\theta = \frac{7}{25}$$

$$\sec\theta = \frac{25}{24} \qquad \csc\theta = \frac{25}{7}$$

$$\tan\theta = \frac{7/25}{24/25} = \frac{7}{24} \qquad \cot\theta = \frac{24}{7}$$

- $\theta = \pi/3$

$$\cos\theta = \frac{1}{2} \qquad \sin\theta = \frac{\sqrt{3}}{2}$$

$$\sec\theta = 2 \qquad \csc\theta = \frac{2}{\sqrt{3}} = \frac{2\sqrt{3}}{3}$$

$$\tan\theta = \frac{\sqrt{3}/2}{1/2} = \frac{\sqrt{3}}{1} = \sqrt{3} \qquad \cot\theta = \frac{1}{\sqrt{3}} = \frac{\sqrt{3}}{3}$$

- $\theta = 5\pi/6$

$$\cos\theta = -\frac{\sqrt{3}}{2} \qquad \sin\theta = \frac{1}{2}$$

$$\sec\theta = -\frac{2}{\sqrt{3}} = -\frac{2\sqrt{3}}{3} \qquad \csc\theta = 2$$

$$\tan\theta = \frac{1/2}{-\sqrt{3}/2} = -\frac{1}{\sqrt{3}} = -\frac{\sqrt{3}}{3} \qquad \cot\theta = -\sqrt{3}$$

- The x-coordinate of θ is 2/5, and θ is in Quadrant IV.

$$\cos\theta = \frac{2}{5} \qquad \sin\theta = -\sqrt{1-\left(\frac{2}{5}\right)^2} = -\frac{\sqrt{21}}{5}$$

$$\sec\theta = \frac{5}{2} \qquad \csc\theta = -\frac{5}{\sqrt{21}} = -\frac{5\sqrt{21}}{21}$$

$$\tan\theta = \frac{-\sqrt{21}/5}{2/5} = -\frac{\sqrt{21}}{2} \qquad \cot\theta = -\frac{2}{\sqrt{21}} = -\frac{2\sqrt{21}}{21}$$

The graph of a trigonometric function is a record of each cycle around the unit circle. For the function $f(x) = \sin x$, x is the angle and $f(x)$ is the y-coordinate of the terminal point determined by the angle x. In the function $g(x) = \cos x$, $g(x)$ is the x-coordinate of the terminal point determined by the angle x. For example, the point determined by the angle $\pi/6$ is $(\sqrt{3}/2, 1/2)$, so $f(\pi/6) = \sin\pi/6 = 1/2$ and $g(\pi/6) = \cos\pi/6 = \sqrt{3}/2$. We will sketch the graph of $f(x) = \sin x$, using the points in Table 13.2.

Table 13.2

x	$\sin x$	Plot this point
-2π	$\sin(-2\pi) = 0$	$(-2\pi, 0)$
$-3\pi/2$	$\sin(-3\pi/2) = 1$	$(-3\pi/2, 1)$
$-\pi$	$\sin(-\pi) = 0$	$(-\pi, 0)$
$-\pi/2$	$\sin(-\pi/2) = -1$	$(-\pi/2, -1)$
0	$\sin 0 = 0$	$(0, 0)$
$\pi/2$	$\sin\pi/2 = 1$	$(\pi/2, 1)$
π	$\sin\pi = 0$	$(\pi, 0)$
$3\pi/2$	$\sin 3\pi/2 = -1$	$(3\pi/2, -1)$
2π	$\sin 2\pi = 0$	$(2\pi, 0)$

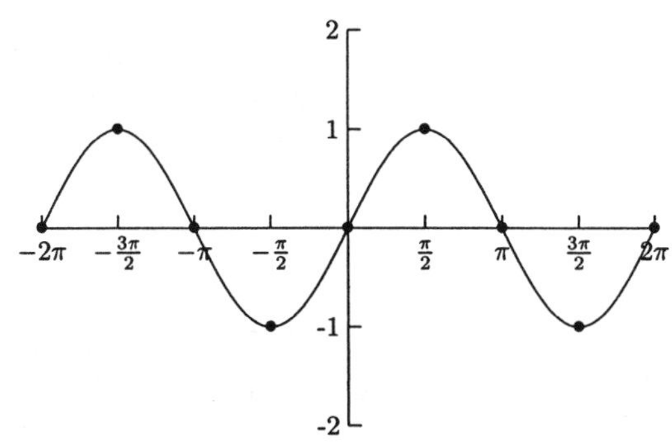

Fig. 13.13.

The graph in Figure 13.13 is two *periods* from the entire graph. This pattern repeats itself in both directions. Each period begins and ends at every multiple of 2π: $\ldots$, $[-2\pi, 0]$, $[0, 2\pi]$, $[2\pi, 4\pi]$, $\ldots$. The graph between 0 and 2π represents sine on the first positive cycle around the unit circle, between 2π and 4π represents the second positive cycle, and between 0 and -2π represents the first negative cycle.

The graph for $g(x) = \cos x$ behaves in the same way. In fact, the graph of $g(x)$ is the graph of $f(x)$ shifted horizontally $\pi/2$ units. (We will see why this is true when we work with right triangles.) The graph for $g(x) = \cos x$ is shown in Figure 13.14.

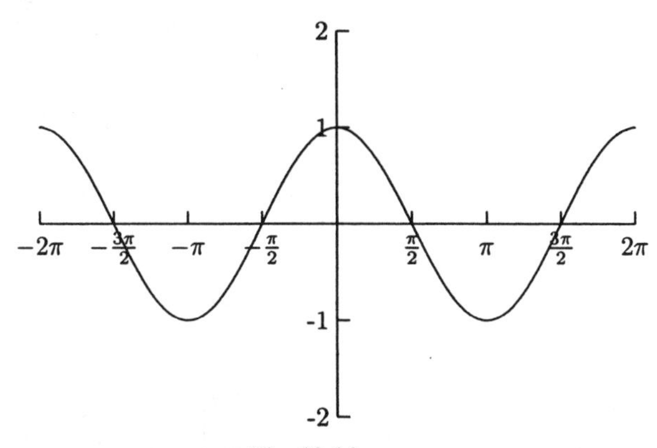

Fig. 13.14.

From their graphs, we can tell that $f(x) = \sin x$ is an odd function ($\sin(-x) = -\sin x$), and $g(x) = \cos x$ is even ($\cos(-x) = \cos x$). We can also see that their domain is all x and their range is all y values between -1 and 1.

The graphs of $f(x) = \sin x$ and $g(x) = \cos x$ can be shifted up or down, left or right, and stretched or compressed in the same way as other graphs. The graphs of $y = c + \sin x$ and $y = c + \cos x$ are shifted up or down c units. The graphs of $y = a \sin x$ and $y = a \cos x$ are vertically stretched or compressed, and the graphs of $y = \sin(x - b)$ and $y = \cos(x - b)$ are shifted horizontally by b units.

EXAMPLES

The dashed graph in Figures 13.15 through 13.18 is one period of the graph of $f(x) = \sin x$, and the solid graphs are transformations. Match the equations below with their graphs.

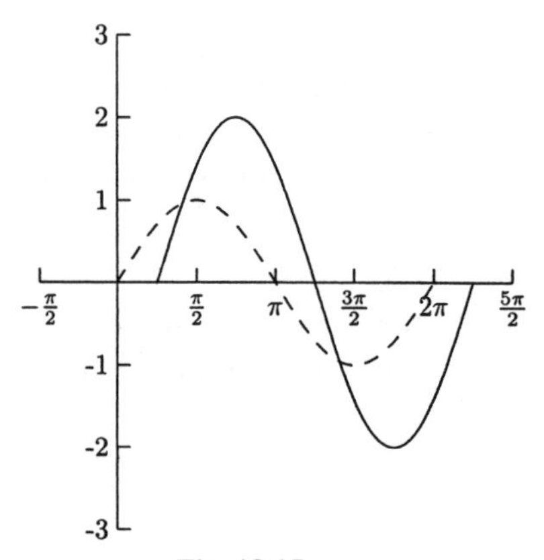

Fig. 13.15.

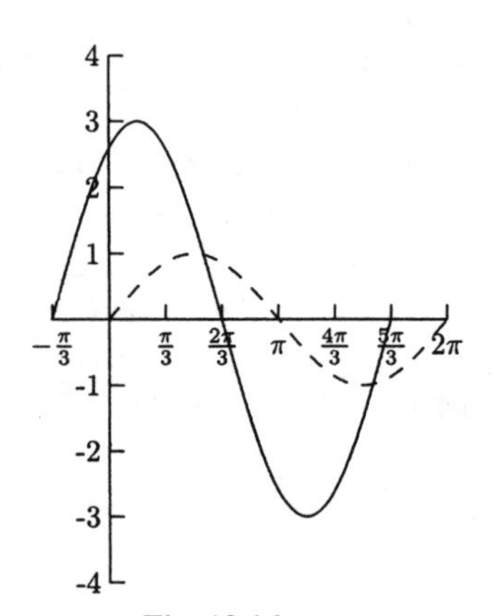

Fig. 13.16.

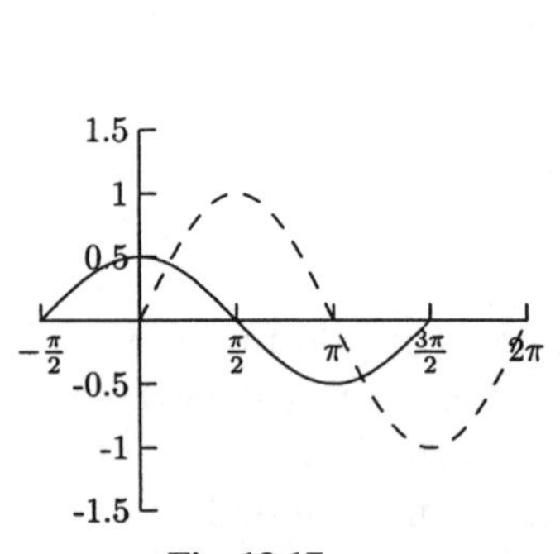

Fig. 13.17.

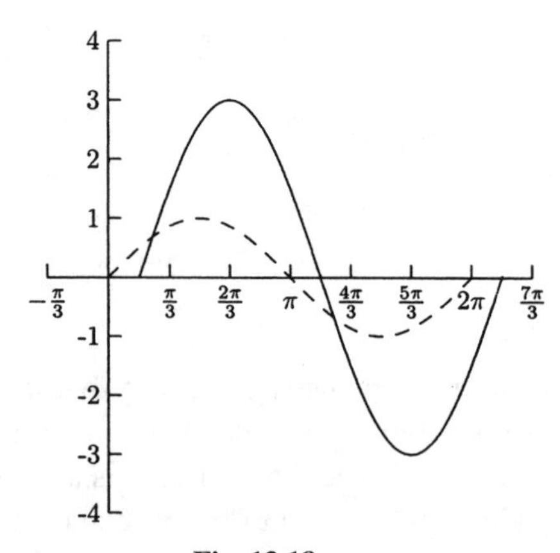

Fig. 13.18.

- $y = 3\sin(x + \pi/3)$

  The graph of this function is vertically stretched by a factor of 3, so we will look for a graph whose y values lie between -3 and 3. The graph will also be shifted to the left by $\pi/3$ units. The graph for this function is shown in Figure 13.16.

- $y = 3\sin(x - \pi/6)$

  The graph of this function is also vertically stretched by a factor of three, but it is shifted to the right by $\pi/6$ units. The graph for this function is shown in Figure 13.18.

- $y = \frac{1}{2}\sin(x + \pi/2)$

  The graph of this function is vertically compressed by a factor of 1/2, so we will look for a graph whose y values are between $-1/2$ and 1/2. The graph will also be shifted to the left by $\pi/2$ units. The graph for this function is shown in Figure 13.17.

- $y = 2\sin(x - \pi/4)$

  The graph of this function is vertically stretched by a factor of 2, so we will look for a graph whose y values are between -2 and 2. It will also be shifted to the right $\pi/4$ units. The graph for this function is shown in Figure 13.15.

Transformations of the graphs of sine and cosine have names. The *amplitude* is the degree of vertical stretching or compressing. The horizontal shift is called the *phase shift*. Horizontal stretching or compressing changes the length of the period. For functions of the form $y = a\sin k(x - b)$ and $y = a\cos k(x - b)$, $|a|$ is the graph's amplitude, b is its phase shift, and $2\pi/k$ is its period.

EXAMPLES

Find the amplitude, period, and phase shift.

- $y = -4\sin 2(x - \pi/3)$

  The amplitude is $|a| = |-4| = 4$, the period is $2\pi/k = 2\pi/2 = \pi$, and the phase shift is $b = \pi/3$.

- $y = -\cos(x + \pi/2)$

  The amplitude is $|a| = |-1| = 1$, the period is $2\pi/k = 2\pi/1 = 2\pi$, and the phase shift is $b = -\pi/2$.

- $y = \frac{1}{2}\cos(2x + 2\pi/3)$

  The amplitude is $|1/2| = 1/2$. In order for us to find k and b for the period and phase shift, we need to write the function in the form $y = a\cos k(x - b)$. We need to factor 2 from $2x + 2\pi/3$.

$$2x + \frac{2\pi}{3} = 2 \cdot x + 2 \cdot \frac{\pi}{3} = 2\left(x + \frac{\pi}{3}\right)$$

  The function can be written as $y = \frac{1}{2}\cos 2(x + \pi/3)$. The period is $2\pi/k = 2\pi/2 = \pi$, and the phase shift is $k = -\pi/3$.

Sketching the Graphs of Sine and Cosine

We can sketch one period of the graphs of sine and cosine or any of their transformations by plotting five key points. These points for $y = \sin x$ and $y = \cos x$ are $x = 0,\ \pi/2,\ \pi,\ 3\pi/2$ and 2π. These points are the x-intercepts and the vertices (where $y = 1$ or -1). For the functions $y = a\sin k(x - b)$ and $y = a\cos k(x - b)$, these points are shifted to $b,\ b + \frac{\pi}{2k},\ b + \frac{\pi}{k},\ b + \frac{3\pi}{2k}$, and $b + \frac{2\pi}{k}$.

EXAMPLES

Sketch one period of the graph for the given function.

- $y = -3\cos\frac{1}{2}x$

Table 13.3

x	$-3\cos\frac{1}{2}x$	Plot this point
$b = 0$	$-3\cos\frac{1}{2}(0) = -3\cos 0 = -3$	$(0, -3)$
$b + \frac{\pi}{2k} = 0 + \frac{\pi}{2(\frac{1}{2})} = 0 + \pi = \pi$	$-3\cos\frac{1}{2}(\pi) = -3\cos\pi/2 = 0$	$(\pi, 0)$
$b + \frac{\pi}{k} = 0 + \frac{\pi}{1/2} = 0 + 2\pi = 2\pi$	$-3\cos\frac{1}{2}(2\pi) = -3\cos\pi = 3$	$(2\pi, 3)$
$b + \frac{3\pi}{2k} = 0 + \frac{3\pi}{2(\frac{1}{2})} = 0 + 3\pi = 3\pi$	$-3\cos\frac{1}{2}(3\pi) = -3\cos 3\pi/2 = 0$	$(3\pi, 0)$
$b + \frac{2\pi}{k} = 0 + \frac{2\pi}{1/2} = 0 + 4\pi = 4\pi$	$-3\cos\frac{1}{2}(4\pi) = -3\cos 2\pi = -3$	$(4\pi, -3)$

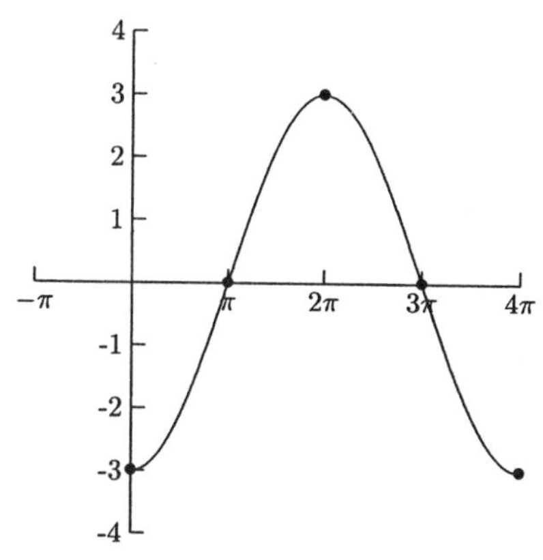

Fig. 13.19.

- $y = 5\sin(3x + \pi/2)$

  We need to write the function in the form $y = a\sin k(x - b)$ so that we can find k and b.

$$3x + \frac{\pi}{2} = 3x + \frac{3}{3} \cdot \frac{\pi}{2} = 3 \cdot x + 3 \cdot \frac{\pi}{6} = 3\left(x + \frac{\pi}{6}\right)$$

Table 13.4

x	$5\sin 3(x + \pi/6)$	Plot this point
$b = -\pi/6$	$5\sin 3(-\pi/6 + \pi/6) = 5\sin 0 = 0$	$(-\pi/6, 0)$
$b + \frac{\pi}{2k} = -\frac{\pi}{6} + \frac{\pi}{2(3)} = 0$	$5\sin 3(0 + \pi/6) = 5\sin \pi/2 = 5$	$(0, 5)$
$b + \frac{\pi}{k} = -\frac{\pi}{6} + \frac{\pi}{3} = \pi/6$	$5\sin 3(\pi/6 + \pi/6) = 5\sin \pi = 0$	$(\pi/6, 0)$
$b + \frac{3\pi}{2k} = -\frac{\pi}{6} + \frac{3\pi}{2(3)} = \pi/3$	$5\sin 3(\pi/3 + \pi/6) = 5\sin 3\pi/2 = -5$	$(\pi/3, -5)$
$b + \frac{2\pi}{k} = -\frac{\pi}{6} + \frac{2\pi}{3} = \pi/2$	$5\sin 3(\pi/2 + \pi/6) = 5\sin 2\pi = 0$	$(\pi/2, 0)$

The points in Table 13.4 are used to construct the graph in Figure 13.20.

PRACTICE

For Problems 1–3, match the function with its graph shown in Figures 13.21–13.23. The dashed graph is the graph of one period of $y = \cos x$. The solid graph is the graph of one period of a transformation.

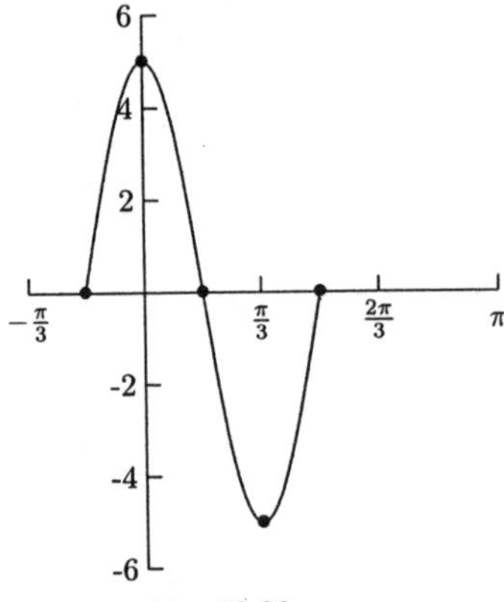

Fig. 13.20.

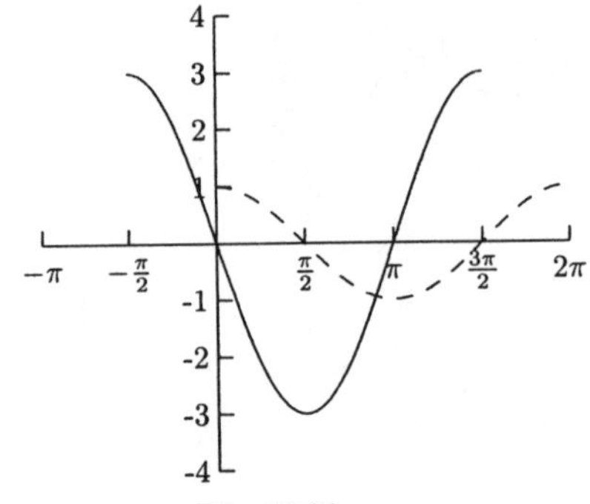

Fig. 13.21.

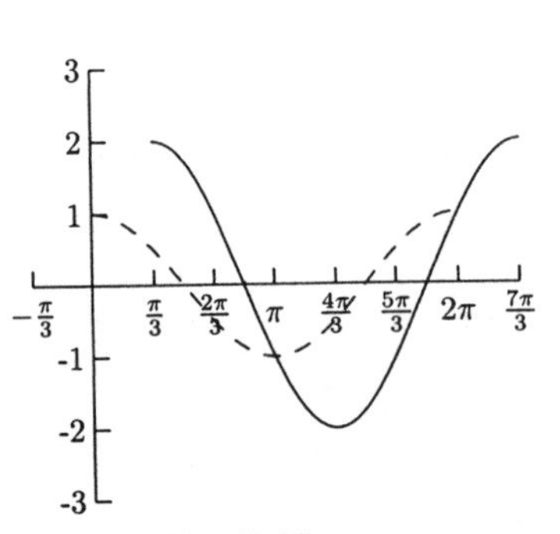

Fig. 13.22.

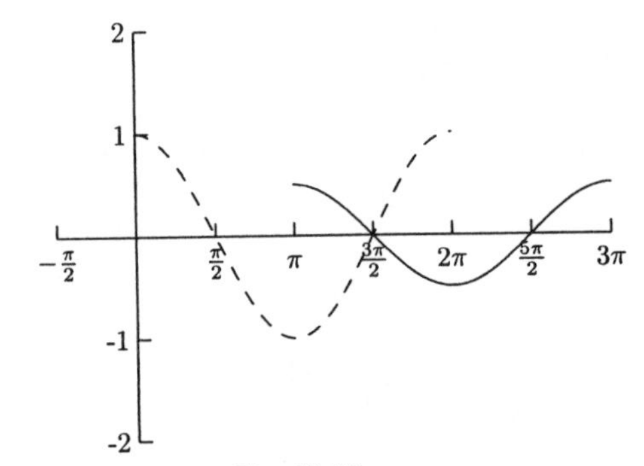

Fig. 13.23.

1. $y = 2\cos(x - \pi/3)$
2. $y = 3\cos(x + \pi/2)$
3. $y = \frac{1}{2}\cos(x - \pi)$
4. Find the amplitude, period, and phase shift for $y = -3\cos\frac{2}{3}(x - \pi/4)$.
5. Find the amplitude, period, and phase shift for $y = 6\sin(2x - \pi/2)$.
6. Sketch one period for the graph of $y = 3\cos\frac{1}{2}(x + \pi/4)$.
7. Sketch one period for the graph of $y = -1 + 2\sin(x - \pi/3)$

SOLUTIONS

1. Figure 13.22
2. Figure 13.21

3. Figure 13.23
4. The amplitude is $|-3| = 3$, the period is $\frac{2\pi}{2/3} = 2\pi \cdot 3/2 = 3\pi$, and the phase shift is $b = \pi/4$.
5. In order to find k and b, we need to write the function in the form $y = a \sin k(x - b)$.

$$2x - \frac{\pi}{2} = 2x - \frac{2}{2} \cdot \frac{\pi}{2} = 2 \cdot x - 2 \cdot \frac{\pi}{4} = 2\left(x - \frac{\pi}{4}\right)$$

The function can be written as $y = 6 \sin 2(x - \pi/4)$. Now we can see that the amplitude is $|6| = 6$, the period is $2\pi/2 = \pi$, and the phase shift is $\pi/4$.

6. Plot points for $x = -\pi/4,\ 3\pi/4,\ 7\pi/4,\ 11\pi/4$, and $15\pi/4$.

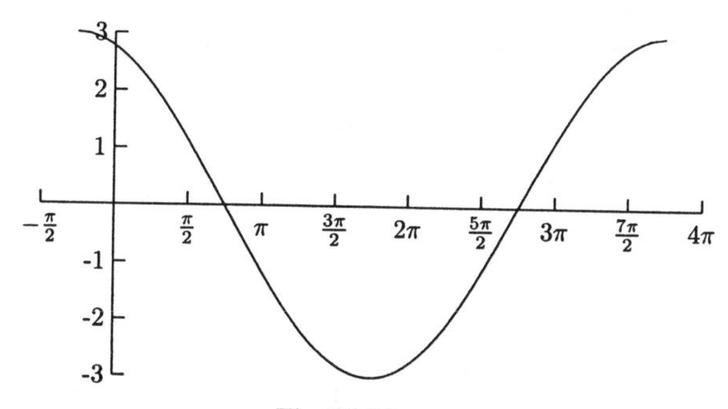

Fig. 13.24.

7. Plot points for $x = \pi/3,\ 5\pi/6,\ 4\pi/3,\ 11\pi/6$, and $7\pi/3$.

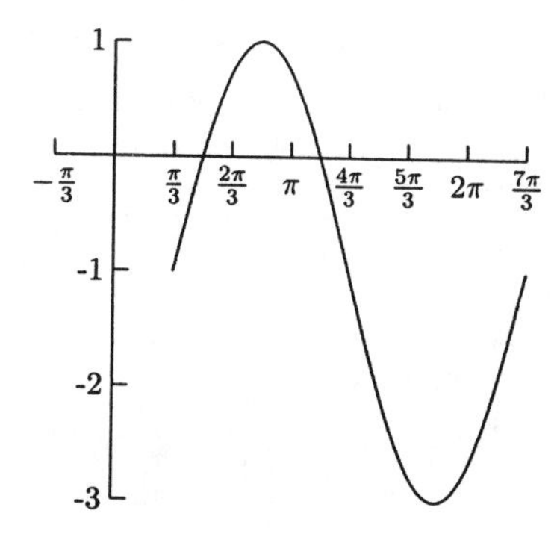

Fig. 13.25.

Graphs for Other Trigonometric Functions

Because $\csc x = 1/\sin x$, the graph of $y = \csc x$ has a vertical asymptote everywhere $y = \sin x$ has an x-intercept (where $\sin x = 0$). Because $\sec x = 1/\cos x$, the graph of $y = \sec x$ has a vertical asymptote everywhere $y = \cos x$ has an x-intercept. The period for $y = \csc x$ and $y = \sec x$ is 2π. The graph for $y = \csc x$ is shown in Figure 13.26, and the graph for $y = \sec x$ is shown in Figure 13.27.

The domain for $y = \csc x$ is all real numbers except for the zeros of $\sin x$, $x \neq \ldots, -2\pi, -\pi, 0, \pi, 2\pi, \ldots$. The range is $(-\infty, -1] \cup [1, \infty)$. The domain for $y = \sec x$ is all real numbers except for the zeros of $\cos x$, $x \neq \ldots, -3\pi/2, -\pi/2, \pi/2, 3\pi/2, \ldots$. The range is $(-\infty, -1] \cup [1, \infty)$. Because

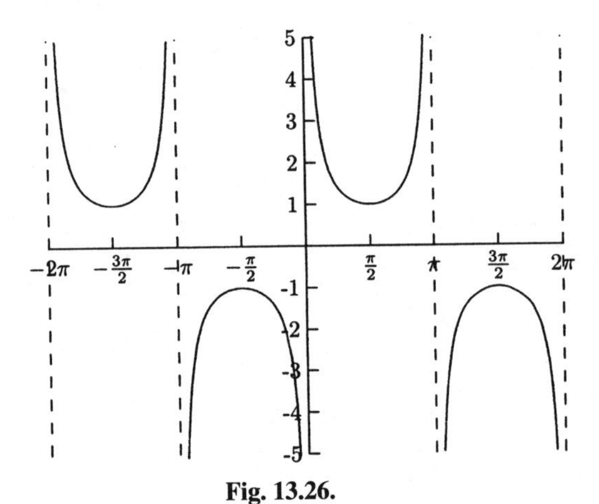

Fig. 13.26.

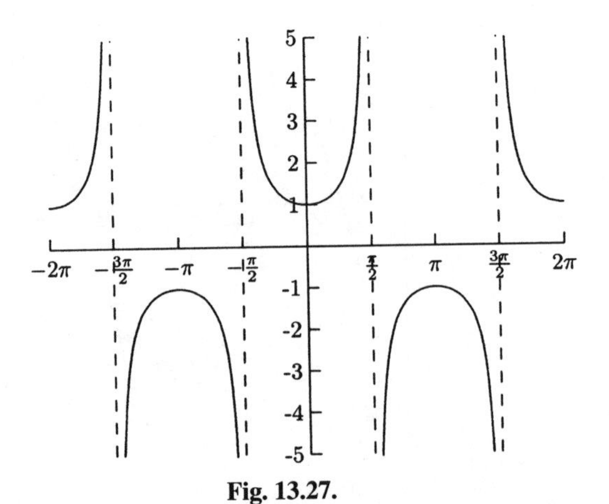

Fig. 13.27.

$y = \sin x$ is an odd function, $y = \csc x$ is also an odd function. Because $y = \cos x$ is an even function, $y = \sec x$ is also an even function.

We can sketch the graphs of $y = \csc x$ and $y = \sec x$ using the graphs of $y = \sin x$ and $y = \cos x$. We will sketch the vertical asymptotes as well as the graphs of $y = \sin x$ or $y = \cos x$ using dashed graphs.

The graph of $y = \sin x$ is given in Figure 13.28. Vertical asymptotes are sketched for every x-intercept.

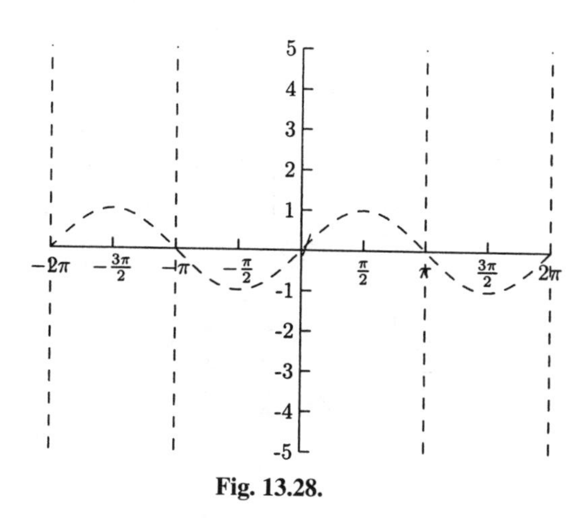

Fig. 13.28.

The vertex for each piece on the graph of $y = \csc x$ is also a vertex for $y = \sin x$.

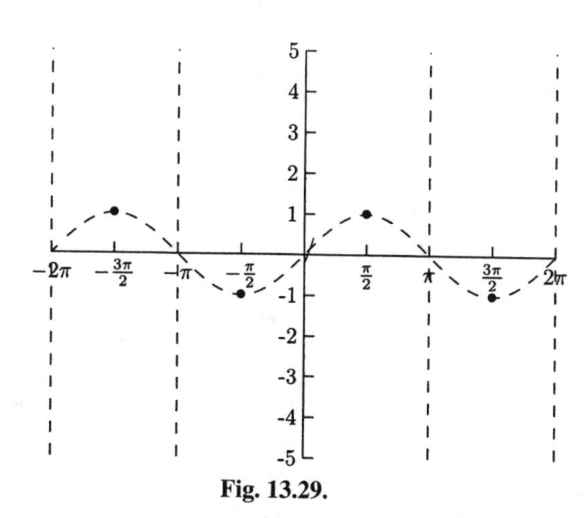

Fig. 13.29.

Then we can plot a point to the left and right of each vertex (staying inside the vertical asymptotes) to show how fast the graph gets close to the vertical asymptotes.

Table 13.5

x	$\csc x$
-1.8π	1.7
-1.2π	1.7
-0.8π	-1.7
-0.2π	-1.7
0.2π	1.7
0.8π	1.7
1.2π	-1.7
1.8π	-1.7

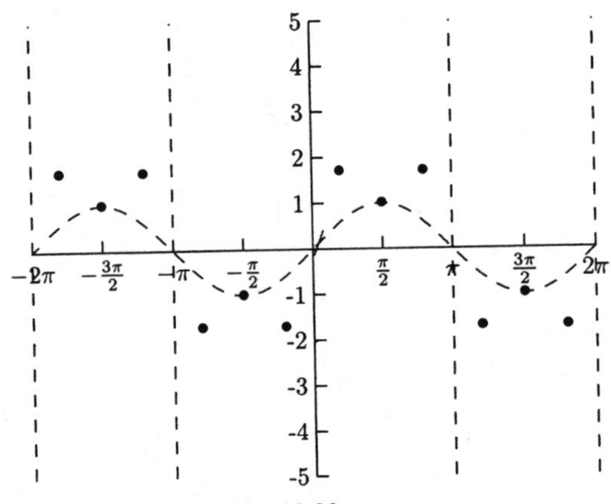

Fig. 13.30.

Now we can draw ⋃ or ⋂ through the points.

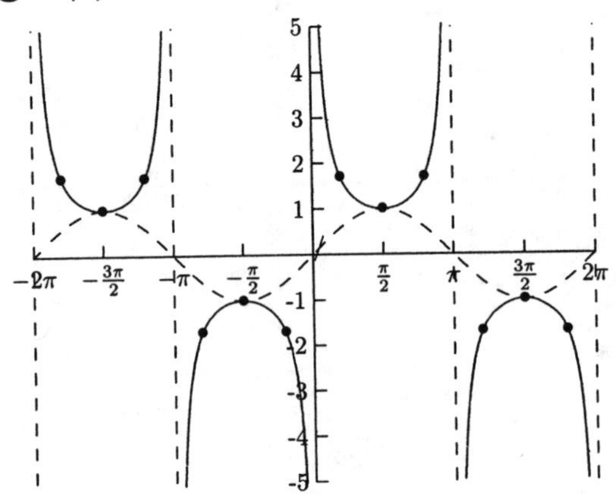

Fig. 13.31.

These steps also work for the graph of $y = \sec x$.

The period for the functions $y = \tan x$ and $y = \cot x$ is π instead of 2π as it is with the other trigonometric functions. These graphs also have vertical asymptotes. The graph of $y = \tan x$ $(= \sin x/\cos x)$ has a vertical asymptote at each zero of $y = \cos x$. The graph of $y = \cot x$ $(= \cos x/\sin x)$ has a vertical asymptote at each zero of $y = \sin x$. The graph of $y = \tan x$ is shown in Figure 13.32, and the graph of $y = \cot x$ is shown in Figure 13.33.

The domain of $y = \tan x$ is all real numbers except the zeros of $y = \cos x$, $x \neq \ldots, -3\pi/2, -\pi/2, \pi/2, 3\pi/2, \ldots$. The domain for $y = \cot x$ is all real numbers except for the zeros of $y = \sin x$, $x \neq \ldots, -2\pi, -\pi, 0, \pi, 2\pi, \ldots$. The range for both $y = \tan x$ and $y = \cot x$ is all real numbers. Both are odd functions.

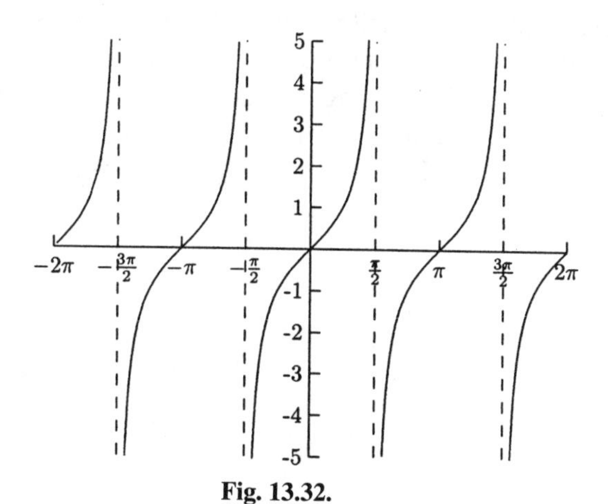

Fig. 13.32.

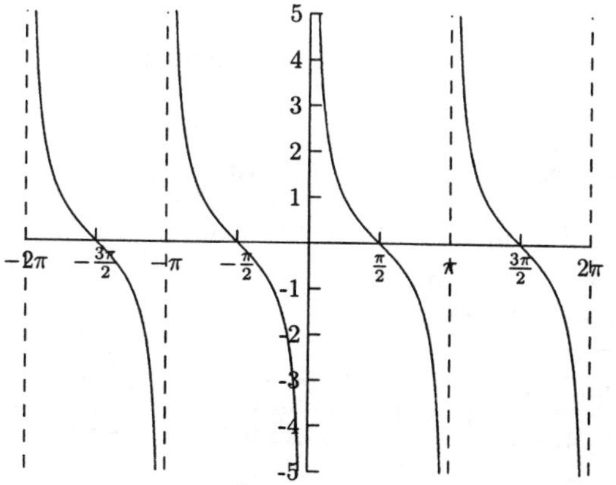

Fig. 13.33.

The transformations of these are similar to those of the other trigonometric functions. For functions of the form $y = a \csc k(x - b)$ and $y = a \sec k(x - b)$, the period is $2\pi/k$, and the phase shift is b. For functions of the form $y = a \tan k(x - b)$ and $y = a \cot k(x - b)$, the period is π/k, and the phase shift is b. The term *amplitude* only applies to the sine and cosine functions.

Right Triangle Trigonometry

Using trigonometry to solve triangles is one of the oldest forms of mathematics. One of its most powerful uses is to measure distances—the height of a tree or building, the distance between earth and the moon, or the dimensions of a plot of land. The trigonometric ratios below are the same as before with the unit circle, only the labels are different. We will begin with right triangles.

In a right triangle, one angle measures 90° and the sum of the other angles is also 90°. The side opposite the 90° angle is the *hypotenuse*. The other sides are the *legs*. If we let θ represent one of the acute angles, then one of the legs is the side opposite θ, and the other side is adjacent to θ. See Figure 13.34.

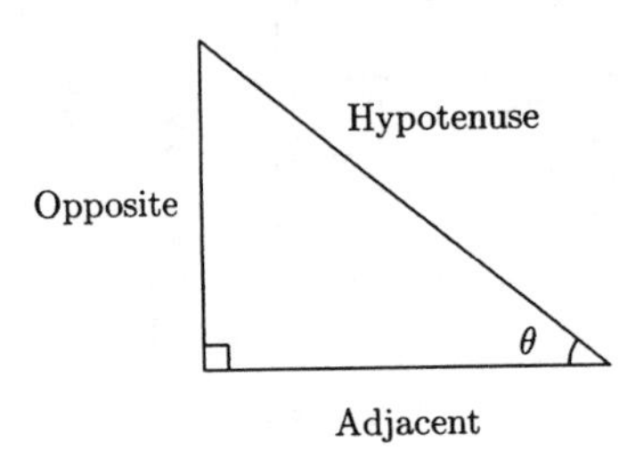

Fig. 13.34.

$$\sin\theta = \frac{\text{Opposite}}{\text{Hypotenuse}} \qquad \cos\theta = \frac{\text{Adjacent}}{\text{Hypotenuse}} \qquad \tan\theta = \frac{\text{Opposite}}{\text{Adjacent}}$$

$$\csc\theta = \frac{\text{Hypotenuse}}{\text{Opposite}} \qquad \sec\theta = \frac{\text{Hypotenuse}}{\text{Adjacent}} \qquad \cot\theta = \frac{\text{Adjacent}}{\text{Opposite}}$$

We can get the identity $\sin^2\theta + \cos^2\theta = 1$ from the Pythagorean Theorem.

$\text{Opposite}^2 + \text{Adjacent}^2 = \text{Hypotenuse}^2$

Divide both sides by Hypotenuse^2.

$$\left(\frac{\text{Opposite}}{\text{Hypotenuse}}\right)^2 + \left(\frac{\text{Adjacent}}{\text{Hypotenuse}}\right)^2 = \left(\frac{\text{Hypotenuse}}{\text{Hypotenuse}}\right)^2$$

$$\sin^2\theta + \cos^2\theta = 1$$

From this equation, we get two others, one from dividing both sides of the equation by $\sin^2\theta$, and the other by dividing both sides by $\cos^2\theta$.

$$\left(\frac{\sin\theta}{\sin\theta}\right)^2 + \left(\frac{\cos\theta}{\sin\theta}\right)^2 = \left(\frac{1}{\sin\theta}\right)^2$$

$$1 + \cot^2\theta = \csc^2\theta$$

$$\left(\frac{\sin\theta}{\cos\theta}\right)^2 + \left(\frac{\cos\theta}{\cos\theta}\right)^2 = \left(\frac{1}{\cos\theta}\right)^2$$

$$\tan^2\theta + 1 = \sec^2\theta$$

EXAMPLES

- Find all six trigonometric ratios for θ.

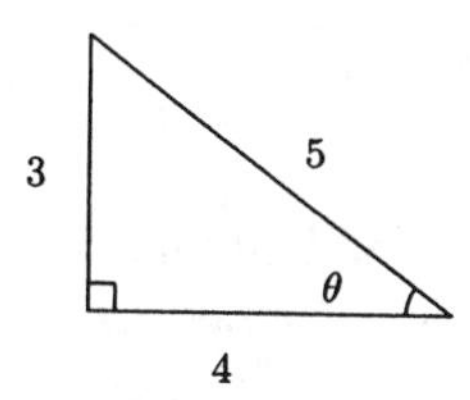

Fig. 13.35.

$$\sin\theta = \frac{\text{Opposite}}{\text{Hypotenuse}} = \frac{3}{5} \qquad \cos\theta = \frac{\text{Adjacent}}{\text{Hypotenuse}} = \frac{4}{5}$$

$$\tan\theta = \frac{\text{Opposite}}{\text{Adjacent}} = \frac{3}{4}$$

$$\csc\theta = \frac{\text{Hypotenuse}}{\text{Opposite}} = \frac{5}{3} \qquad \sec\theta = \frac{\text{Hypotenuse}}{\text{Adjacent}} = \frac{5}{4}$$

$$\cot\theta = \frac{\text{Adjacent}}{\text{Opposite}} = \frac{4}{3}$$

- Find $\sin A$, $\cos B$, $\sec A$, $\csc B$, $\tan A$, and $\cot B$.

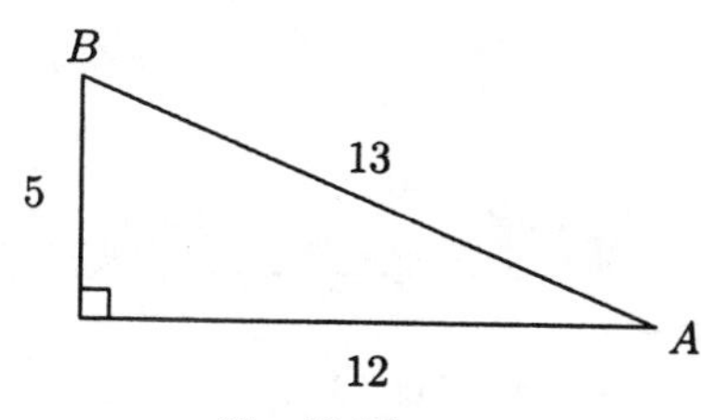

Fig. 13.36.

The hypotenuse is 13, the side opposite $\angle A$ is 5, so $\sin A = 5/13$. The side adjacent to $\angle B$ is 5, so $\cos B = 5/13$. The other ratios are $\sec A = 13/12$, $\csc B = 13/12$, $\tan A = 5/12$, and $\cot B = 5/12$.

The side opposite $\angle A$ is the side adjacent to $\angle B$, and the side adjacent to $\angle A$ is opposite $\angle B$. This is why sine and cosine, secant and cosecant, and tangent and cotangent are co-functions. Because $\angle A + \angle B = 90°$, we have $\angle B = 90° - \angle A$. These facts give us the following important relationships.

$$\sin A = \cos B = \cos(90° - A) \qquad \cos A = \sin B = \sin(90° - A)$$

$$\tan A = \cot B = \cot(90° - A)$$

$$\csc A = \sec B = \sec(90° - A) \qquad \sec A = \csc B = \csc(90° - A)$$

$$\cot A = \tan B = \tan(90° - A)$$

To "solve a triangle" means to find all three angles and the lengths of all three sides. For now, we will solve right triangles. Later, after covering inverse trigonometric functions, we can solve other triangles. When solving right triangles, we will use the Pythagorean Theorem as well as the fact that the sum of the two acute angles is 90°. Except for the angles 30°, 45°, and 60°, we need a calculator. The calculator should be in degree mode. Also, there are probably no keys for secant, cosecant, and cotangent. You will need to use the reciprocal key, marked either $\frac{1}{x}$ or x^{-1}. The keys marked $\sin^{-1}$, $\cos^{-1}$, and $\tan^{-1}$ are used to evaluate the functions covered in the next section.

EXAMPLES

- Solve the triangle.

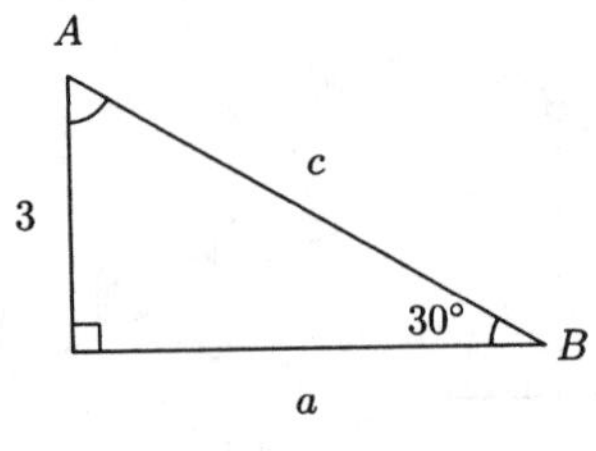

Fig. 13.37.

The side opposite the angle 30° is 3, so $\sin 30° = \frac{3}{c}$. We know that $\sin 30° = 1/2$. This gives us an equation to solve.

$$\frac{1}{2} = \frac{3}{c}$$

$$c = 6$$

We could use trigonometry to find the third side, but it is usually easier to use the Pythagorean Theorem.

$$a^2 + 3^2 = 6^2$$

$$a^2 = 36 - 9 = 27$$

$$a = \sqrt{27} = 3\sqrt{3}$$

$A = 90° - B = 90° - 30° = 60°$.

In some applications of right triangles, we are given the angle of *elevation* or *depression* to an object. The angle of elevation is the measure of upward rotation. The angle of depression is the measure of the downward rotation. See Figure 13.38.

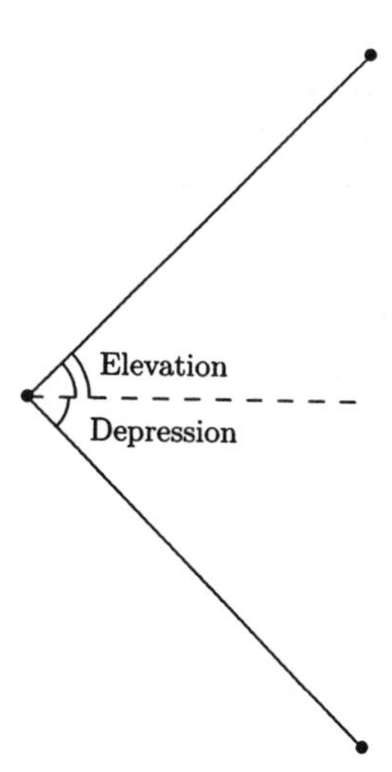

Fig. 13.38.

- A person is standing 300 feet from the base of a five-story building. He estimates that the angle of elevation to the top of the building is 63°. Approximately how tall is the building?

We need to find b in the following triangle.

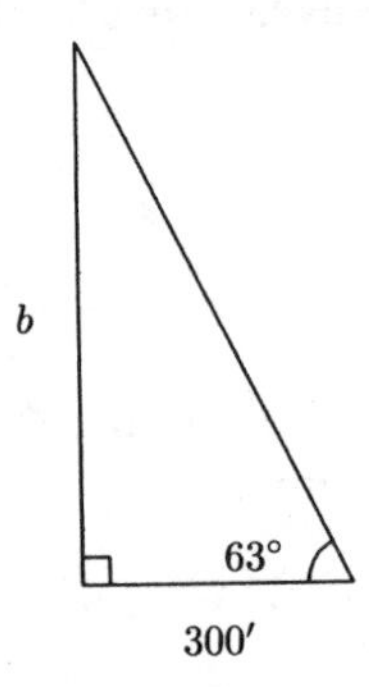

Fig. 13.39.

We could use either of the ratios that use the opposite and adjacent sides, tangent (opposite/adjacent) and cotangent (adjacent/opposite). We will use tangent.

$$\tan 63^\circ = \frac{\text{Opposite}}{\text{Adjacent}} = \frac{b}{300}$$

This gives us the equation $\tan 63^\circ = b/300$. When we solve for b, we have $b = 300 \tan 63^\circ \approx (300)1.9626 \approx 588.78$. The building is about 589 feet tall.

- A guy wire is 60 feet from the base of a tower. The angle of elevation from the top of the tower along the wire is 73°. How long is the wire?
  We need to find c in the following triangle.

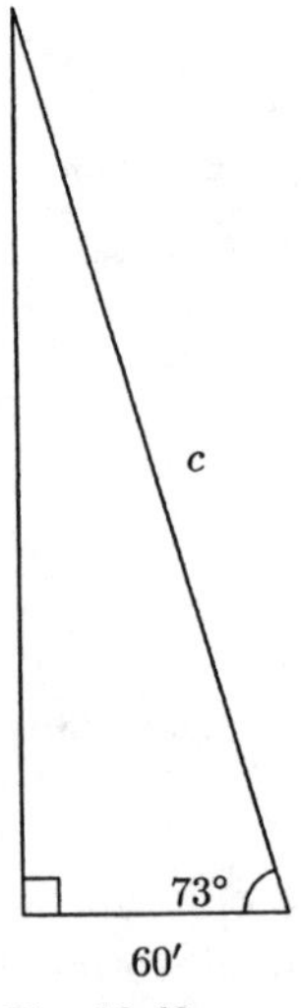

Fig. 13.40.

We could use either cosine (adjacent/hypotenuse) or secant (hypotenuse/adjacent). Using cosine, we have $\cos 73° = 60/c$. Solving this equation for c gives us $c = 60/\cos 73° \approx 60/0.2924 \approx 205$. The wire is about 205 feet long.

PRACTICE

1. Find all six trigonometric ratios for θ.

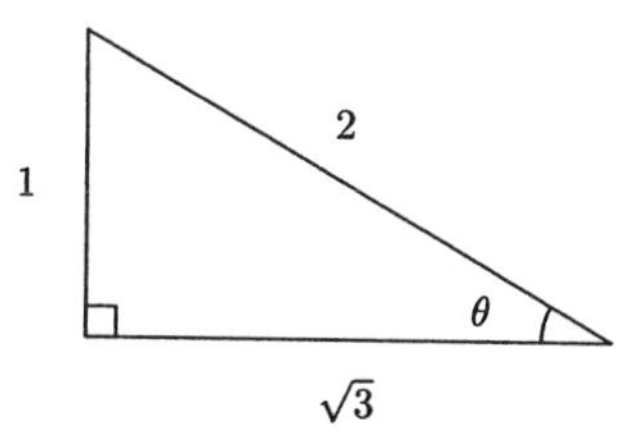

Fig. 13.41.

2. Solve the triangle.

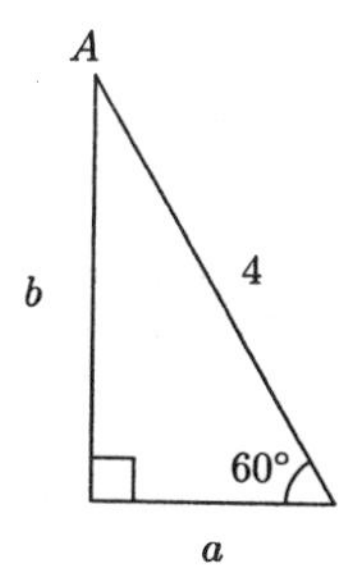

Fig. 13.42.

3. A plane is flying at an altitude of 5000 feet. The angle of elevation to the plane from a car traveling on a highway is about 38.7°. How far apart are the plane and car?

SOLUTIONS

1. $\sin\theta = \dfrac{1}{2}$ $\quad \cos\theta = \dfrac{\sqrt{3}}{2}$ $\quad \tan\theta = \dfrac{1}{\sqrt{3}} = \dfrac{\sqrt{3}}{3}$

   $\csc\theta = 2$ $\quad \sec\theta = \dfrac{2}{\sqrt{3}} = \dfrac{2\sqrt{3}}{3}$ $\quad \cot\theta = \sqrt{3}$

2. We could use any of the ratios involving the hypotenuse. We will use cosine: $\cos 60^\circ = a/4$. Since $\cos 60^\circ = 1/2$, we have $1/2 = a/4$. Solving for a gives us $a = 2$.

$$2^2 + b^2 = 4^2$$

$$b = \sqrt{4^2 - 2^2} = \sqrt{12} = 2\sqrt{3}$$

$\angle A = 90^\circ - 60^\circ = 30^\circ$

3. We need to find c in the following triangle.

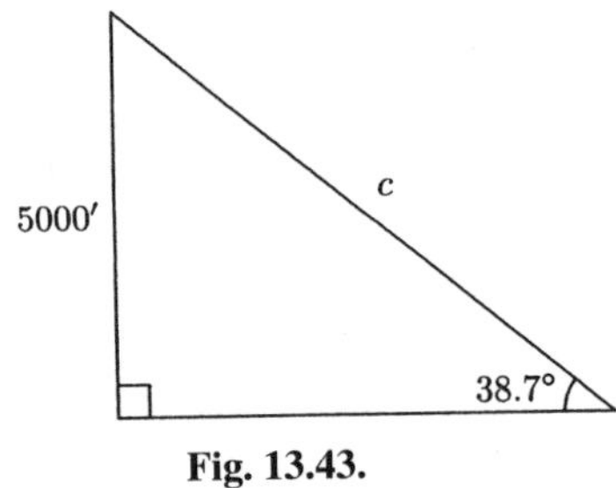

Fig. 13.43.

$$\sin 38.7^\circ = \frac{5000}{c}$$

$$c = \frac{5000}{\sin 38.7^\circ} \approx \frac{5000}{0.6252} \approx 7997$$

The plane and car are about 8000 feet apart.

Inverse Trigonometric Functions

Only one-to-one functions can have inverses, and the trigonometric functions are certainly not one to one. But we can limit their domains and force them to be one to one. Limiting the sine function to the interval from $x = -\pi/2$ to $x = \pi/2$ makes $f(x) = \sin x$ one to one. The graph in Figure 13.44 passes the Horizontal Line Test.

The domain of this function is $[-\pi/2, \pi/2]$, and the range is $[-1, 1]$. If we limit the cosine function to the interval from $x = 0$ to $x = \pi$, we have another one-to-one function. Its graph is shown in Figure 13.45. The domain of this function is $[0, \pi]$ and the range is $[-1, 1]$.

By limiting the tangent function from $x = -\pi/2$ to $x = \pi/2$, $f(x) = \tan x$ is one to one. Its domain is $(-\pi/2, \pi/2)$, and its range is all real numbers.

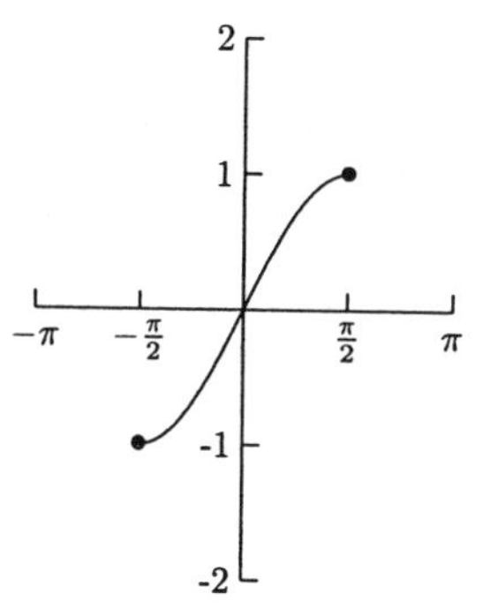

Fig. 13.44.

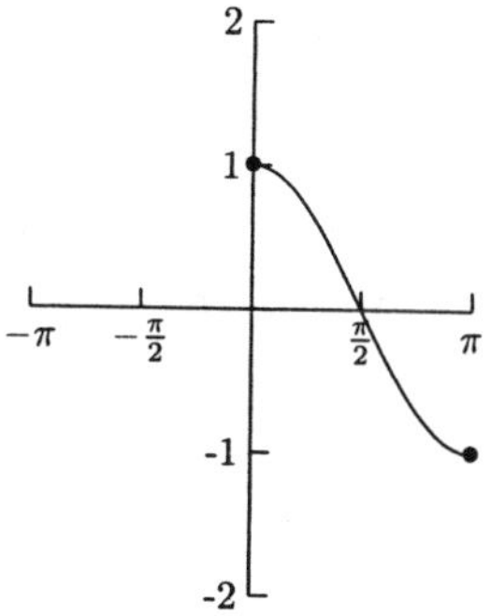

Fig. 13.45.

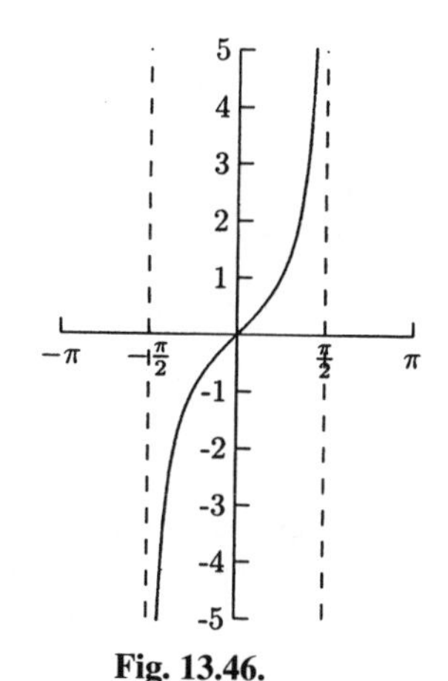

Fig. 13.46.

There are two notations for inverse trigonometric functions. One uses "-1," and the other uses the letters *arc*. For example, the inverse sine function is noted as $\sin^{-1}$ or arcsin. Remember that for any function $f(x)$ and its inverse $f^{-1}(x)$, $f(f^{-1}(x)) = x$ and $f^{-1}(f(x)) = x$. In other words, a function evaluated at its inverse "cancels" itself.

$$\cos^{-1}(\cos \pi/3) = \pi/3 \qquad \sin(\sin^{-1} 1/4) = 1/4$$

$$\tan(\tan^{-1} 1) = 1 \qquad \tan^{-1}(\tan \theta) = \theta$$

The x and y values are reversed for inverse functions. For example, if $(4, 9)$ is a point on the graph of $f(x)$, then $(9, 4)$ is a point on the graph of $f^{-1}(x)$. This means that the y-values for the inverse trigonometric functions are angles. Though we will need to use a calculator to evaluate most of these functions, we can find a few of them without a calculator. For $\cos^{-1} \frac{1}{2}$, ask yourself what angle (between 0 and π) has a cosine of 1/2? Because $\cos \pi/3 = 1/2$, $\cos^{-1} \frac{1}{2} = \frac{\pi}{3}$. When evaluating inverse trigonometric functions, we need to keep in mind what their range is. The domain of $f(x) = \sin(x)$ is $[-\pi/2, \pi/2]$ (Quadrants I and IV), so the range of $y = \sin^{-1} x$ is $[-\pi/2, \pi/2]$. The domain of $f(x) = \cos x$ is $[0, \pi]$, so the range of $y = \cos^{-1} x$ is $[0, \pi]$ (Quadrants I and II). And the domain of $f(x) = \tan x$ is $(-\pi/2, \pi, 2)$, so the range of $y = \tan^{-1} x$ is $(-\pi/2, \pi/2)$ (Quadrants I and IV).

EXAMPLES

- $\sin^{-1} \sqrt{2}/2$

  Because $\sin \pi/4 = \sqrt{2}/2$, $\sin^{-1} \sqrt{2}/2 = \pi/4$.

- $\tan^{-1} \sqrt{3}$

  Because $\tan \pi/3 = \sqrt{3}$, $\tan^{-1} \sqrt{3} = \pi/3$.

- $\cos^{-1}(-1)$

  $\cos^{-1}(-1) = \pi$ because $\cos \pi = -1$.

- $\tan^{-1}(1/3)$

  None of the important angles between $-\pi/2$ and $\pi/2$ has a tangent of 1/3, so we need to use a calculator to get an approximation: $\tan^{-1}(1/3) \approx 0.32175$.

- $\sin^{-1}(\cos \pi/6)$

  $\cos \pi/6 = \sqrt{3}/2$, so we need to replace $\cos \pi/6$ with $\sqrt{3}/2$. This gives us $\sin^{-1} \sqrt{3}/2$. Because $\sin \pi/3 = \sqrt{3}/2$, $\sin^{-1} \sqrt{3}/2 = \pi/3$.

- $\cos(\tan^{-1}(-1))$

  What angle in the interval $(-\pi/2, \pi/2)$ has a tangent of -1? That would be $-\pi/4$, so $\tan^{-1}(-1) = -\pi/4$.

$$\cos(\tan^{-1}(-1)) = \cos\left(-\frac{\pi}{4}\right) = \frac{\sqrt{2}}{2}$$

In the next set of problems, we will use right triangles to find the exact value of expressions like $\cos(\sin^{-1} 2/3)$. We will begin by letting $\sin^{-1} 2/3 = \theta$. We can think of $\sin^{-1} 2/3 = \theta$ as $\sin \theta = 2/3$. This allows us to use (Opposite/Hypotenuse) to represent 2/3. We will create a right triangle with acute angle θ, where the side opposite θ is 2, and the hypotenuse is 3.

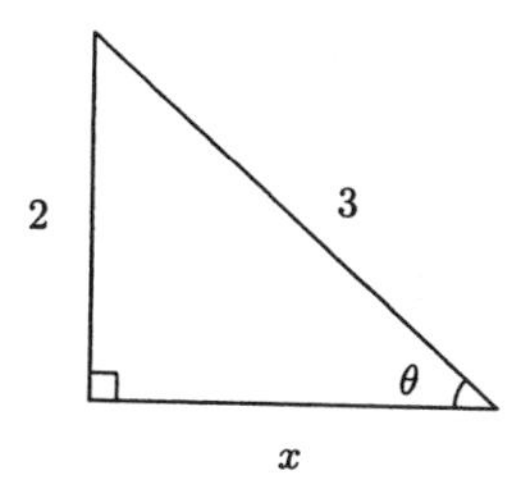

Fig. 13.47.

We want $\cos\theta$. We have the hypotenuse. We will use the Pythagorean Theorem to find x: $x^2 + 2^2 = 3^2$. This gives us $x = \sqrt{5}$ and $\cos\theta = \sqrt{5}/3$. Now we have $\cos(\sin^{-1} 2/3) = \cos\theta = \sqrt{5}/3$.

EXAMPLE

- $\sin(\tan^{-1} 4/5)$
  Let $\tan^{-1} 4/5 = \theta$, so $\tan\theta = 4/5$. We want a right triangle where the side opposite θ is 4 and the side adjacent to θ is 5.

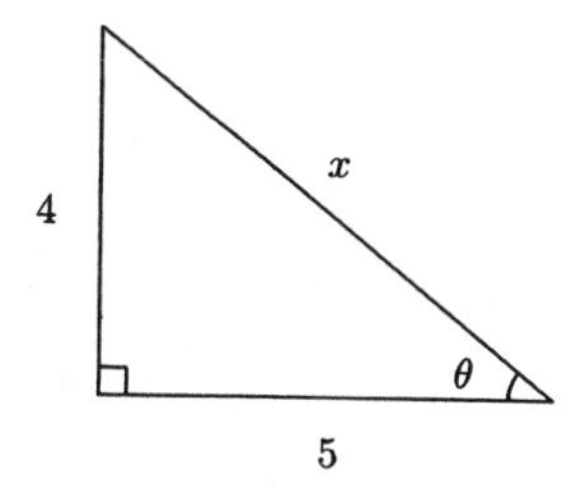

Fig. 13.48.

Solving $4^2 + 5^2 = x^2$ gives us $x = \sqrt{16 + 25} = \sqrt{41}$.

$$\sin\theta = \frac{4}{\sqrt{41}} = \frac{4\sqrt{41}}{41} \text{ so, } \sin\left(\tan^{-1}\frac{4}{5}\right) = \sin\theta = \frac{4\sqrt{41}}{41}$$

We will use inverse trigonometric functions to solve right triangles when we are given one acute angle and the length of one side. We can also use them to solve right triangles when we only have the lengths of two sides.

EXAMPLES

- Solve the triangle.

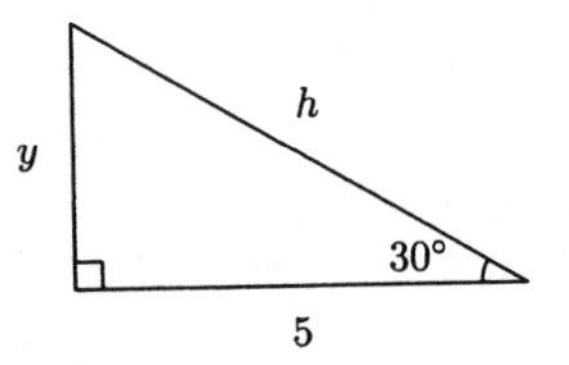

Fig. 13.49.

We need to find the side opposite θ or the hypotenuse. If we want to find the side opposite θ, we can use $\tan 30° = 1/\sqrt{3}$. If we want to find the hypotenuse, we can use $\cos 30° = \sqrt{3}/2$.

$$\cos 30° = \frac{5}{h}$$

$$\frac{\sqrt{3}}{2} = \frac{5}{h}$$

$$h = 5 \cdot \frac{2}{\sqrt{3}} = \frac{10\sqrt{3}}{3}$$

$$\tan 30° = \frac{y}{5}$$

$$\frac{1}{\sqrt{3}} = \frac{y}{5}$$

$$y = \frac{5}{\sqrt{3}} = \frac{5\sqrt{3}}{3}$$

The third angle is $90° - 30° = 60°$.

- Solve the triangle. When rounding is necessary, give your solutions accurate to one decimal place.

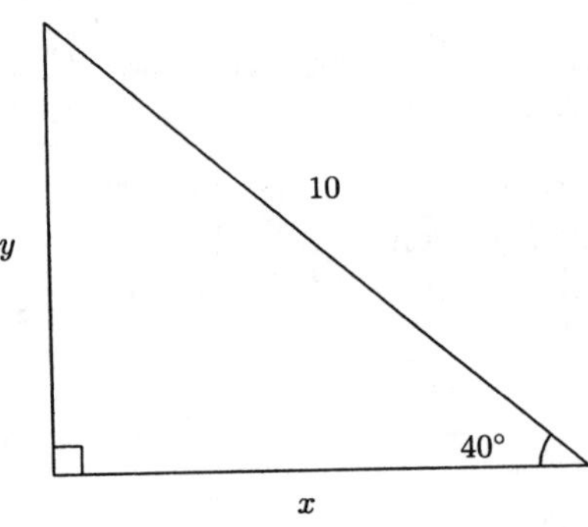

Fig. 13.50.

$$\sin 40° = \frac{y}{10} \qquad \cos 40° = \frac{x}{10}$$

$$y = 10\sin 40° \approx 6.4 \qquad x = 10\cos 40° \approx 7.7$$

The third angle is $90° - 40° = 50°$.

- Solve the triangle. When rounding is necessary, give your solutions accurate to one decimal place.

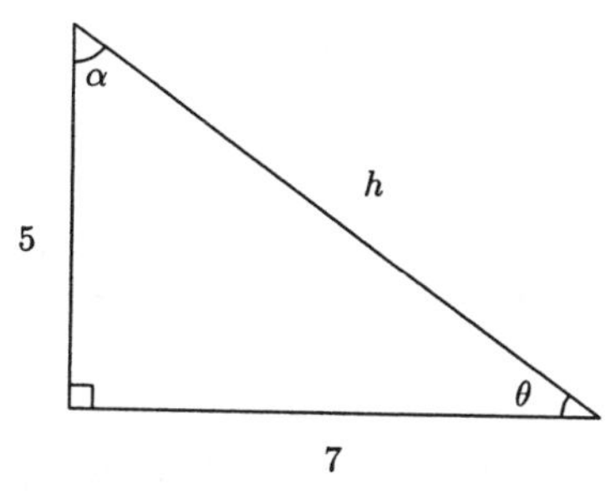

Fig. 13.51.

$$5^2 + 7^2 = h^2$$

$$h = \sqrt{25 + 49} = \sqrt{74}$$

$$\tan\theta = \frac{5}{7}$$

$$\theta = \tan^{-1}\frac{5}{7} \approx 35.5°$$

$$\alpha \approx 90° - 35.5° \approx 54.5°$$

- A 30-foot ladder is leaning against a wall. The top of the ladder is 24 feet above the ground. What angle does the ladder make with the ground?

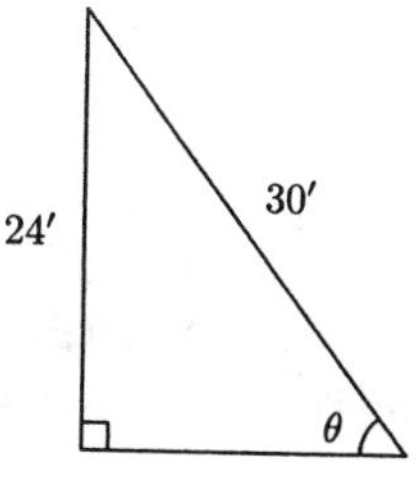

Fig. 13.52.

$$\sin\theta = \frac{24}{30}$$

$$\theta = \sin^{-1}\frac{24}{30} \approx 53.1°$$

- Find x, the height of the triangle.

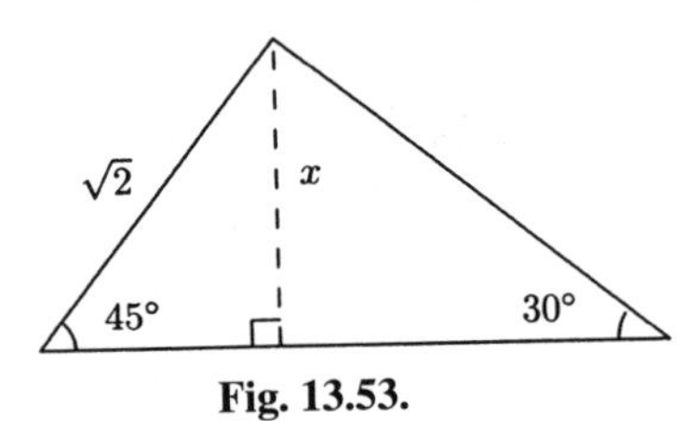

Fig. 13.53.

By viewing the triangle as two separate right triangles, the height of the triangle is the length of one of the legs of the separate triangles. We only need to use one of them.

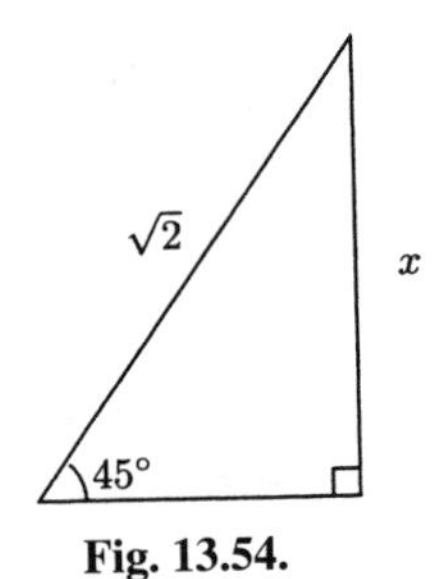

Fig. 13.54.

$$\sin 45° = \frac{x}{\sqrt{2}}$$

$$x = \sqrt{2}\sin 45° = \sqrt{2}\left(\frac{1}{\sqrt{2}}\right) = 1$$

We can solve other triangles using inverse trigonometric functions and the Law of Sines and/or the Law of Cosines. Although all triangles can be solved, sometimes we are given information that is true about more than one triangle or about a triangle that cannot exist. In the following problems, we will use the labels in the following triangles.

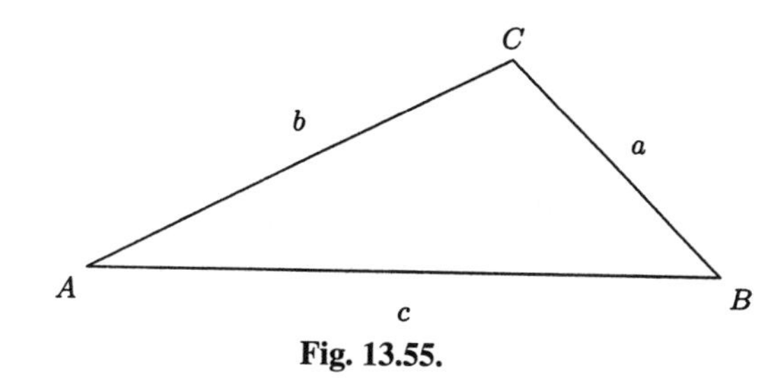

Fig. 13.55.

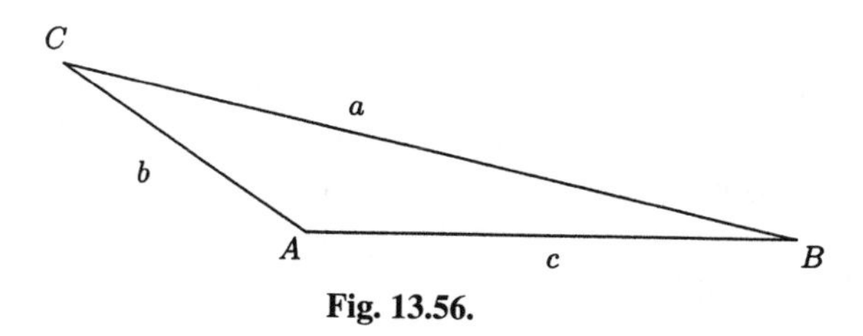

Fig. 13.56.

The angles are A, B, and C. The sides opposite these angles are a, b, and c, respectively.

We cannot solve a triangle if all we know are all three angles. Two triangles can be different sizes but have the same angles. Also, we might be given an angle with the side opposite the angle and another side that makes two triangles true. For example, suppose we are told to find a triangle where $\angle A = 21°$, $a = 3$, and $b = 8$. There are *two* triangles that satisfy these conditions.

Triangle 1	Triangle 2
$\angle A = 21°$	$\angle A = 21°$
$\angle B \approx 72.9°$	$\angle B \approx 107.1°$
$\angle C \approx 86.1°$	$\angle C \approx 52°$
$a = 3$	$a = 3$
$b = 8$	$b = 8$
$c \approx 8.4$	$c \approx 6.6$

There are two triangles when $b \sin A < a < b$. If we have another number in addition to A, a, and b, then there will only be one triangle.

As an example of a triangle that cannot exist, let $\angle A = 20°$, $b = 10$, and $a = 2$. As you can see in Figure 13.57, a is too short to close the triangle. This happens when $a < b \sin A$.

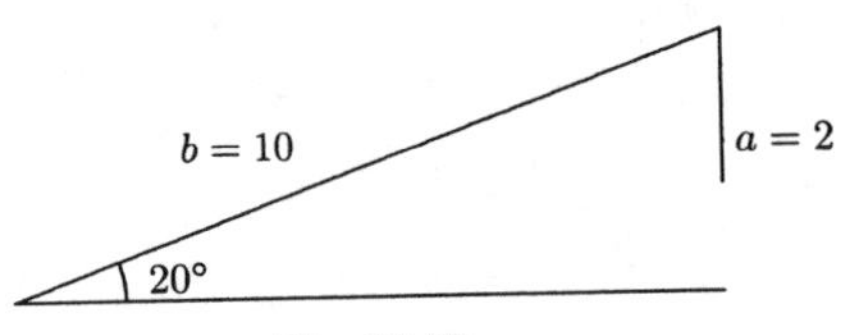

Fig. 13.57.

We can use the Law of Sines to solve a triangle if we know two sides and one of the angles opposite these sides or two angles and one side (if we know two angles, then we know all three because their sum is 180°). If do not have this information, the Law of Cosines works. We can use the Law of Cosines when we have two sides and any angle or when we have all three sides.

Here is the Law of Sines.

$$\frac{\sin A}{a} = \frac{\sin B}{b} = \frac{\sin C}{c}$$

This is really three separate equations.

$$\frac{\sin A}{a} = \frac{\sin B}{b} \qquad \frac{\sin B}{b} = \frac{\sin C}{c} \qquad \frac{\sin A}{a} = \frac{\sin C}{c}$$

Here is the Law of Cosines.

$$a^2 = b^2 + c^2 - 2bc \cos A$$

$$b^2 = a^2 + c^2 - 2ac \cos B$$

$$c^2 = a^2 + b^2 - 2ab \cos C$$

EXAMPLES

Solve the triangle. When rounding is necessary, give your solutions accurate to one decimal place.

- $\angle A = 30°$, $\angle B = 70°$, and $a = 5$

  We will use the Law of Sines because we know an angle, A, and the side opposite it, a.

$$\frac{\sin A}{a} = \frac{\sin B}{b} \text{ becomes } \frac{\sin 30°}{5} = \frac{\sin 70°}{b}$$

$$\frac{\sin 30°}{5} = \frac{\sin 70°}{b}$$

$$\frac{1/2}{5} \approx \frac{0.9397}{b} \qquad (\sin 30° = 1/2,\ \sin 70° \approx 0.9397)$$

$$b \approx 10(0.9397) \approx 9.4$$

Now we will use $(\sin A)/a = (\sin C)/c$ to find c. ($\angle C = 180° - 30° - 70° = 80°$)

$$\frac{\sin 30°}{5} = \frac{\sin 80°}{c}$$

$$\frac{1/2}{5} \approx \frac{0.9848}{c}$$

$$c \approx 10(0.9848) \approx 9.8$$

- $a = 5,\ b = 8$, and $c = 12$

  There is not enough information to get one equation with one variable using the Law of Sines, so we will use the Law of Cosines.

$$a^2 = b^2 + c^2 - 2bc \cos A$$

$$5^2 = 8^2 + 12^2 - 2(8)(12) \cos A$$

$$-183 = -192 \cos A$$

$$\frac{61}{64} = \cos A$$

$$A = \cos^{-1} \frac{61}{64}$$

$$A \approx 17.6°$$

We can use either the Law of Sines or the Law of Cosines to find $\angle B$. The Law of Sines is a little easier.

$$\frac{\sin A}{a} = \frac{\sin B}{b}$$

$$\frac{\sin 17.6°}{5} = \frac{\sin B}{8}$$

$$\sin B = \frac{8 \sin 17.6°}{5} \approx 0.484$$

$$B \approx \sin^{-1} 0.484 \approx 28.9°$$

$\angle C \approx 180° - 17.6° - 28.9° \approx 133.5°$

•

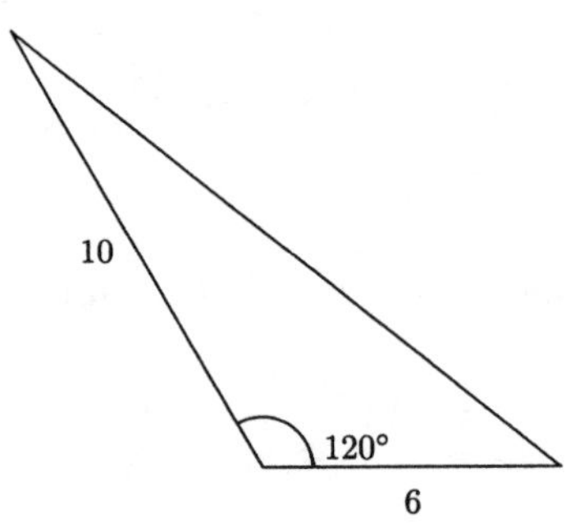

Fig. 13.58.

We will call the 120° angle A, then $b = 10$ and $c = 6$. (It does not matter which side is b and which side is c, as long as we do not label either one of them a.) There is not enough information to use the Law of Sines, so we will use the Law of Cosines.

$$a^2 = b^2 + c^2 - 2bc \cos A$$

$$a^2 = 10^2 + 6^2 - 2(10)(6) \cos 120°$$

$$a^2 = 136 - 120(-0.5)$$

$$a^2 = \sqrt{196}$$

$$a = 14$$

We can use either the Law of Sines or the Law of Cosines to find $\angle B$ or $\angle C$. We will use the Law of Sines to find $\angle B$.

$$\frac{\sin 120°}{14} = \frac{\sin B}{10}$$

$$\frac{\sqrt{3}}{2} \cdot \frac{10}{14} = \sin B \qquad \left(\sin 120° = \frac{\sqrt{3}}{2}\right)$$

$$B = \sin^{-1}\left(\frac{10\sqrt{3}}{28}\right) \approx \sin^{-1} 0.6186 \approx 38.2°$$

$\angle C \approx 180° - 120° - 38.2° \approx 21.8°$

PRACTICE

When rounding is necessary, please give your solutions accurate to one decimal place. The angles for Problems 1–6 are in radians.

1. $\cos^{-1}(\cos \pi/8)$
2. $\tan(\tan^{-1} -1)$
3. $\cos^{-1} 1/2$
4. $\sin^{-1} 1/2$
5. $\tan^{-1} 0$
6. $\sin^{-1} 0.9$
7. Solve the triangle.

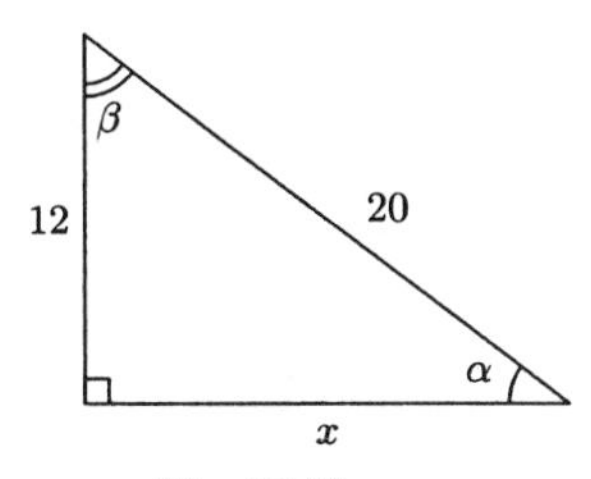

Fig. 13.59.

8. A 20-foot ladder is leaning against a wall. The base of the ladder is four feet from the wall. What angle is formed by the ground and the ladder?
9. Solve the triangle: $\angle A = 42°$, $a = 11$, and $b = 6$.
10. Find all three angles for the triangle whose sides are 6, 8, and 10.
11. A plane is flying over a highway at an altitude of 6000 feet. A blue car is traveling on a highway in front of the plane and a white car is on the highway behind the plane. The angle of elevation from the blue car to the plane is 45°. If the cars are two miles apart, how far is the plane from each car? (Hint: Consider the triangle formed by the cars and plane as two right triangles that share a leg.)

SOLUTIONS

1. $\pi/8$ radians
2. -1 radians

3. $\pi/3$ radians
4. $\pi/6$ radians
5. 0 radians
6. Approximately 1.1 radians
7. $\sin\alpha = \frac{12}{20} = \frac{3}{5}$

$$\alpha = \sin^{-1}\frac{3}{5} \approx 36.9^\circ$$

$$\beta \approx 90^\circ - 36.9^\circ \approx 53.1^\circ$$

$$x^2 + 12^2 = 20^2$$

$$x^2 = 400 - 144$$

$$x = \sqrt{256} = 16$$

8.

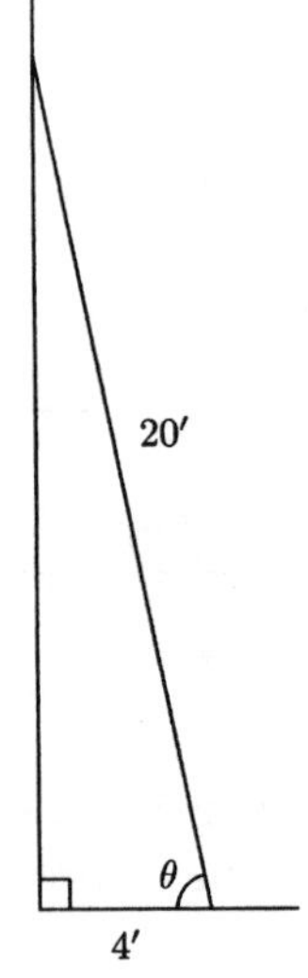

Fig. 13.60.

$$\cos\theta = \frac{4}{20} = \frac{1}{5}$$

$$\theta = \cos^{-1}\frac{1}{5} \approx 78.5^\circ$$

9. We will use the Law of Sines twice.

$$\frac{\sin 42^\circ}{11} = \frac{\sin B}{6}$$

$$\sin B = \frac{6}{11}\sin 42^\circ \approx 0.365$$

$$B \approx \sin^{-1} 0.365 \approx 21.4^\circ$$

$$C \approx 180° - 21.4° - 42° \approx 116.6°$$

$$\frac{\sin 42°}{11} = \frac{\sin 116.6°}{c}$$

$$c \approx \frac{11 \sin 116.6°}{\sin 42°} \approx 14.7$$

10. Let $a = 6$, $b = 8$, and $c = 10$. We will first use the Law of Cosines to find $\angle A$. Then we will use the Law of Sines to find $\angle B$.

$$a^2 = b^2 + c^2 - 2bc \cos A$$

$$6^2 = 8^2 + 10^2 - 2(8)(10) \cos A$$

$$-128 = -160 \cos A$$

$$\frac{4}{5} = \cos A$$

$$A = \cos^{-1} \frac{4}{5} \approx 36.9°$$

$$\frac{\sin 36.9°}{6} = \frac{\sin B}{8}$$

$$\frac{8 \sin 36.9°}{6} = \sin B$$

$$B = \sin^{-1} 0.8 \approx 53.1° \qquad \left(\frac{8}{6} \sin 36.9° \approx 0.8\right)$$

$$C \approx 180° - 36.9° - 53.1° \approx 90°$$

11.

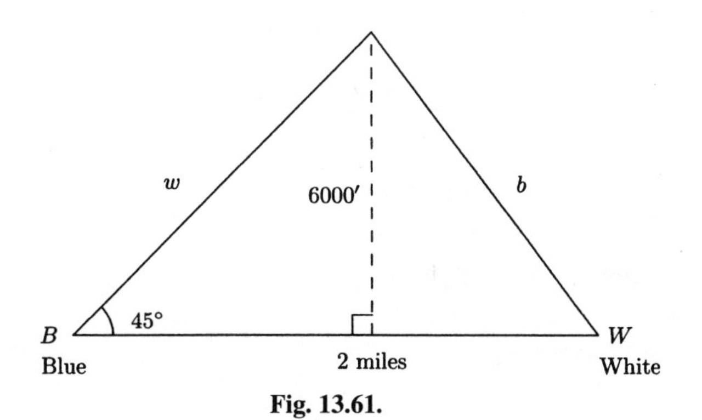

Fig. 13.61.

Let b represent the side of the original triangle that is opposite the angle 45°. Let w represent the side opposite $\angle W$, which is also the distance from the plane to the blue car. Two miles is 2(5280) = 10,560 feet.

$$\sin 45° = \frac{6000}{w}$$

$$w = \frac{6000}{\sin 45°} = \frac{6000}{1/\sqrt{2}} = \sqrt{2}(6000) \approx 8485.3$$

$$b^2 = 8485.3^2 + 10{,}560^2 - 2(8485.3)(10{,}560)\cos 45°$$

$$b^2 \approx 56{,}793{,}637.9$$

$$b \approx \sqrt{56{,}793{,}637.9} \approx 7536.2$$

The plane is about 8485 feet from the blue car and about 7536 feet from the white car.

Miscellaneous Formulas

The formulas in this section are used to find the exact value for more trigonometric ratios than the main angles—0, $\pi/6$, $\pi/4$, $\pi/3$, $\pi/2$. We will find angles that are half, double, or the sum or difference of these angles. These formulas are also used to rewrite functions in a form that fits a calculus formula.

1. Addition and Subtraction Formulas

   (a) $\sin(s+t) = \sin s \cos t + \cos s \sin t$
   (b) $\sin(s-t) = \sin s \cos t - \cos s \sin t$
   (c) $\cos(s+t) = \cos s \cos t - \sin s \sin t$
   (d) $\cos(s-t) = \cos s \cos t + \sin s \sin t$
   (e) $\tan(s+t) = \dfrac{\tan s + \tan t}{1 - \tan s \tan t}$
   (f) $\tan(s-t) = \dfrac{\tan s - \tan t}{1 + \tan s \tan t}$

2. Power Reduction Formulas

   (a) $\sin^2 s = \dfrac{1 - \cos 2s}{2}$

(b) $\cos^2 s = \dfrac{1 + \cos 2s}{2}$

(c) $\tan^2 s = \dfrac{1 - \cos 2s}{1 + \cos 2s}$

3. Half-Angle and Double Angle Formulas

(a) $\sin\left(\dfrac{s}{2}\right) = \pm\sqrt{\dfrac{1 - \cos s}{2}}$

(b) $\cos\left(\dfrac{s}{2}\right) = \pm\sqrt{\dfrac{1 + \cos s}{2}}$

(c) $\tan\left(\dfrac{s}{2}\right) = \dfrac{1 - \cos s}{\sin s} = \dfrac{\sin s}{1 + \cos s}$

The sign of + or − depends on where the angle $s/2$ lies.

(d) $\sin 2s = 2 \sin s \cos s$

(e) $\cos 2s = \cos^2 s - \sin^2 s = 1 - 2\sin^2 s = 2\cos^2 s - 1$

(f) $\tan 2s = \dfrac{2 \tan s}{1 - \tan^2 s}$

4. Product-to-Sum and Sum-to-Product Formulas

(a) $\sin s \cos t = \dfrac{1}{2}[\sin(s + t) + \sin(s - t)]$

(b) $\cos s \sin t = \dfrac{1}{2}[\sin(s + t) - \sin(s - t)]$

(c) $\cos s \cos t = \dfrac{1}{2}[\cos(s + t) + \cos(s - t)]$

(d) $\sin s \sin t = \dfrac{1}{2}[\cos(s - t) - \cos(s + t)]$

(e) $\sin s + \sin t = 2 \sin\left(\dfrac{s + t}{2}\right) \cos\left(\dfrac{s - t}{2}\right)$

(f) $\sin s - \sin t = 2 \cos\left(\dfrac{s + t}{2}\right) \sin\left(\dfrac{s - t}{2}\right)$

(g) $\cos s + \cos t = 2\cos\left(\frac{s+t}{2}\right)\cos\left(\frac{s-t}{2}\right)$

(h) $\cos s - \cos t = -2\sin\left(\frac{s+t}{2}\right)\sin\left(\frac{s-t}{2}\right)$

EXAMPLES

- $\sin 75°$

  We can think of 75° as 45° + 30°. This lets us use formula 1(a).

$$\sin(s+t) = \sin s \cos t + \cos s \sin t$$

$$\sin 75° = \sin(45° + 30°) = \sin 30° \cos 45° + \cos 30° \sin 45°$$

$$= \frac{1}{2} \cdot \frac{\sqrt{2}}{2} + \frac{\sqrt{3}}{2} \cdot \frac{\sqrt{2}}{2} = \frac{\sqrt{2}}{4} + \frac{\sqrt{6}}{4}$$

$$= \frac{\sqrt{2}+\sqrt{6}}{4}$$

- $\cos 15°$

  We will use formula 3(b) because $15° = \frac{30°}{2}$.

$$\cos\frac{s}{2} = \sqrt{\frac{1+\cos s}{2}}$$

$$\cos 15° = \cos\left(\frac{30°}{2}\right) = \sqrt{\frac{1+\cos 30°}{2}}$$

$$= \sqrt{\frac{1+\sqrt{3}/2}{2}} = \sqrt{\frac{\frac{2}{2}+\frac{\sqrt{3}}{2}}{2}} = \sqrt{\frac{\frac{2+\sqrt{3}}{2}}{2}}$$

$$= \sqrt{\frac{2+\sqrt{3}}{4}} = \frac{\sqrt{2+\sqrt{3}}}{\sqrt{4}} = \frac{\sqrt{2+\sqrt{3}}}{2}$$

- $\tan 7\pi/12$

  Because $\frac{7\pi}{12} = \frac{\pi}{4} + \frac{\pi}{3}$, we can use formula 1(e).

$$\tan(s+t) = \frac{\tan s + \tan t}{1 - \tan s \tan t}$$

$$\begin{aligned}
\tan \frac{7\pi}{12} = \tan\left(\frac{\pi}{4} + \frac{\pi}{3}\right) &= \frac{\tan \pi/4 + \tan \pi/3}{1 - \tan \pi/4 \tan \pi/3} = \frac{1+\sqrt{3}}{1-1(\sqrt{3})} \\
&= \frac{1+\sqrt{3}}{1-\sqrt{3}} = \frac{(1+\sqrt{3})(1+\sqrt{3})}{(1-\sqrt{3})(1+\sqrt{3})} \\
&= \frac{1+2\sqrt{3}+(\sqrt{3})^2}{1-(\sqrt{3})^2} \\
&= \frac{1+2\sqrt{3}+3}{1-3} = \frac{4+2\sqrt{3}}{-2} = -\frac{2(2+\sqrt{3})}{2} \\
&= -(2+\sqrt{3})
\end{aligned}$$

- If $\cos\theta = 3/5$ and θ is in Quadrant I, find $\sin 2\theta$.

  By formula 3(d), $\sin 2\theta = 2\sin\theta\cos\theta$. We need to find $\sin\theta$ so that we can use the formula.

$$\sin^2\theta + \cos^2\theta = 1$$

$$\sin^2\theta + \left(\frac{3}{5}\right)^2 = 1$$

$$\sin\theta = \sqrt{1 - \left(\frac{3}{5}\right)^2} = \frac{4}{5}$$

$$\sin 2\theta = 2\sin\theta\cos\theta = 2\left(\frac{4}{5}\right)\left(\frac{3}{5}\right) = \frac{24}{25}$$

- $\cos^2 \pi/12 - \sin^2 \pi/12$

  The expression looks like formula 3(e), where $s = \pi/12$.

$$\cos^2 s - \sin^2 s = \cos 2s$$

$$\cos^2 \frac{\pi}{12} - \sin^2 \frac{\pi}{12} = \cos\left(2 \cdot \frac{\pi}{12}\right)$$

$$= \cos \frac{\pi}{6} = \frac{\sqrt{3}}{2}$$

- Suppose $\cos 2\theta = 1/4$. Find $\sin^2 \theta$.

  We will use formula 2(a).

$$\sin^2 \theta = \frac{1 - \cos 2\theta}{2} = \frac{1 - \frac{1}{4}}{2} = \frac{\frac{4}{4} - \frac{1}{4}}{2} = \frac{\frac{3}{4}}{2} = \frac{3}{4 \cdot 2} = \frac{3}{8}$$

- Write $\cos^4 x$ without squaring any trigonometric functions.
  We will use formula 2(b) twice.

$$\cos^4 x = (\cos^2 x)(\cos^2 x)$$

$$= \left(\frac{1 + \cos 2x}{2}\right)\left(\frac{1 + \cos 2x}{2}\right)$$

$$= \frac{1}{2}(1 + \cos 2x) \cdot \frac{1}{2}(1 + \cos 2x)$$

$$= \frac{1}{4}(1 + \cos 2x)(1 + \cos 2x)$$

$$= \frac{1}{4}(1 + 2\cos 2x + \cos^2 2x) \qquad \text{Use the formula for } s = 2x.$$

$$= \frac{1}{4}\left[1 + 2\cos 2x + \left(\frac{1 + \cos 2 \cdot 2x}{2}\right)\right]$$

$$= \frac{1}{4}\left[1 + 2\cos 2x + \frac{1}{2}(1 + \cos 4x)\right]$$

- Rewrite $\cos 2x \cos 5x$ as a sum or difference.
  Formula 4(c) tells us how to write the product of two cosines as a sum.

$$\cos 2x \cos 5x = \frac{1}{2}\left[\cos(2x + 5x) + \cos(2x - 5x)\right]$$
$$= \frac{1}{2}\left[\cos(7x) + \cos(-3x)\right]$$
$$= \frac{1}{2}(\cos 7x + \cos 3x) \qquad \text{(Because cosine is even, } \cos 3x = \cos(-3x).\text{)}$$

- Rewrite $\sin 3x - \sin 2x$ as a product.
  This fits formula 4(f).

$$\sin 3x - \sin 2x = 2\cos\frac{3x + 2x}{2}\sin\frac{3x - 2x}{2} = 2\cos\frac{5x}{2}\sin\frac{x}{2}$$

PRACTICE

1. Find $\tan 15°$ using the half-angle formula.
2. If $\sin\theta = 2/3$ and θ is in Quadrant II, find $\sin 2\theta$.
3. Write $\sin^4 x$ using only the first powers of trigonometric functions.
4. Write $\cos 4x \sin 6x$ as a sum.

SOLUTIONS

1. Use formula 3(c).

$$\tan 15° = \tan\left(\frac{30°}{2}\right) = \frac{1 - \cos 30°}{\sin 30°} = \frac{1 - \sqrt{3}/2}{1/2} = \left(1 - \frac{\sqrt{3}}{2}\right) \div \frac{1}{2}$$
$$= \left(1 - \frac{\sqrt{3}}{2}\right) \cdot 2 = 2 - \sqrt{3}$$

2. Because θ is in Quadrant II, cosine will be negative.

$$\sin^2\theta + \cos^2\theta = 1$$

$$\left(\frac{2}{3}\right)^2 + \cos^2\theta = 1$$

$$\cos\theta = -\sqrt{1-\left(\frac{2}{3}\right)^2} = -\sqrt{\frac{5}{9}} = -\frac{\sqrt{5}}{3}$$

$$\sin 2\theta = 2\sin\theta\cos\theta \qquad \text{Formula 3(d)}$$

$$= 2\left(\frac{2}{3}\right)\left(\frac{-\sqrt{5}}{3}\right) = -\frac{4\sqrt{5}}{9}$$

3. $\sin^4 x = (\sin^2 x)(\sin^2 x)$

$$(\sin^2 x)(\sin^2 x) = \frac{1-\cos 2x}{2}\cdot\frac{1-\cos 2x}{2} \qquad \text{Formula 2(a)}$$

$$= \frac{1}{2}(1-\cos 2x)\cdot\frac{1}{2}(1-\cos 2x)$$

$$= \frac{1}{4}(1-\cos 2x)(1-\cos 2x)$$

$$= \frac{1}{4}(1-2\cos 2x+\cos^2 2x)$$

$$= \frac{1}{4}\left[1-2\cos 2x+\left(\frac{1+\cos 2\cdot 2x}{2}\right)\right] \qquad \text{Formula 2(b)}$$

$$= \frac{1}{4}\left[1-2\cos 2x+\frac{1}{2}(1+\cos 4x)\right]$$

4. We will use formula 4(b).

$$\cos 4x\sin 6x = \frac{1}{2}[\sin(4x+6x)-\sin(4x-6x)]$$

$$= \frac{1}{2}[\sin(10x)-\sin(-2x)]$$

$$= \frac{1}{2}[\sin 10x+\sin 2x]$$

Because sine is odd, $\sin(-2x) = -\sin 2x$.

CHAPTER 13 REVIEW

1. Find $\sin\theta$ if $\cos\theta = -\frac{1}{4}$ and θ is in Quadrant II.

(a) $\frac{\sqrt{15}}{4}$ (b) $-\frac{\sqrt{15}}{4}$ (c) $\frac{\sqrt{3}}{2}$ (d) $-\frac{\sqrt{3}}{2}$

2. What is the phase shift for $f(x) = 2\cos(3x + \pi/2)$?
(a) $-\frac{\pi}{2}$ (b) $\frac{\pi}{2}$ (c) $-\frac{\pi}{6}$ (d) $\frac{\pi}{6}$

3. What is the period for $f(x) = 2\cos(3x + \pi/2)$?
(a) $\frac{2\pi}{3}$ (b) 6π (c) $\frac{2}{3}$ (d) $\frac{\pi}{3}$

4. From the top of a 200-foot lighthouse, the angle of depression to a ship on the ocean is 20°. How far is the ship from the base of the lighthouse?
(a) About 400 feet (b) About 490 feet
(c) About 550 feet (d) About 690 feet

5. $\cos 15^\circ \cos 10^\circ + \sin 15^\circ \sin 10^\circ =$
(a) $\cos 5^\circ$ (b) $\cos 25^\circ$ (c) $\sin 5^\circ$ (d) $\sin 25^\circ$

6. Find the reference angle for $7\pi/9$.
(a) $\frac{2\pi}{9}$ (b) $-\frac{2\pi}{9}$ (c) $\frac{16\pi}{9}$ (d) $\frac{7\pi}{9}$

7. The terminal point for θ is $(-3/5, 4/5)$. What is $\tan\theta$?
(a) $-\frac{4}{3}$ (b) $-\frac{3}{4}$ (c) $-\frac{5}{3}$ (d) $\frac{5}{4}$

8. $\tan(\cos^{-1} 3/4) =$
(a) $\frac{3\sqrt{7}}{7}$ (b) $\frac{\sqrt{7}}{3}$ (c) $\frac{4\sqrt{7}}{7}$ (d) $\frac{\sqrt{7}}{4}$

9. The graph in Figure 13.62 is the graph of one period of which function?
(a) $y = 2\cos(x + \pi/3)$ (b) $y = 2\cos(x - \pi/3)$
(c) $y = \cos 2(x + \pi/3)$ (d) $y = \cos 2(x - \pi/3)$

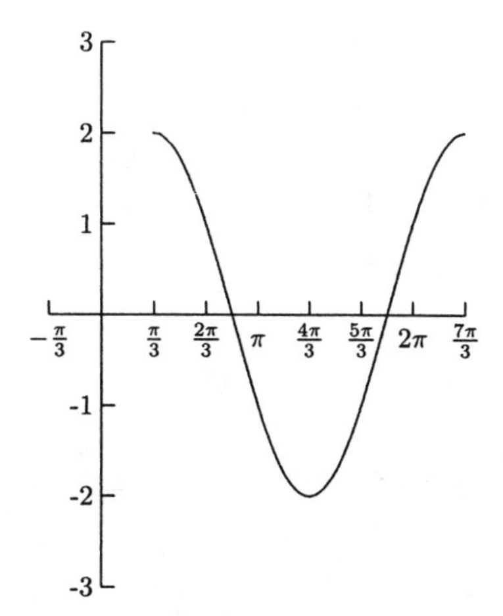

Fig. 13.62.

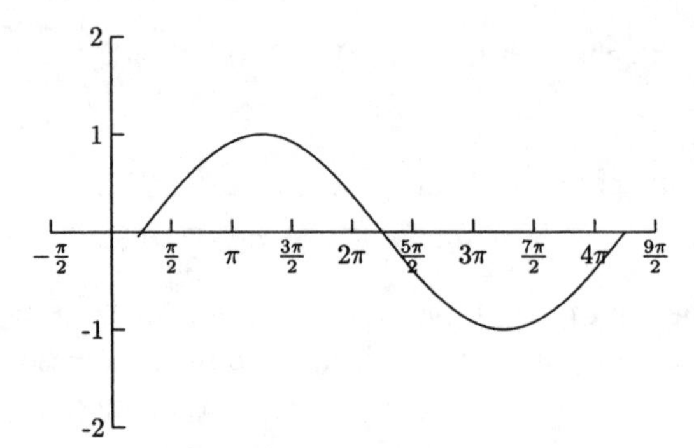

Fig. 13.63.

10. The graph in Figure 13.63 in the graph of one period of which function?
    (a) $y = \sin\frac{1}{2}(x - \pi/4)$ (b) $y = \frac{1}{2}\sin(x - \pi/4)$
    (c) $y = \sin\frac{1}{2}(x + \pi/4)$ (d) $y = \frac{1}{2}\sin(x + \pi/4)$

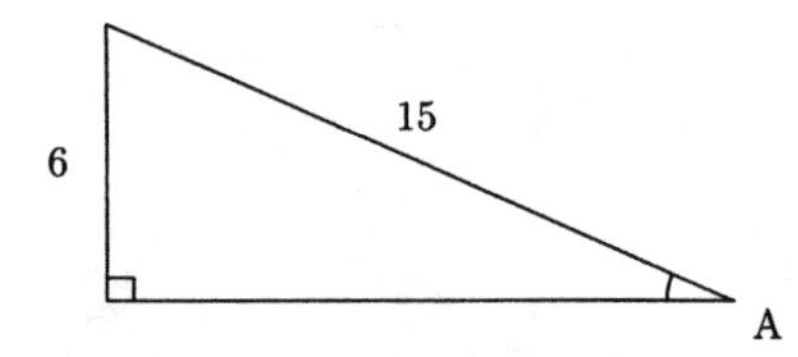

Fig. 13.64.

11. Find $\angle A$.
    (a) About 68.2° (b) About 21.8°
    (c) About 66.4° (d) About 23.6°

SOLUTIONS

1. A 2. C 3. A 4. C 5. A 6. A
7. A 8. B 9. B 10. A 11. D

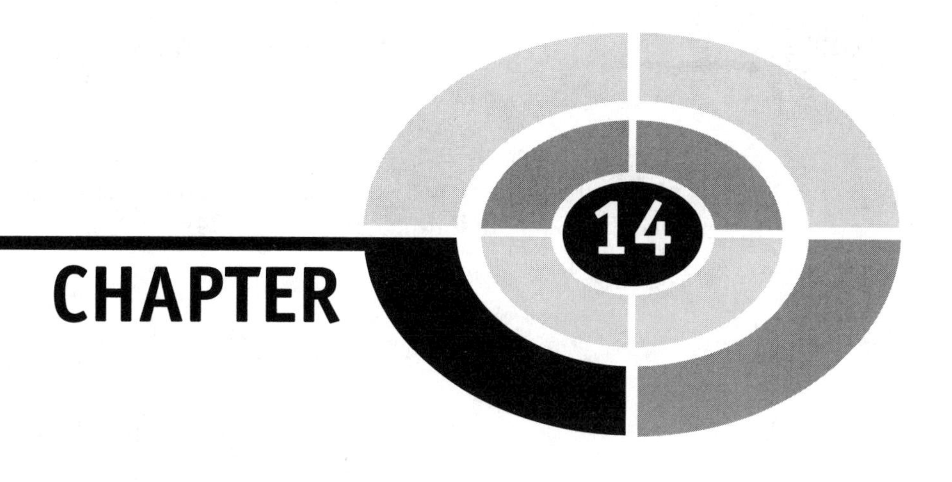

Sequences and Series

A *sequence* is an ordered list of numbers. Although they list the same numbers, the sequence 1, 2, 3, 4, 5, 6, ... is different from the sequence 2, 1, 4, 3, 6, 5, ... Usually a sequence is infinite. This means that there is no last term in the sequence. A *series* is the sum (if it exists) of a sequence. Although a sequence can be any list of numbers, we will work with sequences that can be found from a formula. Formulas describe how to compute the nth term, a_n. For example, the formula $a_n = 2n + 1$ gives us this sequence.

$$\underset{2(1)+1}{3,} \quad \underset{2(2)+1}{5,} \quad \underset{2(3)+1}{7,} \quad \underset{2(4)+1}{9,} \quad \ldots$$

EXAMPLES

Find the first four terms and the 50th term of the sequence.

- $a_n = n^2 - 10$

The first term is $a_1 = 1^2 - 10 = -9$; the second term is $a_2 = 2^2 - 10 = -6$; the third term is $a_3 = 3^2 - 10 = -1$; the fourth term is $a_4 = 4^2 - 10 = 6$; and the 50th term is $a_{50} = 50^2 - 10 = 2490$.

- $a_n = \frac{n-1}{n+1}$

$$a_1 = \frac{1-1}{1+2} = 0 \qquad a_2 = \frac{2-1}{2+1} = \frac{1}{3} \qquad a_3 = \frac{3-1}{3+1} = \frac{1}{2}$$

$$a_4 = \frac{4-1}{4+1} = \frac{3}{5} \qquad a_{50} = \frac{50-1}{50+1} = \frac{49}{51}$$

- $a_n = (-1)^n$

$$a_1 = (-1)^1 = -1 \qquad a_2 = (-1)^2 = 1 \qquad a_3 = (-1)^3 = -1$$

$$a_4 = (-1)^4 = 1 \qquad a_{50} = (-1)^{50} = 1$$

Finding the terms of a sequence is the same function evaluation we did earlier. Sequences are special kinds of functions whose domain is the natural numbers (instead of intervals of real numbers).

We can write the formulas for many sequences using the previous term. For example, the next term of the sequence 3, 5, 7, 9, ... can be found by adding 2 to the previous term. In other words, we could use the formula $a_n = a_{n-1} + 2$. This is a *recursive* formula. This formula is not of much use unless we know how to start. For this reason, the value of a_1 is usually given with recursively defined sequences. A complete recursive definition for this sequence is $a_n = a_{n-1} + 2$, $a_1 = 3$. Now we can compute the terms of the sequence.

$$3, \quad \underset{3+2}{5}, \quad \underset{5+2}{7}, \quad \underset{7+2}{9}, \quad \ldots$$

EXAMPLES

Find the first four terms of the sequence.

- $a_n = 3a_{n-1} + 5$, $a_1 = -4$

  Think of $3a_{n-1} + 5$ as "3 times the previous term plus 5."

$$a_1 = -4 \qquad a_2 = 3(-4) + 5 = -7$$

$$a_3 = 3(-7) + 5 = -16 \qquad a_4 = 3(-16) + 5 = -43$$

- $a_n = \frac{a_{n-1}}{a_{n-2}}$, $a_1 = 2$, $a_2 = 4$

  The terms of this sequence are found by taking the quotient of the previous two terms.

  $$a_1 = 2 \qquad a_2 = 4 \qquad a_3 = \frac{a_{3-1}}{a_{3-2}} = \frac{a_2}{a_1} = \frac{4}{2} = 2 \qquad a_4 = \frac{a_3}{a_2} = \frac{2}{4} = \frac{1}{2}$$

A famous recursively defined sequence is the Fibonacci Sequence. Entire books are written about it! The nth term of the Fibonacci Sequence is $a_n = a_{n-1} + a_{n-2}$ and $a_1 = 1$ and $a_2 = 1$. From the third term on, each term is the sum of the previous two terms. The first few terms are 1, 1, 2, 3, 5, 8, 13, ...

Instead of using a formula to describe a sequence, we might be given the first few terms. From these terms we should be able to see enough of a pattern to write a formula for the nth term.

EXAMPLES

Find the next term in the sequence.

- 2, 6, 18, 54, . . .

  The next term is 3(54) = 162.

- $1, \frac{1}{2}, \frac{1}{3}, \frac{1}{4}, \ldots$

  The next term is $\frac{1}{5}$

- $1, -2, 4, -8, 16, \ldots$

  The next term is $-2(16) = -32$.

  Find a formula for the nth term for the next four examples. Do not use a recursive definition.

- 3, 9, 27, 81, . . .

  $3 = 3^1,\ 9 = 3^2,\ 27 = 3^3,\ 81 = 3^4$

  The nth term is $a_n = 3^n$.

- $-2, -4, -6, -8, -10, \ldots$

  $-2 = -2(1),\ -4 = -2(2),\ -6 = -2(3),\ -8 = -2(4),\ -10 = -2(5)$

  The nth term is $a_n = -2n$.

- $-1, 4, -9, 16, -25, \ldots$

  $-1 = -1^2,\ 4 = 2^2,\ -9 = -3^2,\ 16 = 4^2,\ -25 = -5^2$

If we want the signs to alternate, we can use the factor $(-1)^n$ (if we want the odd-numbered terms to be negative) or $(-1)^{n+1}$ (if we want the even-numbered terms to be negative). The nth term of this sequence is $a_n = (-1)^n n^2$.

- $\frac{1}{2}, \frac{2}{3}, \frac{3}{4}, \frac{4}{5}, \ldots$

$$\frac{1}{2} = \frac{1}{1+1}, \quad \frac{2}{3} = \frac{2}{2+1}, \quad \frac{3}{4} = \frac{3}{3+1}, \quad \frac{4}{5} = \frac{4}{4+1}$$

The nth term is $a_n = \frac{n}{n+1}$.

There are times when we want to add the first n terms of a sequence. The sum

$$a_1 + a_2 + a_3 + \cdots + a_n$$

is called the nth partial sum of the sequence. Its notation is S_n.

$$S_1 = a_1 \qquad S_2 = a_1 + a_2$$

$$S_3 = a_1 + a_2 + a_3 \qquad S_4 = a_1 + a_2 + a_3 + a_4$$

Another common notation for the nth partial sum uses the capital Greek letter sigma, "Σ." This notation also makes use of a_n or a formula for a_n. "$\sum_{n=1}^{5} a_n$" means "add the a_ns beginning with a_1 and ending with a_5.

$$\sum_{n=1}^{5} a_n = a_1 + a_2 + a_3 + a_4 + a_5$$

The subscript n is called the *index of summation*. Other common indices are i, j, and k.

EXAMPLES

Write the sum.

- $\sum_{n=1}^{6} \frac{n^2}{4}$

$$\underset{\frac{1^2}{4}}{\frac{1}{4}} + \underset{\frac{2^2}{4}}{1} + \underset{\frac{3^2}{4}}{\frac{9}{4}} + \underset{\frac{4^2}{4}}{4} + \underset{\frac{5^2}{4}}{\frac{25}{4}} + \underset{\frac{6^2}{4}}{9}$$

- $\sum_{n=1}^{5}(-1)^{n+1}(3n-4)$

$$\underset{(-1)^{1+1}(3\cdot1-4)}{-1} \quad - \quad \underset{(-1)^{2+1}(3\cdot2-4)}{2} \quad + \quad \underset{(-1)^{3+1}(3\cdot3-4)}{5} \quad - \quad \underset{(-1)^{4+1}(3\cdot4-4)}{8} \quad + \quad \underset{(-1)^{5+1}(3\cdot5-4)}{11}$$

Write the sum using summation notation.

- $1+\frac{1}{2}+\frac{1}{3}+\frac{1}{4}+\cdots+\frac{1}{20}$

  This is the sum of the first 20 terms in a sequence, so we will begin by writing "$\sum_{n=1}^{20}$." The nth term of the sequence is $a_n=\frac{1}{n}$, and the summation notation for this sum is

$$\sum_{n=1}^{20}\frac{1}{n}.$$

- $2+4+6+8+10+12$

  This is the sum of the first six terms of the sequence whose nth term is $a_n=2n$. The summation notation is $\sum_{n=1}^{6}2n$.

- $\frac{1}{2}-\frac{1}{4}+\frac{1}{6}-\frac{1}{8}+\frac{1}{10}-\cdots+\frac{1}{18}$

  This is the sum of the first nine terms of the sequence whose nth term is $a_n=(-1)^{n+1}\frac{1}{2n}$. The summation notation is

$$\sum_{n=1}^{9}(-1)^{n+1}\frac{1}{2n}.$$

There are formulas for finding the nth partial sum for special sequences. Using these formulas, we can add many terms of a sequence with little work. We will learn the formulas for the sums of two important sequences, *arithmetic sequences* and *geometric sequences*, later.

PRACTICE

1. Find the first four terms and the 100th term of the sequence whose nth term is $a_n=\frac{2n-1}{n+1}$.
2. Find the first four terms and the 100th term of the sequence whose nth term is $a_n=(-1)^{n+1}\frac{n^2}{2}$.

3. Find the first four terms of the sequence whose nth term is $a_n = \sqrt{a_{n-1}}$ and $a_1 = 256$.

4. Without using a recursive definition, find the nth term for the sequence

$$10,\ 5,\ \frac{5}{2},\ \frac{5}{4},\ \frac{5}{8},\ \frac{5}{16},\ldots.$$

5. Without using a recursive definition, find the nth term for the sequence

$$\frac{0}{3},\ \frac{1}{4},\ \frac{2}{5},\ \frac{3}{6},\ \frac{4}{7},\ldots.$$

6. Write the sum for $\sum_{n=1}^{6} \frac{5}{2n}$.

7. Write the sum using summation notation.

$$\frac{1}{3} - \frac{1}{9} + \frac{1}{27} - \frac{1}{81} + \frac{1}{243}$$

SOLUTIONS

1. $a_1 = \dfrac{2(1)-1}{1+1} = \dfrac{1}{2}$ $\quad a_2 = \dfrac{2(2)-1}{2+1} = 1$ $\quad a_3 = \dfrac{2(3)-1}{3+1} = \dfrac{5}{4}$

   $a_4 = \dfrac{2(4)-1}{4+1} = \dfrac{7}{5}$ $\quad a_{100} = \dfrac{2(100)-1}{100+1} = \dfrac{199}{101}$

2. $a_1 = (-1)^{1+1}\dfrac{1^2}{2} = \dfrac{1}{2}$ $\quad a_2 = (-1)^{2+1}\dfrac{2^2}{2} = -2$

   $a_3 = (-1)^{3+1}\dfrac{3^2}{2} = \dfrac{9}{2}$ $\quad a_4 = (-1)^{4+1}\dfrac{4^2}{2} = -8$

   $a_{100} = (-1)^{100+1}\dfrac{100^2}{2} = -5000$

3. $a_1 = 256$ $\quad a_2 = \sqrt{256} = 16$

   $a_3 = \sqrt{16} = 4$ $\quad a_4 = \sqrt{4} = 2$

4. $a_n = 20(\frac{1}{2})^n$ or $a_n = 10(\frac{1}{2})^{n-1}$

5. $a_n = \dfrac{n-1}{n+2}$

6. $\frac{5}{2}+\frac{5}{4}+\frac{5}{6}+\frac{5}{8}+\frac{5}{10}+\frac{5}{12}$

7. $\sum_{n=1}^{5}(-1)^{n+1}\left(\frac{1}{3}\right)^n$ or $\sum_{n=1}^{5}(-1)^{n+1}\frac{1}{3^n}$

Arithmetic Sequences

A term in an arithmetic sequence is computed by adding a fixed number to the previous term. For example, 3, 7, 11, 15, 19, ... is an arithmetic sequence because we can add 4 to any term to find the following term. We can define the nth term recursively as $a_n = a_{n-1} + d$ or, in more general terms, $a_n = a_1 + (n-1)d$. In the sequence above, $a_1 = 3$ and $d = 4$.

EXAMPLES

Find the first four terms and the 100th term.

- $a_n = 28 + (n-1)1.5$

$$a_1 = 28 \qquad a_2 = 28 + (2-1)1.5 = 29.5$$

$$a_3 = 28 + (3-1)1.5 = 31 \qquad a_4 = 28 + (4-1)1.5 = 32.5$$

$$a_{100} = 28 + (100-1)1.5 = 176.5$$

- $a_n = -2 + (n-1)(-6)$

$$a_1 = -2 \qquad a_2 = -2 + (2-1)(-6) = -8$$

$$a_3 = -2 + (3-1)(-6) = -14 \qquad a_4 = -2 + (4-1)(-6) = -20$$

$$a_{100} = -2 + (100-1)(-6) = -596$$

When asked whether or not a sequence is arithmetic, we will find the difference between consecutive terms. If the difference is the same, the sequence is arithmetic.

EXAMPLES

Determine if the sequence is arithmetic. If it is, find the common difference.

- $-8, -1, 6, 13, 20, \ldots$

  $20 - 13 = 7,\ 13 - 6 = 7,\ 6 - (-1) = 7,\ -1 - (-8) = 7$

The sequence is arithmetic. The common difference is 7.

- $29, 17, 5, -7, -19, \ldots$

  $-19-(-7)=-12,\ -7-5=-12,\ 5-17=-12,\ 17-29=-12$

  The sequence is arithmetic, and the common difference is -12.

- $\frac{5}{3}, \frac{5}{6}, \frac{5}{12}, \frac{5}{24}, \ldots$

  $\frac{5}{24}-\frac{5}{12}=-\frac{5}{24},\ \frac{5}{12}-\frac{5}{6}=-\frac{5}{12}$

  Because the differences are not the same, the sequence is not arithmetic.

We can find any term in an arithmetic sequence if we know either one term and the common difference or two terms. We need to use the formula $a_n=a_1+(n-1)d$ and, if necessary, a little algebra. For example, if we are told the common difference is 6 and the tenth term is 141, then we can put $a_n=141$, $n=10$, and $d=6$ in the formula to find a_1.

$$141=a_1+(10-1)6$$

$$87=a_1$$

The nth term is $a_n=87+(n-1)6$.

EXAMPLES

Find the nth term for the arithmetic sequence.

- The common difference is 2/3 and the seventh term is -10.
  Using $d=\frac{2}{3}$, $n=7$, and $a_n=-10$, the formula $a_n=a_1+(n-1)d$ becomes $-10=a_1+(7-1)\frac{2}{3}$.

  $$-10=a_1+(7-1)\frac{2}{3}$$

  $$-10=a_1+4$$

  $$-14=a_1$$

  The nth term is $a_n=-14+(n-1)\frac{2}{3}$.

- The twelfth term is 8, and the twentieth term is 32.
  The information gives us a system of two equations with two variables. In this example and the rest of the problems in this section, we will add -1

times the first equation to the second. Substitution and matrices would work, too. The equations are $8 = a_1 + (12 - 1)d$ and $32 = a_1 + (20 - 1)d$.

$$\begin{aligned}
-a_1 - 11d &= -8 \\
\underline{a_1 + 19d} &\underline{= 32} \\
8d &= 24 \\
d &= 3 \\
a_1 + 11(3) &= 8 \qquad \text{Let } d = 3 \text{ in } a_1 + 11d = 8 \\
a_1 &= -25
\end{aligned}$$

The nth term is $a_n = -25 + (n - 1)3$.

- The eighth term is 4, and the twentieth term is -38.
  The information in these two terms gives us the system of equations $4 = a_1 + (8 - 1)d$ and $-38 = a_1 + (20 - 1)d$.

$$\begin{aligned}
-a_1 - 7d &= -4 \\
\underline{a_1 + 19d} &\underline{= -38} \\
12d &= -42 \\
d &= -\frac{7}{2} \\
a_1 + 7\left(-\frac{7}{2}\right) &= 4 \qquad \text{Let } d = -\frac{7}{2} \text{ in } a_1 + 7d = 4 \\
a_1 &= \frac{57}{2}
\end{aligned}$$

The nth term is $a_1 = \frac{57}{2} + (n - 1)(-\frac{7}{2})$.

We can add the first n terms of an arithmetic sequence using one of the following two formulas.

$$S_n = \frac{n}{2}(a_1 + a_n) \quad \text{or} \quad S_n = \frac{n}{2}[2a_1 + (n - 1)d]$$

We will use the first formula if we know all of a_1, a_n, and n, and the second if we do not know a_n.

EXAMPLES

- Find the sum.

$$2 + \frac{13}{5} + \frac{16}{5} + \frac{19}{5} + \frac{22}{5} + 5$$

$a_1 = 2,\ a_6 = 5$, and $n = 6$ (because there are six terms)

$$2 + \frac{13}{5} + \frac{16}{5} + \frac{19}{5} + \frac{22}{5} + 5 = \frac{6}{2}(2 + 5) = 21$$

- Find the sum of the first 20 terms of the sequence $-5,\ -1,\ 3,\ 7,\ 11, \ldots$.
$a_1 = -5,\ d = 4$, and $n = 20$.

$$S_{20} = \frac{20}{2}[2(-5) + (20 - 1)4] = 660$$

- $6 + (-2) + (-10) + (-18) + \cdots + (-58)$
We know $a_1 = 6,\ d = -8$ and $a_n = -58$ but not n. We can find n by solving $-58 = 6 + (n - 1)(-8)$.

$$-58 = 6 + (n - 1)(-8)$$

$$-64 = -8(n - 1)$$

$$8 = n - 1$$

$$9 = n$$

$$6 + (-2) + (-10) + (-18) + \cdots + (-58) = \frac{9}{2}[6 + (-58)] = -234$$

- Find the sum of the first thirty terms of the arithmetic sequence whose fifth term is 19 and whose tenth term is 31.5.
In order to use the second sum formula, we need to find a_1 and d. If we were to use the first formula, we would have to find a_{30}, which is a little more work. Because $a_5 = 19$ and $a_{10} = 31.5$, we have the system of equations $19 = a_1 + (5 - 1)d$ and $31.5 = a_1 + (10 - 1)d$.

$$-a_1 - 4d = -19$$

$$\underline{a_1 + 9d = 31.5}$$

$$5d = 12.5$$

$$d = 2.5$$

$$a_1 + 4(2.5) = 19 \qquad \text{Let } d = 2.5 \text{ in } a_1 + 4d = 19$$

$$a_1 = 9$$

$$S_{30} = \frac{30}{2}[2(9) + (30 - 1)(2.5)] = 1357.5$$

PRACTICE

1. Find the first four terms and the 40th term of the arithmetic sequence whose nth term is $a_n = 14 + (n - 1)4$.
2. Determine if the sequence $0.03, 0.33, 0.63, 0.93, \ldots$ is arithmetic.
3. Determine if the sequence $0.4, 0.04, 0.004, 0.0004, \ldots$ is arithmetic.
4. Find the nth term of the arithmetic sequence whose first term is 16 and whose ninth term is 54.
5. Find the nth term of the arithmetic sequence whose sixth term is 12 and whose tenth term is 36.
6. Compute the sum.

$$-8 + \left(-\frac{35}{4}\right) + \left(-\frac{38}{4}\right) + \left(-\frac{41}{4}\right) + (-11) + \left(-\frac{47}{4}\right)$$

7. Compute the sum. $10 + 17 + 24 + 31 + \cdots + 108$
8. Find the sum of the first twelve terms of the arithmetic sequence whose fourth term is 8 and whose tenth term is 56.

SOLUTIONS

1. $a_1 = 14$, $a_2 = 14 + (2 - 1)4 = 18$, $a_3 = 14 + (3 - 1)4 = 22$, $a_4 = 14 + (4 - 1)4 = 26$ and $a_{40} = 14 + (40 - 1)4 = 170$.
2. $0.93 - 0.63 = 0.3$, $0.63 - 0.33 = 0.3$, $0.33 - 0.03 = 0.3$
   The differences are the same, so the sequence is arithmetic.
3. $0.0004 - 0.004 = -0.0036$, $0.004 - 0.04 = -0.036$
   The differences are not the same, so the sequence is not arithmetic.
4. Because $a_1 = 16$, we have $a_n = 16 + (n - 1)d$. Using $a_9 = 54$ in this formula, we have $54 = 16 + (9 - 1)d$. Solving this equation for d gives us $d = \frac{19}{4}$. The nth term is $a_n = 16 + (n - 1)\frac{19}{4}$.

5. From the information in the problem, we have the system $12 = a_1 + (6-1)d$ and $36 = a_1 + (10-1)d$.

$$-a_1 - 5d = -12$$
$$\underline{a_1 + 9d = 36}$$
$$4d = 24$$
$$d = 6$$
$$a_1 + 5(6) = 12 \qquad \text{Let } d = 6 \text{ in } a_1 + 5d = 12$$
$$a_1 = -18$$

The nth term is $a_n = -18 + (n-1)6$.

6. $a_1 = -8$, $a_6 = -\frac{47}{4}$, and $n = 6$. $S_n = \frac{n}{2}(a_1 + a_n)$ becomes $S_6 = \frac{6}{2}(-8 + (-\frac{47}{4})) = -\frac{237}{4}$.

7. $a_1 = 10$, $d = 7$, and $a_n = 108$. We can find n by solving $108 = 10 + (n-1)7$. This gives us $n = 15$. $S_{15} = \frac{15}{2}(10 + 108) = 885$.

8. We will find a_1 and d so that we can use the formula $S_n = \frac{n}{2}[2a_1 + (n-1)d]$. The information in the problem gives the system $8 = a_1 + (4-1)d$ and $56 = a_1 + (10-1)d$.

$$-a_1 - 3d = -8$$
$$\underline{a_1 + 9d = 56}$$
$$6d = 48$$
$$d = 8$$
$$a_1 + 3(8) = 8 \qquad \text{Let } d = 8 \text{ in } a_1 + 3d = 8$$
$$a_1 = -16$$

$$S_{12} = \frac{12}{2}[2(-16) + (12-1)8] = 336$$

Geometric Sequences

In an arithmetic sequence, the difference of any two consecutive terms is the same, and in a geometric sequence, the quotient of any two consecutive terms is the same. A term in a geometric sequence can be found by multiplying the previous term by a fixed number. For example, the next term in the sequence $1, 2, 4, 8, 16, \ldots$ is 2(16)=32, and the term after that is 2(32)=64. This fixed number is called the *common ratio*. We can define the nth term of a geometric sequence recursively by $a_n = ra_{n-1}$. The general formula is $a_n = a_1 r^{n-1}$.

EXAMPLES

- Determine if the sequence 5, 15, 45, 135, 405, ... is geometric.

  We need to see if the ratio of each consecutive pair of numbers is the same.

  $$\frac{405}{135} = 3, \quad \frac{135}{45} = 3, \quad \frac{45}{15} = 3, \quad \text{and} \quad \frac{15}{5} = 3$$

  The ratio is the same number, so the sequence is geometric.

- Determine if the sequence $-8,\ 4,\ -2,\ 1,\ -\frac{1}{2}, \ldots$ is geometric.

  $$\frac{-1/2}{1} = -\frac{1}{2}, \quad \frac{1}{-2} = -\frac{1}{2}, \quad \frac{-2}{4} = -\frac{1}{2}, \quad \text{and} \quad \frac{4}{-8} = -\frac{1}{2}$$

  The ratio is the same number, so the sequence is geometric.

- Determine if the sequence 2430, 729, 240.57, 80.10981, ... is geometric.

  $$\frac{80.10981}{240.57} = 0.333 \quad \text{and} \quad \frac{240.57}{729} = 0.33$$

  The ratios are different, so this is not a geometric sequence.

- Find the first four terms and the tenth term of the sequence $a_n = \frac{1}{100}(-5)^{n-1}$.

  $$a_1 = \frac{1}{100} \qquad a_2 = \frac{1}{100}(-5)^{2-1} = -\frac{1}{20}$$

  $$a_3 = \frac{1}{100}(-5)^{3-1} = \frac{1}{4} \qquad a_4 = \frac{1}{100}(-5)^{4-1} = -\frac{5}{4}$$

  $$a_{10} = \frac{1}{100}(-5)^{10-1} = -\frac{78125}{4}$$

We can find the nth term of a geometric sequence by either knowing one term and the common ratio or by knowing two terms. This is similar to what we did to find the nth term of an arithmetic sequence.

EXAMPLES

Find the nth term of the geometric sequence.

- The common ratio is 3 and the fourth term is 54.
  $a_4 = 54$ and $r = 3$, so $a_n = a_1 r^{n-1}$ becomes $54 = a_1 3^{4-1}$. This gives us $a_1 = 2$. The nth term is $a_n = 2(3)^{n-1}$.

- The third term is 320, and the fifth term is 204.8.
  $a_3 = 320$ and $a_5 = 204.8$ give us the system of equations $320 = a_1r^{3-1}$ and $204.8 = a_1r^{5-1}$. Elimination by addition will not work for the systems in this section, so we will use substitution. Solving for a_1 in $a_1r^2 = 320$ gives us $a_1 = 320/r^2$. Substituting this in $a_1r^4 = 204.8$ gives us the following.

$$a_1r^4 = 204.8$$
$$\frac{320}{r^2} \cdot r^4 = 204.8$$
$$320r^2 = 204.8$$
$$r^2 = 0.64$$
$$r = \pm 0.8$$

  There are two geometric sequences whose third term is 320 and whose fifth term is 204.8, one has a common ratio of 0.8 and the other, -0.8. a_1 for both the sequences is the same.

$$a_1 = \frac{320}{0.8^2} = 500 \quad \text{and} \quad a_1 = \frac{320}{(-0.8)^2} = 500$$

  The nth term for one sequence is $a_n = 500(0.8)^{n-1}$, and the other is $a_n = 500(-0.8)^{n-1}$.

- The third term is 20, and the sixth term is 81.92.
  From $a_3 = 20$ and $a_6 = 81.92$ we have the system of equations $20 = a_1r^{3-1}$ and $81.92 = a_1r^{6-1}$. We will solve for a_1 in $20 = a_1r^2$. Now we will substitute $a_1 = 20/r^2$ for a_1 in $81.92 = a_1r^5$.

$$81.92 = a_1r^5$$
$$81.92 = \frac{20}{r^2} \cdot r^5$$
$$81.92 = 20r^3$$
$$4.096 = r^3$$
$$\sqrt[3]{4.096} = r$$
$$1.6 = r$$
$$a_1 = \frac{20}{1.6^2} = 7.8125$$

  The nth term is $a_n = 7.8125(1.6)^{n-1}$.

We can add the first n terms of a geometric sequence using the following formula (except for $r = 1$).

$$S_n = a_1 \frac{1 - r^n}{1 - r}$$

EXAMPLES

- Find the sum of the first five terms of the geometric sequence whose nth term is $a_n = 3(2)^{n-1}$, $a_1 = 3$ and $r = 2$.

$$S_5 = 3 \cdot \frac{1 - 2^5}{1 - 2} = 3 \cdot \frac{-31}{-1} = 93$$

- Compute $16 + 8 + 4 + 2 + 1 + \frac{1}{2} + \frac{1}{4} + \frac{1}{8} + \frac{1}{16}$

$a_1 = 16, \quad r = \dfrac{1}{2} \quad$ and $\quad n = 9.$

$$S_9 = 16 \frac{1 - (\frac{1}{2})^9}{1 - \frac{1}{2}} = 16 \frac{\frac{512}{512} - \frac{1}{512}}{\frac{1}{2}}$$

$$= 16 \frac{\frac{511}{512}}{\frac{1}{2}} = 16 \left[\frac{511}{512} \div \frac{1}{2} \right] = 16 \left[\frac{511}{512} \cdot 2 \right] = \frac{511}{16}$$

- Find the sum of the first five terms of the geometric sequence whose fourth term is 1.3824 and whose seventh term is 2.3887872.
  We need to find a_1 and r. The terms $a_4 = 1.3824$ and $a_7 = 2.3887872$ give us the system of equations $1.3824 = a_1 r^3$ and $2.3887872 = a_1 r^6$. We will solve for a_1 in the first equation and substitute this for a_1 in the second equation.

$$a_1 = \frac{1.3824}{r^3}$$

$$2.3887872 = a_1 r^6$$

$$2.3887872 = \frac{1.3824}{r^3} r^6$$

$$2.3887872 = 1.3824 r^3$$

$$1.728 = r^3$$

$$\sqrt[3]{1.728} = r$$

$$1.2 = r$$

$$a_1 = \frac{1.3824}{1.2^3} = 0.8$$

We have enough information to compute S_5.

$$S_5 = 0.8\frac{1 - 1.2^5}{1 - 1.2} = 5.95328$$

- $\sum_{i=1}^{6} 6.4(1.5)^{i-1}$

  We are adding the first six terms of the geometric sequence whose nth term is $a_n = 6.4(1.5)^{n-1}$.

$$\sum_{i=1}^{6} 6.4(1.5)^{i-1} = S_6 = 6.4\frac{1 - 1.5^6}{1 - 1.5} = 133$$

- $\sum_{k=0}^{7} 2(3)^{k-1}$

  This problem is tricky because the sum begins with $k=0$ instead of $k=1$. These terms are the first *eight* terms of the geometric sequence $\frac{2}{3}$, 2, 6, 18, 54, 162, 486, 1458, ... Now we can see that $n = 8$, $a_1 = \frac{2}{3}$ and $r = 3$.

$$\sum_{k=0}^{7} 2(3)^{k-1} = S_8 = \frac{2}{3} \cdot \frac{1 - 3^8}{1 - 3} = \frac{6560}{3}$$

- $54 + 18 + 6 + 2 + \frac{2}{3} + \cdots + \frac{2}{81}$

  We have $a_1 = 54$ and $r = \frac{1}{3}$. We need n for $a_n = \frac{2}{81}$.

$$\frac{2}{81} = 54\left(\frac{1}{3}\right)^{n-1}$$

$$\frac{1}{2187} = \left(\frac{1}{3}\right)^{n-1}$$

Because $3^7 = 2187$, $n - 1 = 7$, so $n = 8$.

$$54 + 18 + 6 + 2 + \frac{2}{3} + \cdots + \frac{2}{81} = 54\left(\frac{1-(\frac{1}{3})^8}{1-\frac{1}{3}}\right) = 54\left(\frac{\frac{3^8}{3^8} - \frac{1}{3^8}}{\frac{2}{3}}\right)$$

$$= 54\left(\frac{\frac{3^8-1}{3^8}}{\frac{2}{3}}\right) = 54\left(\frac{6560}{6561} \div \frac{2}{3}\right)$$

$$= 54\left(\frac{6560}{6561} \cdot \frac{3}{2}\right) = \frac{6560}{81}$$

When the common ratio is small enough ($-1 < r < 1$ and $r \neq 0$), the sum of *all* terms in a geometric sequence is a number. In the finite sum $S_n = a_1\frac{1-r^n}{1-r}$, r^n is very small when the ratio is a fraction, so $1 - r^n$ is very close to 1. Using this fact and calculus techniques (usually learned in a later calculus course), it can be shown that the sum of all terms of this kind of geometric sequence is

$$S = a_1\frac{1}{1-r}.$$

The only difference between the infinite sum formula and the partial sum formula is that $1 - r^n$ is replaced by 1. If n is large enough, there is very little difference between the partial sum and the entire sum. We will compare the sum of the first 20 terms of the sequence whose nth term is $a_n = (\frac{1}{2})^{n-1}$ with the sum of all terms.

$$S_{20} = \sum_{n=1}^{20} 1 \cdot \left(\frac{1}{2}\right)^{n-1} = 1 \cdot \frac{1-(\frac{1}{2})^{20}}{1-\frac{1}{2}} \approx 1.999998093 \quad \text{and} \quad S = 1 \cdot \frac{1}{1-\frac{1}{2}} = 2$$

EXAMPLES

- $\sum_{i=1}^{\infty} 6\left(\frac{2}{3}\right)^{i-1}$

  $a_1 = 6,\ r = \frac{2}{3}$

$$\sum_{i=1}^{\infty} 6\left(\frac{2}{3}\right)^{i-1} = S = 6 \cdot \frac{1}{1-\frac{2}{3}} = 6 \cdot \frac{1}{\frac{1}{3}} = 6\left[1 \div \frac{1}{3}\right] = 6\left[1 \cdot \frac{3}{1}\right] = 18$$

- $\sum_{k=0}^{\infty} 15\left(\frac{3}{4}\right)^{k-1}$

  We need to be careful with this sum because the sum begins with $k = 0$ instead of $k = 1$. This means that a_1 is not 15 but

$$a_1 = 15\left(\frac{3}{4}\right)^{0-1} = 15\left(\frac{3}{4}\right)^{-1} = 15\left(\frac{4}{3}\right) = 20.$$

  The common ratio is $\frac{3}{4}$.

$$\sum_{k=0}^{\infty} 15\left(\frac{3}{4}\right)^{k-1} = S = 20 \cdot \frac{1}{1-\frac{3}{4}} = 20 \cdot \frac{1}{\frac{1}{4}} = 20 \cdot 4 = 80$$

PRACTICE

1. What term comes after 18 in the sequence $\frac{2}{9}$, $\frac{2}{3}$, 2, 6, 18, ...?
2. Find the first four terms and the tenth term of the geometric sequence whose nth term is $a_n = -2(4)^{n-1}$.
3. Determine if the sequence 900, 90, 9, 0.9, 0.09, ... is geometric.
4. Determine if the sequence 9, 99, 999, 9999, ... is geometric.
5. Find the nth term of the geometric sequence(s) whose first term is 9 and whose fifth term is $\frac{729}{16}$.
6. Find the nth term of the geometric sequence whose common ratio is -3 and whose sixth term is -1701.
7. Find the nth term of the geometric sequence whose third term is 1 and whose sixth term is $\frac{27}{8}$.
8. Compute the sum.

$$\frac{3}{4} + \frac{3}{8} + \frac{3}{16} + \cdots + \frac{3}{256}$$

9. $\sum_{i=1}^{\infty} \frac{3}{4}\left(\frac{1}{2}\right)^{i-1}$
10. $\sum_{n=0}^{\infty} -4\left(\frac{3}{5}\right)^{n-1}$

SOLUTIONS

1. $3(18) = 54$
2. $a_1 = -2,\ a_2 = -8,\ a_3 = -32,\ a_4 = -128$ and $a_{10} = -524,288$
3. The sequence is geometric because the following ratios are the same.

$$\frac{0.09}{0.9} = 0.1, \quad \frac{0.9}{9} = 0.1, \quad \frac{9}{90} = 0.1, \quad \frac{90}{900} = 0.1$$

4. The sequence is not geometric because the ratios are not the same.

$$\frac{9999}{999} = \frac{1111}{111} \text{ and } \frac{999}{99} = \frac{111}{11}$$

5. Because the fifth term of the sequence is $\frac{729}{16}$, we have the equation $\frac{729}{16} = 9r^{5-1}$. Once we have solved this equation for r, we will be done.

$$\frac{729}{16} = 9r^4$$

$$\frac{1}{9} \cdot \frac{729}{16} = r^4$$

$$\frac{81}{16} = r^4$$

$$\pm\frac{3}{2} = r \qquad \sqrt[4]{\frac{81}{16}} = \frac{3}{2}$$

There are two sequences. The nth term for one of them is $a_n = 9(\frac{3}{2})^{n-1}$ and the other is $a_n = 9(-\frac{3}{2})^{n-1}$.

6. The sixth term is -1701 and $r = -3$, which gives us the equation $-1701 = a_1(-3)^{6-1}$.

$$-1701 = a_1(-3)^5$$

$$-1701 = -243a_1$$

$$7 = a_1$$

The nth term is $a_n = 7(-3)^{n-1}$.

7. The third term is 1 and the sixth term is $\frac{27}{8}$, which gives us the system of equations $1 = a_1 r^{3-1}$ and $\frac{27}{8} = a_1 r^{6-1}$. Solving $1 = a_1 r^2$ for a_1, we get

$a_1 = 1/r^2$. We will substitute this in $\frac{27}{8} = a_1 r^5$.

$$\frac{27}{8} = \left(\frac{1}{r^2}\right) r^5$$

$$\frac{27}{8} = r^3$$

$$\sqrt[3]{\frac{27}{8}} = r$$

$$\frac{3}{2} = r$$

$$a_1 = \frac{1}{(\frac{3}{2})^2} = \left(\frac{2}{3}\right)^2 = \frac{4}{9}$$

The nth term is $a_n = \frac{4}{9}(\frac{3}{2})^{n-1}$.

8. $a_1 = \frac{3}{4}$ and $r = \frac{1}{2}$. We know $a_n = \frac{3}{256}$ but we need n. We will solve $\frac{3}{256} = \frac{3}{4}(\frac{1}{2})^{n-1}$ for n.

$$\frac{3}{256} = \frac{3}{4}\left(\frac{1}{2}\right)^{n-1}$$

$$\frac{4}{3} \cdot \frac{3}{256} = \left(\frac{1}{2}\right)^{n-1}$$

$$\frac{1}{64} = \left(\frac{1}{2}\right)^{n-1}$$

Because $2^6 = 64$, $n - 1 = 6$, so $n = 7$. Now we can find the sum.

$$S_7 = \frac{3}{4} \cdot \frac{1 - (\frac{1}{2})^7}{1 - \frac{1}{2}} = \frac{3}{4} \cdot \frac{\frac{128-1}{128}}{\frac{1}{2}} = \frac{3}{4}\left(\frac{127}{128} \div \frac{1}{2}\right)$$

$$= \frac{3}{4}\left(\frac{127}{128} \cdot \frac{2}{1}\right) = \frac{381}{256}$$

9. $a_1 = \frac{3}{4}$ and $r = \frac{1}{2}$. This is all we need for the infinite sum formula.

$$S = \frac{3}{4} \cdot \frac{1}{1 - \frac{1}{2}} = \frac{3}{4} \cdot \frac{1}{\frac{1}{2}} = \frac{3}{4}\left(1 \div \frac{1}{2}\right) = \frac{3}{4} \cdot (1 \cdot 2) = \frac{3}{2}$$

10. a_1 is not -4 because the sum begins at $n = 0$ instead of $n = 1$.

$$a_1 = -4\left(\frac{3}{5}\right)^{0-1} = -4\left(\frac{3}{5}\right)^{-1} = -4\left(\frac{5}{3}\right) = -\frac{20}{3}$$

Now we can add all of the terms of the geometric sequence whose nth term is $a_n = -\frac{20}{3}(\frac{3}{5})^{n-1}$.

$$S = -\frac{20}{3} \cdot \frac{1}{1-\frac{3}{5}} = -\frac{20}{3} \cdot \frac{1}{\frac{2}{5}} = -\frac{20}{3}\left(1 \div \frac{2}{5}\right) = -\frac{20}{3}\left(1 \cdot \frac{5}{2}\right) = -\frac{50}{3}$$

When regular payments are made to a savings account or to a lottery winner, the annual balances act like terms in a geometric sequence. The common ratio is either $1 + i$ (for savings payments) or $(1 + i)^{-1}$ (for lottery payments), where i is the interest rate per payment period. We learned in Chapter 9 that if we leave P dollars in an account, earning annual interest r, compounded n times per year, for t years, then this will grow to A dollars where $A = P(1 + r/n)^{nt}$. (This is why i replaces r/n.)

We will see what happens to the balance of an account if \$2000 is deposited on January 1 every year for 5 years, earning 10% per year, compounded annually. The first \$2000 will earn interest for the entire 5 years, so it will grow to 2000 $(1 + 0.10/1)^5 = 2000(1.10)^5$. The second \$2000 will earn interest for 4 years, so it will grow to $2000(1.10)^4$. The third \$2000 will earn interest for 3 years, so it will grow to $2000(1.10)^3$. The fourth \$2000 will earn interest for 2 years, so it will grow to $2000(1.10)^2$. And the fifth \$2000 will earn interest for 1 year, so it will grow to $2000(1.10)^1$. The balance after five years is

$$2000(1.10)^5 + 2000(1.10)^4 + 2000(1.10)^3 + 2000(1.10)^2 + 2000(1.10)^1.$$

This is the sum of the first five terms of the geometric sequence whose nth term is $a_n = 2000(1.10)^n$. If we want to use the partial sum formula, we need to rewrite the nth term in the form $a_n = a_1 r^{n-1}$. We will use exponent properties to change $2000(1.10)^n$ to $a_1(1.10)^{n-1}$. We will also use the fact that $n = 1 + n - 1$.

$$2000(1.10)^n = 2000(1.10)^{1+n-1} = 2000(1.10)^1(1.10)^{n-1}$$

$$= [2000(1.10)](1.10)^{n-1} = 2200(1.10)^{n-1}$$

Now we can use the partial sum formula.

$$S_5 = 2200 \cdot \frac{1 - 1.10^5}{1 - 1.10} = 13,431.22$$

The balance in the account will be \$13,431.22.

When a lottery winner wins a \$1,000,000 jackpot, the money is likely to be paid out in \$50,000 annual payments for 20 years. Some states allow the winner to take the *cash value* as a lump sum payment instead. The cash value is the present value of \$1,000,000 to be paid in annual payments over 20 years. The formula for the present value of A dollars, due in t years, earning annual interest r, compounded n times per year is $A(1 + r/n)^{-nt}$. Assume that the money is expected to earn 5% per year. Then the cash value of the jackpot will need to be enough money so that at the beginning of the year (for a payment at the end of the year), they have $50{,}000(1.05)^{-1}$. For a payment at the end of two years, they need $50{,}000(1.05)^{-2}$; at the end of three years, they need $50{,}000(1.05)^{-3}$, and so on until they reach the last payment after 20 years, $50{,}000(1.05)^{-20}$. In other words, the cash value of a \$1,000,000 jackpot with a 20-year payout (assuming 5% interest) is

$$50{,}000(1.05)^{-1} + 50{,}000(1.05)^{-2} + 50{,}000(1.05)^{-3} + \cdots + 50{,}000(1.05)^{-20}.$$

This is the sum of the first 20 terms of the geometric sequence whose nth term is $a_n = 50{,}000(1.05)^{-n}$. We need to use exponent properties to rewrite the nth term in the form $a_n = a_1 r^{n-1}$. We will use the fact that $-n = -n - 1 + 1$ and the exponent facts that $x^{m+n} = x^m x^n$ and $x^{mn} = (x^m)^n$.

$$1.05^{-n} = 1.05^{-n+1-1} = 1.05^{-1} \cdot 1.05^{-n+1} = 1.05^{-1} \cdot 1.05^{-1(n-1)}$$

$$= 1.05^{-1} \cdot (1.05^{-1})^{n-1}$$

Now the nth term can be written as $a_n = [50{,}000(1.05)^{-1}](1.05^{-1})^{n-1}$, where $a_1 = 50{,}000(1.05^{-1})$. Now we can use the partial sum formula.

$$S_{20} = [50{,}000(1.05^{-1})] \cdot \frac{1 - (1.05^{-1})^{20}}{1 - 1.05^{-1}} \approx 623{,}110.51$$

The cash value is \$623,110.51.

CHAPTER 14 REVIEW

1. What term comes next in the sequence?

$$\frac{2}{5}, \frac{3}{6}, \frac{4}{7}, \frac{5}{8}, \frac{6}{9}, \ldots$$

(a) $\frac{7}{10}$ (b) $\frac{7}{11}$ (c) $\frac{8}{11}$ (d) $\frac{8}{10}$

2. What is the fourth term of the sequence whose nth term is $a_n = (-1)^{n+1}(\frac{2}{3})^n$?
   (a) $\frac{16}{81}$ (b) $-\frac{16}{81}$ (c) $\frac{8}{3}$ (d) $-\frac{8}{3}$

3. The terms in the sequence 6, 2, −4, −6, −2, 4, ... can be found using which formula?
   (a) $a_n = a_{n-2} - a_{n-1}$, $a_1 = 6$ and $a_2 = 2$
   (b) $a_n = 6 + (n-1)4$
   (c) $a_n = a_{n-1} - a_{n-2}$, $a_1 = 6$ and $a_2 = 2$
   (d) There is no formula that works.

4. Is the sequence in Problem 3 arithmetic, geometric, or neither?
   (a) Arithmetic
   (b) Geometric
   (c) Neither
   (d) There are not enough terms to tell.

5. Is the sequence $\frac{3}{4}, \frac{1}{4}, \frac{1}{12}, \frac{1}{36}, \ldots$ arithmetic, geometric, or neither?
   (a) Arithmetic
   (b) Geometric
   (c) Neither
   (d) There are not enough terms to tell.

6. What is the third term of the arithmetic sequence whose 17th terms is 9 and whose 21st term is 12?
   (a) $-\frac{3}{2}$ (b) $-\frac{4}{5}$ (c) $\frac{5}{2}$ (d) $-\frac{2}{5}$

7. What is the eighth term of the geometric sequence whose third term is $\frac{5}{4}$ and whose sixth term is 10?
   (a) 36 (b) 40 (c) 45 (d) 49

8. Find the sum.

$$-\frac{2}{3} + \frac{5}{6} + \frac{7}{3} + \frac{23}{6} + \cdots + \frac{59}{6}$$

   (a) $\frac{116}{3}$ (b) $\frac{110}{3}$ (c) $\frac{58}{3}$
   (d) Too many terms are missing to find the sum.

9. Find the sum.

$$\frac{4}{9}+\frac{2}{9}+\frac{1}{9}+\frac{1}{18}+\cdots+\frac{1}{144}$$

(a) $\frac{127}{36}$ (b) $\frac{127}{81}$ (c) $\frac{127}{144}$

(d) Too many terms are missing to find the sum.

10. Find the sum.

$$\sum_{i=1}^{\infty} 5\left(\frac{3}{5}\right)^{i-1}$$

(a) $\frac{25}{2}$ (b) $\frac{75}{2}$ (c) $\frac{5}{2}$

(d) There are too many numbers to add.

SOLUTIONS

1. A	2. B	3. C	4. C	5. B
6. A	7. B	8. B	9. C	10. A

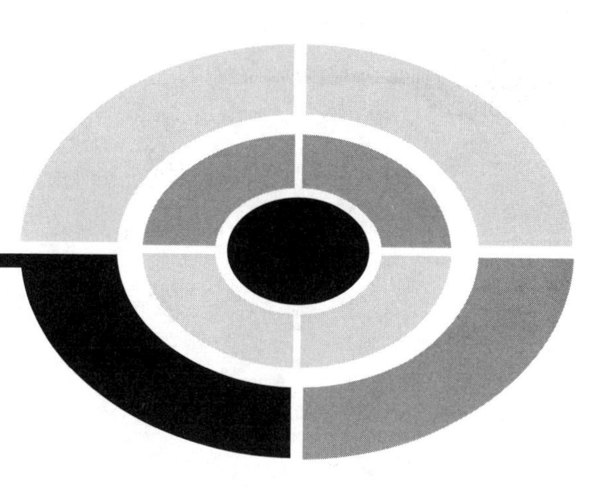

Appendix

Solving Equations and Inequalities

Using algebra to solve equations and inequalities is important in precalculus and calculus. Usually the solution to an equation is a number or numbers. Sometimes, the solution to an equation is simply the equation written another way. To solve for x means to have x, and x only, on one side of the equation. The equation $x = \frac{y-5}{y^2+1}$ is solved for x but $x = \frac{y-5}{x^2+1}$ is not solved for x because x is on both sides of the equation. Solving for x when the equation contains more than one variable is very much like solving for x when the equation has only one variable. We move quantities from one side of the equation by adding, subtracting, multiplying, and dividing.

- Solve for x in the equation $a(x+4) - 2a(x-1) = 5(a+x)$.

$a(x+4) - 2a(x-1) = 5(a+x)$ Simplify both sides of the equation.

$ax + 4a - 2ax + 2a = 5a + 5x$

$ax - 2ax + 6a = 5a + 5x$ Move x terms to one side of the equation.

$$ax - 2ax - 5x + 6a = 5a \quad \text{Move terms without } x \text{ to the other side.}$$

$$ax - 2ax - 5x = -6a + 5a \quad \text{Simplify both sides.}$$

$$-ax - 5x = -a \quad \text{Factor } x.$$

$$x(-a - 5) = -a \quad \text{Divide both sides by } -a - 5$$

$$x = \frac{-a}{-a-5} \text{ or } \frac{a}{a+5}$$

Quadratic Equations

Equations of the form $ax^2 + bx + c = 0$ (where $a \neq 0$) are *quadratic equations.* There are several techniques we can use to solve them, factoring, completing the square, and the quadratic formula. The simplest quadratic equations are in the form $x^2 =$ number. This equation has solutions $x = \sqrt{\text{number}}$ and $x = -\sqrt{\text{number}}$, or simply, $x = \pm\sqrt{\text{number}}$. For example, the solutions for $x^2 = 36$ are $x = 6$ and $x = -6$, or $x = \pm 6$.

Many quadratic equations can be solved by factoring. When there is a zero on one side of the equation, we factor the other side, set each factor equal to zero and solve both equations. This method comes from the fact that $ab = 0$ implies $a = 0$ or $b = 0$.

- $x^2 + x - 6 = 0$

  $x^2 + x - 6$ factors as $(x + 3)(x - 2)$. Set each of $x + 3$ and $x - 2$ equal to 0 and solve for x.

$$x + 3 = 0 \qquad x - 2 = 0$$

$$x = -3 \qquad x = 2$$

- $3x^2 + 24 = -18x$

  We need a zero on one side of the equation, so we will move $18x$ to the other side.

$$3x^2 + 18x + 24 = 0$$

$$3(x^2 + 6x + 8) = 0$$

$$3(x + 2)(x + 4) = 0$$

$$x + 2 = 0 \qquad x + 4 = 0$$

$$x = -2 \qquad x = -4$$

Some quadratic equations are difficult to factor. The quadratic formula can solve *every* quadratic equation. If $a \neq 0$ and $ax^2 + bx + c = 0$, then

$$x = \frac{-b \pm \sqrt{b^2 - 4ac}}{2a}.$$

- $3x^2 - x - 4 = 0$

  $a = 3$, $b = -1$, and $c = -4$

$$x = \frac{-(-1) \pm \sqrt{(-1)^2 - 4(3)(-4)}}{2(3)} = \frac{1 \pm \sqrt{49}}{6} = \frac{1 \pm 7}{6}$$

$$= \frac{8}{6}, \frac{-6}{6} = \frac{4}{3}, -1$$

- $x^2 - 1 = 0$

  $a = 1$, $b = 0$, and $c = -1$

$$x = \frac{-0 \pm \sqrt{0^2 - 4(1)(-1)}}{2(1)} = \frac{\pm\sqrt{4}}{2} = \frac{\pm 2}{2} = \pm 1$$

- $4x^2 + x = 1$

  We need 0 on one side of the equation. Once we move 1 to the other side, we have $4x^2 + x - 1 = 0$.

$$x = \frac{-1 \pm \sqrt{1^2 - 4(4)(-1)}}{2(4)} = \frac{-1 \pm \sqrt{17}}{8}$$

A quadratic equation can have square roots of numbers as solutions that need to be simplified. The square root of a number is simplified when it does not have any perfect squares as factors. For example, $\sqrt{24}$ is not simplified because $24 = 2^2 \times 6$. We can use the exponent properties $\sqrt{ab} = \sqrt{a} \cdot \sqrt{b}$ and $\sqrt[n]{a^n} = a$ to simplify $\sqrt{24}$.

$$\sqrt{24} = \sqrt{4 \cdot 6} = \sqrt{4} \cdot \sqrt{6} = 2\sqrt{6}$$

Square roots of fractions and square roots in denominators are also not considered simplified. These numbers often come up in trigonometry. Sometimes we can multiply the fraction by the denominator over itself.

- $\sqrt{\frac{1}{3}} = \frac{\sqrt{1}}{\sqrt{3}} = \frac{1}{\sqrt{3}} = \frac{1}{\sqrt{3}} \cdot \frac{\sqrt{3}}{\sqrt{3}} = \frac{\sqrt{3}}{(\sqrt{3})^2} = \frac{\sqrt{3}}{3}$

- $\frac{10}{\sqrt{5}} = \frac{10}{\sqrt{5}} \cdot \frac{\sqrt{5}}{\sqrt{5}} = \frac{10\sqrt{5}}{(\sqrt{5})^2} = \frac{10\sqrt{5}}{5} = 2\sqrt{5}$

  This trick will not work for expressions such as $\frac{2}{\sqrt{3}+1}$. To simplify these fractions, we will use the fact that $(a-b)(a+b) = a^2 - b^2$. This allows us to square each term in the denominator individually. The denominator is in the form $a+b$ (where $a = \sqrt{3}$ and $b = 1$). We will multiply the fraction by $a-b$ over itself.

$$\begin{aligned}\frac{2}{\sqrt{3}+1} &= \frac{2}{\sqrt{3}+1} \cdot \frac{\sqrt{3}-1}{\sqrt{3}-1} \\ &= \frac{2(\sqrt{3}-1)}{(\sqrt{3})^2 - 1^2} = \frac{2(\sqrt{3}-1)}{3-1} \\ &= \frac{2(\sqrt{3}-1)}{2} = \sqrt{3}-1\end{aligned}$$

- $\frac{-8}{2-\sqrt{5}}$

  The denominator is in the form $a-b$ (with $a = 2$ and $b = \sqrt{5}$). We will multiply the fraction by $a+b$ over itself.

$$\begin{aligned}\frac{-8}{2-\sqrt{5}} &= \frac{-8}{2-\sqrt{5}} \cdot \frac{2+\sqrt{5}}{2+\sqrt{5}} \\ &= \frac{-8(2+\sqrt{5})}{2^2 - (\sqrt{5})^2} = \frac{-8(2+\sqrt{5})}{4-5} \\ &= \frac{-8(2+\sqrt{5})}{-1} = 8(2+\sqrt{5}) = 16 + 8\sqrt{5}\end{aligned}$$

Factoring by Grouping

Some expressions of the form ax^3+bx^2+cx+d can be factored using a technique called *factoring by grouping*. This technique takes two steps. The first step is to factor the first two terms and the second two terms so that each pair of terms has a common factor. The second step is to factor this common factor. For example, if we factor x^2 from the first two terms of $x^3 + 2x^2 + 3x + 6$, we are left with $x^2(x+2) + 3x + 6$. Now we look at the second two terms, $3x + 6$, and factor it so that $x + 2$ is a factor. If we factor 3 from $3x + 6$, we are left with $x + 2$ as a factor: $3x + 6 = 3(x + 2)$. This leaves us with $x^2(x + 2) + 3(x + 2)$. In the last step, we factor $x + 2$ from each term, leaving x^2 and 3.

$$\begin{aligned} x^3 + 2x^2 + 3x + 6 &= x^2(x + 2) + 3(x + 2) \\ &= (x + 2)(x^2 + 3) \end{aligned}$$

We can use this technique to solve equations.

- $4x^3 - 5x^2 - 36x + 45 = 0$
  Once we have factored $4x^3 - 5x^2 - 36x + 45$, we will set each factor equal to 0 and solve for x. If we factor x^2 from the first two terms, we have $4x^3 - 5x^2 = x^2(4x - 5)$. If we factor -9 from the second two terms, we have $-36x + 45 = -9(4x - 5)$.

$$\begin{aligned} 4x^3 - 5x^2 - 36x + 45 &= 0 \\ x^2(4x - 5) - 9(4x - 5) &= 0 \\ (4x - 5)(x^2 - 9) &= 0 \end{aligned}$$

$$\begin{aligned} 4x - 5 &= 0 & x^2 - 9 &= 0 \\ 4x &= 5 & x^2 &= 9 \\ x &= \frac{5}{4} & x &= \pm 3 \end{aligned}$$

Solving $ax^n = b$ and $a\sqrt[n]{x} = b$

Solve equations of the form $ax^n = b$ by first dividing both sides of the equation by a, then by taking the nth root of both sides. If n is even, use a $\pm$ symbol on one side of the equation to get both solutions.

- $4x^2 = 9$

  $x^2 = \frac{9}{4}$

  $x = \pm\sqrt{\frac{9}{4}} = \pm\frac{3}{2}$

- $8x^3 = -1$

$$x^3 = -\frac{1}{8}$$

$$x = \sqrt[3]{-\frac{1}{8}} = -\frac{1}{2}$$

Solve equations of the form $a\sqrt[n]{x} = b$ by first dividing both sides of the equation by a, then by raising both sides to the nth power.

- $4\sqrt{x} = 5$

$$4\sqrt{x} = 5$$

$$\sqrt{x} = \frac{5}{4}$$

$$(\sqrt{x})^2 = \left(\frac{5}{4}\right)^2$$

$$x = \frac{25}{16}$$

- $4\sqrt{x} - 3 = 0$

  This equation needs to be in the form $a\sqrt{x} = b$ before we square both sides of the equation.

$$4\sqrt{x} - 3 = 0$$

$$4\sqrt{x} = 3$$

$$\sqrt{x} = \frac{3}{4}$$

$$(\sqrt{x})^2 = \left(\frac{3}{4}\right)^2$$

$$x = \frac{9}{16}$$

Inequalities

Solving linear inequalities is much like solving linear equations *except* when multiplying or dividing both sides of the inequality by a negative number, when we must reverse the inequality symbol. Solutions to inequalities are usually given in interval notation. The last page of the appendix has a review of interval notation.

- $5x - 8 > 3x + 10$

$$5x - 8 > 3x + 10$$
$$2x > 18$$
$$x > 9$$

The solution is $(9, \infty)$.

- $3x + 7 \leq 5x - 9$

$$3x + 7 \leq 5x - 9$$
$$-2x \leq -16$$
$$\frac{-2x}{-2} \geq \frac{-16}{-2} \qquad \text{Reverse the sign at this step.}$$
$$x \geq 8$$

The solution is $[8, \infty)$.

A double inequality is notation for two separate inequalities. They are solved the same way as single inequalities.

- $-3 \leq \frac{4x+7}{2} \leq 5$
  This inequality means $-3 \leq \frac{4x+7}{2}$ and $\frac{4x+7}{2} \leq 5$.

$$-3 \leq \frac{4x+7}{2} \leq 5 \qquad \text{Clear the fraction by multiplying all three quantities by 2.}$$
$$-6 \leq 4x + 7 \leq 10 \qquad \text{Subtract 7 from all three quantities.}$$
$$-13 \leq 4x \leq 3 \qquad \text{Divide all three quantities by 4.}$$
$$\frac{-13}{4} \leq x \leq \frac{3}{4}$$

The solution is $[-13/4, \ 3/4]$.

Nonlinear inequalities are solved in a different way. Below is a list of steps we will take to solve polynomial inequalities.

1. Rewrite the expression with 0 on one side.
2. Factor the nonzero side.
3. Set each factor equal to 0 and solve for x.

4. Put these solutions from Step 3 on a number line.
5. Pick a number to the left of the smallest solution (from Step 3), a number between consecutive solutions, and a number to the right of the largest solution.
6. Put these numbers in for x in the original inequality.
7. If a number makes the inequality true, mark "True" over the interval. If a number makes the inequality false, mark "False" over the interval.
8. Write the interval notation for the "True" intervals.

- $2x^2 - x \geq 3$

$$2x^2 - x - 3 \geq 0 \qquad \textbf{Step 1}$$

$$(2x - 3)(x + 1) \geq 0 \qquad \textbf{Step 2}$$

$$2x - 3 = 0 \qquad x + 1 = 0 \qquad \textbf{Step 3}$$

$$x = \frac{3}{2} \qquad x = -1$$

Step 4 Put -1 and $3/2$ on a number line.

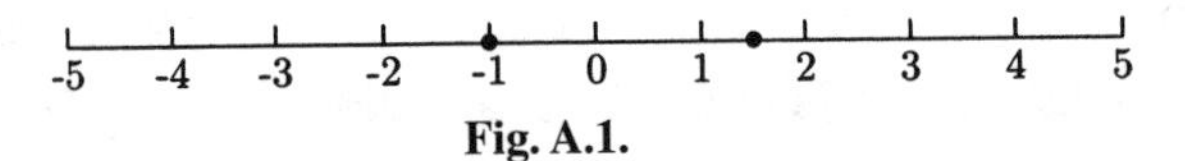

Fig. A.1.

Step 5 We will use $x = -2$ for the number to the left of -1, $x = 0$ for the number between -1 and 3/2, and $x = 2$ for the number to the right of 3/2.

Step 6 We will test these numbers in $2x^2 - x \geq 3$.

Let $x = -2$	$2(-2)^2 - (-2) \geq 3?$	True
Let $x = 0$	$2(0)^2 - 0 \geq 3?$	False
Let $x = 2$	$2(2)^2 - 2 \geq 3?$	True

Step 7 We will mark the interval to the left of -1 "True," the interval between -1 and 3/2 "False," and the interval to the right of 3/2, "True."

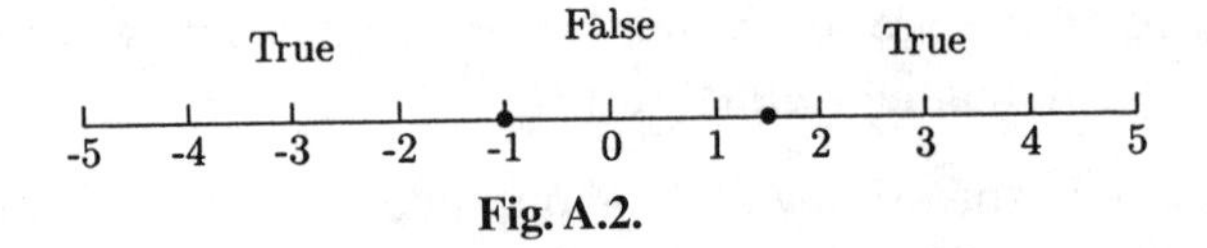

Fig. A.2.

Step 8 The intervals that make the inequality true are $x \leq -1$ and $x \geq 3/2$. The interval notation is $(-\infty, -1] \cup [3/2, \infty)$.

If there is an x in a denominator, the steps change slightly.

1. Get 0 on one side of the inequality.
2. Write the nonzero side as one fraction.
3. Factor the numerator and the denominator.
4. Set each factor equal to 0 and solve for x.
5. Put these solutions from Step 4 on a number line.
6. Pick a number to the left of the smallest solution (from Step 4), a number between consecutive solutions, and a number to the right of the largest solution.
7. Put these numbers in for x in the original inequality.
8. If a number makes the inequality true, mark "True" over the interval. If a number makes the inequality false, mark "False" over the interval.
9. Write the interval notation for the "True" intervals—make sure that the solution does not include any x-value that makes a denominator 0.

- $\frac{x-4}{x+5} > 2$

$$\frac{x-4}{x+5} > 2$$

$$\frac{x-4}{x+5} - 2 > 0 \qquad \textbf{Step 1}$$

$$\frac{x-4}{x+5} - 2\left(\frac{x+5}{x+5}\right) > 0 \qquad \textbf{Step 2}$$

$$\frac{x-4-2(x+5)}{x+5} > 0$$

$$\frac{-x-14}{x+5} > 0$$

$$-x-14=0 \qquad x+5=0 \qquad \textbf{Step 4}$$

$$x=-14 \qquad x=-5$$

Step 5

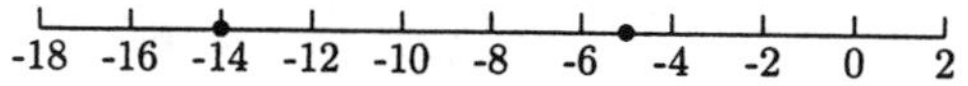

-18 -16 -14 -12 -10 -8 -6 -4 -2 0 2

Fig. A.3.

Step 6 We will use $x=-15$ for the number to the left of -14, $x=-10$ for the number between -14 and -5, and $x=0$ for the number to the right of -5.

Step 7

$$\frac{-15-4}{-15+5} > 2? \qquad \text{False}$$

$$\frac{-10-4}{-10+5} > 2? \qquad \text{True}$$

$$\frac{0-4}{0+5} > 2? \qquad \text{False}$$

We will write "False" over the interval to the left of -14, "True" over the interval between -14 and -5, and "False" over the interval to the right of -5.

Step 8

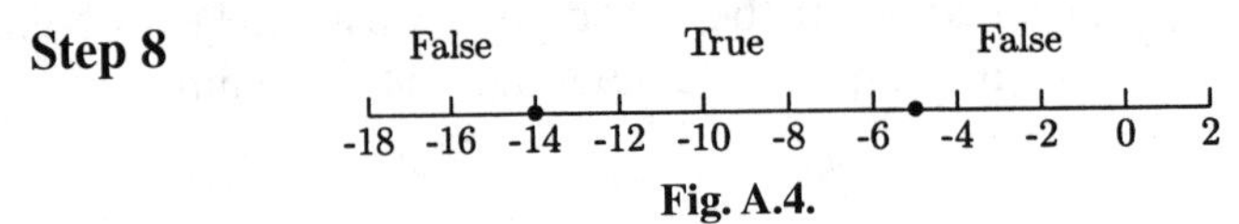

Fig. A.4.

The solution is the interval $(-14, -5)$.

- $\frac{x^2-3x}{x+1} \leq -1$

$$\frac{x^2-3x}{x+1} \leq -1$$

$$\frac{x^2-3x}{x+1} + 1 \leq 0$$

$$\frac{x^2-3x}{x+1} + 1 \cdot \frac{x+1}{x+1} \leq 0$$

$$\frac{x^2-3x+x+1}{x+1} \leq 0$$

$$\frac{x^2-2x+1}{x+1} \leq 0$$

$$\frac{(x-1)(x-1)}{x+1} \leq 0$$

$$x-1=0 \qquad x+1=0$$

$$x=1 \qquad x=-1$$

$$\frac{(-2)^2 - 3(-2)}{-2+1} \leq -1? \qquad \text{True}$$

$$\frac{0^2 - 3(0)}{0+1} \leq -1? \qquad \text{False}$$

$$\frac{2^2 - 3(2)}{2+1} \leq -1? \qquad \text{False}$$

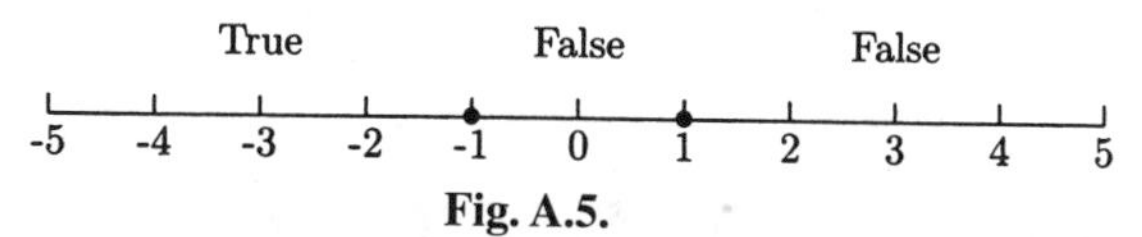

Fig. A.5.

The solution is $(-\infty, -1)$. The solution is *not* $(-\infty, -1]$ because a bracket next to -1 indicates that -1 is part of the solution. We cannot allow $x = -1$ because we would have a zero in a denominator.

Table. A.1

Inequality	Number Line	Interval
$x < a$	**Fig. A.6**	$(-\infty, a)$
$x \leq a$	**Fig. A.7**	$(-\infty, a]$
$x > a$	**Fig. A.8**	(a, ∞)
$x \geq a$	**Fig. A.9**	$[a, \infty)$
$a < x < b$	**Fig. A.10**	(a, b)
$a \leq x \leq b$	**Fig. A.11**	$[a, b]$
$x < a$ or $x > b$	**Fig. A.12**	$(-\infty, a) \cup (b, \infty)$
$x \leq a$ or $x \geq b$	**Fig. A.13**	$(-\infty, a] \cup [b, \infty)$
All x	All real numbers	$(-\infty, \infty)$

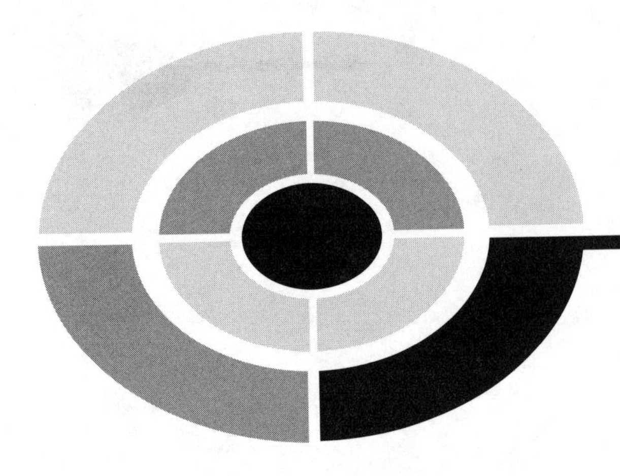

Final Exam

1. What is the maximum or minimum functional value for $f(x) = -(x-5)^2 + 12$?
   (a) The maximum functional value is 12.
   (b) The maximum functional value is 5.
   (c) The minimum functional value is 12.
   (d) The maximum functional value is 5.

2. Find an equation of the line containing the points $(1, 9/2)$ and $(-2, 6)$.
   (a) $y = \frac{1}{2}x + 8$
   (b) $y = \frac{7}{16}x + \frac{9}{2}$
   (c) $y = -\frac{1}{2}x + 5$
   (d) $y = \frac{7}{16}x + 6$

3. What is (are) the vertical asymptote(s) for the graph of

$$f(x) = \frac{x+5}{x-3}?$$

   (a) $x = 3$
   (b) $x = -5$
   (c) $x = 3$ and $x = -5$
   (d) There are no vertical asymptotes.

4. Find the product.

$$\begin{bmatrix} 1 & -1 \\ 0 & 2 \end{bmatrix} \cdot \begin{bmatrix} -5 & 1 & 3 \\ 2 & 2 & 1 \end{bmatrix}$$

(a) $\begin{bmatrix} -3 & 3 & 4 \\ -4 & -4 & -2 \end{bmatrix}$ (b) $\begin{bmatrix} -7 & -1 & 2 \\ 4 & 4 & 2 \end{bmatrix}$

(c) $\begin{bmatrix} -5 & -1 \\ 0 & 4 \end{bmatrix}$ (d) The product does not exist.

5. $\cos 7\pi/6 =$

(a) 1/2 (b) $-1/2$ (c) $\sqrt{3}/2$ (d) $-\sqrt{3}/2$

6. What are the foci for the hyperbola

$$\frac{(y-1)^2}{16} - \frac{(x+1)^2}{9} = 1?$$

(a) $(-1, -4)$ and $(-1, 6)$ (b) $(-6, 1)$ and $(4, 1)$
(c) $(-4, -1)$ and $(6, -1)$ (d) $(1, -6)$ and $(1, 4)$

7. Find $x + y$ for the system.

$$\begin{cases} x - 2y & = 1 \\ 2x + y & = 7 \end{cases}$$

(a) 4 (b) 5 (c) 6 (d) 7

8. Find the fourth term of the arithmetic sequence whose 30th term is -180 and whose 45th term is -300.

(a) 20 (b) 28 (c) 35 (d) 46

9. If \$2000 is deposited into an account earning 9% annual interest, compounded monthly, what is it worth after 10 years?

(a) \$4902.71 (b) \$4734.73 (c) \$2155.17 (d) \$4870.38

10. For $f(x) = 1 - x^2$ and $g(x) = 2x + 5$, what is $f \circ g(x)$?

(a) $-2x^2 + 3$ (b) $\frac{1-x^2}{2x+5}$
(c) $-4x^2 - 20x - 24$ (d) $-2x^3 - 5x^2 + 2x + 5$

11. Evaluate $f(2)$ for $f(x) = 7$.

(a) 2 (b) 7 (c) 14 (d) 2, 7

12. Find $\cos\theta$ if $\sin\theta = -4/5$ and θ is in Quadrant IV.

(a) 3/5 (b) $-3/5$ (c) 9/25 (d) $-9/25$

13. What is the directrix for the parabola $y^2 = 12(x-3)$?
   (a) $y=-3$ (b) $y=3$ (c) $x=0$ (d) $x=3$
14. Are the lines $2x - y = 5$ and $4x - 8y = 9$ parallel, perpendicular, or neither?
   (a) Parallel (b) Perpendicular
   (c) Neither (d) Cannot be determined
15. Find the second term of the arithmetic sequence whose fifth term is $\frac{122}{3}$ and whose tenth term is $\frac{272}{3}$.
   (a) $\frac{5}{6}$ (b) $\frac{14}{3}$ (c) $\frac{25}{16}$ (d) $\frac{32}{3}$
16. What is the domain for $f(x) = \sqrt[3]{x+1}$?
   (a) $(-\infty,-1)\cup(-1,\infty)$ (b) $(-1,\infty)$
   (c) $[-1,\infty)$ (d) $(-\infty,\infty)$
17. Find $f \circ g(-2)$ for $f(x) = x^2 + x$ and $g(x) = 3x + 9$.
   (a) 12 (b) 15 (c) 3 (d) 2
18. What is the vertex for $y = 4x^2 - 6x + 5$?
   (a) $(\frac{3}{2}, 5)$ (b) $(\frac{3}{4}, \frac{11}{4})$ (c) $(3, -4)$ (d) $(3, 14)$
19. What is the inverse of $\begin{bmatrix} -10 & 1 \\ 5 & 2 \end{bmatrix}$?
   (a) $\begin{bmatrix} -\frac{2}{25} & \frac{1}{25} \\ \frac{1}{5} & \frac{2}{5} \end{bmatrix}$ (b) $\begin{bmatrix} \frac{1}{25} & -\frac{2}{25} \\ \frac{2}{5} & \frac{1}{5} \end{bmatrix}$
   (c) $\begin{bmatrix} -\frac{1}{25} & \frac{2}{25} \\ -\frac{2}{5} & -\frac{1}{5} \end{bmatrix}$ (d) $\begin{bmatrix} \frac{1}{25} & \frac{2}{25} \\ \frac{2}{5} & -\frac{1}{5} \end{bmatrix}$
20. What is the fifth term of the sequence where $a_n = n^2$?
   (a) 10 (b) 15 (c) 20 (d) 25
21. What is an asymptote for the hyperbola

$$\frac{(x+1)^2}{9} - \frac{(y-1)^2}{16} = 1?$$

   (a) $y = \frac{3}{4}x + \frac{7}{4}$ (b) $y = \frac{4}{3}x + \frac{1}{3}$
   (c) $y = \frac{4}{3}x + \frac{7}{3}$ (d) $y = \frac{3}{4}x - \frac{1}{4}$
22. What is the phase shift for $y = 3\sin(2x - \pi/3)$?
   (a) $-\pi/3$ (b) $\pi/3$ (c) $-\pi/6$ (d) $\pi/6$

23. What is the period for $y = 3\sin(2x - \pi/3)$?

(a) π (b) $\pi/4$ (c) 4π (d) $\pi/2$

24. Find $x + y$ for the system.

$$\begin{cases} y &= x^2 - x - 8 \\ y &= 2x + 10 \end{cases}$$

(a) 3 and 16 (b) 5 and 21 (c) 1 and 28 (d) 3 and 9

25. $\log_5 5^{2t} =$

(a) t (b) $2t$ (c) 2 (d) $10t$

26. What is the horizontal asymptote for the graph of

$$f(x) = \frac{3x^2 + 2x + 1}{6x^2 + 3x + 4}?$$

(a) $y = 0$ (b) $y = \frac{1}{2}$ (c) There is no horizontal asymptote.

27. Is the sequence $-\frac{8}{3}, -\frac{13}{6}, -\frac{5}{3}, -\frac{7}{6}, -\frac{2}{3}, \ldots$ arithmetic, geometric, or neither?

(a) Arithmetic (b) Geometric
(c) Neither (d) Cannot be determined

28. What is the inverse of $\begin{bmatrix} 1 & 1 & 1 \\ 2 & -1 & 1 \\ -1 & 1 & -3 \end{bmatrix}$?

(a) $\begin{bmatrix} \frac{5}{8} & -\frac{1}{4} & \frac{1}{8} \\ \frac{1}{4} & \frac{1}{2} & \frac{1}{4} \\ \frac{1}{8} & -\frac{1}{4} & -\frac{3}{8} \end{bmatrix}$ (b) $\begin{bmatrix} -\frac{5}{8} & \frac{1}{4} & -\frac{1}{8} \\ \frac{1}{4} & \frac{1}{2} & \frac{1}{4} \\ -\frac{1}{8} & \frac{1}{4} & \frac{3}{8} \end{bmatrix}$

(c) $\begin{bmatrix} \frac{1}{4} & \frac{1}{2} & \frac{1}{4} \\ \frac{1}{8} & -\frac{1}{4} & -\frac{3}{8} \\ \frac{5}{8} & -\frac{1}{4} & \frac{1}{8} \end{bmatrix}$ (d) $\begin{bmatrix} \frac{1}{4} & \frac{1}{2} & \frac{1}{4} \\ \frac{5}{8} & -\frac{1}{4} & \frac{1}{8} \\ \frac{1}{8} & -\frac{1}{4} & -\frac{3}{8} \end{bmatrix}$

29. Write the product as a sum: $\sin 4x \sin x$.

(a) $\frac{1}{2}(\cos 5x + \cos 3x)$ (b) $\frac{1}{2}(\cos 3x - \cos 5x)$
(c) $\frac{1}{2}(\sin 5x + \sin 3x)$ (d) $\frac{1}{2}(\sin 5x - \sin 3x)$

Problems 30–35 refer to the graph in Figure A.14.

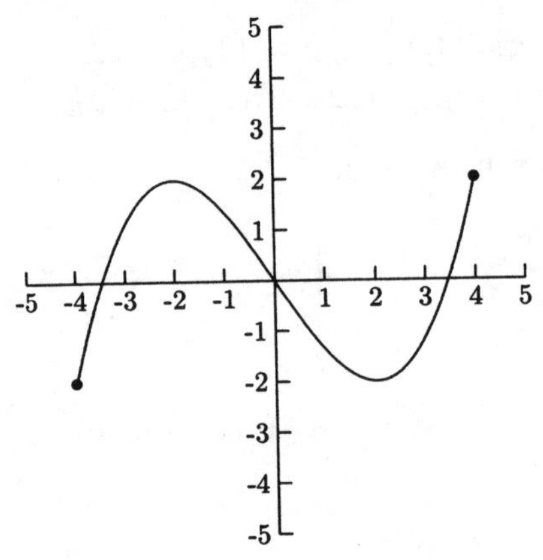

Fig. A.14.

30. Is the graph in Figure A.14 symmetric?
 (a) Yes, with respect to the x-axis (b) Yes, with respect to the y-axis
 (c) Yes, with respect to the origin (d) No

31. What is $f(-4)$?
 (a) -2 (b) 0 (c) 2 (d) 4

32. What is the y-intercept?
 (a) -2 (b) 0 (c) 2 (d) 4

33. Does the function have an inverse?
 (a) Yes (b) No
 (c) Cannot be determined

34. What is the range?
 (a) $[-4, 4]$ (b) $[-2, 2]$ (c) $[-2, -4]$ (d) $[2, 4]$

35. What is the increasing interval(s)?
 (a) $(-4, -2) \cup (2, 4)$ (b) $(-4, 4)$
 (c) $(2, 2)$ (d) $(-2, 2)$

36. Is the function $f(x) = x^2 - x + 2$ even, odd, or neither?
 (a) Even (b) Odd (c) Neither
 (d) Cannot be determined without the graph

37. Are the points $(-4, 2)$, $(1, 3)$, $(-1, 0)$, and $(-2, 5)$ the vertices of a parallelogram?
 (a) Yes (b) No
 (c) Cannot be determined

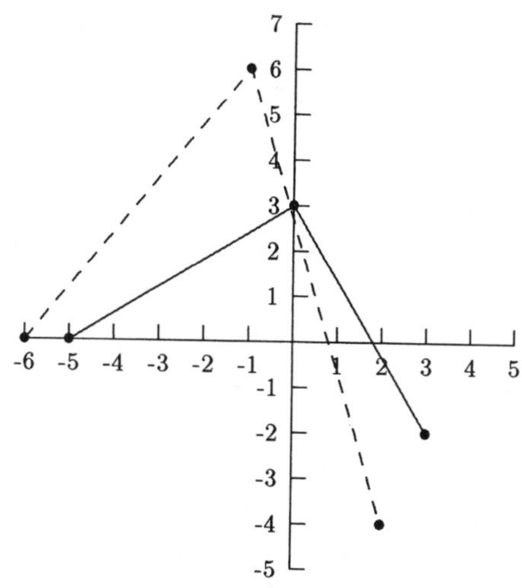

Fig. A.15.

38. The solid graph in Figure A.15 is the graph of $f(x)$, and the dashed graph is the graph of a transformation. What is the transformation?
(a) $f(x+1)+3$ (b) $f(x-1)+3$ (c) $2f(x+1)$ (d) $2f(x-1)$

39. Find the sum.

$$-6+(-2)+2+6+10+\cdots+50$$

(a) 660 (b) 260 (c) 330
(d) Too many terms are missing to find the sum.

40. What is $x+y$ for the system? Use a matrix method.

$$\begin{cases} -x+4y &= 11 \\ 2x+3y &= 22 \end{cases}$$

(a) 7 (b) 8 (c) 9 (d) 10

41. Are the angles 65° and −295° coterminal?
(a) Yes (b) No (c) Cannot be determined

42. Find $f^{-1}(x)$ for $f(x)=\frac{x-3}{x+4}$.
(a) $\dfrac{x+4}{x-3}$ (b) $\dfrac{1}{x+4}$
(c) $\dfrac{-4x-3}{x-1}$ (d) $\dfrac{4x+3}{x-1}$

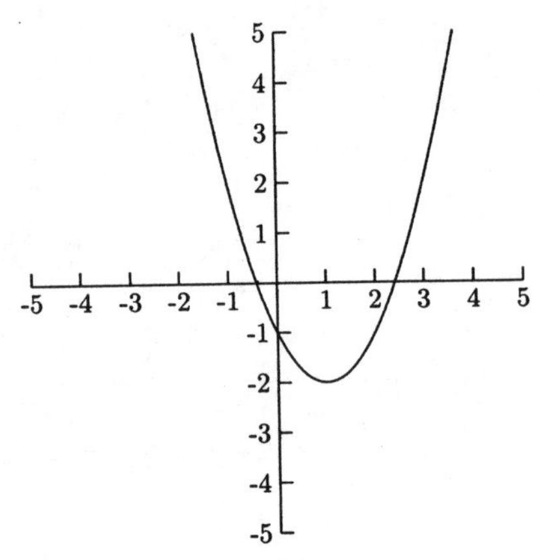

Fig. A.16.

43. The graph in Figure A.16 is the graph of what function?
    (a) $f(x) = x^2 - 2x - 1$ (b) $f(x) = x^2 + 2x - 1$
    (c) $f(x) = x^2 - 3x - 1$ (d) $f(x) = x^2 + 3x - 1$

44. A biscuit recipe calls for 2/3 of a cup of milk for each cup of mix. Find an equation that gives the amount of milk in terms of the amount of mix.
    (a) $y = \frac{3}{2}x$ (b) $y = \frac{2}{3}x$
    (c) $y = \frac{5}{3}x$ (d) Cannot be determined

45. Rewrite $m^r = n$ as a logarithmic equation.
    (a) $\log_n r = m$ (b) $\log_m r = n$
    (c) $\log_n m = r$ (d) $\log_m n = r$

46. A museum offers group discounts for groups of 25 or more. For a group of 25, the ticket price is \$13.50. For each additional person attending, the price drops \$0.50. What group size maximizes the museum's revenue?
    (a) 26 (b) 27 (c) 28 (d) 29

47. The graph in Figure A.17 is the graph of which function?
    (a) $P(x) = (x + 3)(x - 1)^2(x + 1) = x^4 + 2x^3 - 4x^2 - 2x + 3$
    (b) $P(x) = -(x + 3)(x - 1)^2(x + 1) = -x^4 - 2x^3 + 4x^2 + 2x - 3$
    (c) $P(x) = (x + 3)(x - 1)(x + 1) = x^3 + 3x^2 - x - 3$
    (d) $P(x) = -(x + 3)(x - 1)(x + 1) = -x^3 - 3x^2 + x + 3$

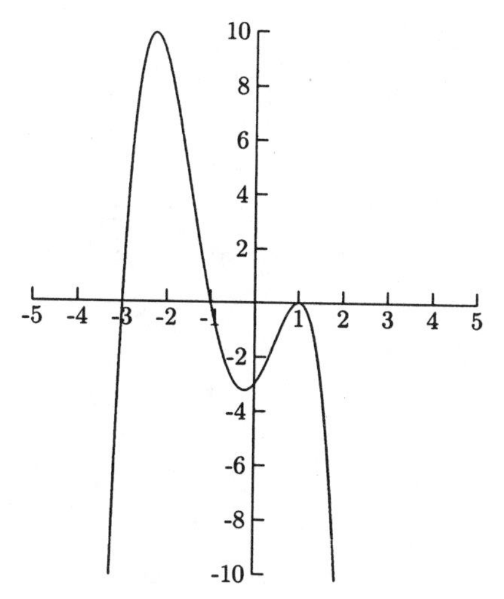

Fig. A.17.

48. What is the determinant for $\begin{bmatrix} 1 & 1 & 0 \\ 1 & 0 & 1 \\ 3 & -2 & 1 \end{bmatrix}$?

(a) 3 (b) 4 (c) 5 (d) 6

49. Find the domain for $f(x) = \frac{6}{x-8}$.

(a) $(8, \infty)$
(b) $(-\infty, 8) \cup (8, \infty)$
(c) $(-\infty, 8] \cup [8, \infty)$
(d) $[8, \infty)$

50. Are $f(x) = 4x^3 + 1$ and $g(x) = \sqrt[3]{\frac{x-1}{4}}$ inverses?

(a) Yes
(b) No
(c) Cannot be determined

51. Solve for x: $\log_5(2x - 7) = 1$.

(a) $x = 4$ (b) $x = 5$ (c) $x = 6$ (d) $x = 7$

52. The graph in Figure A.18 is the graph of which function?

(a) $f(x) = \frac{x}{x+1}$
(b) $f(x) = \frac{x^2}{x+1}$
(c) $f(x) = \frac{x}{x-1}$
(d) $f(x) = \frac{x^2}{x-1}$

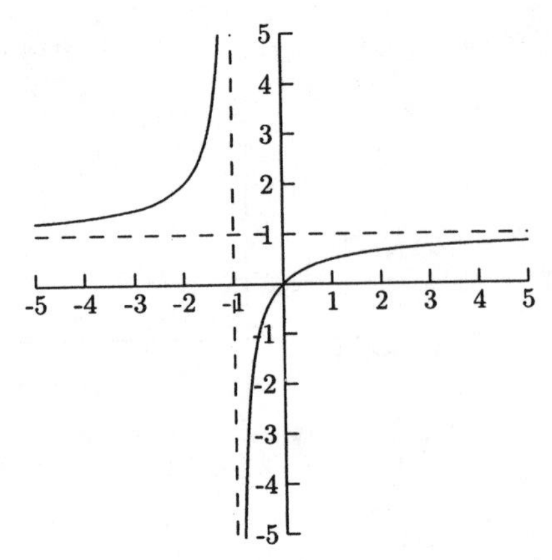

Fig. A.18.

53. According to the Rational Zero Theorem, which of the following is *NOT* a possible rational zero for $f(x) = 6x^4 - x^3 - 3x^2 + x - 10$?
(a) $-\frac{1}{3}$ (b) -10 (c) 3 (d) $\frac{5}{6}$

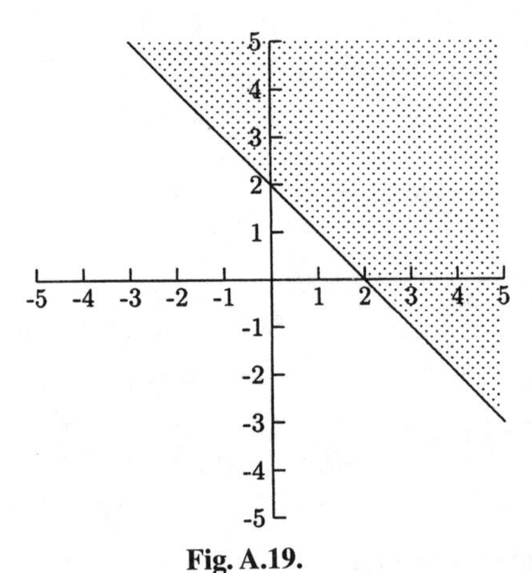

Fig. A.19.

54. The graph in Figure A.19 is the graph of which inequality?
(a) $x + y \geq 2$ (b) $x + y > 2$ (c) $x + y \leq 2$ (d) $x + y < 2$

55. Find the quadratic function with vertex $(-1, 3)$ with the point $(2, -15)$ on its graph.
(a) $f(x) = -18(x-1)^2 + 3$ (b) $f(x) = 18(x-1)^2 + 3$
(c) $f(x) = -2(x+1)^2 + 3$ (d) $f(x) = 2(x+1)^2 + 3$

56. Rewrite as a single logarithm: $\ln x - 3\ln y - \ln z$.

(a) $\ln \frac{x}{3yz}$

(b) $\ln \frac{x}{y^3 z}$

(c) $\ln \frac{xz}{y^3}$

(d) $\frac{\ln x}{3\ln yz}$

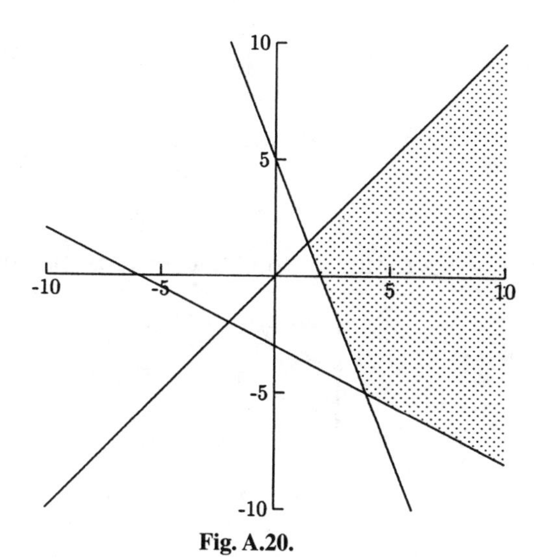

Fig. A.20.

57. The graph in Figure A.20 is the graph of which system?

(a) $\begin{cases} y & \leq x \\ y & \leq -\frac{5}{2}x + 5 \\ y & \leq -\frac{1}{2}x - 3 \end{cases}$

(b) $\begin{cases} y & \leq x \\ y & \geq -\frac{5}{2}x + 5 \\ y & \leq -\frac{1}{2}x - 3 \end{cases}$

(c) $\begin{cases} y & \geq x \\ y & \leq -\frac{5}{2}x + 5 \\ y & \leq -\frac{1}{2}x - 3 \end{cases}$

(d) $\begin{cases} y & \leq x \\ y & \geq -\frac{5}{2}x + 5 \\ y & \geq -\frac{1}{2}x - 3 \end{cases}$

58. The graph in Figure A.21 is the graph of which equation?

(a) $x^2 - \frac{y^2}{4} = 1$

(b) $\frac{x^2}{4} - y^2 = 1$

(c) $x^2 + \frac{y^2}{4} = 1$

(d) $\frac{x^2}{4} + y^2 = 1$

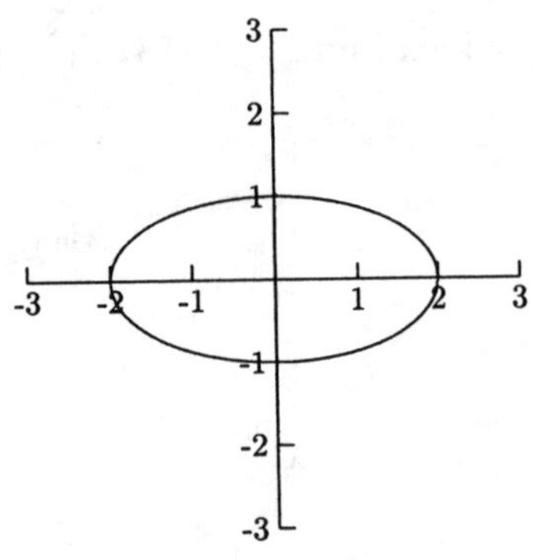

Fig. A.21.

59. What is $x + y + z$ for the system? Use a matrix method.

$$\begin{cases} x + y & = 11 \\ x \quad + z & = -1 \\ 3x - 2y + z & = -11 \end{cases}$$

(a) 6 (b) 7 (c) 8 (d) 9

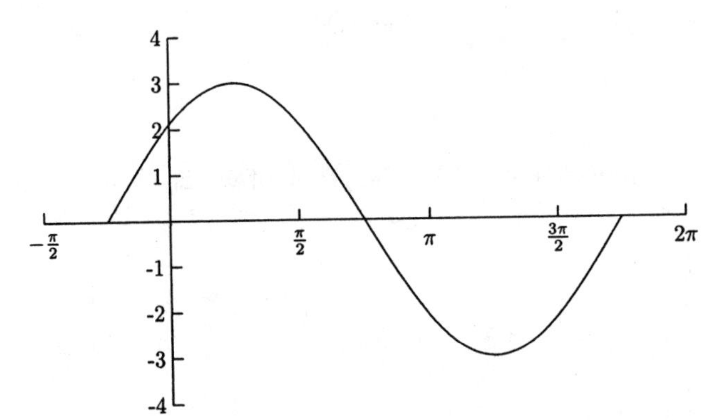

Fig. A.22.

60. The graph in Figure A.22 is the graph of one period of which function?
(a) $y = 3\sin(x - \pi/4)$ (b) $y = 3\sin(x + \pi/4)$
(c) $y = \sin 3(x - \pi/4)$ (d) $y = \sin 3(x + \pi/4)$

61. Evaluate $\frac{f(a+h)-f(a)}{h}$ for $f(x) = 2x^2 - 1$.
(a) $4a + 2h^2 - 2$ (b) $4a + 2h^2 - 1$ (c) $4a + 2h$ (d) $4a + 2h^2$

62. According to Descartes' Rule of Signs, how many possible positive zeros are there for $f(x) = 6x^4 - x^3 - 3x^2 + x - 10$?

(a) 3 or 1 (b) 3 (c) 2 or 0 (d) 2

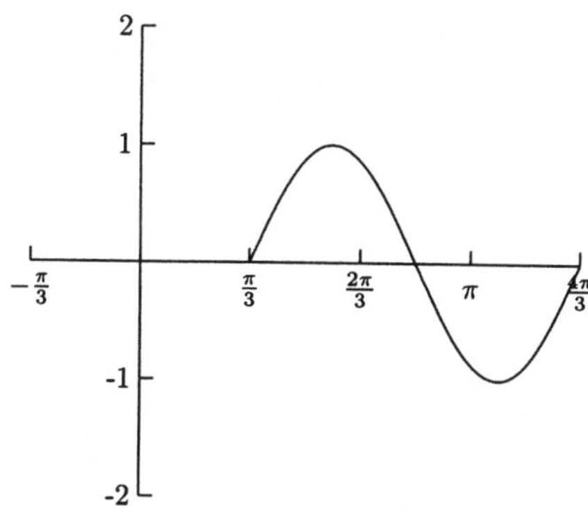

Fig. A.23.

63. The graph in Figure A.23 is the graph of one period of which function?

(a) $y = 2\sin(x - \pi/3)$ (b) $y = 2\sin(x + \pi/3)$

(c) $y = \sin 2(x - \pi/3)$ (d) $y = \sin 2(x + \pi/3)$

64. The graph of $3f(x - 4)$ is the graph of $f(x)$

(a) shifted to the right four units and vertically stretched.

(b) shifted to the left four units and vertically stretched.

(c) shifted to the right four units and vertically compressed.

(d) shifted to the left four units and vertically compressed.

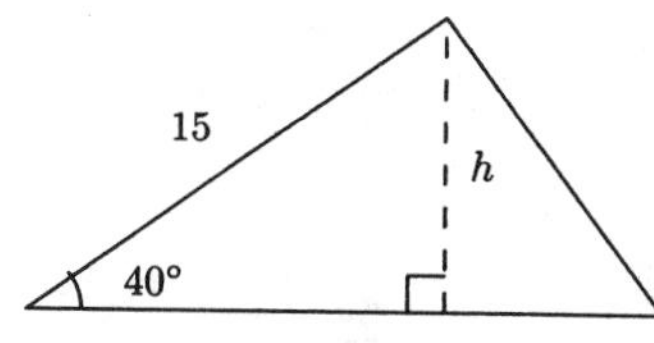

Fig. A.24.

65. Find the height of the triangle in Figure A.24.

(a) About 11.5 (b) About 0.04 (c) About 9.6 (d) About 0.05

66. Find all zeros for $f(x) = 3x^3 - 7x^2 + 8x - 2$.

(a) $-\frac{1}{3}$, $1 \pm i$ (b) $\frac{1}{3}$, $1 \pm i$ (c) $-\frac{1}{3}$, $-1 \pm i$ (d) $-\frac{1}{3}$, $-1 \pm i$

67. What are the intercepts for $f(x) = -x^2 + x + 2$?

    (a) The x-intercepts are 1 and 2, and the y-intercept is 2.

    (b) The x-intercepts are -1 and 2, and the y-intercept is 2.

    (c) The x-intercepts are 1 and -2, and the y-intercept is -2.

    (d) The x-intercepts are -1 and 2, and the y-intercept is -2.

68. $\cos(\tan^{-1} 1/5) =$

    (a) $\frac{5\sqrt{26}}{26}$ (b) $\frac{\sqrt{26}}{5}$ (c) $\frac{\sqrt{26}}{26}$ (d) $\sqrt{26}$

69. What is the slant asymptote for the graph of

$$f(x) = \frac{x^2 - 9}{x + 2}?$$

    (a) $y = x - 11$ (b) $y = x + 2$ (c) $y = x - 7$ (d) $y = x - 2$

70. The population of a town grew from 2000 in the year 1980 to 10,000 in the year 2000. Assuming exponential growth, what is the town's annual growth rate?

    (a) About 6% (b) About 7% (c) About 8% (d) About 9%

71. What is the quotient for $\frac{4+5i}{2-i}$?

    (a) $\frac{13}{3} + 2i$ (b) $\frac{3}{5} + \frac{14}{5}i$ (c) $\frac{13}{5} - \frac{6}{5}i$ (d) $\frac{13}{5} + \frac{6}{5}i$

72. What is the quotient for $(2x^3 - x^2 + 2) \div (x + 3)$?

    (a) $2x^2 - 7x - 21$ (b) $2x^2 + 5x + 15$

    (c) $2x^2 - 7x + 23$ (d) $2x^2 - 7x + 21$

73. What are the vertices for the ellipse

$$\frac{x^2}{25} + \frac{y^2}{16} = 1?$$

    (a) $(-4, 0)$ and $(4, 0)$ (b) $(0, -4)$ and $(0, 4)$

    (c) $(-5, 0)$ and $(5, 0)$ (d) $(0, 5)$ and $(0, -5)$

74. A triangle has sides of length 8, 15, and 20. Which of the following is an approximate angle in this triangle?

    (a) 48.3° (b) 41.7° (c) 50.6° (d) 23.6°

75. Which one of the following statements is NOT true about the polynomial function $f(x) = x^3(x - 4)^2(x + 1)$?

    (a) $x = 0$ is a zero of multiplicity 3.
    (b) $x = 4$ is a zero of multiplicity 2.
    (c) $x = 1$ is a zero of multiplicity 1.
    (d) $x = -1$ is a zero of multiplicity 1.

SOLUTIONS

1. A	2. C	3. A	4. B	5. D	6. A	7. A	8. B
9. A	10. C	11. B	12. A	13. C	14. C	15. D	16. D
17. A	18. B	19. A	20. D	21. C	22. D	23. A	24. C
25. B	26. B	27. A	28. D	29. B	30. C	31. A	32. B
33. B	34. B	35. A	36. C	37. A	38. C	39. C	40. C
41. A	42. C	43. A	44. B	45. D	46. A	47. B	48. B
49. B	50. A	51. C	52. A	53. C	54. A	55. C	56. B
57. D	58. D	59. B	60. B	61. C	62. A	63. C	64. A
65. C	66. B	67. B	68. A	69. D	70. C	71. B	72. D
73. C	74. B	75. C					

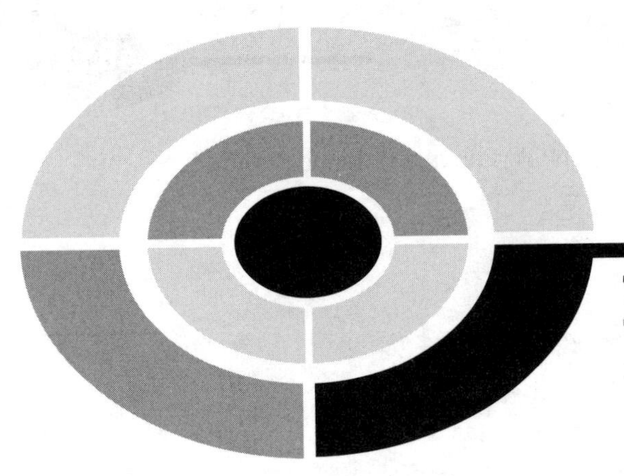

INDEX

ABOUT THE AUTHOR

Rhonda Huettenmueller is the author of the best-selling book in the *Demystified* series, *Algebra Demystified*. Popular with students for her ability to make higher math understandable and even enjoyable, she incorporates many of her teaching techniques in this book. Ms. Huettenmueller has taught mathematics at the college level for almost 15 years. She received her Ph.D. in mathematics from the University of North Texas.

ROBOT BUILDER'S BONANZA

GORDON McCOMB

MYKE PREDKO

THIRD EDITION

McGraw-Hill

New York Chicago San Francisco Lisbon London Madrid Mexico City Milan New Delhi San Juan Seoul Singapore Sydney Toronto

The McGraw-Hill Companies

Cataloging-in-Publication Data is on file with the Library of Congress.

1 2 3 4 5 6 7 8 9 0 DOC/DOC 0 1 0 9 8 7 6

ISBN 0-07-146893-5

The sponsoring editor for this book was Judy Bass and the production supervisor was Pamela A. Pelton. It was set in Souvenir by North Market Street Graphics. The art director for the cover was Anthony Landi.

Printed and bound by RR Donnelley.

This book was printed on recycled, acid-free paper containing a minimum of 50% recycled, de-inked fiber.

CONTENTS

ACKNOWLEDGMENTS

Gordon McComb's Acknowledgments for the Third Edition

Only until you've climbed the mountain can you look behind and see the vast distance that you've covered, and remember those you've met along the way who made your trek a little easier. Now that this book is finally finished, after the many miles of weary travel, I look back to those who helped me turn it into a reality and offer my heartfelt thanks. To the gang on comp.robotics.misc, for the great ideas, wisdom, and support; to Scott Savage, designer of the OOPic; to Frank Manning and Jack Schoof of NetMedia for their help with the BasicX; to Tony Ellis, a real-life "Q" if I ever met one; to Scott Grillo and the editors at McGraw-Hill; to my agents Matt Wagner and Bill Gladstone; and last and certainly not least, to my wife, Jennifer.

Myke Predko's Acknowledgments for the Third Edition

I had no small measure of concern when I was offered the opportunity to work on the third edition of what is affectionately known as *RBB.* The book is a staple for both beginners and experts alike and is crammed with material and knowledge that come from a large number of disciplines. Undertaking this effort required a lot of support from a variety of different individuals.

My editor, Judy Bass, who had the confidence that I could update *RBB* and do a credible job of it, and who kept her sense of humor and interest despite all the emails, questions, and ideas that are generated in a project like this. Judy always gives the confidence that all of McGraw-Hill is behind me.

For technical information and ideas regarding the material in the book, I would like to recognize Ben Wirz, who has an amazing amount of background in robotics and has been the co-designer on the Tab Electronics robot kits; Joe Jones, a very special robot designer and author; and Ken Gracey of Parallax who seems to have dedicated himself to making robots easy for everyone. I want to thank all of you for your time, ideas, and energy toward the development of the third edition of this book.

Over the past few years, I have been involved with the Ontario Science Centre along with Celestica, my daytime employer, helping to put on robot workshops for local families.

Through these workshops, I have learned a lot about what people want to get from robots and have seen their ideas take wing. All the volunteers involved, both Celestica and Ontario Science Centre employees, continue to give excellent suggestions, feedback, and support. I would especially like to thank Blair Clarkson, the special events coordinator of the Ontario Science Centre for his friendship, help, and prodding over the years to help create something that is truly unique.

For those of you following my progress as a writer, you will know that I frequently consult my daughter, Marya, for ideas and a different perspective. Her support started out from just pressing buttons to watch lights flash and now has progressed to trying out a few of the projects and critiquing different aspects of the book.

My wife Patience's continual support and love are necessary ingredients of every book I have written. Her enthusiasm for my hobby despite the mess, time taken from the family, and occasional flames accompanied by loud obscenities is nothing short of wonderful. Nothing that I do would be possible without you.

Lastly, I am indebted to Gordon McComb for all his hard work establishing the framework for *Robot Builder's Bonanza* and the countless hours he has spent making sure this book is the best introduction to robotics there is.

Thank you and I hope I have been able to pass along a bit of what you've given me.

INTRODUCTION

To the robotics experimenter, *robot* has a completely different meaning than what most people think of when they hear the word. A robot is a special brew of motors, solenoids, wires, and assorted electronic odds and ends, a marriage of mechanical and electronic gizmos. Taken together, the parts make a half-living but wholly personable creature that can vacuum the floor, serve drinks, protect the family against intruders and fire, entertain, educate, and lots more. In fact, there's almost no limit to what a well-designed robot can do.

In just about any science, it is the independent experimenter who first establishes the pioneering ideas and technologies. At the turn of the last century, two bicycle mechanics experimenting with strange kites were able to explain the basics of controlled flight. Robert Goddard experimented with liquid-fuel rockets before World War II; his discoveries paved the way for modern-day space flight. Alan Turning, tasked to create logic equipment to decrypt coded radio transmissions during the Second World War also worked at designing the basic architecture for the digital computer. In the 1950s a psychologist, Dr. W. Grey Walter, created the first mobile robots as part of an experiment into the operation of nerves as part of the decision processes in animals.

Robotics—like flight, rocketry, computers, and countless other technology-based endeavors—started small. Today, robotics is well on its way to becoming a necessary part of everyday life; not only are they used in automotive manufacturing, but they are exploring the solar system and prototype robot servants are walking upright, just like humans, as they learn to navigate and interact with our world.

What does this mean for the robotics experimenter? There is plenty of room for growth, with a lot of discoveries yet to be made—perhaps more so than in any other high-tech discipline.

I.1 Inside *Robot Builder's Bonanza*

Robot Builder's Bonanza, Third Edition takes an educational but fun approach to designing working robots. Its modular projects will provide the knowledge to take you from building basic motorized platforms to giving the machine a brain—and teaching it to walk, move about, sense what is going on around it, and obey commands.

If you are interested in mechanics, electronics, or robotics, you'll find this book a treasure chest of information and ideas on making thinking machines. The projects in *Robot Builder's Bonanza* include all the necessary information on how to construct the essential building blocks of a number of different personal robots. Suggested alternative approaches, parts lists, and sources of electronic and mechanical components are also provided where appropriate.

There are quite a few excellent books that have been written on how to design and build robots. But most have been aimed at making just one or two fairly sophisticated automatons, and at a fairly high price. Because of the complexity of the robots detailed in these other books, they require a fairly high level of expertise and pocket money on your part.

Robot Builder's Bonanza is different. Its modular "cookbook" approach offers a mountain of practical, easy to follow, and inexpensive robot experiments and projects. Integrated together, the various projects presented in the book, along with ones you come up with on your own, can be combined to create several different types of highly intelligent and workable robots of all shapes and sizes—rolling robots, walking robots, talking robots, you name it.

I.2 About the Third Edition

This new edition features a new author, Myke Predko, who has revised the second edition (published in 2001 and the original edition in 1987) from the perspective of an electrical engineer. Myke brings his experience as an electrical engineer that has worked with a wide variety of different computer systems as well as the development of low-level software designed for hardware interfacing. Many of the circuits presented in the earlier editions have been redesigned to both simplify them as well as make them more robust in robot applications. In the previous edition of this book, a number of different computer controls were presented while in this edition the projects have been consolidated on the Parallax BASIC Stamp 2, which is an excellent tool for new roboticists. The examples can also inspire the more experienced robot designers who already work with their favorite control hardware. The book has also been updated with new material on such topics as commercially available robots for the home as well as how to organize your own robot competitions.

I.3 What You Will Learn

In the more than three dozen chapters in this book you will learn about a sweeping variety of technologies, all aimed at helping you learn robot design, construction, and application. You'll learn about:

- *Robot-building fundamentals.* How a robot is put together using commonly available parts such as plastic, wood, and aluminum.
- *Locomotion engineering.* How motors, gears, wheels, and legs are used to propel your robot over the ground.
- *Constructing robotic arms and hands.* How to use mechanical linkages to grasp and pick up objects.
- *Sensor design.* How sensors are used to detect objects, measure distance, and navigate open space.

- *Adding sound capabilities.* Giving your robot creation the power of voice and sound effects so that it can talk to you, and you can talk back.
- *Remote control.* How to operate and train your robot using wired and wireless remote control.
- *Computer control.* How to use and program a computer or microcontroller for operating a robot.

Most important, you will gain new insights into problem solving and looking at devices, parts, and materials from a different perspective. No longer will you look at an old CD-player or toy as just junk, but as the potential starting point or parts source for your own creations.

I.4 How to Use This Book

Robot Builder's Bonanza is divided into seven main parts. Each section covers a major component of the common personal or hobby (as opposed to commercial or industrial) robot. The sections are as follows:

1. *Robot Basics.* What you need to get started; setting up shop; how and where to buy robot parts.
2. *Robot Platform Construction.* Robots made of plastic, wood, and metal; working with common metal stock; converting toys into robots or using other mechanical odds and ends to create robots.
3. *Computers and Electronic Control.* An explanation of computer operation; introduction to programming; interfacing computers and controllers to electronic devices.
4. *Power, Motors, and Locomotion.* Using batteries; powering the robot; working with DC, stepper, and servo motors; gear trains; walking robot systems; special robot locomotion systems.
5. *Practical Robotics Projects.* Over a half-dozen step-by-step projects for building wheels and legged robot platforms; arm systems; gripper design.
6. *Sensors and Navigation.* Speech synthesis and recognition; sound detection; robot eyes; smoke, flame, and heat detection; collision detection and avoidance; ultrasonic and infrared ranging; infrared beacon systems; track guidance navigation.
7. *Putting It All Together.* Discussion on the techniques for integrating different parts together into a single robot; finding and efficiently fixing the problems you encounter along the way; putting on a robot competition.

Many chapters present one or more projects that you can duplicate for your own robot creations. Whenever practical, the components were designed as discrete building blocks, so that you can combine the blocks in just about any configuration you desire. The robot you create will be uniquely yours and yours alone.

The *Robot Builder's Bonanza* is not so much a textbook on how to build robots but a treasure map. The trails and paths provided between these covers lead you on your way to building one or more complete and fully functional robots. You decide how you want your robots to appear and what you want your robots to do.

I.5 Expertise You Need

Robot Builder's Bonanza doesn't contain a lot of hard-to-decipher formulas, unrealistic assumptions about your level of electronic or mechanical expertise, or complex designs that only a seasoned professional can tackle. This book was written so that just about anyone can enjoy the thrill and excitement of building a robot. Most of the projects can be duplicated without expensive lab equipment, precision tools, or specialized materials, and at a cost that won't wear the numbers off your credit cards.

If you have some experience in electronics, mechanics, or robot building in general, you can skip around and read only those chapters that provide the information you're looking for. Like the robot designs presented, the chapters are very much stand-alone modules. This allows you to pick and choose, using your time to its best advantage.

However, if you're new to robot building, and the varied disciplines that go into it, you should take a more pedestrian approach and read as much of the book as possible. In this way, you'll get a thorough understanding of how robots tick. When you finish the book, you'll know the kind of robot(s) you'll want to make, and how you'll make them.

I.6 Conventions Used in This Book

Mechanical drawings, schematics, and other diagrams have been created using standard conventions and should not look significantly different from other graphics found in different sources. The basic symbols used in the diagrams will be explained as you read through the book.

If there continue to be symbols or components that are confusing to you, please look at the different reference material listed in the appendices.

Integrated circuits are referenced by their part number. Remember that the part number and the operation of the part can vary when different technologies are used. This means that when you are given a TTL chip of a specific technology (i.e., LS) do not assume that other chips with the same part number, but different technology, can be used.

Details on the specific parts used in the circuits are provided in the parts list tables that accompany the schematic. Refer to the parts list for information on resistor and capacitor type, tolerance, and wattage or voltage rating.

In all full-circuit schematics, the parts are referenced by component type and number.

- IC means an integrated circuit (IC). Some integrated circuits will be referenced by their part number or function if this simplifies the explanation of the circuit and there are many different substitute parts available.
- R means a resistor or potentiometer (variable resistor). All resistors are 1/4 W, 5% tolerance, unless otherwise specified.
- C means a capacitor. Capacitors can be of any type unless specified.
- D means a diode, a zener diode, and, sometimes a light-sensitive photodiode.
- Q means a transistor and, sometimes, a light-sensitive phototransistor.

- LED means a light-emitting diode (most any visible LED will do unless the parts list specifically calls for an infrared or other special-purpose LED).
- XTAL means a crystal or ceramic resonator.
- Finally, S or SW means a switch; RL means a relay; SPKR, a speaker; TR, a transducer (usually ultrasonic); and MIC, a microphone.

Enough talk. Turn the page and open your map. The treasure awaits you.

PART 1

ROBOT BASICS

CHAPTER 1

THE ROBOT EXPERIMENTER

Alone he sits in a dank and musty basement, as he's done countless long nights before; pouring over plans, making endless calculations, and then pounding his creation into being. With each strike of his ball-peen hammer, an ear-shattering bong and echoes ring through the house. Slowly, his work takes shape and form—it started as an unrecognizable blob of metal and plastic, then became an eerie silhouette, then . . .

Brilliant and talented, but perhaps a bit crazed, he is before his time—an adventurer who belongs neither to science nor fiction. He is the robot experimenter, and all he wants to do is make a mechanical creature that will ultimately become his servant and companion. The future hides not what he will ultimately do with his creation, but what his creation will do with him.

Okay, maybe this is a rather dark view of the present-day hobby robotics experimenter. But though you may find a dash of the melodramatic in it, the picture is not entirely unrealistic. It's a view held by many outsiders to the robot-building craft. It's a view that's over 100 years old, from the time when the prospects of building a humanlike machine first came within technology's grasp. It's a view that will continue for another 100 years, perhaps beyond.

Like it or not, if you're a robot experimenter, you are considered to be on society's fringes: an oddball, an egghead, and—yes, let's get it all out—possibly someone looking for a kind of malevolent power!

As a robot experimenter, you're not unlike Victor Frankenstein, the old-world doctor from Mary Wollstonecraft Shelley's immortal 1818 horror thriller. Instead of robbing graves in the still of night, you "rob" electronic stores, flea markets, surplus outlets, and other specialty shops in your unrelenting quest—your thirst—for all kinds and sizes of motors, batteries, gears, wires, switches, and other odds and ends. Like Dr. Frankenstein, you galvanize life from these "dead" parts.

If you have not yet built your first robot, you're in for a wonderful experience. Watching your creation scoot around the floor or table can be exhilarating. Those around you may not immediately share your excitement, but you know that you've built something—however humble—with your own hands and ingenuity.

And yet if you have built a robot, you also know of the heartache and frustration inherent in the process. You know that not every design works and that even a simple engineering flaw can cost weeks of work, not to mention ruined parts. This book will help you—beginner and experienced robot maker alike—get the most out of your robotics hobby.

1.1 The Building-Block Approach

One of the best ways to experiment with and learn about hobby robots is to construct individual robot components, then combine the completed modules to make a finished, fully functional machine. For maximum flexibility, these modules should be interchangeable whenever possible. You should be able to choose locomotion system "A" to work with appendage system "B," and operate the mixture with control system "C"—or any variation thereof.

As you start trying to create your own robots, using a building-blocks approach allows you to make relatively simple and straightforward changes and updates. When designed and constructed properly, the different building blocks, as shown in diagram form in Fig. 1-1, may be shared among a variety of robots.

Most of the building-block designs presented in the following chapters are complete, working subsystems. Some operate without ever being attached to a robot or control computer. The way you interface the modules is up to you and will require some forethought and attention on your part (this book does not provide all the answers!). Feel free to experiment with each subsystem, altering it and improving upon it as you see fit. When it works the way you want, incorporate it into your robot, or save it for a future project.

1.2 Basic Skills

What skills do you need as a robot experimenter? Certainly, if you are already well versed in electronics, programming, and mechanical design, you are on your way to becoming a robot experimenter. But intimate knowledge of these fields is not absolutely necessary; all you really need to start in the right direction as a robot experimenter is a basic familiarity with electronic theory, programming concepts, and mechanics (or time and interest to study the craft). The rest you can learn as you go. If you feel that you're lacking in either beginning electronics or mechanics, pick up a book or two on these subjects at the bookstore or library (see Appendix A, "Further Reading," for a selected list of suggested books and mag-

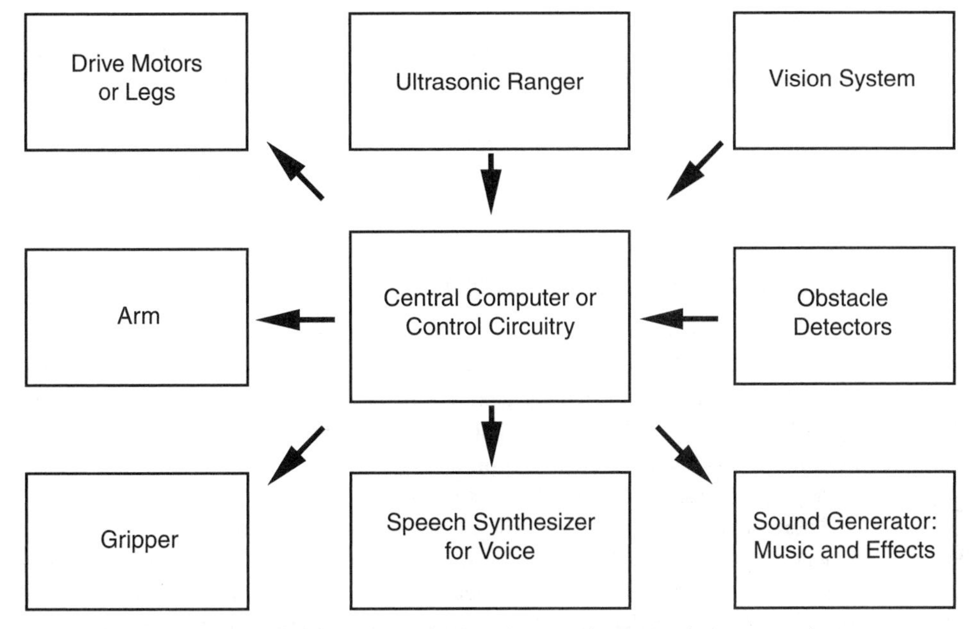

FIGURE 1-1 The basic building blocks of a fully functional robot, including central processor (brain), locomotion (motors), and sensors (switches, sonar, etc.).

azines). In addition, you may wish to read through the seven chapters in Part 1 of this book to learn more about the fundamentals of electronics and computer programming.

1.2.1 ELECTRONICS BACKGROUND

Start by studying analog and digital electronic theory, and learn the function of resistors, capacitors, transistors, and other common electronic components. Your knowledge need not be extensive, just enough so that you can build and troubleshoot electronic circuits for your robot. You'll start out with simple circuits with a minimum of parts, and go from there. As your skills increase, you'll be able to design your own circuits from scratch, or at the very least, customize existing circuits to match your needs.

Schematic diagrams are a kind of recipe for electronic circuits. The designs in this book, as well as those in most any book that deals with electronics, are in schematic form. You owe it to yourself to learn how to read a schematic as there are really only a few dozen common schematic symbols you will have to familiarize yourself with. Several books have been written on how to read schematic diagrams, and the basics are also covered in Chapter 5, "Electronic Components." See also Appendix A for a list of suggested books on robotics.

1.2.2 PROGRAMMING BACKGROUND

Sophisticated robots use a computer or microcontroller to manage their actions. In this book you'll find plenty of projects, plans, and solutions for connecting the hardware of your robot to any of several kinds of robot "brains." Like all computers, the ones for robot control need to be programmed. If you are new or relatively new to computers and programming, start with a beginners' computer book, then move up to more advanced texts.

Chapter 13, "Programming Fundamentals," covers the programming basics. If you've never programmed before, you are probably expecting that there is a lot of knowledge that you must have to successfully program a computer. Actually there is about a half dozen basic programming concepts that once you understand completely you will be able to program just about any computer system in just about any programming language.

1.2.3 MECHANICAL BACKGROUND

Some robot builders are more comfortable with the mechanical side of robot building than the electronic and programming sides—they can see gears meshing and pulleys moving. Regardless of your comfort level with mechanical design, you do not need to possess an extensive knowledge of mechanical and engineering theory to build robots. This book provides some mechanical theory as it pertains to robot building, but you may want to supplement your knowledge with books or study aids.

There are a wealth of books, articles, and online reading materials on mechanical design equations and engineering formulas for you to draw upon when you are designing and building robots. This eliminates the need for this book to repeat this information, but like the information provided in electronics and programming, this book gives you many of the basics required to cobble together the robot's mechanical systems.

1.2.4 WORKSHOP APTITUDE

To be a successful robot builder, you must be comfortable working with your hands and thinking problems through from start to finish. You should know how to use common shop tools, and all related safety procedures, and have some basic familiarity with working with wood, lightweight metals (mostly aluminum), and plastic. Once more, if you feel your skills aren't up to par, read up on the subject and try your hand at a simple project or two first.

You'll find construction tips and techniques throughout this book, but nothing beats hands-on shop experience. With experience comes confidence, and with both comes more professional results. Work at it long enough, and the robots you build may be indistinguishable from store-bought models (in appearance, not capability; yours will undoubtedly be far more sophisticated!).

1.2.5 THE TWO MOST IMPORTANT SKILLS

Two important skills that you can't develop from reading books are *patience* and the *willingness to learn.* Both are absolutely essential if you want to build your own working robots. Give yourself time to experiment with your projects. Don't rush into things because you are bound to make mistakes if you do. If a problem continues to nag at you, put the project aside and let it sit for a few days. Keep a small notebook handy and jot down your ideas so you won't forget them.

If trouble persists, perhaps you need to bone up on the subject before you can adequately tackle the problem. Take the time to learn more about the various sciences and disciplines involved. While you are looking for ways to combat your current dilemma, you are increasing your general robot-building knowledge. Research is never in vain.

1.3 Ready-Made, Kits, or Do-It-Yourself?

This is a wonderful time to be an amateur robot builder. Not only can you construct robots from scratch, you can buy any of several dozen robot kits and assemble them using a screwdriver and other common tools. If you don't particularly like the construction aspects of robotics, you can even purchase ready-made robots—no assembly required. With a ready-made robot you can spend all your time connecting sensors and other apparatuses to it and figuring out new and better ways to program it.

You might go for a third option: hacking an existing platform. With a bit of imagination and luck you can find a toy or some hardware that provides you with an excellent starting place for your own robot. The only limitation is to remember not to take something apart that somebody else values!

Whether you choose to buy a robot in ready-made or kit form, or build your own from basic parts or the ground up, it's important that you match your skills to the project. This is especially true if you are just starting out. While you may seek the challenge of a complex project, if it's beyond your present skills and knowledge level you'll likely become frustrated and abandon robotics before you've given it a fair chance. If you want to build your own robot, start with a simple design—a small rover, like those in Chapters 8 through 11. For now, stay away from the more complex walking and heavy-duty robots.

1.4 The Mind of the Robot Experimenter

Robot experimenters have a unique way of looking at things. They take nothing for granted.

For example, at a restaurant it's the robot experimenter who collects the carcasses of lobsters and crabs to learn how these ocean creatures use articulated joints, in which the muscles and tendons are inside the bone. Perhaps the articulation and structure of a lobster leg can be duplicated in the design of a robotic arm . . .

- At a county fair, it's the robot experimenter who studies the way the egg-beater ride works, watching the various gears spin in perfect unison, perhaps asking themselves if the gear train can be duplicated in an unusual robot locomotion system.
- At a phone booth, it's the robot experimenter who listens to the tones emitted when the buttons are pressed. These tones, the experimenter knows, trigger circuitry at the phone company office to call a specific telephone out of all the millions in the world. Perhaps these or similar tones can be used to remotely control a robot.
- At work on the computer, it's the robot experimenter who rightly assumes that if a computer can control a printer or plotter through an interface port, the same computer and interface can be used to control a robot.

- In the toy store, while children are looking at the new movie-based toys, the robot experimenter sees a vast array of different bases that can be used as the basis for a brand-new robot.
- When taking a snapshot at a family gathering, it's the robot experimenter who studies the inner workings of the automatic focus system of the camera. The camera uses ultrasonic sound waves to measure distance and automatically adjusts its lens to keep things in focus. The same system should be adaptable to a robot, enabling it to judge distances and "see" with sound.

The list could go on and on.

All around us, from nature's designs to the latest electronic gadgets, is an infinite number of ways to make better and increasingly clever robots. Uncovering these solutions requires extrapolation—figuring out how to apply one design and make it work in another application, then experimenting with the contraption until everything works.

1.5 From Here

To learn more about . . .	***Read***
Fundamentals of electronics and basics on how to read a schematic	Chapter 5, "Electronic Components"
Electronics construction techniques	Chapter 7, "Electronic Construction Techniques"
Computer programming fundamentals	Chapter 13, "Programming Fundamentals"
Robot construction using wood, plastic, and metal	Chapters 8 to 10
Making robots from old toys	Chapter 11, "Hacking Toys"
What your robot should do	Chapter 37, "Robot Tasks, Operations, and Behaviors"

CHAPTER 2

ANATOMY OF A ROBOT

We humans are fortunate. The human body is, all things considered, a nearly perfect machine: it is (usually) intelligent, it can lift heavy loads, it can move itself around, and it has built-in protective mechanisms to feed itself when hungry or to run away when threatened. Other living creatures on this earth possess similar functions, though not always in the same form.

Robots are often modeled after humans, if not in form then at least in function. For decades, scientists and experimenters have tried to duplicate the human body, to create machines with intelligence, strength, mobility, and auto-sensory mechanisms. That goal has not yet been realized, but perhaps some day it will.

Nature provides a striking model for robot experimenters to mimic, and it is up to us to take the challenge. Some, but by no means all, of nature's mechanisms—human or otherwise—can be duplicated to some extent in the robot shop. Robots can be built with eyes to see, ears to hear, a mouth to speak, and appendages and locomotion systems of one kind or another to manipulate the environment and explore surroundings.

This is fine theory; what about real life? Exactly what constitutes a real hobby robot? What basic parts must a machine have before it can be given the title *robot*? Let's take a close look in this chapter at the anatomy of robots and the kinds of materials hobbyists use to construct them. For the sake of simplicity, not every robot subsystem in existence will be covered, just the components that are most often found in amateur and hobby robots.

2.1 Tethered versus Self-Contained

People like to debate what makes a machine a real robot. One side says that a robot is a completely *self-contained, autonomous* (self-governed) machine that needs only occasional instructions from its master to set it about its various tasks. A self-contained robot has its own power system, brain, wheels (or legs or tracks), and manipulating devices such as claws or hands. This robot does not depend on any other mechanism or system to perform its tasks. It is complete in and of itself.

The other side says that a robot is anything that moves under its own power for the purpose of performing near-human tasks (this is, in fact, the definition of the word *robot* in many dictionaries). The mechanism that does the actual task is the robot itself; the support electronics or components may be separate. The link between the robot and its control components might be a wire, a beam of infrared light, or a radio signal.

In an experimental robot from 1969 a man sat inside the mechanism and operated it, almost as if driving a car. The purpose of this four-legged lorry was not to create a self-contained robot but to further the development of *cybernetic anthropomorphous machines.* These were otherwise known as *cyborgs,* a concept further popularized by writer Martin Caidin in his 1973 novel *Cyborg* (which served as the inspiration for the 1970s television series, *The Six Million Dollar Man*).

The semantics of robot design won't be argued here (this book is a treasure map after all, not a textbook on theory), but it's still necessary to establish some of the basic characteristics of robots. What makes a robot a robot and not just another machine? For the purposes of this book, let's consider a robot as any device that—in one way or another—mimics human or animal functions. How the robot does this is of no concern; the fact that it does it at all is enough.

The functions that are of interest to the robot builder run a wide gamut: from listening to sounds and acting on them, to talking and walking or moving across the floor, to picking up objects and sensing special conditions such as heat, flames, or light. Therefore, when we talk about a robot it could very well be a self-contained automaton that takes care of itself, perhaps even programming its own brain and learning from its surroundings and environment. Or it could be a small motorized cart operated by a strict set of predetermined instructions that repeats the same task over and over again until its batteries wear out. Or it could be a radio-controlled arm operated manually from a control panel. Each is no less a robot than the others, though some are more useful and flexible. As you'll discover in this chapter and those that follow, how complex your robot creations are is completely up to you.

2.2 Mobile versus Stationary

Not all robots are meant to scoot around the floor. Some are designed to stay put and manipulate some object placed before them. In fact, outside of the research lab and hobbyist garage, the most common types of robots, those used in manufacturing, are *stationary.* Such robots assist in making cars, appliances, and even other robots!

Other common kinds of stationary robots act as shields between a human operator or supervisor and some dangerous material, such as radioactive isotopes or caustic chemicals. Stationary robots are armlike contraptions equipped with grippers or special tools. For example, a robot designed for welding the parts of a car is equipped with a welding torch on the end of its arm. The arm itself moves into position for the weld, while the car slowly passes in front of the robot on a conveyor belt.

Conversely, *mobile* robots are designed to move from one place to another. Wheels, tracks, or legs allow the robot to traverse a terrain. Mobile robots may also feature an armlike appendage that allows them to manipulate objects. Of the two—stationary or mobile—the mobile robot is probably the more popular project for hobbyists to build. There's something endearing about a robot that scampers across the floor, either chasing or being chased by the cat.

As a serious robot experimenter, you should not overlook the challenge and education you can gain from building both types of robots. Stationary robots typically require greater precision, power, and balance, since they are designed to grasp and lift objects—hopefully not destroying the objects they handle in the process. Likewise, mobile robots present their own difficulties, such as maneuverability, adequate power supply, and avoiding collisions.

2.3 Autonomous versus Teleoperated

Among the first robots ever demonstrated for a live audience were the "automatons" of the Middle Ages. These robots were actually machines either performing a preset series of motions or remotely controlled by a person off stage. No matter. People thrilled at the concept of the robot, which many anticipated would be an integral part of their near futures.

These days, the classic view of the robot is a fully autonomous machine, like Robby from *Forbidden Planet,* Robot B-9 from *Lost in Space,* or R2-D2 from *Star Wars.* With these robots (or at least the make-believe fictional versions), there's no human operator, no remote control, no "man behind the curtain." While many actual robots are indeed fully autonomous, many of the most important robots of the past few decades have been *teleoperated.* A teleoperated robot is one that is commanded by a human and operated by remote control. The typical tele-robot uses a video camera that serves as the eyes for the human operator. From some distance—perhaps as near as a few feet to as distant as several million miles—the operator views the scene before the robot and commands it accordingly.

The teleoperated robot of today is a far cry from the radio-controlled robots of the world's fairs of the 1930s and 1940s. Many tele-robots, like the world-famous Mars Rovers Sojourner, Spirit, and Opportunity, are actually half remote controlled and half autonomous. The low-level functions of the robot are handled by microprocessors onboard the machines. The human intervenes to give general-purpose commands, such as "go forward 10 feet" or "hide, here comes a Martian!" The robot is able to carry out basic instructions on its own, freeing the human operator from the need to control every small aspect of the machine's behavior.

The notion of tele-robotics is certainly not new—it goes back to at least the 1940s and the short story "Waldo" by noted science fiction author Robert Heinlein. It was a fantastic idea at the time, but today modern science makes it eminently possible. Stereo video cameras give a human operator 3-D depth perception. Sensors on motors and robotic arms provide feedback to the human operator, who can then feel the motion of the machine or the strain caused by some obstacle. Virtual reality helmets, gloves, and motion platforms literally put the operator "in the driver's seat."

This book doesn't discuss tele-robotics in any extended way, but if the concept interests you, read more about it and perhaps construct a simple tele-robot using a radio or infrared link and a video camera. See Appendix A, "Further Reading," for more information.

2.4 The Body of the Robot

Like the human body, the body of a robot—at least a self-contained one—holds all its vital parts. The body is the superstructure that prevents its electronic and electromechanical guts from spilling out. Robot bodies go by many names, including frame and chassis, but the idea is the same.

2.4.1 SKELETAL STRUCTURES

In nature and in robotics, there are two general types of support frames: endoskeleton and exoskeleton. Which is better? Both: in nature, the living conditions of the animal and its eating and survival tactics determine which skeleton is best. The same is true of robots.

- Endoskeleton support frames are found in many critters, including humans, mammals, reptiles, and most fish. The skeletal structure is on the inside; the organs, muscles, body tissues, and skin are on the outside of the bones. The endoskeleton is a characteristic of vertebrates.
- Exoskeleton support frames have the "bones" on the outside of the organs and muscles. Common creatures with exoskeletons are spiders, all shellfish such as lobsters and crabs, and an endless variety of insects.

2.4.2 FRAME CONSTRUCTION

The main structure of the robot is generally a wood, plastic, or metal frame, which is constructed a little like the frame of a house—with a bottom, top, and sides. This gives the automaton a boxy or cylindrical shape, though any shape is possible. It could even emulate the human form, like the robot Cylon Centurions in *Battlestar Galactica*.

Onto the frame of the robot are attached motors, batteries, electronic circuit boards, and other necessary components. In this design, the main support structure of the robot can be considered an exoskeleton because it is outside the "major organs." Further, this design lacks a central "spine," a characteristic of endoskeletal systems and one of the first things most of us think about when we try to model robots after humans. In many cases, a shell is sometimes placed over these robots, but the "skin" is for looks only (and sometimes the protection of the internal components), not support. Of course, some robots are designed with endoskeletal structures, but most such creatures are reserved for high-tech research and development projects and science fiction films. For the most part, the main bodies of your robots will have an exoskeleton support structure because they are cheaper to build, stronger, and less prone to problems.

2.4.3 SIZE AND SHAPE

The size and shape of the robot can vary greatly, and size alone does not determine the intelligence of the machine or its capabilities. Home-brew robots are generally the size of a small dog, although some are as compact as an aquarium turtle and a few as large as Arnold Schwarzenegger. The overall shape of the robot is generally dictated by the internal components that make up the machine, but most designs fall into one of the following categories:

- *Turtle.* Turtle robots are simple and compact, designed primarily for tabletop robotics. Turtlebots get their name from the fact that their bodies somewhat resemble the shell of a turtle and also from the Logo programming language, which incorporated turtle graphics, adapted for robotics use in the 1970s.
- *Vehicle.* These scooter-type robots are small automatons with wheels. In hobby robotics, they are often built using odds and ends like used compact discs, extra LEGO parts, or the chassis of a radio-controlled car. The small vehicular robot is also used in science and industry: the Mars Rovers, built by NASA, to explore the surface of Mars are examples of this type of robot.
- *Rover.* Greatly resembling the famous R2-D2 of *Star Wars,* rovers tend to be short and stout and are typically built with at least some humanlike capabilities, such as firefighting or intruder detection. Some closely resemble a garbage can—in fact, not a few hobby robots are actually built from metal and plastic trash cans! Despite the euphemistic title, garbage can robots represent an extremely workable design approach.
- *Walker.* A walking robot uses legs, not wheels or tracks, to move about. Most walker 'bots have six legs, like an insect, because they provide excellent support and balance. However, robots with as few as one leg (hoppers) and as many as 8 to 10 legs have been successfully built and demonstrated.
- *Appendage.* Appendage designs are used specifically with robotic arms, whether the arm is attached to a robot or is a stand-alone mechanism.
- *Android.* Android robots are specifically modeled after the human form and are the type most people picture when talk turns to robots. Realistically, android designs are the most restrictive and least workable, inside or outside the robot lab.

This book provides designs and construction details for at least one robot in each of the preceding types except android. That will be left to another book.

2.4.4 FLESH AND BONE

In the 1927 movie classic *Metropolis,* an evil scientist, Dr. Rotwang, transforms a cold and calculating robot into the body of a beautiful woman. This film, generally considered to be the first science fiction cinema epic, also set the psychological stage for later movies, particularly those of the 1950s and 1960s. The shallow and stereotypical character of Dr. Rot-

FIGURE 2-1 The evil Dr. Rotwang and the robot, from the classic motion picture *Metropolis.*

wang, shown in the movie still in Fig. 2-1, proved to be a common theme in countless movies. The shapely robotrix changed form in these other films, but not its evil character. Robots have often been depicted as metal creatures with hearts as cold as their steel bodies.

Which brings us to an interesting question: are all "real" robots made of heavy-gauge steel, stuff so thick that bullets, disinto-ray guns, even atomic bombs can't penetrate? Indeed, while metal of one kind or another is a major component of robot bodies, the list of materials you can use is much larger and diverse. Hobby robots can be easily constructed from aluminum, steel, tin, wood, plastic, paper, foam, or a combination of them all:

- *Aluminum.* Aluminum is the best all-around robot-building material for medium and large machines because it is exceptionally strong for its weight. Aluminum is easy to cut and bend using ordinary shop tools. It is commonly available in long lengths of various shapes, but it is somewhat expensive.
- *Steel.* Although sometimes used in the structural frame of a robot because of its strength, steel is difficult to cut and shape without special tools. Stainless steel, although expensive, is sometimes used for precision components, like arms and hands, and also for parts that require more strength than a lightweight metal (such as aluminum) can provide.
- *Tin, iron, and brass.* Tin and iron are common hardware metals that are often used to make angle brackets, sheet metal (various thicknesses from 1⁄32 in on up), and (when galvanized) nail plates for house framing. Brass is often found in decorative trim for home construction projects and as raw construction material for hobby models. All three metals are stronger and heavier than aluminum. Cost: fairly cheap.
- *Wood.* Wood is an excellent material for robot bodies, although you may not want to use it in all your designs. Wood is easy to work with, can be sanded and sawed to any shape, doesn't conduct electricity (avoids short circuits), and is available everywhere. Disadvantage: ordinary construction plywood is rather weak for its weight, so you need fairly large pieces to provide stability. Better yet, use the more dense (and expensive) multi-ply hardwoods for model airplane and sailboat construction. Common thicknesses are 1⁄4 to 1⁄2 in—perfect for most robot projects.

- *Plastic.* Everything is going plastic these days, including robots. Pound for pound, plastic has more strength than many metals, yet is easier to work with. You can cut it, shape it, drill it, and even glue it. To use plastic effectively you must have some special tools, and extruded pieces may be hard to find unless you live near a well-stocked plastic specialty store. Mail order is an alternative.
- *Rigid expanded plastic sheet.* Expanded sheet plastics are often constructed like a sandwich, with thin outer sheets on the top and bottom and a thicker expanded (air-filled) center section. When cut, the expanded center section often has a kind of foamlike appearance, but the plastic itself is stiff. Rigid expanded plastic sheets are remarkably lightweight for their thickness, making them ideal for small robots. These sheets are known by various trade names such as Sintra and are available at industrial plastics supply outlets.
- *Foamboard.* Art supply stores stock what's known as foamboard (or foam core), a special construction material typically used for building models. Foamboard is a sandwich of paper or plastic glued to both sides of a layer of densely compressed foam. The material comes in sizes from ⅛ in to over ½ in, with ¼ to ⅓ in being fairly common. The board can be readily cut with a small hobby saw (paper-laminated foamboard can be cut with a sharp knife; plastic-laminated foamboard should be cut with a saw). Foamboard is especially well suited for small robots where light weight is of extreme importance.

2.5 Power Systems

We eat food that is processed by the stomach and intestines to make fuel for our muscles, bones, skin, and the rest of our body. While you could probably design a digestive system for a robot and feed it hamburgers, french fries, and other foods, an easier way to generate the power to make your robot go is to use commercially available batteries, connect the batteries to the robot's motors, circuits, and other parts, and you're all set.

2.5.1 TYPES OF BATTERIES

There are several different types of batteries, and Chapter 17, "Batteries and Robot Power Supplies," goes into more detail about them. Here are a few quick facts to start you off.

Batteries generate DC current and come in two distinct categories: rechargeable and nonrechargeable (for now, let's forget the nondescriptive terms like storage, primary, and secondary). Nonrechargeable batteries include the standard zinc and alkaline cells you buy at the supermarket, as well as special-purpose lithium and mercury cells for calculators, smoke detectors, watches, and hearing aids. A few of these (namely, lithium) have practical uses in hobby robotics.

Rechargeable batteries include nickle-metal hydride (NiMH), nickel-cadmium (Ni-Cad), gelled electrolyte, sealed lead-acid cells, and special alkaline. NiMH batteries are a popular choice because they are relatively easy to find, come in popular household sizes (D, C, etc.), can be recharged many hundreds of times using an inexpensive recharger, and are safer for the environment than the other options. Gelled electrolyte (gel-cell) and lead-acid batteries provide longer-lasting power, but they are heavy and bulky.

2.5.2 ALTERNATIVE POWER SOURCES

Batteries are required in most fully self-contained mobile robots because the only automatons connected by power cord to an electrical socket are found in cartoons. That doesn't mean other power sources, including AC or even solar, can't be used in some of your robot designs. On the contrary, stationary robot arms don't have to be capable of moving around the room; they are designed to be placed about the perimeter of the workplace and perform within this predefined area. The motors and control circuits may very well run off AC power, thus freeing you from replacing batteries and worrying about operating times and recharging periods.

This doesn't mean that AC power is necessarily the preferred method. High-voltage AC poses greater shock hazards. Electronic logic circuits ultimately run off DC power, even when the equipment is plugged into an AC outlet, which makes DC power a logical choice.

One alternative to batteries in an all-DC robot system is to construct an AC-operated power station that provides your robot with regulated DC. The power station converts the AC to DC and provides a number of different voltage levels for the various components in your robot, including the motors. This saves you from having to buy new batteries or recharge the robot's batteries all the time.

Small robots can be powered by solar energy when they are equipped with suitable solar cells. Solar-powered robots can tap their motive energy directly from the cells, or the cells can charge up a battery over time. Solar-powered 'bots are a favorite of those designers using the BEAM philosophy—a type of robot design that stresses simplicity, including the power supply of the machine.

2.5.3 PRESSURE SYSTEMS

Two other forms of robotic power, which will not be discussed in depth in this book, are hydraulic and pneumatic. Hydraulic power uses oil or fluid pressure to move linkages. You've seen hydraulic power at work if you've ever watched a bulldozer move dirt from place to place. And while you drive you use it every day when you press down on the brake pedal. Similarly, pneumatic power uses air pressure to move linkages. Pneumatic systems are cleaner than hydraulic systems, but all things considered they aren't as powerful.

Both hydraulic and pneumatic systems must be pressurized to work, and this pressurization is most often performed by a pump. The pump is driven by an electric motor, so in a way robots that use hydraulics or pneumatics are fundamentally electrical. The exception to this is when a pressurized tank, like a scuba tank, is used to provide air pressure in a pneumatic robot system. Eventually, the tank becomes depleted and must either be recharged using some pump on the robot or removed and refilled using a compressor.

Hydraulic and pneumatic systems are rather difficult to implement effectively, but they provide an extra measure of power in comparison to DC and AC motors. With a few hundred dollars in surplus pneumatic cylinders, hoses, fittings, solenoid valves, and a pressure supply (battery-powered pump, air tank, regulator), you could conceivably build a hobby robot that picks up chairs, bicycles, even people!

2.6 Locomotion Systems

As previously mentioned, some robots aren't designed to move around. These include robotic arms, which manipulate objects placed within a work area. But these are exceptions rather than the rule for hobby robots, which are typically designed to get around in this world. They do so in a variety of ways, from using wheels to legs to tank tracks. In each case, the locomotion system is driven by a motor, which turns a shaft, cam, or lever. This motive force affects forward or backward movement.

2.6.1 WHEELS

Wheels are the most popular method for providing robots with mobility. There may be no animals on this earth that use wheels to get around, but for us robot builders it's the simple and foolproof choice. Robot wheels can be just about any size, limited only by the dimensions of the robot and your outlandish imagination. Turtle robots usually have small wheels, less than 2 or 3 in in diameter. Medium-sized rover-type robots use wheels with diameters up to 7 or 8 in. A few unusual designs call for bicycle wheels, which despite their size are lightweight but very sturdy.

Robots can have just about any number of wheels, although two is the most common, creating a *differentially driven robot* (see Fig. 2-2). In this case, the robot is balanced on the two wheels by one or two free-rolling casters, or perhaps even a third swivel wheel. Larger, more powerful four- and six-wheel differentially driven robots have also been built. In these cases all the wheels on a side turn together and provide the robot with better stability and traction than just two wheels. There is a great deal of friction to be overcome, which necessitates powerful drive motors.

Other common wheeled robots use a layout similar to a car or a tricycle. These robot chassis do not have the agility or stability of the differentially driven robot, but they can often be easily adapted from commercially available products such as toys.

2.6.2 LEGS

A minority of robots—particularly the hobby kind—are designed with legs, and such robots can be conversation pieces all their own. You must overcome many difficulties to design and

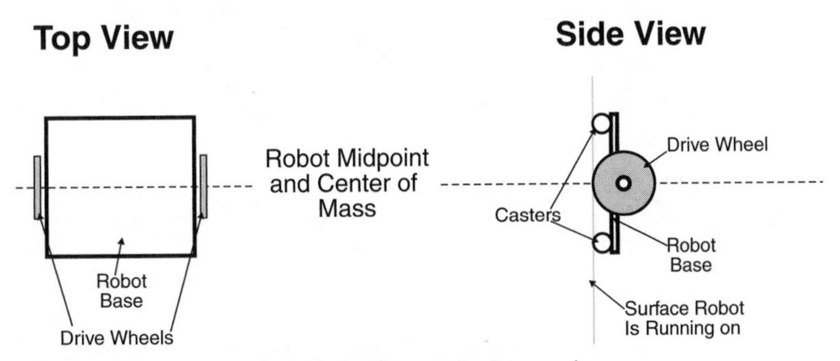

FIGURE 2-2 Design of an ideal differentially driven robot.

construct a legged robot. First, there is the question of the number of legs and how the legs provide stability when the robot is in motion or when it's standing still. Then there is the question of how the legs propel the robot forward or backward—and more difficult still the question of how to turn the robot so it can navigate a corner.

Legged robots create some tough challenges, but they are not insurmountable. Legged robots are a challenge to design and build, but they provide you with an extra level of mobility that wheeled robots do not. Wheel-based robots may have a difficult time navigating through rough terrain, but legged robots can easily walk right over small ditches and obstacles.

A few daring robot experimenters have come out with two-legged robots, but the challenges in assuring balance and control render these designs largely impractical for most robot hobbyists. Four-legged robots (quadrapods) are easier to balance, but good locomotion and steering can be difficult to achieve. Robots with six legs (called hexapods) are able to walk at brisk speeds without falling and are more than capable of turning corners, bounding over uneven terrain, and making the neighborhood dogs and cats run for cover.

2.6.3 TRACKS

The basic design of track-driven robots (as shown in Fig. 2-3) is pretty simple and is based on the differentially driven principle used with wheeled robots. Two tracks, one on each side of the robot, act as giant wheels. The tracks turn, like wheels, and the robot lurches forward or backward. For maximum traction, each track is about as long as the robot itself.

Track drive is preferable for many reasons, including the fact that it makes it possible to mow through all sorts of obstacles, like rocks, ditches, and potholes. Given the right track

FIGURE 2-3 The TAB SumoBot is a tracked, differentially driven robot.

material, track drive provides excellent traction, even on slippery surfaces like snow, wet concrete, or a clean kitchen floor. Track-based robots can be challenging to design and build, but with the proper track material along with powerful motors they are an excellent base for robots and offer the advantages of two-wheeled differentially driven robots with greater stability and the ability to traverse uneven terrain.

2.7 Arms and Hands

The ability to handle objects is a trait that has enabled humans, as well as a few other creatures in the animal kingdom, to manipulate the environment. Without our arms and hands, we wouldn't be able to use tools, and without tools we wouldn't be able to build houses, cars, and—hmmm—robots. It makes sense, then, to provide arms and hands to our robot creations so they can manipulate objects and use tools. A commercial industrial robot arm is shown in Fig. 2-4. Chapters 26 through 28 in Part 5 of this book are devoted entirely to robot arms and hands.

FIGURE 2-4 A robotic arm from General Electric is designed for precision manufacturing. (Photo courtesy General Electric.)

You can duplicate human arms in a robot with just a couple of motors, some metal rods, and a few ball bearings. Add a gripper to the end of the robot arm and you've created a complete arm–hand module. Of course, not all robot arms are modeled after the human appendage. Some look more like forklifts than arms, and a few use retractable push rods to move a hand or gripper toward or away from the robot. See Chapter 26, "Reaching Out with Robot Arms," for a more complete discussion of robot arm design. Chapter 27 concentrates on how to build a popular type of robot arm using a variety of construction techniques.

2.7.1 STAND-ALONE OR BUILT-ON MANIPULATORS

Some arms are complete robots in themselves. Car manufacturing robots are really arms that can reach in just about every possible direction with incredible speed and accuracy. You can build a stand-alone robotic arm trainer, which can be used to manipulate objects within a defined workspace. Or you can build an arm and attach it to your robot. Some arm-robot designs concentrate on the arm part much more than the robot part. They are, in fact, little more than arms on wheels.

2.7.2 GRIPPERS

Robot hands are commonly referred to as grippers or end effectors. We'll stick with the simpler sounding *hands* and *grippers* in this book. Robot grippers come in a variety of styles; few are designed to emulate the human counterpart. A functional robot claw can be built that has just two fingers. The fingers close like a vise and can exert, if desired, a surprising amount of pressure. See Chapter 28, "Experimenting with Gripper Designs," for more information.

2.8 Sensory Devices

Imagine a world without sight, sound, touch, smell, or taste. Without these sense *inputs,* we'd be nothing more than an inanimate machine, like the family car or the living room television, waiting for something to command us to do something. Our senses are an integral part of our lives—if not life itself.

It makes good sense (pardon the pun) to provide at least one type of sense into your robot designs. The more senses a robot has, the more it can interact with its environment and respond to it. The capacity for interaction will make the robot better able to go about its business on its own, which makes possible more sophisticated tasks.

Detecting *objects* around the robot is a sensory system commonly given to robots and helps prevent the robot from running into objects, potentially damaging them or the robots themselves, or just pushing against them and running down their batteries. There are a number of different ways of detecting objects that range from being very simple to very sophisticated. See Chapters 29 and 30 for more details regarding different ways objects are detected.

External *sounds* are easy to detect, and unless you're trying to listen for a specific kind of sound, circuits for sound detection are simple and straightforward. Sounds can be control signals (such as clapping to change the motion of the robot) or could even be a type of object detection (the crunch of the collision can be heard and responded to).

Sensitivity to *light* is also common, and trying to follow a light beam is a classic early robot control application. Sometimes the light sensed is restricted to a slender band of infrared for the purpose of sensing the heat of a fire or navigating through a room using an invisible infrared light beam.

Robot eyesight is a completely different matter. The visual scene surrounding the robot must be electronically rendered into a form the circuits on the robot can accept, and the machine must be programmed to understand and act on the shapes it sees. A great deal of experimental work is underway to allow robots to distinguish objects, but true robot vision is limited to well-funded research teams. Chapter 32, "Robot Vision," provides the basics on how to give crude sight to a robot.

Simple pressure sensors can be constructed cheaply and quickly, however, and though they aren't as accurate as commercially manufactured pressure sensors, they are more than adequate for hobby robotics.

The senses of smell and taste aren't generally implemented in robot systems, though some security robots designed for industrial use are outfitted with a gas sensor that, in effect, smells the presence of dangerous toxic gas.

2.9 Output Devices

Output devices are components that relay information from the robot to the outside world. Common output devices in computer-controlled robots include audio outputs, multiple LEDs, the video screen or (liquid crystal display) panel. As with a personal computer, the robot communicates with its master by flashing messages on a screen or panel.

Another popular robotic output device is the speech synthesizer. In the 1968 movie *2001: A Space Odyssey,* Hal the computer talks to its shipmates in a soothing but electronic voice. The idea of a talking computer was a rather novel concept at the time of the movie, but today voice synthesis is commonplace.

Many hobbyists build robots that contain sound and music generators. These generators are commonly used as warning signals, but by far the most frequent application of speech, music, and sound is for entertainment purposes. Somehow, a robot that wakes you up to an electronic rendition of Bach seems a little more human. Projects in robot sound-making circuits are provided in Chapter 31, "Sound Output and Input."

2.10 Smart versus Dumb Robots

There are smart robots and there are dumb robots, but the difference really has nothing to do with intelligence. Even taking into consideration the science of artificial intelligence, all self-contained autonomous robots are fairly unintelligent, no matter how sophisticated the electronic brain that controls it. Intelligence is not a measurement of computing capacity but the ability to reason, to figure out how to do something by examining all the variables and choosing the best course of action, perhaps even coming up with a course that is entirely new.

In this book, the difference between dumb and smart is defined as the ability to take two or more pieces of data and decide on a preprogrammed course of action. Usually, a smart robot is one that is controlled by a computer. However, some amazingly sophisticated actions can be built into an automaton that contains no computer; instead it relies on simple electronics to provide the robot with some known behavior (such is the concept of BEAM robotics). A dumb robot is one that blindly goes about its task, never taking the time to analyze its actions and what impact they may have.

Using a computer as the brains of a robot will provide you with a great deal of operating flexibility. Unlike a control circuit, which is wired according to a schematic plan and performs a specified task, a computer can be electronically rewired using software instructions—that is, programs. To be effective, the electronics must be connected to all the input and output devices as the *feedback* and *control* subsystems, respectively. This includes the drive motors, the motors that control the arm, the speech synthesizer, the pressure sensors, and so forth. This book presents the theory behind computer control and some sample projects in later chapters.

By following some basic rules, using standard components and code templates, it is not terribly difficult to provide computer control in your robot. While most robot controllers are based on small, inexpensive microcontrollers, you can permanently integrate some computers, particularly laptops, into your larger robot projects.

2.11 The Concept of Robot Work

In Czech, the term *robota* means "compulsory worker," a kind of machine slave like that used by Karel Capek in his now classic play *R.U.R.* (*Rossum's Universal Robots*). In many other Baltic languages the term simply means *work*. It is the work aspect of robotics that is often forgotten, but it defines a robot more than anything else. A robot that is not meant to do something useful is not a robot at all but merely a complicated toy or display piece.

That said, designing and building lightweight demonstrator robots provides a perfectly valid way to learn about the robot-building craft. Still, it should not be the end-all of your robot studies. Never lose sight of the fact that a robot is meant to do something—the more, the better! Once you perfect the little tabletop robot you've been working on the past several months, think of ways to apply your improved robot skills to building a more substantial robot that actually performs some job. The job does not need to be labor saving. We'd all like to have a robot maid like Rosie the Robot on the *Jetsons* cartoon series, but, realistically, it's a pretty sophisticated robot that knows the difference between a clean and dirty pair of socks left on the floor.

2.12 From Here

To learn more about . . .	***Read***
Kinds of batteries for robots	Chapter 17, "Batteries and Robot Power Supplies"
Building mobile robots	Part 2, "Robot Platform Construction"
Building a robot with legs	Chapter 24, "Build a Heavy-Duty Six-Legged Walking Robot"
More on robot arms	Chapters 26 to 28
Robotic sensors	Part 6, "Sensors and Navigation"

CHAPTER 3

STRUCTURAL MATERIALS

There is a thrill in looking in a hardware store, hobby shop, parts catalog, or even your basement workshop and contemplating the ways in which different things that catch your eye can be used in creating your own robot. New materials, while sometimes costly, can give your robot a professional look, especially when the time is taken to paint and finish the structure. At the other end of the scale, with a bit of imagination, some pieces of pipe could be built into the chassis of a robot, or some old steel shelving could be used as the basis for a gripper with the throttle linkage from an old lawnmower as the gripper's actuator. Most robots end up being a combination of newly built and reclaimed parts that give them a Frankenstein look that, surprisingly enough, is quite endearing.

In this chapter, you will be introduced to many of the different materials that are used in building robots along with some comments on attaching the various pieces together and finishing the final product. When you are starting out, remember to start small and don't invest heavily in any one type of material or fastener. Each time you begin a new robot or feature, try a different material and see what works best for you.

In Table 3-1, different materials that are often used for robots are listed along with their unique characteristics. This chart will be referred to throughout the chapter and is a good one to go back to when you are trying to decide what material would be best for your robot. *Availability* is how easy it is for you to get the materials. Although some materials obviously are superior, they can be difficult for you to find. The *strength* rating is relative; depending on the actual material purchased, it can vary considerably. *Cutting* indicates how easy it is to cut a material precisely and end up with a smooth edge. The measurement of how little

TABLE 3-1 Robot Structural Materials Comparison Chart

MATERIAL	AVAILABILITY	COST	STRENGTH	CUTTING	STABILITY	VIBRATION
Wood	Excellent	Good	Poor–Excellent	Poor–Excellent	Poor–Excellent	Good
Plywood	Excellent	Fair	Excellent	Fair	Good–Excellent	Excellent
Steel	Good	Good	Excellent	Poor–Fair	Excellent	Good
Aluminum	Good	Fair	Good	Fair–Good	Excellent	Fair
G10FR4	Fair	Poor	Excellent	Poor	Excellent	Excellent
Particle Board	Excellent	Good	Fair–Good	Fair–Good	Poor–Fair	Poor
Cardboard	Excellent	Excellent	Poor–Fair	Excellent	Poor	Poor
Foamboard	Good	Good	Fair	Excellent	Poor	Excellent
Plexiglas	Good	Good	Fair	Poor–Fair	Good	Poor
Polystyrene	Good	Fair	Poor–Fair	Good–Excellent	Good	Poor

the material's dimensions change over time or temperature is *stability,* and *vibration* lists how well a material will stand up to the vibration of working on a robot.

3.1 Paper

Paper probably does not seem like a very likely material to use in the development of robots, but it, along with related products, are often overlooked as materials that are easy to work with and allow fast results. There are a variety of different paper products to choose from and by spending some time in an artists' supply store, you will probably discover some products that you never thought existed. There are a number of different papers, cardboards, foam-backed boards, and so on that could be used as part of a robot's structure. Like anything, it just takes a bit of imagination.

You may not have thought of this, but paper and cardboard are excellent materials for building prototypes of your robots or different parts of them. Even if you have the correct tools for cutting and shaping wood and metal, you can still cut sample parts from a piece of cardboard much faster to ensure proper fit and clearances to other parts. Many people start by building a prototype structure out of the material they are going to use in the robot with the idea that they can cut, saw, drill, or sand the piece into shape as necessary.

Experimenting on the actual robot material is a lot more work (and potentially more expensive) than first cutting out a piece of cardboard or foam-backed paper and trimming it (using either scissors or a hobby knife) into the correct shape. Once the cardboard has been trimmed to the correct shape, it can be pulled out and measured to find the exact changes, which are then transferred to the drawings used for cutting the real material into shape.

Along with scissors and a hobby knife, you might want to also invest in a cutting board (which can be found at the artists' supply store) designed to be impervious to the sharp edge of the knife and some white glue. An obvious way to ensure that the cardboard or foam-backed paper is at the correct dimensions is to glue a printout of the piece that is going to be used in the robot. There is one area that can be a concern if paper products are used for prototyping robot structures and that is the thickness of the material. Only in very rare cases will the cardboard or whatever material you are using for prototyping be the same thickness as the final material. This isn't a major concern because with a little bit of thinking ahead there will be no problem, but it is something to keep in the back of your mind.

3.2 Wood

There are many different wood products to choose from for use in a robot's structure and each one has its own characteristics that affect its suitability in different areas of the robot's structure. The biggest advantages of wood are its ability to be worked using inexpensive hand tools and its fairly low cost. Even for more exotic hardwoods, which can be quite difficult to work and are relatively expensive, you will find that they can be shaped with patience along with care, and even if the piece ends up being ruined replacement pieces can be purchased for just a few dollars.

When using woods as structural pieces in your robot, you are advised to use the hardest woods that you can find (such as maple, cherry, and oak). Very hard woods will obviously handle the greatest loads and, more importantly, will resist splitting at bolt attachment points. Soft woods like spruce and balsa should be avoided for obvious reasons. When fastening pieces of wood to other pieces, nuts and bolts should be used (and not screws which cut their own threads) to minimize the chance that the wood will be damaged and split or the holes will open up due to vibration.

Along with solid pieces of wood, there are a number of composite products that you might want to consider. The most commonly used composite wood product is plywood (Fig. 3-1), which consists of several sheets of wood glued together in such a way that its strength is maximized. Plywood is often manufactured from softer woods (such as spruce) with the final product being much stronger than the sum of its parts. Home construction plywoods should be avoided; instead you should look at aircraft quality plywood (which can be found in small sheets at hobby stores). Aircraft quality plywood is manufactured from better quality woods to more exacting quality levels to ensure that they can withstand large forces and vibration at a fairly light weight—just what is desired in a robot.

Other composite wood products include chipboard and particle board in which varying sizes of wood cuttings are glued together to form a sheet of material. These materials tend to be very heavy and do not have very good structural qualities in terms of mechanical

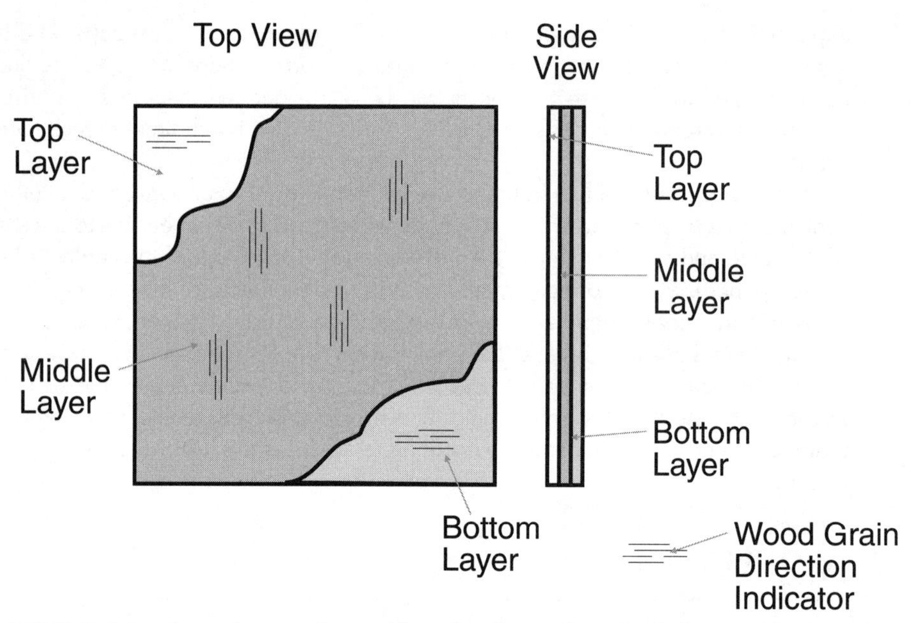

FIGURE 3-1 Plywood is manufactured from thin sheets of wood called veneer glued together with their grains pointing in alternating directions to maximize the strength of the final piece of wood.

strength and vibration resistance. Some of the better quality particle boards may be suitable for use in large robots, but for the most part they should be avoided.

3.3 Plastics

There are probably more different types of plastics than of paper products, woods, and metals combined. Determining the correct type of plastic for a specific application can be a very intensive task and take an unreasonably long period of time. Instead, you more likely will pick up a piece of plastic that you either find as raw material for sale or want to modify an existing product—which for all intents and purposes is the level of consideration required for most hobby robot projects.

There are three major types of plastics you can consider:

Thermoset plastics are hard and have a tightly meshed molecular structure. They generally cure or harden in a mold using chemicals, and during the process they tend to put out a lot of heat. Thermoset plastics can only be shaped by machining as heating them up will destroy the molecular bonds and render the material useless. Thermosets are used in hard plastic toys, appliances, and other products requiring structural strength.

You have most likely encountered elastomer plastics and thought that they were rubber bands. Elastomers have a similar molecular structure to thermosets, but they are much looser, resulting in elastic characteristics. Like thermosets, elastomers cannot be shaped by heating.

Thermoplastics tend to be manufactured as sheets of material of varying thicknesses for a variety of different purposes. At one thickness extreme they are used for plastic bags; and at the other, Plexiglas. Thermoplastics have a long, linear molecular structure that allows them to change shape when heated (unlike the other two types of plastics). Along with Plexiglas, which could be used as a structural material for a robot, sheets of polystyrene (the same material used to make plastic toys and models) can be formed into covers and bodies for a robot using heat and a vacuum-forming process.

3.4 Metal Stock

Metal stock is available from a variety of sources. Your local home improvement store is the best place to start. However, some stock may only be available at specialized metal distributors or specialized product outlets, like hobby stores. Look around in different stores and in the Yellow Pages and you're sure to find what you need.

3.4.1 EXTRUDED ALUMINUM

Extruded stock is made by pushing molten metal out of a shaped orifice. As the metal exits it cools, retaining the exact shape of the orifice. Extruded aluminum stock is readily available at most hardware and home improvement stores. It generally comes in 12-ft sections, but many hardware stores will let you buy cut pieces if you don't need all 12 ft. Even if you have to buy the full 12-ft piece, most hardware stores will cut the pieces to length, saving you the trouble of doing it yourself.

Extruded aluminum is available in more than two dozen common styles, from thin bars to pipes to square posts. Although you can use any of it as you see fit, the following standard sizes may prove to be particularly beneficial in your robot-building endeavors:

- 1-by-1-by-1/16-in angle stock
- 57/64-by-9/16-by-1/16-in channel stock
- 41/64-by-1/2-by-1/16-in channel stock
- Bar stock, in widths from 1 to 3 in and thicknesses of 1/16 to 1/4 in

3.4.2 SHELVING STANDARDS

You've no doubt seen those shelving products where you nail two U-shaped metal rails on the wall and then attach brackets and shelves to them. The rails are referred to as standards, and they are well suited to be girders in robot frames. The standards come in either aluminum or steel and measure 41/64 by 1/2 by 1/16 in. The steel stock is cheaper but considerably heavier, a disadvantage you will want to carefully consider. Limit its use to structural points in your robot that need extra strength. Another disadvantage of using shelving standards instead of extruded aluminum is all the holes and slots you'll find on the standards. The holes are for mounting the standards to a wall; the slots are for attaching shelving brackets. If you are going to use shelving standards, plan to drill into the sides of the rails rather than the base with the holes and slots because this will make integrating them into robots much easier.

3.4.3 MENDING PLATES

Galvanized mending plates are designed to strengthen the joint of two or more pieces of lumber. Most of these plates come preformed in all sorts of weird shapes and so are pretty much unusable for building robots. But flat plates are available in several widths and lengths. You can use the plates as-is or cut them to size. The plates are made of galvanized iron and have numerous predrilled holes to help you hammer in nails. The material is soft enough so you can drill new holes, but if you do so only use sharp drill bits.

Mending plates are available in lengths of about 4, 6, and 12 in. Widths are not as standardized, but 2, 4, 6, and 12 in seem common. You can usually find mending plates near the rain gutter and roofing section in the hardware store. Note that mending plates are heavy, so don't use them for small, lightweight robot designs. Reserve them for medium to large robots where the plate can provide added structural support and strength.

3.4.4 RODS AND SQUARES

Most hardware stores carry a limited quantity of short extruded steel or zinc rods and squares. These are solid and somewhat heavy items and are perfect for use in some advanced projects, such as robotic arms. Lengths are typically limited to 12 or 24 in, and thicknesses range from 1/16 to about 1/2 in.

3.4.5 IRON ANGLE BRACKETS

You will need a way to connect all your metal pieces together. The easiest method is to use galvanized iron brackets. These come in a variety of sizes and shapes and have predrilled holes to facilitate construction. The 3/8-in-wide brackets fit easily into the two sizes of channel stock mentioned at the beginning of the chapter: 57/64 by 9/16 by 1/16 in and 41/64 by 1/2 by 1/16 in. You need only drill a corresponding hole in the channel stock and attach the pieces together with nuts and bolts. The result is a very sturdy and clean-looking frame. You'll find the flat corner angle iron, corner angle (L), and flat mending iron to be particularly useful.

3.5 Quick Turn Mechanical Prototypes

It is very common to design and contract a quick turn, printed circuit assembly house to create custom PCBs for your robots, but up until just a few years ago it wasn't possible to get the same service for mechanical parts built from metal or plastic. Today, there are a number of companies that will take your mechanical designs (some will even provide you with design software free of charge) and turn them into prototypes in just a few weeks (like the one in Fig. 3-2). Using a quick turn mechanical prototype shop for preparing plastic and

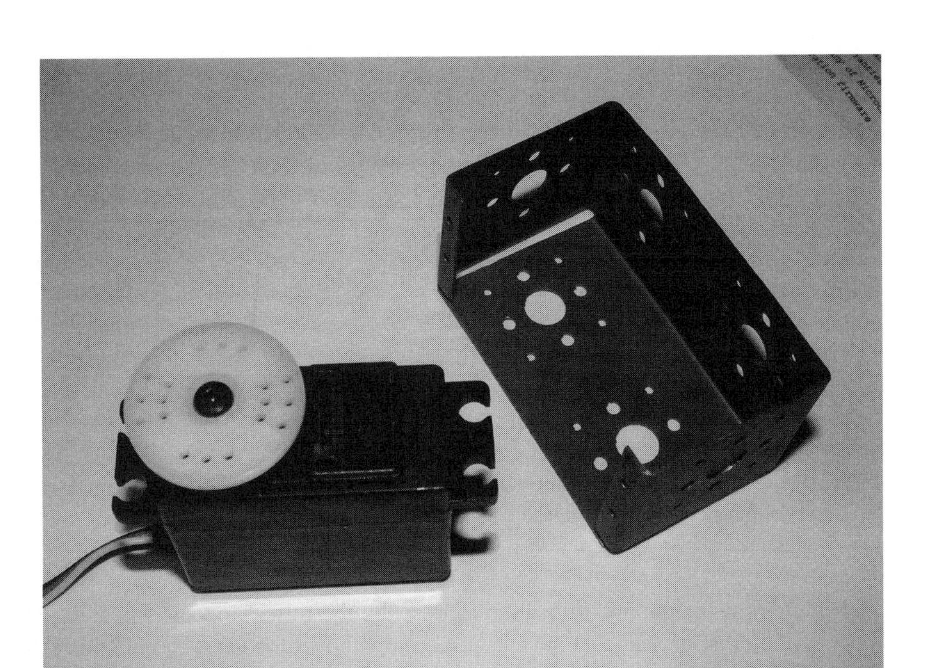

FIGURE 3-2 A servo mounting bracket manufactured at a quick turn prototyping shop.

metal parts will save you the time and money needed to buy the comparable equipment and learn how to use it.

The quick turn mechanical prototype process is still new and you can expect there to be changes over time as it matures. You may find that when you use the quick turn prototypes the choices for materials, finishes, and manufacturing processes are overwhelming. To learn about your options, you should go to your local library and look for references on mechanical design and structural metals and plastics. When you order your first prototypes, keep the order small to minimize costs and maximize the speed with which you receive the parts.

3.6 Fasteners

Once you have decided upon the materials used to build your robot, you will now be left with the task of deciding how to hold them together. What do you think is the number one problem most robots have when they are brought out for a competition? Most people would think of things like dead batteries or codes that can't work in the actual environment (problems with light background noise or the running surface), but it is very common for robots to fall apart or break because the different parts are not held together very well. Part of the problem is the use of an unsuitable material for the structure (like one that breaks during use), but the overwhelming problem is the use of inappropriate adhesives (glues) and fasteners for the robot's structural parts.

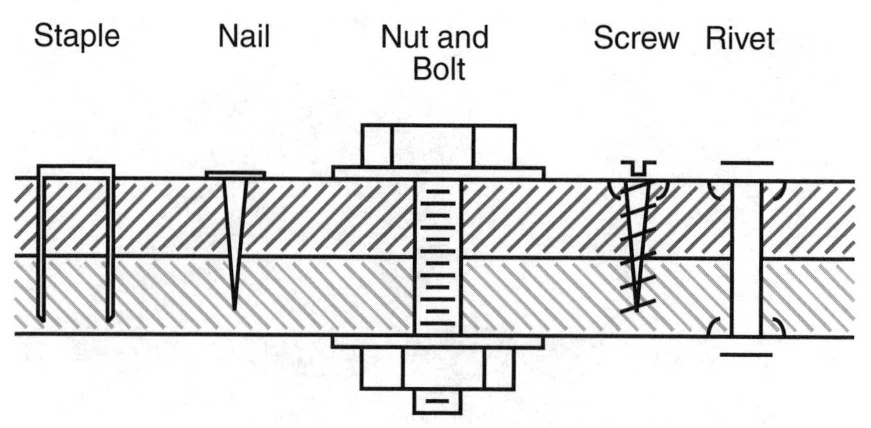

FIGURE 3-3 Some of the different mechanical fasteners that can be used to hold different pieces of a robot's structure together.

Fasteners is the generic term used to describe the miscellaneous nuts and bolts, nails, screws, and other devices that have been developed over the years to hold things together. Fig. 3-3 shows the operation of some common fasteners, and the more commonly used items are listed in the following sections.

3.6.1 NUTS AND BOLTS

Number 6, 8, and 10 nuts and pan-head stove bolts (6/32, 8/32, and 10/24, respectively) are good for all-around construction. Get a variety of bolts in ½, ¾, 1, 1¼, and 1½ in lengths. You may also want to get some 2 and 3-in-long bolts for special applications.

Motor shafts and other heavy-duty applications require ¼-in 20 or 5/16-in hardware. Pan-head stove bolts are the best choice; you don't need hex-head carriage bolts unless you have a specific requirement for them. You can use number 6 (6/32) nuts and bolts for small, lightweight applications. Or for even smaller work, use the miniature hardware available at hobby stores, where you can get screws in standard 5/56, 4/40, and 2/20 sizes.

3.6.2 WASHERS

While you're at the store, stock up on flat washers, fender washers (large washers with small holes), tooth lock washers, and split lock washers. Get an assortment so you have a variety of nut and bolt sizes. Split lock washers are good for heavy-duty applications because they provide more compression locking power. You usually use them with bolt sizes of ¼ in and above.

3.6.3 ALL-THREAD ROD

All-thread comes in varying lengths of stock. It comes in standard thread sizes and pitches. All-thread is good for shafts and linear motion actuators. Get one of each in 8/32, 10/24, and

¼-in 20 threads to start. If you need small sizes, hobby stores provide all-thread rod (typically used for push-pull rods in model airplanes) in a variety of diameters and threads.

3.6.4 SPECIAL NUTS

Coupling nuts are just like regular nuts except that they have been stretched out. They are designed to couple two bolts or pieces of all-thread together, end to end. In robotics, you might use them for everything from linear motion actuators to grippers. Locking nuts have a piece of nylon built into them that provides a locking bite when threaded onto a bolt. It is preferable to use locking nuts over two nuts tightened together.

3.6.5 RIVETS

An often overlooked method of fastening structures together is the blind or pop rivet. This fastener consists of a flanged aluminum tube with a steel rod running through it and a ball at the end. When the ball is pulled into the tube, the tube distorts and flares outward, providing two wide ends to hold together pieces of material. When a certain amount of tension has been placed on the steel rod, it "pops" off, leaving a permanent fastener that is quite a bit shorter over the surface of the material than nuts and bolts. Rivets are quite vibration resistant, and despite being labeled as permanent, they can be drilled out in a few seconds.

3.6.6 ADHESIVES

Adhesives is a ten-dollar word to describe glues. While many people dismiss glues as not being appropriate for use in robots, by following a few simple rules (namely keep the surfaces to be glued together clean, and rough them up with sandpaper to give more surface area for the glue to hold onto), they can be as effective as any of the other methods presented in this chapter and can be a lot easier to work with. Table 3-2 lists a number of the most commonly used adhesives and some of their characteristics and uses.

3.6.7 MISCELLANEOUS METHODS

While the list of fastening methods in the previous sections seems comprehensive, there are still a number of different methods that you can use to hold your structural parts together that are useful in a variety of different applications. The following list describes methods used on different robots that have resulted in structures that are stronger, lighter, and easier to build than using the traditional methods previously described. Some of these suggested fastening methods may seem fanciful, but remember to "never say never"—there are situations where each one of these solutions will be optimal.

- Welding is useful for large heavy robots built on a steel frame or chassis. With the proper tools and training, robot structures can be precisely assembled quite quickly. Training is critical as all types of welding can be dangerous. Oxyacetylene torches can also be used for heating up and bending steel parts—but only if you know what you are doing.
- The next time you open a car's hood to look in the engine compartment, take a look at

TABLE 3-2 Adhesives and the Materials They Are Designed For

ADHESIVE	MATERIALS	COMMENTS
Weldbond	Wood/PCB	Excellent for tying down loose wires and insulating PCBs
Solvents	Plastics	Melts plastics together
Krazy Glue/Locktite	Metal/Plastic	Best for locking nuts
Carpenter's glue	Wood	Works best on unfinished wood
5-minute epoxy	Everything	Very permanent
Contact cement	Flat, Porous	Good for bonding paper/laminates to wood or each other
Two-sided tape	Smooth surface	Good for holding components onto robot structure. Can leave residue during removal.
Hot glue gun	Everything	Not recommended due to poor vibration tolerance

the myriad of fasteners in there. Cable ties and hose clamps are generally specialty items, but chances are you will run across a few applications that are helpful to the robots you are working on.

- Nothing has been written to say that robot structures have to be permanently fastened together. If you are unsure about the best configuration for the parts of your robot or if different parts are needed for multiple robots, why don't you mount them with velcro, magnets, or in a way that allows them to be removed and replaced quickly and easily.
- There are a number of robots that are held together by steel cables and turnbuckles (threaded cable connectors that can be used to adjust the tension on a cable). These robots definitely have a unique look and can be made very light and very rigid.
- Finally, why do the parts have to be fastened together at all? A robot built from interlocking parts could be particularly fascinating and an interesting response to the need for coming up with robot components that can be taken apart and put back together in different ways. Electrical connections made when the parts are assembled together could provide wiring for the robot and prevent invalid configurations from being created.

3.7 Scavenging: Making Do with What You Already Have

You don't always need to buy new (or used or surplus) to get worthwhile robot parts. In fact, some of the best parts for hobby robots may already be in your garage or attic. Consider the typical used VCR, for example. It'll contain at least one motor (and possibly as many as five), numerous gears, and other electronic and mechanical odds and ends. Depending on the brand and when it was made, it could also contain belts and pulleys, appropriate motor drivers, digital electronics chips, infrared receiver modules, miniature push buttons, infrared light-emitting diodes and detectors, and even wire harnesses with multipin connectors. Any and all of these can be salvaged to help build your robot. All told, the typical VCR may have over $50 worth of parts in it.

Never throw away small appliances or mechanical devices without taking them apart and looking for usable parts. If you don't have time to disassemble that CD player that's skipping on all of your compact discs, throw it into a pile for a rainy day when you do have a free moment. Ask friends and neighbors to save their discards for you. You'd be amazed how many people simply toss old VCRs, clock radios, and other items into the trash when they no longer work.

Likewise, make a point of visiting garage sales and thrift stores from time to time, and look for parts bonanzas in used—and perhaps nonfunctioning—goods. Scout the local thrift stores (Goodwill, Disabled American Veterans, Salvation Army, Amvets, etc.) and for very little money you can come away with a trunk full of valuable items that can be salvaged for parts. Goods that are still in functioning order tend to cost more than the broken stuff, but for robot building the broken stuff is just as good. Be sure to ask the store personnel if they have any nonworking items they will sell you at a reasonable cost.

Here is just a short list of the electronic and mechanical items you'll want to be on the lookout for and the primary robot-building components they have inside.

- VCRs are perhaps the best single source for parts, and they are in plentiful supply (hundreds of millions of them have been built since the mid-1970s). As previously discussed, you'll find motors (and driver circuits), switches, LEDs, cable harnesses, and IR receiver modules on many models.
- CD players have optical systems you can gut out if your robot uses a specialty vision system. Apart from the laser diode, CD players have focusing lenses, miniature multicell photodiode arrays, diffraction gratings, and beam splitters, as well as micro-miniature motors and a precision lead-screw positioning device (used by the laser system to read the surface of the CD).
- Old disk drives (floppy and hard drives) also have a number of components that are very useful in robots. Along with the motor that turns the disk, the stepper motor that moves the head is well suited for use in robot arms or even small walking robots. Later in the book, opto-interrupters will be discussed and the typical disk drive has at least two of these that could be used in a robot.
- Fax machines contain numerous motors, gears, miniature leaf switches, and other mechanical parts. These machines also contain an imaging array (it reads the page to fax it) that you might be able to adapt for use as robotic sensors.

- Mice, printers, old scanners, and other discarded computer peripherals contain valuable optical and mechanical parts. Mice contain optical encoders that you can use to count the rotations of your robot's wheels, printers and scanners contain motors and gears, and scanners contain optics you can use for vision systems and other sensors on your robot.
- Mechanical toys, especially the motorized variety, can be used either for parts or as a robot base. Remember to keep the motor drivers (as will be discussed later in the book). When looking at motorized vehicles, favor those that use separate motors for each drive wheel (as opposed to a single motor for both wheels), although other drive configurations can make for interesting and unique robots. Don't limit yourself!

3.8 Finishing Your Robot's Structure

You can change your robot from appearing like something that you cobbled together in your garage over a weekend to a much more professional looking and polished piece of machinery by spending a few minutes over a few days putting on a paint finish. Few people think about finishing a robot's structure, whether it's made of wood or metal, but there are a number of tangible benefits that you should be aware of:

- Eliminating the dust on the surface of the material, allowing for effective two-sided tape attachment (and removal).
- Smoothing wood surface and reducing the lifted fibers that appear when the wood gets moist or wet over time.
- Eliminating wood fibers will also produce more effective glue bonds.
- Eliminating splinters and minimizing surface splintering when drilling into wood.
- Allowing for pencil and ink marking of the surface to be easily wiped off for corrections. This is not possible with bare wood and many metals.

You should consider using aerosol paint to finish the structural pieces of your robot. When properly used there is very little mess and no brushes to clean up. From an auto-body supply house, you should buy an aerosol can of primer (gray is always a good choice) and from a hardware store buy an aerosol can of indoor/outdoor (or marine) acrylic paint (such as Krylon brand) in your favorite color. Red catches the eye, isn't overwhelming, and if there is a blemish in the material or your work, it will become hidden quite nicely. The process outlined here is usable for metal and wood, although wood will tend to have more blemishes and its grain will be more visible than for the metal.

Set up a painting area in a garage or some other well-ventilated space by laying down newspaper both on a flat surface as well as a vertical surface. Next, lay down some bottle caps or other supports for the materials you are going to paint. You will want to finish the ends of the material and don't want to end up with paint flowing between the material and the support (or the newspaper for that matter).

Lightly sand the surfaces that you are going to paint using a fine-grain sandpaper (200 grit at least). You may also want to sand the edges of the material more aggressively to take off any loose wood or metal burrs that could become problems later. Once you have finished this, moisten the rag and wipe it over the surface you have sanded to pick up any loose dust.

Shake the can of primer using the instructions printed on the can. Usually there is a small metal ball inside the can and you will be instructed to shake it until the ball rattles easily inside. Apply a light coat of primer; just enough to change the color of the material. Most primers take 30 minutes or so to dry. Check the instructions on the can before going on to the next step of sanding and putting on new coats of paint or primer.

After the first application of primer, you will probably find that a wood surface is very rough. This is due to the cut fibers in wood standing on end after being moistened from the primer. Repeat the sanding step (along with sanding the ends of the wood and then wiping down with a damp rag) before applying another coat of primer. After the second coat is put down, let it dry, sand very lightly, and wipe down again.

Now you are ready to apply the paint. Shake according to the instructions on the aerosol can and spray the material again, putting on a thin, even coat. You will probably find that the paint will seem to be sucked into the wood and the surface will not be that shiny. This is normal. Once the paint has dried, lightly sand again, wipe down with a wet cloth, and apply a thicker coat of paint.

When this coat has dried, you'll find that the surface is very smooth and shiny. Figure 3-4 shows what can be done with some strips of plywood over a couple of nights. Some of the grain of the wood will still be visible, but it will not be noticeable. You do not have to sand the paint again. The material is now ready to be used in a robot.

FIGURE 3-4 Finished plywood strips for use in a robot base.

3.9 From Here

To learn more about . . .	***Read***
How to solder	Chapter 7, "Electronic Construction Techniques"
Building electronic circuits	Chapter 7, "Electronic Construction Techniques"
Building mechanical apparatus	Part II, "Robot Construction"

CHAPTER 4

BUYING PARTS

Building a robot from scratch can be hard or easy. It's up to you. As a recommendation, when you are starting out, go for the easy route; life is too demanding as it is. The best way to simplify the construction of a robot is to use standard, off-the-shelf parts—things you can get at the neighborhood hardware, auto parts, and electronics store.

Exactly where can you find robot parts? The neighborhood robot store would be the logical place to start—if only such a store existed! Not yet, anyway. Fortunately, other local retail stores are available to fill in the gaps. Moreover, there's a veritable world of places that sell robot junk, probably close to where you live and also on the Internet.

4.1 Hobby and Model Stores

Hobby and model stores are the ideal sources for small parts, including lightweight plastic, brass rod, servo motors for radio control (R/C) cars and airplanes, gears, and construction hardware. Most of the products available at hobby stores are designed for building specific kinds of models and toys. But that shouldn't stop you from raiding the place with an eye to converting the parts for robot use.

Most hobby store owners and salespeople have little knowledge about how to use their line of products for anything but their intended purpose. So you'll likely receive little substantive help in solving your robot construction problem. Your best bet is to browse the store and look for parts that you can put together to build a robot. Some of the parts, particularly those for R/C models, will be behind a counter, but they should still be visible enough for you to conceptualize how you might use them. If you don't have a well-stocked hobby and model store in your area, there's always the Internet.

4.2 Craft Stores

Craft stores sell supplies for home crafts and arts. As a robot builder, you'll be interested in just a few of the aisles at most craft stores, but what's in those aisles will be a veritable gold mine! Look for these useful items:

- *Foam rubber sheets.* These come in various colors and thicknesses and can be used for pads, bumpers, nonslip surfaces, tank treads, and lots more. The foam is very dense; use sharp scissors or a knife to cut it.
- *Foamboard.* Constructed of foam sandwiched between two heavy sheets of paper, foamboard can be used for small, lightweight robots. Foamboard can be cut with a hobby knife and glued with paper glue or hot-melt glue. Look for it in different colors and thicknesses.
- *Parts from dolls and teddy bears.* These can often be used in robots. Fancier dolls use articulations—movable and adjustable joints—that can be used in your robot creations. Look also for linkages, bendable posing wire, and eyes (great for building robots with personality!).
- *Electronic light and sound buttons.* These are designed to make Christmas ornaments and custom greeting cards, but they work just as well in robots. The electric light kits come with low-voltage LEDs or incandescent lights, often in several bright colors. Some flash at random, some sequentially. Sound buttons have a built-in song that plays when you depress a switch. Don't expect high sound quality with these devices. You could use these buttons as touch sensors, for example, or as a tummy switch in an animal-like robot.
- *Plastic crafts construction material.* This can be used in lieu of more expensive building kits, such as LEGO or Erector Set. For example, many stores carry the plastic equivalent of that old favorite, the wooden Popsicle sticks (the wooden variety is also available, but these aren't as strong). The plastic sticks have notches in them so they can be assembled to create frames and structures.
- *Model building supplies.* Many craft stores have these, sometimes at lower prices than the average hobby-model store. Look for assortments of wood and metal pieces, adhesives, and construction tools.

There are, of course, many other interesting products of interest at craft stores. Visit one and take a stroll down its aisles.

4.3 Hardware Stores

Hardware stores and builder's supply outlets (usually open to the public) are the best sources for the wide variety of tools and parts you will need for robot experimentation. Items like nuts and bolts are generally available in bulk, so you can save money. As you tour the hardware stores, keep a notebook handy and jot down the lines each outlet carries. Then, when you find yourself needing a specific item, you have only to refer to your notes. Take an idle stroll through your regular hardware store haunts on a regular basis. You'll always find something new and laughably useful for robot design each time you visit.

4.4 Electronic Stores

Twenty or 30 years ago electronic parts stores were plentiful. Even some automotive outlets carried a full range of tubes, specialty transistors, and other electronic gadgets. Now, Radio Shack remains the only U.S. national electronics store chain. Radio Shack continues to support electronics experimenters and in recent years has improved the selection of parts available primarily through the Internet (you can buy the Parallax BASIC Stamp 2 along with some low-end Microchip PIC MCUs), but for the most part they stock only very common and basic components.

If your needs extend beyond resistors, capacitors, and a few integrated circuits, you must turn to other sources. Locally, you can check the Yellow Pages under Electronics—Retail for a list of electronic parts shops near you.

4.5 Electronics Wholesalers and Distributors

Most electronic stores carry a limited selection, especially if they serve the consumer or hobby market. Most larger cities across the United States—and in other countries throughout the world, for that matter—host one or more electronics wholesalers or distributors. These companies specialize in providing parts for industry.

Wholesalers and distributors are two different kinds of businesses, and it's worthwhile to know how they differ so you can approach them effectively. Wholesalers are accustomed to providing parts in quantity; they offer attractive discounts because they can make up for them with higher volume. Unless you are planning to buy components in the hundreds of thousands, a wholesaler is likely not your best choice.

Distributors may also sell in bulk, but many of them are also set up to sell parts in "onesies and twosies." Cost per item is understandably higher, and not all distributors are willing to sell to the general public. Rather, they prefer to establish relationships with companies and organizations that may purchase thousands of dollars' worth of parts over the course of a year. Still, some electronics parts distributors, particularly those with catalogs on the Internet (see "Finding Parts on the Internet" later in this chapter) are more than happy to work with individuals, though minimum-order requirements may apply. Check with the companies near you and ask for their terms of service. When buying through a distributor, keep in mind that you are seldom able to browse the warehouse to look for goodies. Most distributors provide a listing of the parts they carry. Some only list the lines they offer. You are required to know the make, model, and part number of what you want to order. Fortunately, virtually all electronics manufacturers provide free information about their products on the Internet. Many such Internet sites offer a search tool that allows you to look up parts by function. Once you find a part you want, jot down its number and use it to order from the local distributor.

If you belong to a local robotics club or user's group, you may find it advantageous to establish a relationship with a local electronics parts distributor through the club. Assuming the club has enough members to justify the quantities of each part you'll need to buy, the same approach can work with electronics wholesalers. You may find that the buying power of the group gets you better service and lower prices.

4.6 Samples from Electronics Manufacturers

Only a few electronics manufacturers are willing to send samples of their products to qualified customers. A *qualified customer* is typically an engineer in the industry and will have potential business for the manufacturer—all others will be given the contact information for a recommended distributor. Ten years ago, this was not the case, and many chip manufacturers were willing to send samples based on simple requests from individuals. Over the years, the cost of providing this service, along with contracts with wholesalers and distributors that prohibit manufacturers from dealing directly with individuals, has made this service obsolete.

If you belong to a school or a robotics club, you may still be able to get some sample parts, development tools, and part documentation from chip manufacturers. To find out what is possible from a specific vendor, contact the local sales or support office (which can be found on the company's web site) either by phone or email. When you are making your request, it is a good idea to have a list of requirements for the application along with the part number you believe best meets these requirements; the customer support representative may be able to help you find a part that is cheaper, better meets your requirements, or is easier to work with.

Finally, it should go without saying that when you make the request, it should be reasonable. It can be incredibly infuriating for manufacturers to get demands for large sample quantities of parts from individuals that just seem to be trying to avoid having to pay for them legitimately. One manufacturer has stated that it got out of the parts sampling business because of the number of requests received for specially programmed parts (the part's unprogrammed retail cost is $1.00) that would allow illegally copied software to be played on home video game machines.

4.7 Specialty Stores

Specialty stores are outlets open to the general public that sell items you won't find in a regular hardware or electronic parts store. They don't include surplus outlets, which are discussed in the next section. What specialty stores are of use to robot builders? Consider these:

- *Sewing machine repair shops.* Ideal for finding small gears, cams, levers, and other precision parts. Some shops will sell broken machines to you. Tear the machine to shreds and use the parts for your robot.

- *Auto parts stores.* The independent stores tend to stock more goodies than the national chains, but both kinds offer surprises on every aisle. Keep an eye out for things like hoses, pumps, and automotive gadgets.
- *Used battery depots.* These are usually a business run out of the home of someone who buys old car and motorcycle batteries and refurbishes them. Selling prices are usually between $15 and $25, or 50 to 75 percent less than a new battery.
- *Junkyards.* Old cars are good sources for powerful DC motors, which are used to drive windshield wipers, electric windows, and automatic adjustable seats (though take note: such motors tend to be terribly inefficient for battery-based 'bots). Or how about the hydraulic brake system on a junked 1969 Ford Falcon? Bring tools to salvage the parts you want. And maybe bring the Falcon home with you, too.
- *Lawn mower sales–service shops.* Lawn mowers use all sorts of nifty control cables, wheel bearings, and assorted odds and ends. Pick up new or used parts for a current project or for your own stock at these shops.
- *Bicycle sales–service shops.* Not the department store that sells bikes, but a real professional bicycle shop. Items of interest: control cables, chains, brake calipers, wheels, sprockets, brake linings, and more.
- *Industrial parts outlets.* Some places sell gears, bearings, shafts, motors, and other industrial hardware on a one-piece-at-a-time basis. The penalty: fairly high prices and often the requirement that you buy a higher quantity of an item than you really need.

4.8 Shopping the Surplus Store

Surplus is a wonderful thing, but most people shy away from it. Why? If it's surplus, the reasoning goes, it must be worthless junk. That's simply not true. Surplus is exactly what its name implies: extra stock. Because the stock is extra, it's generally priced accordingly—to move it out the door.

Surplus stores that specialize in new and used mechanical and electronic parts or military surplus (not to be confused with surplus clothing and camping) are a pleasure to find. Most urban areas have at least one such surplus store; some as many as three or four. Get to know each and compare their prices. Bear in mind that surplus stores don't have mass-market appeal, so finding them is not always easy. Start by looking in the phone company's Yellow Pages under Electronics and also under Surplus.

While surplus is a great way to stock up on DC motors, gears, roller chain, sprockets, and other odds and ends, you must shop wisely. Just because the company calls the stuff surplus doesn't mean that it's cheap or even reasonably priced. A popular item in a catalog or advertised on the Internet may sell for top dollar. Always compare the prices of similar items offered by several surplus outlets before buying. Consider all the variables, such as the added cost of insurance, postage and handling, and COD fees if the part is going to be shipped to you.

Remember that most surplus stores sell as-is (often contracted into the single word *asis*) and do not allow returns of any kind. *As-is* means just that, the parts are sold as they are sitting there; this does not mean they work nor does it mean that they are free from scratches, cracks, leaks, or other problems that may limit their usefulness and appeal. Trying to return something to a surplus store is often simply an invitation to be abused. If the item costs more than you are comfortable with losing if it doesn't work properly, then you shouldn't buy it.

4.8.1 WHAT YOU CAN GET SURPLUS

Shopping surplus can be a tough proposition because it's hard to know what you'll need before you need it. And when you need it, there's only a slight chance that the store will have what you want. Still, certain items are almost always in demand by the robotics experimenter. If the price is right (especially on assortments or sets), stock up on the following.

- *Gears.* Small gears between ½ and 3 in are extremely useful. Stick with standard tooth pitches of 24, 32, and 48. Try to get an assortment of sizes with similar pitches. Avoid grab bag collections of gears because you'll find no mates. Plastic and nylon gears are fine for most jobs, but you should use larger metal gears for the main drive systems of your robots.
- *Roller chain and sprockets.* Robotics applications generally call for ¼-in (#25) roller chain, which is smaller and lighter than bicycle chain. When you see this stuff, snatch it up, but make sure you have the master links if the chain isn't permanently riveted together. Sprockets come in various sizes, which are expressed as the number of teeth on the outside of the sprocket. Buy a selection. Plastic and nylon roller chain and sprockets are fine for general use; steel is preferred for main drives.
- *Bushings.* You can use bushings as a kind of ball bearing or to reduce the hub size of gears and sprockets so they fit smaller shafts. Common motor shaft sizes are ⅛ in for small motors and ¼ in for larger motors. Gears and sprockets generally have ⅜-in, ½-in, and ⅝-in hubs. Oil-impregnated Oilite bushings are among the best, but they cost more than regular bushings.
- *Spacers.* These are made of aluminum, brass, or stainless steel. The best kind to get have an inside diameter that accepts 10⁄32 and ¼-in shafts.
- *Motors.* Particularly useful are the 6-V and 12-V DC variety. Most motors turn too fast for robotics applications but you can often luck out and find some geared motors. Final speeds of 20 to 100 r/min at the output of the gear reduction train are ideal. If gear motors aren't available, be on the lookout for gearboxes that you can attach to your motors. Stepper motors are handy, too, but make sure you know what you are buying.
- *Rechargeable batteries.* The sealed lead-acid and gel-cell varieties are common in surplus outlets. Test the battery immediately to make sure it takes a charge and delivers its full capacity (test it under a load, like a heavy-duty motor). These batteries come in 6-V and 12-V capacities, both of which are ideal for robotics. Surplus nickel-cadmium and nickel-metal hydride batteries are available, too, in either single 1.2-V cells or in combination battery packs. Be sure to check these batteries thoroughly.

4.9 Finding Parts on the Internet

The Internet has given a tremendous boost to the art and science of robot building. Through the Internet—and more specifically the World Wide Web—you can now search for and find the most elusive part for your robot. Most of the major surplus and electronics mail-order companies provide online electronic catalogs. You can visit the retailer at their web site and either browse their offerings by category or use a search feature to quickly locate exactly what you want. Moreover, with the help of web search engines such as Google (www.google.com), you can search for items of interest from among the millions of web sites throughout the world. Search engines provide you with a list of web pages that may match your search query. You can then visit the web pages to see if they offer what you're looking for.

Of course, don't limit your use of the Internet and the World Wide Web to just finding parts. You can also use them to find a plethora of useful information on robot building. See Appendix B, "Sources," and Appendix C, "Robot Information on the Internet," for categorized lists of useful robotics destinations on the Internet. These lists are periodically updated at www.robotoid.com.

A number of web sites offer individuals the ability to buy and sell merchandise. Most of these sites are set up as auctions: someone posts an item to sell and then waits for people to make bids on it. Robotics toys, books, kits, and other products are common finds on web auction sites like Ebay (www.ebay.com) and Amazon (www.amazon.com). If your design requires you to pull the guts out of a certain toy that's no longer made, try finding a used one at a web auction site. The price should be reasonable as long as the toy is not a collector's item.

Keep in mind that the World Wide Web is indeed worldwide. Some of the sites you find may not be located in your country. Though many web businesses ship internationally, not all will. Check the web site's fine print to determine if the company will ship to your country, and note any specific payment requirements. If they accept checks or money orders, the denomination of each must be in the company's native currency.

4.10 From Here

To learn more about . . .	*Read*
Tools for robot building	Chapter 6, "Tools"
Description of electronic components	Chapter 5, "Electronic Components"
Details on batteries and battery types	Chapter 17, "Batteries and Robot Power Supplies"
Common motor types used in robotics	Chapter 19, "Choosing the Right Motor"

CHAPTER 5

ELECTRONIC COMPONENTS

Electronics are the central nervous system of your robot and will be responsible for passing information to and from peripheral functions as well as processing inputs and turning them into the output functions the robot performs. Any given hobby robot project might contain a dozen or more electronic components of varying types, including resistors, capacitors, integrated circuits, and light-emitting diodes. In this chapter, you'll read about the components commonly found in hobby robot projects and their many specific varieties. You'll also learn their functions and how they are used.

5.1 Cram Course in Electrical Theory

Understanding basic electronics is a keystone to being able to design and build your own robots. The knowledge required is not all that difficult—in fact the basic theories with diagrams can fit on two sheets of paper (following) which you are encouraged to photocopy and hang up as a quick reference.

Electricity always travels in a circle, or circuit, like the one in Fig. 5-1. If the circuit is broken, or opened, then the electricity flow stops and the circuit stops working.

Electricity consists of electrons, which are easily moved from the atoms of metal conductors. There are two components of electricity that can be measured. *Voltage* is the pressure applied to the electrons to force them to move through the metal wires as well as the

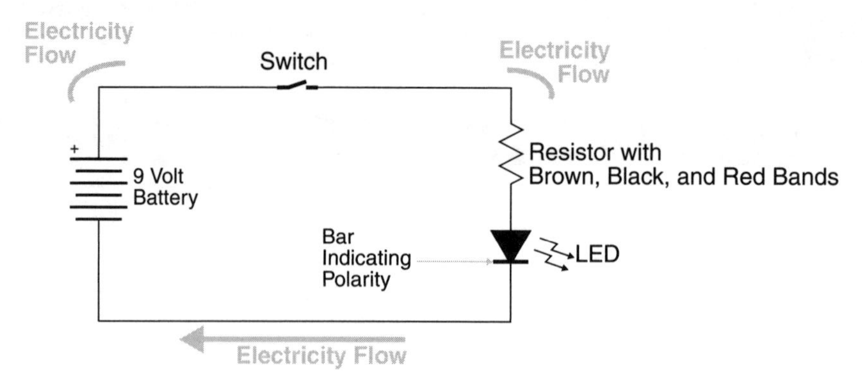

FIGURE 5-1 Electricity flows in a circle, or circuit, from positive (+) to negative. If the circuit is broken (as when the switch is open), electricity stops flowing and the circuit stops working.

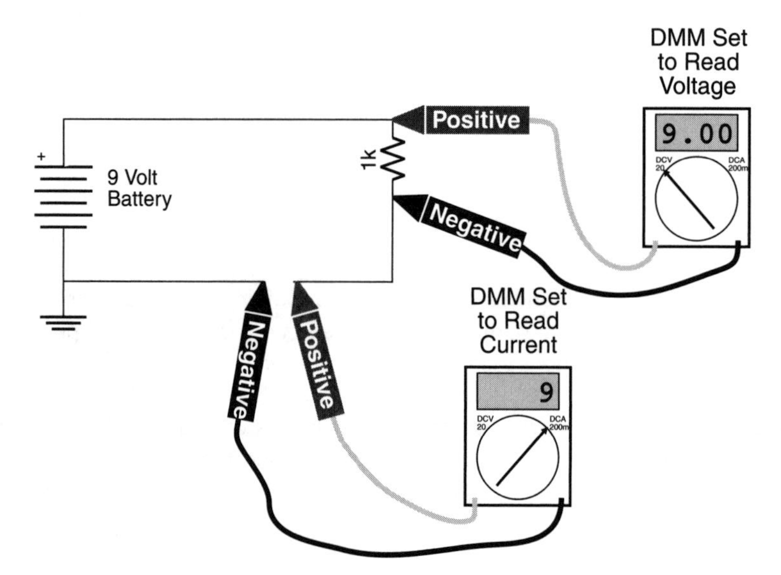

FIGURE 5-2 A digital multimeter can be used to measure the voltage *across* a component as well as the current *through* it.

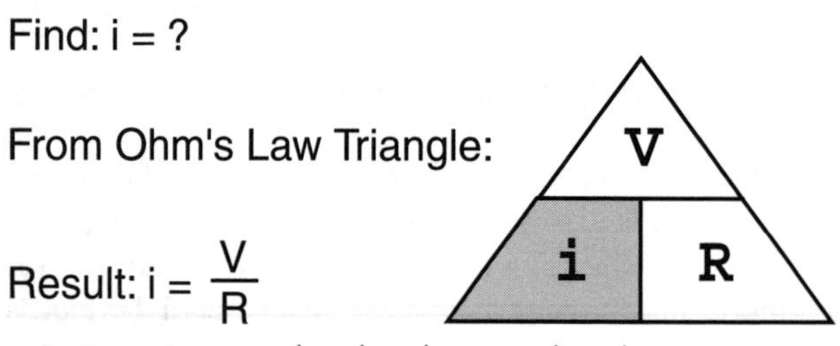

FIGURE 5-3 Using the Ohm's law triangle, voltage, current, or resistance can be calculated when the other two values are known.

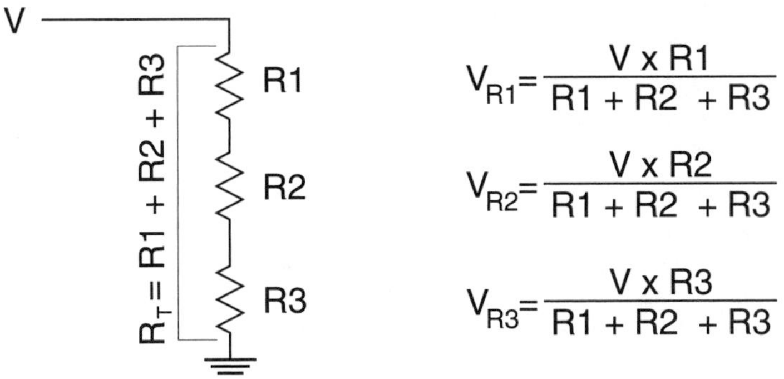

V
$R_T = R1 + R2 + R3$
R1
R2
R3
$V_{R1} = \frac{V \times R1}{R1 + R2 + R3}$
$V_{R2} = \frac{V \times R2}{R1 + R2 + R3}$
$V_{R3} = \frac{V \times R3}{R1 + R2 + R3}$

FIGURE 5-4 When resistors are wired in series, the total resistance of the circuit is proportional to the sum of the resistances.

Vin
+
R1
R2
Vout
$Vout = \frac{Vin \times R2}{R1 + R2}$

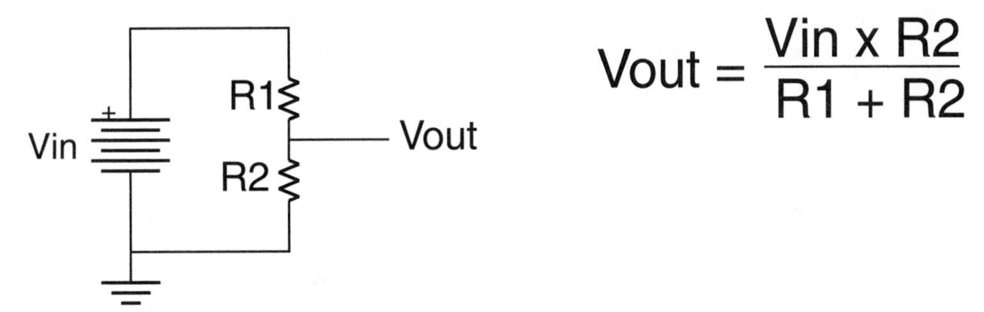

FIGURE 5-5 Two resistors can be used to change an input voltage to another, lower value.

R1
R2
Rn
$R = \frac{1}{\frac{1}{R1} + \frac{1}{R2} \cdots + \frac{1}{Rn}}$

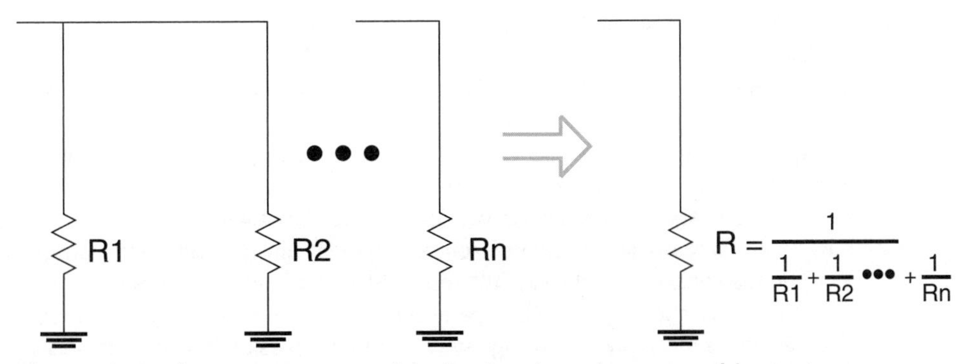

FIGURE 5-6 Placing resistors in parallel will reduce the total resistance of the circuit.

different components in the circuit. As the electrons pass through a component they lose some of the pressure, just as water loses pressure due to friction when it moves through a pipe. The initial voltage applied to the electrons is measured with a volt meter or a multimeter set to measure voltage and is equal to the voltage drops through components in the circuit. The label given to voltage is V.

The second measurement that can be applied to electricity is *current,* which is the number of electrons passing by a point in a given time. There are literally several billion, billion, billion, billion, billion, billion electrons flowing past a point at a given time. For convenience, the unit Coulomb (C) was specified, which is 6.25×10^{18} electrons and is the basis for the ampere (A), which is the number of electrons moving past a point every second. The label given to current is the nonintuitive i.

The voltage across a component and the current through it can be measured using a digital multimeter as shown in Fig. 5-2. It is important to remember that *voltage* is the pressure change across a component, so to measure it you have to put a test lead on either side of the component. *Current* is the volume of electrons moving past a certain point every second, and to measure it, the circuit must be broken and the tester put in line, or in series, with the component being measured.

The current flowing through a component can be calculated if the voltage change, or drop, is known along with the resistance of the component using Ohm's law. This law states that the voltage drop across a resistance is equal to the product of the resistance value and the current flowing through it. Put mathematically, Ohm's law is:

$$V = i \times R$$

Where V is voltage across the component measured in volts, i is the current through the component measured in amperes or amps, and R is the resistance measured in ohms, which has the symbol Ω. Using algebra, when any two of the three values are known, the third can be calculated. If you are not comfortable using algebra to find the missing value, you can use Ohm's law triangle (Fig. 5-3). This tool is quite simple to use. Just place your finger over the value you want to find, and the remaining two values along with how they are located relative to one another shows you the calculation that you must do to find the missing value. For the example in Fig. 5-3, to find the formula to calculate current, put your finger over i and the resulting two values V over R is the formula for finding i (divide the voltage drop by the resistance of the component).

Resistances can be combined, which changes the electrical parameters of the entire circuit. For example, in Fig. 5-4 resistances are shown placed in line or in series and the total resistance is the sum of the resistances. Along with this, the voltage drop across each resistor is proportional to the value of the individual resistors relative to the total resistance of the circuit.

The ratio of voltages in a series circuit can be used to produce a fractional value of the total voltage applied to a circuit. Fig. 5-5 shows a voltage divider, which is built from two series resistors and outputs a lower voltage than was input into the circuit. It is important to remember that this circuit cannot source (provide) any current—any current draw will increase the voltage drop through the top resistor and lower the voltage of the output.

Finally, resistances can be wired parallel to one another as in Fig. 5-6. In this case, the total resistance drops and the voltage stays constant across each resistor (increasing the total amount of current flowing through the circuit). It is important to remember that the equivalent resistance will always be less than the value of the lowest resistance. The general case formula given in Fig. 5-6 probably seems very cumbersome but is quite elegant when applied to two resistors in parallel. The equivalent resistance is calculated using:

$$R_{equivalent} = (R1 \times R2) / (R1 + R2)$$

Whew! This is all there is to it with regards to basic electronics. The diagrams have all been placed in the following to allow you to photocopy them, study while you have a free moment, and pin up over your workbench so you always have them handy.

5.2 Wire Gauge

The most basic component that you will be working with is the wire. There are essentially two types that you should be aware of: solid core and multi-stranded. Solid core wire is exactly what is implied: there is a single conductor within the insulation. Multi-strand wire consists of many small strands of wire, each carrying a fraction of the total current passed through the wire. Solid core wire is best for high current applications, while multi-stranded wire is best for situations where the wire will change shape because the thin strands bend more easily than one large one.

It should not surprise you that the thicker the conductor within the wire, the more current it can carry. Wires must not be overloaded or their internal resistance will cause the wire to overheat, potentially melting the insulation and causing a fire. Table 5-1 lists different American wire gauge (AWG) wire sizes and their specified current-carrying capacity. As a rule of thumb you should halve the amount of current you pass through these different size wires—the current specified in Table 5-1 is the maximum amount of current with a 20 C increase in temperature. By carrying half this amount, there will be minimal heating and power loss within the wires.

TABLE 5-1 Single Conductor Wire Current-Carrying Capacity

CONDUCTOR SIZE	MAXIMUM CURRENT-CARRYING CAPACITY	TYPICAL APPLICATION
12 AWG	36 Amps	Large Motor Power/Robot Battery
14 AWG	27 Amps	Large Motor Power
16 AWG	19 Amps	Medium Motor Power
18 AWG	15 Amps	Medium Motor Power
20 AWG	10 Amps	Logic Power, Small Motor Power
22 AWG	8 Amps	Logic Power, Small Motor Power
30 AWG	2 Amps	Wire Wrapping/Small Signals

Along with the size of the conductor and the amount of current it can carry, a plethora of different options are available when choosing wire for a specific application. There are a variety of ways of providing multiple conductors (each one separated from each other to allow multiple signals or voltage to pass through them) in a single wire; there are different insulations for different applications and different methods of molding the wires so they can be attached to different connectors easily.

Many books this size and larger detail the options regarding wiring for different applications. As you start out with your robot applications, try to use 20 AWG single conductor, multi-stranded wiring for everything (buy it in several colors, including black and red so you can easily determine what is the wire's function). As you build larger robots and understand the electrical requirements of the different parts, you can start experimenting with wire sizes and connecting systems.

5.3 Fixed Resistors

A fixed resistor supplies a predetermined resistance to a circuit. The standard unit of value of a resistor is the ohm (with units in volts per ampere, according to Ohm's law), represented by the symbol Ω. The higher the ohm value, the more resistance the component provides to the circuit. The value on most fixed resistors is identified by color coding, as shown in Fig. 5-7. The color coding starts near the edge of the resistor and comprises four, five, and sometimes six bands of different colors. Most off-the-shelf resistors for hobby projects use standard four-band color coding. The values of each band are listed in Table 5-2, and the formula for determining the resistance from the bands is:

$$\text{Resistance} = ((\text{Band 1 Color Value} \times 10) + (\text{Band 2 Color Value})) \times 10^{\text{Band 3 Color Value}} \text{ ohms}$$

If you are not sure what the resistance is for a particular resistor, use a digital multimeter to check it. Position the test leads on either end of the resistor. If the meter is not autorang-

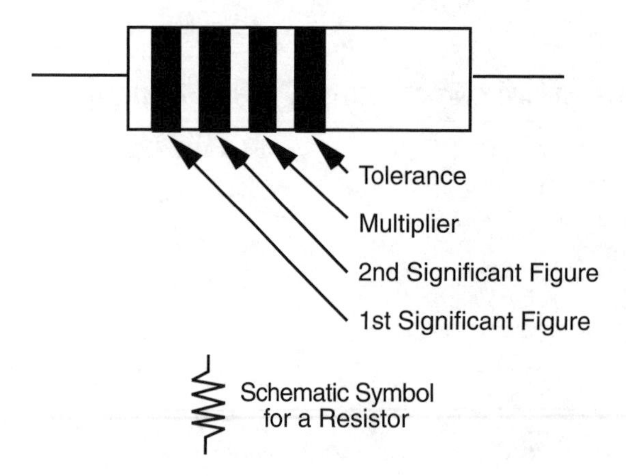

FIGURE 5-7 Resistors use color coding to denote their value. Start from the color band nearest the end. Most resistors have four bands: three for the value and one for the tolerance.

TABLE 5-2 Resistor Band Values

COLOR	BAND COLOR VALUE	TOLERANCE
Black	0	N/A
Brown	1	1%
Red	2	2%
Orange	3	N/A
Yellow	4	N/A
Green	5	0.5%
Blue	6	0.25%
Violet	7	0.1%
Gray	8	0.05%
White	9	N/A
Gold	N/A	5%
Silver	N/A	10%

ing, start at a high range and work down. Be sure you don't touch the test leads or the leads of the resistor; if you do, you'll add the natural resistance of your own body to the reading.

Resistors are also rated by their wattage. The wattage of a resistor indicates the amount of power it can safely dissipate. Resistors used in high-load applications, like motor control, require higher wattages than those used in low-current applications. The majority of resistors you'll use for hobby electronics will be rated at ¼ or even ⅛ of a watt. The wattage of a resistor is not marked on the body of the component; instead, you must infer it from the size of the resistor.

5.4 Variable Resistors

Variable resistors let you dial in a specific resistance. The actual range of resistance is determined by the upward value of the potentiometer. Potentiometers are thus marked with this upward value, such as 10K, 50K, 100K, 1M, and so forth. For example, a 50K potentiometer will let you dial in any resistance from 0 to 50,000 ohms. Note that the range is approximate only.

Potentiometers are of either the dial or slide type, as shown in Fig. 5-8. The dial type is the most familiar and is used in such applications as television volume controls and electric blanket thermostat controls. The rotation of the dial is nearly 360°, depending on which potentiometer you use. In one extreme, the resistance through the potentiometer (or pot) is zero; in the other extreme, the resistance is the maximum value of the component.

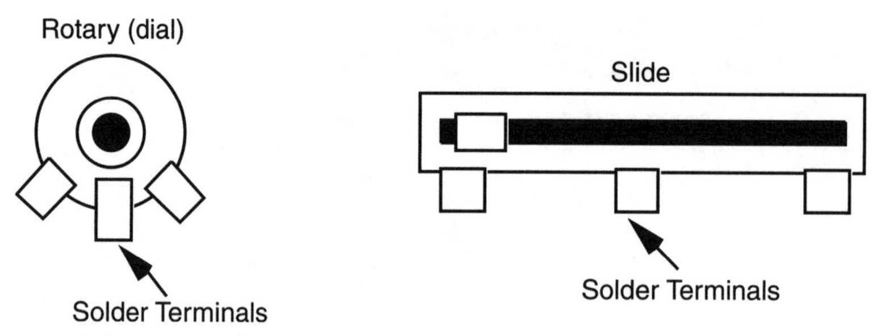

FIGURE 5-8 Potentiometers are variable resistors. You'll find them in rotary or slide versions; rotary potentiometers are the easiest to use in hobby circuits.

Some projects require precision potentiometers. These are referred to as multiturn pots or trimmers. Instead of turning the dial one complete rotation to change the resistance from, say, 0 to 10,000 ohms, a multiturn pot requires you to rotate the knob 3, 5, 10, even 15 times to span the same range. Most are designed to be mounted directly on the printed circuit board. If you have to adjust them, you will need a screwdriver or plastic tool.

5.5 Capacitors

After resistors, capacitors are the second most common component found in the average electronic project. Capacitors serve many purposes. They can be used to remove traces of transient (changing) current ripple in a power supply, to delay the action of some portion of the circuit, or to perform an integration or differentiation of a repeating signal. All these applications depend on the ability of the capacitor to hold an electrical charge for a predetermined time.

Capacitors come in many more sizes, shapes, and varieties than resistors, though only a small handful are truly common. However, most capacitors are made of the same basic stuff: a pair of conductive elements separated by an insulating dielectric (see Fig. 5-9). This dielectric can be composed of many materials, including air (in the case of a variable capacitor, as detailed in the next section), paper, epoxy, plastic, and even oil. Most capacitors

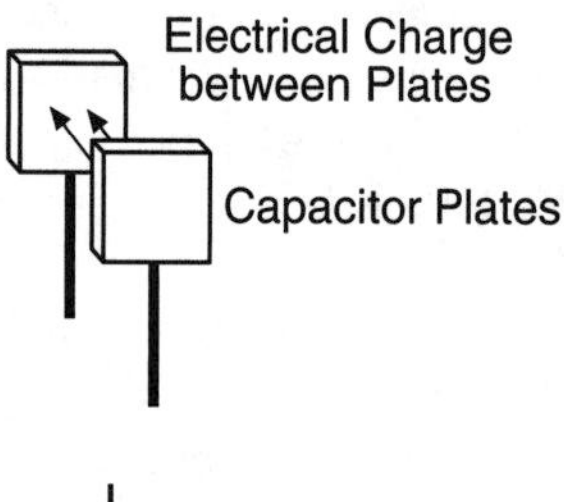

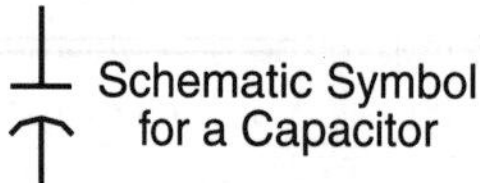

FIGURE 5-9 Capacitors store an electrical charge for a limited time. Along with the resistor, they are critical to the proper functioning of many electronic circuits.

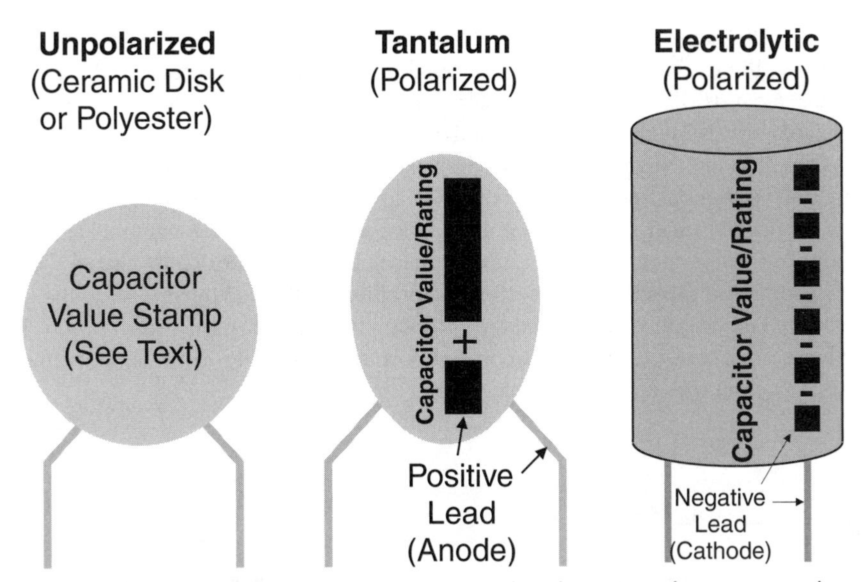

FIGURE 5-10 Different capacitor appearance and markings. Note that in some packages, the positive lead (anode) is indicated while in others it is the negative lead (cathode).

actually have many layers of conducting elements and dielectric. When you select a capacitor for a particular job, you must generally also indicate the type, such as ceramic, mica, or Mylar.

Capacitors are rated by their capacitance, in farads, and by the breakdown voltage of their dielectric. The *farad* is a rather large unit of measurement, so the bulk of capacitors available today are rated in microfarads, or a millionth of a farad. An even smaller rating is the picofarad, or a millionth of a millionth of a farad. The *micro* in the term *microfarad* is most often represented by the Greek mu (μ) character, as in 10 μF. The picofarad is simply shortened to pF. The voltage rating is the highest voltage the capacitor can withstand before the dielectric layers in the component are damaged.

For the most part, capacitors are classified by the dielectric material they use. The most common dielectric materials are aluminum electrolytic, tantalum electrolytic, ceramic, mica, polypropylene, polyester (or Mylar), paper, and polystyrene. The dielectric material used in a capacitor partly determines which applications it should be used for. The larger electrolytic capacitors, which use an aluminum electrolyte, are suited for such chores as power supply filtering, where large values are needed. The values for many capacitors are printed directly on the component. This is especially true with the larger aluminum electrolytic, where the large size of the capacitor provides ample room for printing the capacitance and voltage. Smaller capacitors, such as 0.1 or 0.01 μF mica disc capacitors, use a common three-digit marking system to denote capacitance and tolerance. The numbering system is easy to use, if you remember it's based on picofarads, not microfarads. A number such as 104 means 10, followed by four zeros, as in

100,000

or 100,000 picofarads. Values over 1000 picofarads are most often stated in microfarads. To make the conversion, move the decimal point to the left six spaces: 0.1 μF. Note that values under 1000 picofarads do not use this numbering system. Instead, the actual value, in picofarads, is listed, such as 10 (for 10 pF).

One mark you will find almost exclusively on larger tantalum and aluminum electrolytic is a polarity symbol, most often a minus (–) sign. The polarity symbol indicates the positive and/or negative lead of a capacitor. If a capacitor is polarized, it is extremely important that you follow the proper orientation when you install the capacitor in the circuit. If you reverse the leads to the capacitor—connecting the positive lead (called the *anode*) to the ground rail instead of the negative lead (called the *cathode*), for example—the capacitor may be ruined. Other components in the circuit could also be damaged. Fig. 5-10 shows some different capacitor packages along with their polarity markings.

5.6 Diodes

The diode is the simplest form of semiconductor. It is available in two basic flavors, germanium and silicon, which indicates the material used to manufacture the active junction within the diode. Diodes are used in a variety of applications, and there are numerous subtypes. Here is a list of the most common.

- *Rectifier.* The average diode, it rectifies AC current to provide DC only.
- *Zener.* It limits voltage to a predetermined level. Zeners are used for low-cost voltage regulation.
- *Light-emitting.* These diodes emit infrared of visible light when current is applied.
- *Silicon-controlled rectifier (SCR).* This is a type of high-power switch used to control AC or DC currents.
- *Bridge rectifier.* This is a collection of four diodes strung together in sequence; it is used to rectify an incoming AC current.

Diodes carry two important ratings: peak inverse voltage (PIV) and current. The PIV rating roughly indicates the maximum working voltage for the diode. Similarly, the current rating is the maximum amount of current the diode can withstand. Assuming a diode is rated for 3 amps, it cannot safely conduct more than 3 amps without overheating and failing.

All diodes have positive and negative terminals (polarity). The positive terminal is the anode, and the negative terminal is the cathode. You can readily identify the cathode end of a diode by looking for a colored stripe near one of the leads. Fig. 5-11 shows a diode that has a stripe at the cathode end. Note how the stripe corresponds with the heavy line in the schematic symbol for the diode.

All diodes emit light when current passes through them. This light is generally only in the infrared region of the electromagnetic spectrum. The light-emitting diode (LED) is a special type of semiconductor that is expressly designed to emit light in human visible wavelengths. LEDs are available to produce any of the basic colors (red, yellow, green, blue, or white) of light as well as infrared. The infrared LEDs are especially useful in robots for a variety of different applications.

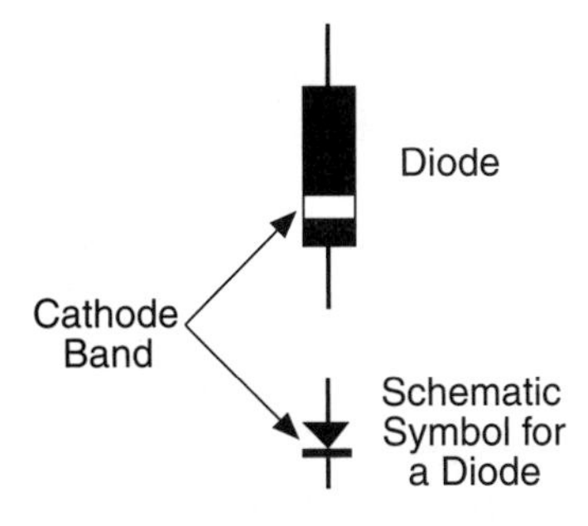

FIGURE 5-11 The polarity of diodes is marked with a stripe. The stripe denotes the cathode (negative) lead.

LEDs carry the same specifications as any other diode. The LED has a PIV rating of about 100 to 150 V, with a maximum current rating of under 40 mA (usually only 5 to 10 mA is applied to the LED). Most LEDs are used in low-power DC circuits and are powered with 12 V or less. Even though this voltage is far below the PIV rating of the LED, the component can still be ruthlessly damaged if you expose it to currents exceeding 40 or 50 mA. A resistor is used to limit the current to the LED.

5.7 Transistors

Transistors were designed as an alternative to the old vacuum tube, and they are used in similar applications, either to amplify a signal by providing a current control or to switch a signal on and off. There are several thousand different transistors available. Besides amplifying or switching a current, transistors are divided into two broad categories:

- *Signal.* These transistors are used with relatively low-current circuits, like radios, telephones, and most other hobby electronics projects.
- *Power.* These transistors are used with high-current circuits, like motor drivers and power supplies.

You can usually tell the difference between the two merely by size. The signal transistor is rarely larger than a pea and uses slender wire leads. The power transistor uses a large metal case to help dissipate heat, and heavy spokelike leads.

Transistors are identified by a unique code, such as 2N2222 or MPS6519. Refer to a data book to ascertain the characteristics and ratings of the particular transistor you are interested in. Transistors are rated by a number of criteria, which are far too extensive for the scope of this book. These ratings include collector-to-base voltage, collector-toe-mitter voltage, maximum collector current, maximum device dissipation, and maximum operating frequency. None of these ratings are printed directly on the transistor.

Signal transistors are available in either plastic or metal cases. The plastic kind is suitable for most uses, but some precision applications require the metal variety. Transistors that use metal cases (or cans) are less susceptible to stray radio frequency interference and they also dissipate heat more readily.

You will probably be using NPN (Fig. 5-12) and PNP (Fig. 5-13) bipolar transistors. These transistors are turned on and off by a control current passing through the base. The current that can pass through the collector is the product of the base current and the constant h_{FE}, which is unique to each transistor.

Bipolar transistors can control the operation and direction of DC motors using fairly simple circuits. Fig. 5-14 shows a simple circuit that will turn a motor on and off using a single NPN bipolar transistor and a diode. When the current passing through coils of a magnetic device changes, the voltage across the device also changes, often in the form of a large spike called *kickback*. These spikes can be a hundred volts or so and can very easily damage the electronic devices they are connected to. By placing a diode across the motor as shown in Fig. 5-15, the spikes produced when the motor is shut off will be shunted through the diode and will not pass along high voltages to the rest of the electronics in the circuit.

The circuit shown in Fig. 5-15 is known as an H-bridge because without the shunt diodes the circuit looks like the letter H. This circuit allows current to pass in either direction through a motor, allowing it to turn in either direction. The motor turns when one of the two connections is made to +V. Both connections can never be connected to +V as this will turn on all the transistors, providing a very low resistance path for current from +V, potentially burning out the driver transistors.

Along with bipolar transistors, which are controlled by current, there are a number of other transistors, some of which are controlled by voltage. For example, the MOSFET (for metal-oxide semiconductor field-effect transistor) is often used in circuits that demand high

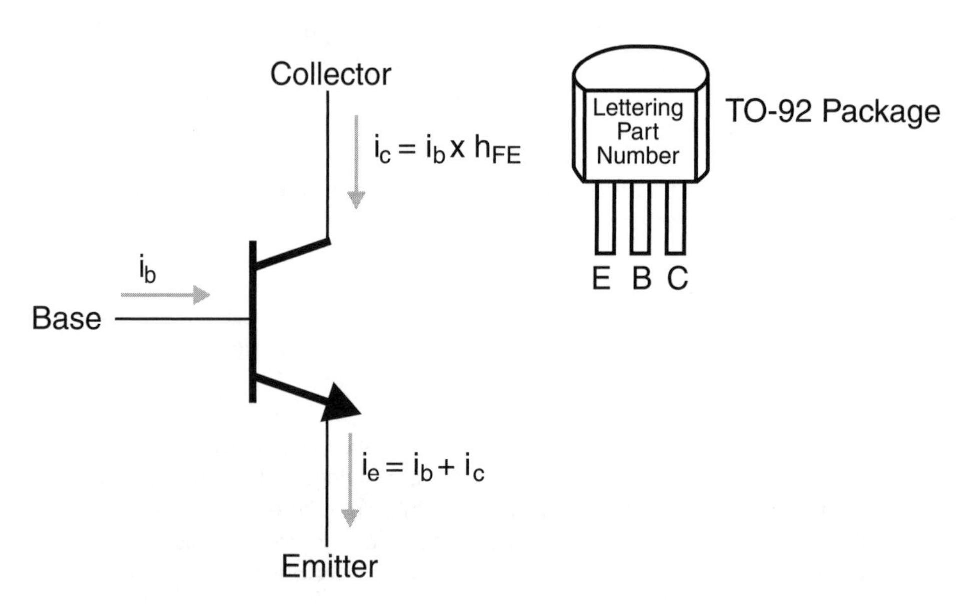

FIGURE 5-12 The NPN bipolar transistor collector current is controlled by current injected into the base.

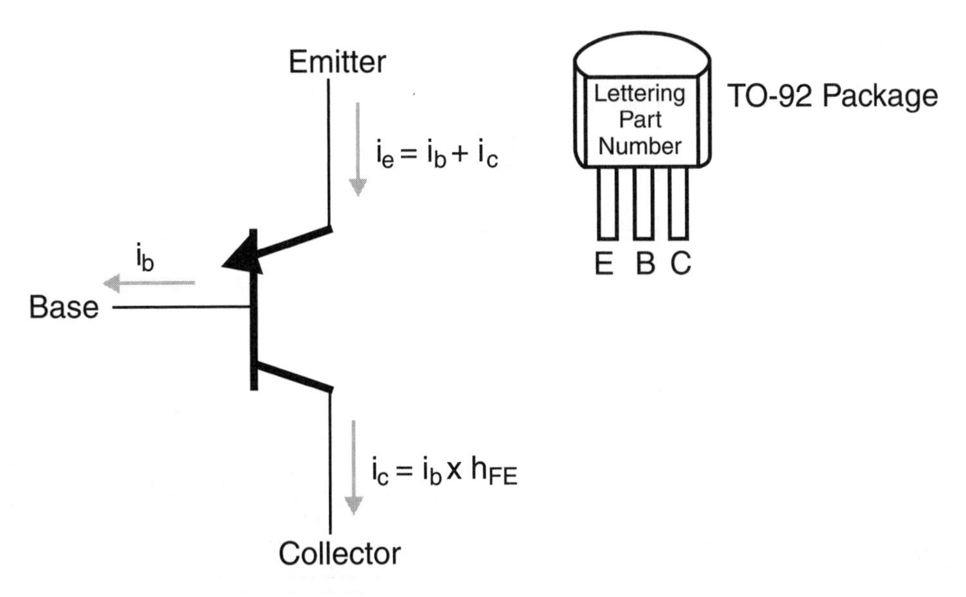

FIGURE 5-13 The PNP bipolar transistor collector current is controlled by current drawn from the base.

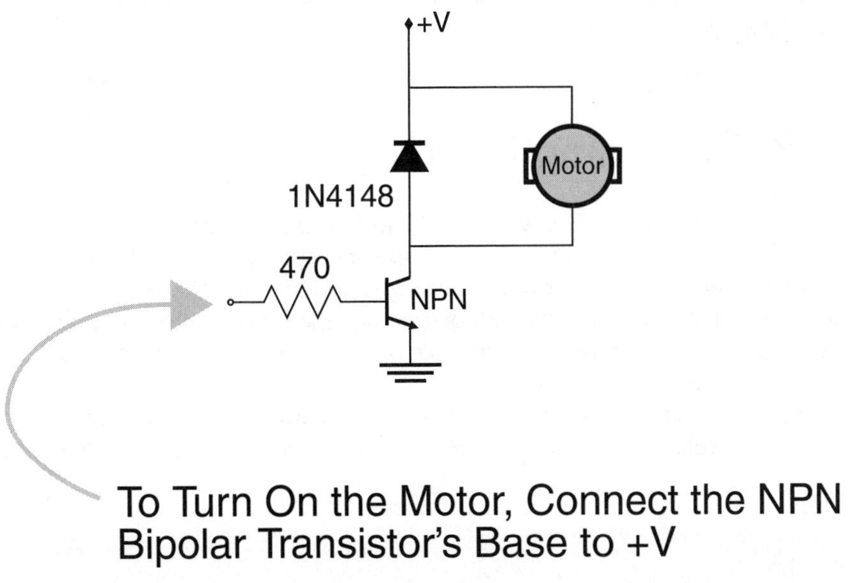

FIGURE 5-14 The complete transistor circuit needed to turn on and off a DC electric motor.

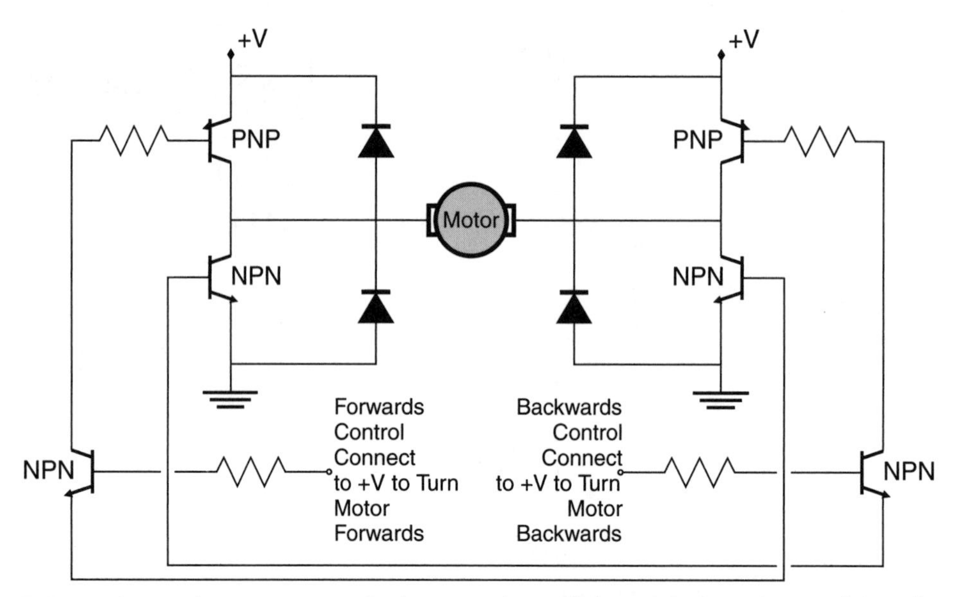

FIGURE 5-15 The six transistor H-bridge motor driver allows a DC motor to be turned on and off and run in either direction. Do not connect both controls to +V.

current and high tolerance. MOSFET transistors don't use the standard base-emitter collector connections. Instead, they call them *gate, drain,* and *source.* The operational differences among the different transistors will become clearer as you become more experienced in creating electronic circuits.

5.8 Grounding Circuitry

When wiring electronic circuits, it is useful to have a large common negative voltage connection or *ground* built into the robot. This connection is normally thought of as being at *earth ground* and is the basic reference for all the components in the circuit. Having a common ground also simplifies the task of drawing schematics; instead of wiring all the negative connections to the negative power supply, all the negative connections are wired to the three bar symbol shown in Fig. 5-16.

Positive voltages are normally indicated with an arrow pointing upward and the label of the positive voltage to be used. These conventions will be used throughout this book.

FIGURE 5-16 This symbol is used for the common negative voltage connection in an electronic circuit.

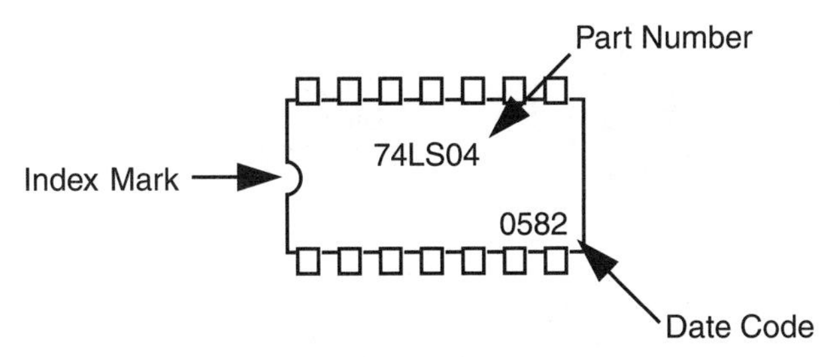

FIGURE 5-17 Integrated circuits (ICs) are common in most any electronic system, including robotics.

5.9 Integrated Circuits

The integrated circuit forms the backbone of the electronics revolution. The typical integrated circuit comprises many transistors, diodes, resistors, and even capacitors. As its name implies, the integrated circuit, or IC, is a discrete and wholly functioning circuit in its own right. ICs are the building blocks of larger circuits. By merely stringing them together you can form just about any project you envision.

Integrated circuits are most often enclosed in dual in-line packages (DIPs), like the one shown in Fig. 5-17. This type of component has a number of pins that can be inserted into holes of a printed circuit board and is also known as a pin through hole (PTH) component. There are numerous types of packages and methods of attaching chips to PCBs but beginners should be working with just PTH DIPs.

As with transistors, ICs are identified by a unique code, such as 7400 or 4017. This code indicates the type of device. You can use this code to look up the specifications and parameters of the IC in a reference book. Many ICs also contain other written information, including manufacturer catalog number and date code. Do not confuse the date code or catalog number with the code used to identify the device.

5.10 Schematics and Electronic Symbols

Electronics use a specialized road map to indicate what components are in a device and how they are connected together. This pictorial road map is the schematic, a kind of blueprint of everything you need to know to build an electronic circuit. Schematics are composed of special symbols that are connected with intersecting lines. The symbols represent individual components, and the lines represent the wires that connect these components together. The language of schematics, while far from universal, is intended to enable most anyone to duplicate the construction of a circuit with little more information than a picture.

The experienced electronics experimenter knows how to read a schematic. This entails recognizing and understanding the symbols used to represent electronic components and how these components are connected. All in all, learning to read a schematic is not difficult. Fig. 5-18 shows many of the most common symbols.

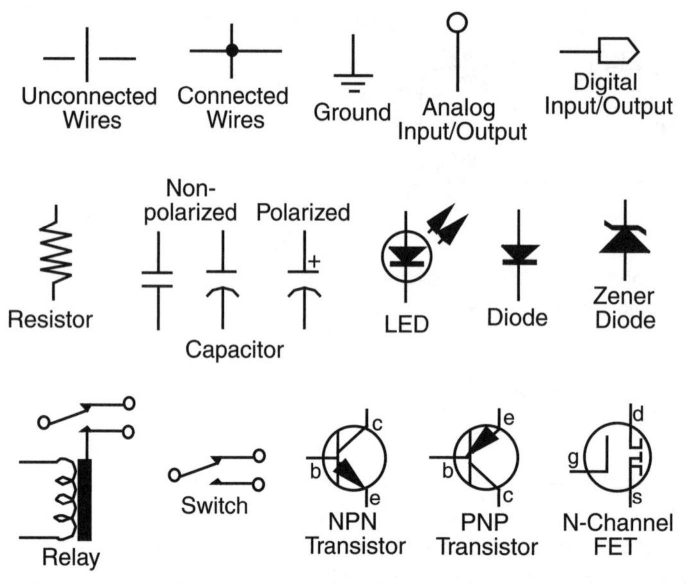

FIGURE 5-18 These symbols are used on schematics to represent different electronic devices.

5.11 From Here

To learn about . . .	*Read*
Finding electronic components	Chapter 4, "Buying Parts"
Working with electronic components	Chapter 6, "Electronic Construction Techniques"
Using electronic components with robot control computers	Chapter 12, "An Overview of Robot 'Brains' "

CHAPTER 6

TOOLS

Take a look at the tools in your garage or workshop. You probably already have all the implements required to build your own robot. Unless your robot designs require a great deal of precision (and most hobby robots don't), a common assortment of hand tools is all that's really needed to construct robot bodies, arms, drive systems, and more. Most of the hardware, parts, and supplies you need are also things you probably already have, left over from old projects around the house. You can readily purchase the pieces you don't have at a hardware store, a few specialty stores around town, or through the mail.

This chapter discusses the basic tools for hobby robot building and how you might use them. You should consider this chapter only as a guide; suggestions for tools are just that—suggestions. By no means should you feel that you must own each tool or have on hand all the parts and supplies mentioned in this chapter. You may have tools that you like to use that aren't listed in this chapter. Once again, the concept behind this book is to provide you with the know-how to build robots from discrete modules. In keeping with that open-ended design, you are free to exchange parts in the modules as you see fit. Some supplies and parts may not be readily available to you, so it's up to you to consider alternatives and how to work them into your design. Ultimately, it will be your task to take a trip to the hardware store, collect the items you need, and hammer out a unique creation that's all your own.

6.1 Safety

When building a robot, there should be one overriding concern and that is the safety of you and the other people building the robot. A momentary distraction or a few seconds of carelessness can lead to you or somebody else being seriously hurt. There are a few simple rules to follow when building robots to make sure that everyone is safe and, even if accidents happen, to minimize the chances for injuries.

1. *Always wear safety glasses.* There are a variety of different safety glasses available. Make sure you get ones that have shatterproof glass and side protection. If you wear glasses, use safety glasses that fit over your regular glasses or have a pair made with impact-resistant lenses and side shields.
2. *Never disable or take off tool safety devices.* This apparatus may seem to make the work more difficult and harder to observe, but they are there for a purpose. If doing the work seems to be particularly onerous due to the safety devices, then chances are you are not using the tool correctly.
3. *Always work in a well-ventilated area.* Some tools or building processes output gases that are not normally toxic, but in a closed environment can be dangerous. If you are working in a garage, it is a good idea to install a bathroom exhaust fan.
4. *Never work without somebody nearby.* There will be times when you work on your own, but make sure there is somebody you can call out to if there is a problem—never work alone in a house.
5. *If something goes wrong, take a few minutes to figure out what was its root cause and how to either fix it or prevent it from happening again later.* Try to avoid getting frustrated or angry at the work and instead go back and fix the reasons for the problem.
6. *Practical jokes have no place in a workshop.* Things that seem funny at the spur of the moment can unleash a catastrophic chain of events.
7. *Keep a fire extinguisher in the work area.* Cutting tools can throw off sparks or raise the temperature of materials to the flashpoint of wood and paper. Soldering irons, by definition, are very hot. Short-circuited batteries can become extremely hot and their casings catch fire.
8. *Make sure there is a telephone close by.* Ideally "911" should be programmed into it.

The most important rule is to never use a tool unless you are trained in its operation and understand all the safety issues. The same goes for chemicals; even apparently benign compounds can become dangerous with the right set of circumstances. Make sure you read and are familiar with all manuals, material safety sheets, and any plan notes before starting work.

Following these simple rules and properly preparing to do the work will greatly minimize the chances of somebody getting hurt.

6.2 Setting Up Shop

You'll need a worktable to construct the mechanisms and electronic circuits of your robots. The garage is an ideal location because it affords you the freedom to cut and drill wood, metal, and plastic without worrying about getting pieces in the carpet or tracking filings and sanding dust throughout the house. The garage is also good because it will minimize any trapped paint or chemical smells, and on those occasions when you burn out an electrical or electronic device, the house won't take on the smell of a store having a fire sale. Your new hobby will be better tolerated and even encouraged if it does not result in extra work or inconvenience for anyone you live with.

In whatever space you choose to set up your robot lab, make sure all your tools are within easy reach. You can keep special tools and supplies in an inexpensive fishing tackle box. It provides lots of small compartments for screws and other parts. For best results, your work space should be an area where the robot-in-progress will not be disturbed if you have to leave it for several hours or several days, as will usually be the case. It should go without saying that the worktable and any power tools should also be off limits or inaccessible to young children.

Good lighting is a must. Both mechanical and electronic assembly requires detail work, and you will need good lighting to see everything properly. Supplement overhead lights with a 60-W desk lamp. You'll be crouched over the worktable for hours at a time, so a comfortable chair or stool is a must. Be sure you adjust the seat for the height of the worktable.

It's not a bad idea to set up your workshop with a networked PC and a phone to be electronically connected to the outside world as well as the systems within your house. Being able to talk to other people (either by phone or the Internet) will be useful, especially if you are having a problem and have the material and tools right in front of you. A PC networked to other PCs in the house could be used for displaying design drawings produced on another computer or allowing you to perform a quick web search for information. Remember to protect the PC and phone from dust and filings that could become airborne when material is being cut.

6.3 Basic Tools

Construction tools are what you use to fashion the frame and other mechanical parts (or structure) of the robot. We will look at the tools needed to assemble the electronics later in this chapter.

The basic tools for creating a robot include:

- *Claw hammer.* These can be used for just about any purpose.
- *Rubber mallet.* For gently bashing together pieces that resist being joined, nothing beats a rubber mallet; it is also useful for forming sheet metal.
- *Measurement tools.* You should have a variety of metal scales, wood and plastic rulers all of varying lengths, as well as a cheap analog dial or digital calipers. You may also want to keep a drill diameter gauge handy along with tools for measuring screw diameters and pitches. Finally, kitchen and fishing scales are useful tools for keeping track of how much the robot is going to weigh.
- *Screwdriver assortment.* Have several sizes of flat-head and Phillips-head screwdrivers. It's also handy to have a few long-blade screwdrivers, as well as a ratchet driver. Get a screwdriver magnetizer/demagnetizer; it lets you magnetize the blade so it attracts and holds screws for easier assembly.

- *Hacksaw.* To cut anything, the hacksaw is the staple of the robot builder. Buy an assortment of blades. Coarse-tooth blades are good for wood and PVC pipe plastic; fine-tooth blades are good for copper, aluminum, and light-gauge steel.
- *Miter box.* To cut straight lines, buy a good miter box and attach it to your worktable (avoid wood miter boxes; they don't last). You'll also use the box to cut stock at near-perfect 45° angles, which is helpful when building robot frames.
- *Wrenches, all types.* Adjustable wrenches are helpful additions to the shop but careless use can strip nuts. The same goes for long-nosed pliers, which are useful for getting at hard-to-reach places. One or two pairs of Vise-Grips will help you hold pieces for cutting and sanding. A set of nut drivers will make it easy to attach nuts to bolts.
- *Measuring tape.* A 6- or 8-ft steel measuring tape is a good length to choose. Also get a cloth measuring tape at a fabric store so you can measure things like chain and cable lengths.
- *Square.* You'll need one to make sure that pieces you cut and assemble from wood, plastic, and metal are square.
- *File assortment.* Files will enable you to smooth the rough edges of cut wood, metal, and plastic (particularly important when you are working with metal because the sharp, unfinished edges can cut you).
- *Motor drill.* Get one that has a variable speed control (reversing is nice but not absolutely necessary). If the drill you have isn't variable speed, buy a variable speed control for it. You need to slow the drill when working with metal and plastic. A fast drill motor is good for wood only. The size of the chuck is not important since most of the drill bits you'll be using will fit a standard ¼-in chuck.
- *Drill bit assortment.* Use good sharp ones only. If yours are dull, have them sharpened (or do it yourself with a drill bit sharpening device), or buy a new set.
- *Vise.* A vise is essential for holding parts while you drill, nail, and otherwise torment them. An extra large vise isn't required, but you should get one that's big enough to handle the size of the pieces you'll be working with. A rule of thumb: a vice that can't close around a 2-in block of metal or wood is too small.
- *Safety goggles.* Wear them when hammering, cutting, and drilling as well as any other time when flying debris could get in your eyes. Be sure you use the goggles. A shred of aluminum sprayed from a drill bit while drilling a hole can rip through your eye, permanently blinding you. No robot project is worth that.

If you plan to build your robots from wood, you may want to consider adding rasps, wood files, coping saws, and other woodworking tools to your toolbox. Working with plastic requires a few extra tools as well, including a burnishing wheel to smooth the edges of the cut plastic (the flame from a cigarette lighter also works but is harder to control), a strip-heater for bending, and special plastic drill bits. These bits have a modified tip that isn't as likely to rip through the plastic material. Small plastic parts can be cut and scored using a sharp razor knife or razor saw, both of which are available at hobby stores.

6.3.1 OPTIONAL TOOLS

There are a number of other tools you can use to make your time in the robot shop more productive and less time consuming. Note that many of these tools are powerful and cause a lot of injury or damage if you are not careful or are inexperienced in their use. If you are unfamiliar with any of the tools, do not buy or use them until you have received training in them!

- A drill press helps you drill better holes because you have more control over the angle and depth of each hole. Be sure to use a drill press vise to hold the pieces. Never use your hands! Along with the drill press, if you are working with metal, it is a good idea to get a spring-loaded center punch, which places an indentation in the material when you press down on it. This indentation will help guide the drill bit to the correct location and is a lot easier to see than a scratch.
- A table saw or circular saw makes it easier to cut through large pieces of wood and plastic. To ensure a straight cut, use a guide fence or fashion one out of wood and clamps. Be sure to use a fine-tooth saw blade if you are cutting through plastic. Using a saw designed for general woodcutting will cause the plastic to shatter.
- A miter saw is a useful tool for precisely cutting wood and plastic parts (making sure that the correct blade for the material is being used). Many of the more recent miter saws have laser guides that will help you precisely line up the cut to marks that you have made in the material.
- An abrasive cutter is a useful tool to have around if you have to cut steel and thick aluminum channel. The cutter looks like a miter saw, but has a silicon carbide cutting disk that chews through the material being cut. The abrasive cutter is reasonably precise, very fast, but will leave a burr that will have to be trimmed off with a file.
- A motorized hobby tool, such as the model shown in Fig. 6-1, is much like a handheld router. The bit spins very fast (25,000 r/min and up), and you can attach a variety of wood, plastic, and metal working bits to it. The better hobby tools, such as those made by Dremel and Weller, have adjustable speed controls. Use the right bit for the job. For example, don't use a wood rasp bit with metal or plastic because the flutes of the rasp will too easily fill with metal and plastic debris.
- A RotoZip tool (that's its trade name) is a larger, more powerful version of a hobby tool. It spins at 30,000 r/min and uses a special cutting bit—it looks like a drill bit, but works like a saw. The RotoZip is commonly used by drywall installers, but it can be used to cut through most any material you'd use for a robot (exception: heavy-gauge steel).
- A Pop Rivet Gun will allow you to quickly fasten two pieces of metal (or metal and plastic) together permanently in just a few seconds. An inexpensive tool (less than $10) can be used.
- Hot-melt glue guns are available at most hardware and hobby stores and come in a variety of sizes. The gun heats up glue from a stick; press the trigger and the glue oozes out the tip. The benefit of hot-melt glue is that it sets very fast—usually under a minute. You can buy glue sticks for normal- or low-temperature guns. Exercise caution when using a hot-melt glue gun: the glue is hot, after all!
- A nibbling tool is a fairly inexpensive accessory (under $20) that lets you "nibble" small chunks from metal and plastic pieces. The maximum thickness depends on the bite of the tool, but it's generally about 1⁄16 in. Use the tool to cut channels and enlarge holes.
- A tap and die set lets you thread holes and shafts to accept standard-sized nuts and bolts. Buy a good set. A cheap assortment is more trouble than it's worth.

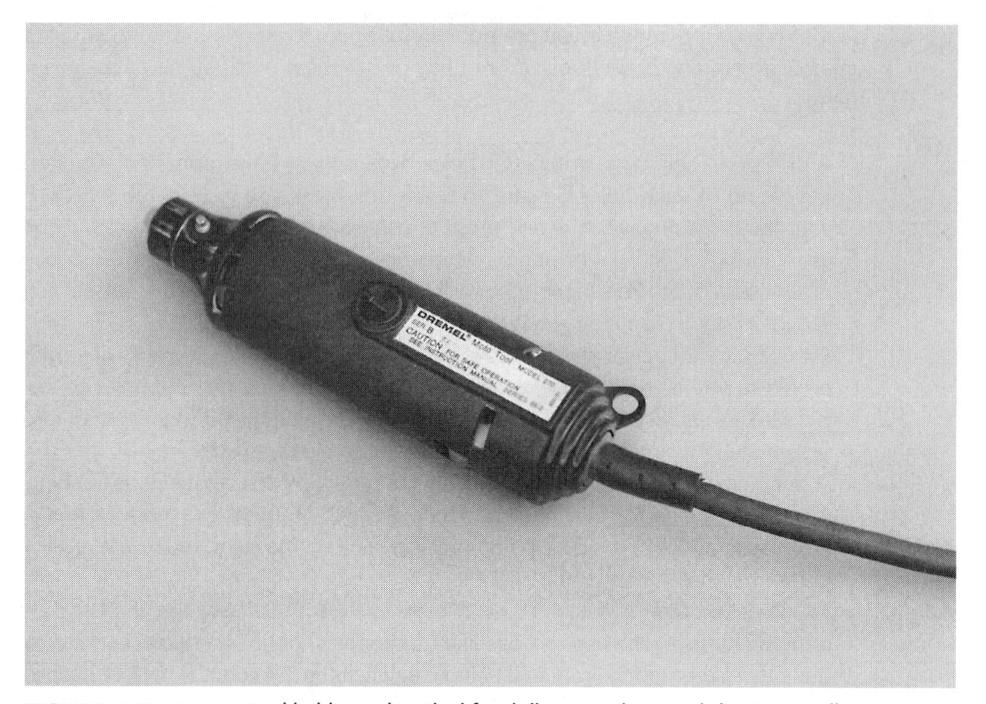

FIGURE 6-1 A motorized hobby tool is ideal for drilling, sanding, and shaping small parts.

- A thread-size gauge, made of stainless steel, may be expensive, but it helps you determine the size of any standard SAE or metric bolt. It's a great accessory for tapping and dieing.
- A brazing tool or small welder lets you spot-weld two metal pieces together. These tools are designed for small pieces only. They don't provide enough heat to adequately weld pieces larger than a few inches in size. Be sure that extra fuel and oxygen cylinders or pellets are readily available for the brazer or welder you buy. There's nothing worse than spending $30 to $40 for a home welding set, only to discover that supplies are not available for it. Be sure to read the instructions that accompany the welder and observe all precautions.

6.4 Electronic Tools

Constructing electronic circuit boards or wiring the power system of your robot requires only a few standard tools. A soldering iron leads the list. For maximum flexibility, invest in a modular soldering pencil, the kind that lets you change the heating element. For routine electronic work, you should get a 25- to 30-W heating element. Anything higher may damage electronic components. You can use a 40- or 50-W element for wiring switches, relays, and power transistors. Stay away from instant-on soldering irons. For any application other than soldering large-gauge wires they put out far too much heat.

As obvious as it seems to most people, do not use soldering iron, solder, or flux that is designed for plumbing applications. These tools and materials are not appropriate for any robot applications and could potentially damage the components being connected together.

- *Soldering stand.* Mandatory for keeping the soldering pencil in a safe, upright position.
- *Soldering tip assortment.* Get one or two small tips for intricate printed circuit board work and a few larger sizes for routine soldering chores.
- *Solder.* Buy resin or flux core type that is designed for electronics. Acid core and silver solder should never be used on electronic components.
- *Solder sponge.* Sponges are useful for cleaning the soldering tip as you use it. Keep the sponge damp, and wipe the tip clean every few joints.
- *Desoldering vacuum tool.* This is useful for soaking up molten solder. Use it to get rid of excess solder, remove components, or redo a wiring job.
- *Solder braid.* Performs a similar function to the desoldering vacuum tool by wicking excess molten solder away from a joint or a component.
- *Dental picks.* These are ideal for probing and separating wires and component leads.
- *Resin cleaner.* Apply the cleaner after soldering is complete to remove excess resin. An ultrasonic jewelry cleaner used with the resin cleaner (or isopropyl alcohol) will get the components very clean with very little work on your part.
- *Solder vise.* This vise serves as a third hand, holding together pieces to be soldered so you are free to work the iron and feed the solder.
- *An illuminated magnifying glass.* Often going by the trade name Dazer, this provides a two or three time magnification of the surface below. It is invaluable for inspecting work or soldering fine components.

6.4.1 STATIC CONTROL

Running robots are terrible environments for electronics. The motors tend to produce large transients on the power lines that come back to the control electronics, running across different surfaces will build up static electrical charges, and different metal parts can produce unexpected voltages and currents that can upset electronics. Fortunately, most modern electronic devices are protected from static electrical discharges, but you should still make some basic concessions to protect them from being damaged during assembly and operation.

Everybody is familiar with the sparks that you can produce by shuffling along a synthetic carpet and touching a metal doorknob. What will probably surprise you is the magnitude of the voltage needed to produce a spark: for you to feel and see the spark, a static charge of 2500 V or more is required. The minute amount of current flow (on the order of microamperes) is why you are not hurt. As you have probably heard, when somebody is electrocuted it is the current that kills, not the voltage.

The amount of static electricity that can damage a silicon chip is significantly less—125 V can ruin a diode's PN junction or a MOSFET's gate oxide layer. Note that 125 V is 20 times less than the 2500-V threshold needed to detect static electricity. This relatively low level is the reason for the concern about static electricity. The term *electrostatic discharge* (ESD) is used to describe the release of static electricity (either from you to the doorknob or into an electronic circuit). Virtually all modern electric devices have built-in protections against ESD, but there are still a number of things that you must do to ensure that the components are not damaged by static electricity, either during assembly or use.

You can buy basic ESD protection kits, which consist of a conducting mat, a wrist strap with cord, and a static bleed line that can be attached to the grounding screw (holding the faceplate on an electrical outlet), for about $20. Before buying this type of kit, there are a number of things that you should be aware of. First, the term *conducting* when applied to the mat, the wrist strap, and the static bleed line is very loosely applied; each of these components should have internal resistances in the mega-ohm range. Do not buy an ESD kit that does not have a mega-ohm resistor in the wrist strap cord or the static bleed line as there could be dangerous voltages passed along them into *you*. Finally, before attaching the static bleed line to the electrical outlet's grounding screw make sure that the socket is wired properly using a socket tester (which costs around $5).

All circuit assembly should take place on the conducting mat while you are wearing the wrist strap. When you buy electronic components, they will either be packaged in conducting plastic tubes or in anti-static bags. When you have completed the assembly operation, the components should be returned to their original packages and any assembled circuitry placed inside an anti-static bag. Operation of the circuitry can have potential problems as was noted at the start of this section. To minimize the chance of the robot's electronics becoming damaged, there are a few precautions that can be taken:

- The metal parts of the robot should all be connected together and connected to the ground (negative) connection of the robot's electronics. This will prevent static electricity from building up within the robot.
- Robot whiskers must be connected to the ground as they can generate static electricity when they run across an object.
- Castor mountings should be attached to the robot's ground connection to avoid buildups of static electricity.
- You might want to let a small metal chain or wire braid run on the ground while the robot is moving to help dissipate static charges from the robot.
- The robot electronics should be within some kind of metal box to protect them from static electricity when the robot is picked up.
- Connectors to PCs (e.g., USB or RS-232) for programming or monitoring must have the outside shells connected to the robot's metal frame and negative connection to make sure any static electricity buildup is not passed through either the robot's or the PC's electronics, damaging them.

6.4.2 DIGITAL MULTIMETER

A *digital multimeter* (DMM and also known as a volt-ohm meter or multitester) is used to test voltage and current levels along with the resistance of different parts of circuits. Along with these basic functions, you can find DMMs that can test transistors and capacitors, and measure signal frequencies and temperature. They can be purchased from under $10 to several thousand, and if you don't already own a volt-ohm meter you should seriously consider buying one immediately. The low cost of a simple unit is disproportionate to the usefulness of the instrument.

There are many DMMs on the market today. For robotics work, a meter of intermediate quality is sufficient and does the job admirably at a price between $30 and $75 (it tends to be on the low side of this range). Meters are available at Radio Shack and most electronics outlets.

While analog (meters with needles) multimeters are still available, you should avoid them. Digital meters employ a numeric display not unlike a digital clock or watch. Analog meters use the older-fashioned mechanical movement with a needle that points to a set of graduated scales. When they first became available, DMMs (like the one shown in Fig. 6-2) used to cost a great deal more than the analog variety, but now they generally cost less than analog meters, are more easily read, and are usually more robust. In fact, it's hard to find a decent analog meter these days.

Most low-cost DMMs require you to select the range before it can make an accurate measurement. For example, if you are measuring the voltage of a 9-V transistor battery, you set the range to the setting closest to, but above, 9 V (with most meters it is the 0 to 20 or 0 to 50-V range). Auto-ranging meters (which cost more than the basic models) don't require you to do this, so they are inherently easier to use. When you want to measure voltage, you set the meter to volts (either AC or DC) and take the measurement. The meter displays the results in the readout panel.

Little of the work you'll do with robot circuits will require a DMM that's superaccurate; when working with electronics, being within a few percentage points of the desired value is normally good enough for a circuit to work properly. The accuracy of a meter is the mini-

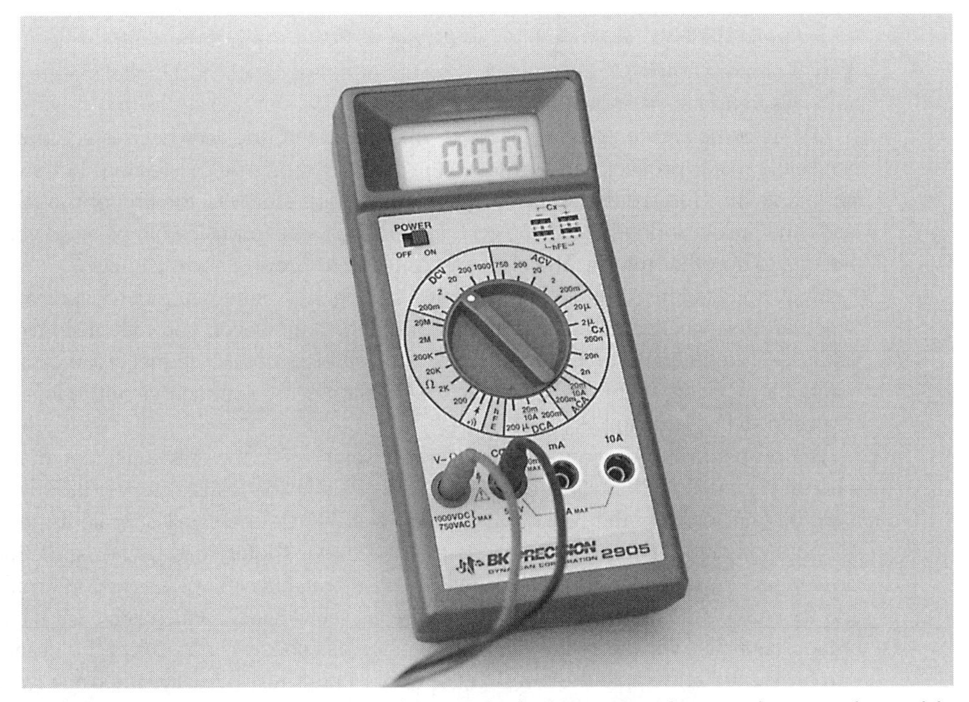

FIGURE 6-2 A volt-ohm meter (or multitester) checks resistance, voltage, and current. This model is digital and has a 3½-digit liquid crystal display (LCD) readout.

mum amount of error that can occur when taking a specific measurement. For example, the meter may be accurate to 2000 V, plus or minus 0.8 percent. A 0.8 percent error at the kinds of voltages used in robots—typically, 5 to 12 V DC—is only 0.096 V.

The number of digits in the DMM display determines the maximum resolution of the measurements. Most DMMs have three and a half digits, so they can display a value as small as 0.001 (the half digit is a 1 on the left side of the display). Anything less than that is not accurately represented and there's little need for accuracy better than this.

DMMs vary greatly in the number and type of functions they provide. At the very least, all standard meters let you measure AC volts, DC volts, milliamps, and ohms. Some also test capacitance and opens or shorts in discrete components like diodes and transistors. These additional functions are not absolutely necessary for building general-purpose robot circuits, but they are handy when troubleshooting a circuit that refuses to work.

The maximum ratings of the meter when measuring volts, milliamps, and resistance also vary. Before buying a specific DMM, make sure you understand what the maximum values are that the meter can handle. Most DMMs have maximum values ratings in the ranges of:

DC voltages to 1000 V

AC voltages to 500 V

DC currents to 200 mA with up to 10 A using a fused input

Resistance 2 MΩ

One exception to this is when you are testing current draw for the entire robot versus just for motors. Many DC motors draw an excess of 200 mA, and the entire robot is likely to draw 2 or more amps. Obviously, this is far out of the range of most digital meters, but there are ways to do it as is shown in Chapter 19.

DMMs come with a pair of test leads, one black and one red. Each is equipped with a needlelike metal probe. Standard leads are fine for most routine testing, but some measurements may require that you use a clip lead. These attach to the end of the regular test leads and have a spring-loaded clip on the end. You can clip the lead in place so your hands are free to do other things. The clips are insulated to prevent short circuits.

Most applications of the DMM involve testing low voltages and resistance, both of which are relatively harmless to humans. Sometimes, however, you may need to test high voltages—like the input to a power supply—and careless use of the meter can cause serious bodily harm. Even when you're not actively testing a high-voltage circuit, dangerous currents can still be exposed.

The proper procedure for using a meter is to set it beside the unit under test, making sure it is close enough so the leads reach the circuit. Plug in the leads, and test the meter operation by first selecting the resistance function setting (use the smallest scale if the meter is not auto-ranging). Touch the leads together: the meter should read 0 Ω or something very close (a half ohm or so). If the meter does not respond, check the leads and internal battery and try again. Analog multimeters often have a "zero adjust," which provides a basic calibration capability for the meter. Once the meter has checked out, select the desired function and range and apply the leads to the circuit under test. Usually, the black lead will be connected to the ground, and the red lead will be connected to the various test points in the circuit.

6.4.3 LOGIC PROBES

Meters are typically used for measuring analog signals. Logic probes test for the presence or absence of low-voltage digital data signals. The 0s and 1s are usually electrically defined as 0 and 5 V, respectively, when used with TTL integrated circuits (ICs). In practice, the actual voltages of the 0 and 1 bits depend entirely on the circuit and the parts used to make it up. You can use a meter to test a logic circuit, but the results aren't always predictable. Further, many logic circuits change states (pulse) quickly, and meters cannot track the voltage switches quickly enough.

Logic probes, such as the model in Fig. 6-3, are designed to give a visual and (usually) audible signal of the logic state of a particular circuit line. One LED (light-emitting diode) on the probe lights up if the logic is 0 (or *low*); another LED lights up if the logic is 1 (or *high*). You should only work with a probe that has a built-in buzzer with different tones for the two logic levels. This feature will allow you to look at the circuitry you are probing rather than having to glance at the probe to see the logic level.

A third LED or tone may indicate a pulsing signal. A good logic probe can detect that a circuit line is pulsing at speeds of up to 10 MHz, which is more than fast enough for robotic applications, even when using computer control. The minimum detectable pulse width (the time the pulse remains at one level) is 50 nanoseconds, which again is more than sufficient.

Another feature that you should be aware of is the ability of the probe to work with different logic families and logic voltage levels. Different CMOS logic families can work at

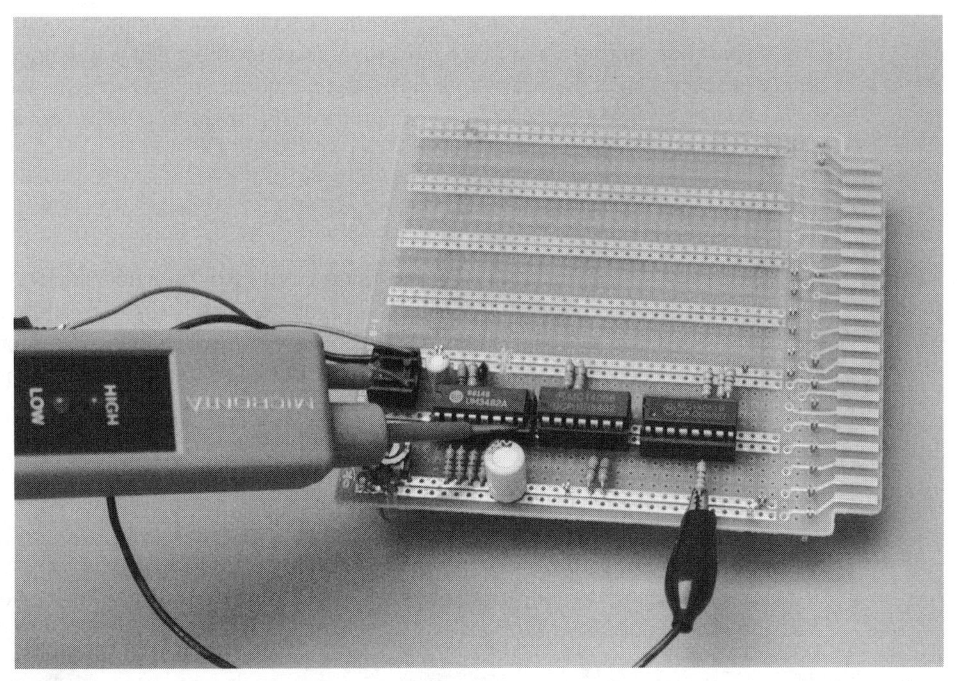

FIGURE 6-3 The logic probe in use. Note that in the photograph the probe derives its power from the circuit under test.

power supply voltages ranging from 3 to 15 V with a logic level transition voltage of one-half the power supply voltage (1.5 to 7.5 V). TTL logic's transition voltage is usually regarded as 1.4 V and is independent of the input power supply (which ranges from 4.75 to 5.25 V). The logic technology that the probe works with is switch selectable.

Most probes are not battery operated; rather, they obtain operating voltage from the circuit under test. This feature allows you to start simply probing your circuit without providing a separate power supply (with a matching ground to the test circuit) for the logic probe and determining the appropriate CMOS logic test level.

Although logic probes may sound complex, they are really simple devices, and their cost reflects this. You can buy a reasonably good logic probe for under $20. The logic probe available from Radio Shack, which has most of the features listed here, can be purchased at this price point. You can also make a logic probe if you wish; it is not recommended as you will be hard pressed for buying the necessary parts for less than the cost of an inexpensive unit.

To use the logic probe successfully you really must have a circuit schematic to refer to. Keep it handy when troubleshooting your projects. It's nearly impossible to blindly use the logic probe on a circuit without knowing what you are testing. And since the probe receives its power from the circuit under test, you need to know where to pick off suitable power. To use the probe, connect the probe's power leads to a voltage source on the board, clip the black ground wire to circuit ground, and touch the tip of the probe against a pin on an integrated circuit or the lead of some other component. For more information on using your probe, consult the manufacturer's instruction sheet.

When designing your robot, it is a good idea to keep your digital logic separate from power supplies and other high-voltage/high-current circuits. When working on an awkward circuit, such as one mounted in a robot, it is not unusual for the metal probe tip to slip and short out other circuits. If the board is all digital logic, then this isn't a problem—but if there are other circuits on the board, you could end up damaging them, your logic probe, and any number of miscellaneous circuits.

6.4.4 OSCILLOSCOPE

An oscilloscope is a pricey tool, but for performing serious work or understanding how the circuitry behaves in your robot, it is invaluable and will save you hours of frustration. Other test equipment will do some of the things you can do with a scope, but oscilloscopes do it all in one box and generally with greater precision. Among the many applications of an oscilloscope, you can do the following:

- Test DC or AC voltage levels
- Analyze the waveforms of digital and analog circuits
- Determine the operating frequency of digital, analog, and RF circuits
- Test logic levels
- Visually check the timing of a circuit to see if things are happening in the correct order and at the prescribed time intervals.

The most common application used to demonstrate the operation of an oscilloscope is converting sound waves into a visual display by passing the output of a microphone into an oscilloscope. This application, while very appealing, does not demonstrate any of the important features of an oscilloscope nor is it representative of the kind of signals that you will probe with it. When you are looking at buying an oscilloscope, you should consider the different features and functions listed in the following.

The resolution of the scope reveals its sensitivity and accuracy. On an oscilloscope, the X (horizontal) axis displays time, and the Y (vertical) axis displays voltage. These values can be measured by the marks, or graticules, on the oscilloscope display. To change the sensitivity, there is usually a knob on the oscilloscope that will make the time between each set of markings larger or smaller. The value between the graticule markings is either displayed on the screen itself electronically or marked on the oscilloscope by the adjustment knob (Fig. 6-4).

There are two different types of oscilloscopes. The analog oscilloscope passes the incoming signal directly from the input probes to the CRT display without any processing. Rather than displaying the signal as it comes in, there is normally a trigger circuit, which starts the display process when the input voltage reaches a specific point. Analog oscilloscopes are best suited for repeating waveforms; they can be used to measure their peak to peak voltages, periods, and timing differences relative to other signals.

The digital storage oscilloscope (DSO) converts the analog voltage to a digital value and then displays it on a computer-like screen. By converting the analog input to digital, the waveform can be saved and displayed after a specific event (also known as the trigger, as in the analog oscilloscope) or processed in some way. Whereas the peak to peak voltage and the waveform's period is measured from the screen in an analog oscilloscope, most digital storage oscilloscopes have the ability to calculate these (and other) values for you. Digital storage oscilloscopes can be very small and flexible; there are a number of products available that connect directly to a PC and avoid the bulk and cost of a display all together. It

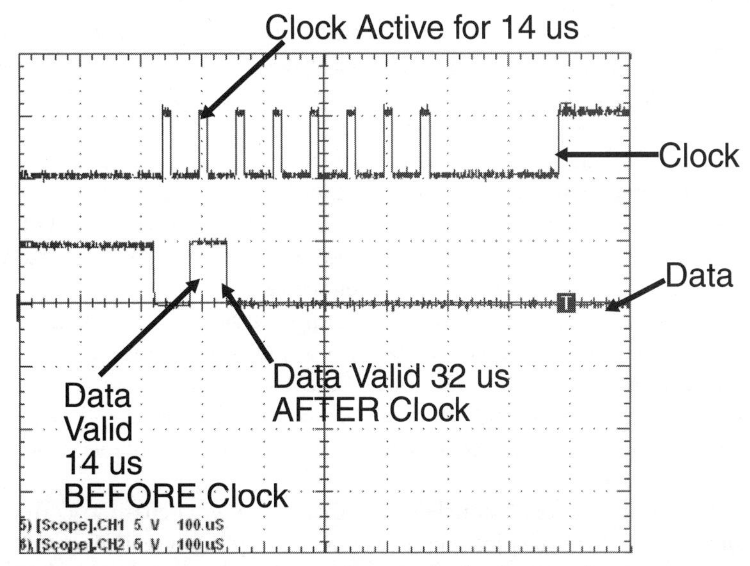

FIGURE 6-4 Digital storage oscilloscope display showing the changing voltage level on the two pins used in the BS2's shiftout statement.

should be noted that the digital storage oscilloscope is capable of displaying the same repeating waveforms as an analog oscilloscope.

One of the most important specifications of an oscilloscope is its *bandwidth,* which is the maximum frequency signal that can be observed accurately. For example, a 20 MHz oscilloscope can accurately display and measure a 20 MHz sine wave. The problem with most signals is that they are not perfect sine waves; they usually consist of much higher frequency harmonics, which make up the signal. To accurately display an arbitrary waveform at a specific frequency, the bandwidth must be significantly higher than the frequency itself; five times the required bandwidth is the minimum that you should settle for, with 10 times being a better value. So, if in your circuit, you have a 20 MHz clock, to accurately observe the signal the oscilloscope's bandwidth should be 100 MHz or more.

Along with the bandwidth measurement in a digital storage oscilloscope, there is also the *sampling* rate of the incoming analog signal. The bandwidth measurement of a digital storage oscilloscope is still relevant; like the analog oscilloscope it specifies the maximum signal frequency that can be input without the internal electronics of the digital storage oscilloscope distorting it. The sampling rate is the number of times per second that the oscilloscope converts the analog signal to a digital value. Most digital storage oscilloscopes will sample at 10 to 50 times the bandwidth and the sampling measurement is in units of samples per second.

Finally, the oscilloscope's trigger is an important feature that many people do not understand how to use properly. As previously noted, the trigger is set to a specific voltage to start displaying (or recording in the case of a digital storage oscilloscope) the incoming analog voltage signal. The trigger allows signals to be displayed without jitter so that the incoming waveform will be displayed as a steady waveform, instead of one that jumps back and forth or appears as a steady blur without any defined start point. The trigger on most oscilloscopes can start the oscilloscope when the signal goes from high to low at a specific voltage level, or from low to high.

Over the years, oscilloscopes have improved dramatically, with many added features and capabilities. Among the most useful features is a delayed sweep, which is helpful when you are analyzing a small portion of a long, complex signal. This feature is not something that you will be comfortable using initially, but as you gain experience with the oscilloscope and debugging you will find that it is an invaluable feature for finding specific problems or observing how the circuitry works after a specific trigger has been executed.

The probes used with oscilloscopes are not just wires with clips on the end of them. To be effective, the better scope probes use low-capacitance/low-resistance shielded wire and a capacitive-compensated tip. These ensure better accuracy.

Most scope probes are passive, meaning they employ a simple circuit of capacitors and resistors to compensate for the effects of capacitive and resistive loading. Many passive probes can be switched between 1X and 10X. At the 1X setting, the probe passes the signal without attenuation (weakening). At the 10X setting, the probe reduces the signal strength by 10 times. This allows you to test a signal that might otherwise overload the scope's circuits.

Active probes use operational amplifiers or other powered circuitry to correct for the effects of capacitive and resistive loading as well as to vary the attenuation of the signal. Table 6-1 shows the typical specifications of passive and active oscilloscope probes.

TABLE 6-1 Specifications For Typical Oscilloscope Probe

PROBE TYPE	FREQUENCY RANGE	RESISTIVE LOAD	CAPACITIVE LOAD
Passive 1X	DC–5 MHz	1 megohm	30 pF
Passive 10X	DC–50 MHz	10 megohms	5 pF
Active	20–500 MHz	10 megohms	2 pF

As an alternative to a stand-alone oscilloscope you may wish to consider a PC-based oscilloscope solution. Such oscilloscopes not only cost less but may provide additional features, such as long-term data storage. A PC-based oscilloscope uses your computer and the software running on it as the active testing component.

Most PC-based oscilloscopes are comprised of an interface card or adapter. The adapter connects to your PC via an expansion board or a serial, parallel, or USB port (different models connect to the PC in different ways). A test probe then connects to the interface. Software running on your PC interprets the data coming through the interface and displays the results on the monitor. Some oscilloscope adapters are designed as probes with simple displays, giving you the capability of the DMM, logic probe, and oscilloscope in a package that you can hold in your hand.

Prices for low-end PC-based oscilloscopes start at about $100. The price goes up the more features and bandwidth you seek. For most robotics work, you don't need the most fancy-dancy model. PC-based oscilloscopes that connect to the parallel, serial, or USB port—rather than internally through an expansion card—can be readily used with a portable computer. This allows you to take your oscilloscope anywhere you happen to be working on your robot.

The designs provided in this book don't absolutely require that you use an oscilloscope, but you'll probably want one if you design your own circuits or want to develop your electronic skills. A basic, no-nonsense model is enough, but don't settle for the cheap, single-trace analog units. A dual-trace (two-channel) digital storage oscilloscope with a 20- to 25-MHz maximum input frequency (with 250,000 samples per second) should do the job nicely. The two channels let you monitor two lines at once (and reference it to a third trigger line), so you can easily compare the input and output signals at the same time. The digital storage features will allow you to capture events and study them at your leisure, allowing you to track the execution progress of the software and its response to different inputs.

Oscilloscopes are not particularly easy to use for the beginner; they have lots of dials and controls for setting operation. Thoroughly familiarize yourself with the operation of your oscilloscope before using it for any construction project or for troubleshooting. Knowing how to set the time per-division knob is as important as knowing how to turn the oscilloscope on and you won't be very efficient at finding problems until you understand exactly how the oscilloscopes trigger. As usual, exercise caution when using the scope with or near high voltages.

6.5 From Here

To learn more about . . .	***Read***
Electronic components	Chapter 5, "Electronic Components"
How to solder	Chapter 6, "Electronic Construction Techniques"
Building electronic circuits	Chapter 6, "Electronic Construction Techniques"
Building mechanical apparatuses	Part 2, "Robot Platform Construction"

CHAPTER 7

ELECTRONIC CONSTRUCTION TECHNIQUES

To operate all but the simplest robots requires an electronic circuit of one type or another. The way you construct these circuits will largely determine how well your robot functions and how long it will last. Poor performance and limited life inevitably result when hobbyists use so-called rat's nest construction techniques such as soldering together the loose leads of components.

Using proper construction techniques will ensure that your robot circuits work well and last as long as you have a use for them. This chapter covers the basics of several types of construction techniques, starting with the basics of soldering, wire-wrapping, and some circuit prototyping techniques. While only the fundamentals are being presented, these methods, techniques, and tools are useful for even very sophisticated circuitry. For more details, consult a book on electronic construction techniques. Appendix A contains a list of suggested sources.

7.1 Soldering Tips and Techniques

Soldering is the process of heating up two pieces of metal together and electrically joining them using a (relatively) low melting temperature metal (normally a lead-tin alloy). Discrete components, chips, printed circuit boards, and wires can all be joined by soldering them together. Soldering is not a construction technique and will not provide a robust mechanical connection that can be used in robot structures. Soldering sounds and looks simple enough, but there is a *lot* of science behind it and care must be taken to get strong, reliable connections (called *joints*) without damaging any of the components being soldered.

Fig. 7-1 shows the important aspects of a solder joint. The two pieces of metal to be connected (which is almost always copper) are joined by another metal (the solder) melted to them. When the solder is melted to the copper, it should *wet* smoothly over the entire copper surface. Even though the solder melts at a lower temperature than the copper, there are very thin interfaces produced that consist of a copper/solder alloy. These interfaces are called the *intermetallic regions* of the solder joints and one of the goals of soldering is to make sure these regions are as thin as possible to avoid alloying of the copper and solder (which raises the melting point of the solder and makes the joint brittle).

The following sections provide an overview of soldering. If you solder sporadically with months in between turning on your soldering iron, it is a good idea to review this material to ensure that the work is carried out efficiently and safely.

7.1.1 SOLDER SAFETY

Keep the following points in mind when soldering:

- Keep your fingers away from the tip of the soldering pencil. A hot soldering iron can seriously burn you.
- Never touch a solder joint until after it has cooled.
- While using the soldering iron, always place it in a properly designed iron stand.
- To avoid inhaling the fumes for any length of time, work only in a well-ventilated area. While the fumes produced during soldering are not particularly offensive, they can be surprisingly toxic.
- Always wear eye protection such as safety glasses or optically clear goggles when clipping leads.
- Keep a fire extinguisher handy, just in case.

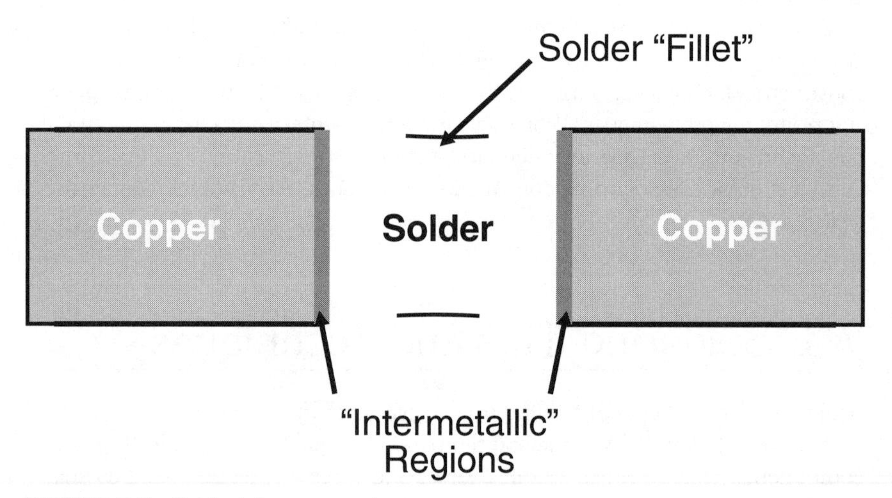

FIGURE 7-1 Solder joint cross-section.

7.1.2 TOOLS AND EQUIPMENT

7.1.2.1 Soldering iron and tip When you are looking for a soldering iron, choose one that is designed for electronics, like the one in Fig. 7-2. The iron should have a fairly low wattage rating (not higher than 30 W). Most soldering irons or *pencils* are designed so you can change the tips as easily as changing a lightbulb. Make sure that you have the smallest pointed tip available for your iron as it will be required for small electronics assembly. It is a good idea to buy a soldering iron that is grounded to keep from running the risk of damaging sensitive electronic components by subjecting them to electrostatic discharge.

Do not use the instant-on type soldering guns favored in the old tube days, and definitely do not use an iron that was designed for plumbing. These types of irons produce far too much unregulated heat and can easily damage the components being soldered.

If your soldering iron has a temperature control and readout, dial it to between 665 and 680°F. This is the typical melting point for solder and will pose the minimum danger of damage to the electronic components. If your iron has just the control and lacks a heat readout, set it to low initially. Wait a few minutes for the iron to heat up, then try one or two test connections. Adjust the heat control so that solder flows onto the connection in under 5 s.

Remember that when you are not using your soldering iron, keep it in an insulated stand.

FIGURE 7-2 Use a low-wattage (25 to 30 W) soldering pencil for all electronics work.

Which *soldering tip* you choose is important. For best results, use a fine tip designed specifically for printed circuit board use (unless you are soldering larger wires; in that case, use a larger tip). Tips are made to fit certain types and brands of soldering irons, so make sure you get ones that are made for your iron.

7.1.2.2 Sponge Keep a damp sponge (just about any type of kitchen sponge will do) by the soldering station and use it to wipe off extra solder. Do not allow globs of solder to remain on the tip. The glob may come off while you're soldering and ruin the connection. The excess solder can also draw away heat from the tip of the pencil, causing a poor soldering job. You'll have to rewet the sponge now and then if you are doing lots of soldering. Never wipe solder off onto a dry sponge, as the sponge could melt or catch on fire.

7.1.2.3 Solder and flux You should use only 63% lead/37% tin *rosin core solder.* It comes in different thicknesses; for best results, use the thin type (0.050 in) for most electronics work, especially if you're working with printed circuit boards. Never use acid core or silver solder on electronic equipment. (Note: certain silver-bearing solders are available for specialty electronics work, and they are acceptable to use although a lot more expensive than tin–lead solders.)

You have probably heard about *lead-free* solders, which, not surprisingly, do not contain any lead in an effort to protect the environment. Unless you are well versed in soldering and are certain that all the components that are going to be soldered have had their leads prepared for lead-free solders, then it is highly recommended that you stay away from them. Lead-free solders melt at a higher temperature than tin–lead solders and lead contamination can cause even higher temperature melting points, which will make it very difficult to remove the components.

Flux is a weak heat-activated acid that cleans the metal surfaces of the components being soldered of oxides and some contaminants. In some cases, such as soldering large wires, you may find it advantageous to apply liquid flux to the surfaces to ensure the larger areas are cleaned during the solder process. To avoid getting into trouble with liquid flux, only buy flux at the store where you bought your solder and make sure that it is the same formulation as the flux in the solder. There are very aggressive fluxes designed for specialized applications that will literally dissolve your components and solder iron tip. If you aren't sure, then refrain from buying any flux and rub the connections with steel wool before soldering to clean off the surfaces to be soldered.

Solder should be kept clean and dry. Avoid tossing your spool of solder into your electronics junk bin. It can collect dust, grime, oil, grease, and other contaminants. Dirty solder requires more heat to melt. In addition, the grime fuses with the solder and melds into the connection. If your solder becomes dirty, wipe it off with a damp paper towel soaked in alcohol, and let it dry.

7.1.2.4 Soldering tools Basic soldering tools include a good pair of small needle-nose pliers, wire strippers, and wire cutters (sometimes called side or diagonal cutters). The stripper should have a dial that lets you select the gauge of wire you are using. A pair of nippy cutters, which cuts wire leads flush to the surface of the board, is also handy.

A *heat sink* should be avoided during soldering—some people may say that it is necessary to avoid heat damage to components, but a heat sink will raise the length of time the iron will have to be applied to the joint in order to melt the solder and could result in poor solder joints as well as an increased amount of heat being passed to the component. As discussed in the following, the iron should only be in contact with the component's leads for a few seconds, much less than the amount of time required to damage the component.

7.1.2.5 Cleaning supplies It is often necessary to clean up before and after you solder. Isopropyl alcohol makes a good, all-around cleaner to remove flux after soldering. After the board has cooled, flux can form a hard surface that is difficult to remove. The best way to clean a circuit board is to use isopropyl alcohol in an ultrasonic cleaner, but you can also use a denture brush (which has much stiffer bristles than an ordinary toothbrush) with isopropyl alcohol as well.

7.1.2.6 Solder vacuum and solder braid These tools are used for removing excess solder and removing components after they have been soldered. *Desoldering* (removing components that have been soldered to a board) is a tricky process that even with a lot of skill can result in a damaged circuit board and components. To minimize the need for desoldering, you should take care to ensure the correct components are being soldered in the correct orientation and use sockets instead of soldering wherever possible.

7.1.3 HOW TO SOLDER

The basis of successful soldering is to use the soldering iron to heat up the work, whether it is a component lead, a wire, or whatever. You then apply the solder to the work. If the solder doesn't flow onto the joint, then check the iron's temperature, add a bit more rosin core solder or even add a bit of liquid flux. Once the solder flows around the joint (and some will flow to the tip), remove the iron and let the joint cool. The joint should look smooth and shiny. If the solder appears dull and crinkly, then you have a cold joint. To fix the joint, apply the soldering iron again to remelt the solder.

Avoid disturbing the solder as it cools; a cold joint might be the result. Do not apply heat any longer than necessary. Prolonged heat can permanently ruin electronic components. A good rule of thumb is that if the iron is on any one spot for more than 5 s, it's too long. If at all possible, you should keep the iron at a 30° to 40° angle for best results. Most tips are beveled for this purpose.

Apply only as much solder to the joint as is required to coat the lead and circuit board pad. A heavy-handed soldering job may lead to soldering bridges, which is when one joint melds with joints around it. At best, solder bridges cause the circuit to cease working; at worst, they cause short circuits that can burn out the entire board.

When soldering on printed circuit boards, you'll need to clip off the excess leads that protrude beyond the solder joint. Use a pair of diagonal or nippy cutters for this task. Be sure to protect your eyes when cutting the lead.

7.1.4 SOLDER TIP MAINTENANCE AND CLEANUP

After soldering, wipe the hot tip on the solder sponge to remove any excess solder, flux, and contaminants. Then let the iron cool. After many hours of use, the soldering tip will become old, pitted, and deformed. This is a good time to replace the tip. Old or damaged tips impair the transfer of heat from the iron to the connection, which can lead to poor soldering joints. Be sure to replace the tip with one made specifically for your soldering iron. Tips are generally not interchangeable between brands.

7.2 Breadboards

The term *breadboard* is used for a variety of experimenter wired circuit products. In this book, the term will be used to describe the temporary prototyping circuit platform shown in Fig. 7-3 in which the holes are connected to adjacent ones by a spring-loaded connector. The typical arrangement is to have the interior holes connected outward while the outside rows of holes are connected together to provide a *bus* structure for power and common signals. Wire (typically 22 gauge) and most electronic components can be pushed into the circuit to make connections and, when the application is finished, to pull out for reuse.

Breadboards are engineered to enable you to experiment with a circuit, without the trouble of soldering. When you are assured that the circuit works, you may use one of the other four construction techniques described in this chapter to make the design permanent. A typical solderless breadboard mounted on a metal carrier is shown in Fig. 7-4. Breadboards are available in many different sizes and styles, but most provide rows of common tie points that are suitable for testing ICs, resistors, capacitors, and most other components that have standard lead diameters.

It is a good idea to first test all the circuits you build on a solderless breadboard. It is important to note that the resistance between adjacent pins is on the order of 5 Ω and the

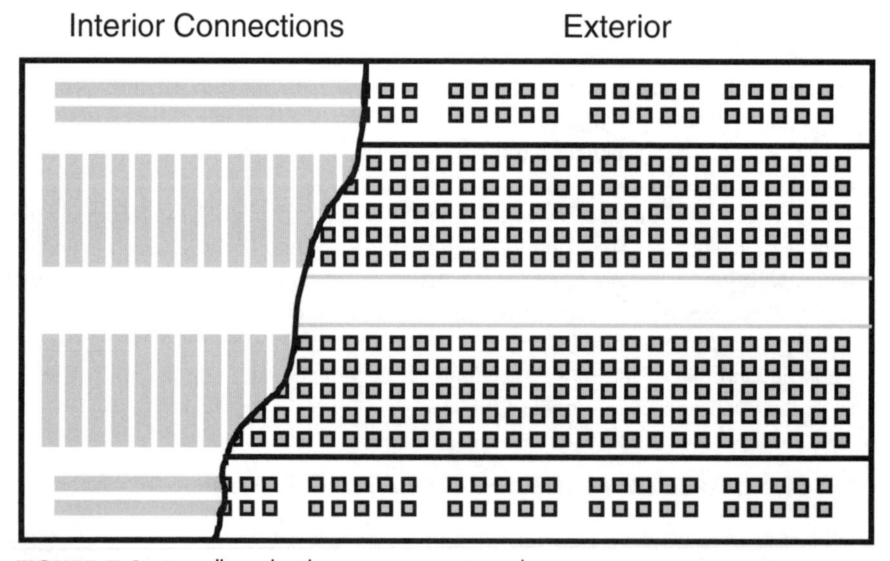

FIGURE 7-3 Breadboard with interior connections shown.

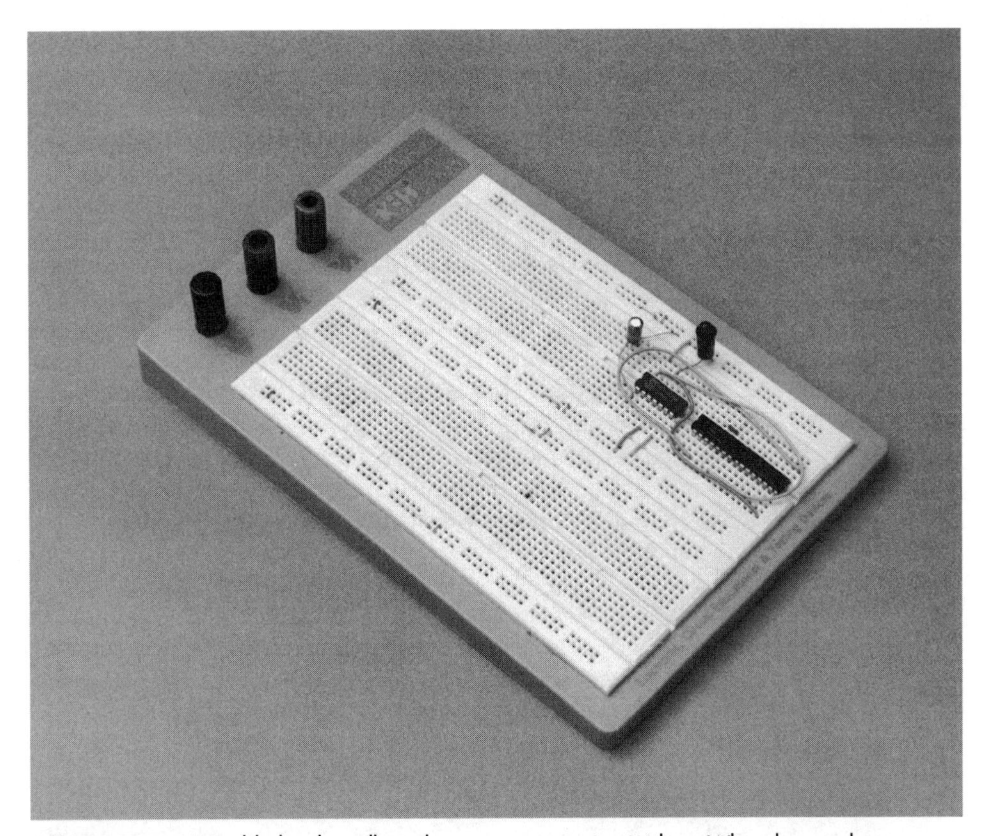

FIGURE 7-4 Use solderless breadboards to test out new circuit ideas. When they work, you can construct a permanent circuit.

capacitance between connectors can be as high as 2 pF. These parasitic impedances mean that your circuit may not perform properly in all cases (especially for currents over 100 mA or signals switching at a few MHz) and you may have to resort to one of the other prototyping methodologies described in the following. It should also be noted that breadboard component connections can come loose when under vibration—such as in a mobile robot.

7.3 Prototyping PCBs

There are many different commercial products that allow you to solder together a custom circuit reasonably quickly and easily. These products may have generic names like breadboard circuit boards, universal solder boards, or experimenters' PC boards. The Vector line of products is commonly used by many robot experimenters. These boards consist of a series of copper strips (which can be cut) to allow you to create custom circuits surprisingly easily using standard PCB soldering techniques. Other prototyping PCBs have sockets and commonly used circuits already etched into the PCBs to further simplify the assembly and wiring of the circuit.

The main disadvantage of universal solder boards is that they don't provide for extremely efficient use of space. It can be difficult to cram the components onto a small space on the board, and you will find that it takes some experience to properly plan your circuits. For this reason, only use prototyping PCBs when wiring a fairly small and simple circuit. Remember to drill mounting holes to secure the board in whatever enclosure or structure you are using.

7.4 Point-to-Point Prototyping Wiring

Point-to-point wiring refers to the practice of mounting electronic components to a prototyping PCB and connecting the leads together directly with solder. This technique was used extensively in the pre-IC days and was often found on commercial products. An example of a circuit built on a prototyping PCB and connected together using point-to-point wiring is shown in Fig. 7-5.

When building point-to-point wired circuits, you must be careful to use insulated wire as point-to-point construction invites short circuits and burnouts. Point-to-point wiring is the most difficult method of creating circuitry discussed in this chapter, but often it is the only

FIGURE 7-5 Circuit created on a prototyping PCB and wired together using point-to-point wiring techniques.

method that allows you to get a robust circuit built quickly. Remember to start small and don't be afraid to experiment with different techniques in order to get the most efficient circuitry.

7.5 Wire-Wrapping

An older method used to wire complex circuitry together is known as *wire-wrapping:* a bared wire is literally wrapped around a long post, which is connected to the pin of an IC or some other component. Wire-wrapping can be used to create very complex circuits, and it is surprisingly robust. However, there are a number of downsides to this wiring method that you should be aware of.

Wire-wrapping requires specialized tools, component sockets (with long, square leads), and prototyping PCBs—each of which can be substantially more expensive than standard parts. It also takes a surprisingly long time. Professional wire-wrappers with automated equipment plan on each wire connection taking 30 s; starting out with hobbyist tools you will be hard pressed to perform (on average) one wire every 3 minutes. It should also be noted that a complex wire-wrapped board quickly becomes a rat's nest and wiring errors are extremely difficult to find. Finally, wiring power and analog components into a wire-wrapped circuit is difficult and can be extremely time consuming.

7.6 Quick Turn Prototype Printed Circuit Boards

Probably the most efficient method of prototyping circuits is to design a simple printed circuit board and have a small number of samples built by a quick turn manufacturer. Depending on your location, you could conceivably have your PCBs within a business day. Quick turn PCBs (like the one shown in Fig. 7-6) are also surprisingly cost efficient—you should plan on the PCBs costing $5 per square inch, approximately the same as what you would pay for PCBs, sockets, and wiring for a wire-wrapped circuit. For a beginner, the time spent designing the PCB is similar to that of wire-wrapping a PCB.

The big advantage of using quick turn PCBs is the ability to change and replicate the circuit very quickly. Using any of the other prototyping techniques previously discussed, the effort in wiring is only applied to one circuit; the time increases linearly with each additional circuit built. A PCB design can be replicated without any additional investment in time and certainly adds to the professionalism of a robot.

An understandable concern about designing printed circuit boards is the time required to learn how to use the design tools, their cost, and learning how to efficiently lay out circuits on a PCB. Currently there are a number of open source software projects and commercial products that allow hobbyists to design PCBs and produce *Gerber files* (which are the design files used to manufacture the PCBs) for free. There are quite a few books devoted to explaining how PCB layout is accomplished and lists many of the tricks that you should be

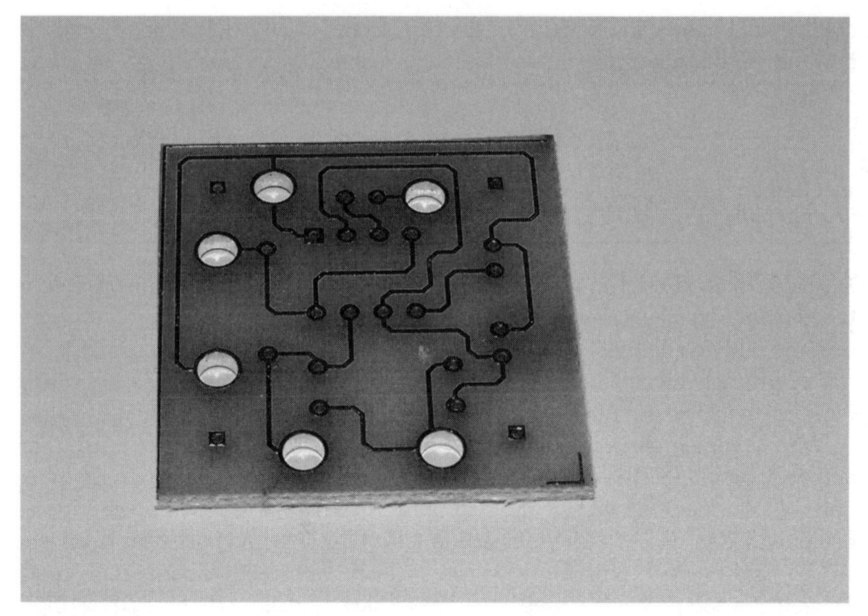

FIGURE 7-6 PCBs, like this one, can be created quickly as well as inexpensively and can be replicated with very little effort.

aware of. Learning how to design PCBs is a marketable skill and one that will certainly pay off in the future.

7.7 Headers and Connectors

Robots are often constructed from subsystems that may not be located on the same circuit board. You must therefore know how to connect subsystems on different circuit boards. Avoid the temptation to directly solder wires between boards. This makes it much harder to work with your robot, including testing variations of your designs with different subsystems.

Instead, use connectors whenever possible, as shown in Fig. 7-7. In this approach you connect the various subsystems of your robot using short lengths of wire. You terminate each wire with a connector of some type or another. The connectors attach to mating pins on each circuit board.

You don't need fancy cables and cable connectors for your robots. In fact, these can add significant weight to your 'bot. Instead, use ordinary 20- to 26-gauge wire, terminated with single- or double-row plastic connectors. You can use ribbon cable for the wire or individual insulated strips of wire. Use plastic ties to bundle the wires together. The plastic connectors are made to mate with single- and double-row headers soldered directly on the circuit board. You can buy connectors and headers that have different numbers of pins or you can salvage them from old parts (the typical VCR is chock-full of them!).

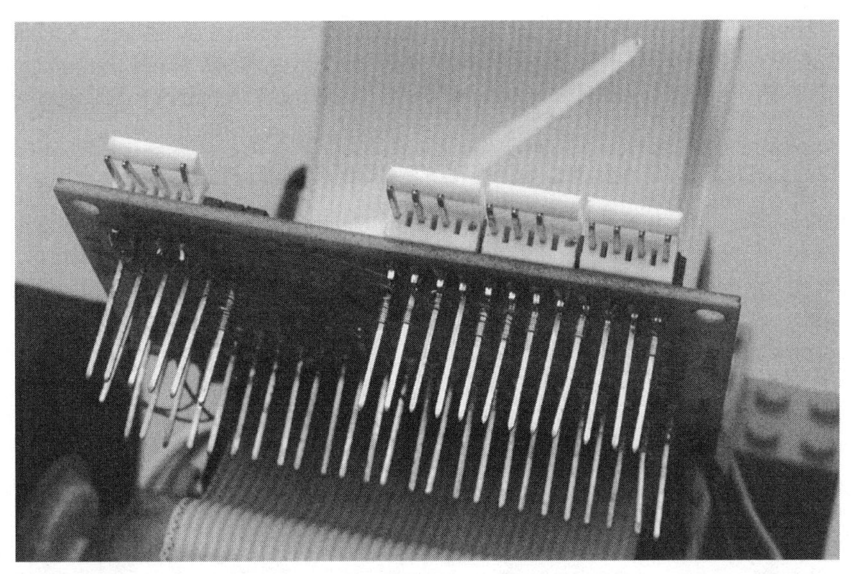

FIGURE 7-7 Using connectors makes for more manageable robots. Use connectors on all subsystems of your robot.

When making interconnecting cables, cut the wires to length so there is a modest amount of slack between subsystems. You don't want, or need, gobs of excess wire. Nor do you want the wire lengths so short that the components are put under stress when you connect them together.

7.8 Eliminating Static Electricity

The ancient Egyptians discovered static electricity when they rubbed animal fur against the smooth surface of amber. Once the materials were rubbed together, they tended to cling to one another. Similarly, two pieces of fur that were rubbed against the amber tended to separate when they were drawn together. While the Egyptians didn't understand this mysterious unseen force—better known now as static electricity—they knew it existed.

Today, you can encounter static electricity by doing nothing more than walking across a carpeted floor. As you walk, your feet rub against the carpet, and your body takes on a static charge. Touch a metal object, like a doorknob or a metal sink, and that static is quickly discharged from your body. You feel the discharge as a shock.

Carpet shock has never been known to kill anyone. The amount of voltage and current is far too low to cause great bodily harm. But the same isn't true of electronic circuits. Considering how your body can develop a 10,000- to 50,000-V charge when you walk across a carpet, try to imagine what that might do to electrical components rated at just 5 or 15 V. The sudden crash of electrical energy can burn holes right through a sensitive transistor or integrated circuit, rendering it completely useless.

Transistors and integrated circuits designed around a metal-oxide substrate can be particularly sensitive to high voltages, regardless of the current level. These components include MOSFET transistors, CMOS integrated circuits, and most computer microprocessors.

7.8.1 STORING STATIC-SENSITIVE COMPONENTS

Plastic is one of the greatest sources of static electricity. Storage and shipping containers are often made of plastic, and it's a great temptation to dump your static-sensitive devices into these containers. Don't do it. Invariably, static electricity will develop, and the component could be damaged. Unfortunately, there's no way to tell if a static-sensitive part has become damaged by electrostatic discharge just by looking at it, so you won't know things are amiss until you actually try to use the component. At first, you'll think the circuit has gone haywire or that your wiring is at fault. If you're like most, you won't blame the transistors and ICs until well after you've torn the rest of the circuit apart.

It's best to store static-sensitive components using one of the following methods. All work by grounding the leads of the IC or transistor together, which diminishes the effect of a strong jolt of static electricity. Note that none of these storage methods is 100 percent foolproof.

- *Anti-static mat.* This mat looks like a black sponge, but it's really conductive foam. You can (and should) test this by placing the leads of a volt-ohm meter on either side of a length of the foam. Dial the meter to ohms. You should get a reading instead of an open circuit. The foam can easily be reused, and large sheets make convenient storage pads for many components.
- *Anti-static pouch or bag.* Anti-static pouches are made of a special plastic (which generates little static) and are coated on the inside with a conductive layer. The bags are available in a variety of forms. Many are a smoky black or gray; others are pink or jet black. As with mats, you should never assume a storage pouch is anti-static just from its color. Check the coating on the inside with a volt-ohm meter—its resistance should be in the 10k to 1 MΩ range.
- *Anti-static tube.* The vast majority of chips are shipped and stored in convenient plastic tubes. These tubes help protect the leads of the IC and are well suited to automatic manufacturing techniques. The construction of the tube is similar to the anti-static pouch: plastic on the outside, a thin layer of conductive material on the inside. You can often get tubes from electronic supply stores by just asking for leftover tubes that would normally be thrown away. It is a good idea to only put one type of chip in each tube, label the tube, and keep the same family of chips together. This will make finding specific components *much* easier.

7.8.2 TIPS TO REDUCE STATIC

Consider using any and all of the following simple techniques to reduce and eliminate the risk of electrostatic discharge.

- *Wear low-static clothing and shoes.* Your choice of clothing can affect the amount of static buildup in your body. Whenever possible, wear natural fabrics such as cotton or wool. Avoid wearing polyester and acetate clothing, as these tend to develop large amounts of static.

- *Use an anti-static wrist strap.* The wrist strap grounds you at all times and prevents static buildup. The strap is one of the most effective means for eliminating electrostatic discharge, and it's one of the least expensive.
- *Ground your soldering iron.* If your soldering pencil operates from AC current, it should be grounded. A grounded iron not only helps prevent damage from electrostatic discharge; it also lessens the chance of receiving a bad shock should you accidentally touch a live wire.
- *Use component sockets.* When you build projects that use ICs, install sockets first. When the entire circuit has been completely wired, you can check your work, then add the chips. Note that some sockets are polarized so the component will fit into them one way only. Be sure to observe this polarity when wiring the socket.

7.9 Good Design Principles

While building circuits for your robots, observe the good design principles described in the following sections, even if the schematic diagrams you are working from don't include them.

7.9.1 PULL-UP RESISTORS

When a digital electronic chip input is left unconnected or floating, the value presented at the input is indeterminate. This is especially true for CMOS chips, which use the gates of MOSFET chips as inputs (TTL inputs are automatically high when inputs are left floating). To ensure that the inputs values are a known value, a pull up should be used or, if the input must be low during operation, then an inverted pull up (shown in Fig. 7-8) should be used. Pull-down resistors on logic inputs should be avoided because of the low resistance (less than 150 Ω) required to ensure TTL inputs are at a low value. To raise the input to a high value, a high-current source must be applied to the input pin to overcome the pull-down resistor. The pull-up resistor can be a relatively high value (10k or more) to minimize any possible power drain and allow simple changes to logic without large current flows.

7.9.2 USE BYPASS CAPACITORS

Some electronic components, especially fast-acting logic chips, generate a lot of noise in the power supply lines. You can reduce or eliminate this noise by putting bypass (so-called decoupling) capacitors between the +V and ground rails of all chips as close to the power supply pins as possible, as shown in Fig. 7-9. Suggested values are 0.01 μF to 0.1 μF. If you are using polarized parts, be sure to properly orient the capacitor.

7.9.3 KEEP LEAD LENGTHS SHORT

Long leads on components can introduce noise in other parts of a circuit. The long leads also act as a virtual antenna, picking up stray signals from the circuit, from overhead light-

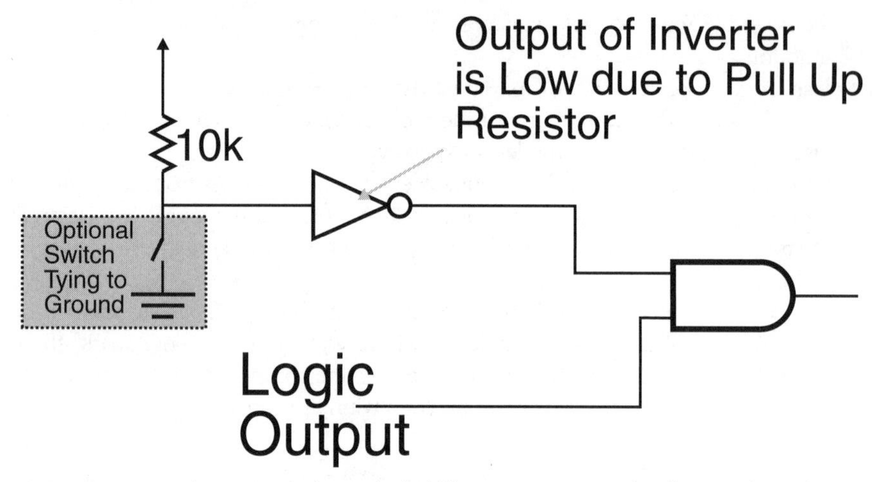

FIGURE 7-8 When an input has to be held low, use an inverted pull up as shown here. This will allow easy logic level changing for circuit test.

ing, and even from your own body. When designing and building circuits, strive for the shortest lead lengths on all components. This means soldering the components close to the board and clipping off any excess lead length, and if you are breaking any connections cut the wires or PCB traces as close to the chip pin as possible.

7.9.4 AVOID GROUND LOOPS

A ground loop is when the ground wire of a circuit comes back and meets itself. The +V and ground of your circuits should always have dead ends to them. Ground loops can cause erratic behavior due to excessive noise in the circuit, which can be very difficult to track down to a root cause.

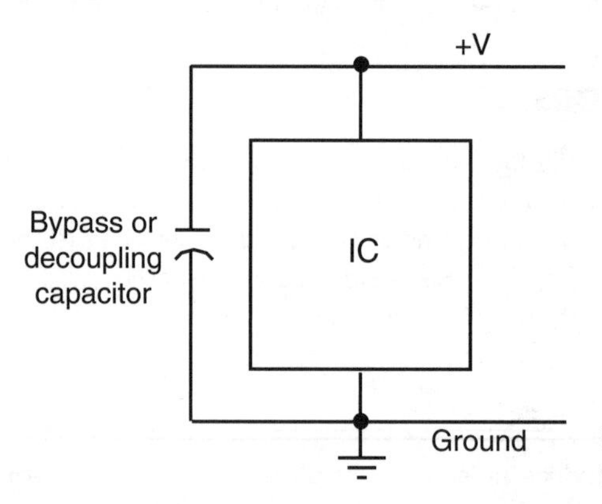

FIGURE 7-9 Add decoupling capacitors near the power and ground pins of integrated circuits.

7.10 From Here

To learn more about . . .	*Read*
Tools for circuit construction	Chapter 6, "Tools"
Where to find electronic components	Chapter 4, "Buying Parts"
Where to find mechanical parts	Chapter 4, "Buying Parts"
Understanding components used in electronic circuitry	Chapter 5, "Electronic Components"

PART 2

ROBOT PLATFORM CONSTRUCTION

CHAPTER 8

PLASTIC PLATFORMS

It all started with billiard balls. A couple of hundred years ago, billiard balls were made from elephant tusks. By the 1850s, the supply of tusk ivory was drying up and its cost had skyrocketed. So, in 1863 Phelan & Collender, a major manufacturer of billiard balls, offered a $10,000 prize for anyone who could come up with a suitable substitute for ivory. A New York printer named John Wesley Hyatt was among several folks who took up the challenge.

Hyatt didn't get the $10,000. His innovation, celluloid, was too brittle to be used for billiard balls. But while Hyatt's name won't go down in the billiard parlor hall of fame, he will be remembered as the man who started the plastics revolution. Hyatt's celluloid was perfect for such things as gentlemen's collars, ladies' combs, containers, and eventually even motion picture film. In the more than 100 years since the introduction of celluloid, plastics have taken over our lives.

Plastic is sometimes the object of ridicule—from plastic money to plastic furniture—yet even its critics are quick to point out its many advantages.

- Plastic is cheaper per square inch than wood, metal, and most other construction materials.
- Certain plastics are extremely strong, approaching the tensile strength of such light metals as copper and aluminum.
- Some plastic is unbreakable.

Plastic is an ideal material for use in hobby robotics. Its properties are well suited for numerous robot designs, from simple frame structures to complete assemblies. Read this chapter to learn more about plastic and how to work with it. At the end of the chapter, we'll show you how to construct an easy-to-build differentially driven robot—the Minibot—from inexpensive and readily available plastic parts.

Included in the robot design is a simple wired (or tethered) remote control that can be used in the other example robots in this section.

8.1 Types of Plastics

Plastics represent a large family of products. Plastics often carry a fancy trade name, like Plexiglas, Lexan, Acrylite, Sintra, or any of a dozen other identifiers. Some plastics are better suited for certain jobs, so it will benefit you to have a basic understanding of the various types of plastics. Here's a short rundown of the plastics you may encounter.

- *ABS.* Short for acrylonitrile butadiene styrene, ABS is most often used in sewer and wastewater plumbing systems. The large black pipes and fittings you see in the hardware store are made of ABS. It is a glossy, translucent plastic that can take on just about any color and texture. It is tough, hard, and yet relatively easy to cut and drill. Besides plumbing fittings, ABS also comes in rods, sheets, and pipes—and as LEGO plastic pieces!
- *Acrylic.* Acrylic is clear and strong, the mainstay of the decorative plastics industry. It can be easily scratched, but if the scratches aren't too deep they can be rubbed out. Acrylic is somewhat tough to cut because it tends to crack, and it must be drilled carefully. The material comes mostly in sheets, but it is also available in extruded tubing, in rods, and in the coating in pour-on plastic laminate.
- *Cellulosics.* Lightweight and flimsy but surprisingly resilient, cellulosic plastics are often used as a sheet covering. Their uses in robotics are minor. One useful application, however, stems from the fact that cellulosics soften at low heat, and thus they can be slowly formed around an object. These plastics come in sheet or film form.
- *Epoxies.* Very durable, clear plastic, epoxies are often used as the binder in fiberglass. Epoxies most often come in liquid form, so they can be poured over something or onto a fiberglass base. The dried material can be cut, drilled, and sanded.
- *Nylon.* Nylon is tough, slippery, self-lubricating stuff that is most often used as a substitute for twine. Plastics distributors also supply nylon in rods and sheets. Nylon is flexible, which makes it moderately hard to cut.
- *Phenolics.* An original plastic, phenolics are usually black or brown, easy to cut and drill, and smell terrible when heated. The material is usually reinforced with wood or cotton bits or laminated with paper or cloth. Even with these additives, phenolic plastics are not unbreakable. They come in rods and sheets and as pour-on coatings. The only application of phenolics in robotics is as circuit board material.
- *Polycarbonate.* Polycarbonate plastic is a close cousin of acrylic but more durable and resistant to breakage. Polycarbonate plastics are slightly cloudy and are easy to mar and scratch. They come in rods, sheets, and tube form. A common, inexpensive window-glazing material, polycarbonates are hard to cut and drill without breakage.
- *Polyethylene.* Polyethylene is lightweight and translucent and is often used to make flexible tubing. It also comes in rod, film, sheet, and pipe form. You can reform the material by applying low heat, and when the material is in tube form you can cut it with a knife.

- *Polypropylene.* Like polyethylene, polypropylene is harder and more resistant to heat.
- *Polystyrene.* Polystyrene is a mainstay of the toy industry. This plastic is hard, clear (though it can be colored with dyes), and cheap. Although often labeled high-impact plastic, polystyrene is brittle and can be damaged by low heat and sunlight. Available in rods, sheets, and foamboard, polystyrene is moderately hard to cut and drill without cracking and breaking.
- *Polyurethane.* These days, polyurethane is most often used as insulation material, but it's also available in rod and sheet form. The plastic is durable, flexible, and relatively easy to cut and drill.
- *PVC.* Short for polyvinyl chloride, PVC is an extremely versatile plastic best known as the material used in freshwater plumbing and in outdoor plastic patio furniture. Usually processed with white pigment, PVC is actually clear and softens in relatively low heat. PVC is extremely easy to cut and drill and almost impervious to breakage. PVC is supplied in film, sheet, rod, tubing, even nut-and-bolt form in addition to being shaped into plumbing fixtures and pipes.
- *Silicone.* Silicone is a large family of plastics in its own right. Because of their elasticity, silicone plastics are most often used in molding compounds. Silicone is slippery and comes in resin form for pouring.

Table 8-1 lists the different types of plastics used for different household applications.

TABLE 8-1 Plastics in Everyday Household Articles

HOUSEHOLD ARTICLE	TYPE OF PLASTIC
Bottles, containers	
• Clear	• Polyester, PVC
• Translucent or opaque	• Polyethylene, polypropylene
Buckets, washtubs	Polyethylene, polypropylene
Foam cushions	Polyurethane foam, PVC foam
Electrical circuit boards	Laminated epoxies, phenolics
Fillers	
• Caulking compounds	• Polyurethane, silicone, PVC
• Grouts	• Silicone, PVC
• Patching compounds	• Polyester, fiberglass
• Putties	• Epoxies, polyester, PVC
Films	
• Art film	• Cellulosics
• Audio tape	• Polyester
• Food wrap	• Polyethylene, polypropylene
• Photographs	• Cellulosics

(continued)

TABLE 8-1 Plastics in Everyday Household Articles (*Continued*)

HOUSEHOLD ARTICLE	TYPE OF PLASTIC
Glasses (drinking)	
• Clear, hard	• Polystyrene
• Flexible	• Polyethylene
• Insulated cups	• Styrofoam (polystyrene foam)
Hoses, garden	PVC
Insulation foam	Polystyrene, polyurethane
Lubricants	Silicones
Plumbing pipes	
• Fresh water	• PVC, polyethylene, ABS
• Gray water	• ABS
Siding and paneling	PVC
Toys	
• Flexible	• Polyethylene, polypropylene
• Rigid	• Polystyrene, ABS
Tubing (clear or translucent)	Polyethylene, PVC

8.2 Working with Plastics

The actions of cutting, drilling, painting, choosing, etc. plastics can be overwhelming if you don't have the basic information. Different plastics have different characteristics that will affect how they are handled and formed and whether or not they are appropriate for an application. In the following, common plastics used in robots are discussed along with information regarding how to work with them.

8.2.1 HOW TO CUT PLASTIC

Soft plastics may be cut with a sharp utility knife. When cutting, place a sheet of cardboard or artboard on the table. This helps keep the knife from cutting into the table, which could ruin the tabletop and dull the knife. Use a carpenter's square or metal rule when you need to cut a straight line. Prolong the blade's life by using the rule against the knife holder, not the blade. Most sheet plastic comes with a protective peel-off plastic on both sides. Keep it on when cutting—not only will it protect the surface from scratches and dings but it can also be marked with a pencil or pen when you are planning your cuts.

Harder plastics can be cut in a variety of ways. When cutting sheet plastic less than ⅛-in thick, use a utility knife and metal carpenter's square to score a cutting line. If necessary, use clamps to hold down the square.

Carefully repeat the scoring two or three times to deepen the cut. Place a ½- or 1-in dowel under the plastic so the score line is on the top of the dowel. With your fingers or the palms of your hands, carefully push down on both sides of the score line. If the sheet is wide, use a piece of 1-by-2 or 2-by-4 lumber to exert even pressure. Breakage and cracking is most likely to occur on the edges, so press on the edges first, then work your way toward the center. Don't force the break. If you can't get the plastic to break off cleanly, deepen the score line with the utility knife. Thicker sheet plastic, as well as extruded tubes, pipes, and bars, must be cut with a saw. If you have a table saw, outfit it with a plywood-paneling blade. Among other applications, this blade can be used to cut plastics. You cut through plastic just as you do with wood, but the feed rate (the speed at which the material is sawed in two) must be slower. Forcing the plastic or using a dull blade heats the plastic, causing it to deform and melt. A band saw is ideal for cutting plastics less than ½-in thick, especially if you need to cut corners. When working with a power saw, use fences or pieces of wood held in place by C-clamps to ensure a straight cut.

You can use a handsaw to cut smaller pieces of plastic. A hacksaw with a medium- or fine-tooth blade (24 or 32 teeth per inch) is a good choice. You can also use a coping saw (with a fine-tooth blade) or a razor saw. These are good choices when cutting angles and corners as well as when doing detail work.

A motorized scroll (or saber) saw can be used to cut plastic, but you must take care to ensure a straight cut. If possible, use a piece of plywood held in place by C-clamps as a guide fence. Routers can be used to cut and score plastic, but unless you are an experienced router user you should not attempt this method.

8.2.2 HOW TO DRILL PLASTIC

Wood drill bits can be used to cut plastics, but bits designed for glass drilling yield better, safer results. If you use wood bits, you should modify them by blunting the tip slightly (otherwise the tip may crack the plastic when it exits the other side). Continue the flute from the cutting lip all the way to the end of the bit (see Fig. 8-1). Blunting the tip of the bit isn't hard

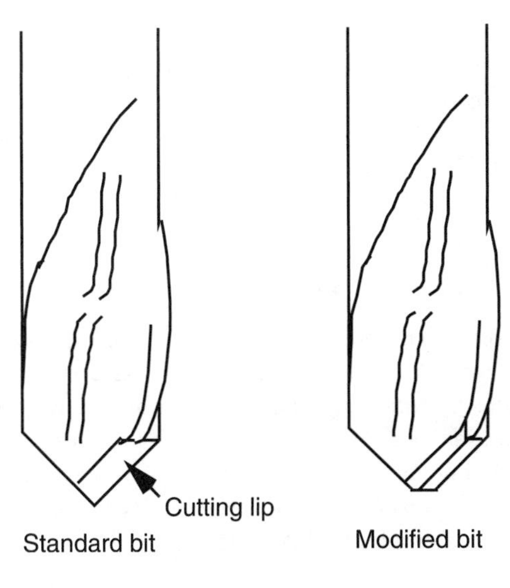

FIGURE 8-1 Suggested modifications for drill bits used with plastic. The end is blunted and the flutes are extended to the end of the cutting tip.

to do, but grinding the flute is a difficult proposition. The best idea is to invest in a few glass or plastic bits, which are already engineered for drilling plastic.

Drilling with a power drill provides the best results. The drill should have a variable speed control. Reduce the speed of the drill to about 500 to 1000 r/min when using twist bits, and to about 1000 to 2000 r/min when using spade bits. When drilling, always back the plastic with a wooden block. Without the block, the plastic is almost guaranteed to crack.

When using spade bits or brad-point bits, drill partially through from one side, then complete the hole by drilling from the other side. As with cutting, don't force the hole and always use sharp bits. Too much friction causes the plastic to melt.

To make holes larger than ¼ in you should first drill a smaller, pilot hole. If the hole is large, over ¼ in in diameter, start with a small drill and work your way up several steps. You should practice drilling on a piece of scrap material until you get the technique right.

8.2.3 HOW TO BEND AND FORM PLASTIC

Most rigid and semirigid plastics can be formed by applying low localized heat. A sure way to bend sheet plastic is to use a strip heater. These are available ready-made at some hardware and plastics supply houses, or you can build your own. A narrow element in the heater applies a regulated amount of heat to the plastic. When the plastic is soft enough, you can bend it into just about any angle you want.

There are two important points to remember when using a strip heater. First, be sure that the plastic is pliable before you try to bend it. Otherwise, you may break it or cause excessive stress at the joint (a stressed joint will look cracked or crazed). Second, bend the plastic past the angle that you want. The plastic will relax a bit when it cools off, so you must anticipate this. Knowing how much to overbend will come with experience, and it will vary depending on the type of plastic and the size of the piece you're working with.

You can mold thinner sheet plastic around shapes by first heating it up with a hair dryer or heat gun, then using your fingers to form the plastic. Be careful that you don't heat up the plastic too much. You don't want it to melt, just conform to the underlying shape. You can soften an entire sheet or piece by placing it into an oven for 10 or so minutes (remove the protective plastic before baking). Set the thermostat to 300°F and be sure to leave the door slightly ajar so any fumes released during the heating can escape. Ventilate the kitchen and avoid breathing the fumes, as they can be noxious. All plastics release gases when they heat up, but the fumes can be downright toxic when the plastic actually ignites. Therefore, avoid heating the plastic so much that it burns. Dripping, molten plastic can also seriously burn you if it drops on your skin.

8.2.4 HOW TO POLISH THE EDGES OF PLASTIC

Plastic that has been cut or scored usually has rough edges. You can file the edges of cut PVC and ABS using a wood or metal file. You should polish the edges of higher-density plastics like acrylics and polycarbonates by sanding, buffing, or burnishing. Try a fine-grit (200 to 300), wet–dry sandpaper and use it wet. Buy an assortment of sandpapers and try several until you find the coarseness that works best with the plastic you're using. You can apply jeweler's rouge, available at many hardware stores in large blocks, by using a polishing wheel. The wheel can be attached to a grinder or drill motor.

Burnishing involves using a very low temperature flame (a match or lighter will do) to melt the plastic slightly. You can also use a propane torch kept some distance from the plastic. Be extremely careful when using a flame to burnish plastic. Don't let the plastic ignite, or you'll end up with an ugly blob that will ruin your project, not to mention filling the room with poisonous gas.

8.2.5 HOW TO GLUE PLASTIC

Most plastics aren't really glued together; they are melted using a solvent that is often called cement. The pieces are fused together—that is, made one. Household adhesives can be used for this, of course, but you get better results when you use specially formulated cements.

Herein lies a problem. The various plastics we have described rarely use the same solvent formulations, so you've got to choose the right adhesive. That means you have to know the type of plastic used in the material you are working with (see Table 8-1). Using Table 8-2, you can select the recommended adhesives for attaching plastics.

When using solvent for PVC or ABS plumbing fixtures, apply the cement in the recommended manner by spreading a thin coat on the pieces to be joined. A cotton applicator is included in the can of cement. Plastic sheet, bars, and other items require more careful cementing, especially if you want the end result to look nice.

With the exception of PVC solvent, the cement for plastics is watery thin and can be applied in a variety of ways. One method is to use a small painter's brush, with a #0 or #1 tip. Joint the pieces to be fused together and paint the cement onto the joint with the brush. Capillary action will draw the cement into the joint, where it will spread out. Another method is to fill a special syringe applicator with cement. With the pieces butted together, squirt a small amount of cement into the joint line.

In all cases, you must be sure that the surfaces at the joint of the two pieces are perfectly flat and that there are no voids where the cement may not make ample contact. If you are

TABLE 8-2 Plastic Bonding Guide

PLASTIC TYPE	CEMENT TO ITSELF, USE	CEMENT TO OTHER PLASTIC, USE	CEMENT TO METAL, USE
ABS	ABS–ABS solvent	Rubber adhesive	Epoxy
Acrylic	Acrylic solvent	Epoxy	Contact cement
Cellulosics	White glue	Rubber adhesive	Contact cement
Polystyrene	Model glue	Epoxy	CA glue*
Polystyrene foam	White glue	Contact cement	Contact cement
Polyurethane	Rubber adhesive	Epoxy, contact cement	Contact cement
PVC	PVC–PVC solvent	PVC–ABS (to ABS)	Contact cement

(*CA stands for cyanoacrylate ester, sometimes known as Super Glue, after a popular brand name.)

joining pieces whose edges you cannot make flush, apply a thicker type of glue, such as contact cement or white household glue. You may find that you can achieve a better bond by first roughing up the joints to be mated. You can use coarse sandpaper or a file for this purpose.

After applying the cement, wait at least 15 minutes for the plastic to fuse and the joint to harden. Disturbing the joint before it has time to set will permanently weaken it. Remember that you cannot apply cement to plastics that have been painted. If you paint the pieces before cementing them, scrape off the paint and refinish the edges so they're smooth again.

8.2.6 USING HOT GLUE WITH PLASTICS

Perhaps the fastest way to glue plastic pieces together is to use hot glue. You heat the glue to a viscous state in a glue gun, then spread it over the area to be bonded. Hot-melt glue and glue guns are available at most hardware, craft, and hobby stores in several different sizes. The glue is available in a normal and a low-temperature form. Low-temperature glue is generally better with most plastics because it avoids the sagging or softening of the plastic sometimes caused by the heat of the glue.

One caveat when working with hot-melt glue and guns (other than the obvious safety warnings) is that you should always rough up the plastic surfaces before trying to bond them. Plastics with a smooth surface will not adhere well when using hot-melt glue, and the joint will be brittle and perhaps break off with only minor pressure. By roughing up the plastic, the glue has more surface area to bond to, resulting in a strong joint. Roughing up plastic before you join pieces is an important step when using any glue or cement (with the exception of CA), but it is particularly important when using hot-melt glue.

8.2.7 HOW TO PAINT PLASTICS

Sheet plastic is available in transparent or opaque colors, and this is the best way to add color to your robot projects. The colors are impregnated in the plastic and can't be scraped or sanded off. However, you can also add a coat of paint to the plastic to add color or to make it opaque. Most plastics accept brush or spray painting.

Spray painting is the preferred method for jobs that don't require extra-fine detail. Carefully select the paint before you use it, and always apply a small amount to a scrap piece of plastic before painting the entire project. Some paints contain solvents that may soften the plastic.

One of the best all-around paints for plastics is the model and hobby spray cans made by Testor. These are specially formulated for styrene model plastic, but the paint adheres well without softening on most plastics. You can purchase this paint in a variety of colors, in either gloss or flat finish. The same colors are available in bottles with self-contained brush applicators. If the plastic is clear, you have the option of painting on the front or back side (or both for that matter). Painting on the front side will produce the paint's standard finish: gloss colors come out gloss; flat colors come out flat. Flat-finish paints tend to scrape off easier, however, so exercise care.

Painting on the back side with gloss or flat paint will only produce a glossy appearance because you look through the clear plastic to the paint on the back side. Moreover, painting imperfections will more or less be hidden, and external scratches won't mar the paint job.

8.2.8 BUYING PLASTIC

Some hardware stores carry plastic, but you'll be sorely frustrated at their selection. The best place to look for plastic—in all its styles, shapes, and chemical compositions—is a plastics specialty store, a sign-making shop, or a hobby shop. Most larger cities have at least one plastic supply store or sign-making shop that's open to the public. Look in the Yellow Pages under Plastics—Retail.

Another useful source is the plastics fabricator. There are actually more of these than retail plastic stores. They are in business to build merchandise, display racks, and other plastic items. Although they don't usually advertise to the general public, most will sell to you. If the fabricator doesn't sell new material, ask to buy the leftover scrap.

8.2.9 PLASTICS AROUND THE HOUSE

You need not purchase plastic for all your robot needs at a hardware or specialty store. You may find all the plastic you really need right in your own home. Here are a few good places to look:

- *Used compact discs (CDs).* CDs, made from polycarbonate plastics, are usually just thrown away and not recycled. With careful drilling and cutting, you can adapt them to serve as body parts and even wheels for your robots. Exercise caution when working with CDs: they can shatter when you drill and cut them, and the pieces are very sharp and dangerous.
- *Old phonograph records.* Found in the local thrift store, records—particularly the thicker 78-r/min variety—can be used in much the same way as CDs and laser discs. The older records made from the 1930s through 1950s used a thicker plastic that is very heavy and durable. Thrift stores are your best bet for old records no one wants anymore (who is that Montovani guy, anyway?). Note that some old records, like the V-Discs made during World War II, are collector's items, so don't wantonly destroy a record unless you're sure it has no value.
- *Salad bowls, serving bowls, and plastic knickknacks.* They can all be revived as robot parts. I regularly prowl garage sales and thrift stores looking for such plastic material.
- *PVC irrigation pipe.* This can be used to construct the frame of a robot. Use the short lengths of pipe left over from a weekend project. You can secure the pieces with glue or hardware or use PVC connector pieces (Ts, Ls, etc.).
- *Old toys.* Not only for structural materials but you can often find a selection of motors, gears, and driver circuitry that can be salvaged and used in your robot projects.

8.3 Build the Minibot

You can use a small piece of scrap sheet acrylic to build the foundation and frame of the Minibot differentially driven robot. The robot is about 6 in^2 and scoots around the floor or table on two small rubber tires. The basic version is meant to be wire-controlled, although in upcoming chapters you'll see how to adapt the Minibot to automatic electronic control,

TABLE 8-3 Minibot Parts List

1	6-in-by-6-in acrylic plastic (1⁄16- or 1⁄8-in thick)
2	Small hobby motors with gear reduction
2	Model airplane wheels
1	3½-in (approx.) 10⁄24 all-thread rod
1	6-in-diameter (approx.) clear plastic dome
1	Four-cell AA battery holder
Misc.	½-in-by-8⁄32 bolts, 8⁄32 nuts, lock washers, ½-in-by-10⁄24 bolts, 10⁄24 nuts, lock washers, cap nuts

even remote control. The power source is a set of four AA flashlight batteries because they are small, lightweight, and provide more driving power than 9-V transistor batteries. The parts list for the Minibot can be found in Tables 8-3 and 8-4.

8.3.1 FOUNDATION OR BASE

The foundation is clear or colored Plexiglas or some similar acrylic sheet plastic. The thickness should be at least ⅛ in, but avoid very thick plastic because of its heavy weight. The prototype Minibot used ⅛-in-thick acrylic, so there was minimum stress caused by bending or flexing.

Cut the plastic as shown in Fig. 8-2. Remember to keep the protective paper cover on the plastic while you cut. File or sand the edges to smooth the cutting and scoring marks. The corners are sharp and can cause injury if the robot is handled by small children. You can easily fix this by rounding off the corners with a file. Find the center and drill a hole with a #10 bit.

Fig. 8-2 also shows the holes for mounting the drive motors. These holes are spaced for a simple clamp mechanism that secures hobby motors commonly available on the market.

8.3.2 MOTOR MOUNT

The small DC motors used in the prototype Minibot were surplus gear motors with an output speed of about 30 r/min. The motors for your Minibot should have a similar speed

TABLE 8-4 Minibot Remote Control Switch

1	Small electronic project enclosure
2	Double-pole, double-throw (DPDT) momentary switches, with center-off return
Misc.	Hookup wire (see text)

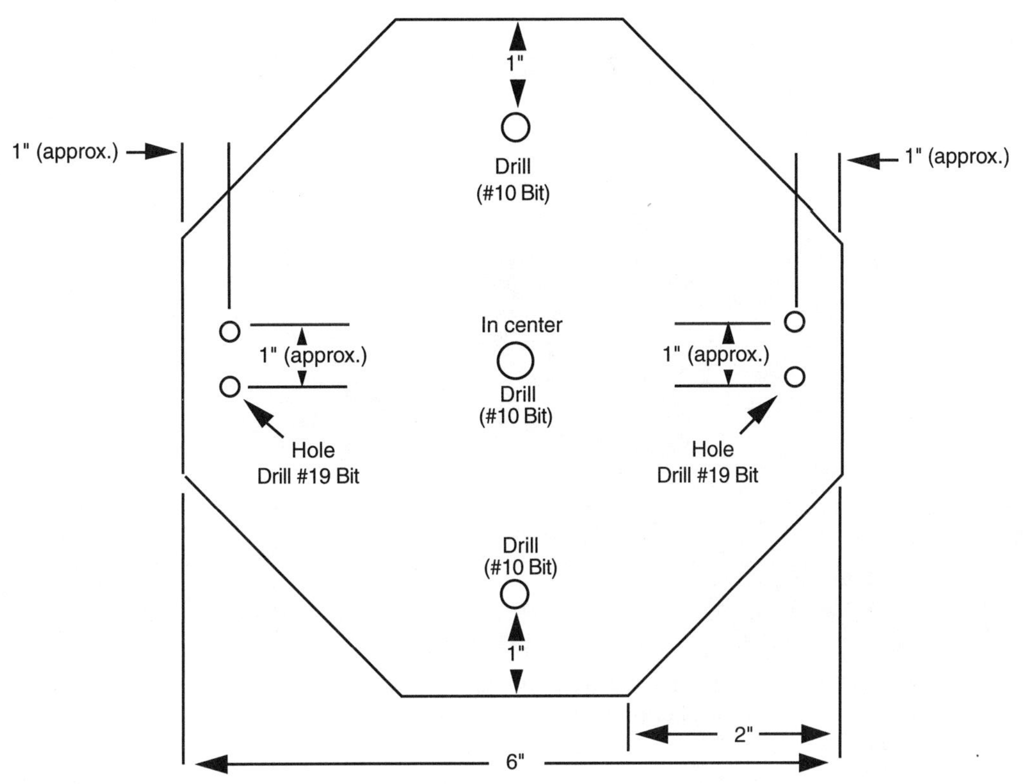

FIGURE 8-2 The cutting guide for the base of the plastic Minibot. The sets of two holes on either side are for the motor mount, and they should be spaced according to the specific mount you are using. Motors of different sizes and types will require different mounting holes.

because even with fairly large wheels, 30 r/min makes the robot scoot around the floor or a table at about 4 to 6 in/s. Choose motors small enough so they don't crowd the base of the robot and add unnecessary weight. Remember that you have other items to add, such as batteries and control electronics.

Use ⅜-in-wide metal mending braces to secure the motor (the prototype used plastic pieces from an old Fastech toy construction kit; you can use these or something similar). You may need to add spacers or extra nuts to balance the motor in the brace. Drill holes for 8⁄32 bolts (#19 bit), spaced to match the holes in the mending plate. Another method is to use small U-bolts, available at any hardware store. Drill the holes for the U-bolts and secure them with a double set of nuts.

Attach the tires to the motor shafts. Tires designed for a radio-controlled airplane or race car are good choices. The tires are well made, and the hubs are threaded in standard screw sizes (the threads may be metric, so watch out!). On the prototype, the motor shaft was threaded and had a 4-40 nut attached on each side of the wheel. Fig. 8-3 shows a mounted motor with a tire attached.

Installing the counterbalances completes the foundation-base plate. These keep the robot from tipping backward and forward along its drive axis. You can use small ball bearings, tiny

FIGURE 8-3 How the drive motors of the Minibot look when mounted. In the prototype Minibot the wheels were threaded directly onto the motor shaft. Note the gear-reduction system built onto the hobby motor.

casters, or—as was used on the prototype—the head of a 10/24 locknut. The locknut is smooth enough to act as a kind of ball bearing and is about the right size for the job. Attach the locknut with a 10/24-by-½-in bolt (if the bolt you have is too long to fit in the locknut, add washers or a 10/24 nut as a spacer).

8.3.3 TOP SHELL

The top shell is optional, and you can leave it off if you choose. The prototype used a round display bowl 6 in in diameter that I purchased from a plastics specialty store. Alternatively, you can use any suitable half sphere for your robot, such as an inverted salad bowl. Feel free to use colored plastic.

Attach the top by measuring the distance from the foundation to the top of the shell, taking into consideration the gap that must be present for the motors and other bulky internal components. Cut a length of 10/24 all-thread rod to size. The length of the prototype shaft was 3½ in. Secure the center shaft to the base using a pair of 10/24 nuts and a tooth lock washer. Secure the center shaft to the top shell with a 10/24 nut and a 10/24 locknut. Use a tooth lock washer on the inside or outside of the shell to keep the shell from spinning loose.

8.3.4 BATTERY HOLDER

You can buy battery holders that hold from one to six dry cells in any of the popular battery sizes. The Minibot motors, like almost all small hobby motors, run off 1.5 to 6 V. A four-cell, AA battery holder does the job nicely. The wiring in the holder connects the batteries in series, so the output is 6 V. Secure the battery holder to the base with 8/32 nuts and bolts. Drill holes to accommodate the hardware. Be sure the nuts and bolts don't extend too far below the base or they may drag when the robot moves. Likewise, be sure the hardware doesn't interfere with the batteries.

8.3.5 WIRING DIAGRAM

The wiring diagram in Fig. 8-4 allows you to control the movement of the Minibot in all directions. This simple two-switch system, which will be used in many other projects in this book, uses double-pole, double-throw (DPDT) switches. The switches called for in the circuit are spring-loaded so they return to a center-off position when you let go of them.

By using the dual pole double-throw switches, the electrical current passed to the motors changes direction as the switches are thrown from one extreme to the other. The actual connections are the same as what is used in an electrical H-bridge, which is discussed elsewhere in the book.

For the hook-up wire used to connect the robot to the remote control box, you might want to try a 6-ft length of Cat-5 network cable; at least four wires must connect the robot to the remote control box (two for power and two for each motor). The color-coded and combined wires are ideal for an application like this.

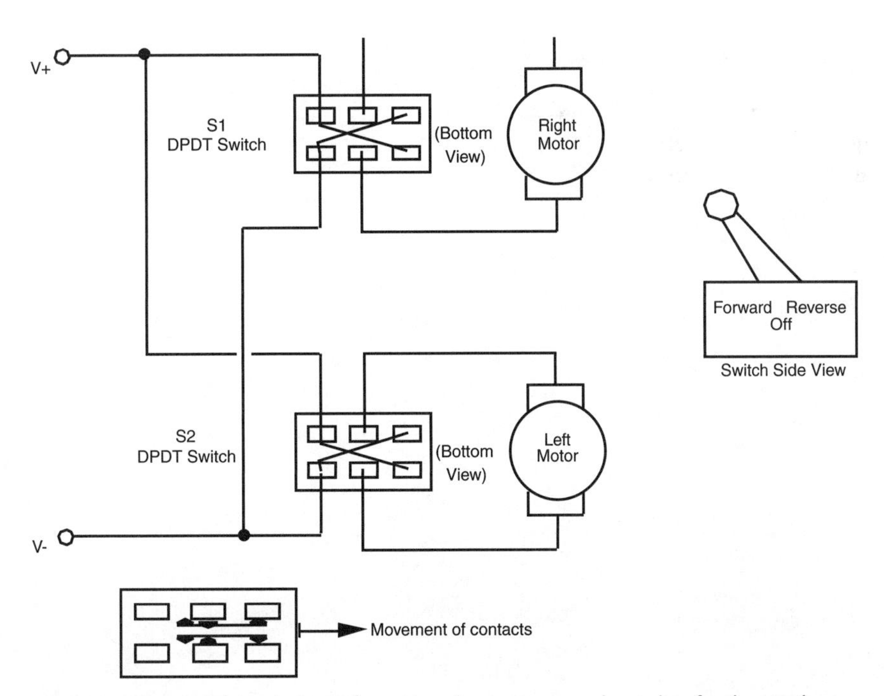

FIGURE 8-4 Use this schematic for wiring the motor control switches for the Minibot. Note that the switches are double-pole, double-throw (DTDP), with a spring return to center-off.

8.4 From Here

To learn more about . . .	***Read***
Wooden robots	Chapter 9, "Wooden Platforms"
Metal robots	Chapter 10, "Metal Platforms"
Using batteries	Chapter 17, "All About Batteries and Robot Power Supplies"
Selecting the right motor	Chapter 19, "Choosing the Right Motor for the Job"
Using a computer or microcontroller	Chapter 12, "An Overview of Robot 'Brains' "

CHAPTER 9

WOODEN PLATFORMS

Wood may not be high-tech, but it's an ideal building material for hobby robots. Wood is available just about everywhere. It's relatively inexpensive, easy to work with, and mistakes can be readily covered up, filled in, or painted over. In this chapter, using wood for robot structures will be presented and how you can apply simple woodworking skills to construct a basic wooden robot platform. This platform can then serve as the foundation for a number of robot designs you may want to explore.

9.1 Choosing the Right Wood

There is good wood and there is bad wood. Obviously, you want the good stuff, but you have to be willing to pay for it. For reasons you'll soon discover, you should buy only the best stock you can get your hands on. The better woods are available at specialty wood stores, particularly the ones that sell mostly hardwoods and exotic woods. Your local lumber and hardware store may have great buys on rough-hewn redwood planking, but it's hardly the stuff of robots.

9.1.1 PLYWOOD

The best overall wood for robotics use, especially for foundation platforms, is plywood. In case you are unfamiliar with plywood (Fig. 9-1), this common building material comes in

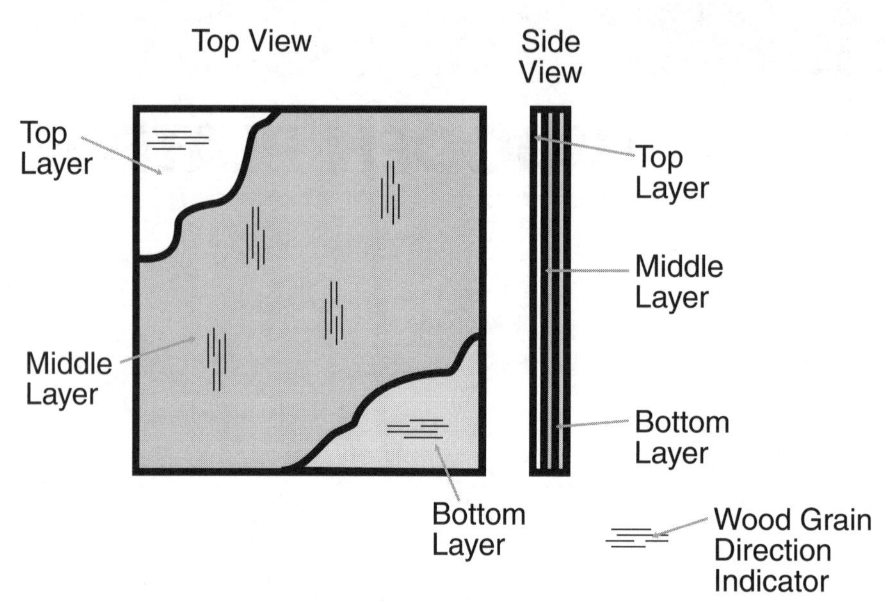

FIGURE 9-1 Plywood is manufactured by laminating (gluing together) thin sheets of wood (called veneer). By changing the angle of the wood's grain, the final plywood is stronger than a single piece of veneer three times thicker than the ones used.

many grades and is made by laminating thin sheets of wood together. The cheapest plywood is called shop grade, and it is the kind often used for flooring and projects where looks aren't too important. The board is full of knots and knotholes, and there may be considerable voids inside the board, all of which detract from its strength.

The remaining grades specify the quality of both sides of the plywood. Grade N is the best and signifies natural finish veneer. The surface quality of grade N really isn't important to us, so we can settle for grade A. Since we want both sides of the board to be in good shape, a plywood with a grade of A-A (grade A on both sides) is desired. Grades B and C are acceptable, but only if better plywoods aren't around. Depending on the availability of these higher grades, you may have to settle for A-C grade plywood (grade A on one side, grade C on the other).

Most plywoods you purchase at the lumber stores are made of softwoods—usually fir and pine. You can get hardwood plywood as well through a specialty wood supplier or from hobby stores (ask for aircraft-quality plywood). Hardwood-based plywood is more desirable because it is more dense and less likely to chip. Don't confuse hardwood plywood with hardboard. The latter is made of sawdust epoxied together under high pressure. Hardboard has a smooth finish; its close cousin, particleboard, does not. Both types are unsuitable for robotics because they are too heavy and brittle.

Plywood comes in various thicknesses starting at about 5⁄16 in and going up to over 1 in. Thinner sheets are acceptable for use in a robotics platform if the plywood is made from hardwoods. When using construction-grade plywoods (the stuff you get at the home improvement store), a thickness in the middle of the range—½ or ⅜ in—is ideal.

Construction plywood generally comes in 4-by-8-ft panels. Hardwood plywoods, particularly material for model building, come in smaller sizes, such as 2 ft by 2 ft. You don't need a large piece of plywood; the smaller the board, the easier it will be to cut to the exact size you need.

Remember that most home improvement stores will cut wood for you—saving you the cost of buying power saws and doing it yourself. The professional-grade saws available at home improvement stores will make a straight, even cut and the staff is usually trained in making the cuts very accurately. Along with cutting the wood to the size you want, they may have leftover pieces for just a dollar or two that can be further cut down to the exact size you want.

9.1.2 PLANKING

An alternative to working with plywood is planking. Use ash, birch, or some other solid hardwood. Stay away from the less meaty softwoods such as fir, pine, and hemlock. Most hardwood planks are available in widths of no more than 12 or 15 in, so you must take this into consideration when designing the platform. You can butt two smaller widths together if absolutely necessary. Use a router to fashion a secure joint, or attach metal mending plates to mate the two pieces together. (This latter option is not recommended; it adds a lot of unnecessary weight to the robot.)

When choosing planked wood, be especially wary of warpage and moisture content. If you have a planer, which cuts down planks of wood into absolutely flat and square pieces and the unit you have can handle the relatively thin and small pieces to be used in robots, then you can select the wood based on its grain and quality. If you don't have a planer, take along a carpenter's square, and check the squareness and levelness of the lumber in every possible direction. Reject any piece that isn't perfectly square; you'll regret it otherwise. Defects in premilled wood go by a variety of colorful names, such as crook, bow, cup, twist, wane, split, shake, and check, but they all mean headache to you.

Wood with excessive moisture may bow and bend as it dries, causing cracks and warpage. These can be devastating in a robot you've just completed and perfected. Buy only seasoned lumber stored inside the lumberyard, not outside. Watch for green specks or grains—these indicate trapped moisture. If the wood is marked, look for an MC specification. An MC-15 rating means that the moisture content doesn't exceed 15 percent, which is acceptable. Good plywoods and hardwood planks meet or exceed this requirement. Don't get anything marked MC-20 or higher or marked S-GREEN—bad stuff for robots.

9.1.3 BALSA

Yes, you've read correctly, *balsa* wood can be useful as part of a robot structure. Balsa is very light (and strong for its density), easily shaped, and readily available in different sizes in virtually all hobby shops throughout the world. It has a number of other properties that make it well suited for robot applications, not only for prototyping complex parts due to the ease in which it can be cut and formed and glued together in other pieces.

You might think that it's understated to say that balsa is a softwood; you have probably pressed your thumbnail into a piece at some time and been amazed how easy it is to put an imprint into the wood. This property along with its ability to absorb vibration make it ideal as a lining for fragile parts mounts. By buying a thick enough piece of balsa, chances are that you can replace the entire hold down with a piece having similar strength, as well as being much lighter.

The major consumer of balsa is not the model airplane industry, as you might have thought, but the liquid natural gas (LNG) industry as an insulator for large LNG tanks. There may be cases where parts of your robot are hot and placed in close proximity to other parts that are temperature sensitive. Before trying to come up with a fan or redesign solution, why don't you try a small piece of balsa, cut to size. You will probably discover that the balsa will help separate the different temperature areas effectively for a small amount of work and cost.

Note that an insulator is not a heat sink or a heat removal device like a fan. If you are using balsa to separate a device that has to stay cool beside one that becomes very hot, make sure that there is a way for the heat from the hot device to leave the robot. Encasing the part in a balsa box to keep heat from affecting other parts will result in the part becoming very hot and eventually burning out or catching fire.

9.1.4 DOWELS

Wood dowels come in every conceivable diameter, from about 1/16 to over 11.2 in. Wood dowels are 3 or 4 ft in length. Most dowels are made of high-quality hardwood, such as birch or ash. The dowel is always cut lengthwise with the grain to increase strength. Other than choosing the proper dimension, there are few considerations to keep in mind when buying dowels.

You should, however, inspect the dowel to make sure it is straight. At the store, roll the dowel on the floor. It should lie flat and roll easily. Warpage is easy to spot. Dowels can be used either to make the frame of the robot or as supports and uprights.

9.2 The Woodcutter's Art

You've cut a piece of wood in two before, haven't you? Sure you have; everyone has. You don't need any special tools or techniques to cut wood for a robot platform. The basic shop cutting tools will suffice: a handsaw, a backsaw, a circular saw, a jigsaw (if the wood is thin enough for the blade), a table saw, a radial arm saw, or—you name it.

Whatever cutting tool you use, make sure the blade is the right one for the wood. For example, the combination blade that probably came with your power saw isn't the right choice for plywood and hardwood. Outfit the saw with a cutoff blade or a plywood-paneling blade. Both have many more teeth per inch. Handsaws generally come in two versions: crosscut and ripsaw. You need the crosscut kind. If you are unfamiliar with the proper use of the tools that you are planning on using, you should go to your local library to look up books on woodworking, or find out when your local home improvement center is having classes on tool usage. Using the right tools and blades along with learning a few tips will make the work go much smoother and faster and result in a better finished product.

9.3 Cutting and Drilling

You can use a hand or motor tools to cut and drill through wood. The choice isn't important and is really a personal one. The most critical parameter, however, is to make sure that you only use sharp cutting tools and drill bits. If your bits or saw teeth are dull, replace them or have them sharpened.

Unless otherwise required, all cuts into wood are perpendicular (90 degrees) to its surface and, in most cases, the edges are also perpendicular to one another. To ensure that the cuts are at the proper angle, you should use either a miter box with your handsaw or a powered miter saw. These tools will help ensure that the cuts you make are at the correct angles and are as accurate as possible.

It's important that you drill straight holes, or your robot will not assemble properly. If your drill press is large enough, you can use it to drill perfectly straight holes in plywood and other large stock. Otherwise, use a portable drill stand. These attach to the drill or work in a number of other ways to guarantee a hole perpendicular to the surface of the material.

Before cutting or drilling, remember the carpenter's adage: "Measure twice and cut once."

9.4 Finishing

You can easily shape wood using rasps and files. If the shaping you need to do is extensive—like creating a circle in the middle of a large plank—you may want to consider getting a handheld rotary cutter. Be careful if you are thinking about buying a simple rotary rasp for your drill; the bearings within a drill are designed for vertical loads, not side-to-side loads and by using a cutter with your drill, you may end up ruining it. After cutting the wood down into the desired shape, use increasingly finer grades of sandpaper followed by the painting process outlined in the following.

The process outlined here will require just a few minutes to accomplish (over a few days, allowing for the paint to dry) and will allow you to create finished pieces of material (wood, metal, and plastic) that will have the functional benefits of:

1. Eliminating the dust on the surface of the material, allowing for effective two-sided tape attachment to (and removal from). Eliminating the fibers in wood also will allow for more effective glue bonds.
2. Smoothing the surface and reducing the lifted fibers that appear when wood gets moist or wet over time.
3. Eliminating splinters and sharp edges when handling the robot and minimizing surface splintering when drilling into wood.
4. Allowing for pencil and ink markings of the surface to be easily wiped off for corrections. This is not possible with bare wood, metal, or plastic.

I tend to just use aerosol paint—when properly used there is very little mess and no brushes to clean up. From an auto-body supply house, you should buy an aerosol can of primer (gray is always the first choice) and from a hardware store buy an aerosol can of indoor/outdoor (or marine) acrylic paint in your favorite color. Red catches the eye, isn't overwhelming, and will hide any blemishes in the wood or your work.

Set up a painting area in a garage or some other well-ventilated area by laying down newspaper both on a flat surface as well as a vertical surface. Next, lay down bottle caps to be used as supports for the materials that you are going to be painting. You don't have to use bottle caps—scraps of wood or other detritus can be just as effective—just make sure that when you are painting something you care about that the supports are smaller than the perimeter of the object being finished. You will want to finish the ends of the plywood and don't want to end up with paint flowing between the plywood and the support.

Lightly sand the surfaces of the material you are going to paint. You may want to sand the edges of the strips more aggressively to take off any loose wood that could become splinters. Once you have finished, moisten a rag and wipe it over the surface you have sanded to pick up any loose dust.

Shake the can of primer using the instructions printed on the can. Usually there is a small metal ball inside the can and you will be instructed to shake it until the ball rattles easily inside. Start with the two plywood strips that have been left over. Place one end on the bottle caps, supporting the surface to be painted and spray about 6 in of the strip, starting at the supported end. Most primers take 30 minutes or so to dry. Check the instructions on the can before going on to the next step of sanding and putting on new coats of paint or primer.

After the first application of primer, you will probably find that the surface of the wood is very rough. This is due to the cut fibers in the wood standing on end after being moistened from the primer. Repeat the sanding step (along with sanding the ends of the wood and wiping down with the damp rag) before applying another coat of primer. After the second coat is put down, let it dry, sand very lightly, and wipe down again.

Now you are ready to apply the paint. Shake according to instructions on the can and spray the plywood strips, putting on a thin, even coat. You will probably find that the paint will seem to be sucked into the wood and the surface will not be that shiny. This is normal. Once the paint has dried, lightly sand again, wipe down with a wet cloth, and apply a thicker coat of paint. When this coat has dried, you'll find that the surface of the plywood is very smooth and shiny. You do not have to sand the paint again. The plywood is now ready to be used in a robot.

9.5 Building a Wooden Motorized Platform

Figs. 9.2 and 9.3 show approaches for constructing a basic square and round motorized wooden platform, respectively. See Table 9-1 for a list of the parts you'll need.

To make the square plywood platform shown in Fig. 9-2, cut a piece of ⅜- or ½-in plywood to 10 by 10 in (the thinner ⅜-in material is acceptable if the plywood is the heavy-duty hardwood variety, such as aircraft-quality plywood). Make sure the cut is square. Notch the wood as shown to make room for the robot's wheels. The notch should be large enough to

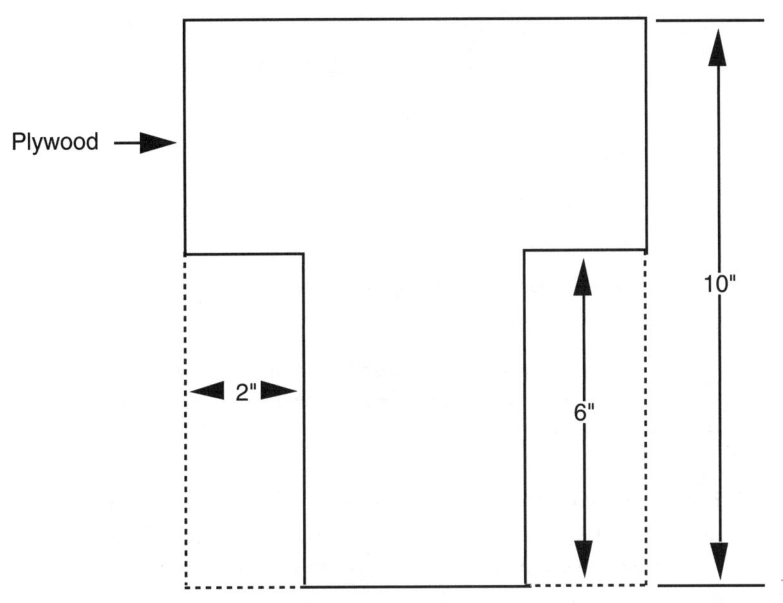

FIGURE 9-2 Cutting plan for a square plywood base.

accommodate the width and diameter of the wheels, with a little breathing room to spare. For example, if the wheels are 6 in diameter and 1.5 in wide, the notch should be about 6.5 by 1.75 in.

To make a motor control switch for this platform, see the parts list in Table 8-4 in Chapter 8.

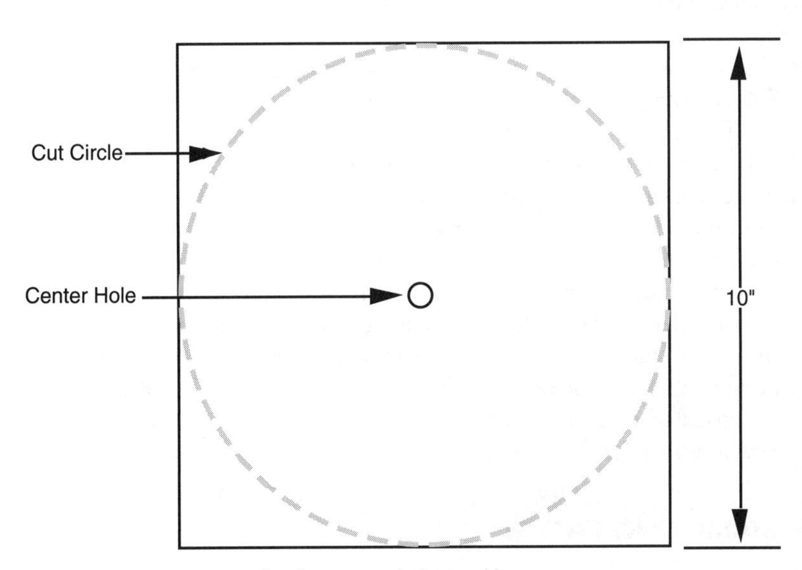

FIGURE 9-3 Cutting plan for a round plywood base.

TABLE 9-1 Parts List For Wood Base

Base	10″ by 10″ ⅜″ to ½″ thick plywood
2	DC gear motors
2	5″ to 7″ rubber wheels
1	1¼″ caster wheel
2	2″ by 4″ lumber (see text)
1	4× "D" cell battery holder
Misc.	1″ by 10/24 stove bolts, 3″ by 10/24 stove bolts, 10/24 nuts and flat washers, 1¼″ by 8/32 stove bolts, 8/32 nuts, flat washers, and lock washers

Fig. 9-3 shows the same 10-in-square piece of ⅜- or ½-in plywood cut into a circle. Use a scroll saw and circle attachment for cutting. As you did for the square platform, make a notch in the center beam of the circle to allow room for the wheels—the larger the wheel, the larger the notch.

9.5.1 ATTACHING THE MOTORS

The wooden platform you have constructed so far is perfect for a fairly sturdy robot, so the motor you choose can be powerful. Use heavy-duty motors, geared down to a top speed of no more than about 75 r/min; 30 to 40 r/min is even better. Anything faster than 75 r/min will cause the robot to dash about at speeds exceeding a few miles per hour, which is unacceptable unless you plan on entering your creation in the sprint category of the Robot Olympics.

Note: you can use electronic controls to reduce the speed of the gear motor by 15 or 20 percent without losing much torque, but you should not slow the motor too much or you'll lose power. The closer the motor operates at its rated speed, the better results you'll have. Throttling motors will be discussed later in the book.

If the motors have mounting flanges and holes on them, attach them using corner brackets. Some motors do not have mounting holes or hardware, so you must fashion a hold-down plate for them. You can make an effective hold-down plate, as shown in Fig. 9-4, out of 2 by 4 lumber. Round out the plate to match the cylindrical body of the motor casing. Then secure the plate to the platform. Last, attach the wheels to the motor shafts. You may need to thread the shafts with a die so you can secure the wheels. Use the proper size nuts and washers on either side of the hub to keep the wheel in place. You'll make your life much easier if you install wheels that have a setscrew. Once they are attached to the shaft, tighten the setscrew to screw the wheels in place.

9.5.2 STABILIZING CASTER

Two motors and a centered caster, attached to the robot's base as depicted in Fig. 9-5, allows you to have full control over the direction your robot travels. You can make the robot

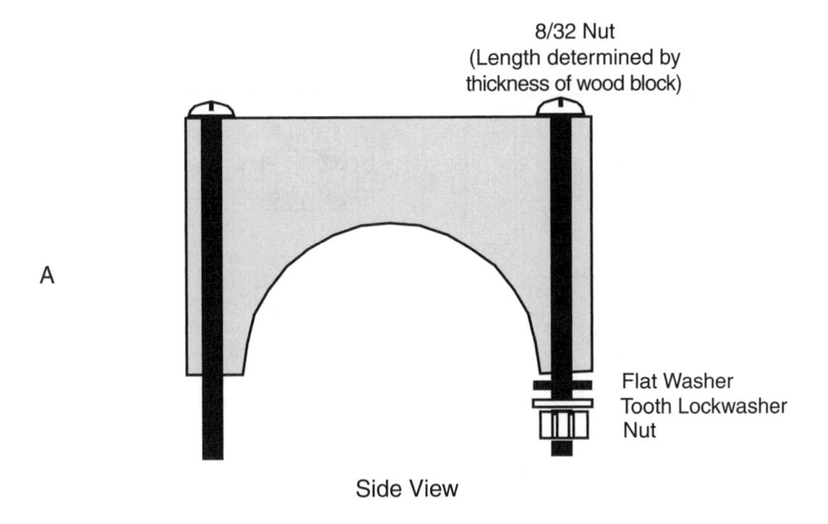

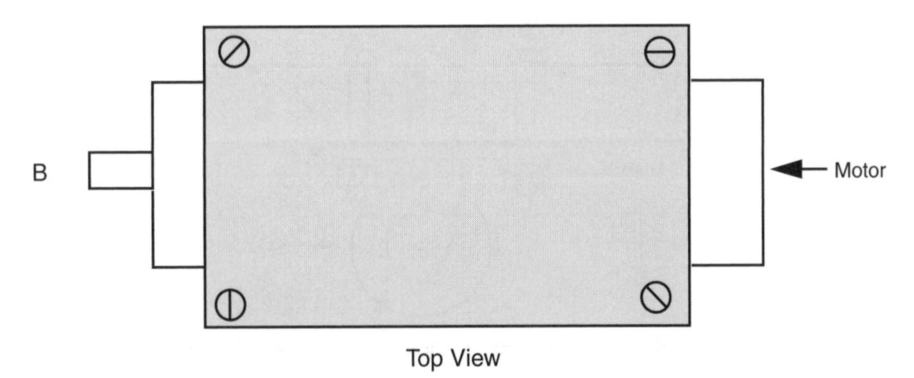

FIGURE 9-4 One way to secure the motors to the base is to use a wood block hollowed out to match the shape of the motor casing. *a.* Side view; *b.* Top view.

turn by stopping or reversing one motor while the other continues turning. Attach the caster using four 8⁄32-by-1-in bolts. Secure the caster with tooth lock washers and 8⁄32 nuts.

The caster should be mounted so that the robot base is level and it can swivel with a minimum of resistance. You may, if necessary, use spacers to increase the distance from the base plate of the caster to the bottom of the platform. If the caster end of the robot is much higher than the wheels, then the motor mounting bolts or the rear of the robot might rub on the ground. If this is the case, you should be looking at using a smaller caster, larger wheels, or mounting the motors on the bottom of the robot base.

9.5.3 BATTERY HOLDER

You can purchase battery holders that contain from one to six dry cells in any of the popular battery sizes. When using 6-V motors, you can use a four-cell D battery holder. You can

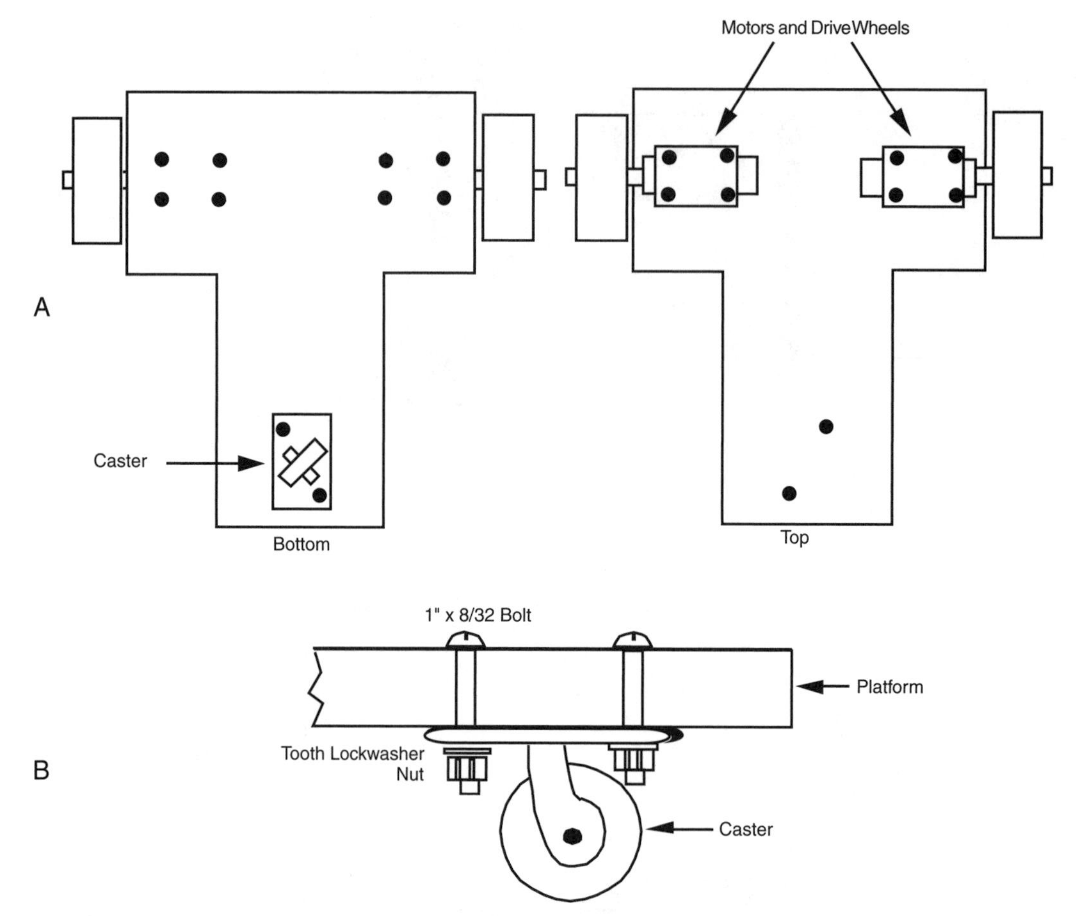

FIGURE 9-5 Attaching the caster to the platform. *a.* Top and bottom view; *b.* caster hardware assembly detail.

also use a single 6-V lantern or rechargeable battery. Motors that require 12 V will need two battery holders, two 6-V batteries, or one 12-V battery (motorcycle batteries are often excellent power sources for robots, which require 12 V and a fair amount of current). For the prototype, 6-V motors and a four-cell D battery holder were used.

Secure the battery holder(s) to the base with 8⁄32 nuts and bolts. Drill holes to accommodate the hardware. Be sure the nuts and bolts don't extend too far below the base or they may drag when the robot moves. Likewise, be sure the hardware doesn't interfere with the batteries.

Wire the batteries and wheels to the DPDT through control switches, as shown in the Minibot project described at the end of Chapter 8, "Plastic Platforms." One switch controls the left motor; the other switch controls the right motor.

9.6 From Here

To learn more about . . .	***Read***
Plastic robots	Chapter 8, "Plastic Platforms"
Metal robots	Chapter 10, "Metal Platforms"
Using batteries	Chapter 17, "All about Batteries and Robot Power Supplies"
Selecting the right motor	Chapter 19, "Choosing the Right Motor for the Job"
Using a computer or microcontroller	Chapter 12, "An Overview of Robot 'Brains' "

CHAPTER 10

METAL PLATFORMS

FIGURE 10-1 The completed Buggybot.

Metal is perhaps the best all-around material for building robots because it offers better strength than other materials. If you've never worked with metal before, you shouldn't worry; there is really nothing to it. The designs outlined in this chapter and the chapters that follow will show you how to construct robots both large and small out of readily available metal stock, without resorting to welding or custom machining.

10.1 Working with Metal

If you have the right tools, working with metal is only slightly harder than working with wood or plastic. You'll have better-than-average results if you always use sharpened, well-made tools. Dull, bargain-basement tools can't effectively cut through aluminum or steel stock. Instead of the tool doing most of the work, you do.

10.1.1 MARKING CUT LINES AND DRILL HOLE CENTERS

Marking metal for cutting and drilling is more difficult than other materials due to its increased hardness and resistance to being permanently marked by inks. Professional metal workers scratch marking lines in metal using a tool called a scratch awl; it can be purchased at hardware stores for just a few dollars.

You'll find that when you drill metal the bit will skate all over the surface until the hole is started. You can eliminate or reduce this skating by using a punch prior to drilling. There are spring-loaded punches that simply require you to press down on them before they snap and pop a small indentation in the material. You can also buy punches that require a hammer to make the indentation, but the spring-loaded ones are easier to work with and do not cost a lot more.

10.1.2 CUTTING

To cut metal, use a hacksaw outfitted with a fine-tooth blade, one with 24 or 32 teeth per inch. Coping saws, keyhole saws, and other handsaws are generally engineered for cutting wood, and their blades aren't fine enough for metal work. You can use a power saw, like a table saw or reciprocating saw, but, again, make sure that you use the right blade. For large aluminum or steel angle stock, an abrasive cutter is invaluable and will make short work of any cuts—remember that the cut ends will be hot for quite a while after the cutting process!

You'll probably do most of your cutting by hand. You can help guarantee straight cuts by using an inexpensive miter box. You don't need anything fancy, but try to stay away from the wooden boxes. They wear out too fast. The hardened plastic and metal boxes are the best buys. Be sure to get a miter box that lets you cut at 45 degrees both vertically and horizontally. Firmly attach the miter box to your workbench using hardware or a large clamp.

10.1.3 DRILLING

Metal requires a slower drilling speed than wood, and you need a power drill that either runs at a low speed or lets you adjust the speed to match the work. Variable speed power drills are available for under $30 these days, and they're a good investment. Be sure to use only sharp drill bits. If your bits are dull, replace them or have them sharpened. Quite often, buying a new set is cheaper than professional sharpening.

When it comes to working with metal, particularly channel and pipe stock, you'll find a drill press is a godsend. It improves accuracy, and you'll find the work goes much faster. Always use a proper vise when working with a drill press. Never hold the work with your hands. Especially with metal, the bit can snag as it's cutting and yank the piece out of your hands. If you can't place the work in the vise, clamp it to the cutting table.

10.1.4 BENDING

One of the biggest challenges you will have when working with metal parts in your robots will be bending them accurately and evenly. When you go to your first robot competitions, you will no doubt see a few robots that look like they were run over by an eighteen-wheeler and hammered back into shape; hopefully you weren't critical; working metal into the desired shape is a challenge. So much so that when you first start working with metal on your own, you probably will want to avoid building robots that require any bent metal.

It should not be surprising that different types of metals require different methods for bending. Steel bar stock, for example, can have a sharp angle bent into it by placing it in a vise at the point of the bend and then hitting it with a hammer to force the bend to the desired angle. Gentle bends in steel require more sophisticated tools; but you could use the equipment designed to bend steel for wrought iron fencing. Aluminum stock, on the other hand, is just about impossible to bend in a home workshop. Using the techniques described for bending aluminum will result in it cracking or even breaking off. If you must use aluminum, then you should either use bar stock cut and fastened together or have a piece of stock milled to the desired shape.

Different *forms* of metal also are treated differently. The techniques previously discussed are not appropriate for sheet metal. Bending sheet metal is best accomplished using a sheet metal brake, which will bend the sheet metal (steel, aluminum, or copper) to any desired angle. Modest brakes can be purchased for about $100 from better machine shop supply houses (which you can find in the Yellow Pages).

10.1.5 FINISHING

Cutting and drilling often leave rough edges, called flashing, in the metal. These edges must be filed down using a medium- or fine-pitch metal file, or else the pieces won't fit together properly. A rotary tool with a carbide wheel will make short work of the flashing. Aluminum flash comes off quickly and easily; you need to work a little harder when removing the flash in steel or zinc stock.

10.2 Building the Buggybot

The Buggybot is a small robot built from a single 6-by-12-in sheet of 1⁄16-in-thick aluminum, nuts, and bolts, and a few other odds and ends. You can use the Buggybot as the foundation and running gear for a very sophisticated petlike robot. As with the robots built with plastic and wood we discussed in the previous two chapters, the basic design of the all-metal Buggybot can be enhanced just about any way you see fit. This chapter details the construction of the framework, locomotion, and power systems for a wired remote control robot. Future chapters will focus on adding more sophisticated features, such as wireless remote control, automatic navigation, and collision avoidance and detection. Refer to Table 10-1 for a list of the parts needed to build the Buggybot.

TABLE 10-1 Parts List for Buggybot (see parts list in Table 8-4 of Chapter 8 for motor control switch)

1	6-by-12-in sheet of $\frac{1}{16}$-in-thick aluminum for the frame
2	Tamiya high-power gearbox motors (from kit—see text)
2	3-in-diameter Lite Flight foam wheels
2	$\frac{5}{56}$ nuts (should be included with the motors)
2	$\frac{3}{16}$-in collars with setscrews
1	Two-cell D battery holder
1	1$\frac{1}{2}$-in swivel caster misc
Misc.	1-in-by-$\frac{6}{32}$ stove bolts, nuts, flat washers, $\frac{1}{2}$-in-by-$\frac{6}{32}$ stove bolts, nuts, tooth lock washers, flat washers (as spacers)

10.2.1 FRAMEWORK

Build the frame of the Buggybot from a single sheet of $\frac{1}{16}$-in-thick aluminum sheet. This sheet, measuring 6 by 12 in, is commonly found at hobby stores. As this is a standard size, there's no need to cut it. Follow the drill-cutting template shown in Fig. 10-2.

After drilling, use a large shop vise or woodblock to bend the aluminum sheet as shown in Fig. 10-3. Accuracy is not all that important. The angled bends are provided to give the Buggybot its unique appearance.

10.2.2 MOTORS AND MOTOR MOUNT

The prototype Buggybot uses two high-power gearbox motor kits from Tamiya, which are available at many hobby stores (as well as Internet sites, such as TowerHobbies.com). These motors come with their own gearbox; choose the 1:64.8 gear ratio. An assembled motor is shown in Fig. 10-4. Note that the output shaft of the motor can be made to protrude a variable distance from the body of the motor. Secure the shaft (using the Allen setscrew that is included) so that only a small portion of the opposite end of the shaft sticks out of the gearbox on the other side, as shown in Fig. 10-4.

You should secure the gearboxes and motors to the aluminum frame of the Buggybot as depicted in Fig. 10-5. Use $\frac{6}{32}$ bolts, flat washers, and nuts. Be sure that the motors are aligned as shown in the drawing. Note that the shaft of each motor protrudes from the side of the Buggybot.

Fig. 10-6 illustrates how to attach the wheels to the shafts of the motors. The wheels used in the prototype were 3-in-diameter foam Lite Flight tires, commonly available at hobby stores. Secure the wheels in place by first threading a $\frac{3}{16}$-in collar (available at hobby stores) over the shaft of the motor. Tighten the collar in place using its Allen setscrew. Then cinch the wheel onto the shaft by tightening a $\frac{5}{56}$ threaded nut to the end of the motor shaft (the nut should be included with the gearbox motor kit). Be sure to tighten down on the nut so the wheel won't slip.

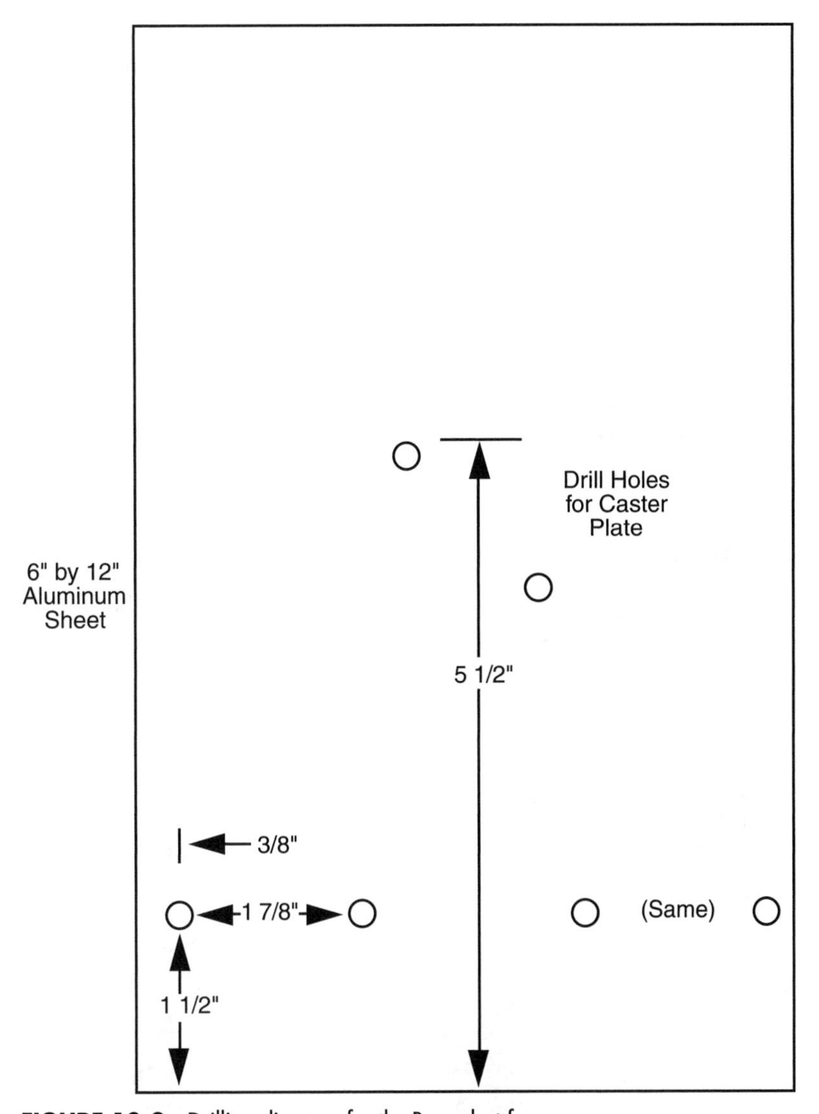

FIGURE 10-2 Drilling diagram for the Buggybot frame.

10.2.3 SUPPORT CASTER

The Buggybot uses the two-wheel-drive tripod arrangement. You need a caster on the other end of the frame to balance the robot and provide a steering swivel. The 1½-in swivel caster is not driven and doesn't do the actual steering. Driving and steering are taken care of by the drive motors. Referring to Fig. 10-7, attach the caster using two 6⁄32 by ½-in bolts and nuts.

Note that the mechanical style of the caster, and indeed the diameter of the caster wheel, is dependent on the diameter of the drive wheels. Larger drive wheels will require either a different mounting or a larger caster. Small drive wheels will likewise require you to adjust the caster mounting and possibly use a smaller-diameter caster wheel.

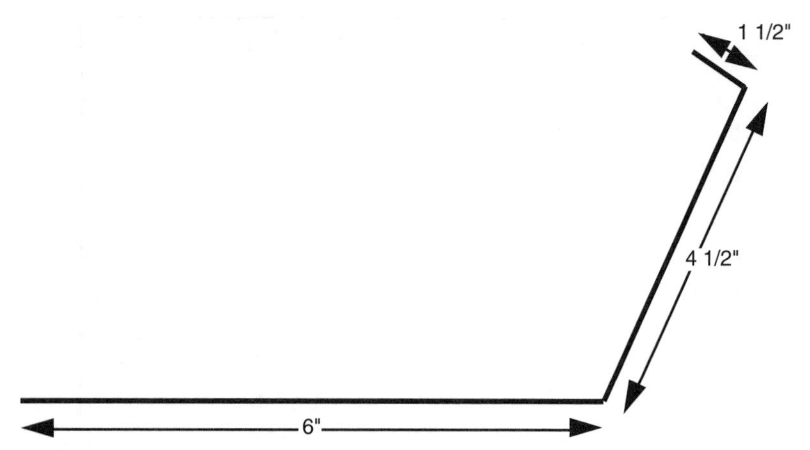

FIGURE 10-3 Bend the aluminum sheet at the approximate angles shown here.

10.2.4 BATTERY HOLDER

The motors require an appreciable amount of current, so the Buggybot really should be powered by heavy-duty C- or D-size cells. The prototype Buggybot used a two-cell D battery holder. The holder fits nicely toward the front end of the robot and acts as a good coun-

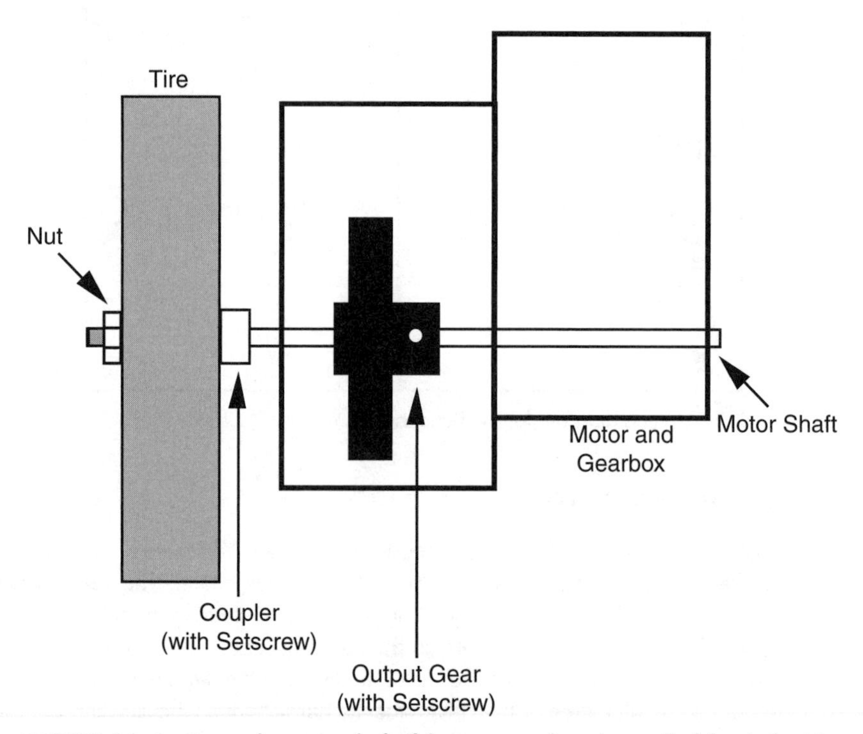

FIGURE 10-4 Secure the output shaft of the motor so that almost all of the shaft sticks out on one side of the motor.

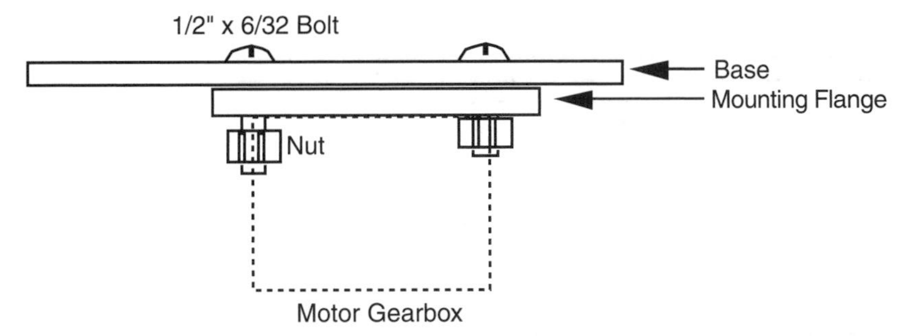

FIGURE 10-5 The gearboxes and motors are attached to the frame of the Buggybot using ordinary hardware.

terweight. You can secure the battery holder to the robot using double-sided tape or hook-and-loop (Velcro) fabric.

10.2.5 WIRING DIAGRAM

The basic Buggybot uses a manual wired switch control. The control is the same one used in the plastic Minibot detailed in Chapter 8, "Plastic Platforms." Refer to the wiring diagram in Fig. 8-4 of that chapter for information on powering the Buggybot.

FIGURE 10-6 Attach the foam wheels (with plastic hubs) for the Buggybot onto the shafts of the motors.

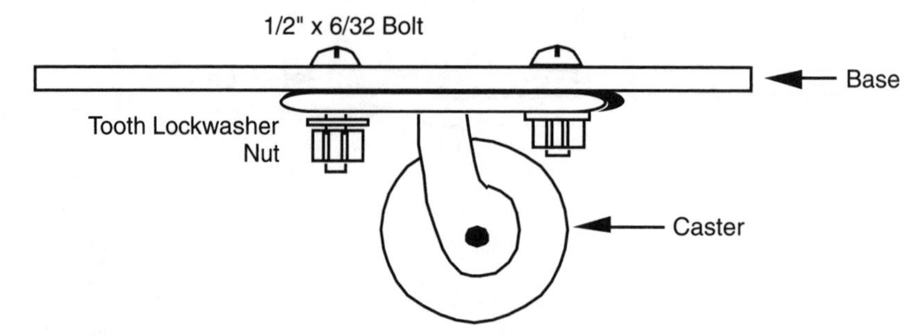

FIGURE 10-7 Mounting the caster to the Buggybot.

To prevent the control wire from interfering with the robot's operation, attach a piece of heavy wire (the bottom rail of a coat hanger will do) to the caster plate and lead the wire up it. Use nylon wire ties to secure the wire. The completed Buggybot is shown in Fig. 10-1.

10.3 Test Run

You'll find that the Buggybot is an amazingly agile robot. The distance it needs to turn is only a little longer than its length, and it has plenty of power to spare. There is room on the robot's front and back to mount additional control circuitry. You can also add control circuits and other enhancements over the battery holder. Just be sure that you can remove the circuit(s) when it comes time to change or recharge the batteries.

10.4 From Here

To learn more about . . .	***Read***
Plastic robots	Chapter 8, "Plastic Platforms"
Metal robots	Chapter 9, "Wooden Platforms"
Using batteries	Chapter 17, "All about Batteries and Robot Power Supplies"
Selecting the right motor	Chapter 19, "Choosing the Right Motor for the Job"
Using a computer or microcontroller	Chapter 12, "An Overview of Robot 'Brains' "

CHAPTER 11

HACKING TOYS

Ready-made toys can be used as the basis for more complex home-brew hobby robots. The toy industry is robot crazy, and you can buy a basic motorized or unmotorized robot for parts, building on it and adding sophistication and features. Snap or screw-together kits, such as the venerable Erector set, let you use premachined parts for your own creations. And some kits, like LEGO and Robotix, are even designed to create futuristic motorized robots and vehicles. You can use the parts in the kits as-is or cannibalize them, modifying them in any way you see fit. Because the parts already come in the exact or approximate shape you need, the construction of your own robots is greatly simplified.

About the only disadvantage to using toys as the basis for more advanced robots is that the plastic and lightweight metal used in the kits and finished products are not suitable for a homemade robot of any significant size or strength. You are pretty much confined to building small minibot or scooterbot-type robots from toy parts. Even so, you can sometimes apply toy parts to robot subsystems, such as a light-duty arm-gripper mechanism installed on a larger automaton.

In the following sections, a number of different toys and building sets are discussed and their appropriateness for use as robots. Let's take a closer look at using toys in your robot designs in this chapter, and examine several simple, cost-effective designs using readily available toy construction kits.

11.1 A Variety of Construction Sets

Toy stores are full of plastic put-together kits and ready-made robot toys that seem to beg you to use them in your own robot designs. Here are some toys you may want to consider for your next project.

11.1.1 ERECTOR SET

The Erector set, now sold by Meccano, has been around since the dawn of time—or so it seems. The kits, once made entirely of metal but now commonly including many plastic pieces, come in various sizes and are generally designed to build a number of different projects. Many kits are engineered for a specific design with perhaps provisions for moderate variations. Among the useful components of the kits are prepunched metal girders, plastic and metal plates, tires, wheels, shafts, and plastic mounting panels. You can use any as you see fit, assembling your robots with the hardware supplied with the kit or with 6/32 or 8/32 nuts and bolts.

Several Erector sets, such as those in the action troopers collection, come with wheels, construction beams, and other assorted parts that you can use to construct a robot base. Motors are typically not included in these kits, but you can readily supply your own. Because Erector set packages regularly come and go, what follows is a general guide to building a robot base. You'll need to adapt and reconfigure based on the Erector set parts you have on hand.

The prepunched metal girders included in the typical Erector set make excellent motor mounts. They are lightweight enough that they can be bent, using a vise, into a U-shaped motor holder. Bend the girder at the ends to create tabs for the bolts, or use the angle stock provided in an Erector set kit. The basic platform is designed for four or more wheels, but the wheel arrangement makes it difficult to steer the robot. The design presented in Fig. 11-1 uses only two wheels. The platform is stabilized using a miniature swivel caster at one end. You'll need to purchase the caster at the hardware store.

Note that the shafts of the motors are not directly linked to the wheels. The shaft of the wheels connects to the base plate as originally designed in the kit. The drive motors are equipped with rollers, which engage against the top of the wheels for traction. You can use a metal or rubber roller, but rubber is better. The pinch roller from a discarded cassette tape player is a good choice, as is a 3/8-in beveled bibb washer, which can be found in the plumbing section of the hardware store. You can easily mount a battery holder on the top of the platform. Position the battery holder in the center of the platform, toward the caster end. This will help distribute the weight of the robot.

The basic platform is now complete. You can attach a dual-switch remote control, as described earlier in the book, or automatic control circuitry as will be outlined in later chapters. Do note that over the years the Erector set brand has gone through many owners. Parts from old Erector sets are unlikely to fit well with new parts. This includes but is not limited to differences in the threads used for the nuts and bolts. If you have a very old Erector set (such as those made and sold by Gilbert), you're probably better off keeping them as collector's items rather than raiding them for robotic parts. The very old Erector sets of the 1930s through 1950s fetch top dollar on the collector's market (when the sets are in good, complete condition, of course).

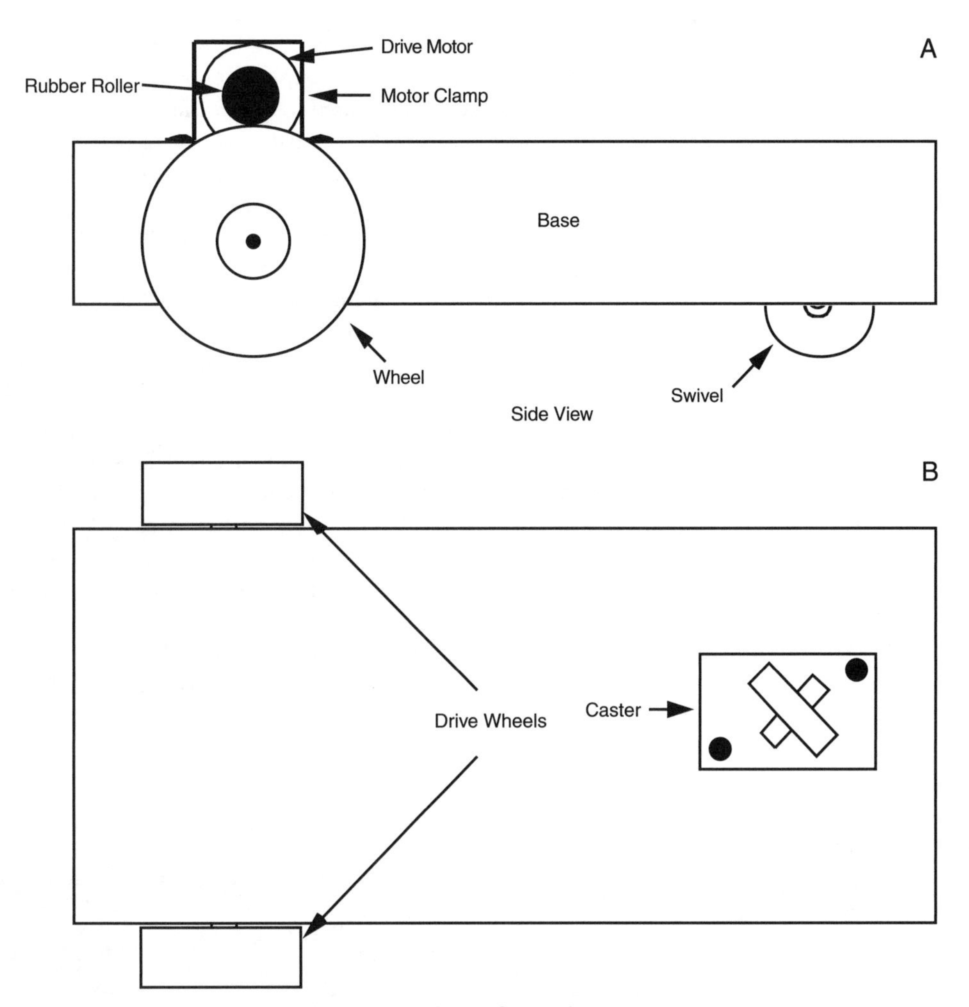

FIGURE 11-1 Constructing the motorized base for a robot using Erector set (Meccano) parts. *a.* Attaching the motor and drive roller over the wheel; *b.* Drive wheel-caster arrangement.

Similarly, today's Meccano sets are only passably compatible with the English-made Meccano sets sold decades ago. Hole spacing and sizes have varied over the years, and mixing and matching is neither practical nor desirable.

11.1.2 ROBOTIX

The Robotix kits, originally manufactured by Milton-Bradley and now sold by Learning Curve, are specially designed to make snap-together walking and rolling robots. Various kits are available, and many of them include at least one motor (additional motors are available separately). You control the motors using a central switch pad. Pushing the switch forward turns the motor in one direction; pushing the switch back turns the motor in the other direction. The output speed of the motors is about 6 r/min, which makes them a bit slow for moving a robot across the room but perfect for arm-gripper designs.

The structural components in the Robotix kits are molded from high-impact plastic. You can connect pieces together to form just about anything. One useful project is to build a robotic arm using several of the motors and structural components. The arm can be used by itself as a robotic trainer or attached to a larger robot. It can lift a reasonable 8 oz or so, and its pincher claw is strong enough to firmly grasp most small objects.

While the Robotix kit allows you to snap the pieces apart when you're experimenting, the design presented here is meant to be permanent. Glue the pieces together using plastic model cement or contact cement. Cementing is optional, of course, and you're free to try other, less permanent methods to secure the parts together, such as small nuts and bolts, screws, or Allen setscrews.

When cemented, the pieces hold together much better, and the arm is considerably stronger. Remember that, once cemented, the parts cannot be easily disassembled, so make sure that your design works properly before you commit to it. When used as a stand-alone arm, you can plug the shoulder motor into the battery holder or base. You don't need to cement this joint.

Refer to Fig. 11-2 as you build the arm. Temporarily attach a motor (call it motor 1) to the Robotix battery holder–base plate. Position the motor so that the drive spindle points straight up. Attach a double plug to the drive spindle and the end connector of another motor (motor 2). Position this motor so that the drive spindle is on one side. Next, attach another double plug and an elbow to the drive spindle of motor 2. Attach the other end of the elbow connector to a beam arm.

FIGURE 11-2 The robot arm constructed with parts from a Robotix construction kit.

Connect a third motor (motor 3) to the large connector on the opposite end of the beam arm. Position this motor so the drive spindle is on the other end of the beam arm. Attach a double plug and an elbow between the drive spindle of motor 3 and the connector opposite the drive spindle of the fourth motor (motor 4). The two claw levers directly attach to the drive spindle of motor 4.

Motorize the joints by plugging in the yellow power cables between the power switch box and the motor connectors. Try each joint, and note the various degrees of freedom. Experiment with picking up various objects with the claw. Make changes now before disassembling the arm and cementing the pieces together.

After the arm is assembled, route the wires around the components, making sure there is sufficient slack to permit free movement. Attach the wires to the arm using nylon wire ties.

11.1.3 LEGO

LEGO has become the premier construction toy, for both children and adults. The LEGO Company, parent company of the LEGO brand, has expanded the line as educational resources, making the ubiquitous LEGO bricks as well as parts suitable for small engineering projects (the Technic line) common in schools across the country and around the world. Along with buying kits, you can buy specific parts directly from LEGO or from a variety of sites on the Internet. LEGO also makes the Mindstorms, a series of sophisticated computerized robots that can be programmed in a variety of different ways. Along with Mindstorms, LEGO makes the Spytech set of robots, which can be used to learn more about behavior-based programming. These tools are excellent for the beginner and can be enhanced with more sophisticated programming environments when you become more comfortable with programming and electronics.

LEGO bricks and parts are excellent for prototyping robot systems, but there are some issues that you should be aware of when you are planning to build a robot from them. First, the bricks do not hold together well in any kind of vibration. You can try gluing them together, but this largely defeats the purpose of having a reconfigurable assembly system. The final issue that you should be aware of is that LEGO bricks and parts are quite heavy for their size, which can cause a problem with some robot designs.

11.1.4 CAPSULA

Capsula is a popular snap-together motorized parts kit that uses unusual tube and sphere shapes. Capsula kits come in different sizes and have one or more gear motors that can be attached to various components. The kits contain unique parts that other put-together toys don't, such as a plastic chain and chain sprockets or gears. Advanced kits come with remote control and computer circuits. All the parts from these kits are interchangeable. The links of the chain snap apart, so you can make any length of chain that you want. Combine the links from many kits and you can make an impressive drive system for an experimental lightweight robot.

11.1.5 FISCHERTECHNIK

The Fischertechnik kits, made in Germany and imported into North America by a few educational companies, are the Rolls-Royces of construction toys. Actually, *toy* isn't the proper term because the Fischertechnik kits are not just designed for use by small children. In fact, many of the kits are meant for high school and college industrial engineering students, and they offer a snap-together approach to making working electromagnetic, hydraulic, pneumatic, static, and robotic mechanisms.

All the Fischertechnik parts are interchangeable and attach to a common plastic base plate. You can extend the lengths of the base plate to just about any size you want, and the base plate can serve as the foundation for your robot, as shown in Fig. 11-3. You can use the motors supplied with the kits or use your own motors with the parts provided. Because of the cost of the Fischertechnik kits, you may not want to cannibalize them for robot components. But if you are interested in learning more about mechanical theory and design, the Fischertechnik kits, used as-is, provide a thorough and programmed method for jumping in with both feet.

11.1.6 K'NEX

K'Nex uses unusual half-round plastic spokes and connector rods (see Fig. 11-4) to build everything from bridges to Ferris wheels to robots. You can build a robot with just K'Nex parts or use the parts in a larger, mixed-component robot. For example, the base of a walking robot may be made from a thin sheet of aluminum, but the legs might be constructed from various K'Nex pieces.

FIGURE 11-3 A sampling of Fischertechnik parts.

FIGURE 11-4 K'Nex sets let you create physically large robots that weigh very little. The plastic pieces form very sturdy structures when properly connected.

A number of K'Nex kits are available, from simple starter sets to rather massive special-purpose collections (many of which are designed to build robots, dinosaurs, or robot dinosaurs). Several of the kits come with small gear motors so you can motorize your creation. Motors that interface to K'Nex parts are also available separately.

11.1.7 ZOOB

Zoob (made by Primordial) is a truly unique form of construction toy. A Zoob piece consists of a stem with a ball or socket on either end. You can create a wide variety of construction projects by linking the balls and sockets together. The balls are dimpled so they connect securely within their sockets. One practical application of Zoob is to create armatures for human- or animal-like robots. The Zoob pieces work in a way similar to bone joints.

11.1.8 CHAOS

Chaos sets are designed for structural construction projects: bridges, buildings, working elevator lifts, and the like. The basic Chaos set provides beams and connectors, along with chutes, pulleys, winches, and other construction pieces. Add-on sets are available that contain parts to build elevators, vortex tubes, and additional beams and connectors.

11.1.9 OTHER CONSTRUCTION TOYS

There are many other construction toys that you may find handy. Check the nearest well-stocked toy store or a toy retailer on the Internet for the following:

- Expandagon Construction System (Hoberman)
- Fiddlestix Gearworks (Toys-N-Things)
- Gears! Gears! Gears! (Learning Resources)
- PowerRings (Fun Source)
- Zome System (Zome System)
- Construx (no longer made, but sets may still be available for sale)

Perhaps the most frequently imitated construction set has been the Meccano/Erector set line. Try finding these imitators, either in new, used, or thrift stores:

- Exacto
- Mek-Struct
- Steel Tec

11.2 Specialty Toys for Robot Hacking

Some toys and kits are just made for hacking (retrofitting, remodeling) into robots. Some are already robots, but you may design them to be controlled manually instead of interfacing to control electronics. The following sections describe some specialty toys you may wish to experiment with.

11.2.1 ROBOSAPIEN

Robosapien is a humanoid robot created by Mark Tilden, the inventor of BEAM robotics. The robot itself is controlled by an infrared remote control and can either perform individual actions (move a leg or an arm, grunt, or grasp) or they can be programmed into a sequence of instructions that the robot will follow. Along with this rudimentary sequenced programming ability, the Robosapien also has a number of sensors (including collision sensors in its feet) that can be used to cause the robot to respond to its environment (again in a very simple manner).

Since its introduction in late 2003, the Robosapien has been the target of many different hacks, including coming up with ways to create sequences on a computer rather than through the remote control and replacing the central processor in the robot with either a programmable microcontroller or even a PDA, which provides a much higher level of control. These hacks are well documented on a variety of different web sites, and there are a number of books detailing how the Robosapien is constructed and what can be done to modify it.

11.2.2 TAMIYA

Tamiya is a manufacturer of a wide range of radio-controlled models. It also sells a small selection of gearboxes in kit form that you can use for your robot creations. One of the most useful is a dual-gear motor, which consists of two small motors and independent drive trains. You can connect the long output shafts to wheels, legs, or tracks.

Tamiya also sells a dual-motor, tracked tractor-shovel kit (see Fig. 11-5) that you control via a switch panel. You can readily substitute the switch panel with computerized control circuitry that will provide for full forward and backward movement of the tank treads as well as the up and down movement of the shovel.

11.2.3 OWIKITS AND MOVITS

The OWIKITS and MOVITS robots are precision-made miniature robots in kit form. A variety of models are available, including manual (switch) and microprocessor-based versions. The robots can be used as-is, or they can be modified for use with your own electronic systems. For example, the OWIKIT Robot Arm Trainer (model OWI007) is normally operated by pressing switches on a wired control pad. With just a bit of work, you can connect the wires from the arm to a computer interface (with relays, for example) and operate the arm via software control. For the most part, the kits can be fairly easily hacked but you will have to design your own motor drivers (which is an important point as will be discussed later in the chapter).

The kit comes with an interesting dual motor that operates the left and right treads. Most of the OWIKITS and MOVITS robots come with preassembled circuit boards; you are expected to assemble the mechanical parts. Some of the robots use extremely small parts and require a keen eye and steady hand. The kits are available in three skill levels: beginner, intermediate, and advanced. If you're just starting out, try one or two kits in the beginner level.

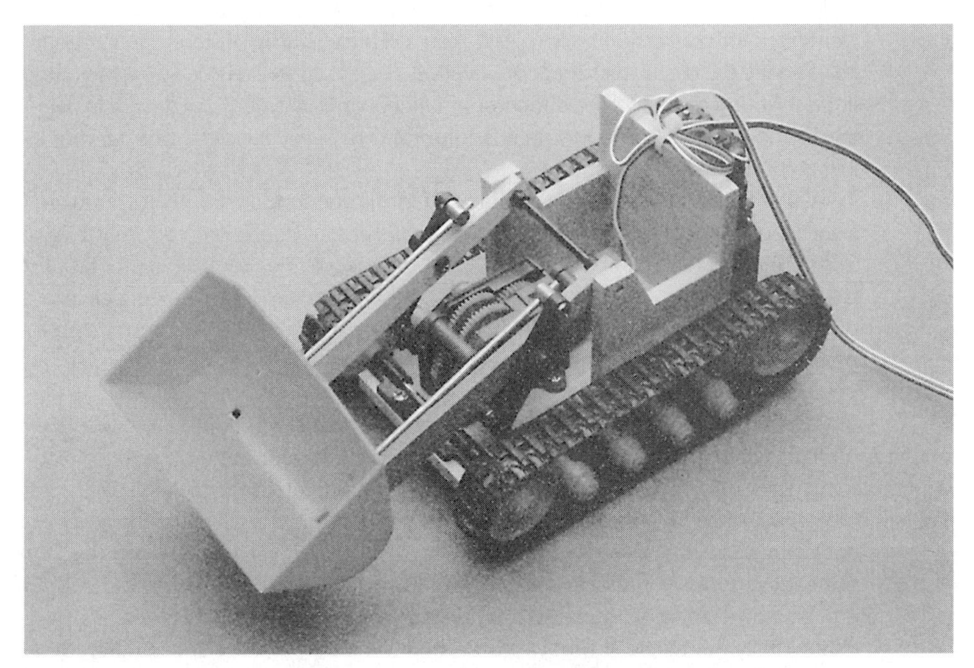

FIGURE 11-5 The Tamiya Bulldozer kit can be used as a lightweight robot platform.

11.2.4 ROKENBOK

Rokenbok toys are radio-controlled construction vehicles that take the form of both wheeled and tracked versions. Despite the European-sounding name, Rokenbok toys are made in the United States and have recently become available through mass-market retailers (the beginner's kit costs around $80). Each vehicle is controlled by a game controller; all the game controllers are connected to a centralized radio transmitter station.

The Sony PlayStation-style controllers can be hacked and connected directly to the input/output (I/O) of a computer or microcontroller (similar to what is shown in Chapter 16). The transmitter station also has a connector for computer control, but as of this writing Rokenbok has not released the specifications or an interface for this port.

11.3 Robots from Converted Vehicles

Motorized toy cars, trucks, and tractors can make ideal robot platforms—with an afternoon's or evening's work they can be modified to run under switch remote control or computer control, as will be shown in this section.

The least expensive radio-controlled cars have a single drive motor and a separate steering servo or solenoid mechanism. Despite what many people think, they can be fairly easily modified into robots although the car-type steering gives them less agility (i.e., the ability to literally turn on a dime) than a tracked vehicle. When hacking these vehicles, care must be taken to ensure you understand how steering is accomplished and you will have to go through the circuitry to ensure that your computer control can drive it.

Most radio- and wire-controlled tractor vehicles or simple toys with differential drives are well suited for conversion into a robot. You will have to strip off the vehicle's body and probably modify the electronics that come with the robot as well as add some of your own. Turning a robot by changing the direction of wheels or tracks does produce a lot of drag; when modifying these toys into robots it is important to keep the weight of the modifications as low as possible.

When considering a vehicle for hacking into a robot, look at how it is powered and consider whether or not it is appropriate—small remote controlled cars that run off a single 1.2 V NiMH battery can be modified into a microcontroller controlled robot, but some work will be required to add a boost power supply for the microcontroller as well as a fairly comprehensive review of the drive electronics to ensure that the correct current is passed to the motor drivers. When you are doing this for the first time, it is easier to select a toy that is powered by three or four AA cells.

Another option is to use two small motorized vehicles (mini four-wheel-drive trucks are perfect), remove the wheels on opposite sides, and mount them on a robot platform. Your robot uses the remaining wheels for traction. Each of the vehicles is driven by a single motor, but since you have two vehicles (see Fig. 11-6), you still gain independent control of both wheel sides. The trick is to make sure that, whatever vehicles you use, they are the same exact type. Variations in design (motor, wheel, etc.) will cause your robot to crab to one side as it attempts to travel a straight line. The reason: the motor in one vehicle will undoubtedly run a little slower or faster than the other, and the speed differential will cause your robot to veer off course.

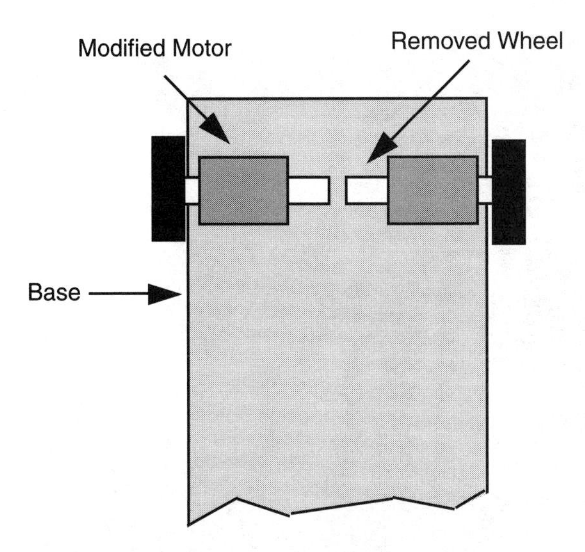

FIGURE 11-6 You can build a motorized robot platform by cannibalizing two small motorized toys and using each "half" of them.

11.3.1 HACKING A TOY INTO A ROBOT

The toy chosen for hacking into a robot was the differentially driven, radio remote control robot shown in Fig. 11-7. The toy was purchased as part of a set of two at a local hardware store for less than $10. This may seem like an incredibly lucky find (a remote control vehicle for only $5), but by keeping your eyes open you will discover that there are many end-of-life toys that can be modified into a robot quite easily.

The body of the toy car was held on by four bolts and was easily removed (Fig. 11-8). The screws holding down the PCB were removed, giving access to the circuitry side of the PCB (Fig. 11-9) as well as the motors, allowing the the snubber circuit built into them to be observed.

When hacking a remote control toy, you are well advised to use as much of the existing circuitry as possible. At one end of the PCB, you will see a number of transistors (at the top of Fig. 11-9) that are wired to the drive motors. These transistors and the resistors associated with them should be used as the motor drivers for your robot. What you should look for are four resistors connected between the transistors and the radio receiver.

You could trace out the circuit used for the motor drivers and compare it to the ones shown in Chapter 20, but it is much easier to find the four resistors leading between the two parts of the PCB and cut the PCB at this point as shown in Fig. 11-10. Before cutting, if you are unsure about the purpose of these transistors, desolder one of the resistors at the receiver side and connect it to the positive power of the toy; one set of wheels should start turning. If it doesn't, then you have either not selected the correct resistors or power is not getting to all the necessary parts of the circuit for the wheels to turn.

Once the PCB is cut, you can attach longer wires to the resistors (which will be used by the robot controller) and put heat shrink tubing over the joints. At this time, the toy's power wiring and the four motor driver leads can be wired to a connector that will be attached to the controller board (Fig. 11-11). This may seem premature, but you have finished the electrical modifications to the toy to turn it into a robot!

FIGURE 11-7 This remote control robot was purchased as a set of two for less than $10. The remote controls work at 27 and 49 MHz and the toys are powered by three AA alkaline batteries.

FIGURE 11-8 The body removed from the remote control car. Note that the component side of the PCB is facing downwards into the vehicle's plastic chassis.

FIGURE 11-9 The remote control receiver PCB with motor drive transistors at the top.

Now, control electronics can be added to the toy to turn it into a robot. For the example toy in this section, a BASIC Stamp 2 (BS2) was added using the circuit shown in Fig. 11-12. The BS2 has a communications/programming interface built in and four I/O pins are used to control the operation of the toy's motors. This circuit was built onto a prototyping PCB along with a mating connector for the toy's motor drivers. Note that the BS2 is powered from the toy's three AA batteries. Don't worry if you don't understand how the BASIC Stamp 2 works and how it can control the robot's motors; the operation of the microcontroller and how it can be used as a robot controller is described in later chapters.

Once the BS2 prototype PCB circuit was assembled, it was mounted to the robot by cutting down some pieces of phenolic PCB material, mounting 6-32 screws on them and epoxying them down to the robot/toy's plastic chassis (Fig. 11-13). The BS2 prototype PCB was screwed down to the chassis using 6-32 acorn nuts, which, along with the six-pin connector to the chassis power supply and motors, allows the PCB to be removed for repair or modification.

The completed robot (Fig. 11-14) is ready for action!

To test the operation of the robot, the following program was used:

```
'  BS2 Robot Move
'
'  This Program Moves the Toy Based Robot Randomly
'
```

FIGURE 11-10 The PCB is cut underneath the current limiting resistors, which pass the receiver's control signals to the toy's motor drivers.

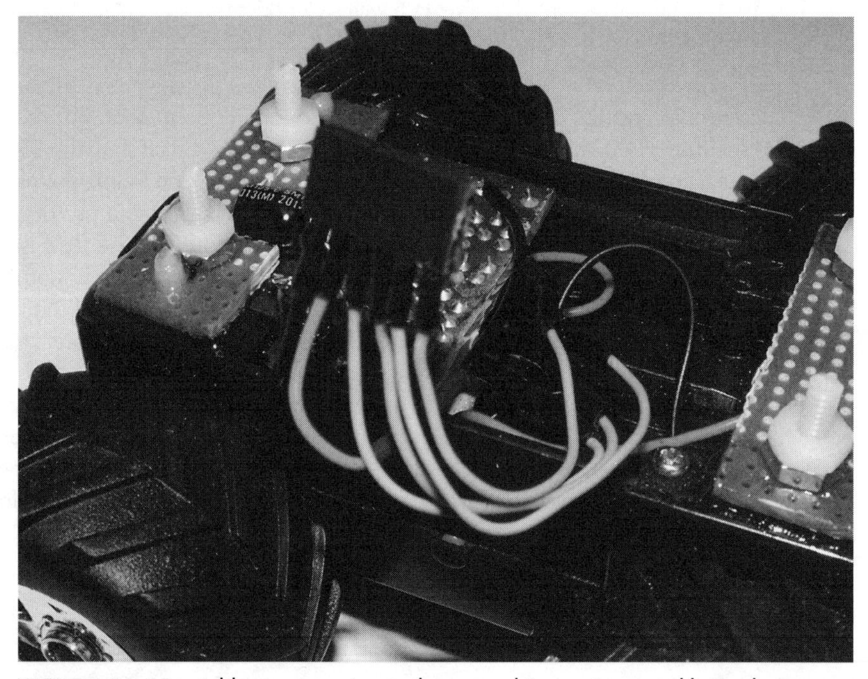

FIGURE 11-11 Add extension wires to the motor driver resistors and bring them out to a connector.

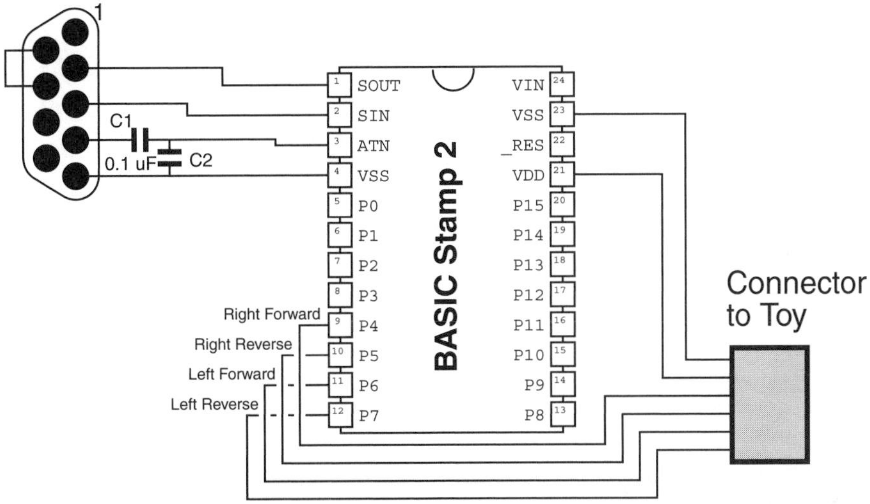

FIGURE 11-12 The BASIC Stamp 2 control circuit for providing computer control of the hacked toy. More information about the BS2 and its operation can be found in Chapter 15.

FIGURE 11-13 Pieces of phenolic PCB material were cut down, drilled for 6-32 nylon bolts, and epoxyed into the toy's plastic chassis to provide mounting for the BS2 PCB.

```
'  myke predko
'
'  05.09.19
'
'{&STAMP BS2}
'{&PBASIC 2.5}

'  I/O Ports
RightForward PIN 4
RightReverse PIN 5
LeftForward  PIN 6
LeftReverse  PIN 7
'  Variables
i VAR Byte

'  Mainline

    LOW RightForward                 '  Make all the Motor Drives Output
    LOW RightReverse
    LOW LeftForward
    LOW LeftReverse

    DO                               '  Loop Forever
```

FIGURE 11-14 Ready to roll! The modifications to the toy are complete and the BS2 has been added and is ready for programming.

```
        HIGH RightForward              ' Move Forward for 1.5 Seconds
        HIGH LeftForward
        PAUSE 1500

        LOW LeftForward                ' Turn Left for 0.75 Seconds
        PAUSE 750

      LOOP
```

This program simply moves the robot forward for a second and a half and then turns left before repeating. The BS2 PCB and mounting hardware weigh quite a bit less than the original toy's chassis, and the toy, which was already quite fast, is extremely fast with the lighter modifications in place. To be able to effectively use the robot, a PWM will have to be put in place to throttle down the motors to get a more realistic and controllable operating speed for the robot.

It should be noted that you will find figuring out the wiring of a hacked toy will rarely be as straightforward as the prototype shown here. Often, toys will use surface mount technology (SMT) components, which are much smaller and harder to trace than the old-fashioned pin through hole (PTH) components found in this toy. Remember that the radio circuitry is always separate from the motor driver circuitry on the PCB and the regular nature of the motor driver circuitry (consisting of multiple transistors and resistors) makes it quite easy to identify. Once you have identified the motor driver circuitry, you should be able to find the current limiting resistors for the motor driver transistors; from here it should be quite easy to work through the robot modification.

11.4 From Here

To learn more about . . .	***Read***
Brains you can add to robots made from toys	Chapter 12, "An Overview of Robot 'Brains' "
Using the BS2 Microcontroller	Chapter 15, "The BASIC Stamp 2 Microcontroller"
Overview of DC Motors	Chapter 20, "Working with DC Motors"

PART 3

COMPUTERS AND ELECTRONIC CONTROL

CHAPTER 12

AN OVERVIEW OF ROBOT "BRAINS"

"Brain, brain, what is brain?" If you're a Trekker, you know this is a line from one of the original *Star Trek* episodes entitled "Spock's Brain." The quality of the story notwithstanding (it is universally regarded as one of the worst, yet paradoxically one of the most popular), the episode was about how Spock's brain was surgically removed by a race of temporarily hyperintelligent women who needed it to run their underground environmental control system. Dr. McCoy was able to create a control mechanism that would allow somebody to operate Spock's brainless body in order for it to be present when the brain was found. Without its brain, Spock's body was not much more than a remotely controlled model car; capable of performing some operations under direct human control, but not able to operate autonomously. The brains of a person or robot process information from the environment; then based on the programming or logic they determine the proper course of action. Without a brain of some type and the ability to respond to different environmental information, a robot is really nothing more than just a motorized toy.

A computer of one type or another is the most common brain found in a robot. A robot control computer is seldom like a PC on your desk, though robots can certainly be operated by most any personal computer. And of course not all robot brains are computerized. A simple assortment of electronic components—a few transistors, resistors, and capacitors—is all that is really needed to make a rather intelligent robot. Endowing your robot with electronic smarts is a huge topic, so additional material is provided in the following chapters to help you understand how electronic sensors and actuators are interfaced to computers and how decisions are made on which actions to take.

12.1 Brains from Discrete Components

You can use the wiring from discrete components (transistors, resistors, etc.) to control a robot. This book contains numerous examples of this type of brain, such as the line-tracing robot circuits in Chapter 33, "Navigation." The line-tracing functionality is provided by just a few common integrated logic circuits and a small assortment of transistors and resistors. Light falling on either or both of two photodetectors causes motor relays to turn on or off. The light is reflected from a piece of tape placed on the ground.

Fig. 12-1 shows another common form of robot brain made from discrete component parts. This brain makes the robot reverse direction when it sees a bright light. The circuit is simple, as is the functionality of the robot: light shining on the photodetector turns on a relay. Variations of this circuit could make the robot stop when it sees a bright light. By using two sensors, each connected to separate motors (much like the line-tracers of Chapter 33), you could make the robot follow a bright light source as it moves. By simply reversing the sensor connections to the motors, you can make the robot behave in the opposite manner as shown in Fig. 12-1, such as steering away from the light source, instead of driving toward it. See Fig. 12-2 for an example.

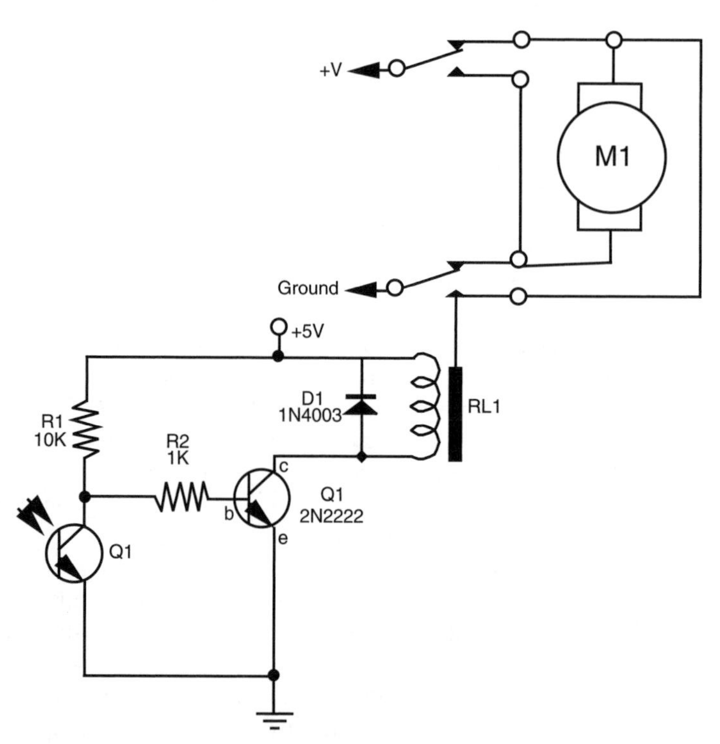

FIGURE 12-1 Only a few electronic components are needed to control a robot using the stimulus of a sensor.

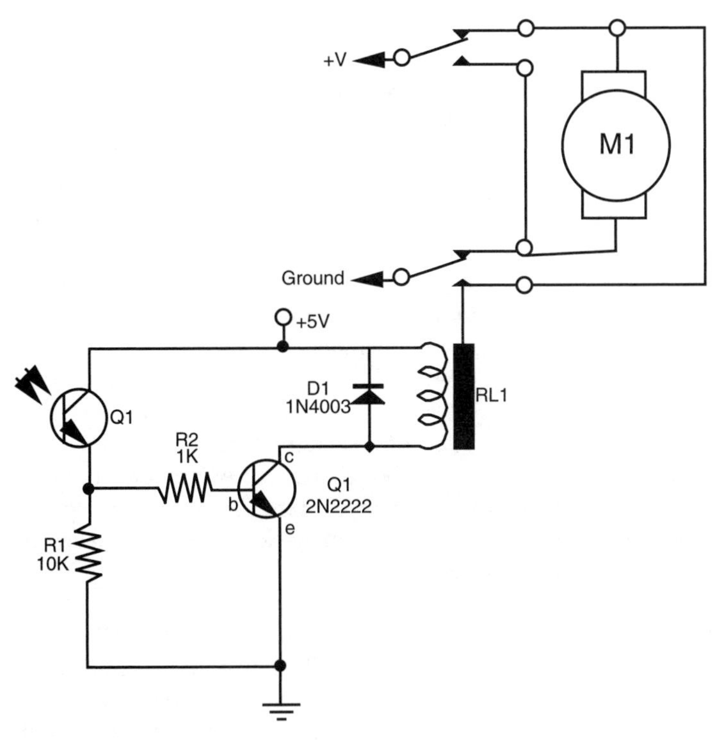

FIGURE 12-2 By connecting the sensors and control electronics differently, a robot can be made to "behave" in different ways.

You could add additional simple circuitry to extend the functionality of robots that use discrete components for brains. For instance, you could use a 555 timer as a time delay: trigger the timer and it runs for 5 or 6 s, then stops. You could wire the 555 to a relay so it applies juice only for a specific amount of time. In a two-motor robot, using two 555 timers with different time delays could make the thing steer around walls and other obstacles.

12.1.1 BEAM TECHNOLOGY

Over the past 15 years, robotics expert Mark Tilden has been designing and testing a class of unique robots that fall into their own classification, which he calls BEAM. These small robots, which can generally be built in an afternoon using mostly reclaimed parts from various electronics devices found around the house, are an excellent way to learn about different robot motors and actuators as well as to see electronic devices working in ways that their designers definitely never intended. Along with the learning aspects of BEAM technology and robots, there are a number of competitive events that have been designed around different basic BEAM operations that are fun and inexpensive to compete in.

BEAM is an acronym for:

- *Biology.* These robots often not only mimic the different structures and behaviors of living organisms, but with their simple construction BEAM robots are meant to evolve through humans continually improving them. It must also be pointed out that an important feature of BEAM technology is the use of renewable energy sources (primarily photocells) for power, which the robots feed from. This need for feeding is often part of the overall operation of the robots.
- *Electronics.* The control and motor drivers used in the robots. As previously noted, many example BEAM robots bend the operation of different components to perform actions that are suited to controlling the operation of the robots.
- *Aesthetics.* Unlike most robots, BEAM robots are designed and constructed with thought applied to what the robot will look like when it is completed. The desire is to have the function of the robot to be immediately obvious from its form.
- *Mechanics.* Instead of the traditional robot shapes, BEAM robots are very efficiently designed, both with an eye toward the aesthetics of the design and making the operation of the robot as efficient as possible by using simple electronic parts that do not have to be tailored to specific motors and sensors.

Normally when you are in a surplus store or pulling apart an old VCR, you are looking at the motors and drive electronics as parts that will be built into a much larger robot. A BEAM roboticist will be looking at the subassembly the motor is built into and how to use that as the basis for an entire robot. This difference in approach is what makes BEAM technology so special.

BEAM robots are a fascinating branch of robots, but the lack of computer control and their typically small size makes them quite difficult to adapt to more than one application or multiple environments. In the appendices, different BEAM resources are listed and it is recommended that you look through them for the technical information and to marvel at the elegance of their designs.

12.2 Brains from Computers and Microcontrollers

The biggest downside of making robot controls from discrete components is the work required to change the hardwired brains to change the behavior of the robot. You either need to change the wires around or add and remove components. Using an experimenter's breadboard makes it easier to try out different designs simply by plugging components and wiring into the board. But this soon becomes tiresome and can lead to errors because parts can work loose from the board.

You can rewire a robot controlled by a computer simply by changing the software running on the computer. For example, if your robot has two light sensors and two motors, you don't need to do much more than change a few lines of programming code to make the robot come toward a light source, rather than move away from it. No changes in hardware

are required. This type of programming functionality is demonstrated in Chapter 15 with a Parallax BASIC Stamp 2 and some of the example applications that it can run.

12.3 Types of Computers for Robots

An almost endless variety of computers can be used as a robot's brain. The most common types that are used are:

- *Microcontroller.* These are programmed either in assembly language or a high-level language such as BASIC or C. There are literally hundreds of different microcontrollers with a plethora of different interfacing capabilities that you can choose from to control your robot.
- *Personal Digital Assistant.* An old Palm Pilot provides a lot of processing power in a fairly small space with a number of features that make it very attractive for use as a robot controller. It can be difficult to interface with.
- *Single-board computer.* A few years ago, complete computer systems built on a PCB were the preferred method of controlling robots. These systems are still used but are much less popular due to the availability of low-cost PC motherboards and more powerful, easy to use microcontrollers.
- *Personal computer motherboards and laptops.* Very small form factor PC motherboards and laptops are common controllers for larger robots. These controllers can be programmed using standard development tools and commercial, digital I/O add-ons for the interfaces needed for the different robot functions.

12.3.1 MICROCONTROLLERS

Microcontrollers are the preferred method for endowing a robot with smarts. The reasons for this include their low costs, simple power requirements (usually 2.5 to 5 V), and ability of most to be programmed using software and a simple hardware interface on your PC. Once programmed, the microcontroller is disconnected from the PC and operates on its own. Microcontrollers can either be downloaded with a program that provides all the functions required of it or execute a *tokenized* program, which provides a set of basic functions for the application and the software already built into the microcontroller handling the various interfaces. Virtually all microcontrollers can be used as either device; the reason for choosing one over the other really comes down to cost, your experience and skills, available features, and ease of use.

You shouldn't be scared at the idea of having to come up with all the code that executes within a microcontroller. If you are familiar with a PC, you know that there are several megabytes of code for the BIOS (basic input/output system) as well as several hundred megabytes of code devoted to the operating system. Going by this standard, it will seem like developing a complete application for a microcontroller is a daunting task. In reality, the operation of the code within the microcontroller is quite simple, and other than requiring a few configuration commands, the software is quite straightforward. The advantage of using a microcontroller that requires a complete application is cost; a microcontroller suitable for use in robots can cost as little as $1.00. A potential drawback to this type of microcontroller is the cost of a programmer, which can be very substantial for some chips.

When source code is *tokenized,* it is passed through a compiler, just like regular application code, but instead of producing a series of instructions, the compiler produces a set of commands that are executed within the microcontroller. This type of microcontroller can have a series of very complex commands programmed into it, which makes them available to new application developers instead of having to puzzle out how to implement them. To further simplify the operation of this type of microcontroller, a bootloader program is typically already burned into them, allowing a simple programming operation that does not require any additional hardware. The Parallax BASIC Stamp 2 discussed later in this book is a bootloader-equipped microcontroller, which has a simple RS-232 programming (and console) interface.

Both kinds of microcontroller are fully programmable, but bootloader-equipped microcontrollers, like the BASIC Stamp 2, are programmed in a high-level language such as BASIC. Stand-alone microcontrollers can usually be programmed in a variety of different high-level languages (BASIC, Java, C) as well as assembly language, giving a lot more flexibility to the application developer.

Microcontrollers are available with 8, 16, or 32-bit processors. While PCs have long since graduated to 32-bit and higher architectures with protect mode and virtual memory operating systems, most applications for microcontrollers do not require more than eight bits.

The exact format and contents of an assembly-language microcontroller program vary between manufacturers. The popular PIC microcontrollers from Microchip follow one language convention. Microcontrollers from Intel, Atmel, Motorola, NEC, Texas Instruments, Philips, Hitachi, Holtek, and other companies all follow their own conventions. While the basic functionality of microcontrollers from these different companies is similar, learning to use each one involves a learning curve. As a result, robot developers tend to fixate on one brand, and even one model, since learning a new assembly language and processor architecture can require a lot of extra work.

Table 12-1 lists a number of different microcontrollers available on the market.

12.3.1.1 The Complete Computer System on a Chip A key benefit of microcontrollers is that they combine a microprocessor component with various inputs/outputs (I/O) that are typically needed to interface with the real world. For example, the 8051 controller sports the following features, many of which are fairly standard among microcontrollers:

- Central processing unit (CPU)
- CPU reset and clocking support circuitry
- Hardware interrupts
- Built-in timer or counter
- Programmable full-duplex serial port
- 32 I/O lines (four eight-bit ports) configurable as an eight-bit RAM and ROM/EPROM bus

Some microcontrollers will have greater or fewer I/O lines, and not all have hardware interrupt inputs. Some will have special-purpose I/O for such things as voltage comparison or analog-to-digital conversion. Just as there is no one car that's perfect for everyone, each microcontroller's design will make it more suitable for certain applications than for others.

TABLE 12-1 Different Microcontrollers

MICROCONTROLLER NAME	MANUFACTURER	COMMENTS
PIC	Microchip	Many different part numbers that are very popular with roboticists. The PIC16F84 has been a traditional favorite, but new microcontrollers are better suited to robot applications. Lots of free development software and example applications available. The BASIC Stamp 2 uses a PIC MCU as its preprogrammed controller.
68HC11	Freescale	Historically a very popular microcontroller for use in robots. Can be somewhat difficult and expensive to work with due to lack of tools available on the Internet. Note: freescale products used to be known as Motorola.
8051	Intel (originally)	Many different varieties and part numbers available from a plethora of manufacturers. The original parts are very simple but different manufacturers have MCUs with sophisticated interfaces. Lots of tools and example applications available on the Internet.
AVR	Atmel	Increasingly popular part for robot applications. Good free tools available for standard and MegaAVR parts.
H8	Hitachi	Used in many commercial robots. Tools available on the Internet but fewer example applications than available for other devices.
8018x	Intel	Microcontroller version of the 8088 used in the original IBM PC. Can be somewhat difficult to find today and requires a BIOS chip for proper operation making it the most difficult to work with of the devices listed here.

It must be pointed out that one of the important features of the microcontroller is the built-in hardware providing reset control and processor clock support. Many modern microcontrollers only require power supplied to them along with a decoupling capacitor and nothing more to run. This makes them an attractive alternative to traditional TTL or CMOS logic chips.

12.3.1.2 Program and Data Storage The typical low-cost microcontroller will only have a few thousand bytes of program storage, which will seem somewhat confining (especially when your PC probably has 512MB or more of main board memory and 40GB or more of disk storage). Despite seeming very small, this amount of memory is usually more than adequate for loading in a robot application; many initial robot programs do not require more than a few dozen lines of program code. If a human-readable display is used, it's typically limited to a small 2-by-16 character LCD, not entire screens of color graphics and text. By using external addressing, advanced microcontrollers may handle more storage:

megabytes of data are not uncommon. For most robot programs, only a few hundred bytes of storage will be required in a microcontroller.

Some microcontrollers—and computers for that matter—stuff programs and data into one lump area and have a single data bus for fetching both program instructions and data. These are said to use the Princeton, or more commonly Von Neumann, architecture. This is the architecture common to the IBM PC compatible and many desktop computers, but is only found in older microcontroller architectures. Rather, most modern microcontrollers use the Harvard architecture, where programs are stored in one place and data in another. Two buses are used: one for program instructions and one for data.

The difference is not trivial. A microprocessor using the Harvard architecture can run faster because it can fetch the next instructions while accessing data. When using the Von Neumann architecture, the processor must constantly switch between going to a data location and a program location on the same bus. The Von Neumann architecture is superior for microcontrollers used in a bootloader configuration and in applications that use real-time operating systems—applications that are beyond the scope of this book.

Because of the clear delineation in program and data space in the Harvard architecture, such microcontrollers have two separate memory areas: EEPROM (electrically erasable programmable read-only memory) for program space and RAM (random access memory) for holding data used while the program runs. A version of EEPROM used in many microcontrollers is known as Flash. You will often see two data storage specifications for microcontrollers.

There are some differences in how the two processor architectures are programmed but learning between the two is not terribly difficult; with the use of high-level languages, the differences become transparent and should not be used to select a microcontroller for a robot. Of much greater importance is the availability of the device itself, the resource materials, development tools, and example applications as well as the ease in which programs can be loaded into the chip.

12.3.1.3 Chip Programming Microcontrollers used in robots are meant to be programmed and erased many times over. A few years ago, finding parts with EEPROM or Flash program memory was not an easy task, and a few microcontrollers (such as the PIC16F84) became very popular devices for robot developers because they could be reprogrammed easily without undergoing any special steps. Older PROM and EEPROM-based microcontrollers were more difficult to work with and required additional tools to erase them before a new program could be burned into them. With the wide availability of EEPROM- and Flash-based microcontrollers, the focus now turns to the cost and overhead of programming applications into microcontrollers.

All microcontroller manufacturers have programmers available for their microcontroller chips and there are third-party tools that can program a wide variety of different chips. The cost for these programmers ranges from $25 to several thousand dollars. Another option for programming microcontroller chips is to build your own programmer, which can cost as little as $2 to $3, depending on the state of your parts drawer.

Another consideration is application debugging. Many microcontrollers are becoming available on the market with built-in debugging features, eliminating the need for an in-circuit emulator and allowing you to debug your applications for just a few hundred dollars. To perform this debugging, several I/O pins of the microcontroller must be dedicated to the debugger hardware connection and cannot be used for application input and output pins. In theory, the ability to debug the program running in a robot seems like an outstanding idea, but in practice, it can be very difficult to implement on a mobile robot.

It may sound ironic, but you should seriously consider a microcontroller that has a debugger interface built in as this could be used as a programming interface, eliminating the need to pull out the chip every time the program is updated. By being able to plug a cable into the robot for programming, the time required to update the program in the microcontroller as well as the opportunity for damaging the microcontroller chip will be greatly reduced.

The ability and cost of a microcontroller programmer are probably the most important considerations that you should make when choosing a microcontroller to work with. While a chip may be ideally suited to the robot application you want to create, if you cannot get a reasonably priced programmer, then you will find that you are spending a disproportionate amount of money on something that does not affect the operation of the robot. A less capable microcontroller with a cheap and simple programming interface will be a lot more useful to you in the long run.

12.3.2 PERSONAL DIGITAL ASSISTANTS

Palm Pilots and other small personal digital assistants (PDAs) can be used as a small robot controller that combines many of the advantages of microcontrollers, larger single-board computers, and PC motherboards and laptops. The built-in power supply and graphic LCD display (with Graffiti stylus input) are further advantages, eliminating the need for supplying power to the PDA and providing a method of entering in parameter data or even modifying the application code without a separate computer. The most significant issue that you will encounter using a PDA as a robot controller is deciding how to interface it to the robot's electronics. PDAs are becoming increasingly popular as robot controllers and there are a variety of products and resources that will make the effort easier for you.

There are two primary types of PDAs to consider based on the operating system used. The Palm O/S was the original, and there are a number of programming languages and native programming environments to consider. One notable product that is very popular with robot developers is the HotPaw (www.hotpaw.com/hotpaw) BASIC programming environment lets you develop, compile, and run BASIC applications on your PDA with a less than $20 license fee. Windows CE is the other type of operating system used by PDAs and offers essentially the same features as the Palm-based devices. At the time of this writing, Microsoft offers a number of eMbedded Visual Tools from their Dveloper's web site (www.microsoft.com) for Windows CE devices including eMbedded Visual BASIC, which can be written and tested on a PC before being downloaded into the PDA.

Once you have decided on the device and the programming environment, you have to decide what is going to be the methodology for connecting the PDA into the robot. You have two choices: the serial port that is built into the robot or the IRDA infrared data port. The serial port is similar to a PC's RS-232 port, but typically only implements two handshaking lines and generally does not produce valid RS-232 voltage levels—neither of which is a significant issue, but things that you should be aware of. The IRDA port is a fairly high speed (from 9.600 to 115.2 kbps) infrared serial port that implements a reasonably complex data protocol. There are a number of IR transceivers and stack protocol chips (such as the Microchip MPC2150) that you can use to implement the interface—the total cost for the stack protocol chip, IR transceiver, and miscellaneous electronics will cost you about $20). It must be pointed out that the IRDA interface will only communicate reliably if the distance between the PDA and other devices is less than 3 ft (1 m).

In either case, an intelligent device will have to be connected to the serial connections and used to interface with the robot's motor drivers and sensors. The advantage of using the serial port is lower parts cost while the IRDA port avoids a direct electrical connection between the robot and the PDA, which eliminates any potential upsets of the PDA due to electrical noise coming from the robot. You can use a prepackaged serial interface chip or you can program a small microcontroller, like a BS2, to provide the serial interface to the robot peripherals.

12.3.3 SINGLE-BOARD COMPUTERS

If you look through older robotics books and magazines, you will see a number of robots that are controlled by single-board computers or SBCs. Like microcontrollers, an SBC can be programmed in either assembly language or in a high-level language such as BASIC or C and contain the processor and memory but also the I/O interfaces necessary to control a robot. SBCs avoid the programming issues of microcontrollers due to the built-in RS-232 or Ethernet interfaces, which allow simple application transfers. Older SBCs used with robots were generally based on the Motorola 68HC11 microcontroller because of the large amount of software and tools available for them.

Another type of SBC that has been popular with roboticists in the past is the PC/104 form factor in which a processor card could be plugged directly into stacking memory, I/O, or video output cards. Despite the apparent bias in its name, many processor cards are available for the PC/104 form factor, providing the ability to use a PC processor, a Motorola 68000, MIPs, or Sparc processor with standard hardware. Applications for PC/104 systems were usually downloaded into on-board EPROM or Flash memory chips. This approach offered tremendous flexibility although at a fairly high cost.

Most modern single-board computers are full PCs with complete systems built on a small circuit board. A very popular form factor is the mini-ITX, which provides a complete PC in a 6.7 in square. These systems can support many hundreds of megabytes of memory and can run Microsoft Windows or Linux operating systems, allowing standard PC development tools to be used for program development.

There are three downsides to using SBCs in robots, especially if you are starting out and are designing your own robot:

- Power required by SBCs is usually more than a single supply and they do not have on-board voltage regulators. A Mini-ITX motherboard requires, +3.3, +5, +12, –5 and –12 V provided to it for proper operation. High-capacity batteries and DC–DC converters can provide appropriate voltages, but their weight and size results in the SBC only being used in large robots.
- Installing an operating system and running an application can be very challenging. Disk drives are very unreliable in a robot environment where the robot vibrates as it moves, and also require a great deal of power. A USB Flash memory thumb drive could be used, but this device would have to be loaded with an image of the operating system along with

application code. Versions of Linux can be found on the Internet that are suitable for loading in and booting from a thumb drive, but this is not a trivial exercise.

- Adding I/O ports would most likely be through the SBC's USB ports. There are a number of suppliers that have digital I/O as well as different advanced function I/O cards available that connect to a PC via USB. Depending on the interface cards, the software interfaces to the I/O functions can be quick, sophisticated, and difficult to program.

Despite the drawbacks, the use of single-board computers, especially those that are PC based, offer some intriguing possibilities. As previously discussed, it would be nice to be able to debug a robot application while it is working in a mobile robot. This is extremely difficult with microcontrollers, but it could be quite easy to do in an SBC-controlled robot; if it were equipped with an 802.11 (or any other wireless) network card, the SBC could run the debugger and make its display available to other PCs on the wireless network via X-Windows. Another possible application would be to use the network interface to stream video from the robot to a base station that was controlling the robot. The code required for these applications is readily available over the Internet and could be added to a robot with very little extra work.

12.3.3.1 Single-Board Computer Kits To handle different kinds of jobs, SBCs are available in larger or smaller sizes than the 4-by-4-in PC/104. And while most SBCs are available in ready-made form, they are also popular as kits. For example, the BotBoard series of single-board computers, designed by robot enthusiast Marvin Green, combine a Motorola 68HC11 microcontroller with outboard interfacing electronics. The Miniboard and HandyBoard, designed by instructors at MIT, are other single-board computers based on the HC11; both are provided in kit and ready-made form from various sources.

12.3.4 PERSONAL COMPUTERS

Having your personal computer control your robot is a good use of available resources because you already have the computer to do the job. Of course it probably also means that your automaton is constantly tethered to your PC, with either a wire or a radio frequency or infrared link.

Just because the average PC is deskbound doesn't mean you can't mount it on your robot and use it in a portable environment. Whether you'd want to is another matter. Certain PCs are more suited for conversion to mobile robot use than others. Here are the qualities to look for if you plan on using your PC as the brains in an untethered robot:

- *Small size.* In this case, small means that the computer can fit in or on your robot. A computer small enough for one robot may be a King Kong to another. Generally speaking, however, a computer larger than about 12 by 12 in is too big for any reasonably sized 'bot.
- *Standard power supply requirements.* A PC will require +5, +12, –5 and –12 V, and many PCs will also require +3.3 V. Depending on the motherboard that you use, you may discover that the +3.3, +5, or +12 V supply is the one that requires the most current as it is used to power most of the components on the PC. A few will function if the

–12 and –5 voltages are absent. It may take a lot of work to figure out which power supplies are required and how much current must be available to each one.

- *Accessibility to the microprocessor system bus or an input/output port.* The computer won't do you much good if you can't access the data, address, and control lines. The IBM PC architecture provides for ready expansion using daughter cards that connect to the motherboard. Standard motherboards also support a variety of standard I/O ports, including parallel (printer), serial, and Universal Serial Bus (USB). There are a number of I/O cards designed for the USB so this may be the most promising avenue.
- *Programmability.* You must be able to program the computer using either assembly language or a higher-level language such as BASIC, C, or Java. The open source community has a number of open source compiler projects available for you to choose from.
- *Mass storage capability.* You need a way to store the programs you write for your robot, or every time the power is removed from the computer you'll have to rekey the program back in. Floppy disks or small, low-power hard disk drives are possible contenders here although the low-cost USB Flash thumb drives are probably your best option. Being solid state, they will be much more reliable in the vibration-filled environment of a robot and they use very little power. Even modest sized (64 MB and less) thumb drives can be loaded with an operating system and application code, eliminating the need for operating systems all together. If these drives are to be used, you will have to make sure that your motherboard can boot from the thumb drive.
- *Availability of technical information.* You can't tinker with a computer unless you have a full technical reference manual. While manufacturers no longer publish these manuals, there are a number of references available for PCs that you can choose from and many of these will discuss alternative applications for PCs (such as building them into robots).

The PC may seem an unlikely computer for robot control, but it offers many worthwhile advantages: expansion slots, large software base, and readily available technical information. Another advantage is that these machines are plentiful on the used market—$20 at some thrift stores. As software for PCs has become more and more sophisticated, older models have to be junked to make room for faster processors and larger memories.

You don't want to put the entire PC on your mobile robot; it would be too heavy. Instead, remove the motherboard from inside the PC, and install that on your 'bot. How successful you are doing this will depend on the design of the motherboard you are using. The supply requirements of older PC-compatible motherboards are rather hefty: you need one or more large batteries to provide power and tight voltage regulation.

Later models of motherboards (those made after about 1990) used large-scale integration chips that dramatically cut down on the number of individual integrated circuits. This reduces the power consumption of the motherboard as well. Favor these "newer" motherboards (sometimes referred to as green motherboards, for their energy-saving qualities), as they will save you the pain and expense of providing extra battery power.

The keyboard is separate and connects to the motherboard by way of a small connector. The BIOS in some motherboards will allow you to run the PC without a keyboard detected. You will want this capability—without it, the motherboard won't boot the operating system, and you'll need to either keep a keyboard connected to the motherboard or rig up some kind of dummy keyboard adapter. The same goes for the video display. Make sure you can operate the motherboard without the display.

12.3.4.1 IBM PC-Compatible Laptops Motherboards from desktop IBM PC-compatible computers can be a pain to use because of the power supply requirements. A laptop on the other hand is an ideal robot-control computer. The built-in battery and display avoid the need for a number of custom power supplies and a display device for the robot. Check online auction sites such as Ebay, a used computer store, or your local classified ads for used units.

You should use the laptop as-is, without removing its parts and mounting them on the robot. That way, you can still access the keyboard and display. Use the parallel, serial, and USB ports on the laptop to connect to the robot. You should not use the laptop's batteries for powering the robot. While they are quite sophisticated and provide a great deal of power, they are designed for the load of the laptop, adding a significant load will cause them to run down faster and be potentially damaged by the extra drain of the robot's motors. You may find that it is impossible to run the robot from the laptop's batteries—protection and monitoring circuitry built into the laptop and the battery packs may detect the additional load and shut down the system fearing that the additional load is a short circuit or a problem with the battery pack itself.

12.3.4.2 Operating Systems If you make the decision to use a PC motherboard (or even a complete PC) in your robot, you will have to make the nontrivial decision on what type of operating system you are going to run and what features you are going to have on it. Fortunately, there are a lot of web sites, books, and other resources that will help you set up the operating system, support libraries, and application code, but you will have to spend some time thinking about how everything should work and what features you want to have available.

The first question you should ask yourself is, what features do you want to have available in the operating system running on your robot's PC? The obvious answer is as few as possible to minimize the amount of space that will be used for the operating system. The reason behind this answer is correct; but the decision point isn't—the operating system that is loaded onto the robot's PC should be as small as possible, but it should have as *many* features built into it as possible.

The reason for wanting as many features as possible is to give you as many options as possible for how the PC will run. While you should avoid running the PC motherboard from a hard disk or CD-ROM, you still want to have the drivers loaded and active in the system in case you want to use them later. Similarly, network interface card (NIC) drivers should be installed along with USB, serial, parallel, and other motherboard interfaces. You should not restrict any flexibility in the PC motherboard's operation until you have finished and perfected the robot (which will probably never happen).

The question you should now be asking is, what else is there to take out? If everything is being left in, there isn't that much space left to save.

Actually there is one big piece that you can take out and that is the graphical user interface—the Windows interface that you work with is at least half the total size of the operating system on disk and in memory. With it taken out, you should have an operating system that will only require 60 MB or so of storage—small enough to fit on a USB Flash thumb or key drive with the application code included. Loading applications or other files is as simple as plugging the USB thumb drive into your PC and copying the files into it, just as you would copy files between folders. If you were not to use a USB thumb drive, you would have to either disconnect the disk drive or eject the CD and update its contents—it's a bit more work but still very doable.

Storing the operating system and application on a USB thumb drive will eliminate the extra power required by a spinning hard disk or CD/DVD drive, as well as avoid the potential reliability issues of rotating devices running on an electrically noisy and vibrating robot. There are numerous web sites devoted to providing information required to set up a thumb drive as a bootable device along with installing an operating system onto it.

Once you have decided where to put the operating system and application code you can decide which operating system you are most comfortable with. All major operating systems can be used (Microsoft Windows (all flavors), MS-DOS, and Linux) after the graphical user interface has been removed from it. There are no significant reasons for choosing one operating system over another; the important points will be which ones you are most comfortable with and which ones have the software development tools that you have and are proficient working with.

Regardless of the operating system used, there are a variety of open source tools that you can download for application development. These can be found using a quick Google search or by looking at www.cygwin.com, which is an X-Windows-based environment for Microsoft Windows that will allow you to simulate the Linux environment and develop and debug applications for any of the three major operating systems. Even if you are unfamiliar with Linux or Unix operating environments and do not consider yourself very technical, it is surprisingly easy to download and work with these tools.

12.4 Inputs and Outputs

The architecture of robots requires inputs, for such things as mode setting or sensors, as well as outputs, for things like motor control or speech. The basic input and output of a computer or microcontroller is a two-state binary voltage level (off and on), usually 0 and 5 V. For example, to place an output of a computer or microcontroller to high, the voltage on that output is brought, under software control, to +5 V.

In addition to standard low/high inputs and outputs, there are several other forms of I/O found on single-board computers and microcontrollers. The more common are listed in the following sections, organized by type. Several of these are discussed in more detail later in the book.

12.4.1 SERIAL COMMUNICATIONS

The most common type of computer interfacing is known as serial, in which multiple bits are sent as a series of bits over time on a single wire. There are a number of different types of serial communications protocols as shown in Fig. 12-3, with asynchronous, synchronous, and Manchester encoding. Each methodology is optimized for different situations.

The two most likely communications methods that you will have to work with when interfacing a robot controller to I/O devices is synchronous serial communications, which consists of a data line and a clock line as shown in Fig. 12-4. There are several different protocols used for synchronous serial communications, but they all have one characteristic in

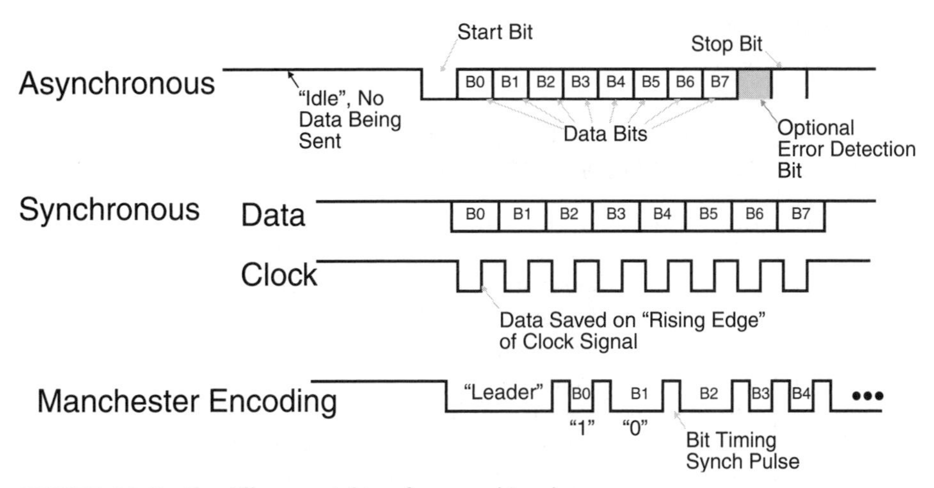

FIGURE 12-3 The different serial interfaces used in robots.

common: the bit on the data line is saved in the receiver when the clock transitions from high to low or low to high. The timing of the clock line does not have to be consistent and in some synchronous serial protocols, there can be multiple devices on the clock and data lines and part of the protocol is used to determine which device is active at any given time.

The three most common types of synchronous serial communication methodologies are:

- *I2C (inter-integrated circuit).* This is a two-wire serial network protocol created by Philips to allow integrated circuits to communicate with one another. With I2C you can install two or more microcontrollers in a robot and have them communicate with one another. One I2C-equipped microcontroller may be the master, while the others are slaves and used for special tasks, such as interrogating sensors or operating the motors.
- *Microwire.* This is a serial synchronous communications protocol developed by National Semiconductor products. Most Microwire-compatible components are used for interfacing with microcontroller or microprocessor support electronics, such as memory and analog-to-digital converters.
- *SPI (serial peripheral interface).* This is a standard used by Motorola and others to communicate between devices. Like Microwire, SPI is most often used to interface with

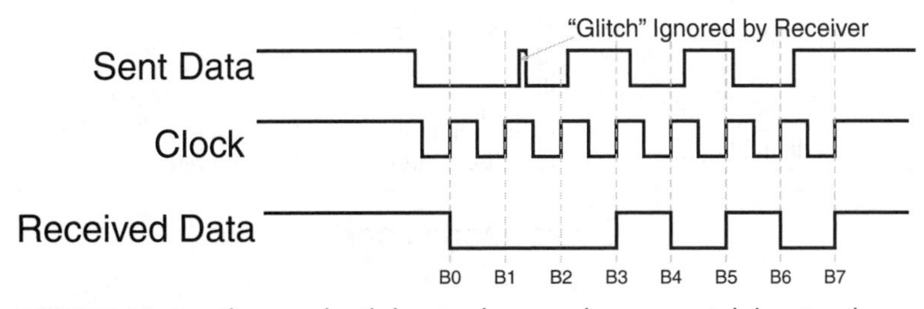

FIGURE 12-4 Close-up detail showing how synchronous serial data is only picked up on the clock edge.

microcontroller or microprocessor support electronics, especially outboard EEPROM memory.

12.4.2 ASYNCHRONOUS SERIAL COMMUNICATIONS (FROM SERIAL COMMUNICATIONS)

Asynchronous serial communications uses a single wire to send a packet that consists of a number of bits, each the same length. The most popular data protocol for asynchronous serial communications is known as non-return to zero (NRZ) and consists of the first bit, the Start Bit is low and is used by the receiver to identify the middle of each bit of the incoming data stream for the most accurate reading as shown in Fig. 12-5. There can be any number of data bits, but for most communications, eight bits, allowing the transmission of a byte, are used. The following data bits have the same period and are read as they are received. The stop bit is a high value (the non-return to zero that resets the data line to a high value so the next start bit will be detected by the receiver) that provides a set amount of time for the sender and receiver to prepare for the next data packet. An error detection bit can be placed at the end of the data bits, but this is rarely done in modern asynchronous serial communications.

The most popular form of asynchronous serial communications and one that you have probably heard of is commonly referred to as RS-232 (although more accurately known as EIA-232) and changes the normal TTL voltage levels of the serial data from 0 to +5 V to +12 (0) and −12 (1) V. There are a number of chips available that will make this voltage conversion simpler, most notably the Maxim MAX232, which generates the positive and negative voltages on the chip from the +5 V supply. Along with RS-232, RS-422, and RS-485

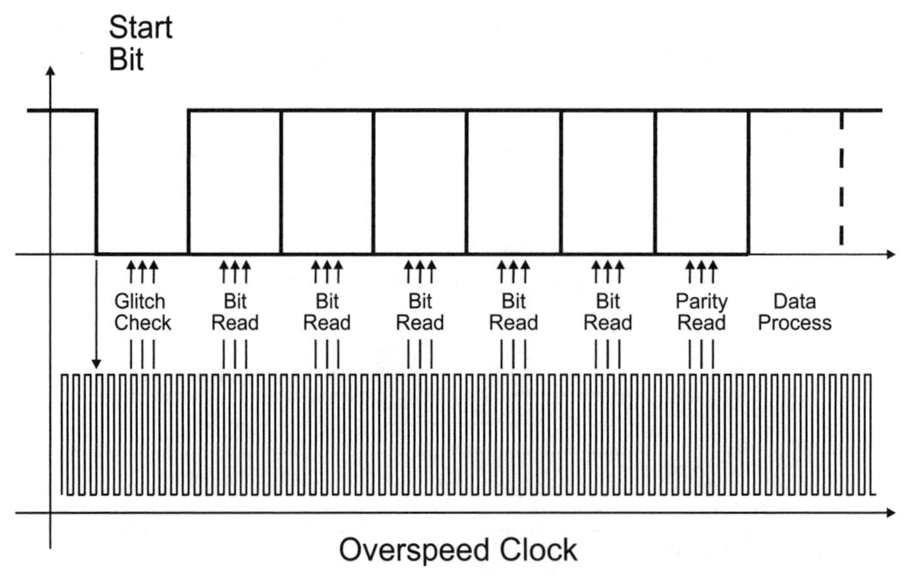

FIGURE 12-5 The asynchronous serial data stream consists of a start bit, which goes low, followed by a number of data bits. The receiver finds the center of the low start bit and then uses this point to read the values of subsequent bits.

are commonly used forms of asynchronous serial communications, and like RS-232, these standards can be implemented using commonly available chips.

Sending and receiving RS-232 data in a computer system may seem like a chore, but it is actually simplified by the universal asynchronous receiver/transmitter (known by its acronym UART) that will send and receive NRZ asynchronous data automatically with the computer system writing to it to start a data send or polling the UART to determine if the last written data byte has been sent or if data has been received. Along with the send and receive data bits, there are a number of other lines that can be used with RS-232 for handshaking (system-to-system communications to indicate that data can be sent or received) but these lines are largely ignored in most modern communications. The UART generally will provide an interface to these bits as well.

12.4.3 DIGITAL-TO-ANALOG CONVERSION

There are two principle types of data conversion:

- *Analog-to-digital conversion* (ADC) transforms analog (linear) voltage changes to binary (digital). ADCs can be outboard, contained in a single integrated circuit, or included as part of a microcontroller. Multiple inputs on an ADC chip allow a single IC to be used with several inputs (4-, 8-, and 16-input ADCs are common).
- *Digital-to-analog conversion* (DAC) transforms binary (digital) signals to analog (linear) voltage levels. DACs are not as commonly employed in robots; rather, they are commonly found on such devices as compact disc players.
- *Comparator.* This is an input that can compare a voltage level against a reference. The value of the input is then lower (0) or higher (1) than the reference. Comparators are most often used as simple analog-to-digital converters where high and low are represented by something other than the normal voltage levels (which can vary, depending on the kind of logic circuit used). For example, a comparator may trigger high at 2.7 v.

12.4.4 PULSE AND FREQUENCY MANAGEMENT

The three major types of pulse and frequency management are the following:

Input capture is an input to a timer that determines the frequency of an incoming digital signal. With this information, for example, a robot could differentiate between inputs, such as two different locator beacons in a room. Input capture is similar in concept to a tunable radio.

Pulse width modulation (PWM) is a digital output that has a square wave of varying duty cycle (e.g., the "on" time for the waveform is longer or shorter than the "off" time). PMW is often used with a simple resistor and capacitor to approximate digital-to-analog conversion, to create sound output, and to control the speed of a DC motor.

Pulse accumulator is an automatic counter that counts the number of pulses received on an input over X period of time. The pulse accumulator is part of the architecture of the microprocessor or microcontroller and can be programmed autonomously. That is, the accumulator can be collecting data even when the rest of the microprocessor or microcontroller is busy running some other program.

12.4.5 SPECIAL FUNCTIONS

There are a variety of other features available in different microcontrollers and computer systems. From a high level they will make your application much more elegant and probably simpler, but they can require specialized knowledge that will take you a while to develop.

- *Hardware interrupts.* Interrupts are special input that provides a means to get the attention of a microprocessor or microcontroller. When the interrupt is triggered, the microprocessor can temporarily suspend normal program execution and run a special subprogram.
- *External reset.* This is an input that resets the computer or microcontroller so it clears any data in RAM and restarts its program (the program stored in EEPROM or elsewhere is not erased).
- *Switch debouncer.* This cleans up the signal transition when a mechanical switch (push button, mercury, magnetic reed, etc.) opens or closes. Without a debouncer, the control electronics may see numerous signal transitions and could interpret each one as a separate switch state. With the debouncer, the control electronics sees just a single transition.
- *Input pull-up.* Pull-up resistors (5 to 10K) are required for many kinds of inputs to control electronics. If the source of the input is not actively generating a signal, the input could float and therefore confuse the robot's brain. These resistors, which can be built into a microcontroller and activated via software, prevent this floating from occurring.

12.5 From Here

To learn more about . . .	***Read***
Connecting computers and other circuits to the outside world	Chapter 14, "Computer Peripherals"
Using the BASIC Stamp microcontroller	Chapter 15, "The BASIC Stamp 2 Microcontroller"

CHAPTER 13

PROGRAMMING FUNDAMENTALS

If you were to watch old movies, you would get the idea that all you needed to build a robot were a couple of motors, a switch or relay, a battery, and some wire. In actuality, most robots, including the hobby variety, are equipped with a computational brain of one type or another that is told what to do through programming. The brain and programming are typically easier and less expensive to implement than are discrete circuitry, which is one reason why it's so popular to use a computer to power a robot.

The nature of the programming depends on what the robot does. If the robot is meant to play tennis, then its programming is designed to help it recognize tennis balls, move in all lateral directions, perform a classic backhand, and maybe jump over the net when it wins.

But no matter what the robot is supposed to do all of the robot's actions come down to a relatively small set of instructions in whatever programming language you are using. If you're new to programming or haven't practiced it in several years, read through this chapter on the basics of programming for controlling your robots. This chapter discusses rudimentary stuff so you can better understand the more technical material in the chapters that follow.

Even if you are familiar with programming, please take a few minutes and read through the chapter. Important concepts needed for programming robots are presented along with some of the fundamental program templates used for programming robots. If you were to skip this chapter, you may become confused at some of the code examples and programming discussions presented later in the book.

13.1 Important Programming Concepts

There are eight critical concepts to understanding programming, whether for robots or otherwise.

In this chapter, we'll talk about each of the following in greater detail:

- Linear program execution
- Variables and I/O ports
- Assignment statements
- Mathematical expressions
- Arrays and character strings
- Decision structures
- Macros, subroutines, and functions
- Console I/O

13.1.1 LINEAR PROGRAM EXECUTION

Modern computer systems, regardless of their sophistication, are really nothing more than electronic circuits that read instructions from their memory in order and execute them as they are received. This may fly in the face of your perception of how a computer program works; especially when you are familiar with working on a PC in which different dialog boxes can be brought up at different times and different buttons or controls can be accessed randomly and not in any predefined sequence. By saying that a computer program is a sequence of instructions that are read and executed may seem to belittle the amount of work that goes into them along with the sophistication of the operations they perform, but this is really all they are.

A sequence of instructions to latch data from a storage location into an external register using two I/O registers (one for data and one for the register clock line) could be:

```
Address   Instruction

   1      I/O Port 1 = All Output
   2      I/O Port 2 = 1 Clock Output Bit
   3      I/O Port 2 Clock Output Bit = Low
   4      Holding Register = Storage Location "A"
   5      I/O Port 1 = Holding Register
   6      I/O Port 2 Clock Output Bit = High
   7      I/O Port 2 Clock Output Bit = Low
```

In this sequence of instructions, the address each instruction is stored in has been included to show that each incrementing address holds the next instruction in sequence for the program.

The first computers could only execute the sequence of instructions as-is and not modify the execution in any way. Modern computers have been given the ability to change which section of instruction sequence is to be executed, either always or conditionally.

The following sequence of instructions will latch the values of one to five into the external register by conditionally changing the section of the instruction sequence to be executed, based on the current value.

```
Address   Instruction
   1      Holding Register = 1
   2      Storage Location "A" = Holding Register
   3      I/O Port 1 = All Output
   4      I/O Port 2 = 1 Clock Output Bit
   5      I/O Port 2 Clock Output Bit = Low
   6      Holding Register = Storage Location "A"
   7      I/O Port 1 = Holding Register
   8      I/O Port 2 Clock Output Bit = High
   9      I/O Port 2 Clock Output Bit = Low
  10      Holding Register = Storage Location "A"
  11      Holding Register = Holding Register + 1
  12      Storage Location "A" = Holding Register
  13      Holding Register = Holding Register - 5
  14      If Holding Register Does Not Equal 0, Next Execution Step is 6
```

In the second program, after the contents of 'Holding Register "A" ' have 1 added to them, the value has 5 subtracted from it and if the result is not equal to zero, execution continues at address 6 rather than ending after address 14. Note that data from a "Storage Location" cannot pass directly to an output port nor can mathematical operations be performed on it without storing the contents in the "Holding Register" first. In this simple program, you can see many of the programming concepts listed in operation, and while they will seem unfamiliar to you, the operation of the program should be reasonably easy to understand.

The sequential execution of different instructions is known as *linear program execution* and is the basis for program execution in computers. To simplify the creation of instruction sequences, you will be writing your software in what is known as a *high-level language* or, more colloquially, a *programming language*. The programming language is designed for you to write your software (or code) in a format similar to the English language, and a computer program known as a *compiler* will convert it into the sequence of instructions needed to carry out the program.

13.1.2 FLOWCHARTS

If you were given instruction in programming before reading this book, you may have been presented with the concept of *flowcharts*—diagrams that show the flow of the application code graphically as passing through different boxes that either perform some function or control the direction the path of execution takes. Fig. 13-1 shows an example flowchart for a robot program in which a robot will go forward until it encounters an object and then reverse and turn left for a half second. The advantage of using flowcharts in teaching programming is how well they describe simple applications graphically and show the operation of high-level programming concepts in a form that is easy to understand.

Flowcharts have four serious drawbacks that limit their usefulness. First, they do not work well for low-level programming concepts. A simple arithmetic statement that multiplies two values together and stores the product in different locations is written as:

```
A = B × C
```

In most programming languages this is easily understood by everyone by virtue that this is a simple equation taught in grade school. The flowcharting standard does not easily handle this operation.

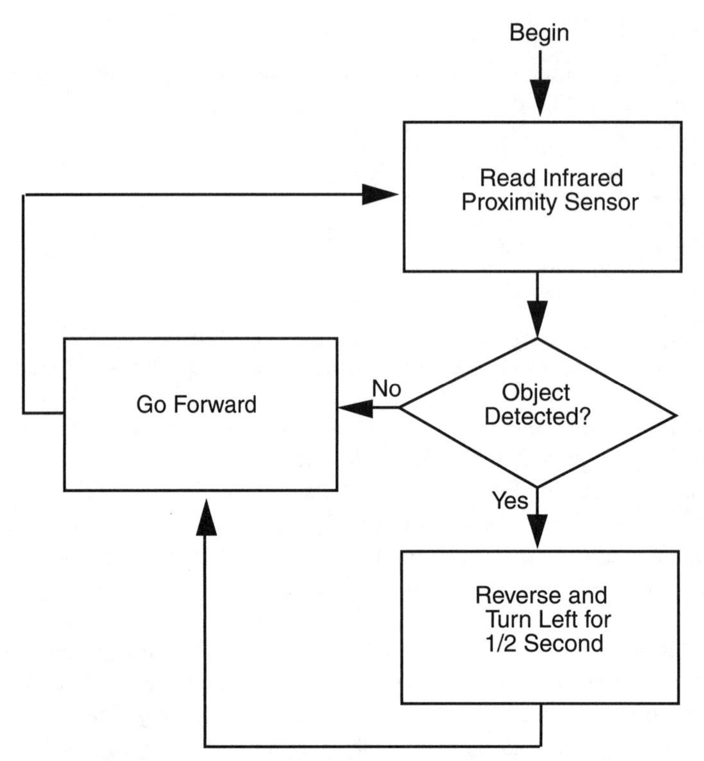

FIGURE 13-1 A flowchart will display high-level program execution.

At the other end of the spectrum, flowcharts cannot easily describe complex programming concepts like the select statement in PBASIC (described in Chapter 15) or sequences of operations.

The third issue with flowcharting is that they are hard to change when a new feature is added to a program or mistakes in the program are corrected. This program is alleviated somewhat with computer-aided graphical programs, but it is still a chore to update a flowchart.

Finally, flowcharts do not show what programs are doing very effectively. Computer programming languages are literally that; a language used by a human to communicate with a computer that is similar in format as a spoken or written language used to communicate ideas and concepts to other humans. While programs may seem difficult to read, their function can usually be determined by applying the same rules of logic used when communicating with other people. This is not true with flowcharting, where the function of the program is difficult to find in a moderately complex diagram.

Flowcharts are effective in explaining certain types of programming operations, but their inability to represent multiple statement types as well as not being very efficient in describing complex programs with different types of statements has pushed them away from the forefront of computer programming education. There are some applications that use a flowcharting format (the programming interface for Lego Mindstorms and

National Instruments' LabView are the most popular), but for the most part computer (and robot) programming is carried out using traditional text-based linear execution programming languages. The use of these programming languages for robotics is discussed in the following.

13.1.3 VARIABLES AND I/O PORTS

The term *variable* comes from the notion that the data is not constant and has to be stored somewhere for later use. Even if you knew what the data value was at some time, it may have changed since the last time it was checked; variable data isn't static. To be able to store and find data in a changeable place is a keystone to programming.

Memory in a computer can be thought of as a series of mailboxes, like the ones shown in Fig. 13-2. Each variable is given a name or a label to allow a program to access it. When data is accessed, this name is used by the computer to retrieve or change the value stored in it. Using the variable names in the mailboxes of Fig. 13-2, an arithmetic assignment statement can be created:

```
i = 7
```

In this statement, the variable i will store the value 7; this is an example of storing a constant value in a variable. Variables can be read and their value can be stored in another variable like

```
j = k
```

in which the contents of k are stored in j. While the statement is described as storing the contents of one variable into another, the original (value source) variable's contents don't change or are taken away—the value is copied from k and the same value put into j. At the end of the statement, both j and k store the same value.

Finally, in

```
i = (j * k) + Src
```

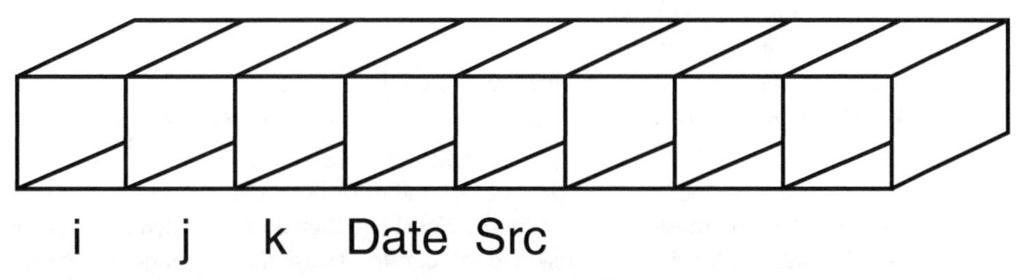

FIGURE 13-2 Variables can be thought of as a series of boxes, each of which is given a name to allow the data to be easily found.

the computer will retrieve the contents of j, k and Src, find the product of j and k, add it to Src, and then save the computed value into i. This is an example of a complex assignment statement, which will be discussed later in the chapter.

I/O ports and hardware special function registers usually use the variable format for reading and writing values into the computer's hardware. So, to output 123 from a computer system I/O port, a statement like this is used:

```
IOPort = 123
```

When you are new to programming, this can be confusing. Hardware that is treated like registers can be difficult to visualize (especially when you retrieve the value of a register and find that it has changed due to a changing input or a counter value).

The data stored in variables come in many forms, and how you choose to store and display data can make your programming much easier to work through or much harder due to many difficult to find errors. There are three different aspects of data, including the amount of space used to store individual values, the format the data is displayed in, and whether relevant pieces of data can be stored together. These three aspects are discussed in the following two sections.

13.1.3.1 Data Types Several different data types are available to all types of programming for storing data in variables and I/O ports. Which one you choose will depend on the data being stored in them. The values might represent a number like 1 or 127, the numeric equivalent of a text character (65 is "A" using the ASCII standard), or a binary value like 00010001, which could mean "run both motors forward" in your robot.

No matter what form the data type is in, most programming languages expect to see data follow predefined types. This is necessary so the data can be properly stored in memory. The most common data type is the eight-bit integer, so-called because the value stores an integer (a whole number) using eight bits. With eight bits, the program can work with a number from 0 to 255 (or –128 to +127, depending on how it uses the eighth bit). The basic data types that you can expect in a programming language are as follows:

- one-bit value, which can hold a 1/0, true/false or high/low value
- eight-bit integer, or byte (can hold a number or a string value)
- eight-bit ASCII character
- 16-bit integer, or word
- 32-bit integer, or long or double word (dword)
- 32-bit floating point, or single (*floating point* means a number with a decimal point)

In many cases, the language provides for either or both signed and unsigned values. The first bit (called the most significant bit, or MSB) is either 0 or 1, which means a positive or negative value. With a 16-bit unsigned integer, for example, the program can store values from 0 to 65535. With a 16-bit signed integer, the program can store values from –32768 to +32767.

13.1.3.2 Number Bases In school you learned that when every number is written out, it is done as part of a base or radix. This is the number of different characters a single digit can have. We are most familiar with the base 10 (known as decimal) due to the number of fingers we have, allowing our ancestors to easily count and add without the need for a pencil and paper.

Values larger than the maximum number of characters allowed in a single digit caused additional, higher value digits to be used. These digits are multiplied by the next power of the number base. To illustrate what this means, consider the number 123.

The value is larger than what a single digit can be so digits multiplied by different powers of the base are used.

The single digits are multiplied by the base to the power 0. "Tens" are the base to the power 1 and "hundreds" are the base to the power 2. The number can be written out mathematically as:

$$
\begin{aligned}
3 \times 10^0 &= 3 \times 1 = 3 \\
2 \times 10^1 &= 2 \times 10 = 20 \\
1 \times 10^2 &= 1 \times 100 = 100 \\
\hline
\text{Total} &= 123
\end{aligned}
$$

Decimal is the most convenient base for humans to work with, but it isn't for computers. Computer circuitry is built from digital logic, which can be either a one (1) or zero (0). Rather than develop circuitry that only works with base 10 numbers, computer designers instead have chosen to have them work in terms of base 2, or binary. The operation of binary is identical to that of decimal, except each digit is a power of 2, not a power of 10.

Using this information, Table 13-1 was developed to show the values for each digit of an eight-bit binary number. Note that the bit number is actually the power of 2 for the digit.

TABLE 13-1 Binary number values. The "%" in front of the binary number is used to indicate the value is binary and not decimal or hexadecimal.

BIT	BINARY VALUE	DECIMAL VALUE
0	%00000001	$1 = 2^0$
1	%00000010	$2 = 2^1$
2	%00000100	$4 = 2^2$
3	%00001000	$8 = 2^3$
4	%00010000	$16 = 2^4$
5	%00100000	$32 = 2^5$
6	%01000000	$64 = 2^6$
7	%10000000	$128 = 2^7$

Knowing the decimal values for each bit, a decimal number can be converted to a binary number by taking away the highest possible value. The decimal number 123 can be converted to binary as:

```
Bit 7 = 128 cannot be taken away from 123 - %0xxxxxxx
Bit 6 =  64 can be taken away from 123    - %01xxxxxx
Bit 5 =  32 can be taken away from 59     - %011xxxxx
Bit 4 =  16 can be taken away from 27     - %0111xxxx
Bit 3 =   8 can be taken away from 11     - %01111xxx
Bit 2 =   4 cannot be taken away from 3   - %011110xx
Bit 1 =   2 can be taken away from 3      - %0111101x
Bit 0 =   1 can be taken away from 1      - %01111011
```

So, the binary number 01111011 is equivalent to the decimal number 123. When binary numbers are written in this book, they are proceeded by a percent (%) character, which is used in BASIC to indicate that a number is binary and not decimal.

Binary numbers can be cumbersome to work with—it should be obvious that trying to remember, say, and record "zero-one-one-one-one-zero-one-one" is tedious at best. To help make working with large binary numbers easier, hexadecimal, or base 16, numbers (which can be represented by four bits) is commonly used. Hexadecimal numbers consist of the first 10 decimal digits followed by the first six letters of the alphabet (A through F) as shown in Table 13-2.

The same methodology is used for converting decimal values to hexadecimal as was used for binary, except that each digit is 16 times the previous digit. The "123" decimal works out to $7B where the "$" character is used in BASIC to indicates it's a hexadecimal number.

When programming, it is recommended that binary values are used for single bits, hexadecimal is used for register values, and decimal for everything else. This will make your program reasonably easy to read and understand and imply what the data is representing in the program without the need for explaining it explicitly.

13.1.4 ASSIGNMENT STATEMENTS

To move data between variables in a computer system, you will have to use what is known as an assignment statement; some value is stored in a specific variable and uses the format

```
DestinationVariable = SourceValue
```

where the SourceValue can be an explicit number (known as a *constant, literal,* or *immediate* value) or a variable. If SourceValue is a variable, its contents are copied from SourceValue and the copy of the value is placed into DestinationVariable. Going back to the previous section, the SourceValue could also be a structure or union element.

The most correct way of saying this is "The contents of SourceValue are stored in DestinationValue," but this implies that the number is physically taken from SourceValue (leaving nothing behind) and storing it in DestinationValue. It has to be remembered that the value stored in SourceValue does not change in an assignment statement.

13.1.4.1 Mathematical Expressions A computer wouldn't be very useful if all it could do is load variables with constants or the value of other variables. They must also

TABLE 13-2 Hexadecimal values with binary and decimal equivalents. Note the "$" in front of the hexadecimal number is used as an indicator like "%" for binary numbers.

HEXADECIMAL NUMBER	BINARY EQUIVALENT	DECIMAL EQUIVALENT
$0	%0000	0
$1	%0001	1
$2	%0010	2
$3	%0011	3
$4	%0100	4
$5	%0101	5
$6	%0110	6
$7	%0111	7
$8	%1000	8
$9	%1001	9
$A	%1010	10
$B	%1011	11
$C	%1100	12
$D	%1101	13
$E	%1110	14
$F	%1111	15

be able to perform some kind of arithmetic manipulation on the values (or data) before storing them into another variable (or even back into the same variable). To do this, a number of different mathematical operations are available in all high-level programming languages to allow you to modify the assignment statement with mathematical functions in the format

```
DestinationVariable = SourceValue1 + SourceValue2
```

in which DestinationVariable is loaded with the sum of SourceValue1 and SourceValue2. DestinationVariable must be a variable, but SourceValue1 and SourceValue2 can be either constants or variables. Their values are read by the computer, stored in a temporary area within the processor, added together, and the result (sum) is stored into the destination variable.

Along with being able to add two values together, programming languages provide a number of different mathematical (addition, subtraction, multiplication, and division) operations that can be performed on variables as well as bitwise (AND, OR, XOR) and comparison (equals, less than, greater than, etc.). Table 13-3 lists the commonly used mathematical operations in computer programming.

The characters used to represent the different operations may have some differences between languages, but for the most part they are quite consistent for the basic mathematical operations. Multiple operations can be performed from a single expression, resulting in a very complex mathematical process being carried out within the computer from a string of characters that were quite easy for you to input into the computer program.

13.1.4.2 Order of Operations All but the oldest or very simple programming languages can handle more than one operator in an expression. This allows you to combine three or more numbers, strings, or variables together to make complex expressions, such as

```
5 + 10 / 2 * 7
```

This feature of multiple operators comes with a penalty, however. You must be careful of the order of precedence or operation, that is, the order in which the program evaluates an expression. Some languages evaluate expressions using a strict left-to-right process (the Parallax BASIC Stamp 2 PBASIC language is one), while others follow a specified pattern where certain operators are dealt with first, then others. This is known as the *order of operations* and ensures that the higher priority operations execute first. The common order of operations for most programming languages is listed in Table 13-3.

The programming language usually does not distinguish between operators that are on the same level or precedence. If it encounters a _ for addition and a - for subtraction, it will evaluate the expression by using the first operator it encounters, going from left to right. You can often specify another calculation order by using parentheses. Values and operators inside the parentheses are evaluated first.

Keeping the order of operations straight for a specific programming language can be difficult, but there is a way of forcing your expressions to execute in a specific order and that

TABLE 13-3 Different Mathematical Operators and Their Order of Operations or Priority

<table>
<tr><th>PRIORITY</th><th>ORDER</th><th>OPERATORS</th></tr>
<tr><td>Lowest</td><td>1</td><td>– (negation), ~(bitwise not), !(logical not)</td></tr>
<tr><td></td><td>2</td><td>+ (addition), –(subtraction)</td></tr>
<tr><td></td><td>3</td><td>* (multiplication), /(division)</td></tr>
<tr><td></td><td>4</td><td>&(bitwise AND), |(bitwise OR), ^(bitwise XOR)</td></tr>
<tr><td></td><td>5</td><td><<(shift left), >>(shift right)</td></tr>
<tr><td></td><td>6</td><td><(less than), <=(less than or equals), =(equals), !=(not equals), >(greater than), >=(greater than or equals)</td></tr>
<tr><td>Highest</td><td>7</td><td>&&(logical AND), ||(logical OR)</td></tr>
</table>

is to place parentheses around the most important parts of the expression. For example, if you wanted to find the product of 4 and the sum of 5 and 3 and store it in a variable, you could write it as

```
A = 4 * (5 + 3)
```

in your program to ensure that the sum of 5 and 3 is found before it is multiplied by 4. If the parentheses were not in place, you would probably have to use a temporary variable to save the sum before calculating the product, like

```
temp = 5 + 3
A = 4 * temp
```

to ensure the calculation would be performed correctly.

13.1.4.3 Using Bitwise Operators Bitwise mathematical operations can be somewhat confusing to new robot programmers primarily because they are not taught in introductory programming courses or explained in introductory programming texts. This is unfortunate because understanding how to manipulate bits in the programming language is critical for robotics programming. Some languages simplify the task of accessing bits by the use of a single bit data type, while others do not provide you with any data types smaller than a byte (eight bits).

The reason why manipulating bits in robots is so important is due to the organization of control bits in the controlling computer systems. Each I/O port register variable address will consist of eight bits, with each bit being effectively a different numeric value (listed in Table 13-4). To look at the contents of an individual bit, you will have to isolate it from all the others using the bitwise AND operator:

```
Bit4 = Register & 16 ' Isolate Bit 4 from the rest of register's bits
```

Looking at this for the first time, it doesn't make a lot of sense although when you consult with Table 13-4, by looking at the binary value you can see that the value of 16 just has one bit of the byte set to 1 and when this is ANDed with the contents of the register, just this one bit's value will be accurate—all the others will be zero.

The set bit value that isolates the single bit of the byte is known as a *mask*. You can either keep a table of mask values in your pocket at all times, or you can use the arithmetic value at the right of Table 13-4. The shift left (<<) operation of one bit is equivalent to finding the set power of 2 or using one of the binary, hex, or decimal values in the table—you just don't have to do any thinking to use it.

To show how the arithmetic value is used to facilitate writing a 1 in a bit of a register, you could use the assignment statement and expression:

```
Register = (Register & ($FF ^ (1 << Bit))) | (1 << Bit)
```

The expression first calculates the byte value with Bit set and then XORs it with all bits set to produce a mask, which will allow the value of each bit of the register to pass unchanged except for Bit, which becomes zero. The register value with Bit zero is then ORed with the

TABLE 13-4 Different Bits in a Byte and Their Numeric Values

BIT NUMBER	BINARY VALUE	HEX VALUE	DECIMAL VALUE	ARITHMETIC VALUE
0	%00000001	$01	1	$2^0 = 1 << 0$
1	%00000010	$02	2	$2^1 = 1 << 1$
2	%00000100	$04	4	$2^2 = 1 << 2$
3	%00001000	$08	8	$2^3 = 1 << 3$
4	%00010000	$10	16	$2^4 = 1 << 4$
5	%00100000	$20	32	$2^5 = 1 << 5$
6	%01000000	$40	64	$2^6 = 1 << 6$
7	%10000000	$80	128	$2^7 = 1 << 7$

bit set and the resulting value is stored in Register. The parentheses, while seemingly complex, force the expression to execute as desired.

13.1.5 ARRAYS AND CHARACTER STRINGS

A variable is normally assigned to a specific memory location within the computer that cannot change. For many kinds of data, this is acceptable, but there will be cases when you would like to access a data address arithmetically rather than having a fixed address. The *array* variable modifier provides you with a number of different variables or *elements,* all based on a location in memory and given the same name, but accessed by their address within the variable.

Fig. 13-3 shows how memory is used to implement an array of bytes. The actual byte (or *array element*) is selected by adding an index to the variable name. Normally the index is added to the address of the variable name, so the first index is zero. To read the third byte in the example array and store it in another variable, the following assignment statement could be used:

```
i = ArrayName (2) ' Read the Third Element (Starting Byte Plus 2)
```

Writing to an array is carried out exactly the same way. Instead of explicitly specifying the array's index, you can calculate it arithmetically using an expression. For example, the value 47 is written to the seventh element if i is equal to three:

```
ArrayName (i * 2) = 47 ' Write to 7th Element (Starting Byte Plus 6)
```

Arrays are often used to implement a string of characters. A string is simply a sequence of alphabetic or numeric characters implemented as a series of incrementing array elements. Most strings are of the ASCIIZ type, which means that they consist of a number of ASCII characters and end with the ASCII NULL character (hex $00).

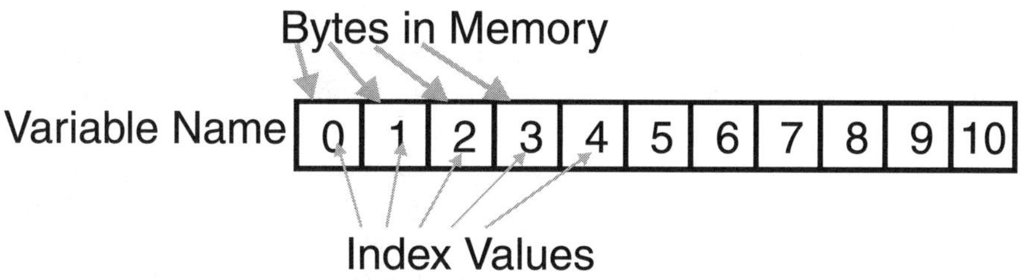

FIGURE 13-3 The memory usage of an array variable consisting of single bytes. Note that this type of array could be a string.

Depending on the programming language, you may be able to access a string like a single element variable or as an array of characters. The BASIC language handles strings as a single element variable, which means you can read and write to them without any special functions or operators. The C language requires the programmer to treat strings of characters as an array and each array element must be handled separately. To simplify the amount of work required to deal with strings in C, there are a number of standard library routines written for it, but new programmers will find working with strings in C challenging.

13.1.6 DECISION STRUCTURES

If you are familiar with programming, when you skim over this section you might think that decision structures is a fancy euphemism for if and other conditional execution statements. That would be an unfortunate conclusion because it simplifies the different structures and what can be done with them. Decision structures are the basis of structured programming and were invented to simplify the task of programming and reading code.

The decision structures all avoid the use of the *goto* statement. This is important because gotos can make code difficult to read and more difficult to write. The issue is that gotos have to have somewhere to go to—either an address, line number, or label. Using only decision structures in your programming avoids this requirement and allows you to concentrate on the program instead of trying to remember if you have used a label before. Even though decision structures and structured programming are considered advanced programming topics, by using them your programs will actually be more simple to create.

The first decision structure to be aware of is the if statement, which is often written out in the form if/else/endif because these are the three statements that make up the decision structure that executes if a set of conditions is true. If a specific number of statements are to execute if the conditions are true, the statements

```
if (Condition)
  Statement1
  Statement2
    :
endif
```

are used. The *condition* statement is normally in the form:

```
Variable|Constant Operator Variable|Constant
```

where the Operator can be one of the six comparison operators listed in Table 13-5. The vertical bar (|) is an OR and indicates that either a variable or a constant can be placed at this part of the condition. Note that the equals and not equals operators can be different depending on the programming language used.

The values on the left and right side of the comparison operator can also be arithmetic expressions (i.e., "i + 4"), but when you first start programming, you should avoid making the comparisons complex as it will make it more difficult for you to debug later.

Multiple comparisons can be put together in a comparison statement using the logical AND and OR operators available in the programming language. Like adding arithmetic expressions to comparisons, you should avoid putting in multiple comparisons in a single statement until you are comfortable with programming.

An optional part of the if decision structure, the *else* statement, can be used to specify statements that execute when the conditions are not true:

```
if (Condition)
  Statement1
  Statement2
    :
else
  Statement1
  Statement2
    :
endif
```

Along with executing some statements if conditions are true, you can also repeatedly execute instructions while a condition is true using the *while* statement in which the condition is exactly the same as the if statement. When the *endwhile* statement is encountered, execution jumps back up to the original while statement and re-evaluates the condition. If it is true, the statements inside while and endwhile repeat, else the statements after the endwhile execute.

```
while (Condition)
   Statement1
   Statement2
     :
endwhile
```

TABLE 13-5 Typical Comparison Operators

OPERATOR	FUNCTION
=, ==	Returns "True" if the values on both sides are equal
<>, !=	Returns "True" if the values are not equal
>	Returns "True" if the value on the left is greater than the one on the right
>=	Returns "True" if the value on the left is greater than or equal to the one on the right
<	Returns "True" if the value on the left is less than the one on the right
<=	Returns "True" if the value on the left is less than or equal to the one on the right

There are many forms of the while statement, including the *until* statement, which continues looping while the condition is not true and stops looping when the condition becomes true. These different types of statements are specific to the programming language.

The final type of decision structure that you should be aware of is the select or switch statements which take the form:

```
select (Variable)
  case Constant1:
    Statement1
    Statement2
      :
  case Constant2:
    Statement1
    Statement2
      :
  :
endselect
```

The *select/switch* statements allow you to avoid multiple if statements for different values of a variable. Usually the case value is a constant, but in some programming languages it can be a variable, a range of values, or a set of conditions.

When you look at these examples and those throughout the book, you should notice that the code is indented to different points depending on whether it is part of a decision statement. This indentation is a programming convention used to indicate that the code is part of a conditional operation. You should make sure that you indent your code in a manner similar to that shown in this book to make your program easier to read and understand when you are trying to figure out what is happening for a specific set of conditions.

13.1.7 SUBROUTINES AND FUNCTIONS

One of the first things a programmer does when starting on a project is to map out the individual segments, or subroutines and functions (sometimes simply called *routines*) that make up the application. A subroutine is a block of code that can be selectively executed by *calling* it. Multiple calls can be placed throughout a program, including calls from other subroutines. Subroutines are meant to replicate repeated code, saving overall space in an application, as well as eliminating the opportunity that repeated code is keyed in incorrectly. Even the longest, most complex program consists of little more than bite-sized subroutines. The program progresses from one subroutine to the next in an orderly and logical fashion.

A subroutine is any self-contained segment of code that performs a basic action. In the context of robot control programs, a subroutine can be a single command or a much larger action. Suppose your program controls a robot that you want to wander around a room, reversing its direction whenever it bumps into something. Such a program could be divided into three distinct subroutines:

- *Subroutine 1.* Drive forward. This is the default action or behavior of the 'bot.
- *Subroutine 2.* Sense bumper collision. This subroutine checks to see if one of the bumper switches on the robot has been activated.
- *Subroutine 3.* Reverse direction. This occurs after the robot has smashed into an object, in response to a bumper collision.

There are two different ways to implement a subroutine in a programming language. The first is to call a label, which executes until a return statement is executed. The second way is to create a procedural subroutine, which is physically separate from the caller. The first method is implemented directly in code like:

```
    :
  call Routine1                 '  Call the Routine
    :        <= Execution Returns here after Call
    :
Routine1:                       '  Routine Label
    :
  return                        '  Return to Statement After call
```

This is the simplest implementation of a subroutine and is used in many beginner programming languages, including PBASIC.

Procedural subroutines are physically separate from the executing code. In the previous code example, if execution continues it will execute the code at Routine1 without calling it. After executing the code of Routine1, it will attempt to return to the statement after the call statement and could end up executing somewhere randomly in the application and will definitely execute in a way that you are not expecting.

A consequence of being physically separate from the caller, a procedural subroutine will have to have different parameters passed to it, which will be used by the subroutine. These parameters are the input data for the subroutine to execute; in the simple call label type of subroutine, the parameters are the program variables that are shared between the main line of the program and the subroutines. The procedural subroutine looks like:

```
    :
  PRoutine1(Parameter1, Parameter2)      '  Call the Routine
    :    <= Execution Returns here after Call
    :
PRoutine1(typedef Parameter1, typedef Parameter2)  '  Entry Point
    :
END PRoutine1                       '  Procedural Routine End/Return Point
```

Looking at this example, the values of Parameter1 and Parameter2 are unique to PRoutine1 and the instance in which it is being called. They are known as local variables and, along with any other variables declared inside the procedural routine, can only be accessed by code within this routine. Global variables are declared outside of any procedure and can be accessed by any subroutine in the application.

While the parameters were described as a consequence of procedural subroutines, they are actually an advantage because they eliminate the need for keeping track of the global variables used as parameters and making sure they are not inadvertently changed in the program. Rather than being a consequence, the parameters and local variables are actually an advantage as they will make your software easier to write.

Changing a variable value in a straight call/return subroutine is quite simple and can be accomplished in a procedural subroutine by accessing a global variable. A better way of changing a variable value in a procedural language is to use a *function,* which is a special type of subroutine that returns a value to the caller. For example:

```
    :
  Variable = Function1(Parameter1, Parameter2)     '  Call the Function
    :      <= Execution Returns here after Call
    :
typedef Function1(typedef Parameter1, typedef Parameter2) ' Entry
    :
  return retValue                  '  Load Return Value
END PRoutine1                      '  Function End/Return Point
```

While execution returns to the statement after the function call, the Variable assignment takes place when execution returns from the function and before it resumes on the statement after the function call. Using functions to return a new value instead of subroutines, which update a common global variable, makes seeing the operation of the code much more obvious and avoids the possibility that the global variable is changed incorrectly by another subroutine.

13.1.8 CONSOLE I/O

Your PC's console is the screen and keyboard used to enter data and display information. At first thought, console I/O is not very useful for robotics developers—the robot is not going to have a direct connection to a computer at all times and the extra work doesn't seem justifiable.

These statements are not completely accurate; knowing how to implement console I/O can be very useful to robot designers as they debug their creations and experiment with different ideas.

One important point to make is that the term *console* is used to describe a monochrome text display with a straight ASCII keyboard. You should not have to worry about displaying graphics and handling mouse inputs and control (unless you want to). The basic console is an 80 character by 24 line display. Data can be formatted specially on the screen, but for the most part, it is a line of text output using the programming language's built-in print function.

Data input can be quite complex and use the cursor control keys on your keyboard, but you should just input data to the console a line at a time with the data being accepted when the enter key is pressed. Waiting for the enter key to be pressed simplifies the data input, allows the use of the backspace key, and allows for the use of an InputString built-in function of the programming language.

The simplest way to implement console I/O on a robot is to use a serial RS-232 connection between the robot and the PC. There are a number of downloadable terminal emulators for the PC, which will provide the console function as a separate window on your PC, allowing you to communicate with your robot or computer circuits with virtually no disruption of the operation of your PC.

13.1.9 COMMENTS

Comments are used by a programmer as remarks or reminders of what a particular line of code or subroutine is for. The comments are especially helpful if the program is shared among many people. When the program is compiled (made ready) for the computer or microcontroller, the comments are ignored. They only appear in the human-ready source code of your programs.

To make a comment using BASIC, use the apostrophe (') character. Any text to the right is treated as a comment and is ignored by the compiler. For example:

```
' this is a comment
```

Note that the symbol used for comments differs between languages. In the C programming language, for instance, the characters // are used (also /* and */ are used to mark a comment block).

13.2 Robotics Programming

When programming is first taught, the concept that a program follows the "input—processing—output" model is used (Fig. 13-4) in which data inputs to the program are passed to a processing block and then passed out of the program in the form of outputs. This model is good for initial programs on a PC in which some number crunching is done to help you learn how the different programming statements work, but they have little relevance when programming robots.

Instead, a programming model like the one shown in Fig. 13-5 should be used. In this model, the program is continually looping (never ending) and reading (or polling) inputs, passing them to a block of code for processing into outputs and then delaying some amount of time before repeating.

The delay block is critical for robot programming as it allows the code to be tuned for the operation of the robot and its parts. If the delay is too short, the robot may vibrate more than move because one set of commands from some input conditions are countermanded by the next set of input conditions, resulting in the original input conditions are true again . . .

Specifying the appropriate amount of delay is something that will have to be found by trial and error. The total loop time should be less than 1 s, but as to the "best" time, that is something that you will have to find by experimenting with your robot. As a rule of thumb, try to keep the total loop time at least 100 ms (0.1 s) so that the motors have time to start up and initiate action before stopping and a new set of inputs polled.

You may also want to leave the motors running during the delay based on the assumption that chances are the processed output will not change radically from loop to loop. This will help make your robot move more smoothly, which is always desirable.

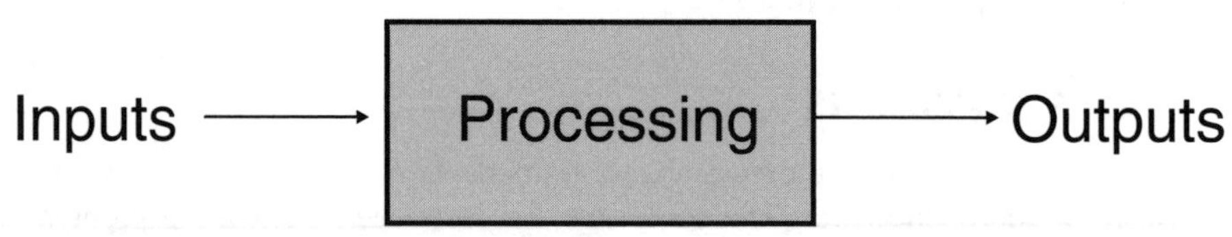

FIGURE 13-4 Typical beginner program flow. Input data is passed to a block of code for processing and then output. This program model cannot be used for robot programming.

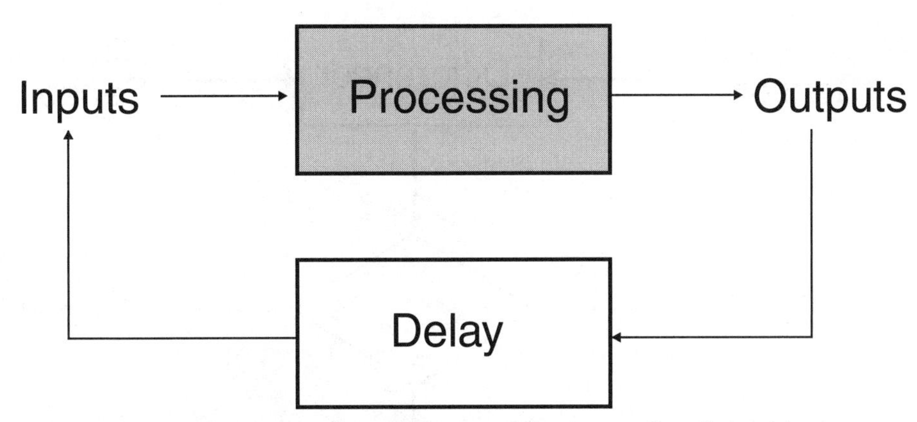

FIGURE 13-5 Robot program flow with the inputs being continually polled and the data processed into outputs. The delay block slows down the execution to make sure that outputs have time to take effect before the environment is polled again for the next set of output commands.

13.3 Graphical Programming

In this chapter, there is a lot of material on text-based programming that probably seems very difficult and complex, especially when you consider that there are many graphical robot programming environments available that could reduce a reasonably complex function like programming a light-seeking robot into the simple chore of drawing a picture like the one in Fig. 13-6.

This program will cause a robot to move toward the brightest point in the room. This is a lot easier to understand than the text-based version, which would look something like

```
while (1)                        '  Loop Forever
    If (Left > Right)            '  Turn Left
        Right Motor = On
        Left Motor = Off
    else
        if (Left < Right)        '  Turn Right
            Right Motor = Off
            Left Motor = On
        else                     '  Left = Right, Go Straight
            Right Motor = On
            Left Motor = On
        endif
    endif
    Dlay(100)  '  Delay 100 ms
endwhile
```

which would probably take you a lot longer to program than moving some blocks and lines around in a graphic editor.

While being more difficult to understand and longer to develop, the text-based program gives you a great deal of flexibility that you don't have with most graphical programming environments. Both programs perform the same operations and work the same way, but what happens if you discover that rather than turning as soon as the left and right light sen-

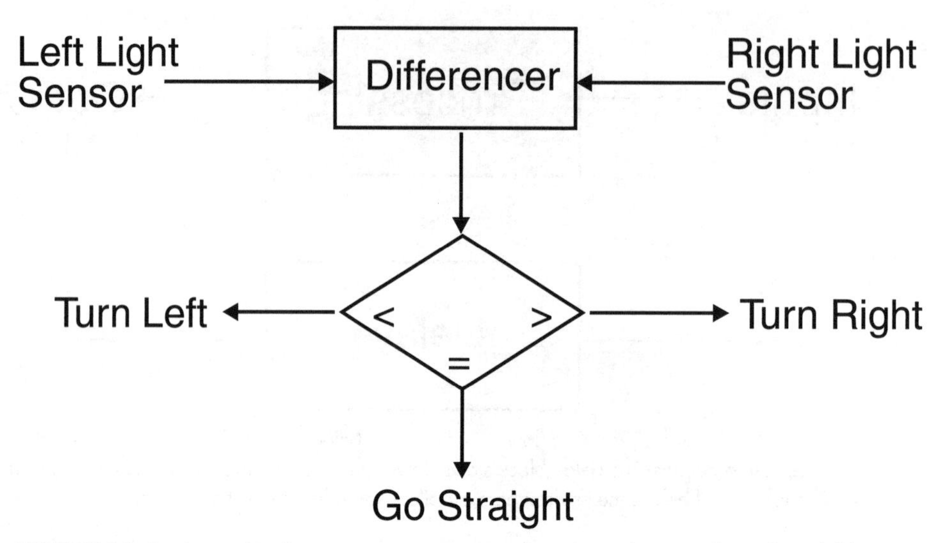

FIGURE 13-6 A graphical programming environment can produce simple and readable programs like the one shown here, but it tends to be difficult to tune with exact parameters as well as enhance with extra logic functions.

sors are different values, the robot could run straight while they were within two values of each other. In the graphical programming environment, this would be impossible, but in the text-based program the two if statements could be changed to

```
if (Left > (Right + 2)

if ((Left + 2) < Right)
```

which would cause the robot to move forward if Left and Right were +/–2 of each other rather than turning if they weren't exactly equal.

There are a lot of other cases where you will find that you need to tweak the program to run optimally, but you find that it is impossible to modify the graphic program for these needs.

Another advantage of text-based programming over graphic programming is the ability to add more complex logic to the operation of the robot. For example, if you wanted the robot to turn right if it has turned left five times in a row you can add the code quite easily to the text program whereas the graphic program does not have that capability (and indeed, very few graphical programming environments even have variables).

It is important to not generalize. While most beginner and hobbyist graphical programming environments and languages do have the limitations listed here, more professional ones do not. For example, National Instruments' LabView programming environment offers the same capabilities and flexibility as a text programming language despite being based on just graphic symbols for operators, variables, and decision structures. Hopefully you realize that these capabilities do not come free; creating a complex graphic program will require at least the same amount of study and practice as developing a text-based one.

13.4 From Here

To learn more about . . .	***Read***
Programming a robot computer or microcontroller	Chapter 12, "An Overview of Robot 'Brains' "
Ideas for working with bitwise values	Chapter 14, "Computer Peripherals"
Programming using infrared remote control	Chapter 16, "Remote Control Systems"

CHAPTER 14

COMPUTER PERIPHERALS

The brains of a robot don't operate in a vacuum. They need to be connected to motors to make the robot move, arms and grippers to pick up things, lights and whistles to let you know what is going on, along with different sensors that give the robot the ability to understand its environment. *Input* is data passed from the environment sensors to the computer, and *output* is the computer's commands to the devices that *do something.* These input and output (I/O) devices are called *peripherals* and usually require some kind of conditioning or programming for the computer or microcontroller in the robot to be able to process information.

This chapter discusses the most common and practical methods for interfacing real-world devices to computers and microcontrollers. Many of the concepts presented in this chapter are quite complex and repeated in other chapters, usually from a different perspective to help you understand them better.

14.1 Sensors as Inputs

By far the most common use for inputs in robotics is sensors. There are a variety of sensors, from the super simple to the amazingly complex. All share a single goal: providing the robot with data it can use to make intelligent decisions. A temperature sensor, for example, might help a robot determine if it's too hot to continue a certain operation. Or an *energy watch* robot might record the temperature as it strolls throughout the house, looking for locations where the temperature varies widely (indicating a possible energy leak).

14.1.1 TYPES OF SENSORS

Broadly speaking, there are two types of sensors (see Fig. 14-1).

- *Digital sensors* provide simple on/off or true/false results or the stepped binary results shown in Fig. 14-1. A switch is a good example of a digital sensor: either the switch is open or it's closed. An ultrasonic ranger, which returns a binary value, with each bit indicating a specific distance, is also a digital sensor.
- *Analog sensors* provide a range of values, usually a voltage. In many cases, the sensor itself provides a varying resistance or current, which is then converted by an external circuit into a voltage. For example, when exposed to light the resistance of a CdS (cadmium sulfide) cell changes dramatically. Built in a simple voltage divider, the voltage output varies with the light striking the CDS cell.

Broadly speaking, in both digital and analog sensors, the input to the computer is a voltage level. In the case of a digital sensor, the robot electronics are only interested in whether the voltage bit value is a logical low (usually 0 V) or a logical high (usually 5 V). Digital sensors can often be directly connected to a robot control computer without any additional interfacing electronics.

In the case of an analog sensor, you need additional robot electronics to convert the varying voltage levels into a form that a control computer can use. This typically involves using an analog-to-digital converter, which is discussed later in this chapter.

14.1.2 EXAMPLES OF SENSORS

One of the joys of building robots is figuring out new ways of making them react to changes in the environment. This is readily done with the wide variety of affordable sensors now

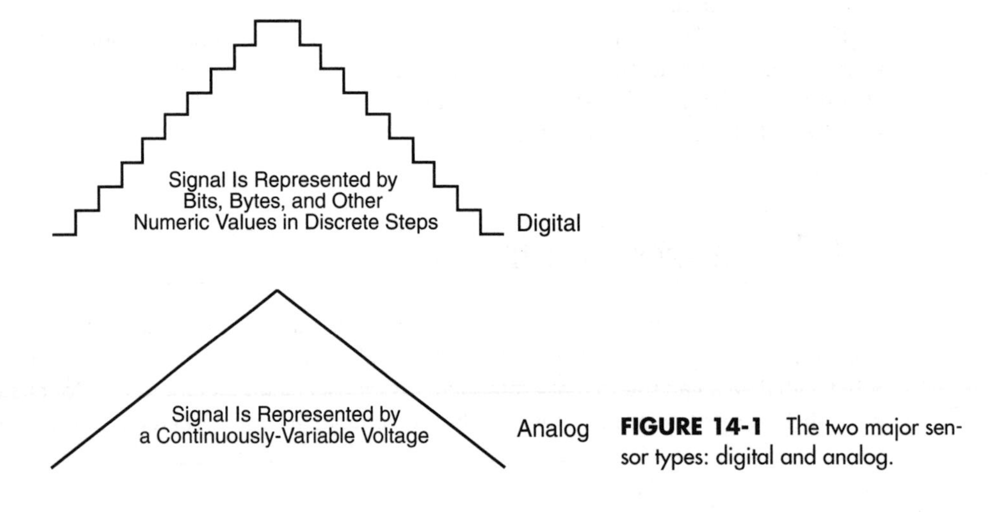

FIGURE 14-1 The two major sensor types: digital and analog.

available. New sensors are constantly being introduced, and it pays to stay abreast of the latest developments. Not all new sensors are affordable for the hobby robot builder, of course—you'll just have to dream about getting that $10,000 vision system. But there are plenty of other sensors that cost much, much less; many are just a few dollars.

Part 6 of this book discusses many different types of sensors commonly available today that are suitable for robotic work. Here is just a short laundry list to whet your appetite:

- *Sonar range finder or proximity detector.* Reflected sound waves are used to judge distances or if a robot is close to an object. The detected range is typically from about a foot to 30 to 40 ft.
- *Infrared range finder or proximity.* Reflected infrared light is used to determine distance and proximity. The detected range is typically from 0 in to 2 or 3 ft.
- *Light sensors.* Various light sensors detect the presence or absence of light. Light sensors can detect patterns when used in groups (called arrays). A sensor with an array of thousands of light-sensitive elements, like a CCD video camera, can be used to construct eyes for a robot.
- *Pyroelectric infrared.* A pyroelectric infrared sensor detects changes in heat patterns and is often used in motion detectors. The detected range is from 0 to 30 ft and beyond.
- *Speech input or recognition.* Your own voice and speech patterns can be used to command the robot.
- *Sound.* Sound sources can be detected by the robot. You can tune the robot to listen to only certain sound wavelengths or to those sounds above a certain volume level.
- *Contact switches.* Used as touch sensors, when activated these switches indicate that the robot has made contact with some object.
- *Accelerometer.* Used to detect changes in speed and/or the pull of the earth's gravity, accelerometers can be used to determine the traveling speed of a robot or whether it's tilted dangerously from center.
- *Gas or smoke.* Gas and smoke sensors detect dangerous levels of noxious or toxic fumes and smoke.
- *Temperature.* A temperature sensor can detect ambient or applied heat. Ambient heat is the heat of the room or air; applied heat is some heat (or cold) source directly applied to the sensor.

14.2 Input and Output Methodologies

The robot's computer controller requires input electrical connections for such things as mode settings or sensors, as well as connections to outputs, for peripherals like motor control or speech. The most basic input and output of a computer or microcontroller are two-state binary voltage levels (off and on), usually between 0 and 5 V. Two types of interfaces are used to transfer these high/low digital signals to the robot's control computer.

14.2.1 PARALLEL INTERFACING

In a parallel interface, multiple bits of data are transferred at one time using (typically) eight separate wires as shown in Fig. 14-2. Parallel interfaces enjoy high speed because more

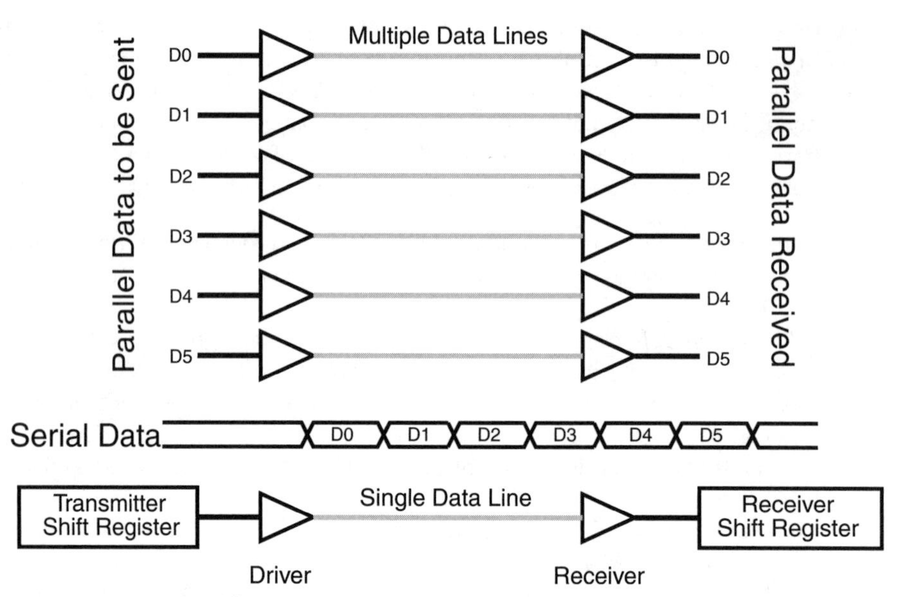

FIGURE 14-2 The differences between passing data serially or in parallel.

information can be shuttled about in less time. A typical parallel interface is the printer port on your personal computer; it sends data to the printer an entire byte (eight bits) at a time. Characters are represented by different combinations of the eight-bit data (using ASCII character codes).

14.2.2 SERIAL INTERFACING

The downside to parallel interfaces is that they consume a large number of input/output lines on the robot computer or microcontroller. There are a limited number of I/O wires or *lines* on the control computer; typically 16 or even fewer. If the robot uses two eight-bit parallel ports, that leaves no I/O lines for anything else.

Serial interfaces, on the other hand, conserve I/O lines because they send data on a single wire (as can be seen in Fig. 14-2). They do this by separating a byte of information into its constituent bits, then sending each bit down the wire at a time, in single-file fashion. There are a variety of serial interface schemes, using one, two, three, or four I/O lines. Additional I/O lines are used for such things as timing and coordination between the data sender and the data recipient.

A number of the sensors you may use with your robot have serial interfaces, and while they appear to be more difficult to interface than parallel connections, they aren't if you use the right combination of hardware and software. The task of working with serial data is made easier when you use a computer or microcontroller because software on the control computer does all the work for you. The BASIC Stamp 2, for example, uses single commands that provides a serial interface on any of its I/O lines.

14.3 Motors and Other Outputs

A robot uses outputs to take some physical action. Most often, one or more motors are attached to the outputs of a robot to allow the machine to move. On a mobile robot, the motors serve to drive wheels, which scoot the 'bot around the floor. On a stationary robot, the motors are attached to arm and gripper mechanisms, allowing the robot to grasp and manipulate objects.

Motors aren't the only ways to provide motility to a robot. Your robot may use solenoids to hop around a table, or pumps and valves to power pneumatic or hydraulic pressure systems. No matter what system the robot uses, the basic concepts are the same: the robot's control circuitry (e.g., a computer) provides a voltage to the output, which controls interface circuitry, allowing motors, solenoids, or pumps to be turned on. When the control voltage changes, the output device stops.

14.3.1 OTHER COMMON TYPES OF OUTPUTS

Along with motors, some other types of outputs are used for the following purposes:

- *Sound.* The robot may use sound to warn you of some impending danger ("Danger, Will Robinson. Danger!") or to scare away intruders. If you've built an R2-D2–like robot (from *Star Wars* fame), your robot might use chirps and bleeps to communicate with you. Hopefully, you'll know what "bebop, pureeep!" means.
- *Voice.* Either synthesized or recorded, a voice lets your robot communicate in more human terms.
- *Visual indication.* Using light-emitting diodes (LEDs), numeric displays, or liquid crystal displays (LCDs), visual indicators help the robot communicate with you in direct ways.

14.4 Sample Output Circuits

The most simple output circuit is the LED driver (Fig. 14-3) in which one of the output lines of the robot controller applies either a high or low voltage to turn on or off an LED. The 470-ohm resistor is used to limit the amount of current passed through the LED to 5 to 6 mA.

Outputs typically drive heavy loads: motors, solenoids, pumps, and even high-volume sound demand lots of current. The typical robotic control computer cannot provide more than 15 to 22 mA (milliamps) of current on any output. That's enough to power one or two LEDs, but not much else. To use an output to drive a load, you need to add a power element that provides adequate current. This can be as simple as one transistor, or it can be a ready-made power driver circuit capable of running large, multi-horsepower motors. One common power driver is the H-bridge, so called because the transistors used inside it are in an H pattern around the motor or other load. The H-bridge can connect directly to the control computer of the robot and provide adequate voltage and current to the load.

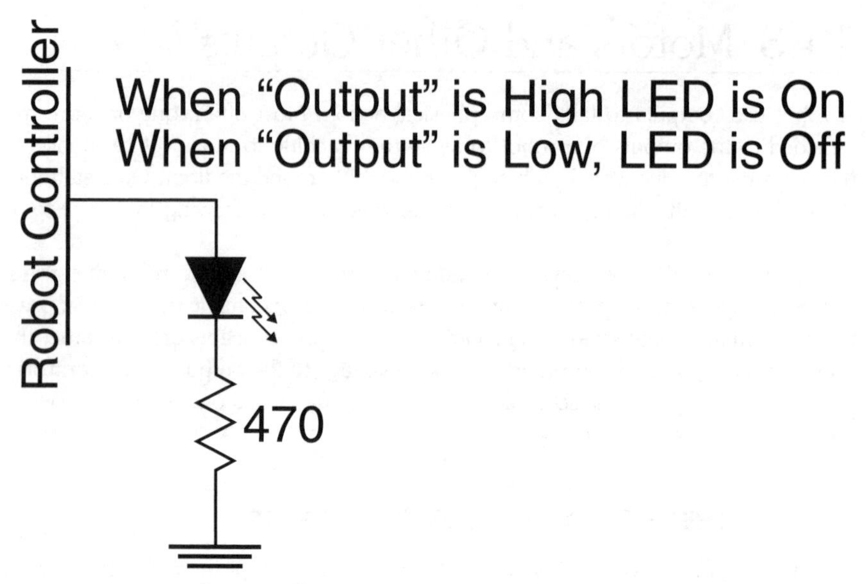

FIGURE 14-3 Robot controller connection to a simple LED.

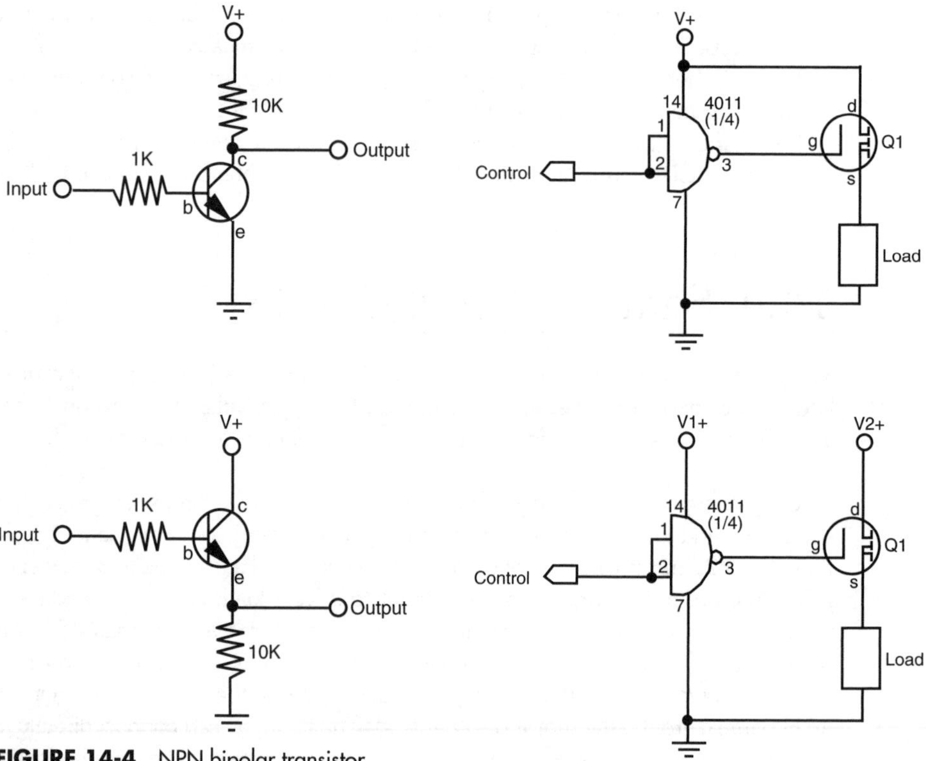

FIGURE 14-4 NPN bipolar transistor drivers.

FIGURE 14-5 Power MOSFET drivers.

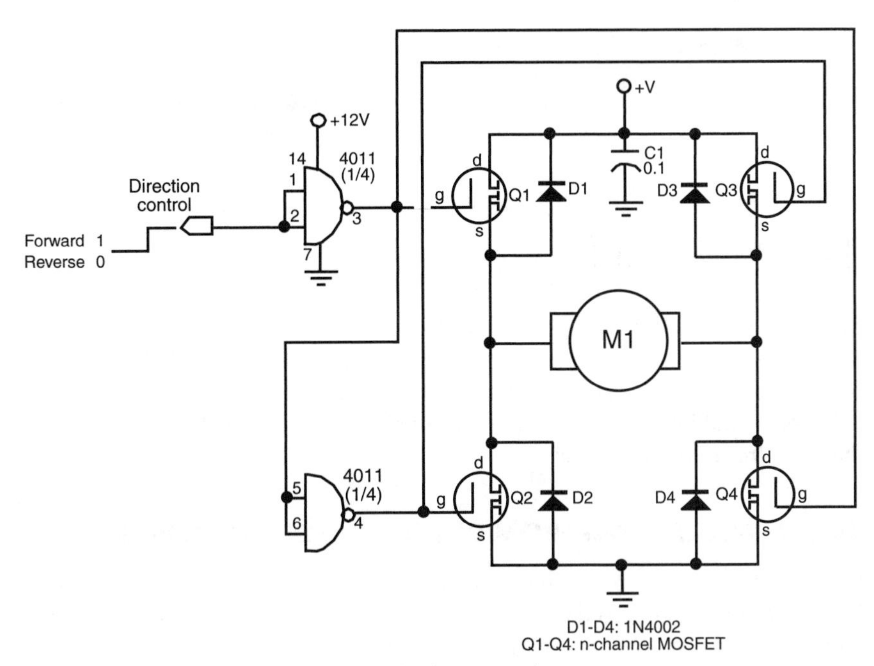

FIGURE 14-6 Discrete component H-bridge motor driver.

Figs. 14-4 through 14-8 show various approaches for doing this, including NPN transistor, power MOSFET, discrete component H-bridge, single-package H-bridge, and buffer circuits. All have their advantages and disadvantages, and they are described in context throughout this book. See especially Chapter 20, "Working with DC Motors," and Chapter 21, "Working with Stepper Motors," for more information on these power drive techniques.

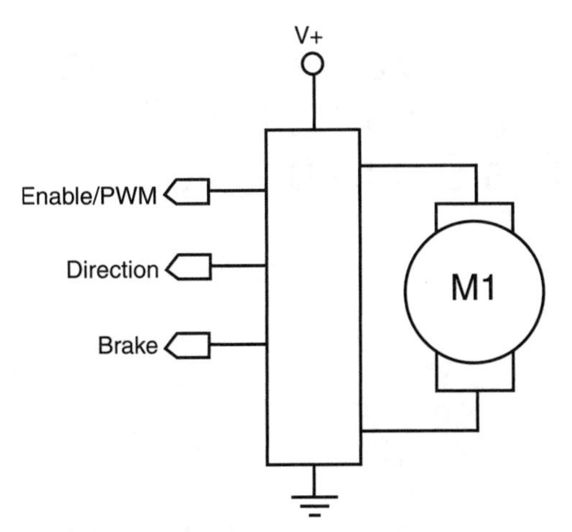

FIGURE 14-7 Packaged H-bridge motor driver.

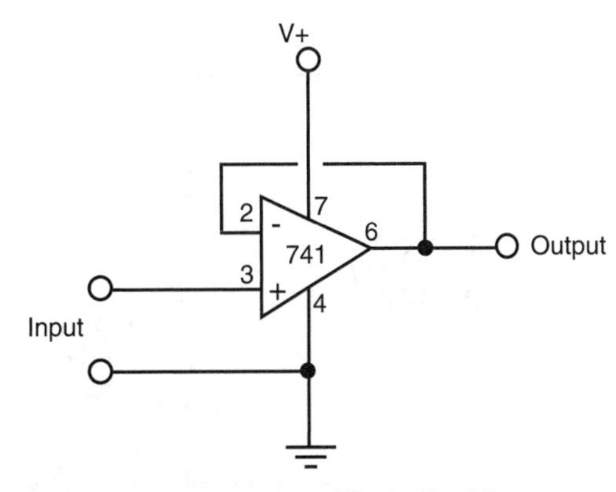

FIGURE 14-8 Non-inverting buffer follower interface.

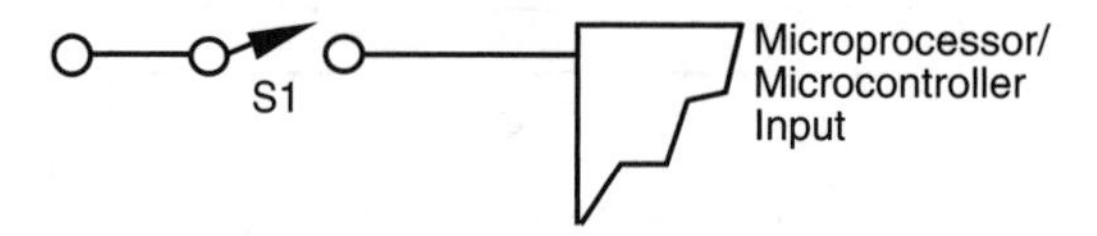

FIGURE 14-9 Switch sensor connected directly to a computer input.

14.5 Digital Inputs

Switches and other strictly digital (on/off) sensors can be readily connected to control electronics. Figs. 14-9 through 14-12 show a variety of techniques, including direct connection of a switch sensor, an LED high/low voltage input and indicator, interface via a switch debouncer, and interface via a buffer. The buffer is recommended to isolate the source of the input from the control electronics.

14.5.1 INTERFACING FROM DIFFERENT VOLTAGE LEVELS

Some digital input devices may operate a voltage that differs from the control electronics. Erratic behavior and even damage to the input device or control electronics could result if you connected components with disparate voltage sources together. So-called logic transla-

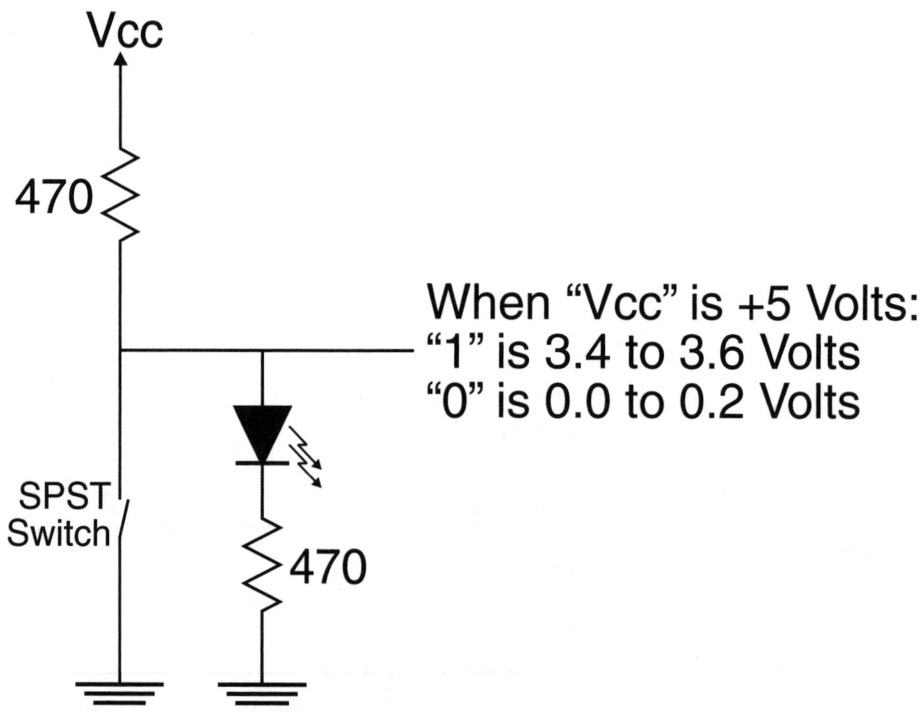

FIGURE 14-10 LED used to indicate the logic level (high or low) or a switch input.

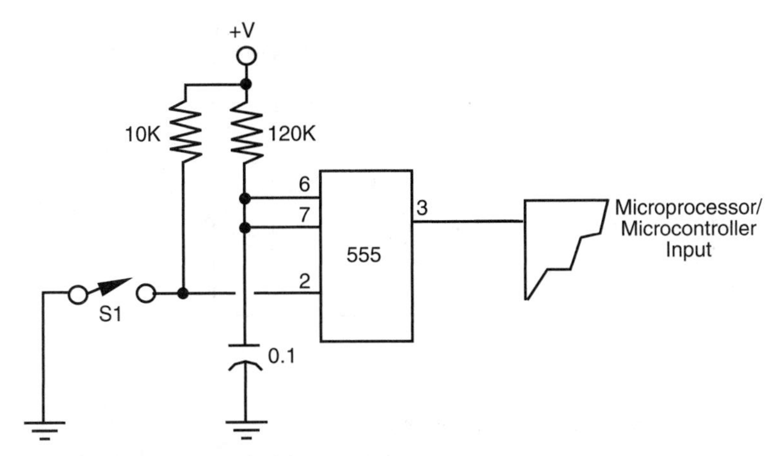

FIGURE 14-11 Switch debouncer input.

tion circuits are needed for these kinds of interfaces. Several integrated circuits provide these functions in off-the-shelf solutions. You can create most of the interfaces you need using standard CMOS and TTL logic chips.

Fig. 14-13 shows how to interface TTL (5 V) to CMOS circuits that use different power sources (use this circuit even if both circuits run under +5 vdc). Fig. 14-14 shows the same concept, but for translating CMOS circuits to TTL circuits that use different power sources.

14.5.2 USING OPTO-ISOLATORS

Note that in both Figs. 14-13 and 14-14, the ground connection is shared. You may wish to keep the power supplies of the inputs and control electronics totally separate. This is most easily done using opto-isolators, which are readily available in IC-like packages. Fig. 14-15 shows the basic concept of the opto-isolator: the source controls a light-emitting diode. The input of the control electronics is connected to a photodetector of the opto-isolator.

Note that since each side of the opto-isolator is governed by its own power supply, you can use these devices for simple level shifting, for example, changing a +5 V signal to +12 V, or vice versa.

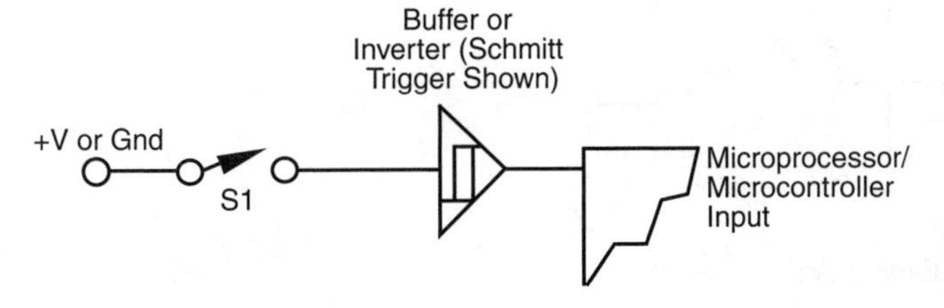

FIGURE 14-12 Buffered input.

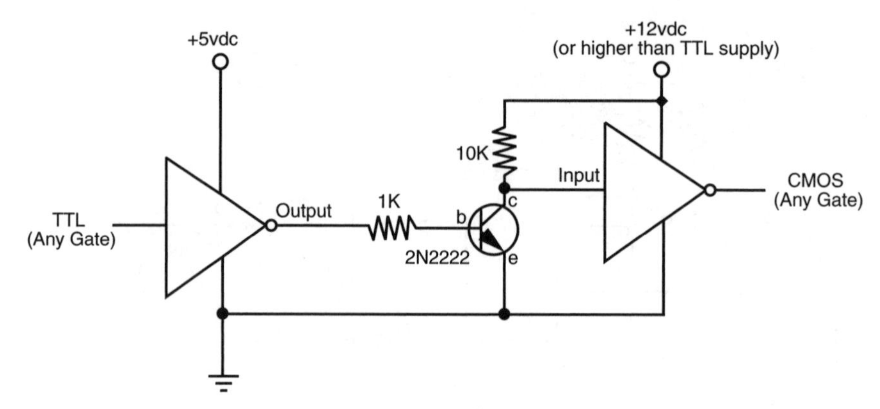

FIGURE 14-13 TTL-to-CMOS translation interface.

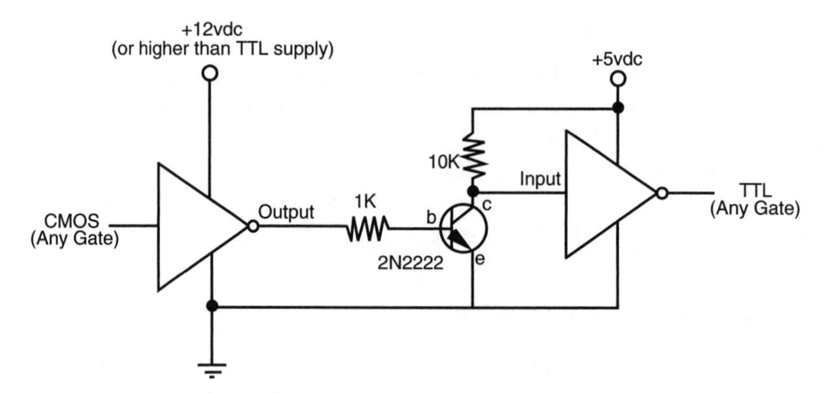

FIGURE 14-14 CMOS-to-TTL translation interface.

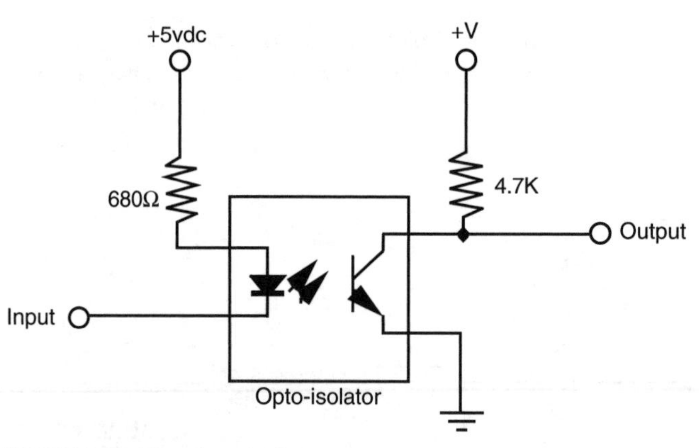

FIGURE 14-15 Opto-isolator.

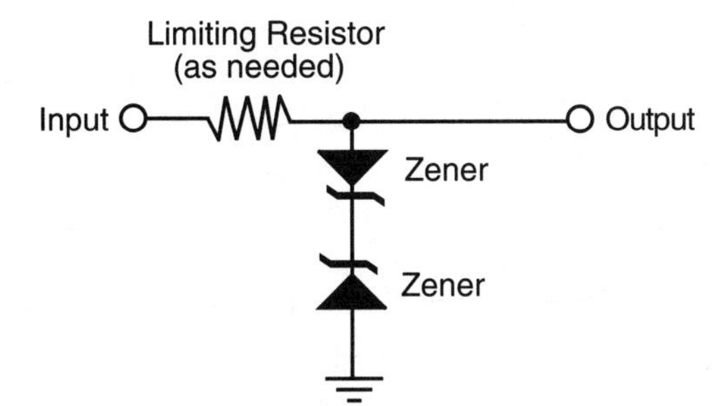

FIGURE 14-16 Zener diode shunt.

14.5.3 ZENER DIODE INPUT PROTECTION

If a signal source may exceed the operating voltage level of the control electronics, you can use a zener diode to clamp the voltage to the input. Zener diodes act like valves that turn on only when a certain voltage level is applied. As shown in Fig. 14-16, by putting a zener diode across the +V and ground of an input, you can basically shunt any excess voltage and prevent it from reaching the control electronics.

Zener diodes are available in different voltages; the 4.7- or 5.1-V zeners are ideal for interfacing to inputs. Use the resistor to limit the current through the zener. The wattage rating of the zener diode you use depends on the maximum voltage presented to the input as well as the current drawn by the input. For most applications where the source signal is no more than 12 to 15 V, a quarter-watt zener should easily suffice. Use a higher wattage resistor for higher current draws.

14.6 Interfacing Analog Input

In most cases, the varying nature of analog inputs means they can't be directly connected to the control circuitry of your robot. If you want to quantify the values from the input, you need to use some form of analog-to-digital conversion (see the section "Using Analog-to-Digital Conversion" later in this chapter for more information).

Additionally, you may need to condition the analog input so its value can be reliably measured. This may include amplifying and buffering the input as described in this section.

14.6.1 VOLTAGE COMPARATOR

Before the robot's controller responds to an input sensor's analog voltage, it must reach a specific threshold. Rather than converting the analog voltage to a binary value using an analog-to-digital converter, the voltage from the input sensor can be *compared* against a reference voltage and a simple binary value, indicating if the analog input is above or below the reference passed to the robot's controller.

The voltage comparator takes a linear, analog voltage and outputs a simple on/off (low/high) signal to the control electronics of your robot. Fig. 14-17 shows a sample voltage comparator circuit. The potentiometer is used to determine the trip point (reference

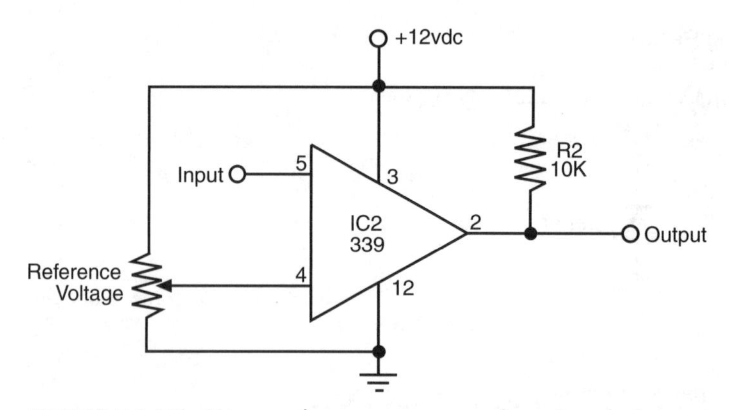

FIGURE 14-17 Using a voltage comparator to determine whether an input voltage is above or below a reference voltage.

voltage) of the comparator. To set the potentiometer, apply the voltage level you want to use as the trip point to the input of the comparator. Adjust the potentiometer so the output of the comparator just changes state.

Note that the pull-up resistor is used on the output of the comparator chip (LM339) used in the circuit. The LM339 uses an open collector output, which means that it can pull the output low, but it cannot pull it high. The pull-up resistor allows the output of the LM339 to pull high.

14.6.2 SIGNAL AMPLIFICATION

Many analog inputs provide on and off signals but not at a voltage high enough to be useful to the control electronics of your robot. In these instances you must amplify the signal, which can be done by using a transistor or an operational amplifier. The op-amp method (Fig. 14-18) is the easiest in most cases, and while the LM741 is probably the most referenced op-amp, you should look around for different devices that have simpler power input requirements (i.e., a single 5 V, which matches the power requirements of your robot's controller).

14.6.3 SIGNAL BUFFERING

The control electronics of your robot may load down the input sources that you use. This is usually caused by a low impedance on the input of the control electronics. When this happens, the electrical characteristics of the sources change, and erratic results can occur. By *buffering* the input you can control the amount of loading and reduce or eliminate any unwanted side effects.

The op-amp circuit, as shown in Fig. 14-19, is a common way of providing high-impedance buffering for inputs to control electronics. R1 sets the input impedance. Note

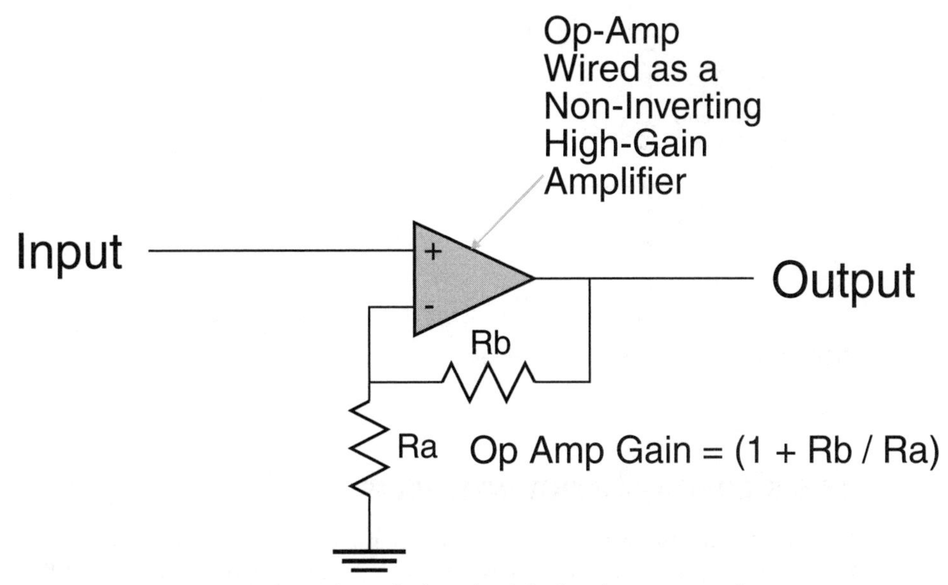

FIGURE 14-18 LED used to indicate the logic level (high or low) or a switch input.

that there are no resistors wired to the op-amp's feedback circuitry, as in Fig. 14-18. In this case, the op-amp is being used in unity gain mode, which does not amplify the signal.

14.6.4 OTHER SIGNAL TECHNIQUES FOR OP-AMPS

There are many other ways to use op-amps for input signal conditioning, and they are too numerous to mention here. A good source for simple, understandable circuits is the *Engineer's Mini-Notebook: Op-Amp IC Circuits,* by Forrest M. Mims III, available through Radio Shack. No robotics lab (or electronics lab, for that matter) should be without Forrest's books.

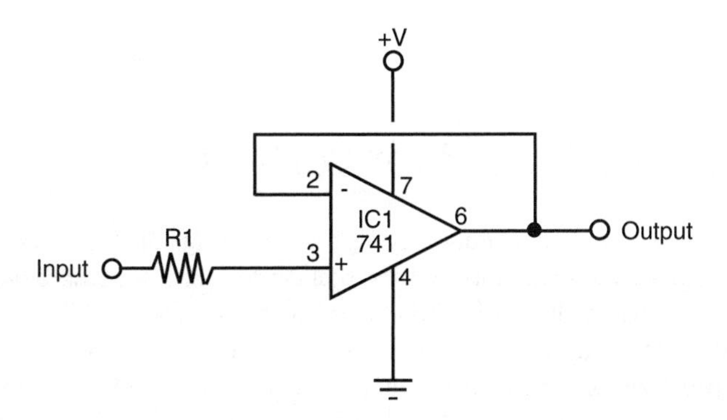

FIGURE 14-19 Op-amp buffer.

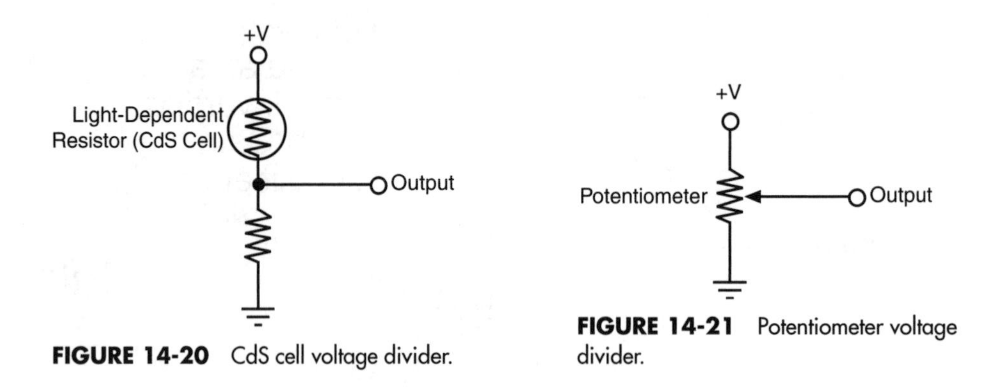

FIGURE 14-20 CdS cell voltage divider.

FIGURE 14-21 Potentiometer voltage divider.

14.6.5 COMMON INPUT INTERFACES

Figs. 14-20 and 14-21 are two basic, common interfaces for analog inputs. These can be connected to analog-to-digital converters (ADC), comparators, buffers, and the like. The most common interfaces are as follows:

- CdS (cadmium-sulfide) cells are, in essence, variable resistors controlled by light and not a mechanical wiper. By putting a CdS cell in series with another resistor between the +V and ground of the circuit (Fig. 14-20), a voltage divider varying voltage is provided that can be read directly into an ADC or comparator. No amplification is typically necessary.
- A potentiometer forms a voltage divider when connected as shown in Fig. 14-21. The voltage varies from ground and +V. No amplification is necessary.
- The output of a phototransistor is a varying current that can be converted to a voltage by using a resistor. (The higher the resistance is, the higher the sensitivity of the device.) The output of a phototransistor is typically ground to close to +V, and therefore no further amplification is necessary.
- Like a phototransistor, the output of a photodiode is a varying current. This output can also be converted into a voltage by using a resistor (see Fig. 14-20). (The higher the resistance, the higher the sensitivity of the device.) This output tends to be fairly weak—on the order of millivolts instead of volts. Therefore, amplification of the output analog voltage is usually required.

14.7 Analog-to-Digital Converters

Computers are binary devices: their digital data is composed of strings of 0s and 1s, strung together to construct meaningful information. But the real world is analog, where data can be almost any value, with literally millions of values between "none" and "lots"!

Analog-to-digital conversion is a system that takes analog information and translates it into a digital, or more precisely binary, format suitable for your robot. Many of the sensors you will connect to the robot are analog in nature. These include temperature sensors, microphones and other audio transducers, variable output tactile feedback (touch) sensors, position potentiometers (the angle of an elbow joint, for example), light detectors, and more. With analog-to-digital conversion you can connect any of them to your robot.

There are a number of ways to construct an analog-to-digital converter, including successive approximation, single slope, delta-sigma, and flash. For the most part, you will not have to understand how the different ADCs work other than being able to read the data sheet and understand how they interface to the analog input and robot controller and how long the analog-to-digital conversion process takes.

14.7.1 HOW ANALOG-TO-DIGITAL CONVERSION WORKS

Analog-to-digital conversion (ADC) works by converting analog values into their binary equivalents. In most cases, low analog values (like a weak light striking a photodetector) might have a low binary equivalent, such as 1 or 2. But a high analog value might have a high binary equivalent, such as 255 or even higher. The ADC circuit will convert small changes in analog values into slightly different binary numbers. The smaller the change in the analog signal required to produce a different binary number, the higher the resolution of the ADC circuit. The resolution of the conversion depends on both the voltage span (0 to 5 V is most common) and the number of bits used for the binary value.

Suppose the signal spans 10 V, and 8 bits (or a byte) are used to represent various levels of that voltage. There are 256 possible combinations of 8 bits, which means the span of 10 V will be represented by 256 different values. Given 10 V and 8 bits of conversion, the ADC system will have a resolution of 0.039 V (39 mV) per step. Obviously, the resolution of the conversion will be finer the smaller the span or the higher the number of bits. With a 10-bit conversion, for instance, there are 1024 possible combinations of bits, or roughly 0.009 V (9 mV) per step.

14.7.2 INSIDE THE SUCCESSIVE APPROXIMATION ADC

Perhaps the most commonly used analog-to-digital converter is the successive approximation approach, which is a form of systematized "20 questions." The ADC arrives at the digital equivalent of any input voltage within the expected range by successively dividing the voltage ranges by two, narrowing the possible result each time. Comparator circuits within the ADC determine if the input value is higher or lower than a built-in reference value. If higher, the ADC branches toward one set of binary values; if lower, the ADC branches to another set.

While this sounds like a roundabout way, the entire process takes just a few microseconds.

One disadvantage of successive approximation (and some other ADC schemes) is that the result may be inaccurate if the input value changes before the conversion is complete. For this reason, most modern analog-to-digital converters employ a built-in sample-and-hold circuit (usually a precision capacitor and resistor) that temporarily stores the value until conversion is complete.

14.7.3 ANALOG-TO-DIGITAL CONVERSION ICS

You can construct analog-to-digital converter circuits using discrete logic chips—basically a string of comparators strung together. An easier approach is a special-purpose ADC integrated circuit. These chips come in a variety of forms besides conversion method (e.g., successive approximation, discussed in the last section).

- *Single or multiplexed input.* Single-input ADC chips, such as the ADC0804, can accept only one analog input. Multiplexed-input ADC chips, like the ADC0809 or the ADC0817, can accept more than one analog input (usually 4, 8, or 16). The control circuitry on the ADC chip allows you to select the input you wish to convert.
- *Bit resolution.* The basic ADC chip has an eight-bit resolution (the ADC08xx ICs discussed earlier are all eight bits). Finer resolution can be achieved with 10-bit and more chips, but they are not widely used or required in robotics.
- *Parallel or serial output.* ADCs with parallel outputs provide separate data lines for each bit. (Ten and higher bit converters may still only have eight data lines; the converted data must be read in two passes.) Serial output ADCs have a single output, and the data is sent one bit at a time. Serial output ADCs are handy when used with microcontrollers and single-board computers, where input/output lines can be scarce. In the most common scheme, a program running on the microcontroller or computer clocks in the data bits one by one in order to reassemble the converted value. The ADC08xx chips have parallel outputs; the 12-bit LTC1298 has a serial output.

14.7.4 INTEGRATED MICROCONTROLLER ADCs

Many microcontrollers and single-board computers come equipped with one or more analog-to-digital converters built in. This saves you the time, trouble, and expense of connecting a stand-alone ADC chip to your robot. You need not worry whether the ADC chip provides data in serial or parallel form since all the data manipulation is done internally. Along with this, you are usually given the option of either using the voltage applied to the microcontroller as the high voltage range or the ADC or an external voltage (generally less than Vdd) from your circuit, giving you the capability of measuring different voltage ranges accurately. You just tell the system to fetch an analog input, and it responds when the conversion is complete.

14.7.5 SAMPLE CIRCUITS

Fig. 14-22 shows a basic circuit for using the ADC0809, which provides eight analog inputs and an eight-bit conversion resolution. The input you want to test is selected using a three-bit control sequence—000 for input 1, 001 for input 2, and so on. Note the ~500 kHz time base, which can come from a ceramic resonator or other clock source or from a resistor/capacitor (RC) time constant. If you need precise analog-to-digital conversion, you should use a more accurate clock than an RC circuit.

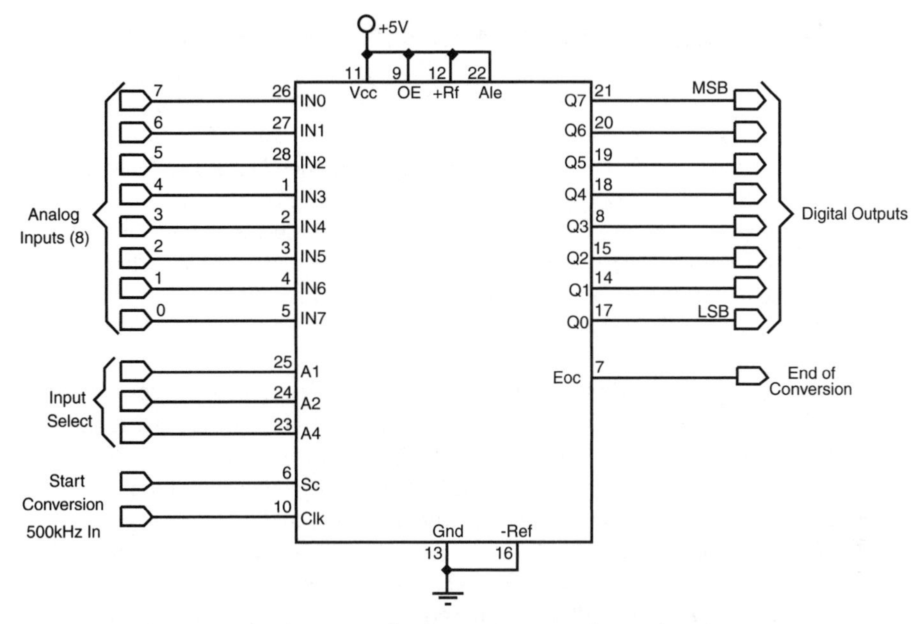

FIGURE 14-22 Basic hookup circuit for the ADC0809 analog-to-digital converter.

14.8 Digital-to-Analog Conversion

Digital-to-analog conversion (DAC) is the inverse of analog-to-digital conversion. With a DAC, a digital signal is converted to a varying analog voltage. DACs are common in some types of products, such as audio compact discs, where the digital signal impressed upon the disc is converted into a melody pleasing to the human ear.

At least in the robotics world, however, DACs are not as commonly used as ADCs, and when they are, simpler approximation circuits are all that's usually necessary. A common technique is to use a capacitor and resistor to form a traditional RC time-constant circuit. A digital device sends periodic pulses through the RC circuit. The capacitor discharges at a more or less specified rate. The more pulses there are during a specific period of time, the higher the voltage that will get stored in the capacitor.

The speed of DC motors is commonly set using a kind of digital-to-analog conversion. Rather than vary the voltage to a motor directly, the most common approach is to use pulse width modulation (PWM), in which a circuit applies a continuous train of pulses to the motor. The longer the pulses are on, the faster the motor will go. This works because motors tend to integrate out the pulses to an average voltage level; no separate digital-to-analog conversion is required. See Chapter 20 for additional information on PWM with DC motors.

You can accomplish digital-to-analog conversion using integrated circuits specially designed for the task. The DAC08, for example, is an inexpensive eight-bit digital-to-analog converter IC that converts an eight-bit digital signal into an analog voltage.

14.9 Expanding Available I/O Lines

A bane of the microcontroller- and computer-controlled robot is the shortage of input/output pins. It always seems that your robot needs one more I/O pin than the computer or microcontroller has. As a result, you think you either need to drop a feature or two from the robot or else add a second computer or microcontroller.

Fortunately, there are alternatives. Perhaps the easiest is to use a data demultiplexer, a handy device that allows you to turn a few I/O lines into many. Demultiplexers are available in a variety of types; a common component offers three input lines and eight output lines. You can individually activate any one of the eight output lines by applying a binary control signal on the three inputs.

Table 14-1 shows which input control signals correspond to which selected outputs.

The demultiplexer includes the venerable 74138 chip, which is designed to bring the selected line low, while all the others stay high. One caveat regarding demultiplexers: only one output can be active at any one time. As soon as you change the input control, the old selected output is deselected, and the new one is selected in its place.

One way around this is to use an addressable latch such as the 74259; another way is to use a serial-to-parallel shift register, such as the 74595. The 74595 chip uses three inputs (and optionally a fourth, but for our purposes it can be ignored) and provides eight outputs. You set the outputs you want to activate by sending the 74595 an eight-bit serial word as shown in Table 14-2.

Fig. 14-23 shows how to interface to the 74595. In operation, software on your robot's computer or microcontroller sends eight clock pulses to the clock line. At each clock pulse, the data line is sent one bit of the serial word you want to use. When all eight pulses have been received, the latch line is activated. The outputs of the 74595 remain active until you change them (or power to the chip is removed, of course).

TABLE 14-1 Binary Demultiplexer Input to Output Device Selection

INPUT CONTROL	SELECTED OUTPUT
000	0
001	1
010	2
011	3
100	4
101	5
110	6
111	7

TABLE 14-2 Using the 74595 Serial-In/Parallel-Out Shift Register to Make Specific Outputs Active

SERIAL WORD IN	SELECTED OUTPUT(s)
00000001	0
00001001	0 and 3
01000110	1, 2, and 6

If this seems like a lot of effort to expend just to turn three I/O lines into eight, many microcontrollers (and some computers) used for robotics include a shiftout command that does all the work for you. This is the case, for example, with the BASIC Stamp 2 (but not the BASIC Stamp I), the BASICX-24, and several others. To use the shiftout command, you indicate the data you want to send and the I/O pins of the microcontroller that are connected to the 74595. You then send a short pulse to the latch line, and you're done! A key benefit of the 74595 is that you can cascade them to expand the I/O options even more.

There are still other ways to expand I/O lines, including serial peripheral interface (SPI), the Dallas 1-Wire protocol, and the 82C55. Several of the more commonly used systems were introduced earlier in this chapter. If your computer or microcontroller supports one or more of these systems, you may wish to investigate using these systems in case you find you are running out of I/O lines for your robot.

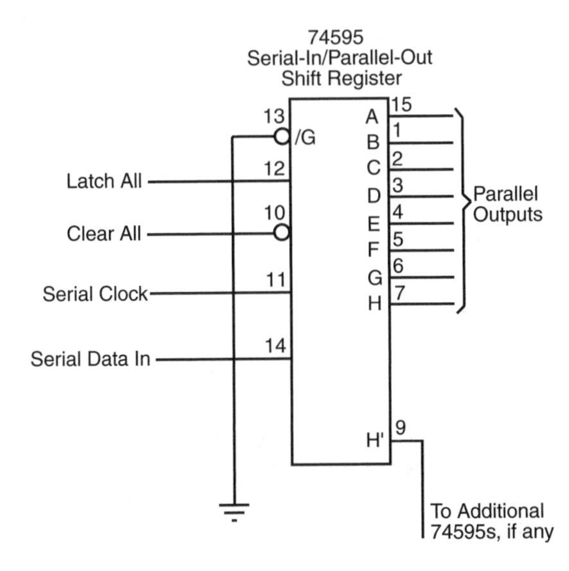

FIGURE 14-23 The 74595 serial-in/parallel-out (SIPO) shift register lets you expand the data lines and select multiple lines at the same time.

14.10 Bitwise Port Programming

Controlling a robot typically involves manipulating one or more input/output lines (bits) on a port attached to a computer or built into a microcontroller. A common layout for an I/O port is eight bits, comprising eight individual connection pins. This is the same general layout as the parallel port found on IBM PC-compatible computers, which provides eight data lines for sending characters to a printer or other device (along with a few additional input and output lines used for control and status).

The design of the typical microcontroller or computer, as well as the usual program tools for it, do not make it easy to directly manipulate the individual bits of a port. Rather, you must manipulate the whole port all at once and, in doing so, hopefully alter only the desired bits. The alternative is to send a whole value—from 0 to 255 for an eight-bit port, and 0 to 15 for a four-bit port—to the port at the same time. This value corresponds to the bits you want to control. For example, given an eight-bit port, the number 54 in binary is 00110110.

Fortunately, with a little bit of programming it's not hard to convert numeric values into their corresponding bits, and vice versa. Each programming language provides a different mechanism for these procedures, and what follows are some simple approaches using PBASIC. Other languages offer more robust bit-handling operators that you can take advantage of. The sample code that follows is meant more to teach you the fundamentals than to be applied directly with a robot. Take the ideas and adapt them to your particular case.

14.10.1 MASKING VALUES BY ANDing

It is not uncommon to have to manipulate the individual bits of a computer or microcontroller port. Quite often, you will find that the bits are not conveniently placed within bytes for easy manipulation. When working with PBASIC (and other microcontroller programming languages), individual bits can be easily manipulated, regardless of which line it comes in on. For example, declaring an input line on pin12 of a BS2 is accomplished by using the statement

```
SwitchIn var INS.bit12
```

and can be manipulated like any other bit in PBASIC

```
A = SwitchIn & 1
```

Quite often, however, it is not possible or practical to address each individual bit. Rather, you must read in an entire register and then mask off the bits that you would like to access as well as shift them down so they are zero-based for easy manipulation. In these cases, you will have to AND the bits with a number of set bits and then shift them up or down depending on the requirements of the application. In the following example program, three bits are input to the BS2, they are incremented by 1 and then output on three different lines:

```
X var word
mask var byte

  mask = %111      '  Just want the value of three bits
do                 '  Repeat Program Forever
  X = INS >> 12    '  Input Bits are 12 to 14, Shift Down
  X = X & mask     '  Mask any other bits
  X = X + 1        '  Increment Output Value
  X = X // 8       '  Make sure "X" is not greater than 7
  X = X << 3       '  Output Bits are 3 to 5, Shift Up
  OUTS = X         '  Output the Incremented Input
loop
```

14.10.2 CONVERTING A VALUE INTO A BINARY-FORMAT STRING

Numeric values can be readily converted into a string of 0s and 1s by testing to see which power of two bits are set in the value.

```
OutputString var byte(8)     '  8 Bit Output String
Power2 var byte              '  Power of 2 Variable
Temp var byte
X var byte                   '  Value to Convert
i var byte                   '  Counter

  X = 47                     '  Initialize the value to convert

  Power2 = 1                 '  Start at 2 to the power 0
  for i = 0 to 7
    Temp = X & Power2        '  Temp uses Power2 to Test Bits
    if (Temp = 0) then       '  If Bit is Zero, put in OutputString
      OutputString(i) = "0"
    else
      OutputString = "1"
      endif
      Power2 = Power2 * 2    '  Increment the Bit to be Tested
    next

  ' Output/Process "OutputString"

  end                        '  Finished, "OutputString" has Bit Values
```

14.10.3 SUMMING BITS INTO A DECIMAL VALUE

You will have plenty of occasions to convert a set of binary digits into a decimal value. This can be accomplished using the reverse of the previous program:

```
InputString var byte (8)     '  8 Bit Input Value String
Power2 var byte              '  Power of 2 Variable
Temp var byte
X var byte                   '  Converted Value
i var byte                   '  Counter

  InputString(0) = '1'       '  Define the Input Data
  InputString(1) = '1'
  InputString(2) = '0'
```

```
  InputString(3) = '1'
  InputString(4) = '1'
  InputString(5) = '1'
  InputString(6) = '0'
  InputString(7) = '0'
  X = 0                        '  Clear the Value
  Power2 = 1                   '  Start at 2 to the power 0
  for i = 0 to 7
    if (InputString(i) = '1') then
      X = X + Power2           '  If Bit is Set, then Add Power to Sum
    endif
    Power2 = Power2 * 2        '  Increment the Bit Value to be Tested
  next

'  Output/Process "X"

  end                          '  Finished, "X" has Value of the Bits
```

14.11 From Here

To learn more about . . .	***Read***
Motor specifications	Chapter 19, "Choosing the Right Motor"
Interfacing circuitry to DC motor loads	Chapter 20, "Working with DC Motors"
Computers and microcontrollers for robotic control	Chapter 12, "An Overview of Robot 'Brains' "
Input and output	Chapter 15, "The BASIC Stamp 2 Microcontroller"
Interfacing sensors	Part 6, "Sensors and Navigation"

CHAPTER 15

THE BASIC STAMP 2 MICROCONTROLLER

If you are familiar with the second edition of this book, you know that the example robot control circuits and programs were written for four different platforms: the PC's parallel port, the Parallax BASIC Stamp, the BasicX microcontroller (MCU), and the OOPic microcontroller. Each of these different platforms provides unique capabilities, but the end result was a book with a bit of a mishmash of different applications, which required the reader to understand each of the different hardware and software development platforms to know what the application was doing and how to convert (or port) it to the one that would be used with the robot. To simplify the book and your work in understanding how each application is to be implemented, it was decided to cut down the number of robot controllers described in the book to one.

Choosing a single example microcontroller with software for the book was not an easy task because there are literally hundreds of different microcontrollers with an equal number of software development tools that go with them. In the end, the decision was made to go with the Parallax BASIC Stamp 2 (abbreviated BS2), one of the devices that was already described in the second edition. The BS2 has the following features, which make it a good choice for new robot developers:

- Low cost for the MCU, and development software
- Programmer interface is built into the MCU
- Built-in +5 V regulator with 100 mA output capability
- Full-featured BASIC derivative language with many enhancements

- Variety of different models with varying performance and features
- RS-232 programming/debug interface
- Literally thousands of example applications to choose from
- Excellent manufacturer (Parallax Inc.) support

As you gain experience and sophistication, the downsides of the BS2 compared to other MCU solutions will become apparent. Many of these issues are a result of trade-offs to provide pin-specific functions or downloadability (in a chip that didn't have this feature when the BS2 was first designed). For example, once you have bought one BS2, the price of subsequent units does not go down because you already have the programmer and development software. As noted, the programmer interface is built into each BS2 rather than buying a single programmer and using that for multiple chips, as is done in most other MCUs. The +5 V regulator is somewhat fragile and cannot be used to power much more than just some LEDs and other basic interfaces. When the BS2 was designed (in the late 1990s), all PCs had RS-232 interfaces, which made sense at the time; now, with RS-232 being phased out of many systems, you will have to buy an USB to RS-232 converter. Finally, the execution speed is about 4000 statements per second, fast enough for most robot requirements but not fast enough for sophisticated interfaces.

Despite these issues, the BS2 is an excellent first MCU to work with and remains a favorite of many robot designers, both beginner and expert. The MCU itself is very reliable (especially after following a few rules that are outlined in this chapter); the software development tools are rock-solid stable (something you will appreciate more as you become more familiar with different systems); and Parallax provides an excellent selection of data sheets and application notes along with a full line of interfacing and support products for you to choose from.

15.1 Choosing the Right Stamp for Your Application

When you see most projects that have a BASIC Stamp in them, they use the standard 24 pin, 0.600-in wide DIP package shown in Fig. 15-1. This is only one of several packages and optional features that are available for the microcontroller. While the 24 pin package is the most popular and the one that you are most likely to use, there are about a dozen different ways that you could use BASIC Stamp packages in your robot. Some of these packages also make the integration of the MCU into the robot a lot easier than using the 24 pin package.

To start off, the Stamps are all designed with a similar architecture; a general-purpose microcontroller is connected to the programming interface to provide a download interface from a PC to the electrically erasable programmable read-only memory (EEPROM) program memory. During program execution, this microcontroller reads the EEPROM, interprets the program instructions, and carries them out using its built-in I/O pins. The I/O pins can be used for a variety of different purposes over and above simple digital input and out-

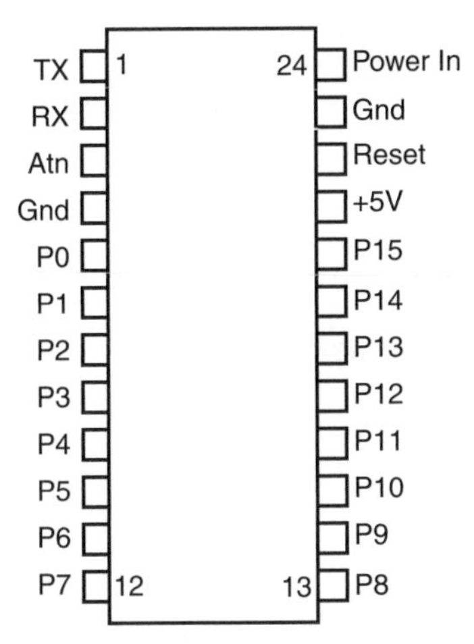

FIGURE 15-1 The connection layout or "pinout" of the BASIC Stamp 2.

put. The microcontroller is usually referred to as the interpreter chip. Along with the microcontroller, EEPROM, and serial programming interface, the Stamp has a power supply that will convert +5 to +15 V DC to 5 V for use by the microcontroller as well as some devices connected to the Stamp, an oscillator for the microcontroller along with a reset circuit. It should be obvious that while the Stamps are referred to as chips (primarily because they plug into a chip socket) they are actually complex modules that integrate a number of different components into a single microcontroller function.

There are three Stamp product families that you should be aware of. The first is the venerable BASIC Stamp 1 (BS1), which first came out in 1992. This microcontroller is built in either a very small *single in-line package* (SIP) (1.4 in long and 0.6 in high) using SMT parts or on a larger PCB (2.5 in by 1.5 in) that includes a small prototyping area. The BS1 has eight I/O pins and runs at a rate of about 2000 instructions per second. The EEPROM, which is used to store the program, will store about 80 instructions. There are a total of 14 variable memory bytes available. The BS1's parallel port programming differs from the other BASIC Stamps and is one of the features that makes it difficult to work with; Windows/2000 and Windows/XP, which are the most commonly used PC operating systems, do not allow programs direct access to the Parallel port. Parallax sells (and has published the schematic so that you can build it yourself) a serial to parallel BS1 programming interface so that the BS1 can be used with the latest versions of Windows (using the latest BASIC Stamp interfacing software).

The BASIC Stamp 2 (BS2), which will be used as the example MCU in this book, is the second-generation MCU and by far the most popular. It is important to realize that there is more than one BS2; Parallax offers a number of different devices (listed in Table 15-1) that offer different features to the original BS2 (which is given the part number BS2-IC by Parallax).

In Table 15-1 the BASIC Stamp's speed is measured in instructions per second; the different BS2s execute 4000 (4k) to 19,000 instructions per second. This is a surprisingly accurate specification for simple operations, and the reciprocal of the speed specification can be used to roughly calculate the operation time of a block of code.

TABLE 15-1 The Features of the Different BS2s

FEATURES	BS2-IC	BS2E-IC	BS2SX-IC	BS2P24-IC	BS2P40-IC	PS2PE-IC	BS2PX-IC
Speed	4k in/s	4k in/s	10k in/s	12k in/s	12k in/s	6k in/s	19k in/s
RAM Bytes	26	26	26	26	26	26	26
EEPROM	2k	16k	16k	16k	16k	16k	16k
I/O Pins	16 + 2s	16 + 2s	16 + 2s	16 + 2s	32 + 2s	16 + 2s	16 + 2s
Source/Sink	20/25	30/30	30/30	30/30	30/30	30/30	30/30

RAM and EEPROM are specified in bytes, and while the EEPROM seems reasonable, the 26 bytes of RAM (used for variables) probably seems to be unusable. Look through the example applications presented in this book and you will probably be surprised to see that very sophisticated applications can be created for the BS2, despite the apparently limited variable memory available to it.

As a rule of thumb, remember that every 2k of EEPROM will store 500 Stamp BASIC statements.

Along with a set number of general-purpose I/O pins, there are two RS-232 serial interface I/O pins available for application use and communicating with another computer. The RS-232 interface consists of a voltage stealing circuit, so a standard –15 to –3 V and +3 to +15 V RS-232 can be used safely. The RS-232 serial port pins double as programming interface and debug communications pins. All I/O pins, except for the two serial interface pins are bi-directional and perform a variety of different functions that are described later in this chapter.

Finally, the amount of current the I/O pins can source (provide to a device) or sink (take away from a device) is measured in milliamps. With the fairly low currents output from the different stamps, it should be noted that they can only control a logic gate, transistor, or LED or two. They do not have sufficient power to drive motors or other magnetic devices.

All the BASIC Stamps have a built-in power supply, but as I will discuss later in this chapter, it should be bypassed for most applications and instead an external regulated power supply should be used. The reason for this recommendation is due to the low current sourcing capability of the power supply on the BS2—if an excessive amount of current is drawn the power supply can be burned out.

The Javelin Stamp executes Java code rather than the Stamp BASIC of the other two types of Stamps. It executes approximately 8.5k instructions per second and has 32k bytes of variable space as well as 32k of EEPROM program memory. The Javelin is generally plug compatible with the BS2, but there will be issues with timing the code to match the application timing of the BS2s.

With all these different Stamp options to choose from, which one is right for you?

For most applications and if you are starting out, the simplest BASIC Stamp 2, the BS2-IC will probably be the best. This part, while seemingly less than the other versions of the

BS2, will be fast enough and have enough EEPROM memory for most applications and it costs as much as 40% less than the other BS2s. The application code written for the BS2-IC can be used with only very minor changes (specifically in the areas of timing) in the other BS2 part numbers, so any investment in software will not be wasted if high-performance BS2s are required later.

Starting at the low end of any device family is always a good idea. With the BASIC Stamp 2, there are 20 to 30 different permutations and combinations or BS2 part number, packaging options, and carrier board options. If you consider the literally thousands of different microcontrollers available in the market that could be used in a robot, the number of options that you have would certainly be greater than 10,000. As you learn more about electronics, programming, and microcontrollers, you will begin to develop preferences for different technical features in the chips you work with; but for now, start with a simple device that is extremely popular (which means there are a number of applications and books that you can reference), can be interfaced to and programmed easily, and is supported by an excellent company with a good variety of data sheets, products, and services you can rely on. When these factors were taken into consideration, the BS2 seemed to be the correct single choice for demonstrating in this book how a microcontroller worked rather than trying to use a variety of different devices for the job.

15.2 Inside the Basic Stamp

When the BASIC Stamp was first designed, low-cost electrically reprogrammable microcontrollers did not exist; this led to the decision to create one by adding a serial EEPROM memory along with a preprogrammed MCU, which resulted in the ability to download and execute applications written on a PC. Fig. 15-2 expands on Fig. 15-1 and shows the circuitry connected to each of the BS2's pins. Instead of building a robot control circuit out of numerous inverters, AND gates, flip-flops, and other hardware, you can use just the BS2 module to provide the same functionality and do everything in software.

Because the Stamp accepts input from the outside world, you can write programs that interact with that input. For instance, it's easy to activate an output line—say, one connected to a motor—when some other input (like a switch) changes logic states. You could use this scheme, for instance, to program your robot to reverse its motors if a bumper switch is activated. Since this is done under program control and not as hardwired circuitry, it's quite simple to change and enhance your robot as you experiment with it.

In operation, your PBASIC program is written on a PC, then downloaded—via a serial connection—to the BASIC Stamp, where it is stored in EEPROM, as shown in Fig. 15-3. The program in the EEPROM is in the form of *tokens,* special instructions that are read, one at a time, by the PBasic interpreter stored in the BASIC Stamp's PROM memory. During program execution, temporary data is kept in RAM. Note that the EEPROM memory of the BASIC Stamp is nonvolatile—remove the power and its contents remain. The same is not true of the RAM. Remove the power from the BASIC Stamp and any data stored in the RAM is gone. Also note that the PBASIC interpreter, which is stored in the programmable read-only memory (PROM) of the interpreter microcontroller, cannot be changed.

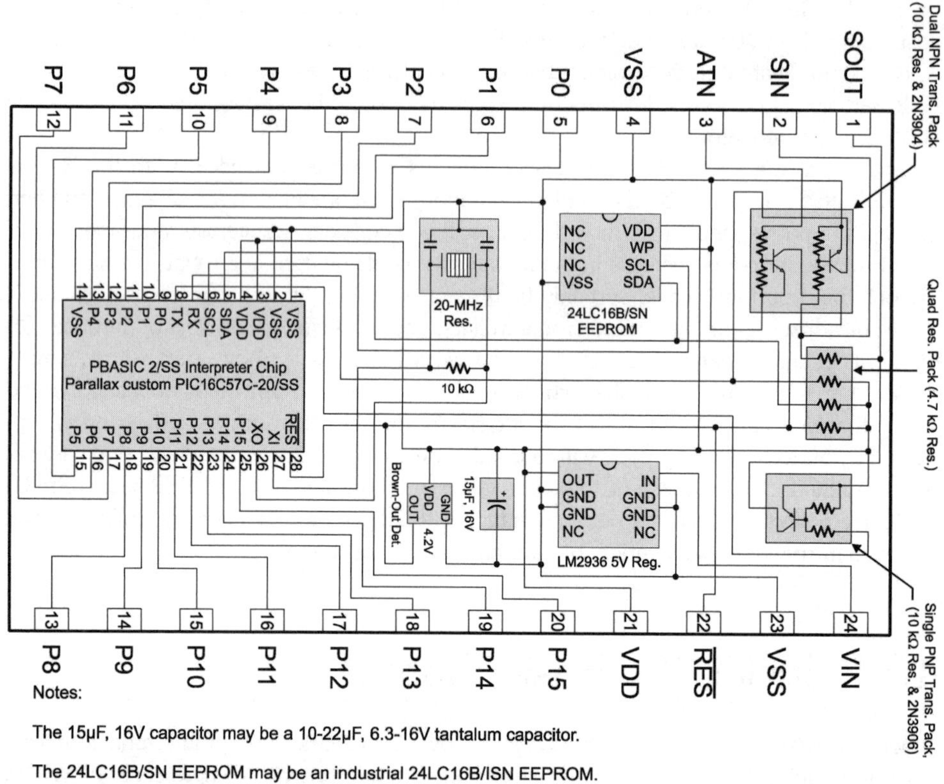

Notes:

The 15μF, 16V capacitor may be a 10-22μF, 6.3-16V tantalum capacitor.

The 24LC16B/SN EEPROM may be an industrial 24LC16B/ISN EEPROM.

The PBASIC2/SS Interpreter may be a PIC16C57C-20/SS or an industrial PIC16C57C-20I/SS.

The 20 MHz resonator is not polarity sensitive and the middle pin can be connected to either VDD or VSS.

FIGURE 15-2 The circuitry behind the BS2's pins. While the circuitry seems quite complex (especially if you are new to electronics), it actually works together to create an easy to use and program device that is well suited to controlling robots.

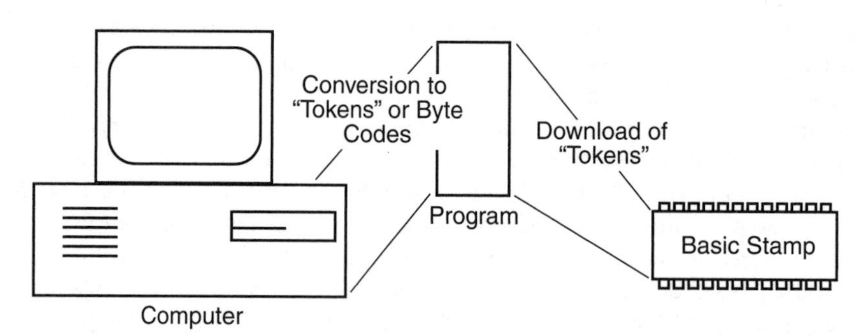

FIGURE 15-3 Programs are originally created on your PC and downloaded to the BS2 where they are stored in "tokenized" format in EEPROM. The PBASIC Interpreter built into the BS2's microcontroller reads and executes these "tokens."

15.3 Developer's Options

The BASIC Stamp 2 is available directly from Parallax, its manufacturer, or from a variety of dealers the world over. In addition to the BS1, BS2, and Javelin Stamp and the variations listed previously, you'll find that there are several assembled development boards that can be used as the complete circuitry for your robots.

- The *Board of Education USB programming board* (Fig. 15-4) is recommended for anyone that has never worked with a microcontroller before. This BS2 carrier board provides a socket for the BS2 along with power supply connectors and control, a breadboarding area, servo connectors, as well as a USB programming interface. The USB programming interface provides a USB to serial programming and debug connection between your PC and the BS2, eliminating the need for a serial port.
- The *Board of Education carrier board—Serial* is similar to the Board of Education USB programming board, except that the programming interface is serial rather than USB.
- The *Parallax Professional Development Board* has virtually all the interfaces built in for you to learn about the different BASIC Stamps. It would be difficult to interface this product into a robot (especially compared to the previously listed products), but it is an excellent tool for learning about BASIC Stamp interfacing and application hardware and software design.
- The *BASIC Stamp HomeWork board* is a stand-alone BS2 built on a PCB along with a breadboard and connectors for a 9-V alkaline (or NiMH) battery. This product is sold

FIGURE 15-4 The USB Board of Education can be used as an experimenter's PCB or even the electronics carrier for your robot.

in groups of 10 or 20 or singly in the WAM Kit from Radio Shack and is an ideal low-cost way of starting out with the BS2. Like the Boards of Education, this PCB could be built into a robot, but additional power would be required for driving servos, which isn't an issue with the Board of Education products.

- The *BASIC Stamp carrier boards* are PCBs that have sockets for BASIC Stamps and allow interface circuitry to be wire wrapped or point to point soldered to them. These boards are useful in situations where a small amount of circuitry is required to interface a Stamp to a robot's circuitry. There are a number of different carrier boards available with different features and able to accept different Stamp modules.

Parallax has designed a number of robots that use the BASIC Stamp 2 as the central processor and use the developer products listed (like the Board of Education PCBs in the case of the Boe-Bot) as carriers. These robots are well designed, consist of high-quality parts, and have excellent documentation. As the creators of PBASIC and the BS2, Parallax has provided additional features to its robots, such as the Gui-Bot software, which provides a graphical programming environment for the BS2 built into the Boe-Bot.

A number of other manufacturers have designed development boards and products around the BS2. The Tab Electronics Sumo-Bot kit from McGraw-Hill is a differentially driven, tracked robot that has a serially programmed BS2 built into it for user programmability.

15.4 Understanding and Using PBASIC

At the heart of the BASIC Stamp series of microcontrollers is the PBASIC programming language. PBASIC has undergone a number of changes over the years and has developed from a very rudimentary language to the current incarnation (2.5 at time of writing), which includes structured programming statements and methods of accessing pins that will make applications easier to read and understand.

PBASIC programs for the BASIC Stamp 2 are developed in the *BASIC Stamp Editor,* a Windows application that can be downloaded from the Parallax web site. The editor lets you write, edit, save, and open PBASIC programs as well as compile and download (run) your finished programs to a BASIC Stamp 2. The download/run step requires that your BASIC Stamp be connected serially to the development PC. Fig. 15-5 shows the BASIC Stamp Editor in operation.

If you do not use Windows, there are Linux and Macintosh (OS 9 and later) versions of the PBASIC software written by Stamp enthusiasts that can be downloaded off the Internet. Along with these interfaces, Parallax has made the *tokenizer* (which is called the *compiler* in this book) for PBASIC available for Windows, Linux, and Mac OS/9+. To find out more about this software, check the Parallax web site for pointers to locations for download.

Like any language, PBASIC is composed of a series of statements strung together in a logical syntax. A statement is *compiled* into a series of instructions that the BS2's interpreter chip carries out or executes. For example, one statement may tell the chip to fetch a value on one of its I/O pins, while another may tell it to wait a certain period of time.

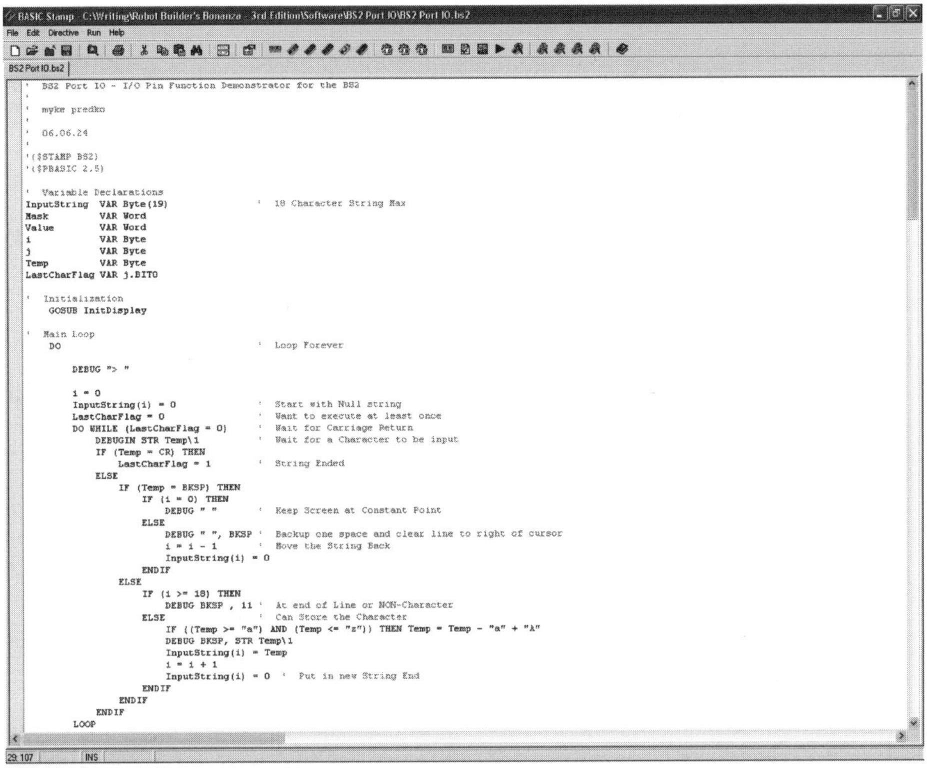
BASIC Stamp - C:\Writing\Robot Builder's Bonanza - 3rd Edition\Software\BS2 Port IO\BS2 Port IO.bs2
File Edit Directive Run Help
BS2 Port IO.bs2
' BS2 Port IO - I/O Pin Function Demonstrator for the BS2
'
' myke predko
'
' 06.06.24
'
'{$STAMP BS2}
'{$PBASIC 2.5}
' Variable Declarations
InputString VAR Byte(19) ' 18 Character String Max
Mask VAR Word
Value VAR Word
i VAR Byte
j VAR Byte
Temp VAR Byte
LastCharFlag VAR j.BIT0
' Initialization
GOSUB InitDisplay
' Main Loop
DO ' Loop Forever
DEBUG "> "
i = 0
InputString(i) = 0 ' Start with Null string
LastCharFlag = 0 ' Want to execute at least once
DO WHILE (LastCharFlag = 0) ' Wait for Carriage Return
DEBUGIN STR Temp\1 ' Wait for a Character to be input
IF (Temp = CR) THEN
LastCharFlag = 1 ' String Ended
ELSE
IF (Temp = BKSP) THEN
IF (i = 0) THEN
DEBUG " " ' Keep Screen at Constant Point
ELSE
DEBUG " ", BKSP ' Backup one space and clear line to right of cursor
i = i - 1 ' Move the String Back
InputString(i) = 0
ENDIF
ELSE
IF (i >= 18) THEN
DEBUG BKSP , 11 ' At end of Line or NON-Character
ELSE ' Can Store the Character
IF ((Temp >= "a") AND (Temp <= "z")) THEN Temp = Temp - "a" + "A"
DEBUG BKSP, STR Temp\1
InputString(i) = Temp
i = i + 1
InputString(i) = 0 ' Put in new String End
ENDIF
ENDIF
ENDIF
LOOP
29:107
INS

FIGURE 15-5 The BASIC Stamp Editor is used to create, compile, and download programs for the BASIC Stamp.

The majority of PBASIC statements can be categorized into four broadly defined groups: variable and pin or port definitions, assignment statements with arithmetic expressions, decision structures, and built-in functions. A PBASIC program will contain elements taken from each group.

The following sections provide an introduction to the PBASIC language with just enough information for you to become dangerous, as the saying goes. For a complete PBASIC language definition, you should go to the Parallax web site and download the PBASIC Language Specification, study it, and experiment with the different features and functions available to you for the BASIC Stamp MCU that you are working with.

15.4.1 VARIABLE AND PIN/PORT DEFINITIONS

PBASIC uses labeled variables as the methodology to store data as well as allow you to read and change the I/O pins of the microcontroller. The variables themselves can be of several different sizes; since there are only 26 bytes of RAM available for use as variables you should always strive to choose the smallest variable size that will accommodate the data you wish to store. The I/O pins can be accessed in a variety of different ways, as will be discussed in this section.

PBASIC provides the following four variable types:

- Bit—1 bit (0 or 1)
- Nibble—4 bits (0 to 15)
- Byte—1 byte (0 to 255)
- Word—2 bytes (0 to 65,535)

Variables must be declared in a PBASIC program before they can be used. This is done using the *var* statement, as in

```
VarName var VarType
```

where VarName is the name (or symbol) of the variable, and VarType is one of the variable types just listed. VarName must start with an alphabetic (a through z) or underscore (_) character and the remaining characters can be alphabetic, underscore, or numeric (0 through 9). Here are some examples:

```
Red  var bit
Blue var byte
```

After reading these statements, the PBASIC compiler will make Red a bit variable and Blue a byte. It should be noted that the PBASIC compiler is pretty good at compressing variables—up to eight bits in a byte, or two nibbles in a byte.

In these examples and in the example code throughout the book, the capitalization of the variables is fairly carefully thought out. This is due to the authors' experience with other high-level languages that do track capitalization whereas PBASIC ignores variable name and other label capitalization. The Stamp Editor will capitalize reserved words to make them stand out visually, but if you were to key in a program in a standard text editor, the capitalization specified in the text file would be unchanged and ignored by the PBASIC compiler. The following declaration statements produce the same result as the previous ones:

```
Red  Var Bit
BLUE VAR BYTE
```

Single dimensional arrays can be specified by simply stating the desired number of elements for the array as part of the type definition:

```
Green Var Byte (10)         '  10 Bytes are used for the Array "Green"
```

The text after the single quote (') is known as a *comment* and is specifically for the use of the human reading the code. Code following the single quote is ignored by the compiler.

Finally, variables can be redefined or smaller parts of variables used to extract specific data information. For example, to look at bit 2 of a returned data variable:

```
RetData var byte            ' Data returned from subroutine
RetFlag var RetData.bit2    ' Variable consisting of bit 2 of
                            ' the returned data
```

Once declared, variables can be used throughout a program. The most rudimentary use for variables is part of the assignment statement, which is discussed in the next section. In later sections, comparisons to different values and use in built-in functions is discussed.

Variables store values that are expected to change as the program runs. PBASIC also supports constants, which are used as a convenience for the programmer. Constants are declared much as variables are, using the *con* statement:

```
MyConstant con 5
```

Any time MyConstant is encountered in the PBASIC program, the string MyConstant is replaced with the value 5.

The BASIC Stamp treats its 16 I/O pins like additional memory that can be accessed like variables or using special functions. There are actually three different ways of accessing I/O pins, which can be confusing for new programmers, especially when the tri-state driver functions are included into the mix.

Each pin has three bits associated with it. The IN bit returns the data value at the output pin. OUT is written to set the output state of the pin (if it is in output mode) and can also be read back. The DIR bit is used to specify whether the pin is in input mode (value 0 and cannot output a logic signal) or output mode (value 1 and can drive a logic signal from the pin).

These three bits are generally thought of as part of three 16-bit variables. INS is the 16-bit variable for each of the 16 IN bits, OUTS is the 16-bit variable for each of the 16 OUT bits, and DIRS is the 16-bit variable for each of the 16 DIR bits. The bits can be accessed as 16 bit (suffix S), 8 bit (suffix L or H), 4 bit (suffix A to D), or 1 bit (suffix 0 to 15). The suffix describes the number of bits and which ones. For example, DIRL will access the lower 8 bits of the DIRS variable. To help understand how these variables operate, look at the I/O Port Simulator at the end of the example applications in this chapter.

Another way to access the I/O pins is to avoid the IN, OUT, and DIR variables all together and use the built-in pin I/O functions of PBASIC, which take a numeric value or a pin variable name (described in the following):

```
High #       '  Put pin in Output Mode and Output a High Voltage ("1")
Low #        '  Put pin in Output Mode and Output a Low Voltage ("1")
Toggle #     '  Change the Output Value (Low to High, High to Low)
Input #      '  Put pin in Input Mode
Reverse #    '  Switch pin mode (Output to Input, Input to Output)
```

The problem with this method is that it does not allow for easy reading of the input pins without using the built-in variables previously specified, which can get confusing when you are new to programming. The pin declaration

```
MyPinName pin 4
```

will create a combination of a variable and constant so that you can access a pin value like a variable or specify its operating mode using one of the five built-in functions previously listed. For example, to create an input pin and poll (read) its value, you could use the code:

```
InputPin pin 4          '  Pin 4 is an input
    Input InputPin      '  Put "InputPin" (Pin 4) into "input mode"
    A = InputPin        '  Read and store the current value of "InputPin"
```

15.4.2 ASSIGNMENT STATEMENTS AND ARITHMETIC EXPRESSIONS

The most fundamental operation that the BS2 (or any computer system for that matter) provides is data movement and manipulation. Without this ability, new outputs cannot be derived from the inputs, and execution changes due to changes in the environment are not possible. The assignment statements and arithmetic expressions used in the BS2 are very similar to their counterparts in other programming languages.

The assignment statement simply evaluates an arithmetic expression and stores the result in a variable. The format of the assignment statement is:

```
VarName = expression
```

While the VarName = part of the assignment statement looks straightforward, the expression part probably seems ominous. There's no need to worry; the term *expression* simply means some number of characters, representing a series of arithmetic operations that evaluate to a numeric value. The assignment statement's expression can be as simple as a constant value like

```
A = 23        '  Save the decimal value '23' in variable 'A'
```

or it can be another variable

```
B = A         '  Save the contents of variable 'A' into variable 'B'
```

Expressions can also be arithmetic operators changing a value in some way. For example, if you wanted to save the value of A multiplied by four in B, you could use the assignment statement and expression combination:

```
B = A * 4     '  Save the product of the contents of 'A' and 4 in 'B'
```

Table 15-2 lists the different expression operators built into PBASIC that can be used with assignment statements. Note that some of them operate with one parameter while others work with two. Only the minus sign (–) can be used with either one parameter or two; when it is before (to the left of) the parameter, it returns the two's complement negative value of the parameter, and when it is between two parameters it subtracts the second from the first.

Multiple operations can be built into an expression, but there is a facet of PBASIC that is different from other programming languages. Normally in a programming language, there is an order of operations that specifies how mathematical operations in an expression execute. In PBASIC, there is no order of operations. Everything executes from left to right unless there are parentheses to help explain the desired order of operations.

For example, if you were to convert a temperature in Fahrenheit to Celsius, you would want to use the arithmetic expression:

$$D_{Fahrenheit} = 9 / 5 * D_{Celsius} + 32$$

This statement would evaluate 9 divided by 5 (1), and multiply it by $D_{Celsius}$ and then add 32 before saving the result in $D_{Fahrenheit}$. The division took place first because it is the left-

TABLE 15-2 PBASIC Assignment Statement Operators

OPERATOR	FUNCTION	PARAMETERS	EXAMPLE USAGE
+	Addition	2	A = B + C
-	Subtraction	2	A = B - C
-	Negation	1	A = -B
*	Multiplication	2	A = B * C
*/	Fractional Multiplication	2	A = B */ C ' Product Middle
**	High Word Multiplication	2	A = B ** C ' High Product Word
/	Division	2	A = B / 2 ' Integer Division
//	Modulus	2	A = B // 2 ' Return Odd/Even
&	Bitwise AND	2	A = B & $F ' Return Low 4 Bits
|	Bitwise OR	2	A = B | %1 ' Return Odd/Even
^	Bitwise XOR	2	A = B ^ 1 ' Switch Lowest Bit
<<	Shift Left n Bits	2	A = B << n ' Shift Left n Bits
>>	Shift Right n Bits	2	A = B >> n ' Shift Right n Bits
ATN	Get arcTangent	2	A = XCord ATN YCord
HYP	Get Hypotenuse	2	A = XCord HYP YCord
MIN	Return Minimum Value	2	A = Value1 MIN Value2
MAX	Return Maximum Value	2	A = Value1 MAX Value2
DIG	Return Specified Digit	2	A = Value DIG Digit
REV	Reverse Set # of Bites	2	A = Value1 REV Value2
ABS	Return Absolute Value	1	A = ABS B
COS	Return Cosine of Value	1	A = COS B
SIN	Return Sine of Value	1	A = SIN(B)
~	Bitwise Invert	1	A = ~B
SQR	Return Square Root	1	A = SQR B
DCD	Set Specified Bit	1	A = DCD 4 ' Same as A = 1 << 4
NCD	Return Highest Bit	1	A = NCD 32 ' Returns '5'

most operator. Dividing constant values can be tricky because numeric values in PBASIC (and other microcontroller programming language values) are integers (with no decimal points) only. Before doing any division, remember to always do the multiplication first.

To force the PBASIC compiler to put your arithmetic operators in the desired order, put parentheses around the higher priority (and first executing) operators and their parameters. Using parameters to fix the degree conversion assignment statement becomes:

```
D<sub>Fahrenheit</sub> = ((D<sub>Celsius</sub> * 9) / 5) + 32
```

When you have more than one operator in an assignment statement, it is usually referred to as a *complex* assignment statement and it is a good idea to use parentheses to ensure the expression is evaluated in the order you want.

15.4.3 EXECUTION FLOW AND DECISION STRUCTURES

Normally, software flows from one statement to the next with each statement performing calculations or inputting or outputting data that are required to meet the application's specifications. In the first computer systems, the programs performed basic calculations and stopped when the answer had been calculated. In robot systems, this is not possible because the robots run continuously; for the programs to execute from line to next line continuously the programs would have to be literally infinitely long. To avoid the chore of writing infinitely long programs, computer scientists came up with the concept of looping a section of code along with being able to change where the program executes algorithmically using decision structures. *Decision structures* is a ten-dollar phrase that describes the programming statements that change the flow of the program's execution based on different inputs and variables. Before you can successfully develop application software programs, you have to understand how a program flows and changes operation using decision structures.

Essentially all robot programs are based on the program template:

```
I/O & Variable Set Up ("Initialization")
Loop Start
    Delay to allow Inputs to Change, Robot to Move
    Input data from the environment
    Process data from the environment
    Output the processed data
Loop End - Jump Back to Loop Start
```

All of the statements in this example program structure can be implemented using the assignment statement except for the Loop Start and Loop End. The Loop End statement causes execution flow of the program to jump back to the start of the loop code, which is indicated by Loop Start. In PBASIC, Loop Start and Loop End are implemented using the DO and LOOP statements, changing the robot program structure to:

```
I/O & Variable Set Up ("Initialization")
DO
    Delay to allow Inputs to Change, Robot to Move
    Input data from the environment
    Process data from the environment
    Output the processed data
LOOP
```

Using the DO/LOOP statements, the steps within them are continuously repeated, eliminating the need to write an infinitely long program. Additional loops can be put inside the main loop. There will be cases, however, in which you want the looping to take place for a certain length of time or while some condition is true.

The DO/LOOP (pronounced "do loop") statements can execute conditionally when WHILE or UNTIL conditions are attached to them. After the WHILE or UNTIL words, a comparison expression is used to indicate when the looping should stop. In Table 15-3, the different types of DO/LOOPs are listed with the WHILE and UNTIL words in place, and Table 15-4 lists the six comparison expressions that can be used with WHILE and UNTIL. Note that the WHILE or UNTIL and comparison expressions can be placed either at the start of the loop (testing before going in the first time) or after working through the loop statements at least once.

The comparison expression is in the format

```
VarName|Constant operator VarName|Constant
```

where VarName I Constant is a variable or constant and operator is the equals, less than, or greater to sign, in one of the six combinations shown in Table 15-4. The complement value is the opposite and lets you use the complement word (UNTIL is the complement of WHILE and vice versa). When WHILE or UNTIL is used with the DO/LOOP, loops execute a set number of times, as in the following example:

```
i = 1
DO WHILE (i <= 7)       ' Loop 7x
    ...                 ' Code that executes 7x
    i = i + 1           ' Increment Loop Counter
LOOP
```

TABLE 15-3 Different Forms of the PBASIC DO/LOOP

DO/LOOP TYPE	DESCRIPTION	THE LOOPING ENDS . . .

`DO`
`... Loop Statements`
`LOOP`
	Infinite Loop, statements between DO and LOOP are executed forever	Never

`DO WHILE Comparison`
`... Loop Statements`
`LOOP`
	Loop while the comparison expression evaluates to true	When comparison expression is false

`DO UNTIL Comparison`
`... Loop Statements`
`LOOP`
	Loop while the comparison expression evaluates to false	When comparison expression is true

`DO`
`... Loop Statements`
`LOOP WHILE Comparison`
	Loop while the comparison expression evaluates to true	When comparison expression is false

`DO`
`... Loop Statements`
`LOOP UNTIL Comparison`
	Loop while the comparison expression evaluates to false	When comparison expression is true

TABLE 15-4 Comparison Expressions With Different Operators

COMPARISON TYPE	OPERATION	COMPLEMENT
A = B	Return true if two arguments are the same value	A <> B
A <> B	Return true if the two arguments are the same	A = B
A > B	Return true if the first argument is greater than the second	A <= B
A >= B	Return true if the first argument is greater than or equal to the second	A < B
A < B	Return true if the first argument is less than the second	A >= B
A <= B	Return true if the first argument is less than or equal to the second	A > B

Note that the code statements inside the DO/LOOP are indented a number of spaces. This optional indentation makes it easier for people to see which code is part of a loop and what is outside it. This convention is used by most programmers and is one that you should follow to make it easier for both yourself and others to read the source code.

In this example code, the code inside the loop executes seven times before execution flows out of the loop and on to the rest of the program. The DO WHILE/LOOP statement can be used for this purpose, but it is more traditional to use the FOR/NEXT statements:

```
FOR VarName = StartValue TO EndValue [STEP StepValue]
    ...                    ' Code Executing inside the "FOR/NEXT" Loop
NEXT
```

The FOR statement initializes a variable with an initial value and then executes until the variable is equal to the final (end) value with the variable being incremented during each loop. If no StepValue is specified, then the incrementing value is one. Using the FOR/NEXT, the seven times loop code shown previously could be reduced to:

```
FOR i = 1 TO 7
    ...                    ' Code that executes 7x
NEXT
```

While the program space gains seem to be marginal over the DO/LOOP, the major advantage of using the FOR/NEXT in this case is how obvious it is to somebody reading the code. Rather than decoding the DO/LOOP statements, the FOR/NEXT is known to be a counting loop.

Before going on, it should be pointed out that the comparison expressions of the DO/LOOP statements can be expanded beyond the simple examples shown so far. First, the comparison values can have arithmetic operators, causing them to calculate values in the DO/LOOP statement as shown below:

```
DO WHILE (A * 4) > 32
    ...                    ' Repeated code (uses "A" in calculations)
    A = A - 4              ' Reduce value of "A"
LOOP                       ' Repeat the Loop
```

Like the complex arithmetic expressions described in the previous section, remember that the evaluation of the expressions takes place left to right with priority given to values and operators in parentheses. When you are starting out, it is probably a good idea to avoid performing calculations in your comparison expressions until you are very comfortable with programming.

The second enhancement to the comparison operators is the ability to AND as well as OR the results of multiple comparison operators together:

```
DO WHILE ((A * 4) > 32) AND (STOPFlag = 0)
    ...                   ' Repeated code (uses "A" in calculations)
    A = A - 4             ' Reduce value of "A"
LOOP                      ' Repeat the loop
```

To make reading the program easier, remember to place each comparison expression inside parentheses (and this is actually a pretty good idea anyway).

Along with looping a program or sections of the program for a set number of times, it is also possible to execute blocks of code conditionally. The traditional way of doing this is to use the IF/ELSE/ENDIF statement which evaluates a conditional expression and if the result is true executes specific code. As shown in the following example code, the condition expression of the IF/ELSE/ENDIF statements is exactly the same as the one used with the DO/LOOP:

```
IF (A = B) THEN
    ...                   ' Code Executes if A = B
ELSE
    ...                   ' Code Executes if A <> B
ENDIF
```

The code after the ELSE statement executes if the comparison expression evaluates to false. Another way of saying this (refer back to Table 15-4), is that the ELSE code executes if the complement comparison expression is true. The ELSE code is optional and no part of it will execute if the comparison is true.

Along with the IF/ELSE/ENDIF statements, if there are multiple constant values for a variable that each result in a different execution path, then the SELECT/CASE/SELECTEND statements can be used to specify blocks of code that execute when the variable is compared to any of these values.

```
SELECT(VarName)
    CASE Constant1
        ...               ' Code Executes if VarName = Constant1
    CASE Constant2
        ...               ' Code Executes if VarName = Constant2
    :
    CASE ELSE
        ...               ' Code Executes if VarName <> ANY Constants
ENDSELECT
```

The SELECT/CASE/ENDSELECT statements could be modeled using IF/ELSE/ENDIF statements arranged as:

```
IF (Varname = Constant1)
    ...                   ' Code Executes if VarName = Constant1
ELSE
        IF (Varname = Constant1)
            ...           ' Code Executes if VarName = Constant2
    ELSE
        ...               ' Code Executes if VarName <> ANY Constants
    ENDIF
ENDIF
```

Note that when the multiple IF/ELSE/ENDIF statements are used together, each line is indented to indicate what previous statement it executes under.

The DO WHILE/LOOP, DO UNTIL/LOOP, DO /LOOP WHILE, DO /LOOP UNTIL, IF/ELSE/ENDIF, and SELECT/CASE/ENDSELECT statements and the code associated with them are known as *decision structures* because they are well-defined blocks of code that execute based on the result of a test (or *decision*). They also go under other names such as *flow control* or *conditional execution* statements. Regardless of their name, they are common to most programming languages that you will be working with.

If you are already familiar with PBASIC or are referencing other books about BS2, you will probably see that a whole class of decision structures, the IF/THEN and BRANCH, statements is not mentioned. Along with this, if you look on the Internet for sample applications, you probably won't see the decision structures listed here at all in the code.

With the introduction of PBASIC 2.5, Parallax introduced these decision structures to allow the application developer to write code using structured programming methodologies. Structured programming eschews the use of GOTO statements as they make a program difficult to read and certain operations, like rewriting the previous IF/ELSE/ENDIF example code.

```
IF (A <> B) THEN NotEquals
    ...                   ' Code Executes if A = B
    GOTO Finished
NotEquals:
    ...                   ' Code Executes if A <> B
Finished:
```

difficult to implement correctly. (The character strings ending in a colon (:) are labels and indicate specific points in the code execution should jump to.) It is considered good programming form to only use structured programming statements, such as the ones presented in this section, and avoid the use of GOTOs or statements that execute GOTOs all together.

Finally, it was implied at the start of the section that all programs should use the program template provided previously as a basis for application programs, which required the use of a DO/LOOP implemented as an infinite loop. There are cases where you will want to execute a simple program that executes a number of statements and stops. This type of program is good for experimenting with the BS2 and learning about different PBASIC statements and functions.

When implementing this type of program, always remember to place an END statement at the bottom of the code. The END statement stops the BS2's interpreter and leaves the

I/O pins in their current state until power or the reset pin is cycled or a new program is loaded into it. Depending on the function of the example program, either a DO/LOOP is used or an END statement.

15.4.4 BUILT-IN FUNCTIONS

The PBASIC language supports several dozen built-in functions that are used to control some activity of the chip, including sounding tones through an I/O pin or waiting for a change of state on an input. The following functions are among the most useful for robotics. You'll want to study these statements more fully in the BASIC Stamp manual, and more information regarding many of them are discussed elsewhere in this book.

- *Button.* The button function momentarily checks the value of an input and then branches to another part of the program if the button is in a low (0) or high (1) state. This function lets you choose which I/O pin to examine, the target state you are looking for (either 0 or 1), and the delay and rate parameters that can be used for such things as switch debouncing. The button function doesn't stop program execution, which allows you to monitor a number of I/O pins at once. This function's operation is somewhat difficult to understand and will be explained in detail later in the book.
- *Debug* and *debugin.* The BASIC Stamp Editor has a built-in terminal that passes data, in different formats, to and from the BASIC Stamp with the programming PC. These functions are highly useful during testing; for example, you can have the debug function display the parameters that were used to calculate the current output state of an I/O pin, so you can determine whether the program is working properly.
- *Freqout.* The freqout function is used to generate tones primarily intended for audio reproduction. You can set the I/O pin, duration, and frequency (in hertz) using this function. An interesting feature of freqout is that you can apply a second frequency, which intermixes with the first. For example, you can combine a straight middle A (440 Hz) with a middle C (523 Hz) to create a kind of chord. Don't expect a symphonic sound, but it works for simple tunes. When freqout is used to drive a speaker you should connect capacitors (and resistors, as required) to build a filter.
- *Pause.* The pause function is used to delay execution by a set amount of time. To use pause you specify the number of milliseconds (thousandths of a second) to wait. For example, pause 1000 pauses for 1 s.
- *Pulsin.* The pulsin function measures the width of a single pulse with a resolution of two microseconds (2 μs). You can specify which I/O pin to use, whether you're looking for a 0-to-1 or 1-to-0 transition, as well as the variable you want to store the result in. Pulsin is handy for measuring time delays in circuits, such as the return ping of an ultrasonic sonar.
- *Pulsout.* Pulsout is the inverse of pulsin; with pulsout you can create a finely measured pulse with a duration of between 2 μs and 131 milliseconds (ms). The pulsout statement is ideal when you need to provide highly accurate waveforms.
- *Rctime.* The rctime statement measures the time it takes for an RC (resistor/capacitor) network to discharge to an opposite logical state. The rctime statement is often used to indirectly measure the capacitance or resistance of a circuit, or simply as a kind of simplified analog-to-digital circuit. Fig. 15-6 shows a sample circuit.

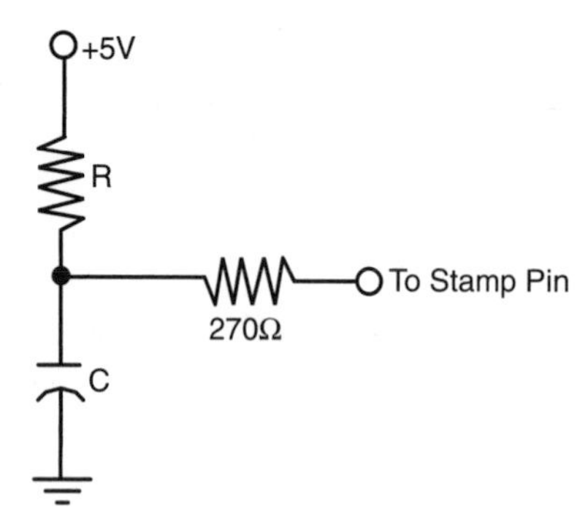

FIGURE 15-6 The rctime statement is used to measure the time it takes for a capacitor to discharge (the timing is accurate to 2 μs intervals). With this information you can indirectly measure capacitance or resistance.

- *Serin* and *serout.* Serin and serout are used to send and receive asynchronous serial communications. They are typically used for RS-232 interfaces, but have a number of formatting options that make them appropriate for a variety of different applications—even communicating between multiple BASIC Stamps.
- *Shiftin* and *shiftout.* The shiftin and shiftout functions are used in two- or three-wire synchronous serial communications. With shiftin/shiftout a separate pin is used for clocking the data between the source and destination. If you're only sending or receiving data, you can use just two pins: one for data and one for clock. If you're both sending and receiving, your best bet is to use three pins: data in, data out, and clock.

These statements are useful when communicating with a variety of external hardware, including serial-to-parallel shift registers and serial analog-to-digital converters.

15.5 Sample Interface Applications

Before going on to interfacing the BS2 (or any other microcontroller) into robots or a final application, you should spend some time learning how to program, create electrical interfaces, and debug applications. In the following sections, some simple applications are presented along with basic wiring information for the BS2 that will come in handy when you are creating your own applications.

A quick Internet search will also yield many different sites with information on robots and microcontrollers. For instance, as this edition of the book was written, a Google search of BS2 and Robots yielded over 10,000 sites. Even if only 1 out of 10 sites had a BS2 application on it, there are over 1000 projects that you can choose from or use as a reference for your robot. Along with links to web pages, you should also consider joining one of the Yahoo! list servers (some listed in the appendices) to share your ideas and ask questions of others.

Please do not consider this book as the ultimate resource on the BS2 and how it is used in robot applications. Go through the appendices for additional introductory how-to reference books on the BS2 and how microcontrollers can be used with robots. The more time spent learning about a specific microcontroller, building your own circuits, and trying different programs as well as seeing a variety of different people's perspectives will minimize the problems that you will have later when you are wiring up and programming your robot.

15.5.1 BASIC BS2 SETUP

At the start of this chapter, it was noted that the BS2 could be programmed cheaply and easily. Unlike other microcontrollers, you can create a breadboard-based development circuit like the one shown in Fig. 15-7 for about $10 (if you use new parts). The circuit consists of a serial interface to your PC along with a small breadboard, a box of precut wires, a breadboard-mountable DPST switch (an EG-1903 is used in Fig. 15-7), and a three AA battery clip attached to the breadboard's backside two-sided tape. The entire package can be built in about 20 minutes and once the BASIC Stamp Editor software has been downloaded into your PC you are ready to go!

The biggest piece of work is soldering five wires and a jumper to a nine-pin female D-shell connector to make the serial communications/programmer interface as shown in Fig. 15-8 (a photograph of an assembled prototype is shown in Fig. 15-9). The BS2 is programmed via RS-232. If your PC does not have a serial port, you can buy a USB to RS-232 adapter. The parts needed to put the interface connector together are listed in Table 15-5. You might be tempted to put the two 0.1 μF capacitors onto the breadboard, but you will find that by spending a few extra minutes soldering them to the nine-pin female D-shell, the effort of having to come up with the most efficient breadboard wiring will be avoided each time the BS2 is used in an application. Also note that the four wires (each a different color to aid in keeping track of them) were cut to the same length with a ½ in (1 cm) or so of bared copper and bent in the same direction to allow them to be pressed into the breadboard.

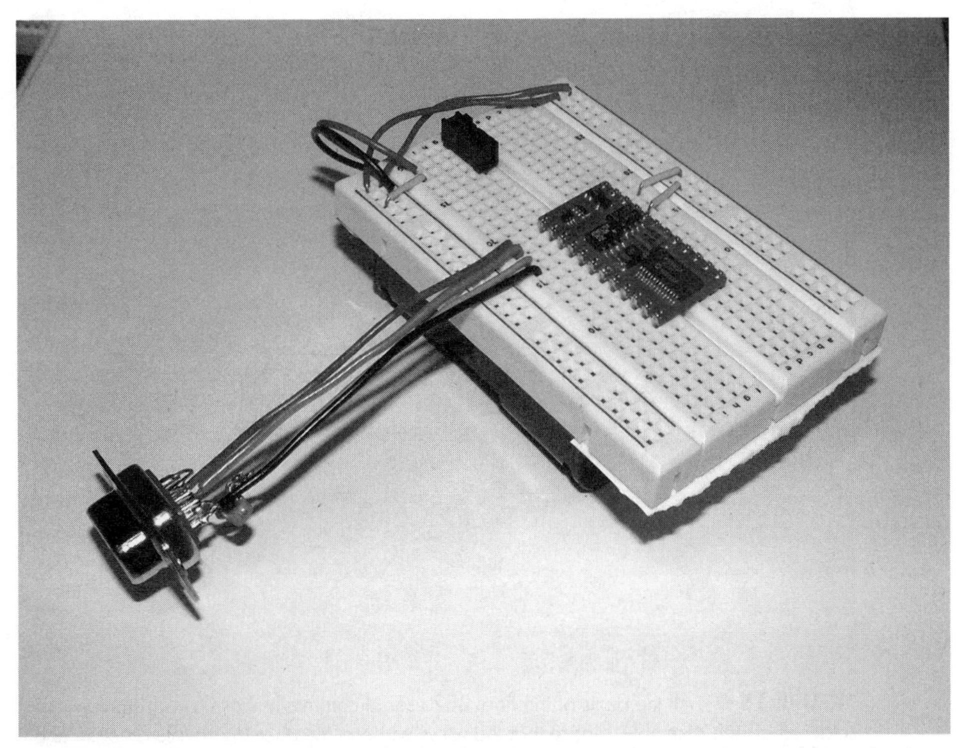

FIGURE 15-7 BS2 on a breadboard with the communications/programmer interface and a three AA battery pack stuck to the back side of the breadboard and providing power to the BS2.

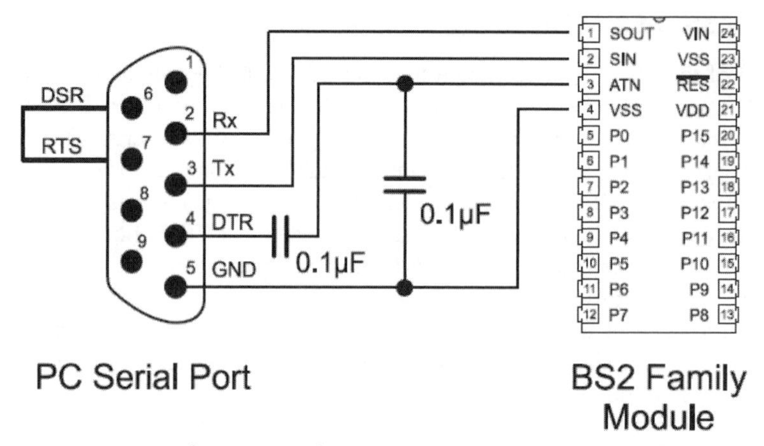

FIGURE 15-8 The BS2's serial communications/programmer interface is quite simple.

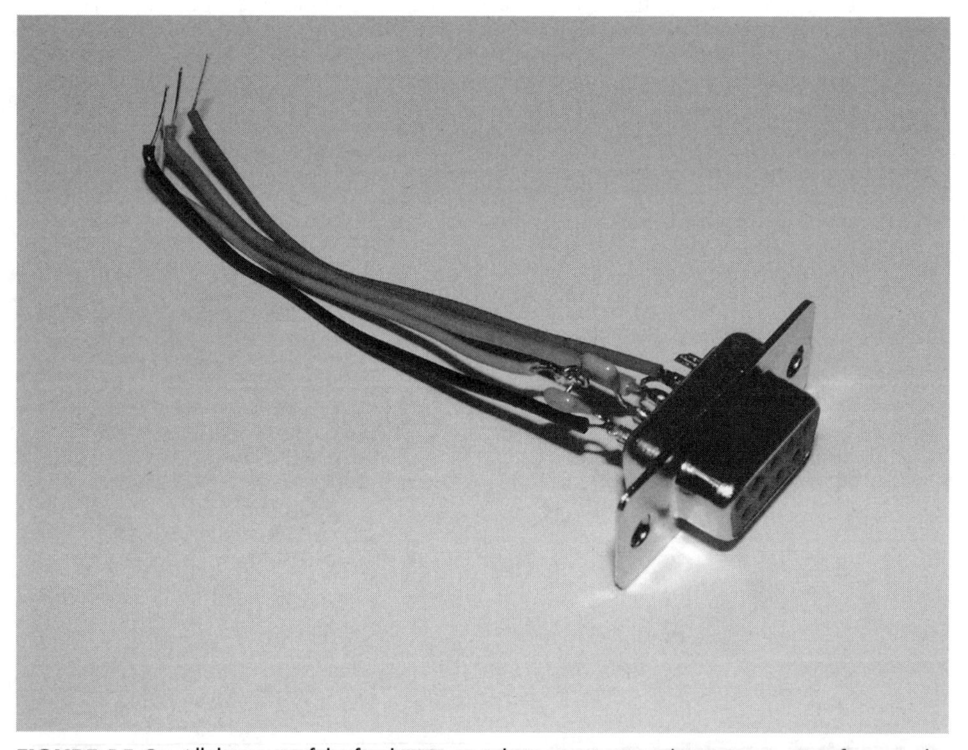

FIGURE 15-9 All the parts of the final BS2's serial communications/programmer interface can be soldered as one assembly. Notice that the wires all bend together to provide an easy and direct interface to the BS2.

TABLE 15-5 Serial Communications/Programming Interface Parts List

Connector	Female nine-pin D-shell connector
Capacitors	0.1 μF capacitors (any type)
Misc.	24 gauge wire, different colors

When you've assembled the serial communications/programming interface, cut a strip of the two-sided tape off the back of a small breadboard and stick on a three or four AA battery clip. For the BS2 to work properly, 4.5 to 6 V must be applied to it. If you are going to use alkaline cells (which produce 1.5 to 1.8 V each), only three batteries are required. If you are going to use NiMH rechargeable batteries (which produce 1.2 V), then four batteries will be required.

You might also want to put a power switch in line with the positive voltage from the battery pack. The positive and negative voltage should be connected to the common strips on the breadboard and connected to VDD (pin 21) and VSS (pin 23), respectively. VIN and the on-board voltage regulator is bypassed as is the _RES (Reset) pin. The BS2 will power up any time power is applied to it.

With the hardware together, you can download the BASIC Software Editor (from www.parallax.com, "Downloads") and install it on your PC. Along with Windows software, there are also Linux and Mac editors available. The editor should install like any other application and once you have connected the serial port to a straight-through nine-pin male to nine-pin female cable, you are ready to try out your first application!

The typical first application of any computer system is the Hello World, which follows. This program will simply print out the welcome phrase to indicate that it is up and running. Key in the program to the BASIC Stamp Editor, save it on your PC's desktop, and then either click on the right-pointing triangle on the toolbar or press Ctrl-R. This will compile, download, and then run the application.

```
'   Hello World
'
'   This Program Prints the String "Hello World!"
'
'
'   Author:  Myke Predko
'
'   Date: 05.06.23
'
'

'{$STAMP BS2}
'{$PBASIC 2.5}

' Variables

' Mainline

  DEBUG "Hello World!", CR    '  Print the String

  END                         '  Stop Program
```

If the program was keyed in correctly and the wiring is correct, a terminal Window should appear with the message "Hello World." If there is an error message indicating there is a problem with the program, check what you've keyed in to make sure the program has been entered correctly. If the PC cannot find a BS2 connected to it, check the wiring of the nine-pin D-shell connector and that at least 4.5 (but not more than 6.0) V is going into the BS2's pin 21 and 23. Additional debugging information can be found in the BASIC Stamp Syntax and Reference manual downloadable from the Parallax web site.

15.5.2 LED OUTPUTS

With the software loaded on your PC and a simple programming/communications interface built and tested for your BS2, you can now start experimenting with the input/output capabilities of the microcontroller. The most fundamental output device that is used with a microcontroller is the light-emitting diode (LED) and with it you can learn a lot about programming and how the BS2 works.

To demonstrate LED interfacing in this and the next section, you should wire the circuit shown in Fig. 15-10, consisting of the BS2, the programming/communications interface, and eight LEDs, wired to the low eight bits of the BS2's I/O port. The parts required for this circuit are listed in Table 15-6. To simplify the wiring in the prototype, eight 5 × 2 mm rectangular LEDs were used instead of square LEDs (Fig. 15-11).

The first task you might want to perform is to flash a single LED (we'll get to the other LEDs later in this section) by performing a delay, turning the LED on, waiting the same delay, and turning the LED off. This process is repeated using the DO/LOOP statements. The first program, listed here, writes directly to the P0 I/O pin to turn the LED on and off with delays in between, as described in this paragraph.

```
'  LED Flash Demonstration 1 - Flash LED on P0 2x per second
'{$STAMP BS2}
'{$PBASIC 2.5}

'  Mainline
    DIR0 = 1                          ' P0 is an output
    DO
        OUT0 = 0                      ' LED On
        PAUSE 250                     ' Delay 1/4 second
        OUT0 = 1                      ' LED off
        PAUSE 250
    LOOP                              ' Repeat
```

When doing any kind of programming for the first time, or if you encounter a problem, it is recommended that you try to come up with three different ways of performing a task. The first way, while it works, might not be the most efficient or easily understood method of performing the task. Writing directly to the I/O ports using assignment statements may be confusing to some people. A second way of flashing the LED on and off could be to use the built-in HIGH and LOW PBASIC functions rather than writing to the I/O ports directly. These functions avoid the need for writing to the bits directly:

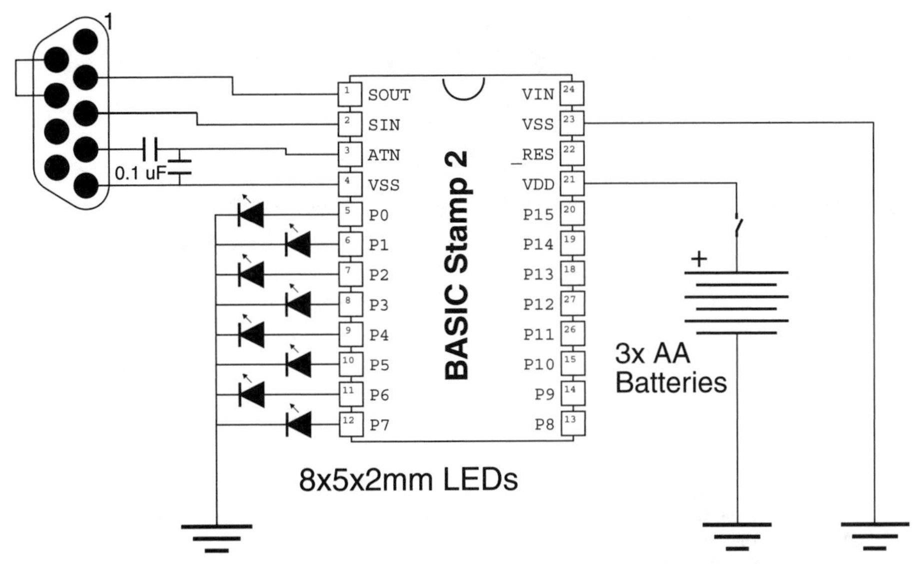

FIGURE 15-10 BS2 with serial communications/programmer interface and eight LEDs.

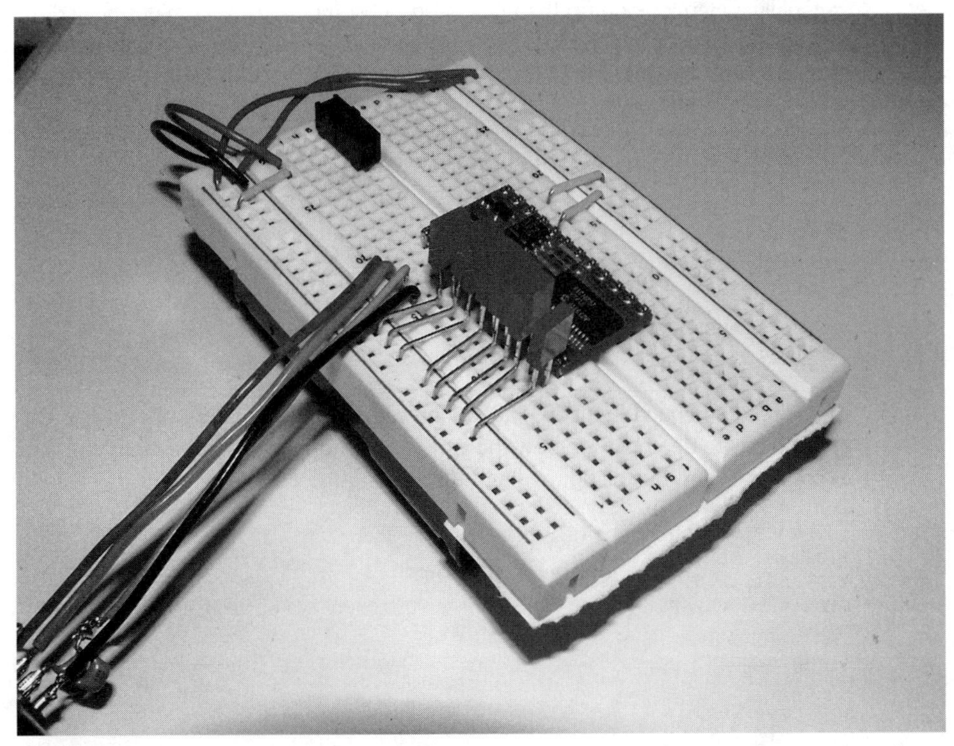

FIGURE 15-11 The circuit is very simple when 5 × 2 mm LEDs are bent into shape as shown here.

TABLE 15-6 BS2 LED Experimentation Circuit Parts List

BS2	Parallax BASIC Stamp 2
LEDs	5 × 2 mm Red LEDs
Programmer	BS2 Programmer/Communications Interface
Misc.	Breadboard, 3x AA Alkaline Battery Clip, Breadboard Wires, Power Switch

```
'  LED Flash Demonstration 2 - Flash LED on P0 2x per second
'{$STAMP BS2}
'{$PBASIC 2.5}

'  Mainline
    DO
        LOW 0                       ' LED On
        PAUSE 250                   ' Delay 1/4 second
        HIGH 0                      ' LED off
        PAUSE 250
    LOOP                            ' Repeat
```

Finally, when looking up the HIGH and LOW functions, you might have discovered the TOGGLE function, which changes the output state of an I/O pin. This feature makes the program even simpler:

```
'  LED Flash Demonstration 3 - Flash LED on P0 2x per second
'{$STAMP BS2}
'{$PBASIC 2.5}

'  Mainline
    OUTPUT 0
    DO
        TOGGLE 0                    ' LED = LED ^ 1
        PAUSE 250                   ' Delay 1/4 second
    LOOP                            ' Repeat
```

It is recommended that you play around with the LEDs and try to come up with different applications to demonstrate how the various functions work. For example, you might want to create a Cylon Eye using the eight LEDs using the code:

```
'  LED Flash Demonstration 4 - Cylon Eye
'{$STAMP BS2}
'{$PBASIC 2.5}
Direction VAR Bit                   '  Up/Down Direction

'  Mainline

    DIRL = $FF                      '  P7:P0 Outputs
    OUTL = 1                        '  P0 is On

    Direction = 0                   '  Go Up
```

```
DO
    PAUSE 100                  '  Delay 1/10 second
    IF (Direction = 0) THEN '  Select Direction
        OUTL = OUTL << 1       '  Shift Up the Turned on LED
        IF (OUT7 = 1) THEN Direction = 1
    ELSE                       '  Go in Opposite Direction
        OUTL = OUTL >> 1
        IF (OUT0 = 1) THEN Direction = 0
    ENDIF
LOOP                           '  Repeat
```

Don't be afraid to try new ideas and if they don't work, spend some time trying to understand what the problem is and fix it. A good rule of thumb is that your first real program will take two weeks to get running properly, your second program a week, the third two days, the fourth four hours, and so on until you are very familiar with programming the BS2 and looking at different ways of approaching the problems. The Cylon Eye program probably seems very impressive and most likely very difficult to implement on your own, but if you work at the problem, try out different things, and always keep in the back of your mind that you want to think of three different ways of approaching a problem, you will become a competent programmer very quickly.

15.5.3 ADDING SWITCHES AND OTHER DIGITAL INPUTS

LEDs, as well as being useful devices for learning how to program, are excellent status indicators for the current state of the program of different inputs. The most basic input device is the momentary on button, which closes the circuit when pressed. This device can be used along with LEDs to demonstrate how digital inputs are passed to the BS2.

The circuit used to test button inputs is shown in Fig. 15-12 and the parts required are listed in Table 15-7. The circuitry was chosen to match that of the previous section to avoid

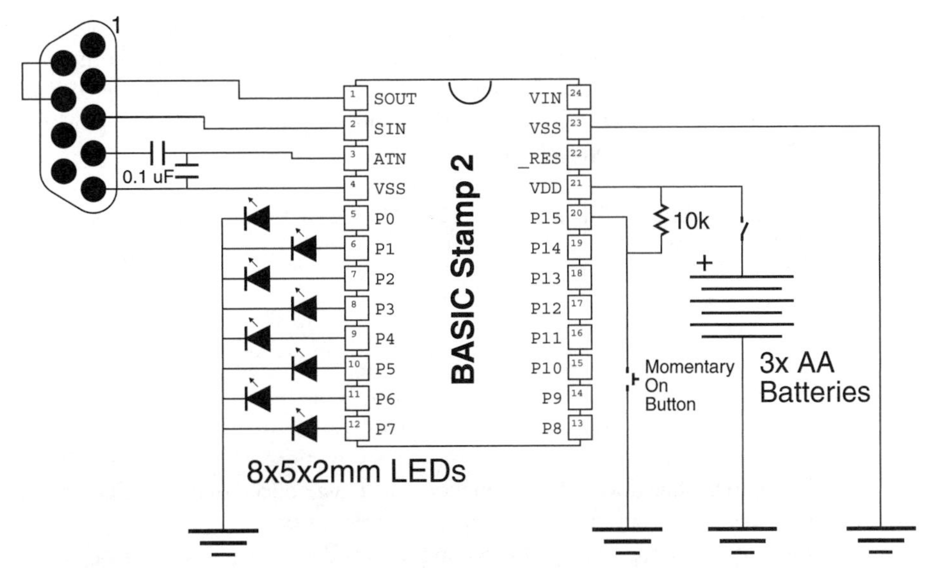

FIGURE 15-12 Momentary on button added to the BS2 eight-LED circuit.

TABLE 15-7 BS2 Button Experimentation Circuit Parts List

BS2	Parallax BASIC Stamp 2
LEDs	5 × 2 mm Red LEDs
10k	10k Resistor
Button	Momentary On Button—with wires soldered on to interface with the breadboard
Programmer	BS2 Programmer/Communications Interface
Misc.	Breadboard, 3x AA Alkaline Battery Clip, Breadboard Wires, Power Switch

the need to tear down and build up a new circuit as you are learning about the BS2. You may be able to find a momentary on button that can be plugged directly into the breadboard, but chances are you will have to solder some wires to a switch and push them into the breadboard to add the switch to the circuit.

The BS2 is built from CMOS technology. As discussed earlier in the book, it does not have its own internal voltage or current source so you must make sure that the input pins are driven either high or low. In this circuit, the 10k resistor pulls up the input pin until the button is pressed and the pin is connected to ground (pulled down). The 10k resistor limits the amount of current that passes through the momentary on switch to ground to about 50 μA. A pull-up circuit, such as this, should always be used with BS2 inputs to ensure that the voltage always transitions from high to low and a very small amount of current passes between them.

To test the circuit, the following program will turn on the LED at P0 any time the input at P15 is low (which is the pulled up momentary on button). Notice that the value at P15 cannot be passed directly to P0—when the button is pressed, the input is low, but to turn on the LED, the output must be high. The XOR (^ operator) with 1 will invert the signal from P15 so it can be used with P0.

```
'  Button Demonstration 1 - Control LED on P0 by Button on P15
'{$STAMP BS2}
'{$PBASIC 2.5}

'  Mainline
  DIR0 = 1                         ' P0 is an output
  DO
        OUT0 = IN15 ^ 1            ' Toggle Button Input
        LOOP                       ' Repeat
```

When you pushed down the button, you may have noticed that the LED flickered on and off. This was due to button bounce, dirty contacts, or your finger getting tired. To try and show this action more clearly, the second button demonstration turns off all eight LEDs and then starts to turn them on (by shifting bits up) when the button is pressed. If the button is lifted, then the program stops until the button is pressed again.

```
'  Button Demonstration 2 - Button Bounce Demonstration
'{$STAMP BS2}
'{$PBASIC 2.5}

'  Mainline
   DIRL = $FF                    ' LEDs on P0-P7
   OUTL = 0
   DO
       DO WHILE (IN15 = 1)       ' Wait for Button Press
       LOOP
       OUTL = 0                  ' Turn off LEDs
       DO WHILE (IN15 = 0)       ' Shift Up while Pressed
           OUTL = (OUTL << 1) + 1
           PAUSE 100
       LOOP
   LOOP                          ' Repeat
```

When you ran this program, you probably saw the LEDs cycle up, but occasionally restart or even just flicker to one on. The reasons for this were listed previously and something that will have to be compensated for in your robot control program. Later in the book, you will be shown how to process this information and figure out exactly what is happening but for now just try to work at understanding how to process simple button inputs.

15.5.4 LCD INTERFACE

The most effective type of display that you can add to your robot is the *liquid crystal display,* best known by its acronym LCD. The LCD display allows you to output data in an arbitrary format in alphanumeric or even graphical format to help you understand what is going on within the robot. LCDs have the reputation for being difficult to work with, but there are products such as the Hitachi 44780 controlled LCDs discussed in this section that are quite easy to add and program to a microcontroller.

The 44780 is a chip that is a bridge between a microcontroller and the LCD hardware and was originally manufactured by Hitachi though now it is made by a wide range of chip manufacturers. It contains the four- or eight-bit interface listed in Table 15-8. In eight-bit mode, all eight bits, D0 through D7, interface to a microcontroller, while in four-bit mode, only the upper four bits (D4 through D7) are used. The other six pins consist of three control and clocking pins, power, and a display contrast voltage.

The RS pin selects between passing characters or commands between the LCD and the microcontroller, and the direction of the data is selected by the RW pin. Normally data are sent from the microcontroller to the LCD and not read, so this line is held low. To indicate that the data on the LCD's pins are correct, the E clock is pulsed from low to high and back to low again. If the LCD is used in four-bit mode, first the four most significant data bits are passed to the LCD, followed by the least significant four data bits.

TABLE 15-8 Hitachi 44780 Controlled LCD Interface

PINS	DESCRIPTION/FUNCTION
1	Ground
2	Vcc
3	Contrast Voltage
4	"RS"-_Instruction/Register Select
5	"RW"-_Write/Read Select
6	"E" Clock
7 to 14	Data I/O Pins

Data sent to the LCD consists of either instructions (low on RS) or characters (high on RS). The instructions specify how the LCD is to operate and can be used to poll the LCD to determine when to send the next character or command. Table 15-9 lists the different pin and data bit values for sending commands and characters to the LCD.

More comprehensive information on LCDs can be found on the Internet at the sites listed in the appendices. To demonstrate how simple it is to add and program an LCD to the BS2, the circuit shown in Fig. 15-13 with the parts listed in Table 15-10 was created. The circuit should be very easy to wire together as it was designed for taking advantage of the natural layout of the BS2 as shown in Fig. 15-14.

The application code to display "Robot Builder's Bonanza" on the two rows of the LCD is:

```
'  BS2 LCD - Display a simple message on an LCD module
'{$STAMP BS2}
'{$PBASIC 2.50}


'  Variables
i VAR Byte
Character VAR Byte             '  Character to Display
LCDData VAR OUTH               '  Define LCD Pins on BS2
LCDE PIN 6
LCDRS PIN 7


'  Initialization
  DIRS = %1111111111000000     '  Make Most Significant 10 Bits Output
  LCDRS = 0: LCDE = 0          '  Initialize LCD interface
  PAUSE 20                     '  Wait for LCD to reset itself
  LCDData = $30: PULSOUT LCDE, 300: PAUSE 5  '  Initialize LCD Module
  PULSOUT LCDE, 300: PULSOUT LCDE, 300       '  Force reset in LCD
  LCDData = $38: PULSOUT LCDE, 300: PAUSE 5  '  Initialize/Set 8 Bit
  LCDData = $10: PULSOUT LCDE, 300  '  No Shifting
  LCDData = $01: PULSOUT LCDE, 300: PAUSE 5  '  Clear LCD
  LCDData = $06: PULSOUT LCDE, 300  '  Specify Cursor Move
  LCDData = $0C: PULSOUT LCDE, 300  '  Enable Display & Cursor Off
```

TABLE 15-9 Hitachi 44780 Command and Data Table

RS	RW	D7	D6	D5	D4	D3	D2	D1	D0	Instruction/Description
4	5	14	13	12	11	10	9	8	7	LCD I/O Pins
0	0	0	0	0	0	0	0	0	1	Clear Display (takes 5 msecs)
0	0	0	0	0	0	0	0	1	0	Move Cursor to "Home" (5 msecs)
0	0	0	0	0	0	0	1	ID	S	ID = 1, Increment Cursor after Write; S = 1, Shift Display when Written to
0	0	0	0	0	0	1	D	C	B	D = 1, Turn Display; C = 1, Cursor On; B = 1, Cursor Blink
0	0	0	0	0	1	SC	RL	0	0	SC = 1, Shift Display after Write; RL = 1, Shift Display Right
0	0	0	0	1	DL	N	F	0	0	Reset the 44780; DL = 1, 8 Data bits; DL = 0, 4 Data bits (D8 - D5); N = 1, 2 Display Lines; F = 1, 5x10 Font (Normally leave at 0)
0	0	0	1	A	A	A	A	A	A	Move Cursor into Character Graphics RAM to "A"ddress
0	0	1	A	A	A	A	A	A	A	Move Cursor to specific position on LCD to "A"ddress
0	1	BF	N/U	N/U	N/U	N/U	N/U	N/U	N/U	Poll the LCD "Busy Flag"
1	0	D	D	D	D	D	D	D	D	Write "D"ata to LCD at Cursor
1	1	D	D	D	D	D	D	D	D	Read "D"ata in LCD at Cursor

TABLE 15-10 BS2 LCD Experimentation Circuit Parts List

BS2	Parallax BASIC Stamp 2
LCDs	16 column by 2 row character Hitachi 44780 controlled LCD
10k	10k Potentiometer
Programmer	BS2 Programmer/Communications Interface
Misc.	Breadboard, 3x AA Alkaline Battery Clip, Breadboard Wires, Power Switch

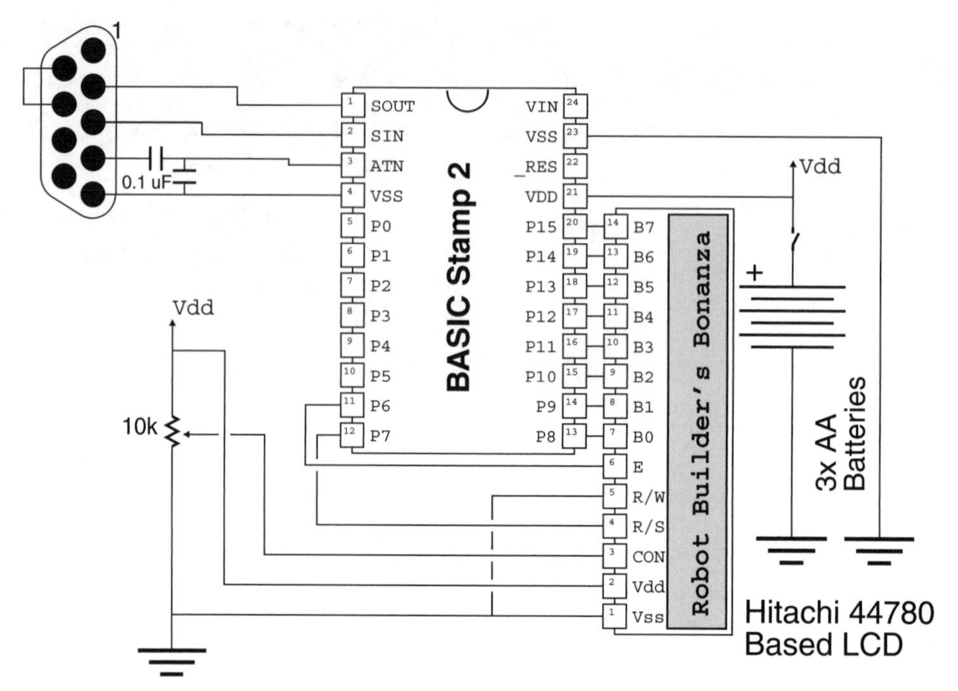

FIGURE 15-13 Circuit for adding an LCD to a BS2.

```
Character = 1: i = 0
LCDRS = 1                       '  Print Characters on Top Line
DO                              '  Read Each Character and Print them
    LOOKUP i, ["Robot Builder's", 0], Character
    IF (Character <> 0) THEN
        LCDData = Character: PULSOUT LCDE, 300
        i = i + 1
    ENDIF
LOOP UNTIL (Character = 0)
LCDRS = 0                       '  Move to Bottom Line
LCDData = $C0: PULSOUT LCDE, 300
Character = 1: i = 0
LCDRS = 1                       '  Print Characters on Bottom Line
DO                              '  Read Each Character and Print them
    LOOKUP i, ["    Bonanza", 0], Character
    IF (Character <> 0) THEN
        LCDData = Character: PULSOUT LCDE, 300
        i = i + 1
    ENDIF
LOOP UNTIL (Character = 0)

END                             '  Finished
```

With the information provided here, you can build and add your own message to an LCD or use it to monitor the progress of a robot. With a bit of research, you can reduce the number of pins required for controlling the LCD as well as demonstrate different capabilities such as changing specific locations on the LCD or adding your own custom characters.

FIGURE 15-14 Wiring for the BS2 LCD demonstration circuit.

15.5.5 I/O PORT SIMULATOR

As was noted previously, the second edition of this book presented four different robot control interfaces and one of the strengths of this approach was to use the different controllers for their respective optimal applications and examples. One of the examples for the PC's parallel port was to provide an I/O pin input/output interface that would allow the reader to see how digital I/O works as well as provide an experimental interface for trying out different hardware interfaces easily.

Along with the usefulness in being able to create a general case I/O interface, there is also the need to better explain all the different ways the BS2's I/O pins and the associated variables are accessed and what happens. To allow you to arbitrarily read and write to different I/O pins and their variables, "BS2 Port IO.bs2" was created:

```
'  BS2 Port IO - I/O Pin Function Demonstrator for the BS2
'
'  myke predko
'
'  06.06.24
'
'{$STAMP BS2}
'{$PBASIC 2.5}
```

```
'  Variable Declarations
InputString VAR Byte(19)              '  18 Character String Max
Mask         VAR Word
Value        VAR Word
i            VAR Byte
j            VAR Byte
Temp         VAR Byte
LastCharFlag  VAR j.BIT0

'  Initialization
    GOSUB InitDisplay

'  Main Loop
    DO                                 '  Loop Forever

        DEBUG "> "
        i = 0
        InputString(i) = 0             '  Start with Null string
        LastCharFlag = 0               '  Want to execute at least once
        DO WHILE (LastCharFlag = 0)    '  Wait for Carriage Return
            DEBUGIN STR Temp\1         '  Wait for Character input
            IF (Temp = CR) THEN
                LastCharFlag = 1       '  String Ended
            ELSE
                IF (Temp = BKSP) THEN
                    IF (i = 0) THEN
                        DEBUG " "      '  Keep Screen at Constant Point
                    ELSE
                        DEBUG " ", BKSP '  Backup one space and clear
                        i = i - 1      '  Move the String Back
                        InputString(i) = 0
                    ENDIF
                ELSE
                    IF (i >= 18) THEN
                        DEBUG BKSP , 11 '  At end of Line or NON-Char
                    ELSE               '  Can Store the Character
                        IF ((Temp >= "a") AND (Temp <= "z")) THEN
                            Temp = Temp - "a" + "A"
                        ENDIF
                        DEBUG BKSP, STR Temp\1
                        InputString(i) = Temp
                        i = i + 1
                        InputString(i) = 0  '  Put in new String End
                    ENDIF
                ENDIF
            ENDIF
        LOOP

        i = 0                          '  Find the Command Start
        DO WHILE (InputString(i) <> 0) AND (InputString(i) = " ")
            i = i + 1
        LOOP
        j = i                          '  Find Register Offset
        DO WHILE (InputString(j) <> 0) AND (InputString(j) <> " ")
            j = j + 1
        LOOP
        DO WHILE (InputString(j) <> 0) AND (InputString(j) = " ")
            j = j + 1                  '  "j" Indexes Start of Register
        LOOP
```

```
        Temp = 96                          '  Code for No Parameters
        IF (inputString(j) <> 0) THEN      '  Decode Register
            IF (InputString(j) = "D") AND (InputString(j+1) = "I")
AND (InputString(j+2) = "R") THEN
                Temp = 0                   '  Code for DIR
                j = j + 3
            ELSE
                IF (InputString(j) = "O") AND (InputString(j+1) = "U")
AND (InputString(j+2) = "T") THEN
                    Temp = 32              '  Code for OUT
                    j = j + 3
                ELSE
                    IF (InputString(j) = "I")
AND (InputString(j+1) = "N") THEN
                        Temp = 64          '  Code for IN
                        j = j + 2
                    ELSE
                        GOSUB UnknownRegister
                        Temp = -1
                    ENDIF
                ENDIF
            ENDIF
            IF (Temp <> -1) THEN
                IF (InputString(j) = "S") THEN
                    Temp = Temp + 31
                    j = j + 1
                ELSE
                    IF (InputString(j) = "H")
OR (InputString(j) = "L") THEN
                        Temp = Temp + 24
                        IF (InputString(j) = "H") THEN Temp = Temp + 1
                        j = j + 1
                    ELSE
                        IF (InputString(j) >= "A")
AND (InputString(j) <= "D") THEN
                            Temp = Temp + 20 + InputString(j) - "A"
                            j = j + 1
                        ELSE
                            IF (InputString(j) >= "0")
AND (InputString(j) <= "9") THEN
                                IF (InputString(j) = "1") THEN
                                    j = j + 1
                                    IF (InputString(j) >= "0")
AND (InputString(j) <= "5") THEN
                                        Temp = 10 + InputString(j) - "0"
                                        j = j + 1
                                    ELSE
                                        Temp = Temp + 1
                                    ENDIF
                                ELSE
                                    Temp = Temp + InputString(j) - "0"
                                    j = j + 1
                                ENDIF
                            ELSE
                                GOSUB UnknownRegister
                                Temp = -1
                            ENDIF
                        ENDIF
                    ENDIF
                ENDIF
            ENDIF
        ENDIF
```

```
        IF (InputString(i) <> 0) AND (Temp <> -1) THEN
            SELECT(InputString(i))          '  Something VALID was Entered
                CASE "H"                    '  Display Help Information
                    GOSUB InitDisplay
                CASE "I"                    '  Get Register Input
                    IF (temp = 96) THEN
                        DEBUG "No Register"
                        GOSUB EnterH
                    ELSE
                        SELECT (Temp & $60)
                            CASE 0
                                Value = DIRS
                            CASE 32
                                Value = OUTS
                            CASE 64
                                Value = INS
                        ENDSELECT
                        IF ((Temp & $1F) = 31) THEN
                            DEBUG IHEX4 Value, "  ", IBIN16 Value, "  ",
DEC Value, CR
                        ELSE
                            IF ((Temp & $1F) >= 24)
AND ((Temp & $1F) <= 25) THEN
                                IF ((Temp & $1F) = 24) THEN
                                    i = Value & $FF
                                ELSE
                                    i = (Value >> 8) & $FF
                                ENDIF
                                DEBUG IHEX2 i, "  ", IBIN8 i, "  ",
DEC i, CR
                            ELSE
                                IF ((Temp & $1F) >= 20)
AND ((Temp & $1F) <= 23) THEN
                                    i = (Temp & %11) * 4
                                    i = (Value >> i) & %1111
                                    DEBUG IHEX1 i, "  ", IBIN4 i, "  ",
DEC i, CR
                                ELSE
                                    i = Temp & %1111
                                    i = (Value >> i) & %1
                                    DEBUG IBIN1 i, CR
                                ENDIF
                            ENDIF
                        ENDIF
                    ENDIF
                CASE "O"                    '  Output Value to Register
                    IF (temp = 96) OR (InputString(j) <> ",") THEN
                        DEBUG "No Register/Value"
                        GOSUB EnterH
                    ELSE
                        j = j + 1        '  Find Value to Write In
                        DO WHILE (InputString(j) <> 0)
AND (InputString(j) = " ")
                            j = j + 1
                        LOOP
                        IF (InputString(j) = 0) THEN
                            DEBUG "No Value"
                            GOSUB EnterH
                        ELSE             '  Read the Value in
                            Value = 0    '  Clear the Value
                            SELECT (InputString(j))
```

```
                              CASE "$"
                                  j = j + 1
                                  DO WHILE ((InputString(j) >= "0")
AND (InputString(j) <= "9")) OR ((InputString(j) >= "A")
AND (InputString(j) <= "F"))
                                      Value = Value * 16
                                      IF (InputString(j)  >= "A") THEN
                                          Value = Value +
InputString(j) + 10 - "A"
                                      ELSE
                                          Value = Value +
InputString(j) - "0"
                                      ENDIF
                                      j = j + 1
                                  LOOP
                              CASE "%"
                                  j = j + 1
                                  DO WHILE (InputString(j) >= "0")
AND (InputString(j) <= "1")
                                      Value = (Value * 2) +
InputString(j) - "0"
                                      j = j + 1
                                  LOOP
                              CASE ELSE
                                  DO WHILE (InputString(j) >= "0")
AND (InputString(j) <= "9")
                                      Value = (Value * 10) +
InputString(j) - "0"
                                      j = j + 1
                                  LOOP
                          ENDSELECT
                          IF (InputString(j) <> 0) THEN
                              DEBUG "Invalid Value"
                              GOSUB EnterH
                          ELSE         '  Write out the Value
                              IF ((Temp & $1F) = 31) THEN
                                  Mask = 0
                              ELSE
                                  IF ((Temp & $1F) >= 24)
AND ((Temp & $1F) <= 25) THEN
                                      IF ((Temp & $1F) = 24) THEN
                                          Mask = $FF00
                                          Value = Value & $FF
                                      ELSE
                                          Mask = $00FF
                                          Value = (Value << 8) & $FF
                                      ENDIF
                                  ELSE
                                      IF ((Temp & $1F) >= 20)
AND ((Temp & $1F) <= 23) THEN
                                          i = (Temp & %11) * 4
                                          Value = (Value & %1111) << i
                                          Mask = ($000F << i) ^ $FFFF
                                      ELSE
                                          i = Temp & %1111
                                          Value = (Value & 1) << i
                                          Mask = ($0001 << i) ^ $FFFF
                                      ENDIF
                                  ENDIF
                              ENDIF
                              SELECT (Temp & $60)
```

```
                                        CASE 0
                                            DIRS = (DIRS & Mask) | Value
                                        CASE 32
                                            OUTS = (OUTS & Mask) | Value
                                        CASE 64
                                            INS = (INS & Mask) | Value
                                  ENDSELECT
                              ENDIF
                          ENDIF
                      ENDIF
                  CASE ELSE
                      DEBUG "Unknown Command"
                      GOSUB EnterH
              ENDSELECT
          ENDIF
      LOOP
EnterH:                                     '  Put in Display Message
    DEBUG " - 'H' for Commands/Regs", CR
    RETURN


UnknownRegister:
    DEBUG "Unknown Register"
    GOSUB EnterH
    DO WHILE (InputString(i) <> 0)
        i = i + 1
    LOOP
    RETURN


InitDisplay:                                '  Display Basic Commands
    DEBUG "Command                         Registers      16Bit  8Bits
1Bit", CR
    DEBUG "H(elp)                          DIR            'S'    'H'/'L'
'#'", CR
    DEBUG "I(nput) Register                OUT   Example: DIRS", CR
    DEBUG "O(utput) Register, Value        IN", CR
    DEBUG "Value: #-Dec/$#-Hex/%#-Bin", CR
    RETURN
```

This program just requires a running BS2 with a serial port interface. After downloading the code (using Ctrl-R), the debug terminal window will come up and display the introductory message:

```
Command                    Registers      16Bit  8Bits   1Bit
H(elp)                     DIR            'S'    'H'/'L' '#'
I(nput)  Register          OUT   Example: DIRS
O(utput) Register, Value   IN
Value: #-Dec/$#-Hex/%#-Bin
```

This menu is a bit cryptic (due to the limitations in the size of the debug terminal display area and the need for additional space for the application code). To net it out, there are three commands that you can enter: "H(elp)" brings up this menu; "I(nput)" will read the contents of a pin variable (which is called a register in this program); and "O(utput)" will write a new value to the pin variable. Only the first character of the command is read and anything else in the command word (ending at a space) is optional.

The pin variables can access the full 16 bits of the pin register (with the variable name ending in S), eight bits (either high or low byte using the L or H postfixes), nibbles, or single bits. When the value in the pin variable is displayed, it is displayed in hexadecimal, binary, and decimal to provide you with the data format that makes the most sense. When writing to the registers, hexadecimal, binary, or decimal values can be specified by leading the value with a $, %, or nothing, respectively.

To access the different registers, add the 16, 8, 1 bit postfix to the register name as shown in Table 15-11 to access the different OUT bits. While all of the 16, 8, and 4 bit labels are shown, only 2 of the possible 16 single bit OUT bit labels are listed.

To change the values of a register, use the "O(utput)" command, like

```
> O DIR0, 1
```

which puts I/O pin 0 into output mode. To read the state of a pin variable, use the "I(nput)" command

```
> I OUT0
```

and the program will display the data.

To demonstrate the operation of the program, wire an LED with its cathode at the BS2's pin 0 and its anode at pin 3. To turn on the LED, enter the following command sequence. The comments following are to explain what each command is doing.

```
> O DIRA, %1001      '  Make BS2 Pins 0 & 3 Outputs
> I DIRS             '  Read Back ALL the Mode Bits to See Changes
> O OUT3, 1          '  Turn on the LED
> I INL              '  Read in the lower 8 Bits of the Input Bits
```

TABLE 15-11 Different Labels to Access the Different BS2 "OUT" Register Bit Combinations

REGISTER BITS	"OUT" LABEL	COMMENTS
0–15	OUTS	All 15 Output Bits
0–7	OUTL	Lower 8 Output Bits
8–15	OUTH	Upper 8 Output Bits
0–3	OUTA	Lowest 4 Output Bits
4–7	OUTB	Second Lowest 4 Output Bits
8–11	OUTC	Second Highest 4 Output Bits
12–15	OUTD	Highest 4 Output Bits
2	OUT2	Bit 2
13	OUT13	Bit 13

Ironically, even though this application was written for beginners, it is really quite advanced. A large part of this is due to the string parsing that the program performs but a large part is due to trying to shoehorn the application into a BS2 (use Ctrl-M in the BASIC Stamp Editor to see what I mean). The program took a bit of work in figuring out which register was being specified and then what was the value to be written to it.

15.6 BS2 Application Design Suggestions

The BS2 will most likely be the easiest microcontroller or computer system that you will work with. It is easy to program, and the built-in functions allow you to do more with the I/O pins than what you would normally expect to be able to do with a simple microcontroller. Electrically, while the BS2 is very robust, there are a few things that you can do to ensure that it will work reliably and safely in your robot application.

The first thing is to provide an externally regulated 5V to the BS2's VDD pin rather than some higher voltage on VIN. The on-board regulator's maximum current output of 100 mA is the reason for suggesting the alternative power input. The voltage regulator has been known to burn out when excessive current is being drawn through the on-board PIC's I/O pins (for turning on LEDs, for example). The regulator can be replaced, but it is a lot easier to plan for providing 5 V directly to the BS2. The voltage of three AA (or even AAA) alkaline batteries in series is sufficient to power the BS2 along with any external peripherals safely.

In some instances, especially when high voltages are involved, the I/O pins of the BS2 can be damaged. The recommended prevention to this problem is to always put a 220 Ω resistor in series with the I/O pin and whatever devices they are connected to. A 220 Ω resistor will limit the maximum current passing through the I/O pins but still provide enough current to light LEDs. The example circuits in this chapter do not have 220 Ω resistors built into them to simplify the wiring of the sample applications, but when you are installing a BS2 into a robot you should always make sure that you have a resistor on every I/O pin.

There is no need for an external reset control. The BS2's "_RES" pin is pulled up, so that if you want to add a momentary on button that can tie the pin to ground and reset the BS2 you can. For the most part, this isn't necessary.

Finally, it is always a good idea to build the BS2's programming interface into your robot. The circuit is simple and will eliminate the need to pull out the BS2 every time the program is changed. This will reduce the wear and tear on the I/O pins as well as the socket connectors and minimize the chance that you will damage your BS2.

Many of these suggestions are applicable to other microcontrollers and computer interfaces. A few extra parts and a few extra minutes can mean that your competitive robot always will be ready for action, not on the bench waiting for its "brains" to be replaced.

15.7 From Here

To learn more about . . .	***Read***
Using DC motors for robot locomotion	Chapter 20, "Working with DC Motors"
Using servo motors for robot control and locomotion	Chapter 22, "Working with Servo Motors"
Choices and alternatives for robot computers and microcontrollers	Chapter 12, "An Overview of Robot 'Brains' "
Attaching real-world hardware computers and microcontrollers	Chapter 14, "Computer Peripherals"

CHAPTER 16

REMOTE CONTROL SYSTEMS

The most basic robot designs—just a step up from motorized toys—use a wired control box on which you flip switches to move the robot around the room or activate the motors in the robotic arm and hand. The wire link can be a nuisance and acts as a tether preventing your robot from freely navigating through the room. You can cut this physical umbilical cord and replace it with a fully electronic one by using a remote control receiver and transmitter.

This chapter details several popular ways to achieve links between you and your robot. You can use the remote controller to activate all of the robot's functions, or, with a suitable on-board computer working as an electronic recorder, you can use the controller as a teaching pendant. You manually program the robot through a series of steps and routines, then play it back under the direction of the computer. Some remote control systems even let you connect your personal computer to your robot. You type on the keyboard, or use a joystick for control, and the invisible link does the rest.

16.1 Controlling Your Robot with a PC Joystick or Control Pad

In previous editions of this book, a joystick taken from an old Atari video game was used as a build-your-own teaching pendant. Over the years, it has become very difficult to find surplus or used Atari video game joysticks and even the ones that are available are generally in poor shape because the plastic boot often cracked or pulled out of the plastic base, making the joystick difficult to use. What should be readily available are PC joysticks or gamepads, which can be easily adapted into wired controllers for robots.

The PC joystick connects to a PC via a 15 pin D-shell connector and is wired according to Fig. 16-1. The pinout of the PC port is given in Table 16-1 and it can accommodate up to two joysticks. For robot applications in which the joystick will be used to control the motion of the robot, only Joystick A along with its two fire buttons are used. For more complex robots, two joysticks could be wired into the port.

The joystick isn't a terrible interface, although it can be difficult to work precisely. Despite the added capability of proportional joysticks, games quickly standardized on a gamepad similar to that used by Nintendo with a cross joystick that was digital (indicates Up, Down, Left, or Right). To interface the gamepad digital joystick with the standard PC game port, the circuit shown in Fig. 16-2 was developed. When the joystick is centered, current flows through one 50k resistor, which is the same resistance as when the analog joystick is centered. If the Right or Down button is pressed, the PNP transistor is turned off and current flows through a second 50k resistor, providing the full 100k resistance of the standard joystick. Finally, if the Left or Up button is pressed, the resistance drops to zero. This circuit works quite well and gives the PC the same interface as a basic game machine.

To demonstrate the operation of a gamepad (or an old PC joystick) and turn on an LED according to whether the joystick is moved to an extreme, the circuit in Fig. 16-3 (and parts

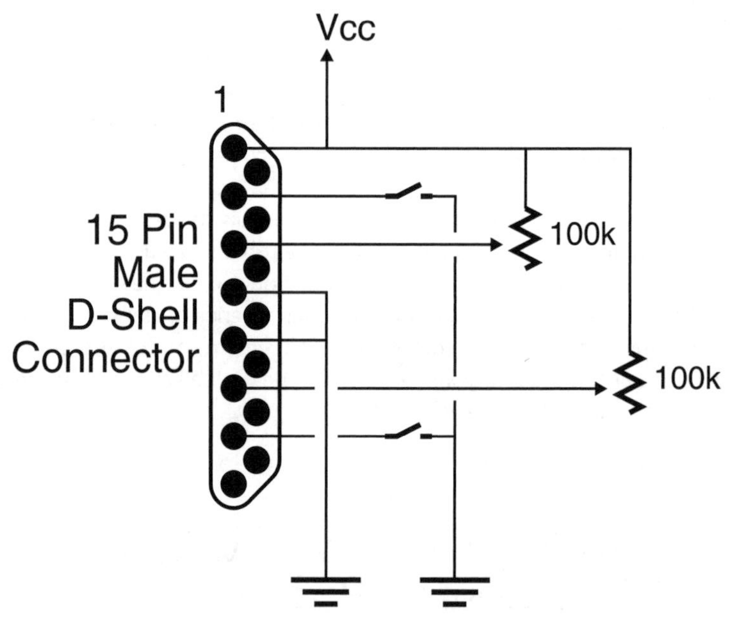

FIGURE 16-1 Internal wiring of PC joystick. Note that the analog outputs are part of a voltage divider tied to Vcc with a resistance of 0 to 100k.

TABLE 16-1 PC Joystick Port Pinout

PIN	FUNCTION
1	+5 V DC
2	B1A—button 1, joystick A
3	XA—*x* axis resistance, joystick A
4	Ground
5	Ground
6	YA—*y* axis resistance, joystick A
7	B2A—button 2, joystick A
8	+5 V DC
9	+5 V DC
10	B1B—button 1, joystick B
11	XB—*x* axis resistance, joystick B
12	GND/MIDI out
13	YB—*y* axis resistance, joystick B
14	B2B—button 2, joystick B
15	+5 V DC/MIDI In

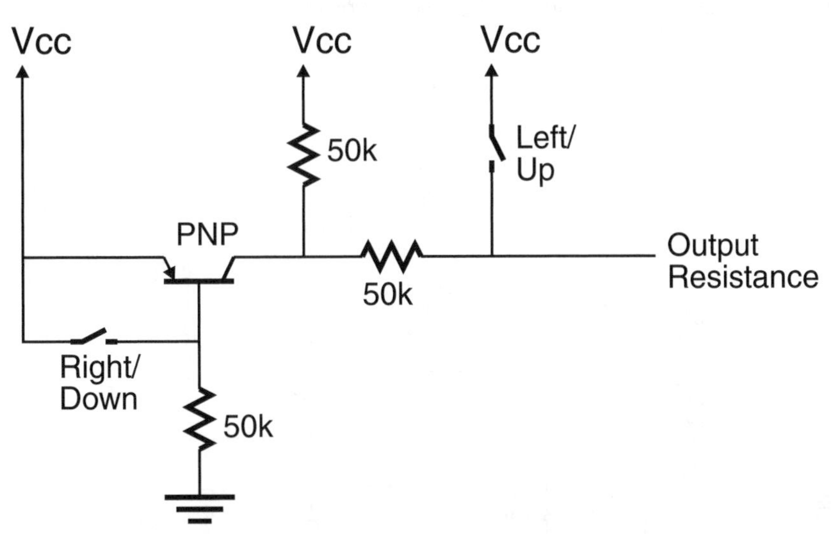

FIGURE 16-2 Internal circuitry for a PC gamepad to provide 100k when the Right/Down button is pressed, 50k when no button is pressed, and 0 Ω when the Left/Up button is pressed.

TABLE 16-2 Parts List for PC Gamepad Interface

15 pin female D-shell	15 pin female D-shell connector
LM339	LM339 quad open-collector output comparator chip
LED1–LED4	Visible light LEDs
R1–R2	10k resistors
R3, R4, R7	100k resistors
R5	22k resistor
R6	47k resistor
R8–R11	470 Ω resistors
Misc.	Breadboard, breadboard wiring, 4x AA battery clip, 4x AA batteries, breadboard mountable switch

list in Table 16-2) was developed. Wires will have to be soldered to the 15 pin female D-shell connector to allow it to be wired to a breadboard.

The circuit uses the two voltages produced by the voltage divider built from R5 through R7 to compare to the voltages produced by the voltage divider, which uses the gamepad/joystick internal resistances along with a 100k resistor to find out if any of the axes have gone to the extreme. The comparator will allow current to flow through the appropriate LED when the gamepad/joystick resistance is at an extreme. This circuit will be used as the basis for a robot motion teaching tool in the next section.

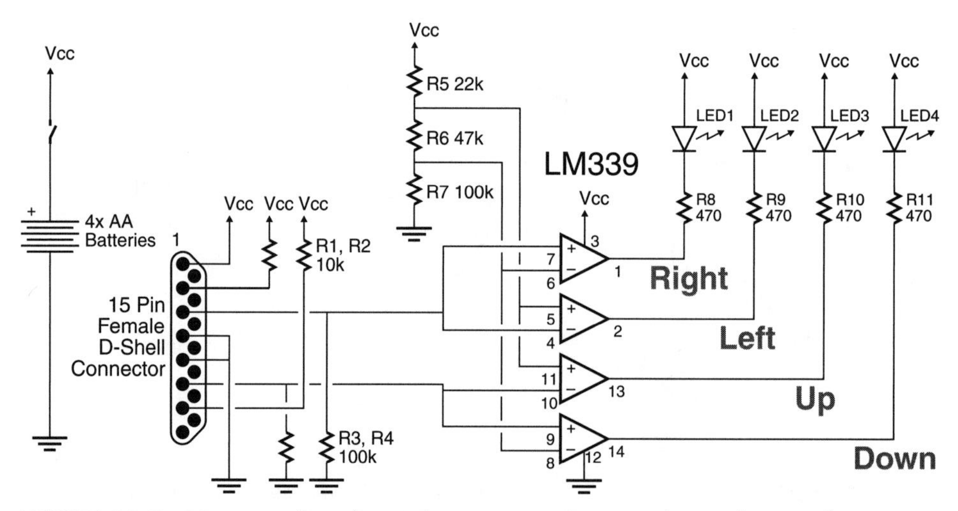

FIGURE 16-3 PC gamepad interface. When a gamepad or joystick control is moved to an extreme, one or two LEDs will light.

16.2 Building a Joystick Teaching Pendant

With the circuit that will convert the position of a PC-compatible gamepad or joystick into four digital signals, you can add a microcontroller that will convert the data bits into motor control values as well as give you the ability to record a set of movements for playback later. This application is often called a *teaching pendant.* In this section, the gamepad/joystick interface circuit will be enhanced to do just that, as well as provide you with a method of saving the recorded positions for use in other applications.

Fig. 16-4 shows how a BS2 can be added to the gamepad/joystick interface along with two LEDs and four motor control bits used to control the motion of a differentially driven robot. Along with the connections for the gamepad/joystick position, the pulled-up A and B buttons are used as control signals to the recording and playback of the teaching pendant application.

The BS2 can be connected to either TTL/CMOS compatible motor drivers or to four LEDs, which will simulate the operation of the motors. For the prototype, the circuit was added to a simple DC driven mobile robot to observe the performance of the robot. After you've built the circuit, you can load the following program into the BS2 to demonstrate the operation of the teaching pendant control.

```
'  BS2 Teaching Pendant - Use BS2 and PC Gamepad for Robot Training
'
'  Along with using the PC Gamepad/Joystick for Tethered Robot Remote
'   Control; it can also be used to record and play back movements
'
'
```

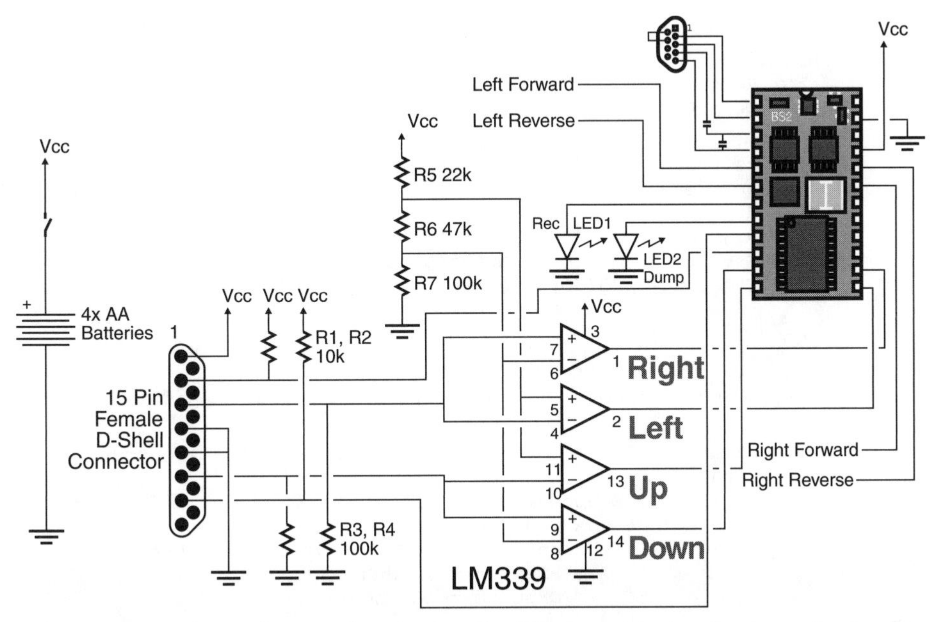

FIGURE 16-4 The teaching pendant circuitry. The BS2 communications/programming interface will be used for saving a series of commands for the robot.

```
'  Author:  Myke Predko
'
'  Date: 05.09.07
'
'

'{$STAMP BS2}
'{$PBASIC 2.5}



'  Pin Declarations
RecLED    PIN 2           '  Indicate Recording
DumpLED   PIN 3           '  Indicate Sending Data to PC
ButtonA   PIN 5           '  Motion Record
ButtonB   PIN 4           '  Record Stop/Play back

JoyLeft   PIN 9           '  Joystick Outputs from LM339
JoyRight  PIN 8           '   - Active Low
JoyUp     PIN 7
JoyDown   PIN 6

LeftFor   PIN 0           '  Motor Control Pins
LeftRev   PIN 1           '   - Active High
RightFor  PIN 15          '   - For Dual H-Bridge
RightRev  PIN 14          '     Differentially Driven Robot

'  Variable Declarations
i         VAR Byte
JValue    VAR Byte        '  Joystick Read in Value
ECount    VAR Word        '  EEPROM Address Counter
DBounce   VAR Word        '  Wait 20ms for debounce
BState    VAR Byte

'  Mainline
    LOW RecLED
    LOW DumpLED

    LOW LeftFor           '  Initialize Motor Outputs to Low
    LOW LeftRev
    LOW RightFor
    LOW RightRev
    WRITE 0, 0            '  Nothing to Start Executing
    DEBUG "PC Gamepad/Joystick Teaching Pendant", CR

    DO

        GOSUB ButtonOff   '  Wait for Both Buttons Raised

        DBounce = 0       '  Wait for Button/2 Button Pressed
        BState = 0
        DO WHILE (DBounce < 50)
            IF (ButtonA = 0) THEN
                IF (ButtonB = 0) THEN   '  Two Buttons Pressed
                    IF (BState = 3) THEN
                        DBounce = DBounce + 1
                    ELSE
                        BState = 3
                        DBounce = 0
```

```
                ENDIF
            ELSE   '  Single Button Pressed
                IF (BState = 1) THEN
                    DBounce = DBounce + 1
                ELSE
                    BState = 1
                    DBounce = 0
                ENDIF
            ENDIF
        ELSE
            IF (ButtonB = 0) THEN
                IF (BState = 2) THEN
                    DBounce = DBounce + 1
                ELSE
                    BState = 2
                    DBounce = 0
                ENDIF
            ELSE   '  Nothing Pressed
                BState = 0
                DBounce = 0
            ENDIF
        ENDIF
    LOOP

    SELECT (BState)
        CASE 1     '  Start Recording
            ECount = 0
            GOSUB ButtonOff
            DEBUG "Start Recording..."
            HIGH RecLED
            DO WHILE (ButtonB = 1) AND (ECount < 1000)
                PAUSE 95     '  Want Roughly 100 ms between Samples
                JValue = (INS >> 6) & $F  '  Get Value to Save
                WRITE ECount, JValue
                ECount = ECount + 1
                GOSUB RobotMove
            LOOP
            WRITE ECount, 0 '  Indicate Recording is Finished
            DEBUG "Finished Recording.", CR
            LOW RecLED
        CASE 2
            ECount = 0
            GOSUB ButtonOff
            DEBUG "Playing Back Recorded..."
            HIGH DumpLED
            JValue = 1
            DO WHILE (ButtonB = 1) AND (ECount < 1000) AND (JValue <> 0)
                PAUSE 95     '  Want Roughly 100 ms between Samples
                READ ECount, JValue
                ECount = ECount + 1
                GOSUB RobotMove
            LOOP
            WRITE ECount, 0 '  Indicate Recording is Finished
            DEBUG "Finished Playing Back.", CR
            LOW DumpLED
        CASE 3
            DEBUG "Dump the Program", CR
            i = 8
            ECount = 0     '  Dump the Program
            READ ECount, JValue
```

```
                DO WHILE (JValue <> 0)
                    IF (i = 8) THEN
                        DEBUG CR, "data ", DEC4 ECount
                        i = 0
                    ENDIF
                    DEBUG ", %", BIN4 JValue
                    ECount = ECount + 1
                    READ ECount, JValue
                    i = i + 1
                LOOP
                DEBUG CR
        ENDSELECT
    LOOP        '  Loop Forever
ButtonOff:      '  Wait for Buttons to be Released
    DBounce = 0
    DO WHILE (DBounce < 100)
        IF (ButtonA = 1) AND (ButtonB = 1) THEN
            DBounce = DBounce + 1
        ELSE
            DBounce = 0
        ENDIF
    LOOP
    RETURN
RobotMove:      '  Using "JValue", Specify Robot
Movement
    SELECT (JValue & %1111)
        CASE %1101  '  Moving Forwards
            HIGH LeftFor
            LOW LeftRev
            HIGH RightFor
            LOW RightRev
        CASE %1001  '  Moving Forwards and to Left
            LOW LeftFor
            LOW LeftRev
            HIGH RightFor
            LOW RightRev
        CASE %0101  '  Moving Forwards and to Right
            HIGH LeftFor
            LOW LeftRev
            LOW RightFor
            LOW RightRev
        CASE %1110  '  Moving in Reverse
            LOW LeftFor
            HIGH LeftRev
            LOW RightFor
            HIGH RightRev
        CASE %1010  '  Moving in Reverse and to Left
            LOW LeftFor
            LOW LeftRev
            LOW RightFor
            HIGH RightRev
        CASE %0110  '  Moving in Reverse and to Right
            LOW LeftFor
            HIGH LeftRev
            LOW RightFor
            LOW RightRev
        CASE %1011  '  Turning to the Left
            LOW LeftFor
            HIGH LeftRev
            HIGH RightFor
            LOW RightRev
```

```
        CASE %0111  '  Turning to the Right
            HIGH LeftFor
            LOW LeftRev
            LOW RightFor
            HIGH RightRev
        CASE %1111  '  Stopped
            LOW LeftFor
            LOW LeftRev
            LOW RightFor
            LOW RightRev
    ENDSELECT
    RETURN
```

To operate the pendant, press the A button on the gamepad/joystick to start the recording operation. Samples of the joystick position are taken every 100 ms and recorded in the BS2's built-in EPROM for up to 100 s (1000 samples). To stop the recording, press the B button. Using the Memory Map function of the BASIC Stamp Editor software, the program was found to take up EEPROM memory locations $474 to $7FF or 2956 bytes, leaving a maximum 1140 bytes available for recording. A total of 1000 samples and bytes was chosen to allow for some expansion of the application later.

To play back the application, press the B button and the robot will perform the actions in the EEPROM memory, exactly as they were recorded. It's actually a lot of fun recording a series of motions and then watching a robot repeat them.

Finally, to dump out the contents of the EEPROM, make sure the robot and BS2 are connected to a PC with the BASIC Stamp Editor software active. Start up a Debug Terminal window and press the A and B buttons simultaneously. When this is done, a series of PBASIC DATA statements are generated, which can be cut and pasted from the Debug Terminal window into an application or another program editor. These statements can be put directly into another application (and you might want to put in the RobotMove subroutine from the previous program as well) for use in it.

16.2.1 POSSIBLE ENHANCEMENTS

There are a number of things that can be done to enhance the teaching pendant; some have tried the program as shown here while others are just being thrown out for your experimentation:

- *Double the memory usage.* One hundred seconds of recorded robot motion is probably more than you will ever need, but in the rare case that you will need more samples, you could double up the memory by placing two four-bit samples into each EEPROM byte rather than the one used in the basic program.
- *Allow control without recording.* Right now, the robot motion is only allowed after the button A is pressed. You might want to modify the program to allow the robot to be controlled at any time with the two buttons controlling recording and playback.
- *Don't erase recorded data on start up.* When the application first powers up, it writes a zero to the first EEPROM location. A zero is interpreted as a data stop. If you would like to save the data from one execution instance to the next, delete the "write 0, 0" statement at the start of the program.

- *Add a second joystick for robot arm control.* This will allow you to control a gripper on the front of a mobile robot or a multi-axis robot arm. Along with this, many gamepads have more than just two buttons—they can have up to four (the second two are used by a second controller) that can be used to enhance the application.
- *Use data as escape maneuvers in combat robots.* One thing you will find difficult to program a robot for is a series of predefined movements that behave exactly as you would like. You will probably find it easier to record the desired sequence and just play it back when required.
- *Use multiple movement recordings in an application.* There's no reason why you wouldn't want to have multiple recordings in an application. If the robot was a combat robot, you might want to randomize your escape maneuvers to make it more difficult for your opponents.
- *Put a B button on a robot to allow playback without the pendant.* It is worth the few seconds to add a button wired to the ground in parallel with the gamepad/joystick's B button to allow the robot to perform the recorded operation without having the pendant attached. Keeping the pendant's wires away from the movement of the robot can be a pain sometimes, especially when you have a sequence preprogrammed in that you like.

16.3 Commanding a Robot with Infrared Remote Control

In the 1960s, the first television remote controls appeared, which had two or three basic functions: on/off, channel change, and sound mute. To change the channel you had to keep depressing the channel button until the desired channel appeared (the channels changed up, from 2 through 13, then started over again). What was more amazing than the remote control itself was how it worked: by ultrasonic sound. Depressing one of the control buttons struck a hammer against a metal bar cut to a specific length and produced a specific high-frequency tone. A microphone in the TV picked up the high-pitched ping and responded accordingly—although it wasn't unusual for women's jewelry and other metal items to produce the same frequencies and cause the channel to change (or worse, the TV to go off) periodically.

These days, remote control of TVs, VCRs, and other electronic devices is taken for granted. Instead of just two functions, the average remote handles dozens—more than the knobs and buttons on the TV or VCR itself. Except for some specialty remotes that use UHF radio signals, today's remote controls operate with infrared (IR) light. Pressing a button on the remote sends a specific signal pattern; the distinctive pattern is deciphered by the unit under control.

You can use the same remote controls to operate a mobile robot. A computer or microcontroller is used to decipher the signal patterns received from the remote via an infrared receiver. Because infrared receiver units are common finds in electronic and surplus stores (they're used heavily in TVs, VCRs, etc.), adapting a remote control for robotics use is actually fairly straightforward. It's mostly a matter of connecting the pieces together. With your infrared remote control you'll be able to command your robot in just about any way you wish—to start, stop, turn, whatever.

Here are the major components of the robot infrared remote control system:

- *Infrared remote.* Most any modern infrared remote control will work, but . . . the signal patterns they use vary considerably. You'll find it most convenient to use a universal remote control (about $10 at a department store). These remote controls can work with just about any make and model of TV set, DVD player, VCR, cable/satellite receiver/decoder available on the market.
- *Infrared receiver module.* The receiver module contains an infrared light detector, along with various electronics to clean up, amplify, and demodulate the signal from the remote control. The remote sends a signal built from a pattern of on/off flashes of infrared light. These flashes are modulated at about 38–40 kHz to differentiate them from other infrared sources in the receiver's environment. The receiver strips out the modulation and provides just the on/off flashing patterns (which will be referred to as the signal or packet in this chapter).
- *Computer or microcontroller.* You need some hardware to decode the light patterns, and a computer or microcontroller, running appropriate software, makes the job straightforward.

Previous editions of the book described the Sharp remote control standard or protocol for the infrared signal, but in this edition, the Sony format will be used. The reason for changing to the Sony standard is the fewer number of bits that are used in the data transmission, the more definite difference between a 1 or 0 of the data, and the greater number of projects available on the Internet and in books that use it.

Receiving the modulated signal is quite simple; there are a number of different remote control receivers available on the market, like the Sharp GP1U57X shown in Fig. 16-5. The receivers are very sensitive to electrical noise, so it is important to provide the 56 Ω resistor, 47 μF, and 0.1 μF filter capacitors to the circuit as shown. Some older receivers are built into metal cans, which may require external grounding as shown in the diagram. The outputs are usually open collector, so a pull-up resistor, like the 10k one shown in Fig. 16-5 is required.

When it comes right down to it, it really doesn't matter which protocol is used; they are all built around a similar data packet using the parts that are shown in Fig. 16-6. The first part is the leader and is used to indicate the start of the packet—for Sony remote controls, this is 2.2 ms in length. The packet's data bits (12 in the case of the Sony protocol) start with a synchronizing pulse with the length of the low pulse being the bit value. The Sony data bits are either 0.55 ms (550 μs) or 1.1 ms in length.

16.3.1 A TYPICAL MICROCONTROLLER INTERFACE

It might be intimidating at first to think about trying to receive and process remote control codes. In this and the next section, three different methods will be demonstrated and there are a number of ways that can be considered for use in a robot. The important

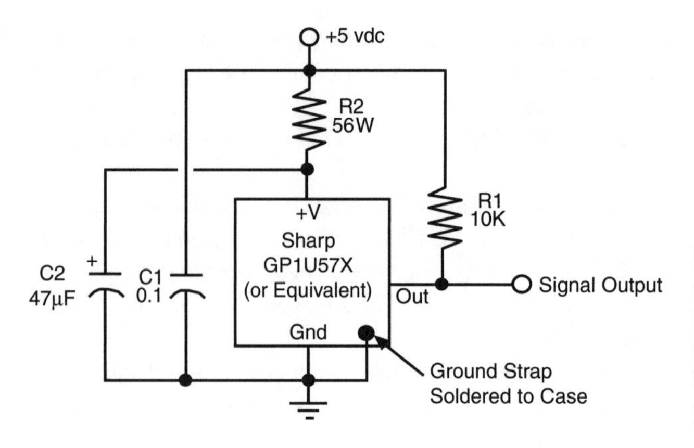

FIGURE 16-5 The infrared receiver-demodulator requires relatively few external components that are primarily used to filter any noise out of the unit's power source.

aspects of the differences are the ability to differentiate between a 1 and a 0 and to determine when an invalid packet (part of it is lost or has been garbled) is received and then reject it.

The first method is a brute force approach of timing the signals coming in.

```
'  Sony Remote Control Receiver Operation Method 1
'   Read the Timing for Each Bit

    DO
        DO WHILE (InputPin = 1) :  LOOP    '  Wait for Signal to go Low
        i = 0: DO WHILE (InputPin = 0) : i = i + 1: LOOP   '  Get Leader
        IF ((i > 2.0ms) AND (i < 2.4ms) THEN   '  Look for Valid Leader
        j = 0                    '  Use "j" to Count the Number of Bits
        RemoteCode = 0           '  Reset the Returned Value
        DO WHILE (j < 12)
            DO WHILE (InputPin = 1) : LOOP  '  Wait for Synch to Finish
            i = 0: DO WHILE (InputPin = 0) : i = i + 1: LOOP  '  Time Bit
            IF ((i < 0.45ms) AND (i > 1.3ms) THEN
                j = 100          '  Indicate Invalid Bit
            ELSE
```

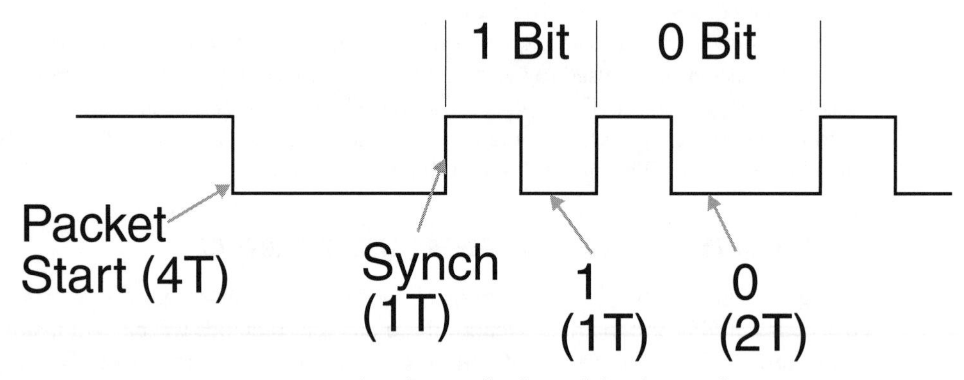

FIGURE 16-6 Sony remote control packet start leader and data bit specification.

```
                IF (i < 0.75ms) THEN  '  "1"
                    RemoteCode = (RemoteCode << 1) + 1
                ELSE  '  "0"
                    RemoteCode = (RemoteCode << 1)
                ENDIF
                j = j + 1     '  Increment Bit Counter
            ENDIF
        LOOP
        IF (j < 100) THEN
'  Use Valid "RemoteCode" to Perform the Next Robot Action
        ENDIF
    LOOP
```

Note that for making the code easier to understand, the counter values are assumed to be directly converted to time values. This is an unreasonable assumption; in a typical application, a free running clock would be reset before each DO WHILE loop and its value used for the comparison afterward.

Timing values explicitly, while working quite well, is difficult to use with a robot. There is a lot of overhead devoted to receiving the signal and it can be very difficult to time the entire application (especially if a PWM is used to control the motor speeds). A better way is to use interrupts to produce a virtual IR receive port using code like:

```
'  Sony Remote Control Receiver Operation Method 2
'   Use Interrupts for IR Receiver Pin Changing State and Timer Overflow


    RemoteCode = 0             '  Initialize Interrupt Handler Variables
    j = 0                      '  Bit Counter
'  Enable Free Running Timer
'  Enable Interrupts for Pin Changing State


    DO                         '  Mainline Code
        DO WHILE (RemoteCode = 0) : LOOP  '  Wait for Remote Signal
        SWITCH (RemoteCode)    '  Respond to the Remote Code Value
            CASE Forwards:
                Motors = Forwards
                RemoteCode = 0'  Clear Remote Code for Next Value
            CASE Reverse:
                :
        ENDSWITCH
    LOOP
'  Interrupt Handler Code
Interrupt PROCEDURE()
    IF (TimerOverFlow) THEN    '  Timer Overflowed, Too Long for Data Bit
        j = 0                  '  Reset Counter
        TimerInterrupt = Disable
    ELSE
        IF (j = 0) THEN        '  Leader Going Low
            j = 13             '  Want to Time the Leader
            TempCode = 0       '  Reset the Temporary Code Value
            Timer = 0          '  Reset the Timer
            TimerInterrupt = Enable  '  Enable Timer Interrupts
        ELSE
            IF (j = 13) THEN   '  Check Leader Timing
                TimerInterrupt = Disable
                IF ((Timer < 2.0ms) OR (Timer > 2.4ms)) THEN
```

```
                j = 0      '  Error, Wait for Next Packet
            ELSE
                j = 12     '  Can Start Reading Bit
            ENDIF
        ELSE               '  Check Data Bits
            IF (InputPin = 0) THEN '  Synch Pulse low
                Timer = 0
                TimerInterrupt = Disable
            ELSE           '  Read Bit Value
                TimerInterrupt = Disable
                IF ((Timer < 0.4ms) OR (Timer > 1.3ms)) THEN
                    j = 0 '  Invalid Bit
                ELSE
                    IF (Timer < 0.75ms) THEN
                        TempCode = (TempCode << 1) + 1
                    ELSE
                        TempCode = (TempCode << 1)
                    ENDIF
                    j = j - 1
                    IF (j = 0) THEN ' Have Read the Packet
                        RemoteCode = TempCode
                    ENDIF
                ENDIF
            ENDIF
        ENDIF
    ENDIF
  ENDIF
END Interrupt PROCEDURE
```

In this interrupt example code, the code does not do anything while waiting for a new RemoteCode value. In actuality, other tasks can be running in the mainline, and RemoteCode can be polled periodically to see if it has changed from zero to a code value. The interrupt code also checks for invalid data bit values, which minimizes the amount of effort the mainline code has to expend determining whether a packet is valid; the mainline just has to compare RemoteCode to the robot's operational values.

16.3.2 BS2 INTERFACE

After reviewing the code in the previous section, you have probably come to the conclusion that the BASIC Stamp 2 cannot read the packets from infrared remote controls. Trying to decode the data using PBASIC statements isn't possible because of the 250 μs statement execution time, but it is possible to add hardware to the BS2 as shown in Fig. 16-7, which performs the decoding for the BS2 and stores the value in a 12 bit serial to parallel decoder that can be read by the BS2. The parts list for the circuit in Fig. 16-7 is in Table 16-3.

The IR receiver (IC1) in the circuit passes the demodulated remote control signals to the trigger input of the 74LS123 as well as the data pin of one of the D-flip flops. When the incoming signal goes low, the 74LS123 single shot produces an 800 μs long pulse, which latches in the data bit along with shifting the contents of all the daisy-chained D-flip flops down. If the data bit is a 1 when the bit is latched into the bit, the next synch pulse will be active and a 1 will be latched in, otherwise the data bit will still be low and a 0 will be latched in. The 2.2 ms leader will be read as a zero and then the data will be shifted through and out of the 12 D-flip flop shift register.

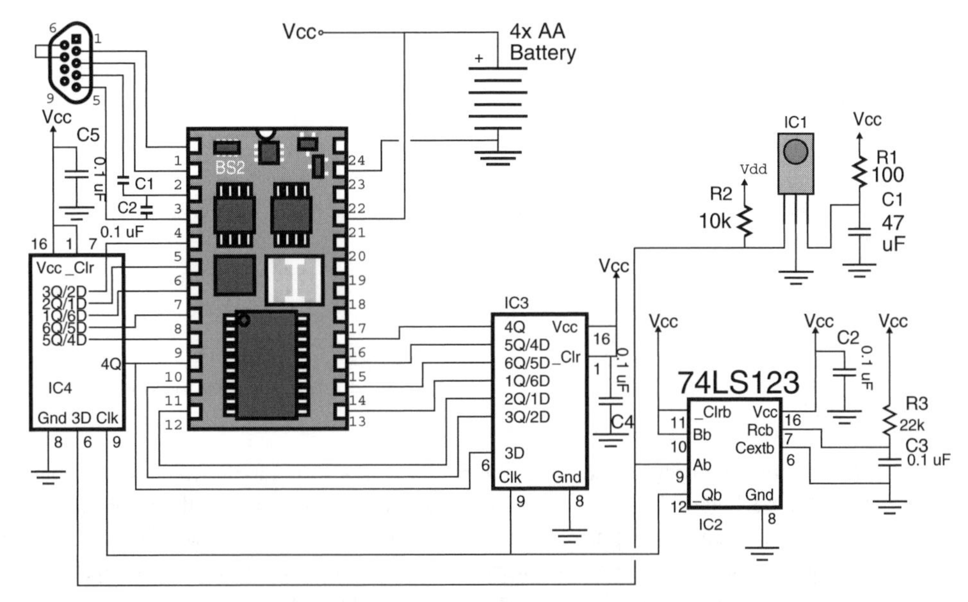

FIGURE 16-7 A single shot delay can produce the clock signal to pass IR serial data to two 74LS174s wired as six-bit shift registers that can be read by the BS2.

To test the operation of the circuit, the following program was developed.

```
'  BS2 Remote Control - Add a Sony Remote Control Read to BS2
'
'  This program reads the 12 bits of a Sony Remote Control and
'   Displays the Result on the Console
'
'  myke predko
```

TABLE 16-3 Parts List for BS2 Sony Remote Control Receiver Hardware

BS2	Parallax BASIC Stamp 2
IC1	38 kHz TV remote control IR receiver (Sharp GP1U57X or equivalent)
IC2	74LA123 dual shot chip
IC3–IC4	74LS174 Hex D-flip flop chips
R1	100 Ω resistor
R2	10k resistor
R3	22k resistor
C1	47 μF electrolytic capacitor
C2–C5	0.1 μF capacitor
Misc.	BS2 communications/programming interface, long breadboard, breadboard wiring kit, 4x AA battery clip, 4x AA alkaline batteries

```
'
'  05.09.04

'{$STAMP BS2}
'{$PBASIC 2.50}

'  Pin/Variable Declarations
RemotePins     VAR INS     '  Use all 16 Bits for the Read
RemoteData     VAR Word

'  Initialization/Mainline
  DO                       '  Repeat forever
    RemoteData = RemotePins & &FFF
    DEBUG "Remote Code is 0x", HEX3 RemoteData, CR
    PAUSE 333              '  Wait 1/3 Second before Repeating
  LOOP
```

Despite the simplicity of the code, the circuit works surprisingly well and data bits are read from the packets with good repeatability. When you build and test the application, there are three things that you will probably notice. The first is that occasionally, when a button is pressed on the remote control, the changing value read from the serial register is different from subsequent values. This is due to the data still being shifted in when the I/O port is executing. The second issue that you might have is the number of BS2 I/O pins that are required for the application—only four are left available for controlling the robot. You may want to shift in the data using the SHIFTIN PBASIC statement with some hardware to arbitrate between the incoming data and reading it. Finally, to indicate that a value has been read, you may want to clear the shift register: this could be accomplished by driving pin 1 of the 74LS174s low, but this would require an additional BS2 pin.

Table 16-4 lists some of the different codes provided by a Sony TV remote control using this circuit and the previous code. By selecting a VCR, DVD, or cable/satellite receiver, you will get slightly different codes for the different functions.

16.3.3 CONTROLLING ROBOT MOTORS WITH A REMOTE CONTROL

As noted earlier in this chapter, the software running in the remote control is very simple; when a button is pressed, a packet of data representing the button is sent out. If it is still pressed 40 or 50 ms later, then a packet is sent again and each and every 40 or 50 ms until the button is released.

The real intelligence is built into the device that is being controlled. Understanding how the remote control works is important to being able to use the remote control to directly control the robot's motors.

The remote control output is so simple because as a communications link infrared data transmission is quite poor, especially compared to other forms of wireless communications. People rarely point the remote control directly at the controlled device's receiver (often, people will point the remote control at the wrong unit altogether); IR communications depend on the ability of the signal to bounce off various objects in the room as well as the receiver's ability to receive and accept very low power signals. Despite this sensitivity and the ability to reflect off of different objects, quite a few packets are normally lost. The receiver compensates for this by continuing the last action some length of time after the last

TABLE 16-4 Sony TV Infrared Remote Control Codes

BUTTON	CODE	BUTTON	CODE
1	0xFEF	Power	0x56F
2	0x7EF	Volume up	0xB6F
3	0xBEF	Volume down	0x36F
4	0x3EF	Channel up	0xF6F
5	0xDEF	Channel down	0x76F
6	0x5EF	Arrow up	0xD0F
7	0x9EF	Arrow down	0x50F
8	0x1EF	Arrow left	0xD2F
9	0xEEF	Arrow right	0x32F
10	0x6EF	OK	0x58F

packet was received, *beyond* the time for the next packet to be received in the expectation that the next was accidentally lost and subsequent ones will continue the operation. This is the mode that is used in robots to control the motors.

To illustrate what is being discussed here, consider a typical situation. The motors move the robot forward when a packet is received and stop when the expected time (50 ms) has gone by without a repeated signal. The motors are turned off. This is a very simple example. For a real-world situation, all the codes would be tested and the motors would command the robot to move forward, reverse, turn left, turn right, etc. In a real robot, motor PWMs would also be active, controlling the speed at which the robot moves.

```
'  Remote Control Motor Example 1
'   Turn Off Motors when Expected Signal Packet is Missed

   DO
       IF (RemoteSignal = RobotForwards) THEN
           Motors = Forwards '  Forwards Signal Received
           RemoteSignal = 0  '  Reset Command
           PAUSE 50          '  Wait for Next Expected Signal
       ELSE                  '  Signal NOT Received
           Motors = Stop     '  Turn off Motors and Wait for Next
       ENDIF
   LOOP
```

The problems with this method of controlling motors are (a) if a packet is missed, the motors will be stopped, periodically resulting in either the robot "shuddering" or a PWM action taking place, resulting in less power to move the robot; and (b) if there is a mistiming between the microcontroller and the remote control, the remote signal will be missed, resulting in the motors being off for some period of time, again producing a PWM action and lowering the amount of power available from the drive motors. Note that the packet received variable RemoteSignal is reset to ensure that any subsequent values of RobotForwards have been newly received and are not saved values from previous data receptions.

To prevent these issues, you can plan for missed packets by running the motors for longer than the time between packets. Many applications use three or four packet delays to ensure the robot is active while it is being commanded.

```
'  Remote Control Motor Example 2
'   Run Motors past expected reception of next Signal Packet

    SignalCount = 0                     '  Reset Timer
    DO
        IF (SignalCount <> 0) THEN      '  Run Motors if Something There
            Motors = Forwards
            SignalCount = SignalCount - 1
        ELSE                            '  Signal NOT Received
          Motors = Stop                 '  Turn off Motors and Wait for Next
        ENDIF

        IF (RemoteSignal = RobotForwards) THEN
            SignalCount = 15            '  Run for three Packet Intervals
            RemoteSignal = 0            '  Reset Command
        ENDIF

        PAUSE 10                        '  Wait 10 ms before repeating

    LOOP
```

In the second example code the SignalCount variable is reset to 15 (three packet intervals in 10 ms increments) every time a RobotForwards packet is received. Along with eliminating the potential shuddering or lowered power operation of the motors, the time available in the PAUSE 10 statement could be put to use for running sensors or expanding the program to accept different inputs and perform different operations. The first example application does not allow itself to be expanded as easily.

16.3.4 GOING FURTHER

Regardless of which microcontroller or computer system you prefer, or maybe you don't want to use the signal patterns for Sony TVs and VCRs, you can adapt the receiver-demodulator interface and the software presented in this chapter for use with a wide variety of controllers, computers, and signal pattern formats. Of course, you'll need to revise the hardware and the program as necessary, and determine the proper bit patterns to use.

You will probably find that the signal patterns used with a great many kind of remote controls will be usable with the circuitry and programs shown in this chapter. Often, you will merely need to test the remote to determine the values to use for each button press. The hardware and software will still work even if the signal from the remote contains more data bits than the 12 provided. For this application, if more bits are sent to the shift registers, the first bits coming through are discarded, which usually doesn't matter—you're not trying to control a TV or VCR but your own robot, and the code values for its control are up to us. The only requirement is that each button on the remote must produce a unique value.

16.4 Using Radio Control Instead of Infrared

If you need to control your robot over longer distances, you should consider using radio signals instead of infrared. You can hack an old pair of walkie-talkies to serve as data transceivers, or even build your own AM or FM transmitter and receiver. An easier (and probably more reliable) method is to use ready-made transmitter/receiver modules. Ming, Abacom, and several other companies make low-cost radio frequency modules that you can use to transmit and receive low-speed (less than 300 bits per second) digital signals. Fig. 16-8 shows transmitter/receiver boards from Ming. Attached to them are daughter boards outfitted with Holtek HT-12E and HT-12D encoder/decoder chips.

The effective maximum range of these units is from 20 to 100 ft, depending on whether you use an external antenna and if there are any obstructions between the transmitter and receiver. More expensive units have increased power outputs, with ranges exceeding 1 mil. You are not limited to using just encoder/decoders like the HT-12. You may wish to construct a remote control system using DTMF (dual-tone multifrequency) systems, the same technology found in Touch-Tone phones. Connect a DTMF encoder to the transmitter and a DTMF decoder to the receiver. Microcontrollers such as the BASIC Stamp can be used as a DTMF encoder, or you can use specialty ICs made for the job.

Another option that you can consider is the use of hobby remote control transmitters and receivers. Interfacing to the servo control signals is somewhat more complex than the

FIGURE 16-8 RF transmitter/receiver modules can be used to remotely control robots from a greater distance than with infrared systems.

digital I/Os of the Holtek chips, but this can be done fairly simply within a microcontroller (a BASIC Stamp 2 can handle this chore with ease; pass each of the servo outputs to a separate BS2 pin and use the Pulsout statement to convert the length of the pulse to a numeric value). If you don't feel comfortable interfacing the receiver to a microcontroller, then servos could be used to throw switches and control the robot mechanically or commercial motor controllers (used for controlling the motors in model airplanes, boats, and cars) can be used.

16.5 From Here

To learn more about . . .	*Read*
Interfacing and controlling DC motors	Chapter 20, "Working with DC Motors"
Connecting to computers and microcontrollers	Chapter 14, "Computer Peripherals"
Using the BASIC Stamp microcontroller	Chapter 15, "The BASIC Stamp 2 Microcontroller"

PART 4

POWER, MOTORS, AND LOCOMOTION

CHAPTER 17

BATTERIES AND ROBOT POWER SUPPLIES

The robots in science fiction films are seldom like the robots in real life. Take the robot power supply. In the movies, robots almost always have some type of advanced nuclear drive or perhaps a space-age solar cell that can soak up the sun's energy, then slowly release it over two or three days. Nuclear power supplies are obviously out of the question. And solar cells don't provide enough power for the typical motorized robot, and as yet they have no power storage capabilities.

Most self-contained real-life robots are powered by batteries, the same kind of batteries used to provide juice to a flashlight, cassette radio, portable television, or other electrical device. Batteries are an integral part of robot design, as important as the frame, motor, and electronic brain—those components we most often think of when the discussion turns to robots. To robots, batteries are lifeblood; without them robots cease to function.

While great strides have been made in electronics during the past 20 years—including entire computers that fit on a chip—battery technology still has a way to go. Today's advanced batteries, used in laptops, cell phones, and remote control vehicles, are able to store significantly more energy than batteries of even five years ago but still don't hold enough energy for their size, weight, and cost. For most robot applications, the most efficient power source is common alkaline and nickel-metal hydride (NiMH) batteries that can provide more than adequate power to all of your robot creations with judicious planning and use.

17.1 Remember: Safety First!

Anytime you are working with electricity, remember to make sure that you understand how much energy is available in the different power supplies and make sure that there is adequate protection—even a AA alkaline radio battery, when shorted out, can produce enough heat to melt the clips that it is in and char the battery's plastic coating, potentially causing a fire.

Be extra careful when wiring power supply circuits that convert household 120 V AC to low DC voltages and triple-check your work. Never operate the supply when the top of the cabinet is off, unless you are testing it. Even then, stay away from the incoming AC. Touching the incoming AC voltage can cause a serious shock, and depending on the circumstances it could kill you. Be extra certain that no wires touch the chassis or front panel. Use an all-plastic enclosure whenever possible.

Do not operate any power supply if it has become wet, has been dropped, or shows signs of visible damage. Fix any problems before powering it up. Understand the fuse requirements for the power supply and robot and do not assemble the supply without a fuse.

17.2 Increasing Robot Performance

It is not widely known or understood, but of the various parameters that are used to choose a battery for an application (such as voltage output, amp-hour rating, recharge rate, etc.) the most important one for robot applications is very rarely considered—the battery's internal resistance. Often people describing themselves as experts will tell you that you should buy the cheapest carbon zinc batteries you can find instead of expensive alkaline or rechargeable batteries because the *ampere hour* (AH) rating of the cheaper batteries is similar to that of the much more expensive battery. This is true, but the inexpensive batteries have a very high internal resistance, which will shorten their lives in your robot and make it difficult for your robot to work reliably.

To understand the effect the internal resistance has in a robot, take a look at Fig. 17-1, which shows three different robot battery power circuits. When you think of electrical circuits, you tend to think of the idealized case (the left circuit diagram) in which there is no resistance in wiring. This is a reasonable approximation when low power and current power sources are being considered. In a robot application, which draws a great deal of current, the internal resistances of the batteries (shown in the middle circuit diagram) have a voltage drop across them which becomes larger as more current is drawn from the batteries. The combined internal resistances can be consolidated into a single resistance as shown in the circuit diagram to the right, the "effective circuit."

This voltage drop is actually a power loss to the robot. The batteries' internal resistances create heat, which is power lost in its most basic form. To make matters worse, most batteries lose the ability to output electrical energy as their internal temperature rises, so this internal resistance not only robs the robot of power, but also reduces the total amount of power available from the batteries.

To summarize, the lower the internal resistance of the battery, the more power is avail-

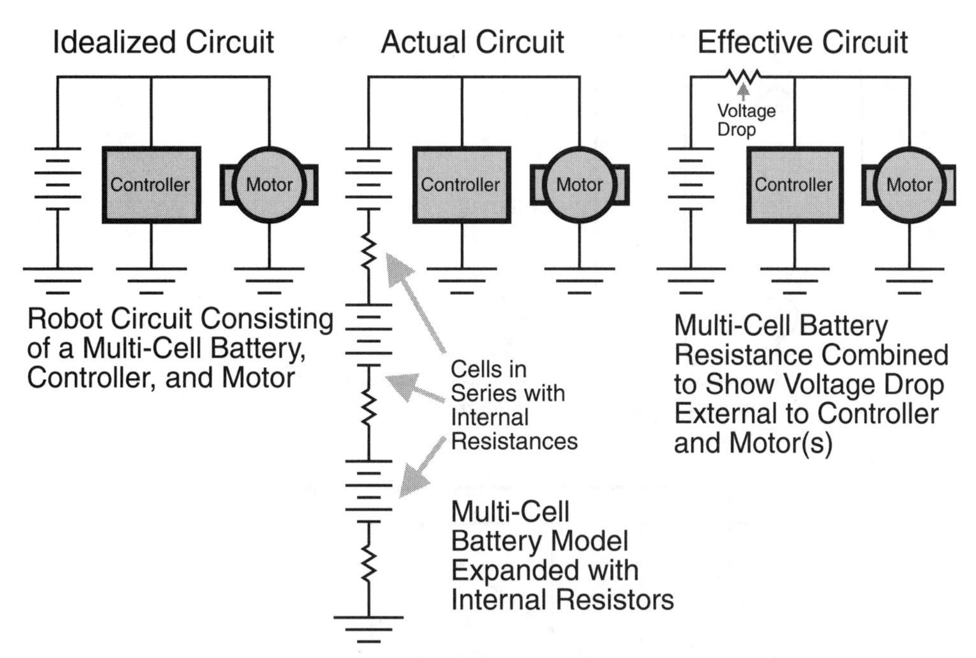

FIGURE 17-1 The effect of the internal resistance of a robot's battery power source on the output voltage available to the robot's motors and electronics.

able to the robot. While you may want to use cheap carbon zinc batteries for testing your robot's electronics, when it comes to moving you should use batteries with the lowest available internal resistance as documented in the battery's data sheet on the manufacturer's web page.

17.3 Combining Batteries

You can obtain higher voltages and current by connecting several cells together, as shown in Fig. 17-2. There are two basic approaches:

- To increase voltage, connect the batteries in series. The resultant voltage is the sum of the voltage outputs of all the cells combined.
- To increase current, connect the batteries in parallel. The resultant current is the sum of the current capacities of all the cells combined.

Take note that when you connect cells together not all cells may be discharged or recharged at the same rate. This is particularly true if you combine two half-used batteries with two new ones. The new ones will do the lion's share of the work and won't last as long. Therefore, you should always replace or recharge all the cells at once. Similarly, if one or more of the cells in a battery pack are permanently damaged and can't deliver or take on a charge like the others, you should replace them.

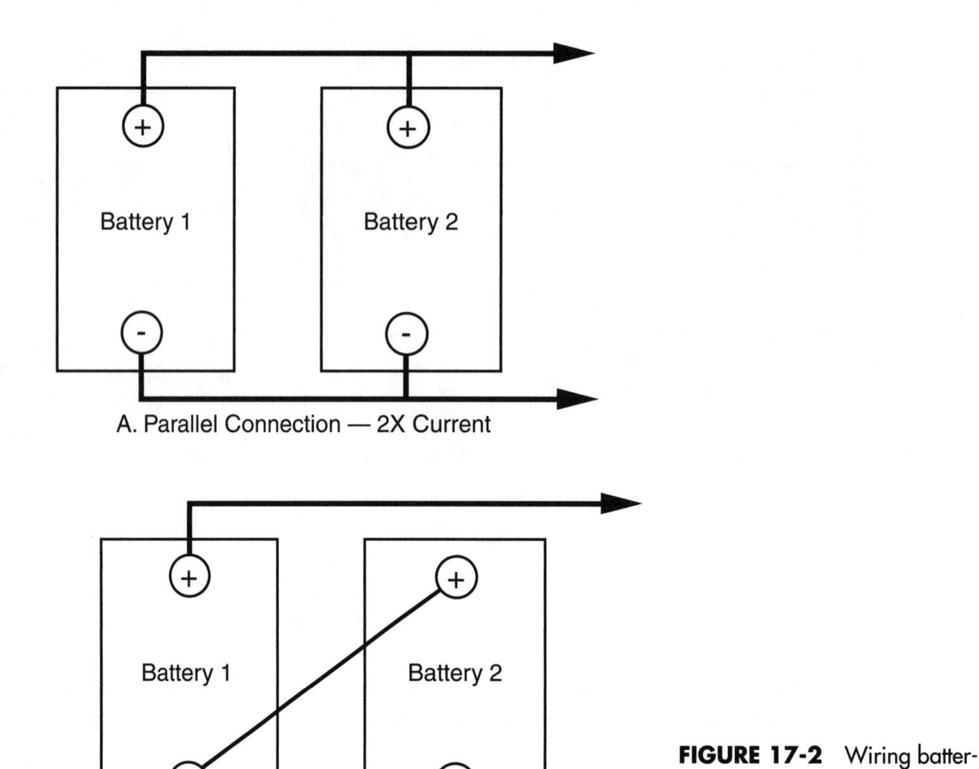

FIGURE 17-2 Wiring batteries to increase ratings. *a.* Parallel connection increases current; *b.* Series connection increases voltage.

17.4 Types of Batteries

There are seven main types of batteries, which come in a variety of shapes, sizes, and configurations.

17.4.1 ZINC

Zinc batteries are the staple of the battery industry and are often referred to simply as flashlight cells. The chemical makeup of zinc batteries takes two forms: carbon zinc and zinc chloride. Carbon zinc, or regular-duty, batteries die out the quickest and are unsuited to robotic applications. Zinc chloride, or heavy-duty, batteries provide a little more power than regular carbon zinc cells and last 25 to 50 percent longer. Despite the added energy, zinc chloride batteries are also unsuitable for most robotics applications.

Both carbon zinc and zinc chloride batteries can be rejuvenated a few times after being drained. Zinc batteries are available in all the standard flashlight (D, C, A, AA, and AAA) and lantern battery sizes.

17.4.2 ALKALINE

Alkaline cells use a special alkaline manganese dioxide formula that lasts up to 800 percent longer than carbon zinc batteries. The actual increase in life expectancy ranges from about 300 to 800 percent, depending on the application. In robotics, where the batteries are driving motors, solenoids, and electronics, the average increase is a reasonable 500 percent or five times the life of carbon zinc batteries.

Alkaline cells, which come in all the standard sizes (as well as 6- and 12-V lantern cells), cost about twice as much as zinc batteries. But the increase in power and service life is worth the cost. Unlike zinc batteries, however, ordinary alkaline batteries cannot be rejuvenated by recharging without risking fire (though some people try it just the same). During recharging, alkaline batteries generate considerable internal heat, which can cause them to explode or catch fire. So when these batteries are dead, just throw them away.

Rechargeable alkaline cells have been on the market for a number of years now. These provide many of the benefits of ordinary alkaline cells but with the added advantage of being rechargeable. A special low-current recharger is required (don't use the recharger on another battery type or you may damage the recharger or the batteries). While rechargeable alkalines cost more than ordinary alkaline cells, over time your savings from reusing the batteries can be considerable.

17.4.3 HIGH-TECH ALKALINE

You've probably seen the advertisements for the very high tech alkaline cells—ones that offer two or three times the life of regular alkaline batteries. While being significantly more expensive than regular alkaline batteries, they do indeed provide 200 to 300 percent of the life of basic alkaline cells. The dramatic improvement in battery life is possible by decreasing the internal resistance of the batteries and improving their ability to respond to changing current demands (such as a motor turning on and off) without damaging the internal structure of the battery.

The much higher price of the high-tech alkaline cells will often preclude their regular use in a robot. They are good to keep in your back pocket for those times when you need a bit more juice in a sumo-competition or the robot is having problems that more power (or less weight) would solve.

17.4.4 NICKEL METAL HYDRIDE

Nickel metal hydride (NiMH) batteries are probably the optimal rechargeable battery technologies for use in robots (and, practically speaking, all other applications that require batteries). NiMH batteries can be recharged 400 or more times and have a low internal resistance, so they can deliver high amounts of current. Nickel metal hydride batteries are about the same size and weight as alkaline cells, but they deliver about 50 percent more operating current than NiCads. In fact, NiMH batteries work best when they are used in very high current situations. Unlike NiCads, NiMH batteries do not exhibit any memory effect, nor do they contain cadmium, a highly toxic material.

NiMH cells are available in all standard sizes, plus special-purpose "sub" sizes for use in sealed battery packs (as in rechargeable handheld vacuum cleaners, photoflash equipment, and so forth). Most sub-size batteries have solder tabs, so you can attach wires directly to the battery instead of placing the cells in a battery holder. NiMH cells don't last quite as long as alkaline cells, but this deficiency is being addressed with the availability of high amp-hour rated batteries at reasonable cost.

While NiMH batteries are discharging, especially at high currents, they can get quite hot. You should consider this when you place the batteries in your robot. If the NiMH pack will be pressed into high-current service, be sure it is located away from any components that may be affected by the heat. This includes any control circuitry or the microcontroller.

NiMH batteries should be recharged using a recharger specially built for them—most modern rechargers have a switch to specify whether the batteries are NiMH or NiCads. According to NiMH battery makers, NiMH batteries should be charged at an aggressive rate. A by-product of this kind of high-current recharging is that NiMH can be recycled back into service more quickly than NiCads. You can deplete your NiMH battery pack, put it under charge for an hour or two, and be back in business.

One disadvantage of NiMH relative to other rechargeable battery technologies is that the battery does not hold the charge well. That is, over time (weeks, even days) the charge in the battery is depleted, even while the battery is in storage. For this reason, it's always a good idea to put NiMH batteries on the recharger at regular intervals, even when they haven't been used. NiMH batteries do not exhibit the memory effect of NiCad batteries.

17.4.5 NICKEL-CADMIUM

When you think "rechargeable battery," you undoubtedly think nickel-cadmium (NiCad for short). In recent years, NiCads have fallen out of favor due to the memory effect in which the battery will tend to last only as long as it is usually asked to; if it is placed in longer service periodically, it will not respond well. Another issue with NiCads is their chemical makeup—cadmium is a highly toxic metal that should not be put into public landfill sites and ideally should be disposed of as toxic waste.

If you decide to use NiCad batteries, remember that they can change output polarity—positive becomes negative and vice versa—under certain circumstances. Polarity reversal is common if the battery is left discharged for too long or if it is discharged below 75 or 80 percent capacity. Excessive discharging can occur if one or more cells in a battery pack wears out. The adjacent cells must work overtime to compensate, and so they discharge themselves too fast and too far.

You can test for polarity reversal by hooking the battery to a volt-ohm meter (remove it from the pack if necessary). If you get a negative reading when the leads are connected properly, the polarity of the cell is reversed. You can sometimes correct polarity reversal by fully charging the battery (connecting it in the recharger in reverse), then shorting it out. Repeat the process a couple of times if necessary. There is about a fifty-fifty chance that the battery will survive this. The alternative is to throw the battery out, so you actually stand to lose very little.

17.4.6 LITHIUM AND LITHIUM-ION

Lithium and rechargeable lithium-ion batteries are popular in laptop computers. They are best used at a steady discharge rate and tend to be expensive. Lithium batteries of various types provide the highest energy density of most any other commercially available battery, and they retain their charge for months, even years. Like other rechargeable battery types, rechargeable lithium-ion batteries require their own special recharging circuitry, or overheating and even fire could result.

17.4.7 LEAD-ACID

The battery in your car is a lead-acid battery. It is made up of not much more than lead plates crammed in a container that's filled with an acid-based electrolyte. These brutes pack a wallop and have an admirable between-charge life. When the battery goes dead, recharge it, just like a NiCad.

Not all lead-acid batteries are as big as the one in your car. You can also get—new or surplus—6-V lead-acid batteries that are about the size of a small radio. The battery is sealed, so the acid doesn't spill out (most automotive batteries are now sealed as well). The sealing isn't complete though: during charging, gases develop inside the battery and are vented out through very small pores. Without proper venting, the battery would be ruined after discharging and recharging. These batteries are often referred to as sealed lead-acid, or SLA.

Lead-acid batteries typically come in self-contained packs. Six-volt packs are the most common, but you can also get 12- and 24-V packs. The packs are actually made by combining several smaller cells. The cells are wired together to provide the rated voltage of the entire pack. Each cell typically provides 2.0 V, so three cells are required to make a 6-V pack. You can, if you wish, take the pack apart, unsolder the cells, and use them separately.

Although lead-acid batteries are powerful, they are heavy. A single 6-V pack can weigh 4 or 5 lb (2 to 2.5 kg). Lead-acid batteries are often used as a backup or emergency power

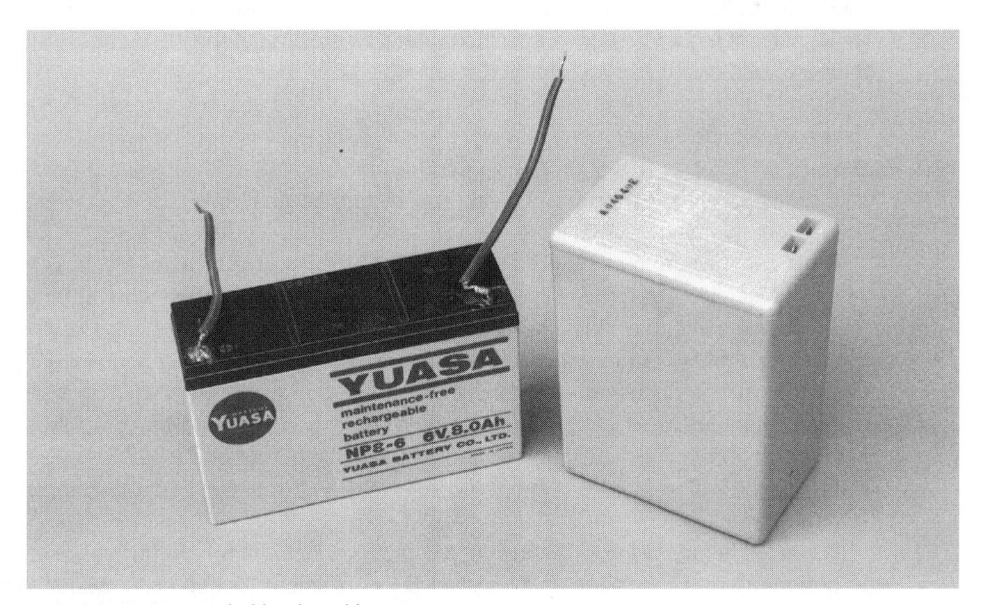

FIGURE 17-3 Sealed lead-acid batteries.

supply for computers, lights, and telephone equipment. The cells are commonly available on the surplus market, and although used they still have many more years of productive life. The retail price of new lead-acid cells is about $25 for a 6-V pack. Surplus prices are 50 to 80 percent lower.

Motorcycle batteries make good power cells for robots. They are easy to get, compact, and relatively lightweight. The batteries come in various amp-hour capacities, so you can choose the best one for your application. New motorcycle batteries are somewhat pricey, although you should be able to find surplus or used batteries for just a few dollars. You can also use car batteries, as long as your robot is large and sturdy enough to support it.

Gelled electrolyte batteries (commonly called gel-cell, after a popular trade name) use a special gelled electrolyte and are the most common form of SLA batteries. They are rechargeable and provide high current for a reasonable time, which makes them perfect for robots. Fig. 17-3 shows typical sealed lead-acid batteries.

17.5 Battery Ratings

Batteries carry all sorts of ratings and specifications. Traditionally, the two most important specifications are per-cell voltage and amp-hour current.

17.5.1 VOLTAGE

The voltage rating of a battery is fairly straightforward. If the cell is rated for 1.5 V, when new, it puts out a bit more. Over time it will drop down to the rate value, give or take. That "give or take" is more important than you may think because few batteries actually deliver their rated voltage throughout their life span. Most rechargeable batteries are recharged 20 to 30 percent higher than their specified rating. For example, the 12-V battery in your car, a type of lead-acid battery, is charged to about 13.8 V.

Standard zinc and alkaline flashlight batteries are rated at 1.5 V per cell. Assuming you have a well-made battery in the first place, the voltage may actually be 1.65 V when the cell is fresh, and drop to 1.3 V or less, at which point the battery is considered dead. The circuit or motor you are powering with the battery must be able to operate sufficiently throughout this range.

Most batteries are considered dead when their power level reaches 80 percent of their rated voltage. That is, if the cell is rated at 6 V, it's considered dead when it puts out only 4.8 V. Some equipment may still function at levels below 80 percent, but the efficiency of the battery is greatly diminished. Below the 80 percent mark, the battery no longer provides the rated current (discussed later), and if it is the rechargeable type, the cell is likely to be damaged and unable to take a new charge. When experimenting with your robot systems, keep a volt-ohm meter handy and periodically test the output of the batteries. Perform the test while the battery is in use. The test results may be erroneous if you do not test the battery under load.

It is often helpful to know the battery's condition when the robot is in use. Using a DMM to periodically test the robot's power plant is inconvenient. But you can build a number of battery monitors into your robot that will sense voltage level. The output of the monitor can be a light-emitting diode (LED), which will allow you to see the relative voltage level, or you can connect the output to a circuit that instructs the robot to seek a recharge or turn off. Several monitor circuits are discussed later in this chapter.

If your robot has an on-board computer, you want to avoid running out of juice midway through some task. Not only will you lose the operating program and have to rekey or reload it, but the robot may damage itself or its surroundings if the power to the computer is suddenly turned off.

17.5.2 CAPACITY

The capacity of a battery is rated as amp-hour current. This is the amount of power, in amps or milliamps, the battery can deliver over a specified period of time. The amp-hour current rating is a little like the current rating of an AC power line, but with a twist. AC power is considered never ending, available night and day, always in the same quantity. But a battery can only store so much energy before it poops out, so the useful service life must be taken into account. The current rating of a battery is at least as important as the voltage rating because a battery that can't provide enough juice won't be able to turn a motor or sufficiently power all the electronic junk you've stuck onto your robot.

What exactly does the term *amp-hour* mean? Basically, the battery will be able to provide the rated current for 1 h before failing. If a battery has a rating of 5 amp-hours (expressed as AH), it can provide up to 5A continuously for 1 h, 1A for 5 h, and so forth, as shown in Fig. 17-4. So far, so good, but the amp-hour rating is not that simple. The 5 AH rating is actually taken at a 10- or 20-h discharge interval. That is, the battery is used for 10 or 20 h, at a low or medium discharge rate. After the specified time, the battery is tested to see how much charge it has left. The rating of the battery is then calculated taking the difference between the discharge rate and the reserve power and multiplying it by the number of hours under test.

What this means is that it's an unusual battery that provides the stated amps in the 1-h

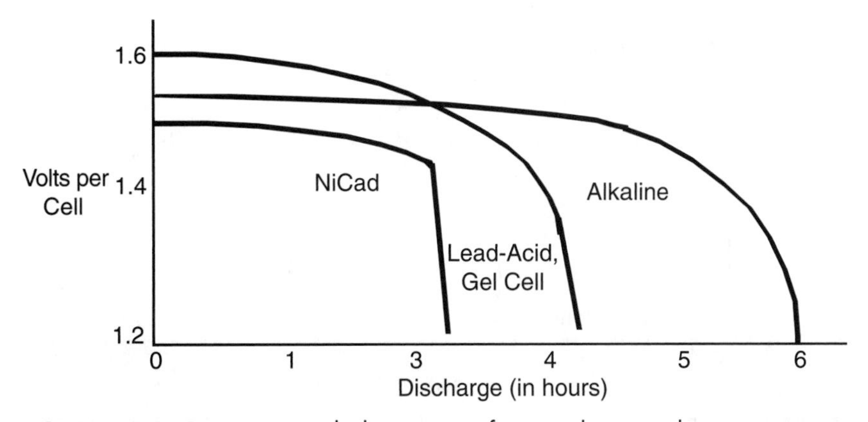

FIGURE 17-4 Representative discharge curves for several common battery types.

period. The battery is much more likely to fail after 30 or 45 min of heavy-duty use and won't be able to supply the specified current for more than about 15 to 20 min. Discharging at or above the amp-hour rating may actually cause damage to the battery. This is especially true of NiCad cells.

The lesson to be learned is that you should always choose a battery that has an amp-hour rating 20 to 40 percent more than what you need to power your robot. Figuring the desired capacity is nearly impossible until the entire robot is designed and built (unless you are very good at computing current consumption). The best advice is to design the robot with the largest battery you think practical. If you find that the battery is way too large for the application, you can always swap it for a smaller one. It's not so easy to do the reverse.

Note that some components in your robot may draw excessive current when first switched on, then settle down to a more reasonable level. Motors are a good example of this. A motor that draws 1 A under load may actually require several amps at startup. The period is very brief, on the order of 100 to 200 ms. No matter; the battery should be able to accommodate the surge. This means that the 20 to 40 percent overhead in using the larger battery is a necessity, not just a design suggestion. A rough comparison of the discharge curve at various discharge times is shown in Fig. 17-5.

17.5.3 RECHARGE RATE

When the first editions of this book were written, most rechargeable batteries required a slow charge process, often taking 12 to 24 h. Modern batteries and charges can perform the task in significantly less time; depending on the battery, this time could be measured in minutes.

Unfortunately, the recharging time is specific to the batteries and the charger used. This means that when you are selecting batteries for use in your robot, you must be cognizant of the recharging time as well as the charging equipment recommended by the battery manu-

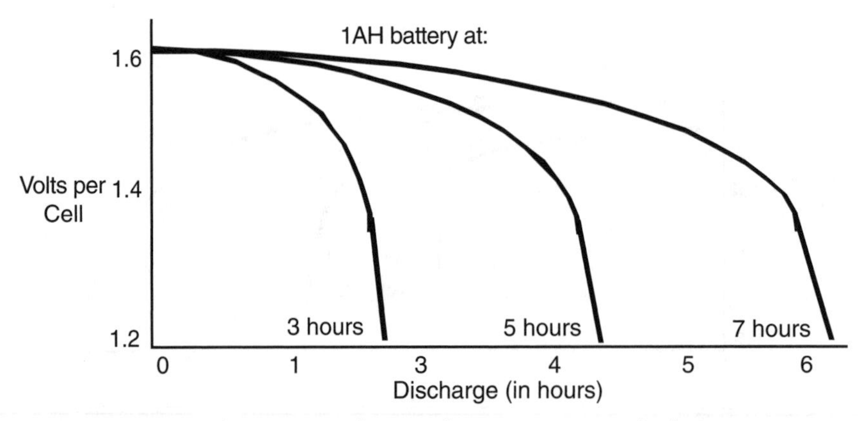

FIGURE 17-5 Discharge curves of a 1-AH battery at 3-, 5-, and 7-h rates.

facturer. For the most part, chargers are quite reasonably priced and can be used with a variety of different manufacturers' batteries.

17.5.4 NOMINAL CELL VOLTAGE

Each battery type generates different nominal (normal, average) output voltages. The traditional rating for zinc, alkaline, and similar nonrechargeable batteries is 1.5 V per cell. The actual voltage delivered by the cell can vary from a high of around 1.7 V (fresh and fully charged) to around 1.2 or 1.3 V (dead). Other battery types, most notably NiCads, provide different nominal cell voltages. Specifically, NiCad and nickel metal hydride batteries provide 1.2 V per cell, and lead-acid batteries provide 2.0 V per cell.

To achieve higher voltages, you can link cells internally or externally (see "Combining Batteries," previously for more information). By internally linking together six 1.5-V cells, for example, the battery will output 9 V.

Nominal cell voltage is important when you are designing the battery power supplies for your robots. If you are using 1.5-V cells, a four-cell battery pack will nominally deliver 6 V, an eight-cell pack will nominally deliver 12 V, and so forth. Conversely, if you are using 1.2-V cells, a four-cell battery pack will nominally deliver 4.8 V and an eight-cell pack will nominally deliver 9.6 V. The lower voltage will have an effect on various robotic subsystems. For example, many microcontrollers used with robots are made to operate at 5 V and will reset (stop working) at 4.5 V. A battery pack that delivers only 4.8 V will likely cause problems with the microcontroller. You will need to add more cells, change the battery type to a kind that provides a higher per-cell voltage, or use a step-up switching voltage regulator (described later in the chapter) to ensure that the microcontroller is powered at the correct voltage.

17.6 Battery Recharging

Most lead-acid and gel-cell batteries can be recharged using a 200- to 800-mA battery charger. Standard NiCad batteries can't withstand recharge rates exceeding 50 to 100 mA, and if you use a charger that supplies too much current you will destroy the cell. Use only a battery charger designed for NiCads. High-capacity NiCad batteries can be charged at higher rates, and there are rechargers designed specially for them.

Nickel metal hydride, rechargeable alkalines, and rechargeable lithium-ion batteries all require special rechargers. Avoid substituting the wrong charger for the battery type you are using, or you run the risk of damaging the charger and/or the battery (and perhaps causing a fire).

You can rejuvenate zinc batteries by placing them in a recharger for a few hours. The process is not true recharging since the battery is not restored to its original charge or voltage level. The rejuvenated battery lasts about 20 to 30 percent as long as it did during its initial use. Most well-built zinc batteries can be rejuvenated two or three times before they are completely depleted.

Rechargeable batteries should be periodically recharged whether they need it or not. Batteries not in regular use should be recharged every two to four months, more frequently for NiMH batteries. Always observe polarity when recharging batteries. Inserting the cells backward in the recharger will destroy the batteries and possibly damage the recharger.

You can purchase ready-made battery chargers for the kind of battery you are using or build your own. The task of building your own is fairly easy because several manufacturers make specialized integrated circuits just for recharging batteries, although you should really look at the time and expense involved in building your own charger. Chances are you will be able to buy a charger for less than what the parts would cost you to build one yourself, and charger manufacturers usually design their chargers to take advantage of the latest charging algorithms for a specific product, allowing for optimized charging cycles. It is unlikely that you will come up with a design that is better or cheaper than commercial units.

17.7 Recharging the Robot

You'll probably want to recharge the batteries while they are inside the robot. This is no problem as long as you install a connector for the charger terminals on the outside of the robot. When the robot is ready for a charge, connect it to the charger.

Ideally, the robot should be turned off during the charge period, or the batteries may never recharge. However, turning off the robot during recharging may not be desirable, as this will end any program currently running in the robot. There are several schemes you can employ that will continue to supply current to the electronics of the robot yet allow the batteries to charge. One way is to use a relay switchout. In this system, the external power plug on your robot consists of four terminals: two for the battery and two for the electronics. When the recharger is plugged in, the batteries are disconnected from the robot. You can use relays to control the changeover or heavy-duty open-circuit jacks and plugs (the ones for audio applications may work). While the batteries are switched out and being recharged, a separate power supply provides power to the robot.

17.8 Battery Care

Batteries are rather sturdy little creatures, but you should follow some simple guidelines when using them. You'll find that your batteries will last much longer, and you'll save yourself some money and grief.

- Store new batteries in the fresh food compartment of your refrigerator (not the freezer) and put them in a plastic bag so if they leak they won't contaminate the food. Remove them from the refrigerator for several hours before using them.
- Avoid using or storing batteries in temperatures above 75° or 80°F. The life of the battery will be severely shortened otherwise. Using a battery above 100° to 125°F causes rapid deterioration.

- Unless you're repairing a misbehaving NiCad, avoid shorting out the terminals of the battery. Besides possibly igniting fumes exhausted by the battery, the sudden and intense current output shortens the life of the cell.
- Keep rechargeable batteries charged. Make a note when the battery was last charged.
- Fully discharge NiCads before charging them again. This prevents memory effect. Other rechargeable battery types (nickel metal hydride, rechargeable alkaline, lead-acid, etc.) don't exhibit a memory effect and can be recharged at your convenience.
- Given the right circumstances all batteries will leak, even the "sealed" variety. When they are not in use, keep batteries in a safe place where leaked electrolyte will not cause damage. Remove batteries from their holder when they are not being used.

17.9 Power Distribution

Now that you know about batteries, you can start using them in your robot designs. The most simple and straightforward arrangement is to use a commercial-made battery holder. Holders are available that contain from two to eight AA, C, or D batteries. The wiring in these holders connects the batteries in series, so a four-cell holder puts out 6 V (1.5 times 4). You attach the leads of the holder (red for positive and black for ground or negative) to the main power supply rail in your robot. If you are using a gel-cell or lead-acid battery, you would follow a similar procedure.

17.9.1 FUSE PROTECTION

Flashlight batteries don't deliver extraordinary current, so fuse protection is not required on the most basic robot designs. Gel-cell, lead-acid, and high-capacity NiCad batteries can deliver a most shocking amount of current. In fact, if the leads of the battery accidentally touch each other or there is a short in the circuit, the wires may melt and a fire could erupt.

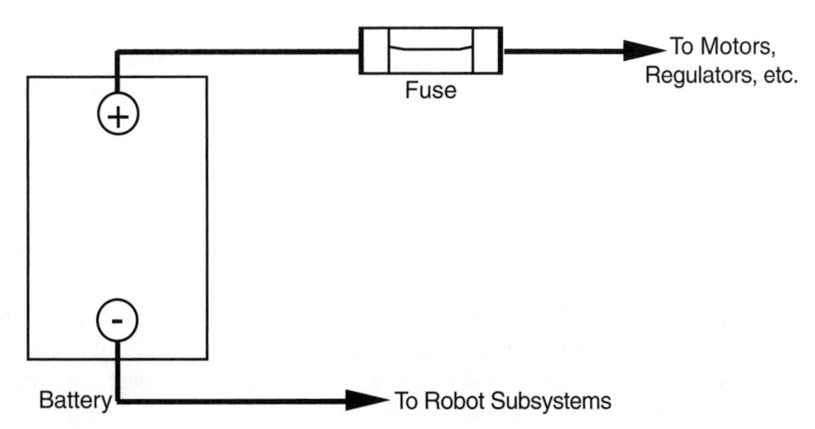

FIGURE 17-6 How to install a fuse in line with the battery and the robot electronics or motor.

Fuse protection helps eliminate the calamity of a short circuit or power overload in your robot. As illustrated in Fig. 17-6, connect the fuse in line with the positive rail of the battery, as near to the battery as possible. You can purchase fuse holders that connect directly to the wire or that mount on a panel or printed circuit board.

Choosing the right value of fuse can be a little tricky, but it is not impossible. It does require that you know how much current your robot draws from the battery during normal and stalled motor operation. You can determine the value of the fuse by adding up the current draw of each separate subsystem, then add 40 to 60 percent contingency for overcurrents above and beyond the measured requirements. These "overcurrents" usually consist of extra motor draws when the robot first starts moving or encounters an obstacle or hill it must climb over—if the motors are stalled, then current draw will be much greater than 50 percent over the nominal operating value, causing the fuse to blow and protecting the robot and its systems.

For example, if the two drive motors in a robot draw 2 A each, the main circuit board draws 1 A, and four other small motors each draw 0.5 A (for a total of, perhaps, 2 A). Add all these up and you get 7 A. Installing a fuse with a rating of 7 A at 125 V will probably end up blowing when the robot starts moving or encounters a bump in the carpet. Putting in a 10- to 12-A fuse will give the extra margin to handle extra current draw during normal operation.

To help prevent the initial motor current draw, you should use a slow-blow fuse. Otherwise, the fuse will burn out every time one of the heavy-duty motors kicks in.

Fuses don't come in every conceivable size. For the sake of standardization, choose the regular 1¼-in-long-by-¼-in-diameter bus fuses. You'll have an easier job finding fuse holders for them and a greater selection of values. Even with a standard fuse size, there is not much to choose from past 8 A, other than 10, 15, and 20 A. For values over 8 A, you may have to go with ceramic fuses, which are used mainly for microwave ovens and kitchen appliances.

17.9.2 MULTIPLE VOLTAGE REQUIREMENTS

Some advanced robot designs require several voltages if they are to operate properly. The drive motors may require 12 V, at perhaps 2 to 4 A, whereas the electronics require +5, and perhaps even –5 V. Multiple voltages can be handled in several ways. The easiest and most straightforward is to use a different set of batteries for each main subsection. The motors operate off one set of large lead-acid or gel-cell batteries; the electronics are driven by smaller capacity NiMH batteries.

This approach is actually preferable when the motors used in the robot draw a lot of current. Motors naturally distribute a lot of electrical noise throughout the power lines, noise that electronic circuitry is extremely sensitive to. The electrical isolation that is provided when you use different batteries virtually eliminates problems caused by noise (the remainder of the noise problems occur when the motor commutators arc, causing RF interference). In addition, when the motors are first started, the excessive current draw from the motors may zap all the juice from the electronics. This sag can cause failed or erratic behavior, and it could cause your robot to lose control.

The other approach to handling multiple voltages is to use one main battery source and a voltage regulator to step it up or down so it can be used with the various components in the system. This is called DC-DC conversion and will be discussed later in this chapter. You can accomplish DC-DC conversion by using circuits of your own design or by purchasing specialty integrated circuit chips that make the job easier. One battery's output voltage can be changed to a wide range of voltages from zero to about 15 V. Normally a single battery will directly drive the motors and, with proper conversion, supply the +5-V power to the circuit boards.

Connecting the batteries judiciously can also yield multiple voltage outputs. By connecting two 6-V batteries in series, as shown in Fig. 17-7, you get +12 V, +6 V, and –6 V. This system isn't nearly as foolproof as it seems, however. More than likely, the two batteries will not be discharged at the same rate. This causes extra current to be drawn from one to the other, and the batteries may not last as long as they might otherwise.

If all of the subsystems in your robot use the same batteries, be sure to add sufficient filtering capacitors across the positive and negative power rails. The capacitors help soak up excessive current spikes and noise, which are most often contributed by motors. Place the capacitors as near to the batteries and the noise source as possible. A good rule of thumb is 100 µF of capacitance for every 250 mA of current drawn by a motor during normal operation.

Be certain the capacitors you use are overrated above the voltage by 50 to 75 percent (e.g., use a 7.5-V rated capacitor for a 5-V circuit; 25 to 35 V is fine). An underrated capacitor will probably burn out or possibly develop a short circuit, which can cause a fire.

You should place smaller value capacitors, such as 0.01 to 0.1 µF, across the positive and negative power rails wherever power enters or exits a circuit board. As a general rule, you should add one of these decoupling capacitors at the power input pins of all ICs. Linear ICs, such as the 555 timer, need decoupling capacitors, or the noise they generate through the power lines can ripple through to other circuits.

17.9.3 SEPARATE BATTERY SUPPLIES

Most hobby robots now contain computer-based control electronics of some type. The computer requires a specific voltage (called regulation, discussed in the next section), and

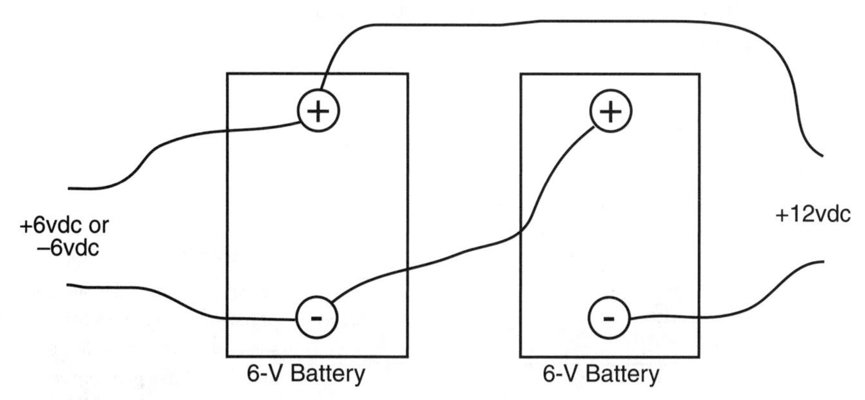

FIGURE 17-7 Various voltage tap-offs from two 6-V batteries. This is not an ideal approach (as the batteries will discharge at different rates).

the chip requires that the voltage be clean and free of noise and other glitches. A common problem in robotic systems is that the motors cause so-called sags and noise in the power supply system, which can affect the operation of the control electronics. You can largely remedy this by using separate battery supplies for the motors and the electronics. The ground connection of the power supplies must be connected together and common throughout the robot.

With this setup, the motors have one unregulated power supply, and the control electronics have their own regulated power supply. Even if the motors turn on and off very rapidly, this approach will minimize sags and noise on the electronics side. It's not always possible to have separate battery supplies, of course. In these cases, use the capacitor filtering techniques described previously. The large capacitors that are needed to achieve good filtering between the electronics and motor sections will increase the size and weight of your robot. A 2200-μF capacitor, for example, may measure ¾ in diameter by over an inch in height. You should plan for this in your design.

17.10 Voltage Regulation

Many types of electronic circuits require a precise voltage or they may be damaged or act erratically. Generally, you provide voltage regulation only to those components and circuit boards in your robot that require it. You can easily add a variety of different solid-state voltage regulators to your electronic circuits. They are easy to obtain, and you can choose from among several styles and output capacities. In this chapter, some of the different types will be described along with their operating characteristics.

17.10.1 ZENER DIODE VOLTAGE REGULATION

A quick and relatively small method for providing regulated voltage is to use *zener* diodes, as shown in Fig. 17-8. With a zener diode, current does not begin to flow through the load circuitry until the voltage exceeds a certain level (called the breakdown voltage). Voltage over this level is then "shunted" through the zener diode, effectively limiting the voltage to the rest of the circuit. Zener diodes are available in a variety of voltages, such as 3.3, 5.1, 6.2, and others.

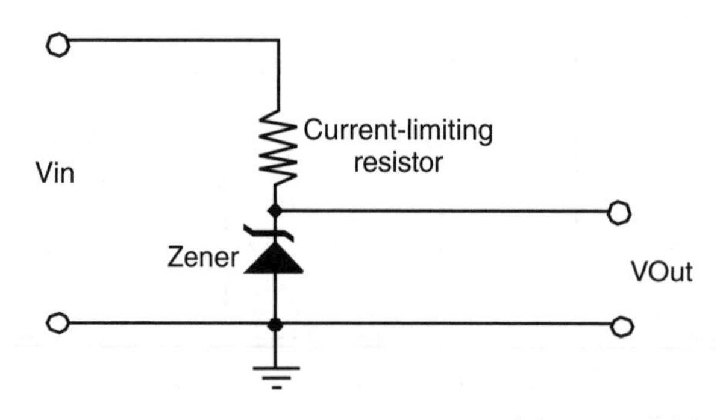

FIGURE 17-8 A zener diode and resistor can make a simple and inexpensive voltage regulator. Be sure to select the proper wattage for the zener and the proper wattage and resistance for the resistor.

Zener diodes are also rated by their tolerance (1 percent and 5 percent are common) and their power rating, in watts. Note the resistor in the schematic shown in Fig. 17-8. This resistor (R1) limits the current through the zener, and its value (and wattage) is determined by the current draw from the load, as well as the input and output voltages.

The process of determining the correct values and ratings for resistor R1 and the zener diode is fairly simple and uses the basic electricity rules presented earlier in the book. The zener voltage rating is, quite obviously, the desired regulated voltage—you may find that the available rated voltages are somewhat awkward (such as 5.1 V), but you should be able to find a value within a few percent of the rated value.

Once you know the voltage rating for the zener diode that you are going to use, you can then calculate the value and ratings for R1. The zener diode regulator shown in Fig. 17-8 is actually a voltage divider, with the lower portion being a set voltage level. To determine the correct resistance of R1, you have to know what the input voltage is and the current that is going to be drawn from the regulator. For example, if you wanted 100 mA at 5.1 V from a 12-V power supply, the resistance of R1 can be calculated as:

$$
\begin{aligned}
V_{R1} &= 12V - 5.1V = 6.9V \\
R &= V / i \\
&= 6.9\ V / 0.100\ A \\
&= 69\ \Omega
\end{aligned}
$$

The closest "standard" resistor value you can get is 68 Ω, which will result in 101 mA being available for the load. With this value in hand, you can now calculate the power being dissipated by the resistor, using the basic power formulas:

$$
\begin{aligned}
\text{Power} &= V \times i \\
&= 6.9\ V \times 0.101\ A \\
&= 0.7\ \text{Watts}
\end{aligned}
$$

Standard resistor power ratings are in ⅛, ¼, ½, 1, 2, and so on watts. A 1-W, 68 Ω resistor would be chosen for this application.

The zener diode will also be dissipating power; how much is something that you should decide. To err on the side of safety, it is recommended that you assume that the zener diode can have 100 percent of the load current passing through it (when the load circuitry is not attached to the power supply). The power rating for the zener diode is calculated exactly the same way as R1:

$$
\begin{aligned}
\text{Power} &= V \times i \\
&= 5.1\ V \times 0.101A \\
&= 0.52\ \text{Watts}
\end{aligned}
$$

For this application, you could probably get away with a ½ W rated zener diode along with the 1 W rated 68 Ω resistor. Working with zener diodes to make power supplies is quite easy, but there is a tremendous price to pay in terms of lost power. In this example, the total power dissipated will be 1.2 W, with 58 percent of it being dissipated through the R1 resistor. This loss may be unacceptable in a battery-powered robot.

17.10.2 LINEAR VOLTAGE REGULATORS

The zener diode regulator can be thought of as a tub of water with a hole at the bottom; the maximum pressure of the water squirting out is dependent on the level of water in the tub. Ideally, there should be more water coming into the tub than will be ever drawn to ensure that the pressure of the water coming out of the hole is constant. This means that a fair amount of water will spill over the edge of the tub. As was shown in the previous section, this is an extremely inefficient method of providing a regulated voltage, as the electrical equivalent of the water pouring over the edge is the power dissipated by R1.

To improve upon the zener diode regulator's inefficiency, a voltage regulator that just lets out enough current at the regulated voltage is desired. The *linear voltage* regulator only allows the required current (at the desired voltage) out and works just like a car's carburetor. In a carburetor, fuel is allowed to flow as required by the engine—if less is required than is available, a valve closes and reduces the amount of fuel that is passed to the engine. The linear voltage regulator works in an identical fashion: an output transistor in the regulator circuit only allows the required amount of current to the load circuit.

The actual circuitry that implements the linear regulator is quite complex, but this is really not a concern of yours as it is usually contained within a single chip like the one shown in Fig. 17-9. The circuit shown here uses one of the most popular linear voltage regulators, the 7805 to regulate a high-voltage source to +5 V for digital electronic circuitry. The parts required for the 7805 based linear regulator are listed in Table 17-1.

Two of the most popular voltage regulators, the 7805 and 7812, provide +5 and +12 V, respectively. Other 7800 series power regulators are designed for +15, +18, +20, and +24 V. The 7900 series provides negative power supply voltages in similar increments. The current capacity of the 7800 and 7900 series that come in the TO-220 style transistor packages (these can often be identified as they have no suffix or use a "T" suffix in their part number) is limited to less than 1 A. As a result, you must use them in circuits that do not draw in an excess of this amount.

Other linear regulators are available in a more traditional TO-3-style transistor package ("K" suffix) that offers current output to several amps. The "L" series regulators come in the small TO-92 transistor packages and are designed for applications that require less than about 500 mA. The different packages limit the amount of heat that can be dissipated by the linear regulators and determine how much current they can source.

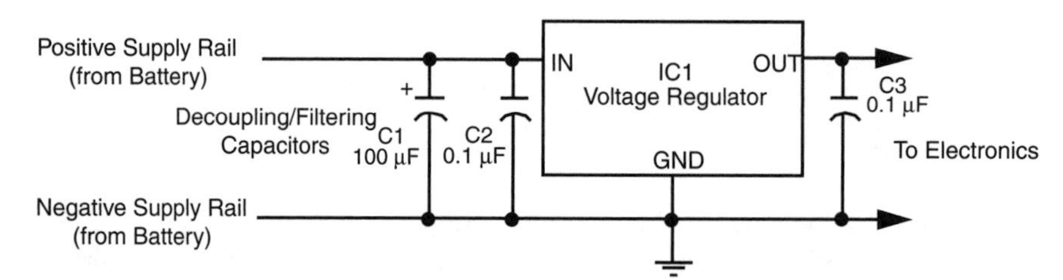

FIGURE 17-9 Three-terminal linear voltage regulators, like the 7805, can be used to provide stable voltages for battery-powered robots. The capacitors help filter (smooth out) the voltage.

TABLE 17-1 Parts List for +5-V Battery Regulator	
IC1	7805 linear voltage regulator
C1	100 µF electrolytic capacitor
C2, C3	0.1 µF capacitor (any type)

Here are some other linear regulators that you may be interested in:

- The 328K provides an adjustable output to 5 V, with a maximum current of 5 A (amperes).
- The 78H05K offers a 5-V output at 5 A.
- The 78H12K offers a 12-V output at 5 A.
- The 78P05K delivers 5 V at 10 A.

An important point to note about linear voltage regulators is their *dropout voltage,* or the minimum voltage that must be provided to operate properly. For the 7800 series, you should provide a minimum of 3 V more than they are rated at. So for a 7805, 8 V or more must be provided to the chip to get 5 V out. This dropout is effectively a voltage drop within the chip. For an 8 V input to the 7805, 37.5 percent of the power passed to the chip will be dissipated as heat; better than the zener diode regulator, but still a significant amount of power loss. If you were to look around, you will find linear regulator chips with much smaller dropout voltages, which will minimize the power lost significantly.

17.10.3 SWITCHING VOLTAGE REGULATION

The regulators described in the previous two sections are not very efficient; they reduce the input voltage in some way, which means they have to dissipate the resulting power lost in the regulator. The switching voltage regulator (more accurately called the *switch mode power supply,* also known by its acronym SMPS) has much higher operating efficiencies and can be configured to raise the incoming voltage or produce a negative voltage output.

The basic circuit for a switching voltage regulator is shown in Fig. 17-10 with its operating waveform shown in Fig. 17-11. This circuit is designed to raise the input voltage from 3 to 5 V by loading and unloading an inductor (or coil) and passing the high voltage to a diode and a filter capacitor. When the transistor attached to the coil is turned off, the coil reacts by producing a large voltage that passes some current through the diode to the load circuit. The VCO, or *voltage controlled oscillator,* controls the rate at which the inductor's transistor is turned on and off to ensure the voltage is regulated reasonably well.

The design of the VCO and the specification of the transistor switching waveform are fairly complex (although not as complex as you might think—many simple microcontrollers on the market can monitor the output voltage and calculate a new transistor switching waveform quite easily). Fortunately, there are a lot of controller chips, such as the

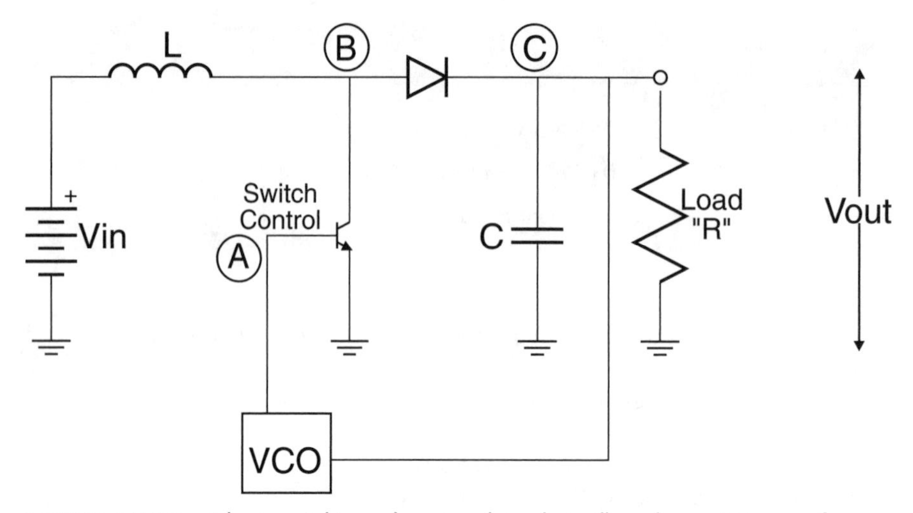

FIGURE 17-10 A basic switching voltage regulator that will produce a 5 V output from 3 V in.

LTC1174CN8-5, shown in Fig. 17-12. This chip will take a voltage from 3 to 15 V and produce a regulated 5 V output with an efficiency well over 90 percent.

The wide input voltage range is part of the advantage of using a switching voltage regulator for robot applications in which a single battery is used for driving the motors as well as the electronics. The switching voltage regulator will produce a remarkably constant output voltage despite the varying input voltage caused by the robot's motors starting, stopping, or stalling. When specifying the circuit to be used in your robot, the nominal voltage output of the batteries while the motors are running should be used.

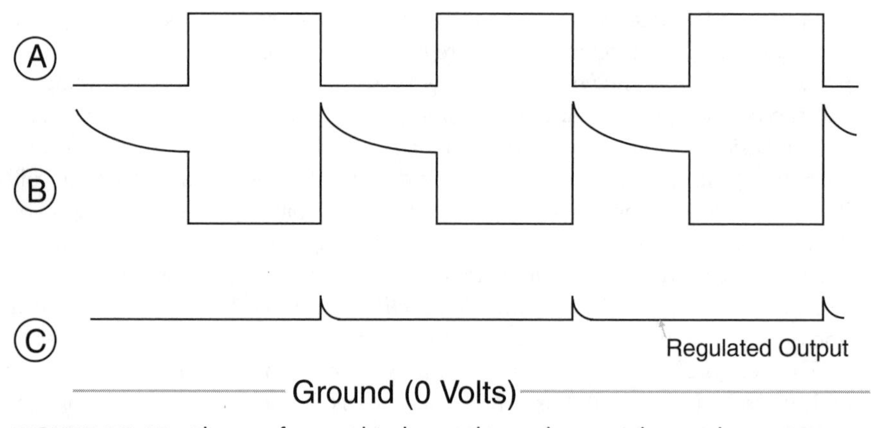

FIGURE 17-11 The waveforms within the switching voltage regulator at the points labeled in Fig. 17-10.

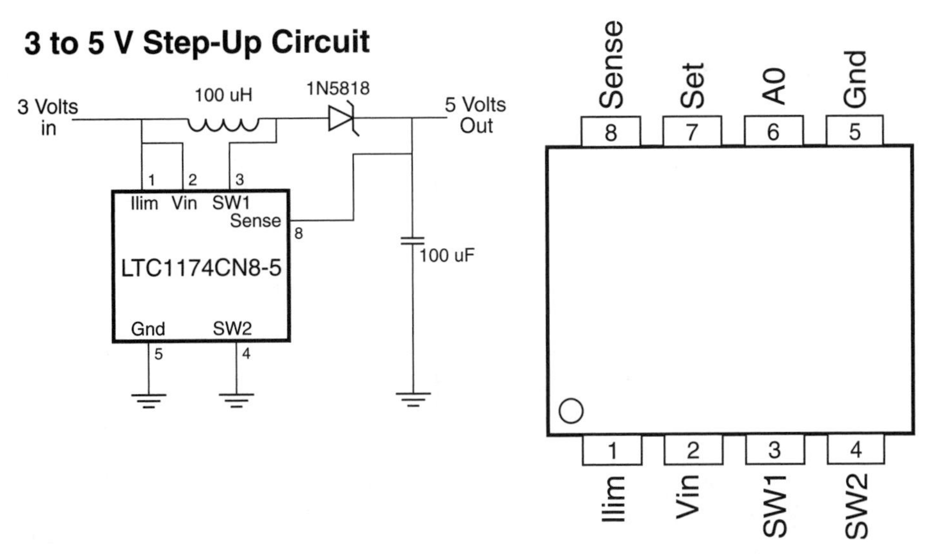

FIGURE 17-12 An LTC1174CN8-5 switching regulator circuit that will regulate an incoming voltage of 3 to 15 V to 5 V with a maximum current of 100 mA.

Switching voltage regulators are a bit more complex to wire, and, depending on the application and the amount of current required for the robot's electronics, will require more than the four components used in the power supply shown in Fig. 17-12. Even if the LTC1174CN8-4 was to be used in a robot application, you would be well advised to pass the incoming electrical power through a diode and filter the diode's output (and input to the voltage regulator) using a 47 µF capacitor or more to ensure that the power would be constant even when there are large voltage transients caused by the motor's changing state.

Switching voltage regulators do not cost much more than the zener diode regulator or linear regulators presented in the previous sections; a big part of the cost of the other regulators is the heat sinking and costs of large current components used in them. The switching voltage regulator does not generate a significant amount of heat due to its high operating efficiency and does not need the same expensive packaging or heat sinking of the other solutions.

17.10.4 POWER DISTRIBUTION

You may choose to place all or most of your robot's electronic components on a single board. You can mount the regulator(s) directly on the board. You can also have several smaller boards share one regulator as long as the boards together don't pull power in excess of what the regulator can supply. Fig. 17-13 shows how to distribute the power from a single battery source to many separate circuit boards. The individual regulators provide power for one or two large boards or a half dozen or so smaller ones.

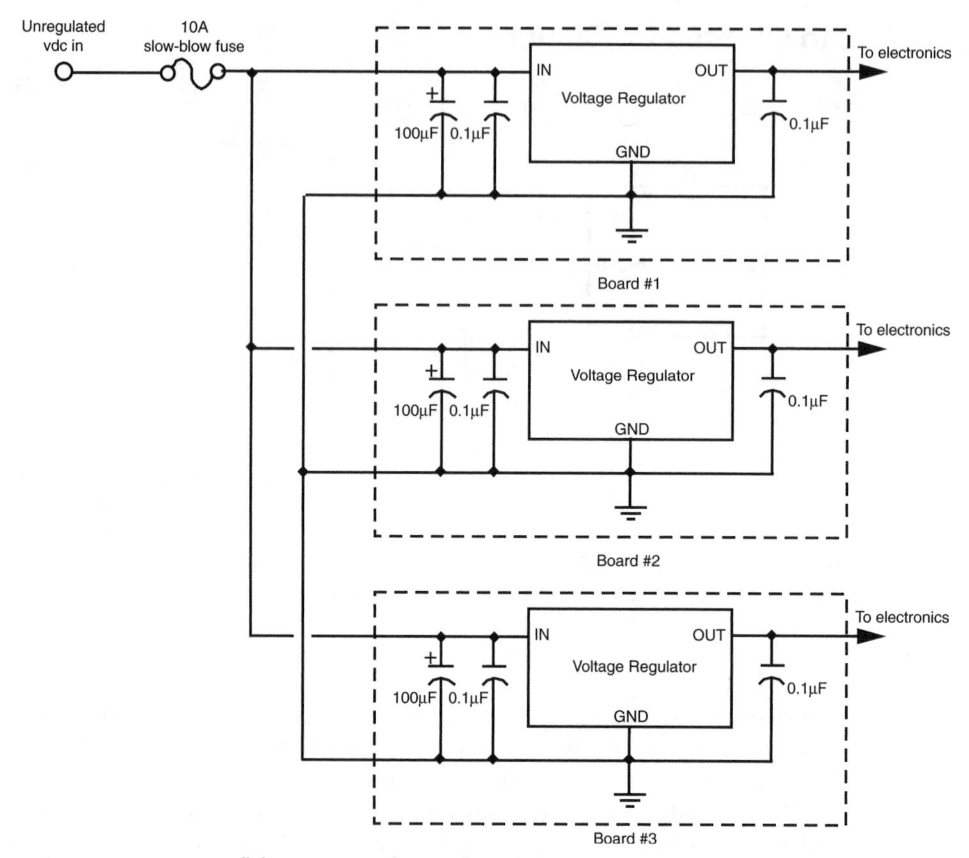

FIGURE 17-13 Parallel connection of circuit boards from a single power source. Each board has its own voltage regulator.

17.11 Battery Monitors

Quick! What's the condition of the battery in your robot? With a battery monitor, you'd know in a flash. A battery monitor continually samples the output voltage of the battery during operation of the robot (the best time to test the battery) and provides a visual or logic output. In this section some of the most common types are described.

17.11.1 4.3 V ZENER BATTERY MONITOR

Fig. 17-14 (refer to the parts list in Table 17-2) shows a simple battery monitor using a 4.3-V quarter-watt zener diode. R1 sets the trip point. When in operation, the LED winks off when the voltage drops below the setpoint. To use the monitor, set R1 (which should be a precision potentiometer, 1 or 3 turn) when the batteries in your robot are low. Adjust the pot carefully until the LED just winks off. Recharge the batteries. The LED should now light.

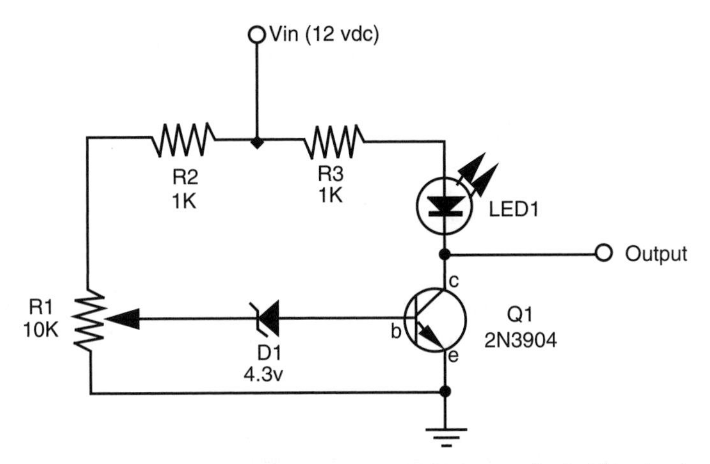

FIGURE 17-14 Battery monitor using 4.3-V zener diode. This circuit is designed to be used with a 12-V battery.

TABLE 17-2 Parts List for 4.3-V Zener Battery Monitor

R1	10K potentiometer (see text)
R2, R3	1K resistor
D1	4.3-V zener diode (¼-W)
Q1	2N3904 NPN transistor
LED1	Light-emitting diode

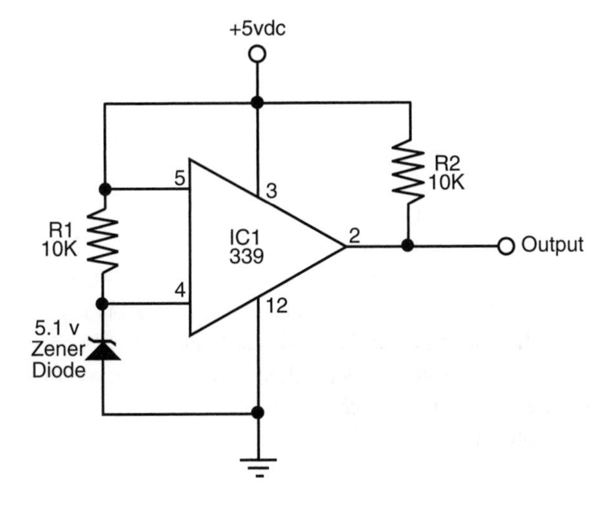

FIGURE 17-15 A zener diode and 339 comparator can be used to construct a fairly accurate 5-V battery monitor.

TABLE 17-3 Parts List for 339 Comparator Battery Monitor	
IC1	339 comparator IC
R1, R2	10K resistor
D1	5.1-V zener diode (¼- or ½-W)

Another, more "scientific" way to adjust R1 is to power the circuit using an adjustable power supply. While watching the voltage output on a meter, set the voltage at the trip point (e.g., for a 12-V robot, set it to about 10 V).

17.11.2 ZENER/COMPARATOR BATTERY MONITOR

A microprocessor-compatible battery monitor is shown in Fig. 17-15 (refer to the parts list in Table 17-3). This monitor uses a 5.1-V quarter-watt zener as a voltage reference for a 339 quad comparator IC. Only one of the comparator circuits in the IC is used; you are free to use any of the remaining three for other applications. The circuit is set to trip when the voltage sags below the (approximate) 5-V threshold of the zener (in my test circuit the comparator tripped when the supply voltage dipped to under 4.5 V). When this happens, the output of the comparator immediately drops to 0 V. Note that the outputs of the 339 are open collector; R2 pulls up the output so a voltage change can be observed.

17.11.3 USING A BATTERY MONITOR WITH A MICROPROCESSOR

You can usually connect battery monitors to a microprocessor or microcontroller input. When in operation, the microprocessor is signaled by the interrupt when the LED is triggered. Software running on the computer interprets the interrupt as "low battery; quick get a recharge." The robot can then place itself into nest mode, where it seeks out its own battery charger. If the charger terminals are constructed properly, it's possible for the robot to plug itself in. Fig. 17-16 shows a simplified flowchart illustrating how this might work.

17.12 A Robot Testing Power Supply

In the previous editions of this book, a great deal of space and effort was devoted to creating a power supply to eliminate the need for supplying the needs of a robot from a battery during development and testing. The circuit rectified and regulated 110 V AC from a household wall socket to +5 V using a 7805. This circuit replicated the capabilities of the logic power supply already built into the robot but probably did not provide enough current to drive the robot's motors. In this edition, a step back is taken with a look at a more practical and easier method of powering a robot on a bench or during development by using commonly available "wall wart" or laptop power supplies.

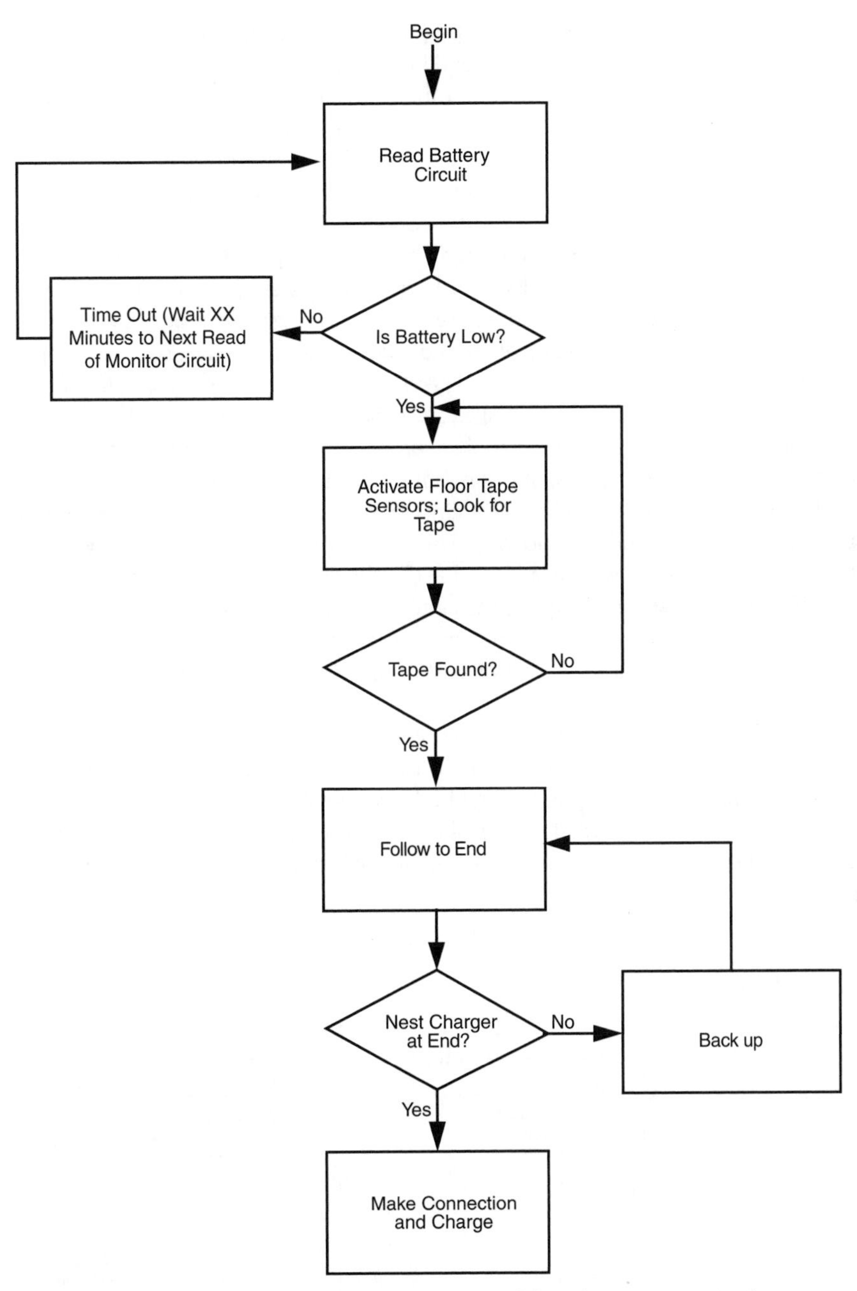

FIGURE 17-16 Software can be used to command the robot to return to its battery recharger nest should the battery exceed a certain low point.

Wall wart is the term used by most people to describe wall-mounted AC/DC power converters to provide power to a variety of different tasks, from powering boom boxes, to toys and game systems, to telephones and other appliances. Chances are you have a number around the house left over from electronic devices that no longer work or have been lost. They can probably be adapted to power your robot's electronics and even motors (if they can produce enough current) with no modifications. Even if you don't, one can be picked up at a local convenience store for just a couple of dollars.

The important feature to look for in these power supplies is that they have a power plug on them rather than a phono plug. A *phono plug* is a single shaft of metal, with multiple connections built into it at different intervals and was originally designed for providing headphone connections to a radio or other audio device. A stereo headset phono plug will have a tip that is separate from a metal band just after it and the main body of the shaft. A *power plug* consists of a hollow barrel that has a power connection on the inside separate from the one on the outside. The phono plug can be used for providing power, but you will find that there is some arcing or sparks when the plug is inserted or withdrawn.

The power plug does not cause arcing or sparks when plugged in or withdrawn from the socket because there is no opportunity for the positive voltage and ground to be shorted together as is shown in Fig. 17-17. When the plug is inserted, the center hole, which provides positive voltage, comes into contact with the central post of the socket and the outside ground connection contacts a metal wiper. The size of a power plug is determined by the diameter or the central post of the plug—2.5 mm is probably the most common size used.

Note that in Fig. 17-17, the negative voltage from the battery passes through a metal connection to the ground wiper, which is disconnected when the plug is inserted. Using this feature will allow the battery to be kept in the robot while external power is being provided.

Looking around, you will find that the most current the typical wall wart power supply can provide is about 800 mA. This is probably not going to be adequate for larger robots.

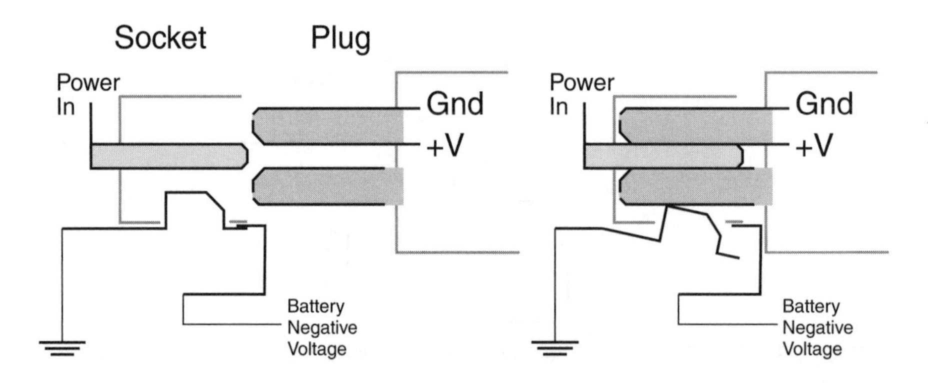

FIGURE 17-17 Power plug and socket combination being shown (a) separate and (b) plug inserted. When the plug is inserted, the wiper connection is broken and the negative voltage connection to the internal battery can be automatically disconnected, eliminating its ability to power the robot.

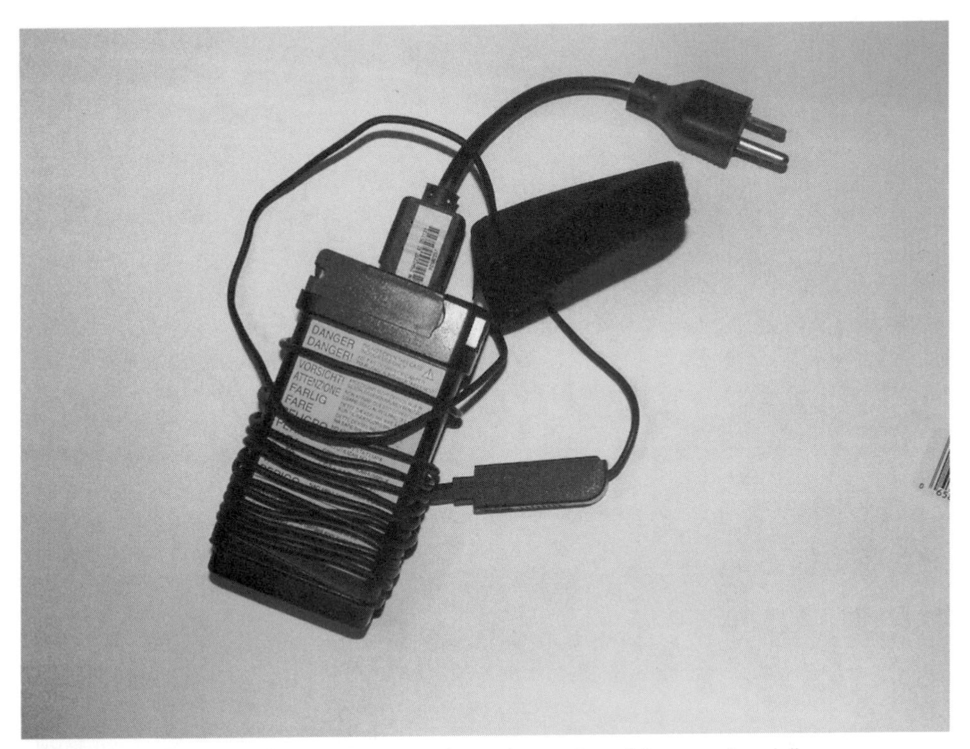

FIGURE 17-18 A used laptop power supply can be purchased for just a few dollars and provides several amps of current.

Instead of the small wall wart, you should be able to find an old laptop power supply, like the one shown in Fig. 17-18, which provides 12 V at up to 5 A at a surplus store for $5 or so. These connectors are in the power plug format but are much larger than the ones typically used for CD players and basic home electronics. Some laptop power supplies provide multiple voltage outputs that may be useful in your robot.

Finally, the DC output from the power supplies cannot be used to charge batteries. NiMH, NiCad, and other battery technologies require a PWM input and often have testing algorithms to determine when they are fully (and not over) charged. Damage or fire could result if a straight DC voltage is passed to them, so only use chargers that are designed for the type of battery you are using.

17.13 From Here

To learn more about . . .	***Read***
Understanding motor current ratings	Chapter 19, "Choosing the Right Motor for the Job"
Other power systems (e.g., hydraulic)	Chapter 26, "Reaching Out with Arms"

CHAPTER 18

PRINCIPLES OF ROBOT LOCOMOTION

As you graduate to building larger mobile robots, you should consider the physical properties of your creations, including their size, weight, and mode of transport. A robot that is too heavy for its frame, or a locomotion mechanism that doesn't provide sufficient stability, will greatly hinder the usefulness of your mechanical invention.

In this chapter you'll find a collection of assorted tips, suggestions, and caveats for designing the locomotion systems for your robots. Because the locomotion system is intimately related to the frame of the robot, we'll cover frames a little bit as well, including their weight and weight distribution. Of course, there's more to the art and science of robot locomotion than we can possibly cover here, but what follows will serve as a good introduction.

18.1 First Things First: Weight

Most hobbyist robots weigh under 20 lbs, and a high percentage of those weigh under 10 lbs. Weight is one of the most important factors affecting the mobility of a robot. A heavy robot requires larger motors and higher capacity batteries—both of which add even more pounds to the machine. At some point, the robot becomes too heavy to even move.

On the other hand, robots designed for heavy-duty work often need some girth and weight. Your own design may call for a robot that needs to weigh a particular amount in order for it to do the work you have envisioned. The parts of a robot that contribute the most to its weight are the following, in (typical) descending order:

- Batteries
- Drive motors
- Frame

A 12-V battery pack can weigh 1 lb; larger-capacity, sealed lead-acid batteries can weigh 5 to 8 lbs. Heavier-duty motors will be needed to move that battery ballast. But bigger and stronger motors weigh more because they must be made of metal and use heavier-duty bushings. And they cost more. Suddenly, your "little robot" is not so little anymore; it has become overweight and expensive.

18.2 Tips for Reducing Weight

If you find that your robot is becoming too heavy, consider putting it on a diet, starting with the batteries. Nickel-cadmium and nickel metal hydride batteries weigh less, volt for volt, than their lead-acid counterparts. While nickel-cadmium and nickel metal hydride batteries may not deliver the amp-hour capacity that a large, sealed lead-acid battery will, your robot will weigh less and therefore may not require the same stringent battery ratings as you had originally thought.

When looking at reducing the weight of your robot or modifying it in any way, remember to try to come up with changes that result in additional benefits. For example, if you were to change your batteries to a lighter set, you will discover that you do not need as powerful a motor. Less powerful motors weigh less than the originally specified motors, further decreasing the weight of the motor. This decrease in the weight of the motor could result in the need for smaller and lighter batteries, which allows you to look at using even smaller and lighter batteries, smaller motors, smaller structure, etcetera. This process can repeat multiple times and it isn't unusual to see a situation where a 10 percent decrease in battery weight results in a 50 percent reduction in overall robot weight. The repeating positive response to a single change is known as a *supereffect,* and you should remember that the reverse is also true: a 10 percent increase in weight in a robot's components could result in a 50 percent increase in weight in the final robot.

If your robot must use a lead-acid battery, consider carefully whether you truly need the capacity of the battery or batteries you have chosen. You may be able to install a smaller battery with a lower amp-hour rating. The battery will weigh less, but, understandably, it will need to be recharged more often. An in-use time of 60 to 120 minutes is reasonable (that is, the robot's batteries must be recharged after an hour or two of continual use).

If you require longer operational times but still need to keep the weight down, consider a replaceable battery system. Mount the battery where it can be easily removed. When the charge on the battery goes down, take it out and replace it with a fully charged one. Place the previously used battery in the charger. The good news is that smaller, lower capacity batteries tend to be significantly less expensive than their larger cousins, so you can probably buy two or three smaller batteries for the price of a single big one.

Drive motors are most often selected because of their availability and cost, not because of their weight or construction. In fact, many robots are designed around the specifications of the selected drive motors. The motors are selected (often they're purchased surplus), and from these the frame of the robot is designed and appropriate batteries are added. Still, it's important to give more thought to the selection of the motors for the robot that you have in mind. Avoid motors that are obviously overpowered in relation to the robot in which they are being used. Motors that are grossly oversized will add unnecessary weight, and they will require larger (and therefore heavier and more expensive) batteries to operate.

18.3 Beware of the Heavy Frame

The frame of the robot can add a surprising amount of weight. An 18-in^2, 2-ft high robot constructed from extruded aluminum and plastic panels might weigh in excess of 20 or 30 lbs, without motors and batteries. The same robot in wood (of sufficient strength and quality) could weigh even more.

Consider ways to lighten your heavy robots, but without sacrificing strength. This can be done by selecting a different construction material and/or by using different construction techniques. For example, instead of building the base of your robot using solid ⅛-in (or thicker) aluminum sheet, consider an aluminum frame with crossbar members for added stability. If you need a surface on which to mount components (the batteries and motors will be mounted to the aluminum frame pieces), add a 1⁄16-in acrylic plastic sheet as a "skin" over the frame. The plastic is strong enough to mount circuit boards, sensors, and other lightweight components on it.

Aluminum and acrylic plastic aren't your only choices for frame materials. Other metals are available as well, but they have a higher weight-to-size ratio. Both steel and brass weigh several times more per square inch than aluminum. Brass sheets, rods, and tubes (both round and square) are commonly available at hobby stores. Unless your robot requires the added strength that brass provides, you may wish to avoid it because of its heavier weight.

Ordinary acrylic plastic is rather dense and therefore fairly heavy, considering its size. Lighter-weight plastics are available but not always easy to find. For example, ABS and PVC plastic—popular for plumbing pipes—can be purchased from larger plastics distributors in rod, tube, and sheet form. There are many special-purpose plastics available that boast both structural strength and light weight. Look for Sintra plastic, for example, which has an expanded core and smooth sides and is therefore lighter than most other plastics.

Check the availability of glues and cements before you purchase or order any material. (See Chapter 10 for more on these and other plastics.)

18.4 Constructing Robots with Multiple Decks

For robots that have additional "decks," like the robot shown in Fig. 18-1, select construction materials that will provide rigidity but the lowest possible weight. One technique, shown

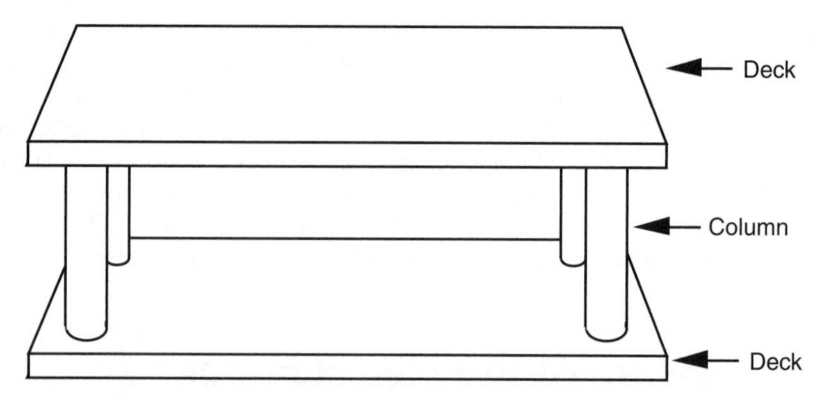

FIGURE 18-1 "Decked" robots provide extra space for batteries and electronics, but they can also add considerably to the weight. Use lightweight construction materials to avoid unduly increasing the weight of the robot.

in the figure, is to use ½-in thin-wall (Schedule 125) PVC pipe for uprights and attach the "decks" using 8⁄32 or 10⁄24 all-thread rod. The PVC pipe encloses the all-thread; both act as a strong support column. You need three such columns for a circular robot, and four columns for a square robot. For small robots, consider electronic circuit board *standoffs,* which are six-sided rods with the ends drilled and tapped for 4-40 screws.

Unless your robot is heavy, be sure to use the thinner-walled Schedule 125 PVC pipe. Schedule 80 pipe, commonly used for irrigation systems, has a heavier wall and may not be needed. Note that PVC pipe is always the same diameter outside, no matter how thick its plastic walls. The thicker the wall, the smaller the inside diameter of the pipe. You can readily cut PVC pipe to length using a PVC pipe cutter or a hacksaw, and you can paint it if you don't like the white color. Use Testor model paints for best results, and be sure to spray lightly. For a bright white look, you can remove the blue marking ink on the outside of the PVC pipe with acetone, which is available in the paint department of your local home improvement store.

18.5 Frame Sagging Caused by Weight

A critical issue in robot frame design is excessive weight that causes the frame to sag in the middle. In a typical robot, a special problem arises when the frame sags: the wheels on either side pivot on the frame and are no longer perpendicular to the ground. Instead, they bow out at the bottom and in at the top (this is called *negative camber*). Depending on which robot tires you use, traction errors can occur because the contact area of the wheel is no longer consistent. As even more weight is added, the robot may have a tendency to veer off to one side or the other.

There are three general fixes for this problem: reduce the weight, strengthen the frame, or add cross-braces to prevent the wheels from cambering. Strengthening the frame usually involves adding even more weight. So if you can, strive for the first solution instead—reduce the weight.

If you can't reduce weight, look for ways to add support beams or braces to prevent sagging. An extra cross-brace along the wheelbase (perhaps stretched between the two motors) may be all that's required to prevent the problem. The cross-brace can be made of lightweight aluminum tubing or even from a wooden dowel. The tubing or dowel does not need to support any weight; it simply needs to act as a brace to prevent compression when the frame sags and the wheels camber.

Yet another method is to apply extreme camber to the wheels, as shown in Fig. 18-2. This minimizes the negative effects of any sagging, and if the tires have a high frictional surface traction is not diminished. However, don't do this with smooth, hard plastic wheels as they don't provide sufficient traction. You can camber the wheels outward or inward. Inward (negative) camber was used in the old Topo and Bob robots made by Nolan Bushnell's failed Androbot company of the mid-1980s. The heavy-duty robot in Fig. 18-2 uses outward (positive) camber. The robot can easily support over 20 lbs in addition to its own weight, which is about 10 lbs, with battery, which is slung under the frame using industrial-strength hook-and-loop (Velcro or similar) fasteners.

FIGURE 18-2 This "Tee-Bot" (so named because it employs the T-braces used for home construction) uses extreme camber to avoid the frame sagging that results from too much weight.

18.6 Horizontal Center of Balance

Your robot's horizontal center of balance (think of it as a balance scale) indicates how well the weight of the robot is distributed on its base. If all the weight of a robot is to one side, for example, then the base will have a lopsided horizontal center of balance. The result is an unstable robot: the robot may not travel in a straight line and it might even tip over.

Ideally, the horizontal center of balance of a robot should be the center of its base (see Fig. 18-3a). Some variation of this theme is allowable, depending on the construction of the robot. For a robot with a single balancing caster, as shown in Fig. 18-3b, it is usually acceptable to place more weight over the drive wheels and less on the caster. This increases traction, and as long as the horizontal center of balance isn't extreme there is no risk that the robot will tip over.

Unequal weight distribution is the most troublesome result if the horizontal center of bal-

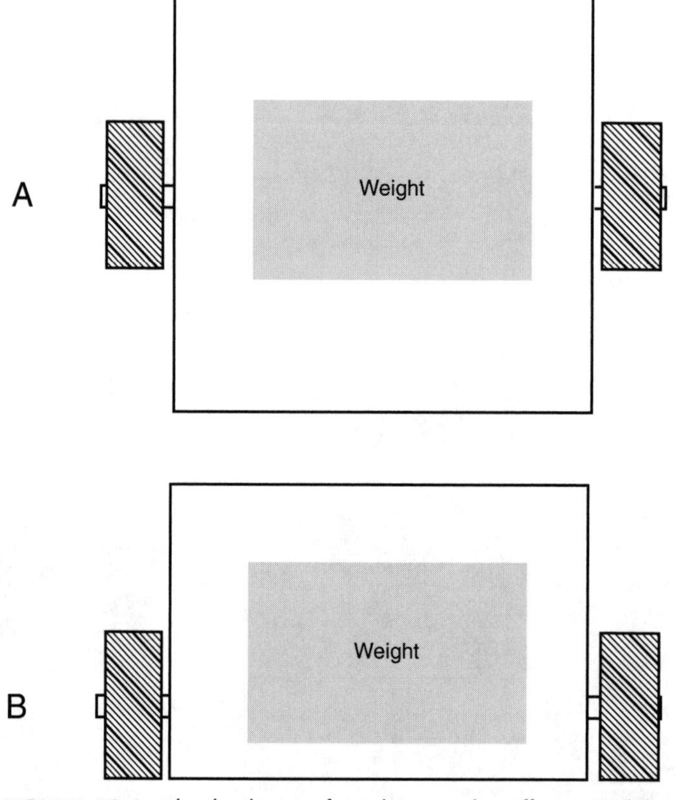

FIGURE 18-3 The distribution of weight on a robot affects its stability and traction. *a.* Centering the weight down the middle in a robot with two balancing casters; *b.* sliding the center of balance toward the drive wheel in a single-caster 'bot.

ance favors one wheel or track over the other—the right side versus the left side, for example. This can cause the robot to continually "crab" toward the heavier side. Since the heavier side has more weight, traction is improved, but motor speed may be impaired because of the extra load.

18.7 Vertical Center of Gravity

City skyscrapers must be rooted firmly in the ground or else there is a risk they will topple over in the slightest wind. The taller an object is, the higher its center of gravity. Of critical importance to vertical center of gravity is the "footprint" or base area of the object—that is, the amount of area in contact with the ground. The ratio between the vertical center of gravity and the area of the base determines how likely it is that the object will fall over. A robot with a small base but high vertical center of gravity risks toppling over. You can correct such a design in either of two ways:

- Reduce the height of the robot to better match the area of the base, or
- Increase the area of the base to compensate for the height of the robot.

(There is also a third method called *dynamic balance*. Here, mechanical weight is dynamically repositioned to keep the robot on even kilter. These systems are difficult to engineer and, in any event, are beyond the scope of this book.)

Which method you choose will largely depend on what you plan to use your robot for. For example, a robot that must interact with people should be at least toddler height. For a pet-size robot, you'll probably not want to reduce the height, but rather increase the base area to prevent the robot from tipping over.

18.8 Locomotion Issues

The way your robot gets from point A to point B is called *locomotion*. Robot locomotion takes many forms, but wheels and tracks are the most common. Legged robots are also popular, especially among hobbyists, as designing them represents a challenge both in construction and weight-balance dynamics.

18.8.1 WHEELS AND TRACKS

Wheels, and to a lesser extent tracks, are the most common means chosen to move robots around. However, some wheels are better for mobile robots than others. Some of the design considerations you may want to keep in mind include the following:

- The wider the wheels, the more the robot will tend to stay on course. With very narrow wheels, the robot may have a tendency to favor one side or the other and will trace a slow curve instead of a straight line. Conversely, if the wheels are too wide, the friction created by the excess wheel area contacting the ground may hinder the robot's ability to make smooth turns.

- Two driven wheels positioned on either side of the robot (and balanced by one or two casters on either end) can provide full mobility. This is the most common drive wheel arrangement and is called a *differential drive.*
- Tracks turn by skidding or slipping, and they are best used on surfaces such as dirt that readily allow low-friction steering.
- Four or more driven wheels, mounted in sets on each side, will function much like tracks. In tight turns, the wheels will experience significant skidding, and they will therefore create friction over any running surface. If you choose this design, position the wheel sets close together.
- You should select wheel and track material to reflect the surface the robot will be used on. Rubber and foam are common choices; both provide adequate grip for most kinds of surfaces. Foam tires are lighter in weight, but they don't skid well on hard surfaces (such as hardwood or tile floors).

18.8.2 LEGS

Thanks to the ready availability of smart microcontrollers, along with the low cost of R/C (radio-controlled) servos, legged automatons are becoming a popular alternative for robot builders. Robots with legs require more precise construction than the average wheeled robot. They also tend to be more expensive. Even a basic six-legged walking robot requires a minimum of two or three servos, with some six- and eight-leg designs requiring 12 or more motors. At about $12 per servo (more for higher-quality ones), the cost can add up quickly!

Obviously, the first design decision is the number of legs. Robots with one leg (hoppers) or two legs are the most difficult to build because of balance issues, and will not be addressed here. Robots with four and six legs are more common. Six legs offer a static balance that ensures the robot won't easily fall over. At any one time, a minimum of three legs touch the ground, forming a stable tripod.

In a four-legged robot, either the robot must move one leg at a time—keeping the other three on the ground for stability—or else employ some kind of dynamic balance when only two of its legs are on the ground at any given time. Dynamic balance is often accomplished by repositioning the robot's center of gravity, typically by moving a weight (such as the robot's head or tail, if it has one). This momentarily redistributes the center of balance to prevent the robot from falling over. The algorithms and mechanisms for achieving dynamic balance are not trivial. Four-legged robots are difficult to steer, unless you add additional degrees of freedom for each leg or articulate the body of the beast like those weird segmented city buses you occasionally see.

The movement of the legs with respect to the robot's body is often neglected in the design of legged robots. The typical six-legged (hexapod) robot uses six identical legs. Yet the crawling insect a hexapod robot attempts to mimic is designed with legs of different lengths and proportions—the legs are made to do different things. The back legs of an insect, for example, are often longer and are positioned near the back for pushing (this is particularly true of insects that burrow through dirt). The front legs may be similarly constructed for digging, carrying food, fighting, and walking. You may wish to replicate this design, or something similar, for your own robots. Watch some documentaries on insects and study how they walk and how their legs are articulated. Remember that the cockroach has been around for over a million years and represents a very advanced form of biological engineering!

18.9 Motor Drives

Next to the batteries, the drive motors are probably the heaviest component in your robot. You'll want to carefully consider where the drive motor(s) are located and how the weight is distributed throughout the base.

One of the most popular mobile robot designs uses two identical motors to spin two wheels on opposite sides of the base (the differentially driven robot). These wheels provide forward and backward locomotion, as shown in Fig. 18-4, as well as left and right steering. If you stop the left motor, the robot turns to the left. By reversing the motors relative to one another, the robot turns by spinning on its wheel axis (turns in place). You use this forward-reverse movement to make hard or sharp right and left turns.

18.9.1 CENTERLINE DRIVE MOTOR MOUNT

You can place the wheels—and hence the motors—just about anywhere along the length of the platform. If they are placed in the middle, as shown in Fig. 18-5, you should add two casters to either end of the platform to provide stability. Since the motors are in the center of the platform, the weight is more evenly distributed across it. You can place the battery or batteries above the centerline of the wheel axis, which will maintain the even distribution.

A benefit of centerline mounting is that the robot has no "front" or "back," at least as far as the drive system is concerned. Therefore, you can create a kind of multidirectional robot that can move forward and backward with the same ease. Of course, this approach also complicates the sensor arrangement of your robot. Instead of having bump switches only in the front of your robot, you'll need to add additional ones in the back in case the robot is reversing direction when it strikes an object.

18.9.2 FRONT-DRIVE MOTOR MOUNT

You can also position the wheels on one end of the platform. In this case, you add one caster on the other end to provide stability and a pivot for turning, as shown in Fig. 18-6. Obviously, the weight is now concentrated more on the motor side of the platform. You should place more weight over the drive wheels, but avoid putting all the weight there since maneuverability and stability may be diminished.

One advantage of front-drive mounting is that it simplifies the construction of the robot. Its *steering circle,* the diameter of the circle in which the robot can be steered, is still the

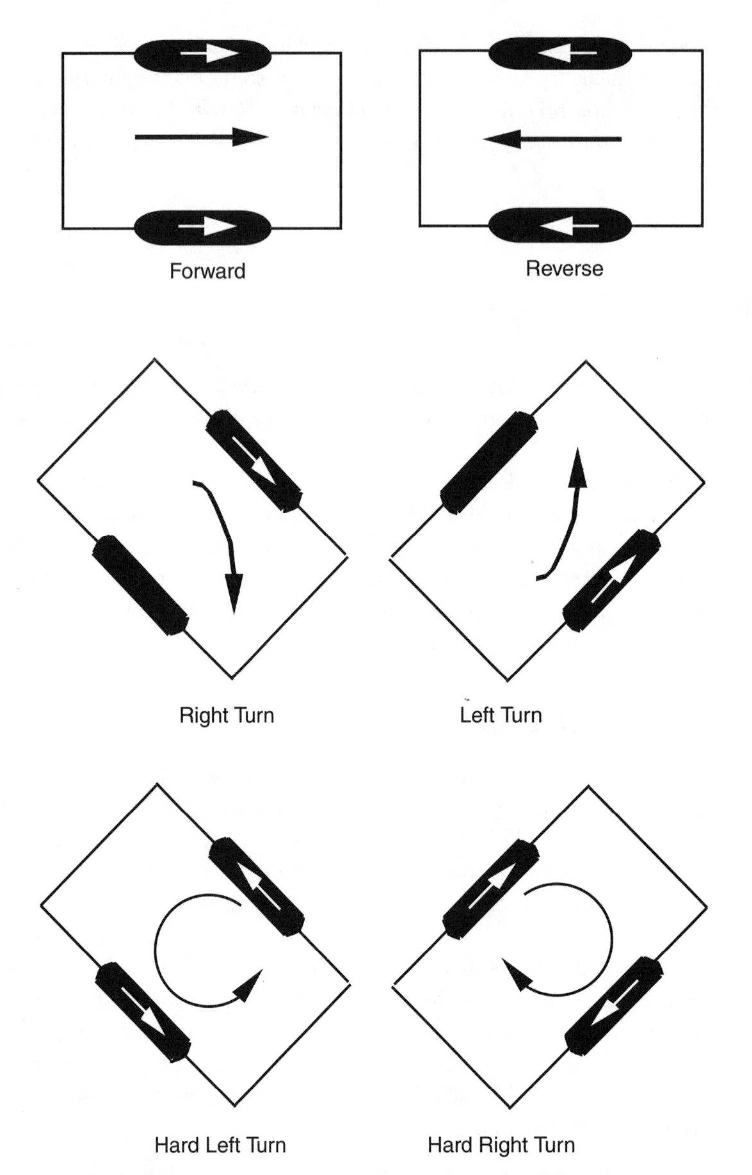

FIGURE 18-4 Two motors mounted on either side of the robot can power two wheels. Casters provide balance. The robot steers by changing the speed and direction of each motor.

same diameter as the centerline drive robot. However, it extends beyond the front/back dimension of the robot (see Fig. 18-7). This may or may not be a problem, depending on the overall size of your robot and how you plan to use it. Any given front-drive robot may be smaller than its centerline drive cousin. Because of the difference in their physical size, the diameter of the steering circle for both may be about the same.

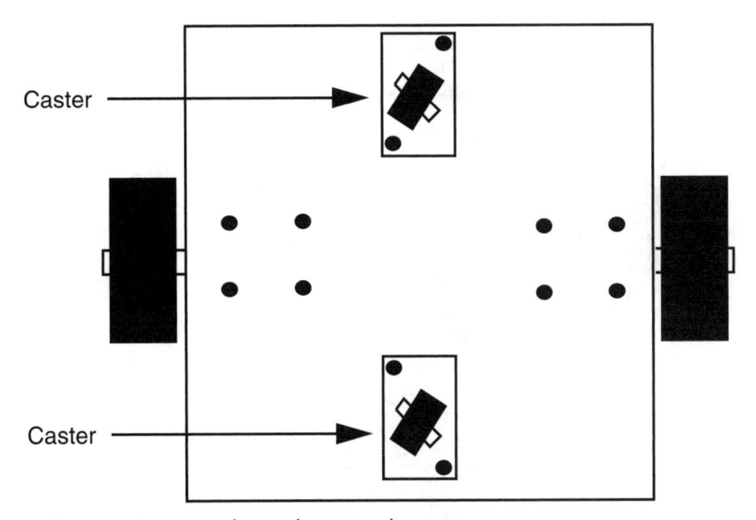

FIGURE 18-5 A robot with a centerline motor mount uses two casters (very occasionally one) for balance. When using one caster, you may need to shift the balance of weight toward the caster end to avoid having the robot tip over.

18.9.3 CASTER CHOICES

As mentioned earlier, most robots employing the two-motor drive system use at least one unpowered caster, which provides support and balance. Two casters are common in robots that use centerline drive-wheel mounting. Each caster is positioned at opposite ends of the robot. When selecting casters it is important to consider the following factors:

- The size of the caster wheel should be in proportion to the drive wheels (see Fig. 18-8).
- When the robot is on the ground, the drive motors must firmly touch terra firma. If the caster wheels are too large, the drive motors may not make adequate contact, and poor

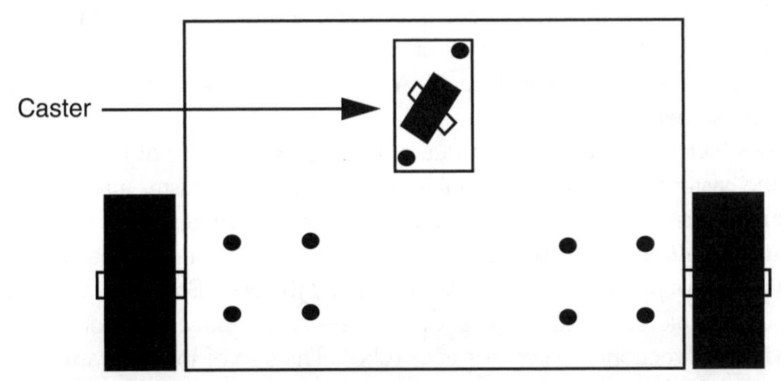

FIGURE 18-6 A robot with a front-drive motor mount uses a single opposing caster for balance. Steering is accomplished using the same technique as a centerline motor mount.

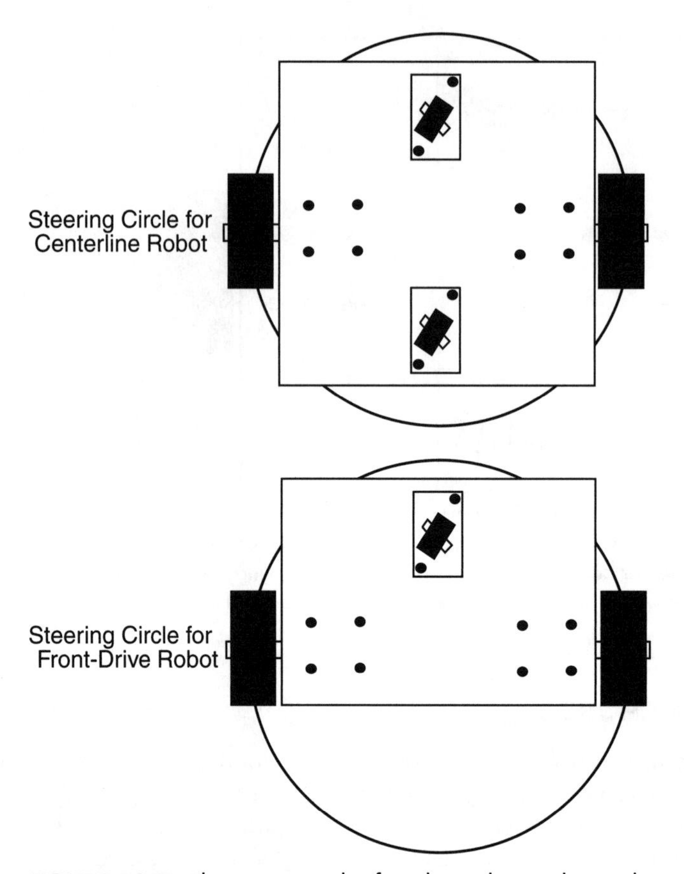

FIGURE 18-7 The steering circle of a robot with centerline and front-drive mounted motors.

traction will result. You might also consider using a suspension system of your own design on the casters to compensate for uneven terrain.

- The casters should spin and swivel freely. A caster that doesn't spin freely will impede the robot's movement.
- In most cases, since the caster is provided only for support and not traction you should construct the caster from a hard material to reduce friction. A caster made of soft rubber will introduce more friction, and it may affect a robot's movements.
- Consider using *ball casters* (also called *ball transfers*), which are primarily designed to be used in materials processing (conveyor chutes and the like). Ball casters (see Fig. 18-9) are made of a single ball—either metal or rubber—held captive in a housing, and they function as omnidirectional casters for your robot. The size of the ball varies from about $^{11}/_{16}$ to over 3 in in diameter. Look for ball casters at mechanical surplus stores and also at industrial supply outlets, such as Grainger and McMaster–Carr.

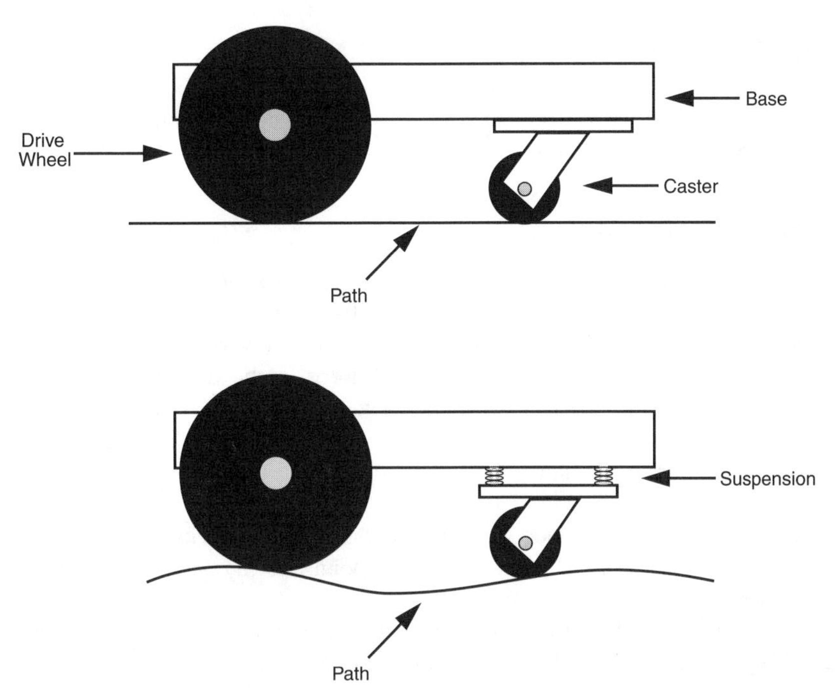

FIGURE 18-8 The height of the caster with respect to the drive wheels will greatly influence the robot's traction and maneuverability. A spring-loaded caster (a kind of suspension) can improve functionality of the robot on semirough terrain.

FIGURE 18-9 Ball casters (or ball transfers) are omnidirectional. For medium- to large-sized robots consider using them instead of wheeled casters.

18.10 Steering Methods

A variety of methods are available to steer your robot. The following sections describe several of the more common approaches.

18.10.1 DIFFERENTIAL

For wheeled and tracked robots, differential steering is the most common method for getting the machine to go in a different direction. The technique is exactly the same as steering a military tank: one side of wheels or treads stops or reverses direction while the other side keeps going. The result is that the robot turns in the direction of the stopped or reversed wheel or tread. Because of friction effects, differential steering is most practical with two-wheel-drive systems. Additional sets of wheels, as well as rubber treads, can increase friction during steering.

- If you are using multiple wheels (dually), position the wheels close together, as shown in Fig. 18-10. The robot will pivot at a virtual point midway between the two wheels on each side.
- If you are using treads, select a relatively low-friction material such as cloth or hard plastic. Very soft rubber treads will not steer well on smooth surfaces. If this cannot be helped, one approach is to always steer by reversing the tread directions. This will reduce the friction.

18.10.2 CAR-TYPE

Pivoting the wheels in the front is yet another method for steering a robot (see Fig. 18-11). Robots with car-type steering are not as maneuverable as differentially steered robots, but they are better suited for outdoor uses, especially over rough terrain. You can obtain somewhat better traction and steering accuracy if the wheel on the inside of the turn pivots more than the wheel on the outside. This technique is called *Ackerman steering* and is found on most cars but not on as many robots.

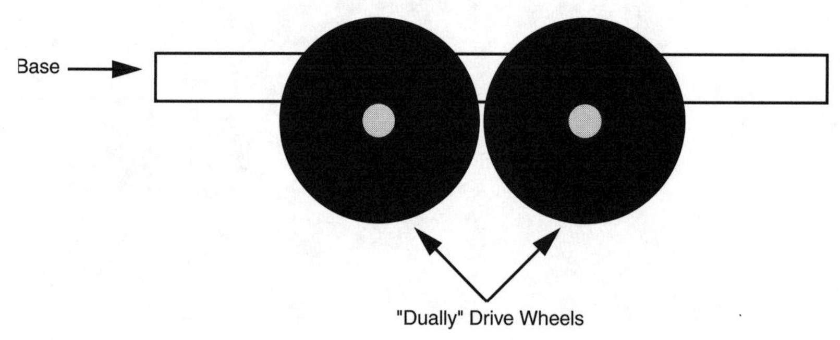

FIGURE 18-10 "Dually" wheels should be placed close to one another. If they are spaced farther apart the robot cannot steer as easily.

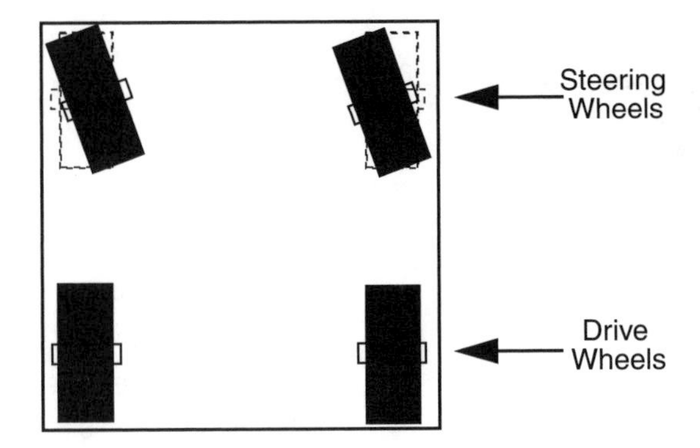

FIGURE 18-11 Car-type steering offers a workable alternative for an outdoors robot, but it is less useful indoors or in places where there are many obstructions that must be steered around.

18.10.3 TRICYCLE

One of the biggest drawbacks of the differentially steered robot is that the robot will veer off course if one motor is even a wee bit slow. You can compensate for this by monitoring the speed of both motors and ensuring that they operate at the same r/min. This typically requires a control computer, as well as added electronics and mechanical parts for sensing the speed of the wheels.

Car-type steering, described in the last section, is one method for avoiding the problem of "crabbing" as a result of differences in motor speed simply because the robot is driven by just one motor. But car-type steering makes for fairly cumbersome indoor mobile robots. A better approach is to use a single drive motor powering two rear wheels and a single steering wheel in the front. This arrangement is just like a child's tricycle, as shown in Fig. 18-12. The robot can be steered in a circle just slightly larger than the width of the machine. Be careful about the wheelbase of the robot (distance from the back wheels to the front steering wheel). A short base will cause instability in turns, and the robot will tip over opposite the direction of the turn.

Tricycle-steered robots must have a very accurate steering motor in the front. The motor must be able to position the front wheel with subdegree accuracy. Otherwise, there is no guarantee the robot will be able to travel a straight line. Most often, the steer-

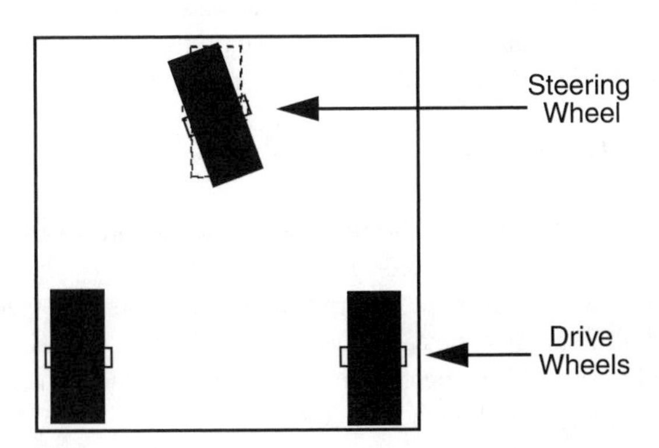

FIGURE 18-12 In tricycle steering, one drive motor powers the robot and a single wheel in front steers the robot. Try and to avoid short wheelbases as this can result in a robot that tips easily when the robot turns.

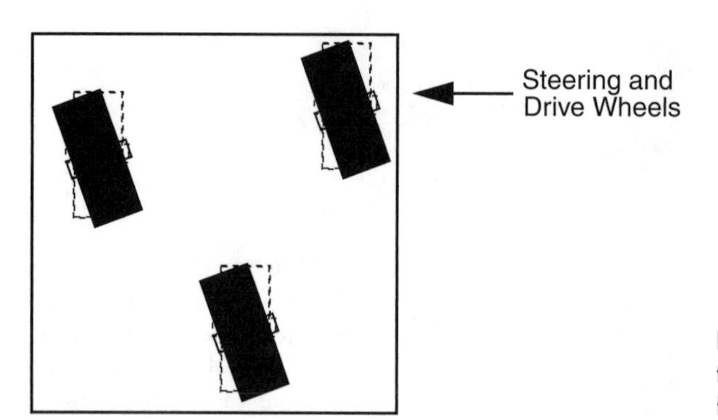

FIGURE 18-13 An omnidirectional robot uses the same wheels for drive and steering.

ing wheel is controlled by a servo motor. Servo motors use a *closed-loop feedback* system that provides a high degree of positional accuracy (depending on the quality of the motor, of course). Read more about servo motors in Chapter 22, "Working with Servo Motors."

18.10.4 OMNIDIRECTIONAL

To have the highest tech of all robots, you may want omnidirectional drive. It uses steerable drive wheels, usually at least three, as shown in Fig. 18-13. The wheels are operated by two motors: one for locomotion and one for steering. In the usual arrangement, the drive/steering wheels are ganged together using gears, rollers, chains, or pulleys. Omnidirectional robots exhibit excellent maneuverability and steering accuracy, but they are technically more difficult to construct.

18.11 Calculating the Speed of Robot Travel

The speed of the drive motors is one of two elements that determines the travel speed of your robot. The other is the diameter of the wheels. For most applications, the speed of the drive motors should be under 130 r/min (under load). With wheels of average size, the resultant travel speed will be approximately 4 ft/s. That's actually pretty fast. A better travel speed is 1 to 2 f/s (approximately 65 r/min), which requires smaller diameter wheels, a slower motor, or both.

How do you calculate the travel speed of your robot? Follow these steps:

1. Divide the r/min speed of the motor by 60. The result is the revolutions of the motor per second (r/s). A 100-r/min motor runs at 1.66 r/s.
2. Multiply the diameter of the drive wheel by pi, or approximately 3.14. This yields the circumference of the wheel. A 7-in wheel has a circumference of about 21.98 in.
3. Multiply the speed of the motor (in r/s) by the circumference of the wheel. The result is the number of linear inches covered by the wheel in 1 s.

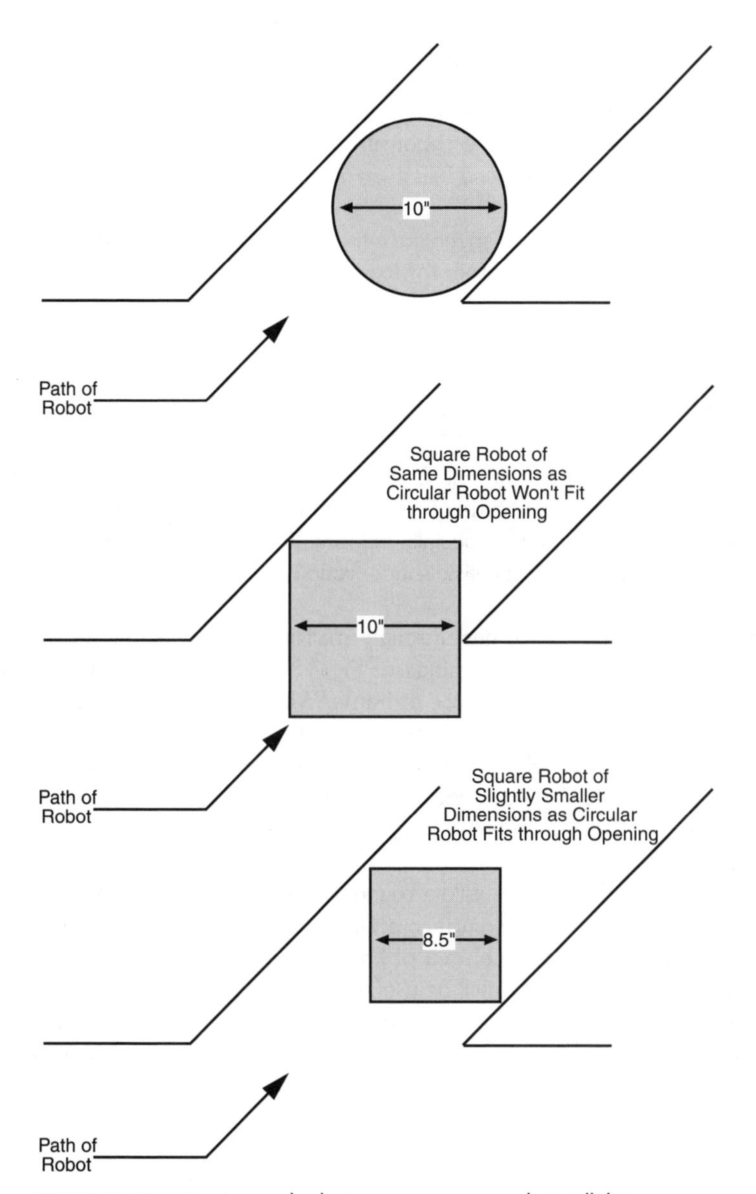

FIGURE 18-14 A round robot versus a square robot. All things being equal, a round robot is better able to navigate through small openings. However, rounded robots also have less usable surface area, so a square-shaped robot can be made smaller and still support the same on-board "real estate."

With a 100-r/min motor and 7-in wheel, the robot will travel at a top speed of 35.168 in/s, or just under 3 ft. That's about 2 mi/h! You can readily see that you can slow down a robot by decreasing the size of the wheel. By reducing the wheel to 5 in instead of 8, the same 100-r/min motor will propel the robot at about 25 in/s. By reducing the motor speed to, say, 75 r/min, the travel speed falls even more, to 19.625 in/s. Now that's more reasonable.

Bear in mind that the actual travel speed once the robot is all put together may be lower than this. The heavier the robot, the larger the load on the motors, so the slower it will turn.

18.12 Round Robots or Square?

Robots can't locomote where they can't fit. Obviously, a robot that's too large to fit through doorways and halls will have a hard time of it. In addition, the overall shape of a robot will also dictate how maneuverable it is, especially indoors. If you want to navigate your robot in tight areas, you should consider its basic shape: round or square.

- A round robot is generally able to pass through smaller openings, no matter what its orientation when going through the opening (see Fig. 18-14). To make a round robot, you must either buy or make a rounded base or frame. Whether you're working with metal, steel, or wood, a round base or frame is not as easy to construct as a square one.
- A square robot must orient itself so that it passes through openings straight ahead rather than at an angle. Square-shaped robot bases and frames are easier to construct than round ones.

While you're deciding whether to build a round- or square-shaped robot, consider that a circle of a given diameter has less surface area than a square of the same width. For example, a 10-in circle has a surface area of about 78 in^2. Moreover, because the surface of the base is circular, less of it will be useful for your robot (unless your printed circuit boards are also circular). Conversely, a 10-by-10-in square robot has a surface area of 100 in. Such a robot could be reduced to about 8.5 in^2, and it would have about the same surface area as a 10-in round robot, and its surface area would be generally more usable.

18.13 From Here

To learn more about . . .	***Read***
Selecting wood, plastic, or metal to construct your robot	Part 2
Choosing a battery for your robot	Chapter 17, "Batteries and Robot Power Supplies"

Selecting motors	Chapter 18, "Choosing the Right Motor"
Building a walking robot	Chapter 24, "Build a Heavy-Duty Six-Legged Walking Robot"
Constructing a treaded robot	Chapter 25, "Advanced Locomotion Systems"
Controlling the speed of a two-motor-driven robot	Chapter 20, "Working with DC Motors"

CHAPTER 19

CHOOSING THE RIGHT MOTOR

Motors are the muscles of robots. Attach a motor to a set of wheels and your robot can scoot around the floor. Attach a motor to a lever, and the shoulder joint for your robot can move up and down. Attach a motor to a roller, and the head of your robot can turn back and forth, scanning its environment. There are many kinds of motors; however, only a select few are truly suitable for home-brew robotics. This chapter will examine the various types of motors and how they are used.

19.1 AC or DC?

Direct current—DC—dominates the field of robotics, either mobile or stationary. DC is used as the main power source for operating the on-board electronics, for opening and closing solenoids, and, yes, for running motors. Few robots use motors designed to operate from AC, even those automatons used in factories. Such robots convert the AC power to DC, then distribute the DC to various subsystems of the machine.

DC motors may be the motors of choice, but that doesn't mean you should use just any DC motor in your robot designs. When looking for suitable motors, be sure the ones you buy are reversible. Few robotic applications call for just unidirectional (one-direction) motors. You must be able to operate the motor in one direction, stop it, and change its direction. DC motors are inherently bidirectional, but some design limitations may prevent reversibility.

The most important factor is the commutator brushes. If the brushes are slanted, the motor probably can't be reversed. In addition, the internal wiring of some DC motors prevents them from going in any but one direction. Spotting the unusual wiring scheme by just looking at the exterior or the motor is difficult, at best, even for a seasoned motor user.

The best and easiest test is to try the motor with a suitable battery or DC power supply. Apply the power leads from the motor to the terminals of the battery or supply. Note the direction of rotation of the motor shaft. Now, reverse the power leads from the motor. The motor shaft should rotate in reverse.

19.2 Continuous or Stepping?

DC motors can be either continuous or stepping. Here is the difference: with a continuous motor, like the ones in Fig. 19-1, the application of power causes the shaft to rotate continually. The shaft stops only when the power is removed or if the motor is stalled because it can no longer drive the load attached to it.

With stepping motors, shown in Fig. 19-2, the application of power causes the shaft to rotate a few degrees, then stop. Continuous rotation of the shaft requires that the power be pulsed to the motor. As with continuous DC motors, there are subtypes of stepping motors. Permanent magnet steppers are the ones you're likely to encounter, and they are also the easiest to use.

The design differences between continuous and stepping DC motors need to be addressed in detail. Chapter 20, "Working with DC Motors," focuses entirely on continuous

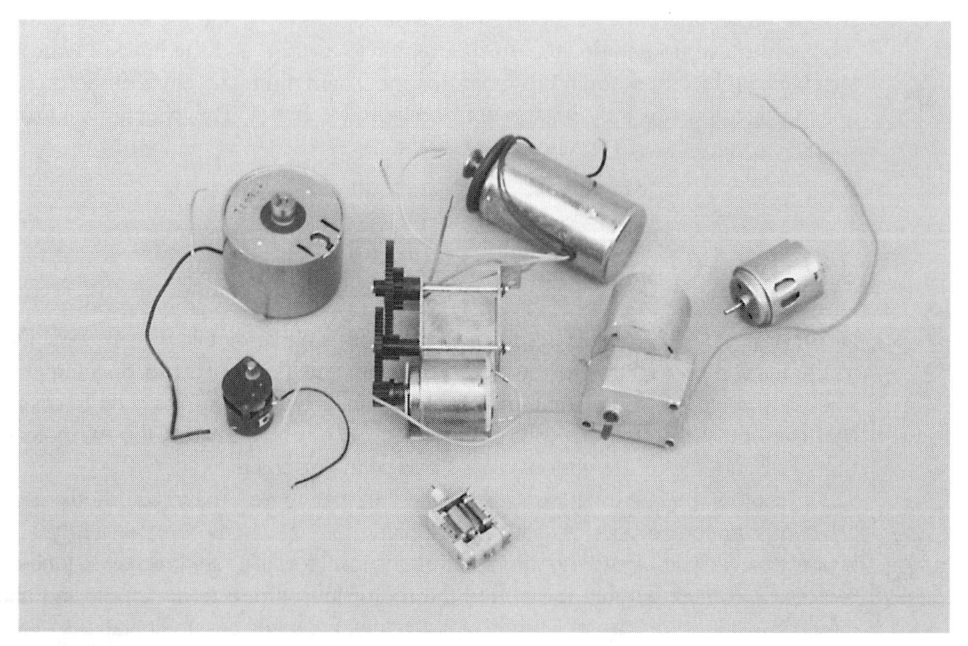

FIGURE 19-1 An assortment of DC motors.

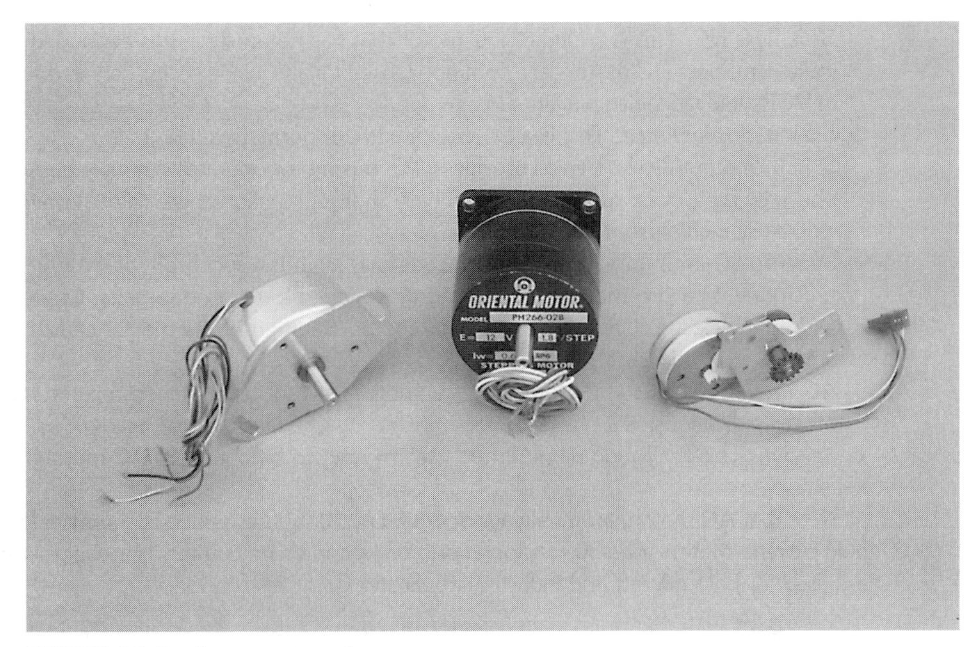

FIGURE 19-2 An assortment of stepper motors.

motors. Chapter 21, "Working with Stepper Motors," focuses entirely on the stepping variety. Although these two chapters focus on the main drive motors of your robot, you can apply the information to motors used for other purposes as well.

19.3 Servo Motors

A special subset of continuous motors is the servo motor, which in typical cases combines a continuous DC motor with a feedback loop to ensure the accurate positioning of the motor. A common form of servo motor is the kind used in model and hobby radio-controlled (R/C) cars and planes.

R/C servos are in plentiful supply, and their cost is reasonable (about $10 to $12 for basic units). Though R/C servos are continuous DC motors at heart, we will devote a separate chapter is devoted just to them. See Chapter 21, "Working with Servo Motors," for more information on using R/C servo motors not only to drive your robot creations across the floor but to operate robot legs, arms, hands, heads, and just about any other appendage.

19.4 Other Motor Types

There are many other types of motors, some of which may be useful in your hobby robot, some of which will not. DC, stepper, and servo motors are the most common, but you may also see references to some of the following:

- *Brushless DC.* This is a kind of DC motor that has no brushes. It is controlled electronically. Brushless DC motors are commonly used in fans inside computers and for motors in VCRs and videodisc players.
- *Switched reluctance.* This is a DC motor without permanent magnets.
- *Synchronous.* Also known as brushless AC, this motor operates synchronously with the phase of the power supply current. These motors function much like stepper motors, which will be discussed in Chapter 21.
- *Synchro.* These motors are considered distinct from the synchronous variety described previously. Synchro motors are commonly designed to be used in pairs, where a "master" motor electrically controls a "slave" motor. Rotation of the master causes an equal amount of rotation in the slave.
- *AC induction.* This is the ordinary AC motor used in fans, kitchen mixers, and many other applications.
- *Sel-Syn.* This is a brand name, often used to refer to synchronous AC motors.

Note that AC motors aren't always operated at 50/60 Hz, which is common for household current. Motors for 400 Hz operation, for example, are common in surplus stores and are used for both aircraft and industrial applications.

19.5 Motor Specifications

Motors come with extensive specifications. The meaning and purpose of some of the specifications are obvious; others aren't. Let's take a look at the primary specifications of motors—voltage, current draw, speed, and torque—and see how they relate to your robot designs.

19.5.1 OPERATING VOLTAGE

All motors are rated by their operating voltage. With small DC hobby motors, the rating is actually a range, usually 1.5 to 6 V. Some high-quality DC motors are designed for a specific voltage, such as 12 or 24 V. The kinds of motors of most interest to robot builders are the low-voltage variety—those that operate at 1.5 to 12 V.

Most motors can be operated satisfactorily at voltages higher or lower than those specified. A 12-V motor is likely to run at 8 V, but it may not be as powerful as it could be and it will run slower (an exception to this is stepper motors; see Chapter 21, "Working with Stepper Motors," for details). You'll find that most motors will refuse to run, or will not run well, at voltages under 50 percent of the specified rating.

Similarly, a 12-V motor is likely to run at 16 V. As you may expect, the speed of the shaft rotation increases, and the motor will exhibit greater power. I do not recommend that you run a motor continuously at more than 30 or 40 percent its rated voltage, however. The windings may overheat, which may cause permanent damage. Motors designed for high-speed operation may turn faster than their ball-bearing construction allows.

If you don't know the voltage rating of a motor, you can take a guess at it by trying various voltages and seeing which one provides the greatest power with the least amount of heat dissipated through the windings (and felt on the outside of the case). You can also listen to the motor. It should not seem as if it is straining under the stress of high speeds.

19.5.2 CURRENT DRAW

Current draw is the amount of current, in milliamps or amps, that the motor requires from the power supply. Current draw is more important when the specification describes motor loading, that is, when the motor is turning something or doing some work. The current draw of a free-running (no-load) motor can be quite low. But have that same motor spin a wheel, which in turn moves a robot across the floor, and the current draw jumps 300, 500, even 1000 percent.

With most permanent magnet motors (the most popular kind), current draw increases with load. You can see this visually in Fig. 19-3. The more the motor has to work to turn the shaft, the more current is required. The load used by the manufacturer when testing the motor isn't standardized, so in your application the current draw may be more or less than that specified.

A point is reached when the motor does all the work it can do, and no more current will flow through it. The shaft stops rotating; the motor has stalled. Some motors, but not many, are rated (by the manufacturer) by the amount of current they draw when stalled.

This is considered the worst-case condition. The motor will never draw more than this current unless it is shorted out, so if the system is designed to handle the stall current it can handle anything. Motors rated by their stall current will be labeled as such. Motors designed for the military, available through surplus stores, are typically rated by their stall current. When providing motors for your robots, you should always know the approximate current draw under load. Most volt-ohm meters can test current. Some special-purpose amp meters are made just for the job.

Be aware that some volt-ohm meters can't handle the kind of current pulled through a motor. Many digital meters can't deal with more than 200 to 400 mA of current in the low-current settings. Small hobby motors can often draw in excess of this. Be sure your meter can accommodate current up to 5 or 10 A and is fuse protected.

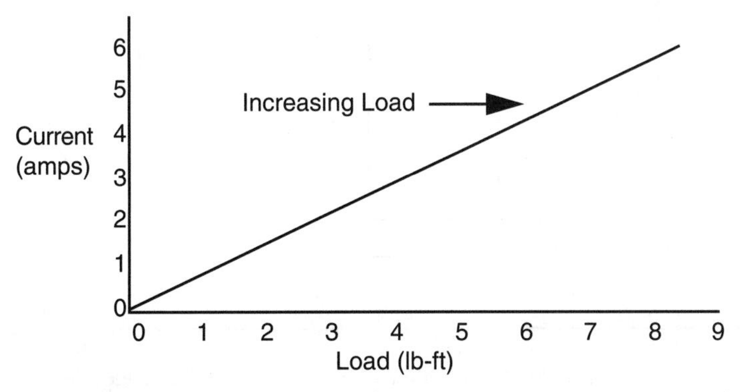

FIGURE 19-3 The current draw of a motor increases in proportion to the load on the motor shaft.

If your meter cannot register this high without popping fuses or burning up, insert a 1- to 10-Ω power resistor (10 to 20 W) between one of the motor terminals and the positive supply rail, as shown in Fig. 19-4. With the meter set on DC voltage, measure the voltage developed across the resistor.

A bit of Ohm's law, I = E/R (I is current, E is voltage, R is resistance) reveals the current draw through the motor. For example, if the resistance is 10 Ω and the voltage is 2.86 V, the current draw is 286 mA. You can watch the voltage go up (and therefore the current, too) by loading the shaft of the motor.

When you are actually measuring voltage across and the current through a motor, you will probably see your readings jump around quite a bit (especially if you are adding a load to your motor). When you are calculating the load of the motor, make sure that you use the worst-case (highest) value for current and the no-load (motor disconnected) voltage of your power supply to ensure that you provide enough power to your application.

19.5.3 SPEED

The rotational speed of a motor is given in revolutions per minute (r/min). Most continuous DC motors have a normal operating speed of 4000 to 7000 r/min. However, some special-purpose motors, such as those used in tape recorders and computer disk drives, operate as slow as 2000 to 3000 r/min. For just about all robotic applications, these speeds are much too high. You must reduce the speed to no more than 150 r/min (even less for motors driving arms and grippers) by using a gear train. You can obtain some reduction by using electronic control, as described in Part 5 of this book, "Computers and Electronic Control."

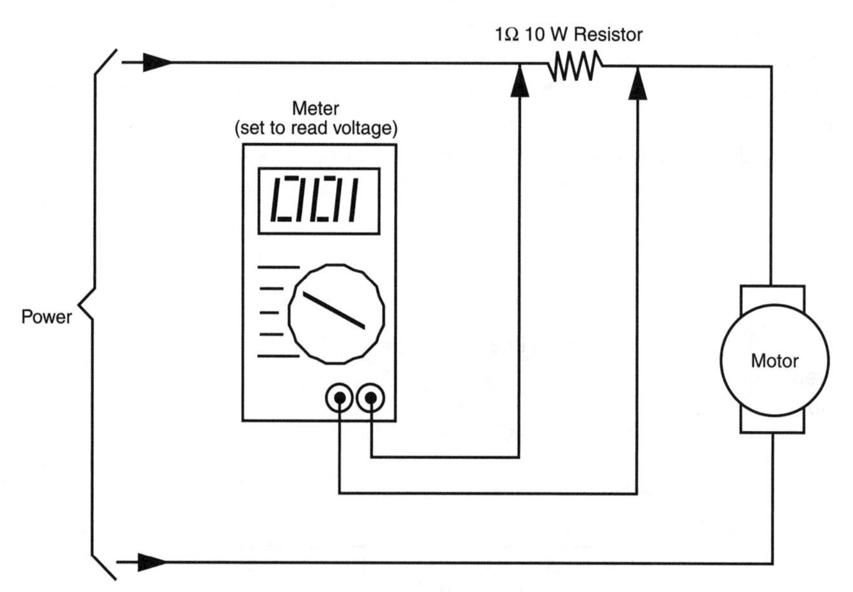

FIGURE 19-4 How to test the current draw of a motor by measuring the voltage developed across an in-line resistor. The actual value of the resistor can vary, but it should be under about 20 Ω. Be sure the resistor is a high-wattage type.

However, such control is designed to make fine-tuned speed adjustments, not reduce the rotation of the motor from 5000 to 50 r/min. Gears, which are explained in later sections of this chapter, are used to provide these large reductions in rotation speeds.

Note that the speed of stepping motors is not rated in r/min but in steps (or pulses) per second. The speed of a stepper motor is a function of the number of steps that are required to make one full revolution plus the number of steps applied to the motor each second. As a comparison, the majority of light- and medium-duty stepper motors operate at the equivalent of 100 to 140 r/min. See Chapter 21, "Working with Stepper Motors," for more information.

19.5.4 TORQUE

Torque is the force the motor exerts upon its load. The higher the torque, the larger the load can be and the faster the motor will spin under that load. Reduce the torque, and the motor slows down, straining under the workload. Reduce the torque even more, and the load may prove too demanding for the motor. The motor will stall to a grinding halt, and in doing so eat up current (and put out a lot of heat).

Torque is perhaps the most confusing design aspect of motors. This is not because there is anything inherently difficult about it but because motor manufacturers have yet to settle on a standard means of measurement. Motors made for industry are rated one way, motors for the military another.

At its most basic level, torque is measured by attaching a lever to the end of the motor shaft and a weight or gauge on the end of that lever, as depicted in Fig. 19-5. The lever can be any number of lengths: 1 cm, 1 in, or 1 ft. Remember this because it plays an important role in torque measurement. The weight can either be a hunk or lead or, more commonly, a spring-loaded scale (as shown in the figure). Turn the motor on and it turns the lever. The amount of weight it lifts is the torque of the motor. There is more to motor testing than this, of course, but it'll do for the moment.

Now for the ratings game. Remember the length of the lever? That length is used in the torque specification. If the lever is 1 in long, and the weight successfully lifted is 2 oz, then the motor is said to have a torque of 2 oz-in. (Some people reverse the "ounce" and "inches" and come up with "inch-ounces.")

The unit of length for the lever usually depends on the unit of measurement given for the weight. When the weight is in grams, the lever is in centimeters (gm-cm). When the weight is in ounces, as already seen, the lever used is in inches (oz-in). Finally, when the weight is in pounds, the lever used is commonly in feet (lb-ft). Like the ounce-inch measurement, gram-centimeter and pound-foot specifications can be reversed—"centimeter-gram" or "foot-pound." Note that these easy-to-follow conventions aren't always used. Some motors may be rated by a mixture of the standards—ounces and feet or pounds and inches.

19.5.5 STALL OR RUNNING TORQUE

Most motors are rated by their running torque, or the force they exert as long as the shaft continues to rotate. For robotic applications, it's the most important rating because it determines how large the load can be and still guarantee that the motor turns. How running torque tests are conducted varies from one motor manufacturer to another, so results can

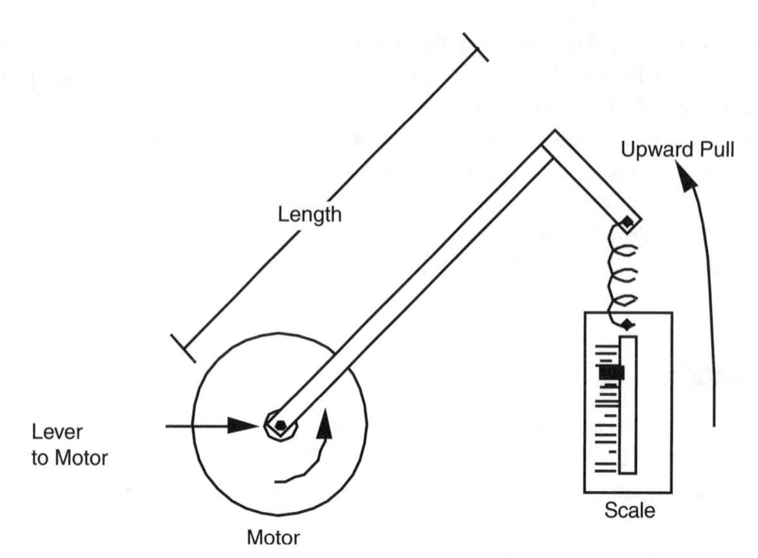

FIGURE 19-5 The torque of a motor is measured by attaching a weight or scale to the end of a lever and mounting the lever of the motor shaft.

differ. The tests are impractical to duplicate in the home shop, unless you have an elaborate slip-clutch test stand, precision scale, and sundry other test jigs.

If the motor(s) you are looking at doesn't have running torque ratings, you must estimate its relative strength. This can be done by mounting it on a makeshift wood or metal platform, attaching wheels, and having it scoot around the floor. If the motor supports the platform, start piling on weights. If the motor continues to operate with, say, 40 or 50 lb of junk on the platform, you've got an excellent motor for driving your robot.

Some motors you may test aren't designed for hauling heavy loads, but they may be suitable for operating arms, grippers, and other mechanical components. You can test the relative strength of these motors by securing them in a vise, then attaching a large pair of Vise-Grips or other lockable pliers to them. Use your own hand as a test jig, or rig one up with fishing weights. Determine the rotational power of the motor by applying juice to the motor and seeing how many weights it can successfully handle.

Such crude tests make more sense if you have a standard by which to judge others. If you've designed a robotic arm before, for example, and are making another one, test the motors that you successfully used in your prototype. If subsequent motors fail to match or exceed the test results of the standard, you know they are unsuitable for the test.

Another torque specification, stall torque, is sometimes provided by the manufacturer instead of or in addition to running torque (this is especially true of stepping motors). Stall torque is the force exerted by the motor when the shaft is clamped tight. There is an indirect relationship between stall torque and running torque, and although it varies from motor to motor you can use the stall torque rating when you select candidate motors for your robot designs.

19.6 Gears and Gear Reduction

We've already discussed the fact that the normal running speed of motors is far too fast for most robotics applications. Locomotion systems need motors with running speeds of 75 to 150 r/min. Any faster than this, and the robot will skim across the floor and bash into walls and people. Arms, gripper mechanisms, and most other mechanical subsystems need even slower motors. The motor for positioning the shoulder joint of an arm needs to have a speed of less than 20 r/min; 5 to 8 r/min is even better.

There are two general ways to decrease motor speed significantly: build a bigger motor (impractical) or add gear reduction. Gear reduction is used in your car, on your bicycle, in the washing machine and dryer, and in countless other motor-operated mechanisms.

19.6.1 GEARS 101

Gears perform two important duties. First, they can make the number of revolutions applied to one gear greater or lesser than the number of revolutions of another gear that is connected to it. They also increase or decrease torque, depending on how the gears are oriented. Gears can also serve to simply transfer force from one place to another.

Gears are actually round levers, and it may help to explain how gears function by first examining the basic mechanical lever. Place a lever on a fulcrum so the majority of the lever is on one side. Push up on the long side, and the short side moves in proportion. Although you may move the lever several feet, the short side is moved only a few inches. Also note that the force available on the short end is proportionately larger than the force applied on the long end. You use this wonderful fact of physics when you dig a rock out of the ground with your shovel or jack up your car to replace a tire.

Now back to gears. Attach a small gear to a large gear, as shown in Fig. 19-6. The small gear is directly driven by a motor. For each revolution of the small gear, the large gear turns one-half a revolution. Expressed another way, if the motor and small gear turn at 1000 r/min, the large gear turns at 500 r/min. The gear ratio is said to be 2:1.

Note that another important thing happens, just as it did with the lever and fulcrum. Decreasing the speed of the motor also increases its torque. The power output is approximately twice the input. Some power is lost in the reduction process due to the friction of the gears. If the drive and driven gears are the same size, the rotation speed is neither increased nor decreased, and the torque is not affected (apart from small frictional losses). You can use same-size gears in robotics design to transfer motive power from one shaft to another, such as driving a set of wheels at the same speed and in the same direction.

19.6.2 ESTABLISHING GEAR REDUCTION

Gears are an old invention, going back to ancient Greece. Today's gears are more refined, and they are available in all sorts of styles and materials. However, they are still based on the old Greek design in which the teeth from the two mating gears mesh with each other. The teeth provide an active physical connection between the two gears, and the force is transferred from one gear to another.

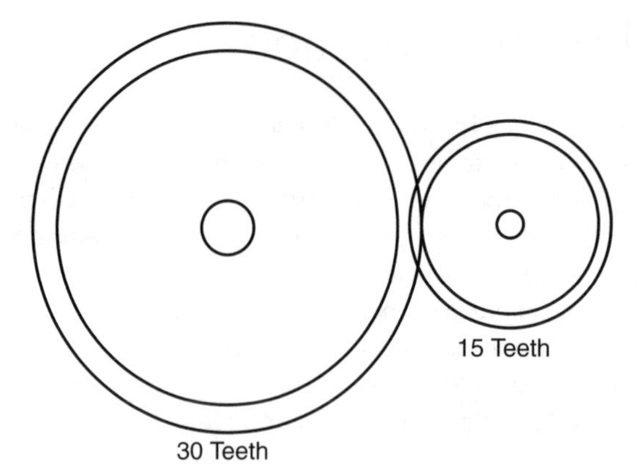

FIGURE 19-6 A representation of a 2:1 gear reduction ratio.

Gears with the same size teeth are usually characterized not by their physical size but by the number of teeth around their circumference. In the example in Fig. 19-6, the small gear contains 15 teeth, the large gear 30 teeth. And, you can string together a number of gears one after the other, all with varying numbers of teeth (see Fig. 19-7). Attach a tachometer to the hub of each gear, and you can measure its speed. You'll discover the following two facts:

- The speed always decreases when going from a small to a large gear.
- The speed always increases when going from a large to a small gear.

There are plenty of times when you need to reduce the speed of a motor from 5000 to 50 r/min. That kind of speed reduction requires a reduction ratio of 100:1. To accomplish that with just two gears you would need, as an example, a drive gear that has 10 teeth and a driven gear that has 1000 teeth. That 1000-tooth gear would be quite large, bigger than the drive motor itself.

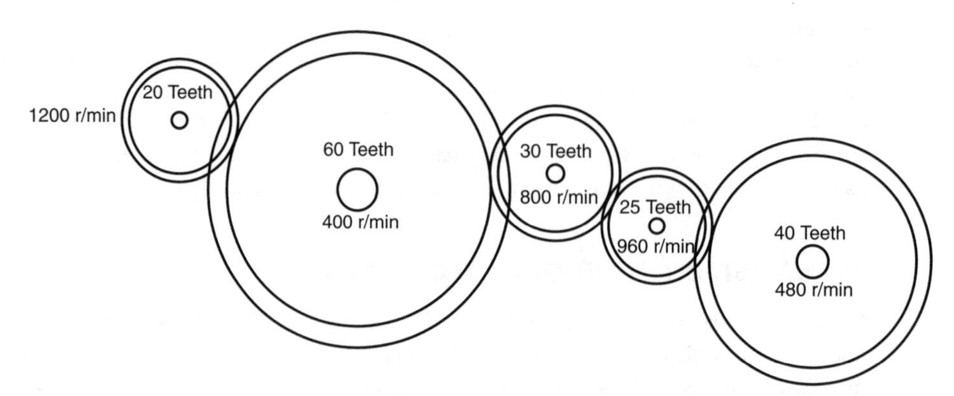

FIGURE 19-7 Gears driven by the 20-tooth gear on the left rotate at different speeds, depending on their diameter.

You can reduce the speed of a motor in steps by using the arrangement shown in Fig. 19-8. Here, the driver gear turns a larger hub gear, which in turn has a smaller gear permanently attached to its shaft. The small hub gear turns the driven gear to produce the final output speed, in this case 50 r/min. You can repeat this process over and over again until the output speed is but a tiny fraction of the input speed. This is the arrangement most often used in motor gear reduction systems.

19.6.3 USING MOTORS WITH GEAR REDUCTION

It's always easiest to use DC motors that already have a gear reduction box built onto them, such as the motor in Fig. 19-9. R/C servo motors already incorporate gear reduction, as do most stepper motors. This fact saves you from having to find a gear reducer that fits the motor and application and attach it yourself. When selecting gear motors, you'll be most interested in the output speed of the gearbox, not the actual running speed of the motor. Note as well that the running and stall torque of the motor will be greatly increased. Make sure that the torque specification on the motor is for the output of the gearbox, not the motor itself.

With most gear reduction systems, the output shaft is opposite the input shaft (but usually off center). With other boxes, the output and input are on the same side of the box. When the shafts are at 90 degrees from one another, the reduction box is said to be a right-angle drive. If you have the option of choosing, select the kind of gear reduction that best suits the design of your robot. You will probably find that shafts on opposite sides is the all-around best choice. Right-angle drives also come in handy, but they usually carry high price tags.

When using motors without built-in gear reduction, you'll need to add reduction boxes, such as the model shown in Fig. 19-10, or make your own. Although it is possible to do both of these yourself, there are many pitfalls:

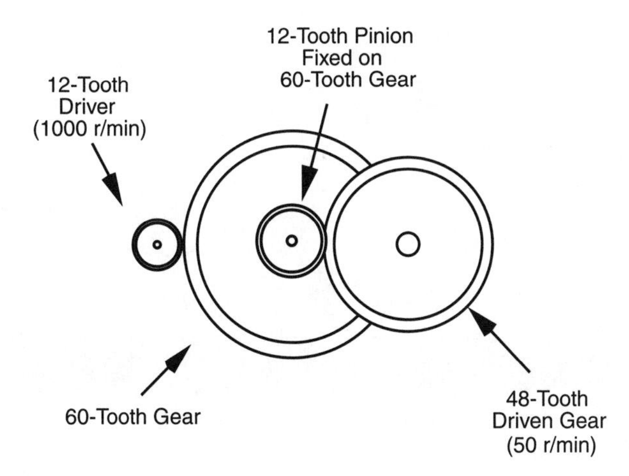

FIGURE 19-8 True gear reduction is achieved by ganging gears on the same shaft.

FIGURE 19-9 A motor with an enclosed gearbox. These are ideal for robotics use.

FIGURE 19-10 A gear reduction box, originally removed from an open-frame AC motor. On this unit, the input and output shafts are on the same side.

- Shaft diameters of motors and ready-made gearboxes may differ, so you must be sure that the motor and gearbox mate.
- Separate gear reduction boxes are hard to find. Most must be cannibalized from salvage motors. Old AC motors are one source of surplus boxes.
- When designing your own gear reduction box, you must take care to ensure that all the gears have the same hub size and that meshing gears exactly match each other.
- Machining the gearbox requires precision, since even a small error can cause the gears to mesh improperly.

19.6.4 ANATOMY OF A GEAR

Gears consist of teeth, but these teeth can come in any number of styles, sizes, and orientations. Spur gears are the most common type. The teeth surround the outside edge of the gear, as shown in Fig. 19-11. Spur gears are used when the drive and driven shafts are parallel. Bevel gears have teeth on the surface of the circle rather than the edge. They are used to transmit power to perpendicular shafts. Miter gears serve a similar function but are designed so that no reduction takes place. Spur, bevel, and miter gears are reversible. That is, unless the gear ratio is very large, you can drive the gears from either end of the gear system, thus increasing or decreasing the input speed.

Worm gears transmit power perpendicularly, like bevel and miter gears, but their design is unique. The worm (or lead screw) resembles a threaded rod. The rod provides the power.

FIGURE 19-11 Spur gears. These particular gears are made of nylon and have aluminum hubs. It's better to use metal hubs in which the gear is secured to the shaft with a setscrew.

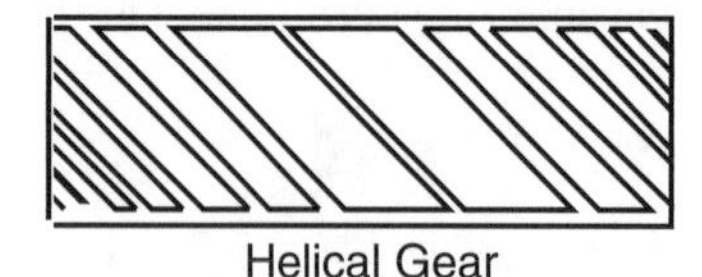

FIGURE 19-12 Standard spur gear versus diagonal helical spur gear. The latter is used to decrease backlash—the play inherent when two gears mesh. Some helical gears are also made for diverting the motion at right angles.

As it turns, the threads engage a modified spur gear (the modification takes into consideration the cylindrical shape of the worm).

Worm gear systems are specifically designed for large-scale reduction. The gearing is not usually reversible; you can't drive the worm by turning the spur gear. This is an important point because it gives worm gear systems a kind of automatic locking capability. Worm gears are particularly well suited for arm mechanisms in which you want the joints to remain where they are. With a traditional gear system, the arm may droop or sink back due to gravity once the power from the drive motor is removed.

Rack gears are like spur gears unrolled into a flat rod. They are primarily intended to transmit rotational motion to linear motion. Racks have a kind of self-locking characteristic as well, but it's not as strong as that found in worm gears.

The size of gear teeth is expressed as pitch, which is roughly calculated by counting the number of teeth on the gear and dividing it by the diameter of the gear. For example, a gear that measures 2 in and has 48 teeth has a tooth pitch of about 24. Common pitches are 12 (large), 24, 32, and 48. Some gears have extra-fine 64-pitch teeth, but these are usually confined to miniature mechanical systems, such as radio-controlled models. Odd-sized pitches exist, of course, as do metric sizes, so you must be careful when matching gears that the pitches are exactly the same. Otherwise, the gears will not mesh properly and may cause excessive wear.

The degree of slope of the face of each tooth is called the pressure angle. The most common pressure angle is 20°, although some gears, particularly high-quality worms and racks, have a 14 ½ in pressure angle. Textbooks claim that you should not mix two gears with different pressure angles even if the pitch is the same, but it can be done. Some excessive wear may result because the teeth aren't meshing fully.

The orientation of the teeth on the gear can differ. The teeth on most spur gears are perpendicular to the edges of the gear. But the teeth can also be angled, as shown in Fig. 19-12, in which case it is called a helical gear.

A number of other unusual tooth geometries are in use. These include double-teeth, where two rows of teeth offset one another, and herringbone, where there are two sets of helical gears at opposite angles. These gears are designed to reduce the backlash phenomenon. The space (or play) between the teeth when meshing can cause the gears to rock back and forth.

19.7 Pulleys, Belts, Sprockets, and Roller Chain

Akin to the gear are pulleys, belts, sprockets, and roller chains. Pulleys are used with belts, and sprockets are used with roller chain. The pulley and sprocket are functionally identical to the gear. The only difference is that pulleys and sprockets use belts and roller chain, respectively, to transfer power. With gears, power is transferred directly.

A benefit of using pulleys-belts or sprockets-chain is that you don't need to be as concerned with the absolute alignment of the mechanical parts of your robot. When using gears you must mount them with high precision. Accuracies to the hundredths of an inch are desirable to avoid slop in the gears as well as the inverse—binding caused by gears that are meshing too tightly. Belts and roller chain are designed to allow for slack; in fact, if there's no slack you run the risk of breaking the pulley or chain!

19.7.1 MORE ABOUT PULLEYS AND BELTS

Pulleys come in a variety of shapes and sizes. You're probably familiar with the pulleys and belts used in automotive applications. These are likely to be too bulky and heavy to be used with a robot. Instead, look for smaller and lighter pulleys and belts used for copiers, fax machines, VCRs, and other electronic equipment. These are available for salvage from whole units or in bits and pieces from surplus outlets.

Pulleys can be either the V type (the pulley wheel has a V-shaped groove in it) or the cog type. Cog pulleys require matching belts. You need to ensure that the belt is not only the proper width for the pulley you are using but also has the same cog pitch.

19.7.2 MORE ABOUT SPROCKETS AND ROLLER CHAIN

Sprockets and roller chain are preferred when you want to ensure synchronism. For large robots you can use ⅜-in bicycle chain. Most smaller robots will do fine with ¼-in roller chain, which can frequently be found in surplus stores. Metal roller chain is commonly available in preset lengths, though you can sometimes shorten or lengthen the chain by adding or removing links. Plastic roller chain, while not as strong, can be adjusted more easily by using snap-on links.

19.8 Mounting the Motor

Every motor requires a different mounting arrangement. It's easier for you when the motor has its own mounting hardware or holes; you can use these to mount the motor in your robot. Remember that Japanese- and European-made motors often have metric threads, so be sure to use the proper-sized bolt.

Other motors may not be as cooperative. Either the mounting holes are in a position where they don't do you much good, or the motor is completely devoid of any means for securing it to your robot. You can still mount these motors successfully by using an assortment of clamps, brackets, woodblocks, and homemade angles.

For example, to secure the motor shown in Fig. 19-13, mounting brackets were fashioned using 6-in galvanized iron mending T plates. A large hole was drilled for the drive shaft and gear to poke through, and the two halves of the mounting bracket were joined together with nuts, bolts, and spacers. The bracket was then attached to the frame of the robot using angle irons and standard hardware. This motor arrangement was made a little more difficult by the addition of a drive gear and sprocket. Construction time for each motor bracket was about 90 min.

Another example is shown in Fig. 19-14. Here, the motor has mounting holes on the end by the shaft, but these holes are in the wrong position for the design of the robot. Two commonly available flat corner irons were used to mount the motor. This is just one approach; a number of other mounting schemes might have worked satisfactorily as well. This design is more thoroughly discussed in Chapter 27, "Build a Revolute Coordinate Arm."

You can also fashion your own mounting brackets using metal or plastic. Cut the bracket to the size you need, and drill mounting holes. This technique works well when you are using servo motors for model radio-controlled cars and airplanes.

FIGURE 19-13 One approach to mounting a large motor in a robot. The motor is sandwiched inside two large hardware plates and is secured to the frame of the motor with angle irons.

FIGURE 19-14 Another approach to mounting a motor to a robot. Flat corner irons secure the motor flange to the frame.

If the motor lacks mounting holes, you can use clamps to hold it in place. U-bolts, available at the hardware store, are excellent solutions. Choose a U-bolt that is large enough to fit around the motor. The rounded shape of the bolt is perfect for motors with round casings. If desired, you can make a holding block out of plastic or wood to keep the bottom of the motor from sliding. Cut the plastic or wood to size, and round it out with a router, rasp, or file so it matches the shape of the motor casing.

19.9 Connecting to the Motor Shaft

Connecting the shaft of the motor to a gear, wheel, lever, or other mechanical part is probably the most difficult task of all. There is one exception to this, however: R/C servo motors are easier to mount, which is one reason they are so popular in hobby robotics. Motor shafts come in many different sizes, and because most—if not all—of the motors you'll use will come from surplus outlets, the shaft may be peculiar to the specific application for which the motor was designed.

Common shaft sizes are $^1/_{16}$- and $^1/_8$-in for small hobby motors and $^1/_4$-, $^3/_8$-, or $^5/_{16}$-in for

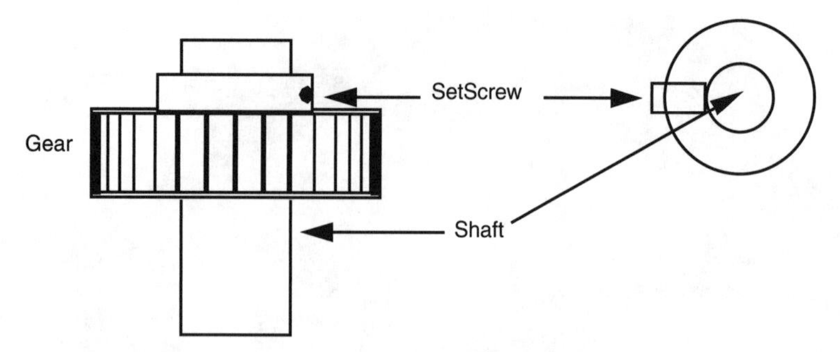

FIGURE 19-15 Use a setscrew to secure the gear to the shaft.

larger motors and gearboxes. Gear hubs are generally ¼-, ½-, or ⅝-in, so you'll need to find reducing bushings at an industrial supply store. Surplus is also a good source. The same goes for wheels, sprockets (for roller chain and timing pulleys), and bearings.

To attach things like gears and sprockets, the gear or sprocket must usually be physically secured to the shaft by way of a setscrew, as depicted in Fig. 19-15. Sometimes a press fit is all that's required. Most better-made gears and sprockets have the setscrews in them or have provisions for inserting them. If the gear or sprocket has no setscrew and there is no hole for one, you'll have to drill and tap the hole for the screw.

There are two common alternatives if you can't use a setscrew. The first method is to add a spline, or key, to secure the gear or sprocket to the shaft. This requires some careful machining, as you must make a slot for the spline in the shaft as well as for the hub of the gear or sprocket. Another method is to thread the gear shaft, and mount the gear or sprocket using nuts and split lock washers (the split in the washer provides compression that keeps the assembly from working loose). Shaft threading is also sometimes necessary when you are attaching wheels. Many people find that threading the shaft is easier. Threading requires you to lock the shaft so it won't turn, which can be a problem with some motors. Also, be careful that the shavings from the threading die do not fall into the motor.

Attaching two shafts to one another is a common, but not insurmountable, problem. The best approach is to use a coupler. You tighten the coupler to the shaft using setscrews. Couplers are available from industrial supply houses and can be expensive, so shop carefully. Some couplers are flexible; that is, they give if the two shafts aren't perfectly aligned. These are the best, considering the not-too-close tolerances inherent in home-built robots. Some couplers are available that accept two shafts of different sizes.

19.10 From Here

To learn more about . . .	***Read***
Robot locomotion systems	Chapter 18, "Principles of Robot Locomotion"

All about DC motors	Chapter 20, "Working with DC Motors"
All about stepper motors	Chapter 21, "Working with Stepper Motors"
All about servo motors	Chapter 22, "Working with Servo Motors"
Operating motors by computer	Chapter 12, "An Overview of Robot 'Brains' "
Interfacing motors to electronic circuitry	Chapter 14, "Computer Peripherals"

CHAPTER 20

WORKING WITH DC MOTORS

DC motors are the mainstay of robotics. A surprisingly small motor, when connected to wheels through a gear reduction system, can power a 25-, 50-, even 100-lb robot with ease. A flick of a switch, a click of a relay, or a tick of a transistor, and the motor stops in its tracks and turns the other way. A simple electronic circuit enables you to gain quick and easy control over speed—from a slow crawl to a fast sprint.

This chapter shows you how to apply open-loop continuous DC motors (as opposed to servo DC motors, which use position feedback) to drive your robots. The emphasis is on using motors to propel a robot across your living room floor, but you can use the same control techniques for any motor application, including gripper closure, elbow flexion, and sensor positioning.

20.1 The Fundamentals of DC Motors

There are many ways to build a DC motor. By their nature, all DC motors are powered by direct current—hence the name DC—rather than the alternating current (AC) used by most motorized household appliances. By and large, AC motors are less expensive to manufacture than DC motors, and because their construction is simpler they tend to last longer than DC motors.

The most common form of DC motor is the permanent magnet type, so-called because it uses two or more permanent magnet pole pieces (called the stator). The turning shaft of

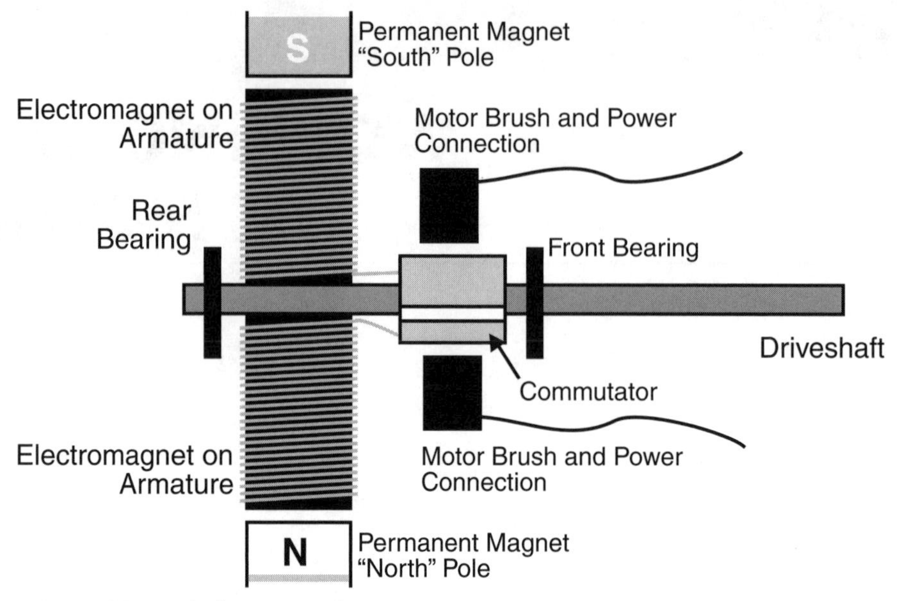

FIGURE 20-1 The basic parts of a DC motor.

the motor, or the rotor, is composed of windings that are connected to a mechanical commutator. Internally, metal brushes (which can wear out!) supply the contact point for the current that turns the motor. Fig. 20-1 provides a side view of the motor while Fig. 20-2 shows the different parts of a motor that has been taken apart.

There are normally three sets of windings to a DC motor as illustrated in Fig. 20-3. The brushes can only come into contact with two of the windings' commutators, resulting in two rotors being magnetized at any given time. These magnetized rotors create an unbalanced force in the DC motor, causing them to be pulled toward their respective permanent magnet poles, and torque on the motor shaft. The placement of the commutators and the brushes causes the motor windings to change polarity as the rotor turns (again shown in Fig. 20-3).

Other types of DC motors exist as well, including the series wound (or universal) and the shunt wound DC motors. These differ from the permanent magnet motor in that no mag-

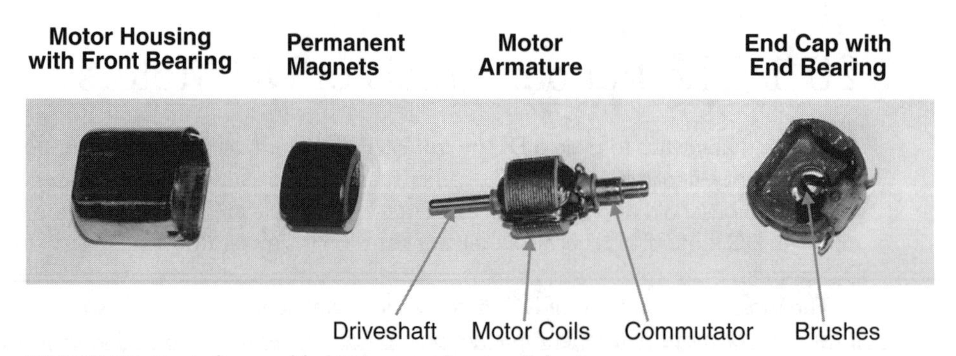

FIGURE 20-2 A disassembled DC motor, showing its basic parts.

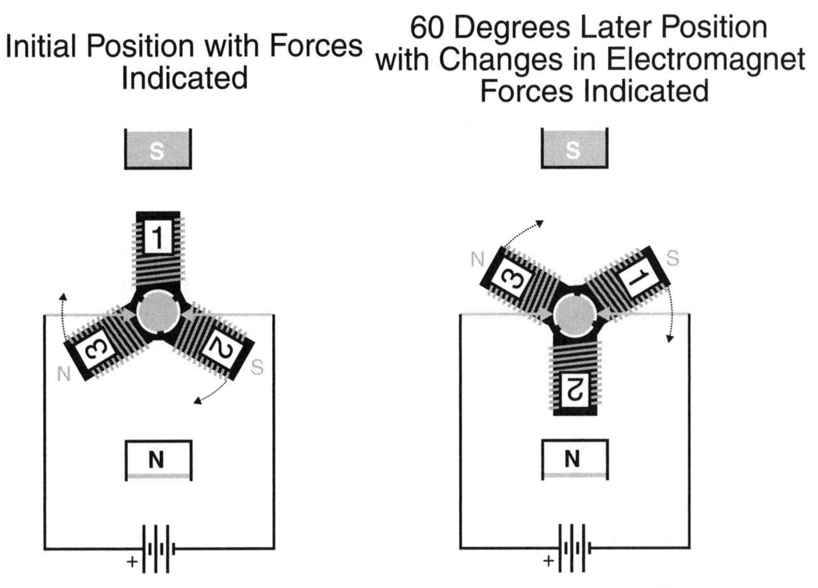

FIGURE 20-3 Two views of the three winding DC motor shaft at different positions. As the shaft turns, the polarity of the different windings change due to the changing position of the brushes relative to the commutators.

nets are used. Instead, the stator is composed of windings that, when supplied with current, become electromagnets.

One of the prime benefits of most, but not all, DC motors is that they are inherently reversible. Apply current in one direction (the "and" on the battery terminals, for example), and the motor may spin clockwise. Apply current in the other direction, and the motor spins counterclockwise. This capability makes DC motors well suited for robotics, where it is often desirable to have the motors reverse direction, such as to back a robot away from an obstacle or to raise or lower a mechanical arm.

Not all DC motors are reversible, and those that are typically exhibit better performance (though often just slightly better) in one direction over the other. For example, the motor may turn a few revolutions per minute faster in one direction. Normally, this is not observable in the typical motor application, but robotics isn't typical. In a robot with the common two-motor drive (see Chapter 18, "Principles of Robot Locomotion"), the motors will be facing opposite directions, so one will turn clockwise while the other turns counterclockwise. If one motor is slightly faster than the other, it can cause the robot to steer off course. Fortunately, this effect isn't usually seen when the robot just travels short distances, and in any case, it can often be corrected by the control circuitry used in the robot.

20.2 Reviewing DC Motor Ratings

Motor ratings, such as voltage and current, were introduced in Chapter 19, "Choosing the Right Motor for the Job." Here are some things to keep in mind when considering a DC motor for your robot:

- DC motors can often be effectively operated at voltages above and below their specified rating. If the motor is rated for 12 V, and you run it at 6 V, the odds are the motor will still turn but at reduced speed and torque. Conversely, if the motor is run at 18 to 24 V, the motor will turn faster and will have increased torque. This does not mean that you should intentionally under- or overdrive the motors you use. Significantly overdriving a motor may cause it to wear out much faster than normal. However, it's usually fairly safe to run a 10-V motor at 12 V or a 6-V motor at 5 V.
- DC motors draw the most current when they are stalled. Stalling occurs if the motor is supplied current, but the shaft does not rotate. If there is no stall detection circuitry built into the motor driver, the battery, control electronics, and drive circuitry you use with the motor must be able to deliver the current at stall, or they (along with the motor) could burn out.
- DC motors vary greatly in efficiency. Many of the least expensive motors you may find are meant to be used in applications (such as automotive) where brute strength, rather than the conservation of electricity, is the most important factor. Since the typical mobile robot is powered by a battery, strive for the most efficient motors you can get. It's best to stay away from automotive starter, windshield wiper, power window, and power seat motors since these are notoriously inefficient.
- The rotational speed of a DC motor is usually too fast to be directly applied in a robot. Gear reduction of some type is necessary to slow down the speed of the motor shaft. Gearing down the output speed has the positive side effect of increasing torque.

20.3 Motor Control

As noted earlier, it's fairly easy to change the rotational direction of a DC motor. Simply switch the power lead connections to the battery, and the motor turns in reverse. The small robots discussed in earlier chapters performed this feat by using a double-pole, double-throw (DPDT) switch. Two such switches were used, one for each of the drive motors. The wiring diagram for these robot motors is duplicated in Fig. 20-4 for your convenience. The DPDT switches used here have a center-off position. When they are in the center position, the motors receive no power so the robot does not move.

You can use the direction control switch for experimenting, but you'll soon want to graduate to more automatic control of your robot. There are a number of ways to accomplish the electronic or electrically assisted direction control of motors. All have their advantages and disadvantages.

20.3.1 RELAY CONTROL

Perhaps the most straightforward approach to the automatic control of DC motors is to use relays. It may seem rather daft to install something as old-fashioned and cumbersome as relays in a high-tech robot, but it is still a useful technique. You'll find that while relays may wear out in time (after a few hundred thousand switchings), they are fairly inexpensive and easy to use.

You can accomplish basic on/off motor control with a single-pole relay. Rig up the relay

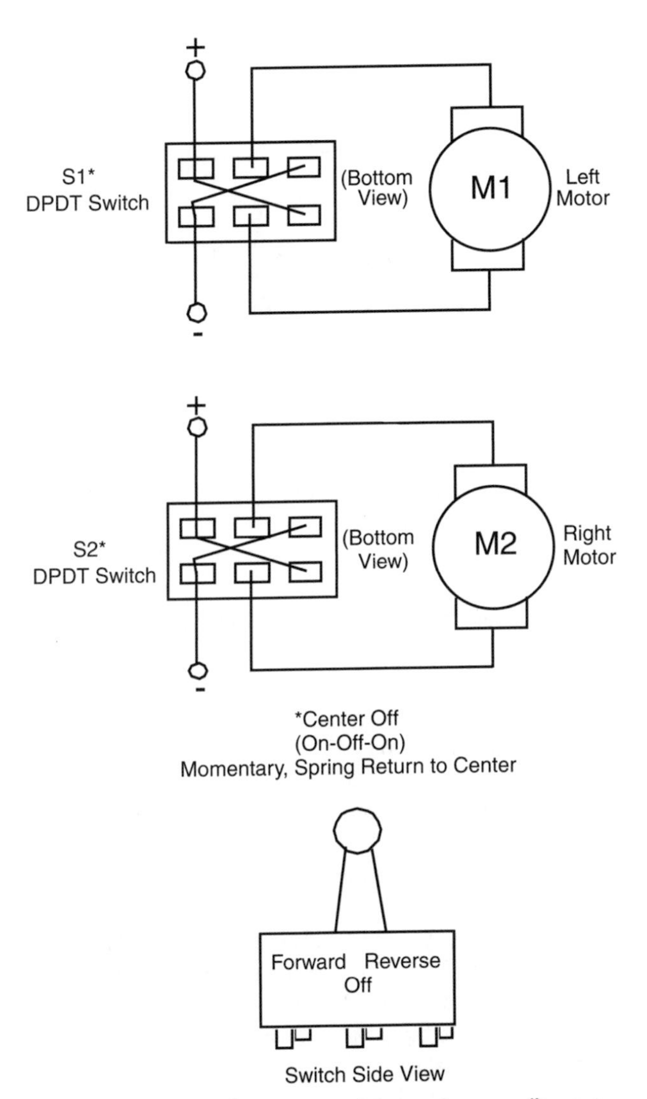

FIGURE 20-4 The basic wiring diagram for controlling twin robot drive motors. Note that the switches are DPDT and the spring return is set to center-off.

so that current is broken when the relay is not activated. Turn on the relay, and the switch closes, thus completing the electrical circuit. The motor turns.

How you activate the relay is something you'll want to consider carefully. You could control it with a push-button switch, but that's no better than the manual switch method just described. Relays can easily be driven by digital signals. Fig. 20-5 shows the complete driver circuit for a relay-controller motor. Logical 0 (LOW) turns the relay off; logical 1 (HIGH) turns it on (refer to the parts list in Table 20-1). The relay can be operated from any digital gate, including a computer or microprocessor port.

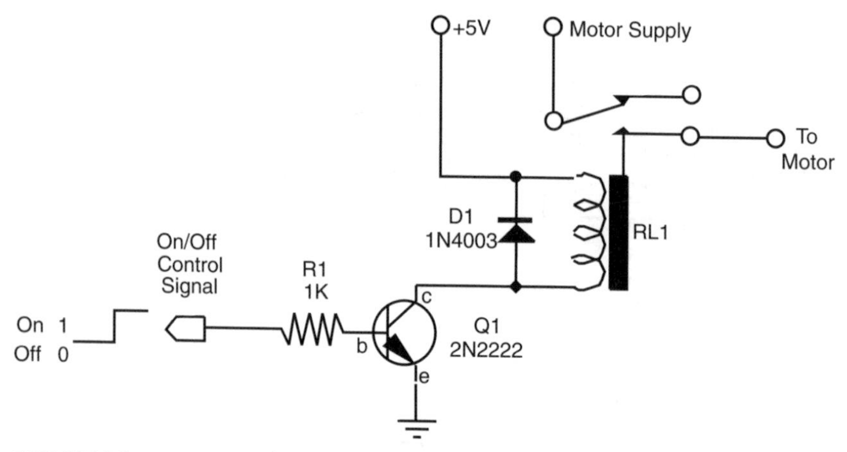

FIGURE 20-5 Using a relay to turn a motor on and off. The input signal is TTL/microprocessor compatible.

Controlling the direction of the motor is only a little more difficult. This requires a double-pole, double-throw (DPDT) relay, wired in series after the on/off relay just described (see Fig. 20-6; refer to the parts list in Table 20-2). With the contacts in the relay in one position, the motor turns clockwise. Activate the relay, and the contacts change positions, turning the motor counterclockwise. Again, you can easily control the direction relay with digital signals. Logical 0 makes the motor turn in one direction (let's say forward), and logical 1 makes the motor turn in the other direction. Both on/off and direction relay controls are shown combined in Fig. 20-7.

You can quickly see how to control the operation and direction of a motor using just two data bits from a computer. Since most robot designs incorporate two drive motors, you can control the movement and direction of your robot with just four data bits. When selecting relays, make sure the contacts are rated for the motors you are using. All relays carry contact ratings, and they vary from a low of about 0.5 A to over 10 A, at 125 V. Higher-capacity relays are larger and may require bigger transistors to trigger them (the very small reed relays can often be triggered by digital control without adding the transistor). For most applications, you don't need a relay rated higher than 2 or 3 A.

TABLE 20-1 Parts List for On/Off Relay Control

RL1	SPDT relay, 5 V coil, contacts rated 2 A or more
Q1	2N2222 NPN transistor
R1	1K resistor
D1	1N4003 diode

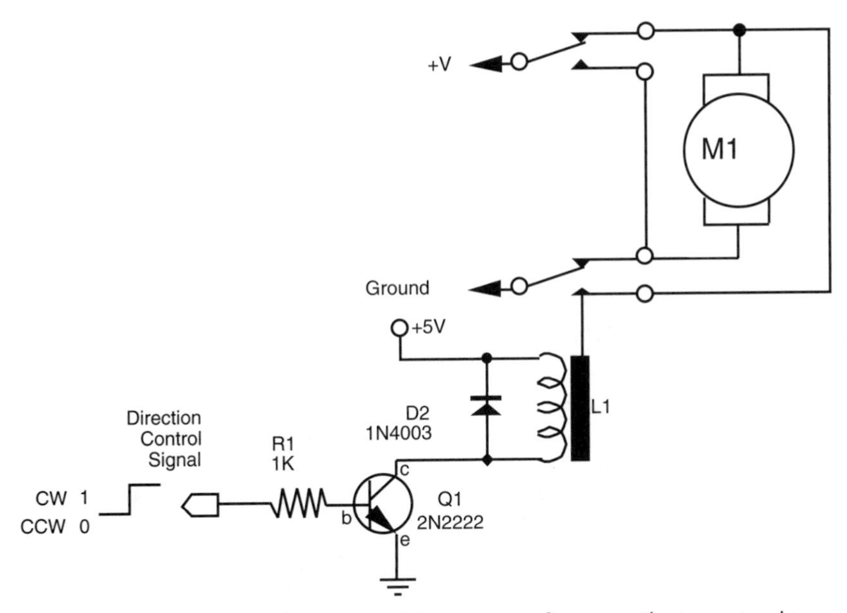

FIGURE 20-6 Using a relay to control the direction of a motor. The input signal is TTL/microprocessor compatible.

20.3.2 BIPOLAR TRANSISTOR CONTROL

Bipolar transistors provide true solid-state control of motors. For the purpose of motor control, you use the bipolar transistor as a simple switch. By the way, note that when a *transistor* is referred to in this section it is a bipolar transistor. There are many kinds of transistors you can use, including the field effect transistor, or FET. In fact, FETs will be discussed in the next section.

There are two common ways to implement the transistor control of motors. One way is shown in Fig. 20-8 (see the parts list in Table 20-3). Here, two transistors do all the work. The motor is connected so that when one transistor is switched on, the shaft turns clockwise. When the other transistor is turned on, the shaft turns counterclockwise. When both transistors are off, the motor stops turning. Notice that this setup requires a dual-polarity

TABLE 20-2 Parts List for Direction Relay Control

RL1	DPDT relay, 5V coil, contacts rated 2 A or more
Q1	2N2222 NPN transistor
R1	1K resistor
D1	1N4003 diode

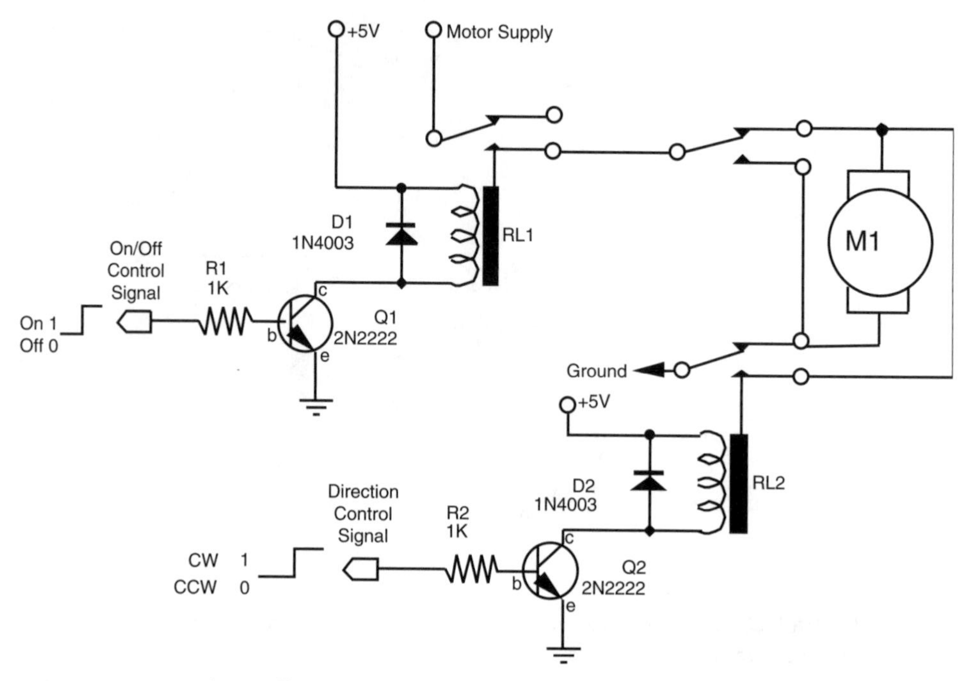

FIGURE 20-7 Both on/off and direction relay controls in one.

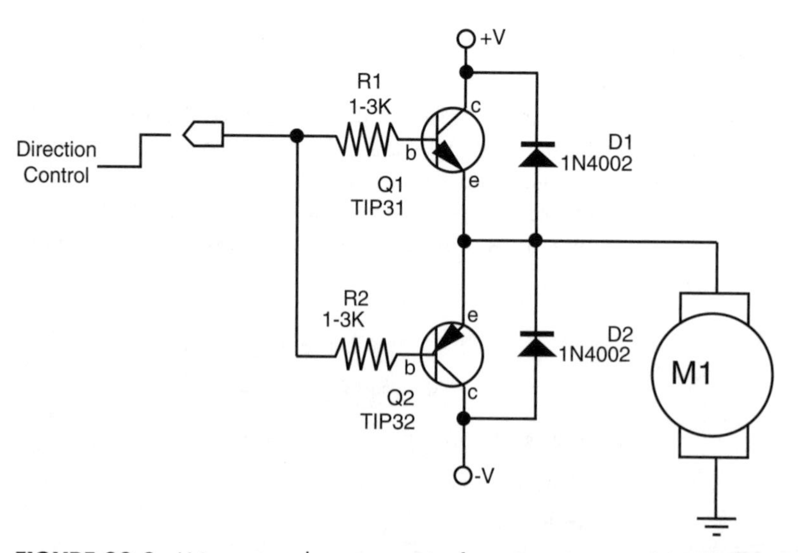

FIGURE 20-8 Using a complementary pair of transistors to control the direction of a motor. Note the double-ended (+ and –) power supply.

TABLE 20-3 Parts List for Two-Transistor Motor Direction Control

Q1	TIP31 NPN power transistor
Q2	TIP32 PNP power transistor
R1, R2	1–3K resistor
D1, D2	1N4002 diode
Misc.	Heat sinks for transistors

power supply. The schematic calls for a 6-V motor and a 6-V and 6-V power source. This is known as a split power supply.

Perhaps the most common way to control DC motors is to use the H-bridge network, as shown in Fig. 20-9 (see the parts list in Table 20-4). The figure shows a simplified H-bridge; some designs get quite complicated. However, this one will do for most hobby robot applications. The H-bridge is wired in such a way that only two transistors are on at a time. When transistors 1 and 4 are on, the motor turns in one direction. When transistors 2 and 3 are on, the motor spins the other way. When all transistors are off, the motor remains still.

Note that the resistor is used to bias the base of each transistor. These are necessary to prevent the transistor from pulling excessive current from the gate controlling it (computer port, logic gate, whatever). Without the resistor, the gate would overheat and be

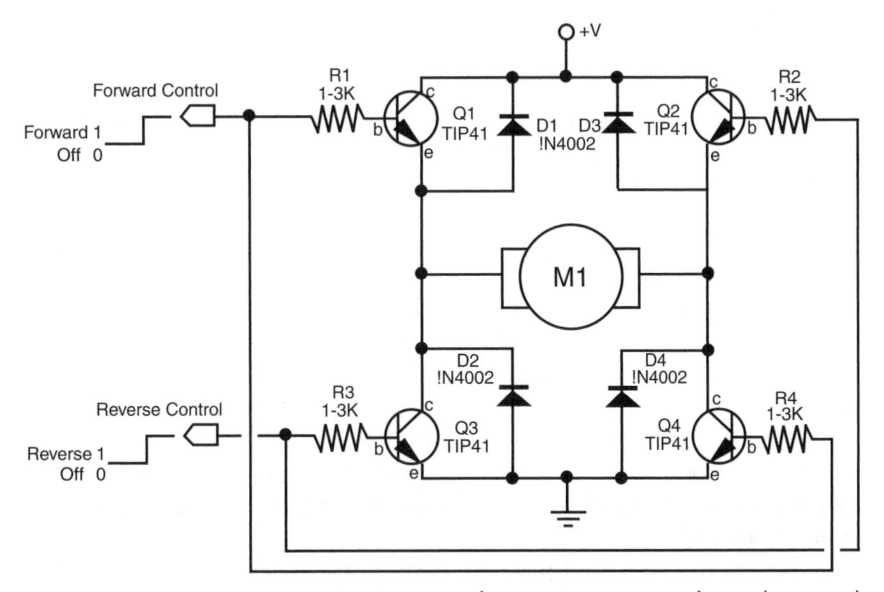

FIGURE 20-9 Four NPN transistors connected in an "H" pattern can be used to control the direction of a motor. The power supply is single ended.

TABLE 20-4 Parts List for H-Bridge Bipolar Transistor Motor Direction Control

Q1–Q4	TIP41 NPN power transistor
R1–R4	1–3K resistor
D1–D4	1N4002 diode
Misc.	Heat sinks for transistors

destroyed. The actual value of the bias resistor depends on the voltage and current draw of the motor, as well as the characteristics of the particular transistors used. For ballpark computations, the resistor is usually in the 1K- to 3K-Ω range. You can calculate the exact value of the resistor using Ohm's law, taking into account the gain and current output of the transistor, or you can experiment until you find a resistor value that works. Start high and work down, noting when the controlling electronics seem to get too hot. Don't go below 1K.

The transistors you choose should comply with some general guidelines. First, they must be capable of handling the current draw demanded by the motors, but which specific transistor you finally choose will largely depend on your application and your design preference. Most large drive motors draw about 1 to 3 A continuously, so the transistors you choose should be able to handle this. This immediately rules out the small signal transistors, which are rated for no more than a few hundred milliamps.

A good NPN transistor for medium-duty applications is the TIP31, which comes in a TO-220 style case. Its PNP counterpart is the TIP32. Both of these transistors are universally available. Use them with suitable heat sinks. For high-power applications, the NPN transistor that's almost universally used is the 2N3055 (get the version in the TO-3 case; it handles more power). Its close PNP counterpart is the MJ2955 (or 2N2955). Both transistors can handle up to 10 A (115 W) when used with a heat sink, such as the one in Fig. 20-10.

Another popular transistor to use in H-bridges is the TIP120, which is known as a Darlington transistor. Internally, it's actually two transistors: a smaller "booster" transistor and a larger power transistor. The TIP120 is preferred because it's often easier to interface it with control electronics. Some transistors, like the 2N3055, require a hefty amount of current in order to switch, and not all computer ports can supply this current. If you're not using a Darlington like the TIP120, it's sometimes necessary to use small-signal transistors (the 2N2222 is common) between the computer port and the power transistor.

The driving transistors should be located off the main circuit board—ideally directly on a large heat sink or at least on a heavy board with clip-on or bolt-on heat sinks attached to the transistors (as in Fig. 20-10). Use the proper mounting hardware when attaching transistors to heat sinks.

Remember that with most power transistors, the case is the collector terminal. This is particularly important when there is more than one transistor on a common heat sink and they aren't supposed to have their collectors connected together. It's also important when

FIGURE 20-10 Power transistors mounted on a heat sink.

that heat sink is connected to the grounded metal frame of the robot. You can avoid any extra hassle by using the insulating washer provided in most transistor mounting kits.

The power leads from the battery and to the motor should be 12- to 16-gauge wire. Use solder lugs or crimp-on connectors to attach the wire to the terminals of T0-3-style transistors. Don't tap off power from the electronics for the driver transistors; get it directly from the battery or main power distribution rail. See Chapter 17, "All about Batteries and Robot Power Supplies," for more detail about robot power distribution systems.

20.3.3 POWER MOSFET CONTROL

Wouldn't it be nice if you could use a transistor without bothering with current limiting resistors? Well, you can as long as you use a special brand of transistor, the power MOSFET. The MOSFET part stands for "metal oxide semiconductor field effect transistor." The "power" part means you can use it for motor control without worrying about it or the controlling circuitry going up in smoke.

Physically, MOSFETs look a lot like bipolar transistors, but there are a few important differences. First, like CMOS ICs, it is entirely possible to damage a MOSFET device by zapping it with static electricity. When handling it, always keep the protective foam around the terminals. Further, the names of the terminals are different from transistors. Instead of base, emitter, and collector, MOSFETs have a gate, source, and drain. You can easily damage a MOSFET by connecting it in the circuit improperly. Always refer to the pinout diagram before wiring the circuit, and double-check your work.

When MOSFETs are on, the resistance between the drain and source (usually referred to as "R_{DS}" in data sheets) is given as a parameter. The lower the drain and source resistance, the higher performing the transistor. This resistance represents power loss and the lower the R_{DS} the lower the power loss in the MOSFETS and the smaller a heat sink that can be used with them.

A commonly available power MOSFET is the IRF-5XX series (such as the IRF-520, IRF-530, etc.) from International Rectifier, one of the world's leading manufacturers of power MOSFET components. These N-channel MOSFETs come in a T0-220-style transistor case and can control several amps of current (when on a suitable heat sink). A current limiting circuit that uses MOSFETs is shown in Fig. 20-11 (see the parts list in Table 20-5). Note the similarity between this design and the transistor design on Fig. 20-9.

An even better H-bridge with power MOSFETs uses two N-channel MOSFETs for the low side of the bridge and two complementary P-channel MOSFETs for the high side. The use of complementary MOSFETs allows all four transistors in the H-bridge to turn completely on, thereby supplying the motor with full voltage. Fig. 20-12 shows a revised schematic (refer to the parts list in Table 20-6).

In both circuits, logic gates provide positive-action control. When the control signal is LOW, the motor turns clockwise. When the control signal is HIGH, the motor turns counterclockwise.

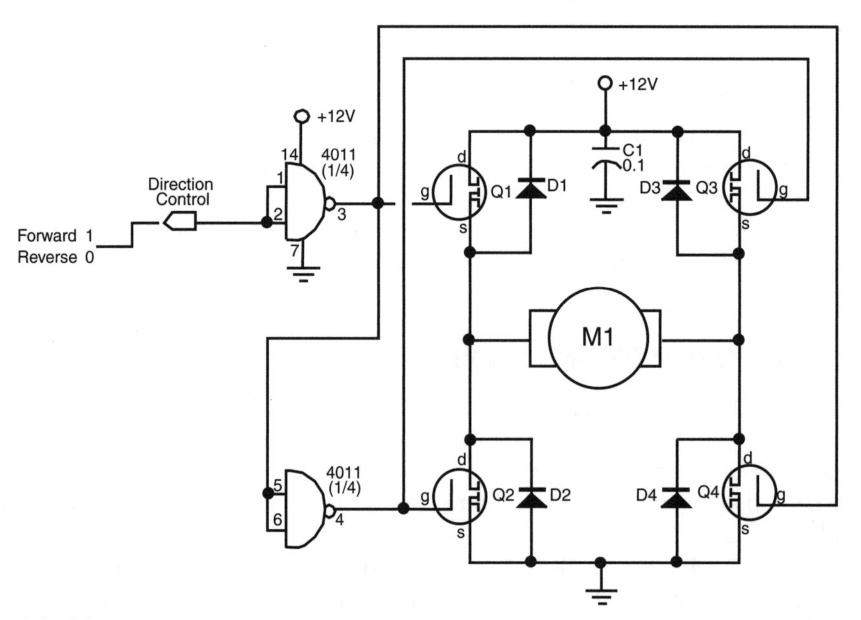

FIGURE 20-11 Four N-channel power MOSFET transistors in an "H" pattern can be used to control the direction of a motor. In a circuit application such as this, MOSFET devices do not strictly require current limiting resistors, as do standard transistors.

TABLE 20-5 Parts List for N-Channel Power MOSFET Motor Control Bridge

IC1	4011 CMOS Quad NAND Gate IC
Q1–Q4	IRF-5XX series (e.g., IRF-530 or equiv.) N-channel power MOSFET
D1–D4	1N4002 diode
Misc.	Heat sinks for transistors

20.3.4 MOTOR BRIDGE CONTROL

The control of motors is big business, and it shouldn't come as a surprise that dozens of companies offer all-in-one solutions for controlling motors through fully electronic means. These products range from inexpensive $2 integrated circuits to sophisticated modules costing tens of thousands of dollars. Of course, the discussion will be confined to the low end of this scale!

The basic motor control is an H-bridge, as discussed earlier—an all-in-one integrated circuit package. Bridges for high-current motors tend to be physically large, and they may come with heat fins or have connections to a heat sink. A good example of a motor bridge is the Allegro Microsystems 3952, which provides in one single package a much improved version of the circuit shown in Fig. 20-13.

Motor control bridges have two or more pins on them for connection to control electronics. Typical functions for the pins are:

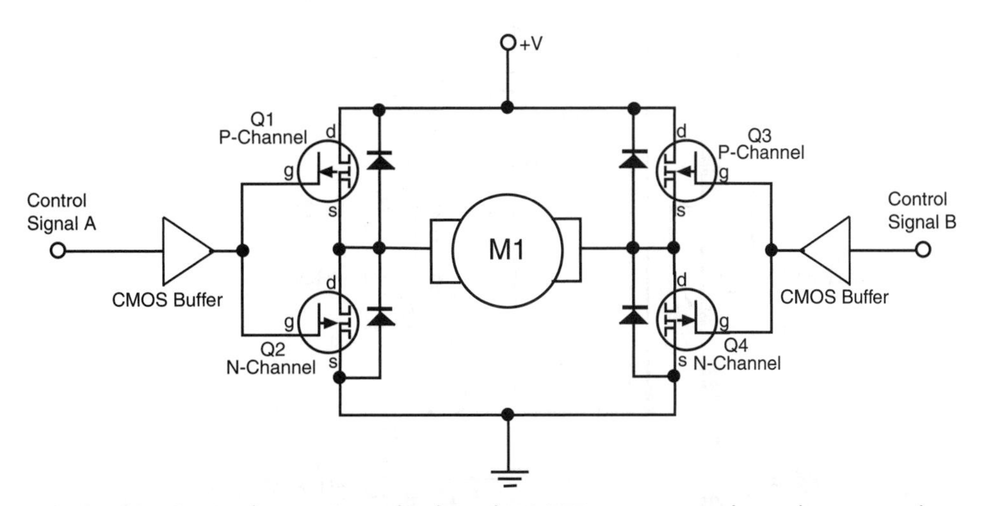

FIGURE 20-12 Combination N- and P-channel MOSFET transistors can be used to increase the voltage flowing to the motors. The MOSFETs should be "complementary pairs" (made to work with one another) that share the same voltage and current ratings. Most makers of MOSFET transistors provide complementary N- and P-channel products.

TABLE 20-6 Parts List for N-Channel Power MOSFET Motor Control Bridge

IC1	4011 CMOS Quad NAND Gate IC
Q1, Q3	IRF-9530 (or equiv.) P-channel power MOSFET
Q2, Q4	IRF-5XX series (e.g., IRF-530 or equiv.) N-channel power MOSFET
D1–D4	1N4002 diode
Misc.	Heat sinks for transistors

- *Motor enable.* When enabled, the motor turns on. When disabled, the motor turns off. Some bridges let the motor float when disabled; that is, the motor coasts to a stop. On other bridges, disabling the motor causes a full or partial short across the motor terminals, which acts as a brake to stop the motor very quickly.
- *Direction.* Setting the direction pin changes the direction of the motor.
- *Brake.* On bridges that allow the motor to float when the enable pin is disengaged, a separate brake input is used to specifically control the braking action of the motor.
- *PWM.* Most H-bridge motor control ICs are used not only to control the direction and power of the motor, but its speed as well. The typical means used to vary the speed of a motor is with pulse width modulation, or PWM. This topic is described more fully in the next section.

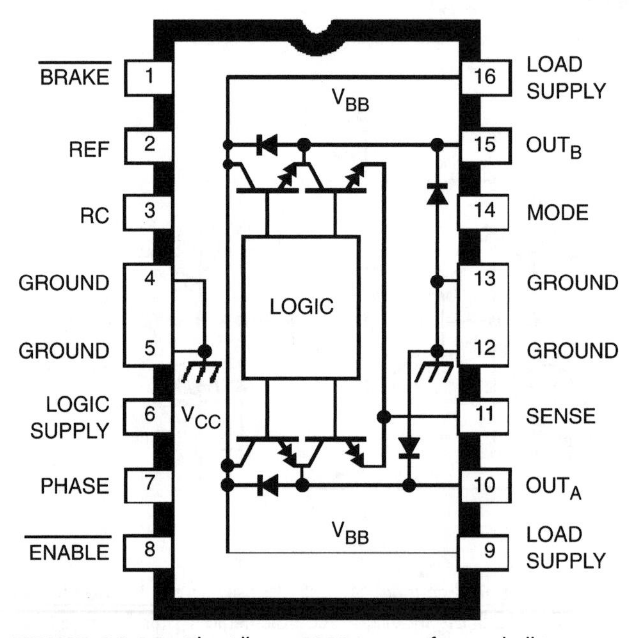

FIGURE 20-13 The Allegro 3952 is one of several all-in-one H-bridge motor control ICs.

The better motor control bridges incorporate overcurrent protection circuitry, which prevents them from being damaged if the motor pulls too much current and overheats the chip. Some even provide for current sense, an output that can be fed back to the control electronics in order to monitor the amount of current being drawn from the motor. This can be useful when you need to determine if the robot has become stuck and the motors have stalled. Recall from earlier in this chapter that DC motors will draw the most current when they are stalled. If the robot gets caught on something and can't budge, the motors will stop, and the current draw will increase.

Some of the available motor control bridges include the L293D and L298N from SGSThomson; the 754410, an improved version of the L293 from Texas Instruments; and the LM18293 from National Semiconductor. Be sure to also check the listings in Appendix B, "Sources," and Appendix C, "Robot Information on the Internet," for additional sources of information on motor control bridges.

20.3.5 RELAY VERSUS BIPOLAR VERSUS FET MOTOR DRIVERS

When making the decision on what type of motor driver to use in a robot, you should consider the advantages of the different types that have been described so far. Some of the comments may be surprising and will hopefully give you some things to think about when you are designing your own robot.

Relay motor control is probably the least efficient and least reliable. Relays are quite costly and large (especially compared to power bipolar and MOSFET transistors) and will fail after a certain amount of time. To make it easier for you, remember to always socket your relays and provide easy access in your robot so that they can be replaced easily. Where relays have an advantage over the transistor motor controllers/drivers is their low motor power loss and voltage drop. It may seem to be a paradox, but relays are often the most effective method of control in small robots that run from radio battery power supplies.

Bipolar transistors are probably the cheapest solution but they can be very inefficient in terms of power usage. Remember that there is a definite voltage drop through the transistor, which translates to a power loss (which is the product of the current flowing through the motor and the voltage drop), which can be very significant in low-voltage situations.

MOSFET transistors tend to be the most efficient type of motor driver in terms of power loss, but they can be fairly expensive depending on the transistors that are chosen for the application. They must always be handled carefully to ensure they are not damaged by ESD. Despite the limitations, the high efficiency of MOSFETs makes them the preferred motor driver for DC motors.

20.4 Motor Speed Control

There will be plenty of times when you'll want the motors in your robot to go a little slower, or perhaps track at a predefined speed. Speed control with continuous DC motors is a science in its own, and there are literally dozens of ways to do it. Some of the more popular methods will be covered in this and later chapters.

20.4.1 NOT THE WAY TO DO IT

Before exploring the right ways to control the speed of motors, it is important to understand how *not* to do it. Many robot experimenters first attempt to vary the speed of a motor by using a potentiometer. While this scheme certainly works, it wastes a lot of energy. Turning up the resistance of the pot decreases the speed of the motor, but it also causes excess current to flow through the pot. That current creates heat and draws off precious battery power.

Another similar approach is shown in Fig. 20-14. Here, the potentiometer controls the base current to a bipolar transistor controlling the operation of the motor. Like the previous case, excess current flows through the transistor, and the energy is dissipated as lost heat. There are, fortunately, far better ways of doing it. Read on.

20.4.2 BASIC PWM SPEED CONTROL

The best way to control the speed of a DC motor is to chop the power being passed to it into different sized chunks. The most common way of doing this is to set up a recurring signal; the time for power to be passed to the motor should be a set fraction of this period. This method is known as *pulse width modulation* or PWM motor control and its basics can be seen in Fig. 20-15. The time the power signal is active is known as the *pulse width,* and the *duty cycle* is the percentage of the pulse width as part of the PWM signal's period.

One issue that you should always be aware of is that when a motor's speed is controlled using a PWM, it will vibrate at the speed of the PWM (due to the coils being turned on and off according to the speed control), which could result in an annoying hum. To avoid this, you should run your PWM either at subaudible (60 Hz or less) or superaudible (18 kHz or more) frequencies. Subaudible frequencies are best for small motors that are usually found in toys, while superaudible frequencies should be used for high-performance or efficiency motors (basically anything that wouldn't be considered in a toy).

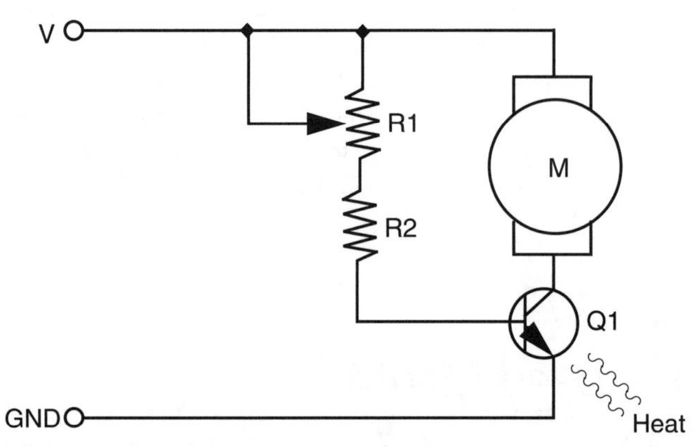

FIGURE 20-14 How not to vary the speed of a motor. This approach is very inefficient as the voltage drop through the transistor along with the current through the motor and transistor will cause a lot of power (heat) to be dissipated.

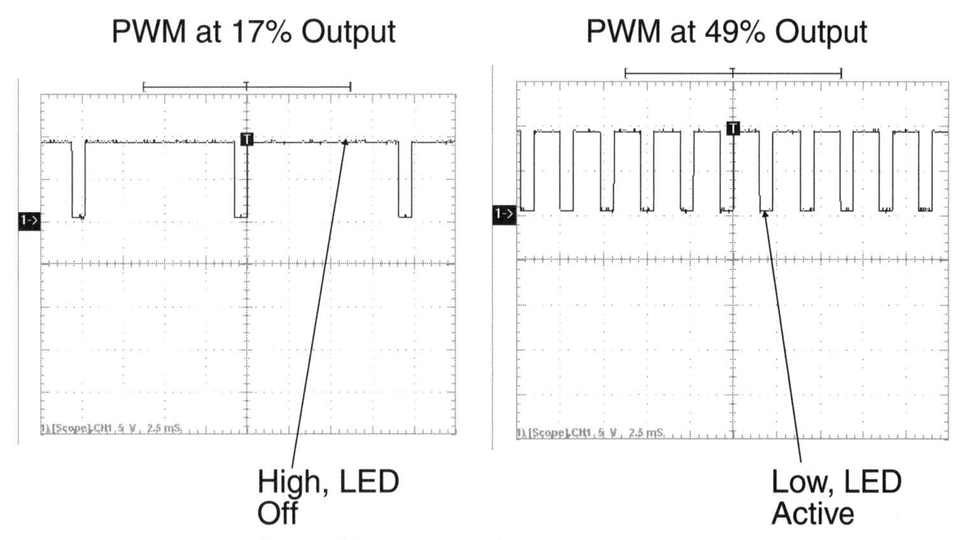

FIGURE 20-15 BASIC features of a PWM signal.

It is interesting to note that the actual power delivered to the motors follows a square law with respect to the duty cycle instead of the expected linear relationship. The reason for power to drop off as the square of the duty cycle can be seen from going back to the DC electricity power equation:

$$\text{Power} = V \times i = V^2/R$$

By halving the time the voltage is applied (a 50 percent duty cycle), the actual power available to the motor is only one-quarter of that available when the duty cycle is 100 percent (or full on). As noted previously, throttling the PWM is not a linear process. When you are setting up your robot, you will have to make sure that the actual power-to-duty cycle relationship is understood as you will discover that if a linear range is used your robot will probably only be active when the PWM duty cycle is 50 percent or more.

Creating a PWM signal is not as easy as you would expect. In the previous versions of this book, a PWM generator circuit was built from a relaxation oscillator, not understanding that while the pulse width would change, so would the period, resulting in a constant PWM duty cycle for a varying frequency. What is desired is a varying PWM pulse width for a constant frequency.

The first demonstration circuit (Fig. 20-16 with parts list in Table 20-7) uses a 555 to allow a potentiometer to brighten or dim an LED.

You will find that the circuit works reasonably well, although it is impossible to completely turn off the LED. This is due to the 555's inability to provide less than a 50 percent duty cycle signal. In Fig. 20-17, the operating signals are shown for different potentiometer settings. The left side of the oscilloscope trace shows the LED dimmed, a 17 percent duty cycle (the time the signal is low, which is when current can flow through the LED) versus the LED at its brightest, a 49 percent duty cycle.

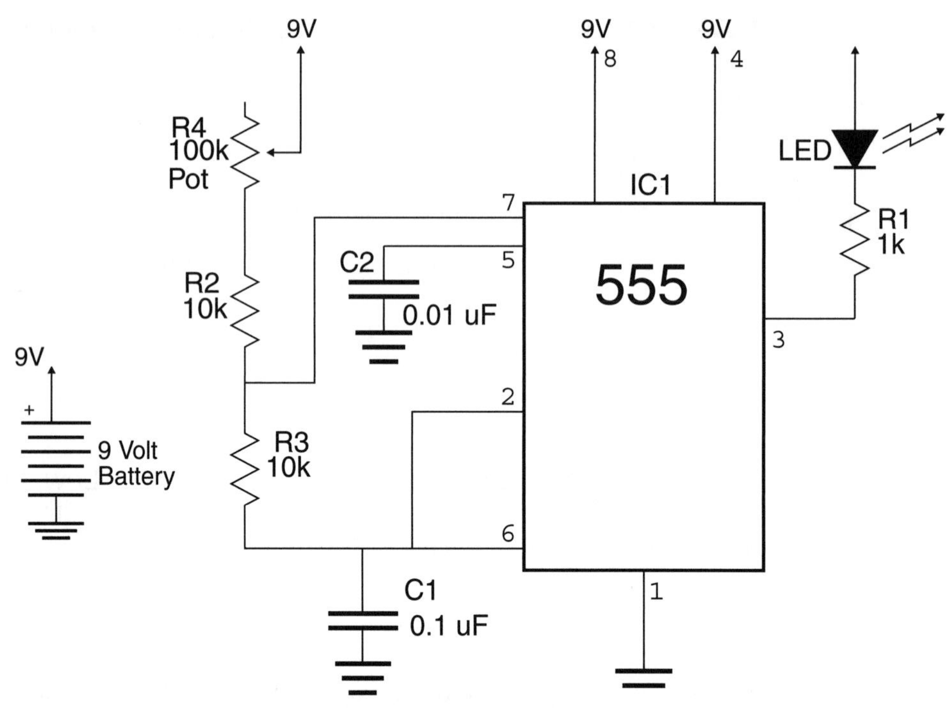

FIGURE 20-16 555-based PWM generator circuit, which can be used to dim an LED.

The oscilloscope pictures in Fig. 20-15 probably do not look like what you would expect for two reasons. First, the LED was turned on when the 555 output was low. This allowed the greatest dynamic duty cycle range, which provided the most visible changes in the LED. Second, the PWM pulse width remained constant while the frequency changed: this is a function of the operation of the 555 timer and must be taken into account if this chip is used as a PWM generator.

TABLE 20-7 Parts List for 555 PWM Circuit

IC1	555 timer chip
LED	Visible light LED
R1	1k resistor
R2, R3	10k resistors
R4	100k potentiometer
C1	0.1 μF capacitor
C2	0.01 μF capacitor
Misc.	Breadboard, breadboard wiring, 9-V battery clip, 9-V battery

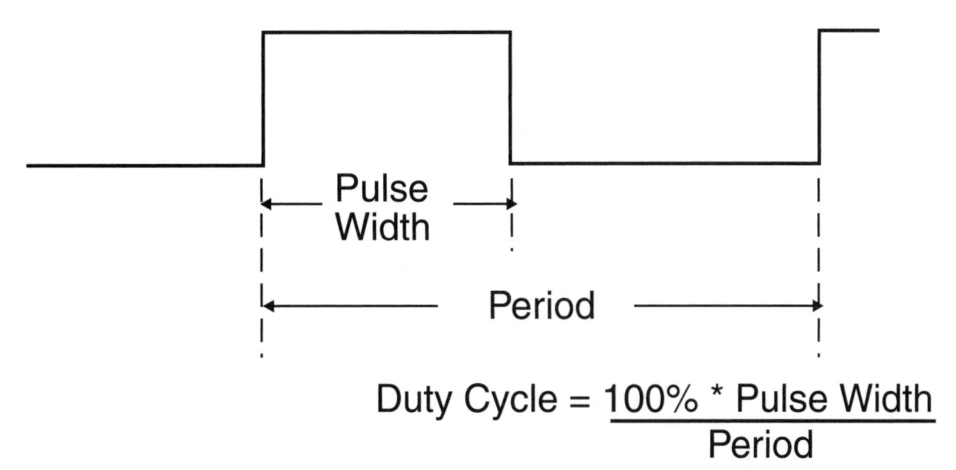

FIGURE 20-17 *555 PWM circuit at the dimmest LED setting and the brightest. Note that when the signal is low, the LED is on and when it is high the LED is off. This PWM circuit keeps the pulse width constant while varying the PWM period.*

While the 555-based PWM generator does not look all that acceptable as a PWM generator, it can be used successfully for motor control. R4's resistance can be changed digitally using a selection of resistors that are allowed to pass current using transistors (as is done in the IBM PC gamepad/joystick interface discussed in Chapter 16). If it is to be used, you may have to switch between using the high and low outputs of the 555 to control the motor drivers, depending on the application and the characteristics of the motor.

The BS2 has a PWM function, which can be used to produce a PWM on each of its pins to control a motor. Fig. 20-18 shows the circuit (Table 20-8 lists the parts) for a BS2 PWM control of the brightness of an LED. The software used for the application is:

```
'  BS2 PWM - Demonstrate BS2 "PWM" Function Operation
'
'  This Program dims then brightens an LED.
'
'  myke predko
'
'  05.09.11

'{$STAMP BS2}
'{$PBASIC 2.5}

'  I/O Pins
LED PIN 14

'  Variables
i         VAR Word

'  Mainline
   DO
```

TABLE 20-8 Parts List for Using A BS2 to Vary the Brightness of an LED

BS2	Parallax BASIC Stamp 2
LED1	Visible light LED
R1	470 ω resistor
Misc.	BS2 communications/programming interface, breadboard, breadboard wiring, 4x AA battery clip, 4x AA alkaline batteries

```
        FOR i = 0 TO 254          '  Dim LED with PWM from
            PWM LED, i, 5         '    100% PWM to 0.4%
        NEXT
        DO WHILE (i <> 1)         '  Brighten LED with PWM
            PWM LED, i, 4
            i = i - 1
        LOOP
    LOOP        '  Repeat
```

The BS2 PWM function is a very effective way to control a motor's speed, and it can also be used to output an analog voltage using a resistor and capacitor, as is shown in the Parallax PBASIC reference manuals. The problem with the PWM function is that it does not run in the background, which is to say that the BS2 cannot do other operations (such as running more than one motor or polling sensors) while the PWM is active. In the next section, a circuit that will allow the BS2 to specify the PWM duty cycle for multiple motors as well as execute other tasks will be presented.

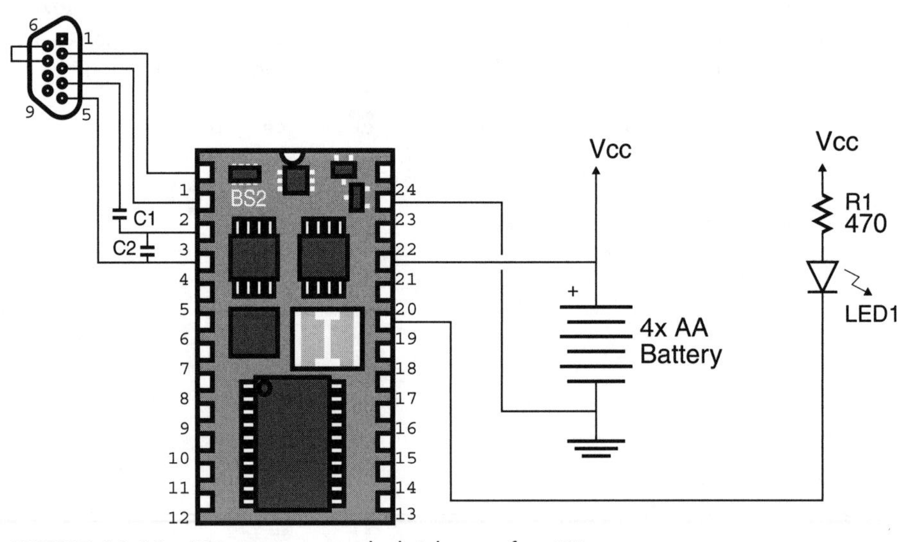

FIGURE 20-18 BS2 circuit to vary the brightness of an LED.

20.4.3 COUNTER-BASED PWM SPEED CONTROL

As mentioned in the previous section, the ideal motor speed control is a PWM in which the period stays constant and the pulse width (and therefore the duty cycle) is variable. Ideally the pulse width would be controlled by a digital value rather than an analog potentiometer or other components, which are difficult to interface digital signals and should not require constant monitoring by the robot's controller. The PWM controller described in this section should meet these criteria and use parts that are readily available.

The PWM controller circuit could be blocked out using the diagram in Fig. 20-19. A clock signal is used to increment a four-bit counter and the output of the counter is continually compared against an expected value. When the counter value is less than the expected value, the output from the digital magnitude comparator is a logic high and when the counter output is equal to or less than the compare value the comparator output is a logic low. When the counter overflows, the process starts over.

The circuit that implements this function is shown in Fig. 20-20 with the parts listed in Table 20-9.

This circuit can be replicated surprisingly easily for controlling multiple motors. You do not have to replicate the entire circuit, you just have to add additional 74LS85 chips for each motor to be controlled. The clock oscillator and counter can be common for multiple PWM circuits. The controller must dedicate four digital outputs for each motor controller for the PWM control as well as any specific outputs for motor direction control. Using this circuit to control two motors, a minimum of 10 lines of the microcontroller must be dedicated to it.

As with any counter-based PWM, the counter clock frequency must be the maximum number value of the counter multiplied by the desired PWM frequency. For the circuit shown in Fig. 20-20, the clock period is 10 kHz; the actual PWM frequency is approxi-

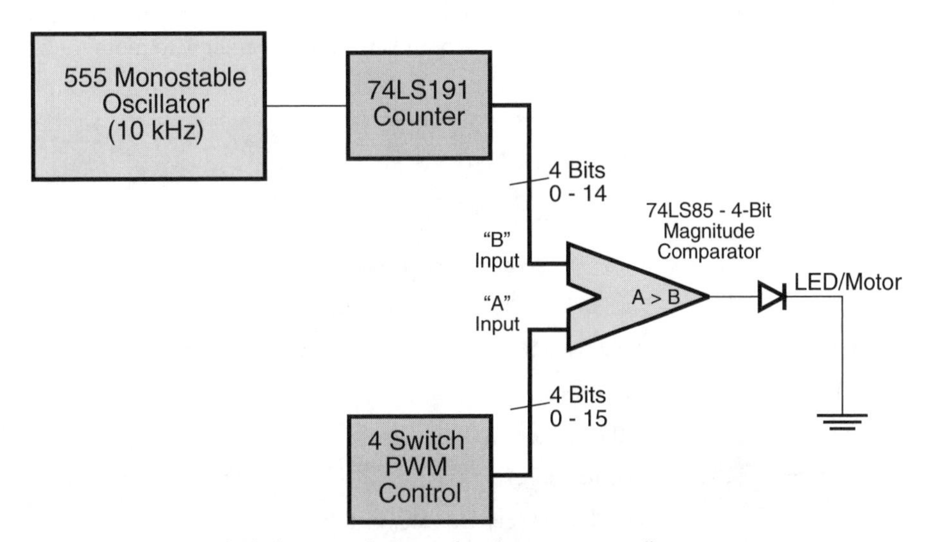

FIGURE 20-19 Block diagram of a basic four-bit PWM controller.

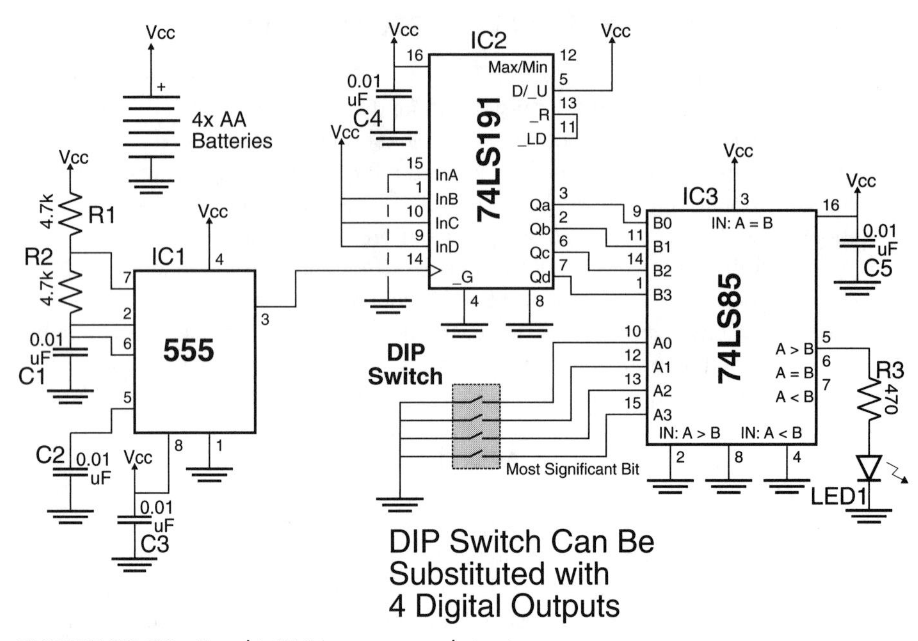

FIGURE 20-20 Four-bit PWM motor control circuit.

mately 667 Hz (10 kHz divided by 15), which is very audible. For a 20 kHz PWM, the PWM clock would have to be 300 kHz. Similarly for a 30 Hz (subaudible) PWM, the PWM clock would have to be 450 Hz.

When more bits are used for the PWM, the clock frequency can become very high. Fortunately for most robot motor applications a four-bit PWM provides adequate resolution.

Finally, this circuit is available in a number of microcontrollers with built-in PWM outputs. Unfortunately, there are very few microcontrollers on the market with more than one built-

TABLE 20-9 Parts List for Four BIT PWM Motor Control Circuit

IC1	555 timer chip
IC2	74LS191 four-bit binary counter chip
IC3	74LS85 four-bit magnitude comparator chip
LED1	Visible light LED
R1–R2	4.7k resistors
R3	470 ω resistor
C1–C5	0.01 μF capacitor
DIP Switch	Breadboard mountable four-bit DIP switch
Misc.	Breadboard, breadboard wiring, 4x AA battery clip, 4x AA batteries

in PWM. Most only have one PWM generation circuit and must either rely on an external circuit, similar to the one demonstrated in Fig. 20-20, or two microcontrollers must be used in the application (one to control each motor's PWM).

20.5 Odometry: Measuring Distance of Travel

Odometry is the science of measuring the movement of a mobile robot by keeping track of the number of revolutions each motor makes and in which direction. The amount of movement is a simple calculation: the number of revolutions multiplied by the diameter of the wheel times pi. As will be explained in this chapter, the odometry measurements are normally carried out using shaft encoders directly connected to the wheels and assume that they can accurately track the position of the robot by how much the wheels have turned.

20.5.1 ANATOMY OF A SHAFT ENCODER

The typical shaft encoder is a disc that has numerous holes or slots along its outside edge. An infrared LED is placed on one side of the disc so that its light shines through the holes. The number of holes or slots is not a consideration here, but for increased speed resolution, there should be as many holes around the outer edge of the disc as possible. An infrared-sensitive phototransistor is positioned directly opposite the LED (see Fig. 20-21) so that when the motor and disc turn, the holes pass the light intermittently. The result, as seen by the phototransistor, is a series of flashing light.

Instead of mounting the shaft encoders on the motor shafts, mount them on the wheel shafts (if they are different). The number of slots in the disc determines the maximum accuracy of the travel circuit. The more slots the disc has, the better the accuracy.

If the encoder disc has 50 slots around its circumference, that represents a minimum sensing angle of 7.2°. As the wheel rotates, it provides a signal to the counting circuit every 7.2°. Stated another way, if the robot is outfitted with a 7-in wheel (circumference = 21.98 in), the maximum travel resolution is approximately 0.44 linear inches. This figure was calculated by taking the circumference of the wheel and dividing it by the number of slots in the shaft encoder.

The outputs of the phototransistor are conditioned by Schmitt triggers. This smooths out the wave shape of the light pulses so only voltage inputs above or below a specific threshold are accepted (this helps prevent spurious triggers). The output of the triggers is applied to the control circuitry of the robot.

20.5.2 THE DISTANCE COUNTER

The pulses from a shaft encoder do not in themselves carry distance measurement. The pulses must be counted and the count converted to distance. Counting and conversion are ideal tasks for a computer. Most single-chip computers and microprocessors, or their interface adapters, are equipped with counters. If your robot lacks a computer or microproces-

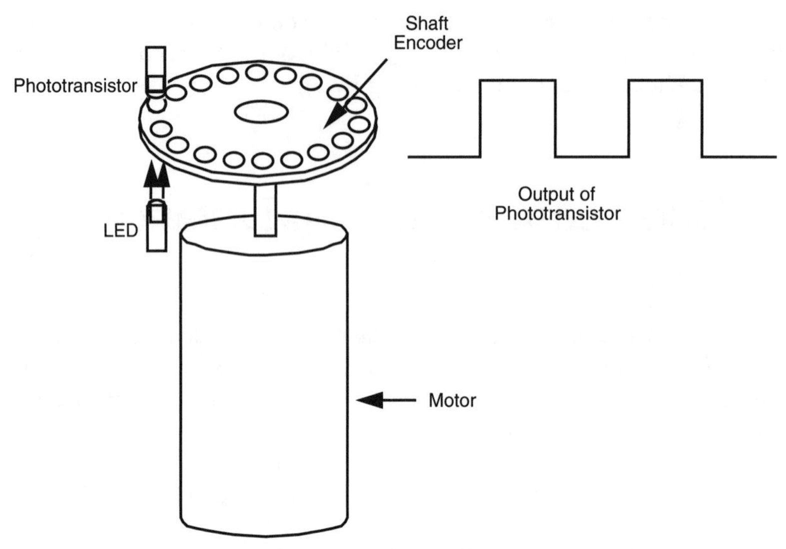

FIGURE 20-21 An optical shaft encoder attached to a motor. Alternatively, you can place a series of reflective strips on a black disc and bounce the LED light into the phototransistor.

sor with a timer, you can add one using a 4040 12-stage binary ripple counter (see Fig. 20-22). This CMOS chip has 12 binary weighted outputs and can count to 4096. It would probably be easiest if you just used the first eight outputs to count to 256.

Any counter with a binary or BCD output can be used with a 7485 magnitude comparator. A pinout of this versatile chip is shown in Fig. 20-23, and a basic hookup diagram in Fig. 20-24. In operation, the chip will compare the binary weighted number at its A and B inputs. One of the three LEDs will then light up, depending on the result of the difference between the two numbers. In a practical circuit, you'd replace the DIP switches (in the dotted box) with a computer port.

You can cascade comparators to count to just about any number. If counting in BCD, three packages can be used to count to 999, which should be enough for most distance recording purposes. Using a disc with 25 slots in it and a 7-in drive wheel, the travel resolution is 0.84 linear inches. Therefore, the counter system will stop the robot within 0.84 in of the desired distance (allowing for coasting and slip between the wheels and ground) up to a maximum working range of 69.93 ft. You can increase the distance by building a counter with more BCD stages or decreasing the number of slots in the encoder disc.

20.5.3 MAKING THE SHAFT ENCODER

By far, the hardest part about odometry is making or adapting the shaft encoders. (You can also buy shaft encoders ready-made.) The shaft encoder you make may not have the fine resolution of a commercially made disc, which often has 256 or 360 slots in it, but the homemade versions will be more than adequate. You may even be able to find already

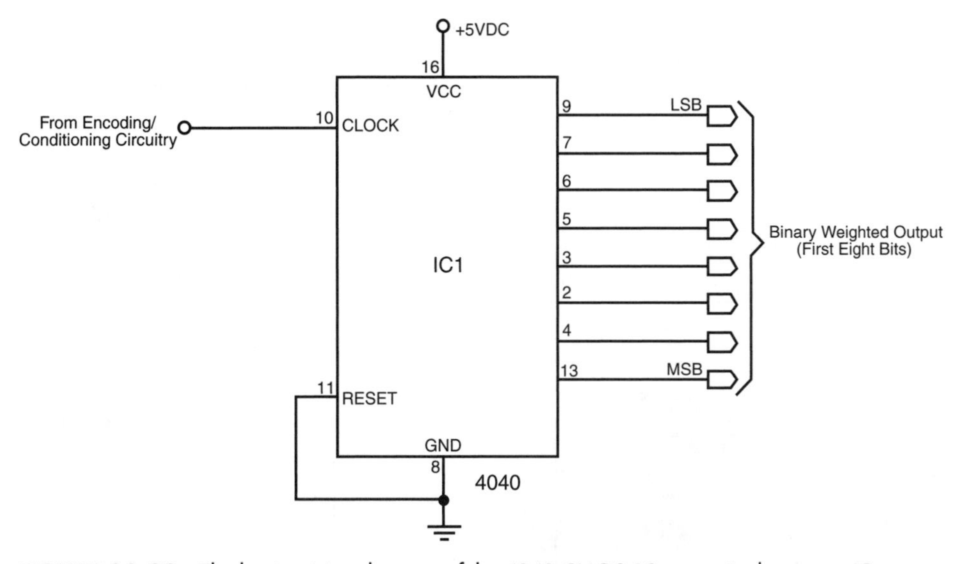

FIGURE 20-22 The basic wiring diagram of the 4040 CMOS 12-stage ripple counter IC.

machined parts that closely fit the bill, such as the encoder wheels in a discarded computer mouse. Fig. 20-25 shows the encoder wheels from a surplus $5 mouse. The mouse contains two encoders, one for each wheel of the robot.

You can also make your own shaft encoder by taking a 1- to 2-in disc of plastic or metal and drilling holes in it. Remember that the disc material must be opaque to infrared light. Some things that may look opaque to you may actually pass infrared light. When in doubt, add a coat or two of flat black or dark blue paint. That should block stray infrared light from reaching the phototransistor. Mark the disc for at least 20 holes, with a minimum size of

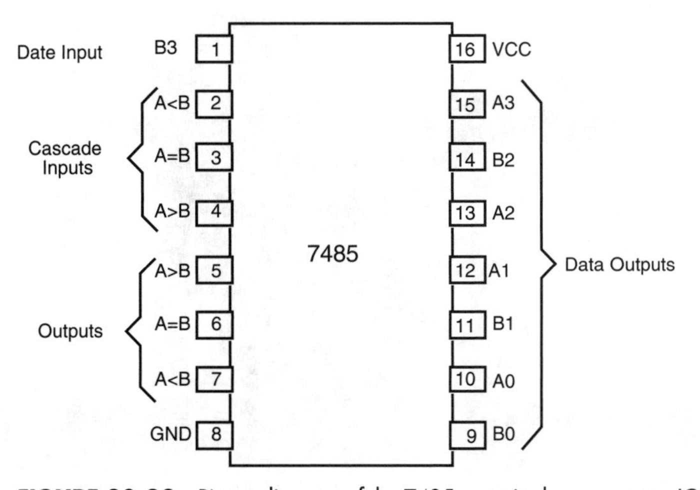

FIGURE 20-23 Pinout diagram of the 7485 magnitude comparator IC.

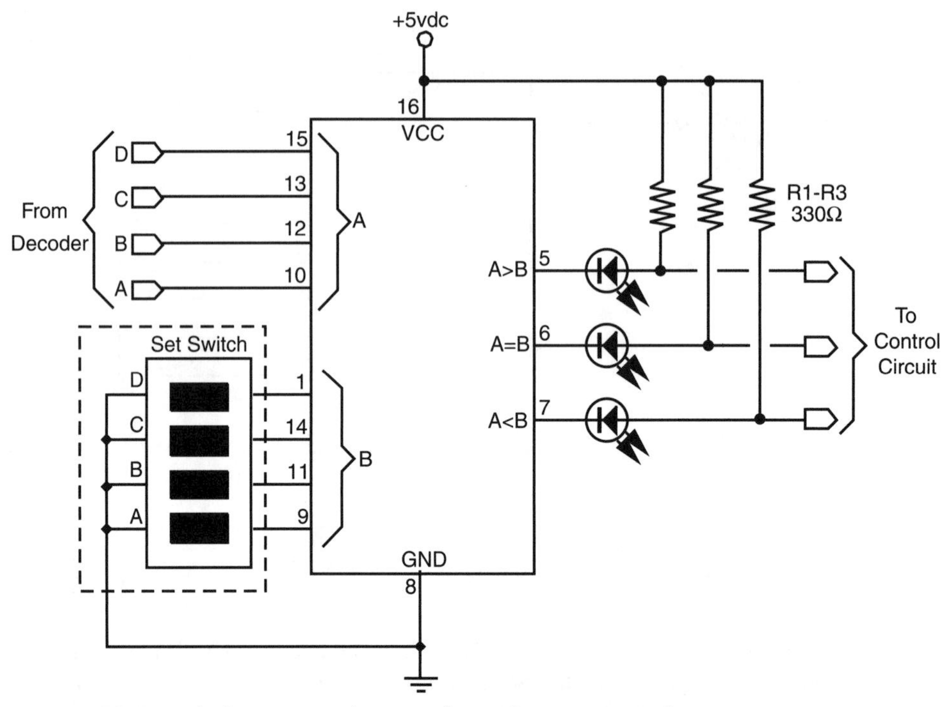

FIGURE 20-24 The basic wiring diagram of a single-state magnitude comparator circuit.

FIGURE 20-25 The typical PC mouse contains two shaft encoder discs. They are about perfect for the average small-or-medium-size robot.

about 1⁄16 in. The more holes the better. Use a compass to scribe an exact circle for drilling. The infrared light will only pass through holes that are on this scribe line.

20.5.4 MOUNTING THE HARDWARE

Secure the shaft encoder to the shaft of the drive motor or wheel. Using brackets, attach the LED so that it fits snugly on the back side of the disc. You can bend the lead of the LED a bit to line it up with the holes. Do the same for the phototransistor. You must mask the phototransistor so it doesn't pick up stray light or reflected light from the LED, as shown in Fig. 20-26. You can increase the effectiveness of the phototransistor placing an infrared filter (a dark red filter will do in a pinch) between the lens of the phototransistor and the disc. You can also use the type of phototransistor that has its own built-in infrared filter.

If you find that the circuit isn't sensitive enough, check whether stray light is hitting the phototransistor. Baffle it with a piece of black construction paper if necessary. Or, if you prefer, you can use a striped disc of alternating white and black spokes as well as a reflectance IR emitter and detector. Reflectance discs are best used when you can control or limit the amount of ambient light that falls on the detector.

20.5.5 QUADRATURE ENCODING

So far we've investigated shaft encoders that have just one output. This output pulses as the shaft encoder turns. By using two LEDs and phototransistors, positioned 90° out of phase (see Fig. 20-27), you can construct a system that not only tells you the amount of travel, but the direction as well. This can be useful if the wheels of your robot may slip. You can determine if the wheels are moving when they aren't supposed to be, and you can determine the

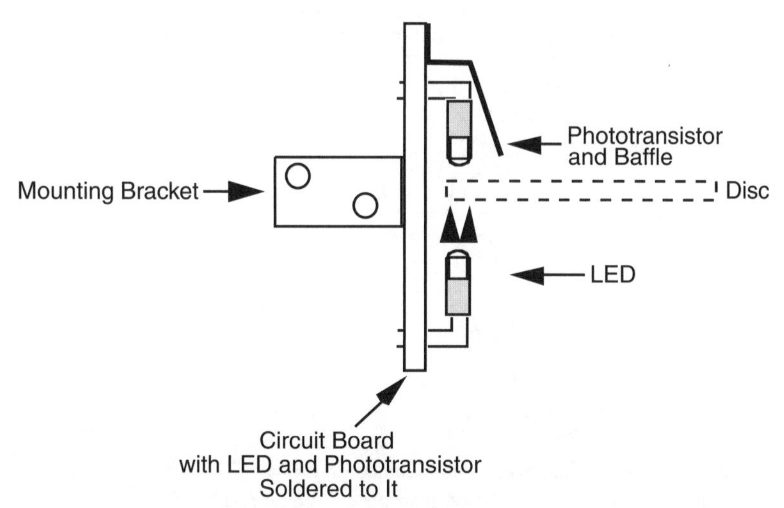

FIGURE 20-26 How to mount an infrared LED and phototransistor on a circuit board for use with an optical shaft encoder disc.

direction of travel. This so-called two-channel system uses quadrature encoding—the channels are out of phase by 90 degrees (one quarter of a circle).

Use the flip-flop circuit in Fig. 20-28 to separate the distance pulses from the direction pulses. Note that this circuit will only work when you are using quadrature encoding, where the pulses are in the following format:

```
off/off

on/off

on/on

off/on (… and repeat.)
```

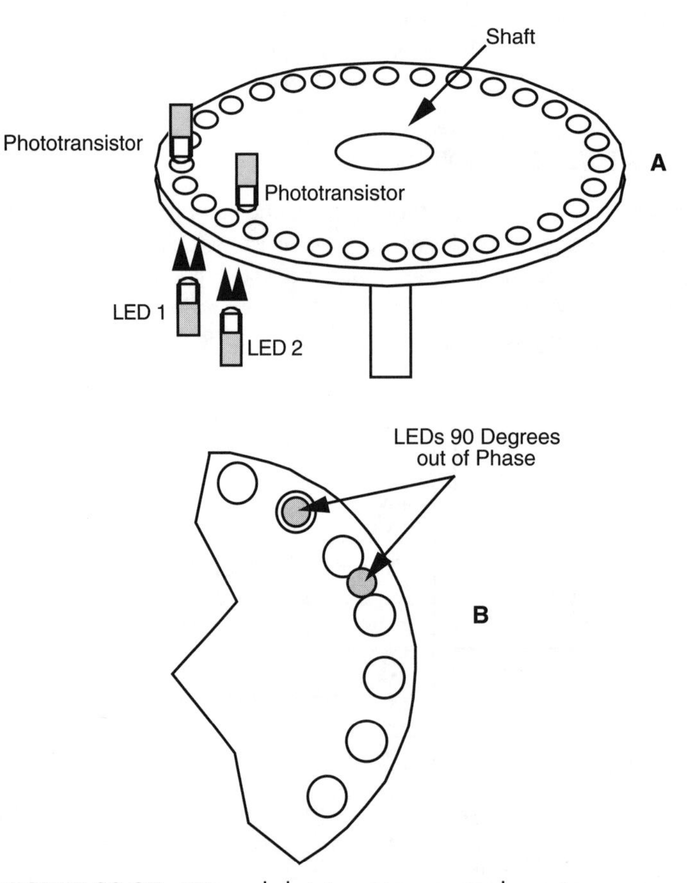

FIGURE 20-27 LEDs and phototransistors mounted on a two-channel optical disc. *a.* The LEDs and phototransistors can be placed anywhere about the circumference of the disc; *b.* the two LEDs and phototransistors must be 90° out of phase.

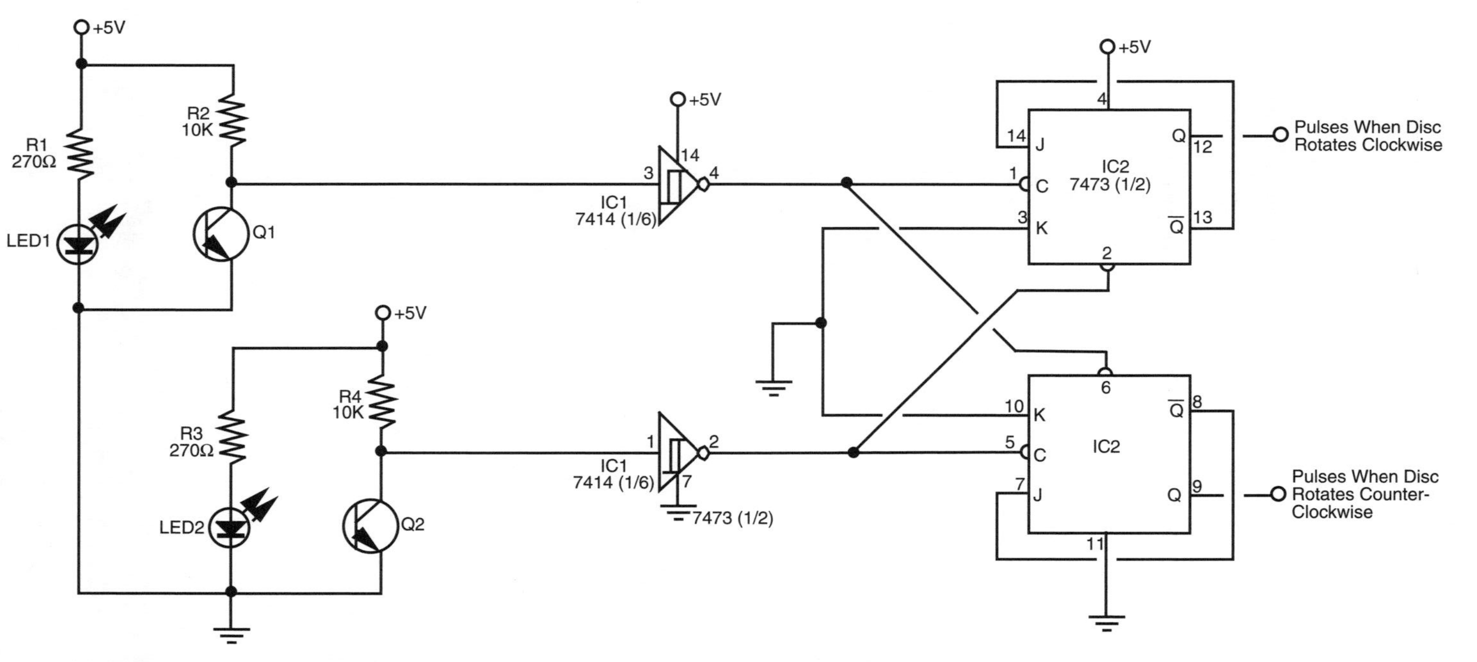

FIGURE 20-28 A two-channel shaft encoder circuit for use with a quadrature (also called two channel, 2-bit Gray code, or sine/cosine) encoder. One of the outputs of the flip-flop indicates distance (or relative speed) and the other the direction of rotation.

20.6 From Here

To learn more about . . .	***Read***
Selecting the right motors for your robot	Chapter 19, "Choosing the Right Motor for the Job"
Using stepper motors	Chapter 21, "Working with Stepper Motors"
Interfacing motors to computers and microcontrollers	Chapter 14, "Computer Peripherals"
More on odometry and measuring the distance of travel of a robot	Chapter 33, "Navigation"

CHAPTER 21

WORKING WITH STEPPER MOTORS

The past chapters have looked at powering robots using everyday continuous DC motors. DC motors are cheap, deliver a lot of torque for their size, and are easily adaptable to a variety of robot designs. By their nature, however, the common DC motors are rather imprecise. Without a servo feedback mechanism or tachometer, there's no telling how fast a DC motor is turning. Furthermore, it's difficult to command the motor to turn a specific number of revolutions, let alone a fraction of a revolution. Yet this is exactly the kind of precision robotics work, particularly arm designs, often requires.

Enter the stepper motor. Stepper motors are, in effect, DC motors with a twist. Instead of being powered by a continuous flow of current, as with regular DC motors, they are driven by pulses of electricity. Each pulse drives the shaft of the motor a little bit. The more pulses that are fed to the motor; the more the shaft turns. As such, stepper motors are inherently digital devices, a fact that will come in handy when you want to control your robot by computer. By the way, there are AC stepper motors as well, but they aren't really suitable for robotics work and so won't be discussed here.

Stepper motors aren't as easy to use as standard DC motors, however, and they're also harder to get and more expensive. But for the applications that require them, stepper motors can solve a lot of problems with a minimum of fuss.

21.1 Inside a Stepper Motor

There are several designs of stepper motors. The most popular variety is the four-phase unipolar stepper, like the one in Fig. 21-1. A unipolar stepper motor is really two motors sandwiched together, as shown in Fig. 21-2. Each motor is composed of two windings. Wires connect to each of the four windings of the motor pair, so there are eight wires coming from the motor. The commons from the windings are often ganged together, which reduces the wire count to five or six instead of eight (see Fig. 21-3).

21.1.1 WAVE STEP SEQUENCE

In operation, the common wires of a unipolar stepper are attached to the positive (sometimes the negative) side of the power supply. Each winding is then energized in turn by grounding it to the power supply for a short time. The motor shaft turns a fraction of a revolution each time a winding is energized. For the shaft to turn properly, the windings must be energized in sequence. For example, energize wires 1, 2, 3, and 4 in sequence and the motor turns clockwise. Reverse the sequence, and the motor turns the other way.

21.1.2 FOUR-STEP SEQUENCE

The wave step sequence is the basic actuation technique of unipolar stepper motors. Another, and far better, approach actuates two windings at once in an on-on/off-off four-

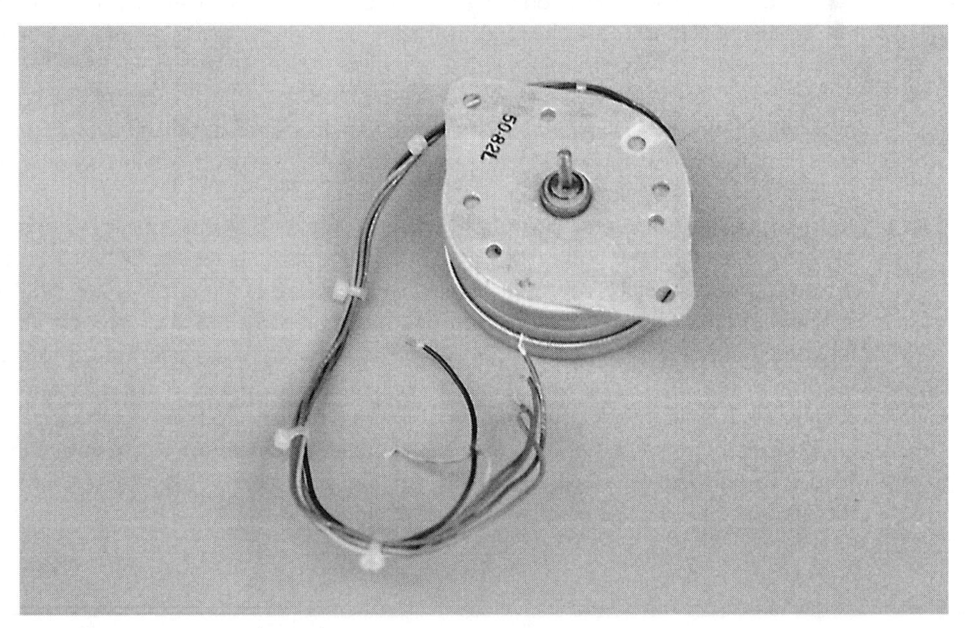

FIGURE 21-1 A typical unipolar stepper motor.

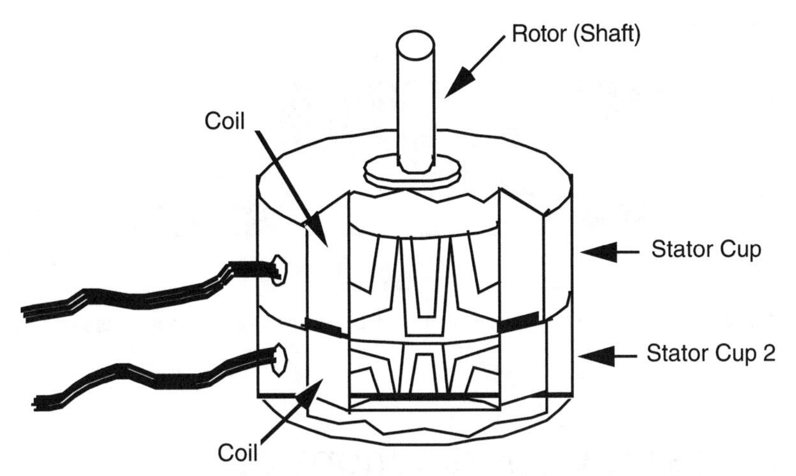

FIGURE 21-2 Inside a unipolar stepper motor. Note the two sets of coils and stators. The unipolar stepper is really two motors sandwiched together.

step sequence, as shown in Fig. 21-4. This enhanced actuation sequence increases the driving power of the motor and provides greater shaft rotation precision.

There are other varieties of stepper motors, and they are actuated in different ways. One you may encounter is bipolar. It has four wires and is pulsed by reversing the polarity of the power supply for each of the four steps. The actuation technique for these motors will be discussed later in this chapter.

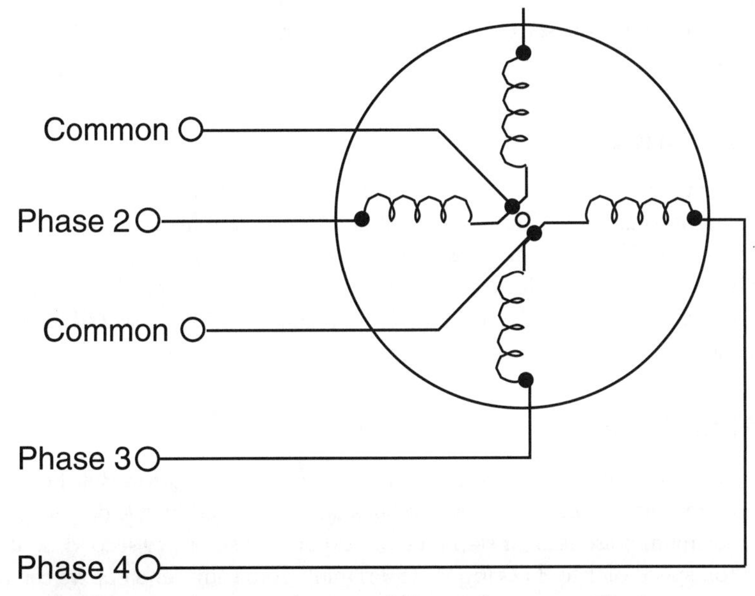

FIGURE 21-3 The wiring diagram of the unipolar stepper. The common connections can be separate or combined.

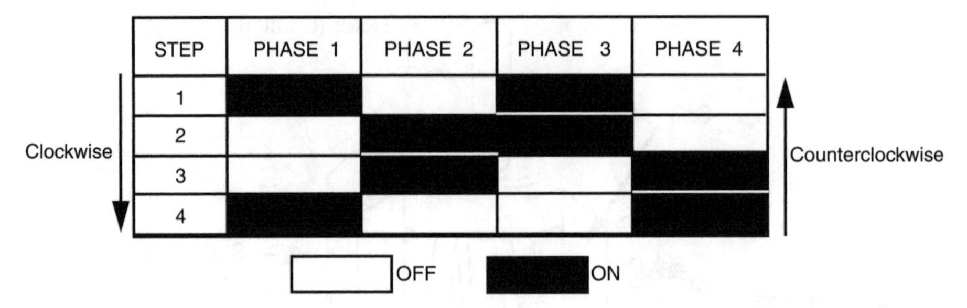

FIGURE 21-4 The enhanced on-on/off-off four-step sequence of a unipolar stepper motor.

21.2 Design Considerations of Stepper Motors

Stepping motors differ in their design characteristics compared with continuous DC motors. The following section discusses the most important design specifications for stepper motors.

21.2.1 STEPPER PHASING

A unipolar stepper requires that a sequence of four pulses be applied to its various windings for it to rotate properly. By their nature, all stepper motors are at least two-phase. Many are four-phase; some are six-phase. Usually, but not always, the more phases in a motor, the more accurate it is.

21.2.2 STEP ANGLE

Stepper motors vary in the amount of rotation of the shaft each time a winding is energized.

The amount of rotation is called the step angle and can vary from as small as 0.9° (1.8° is more common) to 90°. The step angle determines the number of steps per revolution. A stepper with a 1.8° step angle, for example, must be pulsed 200 times for the shaft to turn one complete revolution. A stepper with a 7.5° step angle must be pulsed 48 times for one revolution, and so on.

21.2.3 PULSE RATE

Obviously, the smaller the step angle is, the more accurate the motor. But the number of pulses stepper motors can accept per second has an upper limit. Heavy-duty steppers usually have a maximum pulse rate (or step rate) of 200 or 300 steps per second, so they have an effective top speed of 1 to 3 r/s (60 to 180 r/min). Some smaller steppers can accept a thousand or more pulses per second, but they don't usually provide very much torque and aren't suitable as driving or steering motors.

Note that stepper motors can't be motivated to run at their top speeds immediately from a dead stop. Applying too many pulses right off the bat simply causes the motor to freeze up. To achieve top speeds, you must gradually accelerate the motor. The acceleration can be quite swift in human terms. The speed can be one-third for the first few milliseconds, two-thirds for the next few milliseconds, then full blast after that.

21.2.4 RUNNING TORQUE

Steppers can't deliver as much running torque as standard DC motors of the same size and weight. A typical 12-V, medium-sized stepper motor may have a running torque of only 25 oz-in. The same 12-V, medium-sized standard DC motor may have a running torque that is three or four times more.

However, steppers are at their best when they are turning slowly. With the typical stepper, the slower the motor revolves, the higher the torque. The reverse is usually true of continuous DC motors. Fig. 21-5 shows a graph of the running torque of a medium-duty, unipolar 12-V stepper. This unit has a top running speed of 550 pulses per second. Since the motor has a step angle of 1.8°, that results in a top speed of 2.75 r/s (165 r/min).

21.2.5 BRAKING EFFECT

Actuating one of the windings in a stepper motor advances the shaft. If you continue to apply current to the winding, the motor won't turn anymore. In fact, the shaft will be locked, as if you've applied brakes. As a result of this interesting locking effect, you never need to add a braking circuit to a stepper motor because it has its own brakes built in.

The amount of braking power a stepper motor has is expressed as holding torque. Small stepper motors have a holding torque of a few oz-in. Larger, heavier-duty models have holding torques exceeding 400 oz-in.

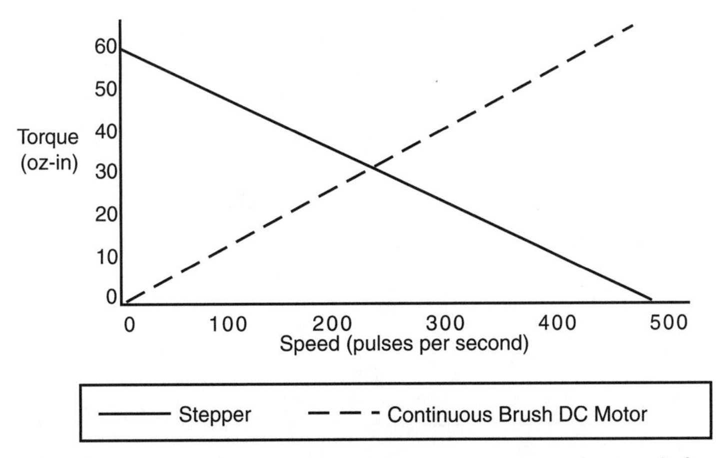

FIGURE 21-5 With a stepper motor, torque increases as the speed of the motor is reduced.

21.2.6 VOLTAGE, CURRENT RATINGS

Like DC motors, stepper motors vary in their voltage and current ratings. Steppers for 5-, 6-, and 12-V operation are not uncommon. But unlike DC motors, if you use a higher voltage than specified for a stepper motor you don't gain faster operation but more running and holding torque. Overpowering a stepper by more than 80 to 100 percent above the rated voltage may eventually burn up the motor.

The current rating of a stepper is expressed in amps (or milliamps) per phase. The power supply driving the motor needs to deliver at least as much as the per-phase specification, preferably more if the motor is driving a heavy load. The four-step actuation sequence powers two phases at a time, which means the power supply must deliver at least twice as much current as the per-phase specification. If, for example, the current per phase is 0.25 A, the power requirement at any one time is 0.50 A.

21.3 Controlling a Stepper Motor

Steppers have been around for a long time. In the old days, stepper motors were actuated by a mechanical switch, a solenoid-driven device that pulsed each of the windings of the motor in the proper sequence. Now, stepper motors are invariably controlled by electronic means. Basic actuation can be accomplished via computer control by pulsing each of the four windings in turn. The computer can't directly power the motor, so transistors must be added to each winding, as shown in Fig. 21-6.

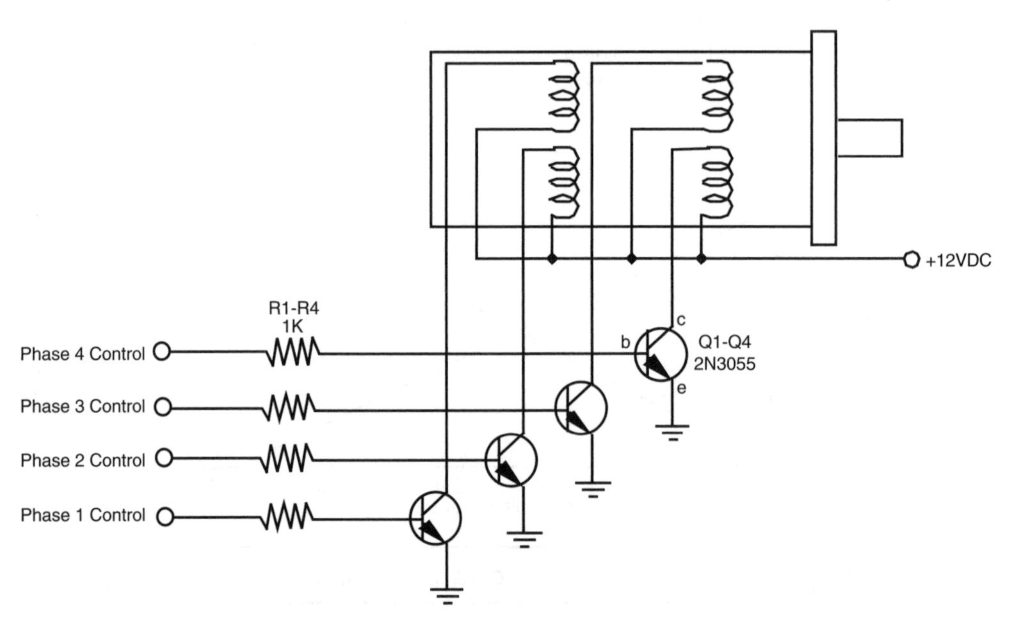

FIGURE 21-6 The basic hookup connection to drive a stepper motor from a computer or other electronic interface. The phasing sequence is provided by software or other means through a port following a four-bit binary sequence: 1010, 0110, 0101, 1001 (reverse the sequence to reverse the motor).

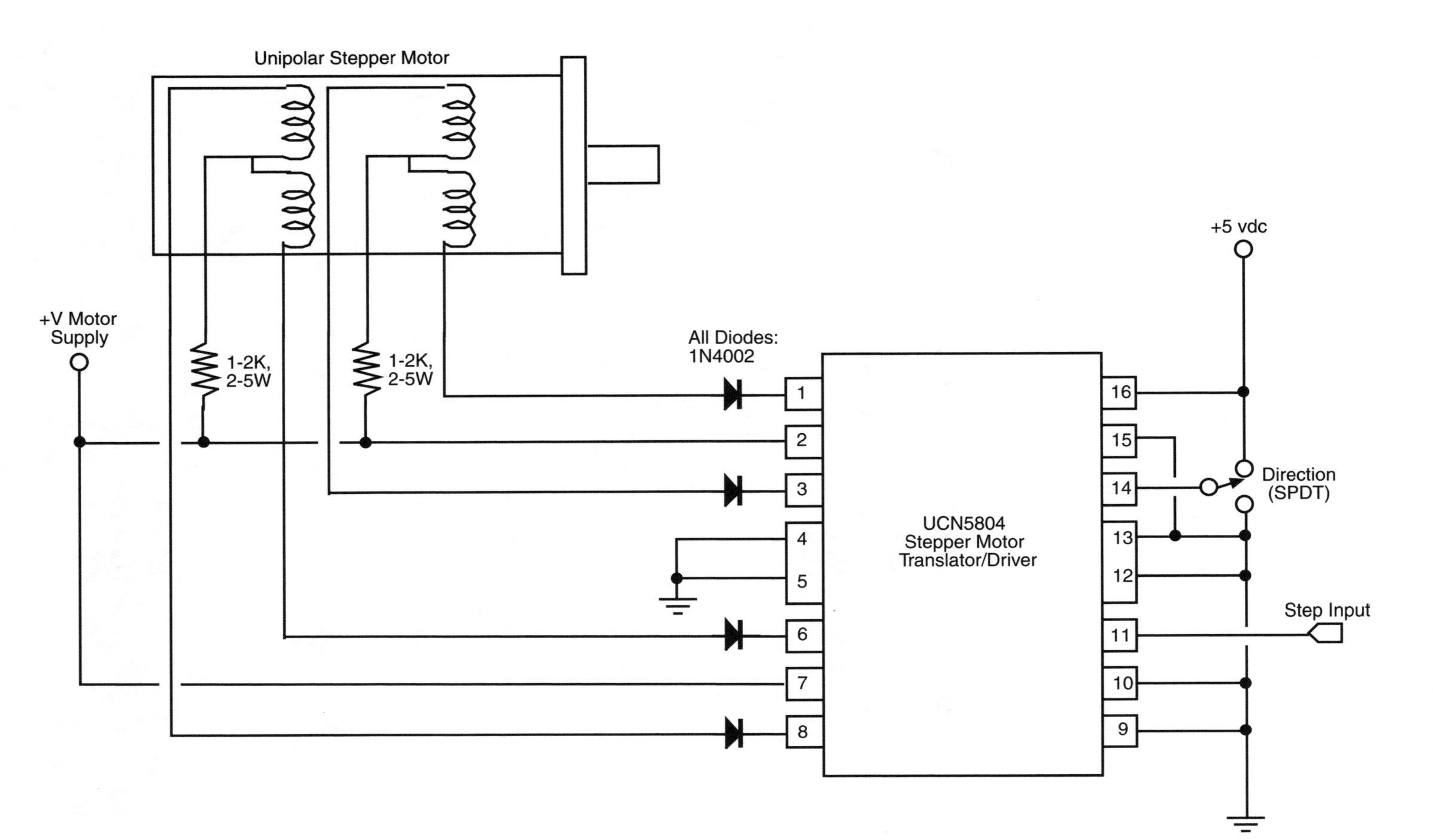

FIGURE 21-7 The basic wiring diagram for the UCN5804.

21.3.1 USING A STEPPER MOTOR CONTROLLER CHIP

In the absence of direct computer control, the easiest way to provide the proper sequence of actuation pulses is to use a custom stepper motor chip, such as the Allegro Microsystems UCN5804. This chip is designed expressly for use with the common unipolar stepper motor and provides a four-step actuation sequence. Stepper motor translator chips tend to be modestly priced, at about $5 to $10, depending on their features and where you buy them.

Fig. 21-7 (refer to the parts list in Table 21-1) shows a typical schematic of the UCN5804. Heavier-duty motors (more than about 1 A per phase) can be driven by adding power transistors to the four outputs of the chips, as shown in the manufacturer's application notes. Note the direction pin. Pulling this pin high or low reverses the rotation of the motor.

21.3.2 USING LOGIC GATES TO CONTROL STEPPER MOTORS

Another approach to operating unipolar stepper motors is to use discrete gates and clock ICs. You can assemble a stepper motor translator circuit using just two IC packages. The circuit can be constructed using TTL or CMOS chips.

The TTL version is shown in Fig. 21-8 (refer to the parts list in Table 21-2). Four exclusive OR gates from a single 7486 IC provide the steering logic. You set the direction by pulling pin 12 HIGH or LOW. The stepping actuation is controlled by a 7476, which contains two JK flip-flops. The Q and $\overline{Q}$ outputs of the flip-flops control the phasing of the motor. Stepping is accomplished by triggering the clock inputs of both flip-flops.

The 7476 can't directly power a stepper motor. You must use power transistors or MOSFETs to drive the windings of the motor. See the section titled "Translator Enhancements" for a complete power driving schematic as well as other options you can add to this circuit.

The CMOS version, shown in Fig. 21-9 (refer to the parts list in Table 21-3), is identical to the TTL version, except that a 4070 chip is used for the exclusive OR gates and a 4027 is used for the flip-flops. The pinouts are slightly different, so follow the correct schematic for the type of chips you use. Note that another CMOS exclusive OR package, the 4030, is also available. Don't use this chip; it behaves erratically in this, as well as other pulsed, circuits.

TABLE 21-1 Parts List for UCN5804 Stepper Motor Translator/Driver

IC1	Allegro UCN5804 Stepper Motor Translator IC
R1, R2	1–2K resistor, 2–5 watts
D1–D4	1N4002 diode
M1	Unipolar stepper motor
Misc.	SPDT switch, heat sinks for UCN5804 (as needed)

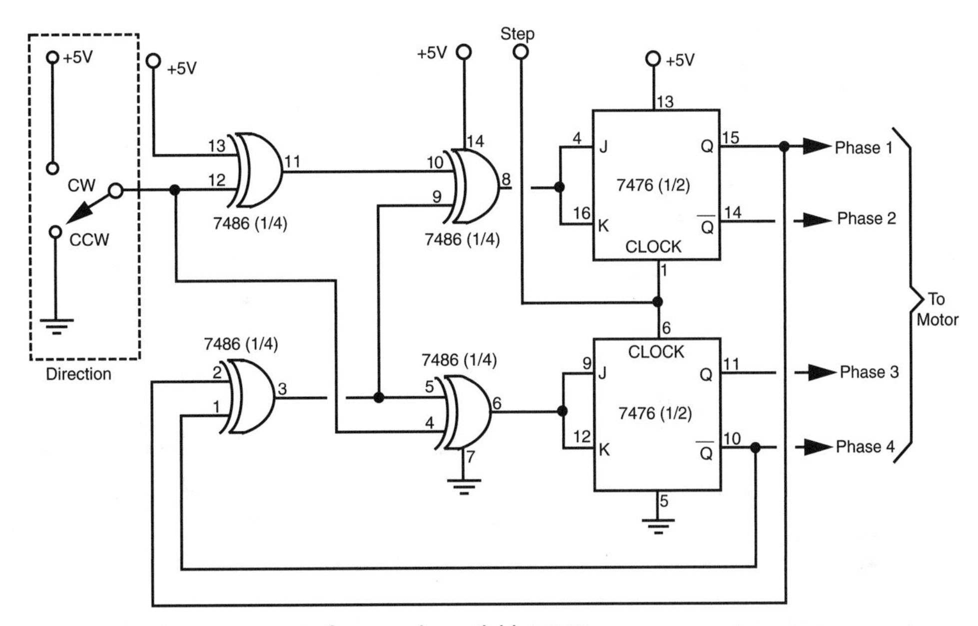

FIGURE 21-8 Using a pair of commonly available TTL ICs to construct your own stepper motor translator circuit.

In both the TTL and CMOS circuits, the stepper motor itself can be operated from a supply voltage that is wholly different from the voltage supplied to the ICs. It should be pointed out that stepper motor translator circuits are excellent first microcontroller applications. Along with providing an intelligent approach to controlling stepper motors, small microcontrollers can cost as little as $0.50—probably less than the cost of one chip used in the circuits shown in Fig. 21-8 and 21-9 and much easier to wire into a circuit.

21.3.3 TRANSLATOR ENHANCEMENTS

Four NPN power transistors, four resistors, and a handful of diodes are all the translator circuits described in the last section need to provide driving power. (You can also use this scheme to increase the driving power of the UCN5804, detailed earlier.) The schematic for the circuit is shown in Fig. 21-10 (refer to the parts list in Table 21-4). Note that you can substitute the bipolar transistors and resistors with power MOSFETs. See Chapter 19, "Choosing the Right Motor," for more information on using power MOSFETs.

TABLE 21-2 Parts List for TTL Stepper Motor Translator

IC1	7485 Quad Exclusive OR Gate IC
IC2	7476 Dual "JK" flip-flop IC

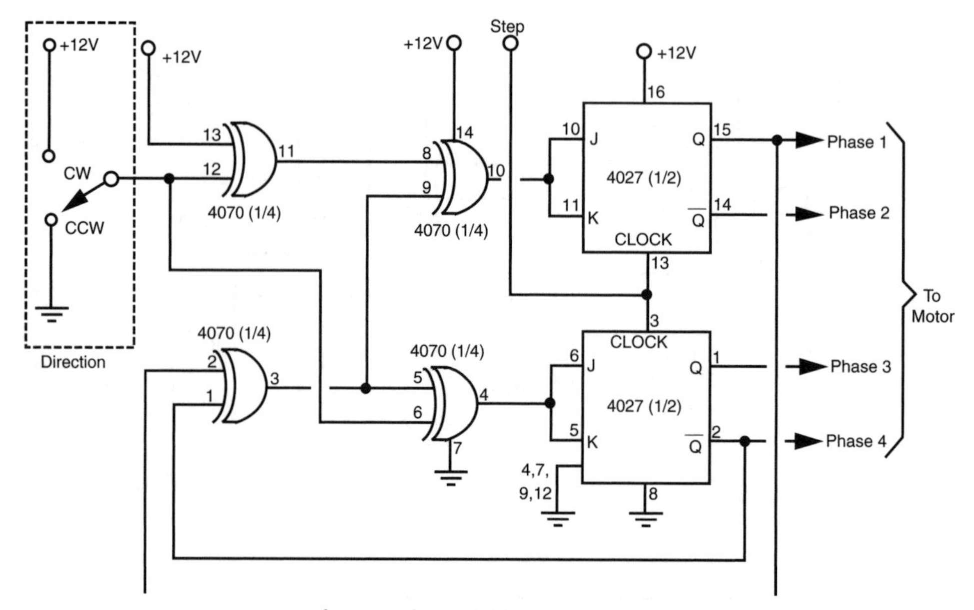

FIGURE 21-9 Using a pair of commonly available CMOS ICs to construct your own stepper motor translator circuit.

You can use just about any NPN power transistor that will handle the motor. The TIP31 is a good choice for applications that require up to 1 A of current. Use the 2N3055 for heavier-duty motors. Mount the drive transistors on a suitable heat sink.

You must insert a bias resistor in series between the outputs of the translation circuit and the base of the transistors. Values between about 1K and 3K should work with most motors and most transistors. Experiment until you find the value that works without causing the flip-flop chips to overheat. You can also apply Ohm's law, figuring in the current draw of the motor and the gain of the transistor, to accurately find the correct value of the resistor. If this is new to you, see Appendix A, "Further Reading," for a list of books on electronic design and theory.

It is sometimes helpful to see a visual representation of the stepping sequence. Adding an LED and current-limiting resistor in parallel with the outputs provides just such a visual indication. See Fig. 21-11 for a wiring diagram (refer to the parts list in Table 21-5). Note the special wiring to the flip-flop outputs. This provides a better visual indication of the stepping action than hooking up the LEDs in the same order as the motor phases.

TABLE 21-3 Parts List for CMOS Stepper Motor Translator

IC1	4070 Quad Exclusive OR Gate IC
IC2	4027 Dual "JK" flip-flop IC

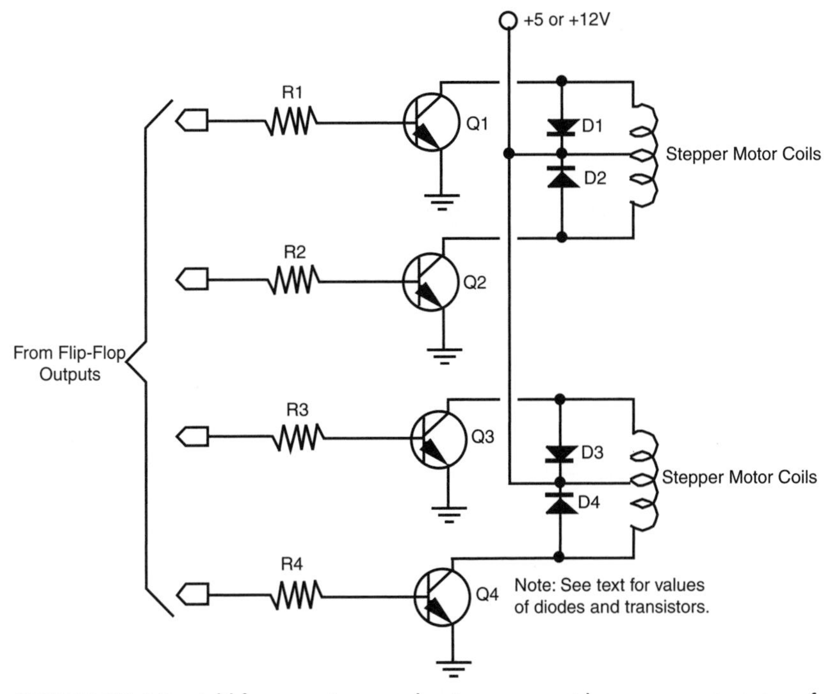

FIGURE 21-10 Add four transistors and resistors to provide a power output stage for the TTL or CMOS stepper motor translator circuits.

Fig. 21-12 shows two stepper motor translator boards. The small board controls up to two stepper motors and is designed using TTL chips. The LED option is used to provide a visual reference of the step sequence. The large board uses CMOS chips and can accommodate up to four motors. The boards were wire-wrapped; the driving transistors are placed on a separate board and heat sink.

21.3.4 TRIGGERING THE TRANSLATOR CIRCUITS

You need a square wave generator to provide the triggering pulses for the motors. You can use the 555 timer wired as an astable multivibrator, or make use of a control line in your

TABLE 21-4 Parts List for Stepper Motor Driver

Q1–Q4	Under 1 A draw per phase: TIP32 NPN transistor 1 to 3 A draw per phase: TIP120 NPN Darlington transistor
R1–R4	1K resistor, 1 W
D1–D4	1N4004 diode
Misc.	Heat sinks for transistors

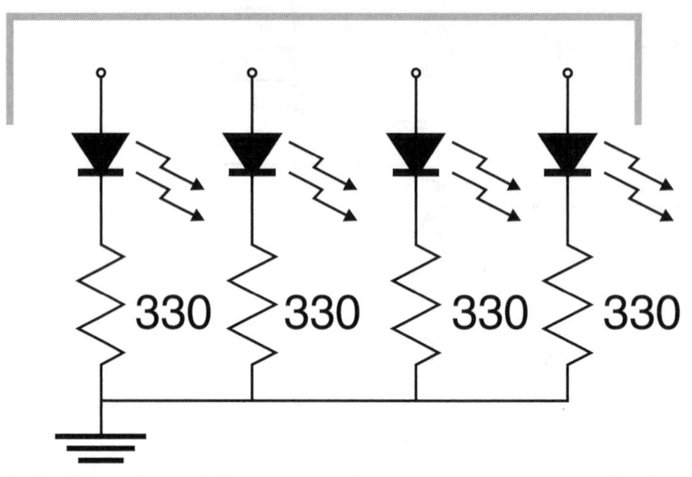

FIGURE 21-11 Add four LEDs and resistors to provide a visual indication of the stepping action.

computer or microcontroller. When using the 555, remember to add the 0.1 μF capacitor across the power pins of the chip. The 555 puts a lot of noise into the power supply, and this noise regularly disturbs the counting logic in the exclusive OR and flip-flop chips. If you are getting erratic results from your circuit, this is probably the cause.

21.3.5 USING BIPOLAR STEPPER MOTORS

As detailed earlier in the chapter, unipolar stepper motors contain four coils in which two of the coils are joined to make a center tap. This center tap is the *common* connection for the motor. Bipolar stepper motors contain two coils, do not use a common connection, and are generally less expensive because they are easier to manufacture. A bipolar stepper motor has four external connection points. An old method for operating a bipolar stepper motor was to use relays to reverse the polarity of a DC voltage to two coils. This caused the motor to inch forward or backward, depending on the phasing sequence.

Today, the more common method for operating a bipolar stepper motor is to use a specialty stepper motor translator, such as the SGS-Thompson L297D (the L297D can also be used to drive unipolar stepper motors). To add more current driving capacity to the L297D

TABLE 21-5 Parts List for Stepper Motor Translator LED Display

R1–R4	330 Ω resistors
LED1–4	Visible light LEDs

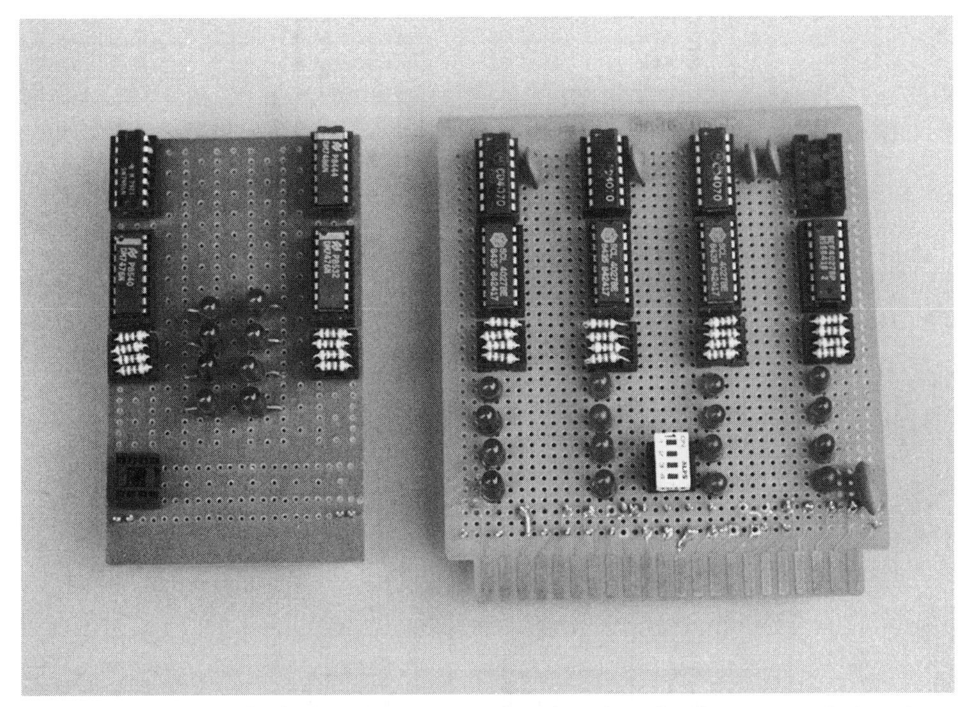

FIGURE 21-12 Two finished stepper motor translator boards, with indicator LEDs. The board on the left controls two stepper motors; the board on the right controls four stepper motors.

you can add a dual H-bridge driver, such as the L298N, which is available from the same company. The truth table for the typical driving sequence of a bipolar stepper motor is shown in Fig. 21-13.

21.3.6 BUYING AND TESTING A STEPPER MOTOR

Spend some time with a stepper motor and you'll invariably come to admire its design and be able to think up all sorts of ways to make it work for you in your robot designs. But to use a stepper, you have to get one. That in itself is not always easy. Then after you have obtained it and taken it home, there's the question of figuring out where all the wires go!

Let's take each problem one at a time.

21.3.7 SOURCES FOR STEPPER MOTORS

Despite their many advantages, stepper motors aren't nearly as common as the trusty DC motor, so they are harder to find. And when you do find them, they're expensive when new. The surplus market is by far the best source for stepper motors for hobby robotics. See Appendix B, "Sources," for a list of selected mail-order surplus companies that regularly carry a variety of stepper motors. They carry most of the name brand steppers: Thompson-

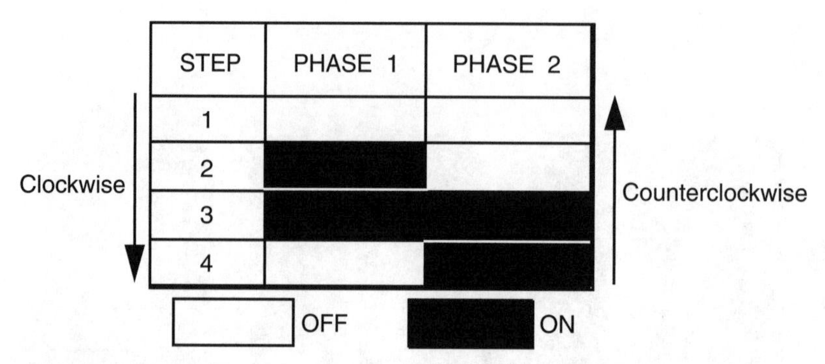

FIGURE 21-13 The phasing sequence for a bipolar stepper motor.

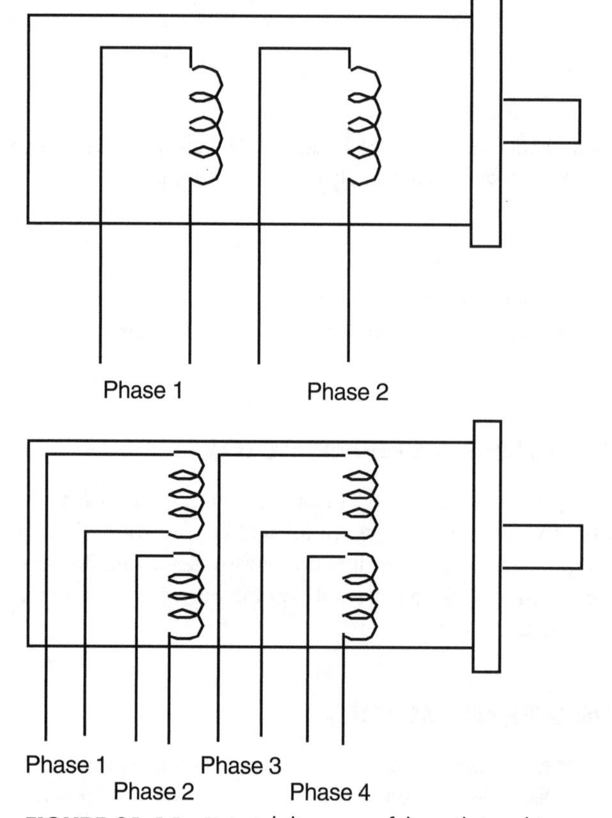

FIGURE 21-14 Pictorial diagrams of the coils in a bipolar and unipolar stepper motor.

Airpax, Molon, Haydon, and Superior Electric. The cost of surplus steppers is often a quarter or fifth of the original list price.

The disadvantage of buying surplus is that you don't always get a hookup diagram or adequate specifications. Purchasing surplus stepper motors is largely a hit-or-miss affair, but most outlets let you return the goods if they aren't what you need. If you like the motor, yet it still lacks a hookup diagram, read the following section on how to decode the wiring.

21.3.8 WIRING DIAGRAM

The internal wiring diagram of both a bipolar and unipolar stepper motor is shown in Fig. 21-14. The wiring in a bipolar stepper is actually easy to decode. A DMM is used to measure the resistance between wire pairs. You can be fairly sure the motor is two-phase if it has only four wires leading to it. You can identify the phases by connecting the leads of the meter to each wire and noting the resistance. Wire pairs that give an open reading (infinite ohms) represent two different coils (phases). You can readily identify mating phases when there is a small resistance through the wire pair.

Unipolar steppers behave the same, but with a slight twist. Let's say, for argument's sake, that the motor has eight wires leading to it. Each winding, then, has a pair of wires. Connect your meter to each wire in turn to identify the mating pairs. As illustrated in Fig. 21-15, no reading (infinite ohms) signifies that the wires do not lead to the same winding; a reading indicates a winding.

If the motor has six wires, then four of the leads go to one side of the windings. The other two are commons and connect to the other side of the windings (see Fig. 21-16). Decoding this wiring scheme takes some patience, but it can be done. First, separate all

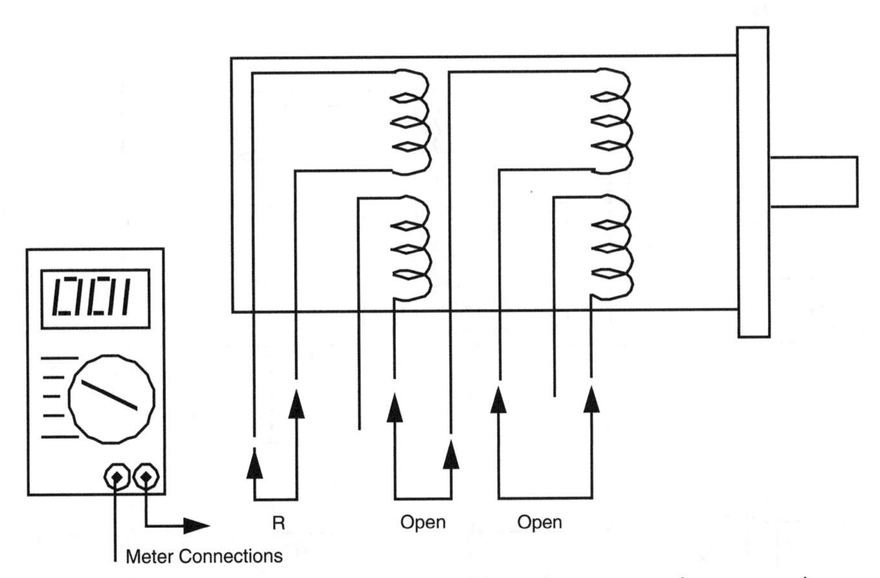

FIGURE 21-15 Connection points and possible readings on an eight-wire unipolar stepper motor.

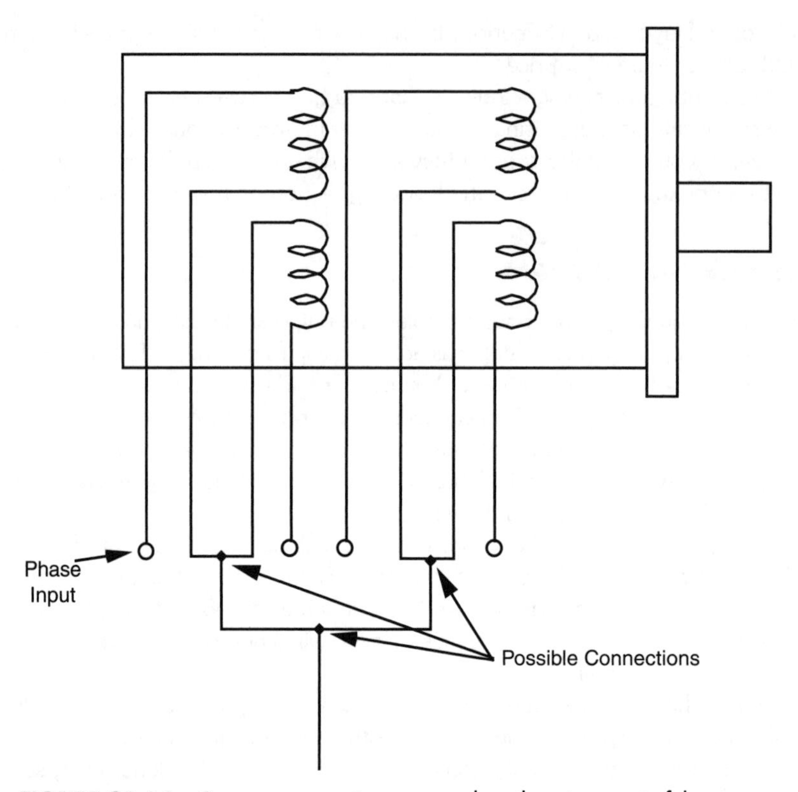

FIGURE 21-16 Common connections may reduce the wire count of the stepper motor to five or six, instead of eight.

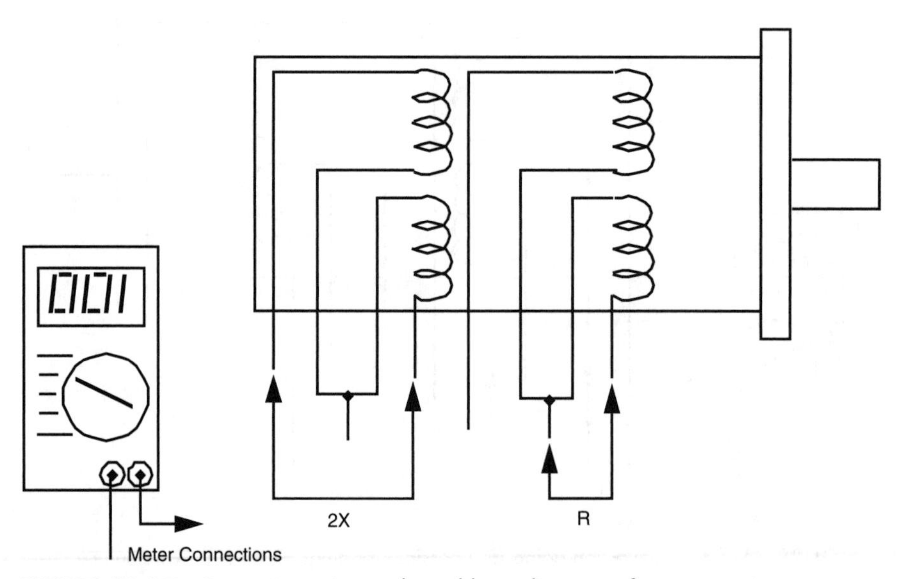

FIGURE 21-17 Connection points and possible readings on a five- or six-wire unipolar stepper motor.

those wires where you get an open reading. At the end of your test, there should be two three-wire sets that provide some reading among each of the leads.

Locate the common wire by following these steps. Take a measurement of each combination of the wires and note the results. You should end up with three measurements: wires 1 and 2, wires 2 and 3, and wires 1 and 3. The meter readings will be the same for two of the sets. For the third set, the resistance should be roughly doubled. These two wires are the main windings. The remaining wire is the common.

Decoding a five-wire motor is the most straightforward procedure. Measure each wire combination, noting the results of each. When you test the leads to one winding, the result will be a specified resistance (let's call it "R"). When you test the leads to two of the windings, the resistance will be double the value of "R," as shown in Fig. 21-17. Isolate this common wire with further testing and you've successfully decoded the wiring.

21.4 From Here

To learn more about . . .	***Read***
Driving a robot	Chapter 18, "Principles of Robot Locomotion"
Selecting a motor for your robot	Chapter 19, "Choosing the Right Motor"
Connecting motors to computers, microcontrollers, and other electronic circuitry	Chapter 14, "Computer Peripherals"

CHAPTER 22

WORKING WITH SERVO MOTORS

DC and stepper motors are inherently *open feedback* systems—you provide electricity and they spin. How much they spin is not always known, not even for a stepper motor, which turns by finite degrees based on the number of pulses it gets. Should something impede the rotation of the motor it may not turn at all, but there's no easy, built-in way that the control electronics would know that.

Servo motors, on the other hand, are designed for *closed feedback* systems. The output of the motor is coupled to a control circuit; as the motor turns, its speed and/or position are relayed to the control circuit. If the rotation of the motor is impeded for whatever reason, the feedback mechanism senses that the output of the motor is not yet in the desired location. The control circuit continues to correct the error until the motor finally reaches its proper point.

Servo motors come in various shapes and sizes. Some are smaller than a walnut, while others are large enough to take up their own seat in your car. They're used for everything from controlling computer-operated lathes to copy machines to model airplanes and cars. It's the last application that is of most interest to hobby robot builders: the same servo motors used with model airplanes and cars can readily be used with your robot.

These servo motors are designed to be operated via a radio-controlled link and so are commonly referred to as radio-controlled (or R/C) servos. But in fact the servo motor itself is not what is radio-controlled; it is merely connected to a radio receiver on the plane or car. The servo takes its signals from the receiver. This means you don't have to control your robot via radio signals just to use an R/C servo—unless you want to, of course. You can control a servo with your PC, a microcontroller such as the BASIC Stamp, or even a simple circuit designed around the familiar 555 timer integrated circuit.

In this chapter R/C servos will be presented along with how they can be put to use in a robot. While there are other types of servo motors, it is the R/C type that is commonly available and reasonably affordable. For simplicity's sake, when you see the term *servo* in the text that follows understand that it specifically means an R/C servo motor, even though there are other types.

22.1 How Servos Work

Fig. 22-1 shows a typical standard-sized R/C servo motor, which is used with flyable model airplanes and model racing cars. The size and mounting of a standard servo is the same regardless of the manufacturer, which means that you have your pick of a variety of makers. Along with the standard-sized servor, there are other common sizes of servo motors also available, which will be discussed later in the chapter.

Inside the servo is a motor, a series of gears to reduce the speed of the motor, a control board, and a potentiometer (see Fig. 22-2). The motor and potentiometer are connected to the control board, all three of which form a closed feedback loop. Both control board and motor are powered by a constant DC voltage (usually between 4.8 and 6.0 V, although many will work with power inputs up to 7.2 V).

To turn the motor, a digital signal is sent to the control board. This activates the motor,

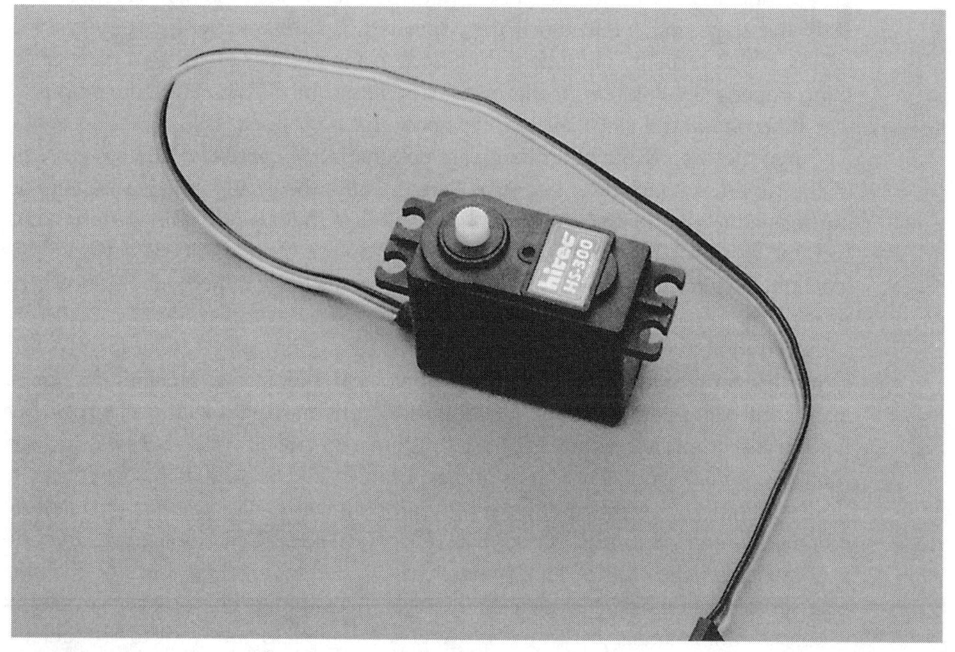

FIGURE 22-1 The typical radio-controlled (R/C) servo motor.

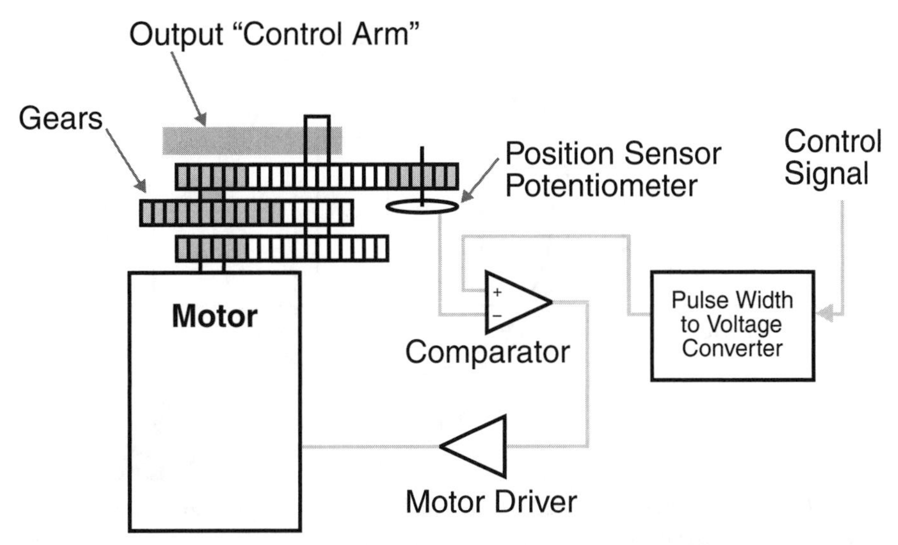

FIGURE 22-2 The internal parts of an R/C servo. The servo consists of a motor, a gear train, a potentiometer, and some control circuit.

which, through a series of gears, is connected to the potentiometer. The position of the potentiometer's shaft indicates the position of the output shaft of the servo. When the potentiometer has reached the desired position, the control board shuts down the motor.

As you can surmise, servo motors are designed for limited rotation rather than for continuous rotation like a DC or stepper motor. While it is possible to modify an R/C servo to rotate continuously (as discussed later in this chapter), the primary use of the R/C servo is to achieve accurate rotational positioning over a range of 90° or 180°. While this may not sound like much, in actuality such control can be used to steer a robot, move legs up and down, rotate a sensor to scan the room, and more. The precise angular rotation of a servo in response to a specific digital signal has enormous uses in all fields of robotics.

22.2 Servos and Pulse Width Modulation

The motor shaft of an R/C servo is positioned by using *pulse width modulation* (PWM). In this system, the servo responds to the duration of a steady stream of digital pulses (Fig. 22-3). Specifically, the control board responds to a digital signal whose pulses vary from about 1 ms (one-thousandth of a second) to about 2 ms. These pulses are sent some 50 times per second. The exact length of the pulse, in fractions of a millisecond, determines the position of the servo. Some servos are very tolerant of varying PWM periods, while others will not work properly or "jitter" if the pulses come at anything other than 50 times per second. To be on the safe side, always try to maintain 20 ms between the start servo control pulses.

At a duration of 1 ms, the servo is commanded to turn all the way in one direction (for

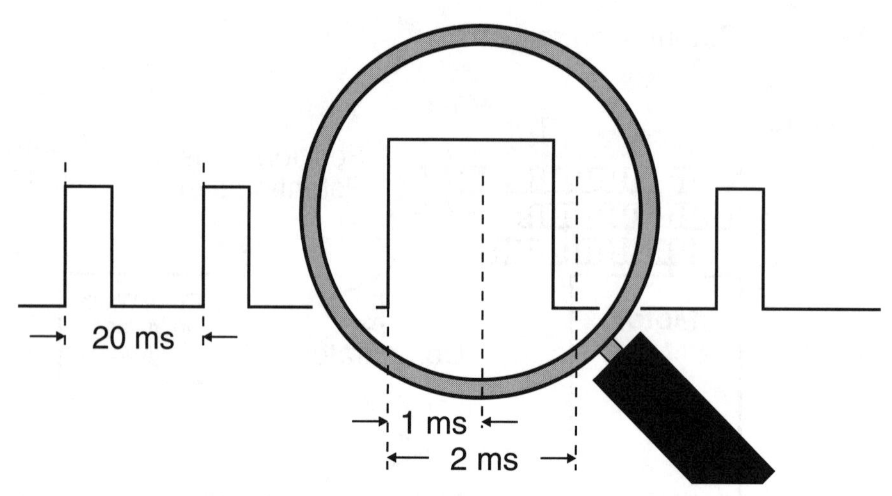

FIGURE 22-3 A pulse, every 20 ms, varying from 1 ms in duration to 2 ms in duration controls the position of the servo.

example counterclockwise). At 2 ms, the servo is commanded to turn all the way in the other direction. Therefore, at 1.5 ms, the servo is commanded to turn to its center (or neutral) position. As mentioned earlier, the angular position of the servo is determined by the width (more precisely, the duration) of the pulse. This technique has gone by many names over the years. One you may have heard of is digital proportional—the movement of the servo is proportional to the digital signal being fed into it.

The power delivered to the motor inside the servo is also proportional to the difference between where the output shaft is and where it's supposed to be. If the servo has only a little way to move to its new location, then the motor is driven at a fairly low speed. This ensures that the motor doesn't overshoot its intended position. But if the servo has a long way to move to its new location, then it's driven at full speed in order to get there as fast as possible. As the output of the servo approaches its desired new position, the motor slows down. What seems like a complicated process actually happens in a very short period of time—the average servo can rotate a full 60° in a quarter to half second.

22.3 The Role of the Potentiometer

The potentiometer of the servo plays a key role in determining when the motor has set the output shaft to the correct position. The potentiometer is physically attached to the output shaft (and in some servo models, the potentiometer is the output shaft). In this way, the position of the potentiometer very accurately reflects the position of the output shaft of the servo. When used in a servo, a potentiometer is wired as a voltage divider and provides a varying voltage to the control circuit when the servo output changes, as shown in Fig. 22-4.

The control circuit in the servo correlates this voltage with the timing of the incoming digital pulses and generates an error signal if the voltage is wrong. This error signal is pro-

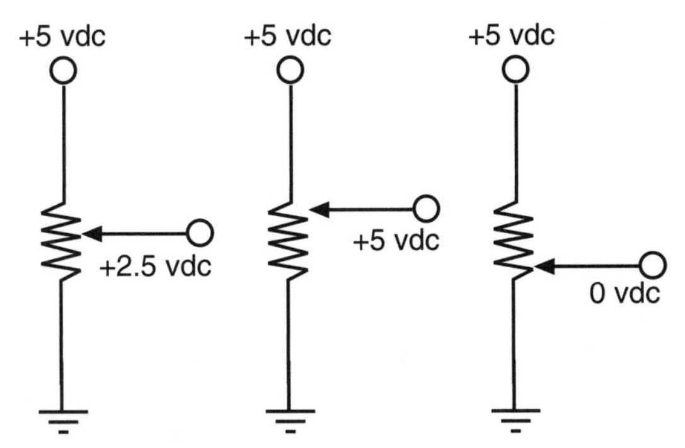

FIGURE 22-4 A potentiometer is often used as a variable voltage divider. As the potentiometer turns, its wiper travels the length of a resistive element. The output of the potentiometer is a varying voltage, from 0 to the +V of the circuit.

portional to the difference between the position of the potentiometer and the timing of the incoming signal. To compensate, the control board applies the error signal to turn the motor. When the voltage from the potentiometer and the timing of the digital pulses match, the error signal is removed, and the motor stops.

22.4 Rotational Limits

Servos also vary by the amount of rotation they will perform for the 1 to 2 ms signal they are provided. Most standard servos are designed to rotate back and forth by 90° to 180°, given the full range of timing pulses. You'll find the majority of servos will be able to turn a full 180°, or very nearly so. Should you attempt to command a servo beyond its mechanical limits, the output shaft of the motor will hit an internal stop. This causes the gears of the servo to grind or chatter. If left this way for more than a few seconds, the gears of the motor or the motor itself and its drivers may be permanently damaged. Therefore, when experimenting with servo motors exercise care to avoid pushing them beyond their natural limits.

The actual length of the pulses is fairly constant at 1 to 2 ms for most manufacturers' products, but it should be noted that the Futaba brand and servos that are compatible have a 1 to 1.5 ms pulse width. There can be a problem with passing a 2 ms pulse to a Futaba servo if the pulse causes the servo to push against its stop (potentially damaging the servo). Rather than experimenting on an unknown servo with 1 to 2 ms pulses, you should start with 1 to 1.5 ms pulses and raise the pulse width to 2 ms if the servo only turns 45 degrees.

22.5 Special-Purpose Servo Types and Sizes

While the standard-sized servo is the one most commonly used in both robotics and radio-controlled models, other R/C servo types, styles, and sizes exist as well.

- *Quarter-scale* (or large-scale) servos are about twice the size of standard servos and are significantly more powerful. Quarter-scale servos are designed to be used in large model airplanes, but they also make perfect power motors for a robot.
- *Mini-micro* servos are about half the size (and smaller!) of standard servos and are designed to be used in tight spaces in a model airplane or car. They aren't as strong as standard servos, however.
- *Sail winch* servos are designed with maximum strength in mind, and are primarily intended to move the jib and mainsail sheets on a model sailboat.
- *Landing-gear retraction* servos are made to retract the landing gear of medium- and large-sized model airplanes. The design of the landing gear often requires the servo to guarantee at least 170° rotation, if not more (i.e., up to and exceeding 360° of motion). It is not uncommon for retraction servos to have a slimmer profile than the standard variety because of the limited space on model airplanes.

22.6 Gear Trains and Power Drives

The motor inside an R/C servo turns at several thousand r/min. This is too fast to be used directly on model airplanes and cars, or on robots. All servos employ a gear train that reduces the output of the motor to the equivalent of about 50 to 100 r/min. Servo gears can be made of plastic, nylon, or metal (usually brass or aluminum).

Metal gears last the longest, but they significantly raise the cost of the servo. Replacement gear sets are available for many servos, particularly the medium- to higher-priced ones ($20+). Should one or more gears fail, the servo can be disassembled and the gears replaced. In some cases, you can upgrade the plastic gears in a less expensive servo to higher-quality metal ones.

Besides the drive gears, the output shaft of the servo receives the most wear and tear. On the least expensive servos this shaft is supported by a plastic bearing, which obviously can wear out very quickly if the servo is used heavily. Actually, this piece is not a bearing at all but a bushing, a sleeve or collar that supports the shaft against the casing of the servo. Metal bushings, typically made from lubricant-impregnated brass, last longer but add to the cost of the servo. The best (and most expensive) servos come equipped with ball bearings, which provide longest life. Ball bearing upgrades are available for some servo models.

22.7 Typical Servo Specs

R/C servo motors enjoy some standardization. This sameness applies primarily to standard-sized servos, which measure approximately 1.6 by 0.8 by 1.4 in. For other servo types the size varies somewhat between makers, as these are designed for specialized tasks.

Table 22-1 outlines typical specifications for several types of servos, including dimensions, weight, torque, and transit time. Of course, except for the size of standard servos, these specifications can vary between brand and model. A few of the terms used in the

TABLE 22-1 Typical Servo Specifications

SERVO TYPE	LENGTH	WIDTH	HEIGHT	WEIGHT	TORQUE	TRANSIT TIME
Standard	1.6″	0.8″	1.4″	1.3 oz	42 oz-in	0.23 sec/60°
¼-scale	2.3″	1.1″	2.0″	3.4 oz	130 oz-in	0.21 sec/60°
Mini-/Nano-	0.85″	0.4″	0.8″	0.3 oz	15 oz-in	0.11 sec/60°
Low profile	1.6″	0.8″	1.0″	1.6 oz	60 oz-in	0.16 sec/60°
Small Sail Winch	1.8″	1.0″	1.7″	2.9 oz	135 oz-in	0.16 sec/60° 1 sec/360°
Large Sail Winch	2.3″	1.1″	2.0″	3.8 oz	195 oz-in	0.22 sec/60° 1.3 sec/360°

specs require extra discussion. As explained in Chapter 19, "Choosing the Right Motor," the torque of the motor is the amount of force it exerts. The standard torque unit of measure for R/C servos is expressed in ounce-inches—or the number of ounces the servo can lift when the weight is extended 1 in from the shaft of the motor. Servos exhibit very high torque thanks to their speed reduction gear trains.

The transit time (also called slew rate) is the approximate time it takes for the servo to rotate the shaft X° (usually specified as 60°). Small servos turn at about a quarter of a second per 60°, while larger servos tend to be a bit slower. The faster the transit time, the faster acting the servo will be.

You can calculate equivalent r/min by multiplying the 60° transit time by 6 (to get full 360° rotation), then dividing the result into 60. For example, if a servo motor has a 60° transit time of 0.20 s, that's one revolution in 1.2 s ($0.2 \times 6 = 1.2$), or 50 r/min ($60 / 1.2 = 50$).

Bear in mind that there are variations on the standard themes for all R/C servo classes. For example, standard servos are available in more expensive high-speed and high-torque versions. Servo manufacturers list the specifications for each model, so you can compare and make the best choice based on your particular needs.

Many R/C servos are designed for use in special applications, and these applications can be adapted to robots. For example, a servo engineered to be used with a model sailboat will be water resistant and therefore useful on a robot that works in or around water.

22.8 Connector Styles and Wiring

While many aspects of servos are standardized, there is much variety between manufacturers in the shape and electrical contacts of the connectors used to attach the servo to a receiver. Despite this diversity, you will find that you can usually connect servos to three

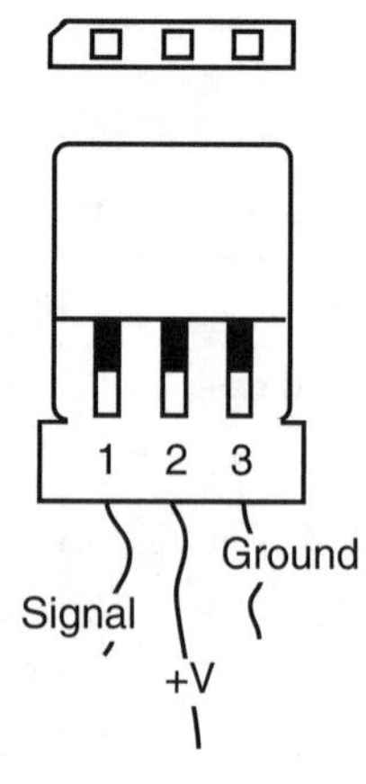

FIGURE 22-5 The standard pinout of servos is pin 1 for signal, pin 2 for +V, and pin 3 for ground. In this configuration damage will not usually occur if you accidentally reverse the connector.

standard 0.100″ connector posts either soldered into a PCB or pressed into a breadboard. You may decide the connector issue isn't worth the hassle, and just cut it off from the servo, hardwiring it to your electronics. This is an acceptable alternative, but hardwiring makes it more difficult to replace the servo should it ever fail.

22.8.1 CONNECTOR TYPE

There are three primary connector types found on R/C servos:

- "J" or Futaba style
- "A" or Airtronics style
- "S" or Hitec/JR style

Servos made by the principle servo manufacturers—Futaba, Airtronics, Hitec, and JR—employ the connector style popularized by that manufacturer. In addition, servos made by competing manufacturers are usually available in a variety of connector styles, and connector adapters are available.

TABLE 22-2 Connector Pinouts of Popular Servo Brands

BRAND	PIN 1 (LEFT)	PIN 2 (CENTER)	PIN 3 (RIGHT)
Airtronics ("new" style)	Signal	+V	Gnd
Airtronics ("old" style)	Signal	Gnd	+V
Futaba	Signal	+V	Gnd
Hitec	Signal	+V	Gnd
JR	Signal	+V	Gnd

TABLE 22-3 Color Coding of Popular Servo Brands

SERVO	+V	GND	SIGNAL
Airtronics	Red	Black	White
	Red stripe Brown	Blue	
Cirrus	Red	Brown	Orange
Daehwah	Red	Black	White
Fleet	Red	Black	White
Futaba	Red	Black	White
Hitec	Red	Black	Yellow
JR	Red	Brown	Orange
KO	Red	Black	Blue
Kraft	Red (4.8 v) White (2.4 v)	Black	Orange Yellow
Sanwa	Red stripe	Black (in center)	Black

22.8.2 PINOUT AND COLOR CODING

The physical shape of the connector is just one consideration. The wiring of the connectors (called the pinout) is also critical. Fortunately, all but the "old-style" Airtronics servos (and the occasional oddball four-wire servo) use the same pinout, as shown in Fig. 22-5. With very few exceptions, R/C servo connectors use three wires, providing DC power, ground, and signal (or control). Table 22-2 lists the pinouts for several popular brands of servos.

Most servos use color coding to indicate the function of each connection wire, but the actual colors used for the wires vary between servo makers. Table 22-3 lists the most common colors used in several popular brands, but chances are you will be right if you assume that red is positive voltage, black is negative, and the third wire is the signal.

22.9 Circuits for Controlling a Servo

Unlike a DC motor, which runs if you simply attach battery power to its leads, a servo motor requires proper interface electronics in order to rotate its output shaft. While the need for interface electronics may complicate to some degree your use of servos, the electronics are actually rather simple. And if you plan on operating your servos with a PC or microcontroller (such as the BASIC Stamp), all you need for the job is a few lines of software.

A DC motor typically needs power transistors, MOSFETs, or relays if it is interfaced to a computer. A servo on the other hand can be directly coupled to a PC or microcontroller with no additional electronics. All of the power-handling needs are taken care of by the control board in the servo, saving you the hassle. This is one of the key benefits of using servos with computer-controlled robots.

22.9.1 CONTROLLING A SERVO VIA A 556 TIMER CHIP

You don't need a computer to control a servo. You can use the venerable 555 timer IC circuit that is doubled in the 556 chip to provide the required pulses to a servo. This circuit uses one of the 555 timers built into the chip to provide the 20 ms pulse period and the second timer to provide the 1 to 2 ms pulse. Fig. 22-6 shows the circuit that can be used to control a servo, and Table 22-4 lists the parts that you will need.

22.9.2 CONTROLLING A SERVO VIA A BASIC STAMP

With a little bit of thought to the software, a BASIC Stamp 2 can be used to control one or more servos as well as have enough time left over to poll inputs and determine what is the appropriate next action. Fig. 22-7 shows the hookup diagram for connecting a standard servo to the BS2. It is very important to note that the power to the servo does not come from the BS2, or any prototyping board it is on. Servos require more current than the power supply on the Stamp can provide. A pack of four AA batteries is sufficient to power

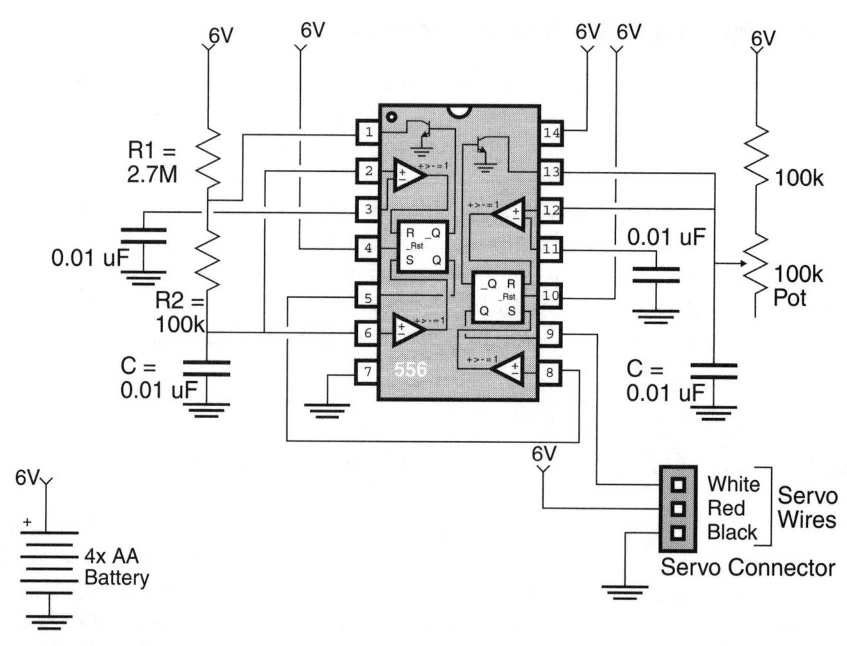

FIGURE 22-6 The left 555 oscillator circuit provides a 20 ms trigger waveform for the monostable oscillator on the right. The monostable oscillator produces a 1 to 2 ms pulse depending on the position of the potentiometer.

TABLE 22-4 Parts Needed to Build the 556 Servo Control Circuit

IC1	556 timer chip
Pot	100k single turn potentiometer
R2, R3	100k resistor
R1	2.7M resistor
C1–C4	0.01 μF capacitors (any type)
Bat	4x AA battery clip
Misc.	Circuit breadboard, wiring, servo connector header

the servo. For proper operation ensure that the grounds are connected between the Stamp and the battery pack. Use a 33 to 47 μF capacitor between the +V and ground of the AA pack to help kill any noise that may be induced into the electronics when the servo turns on and off.

The following application sends the appropriate pulses to a servo for 1 s and then polls the user (using the DEBUG and DEBUGIN console interface statements) to enter in a value that is passed to the servo. The reason why the application is called *calibrate* is due to its use as a method to find the correct stop pulse length for a servo modified for continuous rotation (described in the following).

```
'  Calibrate - Find the Center/Not Moving Point for Servo
'{$STAMP BS2}
'{$PBASIC 2.50}
```

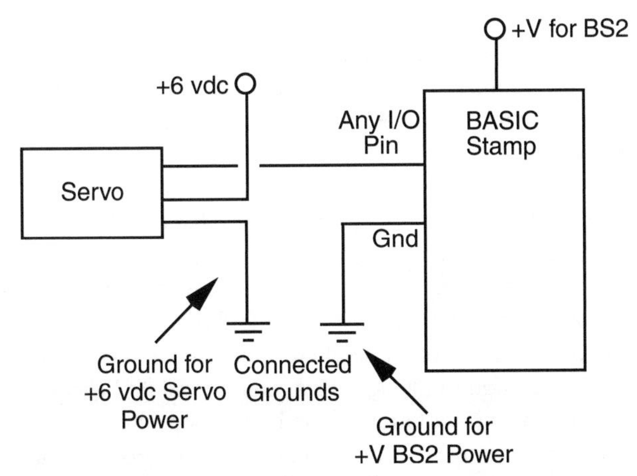

FIGURE 22-7 Hookup diagram for connecting a servo to a BASIC Stamp 2.

```
' Variables
Servo           pin 15
CurrentDelay    var word        ' Servo Center Point
i               var byte

' Initialization/Mainline
  low Servo                     ' Set Servo Pin Low
  CurrentDelay = 750            ' Start at 1.5 ms
  do                            ' Repeat forever
    debug "Current Servo Delay Value = ",dec CurrentDelay,cr
    for i = 0 to 50             ' Output servo value for 1 s
      pulsout Servo, CurrentDelay
      pause 18                  ' 20 msec Cycle Time
    next
    debug "Enter in New Delay Value "
    debugin dec CurrentDelay
    do while (CurrentDelay < 500) or (CurrentDelay > 1000)
      debug "Invalid Value, Must be between 500 and 1000",cr
      debug "Enter in New Delay Value "
      debugin dec CurrentDelay
    loop
  loop
```

22.9.3 USING A DEDICATED CONTROLLER

R/C receivers are designed with a maximum of eight servos in mind. The receiver gets a digital pulse train from the transmitter, beginning with a long sync pulse, followed by as many as eight servo pulses. Each pulse is meant for a given servo attached to the receiver: pulse 1 goes to servo 1, pulse 2 goes to servo 2, and so on. The eight pulses plus the sync pulse take about 20 ms. This means the pulse train can be repeated 50 times each second, which is its refresh rate. As the refresh rate gets slower the servos aren't updated as quickly and can throb or lose position as a result.

Depending on your electronics and programming ability, you may find it very difficult to simultaneously supply pulses to multiple servos at one time (a quick Google search on the Internet will find applications that control up to four servos simultaneously using a single BS2 and more using a Microchip PIC MCU). Rather than invest the time and effort developing this capability, you can buy inexpensive servo controllers from a number of sources (see Appendix B, "Sources," for addresses and web sites) that can control five, eight, or even more servos autonomously, which reduces the program overhead of the microcontroller or computer you are using.

The main benefit of dedicated servo controllers is that a great number of servos can be commanded simultaneously, even if your computer, microcontroller, or other circuitry is not multitasking. For example, suppose your robot requires 24 servos. Say it's an eight-legged spider, and each leg has three servos; each servo controls a different "degree of freedom" of the leg. One approach would be to divide the work among three servo controllers, each capable of handling eight servos. Each controller would be responsible for a given degree of freedom. One might handle the rotation of all eight legs; another might handle the flexion of the legs; and the third might be for the rotation of the bottom leg segment.

Dedicated servo controllers must be used with a computer or microcontroller, as they need to be provided with real-time data in order to operate (commonly sent in a serial data format). A sequence of bytes sent from the computer or microcontroller is decoded by the servo controller, with each byte corresponding to a servo attached to it. Servo controllers typically come with application notes and sample programs for popular computers and microcontrollers, but to make sure things work it's very helpful to have a knowledge of programming and serial communications.

22.9.4 SERVO VOLTAGE MARGINS

Servos are designed to be used with rechargeable model R/C battery packs, which normally put out 4.8 V or four AA alkaline battery packs (which put out approximately 6 V). As the batteries drain the voltage will drop, and you will notice your servos won't be as fast as they used to be. Somewhere below about 4.5 V the servos stop being responsive. Similarly, while servos may work with power supplies greater than 6 V, you cannot count on it and you will find that at some point you will burn out their control electronics or even their motors.

Ideally, your servo power supply should be monitored to ensure that the voltage stays within the range of 4.5 and 6 V. This may mean that the servos in a particular robot will require their own power supply or battery pack, but by ensuring the correct voltage is applied to them, they can be assumed to work properly for their full lives.

22.9.5 WORKING WITH AND AVOIDING THE DEAD BAND

References to the Grateful Dead notwithstanding, all servos exhibit what's known as a *dead band*. The dead band of a servo is the maximum time differential between the incoming control signal and the internal reference signal produced by the position of the potentiometer. If the time difference equates to less than the dead band—say, 5 or 6 ms—the servo will not bother trying to nudge the motor to correct for the error.

Without the dead band, the servo would constantly hunt back and forth to find the exact match between the incoming signal and its own internal reference signal. The dead band allows the servo to minimize this hunting so it will settle down to a position close to, though maybe not exactly, where it's supposed to be.

The dead band varies between servos and is often listed as part of the servo's specifications. A typical dead band is 5 μs. If the servo has a full travel of 180° over a 1000 μs (1 to 2 ms) range, then the 5 μs dead band equates to 1 part in 200. You probably won't even notice the effects of dead band if your control circuitry has a resolution lower than the dead band.

However, if your control circuitry has a resolution higher than the dead band (which is the case with a microcontroller such as the BASIC Stamp 2 or the Motorola MC68HC11) then small changes in the pulse width values may not produce any effect. For instance, if the controller has a resolution of 2 μs and if the servo has a dead band of 5 μs, then a change of just one or even two values—equal to a change of 2 or 4 μs in the pulse width—may not have an effect on the servo.

The bottom line: choose a servo that has a narrow dead band if you need accuracy and if your control circuitry or programming environment has sufficient resolution. Otherwise, ignore the dead band since it probably won't matter one way or another. The trade-off here is that with a narrow dead band the servo will be more prone to hunt to its position and may even buzz after it has found it.

22.9.6 GOING BEYOND THE 1 TO 2 MILLISECOND PULSE RANGE

You've already read that the typical servo responds to signals from 1 to 2 ms. While this is true in theory, in actual practice many servos can be fed higher and lower pulse values in order to maximize their rotational limits. The 1 to 2 ms range may indeed turn a servo one direction or another, but it may not turn it all the way in both directions. However, you won't know the absolute minimums and maximums for a given servo until you experiment with it. Take fair warning: performing this experiment can be risky because operating a servo to its extreme can cause the mechanism to hit its internal stops. As noted earlier in this chapter, if left in this state for any period of time, the gears and electronics of the servo can become damaged.

If you just must have maximum rotation from your servo, connect it to your choice of control circuitry. Start by varying the pulse width in small increments below 1 ms (1000 μs), say in 10 μs chunks. After each additional increment, have your control program swing the servo back to its center or neutral position. When during your testing you hear the servo hit its internal stop (the servo will chatter as the gears slip), you've found the absolute lower-bound value for that servo. Repeat the process for the upper bound. It's not unusual for some servos to have a lower bound of perhaps 250 μs and an upper bound of over 2200 μs. Yet other servos may be so restricted that they cannot even operate over the normal 1 to 2 ms range.

Keep a notebook of the upper and lower operating bounds for each servo in your robot or parts storehouse. Since there can be mechanical differences between servos of the same brand and model, number your servos so you can tell them apart. When it comes time to program them, you can refer to your notes for the lower and upper bounds for that particular servo.

22.10 Modifying a Servo for Continuous Rotation

Many brands and models of R/C servos can be readily modified to allow them to rotate continuously, like a regular DC motor. Such modified servos can be used as drive motors for your robot. Many modern servos also come with the ability to turn continuously with just the flick of a switch. Servos that can turn continuously can be easier to use than regular DC motors since they already have the power drive electronics built in, they come already geared down, and they are easy to mount on your robot.

22.10.1 BASIC MODIFICATION INSTRUCTIONS

Servo modification varies somewhat between makes and models, but the basic steps are the same:

1. Remove the case of the servo to expose the gear train, motor, and potentiometer. This is accomplished by removing the four screws on the back of the servo case and separating the top and bottom.

2. File or cut off the nub on the underside of the output gear that prevents full rotation. This typically means removing one or more gears, so you should be careful not to misplace any parts. If necessary, make a drawing of the gear layout so you can replace things in their proper location!
3. Remove the potentiometer and replace it with two 2.7K-Ω 1 percent tolerance (precision) resistors, wired as shown in Fig. 22-8. This voltage divider circuit "fools" the servo's control circuitry into thinking it's always in the center position and will respond by turning the motor any time the incoming PWM is outside the center position (which is found using the calibrate program previously listed). Another approach is to relocate the potentiometer to the outside of the servo case, so that you can make fine-tune adjustments to the center position. If you do this, it is suggested that a multi-turn trimmer potentiometer is used.
4. Reassemble the case.

22.10.2 APPLYING NEW GREASE

The gears in a servo are lubricated with a white or clear grease. As you remove and replace the gears during your modification surgery, it's inevitable that some of the grease will come off on your fingers. If you feel too much of the grease has come off, you'll want to apply more. Most any viscous synthetic grease suitable for electronics equipment will work, though you can also splurge and buy a small tube of grease especially made for servo gears and other mechanical parts in model cars and airplanes.

When applying grease be sure to spread it around so that it gets onto all the mechanical parts of the servo that mesh or rub. However, avoid getting any of it inside the motor or on the electrical parts. Wipe off any excess.

While it may be tempting, don't apply petroleum-based oil to the gears, such as three-in-one oil or a spray lubricant like WD-40. Some oils may not be compatible with the plastics used in the servo, and spray lubricants aren't permanent enough.

22.10.3 TESTING THE MODIFIED SERVO

After reassembly but before connecting the servo to a control circuit, you'll want to test your handiwork to make sure the output shaft of the servo rotates smoothly. Do this by attach-

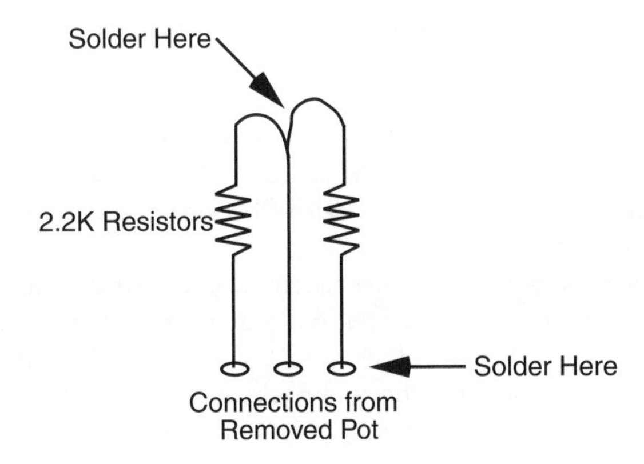

FIGURE 22-8 To modify a servo you must replace the internal potentiometer with two 2.7K resistors, wired as a voltage divider as shown here.

ing a control disc or control horn to the output shaft of the servo. Slowly and carefully rotate the disc or horn and note any snags. Don't spin too quickly, as this will put undo stress on the gears. If you notice any binding while you're turning the disc or horn, it could mean you didn't remove enough of the mechanical stop on the output gear. Disassemble the servo just enough to gain access to the output gear and clip or file off some more.

Once you are comfortable with the servo's ability to turn, you can see how it works using the 556 or BS2 circuits (and software) presented earlier in the chapter.

22.10.4 A CAUTION ON MODIFYING SERVOS

Modifying a servo typically entails removing or gutting the potentiometer and clipping off any mechanical stops or nubs on the output gear. For all practical purposes, this renders the servo unusable for its intended use, that is, to precisely control the angular position of its output shaft. So, before modifying a servo, be sure it's what you want to do. It'll be difficult to reverse the process, and in any case you are voiding its warranty.

22.10.5 SOFTWARE FOR RUNNING MODIFIED SERVOS

Even though a servo has been modified for continuous rotation, the same digital pulses are used to control the motor. Keep the following points in mind when running modified servos:

- If you've used fixed resistors in place of the original potentiometer inside the servo, sending a pulse of about 1.5 ms will stop the motor. Decreasing the pulse width will turn the motor in one direction; increasing the pulse width will turn the motor in the other direction. You will need to experiment (using the calibrate program) with the exact pulse width to find the value that will cause the motor to stop.
- If you've used a replacement 5K potentiometer instead of the original that was inside the servo, you have the ability to set the precise center point that will cause the motor to stop. In your software, you can send a precise 1.5 ms pulse, then adjust the potentiometer until the servo stops. As with fixed resistors, values higher or lower than 1.5 ms will cause the motor to turn one way or another.

22.10.6 LIMITATIONS OF MODIFIED SERVOS

Modifying a servo for continual rotation carries with it a few limitations, exceptions, and "gotchas" that you'll want to keep in mind:

- The average servo is not engineered for lots and lots of continual use. The mechanics of the servo are likely to wear out after perhaps as little as 25 hours (that's elapsed time), depending on the amount of load on the servo. Models with metal gears and/or brass bushing or ball bearings will last longer.
- The control electronics of a servo are made for intermittent duty. Servos used to power a robot across the floor may be used minutes or even hours at a time, and they tend to be under additional mechanical stress because of the weight of the robot. Though this is not exactly common, it is possible to burn out the control circuitry in the servo by overdriving it.

- Standard-sized servos are not particularly strong in comparison to many other DC motors with gear heads. Don't expect a servo to move a 5- or 10-lb robot. If your robot is heavy, consider using either larger, higher-output servos (such as ¼-scale or sail winch), or DC motors with built-in gear heads.
- Last and certainly not least, remember that modifying a servo voids its warranty. You'll want to test the servo before you modify it to ensure that it works.

22.10.7 MODIFYING BY REMOVING THE SERVO CONTROL BOARD

Another way to modify a servo for continuous rotation is to follow the steps outlined previously and also remove the control circuit board. Your robot then connects directly to the servo motor. You'd use this approach if you don't want to bother with the pulse width schema. You get a nice, compact DC motor with gearbox attached.

However, since you've removed the control board, you will also need to provide adequate power output circuitry to drive the motor. The circuitry built onto the servo PCB was designed for the motor; ideally, you might want to do a bit of probing on the servo PCB to try and figure out what are the components and circuit used in the motor driver so that you can replicate it rather than having to come up with your own circuit.

22.11 Attaching Mechanical Linkages to Servos

One of the benefits of using R/C servos with robots is the variety of ways it offers you to connect stuff to the servos. In model airplane and car applications, servos are typically connected to a push/pull linkage of some type. For example, in a plane, a servo for controlling the rudder would connect to a push/pull linkage directly attached to the rudder. As the servo rotates, the linkage draws back and forth, as shown in Fig. 22-9. The rudder is attached to the body of the plane using a hinge, so when the linkage moves, the rudder flaps back and forth.

You can use the exact same hardware designed for model cars and airplanes with your servo-equipped robots. Visit the neighborhood hobby store and scout for possible parts you can use. Collect and read through web sites and catalogs of companies that manufacture and sell servo linkages and other mechanics. Appendix B, "Sources," lists several such companies.

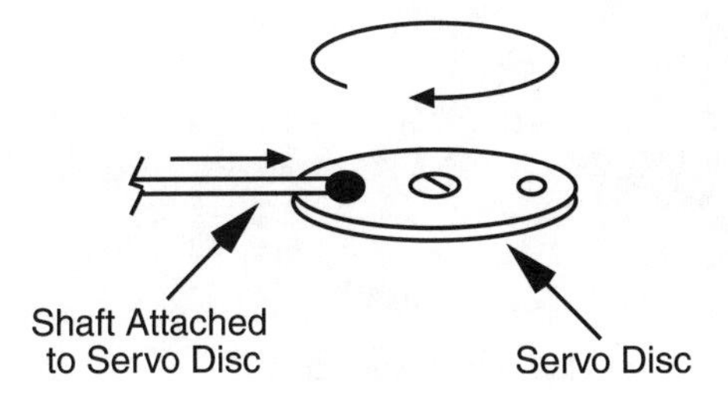

FIGURE 22-9 Servos can be used to transform rotational motion to linear motion.

22.12 Attaching Wheels to Servos

Servos reengineered for full rotation are most often used for robot locomotion and are outfitted with wheels. Since servos are best suited for small- to medium-sized robots (under about 3 lb), the wheels for the robot should ideally be between 2 and 5 in diameter. Larger-diameter wheels make the robot travel faster, but they can weigh more. You won't want to put extra large 7- or 10-in wheels on your robot if each wheel weighs 1.5 lb. There's your 3-lb practical limit right there.

The general approach for attaching wheels to servos is to use the round control disc that comes with the servo (see Fig. 22-10). The underside of the disc fits snugly over the output shaft of the servo. You can glue or screw the wheel to the front of the disc. Here are some ideas:

- Large LEGO balloon tires have a recessed hub that exactly fits the control disc included with Hitec and many other servos. You can simply glue the disc into the rim of the tire.
- Lightweight foam tires, popular with model airplanes, can be glued or screwed to the control disc. The tires are available in a variety of diameters. If you wish, you can grind down the hub of the tire so it fits smoothly against the control disc.

FIGURE 22-10 Attaching a round control disc to the hub of a wheel.

- A gear glued or screwed into the control disc can be used as an ersatz wheel or as a gear that drives a wheel mounted on another shaft.

In all these cases, it's important to maintain access to the screw used to secure the control disc to the servo. When you are attaching a wheel or tire be sure not to block the screw hole. If necessary, insert the screw into the control disc first, then glue or otherwise attach the tire. Make sure the hub of the wheel is large enough to accept the diameter of your screwdriver, so you can tighten the screw over the output shaft of the servo.

22.13 Mounting Servos on the Body of the Robot

Servos should be securely mounted to the robot so the motors don't fall off while the robot is in motion. The following methods do not work well, though they are commonly used:

- *Duct tape or electrical tape.* The "goo" on the tape is elastic, and eventually the servo works itself loose. The tape can also leave a sticky residue.
- *Hook-and-loop, otherwise known as Velcro.* Accurate alignment of the hook-and-loop halves can be tricky, meaning that every time you remove and replace the servos the wheels are at a slightly different angle with respect to the body of the robot. This makes it harder to program repeatable actions.
- *Tie-wraps.* You must cinch the tie-wrap tightly in order to adequately hold the servo in place. Unless your robot is made of metal or strong plastic, you're bound to distort whatever part of the robot you've cinched the wrap against.

Experience shows that hard mounting—gluing, screwing, or bolting—the servos onto the robot body is the best overall solution, and it greatly reduces the frustration level of hobby robotics.

22.13.1 ATTACHING SERVOS WITH GLUE

Gluing is a quick and easy way to mount servos on most any robot body material, including heavy cardboard and plastic. Use only a strong glue, such as two-part epoxy or hot-melt glue. When gluing it is important that all surfaces be clean. Rough up the surfaces with a file or heavy-duty sandpaper for better adhesion. If you're gluing servos to LEGO parts, apply a generous amount so the extra adequately fills between the nubs. LEGO plastic is hard and smooth, so be sure to rough it up first.

22.13.2 ATTACHING SERVOS WITH SCREWS OR BOLTS

A disadvantage of mounting servos with glue is that it's more or less permanent (and, according to Murphy's Law, more permanent than you'd like if you want to remove the servo, less permanent if you want the servo to stay in place!). For the greatest measure of

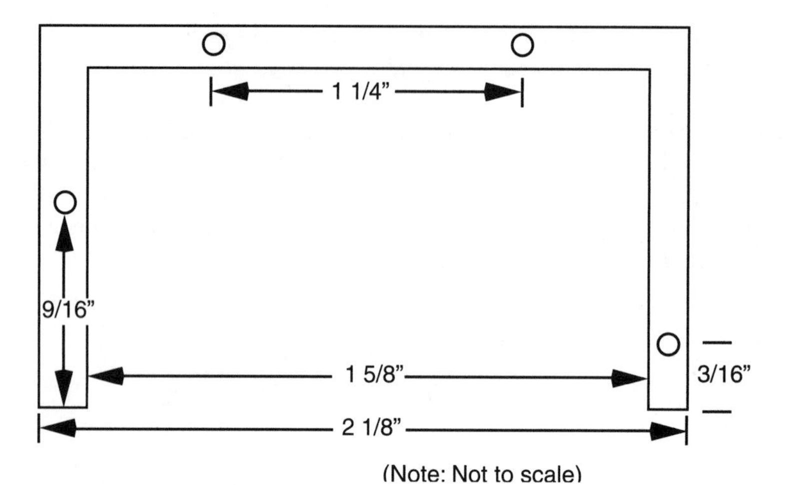

FIGURE 22-11 Use this template to construct a servo mounting bracket. The template may not be reproduced in 1:1 size, so be sure to measure before cutting your metal or plastic.

FIGURE 22-12 A servo mounted on a homemade servo bracket.

flexibility, use screws or bolts to mount your servos to your robot body. All servos have mounting holes in their cases; it's simply a matter of finding or drilling matching holes in the body of your robot.

Servo mounts are included in many R/C radio transmitters and separately available servo sets. You can also buy them separately from the better-stocked hobby stores. The servo mount has space for one, two, or three servos. The mount has additional mounting holes that you can use to secure it to the side or bottom of your robot. Most servo mounts are made of plastic, so if you need to make additional mounting holes they are easy to drill.

You can also construct your own servo mounting brackets using $1/8$-in-thick aluminum or plastic. A template is shown in Fig. 22-11. (Note: the template is not to scale, so don't trace it to make your mount. Use the dimensions to fashion your mount to the proper size.)

The first step in constructing your own servo mounting brackets is to cut and drill the aluminum or plastic, as shown in Fig. 22-12. Use a small hobby file to smooth off the edges and corners. The mounting hole centers provided in the template are designed to line up with the holes in LEGO Technic beams. This allows you to directly attach the servo mounts to LEGO pieces. Use $3/32$ or $4/40$ nuts and bolts, or $4/40$ self-tapping screws, to attach the servo mount to the LEGO beam.

Fig. 22-12 shows a servo mounted on a bracket and attached to a LEGO beam. If necessary, the servos can be easily removed for repair or replacement.

22.14 From Here

To learn more about . . .	***Read***
Using batteries to power your robot	Chapter 17, "Batteries and Robot Power Supplies"
Fundamentals of robot locomotion	Chapter 18, "Principles of Robot Locomotion"
Choosing the best motor for your robot	Chapter 19, "Choosing the Right Motor"
Ways to implement computers and microcontrollers in your robots	Chapter 12, "An Overview of Robot 'Brains' "
Interfacing servos and other motors to control circuitry	Chapter 14, "Computer Peripherals"

PART 5

PRACTICAL ROBOTICS PROJECTS

CHAPTER 23

BUILDING A ROVERBOT

Imagine a robot that can vacuum the floor for you, relieving you of that time-consuming household drudgery and freeing you to do other, more dignified tasks. Imagine a robot that patrols your house, inside or out, listening and watching for the slightest trouble and sounding the alarm if anything goes amiss. Imagine a robot that knows how to look for fire, and when it finds one, puts it out.

Think again. The compact and versatile Roverbot introduced in this chapter can serve as the foundation for building any of these more advanced robots. You can easily add a small DC-operated vacuum cleaner to the robot, then set it free in your living room. Only the sophistication of the control circuit or computer running the robot limits its effectiveness at actually cleaning the rug.

You can attach light and sound sensors to the robot, and provide it with eyes that help it detect potential problems. These sensors, as it turns out, can be the same kind used in household burglar alarm systems. Your only job is to connect them to the robot's other circuits. Similar sensors can be added so your Roverbot actively roams the house, barn, office, or other enclosed area looking for the heat, light, and smoke of fire. An electronically actuated fire extinguisher is used to put out the fire.

The Roverbot described on the following pages represents the base model only (see Fig. 23-1). The other chapters in this book will show you how to add onto the basic framework to create a more sophisticated automaton. The Roverbot borrows from techniques described in Chapter 10, "Metal Platforms." If you haven't yet read that chapter, do so now; it will help you get more out of this one.

FIGURE 23-1 The finished Roverbot (minus the batteries), ready for just about any enhancement you see fit.

23.1 Building the Base

Construct the base of the Roverbot using shelving standards or extruded aluminum channel stock. The prototype Roverbot for this book used aluminum shelving standards because aluminum minimized the weight of the robot. The size of the machine didn't require the heavier-duty steel shelving standards.

The base measures 12⅝ by 9⅛ in. These unusual dimensions make it possible to accommodate the galvanized nailing (mending) plates, which are discussed later in this

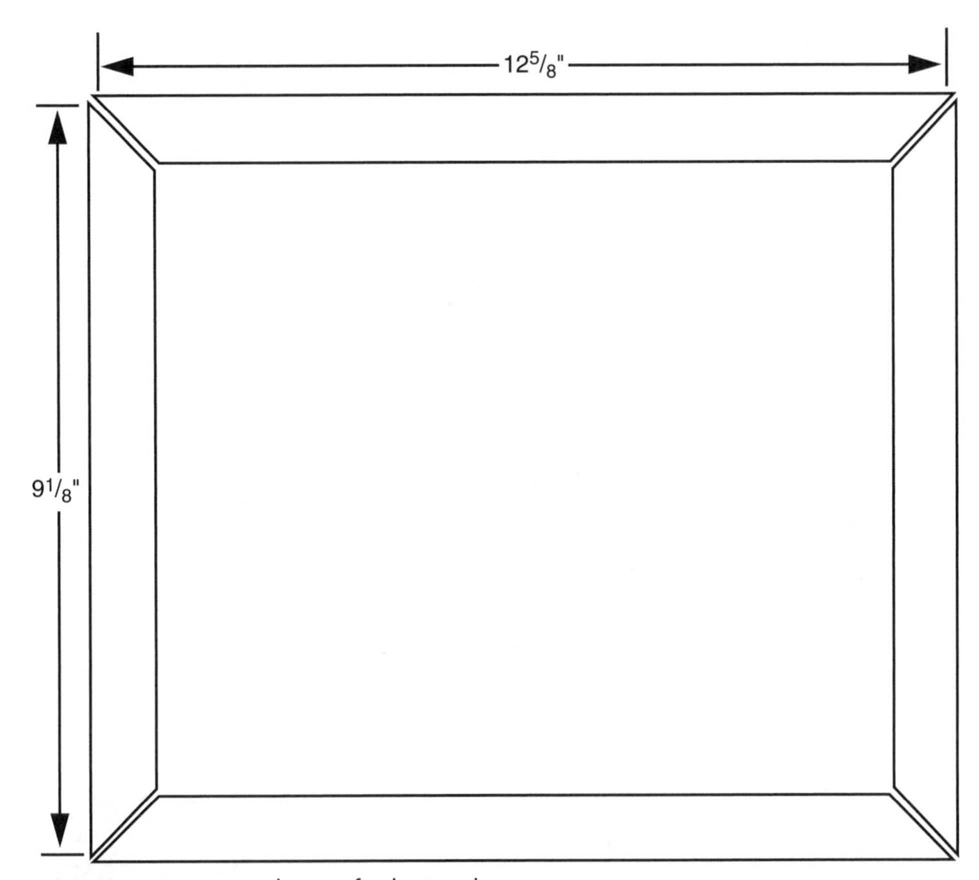

FIGURE 23-2 Cutting diagram for the Roverbot.

chapter. Cut two pieces each of 12⅝-in stock, with 45° miter edges on both sides, as shown in Fig. 23-2 (refer to the parts list in Table 23-1). Do the same with the 9⅛-in stock. Assemble the pieces using 1¼-by-⅜-in flat corner irons and 8⁄32-by-½-in nuts and bolts. Be sure the dimensions are as precise as possible and that the cuts are straight and even. Because you are using the mending plates as a platform, it's doubly important with this design that you have a perfectly square frame. Don't bother to tighten the nuts and bolts at this point.

Attach one 4 3⁄16-by-9-in mending plate to the left third of the base. Temporarily undo the nuts and bolts on the corners to accommodate the plate. Drill new holes for the bolts in the plate if necessary. Repeat the process for the center and left mending plates. When the three plates are in place, tighten all the hardware. Make sure the plates are secure on the frame by drilling additional holes near the inside corners (don't bother if the corner already has a bolt securing it to the frame). Use 8⁄32-by-½-in bolts and nuts to attach the plates into place. The finished frame should look something like the one depicted in Fig. 23-3. The underside should look like Fig. 23-4.

TABLE 23-1 Parts List for Roverbot

	FRAME
2	$12\frac{5}{8}$-in length aluminum or steel shelving standard
2	$9\frac{1}{8}$-in length aluminum or steel shelving standard
3	$4\frac{3}{16}$-by-9-in galvanized nailing (mending) plate
4	$1\frac{1}{4}$-by-$\frac{3}{8}$-in flat corner iron
	RISER
4	15-in length aluminum or steel shelving standard
2	7-in length aluminum or steel shelving standard
2	$10\frac{1}{2}$-in length aluminum or steel shelving standard
4	1-by-$\frac{3}{8}$-in corner angle iron
	MOTORS AND CASTER
2	Gear reduced output 6 or 12 V DC motors
4	$2\frac{1}{2}$-by-$\frac{3}{8}$-in corner angle iron
2	5- to 7-in diameter rubber wheels
2	$1\frac{1}{4}$-in swivel caster
Misc.	Nuts, bolts, fender washers, tooth lock washers, etc. (see text)
	POWER
2	6 or 12 V, 1 or 2 A-h batteries (voltage depending on motor)
2	Battery clamps

23.2 Motors

The Roverbot uses two drive motors for propulsion and steering. These motors, shown in Fig. 23-5, are attached in the center of the frame. The center of the robot was chosen to help distribute the weight evenly across the platform. The robot is less likely to tip over if you keep the center of gravity as close as possible to the center column of the robot.

The 12-V motors used in the prototype were found surplus, and you can use just about any other motor you find as a substitute. The motors used in the prototype Roverbot come with a built-in gearbox that reduces the speed to about 38 r/min. The shafts are $\frac{1}{4}$ in. Each shaft was threaded using a $\frac{1}{4}$-in 20 die to secure the 6-in-diameter lawn mower wheels in place. You can skip the threading if the wheels you use have a setscrew or can

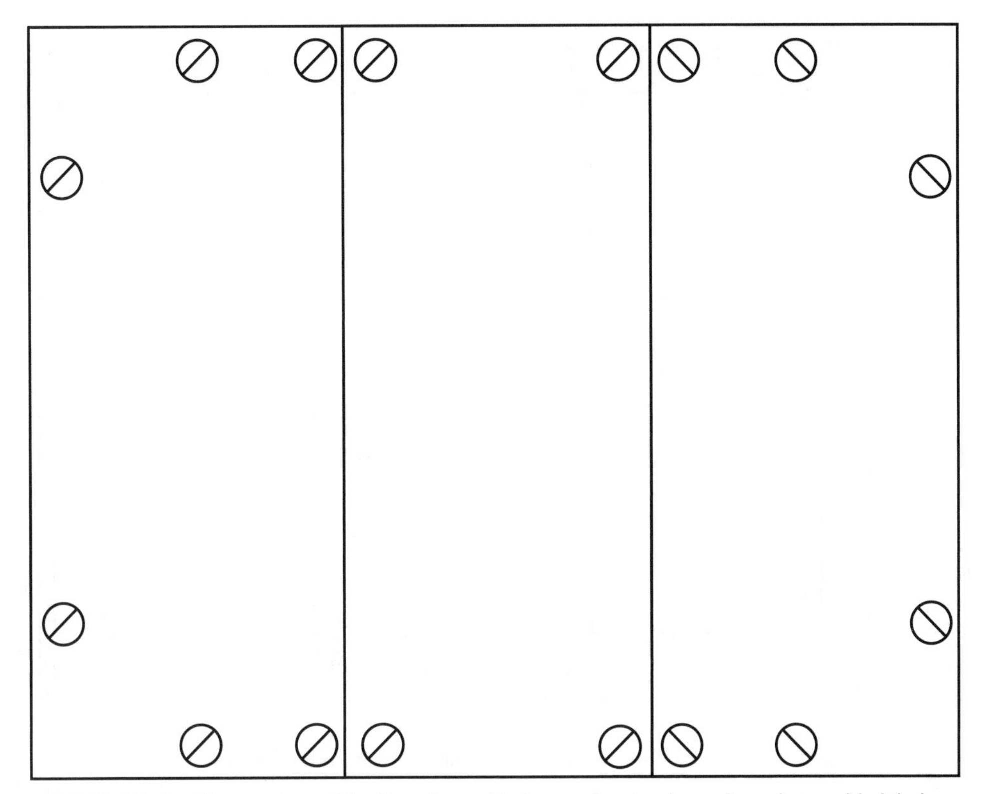

FIGURE 23-3 The top view of the Roverbot, with three galvanized mending plates added (holes in the plates not shown).

be drilled to accept a setscrew. Either way, make sure that the wheels aren't too thick for the shaft. The wheels used in the prototype were 1½ in wide, perfect for the 2-in-long motor shafts.

Mount the motors using two 2½-by-⅜-in corner irons, as illustrated in Fig. 23-6. Cut about 1 in off one leg of the iron so it will fit against the frame of the motor. Secure the irons to the motor using 8⁄32-by-½-in bolts (yes, these motors have pretapped mounting holes!). Finally, secure the motors in the center of the platform using 8⁄32-by-½-in bolts and matching nuts. Be sure that the shafts of the motors are perpendicular to the side of the frame. If either motor is on crooked, the robot will crab to one side when it rolls on the floor. There is generally enough play in the mounting holes on the frame to adjust the motors for proper alignment.

Now attach the wheels. Use reducing bushings if the hub of the wheel is too large for the shaft. If the shaft has been threaded, twist a ¼-in 20 nut onto it, all the way to the base. Install the wheel using the hardware shown in Fig. 23-7. Be sure to use the tooth lock washer. The wheels may loosen and work themselves free otherwise. Repeat the process for the other motor.

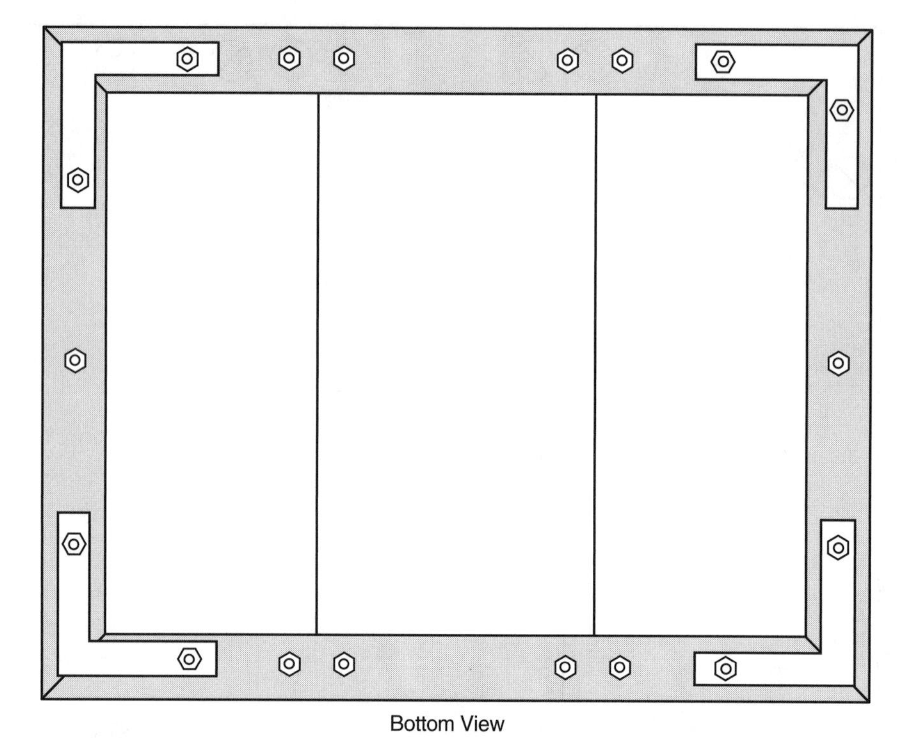

FIGURE 23-4 Hardware detail for the frame of the Roverbot (bottom view).

FIGURE 23-5 One of the drive motors, with wheel, attached to the base of the Roverbot.

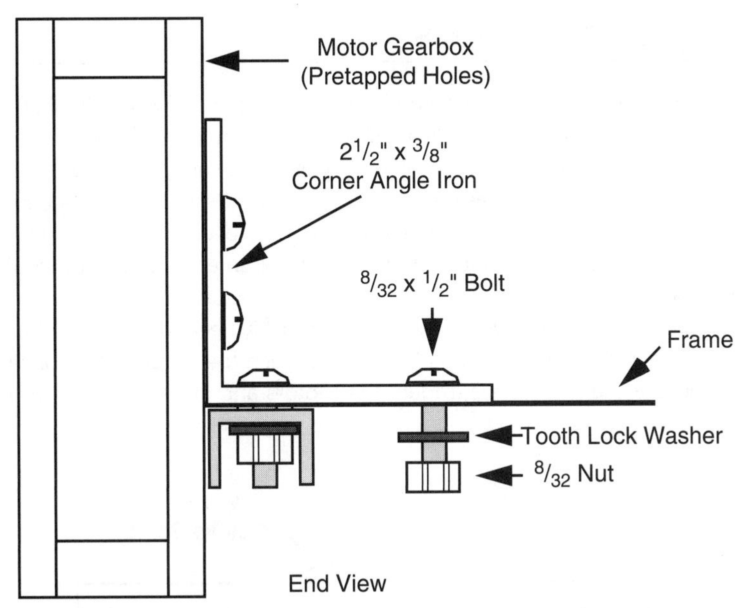

FIGURE 23-6 Hardware detail for the motor mount. Cut the angle iron, if necessary, to accommodate the motor.

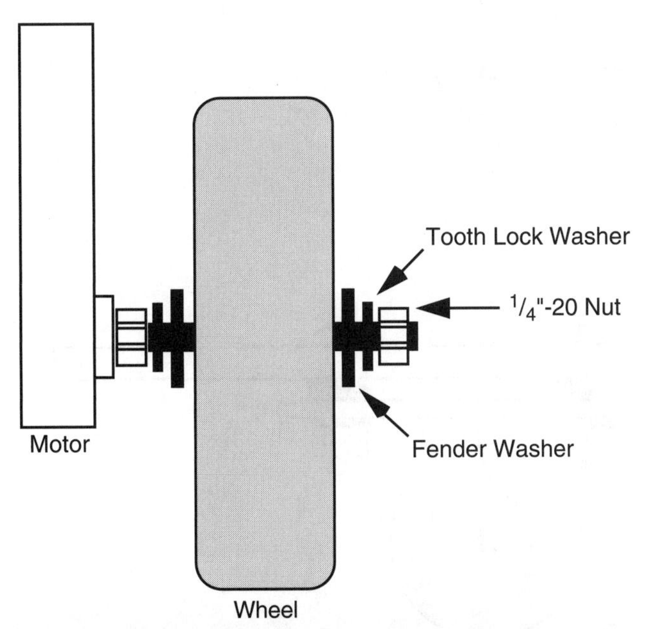

FIGURE 23-7 Hardware detail for attaching the wheels to the motor shafts. The wheels can be secured by threading the shaft and using ¼-in 20 hardware, as shown, or secured to the shaft using a setscrew or collar.

23.3 Support Casters

The ends of the Roverbot must be supported by swivel casters. Use a 2-in-diameter ball-bearing swivel caster, available at the hardware store. Attach the caster by marking holes for drilling on the bottom of the left and right mending plate. You can use the base plate of the caster as a drilling guide. Attach the casters using 8⁄32-by-½-in bolts and 8⁄32 nuts (see Fig. 23-8). You may need to add a few washers between the caster base plate and the mending plate to bring the caster level with the drive wheels (the prototype used a 5⁄16-in spacer). Do the same for the opposite caster.

If you use different motors or drive wheels, you'll probably need to choose a different size caster to match. Otherwise, the four wheels may not touch the ground all at once as they should. Before purchasing the casters, mount the motors and drive wheels, then measure

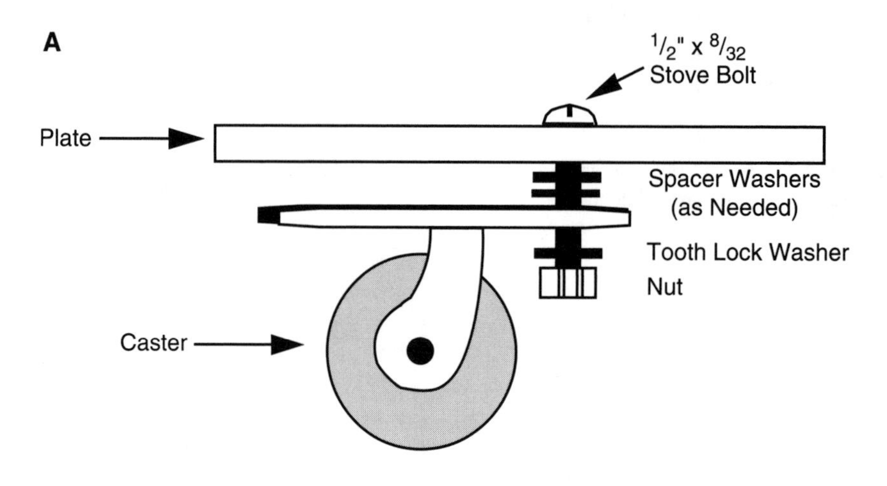

FIGURE 23-8 Adding the casters to the Roverbot. There is one caster on each end, and both must match the depth of the drive wheels (a little short is even better). *a.* Hardware detail; *b.* caster mounted on mending plate.

the distance from the bottom of the mending plate to the ground. Buy casters to match. Again, add washers to increase the depth, if necessary.

23.4 Batteries

Each of the drive motors in the Roverbot consumes ½ A (500 mA) of continuous current with a moderate load. The batteries chosen for the robot, then, need to easily deliver 2 A for a reasonable length of time, say 1 or 2 h of continuous use of the motors. A set of high-capacity NiCads would fit the bill. But the Roverbot is designed so that subsystems can be added to it. Those subsystems haven't been planned yet, so it's impossible to know how much current they will consume. The best approach to take is to overspecify the batteries, allowing for more current than is probably necessary.

Six- and 8-A-H lead-acid batteries are somewhat common on the surplus market. As it happens, 6 or 8 A are about the capacity that would handle intermittent use of the drive motors. (The various electronic subsystems, such as an on-board computer and alarm sensors, should use their own battery.) These heavy-duty batteries are typically available in 6-V packs, so two are required to supply the 12 V needed by the motors. Supplementary power, for some of the linear ICs, like op amps, can come from separate batteries, such as a NiCad pack. A set of C NiCads don't take up much room, but it's a good idea to leave space for them now, instead of redesigning the robot later on to accommodate them.

The main batteries are rechargeable, so they don't need to be immediately accessible in order to be replaced. But you'll want to use a mounting system that allows you to remove the batteries should the need arise. The clamps shown in Fig. 23-9 allow such accessibility.

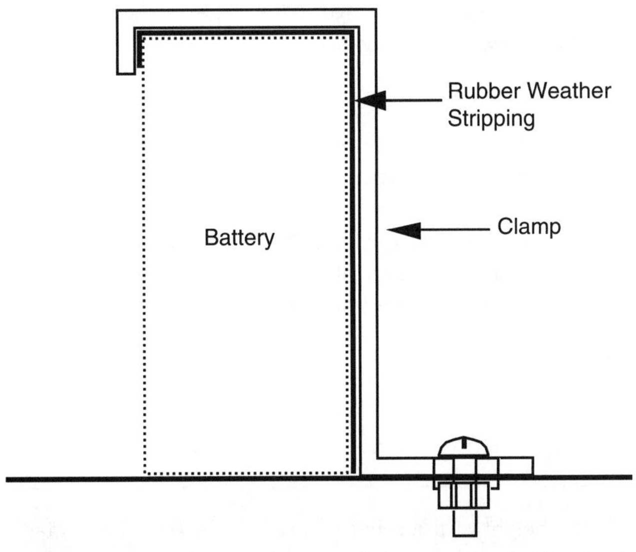

FIGURE 23-9 A battery clamp made from a strip of galvanized plate, bent to the contours of the battery. Line up the metal with weather stripping for a positive grip.

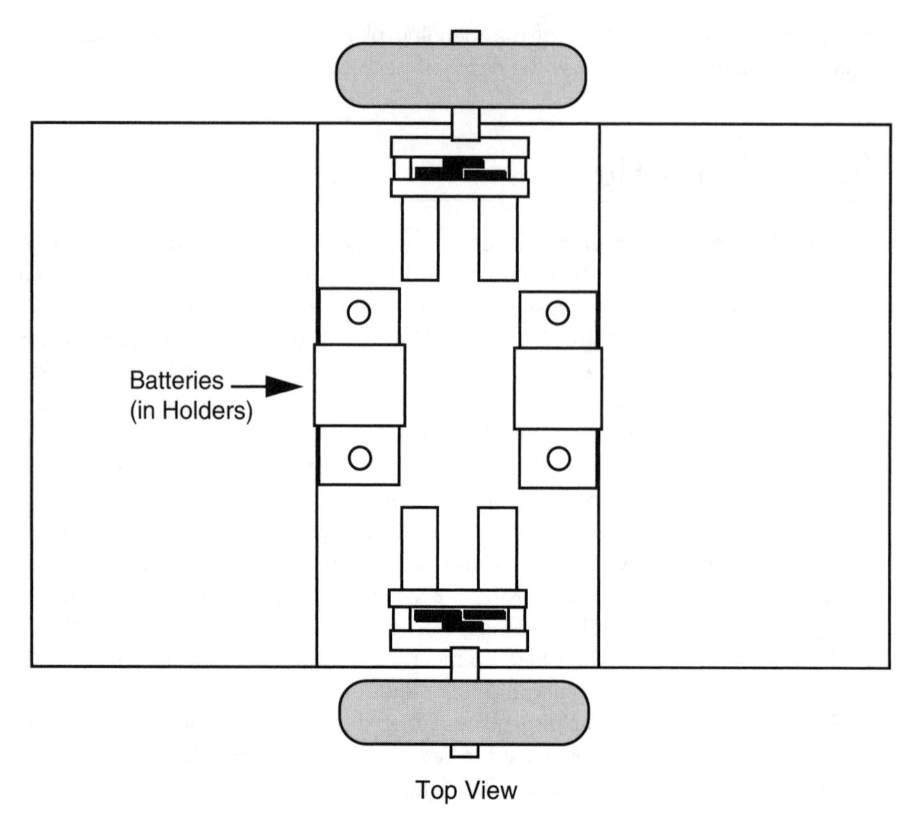

FIGURE 23-10 Top view of the Roverbot, showing the mounted motors and batteries. Note the even distribution of weight across the drive axis. This promotes stability and keeps the robot from tipping over. The wide wheelbase also helps.

The clamps are made from a 1¼-in-wide galvanized mending plate, bent to match the contours of the battery. Rubber weather strip is used on the inside of the clamp to hold the battery firmly in place.

The batteries are positioned off to either side of the drive wheel axis, as shown in Fig. 23-10. This arrangement maintains the center of gravity to the inside center of the robot. The gap also allows for the placement of one or two four-cell C battery packs, should they be necessary.

23.5 Riser Frame

The riser frame extends the height of the robot by approximately 15 in. Attached to this frame will be the sundry circuit boards and support electronics, sensors, fire extinguisher, vacuum cleaner motor, or anything else you care to add. The dimensions are large enough to assure easy placement of at least a couple of full-size circuit boards, a 2½-lb fire extinguisher, and a Black & Decker DustBuster. You can alter the dimensions of the frame, if desired, to accommodate other add-ons.

Make the riser by cutting four 15-in lengths of channel stock. One end of each length should be cut at 90°, the other end at 45°. Cut the mitered corners to make pairs, as shown in Fig. 23-11.

Make the crosspiece by cutting a length of channel stock to exactly 7 in. Miter the ends as shown in the figure.

Connect the two sidepieces and crosspiece using a 1½-by-⅜-in flat angle iron. Secure the angle iron by drilling matching holes in the channel stock. Attach the stock to the angle iron

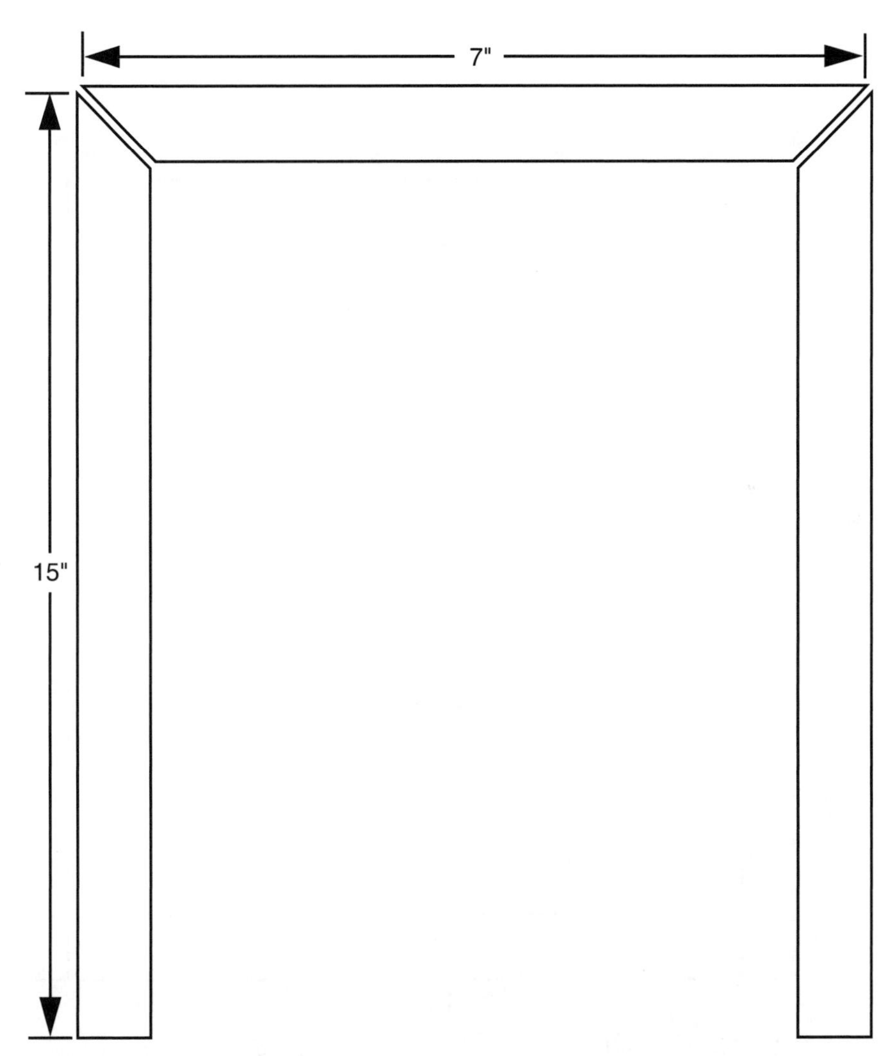

FIGURE 23-11 Cutting diagram for the Roverbot riser pieces (two sets).

by using 8/32-by-½-in bolts on the crosspieces and 8/32-by-1½-in bolts on the riser pieces. Don't tighten the screws yet. Repeat the process for the other riser.

Construct two beams by cutting the angle stock to 10½ in, as illustrated in Fig. 23-12. Do not miter the ends. Secure the beams to the top corners of the risers by using 1-by-⅜-in corner angle irons. Use 8/32-by-½-in bolts to attach the iron to the beam. Connect the angle irons to the risers using the 8/32-by-1½-in bolts installed earlier. Add a spacer between the inside of the channel stock and the angle iron if necessary, as shown in Fig. 23-13. Use 8/32 nuts to tighten everything in place.

Attach the riser to the base plate of the robot using 1-by-⅜-in corner angle irons. As usual, use 8/32-by-½-in bolts and nuts to secure the riser into place. The finished Roverbot body and frame should look at least something like the one in Fig. 23-1.

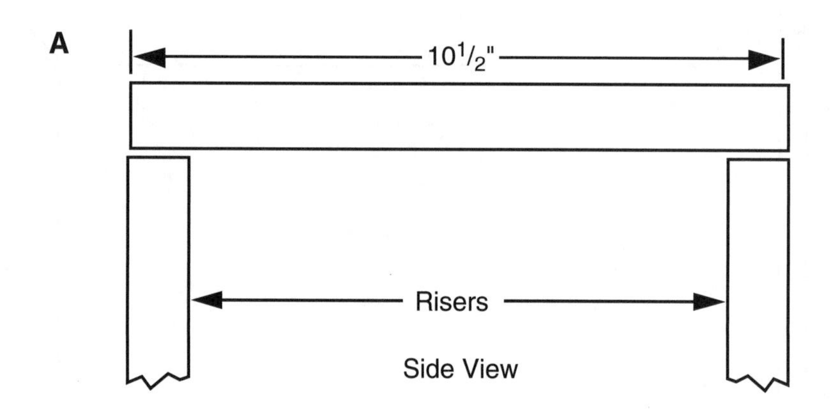

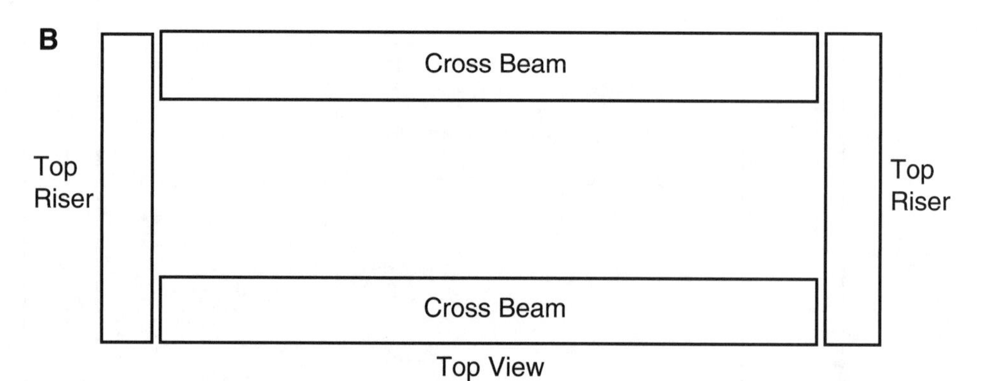

FIGURE 23-12 Construction details for the top of the riser. *a*. Side view showing the crosspiece joining the two riser sides; *b*. top view showing the cross beams and the tops of the risers.

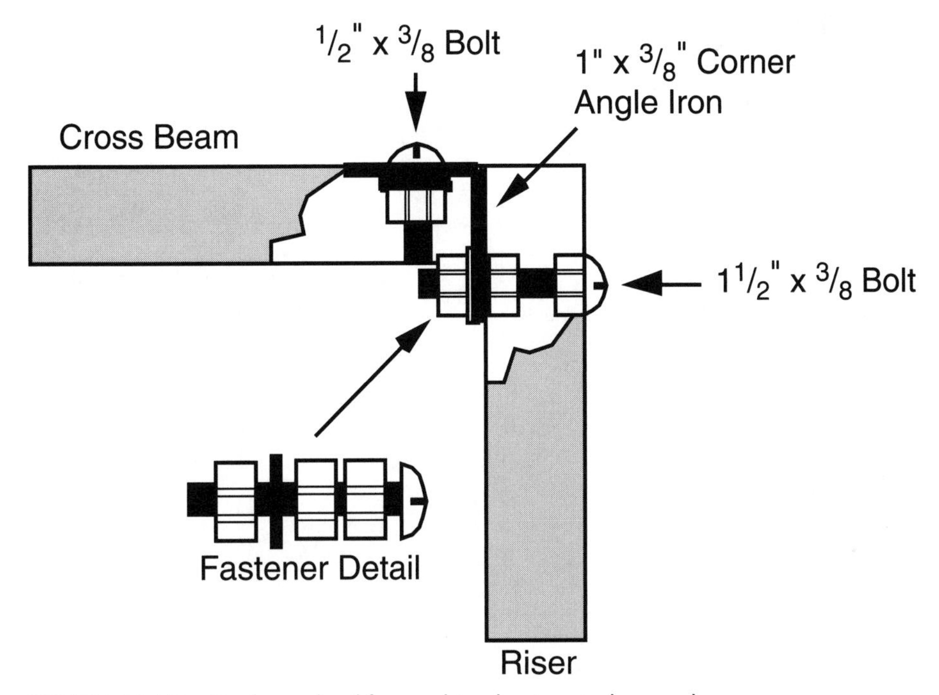

FIGURE 23-13 Hardware detail for attaching the risers to the cross beams.

23.6 Street Test

You can test the operation of the robot by connecting the motors and battery to a temporary control switch. See Chapter 8, "Plastic Platforms," for a wiring diagram. With the components listed in Table 23-1, the robot should travel at a speed of about 1 ft/s. The actual speed will probably be under that because of the weight of the robot. Fully loaded, the Roverbot will probably travel at a moderate speed of about 8 or 9 in/s. That's just right for a robot that vacuums the floor, roams the house for fires, and protects against burglaries. If you need your Roverbot to go a bit faster, the easiest (and cheapest) solution is to use larger wheels. Using 8-in wheels will make the robot travel at a top speed of 15 in/s.

One problem with using larger wheels, however, is that they raise the center of gravity of the robot. Right now, the center of gravity is kept rather low, thanks to the low position of the two heaviest objects, the batteries and motors. Jacking up the robot using larger wheels puts the center of gravity higher, so there is a somewhat greater chance of the robot tipping over. You can minimize any instability by making sure that subsystems are added to the robot from the bottom of the riser and that the heaviest parts are positioned closest to the base. You can also mount the motor on the bottom of the frame instead of on top.

23.7 From Here

To learn more about . . .	***Read***
Constructing robots using metal parts and pieces	Chapter 10, "Building a Metal Platform"
Powering your robot using batteries	Chapter 17, "Batteries and Robot Power Supplies"
Selecting a motor for your robot	Chapter 19, "Choosing the Right Motor"
Operating your robot with a computer	Chapter 12, "An Overview of Robot or Microcontroller 'Brains' "

CHAPTER 24

BUILDING A HEAVY-DUTY SIX-LEGGED WALKING ROBOT

The strange and unique contraption shown in Fig. 24-1 walks on six legs and turns corners with an ease and grace that belies its rather simple design. The Walkerbot design described in this chapter is for the basic frame, motor, battery system, running gear, and legs. You can embellish the robot with additional components, such as arms, a head, as well as computer control. The frame is oversized (in fact, it's too large to fit through some inside doors!), and there's plenty of room to add new subsystems.

The only requirement is that the weight doesn't exceed the driving capacity of the motors and batteries and that the legs and axles don't bend. The prototype Walkerbot weighs about 50 lb. It moves along swiftly and no structural problems have yet occurred. Another 10 or 15 lb could be added without worry.

24.1 Frame

The completed Walkerbot frame measures 18 in wide by 24 in long by 12 in deep. Construction is all aluminum, using a combination of $\frac{41}{64}$-by-$\frac{1}{2}$-by-$\frac{1}{16}$-in channel stock and 1-by-1-by-$\frac{1}{16}$-in angle stock.

Build the bottom of the frame by cutting two 18-in lengths of channel stock and two 24-in lengths of channel stock, as shown in Fig. 24-2 (refer to the parts list in Table 24-1). Miter the ends with a 45° angle. Attach the four pieces using $1\frac{1}{2}$-by-$\frac{3}{8}$-in flat angle irons

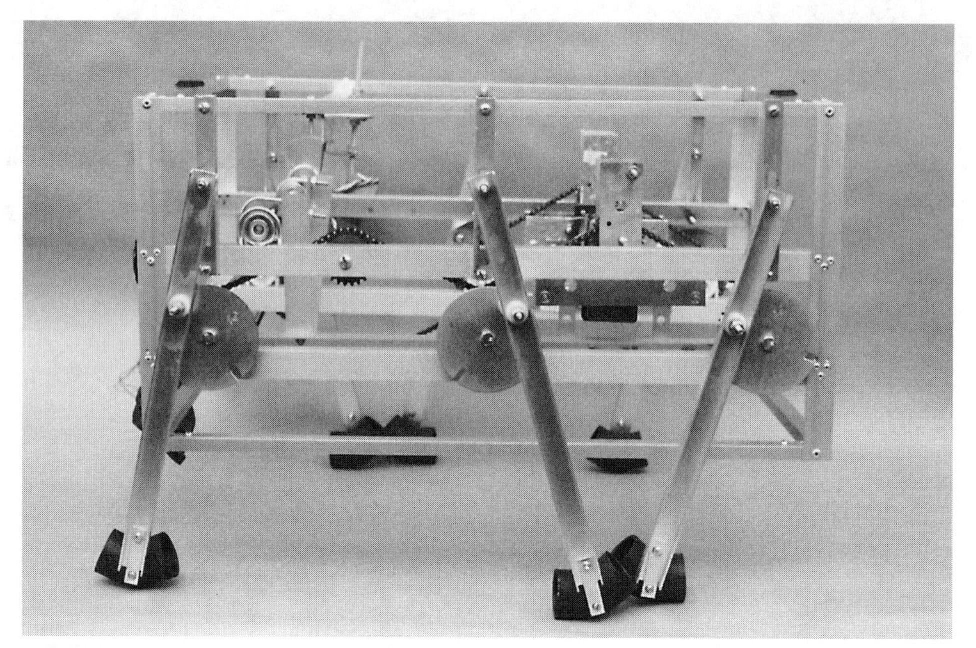

FIGURE 24-1 The completed Walkerbot.

and secure them with 3-by-½-in bolts and nuts. For added strength, use four bolts on each corner.

In the prototype Walkerbot, many of the nuts and bolts were replaced with aluminum pop rivets in order to reduce the weight. Until the entire frame is assembled, however, use the bolts as temporary fasteners. Then, when the frame is assembled, square it up and replace the bolts and nuts with rivets one at a time. Construct the top of the frame in the same manner.

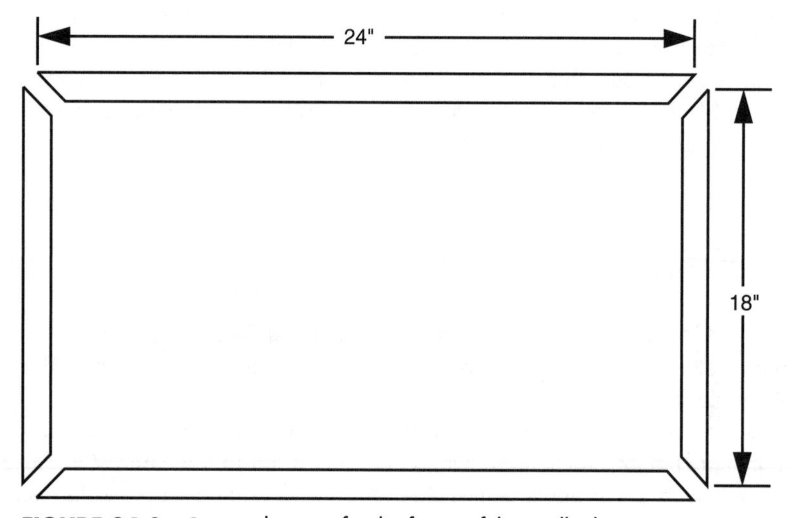

FIGURE 24-2 Cutting diagram for the frame of the Walkerbot (two sets).

TABLE 24-1 Parts List for Walkerbot Frame	
4	24-in lengths $^{41}/_{64}$-by-½-by-$^{1}/_{16}$-in aluminum channel stock
4	18-in lengths $^{41}/_{64}$-by-½-by-$^{1}/_{16}$-in aluminum channel stock
4	12-in lengths 1-by-1-by-$^{1}/_{16}$-in aluminum angle stock
8	1½-by-⅜-in flat angle iron
4	24-in lengths 1-by-1-by-$^{1}/_{16}$-in aluminum angle stock
2	17⅝-in lengths 1-by-1-by-$^{1}/_{16}$-in aluminum angle stock
Misc.	$^{8}/_{32}$ stove bolts, nuts, tooth lock washers, as needed

Connect the two halves with four 12-in lengths of angle stock, as shown in Fig. 24-3. Secure the angle stock to the frame pieces by drilling holes at the corners. Use $^{8}/_{32}$-by-½-in bolts and nuts initially; exchange for pop rivets after the frame is complete. The finished frame should look like the one diagrammed in Fig. 24-4.

Complete the basic frame by adding the running gear mounting rails. Cut four 24-in lengths of 1-by-1-by-$^{1}/_{16}$-in angle stock and two 17⅝-in lengths of the same angle stock. Drill

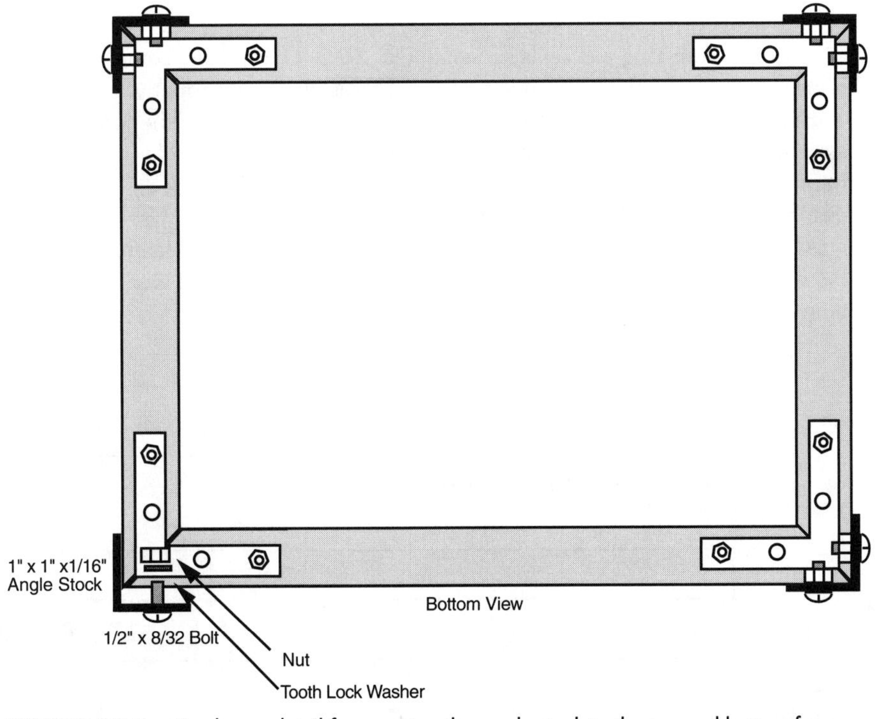

FIGURE 24-3 Hardware detail for securing the angle stock to the top and bottom frame pieces.

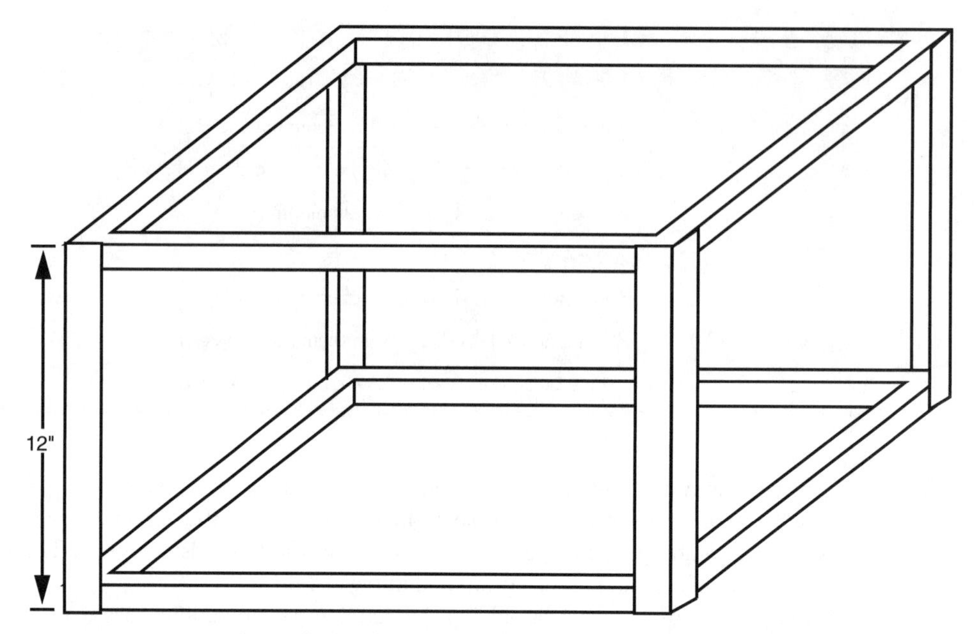

FIGURE 24-4 How the Walkerbot frame should look so far.

¼-in holes in four long pieces as shown in Fig. 24-5. The spacing between the sets of holes is important. If the spacing is incorrect, the U-bolts won't fit properly.

Refer to Fig. 24-6. When the holes are drilled, mount two of the long lengths of angle stock as shown. The holes should point up, with the side of the angle stock flush against the frame of the robot. Mount the two short lengths on the ends. Tuck the short lengths immediately under the two long pieces of angle stock you just secured. Use 8⁄32-by-½-in bolts and nuts to secure the pieces together. Dimensions, drilling, and placement are critical with these components. Put the remaining two long lengths of drilled angle stock aside for the time being.

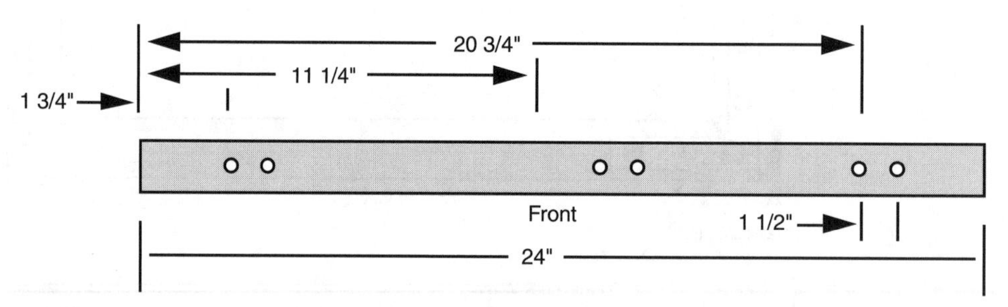

FIGURE 24-5 Cutting and drilling guide for the motor mount rails (four).

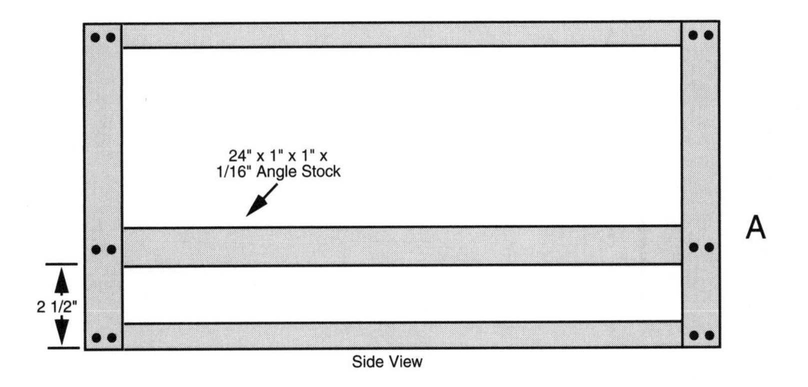

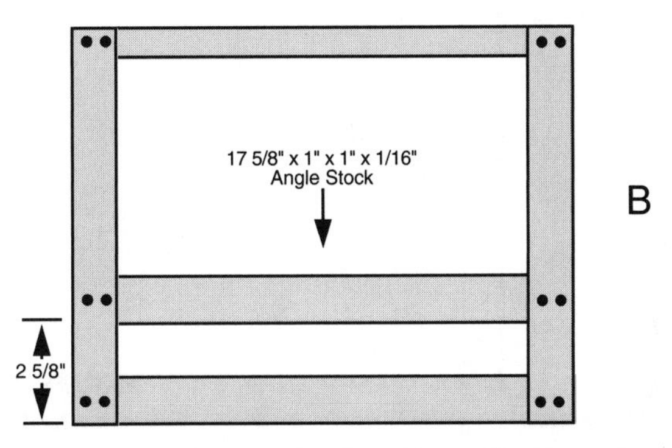

FIGURE 24-6 The motor mount rails secured to the robot. *a.* The long rail mounts 2½ in from the bottom of the frame (the holes drilled earlier point up); *b.* the short end crosspiece rail mounts 2⅝ in from the bottom of the frame.

24.2 Legs

You're now ready to construct and attach the legs. This is probably the hardest part of the project, so take your time and measure everything twice to assure accuracy. Cut six 14-in lengths of 57⁄64-by-9⁄16-by-1⁄16-in aluminum channel stock. Do not miter the ends. Drill a hole with a #19 bit ½ in from one end (the top); drill a ¼-in hole 4¾ in from the top (see Fig. 24-7; refer to the parts list in Table 24-2). Make sure the holes are in the center of the channel stock.

With a ¼-in bit, drill out the center of six 4⅝-in-diameter circular electric receptacle plate covers. The plate cover should have a notched hole near the outside, which is used to secure it to the receptacle box. If the cover doesn't have the hole, drill one with a ¼-in bit ⅜ in from the outside edge. The finished plate cover becomes a cam for operating the up and down movement of the legs.

Assemble four legs as follows: attach the 14-in-long leg piece to the cam using a ½-in length of ½ Schedule 40 PVC pipe and hardware, as shown in Fig. 24-8. Be sure the ends

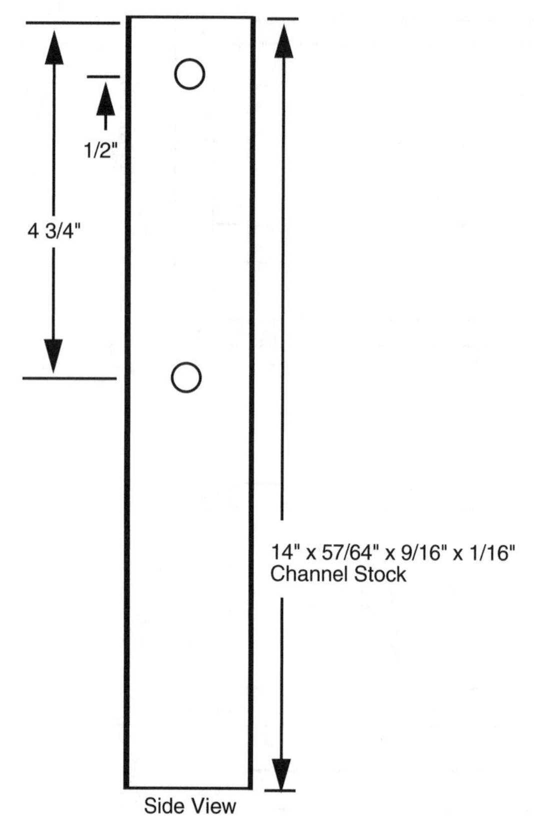

FIGURE 24-7 Cutting and drilling guide for the six legs.

TABLE 24-2 Parts List for Walkerbot Legs

6	14-in lengths 57/64-by-9/16-by-1/16-in aluminum channel stock
6	6-in lengths 41/64-by-1/2-by-1/16-in aluminum channel stock
6	Roller bearings
6	Steel electrical covers (4 5/8-in diameter)
6	5-in hex-head carriage bolt
6	2-by-3/8-in flat mending iron
6	1 1/4-in 45° "Ell" Schedule 40 PCV pipe fitting
Misc.	10/24 and 8/32 stove bolts, nuts, tooth lock washers, locking nuts, flat washers, as needed. 1/2-in Schedule 40 PVC cut to length (see text)

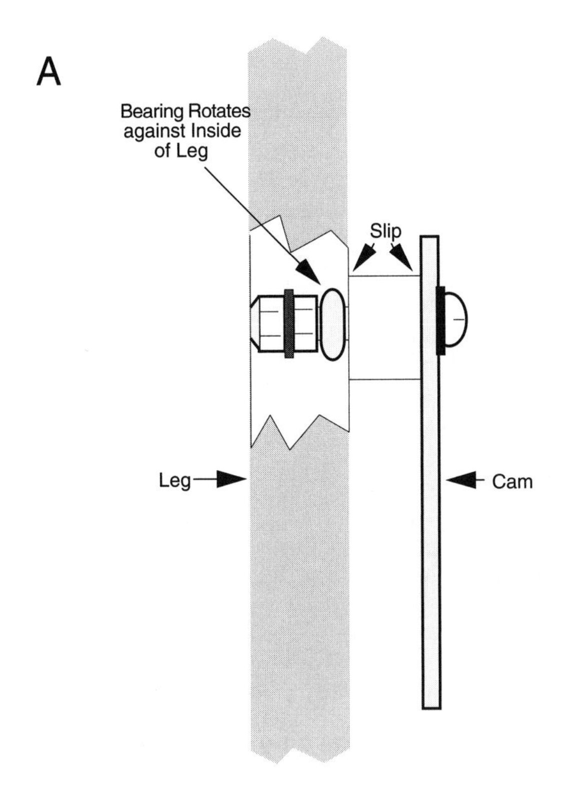

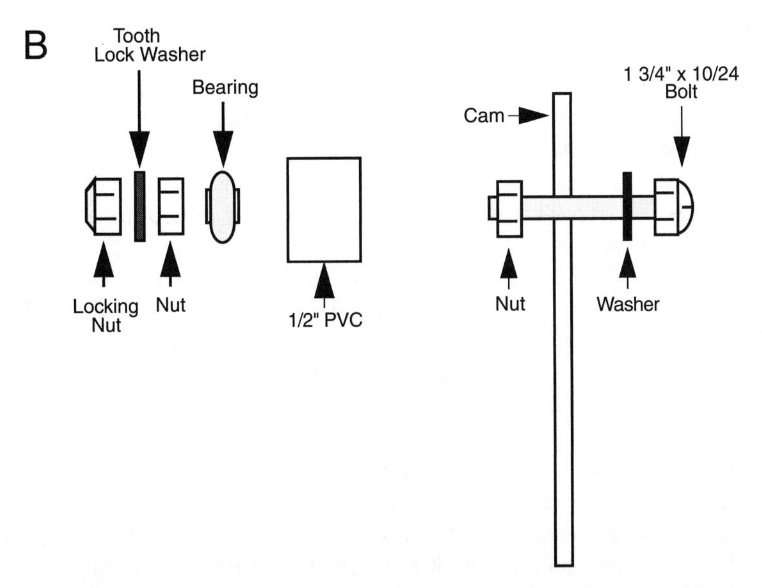

FIGURE 24-8 Hardware detail for the leg cam. *a.* Complete cam and leg; *b.* exploded view. Note that two of the legs use a 2-in piece of PVC and a 3-in bolt.

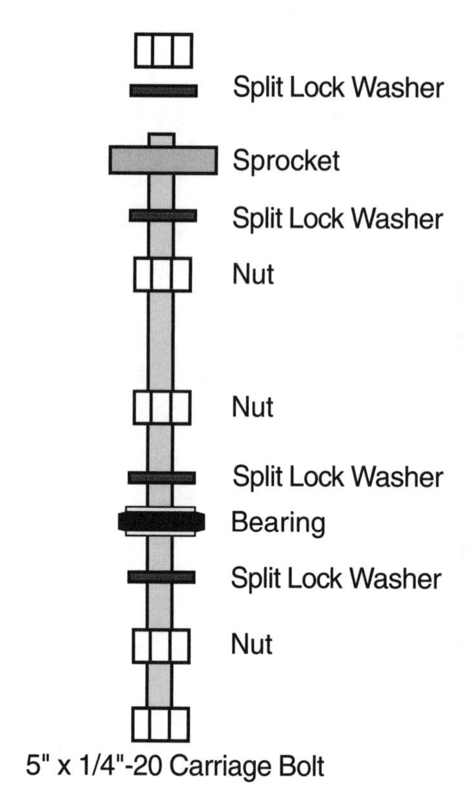

FIGURE 24-9 Hardware detail of the leg shafts.

of the pipe are filed clean and that the cut is as square as possible. The bolt should be tightened against the cam but should freely rotate within the leg hole.

Assemble the remaining two legs in a similar fashion, but use a 2-in length of PVC pipe and a 3-in stove bolt. These two legs will be placed in the center of the robot and will stick out from the others. This allows the legs to cross one another without interfering with the gait of the robot. The bearings used in the prototype were ½-in-diameter closet door rollers.

Now refer to Fig. 24-9. Thread a 5-by-¼-in 20 carriage bolt through the center of the cam, using the hardware shown. Next, install the wheel bearings to the shafts, 1-in from the cam. The 1¼-in-diameter bearings are the kind commonly used in lawn mowers and are readily available. The bearings used in the prototype had ½-in hubs. A ½-to ¼-in reducing bushing was used to make the bearings compatible with the diameter of the shaft.

Install 3½-in-diameter 30 tooth #25 chain sprocket (another size will also do, as long as all the leg mechanism sprockets in the robot are the same size). Like the bearings, a reducing bushing was used to make the ½-in ID hubs of the sprockets fit on the shaft. The exact positioning of the sprockets on the shaft is not important at this time, but follow the spacing diagram shown in Fig. 24-10 as a guide. You'll have to fine-tune the sprockets on the shaft as a final alignment procedure anyway.

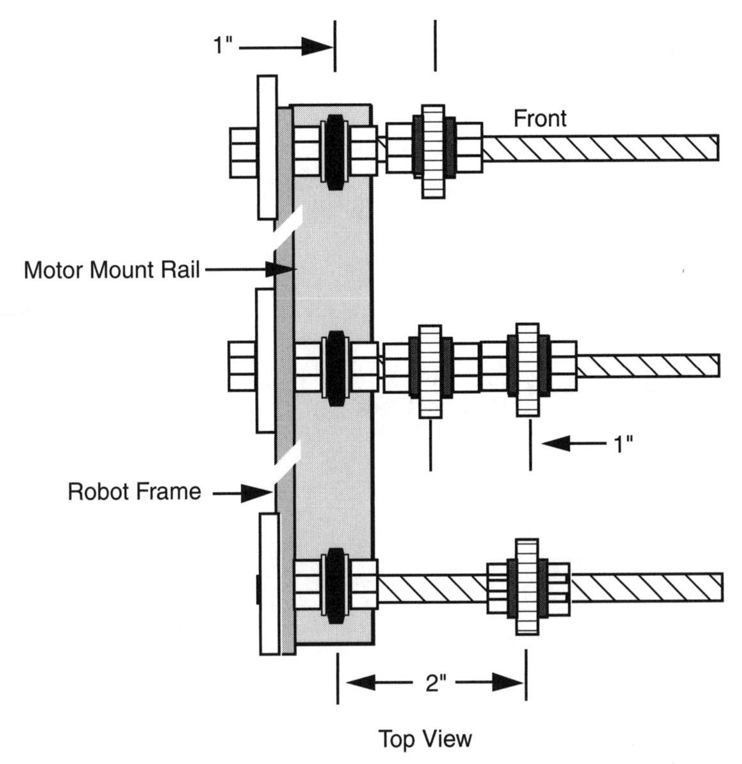

FIGURE 24-10 The leg shafts attached to the motor mount rails (left side shown only).

Once all the legs are complete, install them on the robot using U-bolts. The 1½-in-wide by 2½-in-long by ¼-in 20 thread U-bolts fit over the bearings perfectly. Secure the U-bolts using the ¼-in 20 nuts supplied.

Refer to Fig. 24-11 for the next step. Cut six 6-in lengths of 41/64-by-½-by-1/16-in aluminum channel stock. With a #19 bit, drill holes ⅜ in from the top and bottom of the rail. With a nibbler tool, cut a 3½-in slot in the center of each rail. The slot should start ½ in from one end.

Alternatively, you can use a router, motorized rasp, or other tool to cut the slot. In any case, make sure the slot is perfectly straight. Once cut, polish the edges with a piece of 300 grit wet-dry Emory paper, used wet. Use your fingers to find any rough edges. There can be none. This is a difficult task to do properly, and you may want to take this portion to a sheet metal shop and have them do it for you (it'll save you an hour or two of blister-producing nibbling!). An alternative method, which requires no slot cutting, is shown in Fig. 24-12. Be sure to mount the double rails exactly parallel to one another.

Mount the rails using 8/32-by-2-in bolts and 8/32 nuts. Make sure the rails are directly above the shaft of each leg or the legs may not operate properly. You'll have to drill through both walls of the channel in the top of the frame.

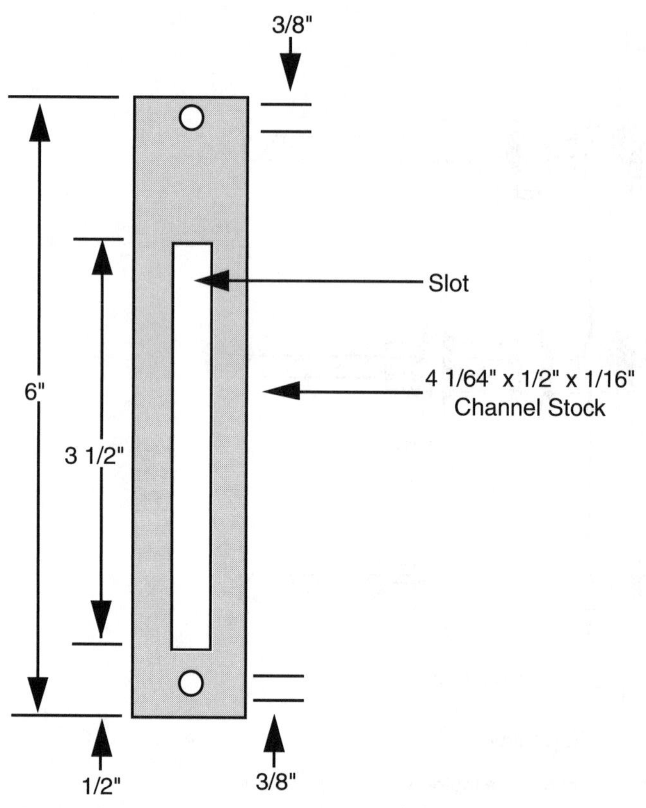

FIGURE 24-11 Cutting and drilling guide for the cam sliders (six required).

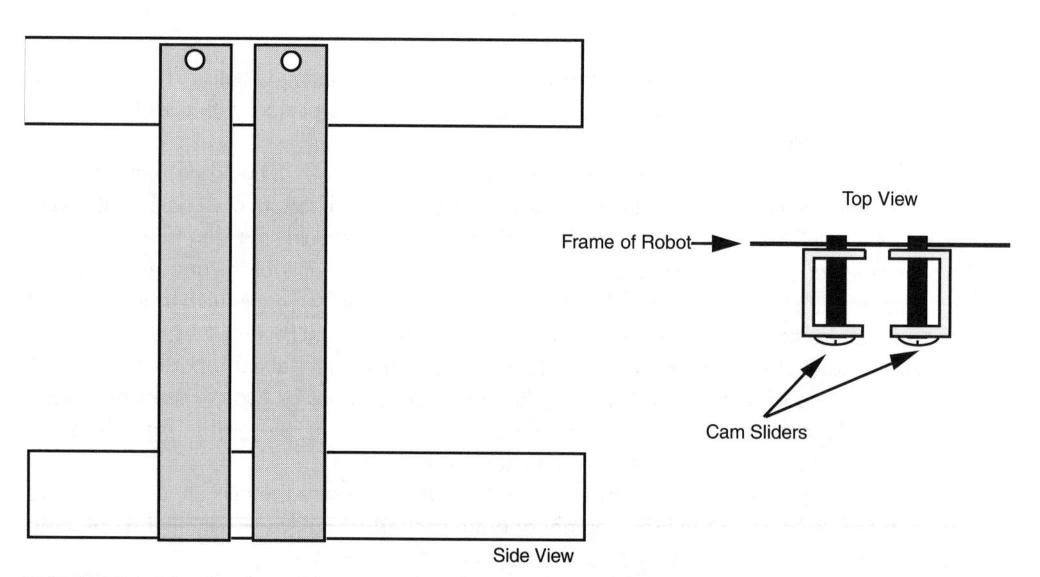

FIGURE 24-12 An alternative approach to the slotted cam sliders.

The rails serve to keep the legs aligned for the up-and-down piston-like stroke of the legs. Attach the legs to the rails using ⅜-by-1½-in bolts. Use nuts and locking nuts fasteners as shown in Fig. 24-13. This finished leg mechanism should look like the one depicted in Fig. 24-14. Use grease or light oil to lubricate the slot. Be sure that there is sufficient play between the slot and the bolt stem. The play cannot be excessive, however, or the leg may bind as the bolt moves up and down inside the slot. Adjust the sliding bolt on all six legs for proper clearance.

Drill small pilot holes in the side of six 45° 1¼-in PVC pipe elbows. These serve as the feet of the legs. Paint the feet at the point if you wish. Using #10 wood screws, attach a 2-by-⅜-in flat mending iron to each of the elbow feet. Drill ¼-in holes 1¼ in from the bottom of the leg. Secure the feet onto the legs using ½-by-¼-in 20 machine bolts, nuts, and lock washers. Apply a 3-in length of rubber weather strip to the bottom of each foot for better traction. The leg should look like the one in Fig. 24-15. The legs should look like the one in Fig. 24-16. A close-up of the cam mechanism is shown in Fig. 24-17.

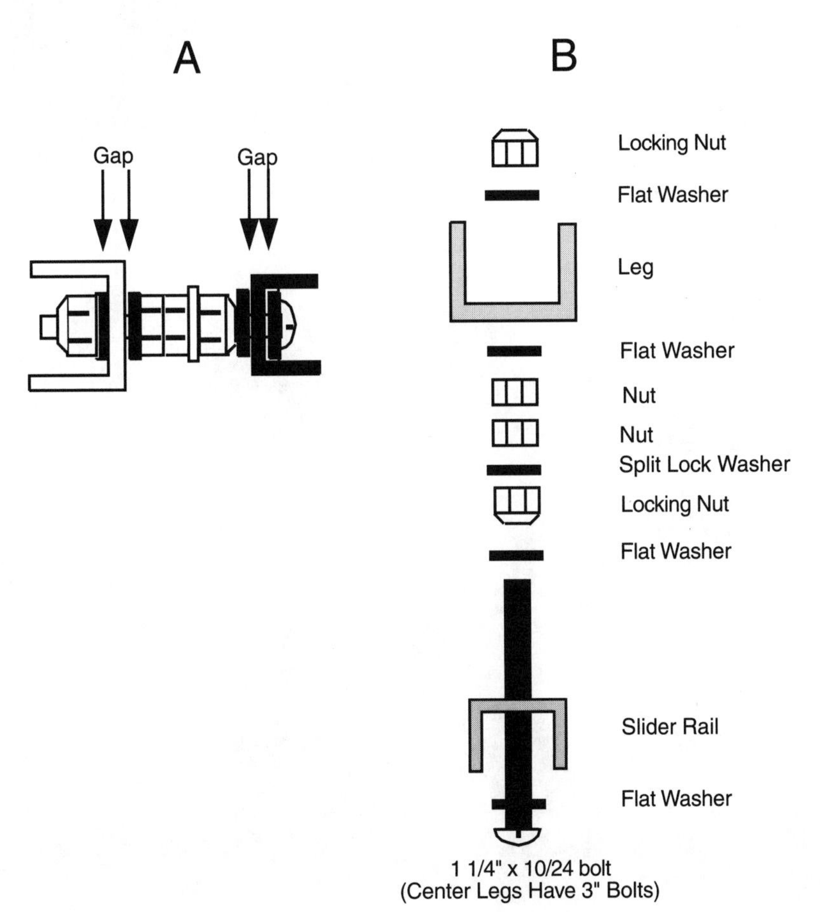

FIGURE 24-13 Cam slider hardware detail. *a*. Complete assembly; *b*. exploded view. Note that the center legs have 2-in bolts.

FIGURE 24-14 The slider cam and hardware. The slot must be smooth and free of burrs, or the leg will snag.

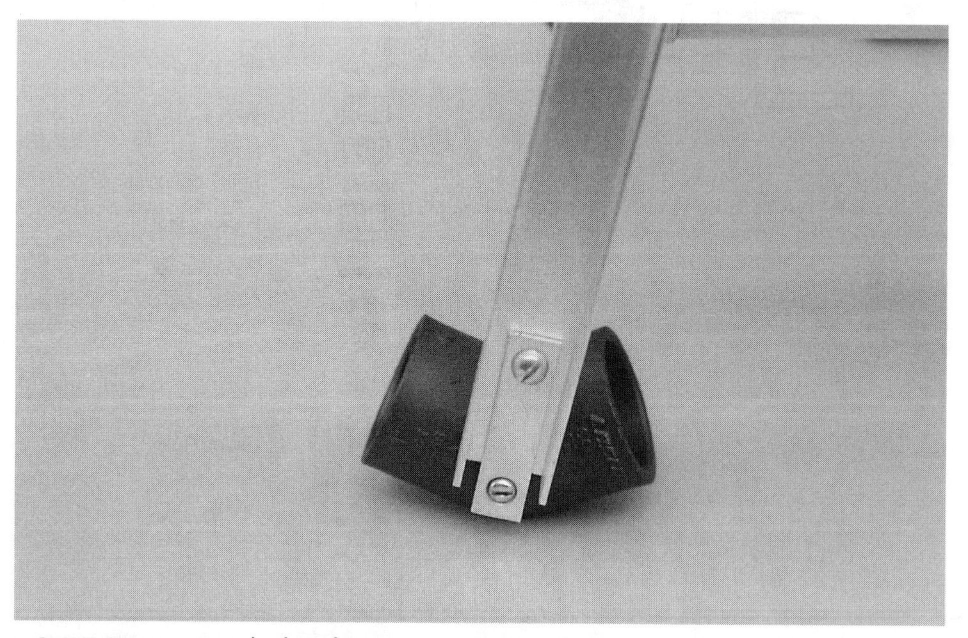

FIGURE 24-15 PVC plumbing fittings used as feet. The feet use a flat mending iron. Add pads or rubber to the bottom of the feet as desired.

FIGURE 24-16 One of six legs, completed (shown already attached to the robot).

FIGURE 24-17 A close-up detail of the leg cam.

24.3 Motors

The motors used in the prototype Walkerbot were surplus finds originally intended as the driving motors in a child's motorized bike or go-cart. The motors have a fairly high torque at 12 V DC and a speed of about 600 r/min. A one-step reduction gear was added to bring the speed down to about 230 r/min. The output speed is further reduced to about 138 r/min by using a drive sprocket. For a walking machine, that's about right, although it could stand to be a bit slower. Electronic speed reduction can be used to slow the motor output down to about 100 r/min. You can use other motors and other driving techniques as long as the motors have a (prereduced) torque of at least 6 lb-ft and a speed that can be reduced to 140 r/min or so.

Mount the motors inside two 6½-by-1½-in mending plate Ts. Drill a large hole, if necessary, for the shaft of the motor to stick through, as shown in Fig. 24-18 (refer to the parts list in Table 24-3). The motors used in the prototype came with a 12-pitch 12-tooth nylon gear. The gear was not removed for assembly, so the hole had to be large enough for it to pass through. The 30-tooth 12-pitch metal gear and 18-tooth ¼-in chain sprocket were also sandwiched between the mending plates.

The ¼-in shaft of the driven gear and sprocket is free running. You can install a bearing on each plate, if you wish, or have the shaft freely rotate in oversize holes. The sprocket and gear have ½-in ID hubs, so reducing bushings were used. The sprocket and gear are held in place with compression. Don't forget the split washers. They provide the necessary compression to keep things from working loose.

Before attaching the two mending plates together, thread a 28½-in length of #25 roller chain over the sprocket. The exact length can be one or two links off; you can correct for any variance later on. Assemble the two plates using 8⁄32 by 3-in bolts and 8⁄32 nuts and lock washers. Separate the plates using 2-in spacers.

Attach the two 17⅞-in lengths of angle bracket on the robot, as shown in Fig. 24-19. The stock mounts directly under the two end pieces. Use ½-in-by-8⁄32 bolts and nuts to secure the crosspieces into place. Secure the leg shafts using 1¼-in bearings and U-bolts.

Mount the motor to the newly added inner mounting rails using 3-by-½-in mending plate Ts. Fasten the plates onto the motor mount, as shown in Fig. 24-20, with 8⁄32 by ½-in bolts and nuts. Position the shaft of the motor approximately 7 in from the back of the robot (you can make any end of the shaft the back; it doesn't matter). Thread the roller chain over the center sprocket and the end sprocket. Position the motor until the roller chain is taut. Mark holes and drill. Secure the motor and mount to the frame using 8⁄32 by ½-in bolts and nuts. Repeat the process for the opposite motor. The final assembly should look like Fig. 24-21.

Thread a 28½-in length of #25 roller chain around sprockets of the center and front legs. Attach an idler sprocket 7½ in from the front of the robot in line with the leg mounts. Use a diameter as close to 2 in as possible for the idler; otherwise, you may need to shorten or lengthen the roller chain. Thread the roller chain around the sprocket, and find a position along the rail until the roller chain is taut (but not overly tight). Make a mark using the center of the sprocket as a guide and drill a ¼-in hole in the rail. Attach the sprocket to the robot. Figs. 24-22 through 24-24 show the motor mount, idler sprocket, and roller chain locations.

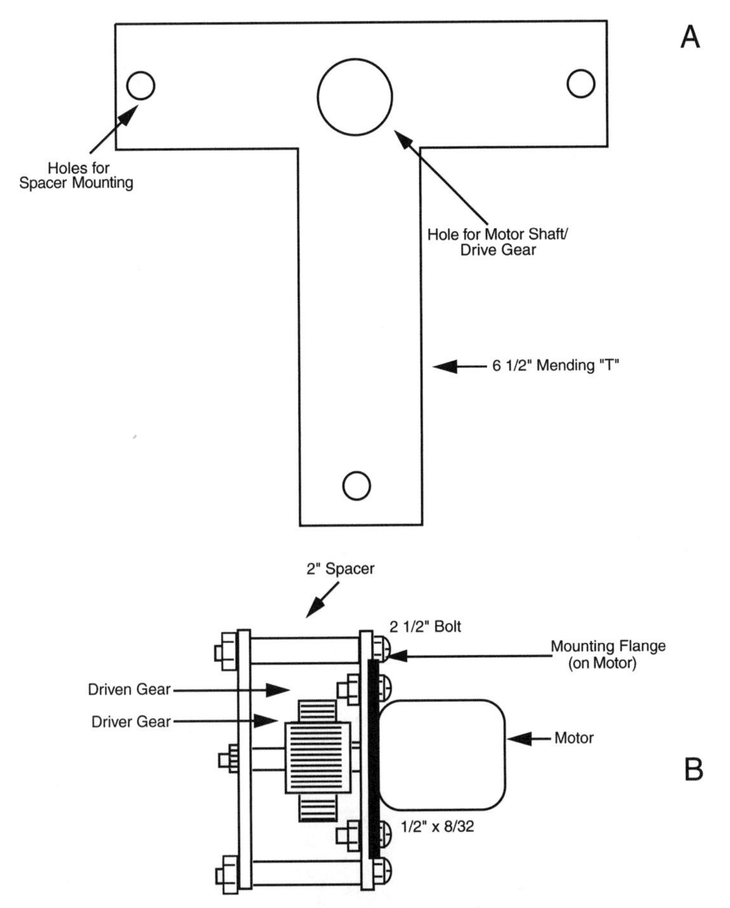

FIGURE 24-18 Motor mount details. *a.* Drilling guide for the mending T; *b.* the motor and drive gear-sprocket mounted with two mending Ts.

TABLE 24-3 Parts List for Walkerbot Mount-Drive System

4	6½-in galvanized mending plate T
4	3-in galvanized mending plate T
2	Heavy-duty gear-reduction DC motors
12	3½-in-diameter 30-tooth #15 chain sprocket
4	28½-in-length #25 roller chain
12	2½-by-1½-by-¼-in 20 U-bolts, with nuts and tooth lock washers
12	1½-in O.D. ¼ to ½-in ID bearing
Misc.	Reducing bushings (see text)

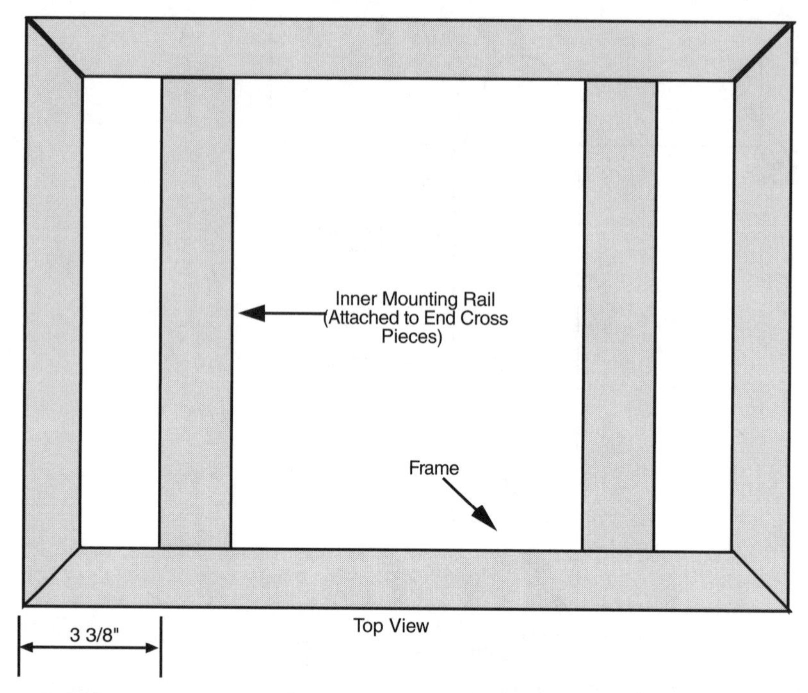

FIGURE 24-19 Mounting location of the inner rails.

FIGURE 24-20 One of the drive motors mounted on the robot using smaller galvanized mending Ts.

FIGURE 24-21 Drive motor attached to the Walkerbot, with drive chain joining the motor to the leg shafts.

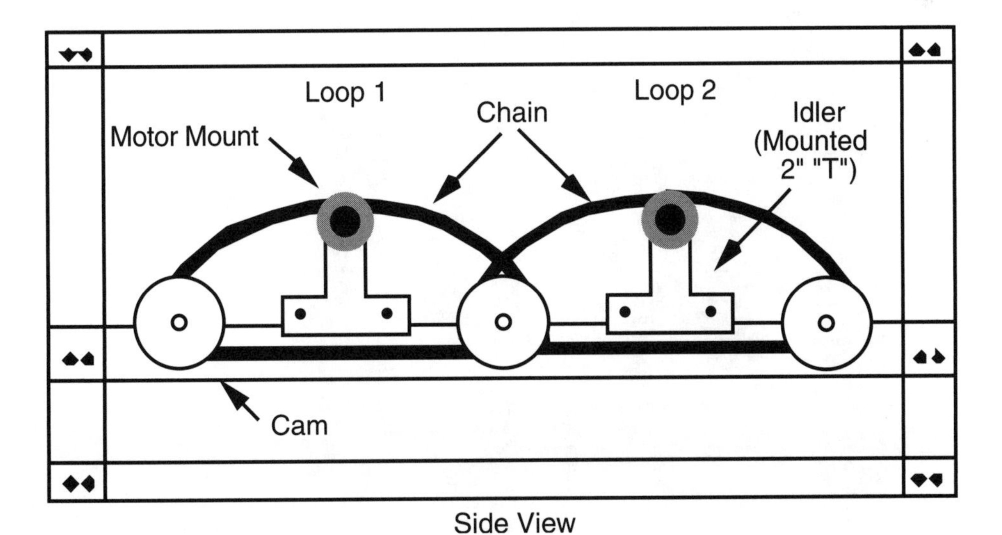

FIGURE 24-22 Mounting locations for idler sprocket.

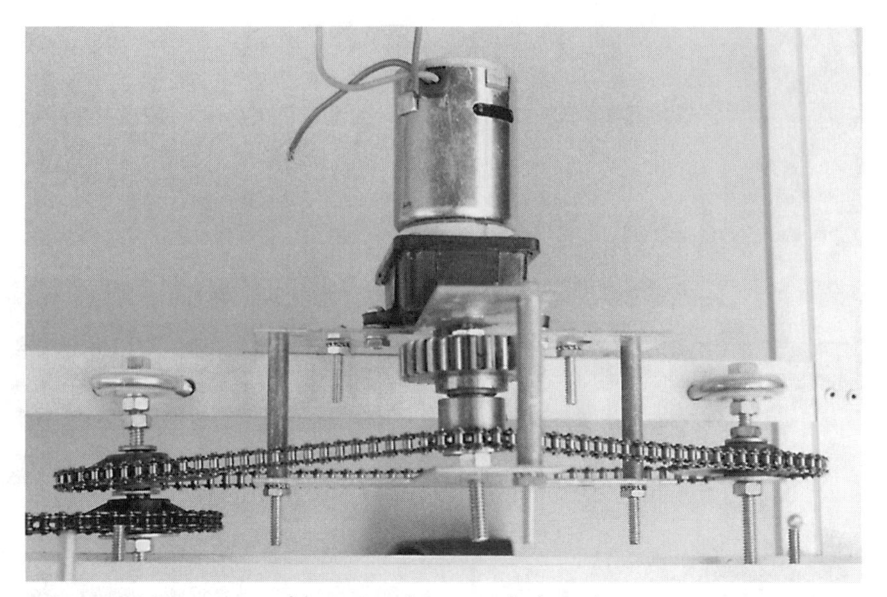

FIGURE 24-23 A view of the mounted motor, with chain drive.

FIGURE 24-24 Left and right motors attached to the robot.

24.4 Batteries

The Walkerbot is not a lightweight robot, and its walking design requires at least 30 percent more power than a wheeled robot. The batteries for the Walkerbot are not trivial. You have a number of alternatives. One workable approach is to use two 6-V motorcycle batteries, each rated at about 30 AH. The two batteries together equal a slimmed-down version of a car battery in size and weight.

You can also use a 12-V motorcycle or dune buggy battery, rated at more than 20 AH. The prototype Walkerbot used 12-AH 6-V gel-cell batteries. The amp-hour capacity is a bit on the low side, considering the 2-A draw from each motor, and the planned heavy use of electronics and support circuits. In tests, the 12-AH batteries provided about 2 h of use before requiring a recharge.

There is plenty of room to mount the batteries. A good spot is slightly behind the center legs. By offsetting the batteries a bit in relation to the drive motors, you restore the center of gravity to the center of the robot. Of course, other components you add to the robot can throw the center of gravity off. Add one or two articulated arms to the robot, and the weight suddenly shifts toward the front. For flexibility, you might want to mount the batteries on a sliding rail, which will allow you to shift their position forward or back depending on the other weight you add to the Walkerbot.

24.5 Testing and Alignment

You can test the operation of the Walkerbot by temporarily installing the wired control box you built earlier for the other, more basic robots that consists of two DPDT switches wired to control the forward and backward motion of the two legs. But before you test the Walkerbot, you need to align its legs. The legs on each side should be positioned so that either the center leg touches the ground or the front and back legs touch the ground. When the two sets of legs are working in tandem, the walking gait should be as shown in Fig. 24-25. This gait is the same as an insect's and provides a great deal of stability. To turn, one set of legs stops (or reverses) while the other set continues. During this time, the tripod arrangement of the gait will be lost, but the robot will still be supported by at least three legs.

An easy way to align the legs is to loosen the chain sprockets (so you can move the legs independently) and position the middle leg all the way forward and the front and back legs all the way back. Retighten the sprockets, and look out for misalignment of the roller chain and sprockets. If a chain bends to mesh with a sprocket, it is likely to pop off when the robot is in motion.

During testing, be on the lookout for things that rub, squeak, and work loose. Keep your wrench handy and adjust gaps and tighten bolts as necessary. Add a dab of oil to those parts that seem to be binding. You may find that a sprocket or gear doesn't stay tightened on a shaft. Look for ways to better secure the component to the shaft, such as by using a setscrew or another split lock washer. It may take several hours of tuning up to get the robot working at top efficiency.

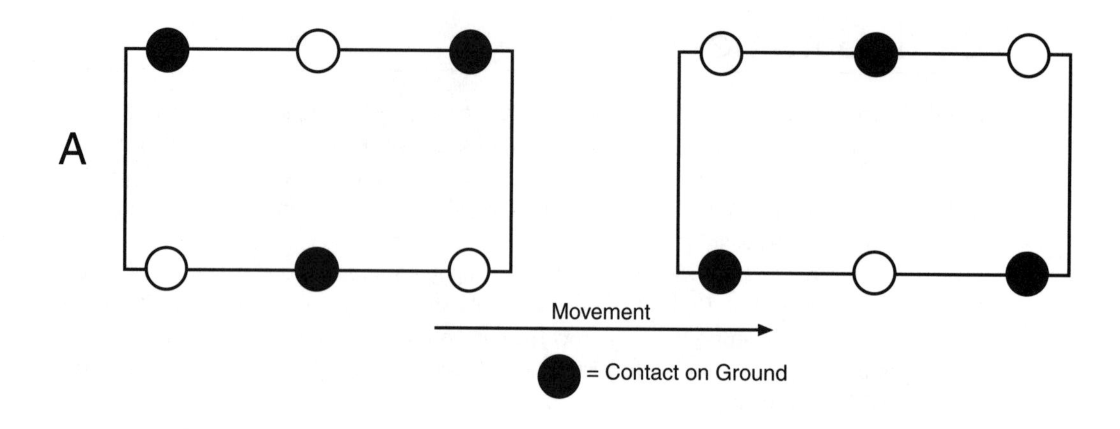

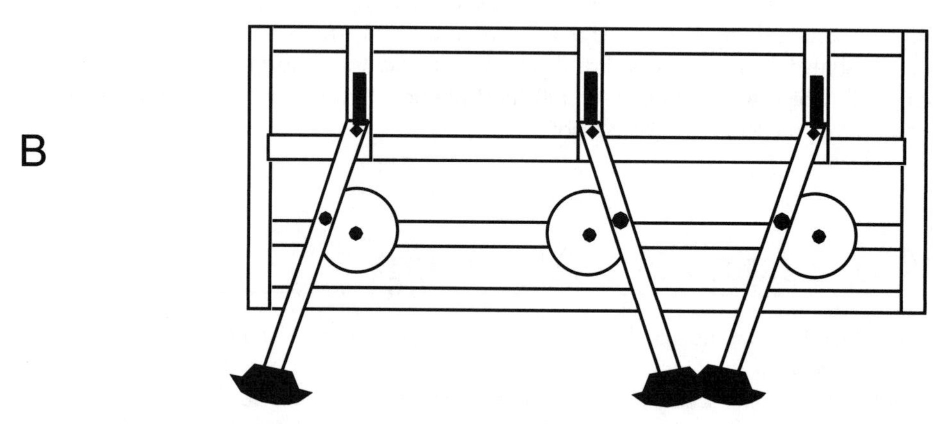

FIGURE 24-25 The walking gait. *a.* The alternating tripod walking style of the Walkerbot, shared with thousands of crawling insects; *b.* the positioning of the legs for proper walking (front and back legs in synchrony; middle leg 180 degrees out of sync). The middle leg doesn't hit either the front or back leg because it is further from the body of the robot.

Once the robot is aligned, run it through its paces by having it walk over level ground, step over small rocks and ditches, and navigate tight corners. Keep an eye on your watch to see how long the batteries provide power. You may need to upgrade the batteries if they cannot provide more than an hour of operation.

The Walkerbot is ideally suited for expansion. Fig. 24-26 shows an arm attached to the front side of the robot. You can add a second arm on the other side for more complete dexterity. Attach a dome on the top of the robot, and you've added a head on which you can attach a video camera, ultrasonic ears and eyes, and lots more. Additional panels can be added to the front and back ends; attach them using hook-and-loop (such as Velcro) strips. That way, you can easily remove the panels should you need quick access to the inside of the robot.

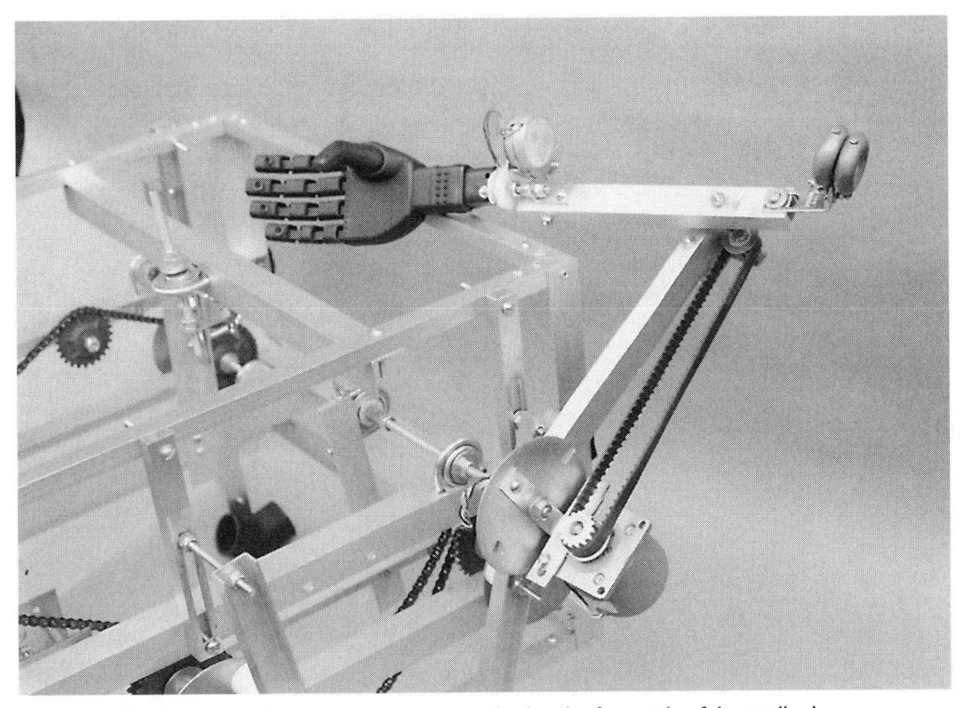

FIGURE 24-26 A revolute coordinate arm attached to the front side of the Walkerbot.

24.6 From Here

To learn more about . . .	***Read***
Working with metal	Chapter 10, "Building a Metal Platform"
Robot locomotion styles, including wheels, treads, and legs	Chapter 18, "Principles of Robot Locomotion"
Using DC motors	Chapter 20, "Working with DC Motors"
Additional locomotion systems based on the Walkerbot frame	Chapter 25, "Advanced Locomotion Systems"
Constructing an arm for the Walkerbot	Chapter 27, "Build a Revolute Coordinate Arm"

CHAPTER 25

ADVANCED ROBOT LOCOMOTION SYSTEMS

Two drive wheels aren't the only way to move a robot across the living room or workshop floor. Here, in this chapter, you'll learn the basics of applying some unique drive systems to propel your robot designs, including a stair-climbing robot, an outdoor tracked robot, and even a six-wheeled Buggybot.

25.1 Making Tracks

There is something exciting about seeing a tank climb embankments, bounding over huge boulders as if they were tiny dirt clods. A robot with tracked drive is a perfect contender for an automaton that's designed for outdoor use. Where a wheeled or legged robot can't go, the tracked robot can roll in with relative ease. Tracked robots, using metal tracks just like tanks, have been designed for the Navy and Army and are even used by many police and fire departments. The all-terrain ability and ruggedness of metal tracked robots made them the design of choice during rescue efforts after the 9/11 attacks.

Using a metal track for your personal robot is decidedly a bad idea. A metal track will be too heavy and much too hard to fabricate. For a home-brew robot, a rubber track is more than adequate. You can use a large timing belt, even an automotive fan belt, for the track or a large rubber O ring (like the one used to drive your vacuum cleaner).

Another alternative that has been used with some success is rubber wetsuit material. Most diving shops have long strips of the rubber lying around that they'll sell or give to you.

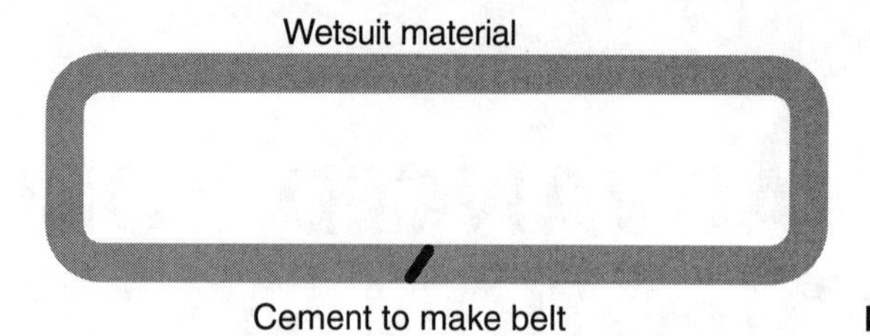

FIGURE 25-1 A wetsuit drive belt.

You can mend the rubber using a special waterproof adhesive. You can glue the strip together to make a band, then glue small rubber cleats onto the band. Fig. 25-1 shows the basic idea.

The drive train for a tracked robot must be engineered for the task of driving a track. A two-wheel differentially driven robot (like the wooden and metal platform robots presented earlier) can be used as a base for a tracked robot. Install three free-turning large pulleys and three small drive pulleys (one driven on each side of the robot), as diagrammed in Fig. 25-2. The track fits inside the groove of the pulleys, so it won't easily slip out.

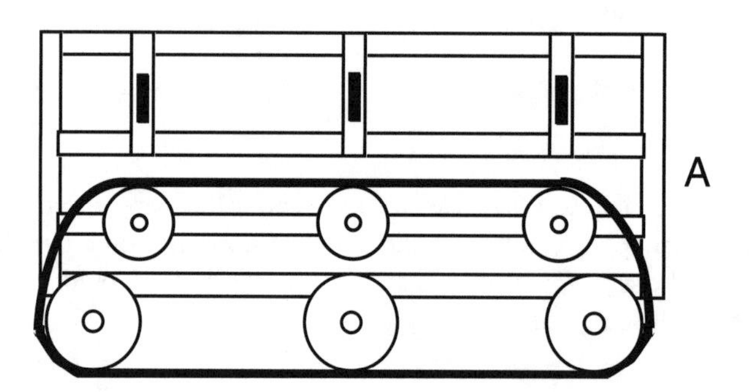

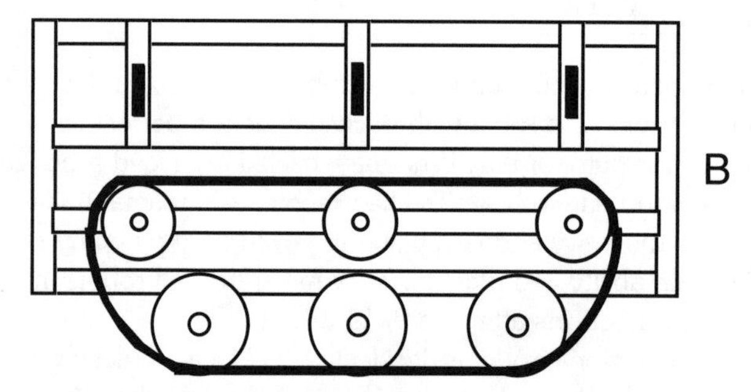

FIGURE 25-2 Two ways to add track drive to the Walkerbot presented in Chapter 22. *a.* Track roller arrangement for good traction and stability but relatively poor turning radius. *b.* Track roller arrangement for good turning radius, but hindered traction and stability.

To propel the robot, you activate both motors so the tracks move in the same direction and at the same speed. To steer, you simply stop or reverse one side (the same as the basic differentially driven robot). For example, to turn left, stop the left track. To make a hard left turn, reverse the left track.

25.2 Steering Wheel Systems

Using dual motors to effect propulsion and steering is just one method for getting your robot around. Another approach is to use a pivoting wheel to steer the robot. The same wheel can provide power, or power can come from two wheels in the rear (the latter is much more common). The arrangement is not unlike golf carts, where the two rear wheels provide power and a single wheel in the front provides steering. See Fig. 25-3 for a diagram of a typical steering-wheel robot. Fig. 25-4 shows a detail of the steering mechanism.

The advantage of a steering-wheel robot is that you need only one powerful drive motor. The motor can power both rear wheels at once as shown in Fig. 25-3, but this isn't recommended for a reason that anyone who is aware of car drivetrains understands. With the

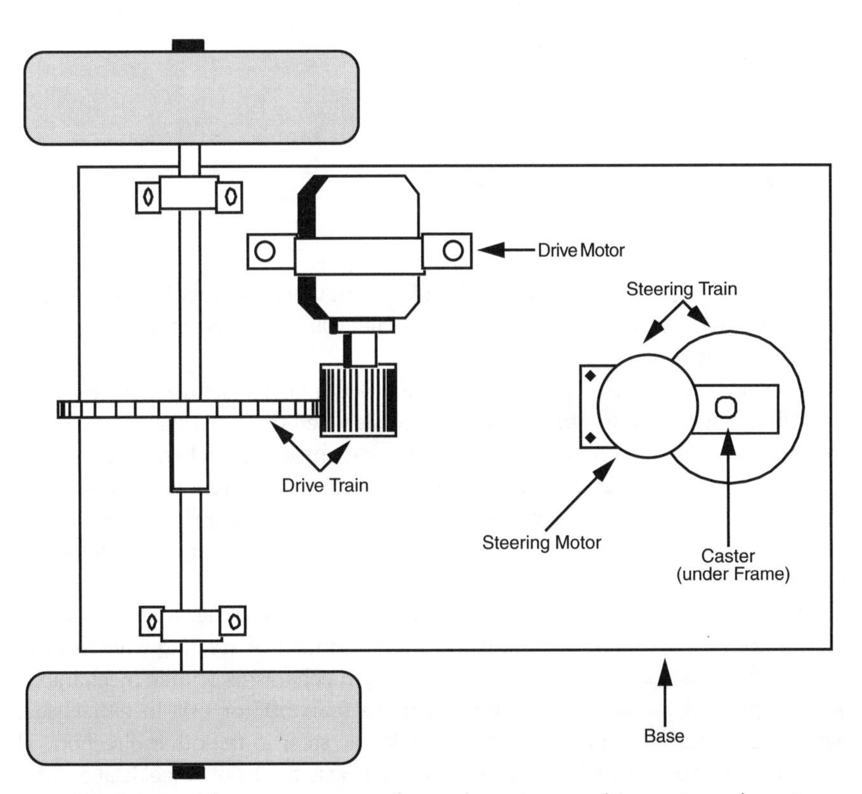

FIGURE 25-3 A basic arrangement for a robot using one drive motor and steering wheel.

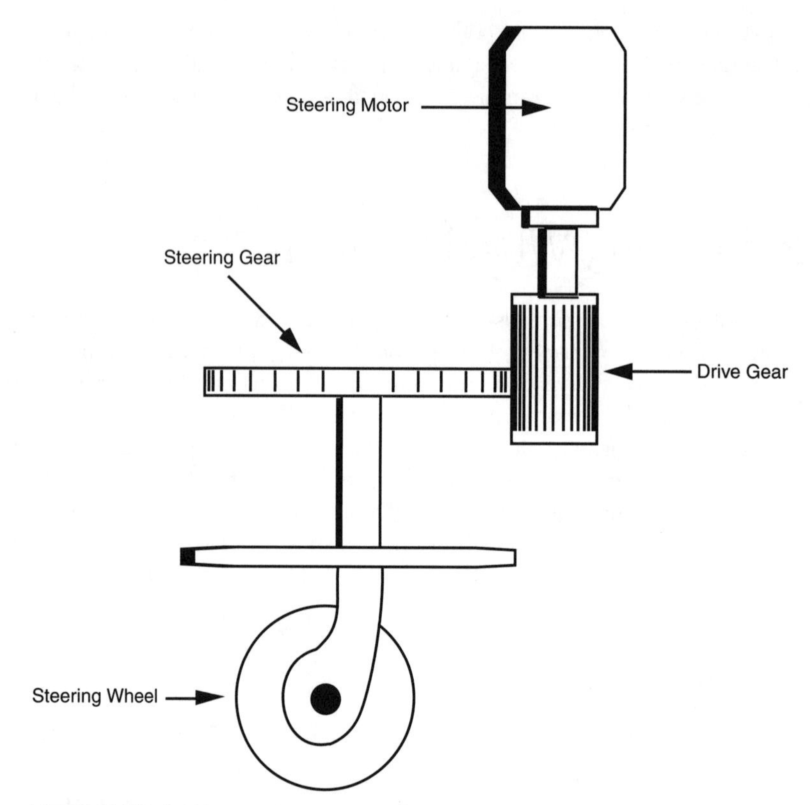

FIGURE 25-4 The steering gear up close.

two wheels turning together, there is a lot of friction when the robot wants to turn because both wheels are locked together even though they will be turning at different speeds as the robot changes direction.

The solution to this dilemma in an automobile is a gear system known as a differential, which allows the wheels to turn at different speeds when the car is changing direction. Finding or making a differential for a robot is a daunting challenge but there are two simple solutions to the dilemma. The first is to drive only one of the two rear wheels and let the other turn freely. This way the other wheel will turn at the appropriate rate for the current motion. The second solution is to let both rear wheels turn freely and independently and drive the *turning wheel.*

The steering-wheel motor needn't be as powerful since all it has to do is swivel the wheel back and forth a few degrees. The biggest disadvantage of steering-wheel systems is the steering! You must build stops into the steering mechanisms (either mechanical or electronic) to prevent the wheel from turning more than 50° or 60° to either side. Angles greater than about 60° cause the robot to suddenly steer in the other direction. They may even cause the robot to lurch to a sudden stop because the front wheel is at a right angle to the rear wheels.

The servo mechanism that controls the steering wheel must know when the wheel is pointing forward. The wheel must return to this exact spot when the robot is commanded to forge straight ahead. Not all servo mechanisms are this accurate. The motor may stop one or more degrees off the center point, and the robot may never actually travel in a straight line. A good steering motor, and a more sophisticated servo mechanism, can reduce this limitation.

A number of robot designs with steering-wheel mechanisms has been described in other robot books and on various web pages. Check out Appendix A, "Further Reading," and Appendix C, "Robot Information on the Internet," for more information.

25.3 Six-Wheeled Robot Cart

A variation on the tracked robot is the six-wheeled rugged terrain cart (also known as a Buggybot), shown in Fig. 25-5. The larger the wheels the better, as long as they aren't greater than the centerline diameter between each drive shaft.

Pneumatic wheels are the best choice because they provide more bounce and handle rough ground better than hard rubber tires. Most hardware stores carry a full assortment of pneumatic tires. Most are designed for things like wheelbarrows and hand dollies. Cost can be high, so you may want to check out the surplus or used industrial supply houses.

Steering is accomplished as with two-wheeled or tracked differentially driven robots. The series of three wheels on each side act as a kind of track tread, so the vehicle behaves much like a tracked vehicle.

The maneuverability isn't as good as with a two-wheeled robot, but you can still turn the robot in a radius a little longer than its length. Sharp turns require you to reverse one set of wheels while applying forward motion to the other.

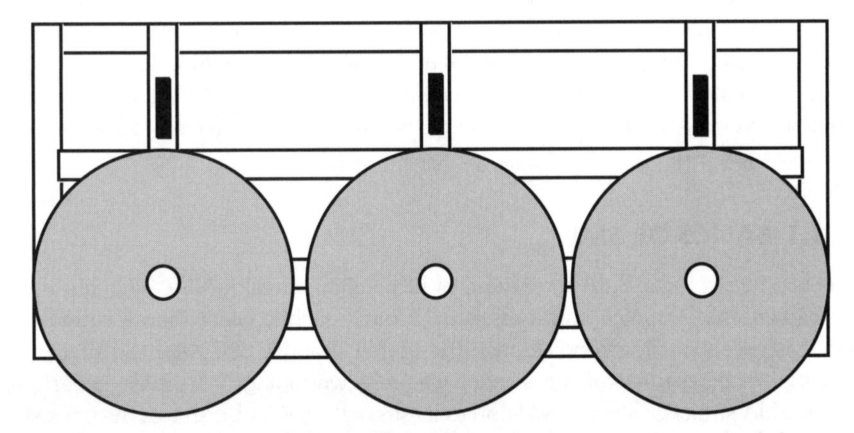

FIGURE 25-5 Converting the Walkerbot (from Chapter 22) into a six-wheeled, all-terrain Buggybot.

25.4 Building Robots with Shape-Memory Alloy

As early as 1938, scientists observed that certain metal alloys, once bent into odd shapes, returned to the original form when heated. This property was considered little more than a laboratory curiosity because the metal alloys were weak, difficult and expensive to manufacture, and they broke apart after just a couple of heating/cooling cycles.

Research into metals with memory took off in 1961, when William Beuhler and his team of researchers at the U.S. Naval Ordnance Laboratory developed a titanium-nickel alloy that repeatedly displayed the memory effect. Beuhler and his cohorts developed the first commercially viable shape-memory alloy, or SMA. They called the stuff Nitinol, a fancy-sounding name derived from Nickel Titanium Naval Ordnance Laboratory.

Since its introduction, Nitinol has been used in a number of commercial products—but not many. For example, several Nitinol engines have been developed that operate with only hot and cold water. In operation, the metal contracts when exposed to hot water and relaxes when exposed to cold water. Combined with various assemblies of springs and cams, the contraction and relaxation (similar to a human muscle) cause the engine to move.

Other commercial applications of Nitinol include pipe fittings that automatically seal when cooled, large antenna arrays that can be bent (using hot water) into most any shape desired, sunglass frames that spring back to their original shape after being bent, and an anti-scald device that shuts off water flow in a shower should the water temperature exceed a certain limit.

Regular Nitinol contracts and relaxes in heat (in air, water, or other liquid). That limits the effectiveness of the metal in many applications where local heat can't be applied. Researchers have attempted to heat the Nitinol metal using electrical current in an effort to exactly control the contraction and relaxation. But because of the molecular construction of Nitinol, hot spots develop along the length of the metal, causing early fatigue and breakage.

In 1985, a Japanese company, Toki, unveiled a new type of shape-memory alloy specially designed to be activated by electrical current. Toki's unique SMA material, tradenamed BioMetal, offers all of the versatility of the original Nitinol, with the added benefit of near-instant electrical actuation. BioMetal and materials like it—Muscle Wire from Mondo-Tronics or Flexinol from Dynalloy—have many uses in robotics, including novel locomotive actuation. From here on out this family of materials will be referred to as shape-memory alloy, or simply SMA.

25.4.1 BASICS OF SMA

At its most basic level, SMA is a strand of nickel titanium alloy wire. Though the material may be very thin (a typical thickness is 0.15 mm—slightly wider than a strand of human hair), it is exceptionally strong. In fact, the tensile strength of SMA rivals that of stainless steel: the breaking point of the slender wire is a whopping 6 lb. Even under this much weight, SMA stretches little. In addition to its strength, SMA also shares the corrosion resistance of stainless steel.

Shape-memory alloys change their internal crystal structure when exposed to certain higher-than-normal temperatures (this includes the induced temperatures caused by passing an electrical current through the wire). The structure changes again when the alloy is allowed to cool. More specifically, during manufacture the SMA wire is heated to a very high temperature, which embosses or memorizes a certain crystal structure. The wire is then cooled and stretched to its practical limits. When the wire is reheated, it contracts because it is returning to the memorized state.

Although most SMA strands are straight, the material can also be manufactured in spring form, usually as an expansion spring. In its normal state, the spring exerts minimum tension, but when current is applied the spring stiffens, exerting greater tension. Used in this fashion, an SMA becomes an "active spring" that can adjust itself to a particular load, pressure, or weight.

Shape-memory alloys have an electrical resistance of about 1 Ω/in. That's more than ordinary hookup wire, so SMAs will heat up more rapidly when an electrical current is passed through them. The more current passes through, the hotter the wire becomes and the more contracted the strand. Under normal conditions, a 2- to 3-in length of SMA is actuated with a current of about 450 mA. That creates an internally generated temperature of about 100 to 130C; 90C is required to achieve the shape-memory change. Most SMAs can be manufactured to change shape at most any temperature, but 90C is the standard value for off-the-shelf material.

Excessive current should be avoided. Why? Extra current causes the wire to overheat, which can greatly degrade its shape-memory characteristics. For best results, current should be as low as necessary to achieve the contraction desired. Shape-memory alloys will contract by 2 to 4 percent of their length, depending on the amount of current applied. The maximum contraction of typical SMA material is 8 percent, but that requires heavy current that can, over a period of just a few seconds, damage the wire.

25.4.2 USING SMA

Shape-memory alloys need little support paraphernalia. Besides the wire itself, you need some type of terminating system, a bias force, and an actuating circuit.

Terminating system The terminators attach the ends of the SMA wires to the support structure or mechanism you are moving. Because SMAs expand as they contract, using glue or other adhesive will not secure the wire to the mechanism. Ordinary soldering is not recommended as the extreme heat of the soldering can permanently damage the wire. The best approach is to use a crimp-on terminator. These and other crimp terminators are available from companies that sell shape-memory alloy wire (either in the experimenter's kit or separately).

You can make your own crimp-on connectors using 18-gauge or smaller solderless crimp connectors (the smaller the better). Although these connectors are rather large for the thin 0.15 mm SMA, you can achieve a fairly secure termination by folding the wire in the connector and pressing firmly with a suitable crimp tool. Be sure to completely flatten the connector. If necessary, place the connector in a vise or use a ball peen hammer to flatten it all the way.

Bias force Apply current to the ends of an SMA wire and it just contracts in air. To be useful, the wire must be attached to one end of the moving mechanism and biased (as

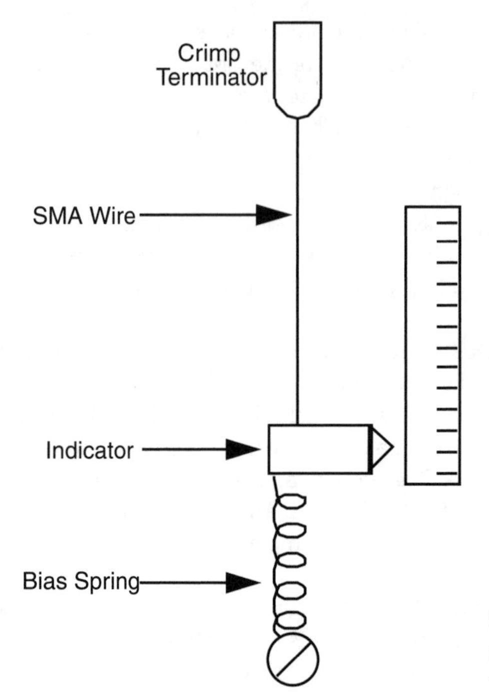

FIGURE 25-6 A bias spring or weight is required as a counterbalance force on the SMA wire.

shown in Fig. 25-6) at the other end. Besides offering physical support, the bias offers the counteracting force that returns the SMA wire to its limber condition once current is removed from the strand.

Actuating circuit SMAs can be actuated with a 1.5-V penlight battery. Because the circuit through the SMA wire is almost a dead short, the battery delivers almost its maximum current capacity. But the average 1.5-V alkaline penlight battery has a maximum current output of only a few hundred milliamps, so the current is limited through the wire. You can connect a simple on/off switch in line with the battery, as detailed in Fig. 25-7, to contract or relax the SMA wire.

The problem with this setup is that it wastes battery power, and if the power switch is left on for too long, it can do some damage to the SMA strand. A more sophisticated approach uses a 555 timer IC that automatically shuts off the current after a short time. The

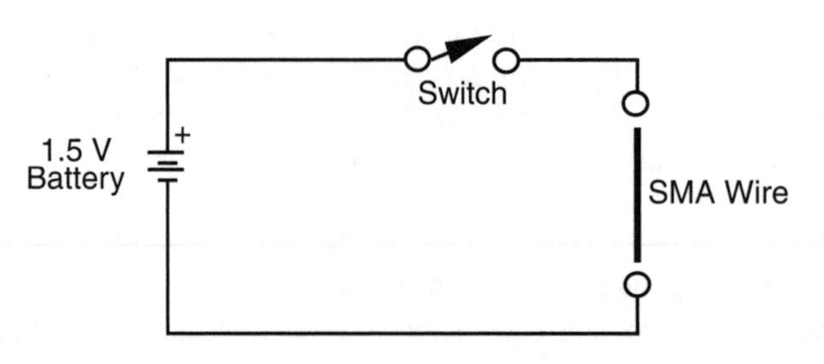

FIGURE 25-7 A simple switch in series with a 1.5-V penlight battery forms a simple SMA driving circuit. The low current delivered by the penlight battery prevents damage to the SMA wire.

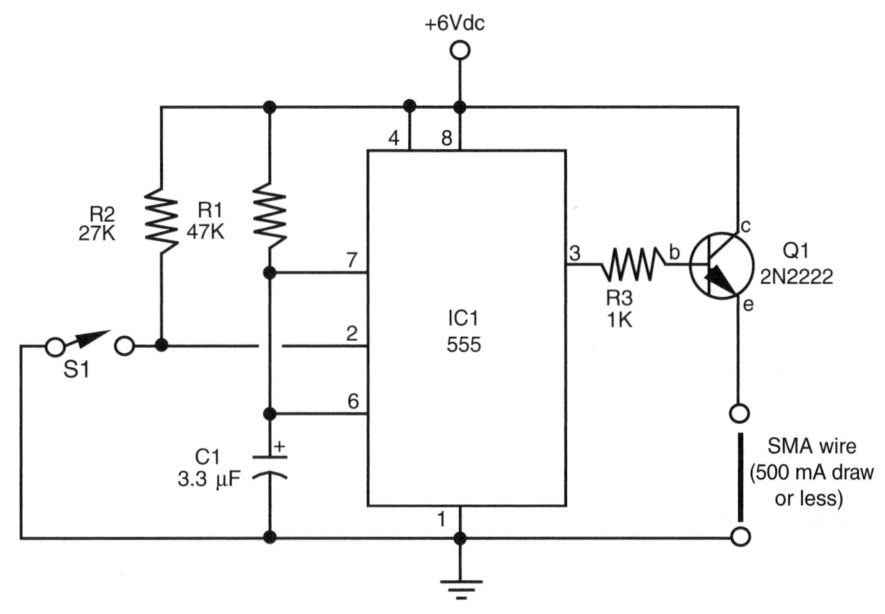

FIGURE 25-8 A 555 timer IC is at the heart of an ideal driving circuit for SMA wire. The 555 removes the current shortly after you release activating switch S1.

schematic in Fig. 25-8 shows one way of connecting a 555 timer IC to turn off power to a length of SMA 0.1 s after the button (S1) is opened. Table 25-1 provides a parts list for the 555 SMA circuit.

In operation, when you press momentary switch S1 current passes through the wire and it contracts. Release S1 immediately, and the SMA stays contracted for an extra fraction of a second, then releases as the 555 timer shuts off. Since the total ON time of the 555 depends on how long you hold S1 down, plus the 1⁄10-of-a-second delay, you should depress the switch only momentarily.

TABLE 25-1 Parts List for 555 SMA Driver

IC1	555 timer
Q1	2N2222 NPN transistor
R1	47K resistor
R2	27K resistor
R3	1K resistor
C1	3.3 µF polarized electrolytic capacitor
Misc.	Momentary SPST switch, SMA wire

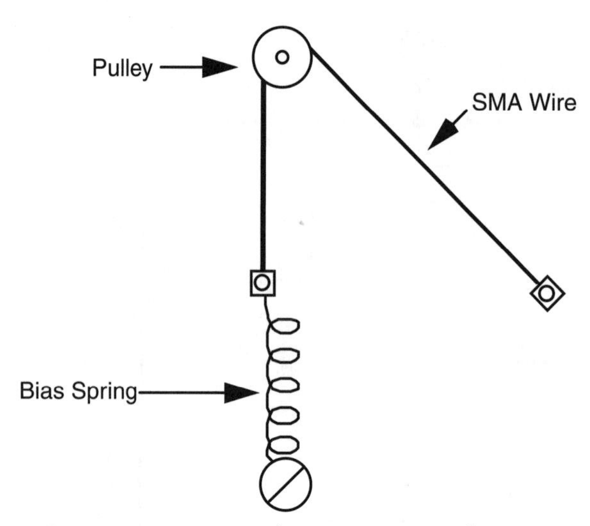

FIGURE 25-9 Concept of using SMA wire with a mechanical pulley.

25.4.3 SHAPE-MEMORY ALLOY MECHANISMS

With the SMA properly terminated and actuated, it's up to you and your own imagination to think of ways to use it in your robots. Fig. 25-9 shows a typical application using an SMA wire in a pulley configuration. Apply current to the wire and the pulley turns, giving you rotational motion. A large-diameter pulley will turn very little when the SMA tenses up, but a small-diameter one will turn an appreciable distance.

Fig. 25-10 shows a length of SMA wire used in a lever arrangement. Here, the metal strand is attached to one end of a bell crank. On the opposite end is a bias spring. Applying juice to the wire causes the bell crank to move. The spot where you attach the drive arm dictates the amount of movement you will obtain when the SMA contracts.

SMA wire is tiny stuff, and you will find that the miniature hardware designed for model R/C airplanes is most useful for constructing mechanisms. Most any well-stocked hobby

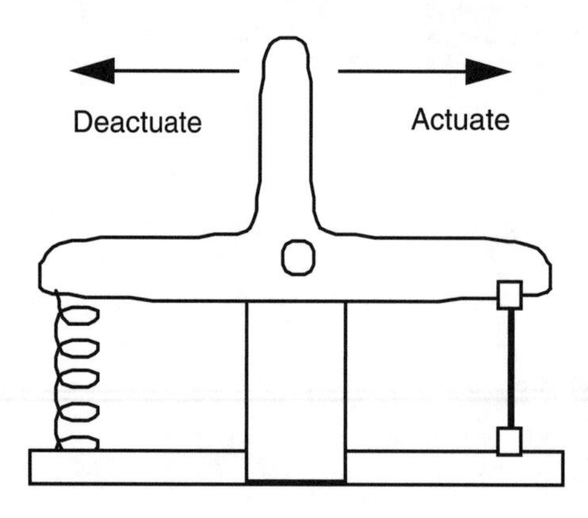

FIGURE 25-10 The bell crank changes the contraction of the SMA for sideways movement of the lever. The spring enables the bell crank to return to its original position after the current is removed from the wire.

store will stock a full variety of bell cranks, levers, pulleys, wheels, gears, springs, and other odds and ends to make your work with SMA more enjoyable.

25.4.4 DESIGNING ROBOTS FOR SMA LIMITATIONS

The operation of shape-memory alloys probably seems like they are the perfect material to simulate the operation of muscles, eliminating the need for motors, hydraulics, and pneumatics in many robots. Before you get visions of building robots that can walk or manipulate objects like humans do, there are a few caveats about SMAs that you should be aware of.

First off, SMAs contract relatively slowly when current is applied and can take a very long time to cool and relax when current is turned off. Coupled with this is the all-or-nothing behavior of SMAs; they are either fully contracted or fully relaxed. It may be possible to specify an amount of contraction, but this would require closed loop control of a power PWM applied to the current and would be difficult to design. These two issues can compound each other, resulting in a robot part that is slow to respond and when it does it goes too far.

SMAs require a disproportionate amount of power when compared to electric motors or R/C servos. Along with this extra power, they cannot provide as much force as an electric motor or servo and when they are used in a robot, you will find that they cannot carry batteries. Any robot you design with SMAs will have a tether wire running to it.

Finally, SMAs are surprisingly difficult to work with. The ideal method of attaching them to structures is to use a crimped or flattened tube, as discussed previously with the SMA being held under light tension when crimping the attachment tube. You will find your first attempts at working with SMA to not work very well, but over time, you will gain the knack and will find that you can work with SMA quickly and efficiently.

These issues will make SMAs seem a lot less attractive than what you might have thought before you read this section, but by keeping these concerns in the back of your mind you can create prototype robots quickly and efficiently. Chances are you had some plans for a very special robot that could take advantage of the different properties of SMAs and you still can, but recognize that your SMA-based robot will have to be built very lightly and probably smaller than the final robot that you want to build. This is actually the advantage of SMA; it can be used to create prototype robot structures quickly and efficiently. Along with allowing you to create prototype structures in just a few moments, its slow operation will allow you to observe the action of the mechanical system and learn what is the correct way to sequence it.

25.5 From Here

To learn more about . . .	***Read***
Selecting a motor for driving a robot	Chapter 19, "Choosing the Right Motor"
Using DC motors for robot locomotion	Chapter 20, "Working with DC Motors"
Building Walkerbot, a six-legged mobile walking robot	Chapter 24, "Build a Heavy-Duty Six-Legged Walking Robot"

CHAPTER 26

REACHING OUT WITH ROBOT ARMS

Robots without arms are limited to rolling or walking about, perhaps noting things that occur around them, but little else. The robot can't, as the slogan goes, "reach out and touch someone," and it certainly can't manipulate its world.

The more sophisticated robots in science, industry, and research and development have at least one arm to grasp, reorient, or move objects. Arms extend the reach of robots and make them more like humans. For all the extra capabilities arms provide a robot, it's interesting that they aren't at all difficult to build. Your arm designs can be used for factory style, stationary "pick-and-place" robots, or they can be attached to a mobile robot as an appendage.

This chapter deals with the concept and design theory of robotic arms. Specific arm projects are presented in the next chapter. Incidentally, when we speak of arms, we will usually mean just the arm mechanism minus the hand (also called the gripper). Chapter 28, "Experimenting with Gripper Designs," talks about how to construct robotic hands and how you can add them to arms to make a complete, functioning appendage.

26.1 The Human Arm

Take a close look at your own arms for a moment. You'll quickly notice a number of important points. First, your arms are amazingly adept mechanisms. They are capable of being maneuvered into just about any position you want. Your arm has two major joints: the shoulder and the elbow (the wrist, as far as robotics is concerned, is usually considered part of the gripper mechanism). Your shoulder can move in two planes, both up and down and back and forth. As well as moving your arm up and down and back and forth, you can rotate your shoulder as well. The elbow joint is capable of moving in one plane; it can be thought of as a simple hinge, and the bone connected to it can be rotated.

The joints in your arm, and your ability to move them, are called degrees of freedom. Your shoulder provides three degrees of freedom in itself; shoulder rotation and two-plane shoulder flexion. The elbow joint adds a fourth and fifth degree of freedom: elbow flexion and elbow rotation.

Robotic arms also have degrees of freedom. But instead of muscles, tendons, ball and socket joints, and bones, robot arms are made from metal, plastic, wood, motors, solenoids, gears, pulleys, and a variety of other mechanical components. Some robot arms provide but one degree of freedom; others provide three, four, and even five separate degrees of freedom.

26.2 Arm Types

Robot arms are classified by the shape of the area that the end of the arm (where the gripper is) can reach. This accessible area is called the *work envelope.* For simplicity's sake, the work envelope does not take into consideration motion by the robot's body, just the arm mechanics.

The human arm has a nearly spherical work envelope. We can reach just about anything, as long as it is within arm's length, within the inside of about three-quarters of a sphere. Imagine being inside a hollowed-out orange. You stand by one edge. When you reach out you can touch the inside walls of about three-quarters of the peel.

In a robot, such a robot arm would be said to have revolute coordinates. The three other main robot arm designs are polar coordinate, cylindrical coordinate, and Cartesian coordinate. You'll note that there are three degrees of freedom in all four basic types of arm designs.

26.2.1 REVOLUTE COORDINATE

Revolute coordinate arms, such as the one depicted in Fig. 26-1, are modeled after the human arm, so they have many of the same capabilities. The typical robotic design is somewhat different, however, because of the complexity of the human shoulder joint.

The shoulder joint of the robotic arm is really two different mechanisms. Shoulder rotation is accomplished by spinning the arm at its base, almost as if the arm were mounted on a record player turntable. Shoulder flexion is accomplished by tilting the upper arm member backward and forward. Elbow flexion works just as it does in the human arm. It moves the forearm up and down. Revolute coordinate arms are a favorite design choice for hobby robots. They provide a great deal of flexibility, and they actually look like arms.

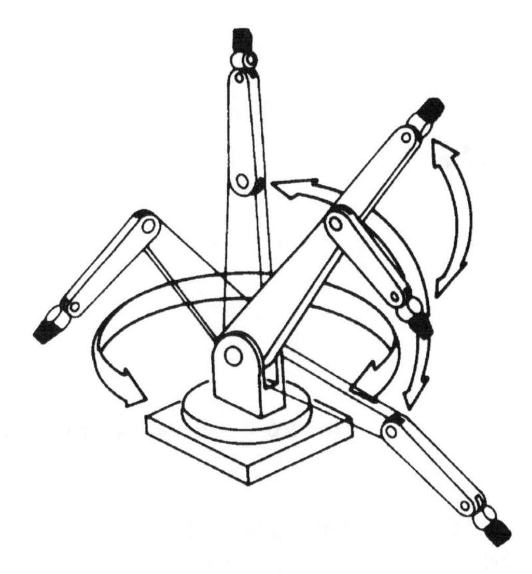

FIGURE 26-1 A revolute coordinate arm.

26.2.2 POLAR COORDINATE

The work envelope of the polar coordinate arm is the shape of a half sphere. Next to the revolute coordinate design, polar coordinate arms are the most flexible in terms of the ability to grasp a variety of objects scattered about the robot. Fig. 26-2 shows a polar coordinate arm and its various degrees of freedom.

A turntable rotates the entire arm, just as it does in a revolute coordinate arm. This function is akin to shoulder rotation. The polar coordinate arm lacks a means for flexing or bending its shoulder, however. The second degree of freedom is the elbow joint, which moves the forearm up and down. The third degree of freedom is accomplished by varying the reach of the forearm. An "inner" forearm extends or retracts to bring the gripper closer to or farther away from the robot. Without the inner forearm, the arm would only be able to grasp objects laid out in a finite two-dimensional circle in front of it. The polar coordinate

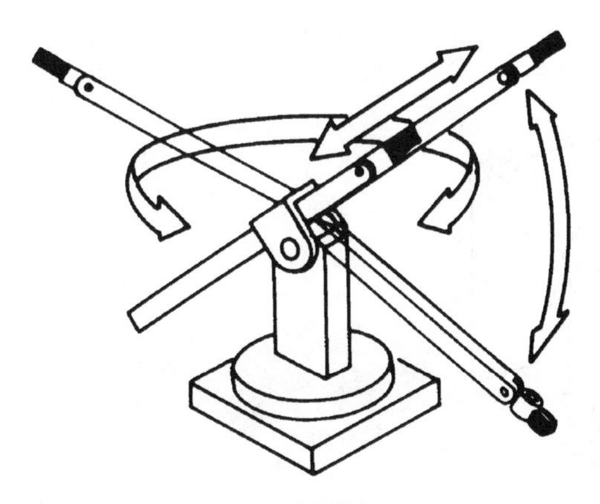

FIGURE 26-2 A polar coordinate arm.

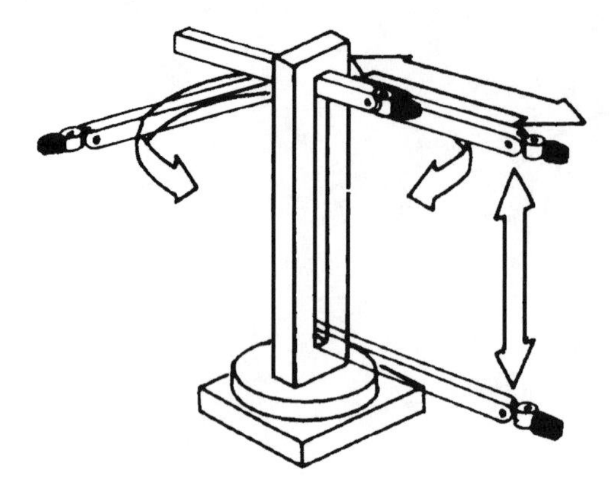

FIGURE 26-3 A cylindrical coordinate arm.

arm is often used in factory robots and finds its greatest application as a stationary device. It can, however, be mounted to a mobile robot for increased flexibility.

26.2.3 CYLINDRICAL COORDINATE

The cylindrical coordinate arm looks a little like a robotic forklift. Its work envelope resembles a thick cylinder, hence its name. Shoulder rotation is accomplished by a revolving base, as in revolute and polar coordinate arms. The forearm is attached to an elevatorlike lift mechanism, as depicted in Fig. 26-3. The forearm moves up and down this column to grasp objects at various heights. To allow the arm to reach objects in three-dimensional space, the forearm is outfitted with an extension mechanism, similar to the one found in a polar coordinate arm.

26.2.4 CARTESIAN COORDINATE

The work envelope of a Cartesian coordinate arm (Fig. 26-4) resembles a box. It is the arm most unlike the human arm and least resembles the other three arm types. It has no rotat-

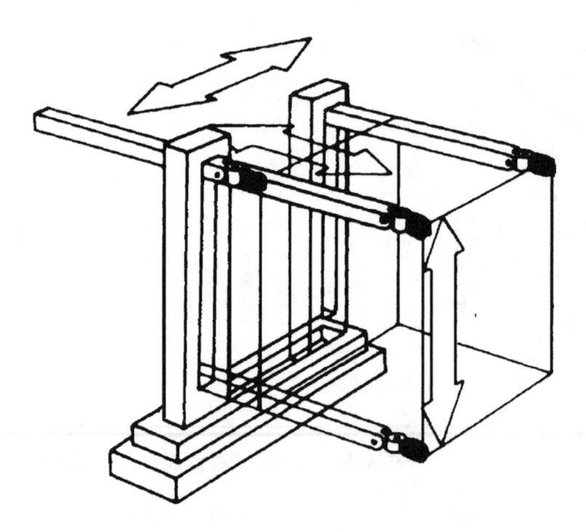

FIGURE 26-4 A Cartesian coordinate arm.

ing parts. The base consists of a conveyer belt-like track. The track moves the elevator column (like the one in a cylindrical coordinate arm) back and forth. The forearm moves up and down the column and has an inner arm that extends the reach closer to or farther away from the robot.

26.3 Activation Techniques

There are three general ways to move the joints in a robot arm:

- Electrical
- Hydraulic
- Pneumatic

Electrical actuation is done with motors, solenoids, and other electromechanical devices. It is the most common and easiest to implement. The motors for elbow flexion, as well as the motors for the gripper mechanism, can be placed in or near the base. Cables, chains, or belts connect the motors to the joints they serve.

Electrical activation doesn't always have to be via an electromechanical device such as a motor or solenoid. Other types of electrically induced activation are possible using a variety of techniques. One of particular interest to hobby robot builders is shape-memory alloy, or SMA, as discussed previously.

Hydraulic actuation uses oil-reservoir pressure cylinders, similar to the kind used in earth-moving equipment and automobile brake systems. The fluid is noncorrosive and inhibits rust: both are the immediate ruin of any hydraulic system. Though water can be used in a hydraulic system, if the parts are made of metal they will no doubt eventually suffer from rust, corrosion, or damage by water deposits. For a simple home-brew robot, however, a water-based hydraulic system using plastic parts is a viable alternative.

Pneumatic actuation is similar to hydraulic, except that pressurized air is used instead of oil or fluid (the air often has a small amount of oil mixed in it for lubrication purposes). Both hydraulic and pneumatic systems provide greater power than electrical actuation, but they are more difficult to use. In addition to the actuation cylinders themselves, such as the one shown in Fig. 26-5, a pump is required to pressurize the air or oil, and values are used to control the retraction or extension of the cylinders. For the best results, you need a holding tank to stabilize the pressurization. For small robot arms, the Lego Technics pneumatic cylinders, controls, and tanks can make the task of actuating an arm quite easy.

An interesting variation on pneumatic actuation is the *Air Muscle,* an ingenious combination of a small rubber tube and black plastic mesh. The rubber tube acts as an expandable bladder, and the plastic mesh forces the tube to inflate in a controllable manner. Air Muscle is available premade in various sizes; it is activated by pumping air into the tube. When filled with air, the tube expands its width but contracts its length (by 25 percent). The result is that the tube and mesh act as a kind of mechanical muscle. The Air Muscle is said to be more efficient than the standard pneumatic cylinder, and according to its makers it has about a 400:1 power-to-weight ratio.

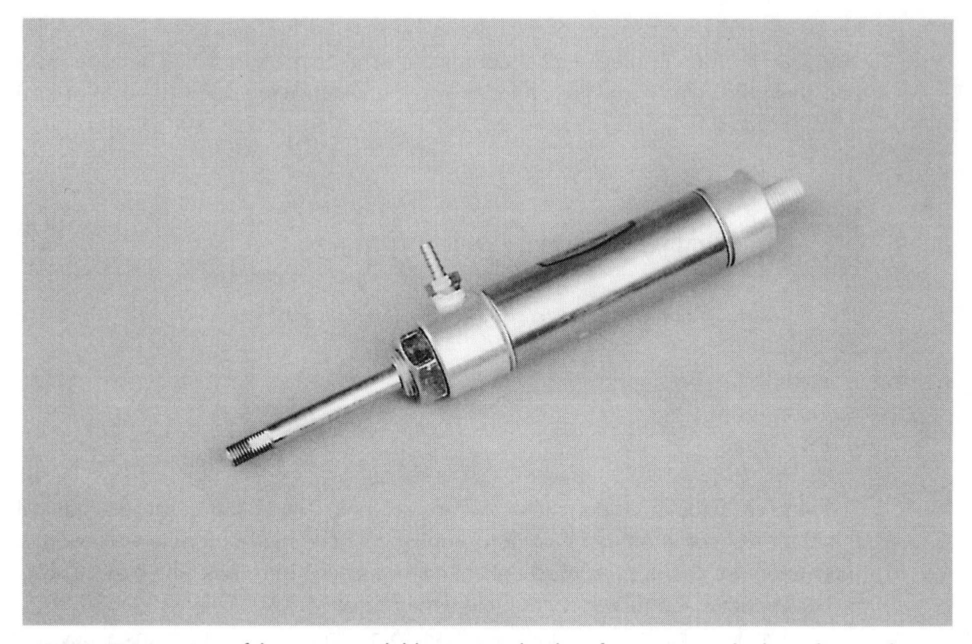

FIGURE 26-5 One of the many available sizes and styles of pneumatic cylinders. This one has a bore of about ½ in and a stroke of three in.

26.4 From Here

To learn more about . . .	*Read*
Building a robotic revolute coordinate arm	Chapter 27, "Build a Revolute Coordinate Arm"
Creating hands for robot arms	Chapter 28, "Experimenting with Gripper Designs"
Endowing robot arms and hands with the sense of touch	Chapter 29, "The Sense of Touch"

CHAPTER 27

BUILDING A REVOLUTE COORDINATE ARM

The revolute coordinate arm design provides a great deal of flexibility, yet requires few components. The arm described in this chapter enjoys only two degrees of freedom. You'll find, however, that even with two degrees of freedom, the arm can do many things. It can be used by itself as a stationary pick-and-place robot, or it can be attached to a mobile platform. The construction details given here are for a left hand; to build a right hand, simply make it a mirror image of the left.

You can use just about any type of gripper with this arm. In Fig. 27-1, the completed arm is shown with a simple gripper built on it. You can design the forearm so it accepts many different grippers interchangeably. See Chapter 27, "Experimenting with Gripper Designs," for more information on robot hands.

27.1 Design Overview

The design of the revolute coordinate arm is modeled after the human arm. A shaft-mounted shoulder joint provides shoulder rotation (degree of freedom #1). A simple swing-arm rotating joint provides the elbow flexion (degree of freedom #2).

You could add a third degree of freedom—shoulder flexion—by providing another joint immediately after the shoulder. Tests have proved that this basic two-degree-of-freedom arm is quite sufficient for most tasks. It is best used, however, on a mobile platform where the

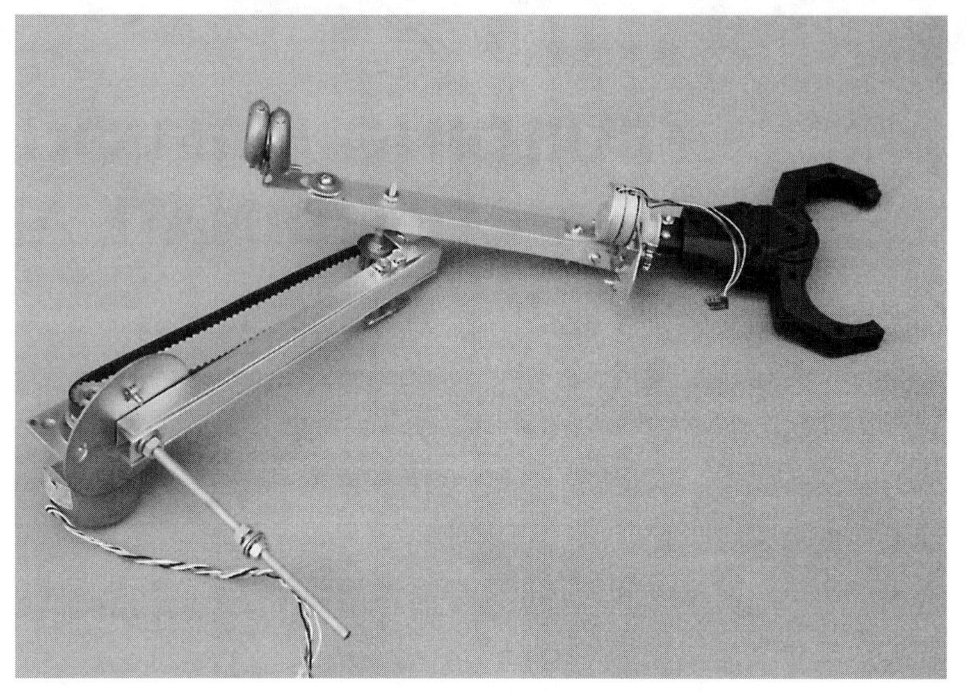

FIGURE 27-1 The completed arm, with gripper (hand) attached.

robot can move closer to or farther away from the object it's grasping. That's cheating, in a way, but it's a lot simpler than adding another joint.

27.2 Shoulder Joint and Upper Arm

The shoulder joint is a shaft that connects to a bearing mounted on the arm base or in the robot. Attached to the shaft is the drive motor for moving the shoulder up and down. The motor is connected by a single-stage gear system, as shown in Fig. 27-2 (refer to the parts list in Table 27-1). In the prototype arm for this book, the output of the motor was approximately 22 r/min, or roughly one-third of a revolution per second.

For a shoulder joint, 22 r/min is a little on the fast side. A gear ratio of 3:1 was chosen to decrease the speed by a factor of three (and increase the torque of the motor roughly by a factor of three). With the gear system, the shoulder joint moves at about one revolution every 8 s. That may seem slow, but remember that the shoulder joint swings in an arc of a little less than 50°, or roughly one-seventh of a complete circle. Thus, the shoulder will go from one extreme to the other in under 2 s.

The upper arm is constructed from a 10-in length of $\frac{57}{64}$-by-$\frac{9}{16}$-by-$\frac{1}{16}$-in aluminum channel stock and a matching 10-in length of $\frac{41}{64}$-by-$\frac{1}{2}$-by-$\frac{1}{16}$-in aluminum channel stock (Fig. 27-3). Sandwich the two stocks together to make a bar. Drill a $\frac{1}{4}$-in hole $\frac{1}{2}$ in from the end

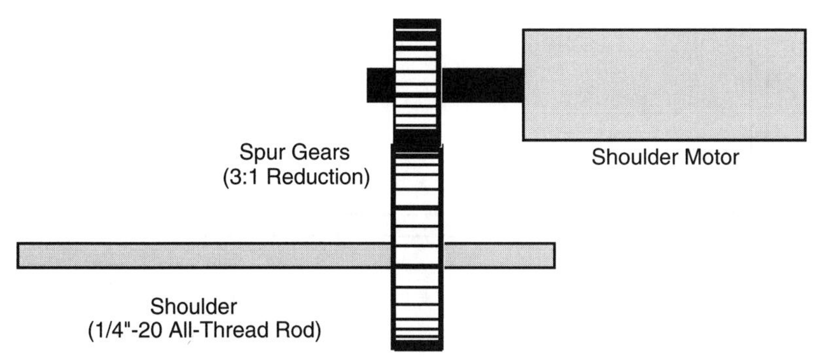

FIGURE 27-2 The gear transfer system used to actuate the shoulder of the revolute arm. You can also use a motor with a built-in reduction gear if the output of the motor is not slow enough for the arm.

of the channel stock pieces. Cut a piece of ¼-in 20 all-thread rod to a length of 7 in (this measurement depends largely on the shoulder motor arrangement, but 7 in gives you room to make changes). Thread a ¼-in 20 nut, flat washer, and locking washer onto one end of the rod. Leave a little extra—about ⅛ to ¼ in—on the outside of the nut. You'll need the room in a bit.

TABLE 27-1 Parts List For Revolute Arm

1	10-in length 57/64-by-9/16-by-1/16-in aluminum channel stock
1	10-in length 41/64-by-½-by-1/16-in aluminum channel stock
1	8-in length 57/64-by-9/16-by-1/16-in aluminum channel stock
1	8-in length 41/64-by-½-by-1/16-in aluminum channel stock
1	7-in length ¼-in 20 all-thread rod
2	1¾-in-by-10/24 stove bolt
2	1½-by-⅜-in flat corner iron
1	3-by-¾-in mending plate "T" (for motor mounting)
2	½-in aluminum spacer
1	¼-in aluminum spacer
2	¾-in-diameter, 5-lugs-per-in timing belt sprocket
1	20½-in-length timing belt (5 lugs per in)
2	Stepper motors (see text)
1	3:1 gear reduction system (such as one 20-tooth 24-pitch spur gear and one 60-tooth 24-pitch spur gear)
Misc.	6/32, 10/24, and ¼-in 20 nuts, washers, tooth lock washers, fishing tackle weights

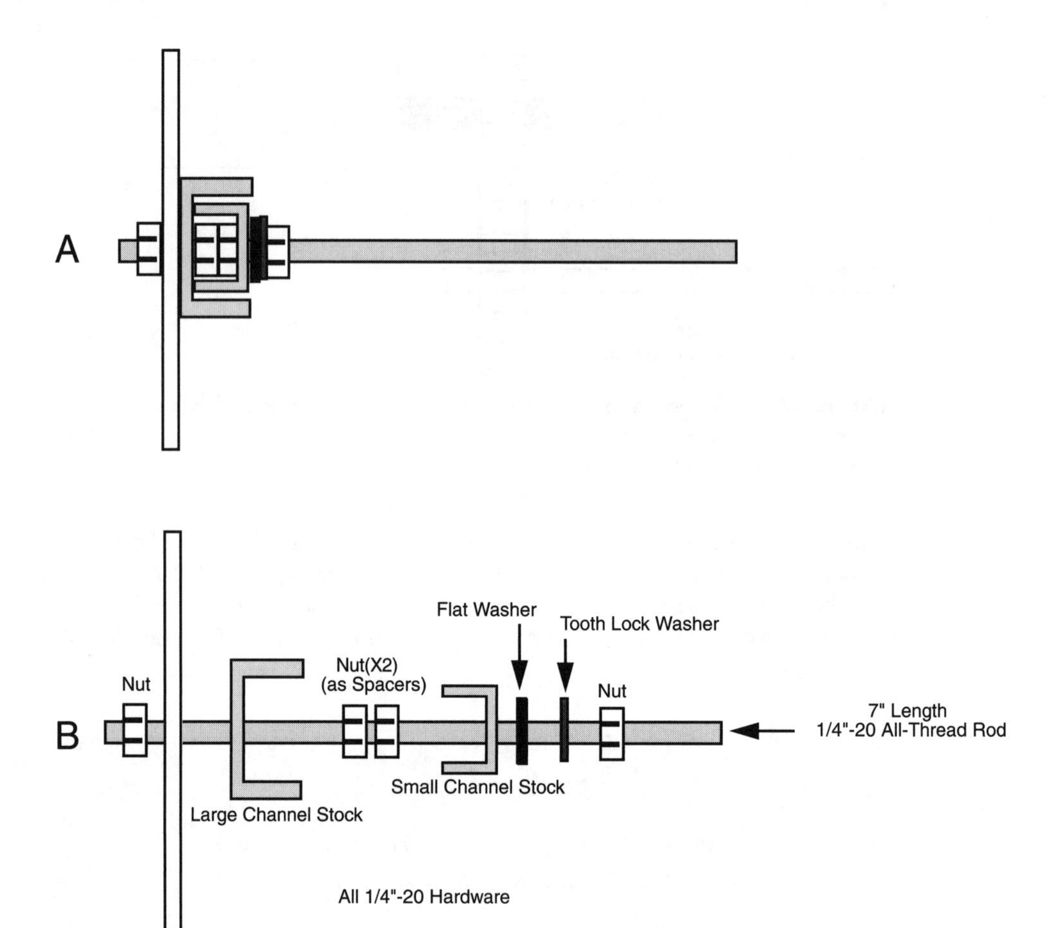

FIGURE 27-3 Shoulder shaft detail. *a.* Completed shaft; *b.* exploded view.

Drill a ¼-in hole in the center of a 3¾-in-diameter metal electrical receptacle cover plate. Insert the rod through it and the hole of the larger channel aluminum. Next, thread two ¼-in 20 nuts onto the rod to act as spacers, then attach the smaller channel aluminum. Lock the pieces together using a flat washer, tooth washer, and ¼-in 20 nut.

The shoulder is now complete.

27.3 Elbow and Forearm

The forearm attaches to the end of the upper arm. The joint there serves as the elbow. The forearm is constructed much like the upper arm: cut the small and large pieces of channel aluminum to 8 in instead of 10 in. Construct the elbow joint as shown in Figs. 27-4 and 27-5, using two 1½-by-⅜-in flat corner angles, ½-in spacers, and 10/24 hardware. The ¾-in

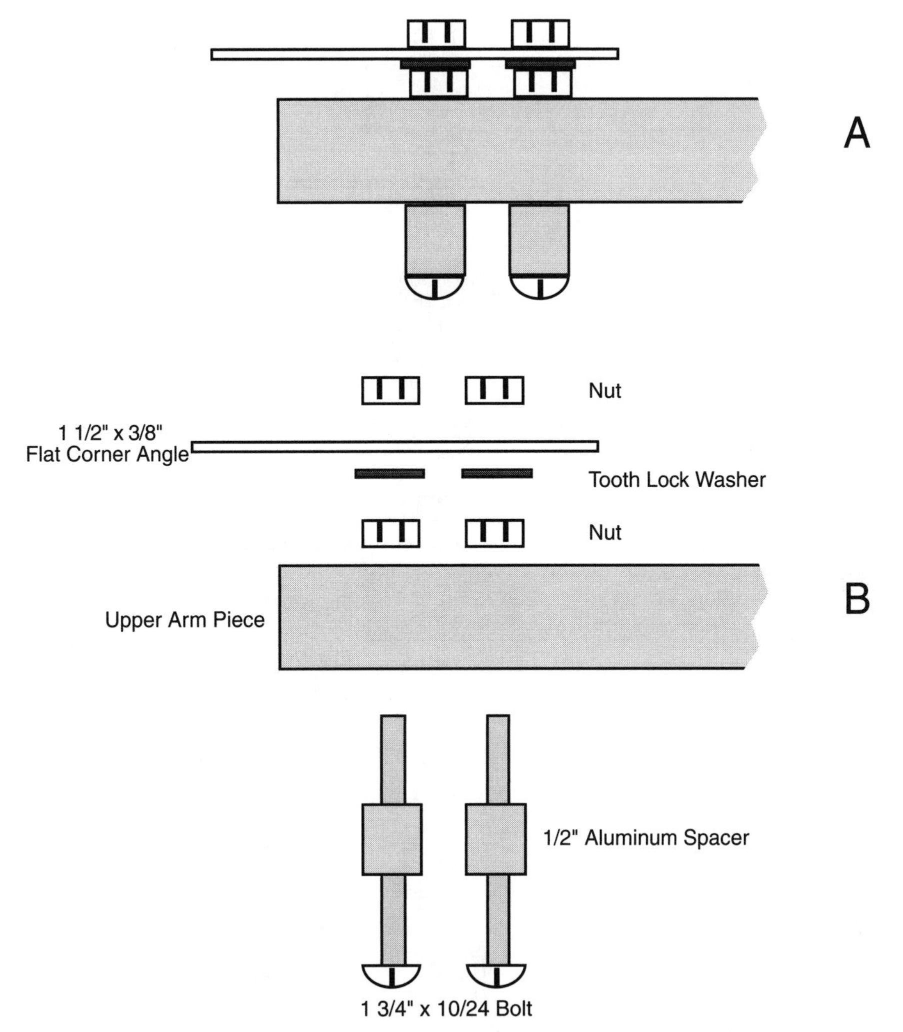

FIGURE 27-4 Upper arm elbow joint detail. *a.* Complete joint; *b.* exploded view.

timing belt sprocket (5 lugs per inch) is used to convey power from the elbow motor, which is mounted at the shoulder. The completed joint is shown in Fig. 27-6.

You can actually use just about any size of timing belt or sprocket. When using the size of sprockets specified in Table 27-1, the timing belt is 20½ in. If you use another size sprocket for the elbow or the motor, you may need to choose another length. You can adjust for some slack by mounting the elbow joint closer to or farther from the end of the upper arm.

You may also use #25 roller chain to power the elbow. Use a sprocket on the elbow and a sprocket on the motor shaft. Connect the two with a #25 roller chain. You'll need to experiment based on the size of sprockets you use to come up with the exact length for the roller chain.

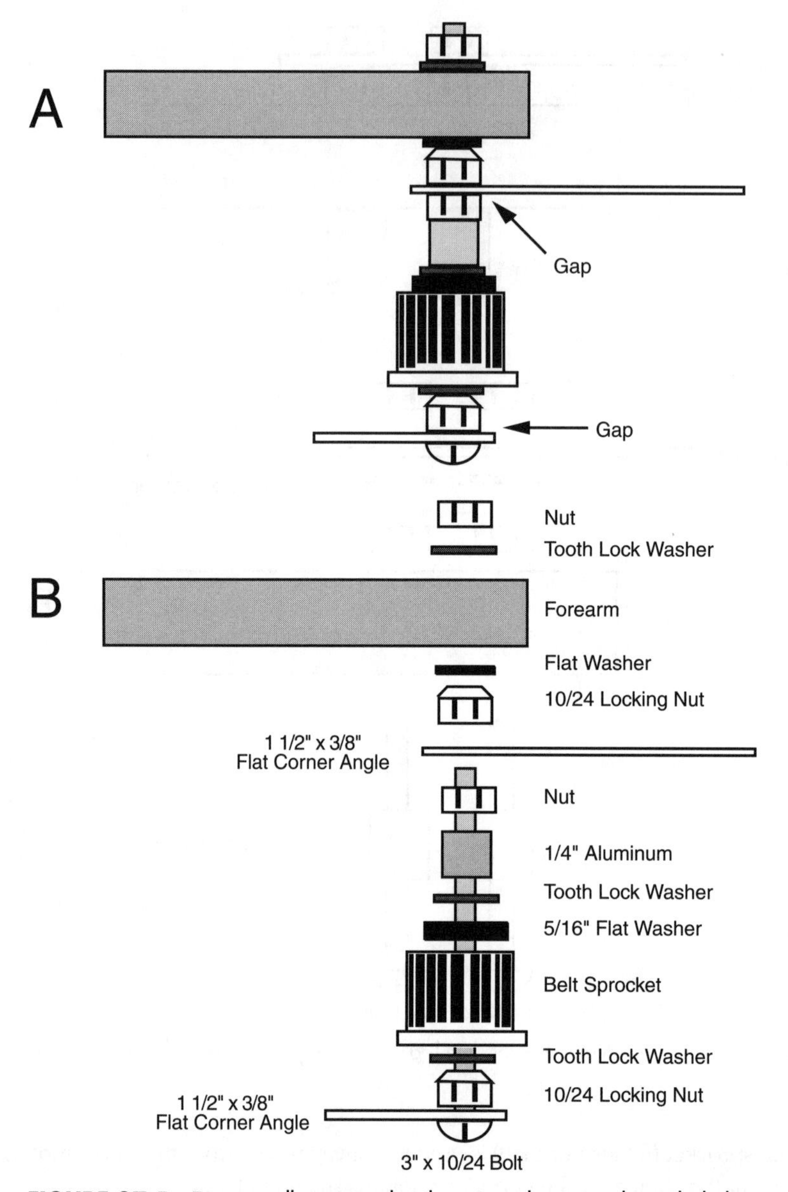

FIGURE 27-5 Forearm elbow joint detail. *a.* Complete joint; *b.* exploded view.

When the elbow and forearm are complete, mount the motor on the shoulder, directly on the plate cover. The motor we chose for the prototype revolute coordinate arm was a 1A medium-duty stepper motor. Predrilled holes on the face of the motor made it easier to mount the arm. A 3-by-¾-in mending plate T was used to secure the motor to the plate, as illustrated in Fig. 27-7. New holes were drilled in the plate to match the holes in the motor (1⅞-in spacing), and the T was bent at the cross.

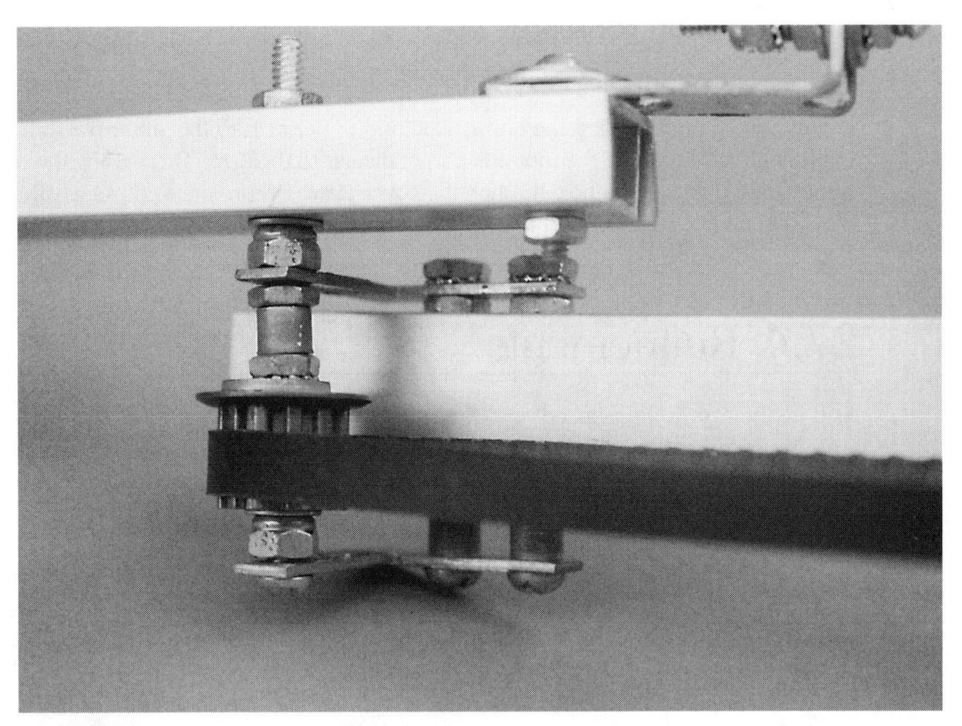

FIGURE 27-6 A close-up view of the elbow joint.

FIGURE 27-7 The motor mounted on the shoulder.

Unscrew the nut holding the cover plate and upper arm to the shaft, place the T on it, and retighten. Make sure the motor is perpendicular to the arm. Then, using the other hole in the T as a guide, drill a hole through the cover plate. Secure the T in place with an 8⁄32-by-½-in bolt and nut. The finished arm, with a gripper attached, is shown in Fig. 27-1.

27.4 Refinements

As it is, the arm is unbalanced, and the shoulder motor must work harder to position the arm. You can help to rebalance the arm by relocating the shoulder rotation shaft and by adding counterweights or springs. Before you do anything hasty, however, you may want to attach a gripper to the end of the forearm. Any attempts to balance the arm now will be severely thwarted when you add the gripper.

The center of gravity for the whole arm, with the elbow drive motor included, is approximately midway along the length of the upper arm (at least this is true of the prototype arm; your arm may be different). Remove the long shaft from the present shoulder joint, and replace it with a short 1½- or 2-in-long ¼-in 20 bolt. Drill a new ¼-in hole through the upper arm at the approximate center of gravity, and thread the shoulder shaft through it. Attach it as before, using ¼-in 20 nuts, flat washers, and toothed lock washers.

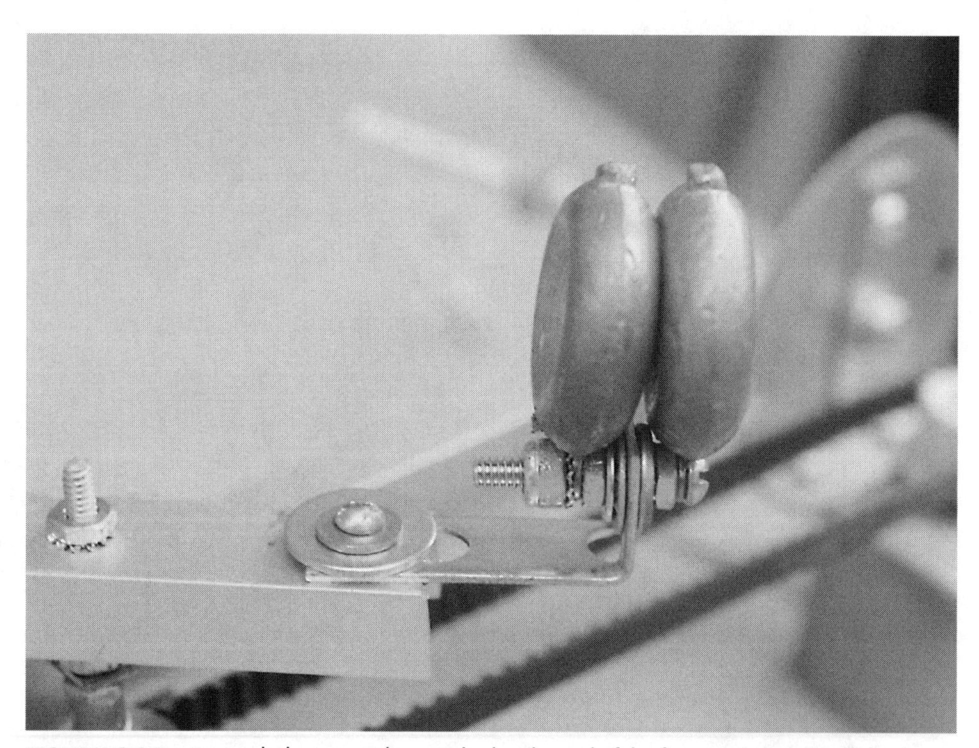

FIGURE 27-8 Counterbalance weights attached to the end of the forearm help redistribute the weight. You can also use springs, which will help reduce the overall weight of the arm.

The forearm is also out of balance, and you can correct it in a similar manner, by attaching the shoulder joint nearer to the center of the arm. This has the unfortunate side effect, however, of shortening the reach of the forearm. One solution is to make the arm longer to compensate. In effect, you'll be keeping the elbow joint where it is, just adding extra length behind it.

This may interfere with the operation of the arm or robot, however, so you may want to opt for counterweights attached to the end of the arm. For the prototype arm, two 4-oz fishing tackle weights were attached to the arm with a 2-by-¾-in corner angle bracket (see Fig. 27-8).

27.5 Position Control

The stepper motors used for the shoulder and elbow joints of the prototype provide a natural control over the position of the arm. Under electronic control, the motors can be commanded to rotate a specific number of steps, which in turn moves the upper arm and forearm a specified amount.

You should supplement the open-loop servo system with limit switches. These switches provide an indication when the arm joints have moved to their extreme positions. The most common limit switches are small leaf switches. You can also construct optical switches using photo-interrupters. A small patch of plastic or metal interrupts the flow of light between an LED and phototransistor, thus signaling the limit of movement. You can build these interrupters by mounting an infrared LED and phototransistor on a small perforated board, or you can purchase ready-made modules (they are common surplus finds). Using an IR LED and phototransistor is actually a simplified version of the limit switches discussed in Chapter 20.

When using continuous DC motors, you need to provide some type of feedback to report the position of the arm. Otherwise, the control electronics (almost always a computer) will never know where the arm is or how far it has moved. There are several ways you can provide this feedback. The most popular methods are a potentiometer and an incremental shaft encoder.

27.5.1 POTENTIOMETER

Attach the shaft of a potentiometer to the shoulder or elbow joint or motor (see Fig. 27-9), and the varying resistance of the pot serves as an indication of the position of the arm. Just about any pot will do, as long as it has a travel rotation the same as or greater than the travel rotation of the joints in the arm. Otherwise, the arm will go past the internal stops of the potentiometer. Travel rotation is usually not a problem in arm systems, where joints seldom move more than 40° or 50°. If your arm design moves more than about 270°, use a multiturn pot. A three-turn pot should suffice.

Another method is to use a slider-pot. You operate a slider-pot by moving the wiper up and down, rather than by turning a shaft. Slider-pots are ideal when you want to measure linear distance, like the amount of travel (distance) of a chain or belt. Fig. 27-10 shows a slider-pot mounted to a cleat in the timing belt used to operate the elbow joint.

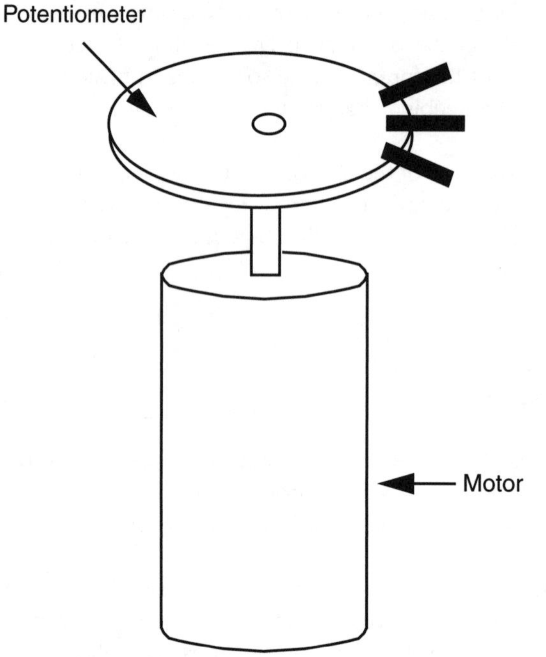

FIGURE 27-9 Using a potentiometer as a position feedback device. Mount the potentiometer on a drive motor or on a joint of the arm.

27.5.2 INCREMENTAL SHAFT ENCODER

The incremental shaft encoder was first introduced in Chapter 20, "Working with DC Motors." The shaft encoder is a disc that has many small holes or slots near its outside circumference. You attach the disc to a motor shaft or the shoulder or elbow joint. The shaft encoder circuit is typically composed of a circuit connected to the phototransistor (the latter of which is baffled to block off ambient light). The phototransistor counts the number of on/off flashes and then converts that number into distance traveled. For example, one on/off flash may equal a 2° movement of the joint. Two flashes may equal a 4° movement, and so forth.

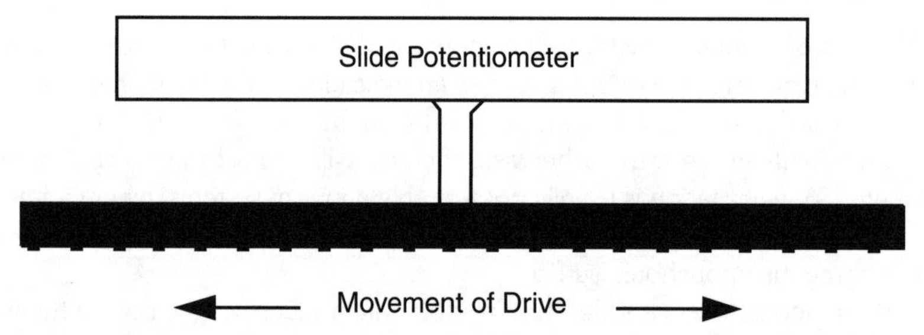

FIGURE 27-10 Using a slide potentiometer to register position feedback. The wiper of the pot can be linked to any mechanical device, like a chain or belt, that moves laterally.

The advantage of the incremental shaft encoder is that its output is inherently digital. You can use a computer, or even a simple counter circuit, to count the number of on/off flashes. The result, when the movement ends, is the new position of the arm.

27.6 From Here

To learn more about . . .	***Read***
Using DC motors and shaft encoders	Chapter 20, "Working with DC Motors"
Using stepper motors to drive robot parts	Chapter 21, "Working with Stepper Motors"
Different robotic arm systems and assemblies	Chapter 24, "An Overview of Arm Systems"
Attaching hands to robotic arms	Chapter 27, "Experimenting with Gripper Designs"
Interfacing feedback sensors to computers and microcontrollers	Chapter 29, "Interfacing with Computers and Microcontrollers"

CHAPTER 28

EXPERIMENTING WITH GRIPPER DESIGNS

The arm system detailed in Chapter 27 isn't much good without hands. In the robotics world, hands are usually called *grippers* (also *end effectors*) because the word more closely describes their function. Few robotic hands can manipulate objects with the fine motor control of a human hand; they simply grasp or grip an object, hence the name gripper. See Fig. 28-1 for an example.

Gripper designs are numerous, and no one single design is ideal for all applications. Each gripper technique has unique advantages over the others, and you must fit the gripper to the application at hand (pun intended). This chapter outlines a number of useful gripper designs for your robots. Most are fairly easy to build; some even make use of inexpensive plastic toys. The gripper designs encompass just the finger or grasping mechanisms. The last section of this chapter details how to add wrist rotation to any of the gripper designs.

28.1 The Clapper

The clapper gripper is a popular design, favored because of its easy construction and simple mechanics. You can build the clapper using metal, plastic, wood, or a combination of all three. The parts list in Table 28-1 is for the parts used to build the metal and plastic clapper shown in Fig. 28-2.

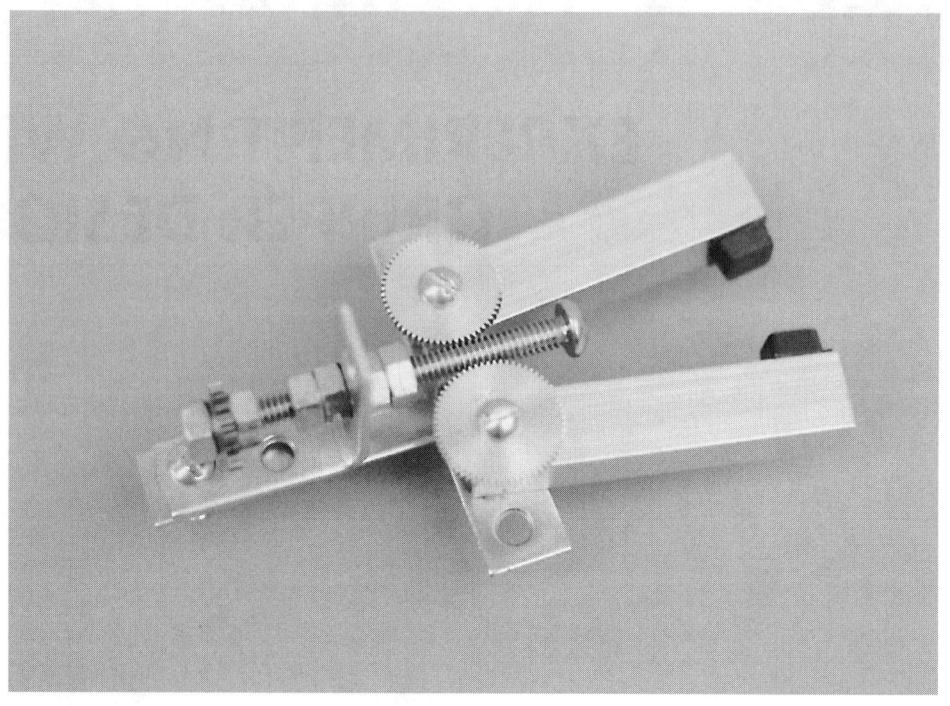

FIGURE 28-1 The two-pincher worm drive gripper.

The clapper consists of a wrist joint (which, for the time being, we'll assume is permanently attached to the forearm of the robot). Connected to the wrist are two plastic plates. The bottom plate is secured to the wrist; the top plate is hinged. A small spring-loaded solenoid is positioned inside, between the two plates. When the solenoid is not activated, the spring pushes the two flaps out, and the gripper is open. When the solenoid is activated, the plunger pulls in, and the gripper closes. The amount of movement at the end of the gripper is minimal—about ½ in with most solenoids. However, that is enough for general gripping tasks.

Cut two 1/16-in-thick acrylic plastic pieces to 1½ by 2⅓ in. Attach the lower flap to two 1-by-⅜-in corner angle brackets. Place the brackets approximately ⅛ in from either side of

TABLE 28-1 Parts List for the Clapper

2	1½-by-2½-by-1/16-in thick acrylic plastic sheet
2	1-by-⅜-in corner angle bracket
1	1½-by-1-in brass or aluminum hinge
1	Small 6- or 12-vdc spring-loaded solenoid
8	½-in-by-6/32 stove bolts, nuts

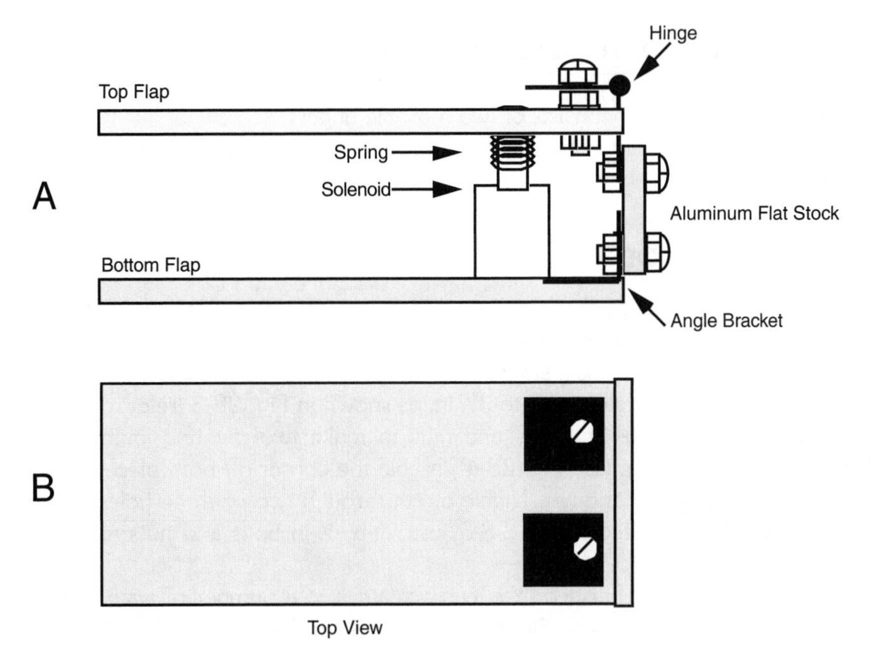

FIGURE 28-2 The clapper gripper. *a.* Assembly detail; *b.* top view.

the flap. Secure the pieces using 6/32-by-½-in bolts and 6/32 nuts. Cut a 1½-in length of 1½-by-⅛-in aluminum bar stock. Mount the two brackets to the bottom of the stock as shown in the figure. Attach the top flap to a 1½-by-1-in (approximately) brass or aluminum miniature hinge. Drill out the holes in the hinge with a #28 drill to accept 6/32 bolts. Secure the hinge using 6/32 bolts and nuts.

The choice of solenoid is important because it must be small enough to fit within the two flaps and it must have a flat bottom to facilitate mounting. It must also operate with the voltage used in your robot, usually 6 or 12 V. Some solenoids have mounting flanges opposite the plunger. If yours does, use the flange to secure the solenoid to the bottom flap. Otherwise, mount the solenoid in the center of the bottom flap, approximately ½ in from the back end (nearest the brackets), with a large glob of household cement. Let it stand to dry.

Align the top flap over the solenoid. Make a mark at the point where the plunger contacts the plastic. Drill a hole just large enough for the plunger; you want a tight fit. Insert the plunger through the hole and push down so that the plunger starts to peek through.

Align the top and bottom flaps so they are parallel to one another.

Using the mounting holes in the hinges as a guide, mark corresponding holes in the aluminum bar. Drill holes and mount the hinge using ½-in-by-6/32 bolts and nuts.

Test the operation of the clapper by activating the solenoid. If the plunger works loose, apply some household cement to keep it in place. You may want to add a short piece of rubber weather stripping to the inside ends of the clappers so they can grasp objects easier. You can also use stick-on rubber feet squares, available at most hardware and electronics stores.

28.2 Two-Pincher Gripper

The two-pincher gripper consists of two movable fingers, somewhat like the claw of a lobster. The steps for constructing one basic and two advanced models are described in this section.

28.2.1 BASIC MODEL

For ease of construction, the basic two-pincher gripper is made from extra Erector set parts (the components from a similar construction kit toy may also be used). Cut two metal girders to 4½ in (since this is a standard Erector set size, you may not have to do any cutting). Cut a length of angle girder to 3½ in, as shown in Fig. 28-3 (refer to the parts list in Table 28-2). Use 6⁄32-by-½-in bolts and nuts to make two pivoting joints. Cut two 3-in lengths and mount them (see Fig. 28-4). Nibble the corner off both pieces to prevent the two from touching one another. Nibble or cut through two or three holes on one end to make a slot. As illustrated in Fig. 28-5, use 6⁄32-by-½-in bolts and nuts to make pivoting joints in the fingers.

The basic gripper is finished. You can actuate it in a number of ways. One way is to mount a small eyelet between the two pivot joints on the angle girder. Thread two small cables or wire through the eyelet and attach the cables. Connect the other end of the cables

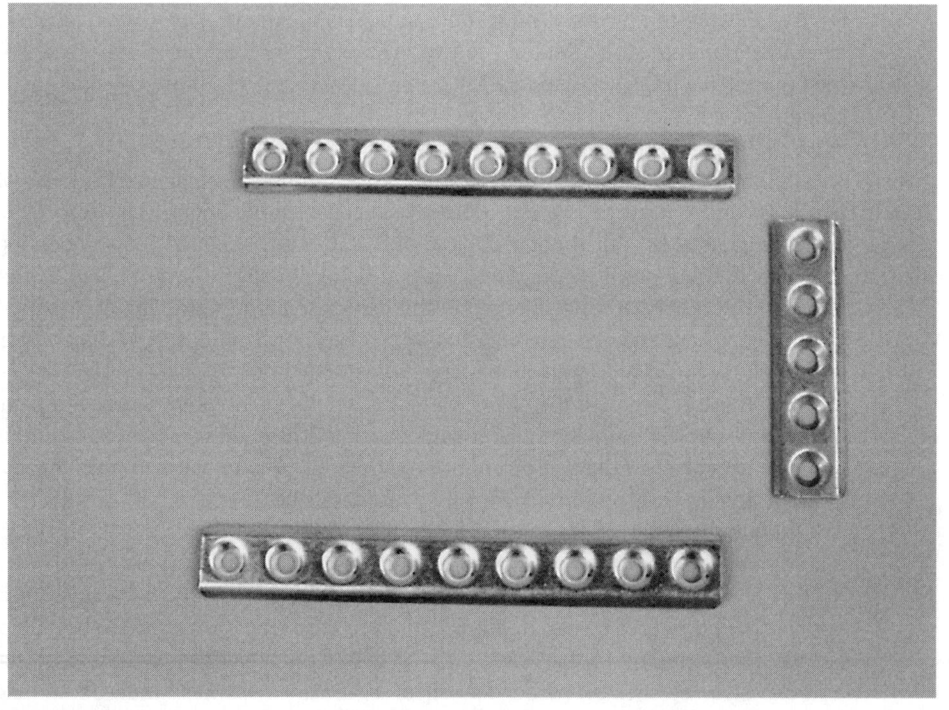

FIGURE 28-3 An assortment of girders from an Erector set toy construction kit.

TABLE 28-2 Parts List For Two-Finger Erector Set Gripper

2	4½-in Erector set girder
1	3½-in-length Erector set girder
4	½-in-by-⁶⁄₃₂ stove bolts, fender washer, tooth lock washer, nuts
Misc.	14- to 16-gauge insulated wire ring lugs, aircraft cable, rubber tabs, ½-by-½-in corner angle brackets (galvanized or from Erector set)

to a solenoid or a motor shaft. Use a light compression spring to force the fingers apart when the solenoid or motor is not actuated.

You can add pads to the fingers by using the corner braces included in most Erector set kits and then attaching weather stripping or rubber feet to the brace. The finished gripper should look like the one depicted in Fig. 28-6.

28.2.2 ADVANCED MODEL NUMBER 1

You can use a readily available plastic toy and convert it into a useful two-pincher gripper for your robot arm. The toy is a plastic extension arm with the pincher claw on one end and a hand gripper on the other (see Fig. 28-7). To close the pincher, you pull on the hand gripper. The contraption is inexpensive—usually under $10—and it is available at many toy stores.

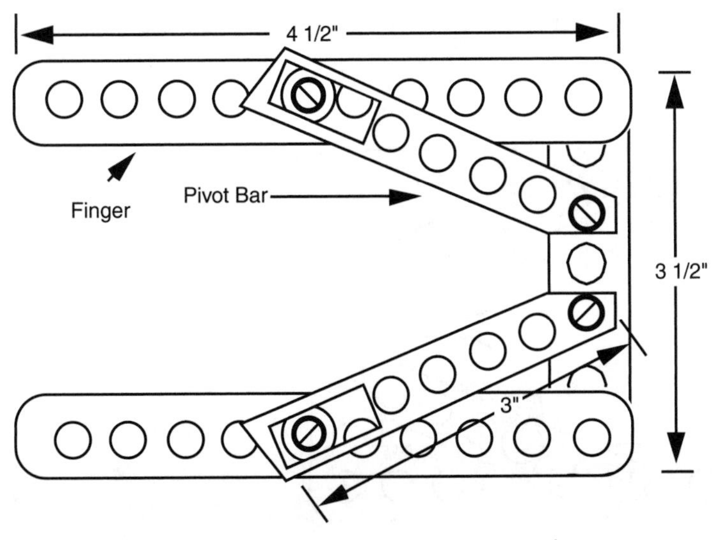

FIGURE 28-4 Construction detail of the basic two-pincher gripper, made with Erector set parts.

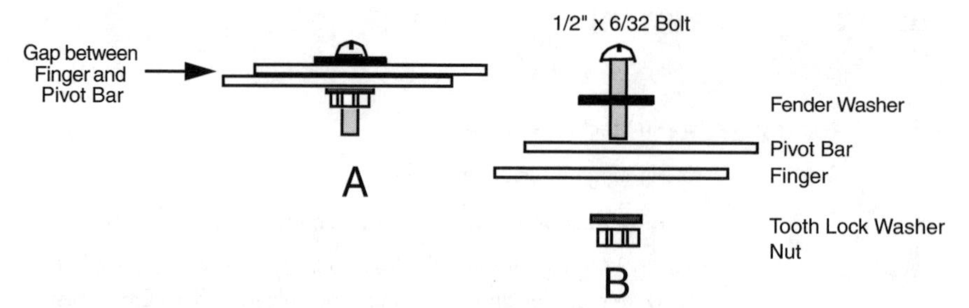

FIGURE 28-5 Hardware assembly detail of the pivot bar and fingers of the two-pincher gripper. *a.* Assembled sliding joint; *b.* exploded view.

Chop off the gripper 3 in below the wrist. You'll cut through an aluminum cable. Now cut off another 1½ in of tubing—just the arm, but not the cable. File off the arm tube until it's straight, then fashion a 1½-in length of ¾-in-diameter dowel to fit into the rectangular arm. Drill a hole for the cable to go through. The cable is off-centered because it attaches to the pull mechanism in the gripper, so allow for this in the hole. Place the cable through the hole, push the dowel at least ½ in into the arm, and then drill two small mounting holes to keep the dowel in place (see Fig. 28-8). Use 6⁄32-by-¾-in bolts and nuts to secure the pieces.

You can now use the dowel to mount the gripper on an arm assembly. You can use a small ¾-in U-bolt or flatten one end of the dowel and attach it directly to the arm. The gripper opens and closes with only a 7⁄16-in pull. Attach the end of the cable to a heavy-duty solenoid that has a stroke of at least 7⁄16 in. You can also attach the gripper cable to a ⅛-in round aircraft cable. Use a crimp-on connector designed for 14- to 16-gauge electrical wire to connect them end to end, as shown in Fig. 28-9. Attach the aircraft cable to a motor or rotary solenoid shaft and activate the motor or solenoid to pull the gripper closed. The

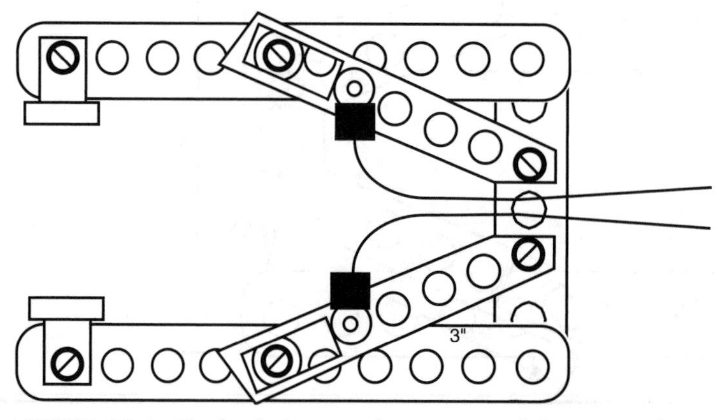

FIGURE 28-6 The finished two-pincher gripper, with fingertip pads and actuating cables.

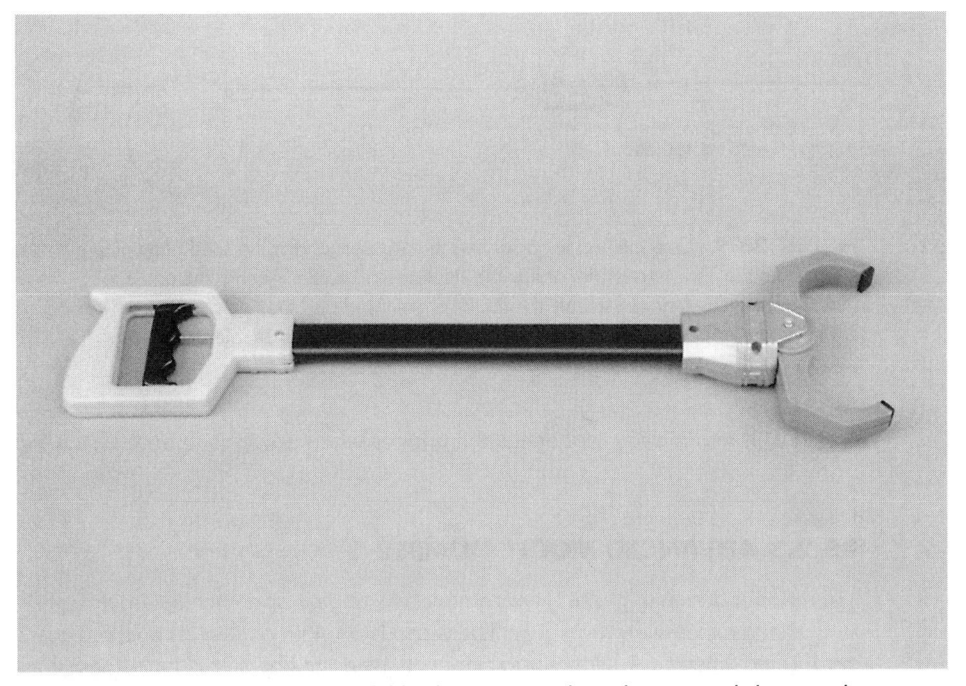

FIGURE 28-7 A commercially available plastic two-pincher robot arm and claw toy. The gripper can be salvaged for use in your own designs.

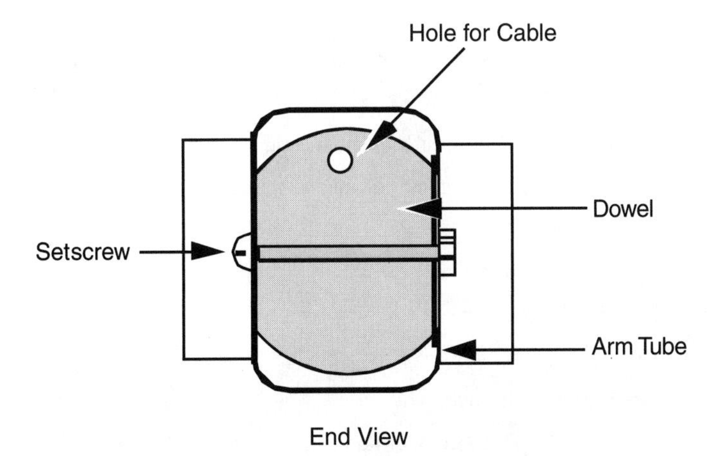

FIGURE 28-8 Assembly detail for the claw gripper and wooden dowel. Drill a hole for the actuating cable to pass through.

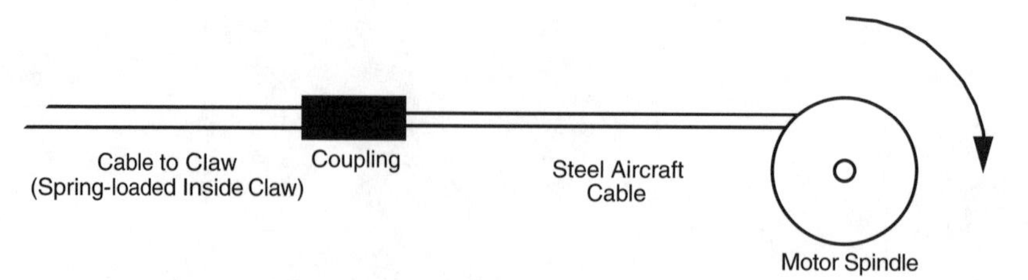

FIGURE 28-9 One method for actuating the gripper: attach the solid aluminum cable from the claw to a length of flexible steel aircraft cable. Anchor the cable to a motor or rotary solenoid. Actuate the motor or solenoid and the gripper closes. The spring in the gripper opens the claw when power to the motor or solenoid is removed.

spring built into the toy arm opens the gripper when power is removed from the solenoid or motor.

28.2.3 ADVANCED MODEL NUMBER 2

This gripper design (Fig. 28-1) uses a novel worm gear approach, without requiring a hard-to-find (and expensive) worm gear. The worm is a length of ¼-in 20 bolt; the gears are standard 1-in-diameter 64-pitch aluminum spur gears (hobby stores have these for about $1 apiece). Turning the bolt opens and closes the two fingers of the gripper. Refer to the parts list in Table 28-3.

Construct the gripper by cutting two 3-in lengths of 41/64-by-½-by-1/16-in aluminum channel stock. Using a 3-in flat mending T plate as a base, attach the fingers and gears to the T as shown in Fig. 28-10. The distance of the holes is critical and depends entirely on the diameter of the gears you have. You may have to experiment with different spacing if you use another gear diameter. Be sure the fingers rotate freely on the base but that the play is not excessive. Too much play will cause the gear mechanism to bind or skip.

TABLE 28-3 Parts List For Worm Drive Gripper

2	3-in lengths 41/64-by-½-by-1/16-in aluminum channel
2	1-in-diameter 64-pitch plastic or aluminum spur gear
1	2-in flat mending T
1	1½-by-½-in corner angle iron
1	3½-by-¼-in 20 stove bolt
2	¼-in 20 locking nuts, nuts, washers, tooth lock washers
2	½-in-by-8/32 stove bolts, nuts, washers
1	1-in-diameter 48-pitch spur gear (to mate with gear on driving motor shaft)

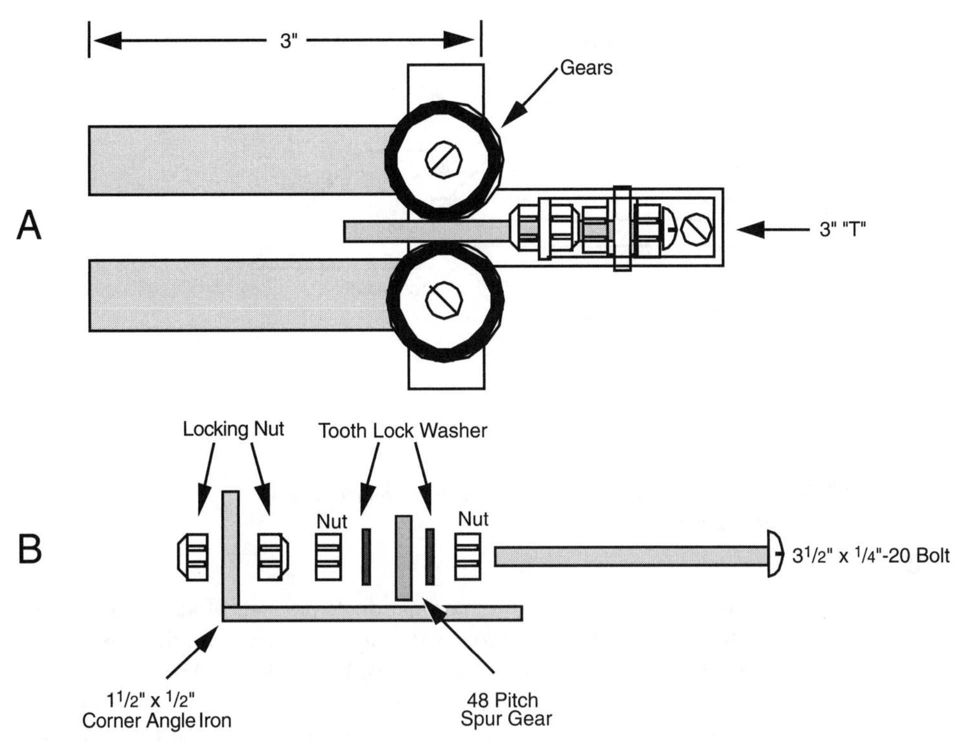

FIGURE 28-10 A two-pincher gripper based on a homemade work drive system. *a.* Assembled gripper; *b.* worm shaft assembly detail.

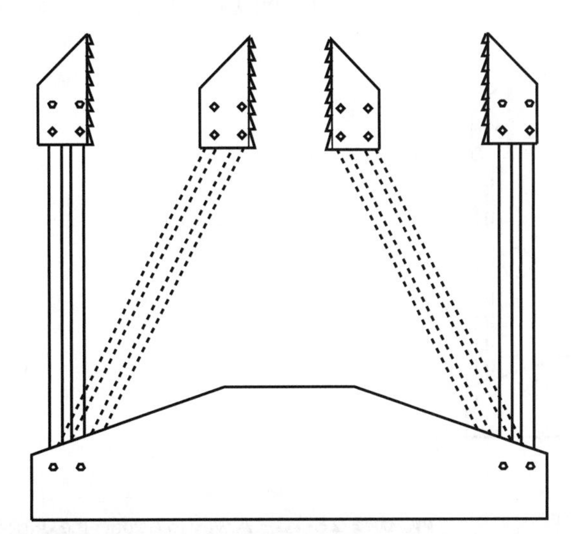

FIGURE 28-11 Adding a second rail to the fingers and allowing the points to freely pivot causes the fingertips to remain parallel to one another.

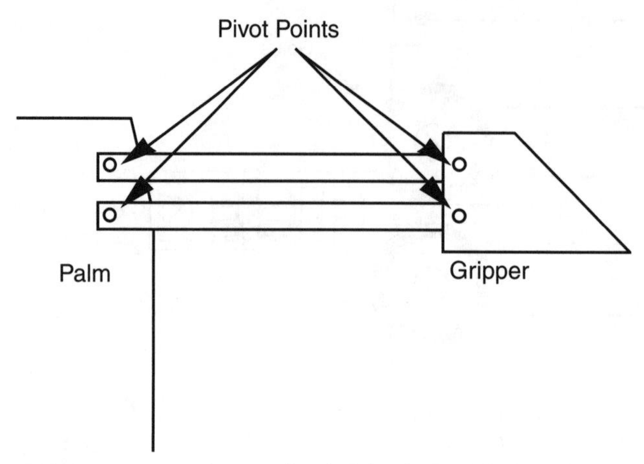

FIGURE 28-12 Close-up detail of the dual-rail finger system. Note the pivot points.

Secure the shaft using a 1½-by-½-in corner angle bracket. Mount it to the stem of the T using an 8/32-by-1-in bolt and nut. Add a #10 flat washer between the T and the bracket to increase the height of the bolt shaft. Mount a 3½-in-long ¼-in 20 machine bolt through the bracket. Use double nuts or locking nuts to form a free-spinning shaft. Reduce the play as much as possible without locking the bolt to the bracket. Align the finger gears to the bolt so they open and close at the same angle.

To actuate the fingers, attach a motor to the base of the bolt shaft. The prototype gripper used a ½-in-diameter 48-pitch spur gear and a matching 1-in 48-pitch spur gear on the

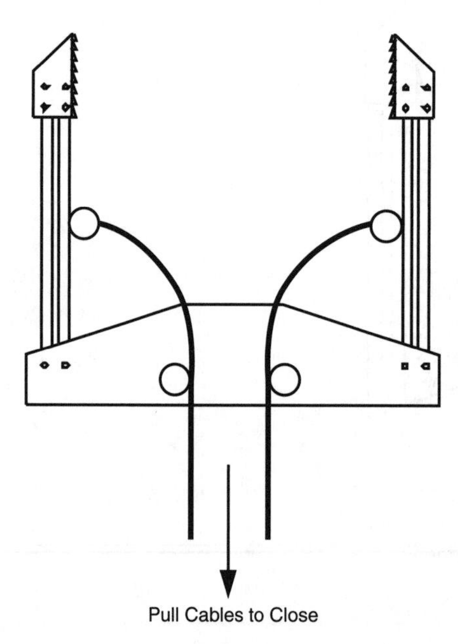

FIGURE 28-13 A way to actuate the gripper. Attach cables to the fingers and pull the cables with a motor or solenoid. Fit a torsion spring along the fingers and palm to open the fingers when power is removed from the motor or solenoid.

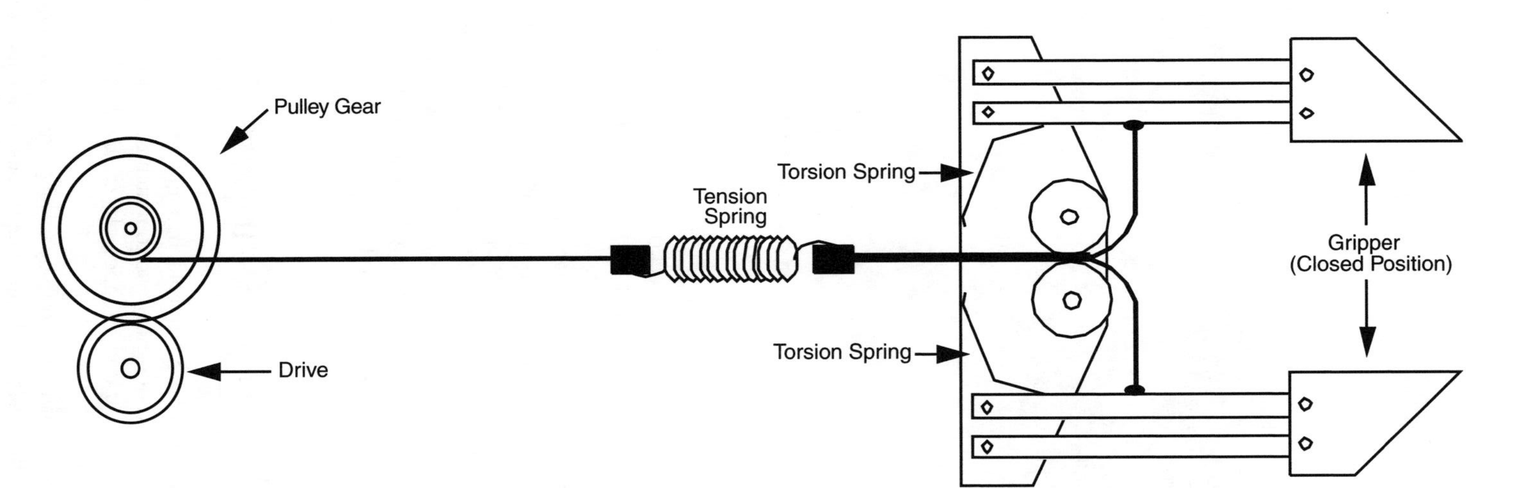

FIGURE 28-14 Actuation detail of a basic two-pincher gripper using a motor. The tension spring prevents undo pressure on the object being grasped. Note the torsion springs in the palm of the gripper.

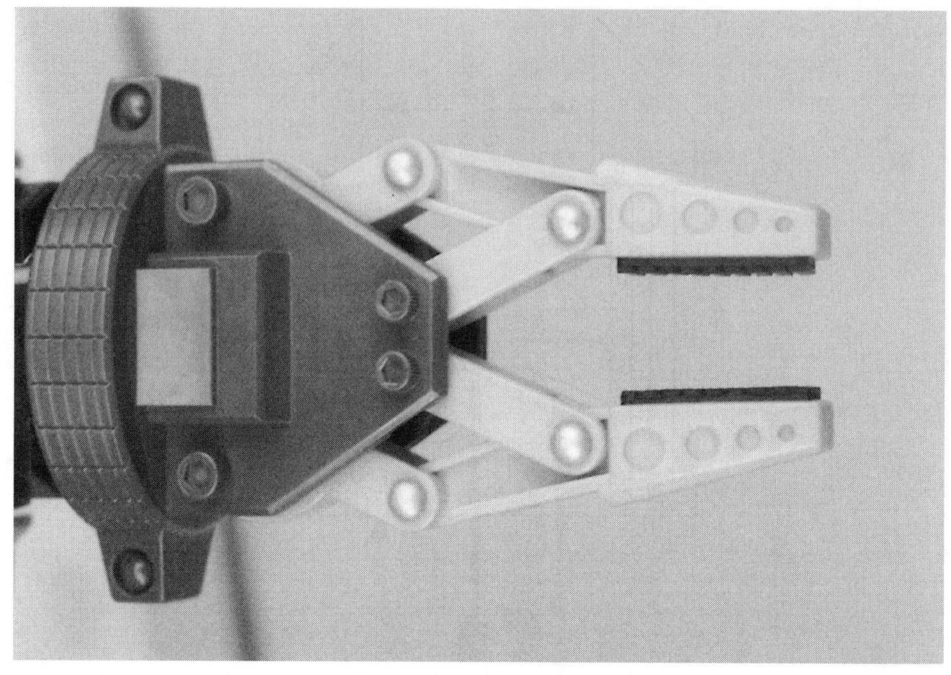

FIGURE 28-15 A close-up view of the Armatron toy gripper. Note the use of the dual-rail finger system to keep the fingertips parallel. The gripper is moderately adaptable to your own designs.

drive motor. Operate the motor in one direction and the fingers close. Operate the motor in the other direction and the fingers open. Apply small rubber feet pads to the inside ends of the grippers to facilitate grasping objects.

Figs. 28-11 through 28-14 show another approach to constructing two-pincher grippers. By adding a second rail to the fingers and allowing a pivot for both, the fingertips remain parallel to one another as the fingers open and close. You can employ several actuation techniques with such a gripper. Fig. 28-15 shows the gripping mechanism of the Radio Shack/Tomy Armatron. Note that it uses double rails to effect parallel closure of the fingers. You can model your own gripper using the design of the Armatron or amputate an Armatron and use its gripper for your own robot.

28.3 Flexible Finger Grippers

Clapper and two-pincher grippers are not like human fingers. One thing they lack is a compliant grip: the capacity to contour the grasp to match the object. The digits in our fingers can wrap around just about any oddly shaped object, which is one of the reasons we are able to use tools successfully.

You can approximate the compliant grip by making articulated fingers for your robot. At least one toy is available that uses this technique; you can use it as a design base. The plastic toy arm described earlier is available with a handlike gripper instead of a claw gripper. Pulling on the handgrip causes the four fingers to close around an object, as shown in Fig. 28-16. The opposing thumb is not articulated, but you can make a thumb that moves in a compliant gripper of your own design.

Make the fingers from hollow tube stock cut at the knuckles. The mitered cuts allow the fingers to fold inward. The fingers are hinged by the remaining plastic on the topside of the tube. Inside the tube fingers is semiflexible plastic, which is attached to the fingertips. Pulling on the handgrip exerts inward force on the fingertips and the fingers collapse at the cut joints.

You can use the ready-made plastic hand for your projects. Mount it as detailed in the previous section on the two-pincher claw arm. You can make your own fingers from a variety of materials. One approach is to use the plastic pieces from some of the toy construction kits. Cut notches into the plastic to make the joints. Attach a length of 20- or 22-gauge stove wire to the fingertip and keep it pressed against the finger using nylon wire ties. Do not make the ties too tight, or the wire won't be able to move. An experimental plastic finger is shown in Fig. 28-17.

You can mount three of four such fingers on a plastic or metal "palm" and connect all the cables from the fingers to a central pull rod. The pull rod is activated by a solenoid or motor. Note that it takes a considerable pull to close the fingers, so the actuating solenoid or motor should be fairly powerful.

The finger opens again when the wire is pushed back out as well as by the natural spring action of the plastic. This springiness may not last forever, and it may vary if you use other materials. One way to guarantee that the fingers open is to attach an expansion spring, or a strip of flexible spring metal, to the tip and base of the finger, on the back side. The spring should give under the inward force of the solenoid or motor, but adequately return the finger to the open position when power is cut.

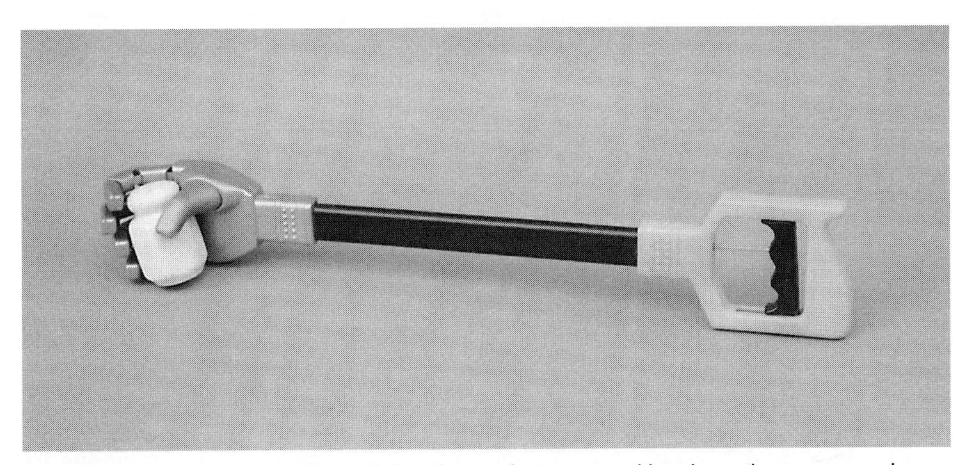

FIGURE 28-16 Commercially available plastic robotic arm and hand toy. The gripper can be salvaged for use in your own designs. The opposing thumb is not articulated, but the fingers have a semicompliant grip.

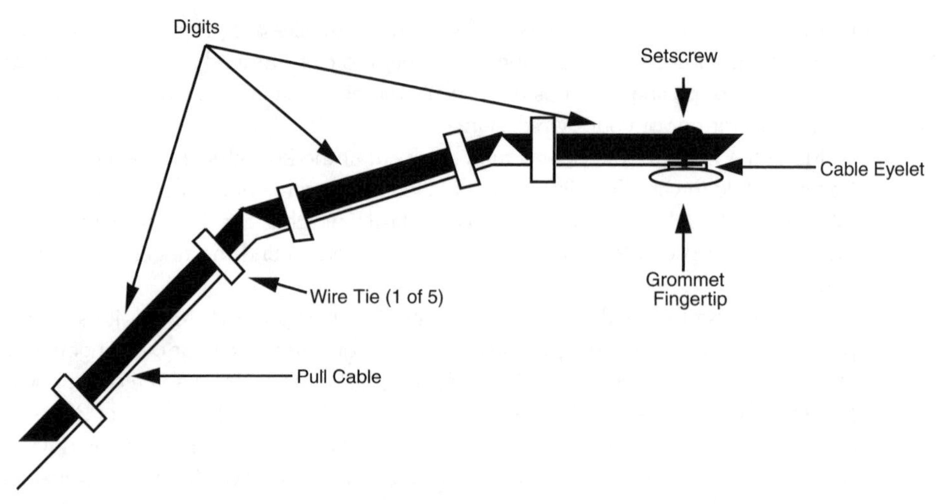

FIGURE 28-17 A design for an experimental compliant finger. Make the finger spring-loaded by attaching a spring to the back of the finger (a strip of lightweight spring metal also works).

28.4 Wrist Rotation

The human wrist has three degrees of freedom: it can twist on the forearm, it can rock up and down, and it can rock from side to side. You can add some or all of these degrees of freedom to a robotic hand. A basic schematic of a three-degree-of-freedom wrist is shown in Fig. 28-18.

With most arm designs, you'll just want to rotate the gripper at the wrist. Wrist rotation is usually performed by a motor attached at the end of the arm or at the base. When the motor is connected at the base (for weight considerations), a cable or chain joins the motor shaft to the wrist. The gripper and motor shaft are outfitted with mating spur gears. You can also use chains (miniature or #25) or timing belts to link the gripper to the drive motor. Fig.

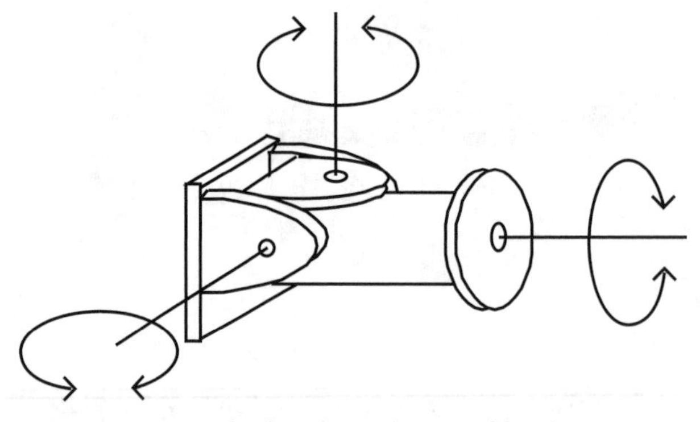

FIGURE 28-18 The three basic degrees of freedom in a human or robotic wrist (wrist rotation in the human arm is actually accomplished by rotating the bones in the forearm).

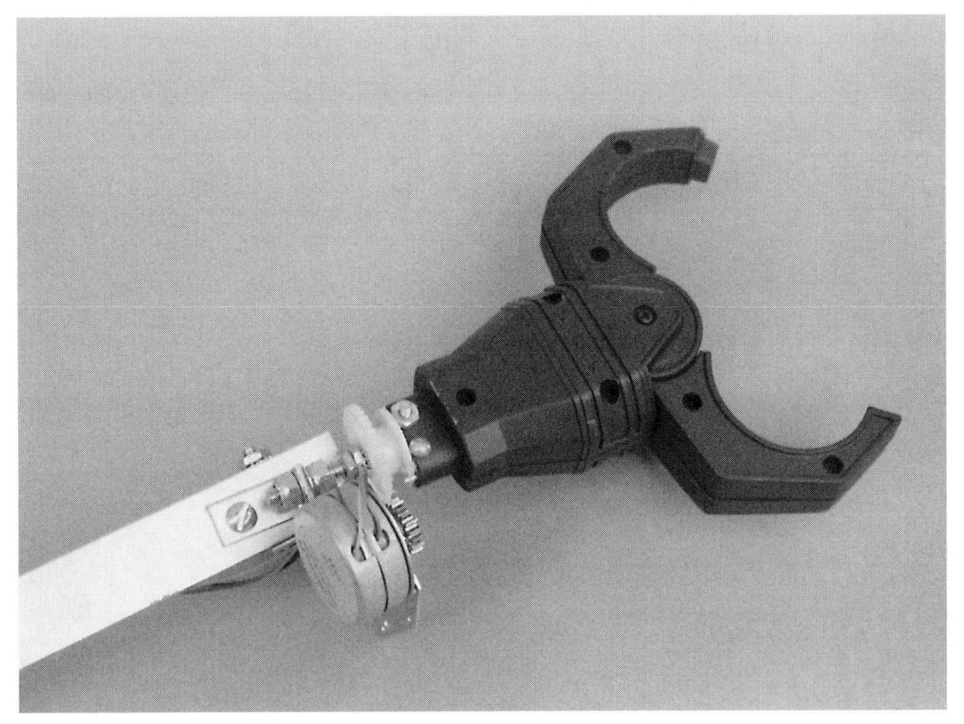

FIGURE 28-19 A two-pincher gripper (from the plastic toy robotic arm detailed earlier in the chapter), attached to the revolute arm described in Chapter 27. A small stepper motor and gear system provide wrist rotation.

28-19 shows the wrist rotation scheme used to add a gripper to the revolute coordinate arm described in Chapter 27.

You can also use a worm gear on the motor shaft. Remember that worm gears introduce a great deal of gear reduction, so take this into account when planning your robot. The wrist should not turn too quickly or too slowly.

Another approach is to use a rotary solenoid. These special-purpose solenoids have a plate that turns 30° to 50° in one direction when power is applied. The plate is spring-loaded, so it returns to its normal position when the power is removed. Mount the solenoid on the arm and attach the plate to the wrist of the gripper.

28.5 From Here

To learn more about . . .	***Read***
Using DC motors and shaft encoders	Chapter 20, "Working with DC Motors"
Using stepper motors to drive robot parts	Chapter 21, "Working with Stepper Motors"
Different robotic arm systems and assemblies	Chapter 26, "Reaching Out with Robot Arms"
Building a robotic revolute coordinate arm	Chapter 27, "Build a Revolute Coordinate Arm"
Interfacing feedback sensors to computers and microcontrollers	Chapter 14, "Computer Peripherals"

PART 6

SENSORS AND NAVIGATION

CHAPTER 29

THE SENSE OF TOUCH

Like the human hand, robotic grippers often need a sense of touch to determine if and when they have something in their grasp. Knowing when to close the gripper to take hold of an object is only part of the story, however. The amount of pressure exerted on the object is also important. Too little pressure and the object may slip out of grasp; too much pressure and the object may be damaged.

The human hand has an immense network of complex nerve endings that serve to sense touch and pressure. Touch sensors in a robot gripper are much cruder with just one or two points, but for most hobby applications these sensors serve their purpose: to provide nominal feedback on the presence of an object and the pressure exerted on the object.

This chapter deals with the fundamental design approaches for several touch sensing systems for use on robot grippers (or should the robot lack hands, elsewhere on the body of the robot). Modify these systems as necessary to match the specific gripper design being used and the control electronics you are using to monitor the sense of touch.

Note that in this chapter the distinction between *touch* and collision is made. Touch is a proactive event, where you specifically wish the robot to determine its environment by making physical contact. Conversely, collision is a reactive event, where (in most cases) you wish the robot to stop what it's doing when a collision is detected and back away from the condition. Chapter 30, "Object Detection," deals with the (potential for) physical contact resulting in a collision.

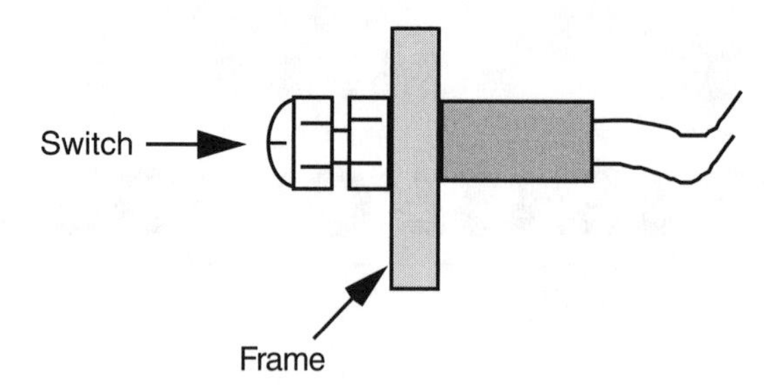

FIGURE 29-1 A mechanical momentary on push button switch makes a perfect touch sensor.

29.1 Mechanical Switch

The lowly mechanical switch is the most common, and simple, form of tactile (touch) feedback. Most any momentary, spring-loaded switch will do; the switch changes state when an object is touched, completing a circuit (or in some cases, the switch opens, breaking the circuit). The switch may be directly connected to a motor or discrete circuit, or it may be connected to a computer or microcontroller, as shown in Fig. 29-1.

You can choose from a wide variety of switch styles when designing contact switches for tactile feedback. Leaf switches (sometimes referred to as Microswitch switches) come with levers of different lengths that enhance the sensitivity of the switch. You can also use miniature contact switches, like those used in keyboards and electronic devices, as touch sensors on your robot.

In all cases, mount the switch so it makes contact with whatever object you wish to sense. In the case of a robotic gripper, you can mount the switch in the hand or finger sections.

In the case of *feelers* for a smaller handless robot, the switch can be mounted fore or aft. It makes contact with an object as it rolls along the ground. By changing the arrangement of the switch from vertical (see Fig. 29-2), you can have the feeler determine if it's reached the edge of a table or a stair landing.

29.1.1 MICROSWITCHES

One of the most convenient components that you can use for sensing objects around a robot or in its grippers is the Microswitch (Fig. 29-3). The name comes from one of the first companies to manufacture switches that can be actuated by external mechanical pressure. Microswitches come in a variety of different shapes and sizes (many more than an inch long, which seems contradictory for something with *micro* in its name).

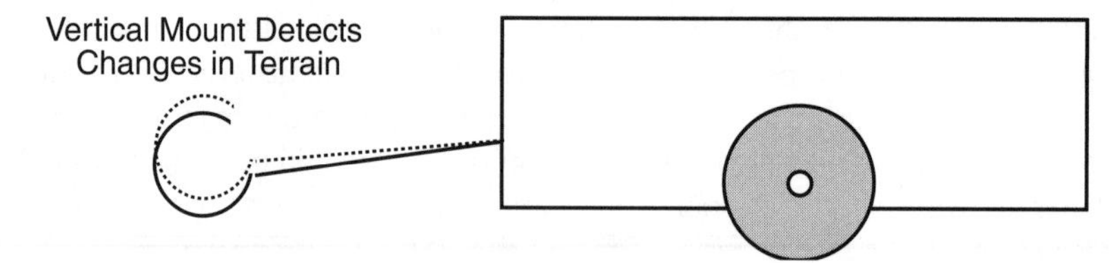

FIGURE 29-2 By orienting the switch to the vertical, your robot can detect changes in topography, such as when it's about to run off the edge of a table.

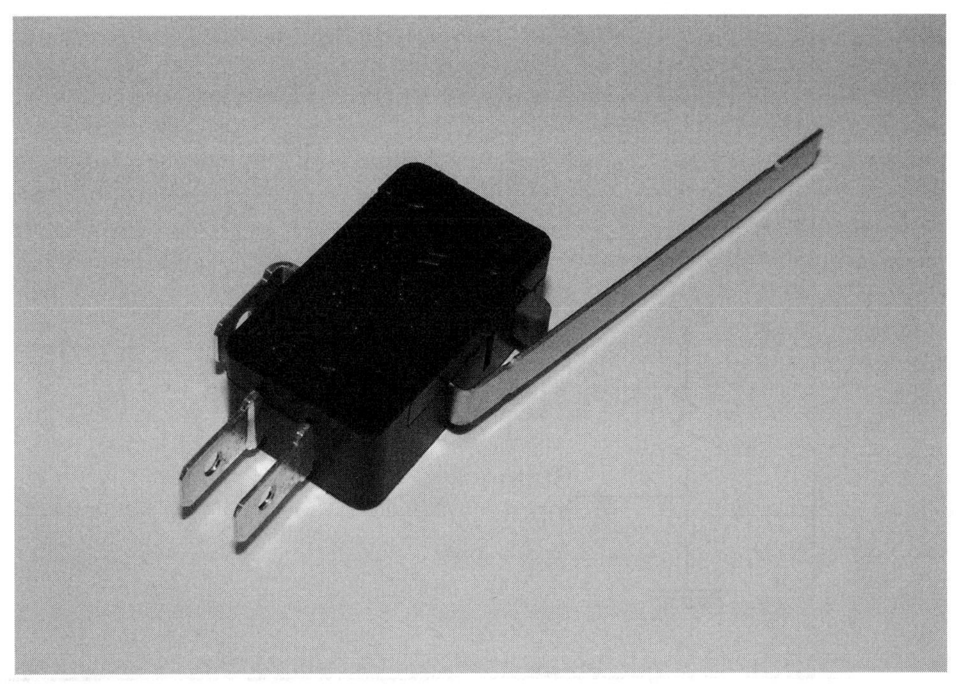

FIGURE 29-3 Microswitch with a long actuator arm built into it. This type of Microswitch is ideally suited for use as a Robot's "whiskers," sensing when the robot collides with another object.

When you start out, you might want to consider a Microswitch with a long metal actuator like the one shown in Fig. 29-3 and have a double throw internal switch (three connections—a common, a normally open, and a normally closed). The actuator can be extended by gluing on a plastic tube (or straw) or soldering on a metal extension. The double throw switch will allow you to use the Microswitch as normally open or normally closed or as part of a debounce circuit that will be discussed later in the chapter. These features will give you a maximum amount of operational flexibility when adding a Microswitch to your robot.

29.2 Switch Bouncing

If you were to read through a catalog of different switches, you would be amazed at the number of options available in terms of number of positions, number of separate switches inside the package (called poles), mounting types, and so on. Despite the plethora of different options, there is one feature that is common to all switches, but is not listed in any data sheet, and that is switch *bounce*—the making and breaking of contacts when a switch changes state. The oscilloscope picture in Fig. 29-4 shows an oscillating signal that takes place when a pulled up momentary on button is pressed.

These oscillations are caused by the contacts within the switch literally bouncing off one another when the switch is closed. Going by Fig. 29-4, you can see that this bouncing takes

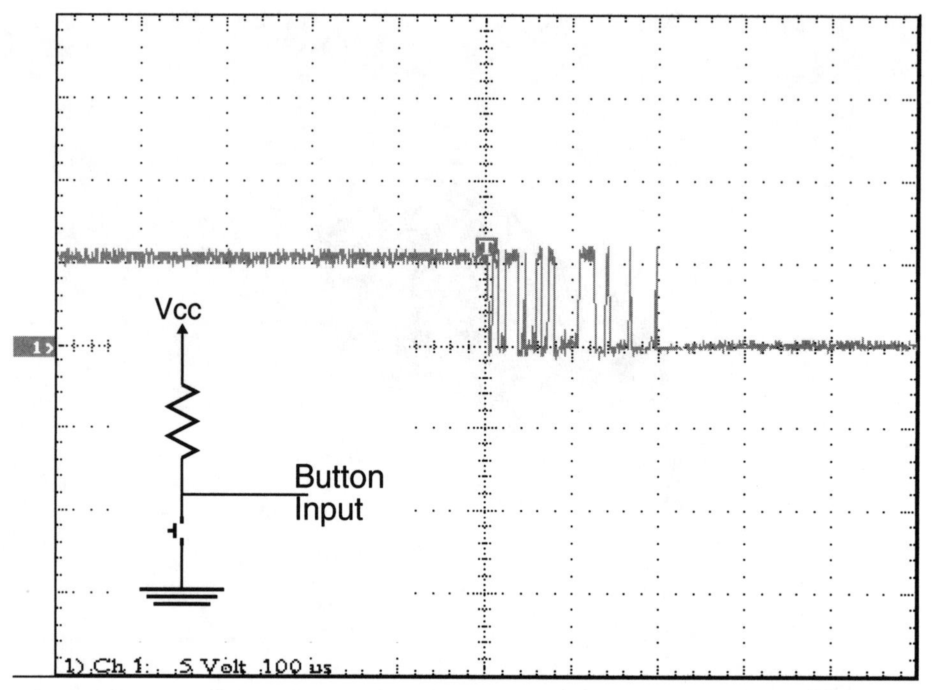

FIGURE 29-4 Oscilloscope picture of the signal produced when a pulled up momentary on switch is pressed. The spikes in the signal indicate that the switches internal contacts are literally bouncing off one another until they make solid contact.

place over a few hundred microseconds—much too fast to be detected by the naked eye. But this bouncing is a problem for computer systems and other electronic devices. Without being properly debounced, each spike may be recognized as an individual event and instead of the robot responding as if the switch changed state once, it will respond to many switch changes, which will potentially make your robot behave erratically.

It is important to note that switch bouncing also takes place when contacts are pulled apart from each other as well as when they come together. You will have to ensure that debouncing takes place for both cases to ensure that no unexpected switch transitions affect the way you expect your robot to operate.

29.2.1 HARDWARE DEBOUNCE

There are some simple hardware approaches to debouncing switch inputs. These methods eliminate the need for software to be developed to process the incoming signals as will be discussed in the next section. Unfortunately, these methods increase the robot's circuitry, so while there is some savings in software complexity, there will be an increase in the overall robot circuitry, which can be difficult to add in a small robot or one that has already been built.

The first method to debounce a circuit uses a double throw switch and a couple of CMOS Schmitt input inverters as shown in Fig. 29-5 (with its parts list in Table 29-1). The two inverters form a flip-flop that only changes state when the switch definitely changes state.

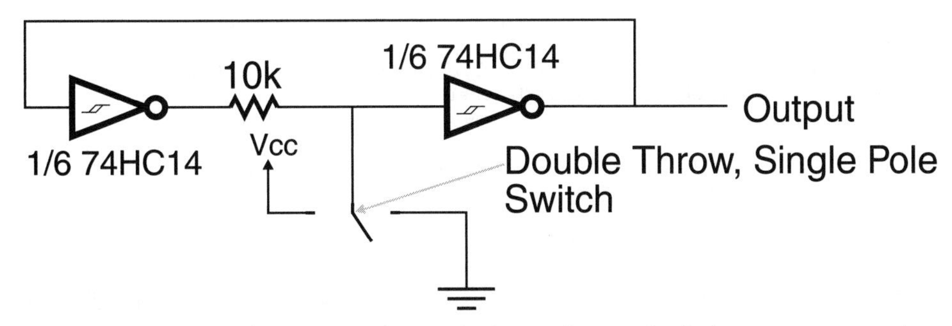

FIGURE 29-5 Two Schmitt triggered inverters along with a couple of other components can be used to make an effective button debounce circuit.

When the common contact is not touching either connector (and is disconnected from Vcc and ground), the flip-flop circuitry remembers the last common contact position. In the circuit given in Fig. 29-5, note that the inverters are Schmitt-triggered CMOS gates—the Schmitt trigger helps ensure that the flip-flop only changes state at a definite voltage level. The CMOS gates along with the 10k current limiting resistor ensures that there is not a large current flow or "backdriving" of an inverter's output.

There are other hardware debounce circuits that you can use in your application as well. One of the most popular is using a 555 set up as a monostable multivibrator with enough of a delay to ensure that any bounces are masked by the 555's output. The circuit shown in Fig. 29-5 is very efficient, both in terms of operation as well as cost; you will be hard pressed to find a circuit that is as simple and effective as this one.

29.2.2 SOFTWARE DEBOUNCE

Debouncing button inputs in software can be tricky; the need for performing other operations in an application means that the debounce code must be integrated with the complete application. This requirement adds a level of complexity to the code that can be difficult to implement for many new programmers. While debouncing button inputs in software is more cost efficient than adding hardware, you may find that the amount of work to get debounce code working in a complete robot application to be much more effort than adding a couple of CMOS Schmitt trigger inverters and a resistor.

The rule of thumb for debouncing a button input is that the line is considered to be debounced and the input stable when it has not changed state for at least 20 ms. The code for debouncing a single button in a stand-alone application is actually quite easy:

TABLE 29-1 Parts List for CMOS Logic Debounce Circuit

74LS14	Hex Schmitt trigger input inverter chip
10k	10k resistor
Switch	Double throw, single pole switch

```
Debounce:                 '  Return when the Button Input is Debounced
    i = 0
    while (i < 20ms)    '  Wait for the Button to be Pressed for 20 ms
        if (Button = Pressed) then
            i = i + 1   '  Increment Pressed Counter
        else
            i = 0       '  Button has been released, reset the counter
        endif
    endwhile
    return              '  Return to caller
```

This code increments a counter each time in a loop that the button is down. If the button is not down then the counter is reset. The counter value is continually checked to see if it has incremented to a value equal to or greater than 20 ms and when it is, the execution of the program leaves the "while" loop and returns to the subroutine's caller.

The problem with debugging a button in software comes when there are additional functions built into the robot. When this happens the time the main robot loop code executes must be known (and becomes the incrementing value for the button counter) and the main loop above becomes distributed. If the robot execution loop was assumed to execute in 1 ms, then the basic code could look like:

```
    ButtonUp = 0                '  Initialize Button Variables
    ButtonDown = 0

    while (1)                   '  Loop Forever
        if (Button = Pressed) then    '  Button State Counter Update
            ButtonUp = 0        '  Button not Up
            ButtonDown = ButtonDown + 1
        else
            ButtonUp = ButtonUp + 1
            ButtonDown = 0
        endif
        if (ButtonUp > 20) then   '  Limit Counter Values
            ButtonUp = 20
        endif
        if (ButtonDown > 20) then
            ButtonDown = 20
        endif

'        Execute Robot Functions   '  Find Robot Outputs to Inputs

      Dlay(1ms - ExecutionTime)   '  Ensure Loop Executes in 1 ms
    endwhile
```

Note that the button "state" variables are limited to a maximum value of 20. This is done to ensure that the variables do not overflow and become zero, which indicates the buttons are not pressed.

While the code in the listing seems complex, you should remember that multiple buttons could be added to the robot application by simply copying this code for each additional button (new counter variables will be required as well). You will discover that the execution time penalty of the button debounce software is negligible, and using this code, you should be able to add a number of buttons to your robot application relatively easily.

29.3 Optical Sensors

Optical sensors use a narrow beam of light to detect when an object is within the grasping area of a gripper. Optical sensors provide the most rudimentary form of touch sensitivity and are often used with other touch sensors, such as mechanical switches.

Building an optical sensor into a gripper is easy. Mount an infrared LED in one finger or pincher; mount an infrared-sensitive phototransistor in another finger or pincher (see Fig. 29-6). Where you place the LED and transistor along the length of the finger or pincher determines the grasping area.

Mounting the infrared pair on the tips of the fingers or pinchers provides little grasping area because the robot is told that an object is within range when only a small portion of it can be grasped. In most gripper designs, two or more LEDs and phototransistors are placed along the length of the grippers or fingers to provide more positive control. Alternatively, you may wish to detect when an object is closest to the palm of the gripper. You'd mount the LED and phototransistor accordingly.

Fig. 29-7 shows the schematic diagram for a single LED-transistor pair. Adjust the value of R2 to increase or decrease the sensitivity of the phototransistor. You may need to place an infrared filter over the phototransistor to prevent it from triggering as a result of ambient light sources (some phototransistors have the filter built into them already). Use an LED-transistor pair equipped with a lens to provide additional rejection of ambient light and to increase sensitivity.

During normal operation, the transistor is on because it is receiving light from the LED. When an object breaks the light path, the transistor switches off. A control circuit connected to the conditioned transistor output detects the change and closes the gripper. In a practical application, using a computer as a controller, you'd write a short software program to control the actuation of the gripper.

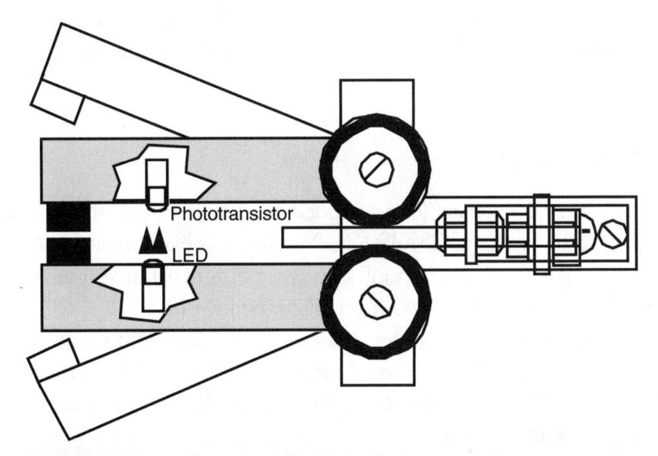

FIGURE 29-6 An infrared LED and phototransistor pair can be added to the fingers of a gripper to provide go/no-go grasp information.

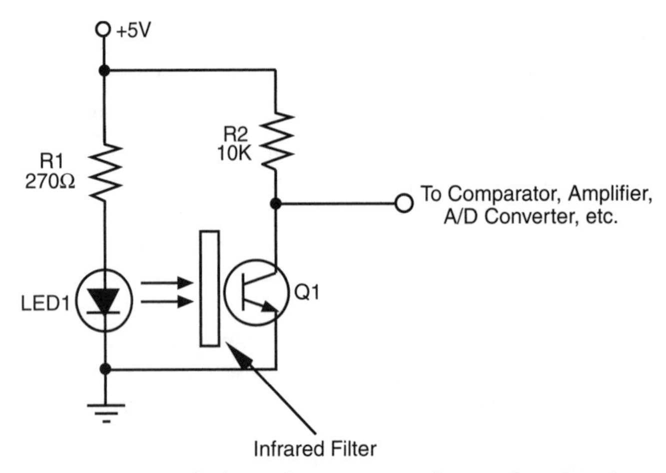

FIGURE 29-7 The basic electronic circuit for an infrared touch sensor. Note the infrared filter; it helps prevent the phototransistor from being activated by ambient light.

29.4 Mechanical Pressure Sensors

An optical sensor is a go/no-go device that can detect only the presence of an object, not the amount of pressure on it. A pressure sensor detects the force exerted by the gripper on the object. The sensor is connected to a converter circuit, or in some cases a servo circuit, to control the amount of pressure applied to the object.

Pressure sensors are best used on grippers where you have incremental control over the position of the fingers or pinchers. A pressure sensor would be of little value when used with a gripper that's actuated by a solenoid. The solenoid is either pulled in or it isn't; there are no in-between states. Grippers actuated by motors are the best choices when you must regulate the amount of pressure exerted on the object.

29.4.1 CONDUCTIVE FOAM

You can make your own pressure sensor (or transducer) out of a piece of discarded conductive foam—the stuff used to package CMOS ICs. The foam is like a resistor. Attach two pieces of wire to either end of a 1-in square hunk and you get a resistance reading on your volt-ohm meter. Press down on the foam, and the resistance lowers.

The foam comes in many thicknesses and densities. Look for semi-stiff foam that regains its shape quickly after it's squeezed. Very dense foams are not useful because they don't quickly spring back to shape. Save the foam from the various ICs you buy and test other types until you find the right stuff for you.

You can make a pressure sensor by sandwiching several pieces of material together, as shown in Fig. 29-8. The conductive foam is placed between two thin sheets of copper or aluminum foil. A short piece of 30 AWG wire-wrapping wire is lightly soldered onto the foil

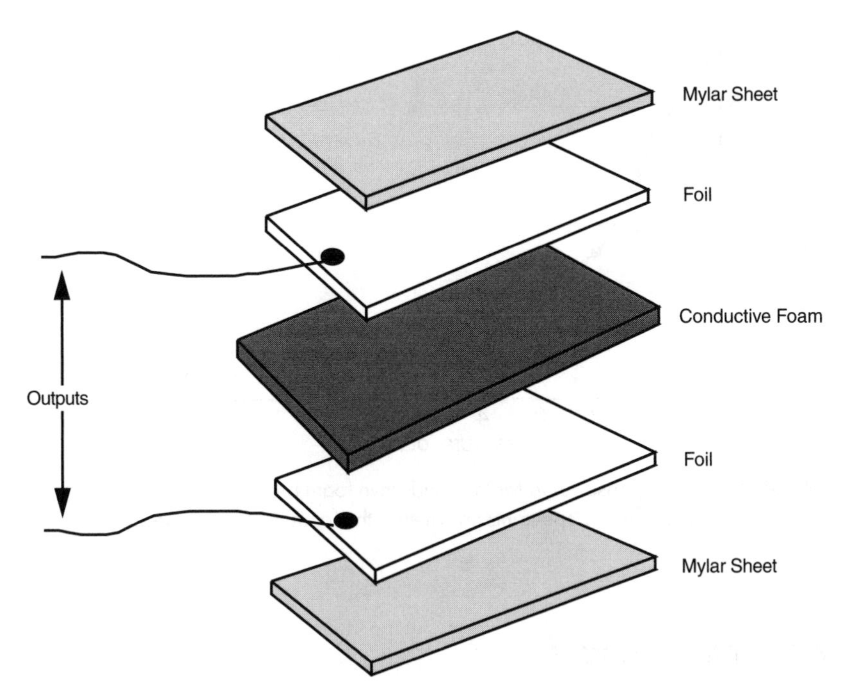

FIGURE 29-8 Construction detail for a pressure sensor using conductive foam. The leads are soldered or attached to foil (copper works best). Choose a foam that has a good "spring" to it.

(when using aluminum foil, the wire is wound around one end). Mylar plastic, like the kind used to make heavy-duty garbage bags, is glued on the outside of the sensor to provide electrical insulation. If the sensor is small and the sense of touch does not need to be too great, you can encase the foam and foil in heat-shrink tubing. There are many sizes and thicknesses of tubing; experiment with a few types until you find one that meets your requirements.

The resistance of the conductive foam pressure transducers changes abruptly when they are compressed. The output may not return to its original resistance value (see Fig. 29-9). So in the control software, you should always reset the transducer just prior to grasping an object.

For example, the transducer may first register an output of 30K Ω (the exact value depends on the foam, the dimensions of the piece, and the distance between wire terminals). The software reads this value and uses it as the set point for a normal (nongrasping) level to 30K. When an object is grasped, the output drops to 5K. The difference—25K—is the amount of pressure. Keep in mind that the resistance value is relative, and you must experiment to find out how much pressure is represented by each 1K of resistance change.

The transducer may not go back to 30K when the object is released. It may spring up to 40K or go only as far as 25K. The software uses this new value as the new set point for the next occasion when the gripper grasps an object.

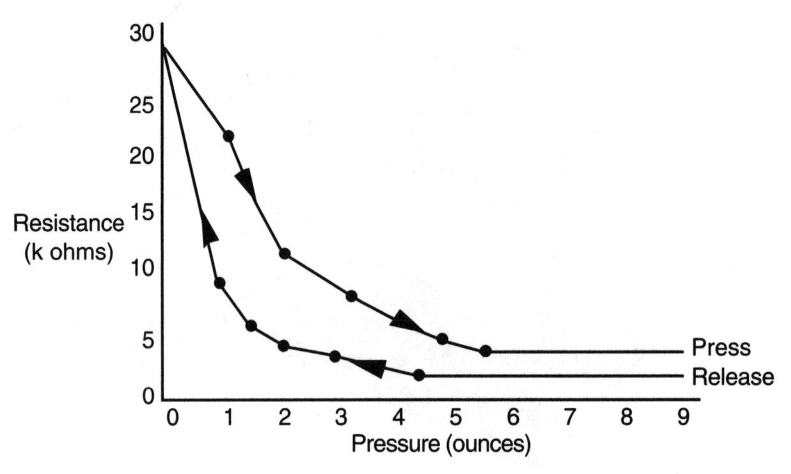

FIGURE 29-9 The response curve for the conductive foam pressure sensor. Note that the resistance varies depending on whether the foam is being pressed or released.

29.4.2 STRAIN GAUGES

Obviously, the home-built pressure sensors described so far leave a lot to be desired in terms of accuracy. If you need greater accuracy, you should consider commercially available strain gauges. These work by registering the amount of strain (the same as pressure) exerted on various points along the surface of the gauge.

Strain gauges are somewhat pricey—about $10 and over in quantities of 5 or 10. The cost may be offset by the increased accuracy the gauges offer. You want a gauge that's as small as possible, and preferably one mounted on a flexible membrane. See Appendix B, "Sources," for a list of companies offering such gauges.

29.4.3 CONVERTING PRESSURE DATA TO COMPUTER DATA

The output of both the homemade conductive foam pressure transducers and the strain gauges is analog—a resistance or voltage. Neither can be directly used by a computer, so the output of these devices must be converted into digital form first.

Both types of sensors are perfect for use with an analog-to-digital converter. You can use one ADC0808 chip (under $5) with up to eight sensors. You select which sensor output you want to convert into digital form. The converted output of the ADC0808 chip is an eight-bit word, which can be fed directly to a microprocessor or computer port. Fig. 29-10a shows the basic wiring diagram for the ADC0808 chip, which can be used with conductive foam transducer; Fig. 29-10b shows how to connect a conductive foam transducer to one of the analog inputs of the ADC0808.

Notice the 10K resistor in Fig. 29-10, placed in series between the pressure sensor and ground. This converts the output of the sensor from resistance to voltage. You can change the value of this resistor to alter the sensitivity of the circuit. For more information on ADCs, see Chapter 14, "Computer Peripherals."

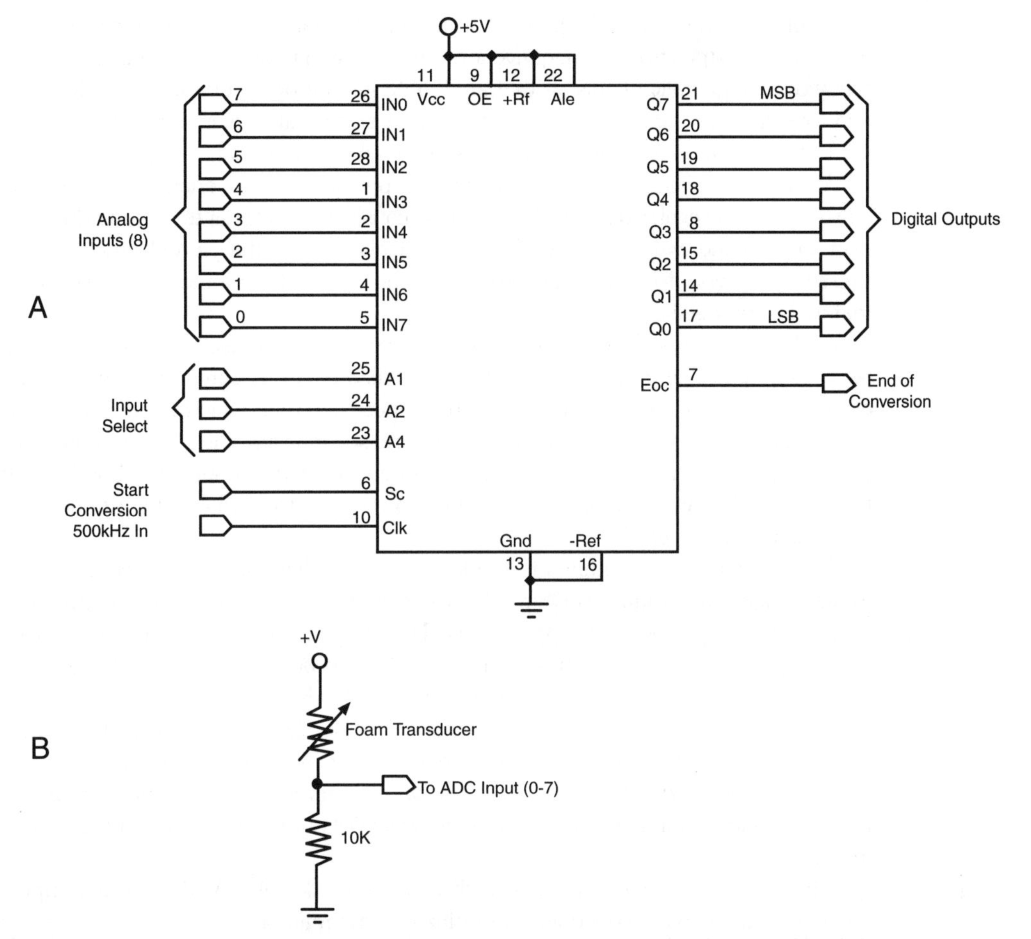

FIGURE 29-10 *a.* The basic wiring diagram for converting pressure data into digital data, using an ADC0808 analog-to-digital converter (ADC) IC. You can connect up to eight pressure sensors to the one chip. *b.* How to interface a conductive foam sensor to the ADC0808 chip.

29.5 Experimenting with Piezoelectric Touch Sensors

A new form of electricity was discovered just a little more than a century ago when the two scientists Pierre and Jacques Curie placed a weight on a certain crystal. The strain on the crystal produced an odd form of electricity—significant amounts of it, in fact. The Curie brothers coined this new electricity *piezoelectricity*; *piezo* is derived from the Greek word meaning *press*. Later, the Curies discovered that the piezoelectric crystals used in their experiments underwent a physical transformation when voltage was applied to them. They also found that the piezoelectric phenomenon is a two-way street. Press the crystals and out comes a voltage; apply a voltage to the crystals and they respond by flexing and contracting.

All piezoelectric materials share a common molecular structure, in which all the movable electric dipoles (positive and negative ions) are oriented in one specific direction. Piezoelectricity occurs naturally in crystals that are highly symmetrical—quartz, Rochelle salt crystals, and tourmaline, for example. The alignment of electric dipoles in a crystal structure is similar to the alignment of magnetic dipoles in a magnetic material.

When an electrical voltage is applied to the piezoelectric material, the physical distances between the molecular dipoles change. This causes the material to contract in one dimension (or axis) and expand in the other. Conversely, placing the piezoelectric material under pressure (in a vise, for example) compresses the dipoles in one more axis. This causes the material to produce a voltage.

While natural crystals were the first piezoelectric materials used, synthetic materials have been developed that greatly demonstrate the piezo effect. A common human-made piezoelectric material is ferroelectric zirconium titanate ceramic, which is often found in piezo buzzers used in smoke alarms, wristwatches, and security systems. The zirconium titanate is evenly deposited on a metal disc. Electrical signals, applied through wires bonded to the surfaces of the disc and ceramic, cause the piezo material to vibrate at high frequencies (usually 4 kHz and above).

Piezo activity is not confined to brittle ceramics. PVDF, or polyvinylidene fluoride (used to make high-temperature PVDF plastic water pipes), is a semicrystalline polymer that lends itself to unusual piezoelectric applications. The plastic is pressed into thin, clear sheets and is given precise piezo properties during manufacture by—among other things—stretching the sheets and exposing them to intense electrical fields.

PVDF piezo film is currently used in many commercial products, including noninductive guitar pickups, microphones, even solid-state fans for computers and other electrical equipment. One PVDF film you can obtain and experiment with is Kynar, available directly from the manufacturer (see Measurement Specialists at www.msiusa.com for more information).

Whether you are experimenting with ceramic or flexible PVDF film, it's important to understand a few basic concepts about piezoelectric materials:

- Piezoelectric materials are voltage sensitive. The higher the voltage is, the more the piezoelectric material changes. Apply 1 V to a ceramic disc and crystal movement will be slight. Apply 100 V and the movement will be much greater.
- Piezoelectric materials act as capacitors. Piezo materials develop and retain an electrical charge.
- Piezoelectric materials are bipolar. Apply a positive voltage and the material expands in one axis. Apply a negative voltage and the material contracts in that axis.

29.5.1 EXPERIMENTING WITH CERAMIC DISCS

The ubiquitous ceramic disc is perhaps the easiest form of piezoelectric transducer to experiment with. A sample disc is shown in Fig. 29-11. The disc is made of nonferrous metal, and the ceramic-based piezo material is applied to one side. Most discs available for purchase have two leads already attached. The black lead is the ground of the disc and is directly attached to the metal itself.

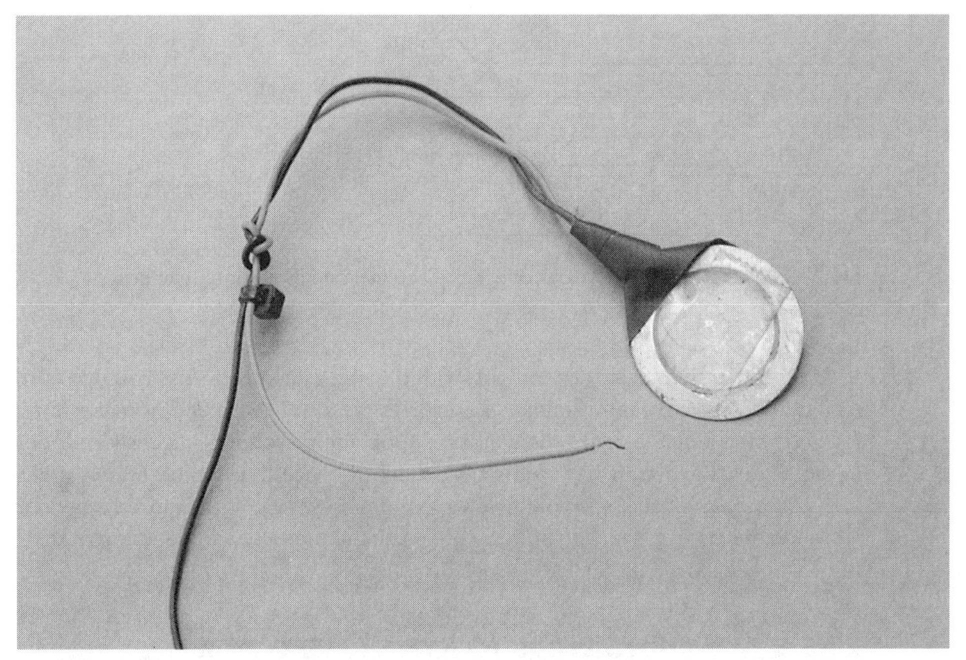

FIGURE 29-11 Piezo ceramic discs are ideally suited to be contact and pressure sensors for robotics.

You can use a ceramic disc as an audio transducer by applying an audio signal to it. Most piezo discs will emit sound in the 1K to 10K region, with a resonant frequency of between 3K and 4K. At this resonant frequency, the output of the disc will be at its highest.

When the piezo material of the disc is under pressure—even a slight amount—the disc outputs a voltage proportional to the amount of pressure. This voltage is short lived: shortly after the initial change in pressure, the voltage output of the disc will return to 0. A negative voltage is created when the pressure is released (see the discussion of the bipolar nature of piezo materials earlier in the chapter).

You can easily interface piezo discs to a computer or microcontroller, either with or without an analog-to-digital converter. Chapter 30, "Object Detection," discusses several interface approaches. See the section "Piezo Disc Touch Bar" in that chapter for more information.

29.5.2 EXPERIMENTING WITH KYNAR PIEZO FILM

Samples of Kynar piezoelectric film are available in a variety of shapes and sizes. The wafers, which are about the same thickness as the paper in this book, have two connection points, as illustrated in Fig. 29-12. Like ceramic discs, these two connection points are used to activate the film with an electrical signal or to relay pressure on the film as an electrical impulse.

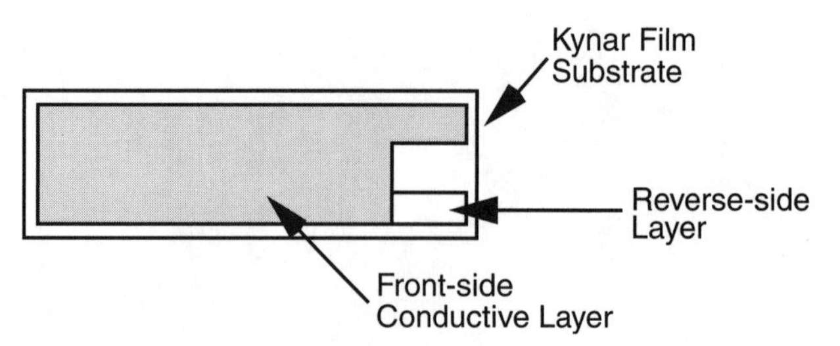

FIGURE 29-12 A close-up look at Kynar piezo film and its electrical contacts.

You can perform basic experiments with the film using just an oscilloscope (preferred) or a high-impedance digital voltmeter. Connect the leads of the scope or meter to the tabs on the end of the film (the connection will be sloppy; there are some suggestions for other ways to attach leads to Kynar film in the next section). Place the film on a table and tap on it. You'll see a fast voltage spike on the scope or an instantaneous rise in voltage on the meter. If the meter isn't auto-ranging and you are using the meter at a low setting, chances are that the voltage spike will exceed the selected range.

29.5.3 ATTACHING LEADS TO KYNAR PIEZO FILM

Unlike piezoelectric ceramic discs, Kynar film doesn't usually come with pre-attached leads (although you can order samples with leads attached, but they are expensive). There are a variety of ways to attach leads to Kynar film. Obviously, soldering the leads onto the film contact areas is out of the question. Acceptable methods include applying conductive ink or paint, self-adhesive copper-foil tape, small metal hardware, and even miniature rivets. In all instances, use small-gauge wire—22 AWG or smaller. You can expect the best results using 28 AWG and 30 AWG solid wire-wrapping wire. The following are the best methods:

- *Conductive ink or paint.* Conductive ink, such as GC Electronics' Nickel-Print paint, bonds thin wire leads directly to the contact points on Kynar film. Apply a small globule of paint to the contact point, and then slide the end of the wire in place. Wait several minutes for the paint to set before handling. Apply a strip of electrical tape to provide physical strength.
- *Self-adhesive copper-foil tape.* You can use copper-foil tape designed for repairing printed circuit boards to attach wires to Kynar film. The tape uses a conductive adhesive and can be applied quickly and simply. As with conductive inks and paints, apply a strip of electrical tape to the joint to give it physical strength.
- *Metal hardware.* Use small 2/56 or 4/40 nuts, washers, and bolts (available at hobby stores) to mechanically attach leads to the Kynar. Poke a small hole in the film, slip the bolt through, add the washer, and wrap the end of a wire around the bolt. Tighten with the nut.
- *Miniature rivets.* Homemade jewelry often uses miniature brass or stainless steel rivets. You can obtain the rivets and the proper riveting tool from many hobby and jewelry-

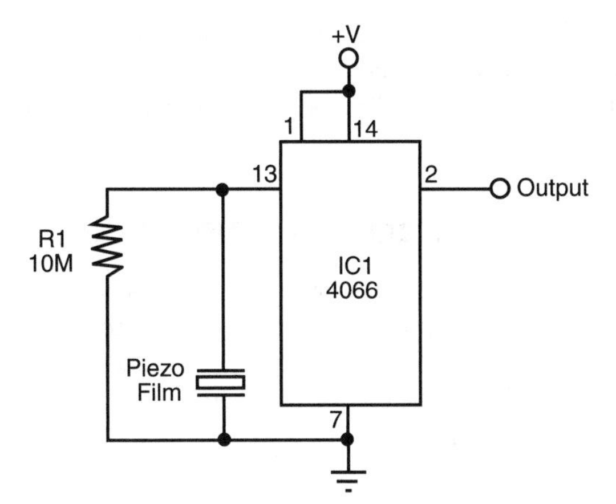

FIGURE 29-13 A strike/vibration indicator using Kynar piezo film.

making stores. To use them, pierce the film to make a small hole, wrap the end of the wire around the rivet post, and squeeze the riveting tool (you may need to use metal washers to keep the wire in place).

29.5.4 USING KYNAR PIEZO FILM AS A MECHANICAL TRANSDUCER

Fig. 29-13 shows a simple demonstrator circuit you can build that indicates each time a piece of Kynar film is struck. Tapping the film produces a voltage output, which is visually

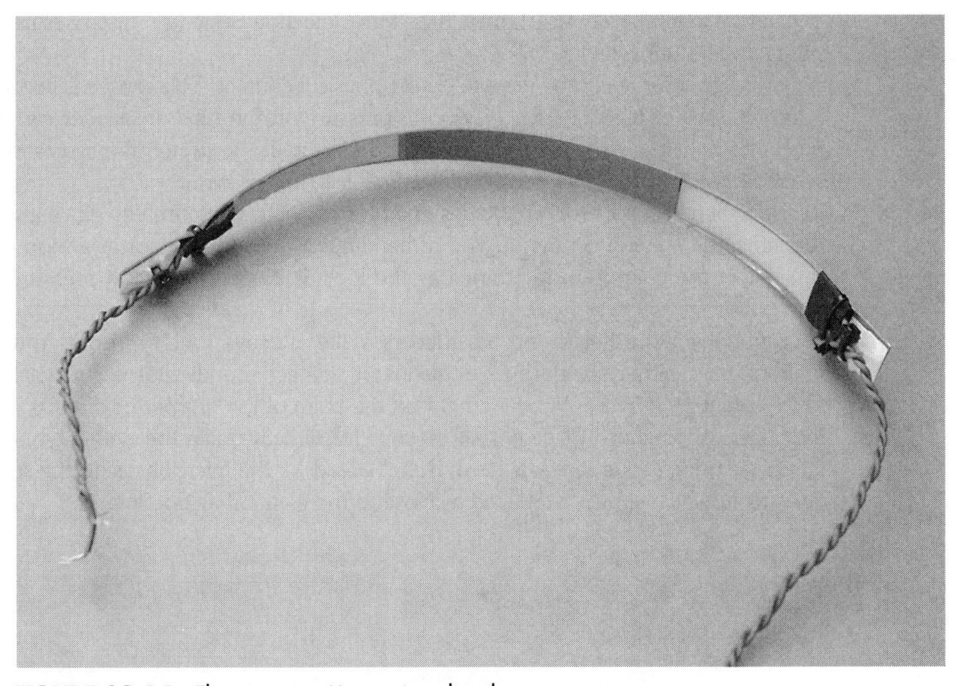

FIGURE 29-14 The prototype Kynar piezo bend sensor.

indicated when the LED flashes. The 4066 IC is an analog switch. When a voltage is applied to pin 3, the connection between pins 1 and 2 is completed and that finishes the electrical circuit to light the LED. For a robotic application, you can connect the output to a computer or microcontroller.

29.5.5 CONSTRUCTING A KYNAR PIEZO FILM BEND SENSOR

You can easily create a workable touch sensor by attaching one or two small Kynar transducers to a thick piece of plastic. The finished prototype sensor is depicted in Fig. 29-14. The plastic membrane could be mounted on the front of a robot, to detect touch contact, or even in the palm of the robot's hand. Any flexing of the membrane causes a voltage change at the output of one or both Kynar film pieces.

29.6 Other Types of Touch Sensors

The human body has many kinds of touch receptors embedded within the skin. Some receptors are sensitive to physical pressure, while others are sensitive to heat. You may wish to endow your robot with some additional touch sensors like:

- Heat sensors can detect changes in the heat of objects within grasp. Heat sensors are available in many forms, including thermisters (resistors that change their value depending on temperature) and solid-state diodes that are specifically made to be ultrasensitive to changes in temperature. Chapter 34, "Fire Detection Systems," discusses using solid-state temperature sensors.
- Air pressure sensors can be used to detect physical contact. The sensor is connected to a flexible tube or bladder (like a balloon); pressure on the tube or bladder causes air to push into or out of the sensor, thereby triggering it. To be useful, the sensor should be sensitive to increases in air pressure to about 1 lb/in^2, or less.
- Resistive bend sensors, originally designed for use with virtual reality gloves, vary their resistance depending on the degree of bending. Mount the sensor in a loop, and you can detect the change in resistance as the loop is deformed by the pressure of contact.
- Microphones and other sound transducers make effective touch sensors. You can use microphones, either standard or ultrasonic, to detect sounds that occur when objects touch. Mount the microphone element on the palm of the gripper or directly on one of the fingers or pinchers. Place a small piece of felt directly under the element, and cement it in place using a glue that sets hard. Run the leads of the microphone to the sound trigger circuit, which should be placed as close to the element as possible.

29.7 From Here

To learn more about . . .	***Read***
Designing and building robot hands	Chapter 28, "Experimenting with Gripper Designs"
Connecting sensors to computers and microcontrollers	Chapter 14, "Computer Peripherals"
Collision detection systems	Chapter 30, "Object Detection"
Building light sensors	Chapter 32, "Robot Vision"
Fire, heat, and smoke detection for robotics	Chapter 34, "Fire Detection Systems"

CHAPTER 30

OBJECT DETECTION

You've spent hundreds of hours designing and building your latest robot creation. It's filled with complex little doodads and precision instrumentation. You bring it into your living room, fire it up, and step back. Promptly, the beautiful new robot smashes into the fireplace and scatters itself over the living room rug. You remembered things like motor speed controls, electronic eyes and ears, even a synthetic voice, but you forgot to provide your robot with the ability to look before it leaps.

Object detection systems take many forms and work in many different ways requiring different interfaces and programming. This chapter presents a number of passive and active object detection systems that are easy to build and use. Some of the systems are designed to detect objects close to the robot (called near-object, or proximity, detection), and some are designed to detect objects at distances of 10 ft or more (called far-object detection). Depending on how the sensor works, your robot may change trajectory to avoid an object far away from it or it could turn hard away or stop to avoid something that was sensed immediately in its path.

The material in this chapter may seem very similar to that of Chapter 29, "the Sense of Touch," and there is some overlap in the material and programming differences. To clear up any confusion, Chapter 29 discusses different methods of detecting whether a gripper has detected an object to pick up, while in this chapter the sensors reviewed are for mobile robots to ensure that they are not damaged by nor do they damage objects they collide with as they are moving about.

30.1 Design Overview

Object detection can be further divided into *collision avoidance* and *collision detection.* With collision avoidance, the robot uses noncontact techniques to determine the proximity of objects around it. It then avoids any objects it detects. Collision detection concerns what happens when the robot has already gone too far, and contact has been made with whatever foreign object was unlucky enough to be in the machine's path.

Collision avoidance can be further broken down into two subtypes: near-object detection and far-object detection. By its nature, all cases of collision detection involve making contact with nearby objects. All of these concepts are discussed in this chapter.

Additionally, robot builders commonly use certain object detection methods to navigate a robot from one spot to the next. Many of these techniques are introduced here because they are relevant to object detection, but we develop them more fully in Chapter 33, "Navigation."

30.1.1 NEAR-OBJECT DETECTION

Near-object detection does just what its name implies: it senses objects that are close by, from perhaps just a breath away to as much as 8 or 10 ft. These are objects that a robot can consider to be in its immediate environment—objects it may have to deal with, and soon. These objects may be people, animals, furniture, or other robots. By detecting them, your robot can take appropriate action, which is defined by the program you give it. Your robot may be programmed to come up to people and ask them their name. Or it might be programmed to run away whenever it sees movement. In either case, it won't be able to accomplish either behavior unless it can detect objects in its immediate area.

There are two ways to effect near-object detection: proximity and distance:

- Proximity sensors care only that some object is within a zone of relevance. That is, if an object is near enough in the physical scene the robot is looking at, the sensor detects it and triggers the appropriate circuit in the robot. Objects beyond the proximal range of a sensor are effectively ignored because they cannot be detected.
- Distance measurement sensors determine the distance between the sensor and to some object within range. Distance measurement techniques vary; almost all have notable minimum and maximum ranges. Few yield accurate data if an object is very close to the robot. Likewise, objects just outside the sensor's effective range can yield inaccurate results. Large objects far away may appear closer than they really are; very close small objects may appear abnormally larger than they really are, and so on. If there are multiple objects within the sensor's field of view, the robot may have difficulty sorting them out and figuring out which one is closest (and is the most likely danger).

Sensors have depth and breadth limitations: depth is the maximum distance an object can be from the robot and still be detected by the sensor. Breadth is the maximum height and width of the sensor detection area. Some sensors see in a relatively narrow zone, typically in a conical pattern, like the beam of a flashlight. Light sensors are a good example. Adding a lens in front of the sensor narrows the pattern even more. Other sensors have specific breadth patterns. The typical passive infrared sensor (the kind used on motion alarms) uses a Fresnel lens that expands the field of coverage on the top but collapses it on the bottom. This makes the sensor better suited for detecting human motion instead of cats, dogs, and other furry creatures (humans being, on average, taller than furry creatures). The detector uses a pyroelectric element to sense changes in heat patterns in front of it.

30.1.2 FAR-OBJECT DETECTION

Far-object detection focuses on objects that are outside the robot's primary area of interest but still within a detection range. A wall 50 ft away is not of critical importance to a robot (conversely, the same wall when it's 1 ft away is very important). Far-object detection is typically used for area and scene mapping to allow the robot to get a sense of its environment. Most hobby robots don't employ far-object detection because it requires fairly sophisticated sensors, such as narrow-beam radar or pulsed lasers.

The difference between near- and far-object detection is relative. As the designer, builder, and master of your robot, you get to decide the threshold between near and far objects. Perhaps your robot is small and travels fairly slowly. In that case, far objects are those 4 to 5 ft away; anything closer is considered "near." With such a robot, you can employ ordinary sonar distance systems for far-object detection, including area mapping.

This chapter will concentrate on near-object detection methods since traditional far-object detection is beyond the reach and riches of most hobby robot makers (with the exception of sonar systems, which have a maximum range of about 30 ft). You may, if you wish, employ near-object techniques to detect objects that are far away relative to the world your robot lives in.

30.1.3 REMEMBERING THE KISS PRINCIPLE

Engineering texts like to tout the concept of KISS: "Keep It Simple, Stupid." If the admonition is intentionally insulting it is to remind all of us that usually the simple techniques are the best. Of course, *simplicity* is relative. An ant is simple compared to a human being, but so far no scientist has ever created the equivalent of a living ant.

KISS certainly applies to using robotic sensors for object detection. Your ultimate goal may be to put "eyes" on your robot to help it see the world the way that you do. In fact, such eyes already exist in the form of CCD and CMOS video imagers. They're relatively cheap, too—less than $50 retail for camera modules or hackable webcams. What's missing in the case of vision systems is the ability to process the incoming signal ways to use the wealth of information provided by the sensor. How do you make a robot differentiate between a can of Dr. Pepper and Mrs. Johnson's slobbering two-year-old—both of which are very wet when tipped over?

When you think about which object detection sensor or system to add to your robot, consider the system's relative complexity in relation to the rest of the project. If all your small 'bot needs is a bumper switch, then avoid going overboard with a $100 sonar system. Conversely, if the context of the robot merits it, skimp with inadequate sensors. Larger, heavier robots require more effective object-detection systems—if for no other reason than to prevent injuring *you*—if it happens to start to run amok.

30.1.4 REDUNDANCY

Two heads are better than one? Maybe. One thing is for sure: two eyes are definitely better than one. The same goes for ears and many other kinds of sensors. This is sensor redundancy at work (having two eyes and ears also provides stereo vision or hearing, which aids in perception). Sensor redundancy—especially for object detection—is not intended primarily to compensate for system failure, the way NASA builds backups into its space projects in case some key system fails 25,000 miles up in space. Rather, sensor redundancy is meant to provide a more complete picture and better situational awareness for the robot by using multiple sensors with different characteristics. If a long-range sensor says an object is 10 ft away and a mid-range sensor that is 8 ft away says it's a foot away, then the robot's control computer has dual conformation that there is an object present and can be confident that it is about 8 ft away.

With only one sensor the robot must blindly (excuse the pun) trust that the sensor data is reliable. This is not a good idea because even for the best sensors data is not 100 percent reliable. There are two kinds of redundancy:

- Same-sensor redundancy relies on two or more sensors of an identical type. Each sensor more or less sees the same scene. You can use sensor data in either (or both) of two ways: through statistical analysis or interpolation. With statistical analysis, the robot's control circuitry combines the input from the sensors and uses a statistical formula to whittle the data to a most likely result. For example, sensors with wildly disparate results may be rejected out of hand, and the values of the remaining sensors may be averaged out. With interpolation, the data of two or more sensors is combined and cross-correlated to triangulate on the object, just like having two eyes and two ears adds depth and direction to our visual and aural senses.
- Complementary-sensor redundancy relies on two or more sensors of different types. Since the sensors are fundamentally different—for example, they use completely different collection methods—the data from the sensors are always interpolated. For instance, if a robot has both a sonar and an infrared distance-measuring system, it uses both because it understands that for some kinds of objects the data from the infrared system will be more reliable, and for other objects the data from the sonar system will be more reliable. The previous example of the object 8 ft away is an example of complementary sensor redundancy.

Budget and time constraints will likely be the limiting factors in whether you employ redundant sensor systems in your robots. So when combining sensors, do so logically: consider which sensors complement others well and if they can be reasonably added. For example, both sonar and infrared proximity sensors can use the same 40 kHz modulation system. If you have one, adding the other need not be difficult, expensive, or time-consuming.

30.2 Noncontact Near-Object Detection

Avoiding a collision is better than detecting it after it has happened. Short of building some elaborate radar distance measurement system, the ways for providing proximity detection to avoid collisions fall into two categories: light and sound. The following sections will take a closer look at several light- and sound-based techniques.

30.2.1 SIMPLE INFRARED LIGHT PROXIMITY SENSOR

Light may always travel in a straight line, but it bounces off nearly everything. You can use this to your advantage to build an infrared collision detection system. You can mount several infrared bumper sensors around the periphery of your robot. They can be linked together to tell the robot that something is out there, or they can provide specific details about the outside environment to a computer or control circuit.

The basic infrared detector is shown in Fig. 30-1 (refer to the parts list in Table 30-1). It provides an output that can be polled by a robot's controller comparator or ADC input. This uses an infrared LED and infrared phototransistor. Fig. 30-2 shows how the LED and phototransistor might be mounted around the base of the robot to detect an obstacle like a wall, chair, or person.

Sensitivity can be adjusted by changing the value of R2; reduce the value to increase sensitivity. An increase in sensitivity means that the robot will be able to detect objects farther away. A decrease in sensitivity means that the robot must be fairly close to the object before it is detected.

Bear in mind that all objects reflect light in different ways. You'll probably want to adjust the sensitivity so the robot behaves itself best in a room with white walls. But that sensitivity may not be as great when the robot comes to a dark brown couch or the coal gray suit of your boss.

The infrared phototransistor should be baffled—blocked—from both ambient room light as well as direct light from the LED. The positioning of the LED and phototransistor is very important, and you must take care to ensure that the two are properly aligned. You may wish to mount the LED–phototransistor pair in a small block of wood. Drill holes for the LED and phototransistor.

TABLE 30-1 Parts List for Infrared Proximity Switch

R1	270 Ω resistor
R2	10K resistor
Q1	Infrared sensitive phototransistor
LED1	Infrared light-emitting diode
Misc.	Infrared filter for phototransistor (if needed)

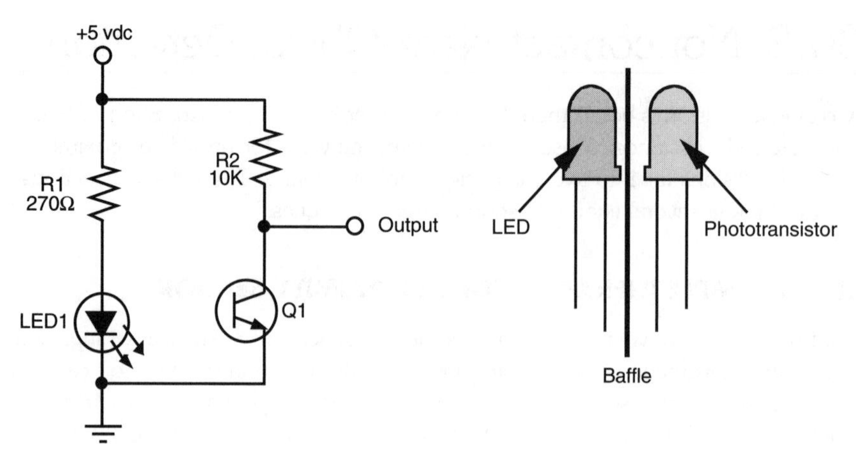

FIGURE 30-1 The basic design of the infrared proximity sensor.

Or, if you prefer, you can buy the detector pair already made up and installed in a similar block. The component shown in Fig. 30-3 is a TIL139 (or equivalent) from Texas Instruments. This particular component was purchased at a surplus store for about $1.

30.2.2 BETTER IR PROXIMITY SENSOR

In Chapter 16, the TV remote control, which receives a modulated infrared light signal, was introduced as a basic method of providing a remote control to robots. Along with receiving commands for robot movement, it can also be used to seek out and find the range to objects around a robot just as a submarine detects objects around it using sonar. The equipment required to provide this capability just costs a few dollars and is surprisingly easy to experiment with as will be shown in the following section.

An IR LED can be used to generate a modulated IR signal using the hardware shown in Fig. 30-4. The modulated IR signal lasts for about 500 μs, which is enough time for the TV remote control receiver to recognize it. If there is no object from which the signal can reflect off of, the TV remote control receiver does not indicate that a signal is present. The modu-

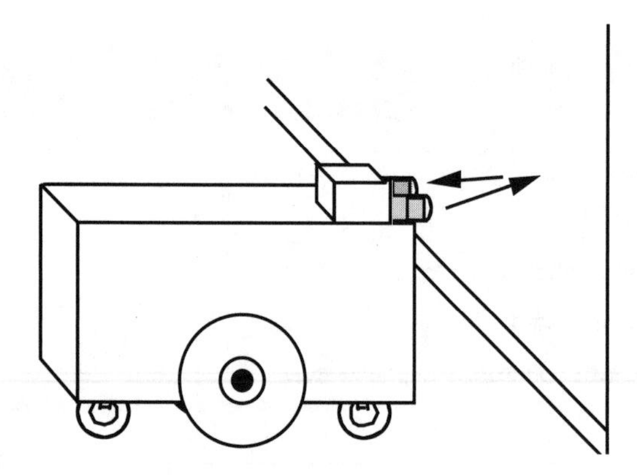

FIGURE 30-2 How the sensor is used to test proximity to a nearby object.

FIGURE 30-3 The Texas Instruments TIL139 infrared emitter/detector sensor unit. These types of units are often available on the surplus market.

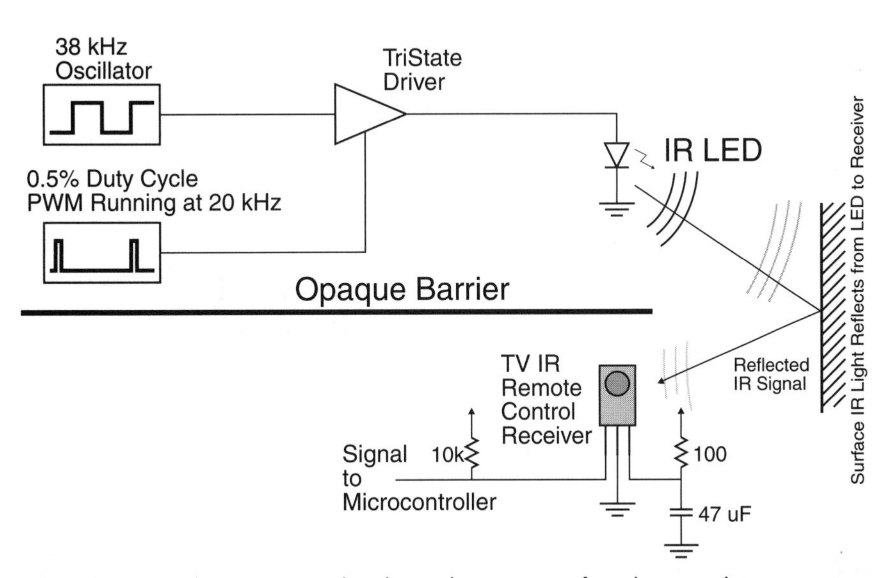

FIGURE 30-4 The circuitry used to detect objects using infrared LEDs and TV remote controls operates on the same theory as sonar; signals sent out from the LED reflect off of objects and are detected using the TV remote control.

lated signal is only active for a few hundred milliseconds to prevent the receiver from treating the signal as part of the background noise that it must filter out—leaving the signal active for more than a few milliseconds will cause the receiver to filter out the signal and look for other more transient signals.

The opaque barrier is to ensure there is not a direct path from the IR LED to the TV remote control. As with a simple IR LED and IR phototransistor, a block of wood drilled with two holes, one for the LED and one for the phototransistor, can be used for this purpose. Other materials such as electrical tape or sheet metal can also be used. Many materials that you would think are opaque to infrared light—such as cardboard—actually are not and you should test the different material you are planning on using in your robot. Probably the most common and easy to use opaque barrier is 5 mm black heat-shrink tubing, cut to ¾ to 1 in long and slipped over the IR LED; this prevents a direct path for the IR light from the LED to the TV remote control receiver and directs the IR light into a fairly narrow beam, allowing you to control where objects can be detected.

The attenuated signal can only be recognized by the TV remote control receiver when enough energy has been reflected into it as shown in Fig. 30-5. The receiver does not immediately respond to the received signal and remains active some time after the signal

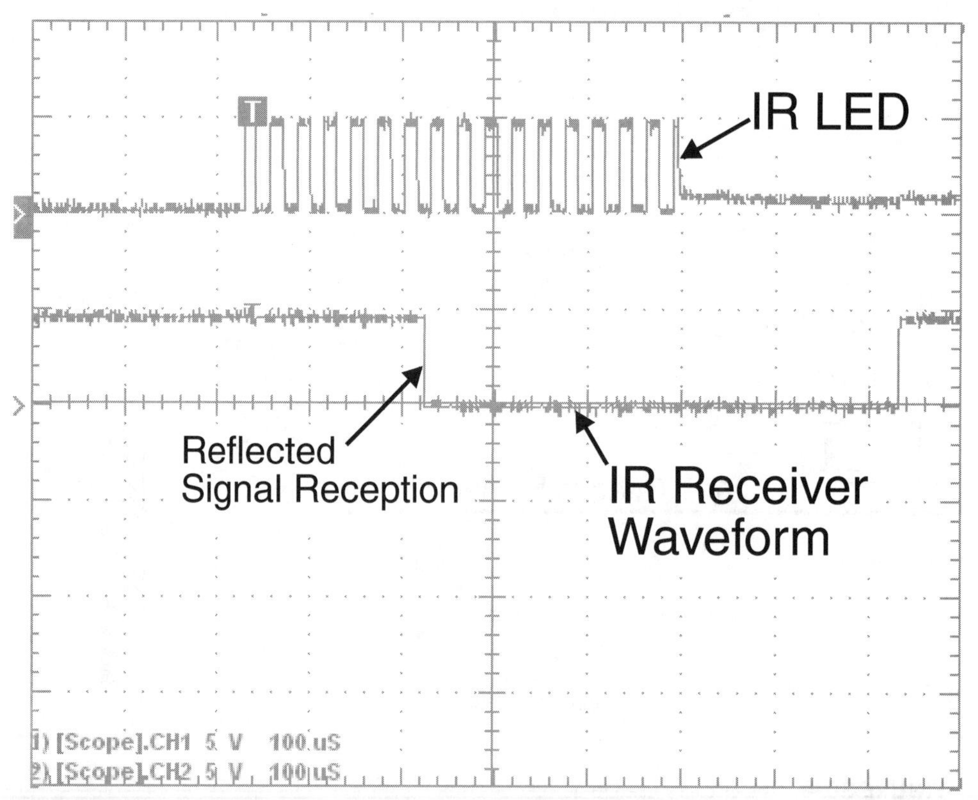

FIGURE 30-5 Oscilloscope waveform of the signal sent to an IR LED and the output of the TV remote control receiver when the signal is reflected from some object.

has stopped. If the LED output was at full power, then the signal would reflect off of other objects around the robot, resulting in reflections coming into the TV remote control receiver at such power that it would be recognized as a reflection. The solution to this problem is to attenuate, or lessen the ability of the TV remote control receiver to receive the incoming signals.

The obvious way to lessen the power of the signal is to dim the LED by allowing less current to pass through it, but a much better way of handling the signal is to shift the modulation frequency. If you were to look at a TV remote control receiver data sheet, you would discover that the ability to detect incoming signals changes as the frequency of the modulated signal changes. The closer to the design frequency of the receiver, the more sensitivity it has. By using a different frequency from the TV remote control receiver's nominal frequency, its ability to detect modulated signals decreases. In Table 30-2, the distance a particular 38 kHz IR TV remote control receiver can receive signals reflected from other objects is given.

These values are for a specific manufacturer's part number receiver; you will find that other manufacturers' products will respond differently.

What is interesting to note about Table 30-2 is that as the modulation frequency changes, the object detection distance decreases. As will be shown in the next section, by changing the modulation frequency of the IR LED, you can estimate the range to an object from the IR LED and TV remote control, which allows you to decide whether to take immediate evasive action or slightly alter the course of the robot.

BS2 Implementation of the Proximity Sensor The BS2 does not have the capability to produce 38 or 40 kHz signals, which are the most commonly used modulation frequencies for TV remote control receivers to drive an LED for detecting objects directly. Instead, the freqout function produces a digital signal that can be filtered into an analog sine wave of a specific frequency. The mathematics behind the operation of the freqout function are not important except to note that harmonics at the IR receiver's frequency are produced which allows it to be used for detecting objects around the BS2 as well as providing a simple range estimate for them.

To test the ability of the BS2 to detect objects around it the circuit shown in Fig. 30-6 was created (its parts list is Table 30-3). A modulated signal can be output on Pin 0 and after

TABLE 30-2 38 kHz TV Remote Control Receiver Object Detection Distance for Different Modulation Frequencies. Modulation Frequencies Generated by a Microchip PIC Microcontroller.

MODULATION FREQUENCY	OBJECT DETECTION DISTANCE
38.5 kHz	30 ft (360 in)
35.7 kHz	2 ft (24 in)
33.3 kHz	10 in
31.3 kHz	6 in

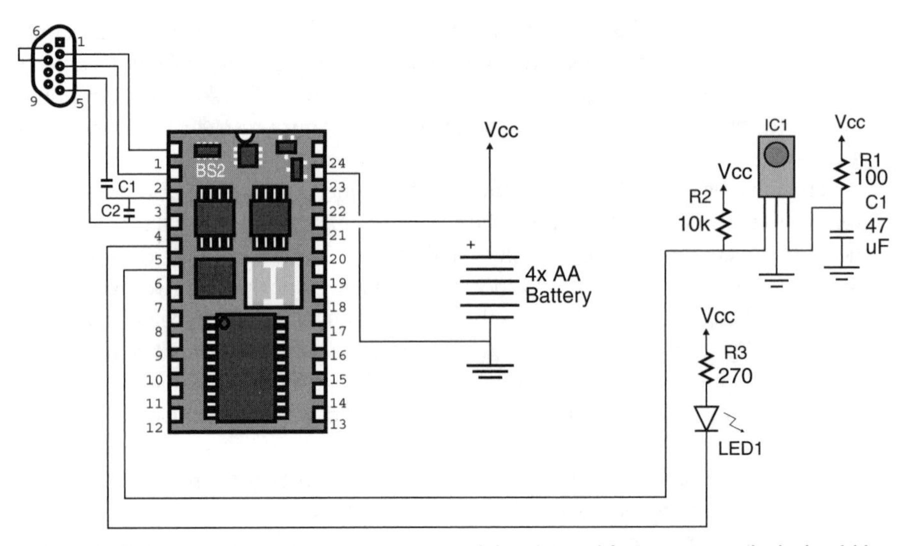

FIGURE 30-6 BS2 circuit that generates a modulated signal for an IR LED which should be detected by a TV remote.

the BS2 has finished sending it, the output of an IR receiver can be polled at Pin 1. The application that tests the operation of the IR receiver follows. Remember to slip a piece of 5 mm heat shrink tubing over the IR LED.

```
'  BS2 Object Detection - Using BS2 "Freqout" Detect Objects
'
'  myke predko
'
'  05.09.10
```

TABLE 30-3 Parts List for BS2 IR Object Detection/Range Finder

BS2	Parallax BASIC Stamp 2
IC1	38 kHz IR TV remote control receiver
LED1	5mm IR LED
R1	100 Ω resistor
R2	10k resistor
R3	270 Ω resistor
C1	47 μF electrolytic capacitor
Misc.	BS2 communications/programming interface, breadboard, breadboard wiring, 4× AA battery clip, 4× AA batteries, ¾ in long piece of 5 mm heat shrink tubing

```
'{$STAMP BS2}
'{$PBASIC 2.5}
'  I/O Pins
IREmitter PIN 0
IRRxr     PIN 1

'  Variables
IRStatus  VAR Bit

'  Mainline
   DO
       FREQOUT IREmitter, 1, 36500  '  Output IR Signal
       IRStatus = (IRRxr ^ 1)   '  High When Received

       DEBUG "Sensor Value = ", BIN1 IRStatus, CR
       PAUSE 250
LOOP                            '  Repeat
```

When the IR LED is arranged to fire perpendicularly from the face of the TV remote control receiver, you should find that the IR receiver will detect objects at about 18 in away. As noted in the previous section, this distance can be reduced by changing the frequency, but instead of changing a single frequency to detect objects at a specific distance from the robot, why not also determine the distance objects are from the robot?

In the next program listed, the frequency is changed to give a three-bit value indicating the relative distance of the object away from the IR LED/TV remote control receiver. When the least significant bit of the return value is set, the object is approximately a foot away from the sensor. The next bit indicates if the object is within 6 in of the sensor, and the most significant bit indicates if the object is within 3 in of the IR LED/TV remote control receiver sensor. This amount of accuracy should be adequate for most robot applications.

```
'  BS2 Object Ranging - Using BS2 "Freqout" Detect Objects
'
'  Different IR Frequencies are used to Determine the
'    (Rough) Distance to an Object.
'
'  Frequency Values Specified by Author.
'
'  myke predko
'
'  05.09.10

'{$STAMP BS2}
'{$PBASIC 2.5}

'  I/O Pins
IREmitter PIN 0
IRRxr     PIN 1

'  Variables
IRStatus  VAR Byte
i         VAR Byte
IRDist    VAR Byte
IRFreq    VAR Word
```

```
'  Mainline
   DO
       IRStatus = 0
       FOR i = 0 TO 2
           LOOKUP i,[48000, 44500, 42000], IRFreq

           FREQOUT IREmitter, 1, IRFreq  '  Output IR Signal
           IRStatus = (IRStatus << 1) + (IRRxr ^ 1)
           PAUSE 100
       NEXT

       DEBUG "Sensor Value = ", DEC IRStatus, CR
       PAUSE 250
   LOOP                         '  Repeat
```

The values passed to freqout were determined empirically; that is to say different values were tested for the TV remote control receiver used in the prototype circuit. You will find that different remote control receivers will behave differently with different freqout values—the important point to recognize here is that the same manufacturer's receivers of a specific part number should be used for a robot or else you will have to come up with frequency values for each receiver used in the robot.

30.2.3 SHARP INFRARED OBJECT SENSORS

The premier maker of infrared object and distance measurement sensors for use in robotics is Japan-based Sharp. One of their infrared distance measurement sensors, the GP2D02, is shown in Fig. 30-7. Actually, Sharp doesn't make these sensors for the robot-

FIGURE 30-7 The Sharp infrared sensors work similarly to the BS2 proximity sensor presented earlier in the book.

ics industry; rather, they are principally intended for use in cars for proximity devices and copiers for paper detection. Depending on the model, the sensors have a range of about 4 in (10 cm) and 31.5 in (80 cm).

The Sharp infrared sensors share better-than-average immunity to ambient light levels, so you can use them under a variety of lighting conditions (except perhaps very bright light outdoors). The sensors use a modulated—as opposed to a continuous—infrared beam that helps reject false triggering. It also makes the system accurate even if the detected object absorbs or scatters infrared light, such as heavy curtains or dark-colored fabrics.

Sharp GP2D12 Analog Output Infrared Ranging Sensor The GP2D12 is probably the most popular infrared object sensor used in robotics. While other Sharp ranging modules provide a digital output (either in the form of a changing signal when an object within a threshold is reached or by a synchronous serial I/O), the GP2D12 outputs an analog signal that can be read by a microcontroller and converted into a distance to object. To demonstrate how this can be done, an ADC0804 was attached to a BS2 (as shown in Fig. 30-8) and the distance to an object was calculated from the analog voltage output from the GP2D12. The parts list for the circuit is listed in Table 30-4.

The program used to read the current analog voltage from the GP2D12 output from the ADC0804 and convert it to a distance follows. To calculate the distance, the voltage levels at 10 cm intervals were recorded and then the slopes and the X intercept values for the lines

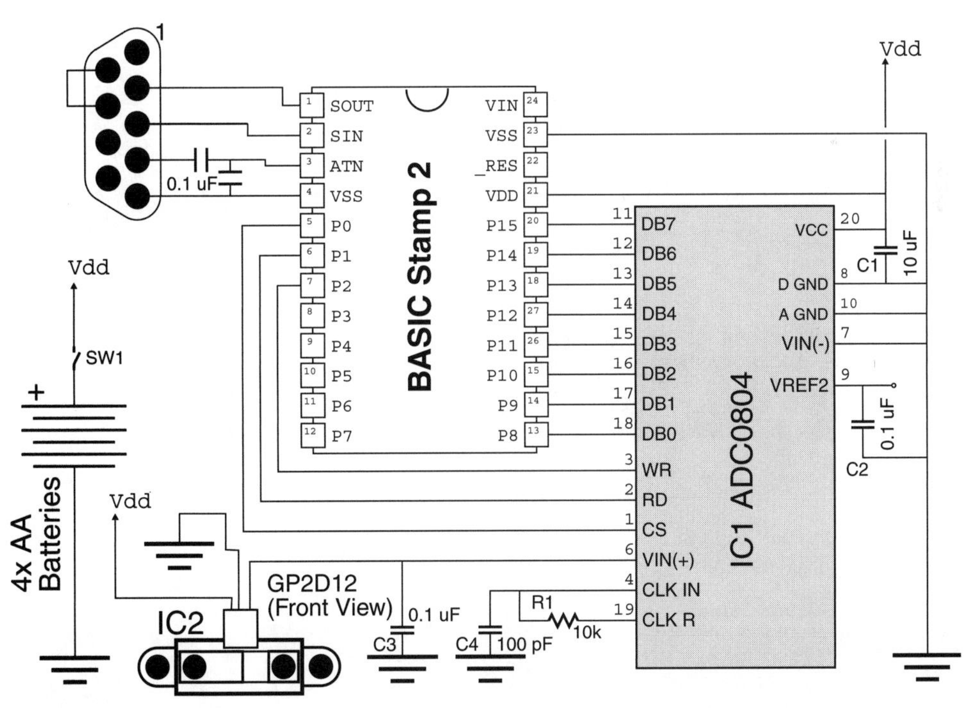

FIGURE 30-8 BS2 Circuit to output the distance from a Sharp GP2D12 infrared ranging sensor to a PC's console.

TABLE 30-4 Parts List for BS2/GP2D12 Object Distance Measuring Circuit

BS2	Parallax BASIC Stamp 2
IC1	ADC0804 ADC converter
IC2	Sharp GP2D12 with breadboard wires attached
R1	10k resistor
C1–C3	0.1 μF capacitor, any type
C4	100 pF capacitor, any type
Misc.	BS2 serial port communications/programming interface, breadboard, breadboard wiring, 4× AA battery clips, 4× AA alkaline batteries.

between 10 cm points were calculated. The voltage value was then interpolated between these points on the line.

```
'  BS2 GP2D12 - Measuring Distance Using a Sharp GP2D12
'
'  Distance Information is taken from Figure 6 of GP2D12
'   Datasheet
'  10cm - 2.40 Volts - ADC = 123
'  20cm - 1.40 Volts - ADC =  72
'  30cm - 1.00 Volts - ADC =  51
'  40cm - 0.75 Volts - ADC =  38
'  50cm - 0.55 Volts - ADC =  28
'  Values In between the ranges will be interpolated using a straight
line.
'  ADC Value is based on a reference voltage of 5.00 Volts
'
'  myke predko
'
'  05.08.25
'
'  Pin 0 - ADC0804 CS Pin (Output)
'  Pin 1 - ADC0804 RD Pin (Output)
'  Pin 2 - ADC0804 WR Pin (Output)
'  Pin 15-8 - ADC0804 Data Pins
'
'{$STAMP BS2}
'{$PBASIC 2.5}

'  Variable Declarations
Temp        VAR Byte
x           VAR Word
Slope       VAR Word
ZeroPoint   VAR Word
y           VAR Word
CS          PIN 0
RD          PIN 1
WR          PIN 2
```

```
'  Initialization
    HIGH CS                          '  No Operation
    HIGH RD
    HIGH WR

'  Main Loop
    DO                               '  Loop Forever

        LOW CS                       '  Start ADC Operation
        LOW WR                       '  Low on WR Starts Operation
        HIGH WR
        HIGH CS

        LOW CS                       '  Enable the chip
        LOW RD                       '  Read ADC Value
        x = INH                      '  Read the High 8 Bits
        HIGH RD
        HIGH CS

        IF (x > 72) THEN             '  Between 20 and 10 cm
            Slope = ABS(72 - 123)    '  Slope is divisor per 10 cm
            ZeroPoint = 10 + ((123 * 10) / Slope)
            y = ZeroPoint - ((x * 10) / Slope)
            DEBUG "Distance = ", DEC y, " cm", CR
        ELSE
            IF (x > 51) THEN         '  Between 30 and 20 cm
                Slope = ABS(51 - 72)         '  Slope is divisor per 10 cm
                ZeroPoint = 20 + ((72 * 10) / Slope)
                y = ZeroPoint - ((x * 10) / Slope)
                DEBUG "Distance = ", DEC y, " cm", CR
            ELSE
                IF (x > 38) THEN          '  Between 40 and 30 cm
                    Slope = ABS(38 - 51)     '  Slope is divisor per 10 cm
                    ZeroPoint = 30 + ((51 * 10) / Slope)
                    y = ZeroPoint - ((x * 10) / Slope)
                    DEBUG "Distance = ", DEC y, " cm", CR
                ELSE
                    IF (x > 28) THEN      '  Between 50 and 40 cm
                        Slope = ABS(28 - 38)
                        ZeroPoint = 40 + ((38 * 10) / Slope)
                        y = ZeroPoint - ((x * 10) / Slope)
                        DEBUG "Distance = ", DEC y, " cm", CR
                    ELSE                  '  Further Away than 50 cm
                        DEBUG "Nothing In front of GP2D12", CR
                    ENDIF
                ENDIF
            ENDIF
        ENDIF

        PAUSE 250                    '  Delay to 4x per second

    LOOP
END
```

The development of the slope and the line was not trivial (although it does not look very complex in the code above) and it is calculated from first principles you learned in high school math. The lines that are produced are actually counterintuitive because the *x* axis is voltage while the *y* axis is distance (most graphs will show this data in reverse). When you are developing a project like this one, make sure that you use lots of DEBUG statements to verify the software calculations against your own. Because the BS2 cannot perform floating point operations, you will find that you will have to do strange things, like make your slopes in units of 10s of mV per decameters.

The distance output is surprisingly accurate although it quickly diminishes when the power supply voltage varies from 5 V. In the schematic (Fig. 30-8) and parts list (Table 30-4), four AA alkaline batteries are specified, which will produce approximately 6 V (20% higher than the nominal voltage). In the prototype circuit, a Parallax BASIC Stamp Homework Board was used, which produces 5 V from a regulator, and the distances output matched measured distances surprisingly well.

30.2.4 PASSIVE INFRARED DETECTION

You can use commonly available passive infrared detection systems to detect the proximity of humans and animals. These systems, popular in both indoor and outdoor security systems, work by detecting the change in infrared thermal heat patterns in front of a sensor. This sensor uses a pair of pyroelectric elements that react to changes in temperature. Instantaneous differences in the output of the two elements are detected as movement, especially movement by a heat-bearing object, such as a human.

You can purchase pyroelectric sensors—commonly referred to as PIR, for passive infrared—new or salvage them from an existing motion detector. When salvaging from an existing detector, you can opt to unsolder the sensor itself and construct an amplification circuit around the removed sensor, or you can attempt to tap into the existing circuit of the detector to locate a suitable signal. Both methods are described next.

30.2.5 USING A NEW OR REMOVED-FROM-CIRCUIT DETECTOR

Using a new PIR sensor is by far the easiest approach since new PIR sensors will come with a data sheet from the manufacturer (or one will be readily available on the Internet). Some sensors—such as the Eltec 422—have built-in amplification, and you can connect them directly to a microcontroller or computer. Others require extra external circuitry, including amplification and signal filtering and conditioning.

If you prefer, you can attempt to salvage a PIR sensor from a discarded motion detector. Disassemble the motion detector, and carefully unsolder the sensor from its circuit board. The sensor will likely be securely soldered to the board so as to reduce the effects of vibration. Therefore, the unsoldered sensor will have fairly short connection leads. You'll want to resolder the sensor onto another board, being careful to avoid applying excessive heat.

Fig. 30-9 shows a typical three-lead PIR device. The pinouts are not industry standard, but the arrangement shown is common. Pin 1 connects to +V (often 5 V); pin 2 is the output, and pin 3 is ground. Physically, PIR sensors look a lot like old-style transistors and come in metal cans with a dark rectangular window on top (see Fig. 30-10). Often, a tab or notch will be located near pin 1. As even unamplified PIR sensors include an internal FET transistor for signal conditioning, the power connect and output of the sensor are commonly referred to by their common FET pinout names of *drain* and *source*.

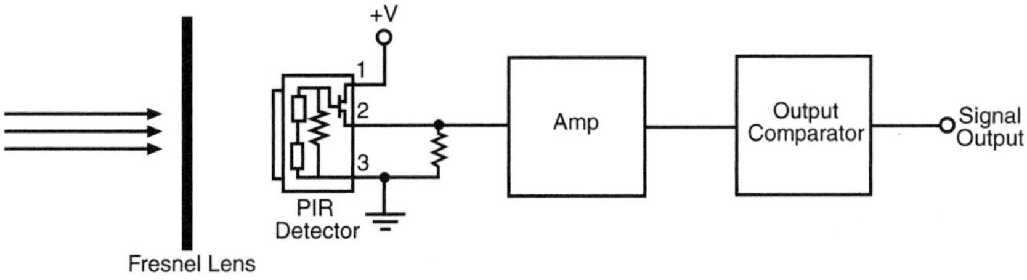

FIGURE 30-9 Most PIR sensors are large, transistorlike devices with a fairly common pinout arrangement. This is a block diagram of how the typical PIR sensor works.

If the sensor incorporates an internal output amplifier and signal conditioner, its output will be suitable for direct connection to a microcontroller or other logic input. A buffer circuit, like that shown in Fig. 30-11, is often recommended to increase input impedance. The circuit uses an op amp in unity gain configuration. If the sensor you are using lacks a preamplifier and signal condition, you can easily add your own with the basic circuit shown in Fig. 30-12.

With both the circuits shown in Figs. 30-11 and 30-12, the ideal interface to a robot computer or microcontroller is via an analog-to-digital converter (ADC). Many microcon-

FIGURE 30-10 The PIR sensor has an infrared window on the top to let in infrared heat radiated by objects. Movement of those objects is what the sensor is made to detect, not just the heat from an object.

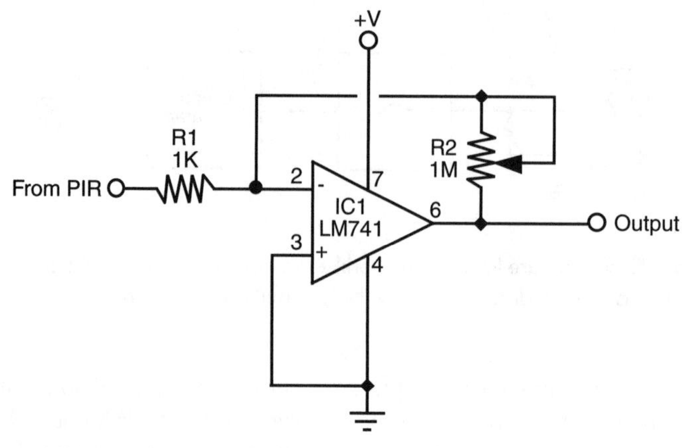

FIGURE 30-11 Use a buffer circuit between the output of the amplified PIR device and the microcontroller or other logic input.

trollers offer these on chip. If your control circuit lacks a built-in ADC, you can add one using one of the approaches outlined in Chapter 14, "Computer Peripherals." The output of the PIR sensor will be a voltage between ground and +V. For example's sake, assume the output will be the full 0 to 5 V, though in practice the actual voltage switch will be more restricted (e.g., 2.2 to 4.3 V, depending on the circuitry you use). Assuming a 0 to 5 vdc output, with no movement detected, the output of the sensor will be 2.5 V. As movement is detected, the output will swing first in one direction, then the other. It's important to keep this action in mind; it is caused by the nature of the pyroelectric element inside the sensor. It is also important to keep in mind that a heat source, even directly in front of the sensor, will not be detected if it doesn't move. For a PIR device to work, the heat source must be in motion. When programming your computer or microcontroller, you can look for variances in the voltage that will indicate a rise or fall in the output of the sensor.

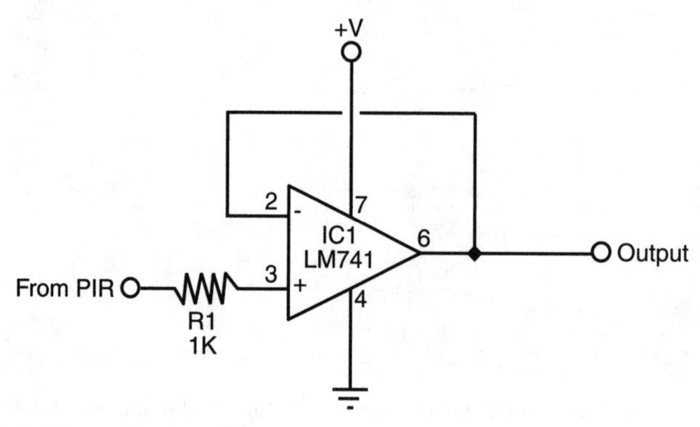

FIGURE 30-12 If the PIR sensor you are using lacks a built-in output amplifier, you can construct one using commonly available op amps.

30.2.6 HACKING A MOTION DETECTOR BOARD

Rather than unsolder the PIR sensor from a motion detector unit, you may be able to hack into the motion detector circuit board to find a suitable output signal. The advantage of this approach is that you don't have to build a new amplifier for the sensor. The disadvantage is that this can be hard to do depending on the make and model of the motion control unit that you use.

For best results, use a motion detector unit that is battery powered. This avoids any possibility that the circuit board in the unit also includes components for rectifying and reducing an incoming AC voltage. After disassembling the motion detector unit, connect +5 vdc power to the board. (Note: some PIR boards operate on higher voltages, usually 9 to 12 V. You may need to increase the supply voltage to properly operate the board.) Using a multitester or oscilloscope (the scope is the preferred method), carefully probe various points on the circuit board and observe the reading on the meter or scope. Wave your hand over the sensor and watch the meter or scope. If you're lucky, you'll find two kinds of useful signals:

- *Digital (on/off) output.* The output will normally be LOW and will go HIGH when movement is detected. After a brief period (less than 1 s), the output will go LOW again when movement is no longer detected. With this output you do not need to connect the sensor to an analog-to-digital converter.
- *Analog output.* The output, which will vary several volts, is the amplified output of the PIR sensor. With this output you will need to connect the sensor to an analog-to-digital converter (or an analog comparator).

You may also locate a timed output, where the output will stay HIGH for a period of time—up to several minutes—after movement is detected. This output is not as useful. Fig. 30-13 shows the innards of a hacked motion detector. If the PIR board you are using operates with a 5 vdc supply, you can connect the wire you added directly to a microprocessor or microcontroller input. If the PIR board operates from a higher voltage, use a logic level translation circuit, or connect the wire you added from the PIR board to the coil terminals of a 9 or 12 volt reed relay.

30.2.7 USING THE FOCUSING LENS

PIR sensors work by detecting electromagnetic radiation in the infrared region, especially about 5 to 15 micrometers (5000 to 15,000 nanometers). Infrared radiation in this part of the spectrum can be focused using optics for visible light. While you can use a PIR device without focusing, you'll find range and sensitivity are greatly enhanced when you use a lens. Most motion detectors use either a specially designed Fresnel lens to focus infrared radiation or a motion detection lens, which causes the amount of light falling on the detector to change with movement. The Fresnel lens is a piece of plastic with grooves, and is made to gather more light at the top than at the bottom. With this geometry, when the sensor is mounted high and pointing down the motion detector is more sensitive to movement farther away than right underneath. The motion detection lens consists of a series of flat sections, which focus light from different areas—when a person moves, the amount of light reaching the detector changes, resulting in a changing signal.

FIGURE 30-13 A hacked PIR detector, showing the DC-operated circuit board.

If you've gotten your PIR sensor by hacking a motion detector, you can use the same lens for your robot. You may wish to invert the lens from its usual orientation (because your robot will likely be near the ground, looking up). Or, you can substitute an ordinary positive diopter lens and mount it in front of the PIR sensor. Chapter 32, "Robot Vision," has more information on the use and proper mounting of lenses. Note that, oddly enough, plastic lenses are probably a better choice than glass lenses. Several kinds of glass actively absorb infrared radiation, as do optical coatings applied to finer quality lenses. You may need to experiment with the lens material you use or else obtain a specialty lens designed for use with PIR devices.

30.2.8 ULTRASONIC SOUND

Like light, sound has a tendency to travel in straight lines and bounce off any object in its path. You can use sound waves for many of the same things that light can be used for, including detecting objects. High-frequency sound beyond human hearing (ultrasonic) can be used to detect both proximity to objects as well as distance.

In operation, ultrasonic sound is transmitted from a transducer, reflected by a nearby object, then received by another transducer. The advantage of using sound is that it is not sensitive to objects of different color and light-reflective properties. Keep in mind, however, that some materials reflect sound better than others and that some even absorb sound completely.

This system is adaptable for use with either a single transmitter/receiver pair or multiple pairs. Ultrasonic transmitter and receiver transducers are common finds in the surplus market and even when new cost under $5 each (depending on make and model). Ultrasonic transducers are available from a number of retail and surplus outlets; see Appendix B, "Sources," for a more complete list of electronics suppliers.

You can also mount a single pair of transducers on a scanning platform (also called a turret or carousel), as shown in Fig. 30-14. The scanner can be operated using a standard RC servo (see Chapter 22, "Working with Servo Motors," for more information).

Figs. 30-15 and 30-16 show a basic circuit you can build that provides ultrasonic proximity detection and has two parts: a transmitter and a receiver (refer to the parts list in Tables 36-2 and 36-3). The transmitter circuit works as follows: a stream of 40 kHz pulses are produced by a 555 timer wired up as an astable multivibrator.

The receiving transducer is positioned two or more inches away from the transmitter transducer. For best results, you may wish to place a piece of foam between the two transducers to eliminate direct interference. The signal from the receiving transducer needs to be amplified; an op amp (such as an LM741, as shown in Fig. 30-16) is more than sufficient for the job. The amplified output of the receiver transducer is directly connected to another 741 op amp wired as a comparator. The ultrasonic receiver is sensitive only to sounds in about the 40 kHz range (± about 4 kHz).

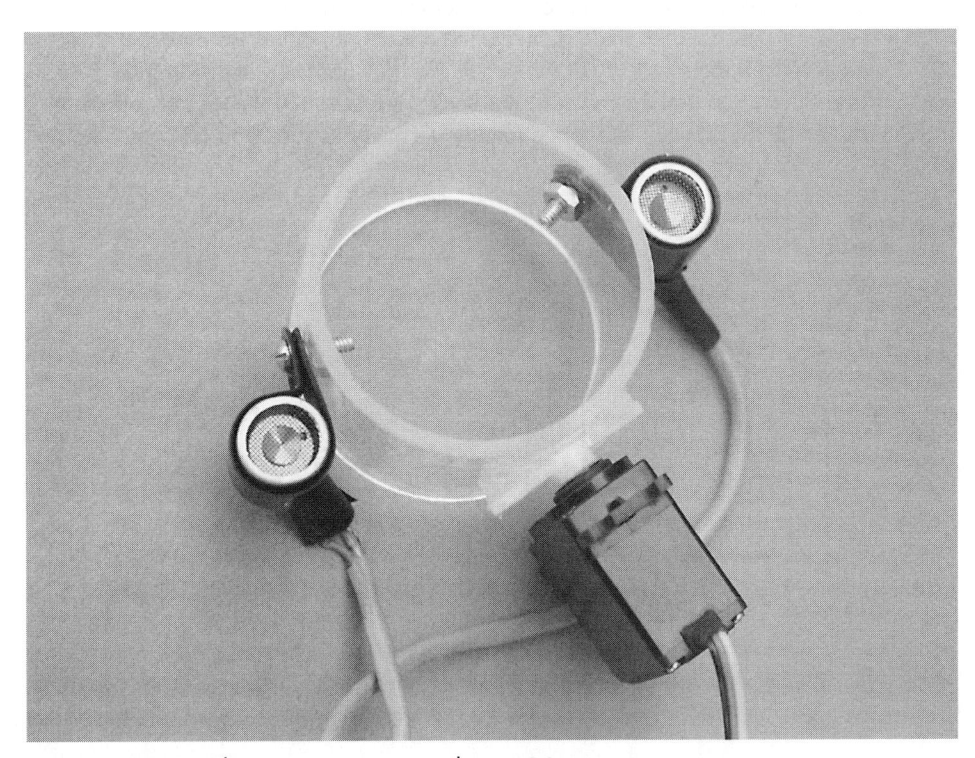

FIGURE 30-14 Ultrasonic sensors mounted on an RC servo scanner turret.

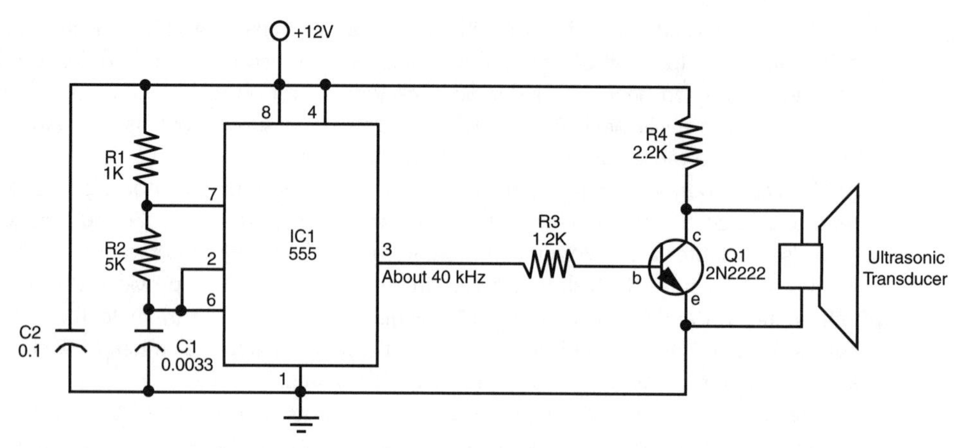

FIGURE 30-15 Schematic diagram for a basic ultrasonic proximity transmitter.

The closer the ultrasonic sensor is to an object, the stronger the reflected sound will be. (Note, too, that the strength of the reflected signal will also vary depending on the material bouncing the sound.) The output of the comparator will change between LOW and HIGH as the sensor is moved closer to or farther away from an object.

Once you get the circuit debugged and working, adjust potentiometer R2, on the op amp, to vary the sensitivity of the circuit. You will find that, depending on the quality of the transducers you use, the range of this sensor is quite large. When the gain of the op amp is turned all the way up, the range may be as much as 6 to 8 ft. (The op amp may ring, or

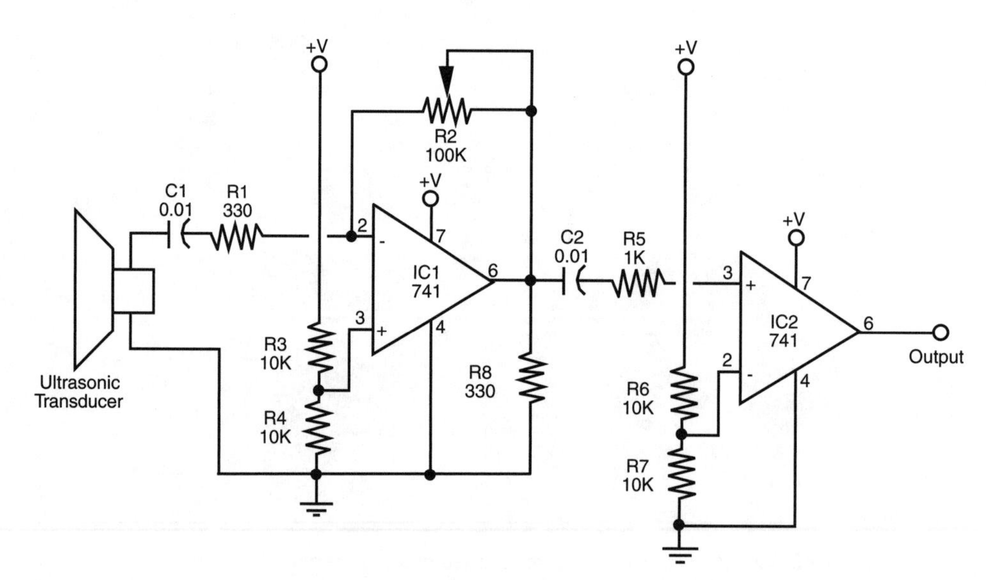

FIGURE 30-16 Schematic diagram for a basic ultrasonic proximity sensor receiver.

oscillate, at very high gain levels, so use your logic probe to choose a sensitivity setting just below the ringing threshold.)

30.3 Contact Detection

A sure way to detect objects is to make physical contact with them. Contact is perhaps the most common form of object detection and is often accomplished by using simple switches. This section will review several contact methods, including soft-contact techniques where the robot can detect contact with an object using just a slight touch.

30.3.1 PHYSICAL CONTACT BUMPER SWITCH

An ordinary switch can be used to detect physical contact with an object. So-called bumper switches are spring-loaded push-button switches mounted on the frame of the robot, as shown in Fig. 30-17. The plunger of the switch is pushed in whenever the robot collides with an object. Obviously, the plunger must extend farther than all other parts of the robot. You may need to mount the switch on a bracket to extend its reach.

The surface area of most push-button switches tends to be very small. You can enlarge the contact area by attaching a metal or plastic plate or a length of wire to the switch plunger. A piece of rigid 1/16-in-thick plastic or aluminum is a good choice for bumper plates. Glue the plate onto the plunger. Low-cost push-button switches are not known for their sensitivity. The robot may have to crash into an object with a fair amount of force before the switch makes positive contact, and for most applications that's obviously not desirable.

Leaf switches require only a small touch before they trigger. The plunger in a leaf switch (often referred to as a Microswitch), is extra small and travels only a few fractions of an inch before its contacts close. A metal strip, or leaf attached to the strip, acts as a lever, further increasing sensitivity. You can mount a plastic or metal plate to the end of the leaf to increase surface area. If the leaf is wide enough, you can use miniature 4/40 or 3/38 hardware to mount the plate in place.

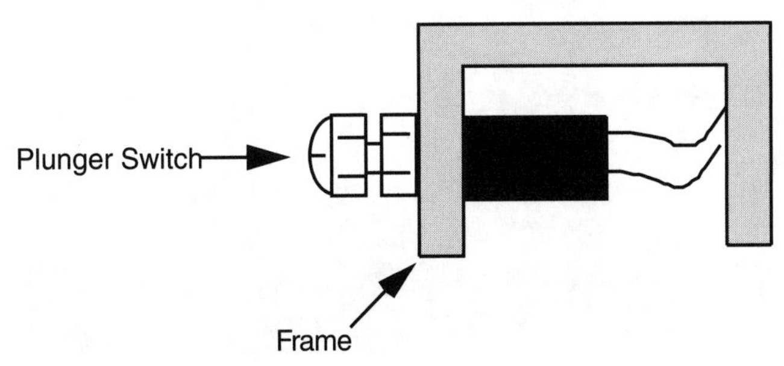

FIGURE 30-17 An SPST spring-loaded plunger switch mounted in the frame or body of the robot, used as a contact sensor.

30.3.2 WHISKER

Many animal experts believe that a cat's whiskers are used to measure space. If the whiskers touch when a cat is trying to get through a hole, it knows there is not enough space for its body. We can apply a similar technique to our robot designs—whether or not a cat actually uses whiskers for this purpose.

You can use thin 20- to 25-gauge piano or stove wire for the whiskers of the robot. Attach the wires to the end of switches, or mount them in a receptacle so the wire is supported by a small rubber grommet.

By bending the whiskers, you can extend their usefulness and application. The commercially made robot shown in Fig. 30-18, the Movit WAO, has two whiskers that can be rotated in their switch sockets. When the whiskers are positioned so the loop is vertical, they can detect changes in topography to watch for such things as the edge of a table, the corner of a rug, and so forth.

A more complex whisker setup is shown in Fig. 30-19. Two different lengths of whiskers are used for the two sides of the robots. The longer-length whiskers represent a space a few inches wider than the robot. If these whiskers are actuated by rubbing against an object but the short whiskers are not, then the robot understands that the pathway is clear to travel but space is tight.

FIGURE 30-18 The Movit WAO robot (one of the older models, but the newer ones are similar). Its two tentacles, or whiskers, allow it to navigate a space.

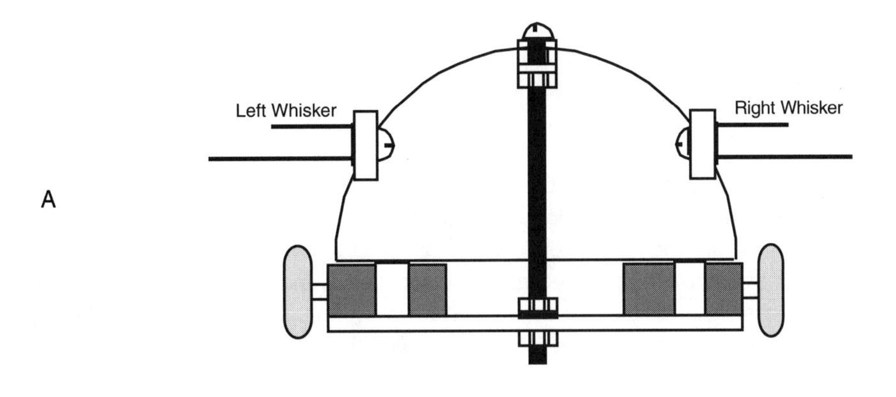

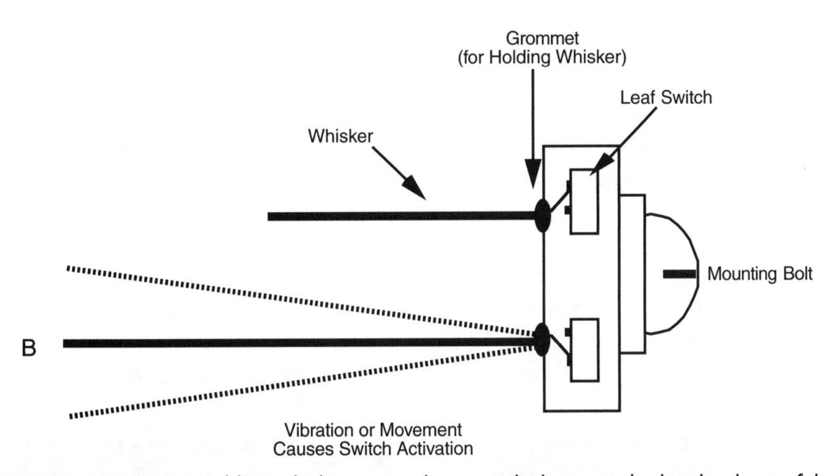

FIGURE 30-19 Adding whiskers to a robot. *a.* Whiskers attached to the dome of the Minibot (see Chap. 8); *b.* construction detail of the whiskers and actuation switches.

The short whiskers are cut to represent the width of the robot. Should the short whiskers on only one side of the robot be triggered, then the robot will turn the opposite direction to avoid the obstacle. If both sides of short whiskers are activated, then the robot knows that it cannot fit through the passageway, and it either stops or turns around.

Before building bumper switches or whiskers into your robot, be aware that most electronic circuits will misbehave when they are triggered by a mechanical switch contact. The contact has a tendency to bounce as it closes and opens, so it needs to be conditioned. See the debouncing circuits in Chapter 29 for ways to clean up the contact closure so switches can directly drive your robot circuits.

30.3.3 SPRING WHISKERS

You can make a very inexpensive and simple whisker sensor for your robots using an old spring and a piece of thick solid core wire as shown in Fig. 30-20. The whisker consists of a spring that has been pulled apart and the two ends straightened with one end being several inches long and acting as the whisker. Instead of using the straightened spring wire for the whisker, a piece of narrow plastic tubing can be used. The other end of the spring is simply soldered to the mounting PCB.

Running through the center of the spring is a piece of bared, solid core (20 gauge is good for this application) that does not touch the spring unless there is a force against the whisker (straightened spring end), in which case, the two pieces of metal form a contact. Ideally, the center contact wire should be a pulled up connection to the robot controller while the spring and its integral whisker is connected to the robot's ground. This wiring arrangement minimizes the chance that static electricity can form on the whisker and cause damage to the robot's controller when it comes into contact with the center wire.

30.3.4 PRESSURE PAD

In Chapter 29 you learned how to give the sense of touch to robot fingers and grippers. One of the materials used as a touch sensor was conductive foam, which is packaged with most CMOS and microprocessor ICs. This foam is available in large sheets and is perfect for use as collision detection pressure pads. Radio Shack sells a nice 5-in square pad that's ideal for the job.

Attach wires to the pad as described in Chapter 29, and glue the pad to the frame or skin of your robot. Unlike fingertip touch, where the amount of pressure is important, the salient ingredient with a collision detector is that contact has been made with something. This makes the interface electronics that much easier to build.

Fig. 30-21 shows a suitable interface for use with the pad (refer to the parts list in Table 30-5). The pad is placed in series with a 3.3K resistor between ground and the positive supply voltage to form a voltage divider. When the pad is pressed down, the voltage at the output of the sensor will vary. The output of the sensor, which is the point between the pad

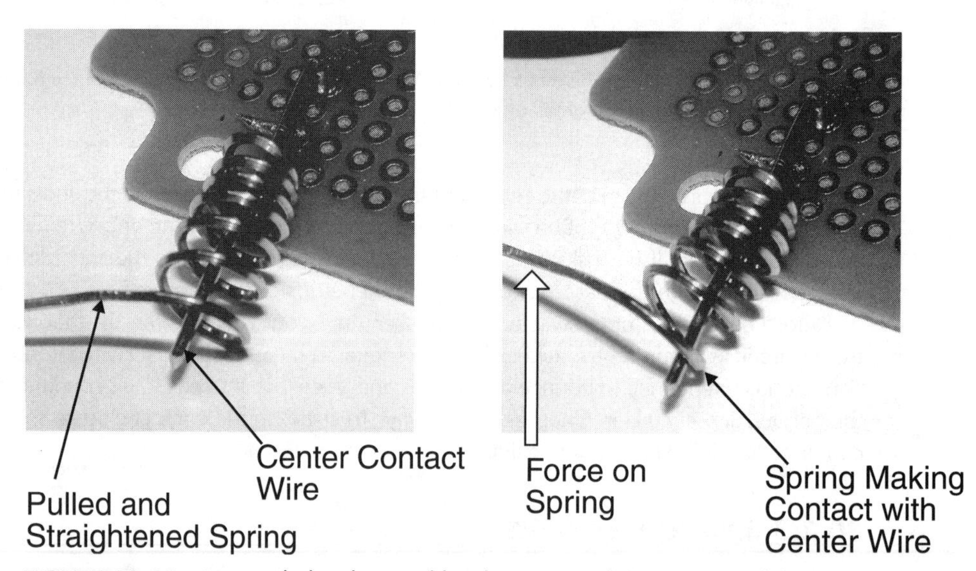

FIGURE 30-20 Spring whisker shown soldered to a PCB with a 20-gauge solid core wire running up through the center. When the whisker has a force applied to it, the spring comes into contact with the center wire.

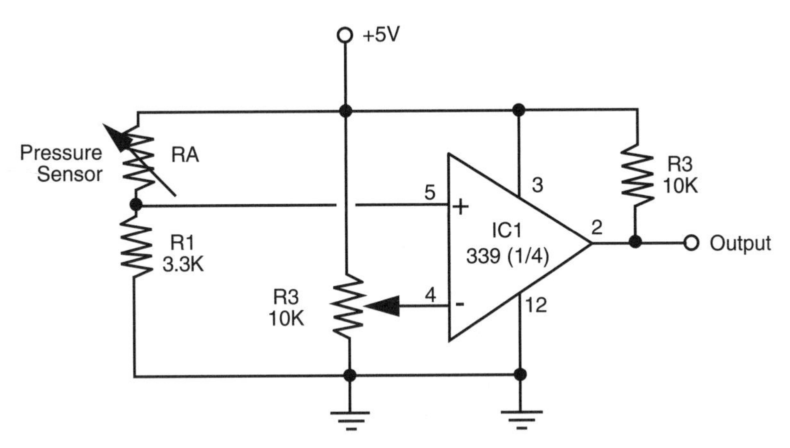

FIGURE 30-21 Converting the output of a conductive foam pressure sensor to an on/off type switch output.

and resistor, is applied to the inverting pin of a 339 comparator. (There are four separate comparators in the 339 package, so one chip can service four pressure pads.) When the voltage from the pad exceeds the reference voltage supplied to the comparator, the comparator changes states, thus indicating a collision.

The comparator output can be used to drive a motor direction control relay or can be tied directly to a microprocessor or computer port. Follow the interface guidelines provided in Chapter 14.

30.3.5 MULTIPLE BUMPER SWITCHES

What happens when you have many switches or proximity devices scattered around the periphery of your robot? You could connect the output of each switch to the computer, but that's a waste of interface ports. A better way is to use a priority encoder or multiplexer. Both schemes allow you to connect several switches to a common control circuit. The robot's microprocessor or computer queries the control circuit instead of the individual switches or proximity devices.

Using a Priority Encoder The circuit in Fig. 30-22 uses a 74148 priority encoder IC. Switches are shown at the inputs of the chip. When a switch is closed, its binary equivalent

TABLE 30-5 Parts List for Pressure Sensor Bumper Switch

IC1	LM339 quad comparator IC
R1	3.3K resistor
R2	10K potentiometer
R3	10K resistor
Misc.	Conductive foam pressure transducer (see text)

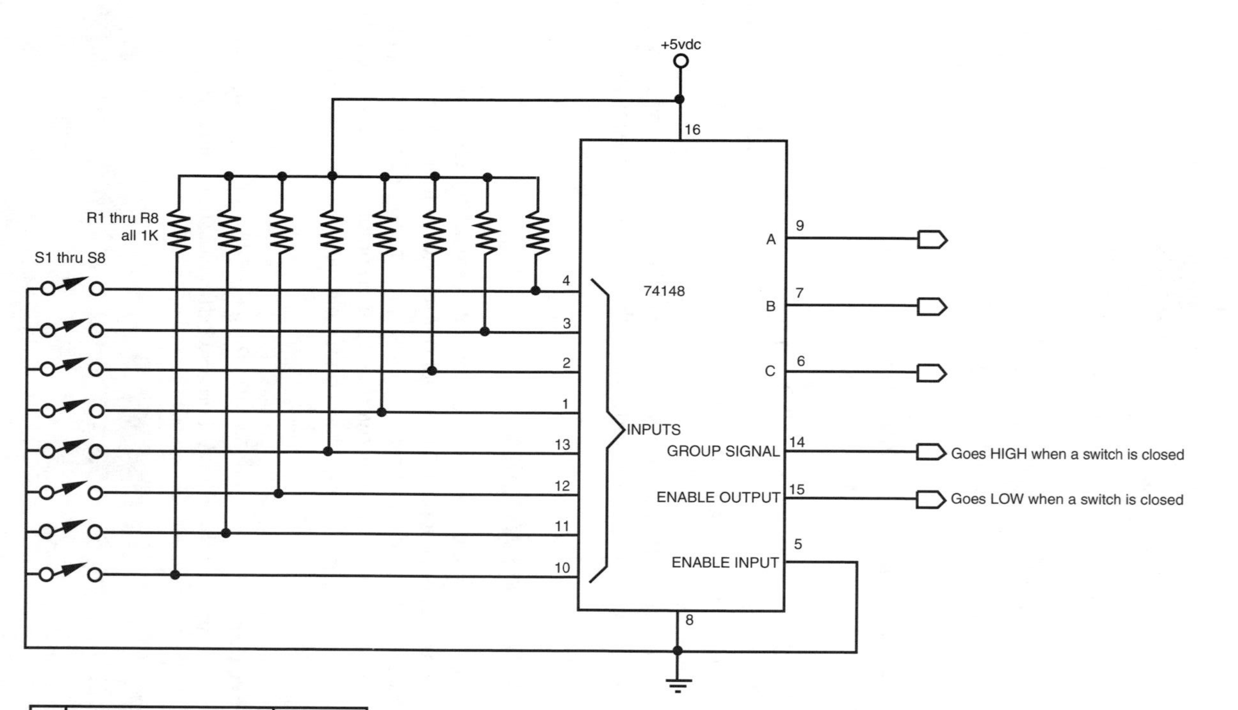

EI	SWITCH 1	2	3	4	5	6	7	8	OUTPUT A	B	C
1	X	X	X	X	X	X	X	X	1	1	1
0	C	O	0	C	C	C	C	C	1	1	1
0	0	C	0	0	0	0	0	0	0	1	1
0	0	0	C	0	0	0	0	0	1	0	1
0	0	0	0	C	0	0	0	0	0	0	1
0	0	0	0	0	C	0	0	0	1	1	0
0	0	0	0	0	0	C	0	0	0	1	0
0	0	0	0	0	0	0	C	0	1	0	0
0	0	0	0	0	0	0	0	0	0	0	0

Truth Table--74148 priority encoder

FIGURE 30-22 Multiple switch detection using the 74148 priority encoder IC.

appears at the A-B-C output pins. With a priority encoder, only the highest value switch is indicated at the output. In other words, if switches 4 and 7 are both closed, the output will only reflect the closure of pin 4.

Another method is shown in Fig. 30-23. Here, a 74150 multiplexer IC is used as a switch selector. To read whether a switch is or not, the computer or microprocessor applies a binary weighted number to the input select pins. The state of the desired input is shown in inverted form at the Out pin (pin 10). The advantage of the 74150 is that the state of any switch can be read at any time, even if several switches are closed.

Using a Resistor Ladder If the computer or microcontroller used in your robot has an analog-to-digital converter (ADC) or you don't mind adding one, you can use another technique for interfacing multiple switches: the resistor ladder. The concept is simple, as Fig. 30-24 shows. Each switch is connected to ground on one side and to positive voltage in series with a resistor on the other side. Multiple switches are connected in parallel to an ADC input, as depicted in the figure. The resistors form a voltage divider. Each resistor has a different value, so when a switch closes the voltage through that switch is uniquely different.

Note that because the resistors are in parallel, you can close more than one switch at one time. An in-between voltage will result. Feel free to experiment with the values of the resistors connected to each switch to obtain maximum flexibility.

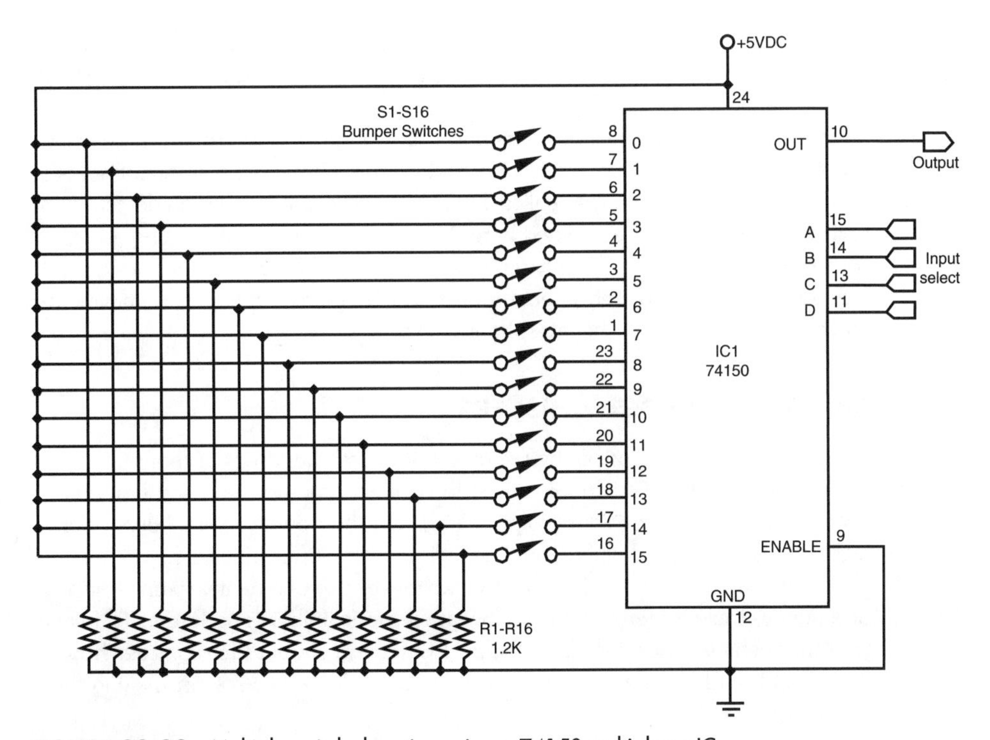

FIGURE 30-23 Multiple switch detection using a 74150 multiplexer IC.

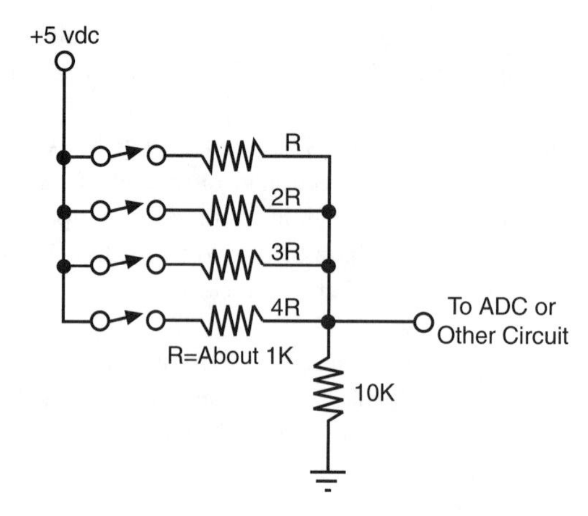

FIGURE 30-24 A resistor ladder provides a variable voltage; the voltage at the output of the ladder is dependent on the switch(es) that are closed.

30.4 Soft Touch and Compliant Collision Detection

The last nickname you'd want for your robot is "Bull in a China Closet," a not too flattering reference to your automaton's habit of crashing into and breaking everything. Unfortunately, even the best behaved robots occasionally bump into obstacles, including walls, furniture, and the cat (your robot can probably survive a head-on collision with a solid wall, but the family feline . . . maybe not!).

Since it's impractical—not to mention darn near impossible—to always prevent your robot from colliding with objects, the next best thing is to make those collisions as "soft" as possible. This is done using so-called soft touch or compliant collision detection means. Several such approaches are outlined here. You can try some or all; mixing and matching sensors on one robot is not only encouraged, it's a good idea. As long as the sensor redundancy does not unduly affect the size, weight, or cost of the robot, having backups can make your robot a better behaved houseguest.

30.4.1 LASER FIBER WHISKERS

You know about fiber optics: they're used to transmit hundreds of thousands of phone calls through a thin wire. They're also used to connect together high-end home entertainment gear and even to make "light sculpture" art. On their own, optical fibers offer a wealth of technical solutions, and when combined with a laser, optical fibers can do even more.

The unique whiskers project that follows makes use of a relatively underappreciated (and often undesirable) synergy between low-grade optical fibers and lasers. To fully understand what happens to laser light in an optical fiber, you should first consider how fiber optics work and then how the properties of laser light play a key role in making the fiber-optic robowhiskers function.

Fiber Optics: An Introduction An optical fiber is to light what PVC pipe is to water. Though the fiber is a solid, it channels light from one end to the other. Even if the fiber is bent, the light follows the path, altering its course at the bend and traveling on. Because light acts as an information carrier, a strand of optical fiber no bigger than a human hair can carry the same amount of data as some 900 copper wires.

The idea for optical fibers is over 100 years old. British physicist John Tyndall once demonstrated how a bright beam of light was internally reflected through a stream of water flowing out of a tank. Serious research into light transmission through solid material started in 1934, when Bell Labs was issued a patent for the light pipe. In the 1950s, the American Optical Corporation developed glass fibers that transmitted light over short distances (a few yards). The technology of fiber optics really took off around 1970 when scientists at Corning Glass Works developed long-distance optical fibers.

Optical fibers are composed of two basic materials, as illustrated in Fig. 30-25: the core and the cladding. The core is a dense glass or plastic material that the light actually passes through as it travels the length of the fiber. The cladding is a less dense sheath, also of plastic or glass, that serves as a refracting medium. An optical fiber may or may not have an outer jacket, a plastic or rubber insulation used as protection.

Optical fibers transmit light by total internal reflection (TIR), as shown in Fig. 30-26. Imagine a ray of light entering the end of an optical fiber strand. If the fiber is perfectly straight, the light will pass through the medium just as it passes through a plate of glass. But if the fiber is bent slightly, the light will eventually strike the outside edge of the fiber. If the angle of incidence is great (more than the so-called critical angle), the light will be reflected internally and will continue its path through the fiber. But if the bend is large and the angle of incidence is small (e.g., less than the critical angle), the light will pass through the fiber and be lost.

Note the cone of acceptance, as shown in Fig. 30-26. The cone represents the degree to which the incoming light can be off axis and still make it into the fiber. The cone of acceptance (usually 30°) of an optical fiber determines how far the light source can be from the optical axis and still manage to make it into the fiber. Though the cone of acceptance may

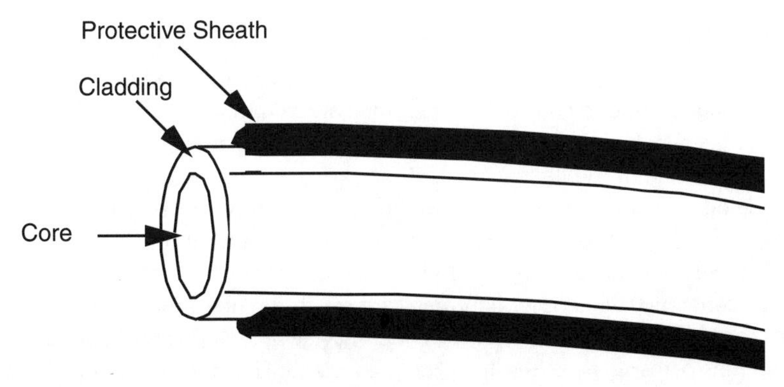

FIGURE 30-25 The physical makeup of an optical fiber, consisting of core and cladding.

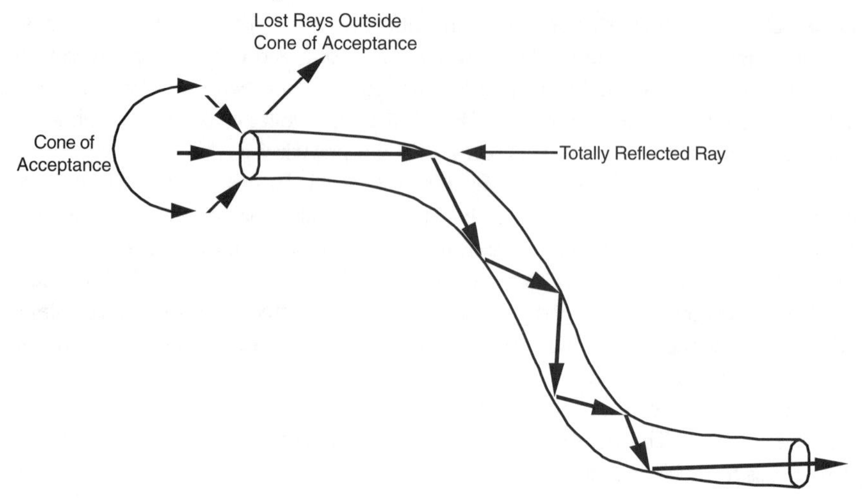

FIGURE 30-26 Light travels through optical fibers due to a process called total internal reflection (TIR).

be great, fiber optics perform best when the light source (and detector) are aligned to the optical axis.

Types of Optical Fibers The classic optical fiber is made of glass, otherwise known as silica (which is plain ol' sand). Glass fibers tend to be expensive and are more brittle than stranded copper wire, but they are excellent conductors of light, especially light in the infrared region between 850 and 1300 nanometers (nm).

Less expensive optical fibers are made of plastic. Though light loss through plastic fibers is greater than through glass fibers, they are more durable. Plastic fibers are best used in communications experiments with near-infrared light sources—the 780 to 950 nm range. This nicely corresponds to the output wavelength and sensitivity of common infrared emitters and detectors.

Optical fiber bundles may be coherent or incoherent. These terms relate to the arrangement of the individual strands in the bundle. If the strands are arranged so that the fibers can transmit an image from one end to the other, they are said to be coherent. The vast majority of optical fibers are incoherent: an image or particular pattern of light is lost when it reaches the other end of the fiber.

The cladding used in optical fibers may be one of two types: step-index and graded-index. Step-index fibers provide a discrete boundary between more dense and less dense regions of core and cladding. They are the easiest to manufacture, but their design causes a loss of ray coherency when laser light passes through the fiber: that is, coherent light in, largely incoherent light out. The loss of coherency, which is due to light rays traveling slightly different paths through the fiber, reduces the efficiency of the laser beam. Still, it offers some very practical benefits.

There is no discrete refractive boundary in graded-index fibers. The core and cladding media slowly blend, like an exotic tropical drink. The grading acts to refract light evenly, at any angle of incidence. This preserves coherency and improves the efficiency of the fiber. As you might have guessed, graded-index optical fibers are the most expensive of the bunch.

Working with Fiber Optics Optical fibers may be cut with wire cutters, nippers, or even a knife. But you must exercise care to avoid injuring yourself from shards of glass that may fly out when the fiber is cut (plastic fibers don't shatter when cut). Wear heavy cotton gloves and eye protection when working with optical fibers. Avoid working with fibers around food-serving or -preparation areas (that means stay out of the kitchen!). The bits of glass may inadvertently settle on food, plates, or eating utensils and could cause bodily harm.

One good way to cut glass fiber is to gently nick it with a sharp knife or razor, then snap it in two. Position the thumb and index finger of both hands as close to the nick as possible, then break the fiber with a swift downward motion (snapping upward increases the chance that glass shards will fly off in your direction).

Building the Laser-Optic Whisker Consider the arrangement in Fig. 30-27. A laser is pointed at one end of a stepped-index optical fiber. The fiber forms one or more loops around the front, side, or back of the robot. At the opposite end of the fiber is an ordinary phototransistor or photodiode. When the laser is turned on, the photodetector registers a certain voltage level from the laser light, say 2.5 V. This is the quiescent level.

When one or more of the loops of the fiber are deformed—the robot has touched a person or thing, for instance—the laser light passing through the fiber is diverted in its path, and this changes the interference patterns at the photodetector end. The change in light level received by the photodetector does not span a very wide range, perhaps 1 V total. But this 1 V is enough to not only determine when the robot has touched an object but the relative intensity of the collision. The more the robot has connected to some object, the more the fibers will deform and the greater the output change of the light as it reaches the photodetector.

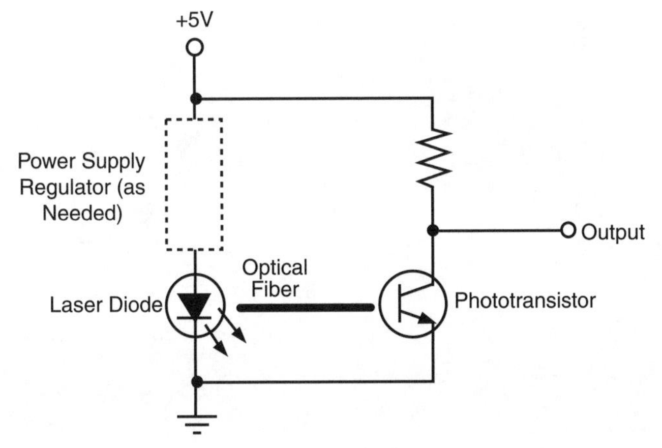

FIGURE 30-27 The basic parts of a laser-optic whisker are a laser, a length of fiber optics, and a photodetector.

The key benefit of the laser-optic whisker system is that a collision can be detected with just a feather touch. In fact, your robot may know when it's bumped into you before you do! Since contact with the robot is through a tiny piece of plastic, there's little chance the machine will damage or hurt anything it bumps into. The whiskers can protrude several inches from the body of the robot and omnidirectionally, if you desire. In this way it will sense contact from any direction.

Fig. 30-28 shows a prototype of this technique that consists of a hacked visible light penlight laser, several strands of cheap (very cheap) stepped-index optical fibers, and a set of three phototransistors. The optical fibers are tied together in a bundle using a small brass collar, electrical tape, and tie-wrap. This bundle is then inserted into the opening of the penlight laser and held in place with a sticky-back tie-wrap connector (available at Radio Shack and many other places).

On the opposite ends of the optical fibers are #18 crimp-type bullet connectors. These are designed to splice two #18 or #20 wires together, end to end. Carefully crimp them onto the ends of the fibers, so they act as plug-in connectors. As shown in Fig. 30-29, these ersatz connectors plug into makeshift optical jacks, which are nothing more than ¼-in-diameter by ⅜-in-aluminum tubing. The tubing is glued over the ends of the phototransistors and the phototransistors are soldered near the edge of the prototyping PCB.

Refer to Fig. 30-30 for a schematic wiring diagram of a power regulator for the penlight laser. Note the zener diode voltage regulator. The laser I used was powered by two AAA batteries, or roughly 3 V. Diode lasers are sensitive to high input voltage, and many will burn

FIGURE 30-28 The prototype laser-optic sensor, showing the loose fibers (on the robot these fibers are neatly looped to create a kind of sensor antenna).

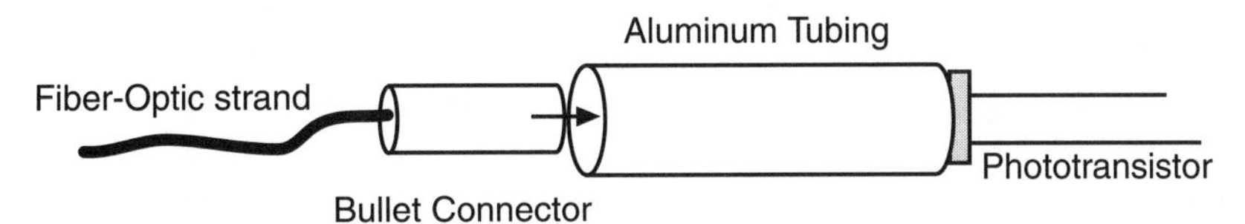

FIGURE 30-29 Use short lengths of aluminum tubing, available at hobby stores, and a crimp-on bullet connector to create "optical jacks" for the laser-optic whisker system.

out if fed a higher voltage than they are designed for. The penlight laser consumes less than about 30 mA. An alternative is to use three signal diodes (e.g., 1N4148) in series between the +V and the input of the laser to drop the 5 vdc voltage to about 2.7 to 3.0 V. The diodes you use should be rated for ¼-W or higher.

Interfacing the Photodetectors The output of a phototransistor is close to the full 0 to 5 V range of the circuit's supply range. You'll want your robot to be able to determine the intensity changes as the whiskers bump against objects. If you're using a computer or microcontroller to operate your robot, this means you'll need to convert the analog signal produced by the detectors into a digital signal suitable for the brains on your 'bot. Most popular microcontroller families have analog-to-digital converter (ADC) ports built in. If your computer or controller doesn't have ADC inputs, you can add an outboard ADC using an ADC0809 or similar chips. See Chapter 14 for more information on interfacing an analog signal to a digital input by way of an analog-to-digital converter.

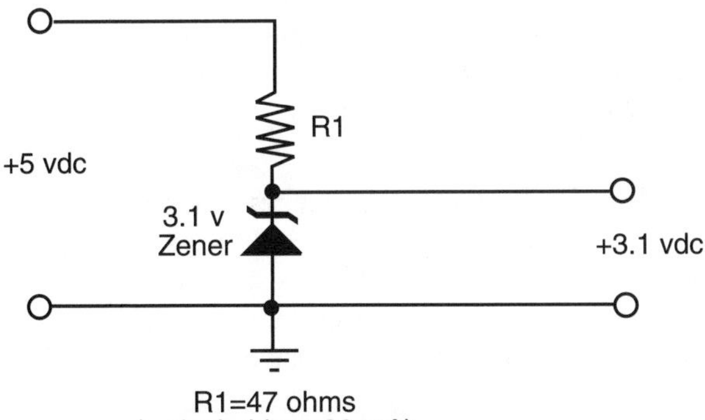

FIGURE 30-30 Most penlight lasers are designed to operate with 3 vdc; use a zener diode or voltage regulator to provide the proper voltage.

Creating the Whisker Loops Okay, so the laser-optic whisker system may not use cat-type whiskers with ends that stick out. Still, the word *whisker* aptly describes the way the system works. If something even so much as brushes lightly against the whisker, the light reaching the photodetector will change, and your robot can react accordingly.

The prototype system for this book used three whiskers, all of which were formed into three small loops around the front and two sides of the test robot. The loops can be held in place with small screws, dabs of glue (don't use hot-melt glue!), or even LEGO parts should your robot be constructed with them. When forming the loops don't make them too tight. The more compliant the loops are, the more they will detect small amounts of pressure. If the loops are very tight, the fibers become rigid and not very compliant. This reduces the effectiveness of the whiskers.

At the same time, the loops should not be so loose that they tend to wobble or flap while the robot is in motion. Should this occur, the natural vibration and movement of the fiber will cause false readings. A loop diameter of from 4 to 6 in should be sufficient given optical fiber pieces of average diameter and stiffness. Experiment with the optical fibers you obtain for the project. Your laser-optic whisker system does not need to use three separate fiber strands. One strand may be enough, especially if the robot is small. The prototype used three so the robot could independently determine in which direction (left, front, right) a collision or bump had occurred.

Getting the Right Kind of Optical Fiber Perhaps the hardest part of constructing this project is finding the right kind of optical fiber. You want to avoid any kind of graded-index fiber (described earlier) because these will not produce the internal interference patterns that the project depends on. In essence, what you want is the cheapest, lousiest fiber-optic strands you can find. The kind designed for "light fountain art" (popular in the early 1970s) is ideal. You do not want to use data communications-grade optical fiber.

Before you buy miles of optical fiber, test a 2-ft strand with a suitable diode laser and phototransistor. Loop the fiber and tape it snugly to your desk or workbench. Connect the phototransistor to a sensitive volt-ohm meter or, better yet, an oscilloscope. Gently touch the fiber loops to deform them. You should observe a definite change of output in the phototransistor. If you do not, examine your setup to rule out a wiring error, and try again. Turn the laser off momentarily and observe the change in output.

Working with Laser Diodes Penlight lasers can be easily hacked for a wide variety of interesting robot projects—the soft-touch fiber-optic whisker is just one of them. Penlight lasers use a semiconductor lasing element. While these elements are fairly hearty, they do require certain handling precautions. And even though they are small, they still emit laser light that can be potentially dangerous to your eyes. So keep the following points in mind:

- Always make sure the terminals of a laser diode are connected properly to the drive circuit.
- Never apply more than the rated voltage to the laser or it will burn up.

- Extend the same care to laser diodes that you do to any static-sensitive device. Wear an anti-static wrist strap while handling the bare laser element, and keep the device in a protective, anti-static bag until it's ready for use.
- Use only a grounded soldering pencil when attaching wires to the laser diode terminals.
- Limit soldering duration to less than 5 s per terminal.
- Never connect the probes of a volt-ohm meter across the terminals of a laser diode. The current from the internal battery of the meter may damage the laser.
- Use only batteries or well-filtered AC power supplies. Laser diodes are susceptible to voltage transients and can be ruined when powered by poorly filtered line-operated supplies.
- Take care not to short the terminals of the laser during operation.
- Avoid looking into the window of the laser while it is operating, even if you can't see any light coming out (is the diode the infrared type?).
- Unless otherwise specified by the manufacturer, clean the output window of the laser diode with a cotton swab dipped in technical-grade isopropyl alcohol (less than 20% water). Alternatively, you can use optics-grade lens cleaning fluid.
- If you are using a laser from a laser penlight, bear in mind that the penlight casing acts as a heat sink. If you remove the laser from the penlight casing, be sure to attach the laser to a suitable heat sink to avoid possible damage. If you keep the laser in the casing, there is usually no need to add the heat sink—the casing should be enough.

30.4.2 PIEZO DISC TOUCH BAR

The laser-optic whisker system described earlier is a great way to detect even your robot's minor collisions. But it may be overkill in some instances, providing too much sensitivity for a zippy little robot always on the go. The soft-touch collision sensor described in this section, which uses commonly available piezo ceramic discs, is a good alternative to the laser-optic whisker system for lower-sensitivity applications. This sensor is constructed with a half-round bar to increase the area of contact.

Construction of the Piezo Disc Touch Bar The main sensing elements of the piezo disc touch bar are two 1-in-diameter bare piezo ceramic discs. These discs are available at Radio Shack and many surplus electronics stores; they typically cost under $1 or $2 each, and you can often find them for even less.

You attach the discs to a 6½-in long support bar, which you can make out of plastic, even a long LEGO Technic beam. As shown in Fig. 30-31, you glue the discs into place with ⅛-in foam (available at most arts and craft stores) so it sticks to the ceramic surface of the disc and acts as a cushion. You then bend a length of ⅛-in-diameter aluminum tubing (approximately 8 to 9 in) into a half-circle; thread through two small grommets, as shown in Fig. 30-32; and glue the grommets to the support bar. You flatten the ends of the tubing and bend them at right angles to create a foot; the foot rests on the foam-padded surface of the discs. The half-round tubing slopes downward slightly on the prototype. This is intentional, so the robot can adequately sense objects directly in front of it near the ground.

To construct the piezo disc touch bar prototype, hot-melt glue was used to attach the discs and grommets to the support bar. You can use most any other adhesive or glue you

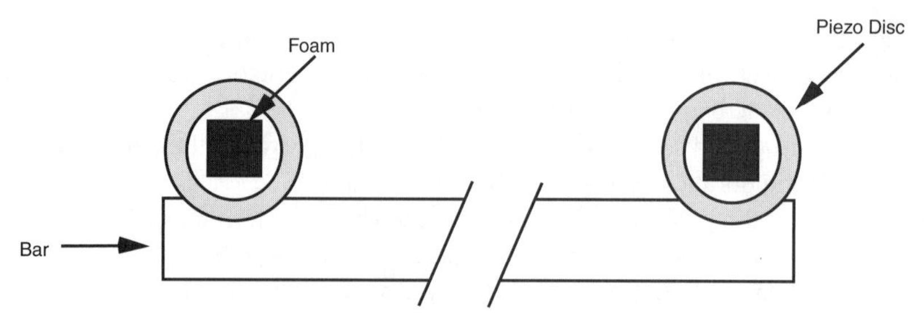

FIGURE 30-31 Glue the piezo discs to a piece of plastic; the plastic is a support bar for the discs that also makes it easier to mount the touch bar sensor onto your robot.

wish, but be sure it provides a good, strong hold for the different materials used in this project (metal, plastic, and rubber).

Constructing the Interface Circuit Piezo discs are curious creatures: when a voltage is applied to them, the crystalline ceramic on the surface of the disc vibrates. It is the nature of piezoelectricity to be both a consumer and a producer of electricity. When the disc is connected to an input, any physical tap or pressure on the disc will produce a voltage. The exact voltage is approximately proportional to the amount of force exerted on the disc:

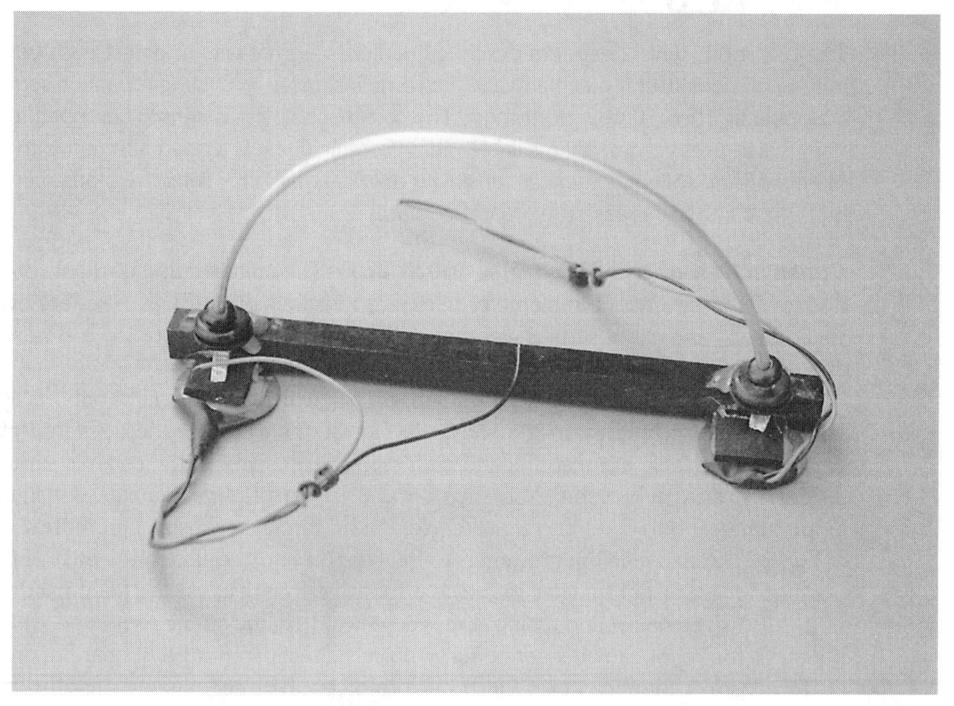

FIGURE 30-32 The finished prototype of the piezo disc touch bar. One variation is to mount the discs a little lower so the metal bar physically deforms the disc rather than pushes against its center.

apply a little pressure or tap and you get a little voltage. Apply a heavier pressure or tap, and you get a bigger voltage.

The piezoelectric material on ceramic piezo discs is so efficient that even a moderately strong force on the disc will produce in excess of 5 or 10 V. That's good in that it makes it easy to interface the discs to a circuit, since there is usually no need to amplify the signal. But it's also bad in that the voltage from the disc can easily exceed the maximum inputs of the computer, microcontroller, or other electronic device you're interfacing with. (Pound on a piezo disc with a hammer, and, though it might be broken when you're done, it will also produce a 1000 V or more.)

To prevent damage to your support electronics, attach two 5.1-V zener diodes as shown in Fig. 30-33, to each disc of the touch bar. The zener diodes limit the output of the disc to 5.1 V, a safe enough level for most interface circuitry. For an extra measure of safety, use 4.7-V zeners instead of 5.1 V.

Note that piezoelectric discs also make great capacitors. This means that over time the disc will take a charge, and the charge will show up as a constantly changing voltage at the output of the disc. To prevent this, insert a resistor across the output of the disc and ground. In the prototype circuits, an 82k resistor eliminated the charge buildup without excessively diminishing the sensitivity of the disc. Experiment with the value of the resistor. A higher value will increase sensitivity, but it could cause an excessive charge buildup. A lower value will reduce the buildup but also reduce the sensitivity of the disc. It is also helpful to route the output of the disc to an op amp, preferably through a 100K or higher resistor.

Mounting the Touch Bar Once you have constructed the piezo disc touch bar and added the voltage-limiting circuitry, you can attach it to the body of the robot. The front of the robot is the likely choice, but you can add additional bars to the sides and rear to obtain a near 360 degree sensing pattern. The width of the bar makes it ideal for any robot that's between about 8 and 14 in wide.

Since the sensing element of the touch bar, the aluminum tube, has a half-round shape, the sensor is also suitable for mounting on a circular robot base. For added compliancy, you may wish to mount the bar using a thick foam pad, spring, or shock absorber (shocks made for model racing cars work well). If the bar is mounting directly to the robot the sensor exhibits relatively little compliancy.

You should mount the bar at a height that is consistent with the kinds of objects the robot is most likely to collide with. For a wall-hugging robot, for example, you may wish to mount the bar low and ensure that the half-round tube slopes downward. That way, the sensor is more likely to strike the baseboard at the bottom of the wall.

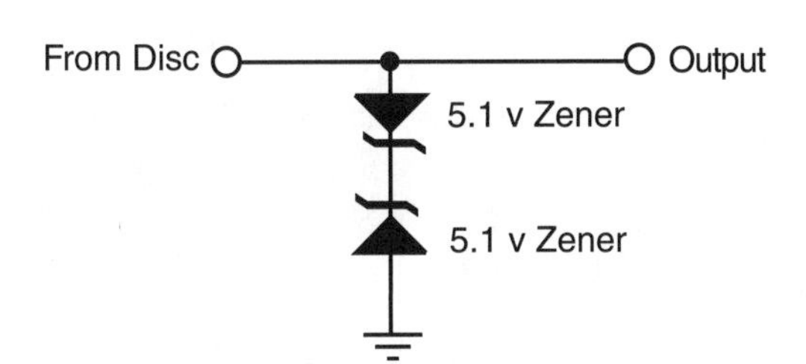

FIGURE 30-33 A suitable interface circuit for connecting a piezo disc to a TTL-compatible input or op amp.

Software for Sensing a Collision The following code is a short sample program for reading the values provided by the piezo disc touch bar. The program is written for the BASIC Stamp 2 microcontroller and requires the addition of one or more serial-output analog-to-digital converter chips (an ADC0831 was used for the prototype). You need only one ADC if it has multiple inputs; you'll need two ADCs if the chips have but a single input. See the comments in the program for hookup information.

```
' For the BASIC Stamp 2
' Uses an ADC081 serial ADC
Adress var byte          ' A-to-D result: one byte
CS con 13                ' Chip select is pin 13
Adata con 14             ' ADC data output is pin 14
CLK con 15               ' Clock is pin 15
Vref con 0               ' VRef

    high Vref
    high CS              ' Deselect ADC to start

    DO
        low CS                                          ' Activate ADC
        shiftin AData, CLK, msbpost, [ADres\9]         ' Shift in the data
        high CS                                         ' Deactivate ADC
        debug ? Adres                                   ' Display result
        pause 100                                       ' Wait 1/10 second
      LOOP               ' Repeat
```

30.4.3 OTHER APPROACHES FOR SOFT-TOUCH SENSORS

There are several other approaches for using soft-touch sensors. For example, the resistive bend sensor changes its resistance the more it is curved or bent. Positioned in the front of your robot in a loop, the bend sensor could be used to detect the deflection caused by running into an object.

If you like the idea of piezoelectric elements but want a more localized touch sensor than the touch bar described in the previous section, you might try mounting piezoelectric material and discs on rubber or felt pads, or even to the bubbles of bubble pack shipping material, to create fingers for your 'bot.

30.5 From Here

To learn more about . . .	*Read*
Connecting analog and digital sensors to computers, microcontrollers, and other circuitry	Chapter 14, "Computer Peripherals"
Building and using sensors for tactile feedback	Chapter 29, "The Sense of Touch"
Giving your robots the gift of sight	Chapter 32, "Robot Vision"
Using sensors to provide navigation assistance to mobile robots	Chapter 33, "Navigation"

CHAPTER 31

SOUND INPUT AND OUTPUT

The robots of science fiction are seldom mute or deaf. They may speak pithy warnings—the most famous probably being: "Danger, Danger, Will Robinson"—or squeak out blips and beeps in some "advanced" language only other robots can understand. Voice and sound input and output make a robot more humanlike, or at least more entertaining. What is a personal robot for if not to entertain?

What's good for robots in novels and in the movies is good enough for us, so this chapter presents a number of useful projects for giving your mechanical creations the ability to make and hear noise. The projects include using recorded sound, generating warning sirens, recognizing and responding to your voice commands, and listening for sound events. Admittedly, this chapter only scratches the surface of what's possible today, especially with technologies like MP3 compressed digitized sound.

31.1 Cassette Recorder Sound Output

Before electronic doodads took over robotics there were mechanical solutions for just about everything. While they may not always have been as small as an electrical circuit, they were often easier to use. Case in point: you can use an ordinary cassette tape and playback mechanism to produce music, voice, or sound effects. Tape players and tape player mechanisms are common finds in the surplus market, and you can often find complete (and still working) portable cassette players-recorders at thrift stores. With just a few

FIGURE 31-1 A surplus cassette deck transport. This model is entirely solenoid driven and so is perfect for robotics.

wires you can rig a cassette tape player in the robot and have the sound played back, on your command.

When looking for a cassette player try to find the kind shown in Fig. 31-1, which is solenoid controlled. These are handy for your robot designs because instead of pressing mechanical buttons, you can actuate solenoids by remote or computer control to play, fast-forward, or rewind the tape. They can be somewhat difficult to find as they are usually built into more expensive soft-touch tape decks and are not readily found in surplus bins.

For most cassette decks you only need to provide power to operate the motor(s) and solenoids (if any) and a connection from the playback head to an amplifier. Since you are not using the deck for recording, you don't have to worry about the erase head or biasing

TABLE 31-1 Parts List for Cassette Tape Head Player Amplifier

IC1	LT1007 low-noise operational amplifier (Linear Technology)
R1	330K resistor
R2	4.9K resistor
R3	100-Ω resistor
C1	0.1 μF ceramic capacitor

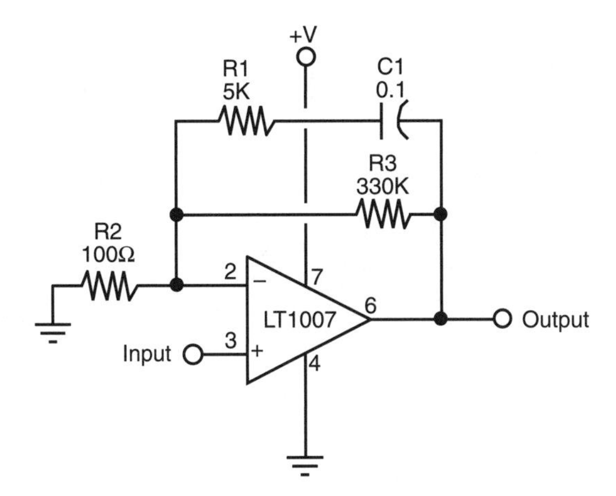

FIGURE 31-2 Preamplifier circuit for use with a magnetic tape playback head.

the record head. If the deck already has a small preamplifier for the playback head, use it. It'll improve the sound quality. If not, you can use the tape head preamplifier shown in Fig. 31-2 (you can use a less expensive op amp than the one specified in the parts list in Table 31-1, but noise can be a problem). Place the preamplifier board as close to the cassette deck as possible to minimize stray pickup.

31.2 Electronically Recorded Sound Output

While mechanical sound playback systems like the cassette recorder are adequate, they lack the response and flexibility of a truly electronic approach. Fortunately, all-electronic reproduction of sound is fairly simple and inexpensive, in large part because of the wide availability of custom-integrated circuits that are designed to record, store, and play back recorded sound. Most of these chips are made for commercial products such as microwave ovens, cellular phones, or car alarms.

31.2.1 HACKING A TOY SOUND RECORDER

You can hack toy sound recorders, such as the Yak Bak, for use in your robot. These units, which can often be found at toy stores for under $10, contain a digital sound-recording chip, microphone, amplifier, and speaker (and sometimes sound effects generator). To use them, you press the Record button and speak into the microphone. Then, stop recording and press the play button and the sound will play back until you make a new recording.

Fig. 31-3 shows a Yak Bak toy that was disassembled and hacked by soldering wires directly to the circuit board. The wires, which connect to a microcontroller or computer, are in lieu of pressing buttons on the toy to record and play back sounds. The buttons on most of these sound recorder toys are made of conductive rubber and are easily removed. To operate the unit via a microcontroller or computer, you bring the button inputs HIGH or LOW. (Which value you choose depends on the design of the circuit; you need to experi-

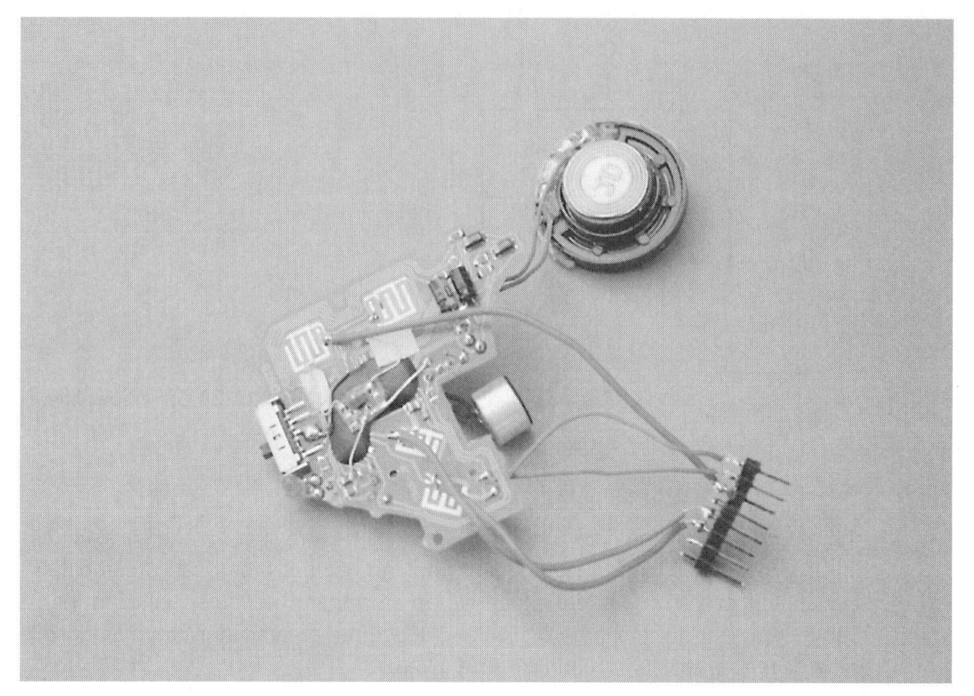

FIGURE 31-3 A hacked Yak Bak can be used to store and play short sound snips. You can record sounds for later playback, which can be via computer control. This model has two extra buttons for sound effects, which are also connected to the robot's microcontroller or computer.

ment to find out which to use.) Connect a 1K to 3K resistor between the I/O pin and the button input.

Suppose you have a Yak Bak or similar toy connected to I/O pin 1 on a Basic Stamp 2. Assume that on the toy you are using, bringing the button input HIGH triggers a previously recorded sound snip. The control program is as simple as this:

```
high 1
pause 10
low 1
```

The program starts by bringing the button input (the input of the toy connected to pin 1 of the BASIC Stamp) HIGH. The pause statement waits 10 ms and then places the button LOW again.

The built-in amplifier of these sound recorder/playback toys isn't very powerful. You may wish to connect the output of the toy to one of the audio output amplifiers described later in the chapter (see "Audio Amplifiers").

31.2.2 USING THE ISD FAMILY OF VOICE-SOUND RECORDERS

Toy sound recorders are limited to playing only a single sample. For truly creative robot yapping, you need a sound chip in which you can control the playback of any of several prerecorded snips. You can do this easily by using the family of sound storage and playback chips produced by Information Storage Devices (ISD). The company has made these ChipCorder ICs readily available to the electronics hobbyist and amateur robot builder.

You can purchase ISD sound recorder chips from a variety of sources, including Jameco and Digikey (see Appendix B). Prices for these chips vary depending on feature and recording time, but most cost under $10. While there are certainly other makers of sound storage/playback integrated circuits, the ISD chips are by far the most widely used and among the most affordable.

Adding a BS2 to a robot is a fairly simple operation, although it can be difficult to work through using just the data sheets. Using the circuit in Fig. 31-4 (with the parts listed in Table 31-2), you can create your own BS2 to an ISD chip interface circuit using BS2 ICD.bs2 (in the following). When wiring the circuit, make sure you use a reasonably large breadboard and that you leave lots of space between the chips and the end of the breadboard for the various resistors and capacitors.

```
'   BS2 ISD2532 - Controlling the Operation of an IS2532
'
'   myke predko
'
'   05.08.27
'
'   Pin 8 - ISD2532 Chip Enable
'   Pin 9 - ISD2532 Play/Record
'
'{$STAMP BS2}
'{$PBASIC 2.5}
```

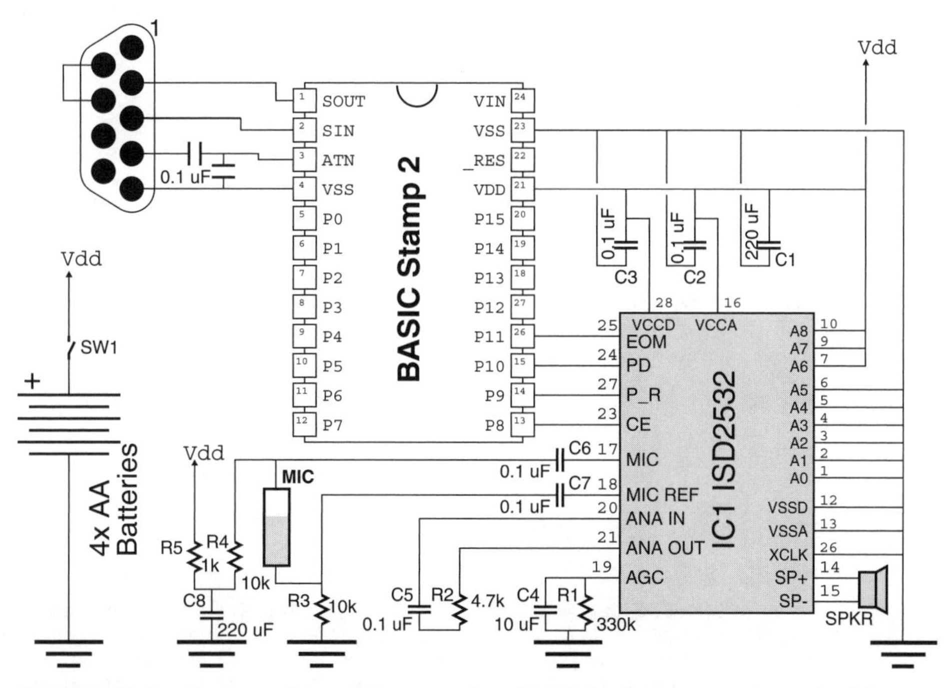

FIGURE 31-4 Circuit to allow a BS2 to control an ISD2532 solid-state sound recorder/player. While this circuit's operation is controlled from a PC console, by changing the control software, it can be easily integrated into a robot.

TABLE 31-2 Parts List for BS2 Control of an ISD Sound Recorder/Player Chip

BS2	Parallax BASIC Stamp 2
IC1	ISD2532 sound recorder/player
R1	330k resistor
R2	4.7k resistor
R3, R4	10k resistor
R5	1k resistor
C1	220 μF electrolytic capacitor
C2, C3	0.1 μF electrolytic capacitor
C4	10 μF electrolytic capacitor
C5–C7	0.1 μF ceramic capacitor
C8	220 μF electrolytic capacitor
SW1	Breadboard mountable SPDT switch
MIC	Electret microphone
SPKR	16 Ω speaker
Misc.	Large breadboard (6 in or longer), breadboard wiring, 4× AA battery clip, 4× AA alkaline batteries, BS2 serial port programmer interface

```
'  Variable Declarations
InputString  VAR Byte(2)
Temp         VAR Byte
LastCharFlag VAR Bit
i            VAR Byte
CE           PIN 8
P_R          PIN 9
PD           PIN 10
EOM          PIN 11

'  Initialization
    HIGH CE                                 '  No Operation
    LOW P_R                                 '  Start with Recording
    HIGH PD                                 '  Hold the Chip High
'  Main Loop
    DO                                      '  Loop Forever

        DEBUG "Command (R/P)> "

        i = 0
        InputString(i) = 0                  '  Start with Null string
        LastCharFlag = 0                    '  Want to execute at least once
        DO WHILE (LastCharFlag = 0)         '  Wait for Carriage Return
            DEBUGIN STR Temp\1              '  Wait for a Char to be input
            IF (Temp = CR) THEN
                LastCharFlag = 1            '  String Ended
```

```
            ELSE
                IF (Temp = BKSP) THEN
                    IF (i = 0) THEN
                        DEBUG " "           '  Keep Screen at Constant Point
                    ELSE
                        DEBUG " ", BKSP '  Backup one space
                        i = i - 1           '  Move the String Back
                        InputString(i) = 0
                    ENDIF
                ELSE
                    IF (i >= 1) THEN
                        DEBUG BKSP, 11    '  At end of Line or NON-Char
                    ELSE                    '  Can Store the Character
                        IF ((Temp >= "a") AND (Temp <= "z")) THEN
                            Temp = Temp - "a" + "A"
                        ENDIF
                        DEBUG BKSP, STR Temp\1
                        InputString(i) = Temp
                        i = i + 1
                        InputString(i) = 0     '  Put in new String End
                    ENDIF
                ENDIF
            ENDIF
        LOOP

        IF (InputString(0) = "R") THEN
            LOW P_R                         '  Specify Record
            LOW PD                          '  Enable the ISD Chip
            PULSOUT CE, 1                   '  Start the Recording Process
            DEBUG "Recording. Press Any Key to Stop..."
            DEBUGIN STR Temp\1
            PULSOUT CE, 1                   '  Start the Recording Process
            IF (Temp <> CR) THEN DEBUG CR
            HIGH PD                         '  Finished
        ELSE
            IF (InputString(0) = "P") THEN
                HIGH P_R                    '  Specify Play
                LOW PD
                PULSOUT CE, 1               '  Start the Recording Process
                DEBUG "Playing..."
                DO WHILE (EOM = 1): LOOP
                HIGH PD                     '  Finished
                DEBUG CR
            ELSE
                DEBUG "Invalid Command", CR
            ENDIF
        ENDIF

    LOOP
```

This program uses the console interface presented earlier in the book for the BS2 I/O port read and write accessing to demonstrate how this function can be used for different applications. In a robot application, just the code that is used to Specify Record and Specify Play is required.

This application runs the ISD chip in push-button control mode, which allows multiple messages to be stored on the ISD chip, using address pins A0 through A5. This will give you up to 32 (short) messages that can be used to show off the capabilities of the robot or give you feedback as to what its input is or what functions it is executing.

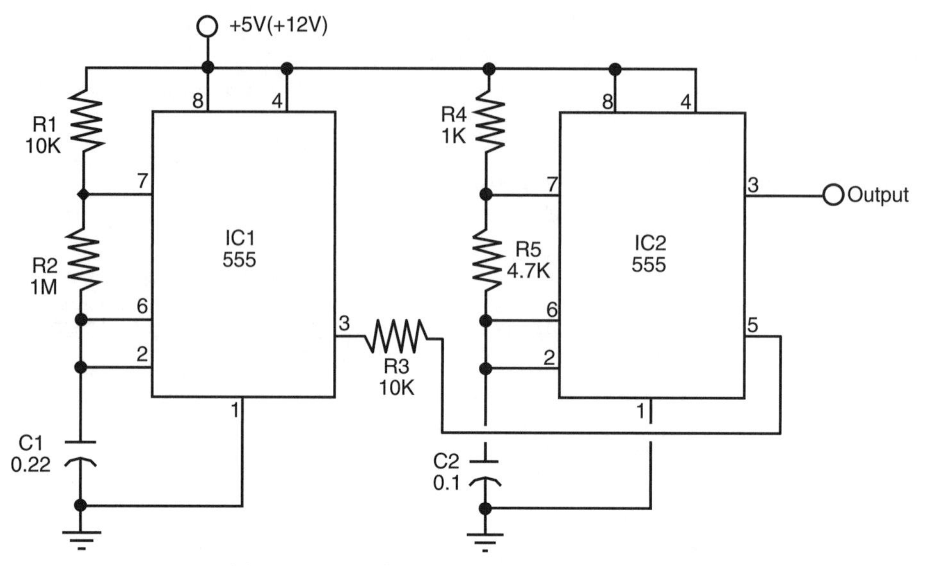

FIGURE 31-5 A warbler siren made from two 555 timer ICs.

31.3 Sirens and Other Warning Sounds

If you use your robot as a security device or to detect intruders, fire, water, or whatever, then you probably want the machine to warn you of immediate or impending danger. The warbling siren shown in Fig. 31-5 will do the trick, assuming it's connected to a strong enough amplifier (refer to the parts list in Table 31-3). The circuit is constructed using two 555 timer chips (alternatively, you can combine the functions into the 556 dual timer chip). To change the speed and pitch of the siren, alter the values or R1 and R4, respectively.

For maximum effectiveness, connect the output of the IC2 to a high-powered amplifier. You can get audio amplifiers with wattages of 8, 16, and more in easy-to-build kit

TABLE 31-3 Parts List for Siren

IC1, IC2	555 timer IC
R1	10K resistor
R2	1 MΩ
R3	10K resistor
R4	1K resistor
R5	4.7K resistor
C1	0.22 μF ceramic capacitor
C2	0.1 μF ceramic capacitor

form. See Appendix B, "Sources," for a list of mail-order companies that also sell electronic kits.

31.4 Sound Control

Unless you have all of the sound-making circuits in your robot hooked up to separate amplifiers and speakers (not a good idea), you'll need a way to select between the sounds. The circuit in Fig. 31-6 uses a 4051 CMOS analog switch and lets you choose from among eight different analog signal sources. You select input by providing a three-bit binary word to the select lines. You can load the selection via computer or set it manually with a switch. A binary-coded-decimal (BCD) thumbwheel switch is a good choice, or you can use a four-bank DIP switch. Table 31-4 is the truth table for selecting any of the eight inputs.

You can route just about any of your sound projects through this chip, just as long as the outlet level doesn't exceed a few milliwatts. Do not pass amplified sound through the chip. Besides in all likelihood destroying the chip, it'll cause excessive cross-talk between the channels. It's also important that each input signal not have a voltage swing that exceeds the supply voltage to the 4051.

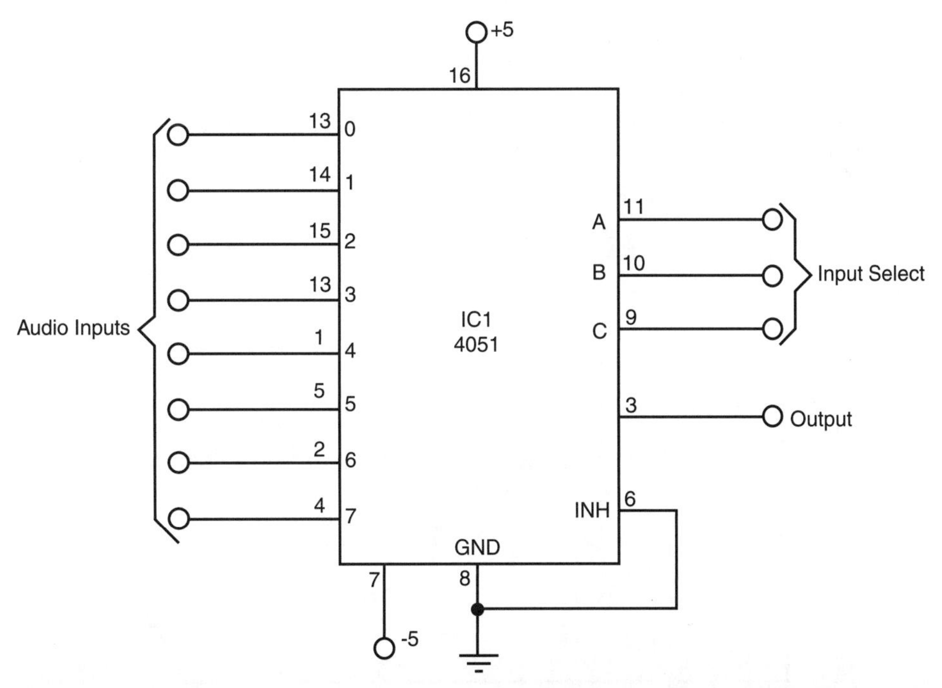

FIGURE 31-6 How to use the 4051 CMOS eight-input analog switch to control the output of the various sound-making circuits in your robot. You can choose the sound source you want routed to the output amplifier by selecting its input with the Input Select lines (they are binary weighted: A = 1, V = 2, C = 4). For best results, the audio inputs should not already be amplified.

TABLE 31-4 4051 Truth Table

C	B	A	Selected Output	Input Pin
0	0	0	0	13
0	0	1	1	14
0	1	0	2	15
0	1	1	3	12
1	0	0	4	1
1	0	1	5	5
1	1	0	6	2
1	1	1	7	4

31.5 Audio Amplifiers

Fig. 31-7 (parts list in Table 31-5) shows a rather straightforward 0.5-W sound amplifier that uses the LM386 integrated amplifier. The sound output is minimal, but the chip is easy to get, cheap, and can be wired up quickly. It's perfect for experimenting with sound projects. The amplifier as shown has a gain of approximately 20, using minimal parts. You can increase the gain to about 200 by making a few wiring changes, as shown in Fig. 31-8 (parts list in Table 31-6). Either amplifier will drive a small (2- or 3-in) 8-Ω speaker.

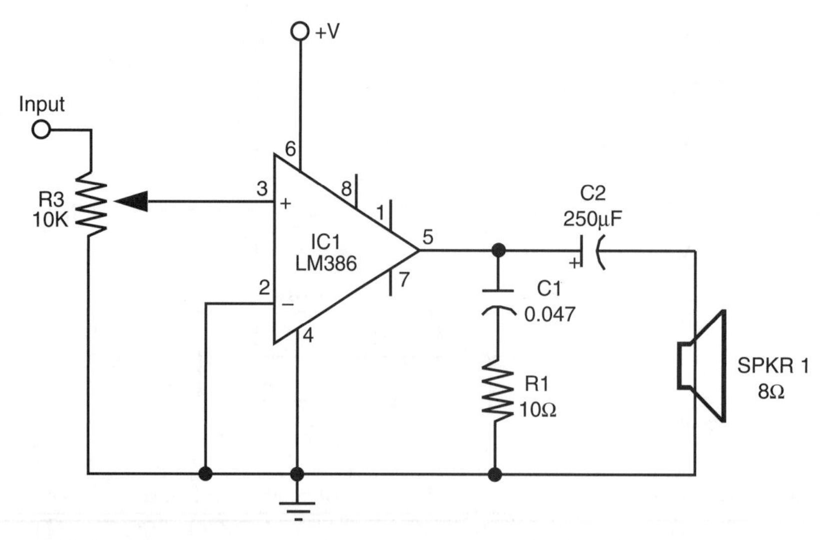

FIGURE 31-7 A simple gain-of-50 integrated amplifier, based on the popular LM386 audio amplifier IC.

TABLE 31-5 Parts List for GAIN-200 Audio Amplifier

IC1	LM386 audio amplifier IC
R1	10 Ω resistor
R2	10K potentiometer
C1	0.047 μF ceramic capacitor
C2	250 μF electrolytic capacitor
SPKR1	8-Ω miniature speaker

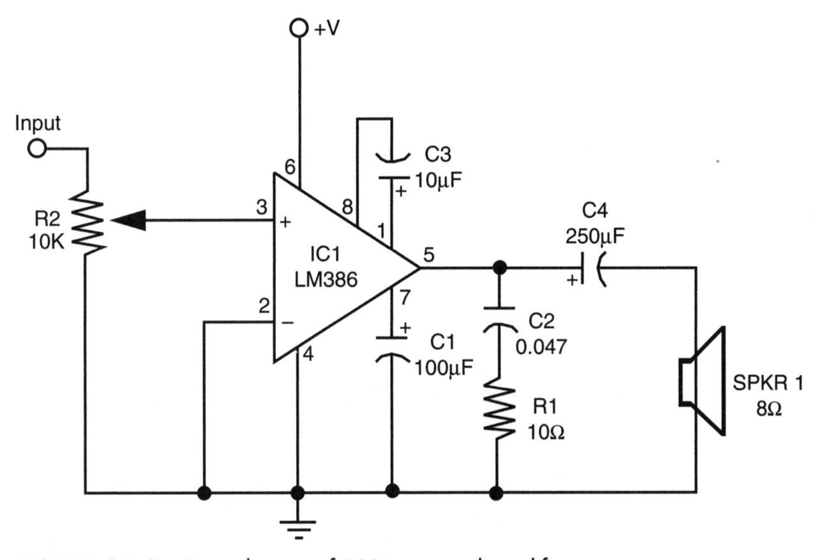

FIGURE 31-8 A simple gain-of-200 integrated amplifier.

TABLE 31-6 Parts List for GAIN-of-200 Audio Amplifier

IC1	LM386 audio amplifier IC
R1	10 Ω resistor
R2	10K potentiometer
C1	100 μF electrolytic capacitor
C2	0.047 μF ceramic capacitor
C3	10 μF electrolytic capacitor
C4	250 μF electrolytic capacitor
SPKR1	8-Ω miniature speaker

31.6 Speech Recognition

Robots that listen to your voice commands and obey? The technology is not only available, it's relatively inexpensive. Several companies, such as Sensory Inc. and Images Company, offer full-featured speech recognition systems for under $100. Both require you to train the system to recognize your voice patterns. Once trained, you simply repeat the command, and the system sets one or more of its outputs accordingly.

The Voice Direct, from Sensory Inc., is relatively easy to set up and use. The unit consists of a small double-sided circuit board that is ready to be connected to a microphone, speaker (for auditory confirmation), battery, and either relays or a microcontroller. The Voice Direct board recognizes up to 15 words or phrases and is said to have a 99 percent or better recognition accuracy. Phrases of up to 3.2 s can be stored, so you can tell your robot to "come here" or "stop, don't do that!" Fig. 31-9 shows a Voice Direct module; the product comes with complete circuit and connection diagrams.

Keep the following in mind when using a voice recognition system:

- You must be reasonably close to the microphone for the system to accurately understand your commands. The better the quality of the microphone you have, the better the accuracy of the speech recognition.

FIGURE 31-9 The Voice Direct voice recognition system from Sensory Inc.

- If you are using a voice recognition system on a mobile robot, you may wish to extend the microphone away from the robot so motor noise is reduced. For best results, you'll need to be fairly close to the robot and speak directly and clearly into the microphone.
- Consider using a good-quality RF or infrared wireless microphone for your voice recognition system. The receiver of the wireless microphone is attached to your robot; you hold the microphone itself in your hand.

31.7 Speech Synthesis

Not long ago, integrated circuits for the reproduction of human-sounding speech were fairly common. Several companies, including National Semiconductor, Votrax, Texas Instruments, and General Instrument, offered ICs that were not only fairly easy to use, even on the hobbyist level, but surprisingly inexpensive. Voice-driven products using these chips included the Speak-and-Spell toys and voice synthesizers for the blind. In most cases, these ICs could create unlimited speech because they reproduced the fundamental sounds of speech (called phonemes). With the proliferation of digitized recorded speech (using chips like the ISD chip presented previously), unlimited speech synthesizers have become an exception instead of the rule. The companies that made stand-alone speech synthesizer chips either stopped manufacturing them or were themselves sold to other firms that no longer carry the old speech parts.

In addition, products such as the sound card for the IBM PC-compatible computers obviated the need for a separate, stand-alone speech circuit. Using only software and a sound card, it is possible to reproduce a male or female voice. In fact, Microsoft provides free speech-making tools and operating system APIs for Windows. If you are using a robot controlled by a laptop running Windows, adding synthesized speech is remarkably easy. See the Microsoft site (www.microsoft.com) for more information.

31.8 Sound Input Sensors

Next to sight, the most important human sense is hearing. And compared to sight, sound detection is far easier to implement in robots. You can build simple "ears" in less than an hour that let your robot listen to the world around it.

Sound detection allows your robot creation to respond to your commands, whether they take the form of a series of tones, an ultrasonic whistle, or a hand clap. It can also listen for the tell-tale sounds of intruders or search out the sounds in the room to look for and follow its master. The remainder of this chapter presents several ways to detect sound. Once detected, the sound can trigger a motor to motivate, a light to go lit, a buzzer to buzz, or a computer to compute.

31.8.1 MICROPHONE

Obviously, your robot needs a microphone (or mike) to pick up the sounds around it. The most sensitive type of microphone is the electret condenser, which is used in most higher quality hi-fi mikes. The trouble with electret condenser elements, unlike crystal element mike, is that they need electricity to operate. Supplying electricity to the microphone element really isn't a problem, however, because the voltage level is low—under 4 or 5 V.

Most electret condenser microphone elements come with a built-in field effect transistor (FET) amplifier stage. As a result, the sound is amplified before it is passed on to the main amplifier. Electret condenser elements are available from a number of sources, including Radio Shack, for under $3 or $4. You should buy the best one you can. A cheap microphone isn't sensitive enough.

The placement of the microphone is important. You should mount the mike element at a location on the robot where vibration from motors is minimal. Otherwise, the robot will do nothing but listen to itself. Depending on the application, such as listening for intruders, you might never be able to place the microphone far enough away from sound sources or make your robot quiet enough. You'll have to program the machine to stop and then listen.

31.8.2 AMPLIFIER INPUT STAGE

The circuit in Fig. 31-10 is a basic amplifier for an electret condenser microphone (refer to parts list in Table 31-7). The circuit is designed around the common LM741 op amp, which is wired to operate from a single-ended power supply. Potentiometer R1 lets you adjust the gain of the op amp, and hence the sensitivity of the circuit to sound. After experimenting

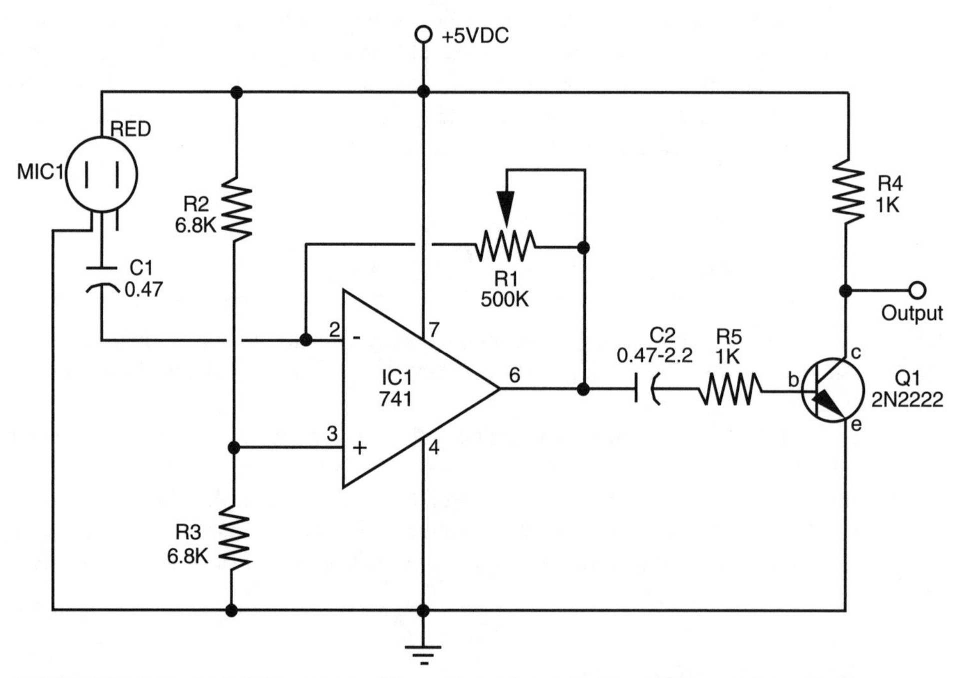

FIGURE 31-10 Sound detector amplifier. Adjust R1 to increase or decrease the sensitivity, or replace the potentiometer with the circuit that appears in Fig. 31-10.

TABLE 31-7 Parts List for Sound Detector

IC1	LM741 op amp IC
Q1	2N2222 transistor
R1	500K potentiometer
R2,R3	6.8K resistor
R4,R5	1K resistor
C1,C2	0.47 μF ceramic capacitor
MIC1	Electret condenser microphone

with the circuit and adjusting R1 for best sensitivity, you can substitute the potentiometer for a fixed-value resistor. Remove R1 from the circuit and check its resistance with a volt-ohm meter. Use the closest standard value of resistor.

By adding the optional circuit in Fig. 31-11, you can choose up to four gain levels via computer control. The resistors, R1 and R2 (you decide on their value based on the gain you wish), are connected to the inverting input of the op amp and the inputs of a 4066 CMOS 1-of-4 analog switch. Select the resistor value by placing a HIGH bit on the switch you want to activate. The manufacturer's specification sheets for this chip recommend that only one switch be closed at a time.

31.8.3 TONE DECODING DETECTION

The 741 op amp is sensitive to sound frequencies in a very wide band and can pick up everything that the microphone has to send it. You may wish to listen for sounds that occur only in a specific frequency range. You can easily add a 567 tone decoder IC to the amplifier input stage to look for these specific sounds.

The 567 is almost like a 555 in reverse. You select a resistor and capacitor to establish an operating frequency, called the center frequency. Additional components are used to establish a bandwidth—or the percentage variance that the decoder will accept as a desired frequency (the variance can be as high as 14 percent). Fig. 31-12 shows how to connect a 567 to listen to and trigger on about a 1 kHz tone (refer to the parts list in Table 31-8).

Before you get too excited about the 567 tone decoder, you should know about a few minor faults. The 567 has a tendency to trigger on harmonics of the desired frequency. You can limit this effect, if you need to, by adjusting the sensitivity of the input amplifier and decreasing the bandwidth of the chip.

Another minor problem is that the 567 requires at least eight wave fronts of the desired sound frequency before it triggers on it. This reduces false alarms, but it also makes detection of very low frequency sounds impractical. Though the 567 has a lower threshold of about 1 hz, it is impractical for most uses at frequencies that low.

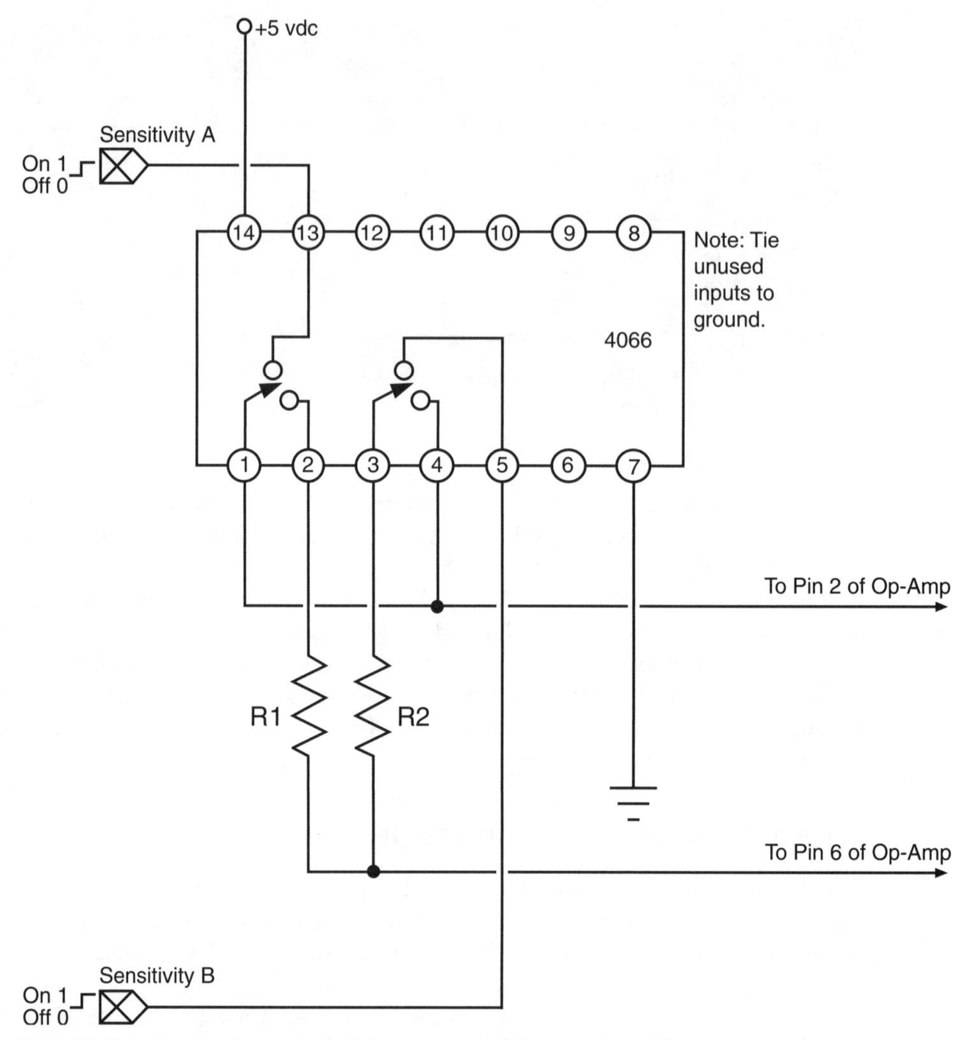

FIGURE 31-11 Remote control of the sensitivity of the sound amplifier circuit. Under computer control, the robot can select the best sensitivity for a given task.

Being an older part, the 567 has been "obsoleted" by several manufacturers that used to make it. For the time being, however, you can still purchase 567 chips from most new and surplus retail and mail-order electronics companies.

Another option is to use a telephone Touch-Tone (known as DTMF) decoder. There are several chips available, such as the MT8880 (available from Parallax) that can decode the dual tones of a Touch-Tone phone and output a digital value. An obvious advantage of using a DTMF decoder over the 567 is the greater number of outputs, allowing you to implement multiple controls for your robot. The remote control sound source for this chip could be an old telephone handset that is powered by external batteries and the tones are amplified, or a microcontroller, like the BS2, which has a DTMFOUT command.

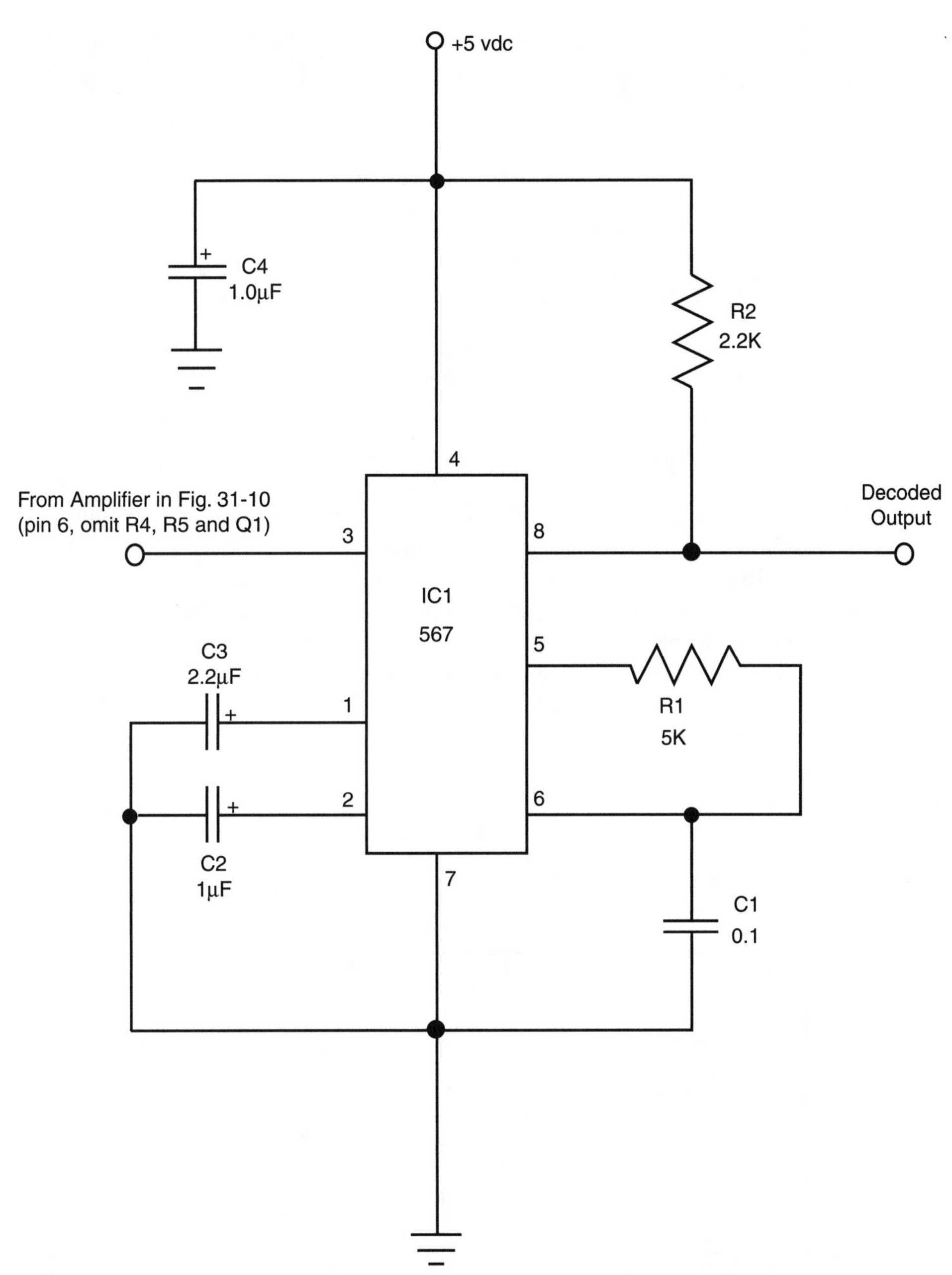

FIGURE 31-12 A 567 tone decoder IC, wired to detect tones at about 1 kHz.

TABLE 31-8 Parts List for Tone Decoder

IC1	567 tone decoder IC
R1	50K 3- to 15-turn precision potentiometer
R2	2.2K resistor
C1	0.1 μF ceramic capacitor
C2	2.2 μF tantalum or electrolytic capacitor
C3, C4	1.0 μF tantalum or electrolytic capacitor

31.8.4 BUILDING A SOUND SOURCE

With the 567 decoder, you'll be able to control your robot using specific tones. With a tone generator, you'll be able to make those tones so you can signal your robot via simple sounds. Such a tone-generator sound source is shown in Fig. 31-13 (the parts list is Table 31-9). The values shown in the circuit generate sounds in the 48-kHz to 144-Hz range. To extend the range higher or lower, substitute a higher or lower value for C1. Basic design formulas and tables for the 555 are provided in the appendices.

For frequencies between about 5 and 15 kHz, use a piezoelectric element as the sound source. Use a miniature speaker for frequencies under 5 kHz and an ultrasonic transducer for frequencies over 30 kHz. Cram all the components in a small box, stick a battery inside,

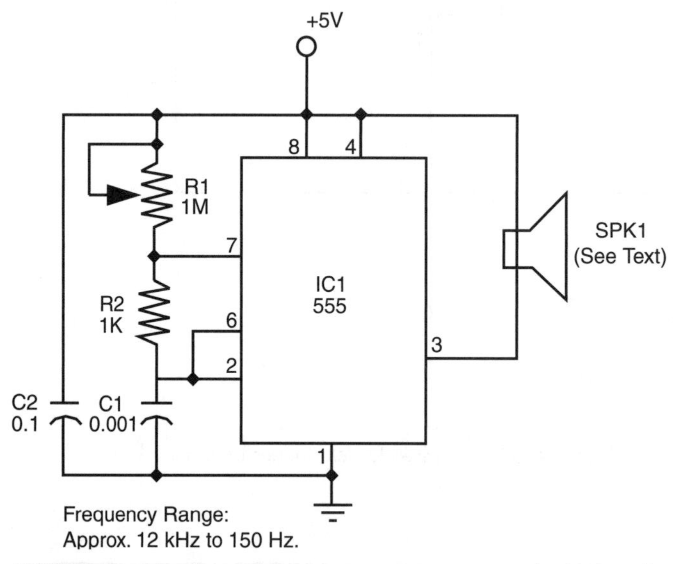

FIGURE 31-13 A variable frequency tone generator, built around the common 555 timer IC. The tone output spans the range of human hearing and beyond.

TABLE 31-9 Parts List for Tone Generator

IC1	555 timer IC
R1	1 MΩ potentiometer
R2	1K resistor
C1	0.001 μF ceramic capacitor
C2	0.1 μF ceramic capacitor
SPKR1	4 or 8 Ω miniature speaker

and push the button to emit the tone. Be aware that the sound level from the speaker and especially the piezoelectric element can be quite high. Do not operate the tone generator close to your ears or anyone else's ears except your robot's.

Do not limit yourself to only using a 555 control. As discussed previously, a microcontroller can output specific frequencies very accurately. Some microcontrollers, such as the BS2, can be used to output telephone DTMF tones, which will give you up to 16 different commands for your robot—along with the basic 12 that are on your telephone's keypad, there are four additional tone combinations available.

31.9 From Here

To learn more about . . .	***Read***
Computer and microcontroller options for robotics	Chapter 12, "An Overview of Robot 'Brains' "
Interfacing sound inputs/outputs to a computer	Chapter 14, "Computer Peripherals"
Sensors to prevent your robot from bumping into things	Chapter 30, "Object Detection"
Eyes to go along with the ears of your robot	Chapter 32, "Robot Vision"

CHAPTER 32

ROBOT VISION

Robotic vision systems can be simple or complex to match your specific requirements and your itch to tinker. Rudimentary Cyclops vision systems are used to detect nothing more than the presence or absence of light. Aside from this rather mundane task, there are plenty of useful applications for an on/off light detector. More advanced vision systems decode relative intensities of light and can even make out patterns and crude shapes.

While the hardware for making robot eyes is rather simple, using the vision information they generate is not. Except for the one-cell light detector, vision systems must be interfaced to a computer to be useful. You can adapt the designs presented in this chapter to just about any computer using a microprocessor data bus or one or more parallel printer ports.

32.1 Simple Sensors for Vision

A number of simple electronic devices can be used as eyes for your robot. These include the following:

- *Photoresistors.* These are typically a cadmium-sulfide (CdS) cell (often referred to simply as a photocell). A CdS cell acts like a light-dependent resistor (also referred to as an LDR): the resistance of the cell varies depending on the intensity of the light striking it. When no light strikes the cell, the device exhibits very high resistance, typically in the high 100 kilohms, or even megohms. Light reduces the resistance, usually significantly (a few hundreds or thousands of ohms). CdS cells are very easy to interface to other electronics, but they react somewhat slowly and are unable to discern when the light level changes more than 20 or 30 times per second. This trait actually comes in handy because it means CdS cells basically ignore the on/off flashes of AC-operated lights.

- *Phototransistors.* These are very much like regular transistors with their metal or plastic tops removed. A glass or plastic cover protects the delicate transistor substrate inside. Unlike CdS cells, phototransistors act very quickly and are able to sense tens of thousands of changes in light level per second. The output of a phototransistor is not linear; that is, there is a disproportionate change in the output of a phototransistor as more and more light strikes it. A phototransistor can become easily swamped with too much light. In this condition, even as more light shines on the device, the phototransistor is not able to detect any change.
- *Photodiodes.* These are the simpler diode versions of phototransistors. Like phototransistors, they are made with a glass or plastic cover to protect the semiconductor material inside them. And like phototransistors, photodiodes act very fast and can become swamped when exposed to a certain threshold of light. One common characteristic of most photodiodes is that their output is rather low, even when fully exposed to bright light. This means that to be effective the output of the photodiode must usually be connected to a small amplifier.

Photoresistors, photodiodes, and phototransistors are connected to other electronics in about the same way: you place a resistor between the device and either +V or ground. The point between the device and the resistor is the output, as shown in Fig. 32-1. With this arrangement, all three devices therefore output a varying voltage. The exact arrangement of the connection determines if the voltage output increases or decreases when more light strikes the sensor.

Light-sensitive devices differ in their spectral response, which is the span of the visible and near-infrared light region of the electromagnetic spectrum that they are most sensitive to. CdS cells exhibit a spectral response very close to that of the human eye, with the greatest degree of sensitivity in the green or yellow-green region (see Fig. 32-2). Both phototransistors and photodiodes have peak spectral responses in the infrared and near-infrared regions. In addition, some phototransistors and photodiodes incorporate optical filtration to decrease their sensitivity to the visible light spectrum. This filtration makes the sensors more sensitive to infrared and near-infrared light.

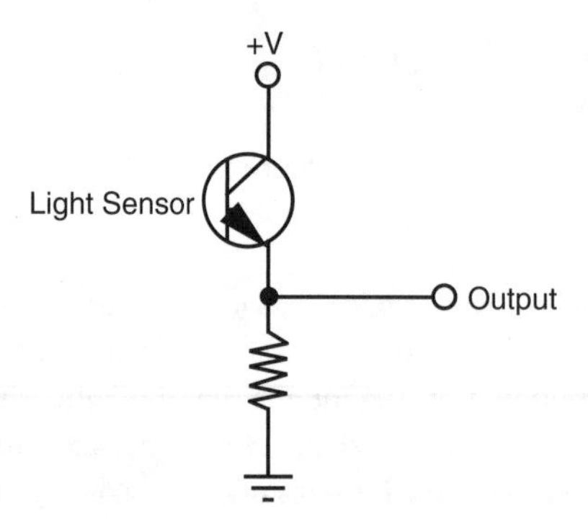

FIGURE 32-1 The basic connection scheme for phototransistors, photodiodes, and photoresistors uses a discrete resistor to form a voltage divider. The output is a varying voltage, which can go from 0 to +V depending on the sensor.

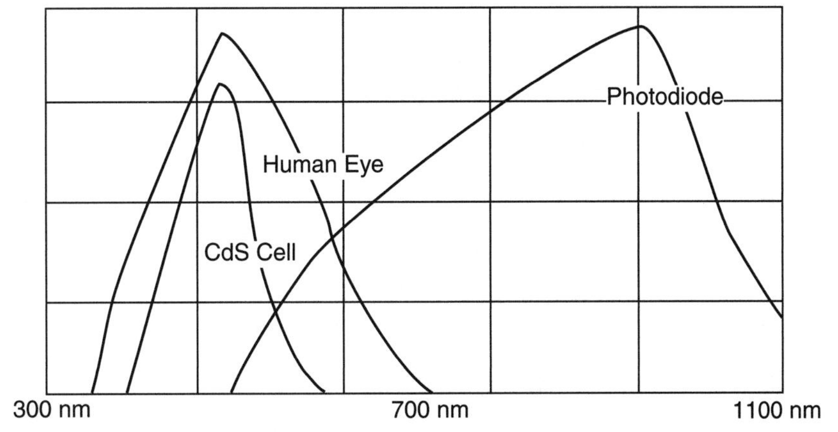

FIGURE 32-2 Light sensors vary in their sensitivity to different colors of the electromagnetic spectrum. The color sensitivity of CdS cells is very similar to that of the human eye.

32.2 One-Cell Cyclops

A single light-sensitive photocell is all your robot needs to sense the presence of light. The photocell is a variable resistor that works much like a potentiometer but has no control shaft. You vary its resistance by increasing or decreasing the light. Connect the photocell as shown in Fig. 32-3. Note that, as explained in the previous section, a resistor is placed in series with the photocell and that the output tap is between the cell and resistor. This converts the output of the photocell from resistance to voltage, the latter of which is easier to use in a practical circuit. The value of the resistor is given at 3.3K Ω but is open to experimentation. You can vary the sensitivity of the cell by substituting a higher or lower value. For experimental purposes, connect a 1K resistor in series with a 50K pot (in place of the 3.3K Ω resistor) and try using the cell at various settings of the wiper. Test the cell output by connecting a volt-ohm meter to the ground and output terminals.

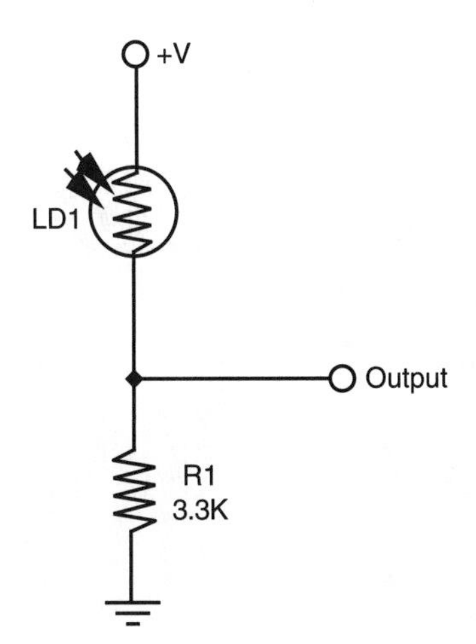

FIGURE 32-3 A one-cell robotic eye, using a CdS photocell as a light sensor.

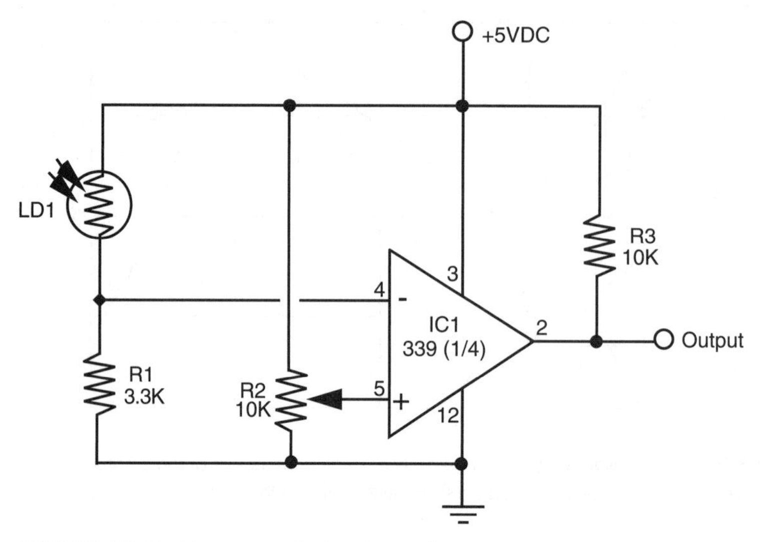

FIGURE 32-4 How to couple the photocell to a comparator.

So far, you have a nice light-to-voltage sensor, and when you think about it there are numerous ways to interface this ultrasimple circuit to a robot. One way is to connect the output of the sensor to the input of a comparator. (The LM339 quad comparator IC is a good choice, but you can use just about any comparator.) The output of the comparator changes state when the voltage at its input goes beyond or below a certain *trip point*. In the circuit shown in Fig. 32-4 (refer to the parts list in Table 32-1), the comparator is hooked up so the noninverting input serves as a voltage reference. Adjust the potentiometer to set the trip point (or *voltage threshold*) by first setting it at the midway point and then adjusting the trip point higher or lower as required. The output of the photocell voltage divider circuit is connected to the inverting input of the comparator, which will change. When the photocell voltage divider voltage passes through the threshold voltage, then output of the comparator changes state.

TABLE 32-1 Parts List for Single-Cell Robotic Eye

IC1	LM339 quad comparator IC
R1	3.3K resistor
R2	10K potentiometer
R3	10K resistor
LD1	Photocell

One practical application of this circuit is to detect light levels that are higher than the ambient light in the room. Doing so enables your robot to ignore the background light level and respond only to the higher intensity light. To begin, set the trip point potentiometer so the circuit just switches HIGH. Use a flashlight to focus a beam directly onto the photocell, and watch the output of the comparator change state. Another application is to use the photocell as a light detector. Set the potentiometer to one extreme so the comparator changes state just after light is applied to the surface of the cell.

32.3 Multiple-Cell Light Sensors

The human eye has millions of tiny light receptacles. Combined, these receptacles allow us to discern shapes, to actually see rather than just detect light levels. A crude but surprisingly useful approximation of human sight is given in Fig. 32-5 (refer to the parts list in Table 32-2). Here, eight photocells are connected to a 16-channel multiplexed analog-to-digital converter (ADC). The ADC, which has enough pins for another eight cells, converts the analog voltages from the outputs of each photocell and one by one converts them into digital data. The eight-bit binary number presented at the output of the ADC represents any of 256 different light levels.

The converter is hooked up in such a way that the outputs of the photocells are converted sequentially, in a row and column pattern, following the suggested mounting scheme shown in Figs. 32-6 and 32-7. A computer hooked up to the A/D converter records the digital value of each cell and creates an image matrix, which can be used to discern crude shapes.

Each photocell is connected in series with a resistor, as with the one-cell eye presented earlier. Initially, use 2.2K resistors, but feel free to substitute higher or lower values to increase or decrease sensitivity. The photocells should be identical, and for the best results, they should be brand-new prime components. Before placing the cells in the circuit, test each one with a volt-ohm meter and a carefully controlled light source. Check the resistance of the photocell in complete darkness, then again with a light shining at it a specific distance away. Reject cells that do not fall within a 5 to 10 percent "pass" range. See Chapter 14, "Computer Peripherals," for more information on using ADCs and connecting them to computer ports and microprocessors.

Note the short pulse that appears at pin 13 of the ADC; the end-of-conversion (EOC) output. This pin signals that the data at the output lines are valid. If you are using a computer or microcontroller, you can connect this pin to an interrupt line (if available). Using an interrupt line lets your computer do other things while it waits for the ADC to signal the end of a conversion.

You can get by without using the EOC pin—the circuit is easier to implement without it—but you must set up a timing delay circuit or routine to do so. The delay routine is probably the easiest route; simply wait long enough for the conversion to take place (a maximum of about 115 µs), then read the data. Even with a delay of 125 µs (to allow for settling, etc.), it takes no more than about 200 ms to read the entire matrix of cells.

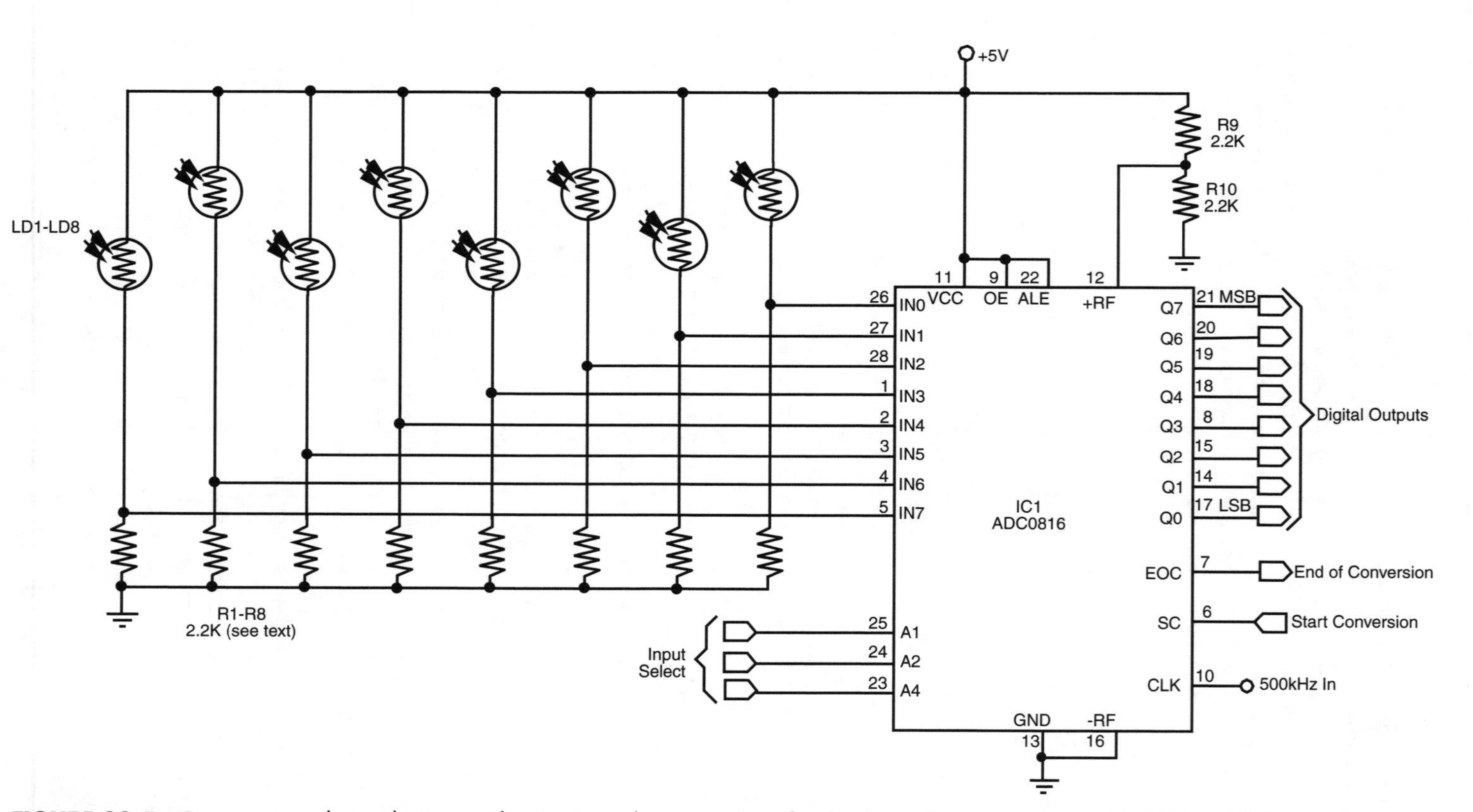

FIGURE 32-5 One way to make a robotic eye. The circuit, as shown, consists of eight photocells connected to an ADC0816 eight-bit, 16-input analog-to-digital converter IC. The output of each photocell is converted when selected at the Input Select lines. The ADC0816 can handle up to 16 inputs, so you can add another eight cells.

TABLE 32-2 Parts List for Multicell Robotic Eye

IC1	ADC0816 eight-bit analog-to-digital converter IC
R1–R8	2.2K resistor (adjust value to gain best response of photocells)
R9, R10	2.2K resistors
LD1–LD8	Photocell

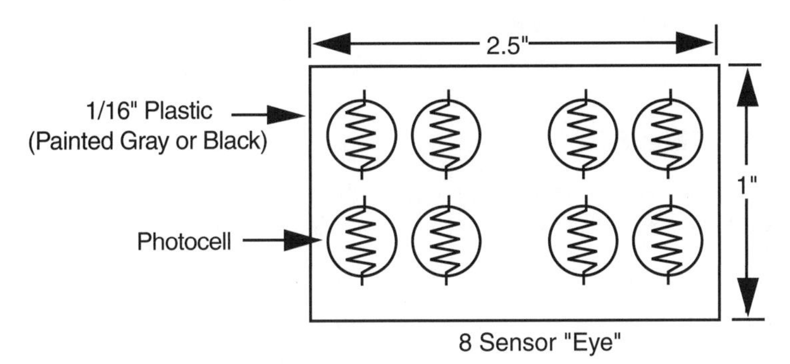

FIGURE 32-6 Mounting the photocells for an eight-cell eye.

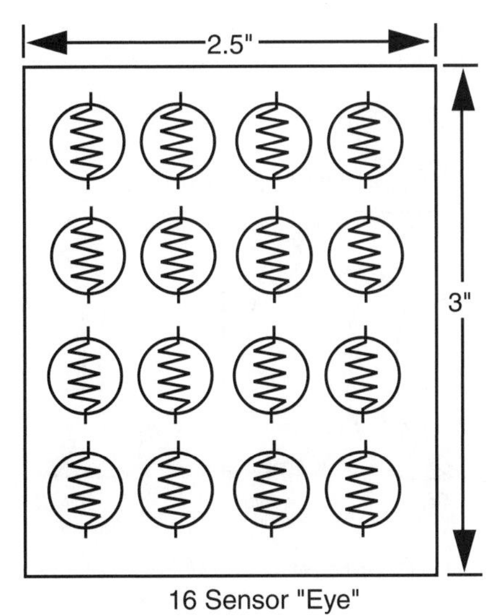

FIGURE 32-7 Mounting an array of four by four photocells.

32.4 Using Lenses and Filters with Light-Sensitive Sensors

Simple lenses and filters can be used to greatly enhance the sensitivity, directionality, and effectiveness of both single- and multicell-vision systems. By placing a lens over a small cluster of light cells, for example, you can concentrate room light to make the cells more sensitive to the movement of humans and other animate objects. The lens need not be complex; an ordinary ½- to 1-in-diameter magnifying lens, purchased new or surplus, is all you need.

You can also use optical filters to enhance the operation of light cells. Optical filters work by allowing only certain wavelengths of light to pass through and blocking the others. CdS photocells tend to be sensitive to a wide range of visible and infrared light. You can readily accentuate the sensitivity of a certain color (and thereby de-accentuate other colors) just by putting a colored gel or other filter over the photocell.

32.4.1 USING LENSES

Lenses are refractive media constructed so that light bends in a particular way. The two most important factors in selecting a lens for a given application are lens focal length and lens diameter:

- *Lens focal length.* Simply stated, the focal length of a lens is the distance between the lens and the spot where rays are brought to a common point. (Actually, this is true of positive lenses only; negative lenses behave in an almost opposite way, as discussed later.)
- *Lens diameter.* The diameter of the lens determines its light-gathering capability. The larger the lens is, the more light it collects.

There are six major types of lenses, shown in Fig. 32-8. Such combinations as plano-convex and bi-concave refer to each side of the lens. A plano-convex lens is flat on one side

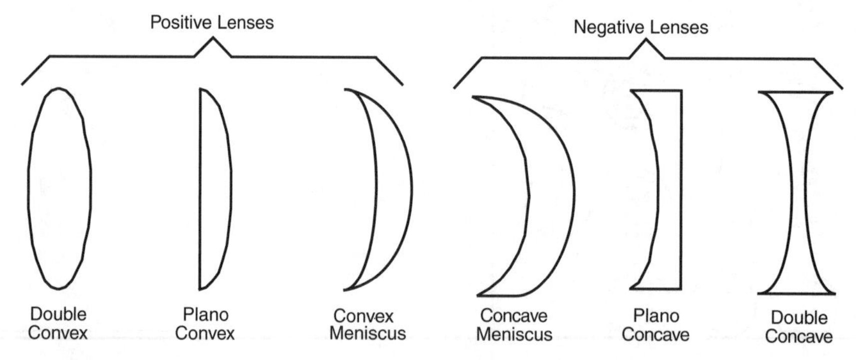

FIGURE 32-8 Lenses come in a variety of forms. Plano-convex and double-convex are among the most common.

and curved outward on the other. A bi-concave lens curves inward on both sides. *Negative* and *positive* refer to the focal point of the lens, as determined by its design.

Lenses form two kinds of images: real and virtual. A real image is one that is focused to a point in front of the lens, such as the image of the sun focused to a small disc on a piece of paper. A virtual image is one that doesn't come to a discrete focus. You see a virtual image behind the lens, as when you are using a lens as a magnifying glass. Positive lenses, which magnify the size of an object, create both real and virtual images. Their focal length is stated as a positive number, such as +1 or +2.5. Negative lenses, which reduce the size of an object, create only virtual images. Their focal length is stated as a negative number.

Lenses are common finds in surplus stores, and you may not have precise control over what you get. For robotics vision applications, plano-convex or double-convex lenses of about 0.5 to 1.25 in diameter are ideal. The focal length should be fairly short, about 1 to 3 in. When you are buying an assortment of lenses, the diameter and focal length of each lens is usually provided, but if not, use a tape to measure the diameter of the lens and its focal length (see Fig. 32-9). Use any point source except the sun—focusing the light of the sun onto a small point can cause a fire!

To use the lens, position it over the light cell(s) using any convenient mounting technique. One approach is to glue the lens to a plastic or wood lens board. Or, if the lens is the correct diameter, you can mount it inside a short length of plastic PVC pipe; attach the other end of the pipe to the light cells. Be sure you block out stray light. You can use black construction paper to create light baffles. This will make the robot see only the light shining through the lens. If desired, attach a filter over the light cells. You can use a dab of glue to secure the filter in place.

Using Fig. 32-6 as a guide, you can create a kind of two-eyed robot by placing a lens over each group of four photocells. The lenses are mounted in front of the photocells,

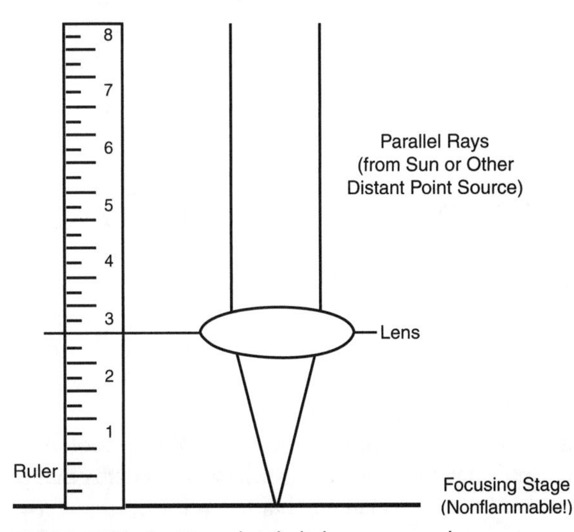

FIGURE 32-9 Use a bright light source, such as an incandescent lamp, and a tape to measure the focal point of a lens.

which are secured to a circuit board in two groups of four. The smaller the photocell is, the easier it will be to mount with the lens (although soldering them may be more difficult). Each of the eight cells is connected to a separate input of an eight-input analog-to-digital converter (ADC) chip. By using an eight-input ADC, the values of all eight cells can be readily sensed without the need for separate ADC chips and extra wiring.

32.4.2 USING FILTERS

Filters accept light at certain wavelengths and block all others. A common filter used in robot design is intended to pass infrared radiation and block visible light. Such filters are commonly used in front of phototransistors and photodiodes to block out unwanted ambient (room) light. Only infrared light—from a laser diode, for instance—is allowed to pass through and strike the sensor. Optical filters come in three general forms: colored gel, interference, and dichroic.

- Colored gel filters are made by mixing dyes into a Mylar or plastic base. Good gel filters use dyes that are precisely controlled during manufacture to make filters that pass only certain colors. Depending on the dye used, the filter is capable of passing only a certain band of wavelengths. A good gel filter may have a bandpass region (the spectrum of light passed) of 40 to 60 nanometers (nm), which equates to nearly one full color of the basic six-color rainbow.
- Interference filters consist of several dielectric and sometimes metallic layers that each block a certain range of wavelengths. One layer may block light under 500 nm, and another layer may block light above 550 nm. The band between 500 and 550 nm is passed by the filter. Interference filters can be made to pass only a very small range of wavelengths.
- Dichroic filters use organic dyes or chemicals to absorb light at certain wavelengths. Some filters are made from crystals that exhibit two or more different colors when viewed at different axes. Color control is maintained by cutting the crystal at a specific axis.

32.5 Introduction to Video Vision Systems

Single- and multicell-vision systems are useful for detecting the absence or presence of light, but they cannot make out the shapes of objects. This greatly limits the environment into which such a robot can be placed. By detecting the shape of an object, a robot might be able to make intelligent assumptions about its surroundings and perhaps be able to navigate those surroundings, recognize its "master," and more.

A few years ago, video vision was an expensive proposition for any robot experimenter. But the advent of inexpensive and small pinhole and lipstick video cameras that can output video data in a variety of different formats makes the hardware for machine vision affordable (while still not a trivial exercise).

A video system for robot vision need not be overly sophisticated. The resolution of the image can be as low as about 100 by 100 pixels (10,000 pixels total), though a resolution of no less than 300 by 200 pixels (60,000 pixels total) is preferred. The higher the resolution is, the better the image and therefore the greater the robot's ability to discern shapes. A color camera is not mandatory and, in some cases, makes it harder to write suitable video interpolating software.

Video systems that provide a digital output are generally easier to work with than those that provide only an analog video output. You can connect digital video systems directly to a PC, such as through a serial, parallel, or USB port. Analog video systems require that a video capture card, a fast analog-to-digital converter, or some other similar device be attached to the robot's computer.

While the hardware for video vision is now affordable to most any robot builder, the job of translating a visual image a robot can use requires high-speed processing and complicated computer programming. Giving robots the ability to recognize shapes has proved to be a difficult task. Consider the static image of a doorway. Our brains easily comprehend the image, adapting to the angle at which we are viewing the doorway; the amount, direction, and contrast of the light falling on it; the size and kind of frame used in the doorway; whether the door is open or closed; and hundreds or even thousands of other variations. Robot vision requires that each of these variables be analyzed, a job that requires computer power and programming complexity beyond the means of most robot experimenters.

32.5.1 ROBOT VIEW DIGITAL CAMERA

If a live full-motion video system seems too much for you, a small digital camera (like the one shown in Fig. 32-10) can be added to a robot very easily. This camera was bought

FIGURE 32-10 A small, inexpensive digital camera can be mounted to and controlled by your robot very easily.

for under $20 at a local Radio Shack and can store 32 pictures at its highest resolution (640 by 480 pixels, or VGA resolution), and the pictures can be downloaded to a PC using a USB port. Depending on the camera and robot, you may have to modify the camera to take pictures under computer control, but this should not be a significant amount of work.

The camera mount on the robot can be a bit tricky. You will have to make sure that the camera lens has an unobstructed view before it, the battery compartment must be accessible as well as the USB interface. You will also have to provide access to the controls and shutter release on the camera. The camera's buttons can either be electronically controlled (the interface follows) or physically pressed using a radio control servo controlled by the robot. In either case, the camera mounting will have to be designed around all these interfaces and apertures.

The prototype camera shown in Fig. 32-10 had a control button to place the camera in a specific mode (depending on the number of times it was pressed). This button did not seem to work reliably enough for an electronic control or a servo actuator—it seemed to take a different number of presses to get the camera into a specific mode (as indicated by an LCD display on the front of the camera). The decision was made to set the camera to a specific mode before sending the robot out on a photographic "mission." Along with this simplification of the control interface, the decision was made to mount the camera to the front of a robot using two-sided tape. There did not seem to be an easy solution to "hard" mounting the camera and allowing access to the various controls and parts of the camera except by mounting the camera by its back to the robot.

The camera's shutter release was controlled electronically using the circuit shown in Fig. 32-11 (parts list in Table 32-3). To access the switch connections, a rotary tool was used to cut away the plastic case on the back side, and the transistor's collector and emitter were soldered across the switch's terminals. To find the negative connection of the switch to wire

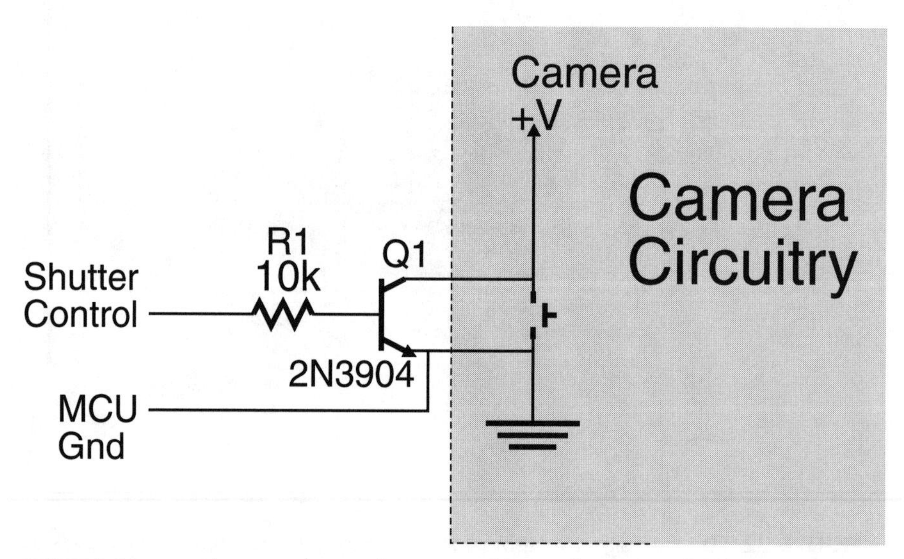

FIGURE 32-11 Electronic shutter release control for a robot-mounted digital camera.

TABLE 32-3 Parts List for Robot Digital Camera

Q1	2N3904 NPN transistor
R1	10k resistor
Misc.	Wiring, soldering iron, solder

the transistor's collector to, use a DMM to find the button lead that is connected directly to the negative battery connection.

When a high voltage is applied to the Shutter Control line, the NPN transistor will turn on and pull the high side of the switch to ground. The PBASIC instruction to take a picture was the simple statement:

```
PULSOUT Pin, 100
```

which holds down the shutter release for 100 ms or 0.1 s, which was adequate for the camera that was used.

If you do not feel comfortable opening up the camera body and probing to identify the negative button lead, you can use a servo with a long arm on it to press the camera's shut-

FIGURE 32-12 Photograph taken from a robot with the camera shown in Fig. 32-10 mounted on its front. The perspective could be described as a mouse's point of view.

ter release button. A servo could also be used for any operating mode controls, although as previously noted, the camera that was used for the prototype did not seem to have a control button that worked reliably enough for a servo (or electronic control) to engage it a set number of times and to put the camera into the desired operating mode.

What makes adding a camera to the robot very interesting is the "mouse eye view" pictures that you will get from your robot. Chances are you will think you know what the robot will see before it, but in actuality, you will get a number of pictures that you will be hard pressed to identify what are in them, like Fig. 32-12. Admittedly, the digital photograph's quality has been impaired by the low resolution (maximum 640 by 480 pixel) of the camera used in the prototype, but the low perspective makes it very difficult to get a good understanding of what the picture is of actually.

In case you are wondering what Fig. 32-12 is a picture of, consider the more traditional picture in Fig. 32-13. The camera took a picture of the edge of a broom and dustpan and a child's bead toy. This will probably be obvious after looking at Fig. 32-13, but when you are just looking at the raw picture, you will be hard pressed to exactly identify what it is you are seeing. The difficulty you had will give you an appreciation as well as some examples of the scenes that a live video system will have to decode before the robot can move.

FIGURE 32-13 Human perspective on the photograph in Fig. 32-12. The scene is much clearer and more obvious.

32.6 Vision by Laser Light

Here's another low-cost method of experimenting with robot vision that you might want to tackle, and it uses only about $30 worth of parts (minus the video camera). You need a simple penlight laser, a red filter, and a small piece of diffraction grating (available from Edmund Scientific Company and other sources for optical components; see Appendix B, "Sources," for additional information).

The system works on a principle similar to the three-beam focusing scheme used in CD players. In a CD player, laser light is broken into sub-beams by the use of a diffraction grating. A single, strong beam appears in the center, flanked by weaker beams on both sides, as shown in Fig. 32-14. The three-beam CD focusing system uses the two closest side beams, ignoring all the others.

The beam spacing increases as the distance from the lens to the surface of the disc increases. Similarly, the beam spacing decreases as the lens-to-CD distance decreases. A multicelled photodetector in the CD players integrates the light reflected by these beams and determines whether the lens should be moved closer to, or farther away from, the disc. For history buffs, the fundamental basis of this focusing technique is over a hundred years old and was pioneered by French physicist Jean Foucault.

CD players use a diffraction grating in which lines are scribed into a piece of plastic in only one plane. This causes the laser beam to break up into several beams along the same plane. With a diffraction grating that has lines scribed both vertically and horizontally, the laser beam is split up into multiple beams that form a grid when projected on a flat surface (see Figs. 32-15 and 32-16). The beams move closer together as the distance from the laser and surface is decreased; the beams move farther apart as the distance from the laser and surface is increased.

As you can guess, when the beams are projected onto a three-dimensional scene, they form a kind of topographical map in which they appear closer or farther apart depending on the distance of the object from the laser.

The red filter placed in front of the camera lens filters out most of the light except for the red beams from the penlight laser. For best results, use a high-quality optical bandpass filter that accepts only the precise wavelength of the diode laser, typically 635 or 680 nm. Check the specifications of the laser you are using so you can get the correct filter. Meredith

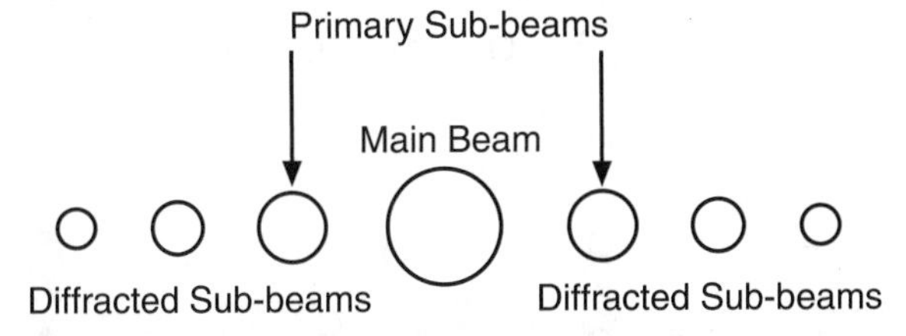

FIGURE 32-14 Most CD players use a diffraction grating to break up the single laser beam into several sub-beams. The sub-beams are used to focus and track the optical system.

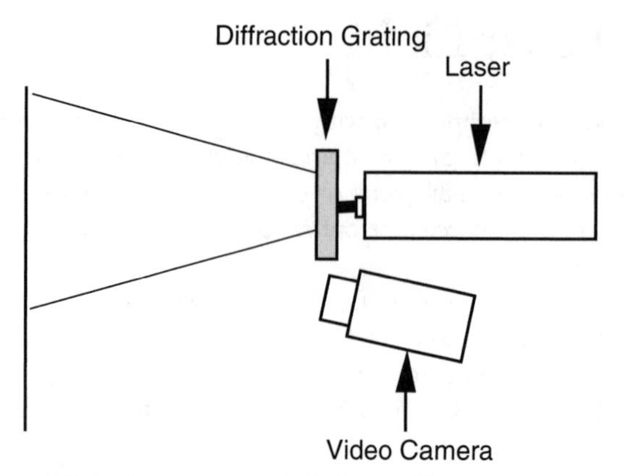

FIGURE 32-15 A penlight laser, diffraction grating, filter, and video camera can be used to create a low-cost machine vision system.

Instruments and Midwest Laser Products, among other sources, provide a variety of penlight lasers and optical filters you can use (see Appendix B).

The main benefit of the laser diffraction system is this: it's easier to write software that measures the distance between pixels than it is to write software that attempts to recognize shapes and patterns. For many machine vision applications, it is not as important for the robot to recognize the actual shape of an object as it is to navigate around or manipulate that shape. As an example, a robot may see a chair in its path, but there is little practical need for it to recognize the chair as an early-eighteenth-century Queen Anne–style two-

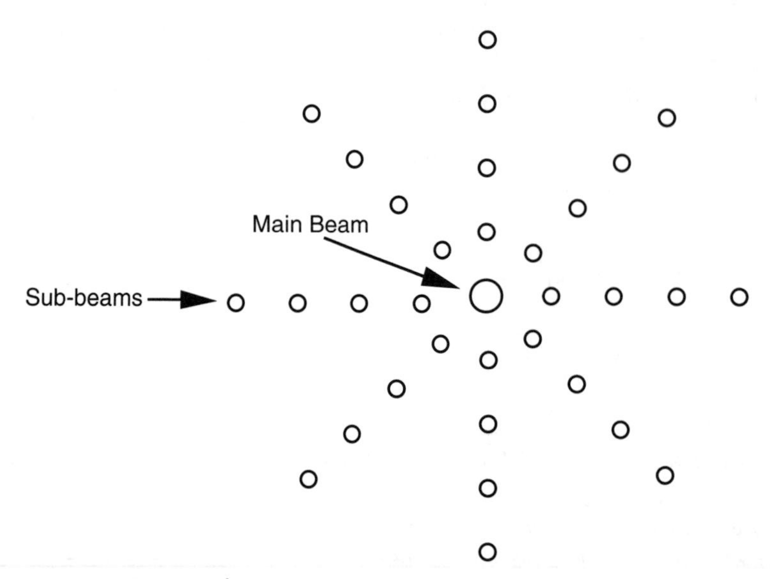

FIGURE 32-16 When projected onto a flat surface, the beams from the diffracted laser light form a regular grid.

seater settee. All it really needs to know is that something is there, and by moving left or right that object can be avoided.

32.7 Going Beyond Light-Sensitive Vision

Sight provides a fast and efficient way for us to determine our surroundings. The eyes take in a wide field, and the brain processes what the eyes see to compose a picture of the immediate environment. Taking a cue from the special senses evolved by some animals, however, visual eyesight is not the only way to see. For instance, bats use high-pitched sound to quickly and efficiently navigate through dark caves. So accurate is their sonar that bats can sense tiny insects flying a dozen or more feet away.

Similarly, robots don't always need light-sensitive vision systems. You may want to consider using an alternative system, either instead of or in addition to light-sensitive vision. The following sections outline some affordable technologies you can easily use.

32.7.1 ULTRASONICS

Like a bat, your robot can use high-frequency sounds to navigate its surroundings. Ultrasonic transducers are common in Polaroid instant cameras, electronic tape-measuring devices, automotive backup alarms, and security systems. All work by sending out a high-frequency burst of sound, then measuring the amount of time it takes to receive the reflected sound.

Ultrasonic systems are designed to determine distance between the transducer and an object in front of it. More accurate versions can map an area to create a type of topographical image, showing the relative distances of several nearby objects along a kind of 3-D plane. Such ultrasonic systems are regularly used in the medical field (e.g., an ultrasound picture of a baby still inside the mother). Some transducers are designed to be used in pairs: one transducer to emit a series of short ultrasonic bursts, another transducer to receive the sound. Other transducers, such as the kind used on Polaroid cameras and electronic tape-measuring devices, combine the transmitter and receiver into one unit. It should be noted that ultrasonics tend to require a great deal of power making them best suited for large robots.

An important aspect of ultrasonic imagery is that high sound frequencies disperse less readily than do low-frequency ones. That is, the sound wave produced by a high-frequency source spreads out much less broadly than the sound wave from a low-frequency source. This phenomenon improves the accuracy of ultrasonic systems. Both DigiKey and All Electronics, among others, have been known to carry new and surplus ultrasonic components suitable for robot experimenters. See Chapter 30 for more information on using ultrasonic sensors to guide your robots.

32.7.2 RADAR

Radar systems work on the same basic principle as ultrasonics, but instead of high-frequency sound they use a high-frequency radio wave. Most people know about the high-powered radar equipment used in aviation, but lower-powered versions are commonly used in security systems, automatic door openers, automotive backup alarms, and of course, speed-measuring devices used by the police.

Radar is less commonly found on robotics systems because it costs more than ultrasonics. This may change in the future with the availability of low-cost and low-power ultra-wideband (UWB) radar. Rather than emitting a radio signal at a single frequency, UWB emits a signal at a wide range of frequencies (from 1 Hz to several GHz). This wide range of frequencies makes UWB radar sensitive to virtually all objects, allowing you to sense effectively all objects in front of the robot, without having to resort to multiple sensors to ensure that nothing is missed. As indicated, UWB radar is not yet available for hobby robotics, but this will change as it becomes available in more and more products and hackable units become available.

32.7.3 PASSIVE INFRARED

A favorite for security systems and automatic outdoor lighting, passive pyroelectric infrared (PIR) sensors detect the natural heat that all objects emit. This heat takes the form of infrared radiation—a form of light that is beyond the limits of human vision. The PIR system merely detects a rapid change in the heat reaching the sensor; such a change usually represents movement.

The typical PIR sensor is equipped with either a Fresnel or motion detection lens. The Fresnel lens focuses infrared light from a fairly wide area onto the pea-sized surface of the detector. In a robotics vision application, you can replace the Fresnel lens with a telephoto lens arrangement that permits the detector to view only a small area at a time. Mounted onto a movable platform, the sensor could detect the instantaneous variations of infrared radiation of whatever objects are in front of the robot. See Chapter 30, "Object Detection," for more information on the use of PIR sensors.

The motion detection lens is the faceted lens that you have probably seen on a burglar alarm object sensor. The facets pass infrared light from different positions to the PIR sensor. When there is a single infrared light source (e.g., an intruder), as they move across the motion detector lens' field of view, the facets that pass the infrared light to the PIR sensor changes. This change results in a change in the amount of infrared light hitting the PIR sensor, allowing it to detect the motion of a hot object in front of it. To compensate for changing temperatures in a room, a new "reference" infrared light level is continually checked against the previously stored value and if the difference is within an acceptable range, the system accepts this as the new reference else it sounds the alarm.

32.7.4 TACTILE FEEDBACK

Many robots can be effective navigators with little more than a switch or two to guide their way. Each switch on the robot is a kind of touch sensor: when a switch is depressed, the robot knows it has touched some object in front of it. Based on this information, the robot can stop and negotiate a different path to its destination.

To be useful, the robot's touch sensors must be mounted where they will come into contact with the objects in their surroundings. For example, you can mount four switches along the bottom periphery of a square-shaped robot so contact with any object will trigger one of the switches. Mechanical switches are triggered only on physical contact; switches that use reflected infrared light or capacitance can be triggered by the proximity of objects. Noncontact switches are useful if the robot might be damaged by running into an object, or vice versa. See Chapter 35, "Adding the Sense of Touch," for more information on tactile sensors.

32.8 From Here

To learn more about . . .	***Read***
Using a computer within your robot	Chapter 12, "An Overview of Robot 'Brains' "
Connecting sensors to a robot computer or microcontroller	Chapter 14, "Computer Peripherals"
Using touch to guide your robot	Chapter 29, "The Sense of Touch"
Getting your robot from point A to point B	Chapter 33, "Navigation"

CHAPTER 33

NAVIGATION

The projects and discussion in this chapter focus on navigating your robot through space—not the outerspace kind, but the space between two chairs in your living room, between your bedroom and the hall bathroom, or outside your home by the pool. Robots suddenly become useful once they can master their surroundings, and being able to wend their way through their surrounds is the first step toward that mastery.

The techniques used to provide such navigation are varied: path-track systems, infrared beacons, ultrasonic rangers, compass bearings, dead reckoning, and more.

33.1 A Game of Goals

A helpful way to look at robot navigation is to think of it as a game, like soccer. The aim of soccer is for the members of one team to kick the ball into a goal. That goal is guarded by a member of the other team, so it's not all that easy to get the ball into the goal. Similarly, for a robot a lot stands between it and its goal of getting from one place to another. Those obstacles include humans, chairs, cats, a puddle of water, an electrical cord—just about anything can prevent a robot from successfully traversing a room or yard.

To go from point A to point B, your robot will consider the following process (as shown in Fig. 33-1):

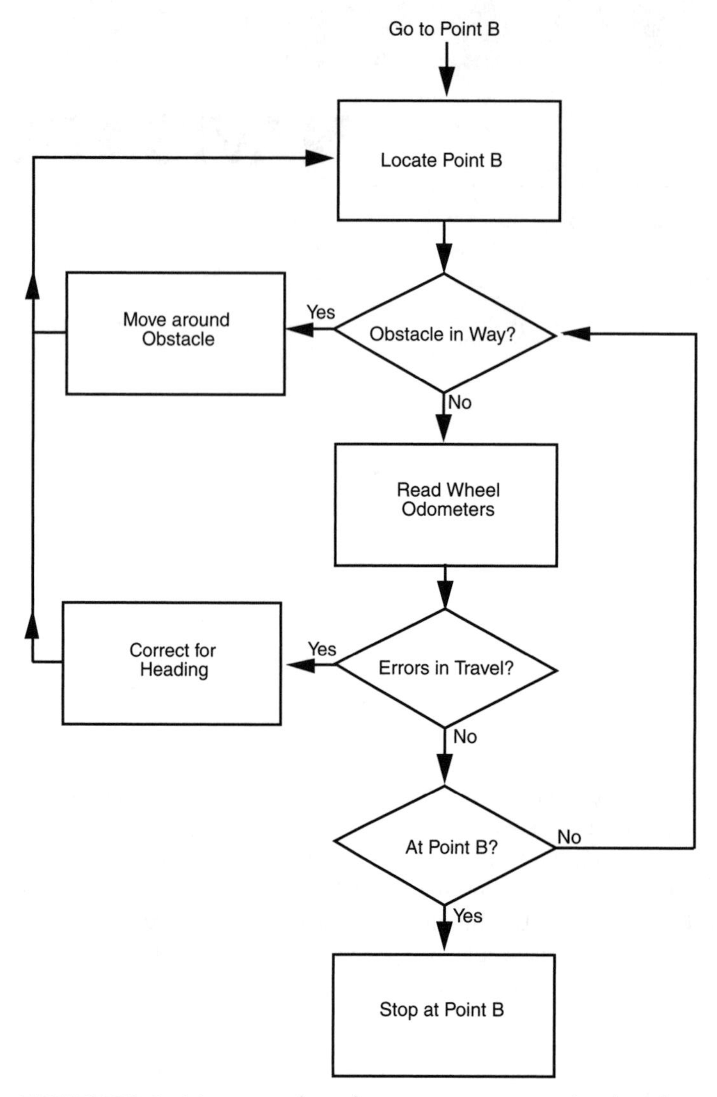

FIGURE 33-1 Navigation through open space requires that the robot be programmed not only to achieve the "goal" of a specific task but to self-correct for possible obstacles.

1. Retrieve instruction of goal: get to point B. This can come from an internal condition (battery is getting low; must get to power recharge station) or from a programmed or external stimulus.
2. Determine where point B is in relation to current position (point A), and determine a path to point B. This requires obtaining the current position using known landmarks or references.

3. Avoid obstacles along the way. If an immovable obstacle is encountered, move around the obstacle and recalculate the path to get to point B.
4. Correct for errors in navigation ("in-path error correction") caused by such things as wheel slippage. This can be accomplished by periodically reassessing current position using known landmarks or references.
5. Optionally, time out (give up) if goal is not reached within a specific period of time or distance traveled.

Notice the intervening issues that can retard or inhibit the robot from reaching its goal. If there are any immovable obstacles in the way, the robot must steer around them. This means its predefined path to get from point A to point B must be recalculated. Position and navigation errors are normal and are to be expected. You can reduce the effects of error by having the robot periodically reassess its position. This can be accomplished by using a number of referencing schemes, such as mapping, active beacons, or landmarks.

People don't like to admit failure, but a robot is just a machine and doesn't know (or care) that it failed to reach its intended destination. You should account for the possibility that the robot may never get to point B. This can be accomplished by using time-outs, which entails either determining the maximum reasonable time to accomplish the goal or, better yet, the maximum reasonable distance that should be traveled to reach the goal.

You can build other fail-safes into the system as well, including a program override if the robot can no longer reassess its current location using known landmarks or references. In such a scenario, this could mean its sensors have gone kaput or that the landmarks or references are no longer functioning or accurate. One course of action is to have the robot shut down and wait to be bailed out by its human master.

33.2 Following a Predefined Path: Line Tracing

Perhaps the simplest navigation system for mobile robots involves following some predefined path that's marked on the ground. The path can be a black or white line painted on a hard-surfaced floor, a wire buried beneath a carpet, a physical track, or any of several other methods. This type of robot navigation is used in some factories. Marking the path with reflective tape is preferred in factories because the track can easily be changed without ripping up or repainting the floor.

You can readily incorporate a tape-track navigation system in your robot. The line-tracing feature can be the robot's only means of semi-intelligent action, or it can be just one part of a more sophisticated machine. You could, for example, use the tape to help guide a robot back to its battery charger nest.

With a line-tracing robot, you place a piece of white or reflective tape on the floor. For the best results, the floor should be hard, like wood, concrete, or linoleum, and not carpeted. One or more optical sensors are placed on the robot. These sensors incorporate an infrared LED and an infrared phototransistor. When the transistor turns on it sees the light from the LED reflected off the tape. Obviously, darker floors provide better contrast against light-colored tape.

In a working robot, mount the LED and phototransistors in a suitable enclosure, as described more fully in Chapter 30, "Object Detection." Or, use a commercially available LED–phototransistor pair (again, see Chapter 30). Mount the detectors on the bottom of the robot, as shown in Fig. 33-2, in which two detectors are set apart twice the width of the tape.

Fig. 33-3 shows the basic sensor circuit and how the LED and phototransistor are wired. Feel free to experiment with the value of R2; it determines the sensitivity of the phototransistor and its voltage range when just the floor or the tape is reflecting the infrared light or a combination of the two. Ideally, you would like the voltage output to be as "binary" as possible: a high voltage level for a black floor (no light being passed to the phototransistor, so no current passes through it) and low for the white tape. Fig. 33-4 shows the sensor output passed to a comparator circuit that forms the basis of the line-tracing system. Refer to this figure often because this circuit can be used in other applications.

You can use the schematics in Figs. 33-5 and 33-6 to build a complete line-tracing system (refer to the parts lists in Tables 33-1 and 33-2). You can build the circuit using just three IC packages: an LM339 quad comparator, a 7486 quad exclusive OR gate, and a 7400 quad NAND gate. Before using the robot, block the phototransistors so they don't receive any light. Rotate the shaft of the set-point pots until the relays kick in, then back off again. You may have to experiment with the settings of the set-point pots as you try out the system.

Depending on which motors you use and the switching speed of the relays, you may find your robot waddling its way down the track, overcorrecting for its errors every time. You can help minimize this by using faster-acting relays. Another approach is to vary the gap between the two sensors. By making it wide, the robot won't be turning back and forth as much to correct for small errors.

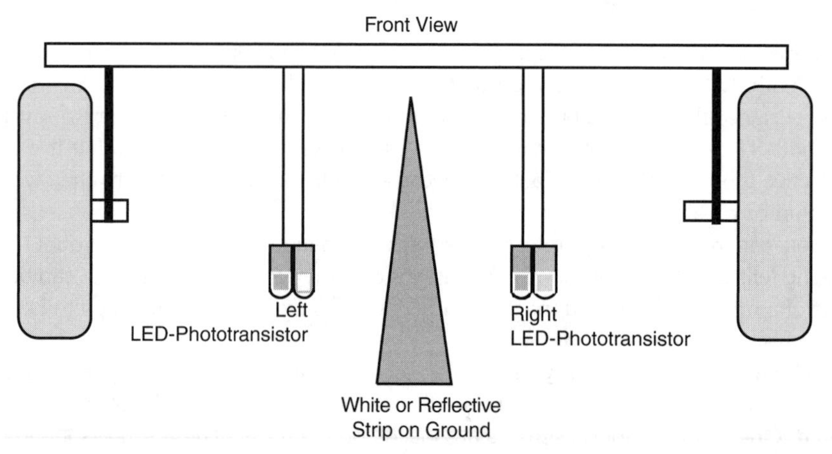

FIGURE 33-2 Placement of the left and right phototransistor–LED pair for the line-tracing robot.

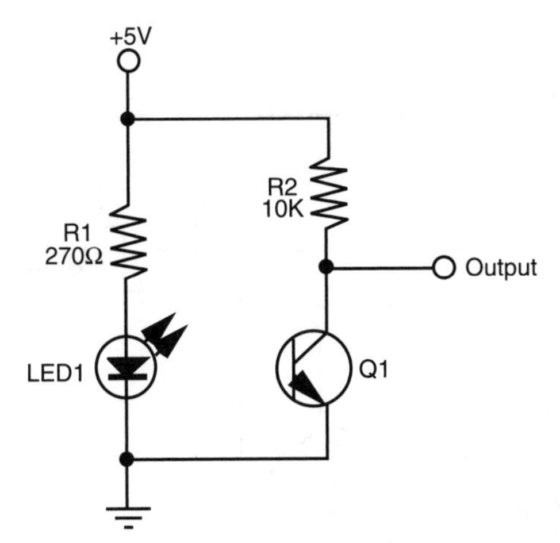

FIGURE 33-3 The basic LED-phototransistor wiring diagram.

The actual turn radius will depend entirely on the robot. If you need your robot to turn very tight, small corners, build it small. If your robot has a brain, whether it is a computer or central microprocessor, you can use it instead of the direct connection to the relays for motor control. The output of the comparators, when used with a +5 V supply, is compatible with computer and microprocessor circuitry, as long as you follow the interface guidelines provided in Chapter 14. The two sensors require only two bits of an eight-bit port.

33.2.1 COMPUTER CONTROLLED LINE FOLLOWING

It is quite easy to add and program computer control for a line-following robot. The following program was written for a three sensor line following robot with the distance between

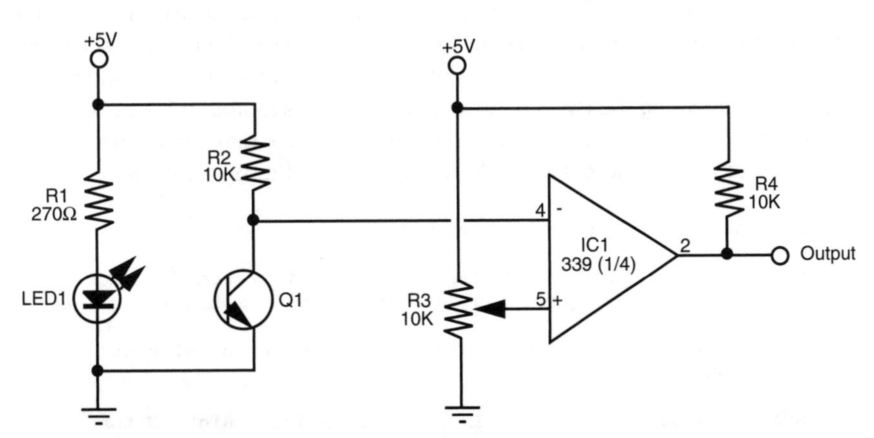

FIGURE 33-4 Connecting the LED and phototransistor to an LM339 quad comparator IC. The output of the comparator switches between HIGH and LOW depends on the amount of light falling on the phototransistor. Note the addition of the 10K pull-up resistor on the output of the comparator. This is needed to assure proper HIGH/LOW action.

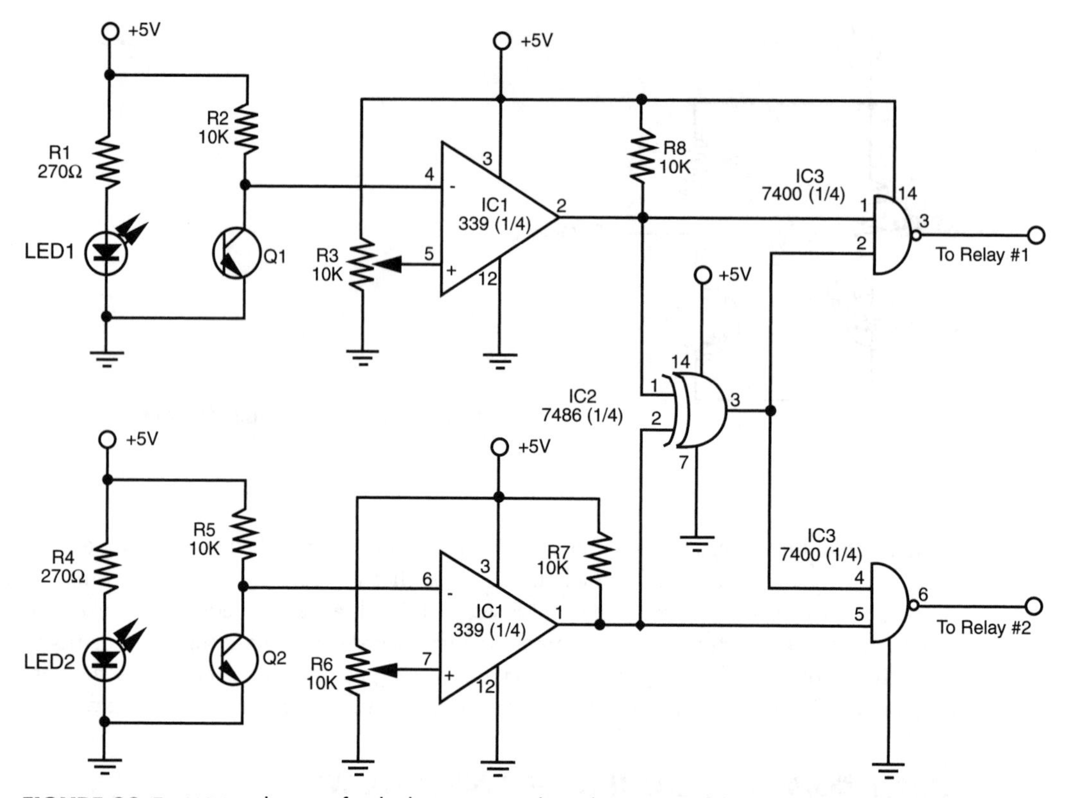

FIGURE 33-5 Wiring diagram for the line-tracing robot. The outputs of the 7400 are routed to the relays in Fig. 33-6.

the sensors being about 75 percent of the width of the line—which means that as the line changes, two sensors are active at any given time, allowing the robot to slow down on that side rather than turn off the motor completely. Slowing down a side of the robot is accomplished by decreasing the PWM duty cycle on that side's motor by 50 percent—you will see that the turn logic is replicated twice in the following program, with differing values for the slow turns to run the motors more slowly than if a hard turn is required.

```
'  BS2 Line Follow - Line Following Software
'
'  This software takes the inputs from three infrared detectors
'   to control a robot following a white line on a black surface.
'
'  A simple 30 Hz PWM has been created for the robot with the following
'   responses to the inputs:
'
'  Left IR  Middle IR   Right IR    Left Motor    Right Motor
'  1        0           1           Full          Full
'  0        0           1           Half          Full
'  0        1           1           Zero          Full
'  1        0           0           Full          Half
```

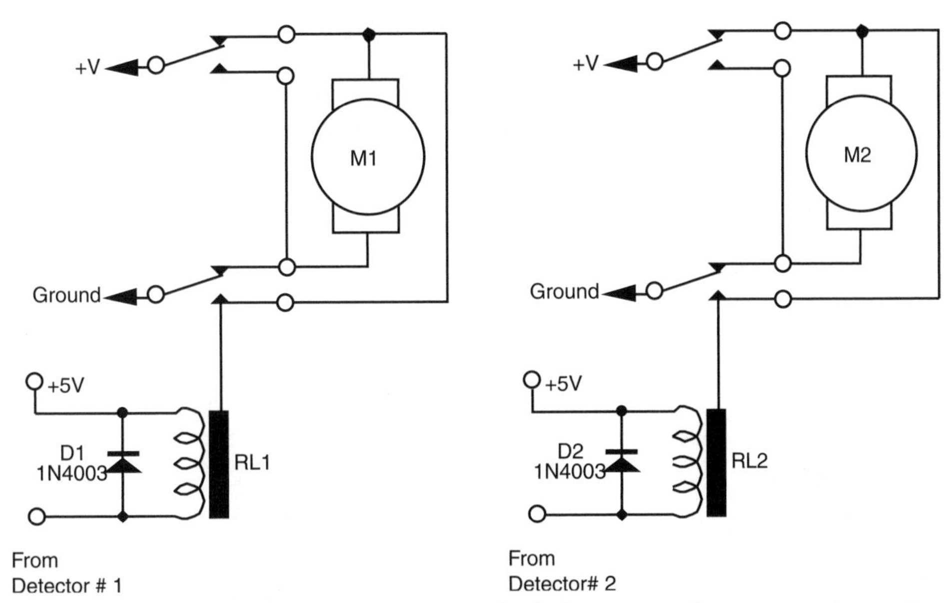

FIGURE 33-6 Motor direction and control relays for the line-tracing robot. You can substitute the relays for purely electronic control; refer to Chap. 18.

TABLE 33-1 Parts List for Line Tracer

IC1	LM339 quad comparator
IC2	7486 quad exclusive OR gate IC
IC3	7400 quad NAND gate IC
Q1, Q2	Infrared-sensitive phototransistors
R1, R4	270 Ω resistors
R2, R5, R7, R8	10k resistors
R3, R6	10k potentiometers
LED1, LED2	Infrared LEDs
Misc.	Infrared filter for phototransistors (if needed)

TABLE 33-2 Parts List for Relay Control

RL1, RL2	DPDT fast-acting relay; contacts rated 2 A or more
D1, D2	1N4003 diodes

```
'  1        1          0          Full          Zero
'  0        0          0          Zero          Zero
'  1        1          1          Zero          Zero
'
'  A white line perpendicular to the following line (all three Sensors
'    0) is the "Stop" line, and the robot will stop at this position.
'
'  When the robot loses the line (all three Sensors 1) the robot stops
'
'  myke predko
'  05.08.29

'{$STAMP BS2}
'{$PBASIC 2.5}

'  Mainline
LeftIR     PIN 0
MiddleIR   PIN 1
RightIR    PIN 2
LeftMotor  PIN 4
RightMotor PIN 5

    LOW LeftMotor                  '  Motors Output/Stop
    LOW RightMotor

    do
        IF (LeftIR = 0) then
            IF (MiddleIR = 0) then
                IF (RightIR = 0) then
                    LeftMotor = 0  '  At Stop Line
                    RightMotor = 0
                else
                    LeftMotor = 0  '  Slow Turn Left
                    RightMotor = 1
                endif
            ELSE                   '  Only Left Sensor on Line
                LeftMotor = 0      '  Hard Turn Left
                RightMotor = 1
            ENDIF
        ELSE
            IF (RightIR = 0) THEN
                IF (MiddleIR = 0) then
                    LeftMotor = 1   '  Slow Turn Right
                    RightMotor = 0
                else
                    LeftMotor = 1   '  Hard Turn Right
                    RightMotor = 0
                endif
            ELSE                    '  Only Left Sensor on Line
                IF (MiddleIR = 0) then
                    LeftMotor = 1   '  Straight Down the Middle
                    RightMotor = 1
                else
                    LeftMotor = 0   '  Lost Line, Stop
                    RightMotor = 0
                ENDIF
            ENDIF
        endif
```

```
    PAUSE 15        '  Hold Motor State for 1/60th s

    IF (LeftIR = 0) THEN
        IF (MiddleIR = 0) THEN
            IF (RightIR = 0) THEN
                LeftMotor = 0  '  At Stop Line
                RightMotor = 0
            ELSE
                LeftMotor = 1  '  Slow Turn Left
                RightMotor = 1
            ENDIF
        ELSE                   '  Only Left Sensor on Line
            LeftMotor = 0      '  Hard Turn Left
            RightMotor = 1
        ENDIF
    ELSE
        IF (RightIR = 0) THEN
            IF (MiddleIR = 0) THEN
                LeftMotor = 1  '  Slow Turn Right
                RightMotor = 1
            ELSE
                LeftMotor = 1  '  Hard Turn Right
                RightMotor = 0
            ENDIF
        ELSE                   '  Only Left Sensor on Line
            IF (MiddleIR = 0) THEN
                LeftMotor = 1   '  Straight Down the Middle
                RightMotor = 1
            ELSE
                LeftMotor = 0  '  Lost Line, Stop
                RightMotor = 0
            ENDIF
        ENDIF
    ENDIF

    PAUSE 15                   '  Hold Motor State for 1/60th Second

LOOP                           '  Repeat
```

The resulting motion is surprisingly smooth and follows the line very accurately with very few excursions to the "hard" turn required. To help you see the operation of the robot, you might want to add some LEDs to the BS2's I/O pins to "parrot" the IR line sensor inputs and the motor control outputs.

33.3 Wall Following

Robots that can follow walls are similar to those that can trace a line. Like the line, the wall is used to provide the robot with navigation orientation. One benefit of wall-following robots is that you can use them without having to paint any lines or lay down tape. Depending on the robot's design, the machine can even maneuver around small obstacles (doorstops, door frame molding, radiator pipes, etc.).

33.3.1 VARIATIONS OF WALL FOLLOWING

Wall following can be accomplished with any of four methods:

- *Contact.* The robot uses a mechanical switch, or a stiff wire connected to a switch, to sense contact with the wall, as shown in Fig. 33-7a. This is by far the simplest method, but the switch is prone to mechanical damage over time.
- *Noncontact, active sensor.* The robot uses active proximity sensors, such as infrared or ultrasonic, to determine its distance from the wall. No physical contact with the wall is needed. In a typical noncontact system, two sensors are used to judge when the robot is parallel to the wall (see Fig. 33-7b).
- *Noncontact, passive sensor.* The robot uses passive sensors, such as linear Hall effect switches, to judge distance from a specially prepared wall (Fig. 33-7c). In the case of Hall effect switches, you could outfit the baseboard or wall with an electrical wire through which a low-voltage alternating current is fed. When the robot is in the proximity of the switches the sensors will pick up the induced magnetic field provided by the alternating current. Or, if the baseboard is metal, the Hall effect sensor (when rigged with a small magnet on its opposite side) could detect proximity to a wall.

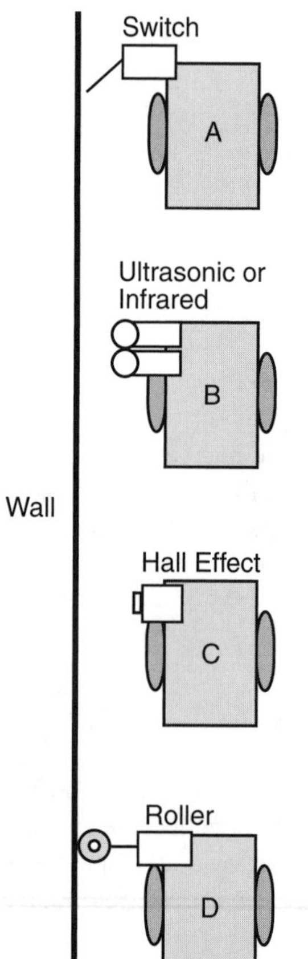

FIGURE 33-7 Ways to follow the wall include: *a.* Contact switch; *b.* noncontact active sensor (such as infrared); *c.* noncontact passive sensor (e.g., Hall effect sensor and magnetic, electromagnetic, or ferrous metal wall/baseboard); and *d.* "soft contact" using pliable material such as foam rollers.

- *Soft contact.* The robot uses mechanical means to detect contact with the wall, but the contact is softened by using pliable materials. For example, you can use a lightweight foam wheel as a wall roller, as shown in Fig. 33-7d. The benefit of soft contact is that mechanical failure is reduced or eliminated because the contact with the wall is made through an elastic or pliable medium.

In all cases, upon encountering a wall the robot goes into a controlled program phase to follow the wall in order to get to its destination. In a simple contact system, the robot may back up a short moment after touching the wall, then swing in a long arc toward the wall again. This process is repeated, and the net effect is that the robot "follows the wall."

With the other methods, the preferred approach is for the robot to maintain proper distance from the wall. Only when proximity to the wall is lost does the robot go into a "find wall" mode. This entails arcing the robot toward the anticipated direction of the wall. When contact is made, the robot alters course slightly and starts a new arc. A typical pattern of movement is shown in Fig. 33-8.

33.3.2 ULTRASONIC WALL FOLLOWING

A simple ultrasonic wall follower can use two ultrasonic transmitter/receiver pairs. Each transmitter and receiver is mounted several inches apart to avoid cross-talk. Two transmitter/receiver pairs are used to help the robot travel parallel to the wall. Suitable ultrasonic transmitter and receiver circuits are detailed in Chapter 30, "Object Detection."

Because the robot will likely be close to the wall (within a few inches), you will want to drive the transmitters at very low power and use only moderate amplification, if any, for the receiver. You can drive the transmitters at very low power by reducing the voltage to the transmitter.

33.3.3 SOFT-CONTACT FOLLOWING WITH FOAM WHEELS

Soft-contact wall following with a roller wheel offers you some interesting possibilities. In fact, you may be able to substantially simplify the sensors and control electronics by placing an idler roller made of soft foam as an outrigger to the robot and then having the robot constantly steer inward toward the wall. This can be done simply by running the inward wheel (the wheel on the side of the wall) a little slower than the other. The foam idler roller will prevent the robot from hitting the wall.

33.3.4 DEALING WITH DOORWAYS AND OBJECTS

Merely following a wall is, in essence, not that difficult. The task becomes more challenging when you want the robot to maneuver around obstacles or skip past doorways. This requires additional sensors, perhaps whiskers or other touch sensors in the forward portion of the robot. These are used to detect corners as well. This is especially important when you are constructing a robot that has a simple inward-arc behavior toward following walls. Without the ability to sense a wall straight ahead, the robot may become hopelessly trapped in a corner.

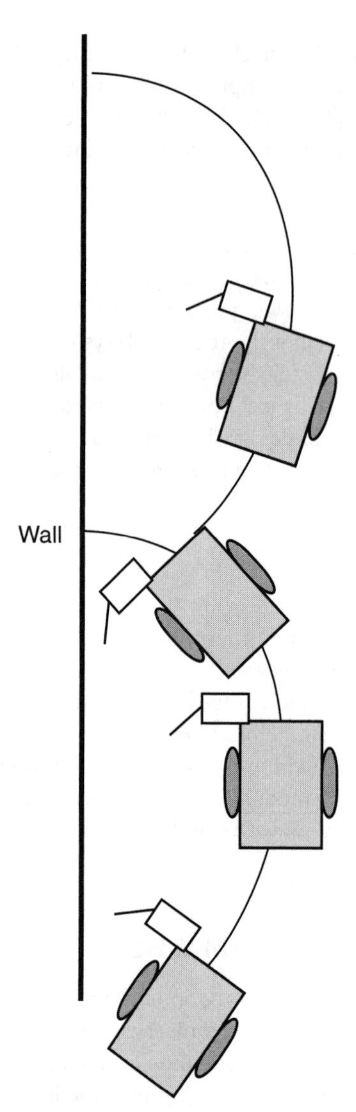

FIGURE 33-8 A wall-following robot that merely "feels" its way around the room might make wide, sweeping arcs. The arc movement is easily accomplished in a typical two-wheeled robot by running one motor slower than the other.

Open doorways that lead into other rooms can be sensed using a longer-range ultrasonic transducer. Here, the long-range ultrasonic detects that the robot is far from any wall and places the machine in a "go straight" mode. Ideally, the robot should time the duration of this mode to account for the maximum distance of an open doorway. If a wall is not detected within *x* seconds, the robot should go into a "look for wall" mode.

33.3.5 CODING YOUR WALL-FOLLOWING ROBOT

Software for a wall-following robot can be accomplished by using similar software as the line-following robot; a PWM can be used to make gentle turns, and hard turns can be made when required (such as when a wall is in front of the robot). The following program is optimized to eliminate some of the decision making for the "slow" versus "hard" turns in the first 15-ms portion of the PWM operation; in the second part, the decision is made whether or not to turn the motors off for the full PWM cycle.

```
'  BS2 Line Follow - Wall Following Software
'
'  This software takes the inputs from two Wall sensors/Whiskers
'   (one at the side and one in front) to control a robot
'   following a wall on the left hand side of the robot.
'
'  A simple 30 Hz PWM has been created for the robot with the following
'   responses to the inputs:
'
'  Left Whisker  Front Wisker     Left Motor     Right Motor
'  Contact        No Contact       Full           Half
'  No Contact     No Contact       Half           Full
'  No Contact     Contact          Full           Zero
'  Contact        Contact          Full           Zero
'
'  When the robot loses contact with the wall, it circles to the left.
'
'  myke predko
'  05.08.29


'{$STAMP BS2}
'{$PBASIC 2.5}


'  Mainline
LeftIR     PIN 15
FrontIR    PIN 14
LeftMotor  PIN 4
RightMotor PIN 5


    LOW LeftMotor                 '  Motors Output/Stop
    LOW RightMotor


    DO


        IF (LeftIR = 0) THEN      '  Left Contact
            LeftMotor = 1
            RightMotor = 0        '  Hard/Slow Turn Right
        ELSE
            LeftMotor = 0         '  Hard/Slow Turn Left
            RightMotor = 1
        ENDIF


        PAUSE 15                  '  Hold Motor State for 1/60th Second


        IF (LeftIR = 0) THEN      '  Left Contact
            IF (FrontIR = 0) THEN
                LeftMotor = 1     '  Something in Front
                RightMotor = 1    '  Slow Turn Right
            ELSE
                LeftMotor = 1     '  Hard Turn Right
                RightMotor = 0
            ENDIF
```

```
        ELSE
            IF (FrontIR = 0) THEN
                LeftMotor = `          '  Slow Turn Left
                RightMotor = 1
            ELSE
                LeftMotor = 0          '  Hard Turn Left
                RightMotor = 1
            ENDIF

        PAUSE 15                       '  Hold Motor State for 1/60th Second

LOOP                                   '  Repeat
```

This application was run on the same robot as the line follower, but used Sharp IR object sensors (see Chapter 30, "Object Detection," for more information on the sensors).

33.4 Odometry: The Art of Dead Reckoning

Hop into your car. Note the reading on the odometer. Now drive straight down the road for exactly one minute, paying no attention to the speedometer or anything else (of course, keep your eyes on the road!). Again note the reading on the odometer. The information on the odometer can be used to tell you where you are. Suppose it says one mile. You know that if you turn the car around exactly 180° and travel back a mile, at whatever speed, you'll reach home again.

This is the essence of odometry; reading the motion of a robot's wheels to determine how far it's gone. Odometry is perhaps the most common method for determining where a robot is at any given time. It's cheap, easy to implement, and fairly accurate over short distances. Odometry is similar to the dead-reckoning navigation used by sea captains and pilots before the age of satellites, radar, and other electronic schemes. Hence, odometry is also referred to in robot literature as *dead reckoning*.

33.4.1 OPTICAL ENCODERS

You can use a small disc fashioned around the hub of a drive wheel, or even the shaft of a drive motor, as an optical shaft encoder (described in "Anatomy of a Shaft Encoder," in Chapter 20). The disc can be either the reflectance or the slotted type:

- With a reflectance disc, infrared light strikes the disc and is reflected back to a photodetector.
- With a slotted disc, infrared light is alternately blocked and passed and is picked up on the other side by a photodetector.

With either method, a pulse is generated each time the photodetector senses the light.

33.4.2 MAGNETIC ENCODERS

You can construct a magnetic encoder using a Hall effect switch (a semiconductor sensitive to magnetic fields) and one or more magnets. A pulse is generated each time a magnet

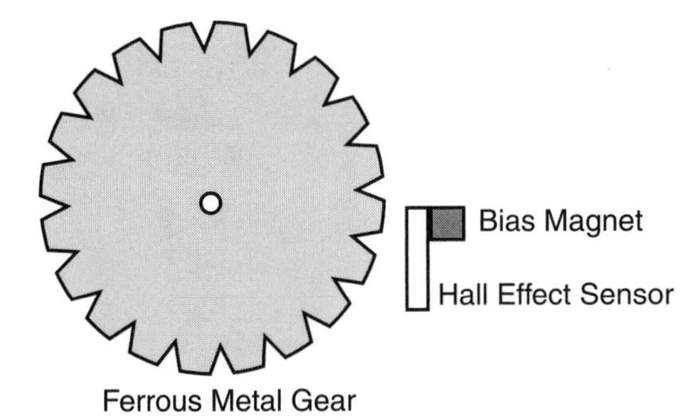

FIGURE 33-9 A Hall effect sensor outfitted with a small "bias" magnet and sensitive to the changes in magnetic flux caused by a rotating ferrous metal gear.

passes by the Hall effect switch. A variation on the theme uses a metal gear and a special Hall effect sensor that is sensitive to the variations in the magnetic influence produced by the gear (see Fig. 33-9).

A bias magnet is placed behind the Hall effect sensor. A pulse is generated each time a tooth of the gear passes in front of the sensor. The technique provides more pulses on each revolution of the wheel or motor shaft, and without having to use separate magnets on the rim of the wheel or wheel shaft.

33.4.3 THE FUNCTION OF ENCODERS IN ODOMETRY

As the wheel or motor shaft turns, the encoder (optical or magnetic) produces a series of pulses relative to the distance the robot travels. Assume the wheel is 3 in diameter (9.42 in in circumference), and the encoder wheel has 32 slots. Each pulse of the encoder represents 0.294 in of travel (9.42/32). If the robot senses 10 pulses, it can calculate the movement to 2.94 in.

It's best to make odometry measurements using a microcontroller that is outfitted with a pulse accumulator or counter input. These kinds of inputs independently count the number of pulses received since the last time they were reset. To take an odometry reading, you clear the accumulator or counter and then start the motors. Your software need not monitor the accumulator or counter. Stop the motors, and then read the value in the accumulator or counter. Multiply the number of pulses by the known distance of travel for each pulse. (This will vary depending on the construction of your robot; consider the diameter of the wheels and the number of pulses of the encoder per revolution.)

33.4.4 ERRORS IN ODOMETRY

If the robot uses the traditional two-wheel-drive approach, you should attach optical encoders to both wheels. This is necessary because the drive wheels of a robot are bound to turn at slightly different speeds over time. By integrating the results of both optical encoders, it's possible to determine where the robot really is as opposed to where it should be (see Fig. 33-10). As well, if one wheel rolls over a cord or other small lump, its rotation will be hindered. This can cause the robot to veer off course, possibly by as much as 3 to 5 degrees or more. Again, the encoders will detect this change.

If the number of pulses from both encoders is the same, you can assume that the robot traveled in a straight line, and you have only to multiply the number of pulses by the dis-

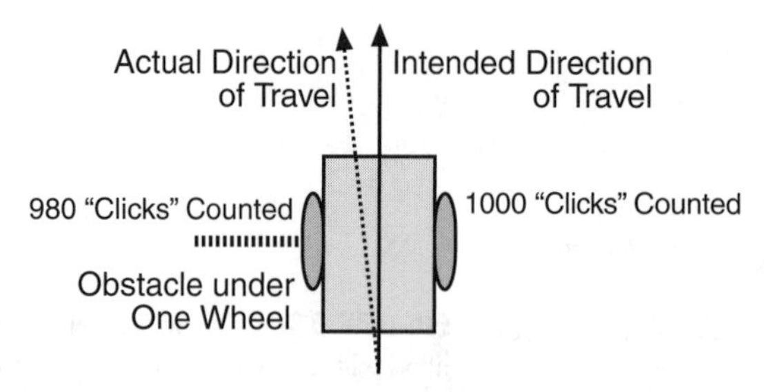

FIGURE 33-10 The relative number of "counts" from each encoder of the typical two-wheeled robot can be used to indicate deviation in travel. If an encoder shows that one wheel turned a fewer number of times than the other wheel, then it can be assumed the robot did not travel in a straight line.

tance per pulse. For example, if there are 1055 pulses in the accumulator-counter, and if each pulse represents 0.294 in of travel, then the robot has moved 310.17 in straight forward.

In a perfect world, robots would not need anything more than a single odometer to determine exactly where they were at any given time. Unfortunately, robots live and work in a world that is far from perfect; as a result, their odometers are far from accurate. Over a 20- to 30-ft range, for example, it's not uncommon for the average odometer to misrepresent the position of the robot by as much as half a foot or more!

Why the discrepancy? First and foremost: wheels slip. As a wheel turns, it is bound to slip, especially if the surface is hard and smooth, like a kitchen floor. Wheels slip even more when they turn. The wheel encoder may register a certain number of pulses, but because of slip the actual distance of travel will be less. Certain robot drive designs are more prone to error than others. Robots with tracks are steered using slip—lots of it. The encoders will register pulses, but the robot will not actually be moving in proportion.

There are less subtle reasons for odometry error. If you're even a hundredth of an inch off when measuring the diameter of the wheel, the error will be compounded over long distances. If the robot is equipped with soft or pneumatic wheels, the weight of the robot can deform the wheels, thereby changing their effective diameter.

Because of odometry errors, it is necessary to combine it with other navigation techniques, such as active beacons, distance mapping, or landmark recognition. All three are detailed later in this chapter.

33.5 Compass Bearings

Besides the stars, the magnetic compass has served as humankind's principal navigation aid over long distances. You know how it works: a needle points to the magnetic North Pole of the earth. Once you know which way is north, you can more easily reorient yourself in your travels.

Robots can use compasses as well, and a number of electronic and electromechanical compasses are available for use in hobby robots. One of the least expensive is the Dinsmore 1490, from Dinsmore Instrument Co. The 1490 looks like an overfed transistor, with 12 leads protruding from its underside. The leads are in four groups of three; each group represents a major compass heading: north, south, east, and west. The three leads in each group are for power, ground, and signal. A Dinsmore 1490, mounted on a circuit board, is shown in Fig. 33-11.

The 1490 provides eight directions of heading information (N, S, E, W, SE, SW, NE, NW) by measuring the earth's magnetic field. It does this by using miniature Hall effect sensors and a rotating compass needle (similar to ordinary compasses). The sensor is said to be internally designed to respond to directional changes much like a liquid-filled compass. It turns to the indicated direction from a 90° displacement in approximately 2.5 s. The manufacturer's specification sheet claims that the unit can operate with up to 12 degrees of tilt with acceptable error, but it is important to note that any tilting from center will cause a corresponding loss in accuracy.

Fig. 33-12 shows the circuit diagram for the 1490, which uses four inputs to a computer or microcontroller. Note the use of pull-up resistors. With this setup, your robot can determine its orientation with an accuracy of about 45 degrees (less if the 1490 compass is tilted). Dinsmore also makes an analog-output compass that exhibits better accuracy.

You could also consider the Vector 2X and 2XG. These units use magneto-inductive sen-

FIGURE 33-11 The Dinsmore 1490 digital compass provides simple bearings for a robot. The sensor is accurate to about 45 degrees.

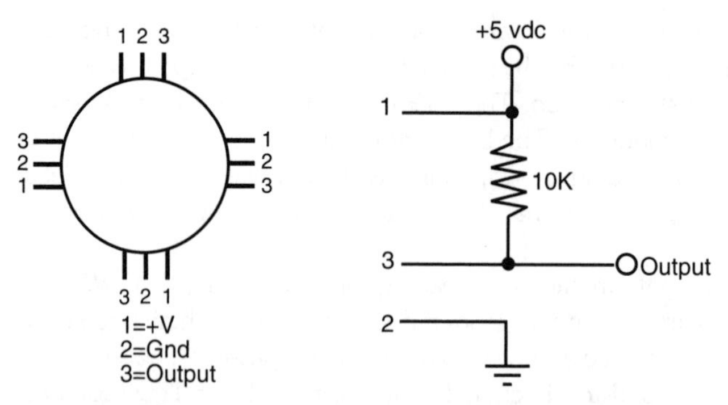

FIGURE 33-12 Circuit diagram for using the Dinsmore 1490 digital compass. When used with a +5 vdc supply, the four outputs can be connected directly to a microcontroller. One or two outputs can be activated at a time; if two are activated, the sensor is reading between the four compass points (e.g., N and W outputs denotes NW position).

sors for sensing magnetic fields. The Vector 2X/2XG provides either compass heading or uncalibrated magnetic field data. This information is output via a three-wire serial format and is compatible with Motorola SPI and National Semiconductor Microwire interface standards. Position data can be provided either 2.5 or 5 times per second.

Vector claims accuracy of ±2 degrees. The 2X is meant to be used in level applications. The more pricey 2XG has a built-in gimbal mechanism that keeps the active magnetic-inductive element level, even when the rest of the unit is tilted. The gimbal allows tilt up to 12 degrees.

A final option to consider is a commercially available GPS unit with a built-in compass and RS-232 serial interface like a Garmin "ETrex" line of outdoor GPS units. RS-232 data are normally sent as a "NMEA" data stream, which can be easily filtered by a microcontroller like a BS2. A handheld GPS unit usually has a very accurate compass built into the unit (to a degree or less) and is very insensitive to tilting. The only downside to using a handheld GPS unit is the price—they can often be several hundred dollars.

33.6 Ultrasonic Distance Measurement

Police radar systems work by sending out a high-frequency radio beam that is reflected off nearby objects, such as your car as you are speeding down the road. Most people believe that the speed is calculated by timing how long it takes for a signal sent from the radar gun to bounce off the car and return; multiple "shots" are timed and the speed is calculated based on the rate of change in the return time of these pulses. Speed is actually calculated using the Doppler effect: the frequency of the reflected signal varies according to how fast you are going and whether you are approaching or going away from the radar unit.

Radar systems are complex and expensive, and most require certification by a government authority, such as the Federal Communications Commission for devices used in the

United States. There is another approach: you can use high-frequency sound instead to measure distance, and with the right circuitry you can even provide a rough indication of speed.

Ultrasonic ranging is, by now, an old science. Polaroid used it for years as an automatic focusing aid on their instant cameras. Other camera manufacturers have used a similar technique, though it is now more common to implement infrared ranging (covered later in the chapter). The Doppler effect that is caused when something moves toward or away from the ultrasonic unit is used in home burglar alarm systems. However, for robotics the more typical application of ultrasonic sound is either to detect proximity to an object (see "Ultrasonic Wall Following," earlier in the chapter) or to measure distance (also called ultrasonic ranging).

To measure distance, a short burst of ultrasonic sound—usually at a frequency of 40 kHz for most ultrasonic ranging systems—is sent out through a transducer (a specially built ultrasonic speaker). The sound bounces off an object, and the echo is received by another transducer (this one a specially built ultrasonic microphone). A circuit then computes the time it took between the transmit pulse and the echo and comes up with distance.

Certainly, the popularity of ultrasonics does not detract from its usefulness in robot design. The system presented here is suited for use with a computer or microcontroller. There are a variety of ways to implement ultrasonic ranging. One method is to use the ultrasonic transducer and driver board from an old Polaroid instant camera, such as the Polaroid Sun 660 or the Polaroid SX-70 One Step. However, the driver board used in these cameras may require some modification to allow more than one "ping" of ultrasonic sound without having to cycle the power to the board off, then back on. More about this in a bit. You can also purchase a new Polaroid ultrasonic transducer and driver board from a number of mail-order sources, including on the Internet. Several of these outlets are listed in Appendix B, "Sources." These units are new, and most come with documentation, including hookup instructions for connecting to popular microcontrollers, such as the BASIC Stamp. Perhaps the most common Polaroid distance-measuring kit is composed of the so-called 600 Series Instrument Grade transducer along with its associated Model 6500 Ranging Module (Fig. 33-13).

The transducer, which is about the size of a silver dollar coin, acts as both ultrasonic transmitter and receiver. Because only a single transducer is used, the Polaroid system as described in this section cannot detect objects closer than about 1.3 ft. This is because of the amount of time required for the transducer to stop oscillating before it sets itself up to receive. The maximum distance of the sensor is about 35 ft when used indoors, and a little less when used outdoors, especially on a windy day. The system is powered by a single 6-vdc battery pack and can be interfaced to any computer or microcontroller.

33.6.1 FACTS AND FIGURES

First some statistics. At sea level, sound travels at a speed of about 1130 f/s (about 344 m/s) or 13,560 in/s. While this time varies depending on atmospheric conditions, including air pressure (which varies by altitude), temperature, and humidity, it is a good value to use when working with an ultrasonic ranging system. The overall time between transmit pulse and echo is divided by two to compensate for the round-trip travel time between the robot and the object. Given a travel time of 13,560 in/s for sound, it takes just 73.7 μs

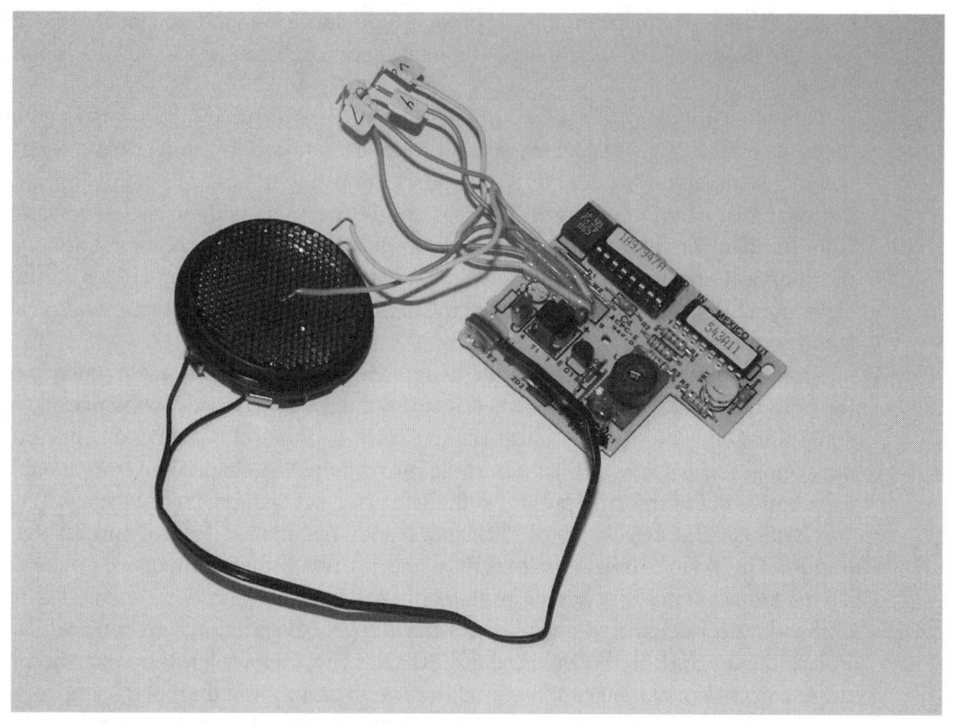

FIGURE 33-13 Polaroid 6500 module with wires attached to simplify adding the circuit to a robot. Note that 1 A of current is required and to minimize noise in the rest of the circuit, a 1000 µF capacitor is placed across the Polaroid 6500's power leads.

(0.0000737 s) for sound to travel 1 in. This means that for every 147.4 µs of time, the distance between the transducer and an object changes by 1 in (2.5 cm).

33.6.2 INTERFACING A POLAROID 6500 ULTRASONIC RANGE FINDER

With the figure of 1 in per 147.4 µs in the back of our minds, let's consider how the Polaroid ranging system works. The Ranging Module is connected to a computer or microcontroller using only two wires: INIT (for INITiate) and ECHO. INIT is an output, and ECHO is an input. The Ranging Module contains other I/O connections, such as BLNK and BINH, but these are not strictly required when you are determining distance to a single object, and so they will not be discussed here.

To trigger the Ranging Module and have it send out a burst of ultrasonic sound, the computer or microcontroller brings the INIT line HIGH. The computer-microcontroller then waits for the ECHO line to change from LOW to HIGH. The time difference, in microseconds, is divided in two, and that gives you distance. To measure the time between the INIT pulse and the return ECHO, the computer or microcontroller uses a timer to precisely count the time interval.

To demonstrate the operation of the Polaroid 6500 with a BS2, some additional circuitry has to be added as shown in Fig. 33-14 (parts list in Table 33-3). The 74LS123 and

74LS74 create a delayed pulse that can be measured by a BASIC Stamp 2's Pulsin statement. Using one of the single 74LS123 monostable multivibrator's single-shot outputs creates two D-flip flop triggering pulses that ensure that the resulting pulse passed to the BS2 has the same timing as the delay between the initialization of the Init and the Echo return as shown in Fig. 33-15.

It is important to note that the Polaroid 6500 uses a massive amount of current when it is operating. To ensure that the BS2 and the TTL chips' operation are not affected by the 6500, the 6500 is given a power supply separate from the other chips. For the prototype circuit, a 6-V lantern battery was used for powering the 6500. The BS2's on-board regulator is used to power both it as well as the two TTL chips.

When the 6500 is active, make sure the transducer is mounted on a nonconducting surface and you are not touching it! The 6500 will give you a nasty shock when it is active and could potentially damage any electronics that have common pins to it.

The software to use the BS2/Polaroid 6500 combination follows and consists of resetting the 74LS74 D-flip flops before sending out the Init signal and waiting for the delayed pulse back.

```
'  BS2 Polaroid 6500 - Use Polaroid 6500 Sonar Ranging Module
'
'  myke predko
'
'  04.01.25
```

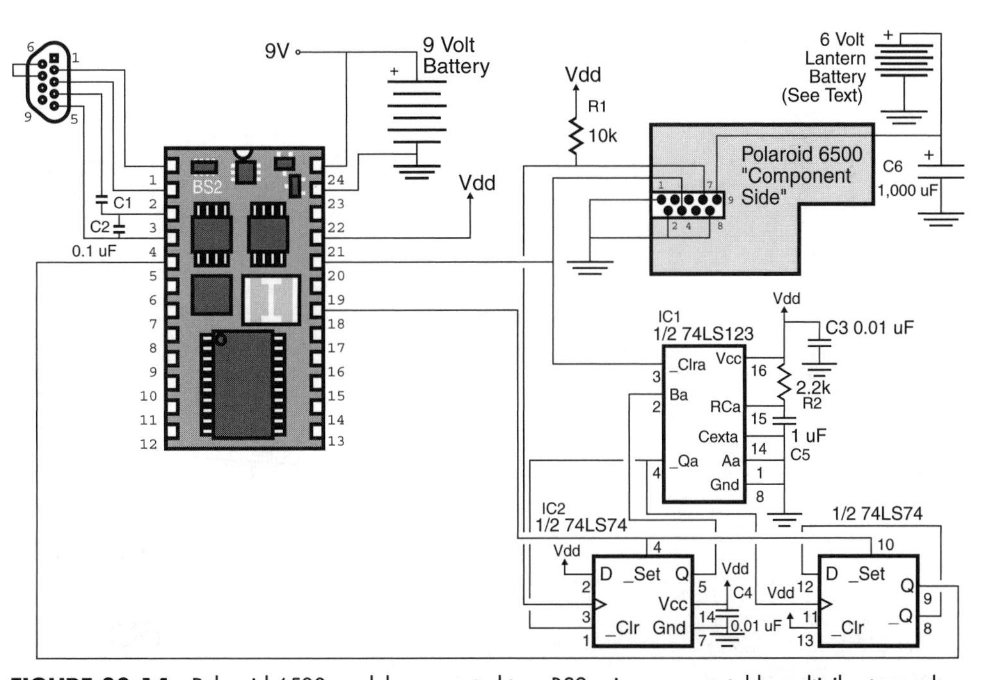

FIGURE 33-14 Polaroid 6500 module connected to a BS2 using a monostable multivibrator and two D-flip flops to ensure the waveform passed to the BS2 is a pulse and available when the BS2 is ready for it.

TABLE 33-3 Polaroid 6500 Interface Circuit

BS2	Parallax BS2
Polaroid 6500	Polaroid 6500 ranging module with wires attached for breadboard interfacing
IC1	74LS123 dual monostable multivibrator chip
IC2	74LS74 dual D-flip flop chip
R1	10k resistor
R2	2.2k resistor
C1, C2	0.1 μF capacitor
C3, C4	0.01 μF capacitor
C5	1 μF capacitor
C6	1000 μF capacitor (see text)
Misc.	Breadboard, BS2 communication/programming interface, 9-V battery, 6-V lantern battery (see text)

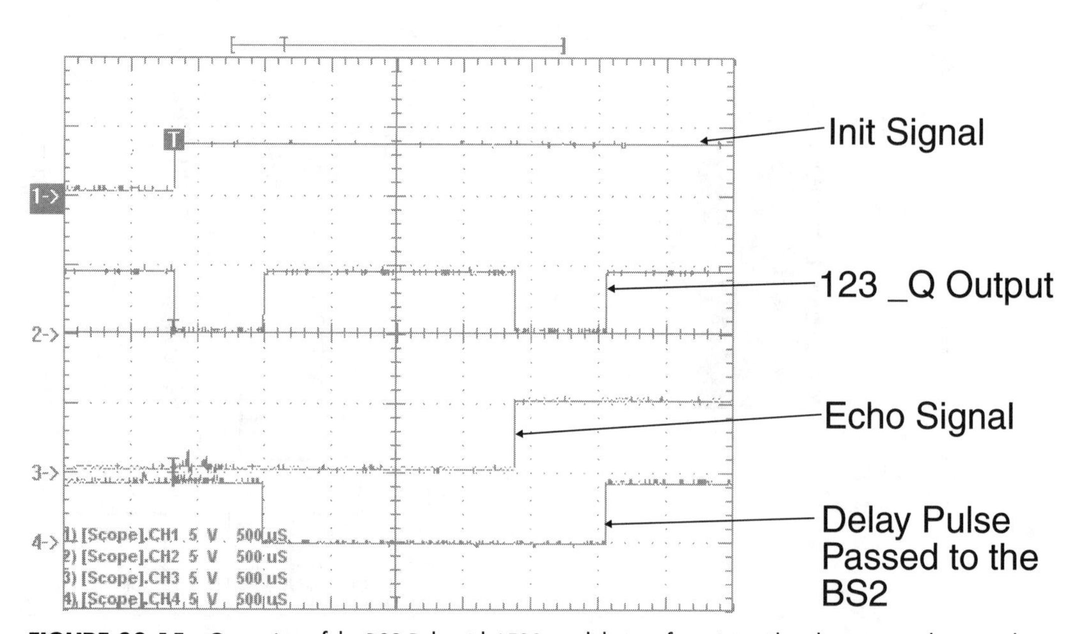

FIGURE 33-15 Operation of the BS2 Polaroid 6500 module interface. Note that the Init signal causes the 74LS123 to trigger, starting a pulse output on the 74LS74 to be measured by the BS2. When the echo has returned, the 74LS123 is triggered again, changing the state of the 74LS74 and completing the pulse.

```
'{$STAMP BS2}
'{$PBASIC 2.50}

'  Pin/Variable Declarations
InitPin        PIN 15
InitSetupPin   PIN 13
FlightPin      PIN 0
SoundFlight    VAR Word
SoundIn        VAR Word
SoundFt        VAR Word

'  Initialization/Mainline
  LOW InitPin                   '  Set Initialization Pin To O/P & Low
HIGH InitSetupPin               '  Pulsed Low to Setup Conditions
INPUT FlightPin

DO                              '  Repeat forever
    PULSOUT InitSetupPin, 10    '  Setup Hardware for Pulse Read
    HIGH InitPin                '  Output to Cause 150 ms Pulse
    PULSIN FlightPin, 0, SoundFlight
    IF (SoundFlight <> 0) THEN'  If "0" Returned, Echo not found
      DEBUG "Time of Flight is ", DEC SoundFlight * 2, " ms", CR
      SoundIn = SoundFlight / 153
      SoundFt = SoundIn / 12: SoundIn = SoundIn // 12
      DEBUG "Distance from Sensor to Object ", DEC SoundFt, "' ", DEC
SoundIn, REP 34\1, CR
    ENDIF
    LOW InitPin
    PAUSE 1000                  '  Wait a Second before Repeating
  LOOP
```

33.7 "Where Am I?": Sighting Landmarks

Explorers rely on landmarks to navigate wide-open areas. It might be an unusual outcropping of rocks or a bend in a river. Or the 7-Eleven down the street. In all cases, a landmark serves to give you general bearings. From these general bearings you can more readily navigate a given locale.

Robots can use the same techniques, though rocks, rivers, and convenience stores are somewhat atypical as useful landmarks. Instead, robots can use such techniques as beacons to determine their absolute position within a known area. The following sections describe some techniques you may wish to consider for your next robot project.

33.7.1 INFRARED BEACON

Unless you confine your robot to playing just within the laboratory, you'll probably want to provide it with a means to distinguish one room in your house from the next. This is particularly important if you've designed the robot with even a rudimentary form of object and area mapping. This mapping can be stored in the robot's memory and used to steer around objects and avoid walls.

For less than a week's worth of groceries, you can construct an infrared beacon system that your robot can use to determine when it has passed from one room to the next. The robot is equipped with a receiver that will detect transmitters placed in each room. These transmitters send out a unique code, which the robot interprets as a specific room. Once it has identified the room, it can retrieve the mapping information previously stored for it and use it to navigate through its surroundings.

The beacon system presented here is designed around a set of television and VCR remote control chips sold by Holtek. The chips are reasonably inexpensive but can be difficult to find. The chips used in this project are HT12D and HT12E, available from Jameco (www.jameco.com, but you should check the Internet for other sources as well).

You can, of course, use just about any wireless remote control system you desire. The only requirements are that you must be able to set up different codes for each transmitter and that the system must work with infrared light.

You can connect the four-bit output of the HT-12D decoder IC to a microcontroller or computer. You will also want to connect the VD (valid data) line to a pin of your microcontroller or computer. When this line "winks" LOW, it means there is valid data on the four data lines. The value at the four data lines will coincide with the setting of the four-position DIP switch on each transmitter.

33.7.2 RADIO FREQUENCY IDENTIFICATION

Radio frequency identification (RFID) is a hot technology that uses small passive devices that radiate a digital signature when exposed to a radio frequency signal. RFID is found in products ranging from toys to trucking, farm animal inventories, automobile manufacturing, and more, including replacing the bar codes used at retail checkouts.

A transmitter/receiver, called the interrogator or reader, radiates a low- or medium-frequency carrier RF signal. If it is within range, a passive (unpowered) or active (powered) detector, called a tag or transponder, re-radiates (or backscatters) the carrier frequency, along with a digital signature that uniquely identifies the device. RFID systems in use today operate on several common RF bands, including a low-speed 100 to 150 kHz band and a higher 13.5 MHz band.

The tag is composed of an antenna coil along with an integrated circuit. The radio signal provides power when used with passive tags, using well-known RF field induction principles. Inside the integrated circuit are decoding electronics and a small memory. A variety of data transmission schemes are used, including non-return-to-zero, frequency shift keying, and phase shift keying. Manufacturers of the RFID devices tend to favor one system over another for specific applications. Some data modulation schemes are better at long distances, for example.

Different RFID tags have different amounts of memory, but a common device might provide for 64 to 128 bits of data. This is more than enough to serve as room-by-room or locale-by-locale beacons. The advantage RFID has over infrared beacons (see earlier in this chapter) is that the coverage of the RF signal is naturally limited. While this limitation can certainly be a disadvantage, when properly deployed it can serve as a convenient way to differentiate between different areas of a house's robotic work space. The average working distance between interrogator and tag is several feet, though this varies greatly depending on the power output of the interrogator. Units with higher RF power can be used over longer distances. For room-by-room robotics use, however, a unit with limited range is preferred, which also means a less expensive system.

While RFID systems are not complex, their cost can be a bit expensive for hobbyists. Demonstration and developers' kits are available from some manufacturers for $100 to $200, and this includes the reader and an assortment of tags. However, once implemented RFID is a low-maintenance, long-term solution for helping your robot know where it is.

33.7.3 LANDMARK RECOGNITION

As mentioned, humans navigate the real world by using landmarks: the red barn on the way to work signals you're getting close to your turnoff. Robots can use the same kind of visual cues to help them navigate a space. Landmarks can be natural—a support pillar in a warehouse for example—or they can be artificial—reflectors, posts, or bar codes positioned just for use by the robot. A key benefit of landmark recognition is that most systems are easy to install, cheap, and when done properly, unmistakable from the robot's point of view.

Wide Field Bar Code One technique to consider is the use of wide-field bar codes, which are commonly used in warehouses for quick and easy inventory. The bar code pattern is printed very large, perhaps as tall as 2 in and as wide as a foot. A traditional laser bar code reader then scans the code. The large size of the bar code makes it possible to use the bar code reader even from a distance—10 to 20 ft or more.

You can adapt the same method to help your robot navigate from room to room, and even within a room. For each location you want to identify, print up a large bar code. Free and low-cost bar code printing software is available over the Internet and in several commercial packages. You can either make or purchase a wide-field bar code scanner and connect it to your robot's computer or microcontroller. As your robot roams about, the scanner can be constantly looking for bar codes. The laser light output from the scanner is very low and, if properly manufactured, is well within safe limits even if the beam should quickly scan past the eyes of people or animals.

Door Frame "Flags" Yet another technique that merits consideration is the use of reflective tape placed around the frames of doors. Doorways are uniquely helpful in robot navigation because in the human world we tend to leave the space around them open and uncluttered. This allows us to enter and exit a room without tripping over something. It also typically means that the line of sight of the door will not be blocked, creating a reliable landmark for a robot.

Imagine vertical strips of reflective tape on either side of the doorway. These strips could reflect the light from a scanning laser mounted on the robot, as shown in Fig. 33-16. The laser light would be reflected from the tape and received by a sensor on the robot. Since the speed of the laser scan is known, the timing between the return pulses of the reflected laser light would indicate the relative distance between the robot and the doorway. You could use additional tape strips to reduce the ambiguity that results when the robot approaches the doorway at an angle.

Or consider using a CCD or CMOS camera. The robot could use several high-output infrared LEDs to illuminate the tape strips. Since the tape is much more reflective than the walls or door frame, it returns the most light. The CCD or CMOS camera is set with a high contrast ratio, so it effectively ignores anything but the bright tapes. Assuming the robot is positioned straight ahead of the door, the tapes will appear to be parallel. The distance between the tapes indicates the distance between the robot and the doorway. Should the

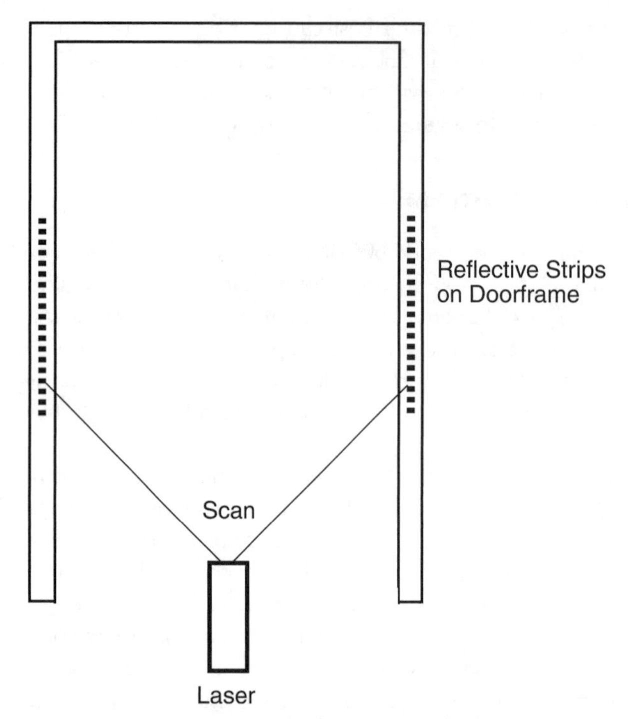

FIGURE 33-16 A scanning laser mounted on your robot can be used to detect the patterns of reflective tape located on or near doorways. Since the speed of the scan is known, electronics on your robot calculate distance (and position, given more strips) from robot to door.

robot be at an angle to the door, the tapes will not be parallel. Their angle, distance, and position can once again be interpolated to provide the robot's position relative to the door.

33.7.4 OTHER TECHNIQUES FOR BEACONS AND LIGHTHOUSES

There are scores of ways to relay position information to a robot. You've already seen two beacon-type systems: infrared and radio frequency. There are plenty more. There isn't enough space in this book to discuss them all, but the following sections outline some techniques you might want to consider. Many of these systems rely on a line of sight between the beacon or lighthouse and the robot. If the line of sight is broken, the robot may very well get lost.

Three-Point Triangulation Traditional three-point triangulation is possible using either of two methods:

- *Active beacon.* A sensor array on the robot determines its location by integrating the relative brightness of the light from three active light sources.
- *Active robot.* The robot sends out a signal that is received by three sensors located

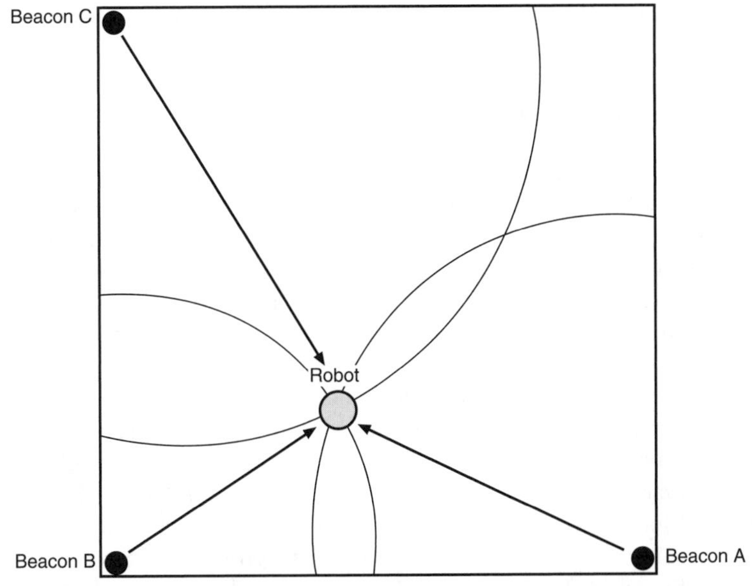

FIGURE 33-17 Three active beacons, connected to fire in sequence, provide both infrared and ultrasonic sound signals. Using a burst of infrared light, the robot times how long it takes for the sonar ping to reach it. Repeated three times—one for each beacon—the robot is able to make an accurate fix within the room.

around the room. The sensors integrate the robot's position, then relay this information back to the robot (via RF or an infrared radio link).

33.7.5 COUPLED SONAR AND IR LIGHT

This technique calculates time of flight using sound, and it offers excellent accuracy. You equip three active beacons with sonar transmitters and high-output infrared light-emitting diodes. You then connect the three beacons electrically so they will fire in sequence. When fired, both the sonar transmitter and IR LEDs emit a short 40 kHz signal. Because light travels much faster than sound, the robot will detect the IR signal first, followed by the sound signal.

The difference in time between the reception of the IR and sound signals represents distance. Each beacon provides a circle path that the robot can be in. All three circles will intersect at only one spot in the room, and that will be the location of the robot. See Fig. 33-17 for a demonstration of how this works.

33.8 Exploring Other Position-Referencing Systems

Over the years a number of worthwhile techniques have been developed to help robots know where they are. We've covered many of the most common techniques here. If your budget and construction skills allow for it, however, you might want to consider any or all of the following.

33.8.1 GLOBAL POSITIONING SATELLITE

Hovering over the earth are some two dozen satellites that provide accurate world-positioning data to vehicles, ships, and aircraft. The satellite network, referred to as global positioning system (GPS), works by triangulation: the signals from three or more satellites are received and their timings are compared. The difference in the timings indicates the relative distances between the satellites and the receiver. This provides a "fix" by which the receiver can determine not only the latitude and longitude most anywhere on the earth, but also elevation.

GPS was primarily developed by the United States government for various defense systems, but it is also regularly used by private commerce and even consumers. Until recently, the signals received by a consumer-level GPS receiver have been intentionally "fuzzied" to decrease the accuracy of the device. (This is called *selective availability,* imposed by the U.S. government for national security reasons.) Instead of the accuracies of a few feet or less that are possible with military-grade GPS receivers, consumer GPS receivers have had a nominal resolution of 100 m, or about 325 ft. In practical use, with selective availability activated in the GPS satellites, the actual error is typically 50 to 100 ft. Selective availability has since been deactivated (but could be re-activated in the event of a national emergency), and the resolution of consumer GPS receivers can be under 20 to 25 ft.

Furthermore, a system called differential GPS, in which the satellite signals are correlated with a second known reference, demonstrably increases the resolution of GPS signals to less than 5 in. When used outdoors (the signal from the satellites is too weak for indoor use), this can provide your robot with highly accurate positioning information, especially if your 'bot wanders hundreds of feet from its base station. Real-time differential GPS systems are still fairly costly, but their outputs can read into the robot's computer in real time. It takes from one to three minutes for the GPS system to lock onto the satellites overhead, however. Every time the lock is broken—the satellite signals are blocked or otherwise lost—it takes another one to three minutes to reestablish a fix.

If you're interested in experimenting with GPS, look for a receiver that has a NMEA-0183 or RS-232 compatible computer interface. A number of amateur radio sites on the Internet discuss how to use software to interpret the signals from a GPS receiver.

33.8.2 INERTIAL NAVIGATION

You can use the same physics that keeps a bicycle upright when its wheels are in motion to provide motion data to a robot. Consider a bicycle wheel spinning in front of you while you hold the axle between your hands. Turn sideways and the wheel tilts. This is the gyroscopic effect in action; the angle of the wheel is directly proportional to the amount and time you are turning. Put a gyroscope in an airplane or ship and you can record even imperceptible changes in movement, assuming you are using a precision gyroscope.

Gyros are still used in airplanes today, even with radar, ground controllers, and radios to guide their way. While many modern aircraft have substituted mechanical gyros with completely electronic ones, the concept is the same. During flight, any changes in direction are recorded by the inertial guidance system in the plane (there are three gyros, for all three axes). At any time during the flight the course of the plane can be scrutinized by looking at the output of the gyroscopes.

Inertial guidance systems for planes, ships, missiles, and other such devices are far, far too expensive for robots. However, there are some low-cost gyros that provide modest accuracies. One reasonably affordable model is the Max Products MX-9100 micro piezo gyro, often used in model helicopters. The MX-9100 uses a piezoelectric transducer to sense motion. This motion is converted into a digital signal whose duty cycle changes in proportion to the rate of change in the gyro.

Laser- and fiber-optic-based gyroscopes offer another navigational possibility, though the price for ready-made systems is still out of the reach of most hobby robot enthusiasts. These devices use interferometry—the subtle changes in the measured wavelength of a light source that travels in opposite directions around the circumference of the gyroscope. The light is recombined onto a photosensor or a photocell array such as a CCD camera. In the traditional laser-based gyroscopes (e.g., the Honeywell ring gyro), the two light beams create a bull's-eye pattern that is analyzed by a computer. In simpler fiber-optic systems, the light beams are mixed and received by a single phototransistor. The wave patterns of the laser light produce sum and difference signals (heterodyning). The difference signals are well within audio frequency ranges, and these can be interpreted using a simple frequency-to-voltage converter. From there, relative motion can be calculated.

A low-cost method to experiment with inertial navigation is to use accelerometers (similar to those described in detail in Chapter 35, "Experimenting with Tilt and Gravity Sensors"). The nature of accelerometers, particularly the less-expensive piezoelectric variety, can make them difficult to employ in an inertial system. Accelerometers are sensitive to the earth's gravity, and tilting on the part of the robot can introduce errors. By using multiple accelerometers—one to measure movement of the robot and one to determine tilt—it is generally possible to reduce (but perhaps not eliminate) these errors.

33.8.3 MAP MATCHING

Maps help us navigate strange towns and roads. By correlating what we see out the windshield with the street names on the map, we can readily determine where we are—or just how lost we are! Likewise, given a map of its environment, a robot could use its various sensors to correlate its position with the information in a map. Map-based positioning, also known as map matching, uses a map you prepare for the robot or that the robot prepares for itself.

During navigation, the robot uses whatever sensors it has at its disposal (infrared, ultrasonic, vision, etc.) to visualize its environment. It checks the results of its sensors against the map previously stored in its memory. From there, it can compute its actual position and orientation. One common technique, developed by robot pioneer Hans Moravec, uses a "certainty grid" that consists of squares drawn inside the mapped environment (think of graph paper used in school). Objects, including obstacles, are placed within the squares. The robot can use this grid map to determine its location by attempting to match what it sees through its sensors with the patterns on the map.

Obviously, map matching requires that a map of the robot's environment be created first. Several consumer robots, like the Cyebot, are designed to do this mapping autonomously by exploring the environment over a period of time. Industrial robots typically require that the map be created using a CAD program and the structure and objects within it very accurately rendered. The introduction of new objects into the environment can drastically decrease the accuracy of map matching, however. The robot may mistake a car for a foot stool, for example, and seriously misjudge its location.

33.9 From Here

To learn more about . . .	***Read***
Using computers and microcontrollers in your robots	Chapter 14, "Computer Peripherals"
Infrared and wireless communications techniques	Chapter 16, "Remote Control Systems"
Keeping your robot from crashing into things	Chapter 30, "Object Detection"
Vision for your robot	Chapter 32, "Robot Vision"
Preventing your robot from falling over	Chapter 35, "Experimenting with Tilt and Gravity Sensors"

CHAPTER 34

FIRE DETECTION SYSTEMS

Everyone complains that a robot is good for nothing—except, perhaps, providing its master with a way to tinker with gadgets in the name of "science"! But here's one useful and potentially life-saving application you can give your robot in short order: fire and smoke detection. As this chapter will show, you can easily attach sensors to your robot to detect flames, heat, and smoke, making your robot into a mobile smoke detector.

34.1 Flame Detection

Flame detection requires little more than a sensor that detects infrared light and a circuit to trigger a motor, siren, computer, or other device when the sensor is activated. As it turns out, almost all phototransistors are specifically designed to be sensitive primarily to infrared or near-infrared light. You need only connect a few components to the phototransistor and you've made a complete flame detection circuit. Interestingly, the detector can "see" flames that we can't. Many gases, including hydrogen and propane, burn with little visible flame. The detector can spot them before you can, or before the flames light something on fire and smoke fills the room.

34.1.1 DETECTING THE INFRARED LIGHT FROM A FIRE

The simple circuit in Fig. 34-1 shows the most straightforward method for detecting flames. (See parts list in Table 34-1.) You mask a phototransistor so it sees only infrared light by

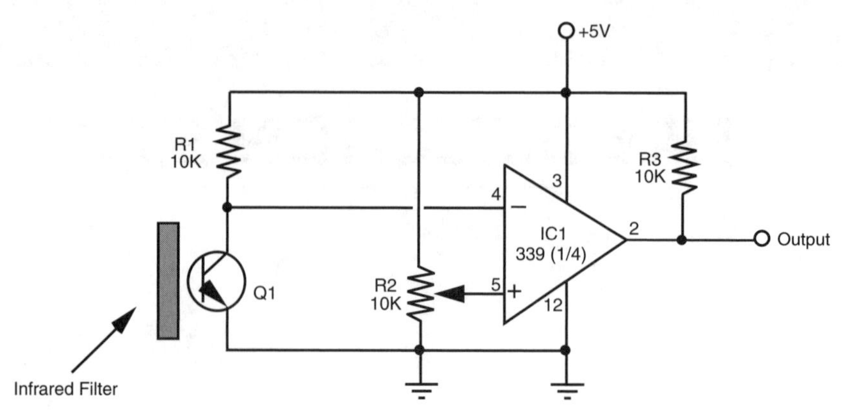

FIGURE 34-1 A flame detector built around the LM399 quad comparator IC.

using an opaque infrared filter. Some phototransistors have the filter built in; with others, you'll have to add the filter yourself. If the phototransistor does not have an infrared-filtered lens, add one for the light to pass through. Place the transistor at the end of a small opaque tube, say one with a ¼- or ½-in I.D. (the black tubing for drip irrigation is a good choice). Glue the filter to the end of the tube. The idea is to block all light to the transistor except that which is passed through the filter.

If you are looking for an infrared filter, a low-cost one that you can use is a piece of overexposed film negative. While digital cameras are becoming very popular, there are still a lot of film cameras being used and you should be able to get a scrap of processed overexposed film negative from a photo developer.

In the circuit, when infrared light hits the phototransistor it triggers on. The brighter the infrared source is, the more voltage is applied to the inverting input of the comparator. If the input voltage exceeds the reference voltage applied to the noninverting input, the output of the comparator changes state.

Potentiometer R2 sets the sensitivity of the circuit. You'll want to turn the sensitivity down so ambient infrared light does not trigger the comparator. You'll find that the circuit does not work when the background light has excessive infrared content. You can't, for example, use the circuit when the sensor is pointed directly at an incandescent light or the sun.

TABLE 34-1 Parts List for Flame Detector

IC1	LM339 quad comparator IC
R1, R3	10K resistor
R2	10K potentiometer
Q1	Infrared-sensitive phototransistor
Misc.	Infrared filter for phototransistor (if needed)

Test the circuit by connecting an LED and 270-Ω resistor from the Vout terminal to ground. Point the sensor at a wall, and note the condition of the LED. Now, wave a match in front of the phototransistor. The LED should blink on and off. You'll notice that the circuit is sensitive to all sources of infrared light, which includes the sun, strong photolamps, and electric burners. If the circuit doesn't seem to be working quite right, look for hidden sources of infrared light. With the resistor values shown, the circuit is fairly sensitive; you can change them by adjusting the value of R1 and R2.

34.1.2 WATCHING FOR THE FLICKER OF FIRE

No doubt you've watched a fire at the beach or in a fireplace and noted that the flame changes color depending on the material being burned. Some materials burn yellow or orange, while others burn green or blue (indeed, this is how those specialty fireplace logs burn in different colors). Just like the color signature given off when different materials burn, the flames of the fire flicker at different but predictable rates.

You can use this so-called *flame modulation* in a robot fire detection system to determine what is a real fire and what is likely just sunlight streaming through a window or light from a nearby incandescent lamp. By detecting the rate of flicker from a fire and referencing it against known values, it is possible to greatly reduce false alarms. The technique is beyond the scope of this book, but you could design a simple flame-flicker system using an op amp, a fast analog-to-digital converter, and a computer or microcontroller. The analog-to-digital converter would translate the instantaneous brightness changes of the fire into digital signals.

The patterns made by those signals could then be referenced against those made by known sources of fire. The closer the patterns match, the greater the likelihood that there is a real fire. In a commercial product of this nature, it is more likely that the device would use more sophisticated digital signal processing.

34.2 Using a Pyroelectric Sensor to Detect Fire

A pyroelectric sensor is sensitive to the infrared radiation emitted by most fires. The most common use of pyroelectric infrared (or PIR) sensors is in burglar alarms and motion detectors. The sensor detects the change in ambient infrared radiation as a person (or animal or other heat-generating object) moves within the field of view of the sensor. The key ingredient here is change: a PIR sensor cannot detect heat per se but the changes in the heat within its field of view. In larger fires, the flickering flames create enough of a change to trigger the PIR detector.

Chapter 30, "Object Detection," discusses how to use PIR sensors to detect the motion of people and animals around a robot. The same sensor, with little or no change, can be employed to detect fires. To be effective as a firefighter, you should ideally reduce the sensor's field of view so the robot can detect smaller fires. The larger the field of view, the more the temperature and/or position of the heat source must change in order for the PIR sensor to detect it.

With a smaller field of view, the magnitude of change can be lower. However, with a small field of view, your robot will likely need to sweep the room, using a servo or stepper motor, in order to observe any possible fires. The sweeping must stop periodically so the robot can take a room reading. Otherwise, the motion of the sensor could trigger false alarms.

34.3 Smoke Detection

"Where there's smoke, there's fire." Statistics show that the majority of fire deaths each year are caused not by burns but by smoke inhalation. For less than $15, you can add smoke detection to your robot's long list of capabilities and with a little bit of programming have it wander through the house checking each room for trouble. You'll probably want to keep it in the most fire-prone rooms, such as the basement, kitchen, laundry room, and robot lab.

You can build your own smoke detector using individually purchased components, but some items, such as the smoke detector cell, are hard to find. It's much easier to use a commercially available smoke detector and modify it for use with your robot. In fact, the process is so simple that you can add one to each of your robots. Tear the smoke detector apart and strip it down to the base circuit board.

The active element used for detecting smoke—the radioactive substance Americium 241—has a half-life of approximately seven years. After about five to seven years, the effectiveness of the alarm is diminished, and you should replace it. Using a very old alarm will render your "Smokey the Robot" fairly ineffectual at detecting the smoke of fires. Remember to properly dispose of old smoke detectors according to the rules and laws where you live.

34.3.1 HACKING A SMOKE ALARM

You can either buy a new smoke detector module for your robot or scavenge one from a commercial smoke alarm unit. The latter tends to be considerably cheaper—you can buy quality smoke alarms for as little as $7 to $10. Hacking a commercial smoke alarm, specifically a Kidde model 0915K, so it can be directly connected to a robot's computer port or microcontroller will be presented in this chapter. Of course, smoke alarms are not all designed the same, but the basic construction is similar to that described here. You should have relatively little trouble hacking most any smoke detector.

However, you should limit your hacking attempts to those smoke alarms that use traditional 9-V batteries. Certain smoke alarm models, particularly older ones, require AC power or specialized batteries (such as 22-V mercury cells). These are harder to salvage and, besides, their age makes them less suitable for sensitive smoke detection.

Start by checking the alarm for proper operation. If it doesn't have one already, insert a fresh battery into the battery compartment. Put plugs in your ears (or cover up the audio transducer hole on the alarm). Press the test button on the alarm; if it is properly functioning the alarm should emit a loud, piercing tone. If everything checks out okay, remove the battery, and disassemble the alarm. Less expensive models will not have screws but will likely use a snap-on construction. Use a small flat-headed screwdriver to unsnap the snaps.

Inside the smoke detector will be a circuit board, like the one in Fig. 34-2, that consists of the drive electronics and the smoke detector chamber.

Either mounted on the board or located elsewhere will be the piezo disc used to make the loud tone. Remove the circuit board, being careful you don't damage it. Examine the board for obvious hack points, and note the wiring to the piezo disc. More than likely, there will be either two or three wires going to the disc.

- *Two wires to the piezo disc.* The wires will provide ground and +V power. This design is typical when you are using all-in-one piezo disc buzzers, in which the disc itself contains the electronics to produce the signal for audible tones.
- *Three wires to the piezo disc.* The wires will provide ground, +V power, and a signal that causes the disc to oscillate with an audible tone.

Using a volt-ohm meter, find the wire that serves as ground. It is probably black or brown, but if no obvious color coding is used, examine the circuit board and determine where the wires are attached. An easy way to find the ground connection is to find the lowest resistance wire to the negative battery lead. Connect the other test lead to the remaining wire, or if the disc has three wires, connect the test lead to one of the remaining wires.

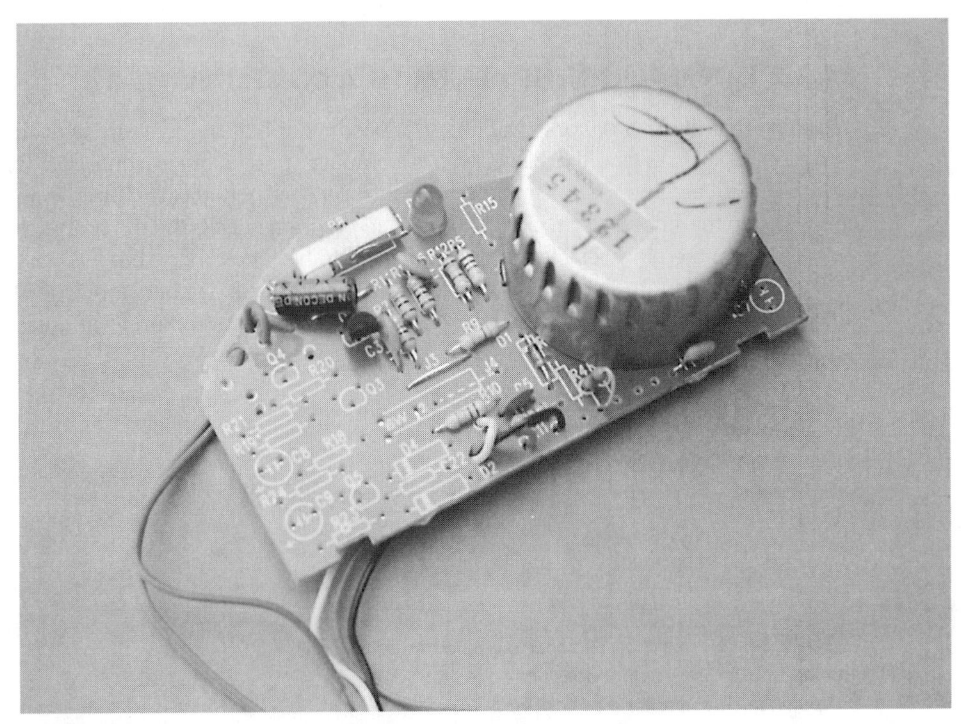

FIGURE 34-2 The guts of a smoke detector.

Replace the battery in the battery compartment, and depress the test button on the alarm. Watch for a change in voltage. For a two-wire disc you should see the voltage change as the tone is produced. For a three-wire disc, try each wire to determine which produces the higher voltage; that is the one you wish to use.

Once you have determined the functions of the wires to the piezo disc, clip off the disc and save it for some other project. Retest the alarm's circuit board to make sure you can still read the voltage changes with your volt-ohm meter. Then clip off the wires to the battery compartment, noting their polarity. Connect the circuit to a +5 V power supply. Depress the test button again. Ideally, the circuit will still function with the lower voltage. If it does not, you'll need to operate the smoke alarm circuit board with +9 V, which can complicate your robot's power supply and interfacing needs.

Remember to note the voltage when the piezo disc is active. It should not be more than +5 V; if it is, that means the circuit board contains circuitry for increasing the drive voltage to the piezo disc. You don't want this when you are interfacing the board to a computer port or microcontroller, so you'll need to limit the voltage by using a circuit such as that shown in Fig. 34-3. Here, the output of the smoke alarm circuit is clamped at no more than 5.1 V, thanks to the 5.1-V zener diode.

There is no effective way to measure the peak voltage by using a volt-ohm meter, because the output of the smoke alarm detector is often an oscillating signal. The meter will only show an average of the voltage provided by the circuit. If you are limited to using only a volt-ohm meter for your testing, for safety's sake add the 5.1-V zener circuit as shown in Fig. 34-3. While this may be unnecessary in some instances, it will help protect your digital interface from possible damage caused by overvoltage from the smoke alarm circuit board.

34.3.2 INTERFACING THE ALARM TO A COMPUTER

Assuming that the board works with +5 V applied, your hacking is basically over, and you can proceed to interface the alarm with a computer port or microcontroller. By way of example, we'll assume that a simple microcontroller that periodically polls the input pin is connected to the smoke alarm circuit board. The program checks the pin several times each second. When the pin goes HIGH, the smoke alarm has been triggered.

If your microcontroller supports interrupts (ideally with latched inputs that can be polled), a better scheme is to connect the smoke alarm circuit board to an interrupt pin.

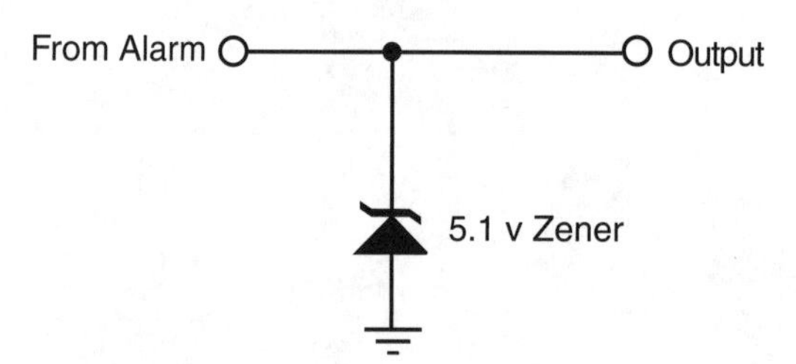

FIGURE 34-3 Use a 5.1 V zener diode to ensure that the smoke alarm output does not drive the computer/microcontroller input above 5 V.

Then write your software so that if the interrupt pin input state changes, a special "I smell smoke" routine is run. The benefit of an interrupt or latched input over polling is that the latter requires your program to constantly branch off to check the condition of the input pin. With an interrupt or latched input, your software program can effectively be ignorant of any smoke detector functionality. If and when the signal pin is active because the smoke alarm circuit was tripped, a special software routine takes over, commanding the robot to do something else.

Rather than connect the output of the smoke alarm circuit board directly to the input pin, use a buffer to protect the microcontroller or computer against possible damage. You can construct a buffer using logic circuits (either TTL or CMOS) or with an op amp wired for unity-gain (with unity-gain, the op amp doesn't amplify anything). Note also that the smoke alarm circuit board derives its power from the robot's main +5 V power supply and not from the microcontroller.

Alternatively, you can use an opto-isolator. The opto-isolator bridges the gap between the detector and the robot. You do not need to condition the output of the opto-isolator if you are connecting it to a computer or microprocessor port or to a microcontroller.

34.3.3 TESTING THE ALARM

Once the smoke alarm circuit board is connected to the microcontroller or computer port, test it and your software by triggering the test button on the smoke alarm. The software should branch off to its "I smell smoke" subroutine. For a final test, light a match, and then blow it out. Wave the smoldering match near the smoke detector chamber. Again, the software runs the "I smell smoke" subroutine.

34.3.4 LIMITATIONS OF ROBOTS DETECTING SMOKE

You should be aware of certain limitations inherent in robot fire detectors. In the early stages of a fire, smoke tends to cling to the ceilings. That's why manufacturers recommend that you place smoke detectors on the ceiling rather than on the wall. Only when the fire gets going and smoke builds up does it start to fill up the rest of the room.

Your robot is probably a rather short creature, and it might not detect smoke that confines itself only to the ceiling. This is not to say that the smoke detector mounted on even a 1-ft-high robot won't detect the smoke from a small fire; just don't count on it. Back up the robot smoke sensor with conventionally mounted smoke detection units, and do not rely only on the robot's smoke alarm.

34.3.5 DETECTING NOXIOUS FUMES

Smoke alarms detect the smoke from fires but not noxious fumes. Some fires emit very little smoke but plenty of toxic fumes, and these are left undetected by the traditional smoke alarm. Moreover, potentially deadly fumes can be produced in the absence of a fire. For example, a malfunctioning gas heater can generate poisonous carbon monoxide gas. This colorless, odorless gas can cause dizziness, headaches, sleepiness, and—if the concentration is high enough—even death.

Just as there are alarms for detecting smoke, so there are alarms for detecting noxious gases, including carbon monoxide. Such gas alarms tend to be a little more expensive than smoke alarms, but they can be hacked in much the same way as a smoke alarm. Deduce the signal wires to the piezo disc and connect them (perhaps via a buffer and zener diode voltage clamp) to a computer port or microcontroller.

Combination units that include both a smoke and gas alarm are also available. You should determine if the all-in-one design will be useful for you. In some combination smoke–gas alarm units, there is no simple way to determine which has been detected. Ideally, you'll want your robot to determine the nature of the alarm, either smoke or gas (or perhaps both).

34.4 Heat Sensing

In a fire, smoke and flames are most often encountered before heat, which isn't felt until the fire is going strong. But what about before the fire gets started in the first place, such as when a kerosene heater is inadvertently left on or an iron has been tipped over and is melting the nylon clothes underneath?

If your robot is on wheels (or legs) and is wandering through the house, perhaps it'll be in the right place at the right time and sense these irregular situations. A fire is brewing, and before the house fills with smoke or flames the air gets a little warm. Equipped with a heat sensor, the robot can actually seek out warmer air, and if the air temperature gets too high it can sound an initial alarm.

Realistically, heat sensors provide the least protection against a fire. But heat sensors are easy to build, and, besides, when the robot isn't sniffing out fires it can be wandering through the house giving it an energy check or reporting on the outside temperature or . . . you get the idea.

Fig. 34-4 shows a basic but workable circuit centered around an LM355 temperature sensor (it and the other parts in the circuit are listed in Table 34-2). This device is relatively easy to find and costs under $1.50. The output of the device, when wired as shown, is a linear voltage. The voltage increases 10 mV for every rise in temperature of 1° Kelvin (K).

Degrees Kelvin uses the same scale as degrees Celsius (C), except that the zero point is absolute zero—about –273C. One degree Centigrade equals 1° Kelvin; only the start points

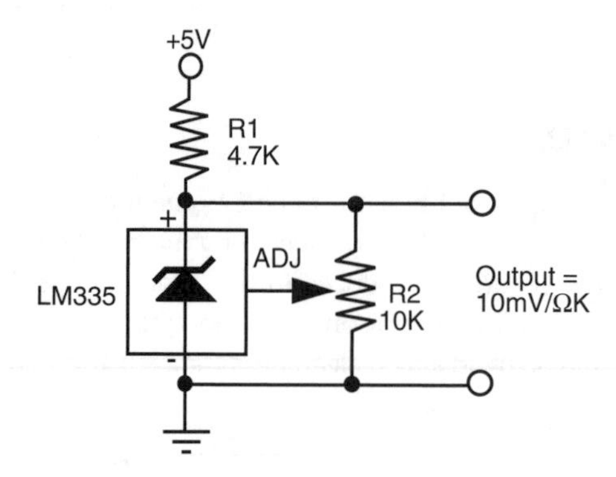

FIGURE 34-4 The basic wiring diagram for the LM355 temperature sensor.

TABLE 34-2 Parts List for the Basic Temperature Transducer

R1	4.7K resistor, 1 percent tolerance
R2	10K 10-turn precision potentiometer
D1	LM335 temperature sensor diode

differ. You can use this to your advantage because it lets you easily convert degrees Kelvin into degrees Celsius. Actually, since your robot will be deciding when hot is hot, and doesn't care what temperature scale is used, conversion really isn't necessary.

You can test the circuit by connecting a volt-ohm meter to the ground and output terminals of the circuit. At room temperature, the output should be about 2.98 V. You can calculate the temperature if you get another reading by subtracting the voltage by 273 (ignore the decimal point but make sure there are two digits to the right of it, even if they are zeros). What's left is the temperature in degrees Celsius. For example, if the reading is 3.10 V, the temperature is 62C (310 – 273 = 62). By the way, that's pretty hot (144° Fahrenheit).

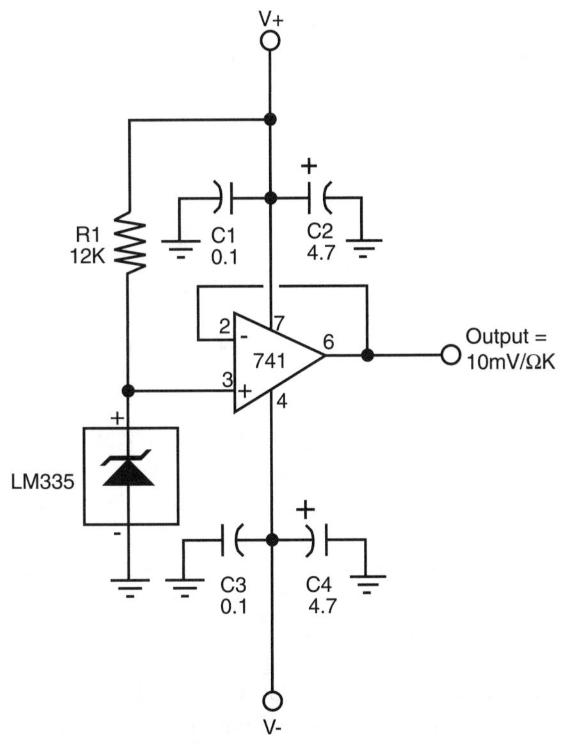

FIGURE 34-5 An enhanced wiring scheme for the LM355 temperature sensor. The load of the output is buffered and does not affect the reading from the LM355.

TABLE 34-3 Parts List for the Buffered Temperature Transducer

R1	12K resistor, 1 percent tolerance
C1, C3	0.1 µF ceramic capacitor
C2, C4	4.7 µF tantalum capacitor
D1	LM335 temperature sensor diode

You can calibrate the circuit, if needed, by using an accurate bulb thermometer as a reference and adjusting R2 for the proper voltage. By knowing the temperature, you can determine what the output voltage should be by adding the temperature (in degrees Celsius) to 273. If the temperature is 20C, then the output voltage should be 2.93 V. For more accuracy, float some ice in a glass of water for 15 to 20 minutes and stick the sensor in it (keep the leads of the testing apparatus dry). Wait 5 to 10 minutes for the sensor to settle and read the voltage. It should be exactly 2.73 V.

The load presented at the outputs of the sensor circuit can throw off the reading. The schematic in Fig. 34-5 and parts list in Table 34-3 provides a buffer circuit so the load does not interfere with the output of the 355 temperature sensor. Note the use of the decoupling capacitors as recommended in the manufacturer's application notes. These aren't essential, but they are a good idea.

34.5 Firefighting

By attaching a small fire extinguisher to your robot, you can have the automaton put out the fires it detects. Obviously, you'll want to make sure that the fire detection scheme you've put into use is relatively free of false alarms and that it doesn't overreact to normal situations. Having your robot rush over to one of your guests and put out a cigarette he just lit is a potential values judgment.

It's a good idea to use a clean fire-extinguishing agent for your firefighting 'bot. Halon is one of the best such agents, but is no longer manufactured for general consumption. Alternatives include inert gases (helium, argon) and noncombustible gases (e.g., nitrogen), and they will not leave a sediment on whatever they are sprayed on.

No matter what you use for the fire extinguisher, be sure to use caution as a guide when building any firefighting robot. Consider limiting your robot for experimental use, and test it only in well-ventilated rooms—or better yet, outside.

The exact mounting and triggering scheme you use depends entirely on the design of the fire extinguisher. The bottle used in the prototype firebot is a Kidde Force-9, 2½ lb Halon extinguisher. It has a diameter of about 3¼ in. You can mount the extinguisher in the robot by using plumber's tape, that flexible metallic strip used by plumbers to mount water and gas pipes. It has lots of holes already drilled into it for easy mounting. Use two strips to hold the bottle securely. Remember that a fully charged extinguisher is heavy—in this case over

3 lb (2½ lb for the Halon chemical and about ½ lb for the bottle). If you add a fire extinguisher to your robot, you must relocate other components to evenly distribute the weight.

The extinguisher used in the prototype system for this book used a standard actuating valve. To release the fire retardant, you squeeze two levers together. Fig. 34-6 shows how to use a heavy-duty solenoid to remotely actuate the valve. You may be able to access the valve plunger itself (you may have to remove the levers to do so). Rig up a heavy-duty solenoid and lever system. A computer or control circuit activates the solenoid.

For best results, the valve should be opened and closed in quick bursts (200 to 300 ms are about right). The body of the robot should also pivot back and forth so the extinguishing agent is spread evenly over the fire. Remember that to be effective, the extinguishing agent must be sprayed at the base of the fire, not at the flames. For most fires, this is not a problem because the typical robot stays close to the floor. If the fire is up high, the robot may not be able to effectively fight it.

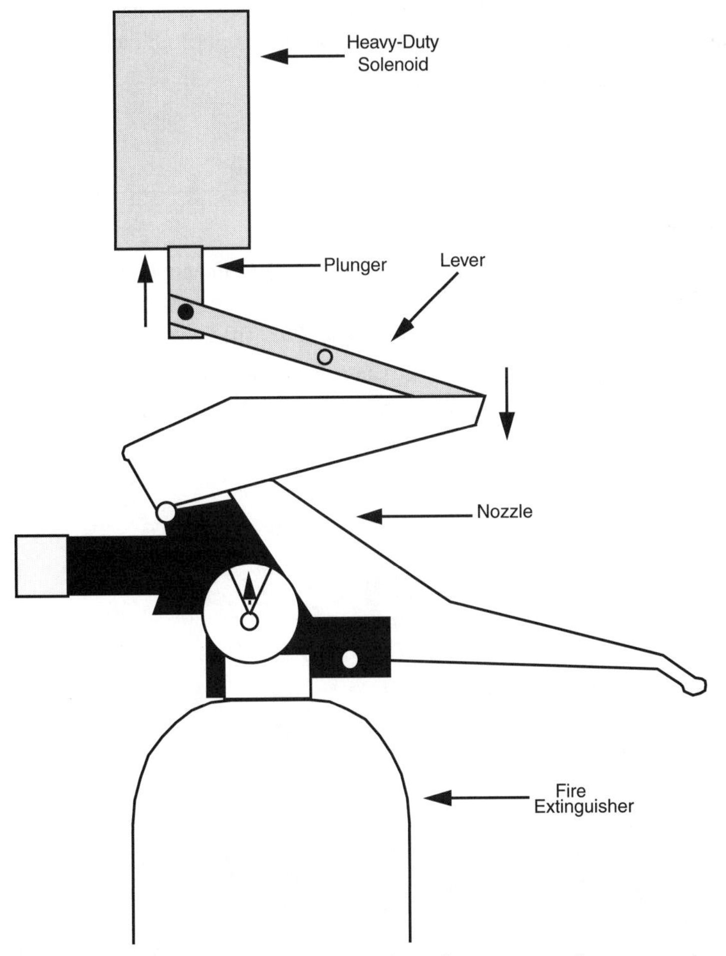

FIGURE 34-6 Using a heavy-duty solenoid to activate a fire extinguisher.

You can test the fire extinguisher a few times before the bottle will need recharging. The fully charged bottle in the prototype was able to squeeze off several dozen short blasts before the built-in pressure gauge registered that a new charge was needed. Don't use your only extinguisher for your robot experiments; keep an extra handy in the unlikely event that you have to fight a fire yourself.

If the firefighting robot bug bites you hard, consider entering your machine in the annual Trinity College Firefighting Home Robot Contest (see www.trincoll.edu/events/robot/ for additional information, including rules and a description of the event). This contest involves timing a robot as it goes from room to room in a houselike test field (the "house" and all its rooms are in a reduced scale). The object is to find the fire of a candle and snuff it out in the least amount of time. Separate competitions involving a junior division (high school and younger) and a senior division (everyone else) help to provide an even playing field for the contestants.

Rather than a fire extinguisher, competition firefighting 'bots usually use a fan (such as the muffin fans used to cool computer systems) to blow out the candle that is used as the fire source. The advantages of the fan over the fire extinguisher is its light weight, fairly low power consumption, and ease in which it can be scanned about to blow out the flame.

34.6 From Here

To learn more about . . .	***Read***
Connecting sensors to computers and microcontrollers	Chapter 14, "Computer Peripherals"
Adding the sensation of "touch"	Chapter 29, "The Sense of Touch"
Optical systems for detecting light	Chapter 32, "Robot Vision"
Enabling the robot to move around in a room or house	Chapter 33, "Navigating through Space"
Adding a siren or other warning device	Chapter 31, "Sound Output and Input"

CHAPTER 35

EXPERIMENTING WITH TILT AND GRAVITY SENSORS

Every schoolchild learns the human body has five senses: sight, hearing, touch, smell, and taste. These are primary developed senses; yet the body is endowed with far more senses, including many we often take for granted. These more "primitive" human senses are typically termed *sixth senses*—a generic phrase for a sense that doesn't otherwise fit within the common five. One of the most important sixth senses is the sense of balance. This sense is made possible by a complex network of nerves throughout the body, including those in the inner ear. The sense of balance helps us to stand upright and to sense when we're falling. When we're off balance, the body naturally attempts to reestablish an equilibrium. The sense of balance is one of the primary prerequisites for two-legged walking.

Our sense of balance combines information about both the body's angle and its motion. At least part of the sense of balance is derived from a sensation of gravity—the pull on our bodies from the earth's mass. Gravity is an extraordinarily strong physical force, but strangely enough it is not often used in hobby robotics because accurate sensors for measuring it have been prohibitively expensive.

But just consider the possibilities if a robot were given the ability to "feel" gravity. The same forces of gravity that help us to stay upright might provide a two-legged robot with the sensation that would keep it upright. Or a rolling robot—on wheels or tracks—might avoid tipping over and damaging something by determining if its angle is too steep. The sense of gravity might enable the robot to avoid traveling over that terrain, or it might tell the robot to shift some internal ballast weight (assuming it were so equipped) to change its center of balance.

35.1 Sensors to Measure Tilt

One of the most common means for providing a robot with a sense of balance is to use a tilt sensor or tilt switch. The sensor or switch measures the relative angle of the robot with respect to the center of the earth. If the robot tips over, the angle of a weight inside the sensor changes, and this change produces an electrical signal that can be detected by electronics in the robot. Tilt sensors and switches come in various forms and packages, but the most common are the following:

- Mercury-filled glass ampoules that form a simple on/off switch. When the tilt switch is in one position (say, horizontal), the liquid-mercury metal touches contacts inside the ampoule, and the switch is closed. But when the switch is rotated to vertical, the mercury no longer touches the contacts, and the switch is open. The major disadvantage of mercury tilt switches is the mercury itself, which is a highly toxic metal.
- The ball-in-cage (see Fig. 35-1) is an all-mechanical switch popular in pinball machines and other devices where small changes in level are required. The switch is a square or round capsule with a metal ball inside. Inside the capsule are two or more electrical con-

FIGURE 35-1 Typical ball-in-cage switches.

tacts. The weight of the ball makes it touch the electrical contacts, which forms a switch. The capsule may have multiple contacts so it can measure tilt in all directions.

- Electronic spirit-level sensors use the common fluid bubble you see on ordinary levels at the hardware store plus some interfacing electronics. A spirit level is merely a glass tube filled, though not to capacity, with water or some other fluid. A bubble forms at the top of the tube since it isn't completely filled. When you tilt the tube, gravity makes the bubble slosh back and forth. An optical sensor—an infrared LED and detector, for example—can be used to measure the relative size and position of the bubble.
- Electrolytic tilt sensors are like mercury switches but more complex and a lot more costly. In an electrolytic tilt sensor, a glass ampoule is filled with a special electrolyte liquid—that is, a liquid that conducts electricity but in very measured amounts. As the switch tilts, the electrolyte in the ampoule sloshes around, changing the capacitance between two (or more) metal contacts.

35.1.1 BUILDING A BALANCE SYSTEM WITH A MERCURY SWITCH

You can construct a simple but practical balance system for your robot using two small mercury switches. You want mercury switches that will open (or close) at fairly minor angles, perhaps 30 to 35 degrees or so—just enough to signal to the robot that it is in danger of tipping over. You may have to purchase the switch with these specifications through a specialty industrial parts store, unless you're lucky enough to find one on the used or surplus market.

Mount the switch in an upright position. If the level of the robot becomes extreme, the switch will trigger. You can directly interface the switch to an I/O (input/output) line on a PC or microcontroller. However, you'll probably want to include a debounce circuit in line with the switch and I/O line since mercury switches can be fairly noisy electrically. A suitable debouncer circuit is shown in Chapter 29.

35.1.2 BUILDING A BALANCE SYSTEM WITH A BALL-IN-CAGE SWITCH

The four-conductor ball-in-cage switch is a rather common find in the surplus market, and it's very inexpensive. If the switch is tilted in any direction by more than about 25 to 30 degrees, at least one of the four contacts in the switch will close, thus indicating that the robot is off level. You can use a debouncer circuit with the ball-in-cage tilt sensor.

Because the ball-in-cage sensor has four contacts (plus a center common), you can either provide independent outputs of the switch or a common output. With independent outputs, a PC or microcontroller on your robot can determine in which direction the robot is tilting (if two contacts are closed, then the ball is straddling two contacts at the same time). However, unless you come up with some fancy interface circuitry, you'll need to dedicate four I/O lines on the PC or microcontroller, one for each switch contact.

Conversely, with the common output approach you can wire all the outputs together in a serial chain. The switch will close if the ball touches any contact. This approach uses only one I/O line, but it deprives the robot of the ability to know in exactly which direction it is off level. A variation on this theme is to use resistors of specific values to form a voltage divider. When you connect the resistors to an I/O line capable of analog input, you can easily determine by the changing voltage at the input which contact switch has been closed.

35.2 Using an Accelerometer to Measure Tilt

One of the most accurate, yet surprisingly low-cost, methods for tilt measurement involves an accelerometer. Once the province only of high-tech aviation and automotive testing labs, accelerometers are quickly becoming common staples in consumer electronics. It's quite possible, for example, that your late-model car contains at least one accelerometer—if not as part of its collision safety system (such as an airbag), then perhaps as an integral part of its burglar alarm. Accelerometers are also increasingly used in laptop computers (used to put the laptop into a suspend mode with the hard drive parked, assuming that if the angle changes drastically, it has tipped off a table), high-end video game controllers, portable electric heaters, and in-home medical equipment.

New techniques for manufacturing accelerometers have made them more sensitive and accurate yet also less expensive. A device that might have cost upwards of $500 a few years ago sells in quantity for under $10 today.

The Analog Devices accelerometers discussed in the following sections are available through a number of retail outlets, and neither requires extensive external circuitry. While the text that follows is specific to the accelerometers from Analog Devices, you may substitute units from other sources after making the appropriate changes in the circuitry and computer interface software.

35.2.1 WHAT IS AN ACCELEROMETER?

The basic accelerometer is a device that measures change in speed. Put an accelerometer in your car, for example, and step on the gas. The device will measure the increase in speed. Most accelerometers only measure acceleration (or deceleration) and not constant speed or velocity. Such is the case with the accelerometers detailed here.

Though accelerometers are designed to measure changes in speed, many types of accelerometers—including the ones detailed in the following sections—are also sensitive to the constant pull of the earth's gravity. It is this latter capability that is of interest to us since it means you can use the accelerometer to measure the tilt, or "attitude," of your robot at any given time. This tilt is represented by a change in the gravitational forces acting on the sensor. The output of the accelerometer is either a linear AC or DC voltage or, more handily, a digital pulse that changes in response to the acceleration or gravity forces.

To measure the tilt of a robot, an accelerometer is placed parallel to the surface a robot runs on. When the robot is stopped on a level surface, the accelerometer will not detect any acceleration (due to gravity). When the robot is tilted, the acceleration due to gravity will be sensed by the accelerometer according to the formula:

$$\text{Tilt Acceleration Due to Gravity} = 9.9 \text{ m/s}^2 \times \sin(\text{Tilt Angle})$$

Note that when you try to measure the tilt angle of a robot from the acceleration an accelerometer measures, you will discover that it is impossible to measure unless the robot has stopped. While Newton's First Law of Motion implies that if the robot is moving constantly, then accelerometer measurements of tilt angle should be accurate, this is not the case in the real world. Motor surges, vibration, and bumps in the surface make it impossible to get an accurate measurement of the tilt angle.

35.2.2 ADDITIONAL USES FOR ACCELEROMETERS

Before going into the details of using accelerometers for tilt and angle measurement in robots, here is a list of some of the different robotics-based sensor applications for these devices. Apart from sensing the angle of tilt, a gravity-sensitive accelerometer can also be used for the following tasks:

- *Shock and vibration.* If the robot bumps into something, the output of the accelerometer will spike instantaneously. Because the output of the accelerometer is proportional to the power of the impact, the harder the robot bumps into something, the larger the voltage spike. You can use this feature for collision detection, obviously, but in ways that far exceed what is possible with simple bumper switches since an accelerometer is sensitive to shock from most any direction.
- *Motion detection.* An accelerometer can detect motion even if the robot's wheels aren't moving. This might be useful for robots that must travel over uneven or unpredictable terrain. Should the robot move (or stop moving) when it's not supposed to, this will show up as a change in speed and will therefore be sensed by the accelerometer.
- *Telerobotic control.* You can use accelerometers mounted on your clothes to transmit your movements to a robot. For instance, accelerometers attached to your feet can detect the motion of your legs. This information could be transmitted (via radio or infrared link) to a legged robot, which could replicate those moves. Or you might construct an "air stick" wireless joystick, which would simply be a pipe with an accelerometer at the top or bottom and some kind of transmitter circuit. As you move the joystick your movements are sent to your robot, which acts in kind.

35.2.3 SINGLE- AND DUAL-AXIS SENSING

The basic accelerometer is single axis, meaning it can detect a change in acceleration (or gravity) in one axis only, as shown in Fig. 35-2. While this is moderately restrictive, you can still use such a device to create a capable and accurate tilt-and-motion sensor for your robot. The first accelerometer project described in this chapter uses such a single-axis device.

A dual-axis accelerometer detects changes in acceleration and gravity in both the x and y planes (see Fig. 35-2). If the sensor is mounted vertically—so that the y axis points straight up and down—the y axis detects up and down changes, and the x axis will detect side-to-side motion. Conversely, if the sensor is mounted horizontally, the y axis detects motion forward and backward, and the x axis detects motion from side to side.

35.2.4 THE ANALOG DEVICES' ADXL ACCELEROMETER FAMILY

Analog Devices is a semiconductor maker primarily for industrial- and military-grade operational amplifiers, digital-to-analog and analog-to-digital converters, and motion control products. One of their key product lines is accelerometers, and for them they use a patented

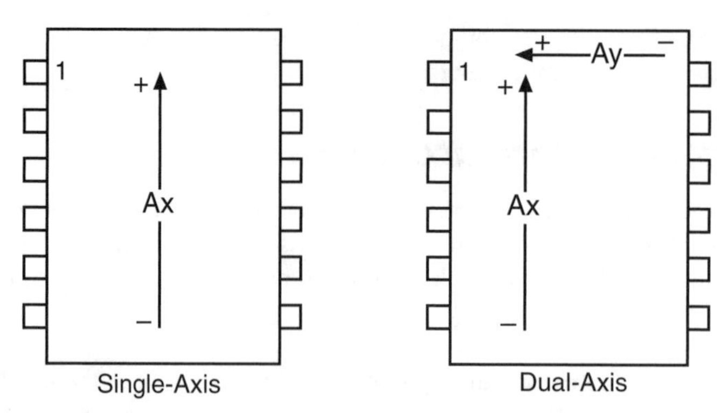

FIGURE 35-2 *a.* A single-axis accelerometer; *b.* a dual-axis accelerometer.

fabrication process to create a series of near-microscopic mechanical beams. This micromachining involves etching material out of a substrate. During acceleration, the beam is distended along its length. This distention changes the capacitance in nearby plates. This change in capacitance is correlated as acceleration.

A thorough discussion of the theory of operation behind the ADXL family of accelerometers is beyond the scope of this chapter, but you can obtain much more on the subject directly from the manufacturer. Check the Analog Devices web site at www.analog.com.

In addition to the mechanical portions of the accelerometer, all the basic interface circuitry is part of the device. In fact, when looking at one of the ADXL accelerometers, you'd think it was just an integrated circuit of some kind. Because the basic circuitry is included as part of the accelerometer, only a minimum number of external parts are needed.

35.3 Constructing a Dual-Axis Accelerometer Robotic Sensor

Analog Devices makes a low-cost line of accelerometers specifically designed for consumer products. Their ADXL202 is a dual-axis device with a ±2-g sensitivity (if you need more g's, check out the ADXL210, which is rated at ±10 g's). Along with being cheaper to buy, the ADXL202 has a simplified digital output that requires very little hardware interfacing. Instead of a linear voltage (as is used in other accelerometers), the output is purely digital. As acceleration changes, the timing of the pulses at the output of the ADXL202 chip changes. This change can be readily determined by a PC or microcontroller, using simple software (see the example for the BASIC Stamp 2 later in the chapter).

The ADXL202 is a surface-mount component. By a long measure, the ready-made ADXL202 evaluation board is the easiest way to use this device. It comes on a small postage-stamp carrier, which can be directly soldered to the BASIC Stamp or other microcontroller.

35.3.1 WIRING DIAGRAM

The basic hookup diagram for the ADXL202 is shown in Fig. 35-3 with the parts list in Table 35-1. Note that except for two filter capacitors and a single resistor, there are no external components. I have specified a rather low bandwidth of 10 Hz for the device. According to the ADXL202 data sheet, the value of C1 and C2 for this bandwidth should be 0.47 μF.

Resistor R1 sets the value of the timing pulse used for the output of the *x* and *y* axes of the accelerometer chip. The 620k resistor produces the relatively modest timing pulse of 5 ms; according to the data sheet this requires a nominal value of about 625K for R1. Note that the exact timing of the pulse is not critical, as any variation will be accounted for in the software. You will want to select a higher or lower timing pulse based on the capabilities of the PC or microcontroller you are using and the resolution you desire.

35.3.2 UNDERSTANDING THE OUTPUT OF THE ADXL202

The ADXL202 delivers a PWM with a varying duty cycle. The duty cycle or timing of the pulses, defined as T2, is set by R1. For this application, the pulses are 5 ms apart. Changes in acceleration change the width of each pulse. For the ADXL202, the pulse width changes 12.5 percent in each way for each g of acceleration. Therefore, the width of these 5-ms pulses will change by 50 percent for the entire ±2-g range of the device. A zero g state has a 50 percent duty cycle, while a 1 g acceleration will have a 25 percent duty cycle (2.5 ms pulse width) or 75 percent duty cycle (7.5 ms pulse width), depending on the direction of the acceleration. The period of the PWM signal is defined as T1. Because the ADXL202 uses a pulse width modulated output, rather than a linear DC output, no analog-to-digital conversion is necessary.

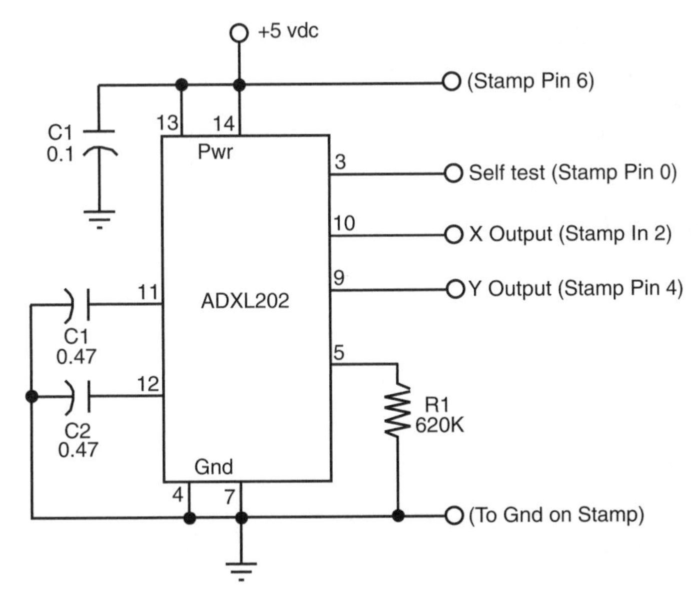

FIGURE 35-3 A basic schematic diagram for using the Analog Devices ADXL150 single-axis accelerometer.

TABLE 35-1 ADXL202 Test Circuit Parts List

IC1	ADXL202 acceleration sensor
R1	620k resistor
C1	0.1 μF capacitor
C2, C3	0.47 μF capacitor
Misc.	BS2 on development carrier board, PC for programming and interfacing

35.3.3 ORIENTING THE ACCELEROMETER

Because the ADXL202 has two axes, it can detect acceleration and gravity changes in two axes at once. You can use the device in vertical or horizontal orientation. As a tilt sensor, orient the device horizontally; any tilt in any direction will therefore be sensed. In this position, the ADXL202 can also be used as a motion detector to determine the speed, direction, and possibly even the distance (given the resolution of the control circuitry you use) of that movement.

35.3.4 CONTROL INTERFACE AND SOFTWARE

The control interface for the ADXL202 is surprisingly simple. Fig. 35-4 shows the hookup diagram for connecting the ADXL202 surface-mount chip and evaluation board to a BASIC Stamp 2. In both cases, power for the '202 comes from one of the Stamp's I/O pins. As mentioned in an application note written by an Analog Devices engineer on the subject of interfacing the ADXL202 to a BASIC Stamp, this isn't the overall best design choice, but for experimenting it's quick and simple.

The following is a short program written in PBASIC for the BASIC Stamp 2 that allows continual reading of the two outputs of the ADXL202. The program works by first determining the period of the T2 basic pulse. It then uses the PULSIN command with both the T1y and T1x axis signals. PULSIN returns the length of the pulse; a longer pulse means higher g; a shorter pulse means lower g.

```
Freq Var Word
T1x Var Word
T1y Var Word
T2 Var Word

Low   0 ' self test, pin 0
Input 2 ' X accel, pin 2
Input 4 ' Y accel, pin 4
High 6  ' V+ pin 6

  Count 4, 500, Freq
  T2 = Freq * 2
  do
    debug cls
    Pulsin 2,1,T1y
```

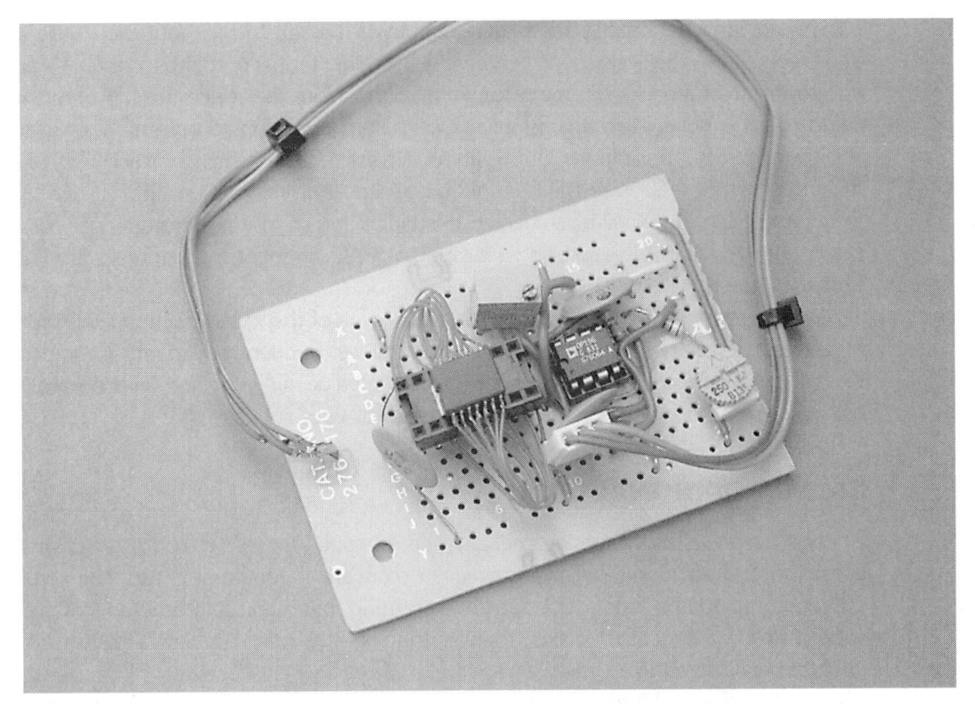

FIGURE 35-4 Connection diagram for hooking up an ADXL202EB (evaluation board) to a BASIC Stamp 2.

```
    T1y = 2 * T1y
    Pulsin 4,1,T1x
    T1x = 2 * T1x
    T1y = 8 * T1y / T2
    T1x = 8 * T1x / T2
    debug dec T1x, tab, dec T1y, tab, cr
    Pause 150
  loop
```

Because the BASIC Stamp 2 has a clock frequency of 2 μs, the actual time of the T1y and T1x pulses are converted to microseconds with the lines

```
T1y = 2 * T1y
T1x = 2 * T1x
```

T1y and T1x are the pulse widths, in microseconds. These widths are then referenced to the T2 value previously obtained by the program with the lines

```
T1y = 8 * T1y / T2
T1x = 8 * T1x / T2
```

The typical results of this program are numbers like 200 and 170, for the *x* and *y* axes, respectively. Note that even on a flat surface, the two outputs of the ADXL202 may not exactly match because of manufacturing tolerances.

The do/loop continually reads the outputs of the sensor. Without the Pause statement and Debug lines, the code loops very fast—just a few tenths of microseconds—which allows you to insert other programming for your robot. Note that once the loop has begun, the value of T2 is never read again (unless you restart the entire program). This is acceptable for low-accuracy applications like basic tilt sensing. But when higher accuracy is required, the timing of the T2 pulse train should be re-read every 5 or 10 minutes, and even more frequently if the robot will be subjected to sudden and sharp temperature changes. The output of the ADXL202 is sensitive to temperature, so changes in temperature will affect the timing of the T2 pulse.

As the program runs you will note that the value of the x and y outputs will change ± 50 to ± 75 just by tilting the accelerometer on its sides. Sudden movement of the accelerometer will produce more drastic changes. Note the values you get and incorporate them into the accelerometer control software you devise for your robot.

35.3.5 ADDITIONAL USES

Though the ADXL202 accelerometer is ideally suited for use as a tilt sensor, it has other uses, too. No additional hardware or even software is required to turn the sensor into a movement, vibration, and shock sensor. Assuming that the accelerometer is oriented so the robot travels in the chip's x axis, then as the robot moves the ADXL202 will register the change in acceleration.

Should the robot hit a wall or other obstacle, it will be sensed as a very high acceleration/deceleration spike. Your control software will need to loop through the code at a high enough rate to catch these momentary changes in output if you want your robot to react to shocks and vibrations. The do/loop in the previous code repeats often enough that your robot should detect most collisions with objects.

If you absolutely must detect all collisions you'll need to devise some kind of hardware interrupt that will trigger the microprocessor or microcontroller when the output of the ADXL202 exceeds a certain threshold. Another approach is to dedicate a fast-acting microcontroller just to the task of monitoring the output of the ADXL202. The low cost of microcontrollers these days makes such dedicated applications a reasonable alternative.

35.4 Alternatives to Store-Bought Accelerometers

While factory-made accelerometers, such as the Analog Devices ADXL150 and ADXL202, are the most convenient for use with robotics, there are some low-cost alternatives you might want to experiment with. You can make your own home-brew accelerometer using a 50-cent piezo ceramic disc and a heavy steel ball or other weight. The home-brew piezo accelerometer isn't as accurate as the ADXL series or other factory-made accelerometers, but it'll do in a pinch and teach you about the physics of motion in the process.

The piezo disc accelerometer works by using a well-known behavior of piezoelectric material: it is both a consumer of energy and a producer of energy. Most applications of piezoelectric materials are in consumer products like speakers and beepers. Apply a voltage to the piezoelectric material, and it vibrates, producing a tone. Conversely, if you vibrate the piezoelectric material using some mechanical means, the output is an electrical signal.

Piezo discs are common finds in electronic and surplus stores. These units are typically used as the elements in low-cost speakers or tone-makers (like smoke alarms, car alarms, etc.). The typical piezo disc is about an inch in diameter and is made of brass or some other nonferrous metal. Deposited on one side of the disc is a ceramic material made of piezo crystals. These crystals vibrate when a voltage is applied to the disc. Most piezo discs already have two wire leads conveniently soldered to them so they can be easily connected to the rest of your circuit.

35.4.1 CONSTRUCTING THE PIEZO DISC ACCELEROMETER

The disc will be used in electricity-producing mode, with the help of a steel ball or other heavy weight to provide mechanical energy. Place the ball or weight on the disc—ceramic side up—and tape the ball in place so it won't roll or fall off the disc. Connect the output of the disc to a fast-acting voltage meter or an oscilloscope. Lift the disc up and down rapidly, and you'll see the voltage output of the disc fluctuate, perhaps as much as a full volt or two. The faster you move the disc, the more the voltage will swing.

Just as important as noting the magnitude of the voltage during a change in movement, note that the polarity of the voltage changes depending on the direction of travel. The output of the disc might be in positive volts when moving up but negative volts when moving down.

To complete the construction, mount the disc either on a separate sensor board or on the robot itself. As an accelerometer that senses lateral motion, the disc can be mounted in a vertical position, though that will reduce its sensitivity since the ball or weight is being pulled off the disc by gravity. Be sure that the tape holding the ball is secure. You may wish to construct a more reliable captive mechanism, perhaps housing the disc and ball or weight in an enclosure. A 35-mm film can cut to size or a plastic "bug case" (like the kind used for prizes in bubble gum machines) are good options.

As with a factory-made accelerometer, you can use the piezo disc accelerometer for vibration and shock detection. Sudden jolts—like when the robot bumps into something—will translate into larger-than-normal variations in the output of the disc. When you connect this accelerometer to the brains of your robot, this information can be used to determine the machine's proper course of action.

35.4.2 LIMITATIONS OF THE PIEZO DISC ACCELEROMETER

While the piezo disc makes for a cheap and easy accelerometer, it's not without its limitations. Here are three you will need to consider:

- The disc will only measure changes in momentum since it is inherently an AC device. Once the momentum of the disc normalizes, the output voltage will fall back to its nominal state. Since gravity acts like a constant DC signal, this means you will not be able to use the piezo disc as a tilt sensor very easily.
- The output of the piezo disc can easily exceed the input voltage of the interfacing elec-

tronics. Should the disc receive the blow of a sharp impact, the voltage output can easily exceed 20, 50, and even 100 V. For this reason, you must always place a zener diode to act as a voltage clamp, as shown in Fig. 29-14 of Chapter 29. Select a zener diode voltage that is compatible with the input voltage for the interface you are using. For example, if the interface voltage is 5 vdc, use a 5.1-V zener.

- The piezo disc is basically a capacitor so it stores a charge over time. You can reduce the effects of the capacitive charge by placing a 50K to 250K resistor across the output leads of the disc (this will help to bleed off the charge). You may also want to feed the output of the disc to an op amp.

35.5 From Here

To learn more about . . .	***Read***
Connecting hardware to a computer or microcontroller	Chapter 29, "Computer Peripherals"
Navigating through an Environment	Chapter 33, "Navigating through Space"

CHAPTER 36

HOME ROBOTS AND HOW NOT TO CHEW UP YOUR FURNITURE

In 2002, the iRobot corporation announced the availability of the Roomba vacuum cleaning robot for the home. This announcement is significant because the Roomba was the first practical robot for home use that could be used by anybody. The robot did not require any specialized programming or operation. The user was responsible for just making sure the Roomba's batteries were charged and to place a couple of electronic markers to prevent the Roomba from leaving the room that it is being tasked to clean. Since its introduction a number of copies of the Roomba have been announced.

While the Roomba is a significant step forward, it is not what people think of when they envision home robots. The Honda Asimo or Sony Qiro, both humanoid-shaped robots that are able to perform a variety of different tasks, better meet people's expectations of what a home robot should look like. Unfortunately, these robots have had thousands of hours of engineering effort and many millions of dollars invested in them to get to the point where they are, and yet they are still unable to perform the most basic task that the public at large expects from a robot—fetching a beverage from the kitchen refrigerator for them while they are watching TV.

This chapter examines some of the more immediate problems of implementing a robot that runs around the house. There are a number of different issues, such as how the robot will find its recharging station automatically and how it will work with objects on surfaces of tables or counters, all of which are at different heights.

While it is unlikely that you will come up with a robot that will rival the functionality of the Asimo or Qiro, you can definitely come up with a robot that can reproduce the capabilities of the Roomba and maybe find a new niche application that is perfect for home robotics, using today's technology. Who knows—you might find your fortune coming up with a unique and practical way of having a robot retrieve a cold beverage from your refrigerator.

36.1 Sensing the Environment: Protecting the Furniture and the Robot

It will take a leap of faith to allow a robot to move about your home without supervision. Chances are, it won't be *your* leap of faith, but of others responsible for the upkeep of the house. You must think about what goes through their minds when you say you want to let a robot loose and it will save them a lot of work—chances are they are thinking about cleaning up all the mess and repairing the damage the robot causes.

A number of things you can do to make the robot safe for operating in a room are shown in the cross-sectional view of a home robot in Fig. 36-1. The robot will have a soft, flexible bumper that will not mar or scratch objects that it runs into or rubs against. Inside the bumper will be contact switches all the way around the robot that are fairly sensitive and will close when the robot encounters an object with more than a few ounces of resistance.

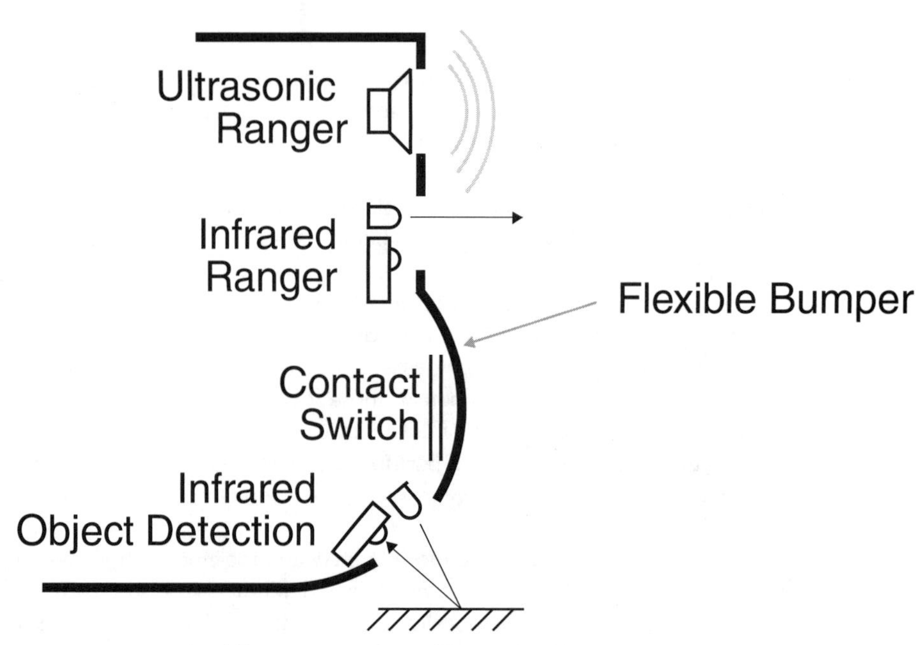

FIGURE 36-1 The different sensors that will be required to ensure that a home robot does not run into and damage objects around a home. These sensors must be located around the robot to ensure that nothing is missed in between the two sensors, resulting in the robot colliding with it.

The bumpers are the last line of defense in protecting a home against a robot. The purpose of the *ultrasonic ranger* and *infrared ranger* is to map out objects around the robot so it can change course before colliding with objects in its path. As you will discover when you work with different object sensors and ranging equipment different methods will detect different objects better than others; the ultrasonic ranger works better on dark or black objects that do not reflect light while the infrared ranger works better with curtains that allow ultrasonic signals to pass through while being attenuated (and will not have enough energy to be detected after reflecting off of hard objects behind the curtains).

The downward-pointing *infrared object detection* sensor is used to detect the surface the robot is running on and to ensure the robot doesn't fall down any stairs. While this may sound funny, it is a very serious concern. It is especially serious if the robot weighs 10 lb or more and could cause some damage or injury falling down a set of stairs.

The bumpers, while being the last line of defense, will probably come into contact with a variety of objects the ultrasonic and infrared ranging sensors will not detect. The most common objects that they will miss are thin table and chair legs. This means that while your robot can go very fast, it shouldn't. It should only move at a few feet per second so that it can stop in a reasonable distance and not harm the object it has collided with.

Cats, dogs, and other pets will have to be accounted for as well. Their fur can be difficult for sensors to detect and they can react unpredictably and violently to objects running into them. Consideration must be made as to how the robot should respond to a collision with an animal (as opposed to an object). While turning around and moving away from the object that has been collided with may be an option, a better one might be to stop and wait to see if the object *attacks* the robot as many pets will do. In this case, it might be best for the robot to stay stopped until it has been determined that the pet has left or gone back to sleep.

36.2 Movement Algorithms

Coming up with an algorithm for your robot to move about the room could be the most challenging aspect of designing a home robot. Humans take for granted the ability to move through rooms, but it is extremely difficult to come up with different ways of programming a robot to do so effectively. The problem gets more complex when the robot has to do something in the room such as find a location within the room or even come up with a way of cleaning it.

The point of cleaning the room is brought up because of the difficulty roboticists at iRobot (the company that designed the Roomba) had in coming up with an algorithm that would maximize the amount of the room that would be vacuumed by their robot. The algorithm chosen is shown graphically in Fig. 36-2. The robot starts out with a spiral pattern until it comes in contact with one of the room's walls. From here, it then follows the wall until it is sure that it has covered the entire perimeter of the room. When the perimeter has been cleaned, the robot then begins moving back and forth throughout the room in a random pattern. After some period of time (many minutes), the robot restarts the spiral pattern and the algorithm repeats itself.

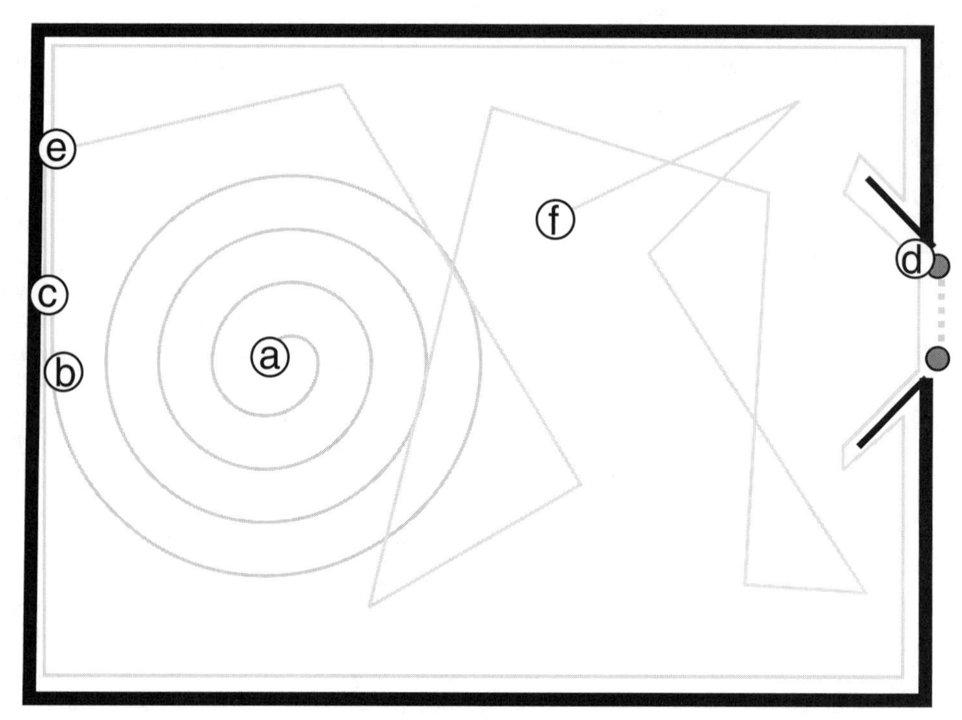

FIGURE 36-2 The iRobot Roomba vacuum robot room-cleaning algorithm. *a.* The robot starts with a spiral pattern until it comes in contact with a wall. *b.* After colliding with the wall, the robot follows the perimeter of the room. *c.* It stays inside the room by not violating an infrared barrier. *d.* It determines that it has cleaned the room's perimeter. *e.* It begins to move randomly throughout the room, vacuuming as it goes. *f.* After moving randomly, the process repeats with the robot starting out at some random point in the room with the robot making another spiral pattern.

This algorithm works quite well for the application and could conceivably clean the entire floor of a home if it were allowed to run long enough. Of course, it could take days for the robot to move randomly into each room and stay there until they are all vacuumed. It must be recognized that there are situations (such as the chair leg noted in the previous section) that will make the Roomba's room-cleaning algorithm much less efficient.

Moving from room to room in a purposeful manner is difficult and will most likely require the IR or RF beacons in each room mentioned elsewhere for the robot to determine where it is and where it must go. With the beacons, the task of navigating becomes more of an IT problem rather than a robotics problem as the path through the house must consist of sorting the beacons of specific rooms along with the adjacent rooms, leading to the final destination. For example, a robot that finds itself in the "Master Bedroom" and has to go to the "Main Bathroom" would have to figure out the complete set of beacons to pass by. Using the beacons defined in Table 36-1, the sequence would be:

3 to 5 to 2 to 6

TABLE 36-1 Sample Beacon Layout for a Home

ROOM	BEACON NUMBER	ADJACENT BEACONS
Kitchen	1	2, 5, 10
Main Hallway	2	1, 5, 6, 7, 8, 9
Master Bedroom	3	5, 4
Ensuite Bathroom	4	5
Living Room	5	1, 3, 2
Main Bathroom	6	2
Bedroom 1	7	2
Bedroom 2	8	2
Bedroom 3	9	2
Family Room	10	1

When the robot finds itself in a room with a different beacon, it will have to reverse course, return to a room that is on the course, and try to find the next one in the sequence.

The final issue that you will have to contend with is stairs. Stairs are very difficult to climb and descend and discovering a way to do so efficiently is one of the Holy Grails of robotics. It is not difficult to design a set of large, knobby wheels that can move up a set of stairs. They generally work best when the robot is exactly perpendicular to the stairs. The problem with climbing stairs is twofold; the first is approaching the stairs at exactly 90 degrees. Even a couple of degrees off will cause the robot to lose its footing and potentially fall down or topple.

To make matters worse, many houses have stairs that are irregular as shown in Fig. 36-3. In each of these cases, either the angle the robot approaches the steps changes or the steps are encountered at different times, which the robot cannot automatically accommodate, leading to it toppling or falling.

36.3 Communicating with the Robot

Communicating with a robot is easy to do when it is in the room with you. A TV remote control can be your interface to the robot and a series of LEDs or a speaker can be the robot's way of responding to your commands and indicating its status. While this is not how most people imagine robots communicate with their owners (speech, both to and from the robot is the expected method), it is efficient and practical. Problems with this method of communication occur when the robot is in another room.

If you want to get the attention of somebody in another room, the first thing that you probably do is shout out that person's name. Unfortunately, current speech recognition sys-

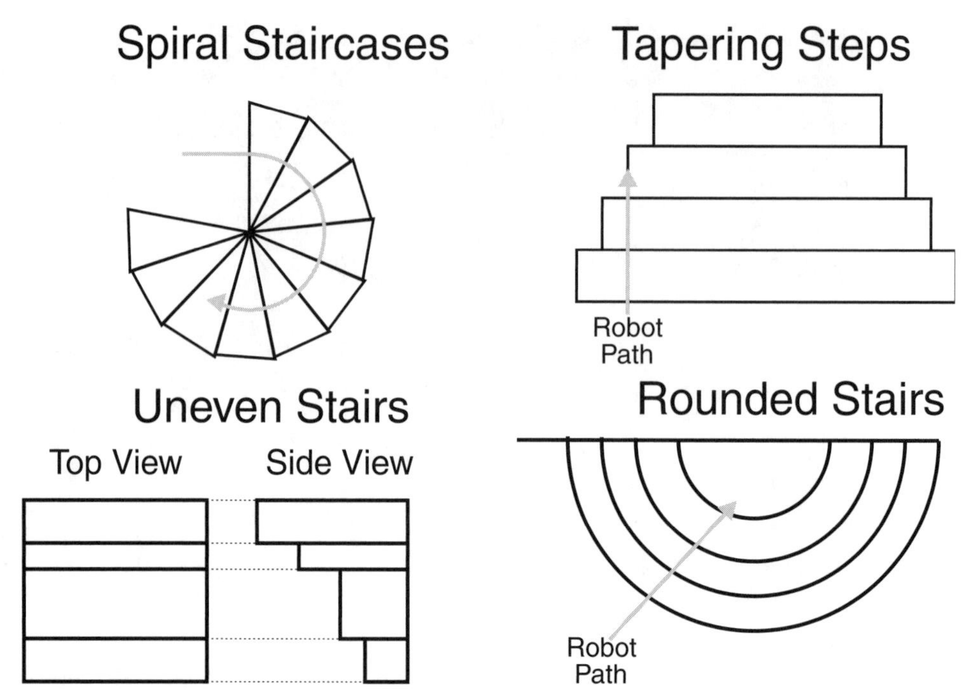

FIGURE 36-3 Some of the different situations a stair-climbing robot will encounter and have to overcome.

tems are nowhere near up to the task of being able to differentiate a shout from any other loud background noise and determine if it is a command that it should respond to. A call of "Robot, get me a cola!" will probably result in the robot staying put and you going thirsty.

Using today's technology, the most effective way to communicate with a robot that could be somewhere within the house is wirelessly using something like a home WiFi network. This does not mean that you will have to go to a computer to communicate with the robot; there are many PDAs and other handheld devices available that communicate with home networks via WiFi. When communicating with the robot, it should be able to respond with where it is and its status.

Status information should not be difficult for the robot to provide, but it will be a challenge for the robot to determine where it is in your house. If you don't want the robot to move randomly about the house until it finds you, you might want to consider:

- Using the infrared or low-power RF beacons in each room discussed in the previous chapter so the robot can query to find its way to you.
- Adding a TV camera to the robot that broadcasts its view of the world to you and lets you command it, by remote control. Pictures could be transmitted via WiFi or to your television set using a low-power (FCC approved) transmitter.

Of the two methods, the beacon method is more technically impressive, while the remote TV camera is much more fun.

36.4 From Here

To learn more about . . .	***Read***
Computers and microcontrollers for robotic control	Chapter 12, "An Overview of Robot 'Brains' "
Interfacing circuitry to DC motor loads	Chapter 20, "Working with DC Motors"
Input and output	Chapter 14, "Computer Peripherals"
Interfacing sensors	Part 6, "Sensors and Navigation"
Building a robotic revolute coordinate arm	Chapter 27, "Build a Revolute Coordinate Arm"
Creating hands for robot arms	Chapter 28, "Experimenting with Gripper Designs"
Endowing robot arms and hands with the sense of touch	Chapter 29, "The Sense of Touch"

PART 7

PUTTING IT ALL TOGETHER

CHAPTER 37

ROBOT TASKS, OPERATIONS, AND BEHAVIORS

All robots are different because their creators have different tasks in mind for their creations to accomplish. A robot designed to find empty soda cans is going to be radically different from one made to roam around a warehouse sniffing out the smoke and flames of a fire.

Consider that a true robot is a machine that not only acts independently within an environment but reacts independently of that environment. In describing what a robot is it's often easier to first consider what it isn't:

- Your car is a machine, but it's not a robot. Unless you outfit it with special gizmos, it has no way of driving itself (okay, so Q can make a self-running car for James Bond). It requires you to control it, steer the wheels, and operate the gas and brake pedals, and roll down the window to talk to the nice police officer.
- Your refrigerator is a machine, but it's not a robot. It may have automatic circuitry that can react to an environment (increase the cold inside if it gets hot outside), but it cannot load or unload its own food, so it still needs you for its most basic function.
- Your dishwasher is a machine, but it's not a robot. Like the refrigerator, the dishwasher is not self-loading, may not adjust itself in response to how dirty the dishes are, and cannot be reprogrammed to accommodate changes in the soap you use, nor can it detect that you've loaded it with $100-a-plate porcelain—so go easy on the rinse cycle, thank you very much.

Other machines around your home and office are the same. Consider your telephone answering machine, your copier, or even your personal computer. All need you to make them work and accomplish their basic tasks.

A real robot, on the other hand, doesn't need you to fulfill its chores. A robot is programmed ahead of time to perform some job, and it goes about doing it. Here, the distinction between a robot and an automatic machine becomes a little blurry because both can run almost indefinitely without human intervention (not counting wear and tear and the availability of power). However, most automatic machines lack the means to interact with their environment and to change that environment if necessary. This feature is often found in more complex robots.

Beyond this broad distinction, the semantics of what is and is not a robot isn't a major concern of this book. The main point is this: once the robot is properly programmed, it should not need your assistance to complete its basic task(s), barring any unforeseen obstacles or a mechanical failure.

37.1 "What Does My Robot Do?": A Design Approach

Before you can build a robot, you must decide what you want the robot to do. That seems obvious, but you'd be surprised how many first-time robot makers neglect this important step. By reducing the tasks to a simple list, you can more easily design the size, shape, and capabilities of your robot. Let's create an imaginary homebuilt robot named RoBuddy, for "Robotic Buddy," and go through the steps of planning its design. We'll start from the standpoint of the jobs it is meant to do. For the sake of simplicity, we'll design RoBuddy so that it's an entertainment 'bot—it's for fun and games and is not built for handling radioactive waste or picking up after your dog Spot.

The first question people ask when seeing a robot is, "So, what does it do?" That's not always an easy question to answer because the function of a robot can't always be summarized in a quick sentence. Yet most people don't have the patience to listen to a complex explanation. Such is the quandary of the robot builder!

37.1.1 AN ITINERARY OF FUNCTIONS

One of the best shortcuts to explaining what a robot can do is to simply give the darned thing a vacuum cleaner. That way, when you don't feel like repeating the whole litany of capabilities, you can merely say, "it cleans the floors." That's almost always guaranteed to elicit a positive response. So this is Basic Requirement #1: RoBuddy must be equipped with a vacuum cleaner. And since RoBuddy is designed to be self-powered from batteries, the vacuum cleaner needs to run under battery power, too. Fortunately, auto parts stores carry a number of 12-V portable vacuum cleaners from which you can choose.

Like the family dog that performs tricks for guests, a robot that mimics some activity amusing to humans is a great source of entertainment. A very useful and effective activity is pouring and serving drinks. That takes at least one arm and gripper, preferably two, and the arms must be strong and powerful enough to lift at least 12 oz of beverage. We now have Basic Requirement #2: RoBuddy must be equipped with at least one appendage that has a gripper designed for drinking glasses and soda cans.

The RoBuddy must also have some kind of mobility so that at the very least it can move around and vacuum the floor. There are a number of ways to provide locomotion to a robot, as described in earlier chapters. But for the sake of description, let's assume we use the common two-wheel-drive approach, which consists of two motorized wheels counterbalanced by one or two nonpowered casters. That's Basic Requirement #3: RoBuddy must have two drive motors and two wheels for moving across the floor.

Since RoBuddy flits about your house all on its own accord, it has to be able to detect obstacles so it can avoid them. Obviously, then, the robot must be endowed with some kind of obstacle detection devices. We're up to Basic Requirement #4: RoBuddy must be equipped with passive and active sensors to detect and avoid objects in its path.

Serving drinks, vacuuming the floor, and avoiding obstacles requires an extensive degree of intelligence and is beyond the convenient capability of hard-wired discrete circuits consisting of some resistors, a few capacitors, and a handful of transistors. A better approach is to use a computer, which is capable of being programmed and reprogrammed at will. This computer is connected to the vacuum cleaner, arm and gripper, sensors, and drive motors. Finally, then, this is Basic Requirement #5: RoBuddy must be equipped with a computer to control the robot's actions.

These five basic requirements may or may not be important to you or applicable to all your robot creations. However, they give you an idea of how you should outline the functions of your robot and match them with a hardware requirement.

37.1.2 ADDITIONAL FEATURES

Depending on your time, budget, and construction skill, you may wish to endow your robot(s) with a number of other useful features, such as:

- Sound output, perhaps combining speech, sound effects, and music.
- Variable-speed motors so your robot can get from room to room in a hurry but slow down when it's around people, pets, and furniture.
- Set-and-forget motor control, so the "brains on board" controlling your 'bot needn't spend all its processing power just running the drive motors.
- Distance sensors for the drive motors so the robot knows how far it has traveled (odometry).
- Infrared and ultrasonic sensors to keep the robot from hitting things.
- Contact bumper switches on the robot so it knows when it's hit something and to stop immediately.
- LCD panels, indicator lights, or multidigit displays to show current operating status.
- Tilt switches, gyroscopes, or accelerometers to indicate when the robot has fallen over, or is about to.
- Voice input, for voice command, voice recognition, and other neat-o things.
- Teaching pendant and remote control so you can move a joystick to control the drive motors and record basic movements.

Of course, all of these are discussed in previous chapters. Review the table of contents or index to locate the relevant text on these subjects.

37.2 Reality versus Fantasy

In building robots it's important to separate the reality from the fantasy. Fantasy is a *Star Wars* R2-D2 robot projecting a hologram of a beautiful princess. Reality is a home-brew robot that scares the dog as it rolls down the hallway—and probably hits the walls as it goes. Fantasy is a giant killer robot that walks on two legs and shoots a death ray from a visor in its head. Reality is foot-tall trash can robot that pours your houseguests a Diet Coke. Okay, so it spills a little every now and then . . . now you know why a robot equipped with a vacuum cleaner comes in handy!

It's easy to get caught up in the romance of designing and building a robot. But it's important to be wary of impossible plans. Don't attempt to give your robot features and capabilities that are beyond your technical expertise, budget, or both (and let's not also forget the limits of modern science). In attempting to do so, you run the risk of becoming frustrated with your inability to make the contraption work, and you miss out on an otherwise rewarding endeavor.

When designing your automaton, you may find it helpful to put the notes away and let them gel in your brain for a week. Quite often, when you review your original design, you will realize that some of the features and capabilities are mere wishful thinking and beyond the scope of your time, finances, or skills. Make it a point to refine, alter, and adjust the design of the robot before, and even during, construction.

37.3 Understanding and Using Robot Behaviors

A current trend in the field of robot building is *behavior-based robotics,* where you program a robot to act in some predictable way based on both internal programming and external input. For example, if the battery of your robot becomes weak, it can be programmed with a "find energy" behavior that will signal the robot to return to its battery charger. Behaviors are a convenient way to describe the core functionality of robots—a kind of component architecture to define what a robot will do given a certain set of conditions.

The concept of behavior-based robotics has been around since the 1980s and was developed as a way to simplify the brain-numbing computational requirements of artificial intelligence systems popular at the time. Behavior-based robotics is a favorite at the Massachusetts Institute of Technology, and Professor Rodney Brooks, a renowned leader in the field of robot intelligence, is one of its major proponents.

Since the introduction of behavior-based robotics, the idea has been discussed in countless books, papers, and magazine articles, and has even found its way into commercial products. The LEGO Mindstorms robot, which is based on original work done at MIT, uses behavior principles. See Appendix A, "Further Reading," for books that contain useful information on behavior-based robotics.

37.3.1 WHEN A BEHAVIOR IS JUST A SIMPLE ACTION

Since the introduction of behavior-based robotics, numerous writers have applied the term *behavior* to cover a wide variety of things—to the point that everything a robot does becomes a behavior. The result is that robot builders can become convinced their creations are really exhibiting human- or animal-like reactions, when all they are doing is carrying out basic instructions from a computer or simple electronic circuit. Delusions aside, this has the larger effect of distracting you from focusing on other useful approaches for dealing with robots.

To help explain how this is a problem, consider this analogy: suppose you see a magic show so many times that you end up believing the disappearing lady is really gone. Not so. It's an optical and psychological trick every time. Sometimes a robot displays a simple action as the result of rudimentary programming, and by calling everything it does a *behavior* we lose a clearer view of how the machine is really operating.

The following sections contain a brief discourse on behavior-based robotics.

37.3.2 WALL FOLLOWING: A COMMON BEHAVIOR?

One common example of behavior-based robotics is the *wall follower,* which is typically a robot that always turns in an arc, waiting to hit a wall. A sensor on the front of the robot detects the wall collision. When the sensor is triggered, the robot will turn away from the wall some amount and repeat the whole process over again.

This is a perfect example of how the term *behavior* has been misplaced: the true behavior of the robot is not to follow a wall but simply to turn in circles until it hits something. When a collision occurs, the robot turns to clear the obstacle and then continues to turn in a circle once again. In the absence of the wall—a reasonable change in environment—the robot would not exhibit its namesake *behavior.* Or conversely, if there were additional objects in the room, the robot would treat them as walls, too. In that case, the robot might be considered useless, misprogrammed, or worse.

If wall following is not a true behavior, then what is it? There really is no industry-standard term for this type of action. The important thing to remember is that a true behavior is independent, or nearly so, of the robot's typical physical environment. That's rule number one to keep in mind. Note that environment is not the same as a condition. A condition is a light shining on the robot that it might move toward or away from; an environment is a room or other area that may or may not have certain attributes. Conditions contribute to the function of the robot, just like batteries or other electric power contributes to the robot's ability to move its motors. Conversely, environments can be ever changing and in many ways unmanageable. Environments consist of physical parameters under which the robot may or may not operate at any given time.

Robotic behaviors are most useful when they encapsulate multiple variables, particularly those that are in response to external input (senses). This is rule number two of true behavior-based robotics. The more the robot is able to integrate and differentiate between different inputs (senses), irrespective of environment—and still carry out its proper programming—the more it can demonstrate its true behaviors.

37.3.3 THE WALT DISNEY EFFECT

It is tempting to endow robots with human- or animal-like emotions and traits, such as hunger (battery power) or affinity/love (a beacon or an operator clicking a clicker). But in my opinion these aren't behaviors at all. They are anthropomorphic qualities that merely appear to result in a human-type response simply because we want them to.

In other words, it's completely made up. Imagine this in the extreme: is a robot suicidal if it has a tendency to drive off the workbench and break as it hits the floor? Or is it that your workbench is too small and crowded, and your concrete floor is too hard? Emotions such as love are extremely complex; as a robot builder, it's easy to get confused about what your creation can really do and feel.

In his seminal book *Vehicles,* Valentino Braitenberg presents a study of synthetic psychology on which fictional vehicles demonstrate certain behavioral traits. For example, Braitenberg's Vehicle 2 has two motors and two sensors (say, light sensors). By connecting the sensors to the motors in different ways the robot is said to exhibit emotions, or at the least actions we humans may interpret as quasi-intelligent or human-like emotional responses. In one configuration, the robot may steer toward the light source, exhibiting "love." In another configuration, the robot may steer away, exhibiting "fear."

Obviously, the robot is feeling neither of these emotions, nor does Braitenberg suggest this. Instead, he gives us vehicles that are fictional representations of human-like traits. It's important not to get caught up in Disneyesque anthropomorphism. A good portion of behavior-based robotics centers around human interpretation of the robot's mechanical actions. We interpret those actions as intelligent, or even as cognition. This is valid up to a point, but consider that only we ourselves experience our own intelligence and cognition (that is, we are self-aware); a robot does not. Human-like machine intelligence and emotions are in the eye of a human beholder, not in the brain of the robot. This, however, may change in the future as new computing models are discovered, invented, and explored.

37.3.4 ROBOTIC FUNCTIONS AND ERROR CORRECTION

When creating behaviors for your robots, keep in mind the function that you wish to accomplish and then consider how that function is negatively affected by variables in the robot's likely environments. For practical reasons (budget, construction skill), you must consider at least some of the limitations of the robot's environment in order to make it reliably demonstrate a given behavior. A line-following robot, which is relatively easy to build and program, will not exhibit its line-following behavior without a line. By itself, such a robot would merely be demonstrating a simple action. But by adding error correction—to compensate for unknown or unexpected changes in environment—the line-following robot begins to demonstrate a useful behavior. This behavior extends beyond the robot's immediate environmental limits. The machine's ability to go into a secondary, error-correcting state to find a line to follow is part of what makes a valid line-following behavior—even more so if in the absence of a line to follow the robot can eventually make its own.

Error correction is rule number three of behavior-based robotics. Without error correction, robots operating in restrictive environments are more likely to exhibit simple, even stupid, actions in response to a single stimulus. Consider the basic wall-following robot again: it requires a room with walls—and, at that, walls that are closer together than its turning radius. Outside or in a larger room, the robot behaves completely different, yet its programming is exactly the same. The problem of the wall-following robot could be fixed either by adding error correction or by renaming the base behavior to more accurately describe what it physically is doing.

37.3.5 ANALYZING SENSOR DATA TO DEFINE BEHAVIORS

By definition, behavior-based robotics is reactive, so it requires some sort of external input by which a behavior can be triggered. Without input (a light sensor, ultrasonic detector, bumper switch, etc.) the robot merely plays out a preprogrammed set of moves—simple actions, like a player piano. More complex behaviors become possible if the following capabilities are added:

- The ability to analyze the data from an analog, as opposed to a digital, sensor. The output of an analog sensor provides more useful information than the simple on/off state of a digital sensor. This data and ability could be called *sensor data parametrics.*
- The ability to analyze the data from multiple sensors, either several sensors of the same type (a gang of light-sensitive resistors, for example) or sensors of different types (a light sensor and an ultrasonic sensor). This is commonly referred to as *sensor fusion.*

Consider sensor parametrics first. Suppose your robot has a temperature sensor connected to its on-board computer. Temperature sensors are analog devices; their output is proportional to the temperature. You use this feature to determine a set or range of preprogrammed actions, depending on the specified temperature. This set of actions constitutes a behavior or, if the actions are distinct at different temperatures, a variety of behaviors. Similarly, a photophilic robot that can discern the brightest light among many lights also exhibits sensor analysis from parametric data.

Sensor fusion analyzes the output of several sensors. Your robot initiates the appropriate behavioral response as a result. For example, your robot may be programmed to follow the brightest light but also detect obstructions in its path. When an obstruction is encountered, the robot is programmed to go around it and then continue—perhaps from a new direction—toward the light source.

Sensor fusion helps provide error correction and allows a robot to continue exhibiting its behavior (one might call it the robot's *prime directive*) even in the face of unpredictable environmental variables. The variety, sophistication, and accuracy of the sensors determine how well the robot will perform in any given circumstance. Obviously, it's not practical, economic or otherwise, to ensure that your robot will work flawlessly under all environments and conditions. But the more you give your robot the ability to overcome common and reasonable environmental variables (such as socks on the floor), the better it will display the behavior you want.

37.4 Multiple Robot Interaction

An exciting field of research is the interaction of several robots working together. Rather than build one big, powerful robot that does everything, multirobot scenarios combine the strengths of two or more smaller, simpler machines to achieve synergy: the whole is greater than the parts. Anyone who has seen the old science fiction film *Silent Running* knows what three diminutive robots (named Huey, Luey, and Dewey, by the way) can do!

Robot *tag teams* or *swarms* are common in college and university robot labs, where groups of robot researchers compete with their robots as the players of a game (robo-soccer is popular). Each robot in the competition has a specific job, and the goal is to have them work together. There are three common types of robot-to-robot interaction:

- *Peer-to-peer.* Each robot is considered equal, though each one may have a different job to do, based on predefined programming. The workload may also be divided based on physical proximity to the work, and whether the other robots in the group are busy doing other things.
- *Queen/drone.* One robot serves as the leader, and one or more additional robots serve as worker drones. Each drone takes its work orders directly from the queen and may interact only peripherally with the other drone 'bots.
- *Convoys.* Combining the first two types, the leader of the convoy is the "queen" robot, and the other robots act as peers among themselves. In convoy fashion, each robot may rely on the one just ahead for important information. This approach is useful when the "queen" is not capable, for computer processing reasons or otherwise, to control a large number of fairly mindless drones.

Why all the fuss with multiple robots? First and foremost because it's generally easier and cheaper to build many small and simple robots than a single big and complex one. Second, the mechanical failure of one robot can be compensated for by the remaining good robots. In many instances, the "queen" or leader robot is no different from the others, it just plays a coordinating role. In this way, should the leader 'bot go down for the count, any other robot can easily take its place. And third, work tends to get done faster with more hands helping.

37.5 The Role of Subsumption Architecture

Subsumption architecture isn't an odd style of building. Rather, it's a technique devised by Dr. Brooks at MIT that has become a common approach for dealing with the complexities of sensor fusion and artificial machine intelligence. With subsumption, sensor inputs are prioritized. Higher-priority sensors override lower-priority ones. In the typical subsumption model, the robot may not even be aware that a low-priority sensor was triggered. The net result is that a robot can demonstrate very sophisticated responses to inputs with relatively simple programs and low-performance microcontrollers.

More complex hybrid systems may employ a form of simple subsumption along with more traditional artificial intelligence programming. The robot's computer may evaluate the relative merits of low-priority sensors and use this information to intuit a unique course of action, perhaps one in which direct programming for the combination of input variables does not yet exist. In some cases, the output of a low-priority sensor may moderate the interpretation of a high-priority one.

As an example, a firefighting robot may have both a smoke detector and a flame detector. The smoke detector is likely to sense smoke before any fire can be identified, since smoke so easily permeates a structure. Therefore, the smoke sensor will likely be given a lower priority to the flame detector, since it is so easily triggered. But consider that flames can exist without a destructive fire (e.g., a fireplace and candlelight, both of which do not emit much smoke under normal circumstances). Rather than have the robot totally ignore its other sensors when the high-priority flame detector is triggered, the robot instead integrates the output of both flame and smoke sensors to determine what is, and isn't, a fire that needs to be put out.

37.6 From Here

To learn more about . . .	***Read***
Deciding on the appropriate design	Chapter 18, "Principles of Robot Locomotion"
Manipulating objects around the robot	Chapter 26, "Reach out with Robot Arms"
Providing sound interfaces	Chapter 31, "Sound Input and Output"
Light interfaces	Chapter 32, "Robot Vision"
Sensing objects around the robot	Chapter 30, "Object Detection"
Making sense of the world around the robot	Chapter 33, "Navigation"

CHAPTER 38

INTEGRATING THE BLOCKS

With the function of the robot defined, you will have to decide how you are going to bring all the pieces together. So far in this book, different building methodologies, electronic circuits for motor driving and control, computer and microcontroller devices, sensors and output devices have been discussed along with example circuits. You probably feel comfortable with the individual building blocks that have been presented although bringing them all together with a single controller probably seems scary and difficult. By following a few simple rules, you will be able to bring all the pieces together into a functional robot.

38.1 Basic Program Structure

The first rule is to try and maintain the basic program format that was discussed in Chapter 13. Written out in pseudo-code, the structure of your first robot programs should always be:

```
'  Initializations
'  Software Variables
'  Hardware Registers and Devices

   WHILE (1)              '  Loop Forever
       Outputs = Processing(Inputs) ' Repeat for Each Output
       Delay(value)       ' Delay 100 ms or so for each loop
   WEND                   ' Jump Back to "WHILE(1)" and repeat
```

where Outputs = Processing(Inputs) is a simple way of saying that each of the robot's outputs is a function of the various inputs available. This will reduce the complexity of your software to just a simple Input—Processing—Output model for each output and will allow you to easily observe the response of some outputs to specific inputs and if they are not working properly they can be changed without affecting other functions in the robots.

For a light-following robot, the LeftMotorProcess function could be

```
MotorControl LeftMotorProcess(LightLeft, LightRight)

    IF (LightRight >= LightLeft) THEN
        MotorControl = Forward
    ELSE
        MotorControl = Stop
    ENDIF

End LeftMotorProcess
```

and used to control the left motor of the robot directly. In this function, the Left Motor will be active if the robot is to move forward (LeftLight equal to RightLight) or turn right (RightLight is greater than LeftLight). If the robot has to move left, then the motor stops and the left wheel becomes a pivot the robot turns around.

The right motor function (RightMotorProcess) would be identical except that the comparison statement would be reversed, becoming;

```
IF (LightLeft >= LightRight) THEN
```

As the application became more complex, then the functions governing the actions of individual outputs can be made more complex without affecting other output functions and additional ones could be added without affecting any of the original output functions.

This approach may seem somewhat inefficient because much of the code will be repeated, but as an approach for your first attempts at getting a robot working, they will be quite effective, easy to code, and (most importantly) easy to debug and get running.

38.2 Allocating Resources

In contrast to a PC, where there seems to be almost limitless resources available, a microcontroller or microprocessor used to control a robot has only a limited amount of resources that can be brought to bear on the task of controlling a robot. Before beginning to design the robot, the resources that will be required should be listed and a controller selected from the chips and system PCBs that have the required resources.

38.2.1 I/O PINS

The resource that will seem to be the most restricted when deciding on which controller to use will be I/O pins: both digital and analog. During planning of the robot features, the different I/O requirements must be cataloged and matched to I/O pins with the required capabilities. This task is not difficult, but forgetting to do it can cause a great deal of stress and rebuilding when you are finishing the robot later.

When you are first starting out, invest in microcontrollers that have quite a few I/O pins available; larger chips can have 30 or so I/O pins that can be used for a variety of different tasks. These chips will be somewhat more expensive and physically larger than those that have a dozen or so I/O pins, but the extra I/O pins will be useful when you want to add another feature to your robot, LEDs, or a speaker to indicate different input conditions from the robot's sensors that will be important for either debugging the application code or better understanding the input data being received by the controller.

38.2.2 INTERNAL FEATURES

Along with I/O pins, most controllers have a number of other features that you will want to take advantage of in your application design. Timers, analog-to-digital converters, voltage comparators, and serial interfaces are all resources that must be managed like I/O pins as they are available only at specific pins and addresses within the devices.

The BASIC Stamp 2, which has been used for demonstrating different functions, is unusual in the flexibility given to each pin. They can be used for digital I/O, serial I/O, PWM inputs and outputs, limited serial interfaces and so on, which are usually devoted to specific pins in a traditional controller. For this reason, you may want to use only a controller like a BS2 because you are not as constrained as you would be with other controllers and chips.

The best way to manage limited internal features and resources is to treat them as central resources—available for the complete robot application rather than only specific functions. One of the best ways to ensure that the resources are available to all the functions in the robot is to continually execute the internal functions and make the results available to all the output procedures as global variables that can have status flags and values read by the output functions as inputs, just like digital I/O pins or register values to prevent any one function from affecting the value for the other output functions which also rely on the resource's data.

38.3 Getting a Program's Attention Via Hardware

Even in systems that lack multitasking capability it's still possible to write a robot control program that doesn't include a repeating loop that constantly scans (polls) the condition of sensors and other input. Two common ways of dealing with unpredictable external events are using a timer (software) interrupt or a hardware (physical connection) interrupt.

When using interrupts, they should set flags or other variable values to indicate that the interrupt has been requested and processed. During the main loop of the control program, the flags and variables should be used as inputs, just like object, sound, speed, and other standard sensors.

38.3.1 TIMER INTERRUPT

A timer built into the computer or microcontroller runs in the background. At predefined intervals—most commonly when the timer overflows its count—the timer requests the attention of the microprocessor, which in turn temporarily suspends the main program, if it can spare the cycles. The microprocessor runs a special timer interrupt handler subroutine, which in the case of a task-based robot would poll the various sensors and other input looking for possible error modes. (Think of the timer as a heartbeat; at every beat the microprocessor pauses to do something special.)

If no error is found, the microprocessor resumes the main program. If an error is found, the microprocessor runs the relevant section in code that deals with the error. Timer interrupts can occur hundreds of times each second. That may seem like a lot in human terms, but it can be trivial to a microprocessor running at several million cycles per second.

38.3.2 HARDWARE INTERRUPT

A hardware interrupt is a mechanism by which to immediately request attention from the microprocessor. It is a physical connection on the microprocessor that can in turn be attached to some sensor or other input device on the robot. With a hardware interrupt the microprocessor can spend 100 percent of its time on the main program and temporarily suspend it if, and only if, the hardware interrupt is triggered.

Hardware interrupts are used extensively in most computers, and their benefits are well established. Your PC has several hardware interrupts. For example, the keyboard is connected to a hardware interrupt, so when a key is pressed the request is sent to the processor to stop executing the current code and devote its attention to processing the data from the keyboard. The standard PC hardware has 16 hardware interrupt sources, which are prioritized by hardware within the PC down to just one interrupt request pin on the microprocessor. You can do something similar in your own robot designs.

38.3.3 GLASS HALF-EMPTY, HALF-FULL

There are two basic ways to deal with error modes in an interrupt-based system. One is to treat them as "exceptions" rather than the rule:

- In the exception model, the program assumes no error mode and only stops to execute some code when an error is explicitly encountered. This is the case with a hardware interrupt, which will stop execution of the current application anytime an error condition is detected and used to cause an interrupt.
- In the opposite model, the program assumes the possibility of an error mode all the time and checks to see if its hunch is correct. This is the case with the timer interrupt in which the handler subroutine will poll all the robot's sensors periodically.

The approach you use will depend on the hardware choices available to you. If you have both a timer and a hardware interrupt at your disposal, the hardware interrupt is probably the more straightforward method because it allows the microprocessor to be used more efficiently.

38.4 Task-Oriented Robot Control

As workers, robots have a task to do. In many books on robotics theory and application, these tasks are often referred to as *goals*. A robot may be given multiple tasks at the same time, such as the following:

1. Get a can of Dr. Pepper.
2. Avoid running into the wall while doing so.
3. Watch out for the cat and other ground-based obstacles.
4. Bring the soda back to the "master."

These tasks form a hierarchy. Task 4 cannot be completed before task 1. Together, these two form the primary directive tasks. Tasks 2 and 3 may or may not occur; these are error mode tasks. Should they occur, they temporarily suspend the processing of the primary directive tasks.

38.4.1 PROGRAMMING FOR TASKS

From a programming standpoint, you can consider most any job you give a robot to be coded something like this:

```
DO
    Task X                     '  The Primary Task
    DO WHILE (ERROR)           '  An Error Condition
        Task Y                 '  The Error Correction Code
    LOOP
LOOP UNTIL Task X complete     '  Continue Task X until it is Complete
```

X is the primary directive task, the thing the robot is expected to do. Y is a special function that gets the robot out of trouble should an error condition—of which there may be many—occurs. Most error modes will prevent the robot from accomplishing its primary directive task. Therefore, it is necessary to clear the error first before resuming the primary directive.

Note that it is entirely possible that the task will be completed without any kind of complication (no errors). In this case, the error condition is never raised, and the Y functionality is not activated. The robot programming is likewise written so that when the error condition is cleared, it can resume its prime directive task.

38.4.2 MULTITASKING ERROR MODES FOR OPTIMAL FLEXIBILITY

For a real-world robot, errors are just as important a consideration as tasks. Your robot programming must deal with problems, both anticipated (walls, chairs, cats) and unanticipated (water on the kitchen floor, no sodas in the fridge). The more your robot can recognize error modes, the better it can get itself out of trouble. And once out of an error mode, the robot can be reasonably expected to complete its task.

How you program various tasks in your robot is up to you and the capabilities of your robot software platform. If your software supports multitasking, try to use this feature whenever possible. By dealing with tasks as discrete units, you can better add and subtract functionality simply by including or removing tasks in your program.

Equally important, you can make your robot automatically enter an error mode task without specifically waiting for it in code. In non-multitasking procedural programming, your code is required to repeatedly check (poll) sensors and other devices that warn the robot of an error mode. If an error mode is detected, the program temporarily branches to a portion of the code written to handle it. Once the error is cleared, the program can resume execution where it left off.

With a multitasking program, each task runs simultaneously. Tasks devoted to error modes can temporarily take over the processing focus to ensure that the error is fixed before continuing. The transfer of execution within the program is all done automatically. To ensure that this transfer occurs in a logical and orderly manner, the program should give priorities to certain tasks. Higher-priority tasks are able to take over (*subsume,* a word now in common parlance) other running tasks when necessary. Once a high-priority task is completed, control can resume with the lower-priority activities, if that's desired.

38.5 From Here

To learn more about . . .	***Read***
Computer capabilities	Chapter 12, "An Overview of Robot 'Brains' "
Programming computer systems	Chapter 13, "Programming Fundamentals"
Connecting computers and other control circuits to the outside world	Chapter 14, "Computer Peripherals"
Sensing objects around the robot	Chapter 30, "Object Detection"

CHAPTER 39

FAILURE ANALYSIS

Anything that can go wrong, will go wrong.
—Murphy's Law

Legend has it that the original "Murphy" was a U.S. Air Force officer that worked on rocket sleds in the late 1940s and early 1950s. These sleds were used to learn about the effects of high accelerations (g forces) on the human body as well as to design appropriate restraint and safety systems for high-performance aircraft and spaceships. Computer simulations and mechanical models of the human body were not available at the time, so the tests were performed on a living person. The sleds were capable of tremendous speeds (approaching the speed of sound) and there was a very high probability for accidents due to mechanical failure. In readying a sled for a test and working through a myriad of problems, Murphy reportedly muttered the comment for which he was to become immortalized. You probably have already heard of Murphy's Law and it has been referenced so many times in books that it has become trite. It was really put in here to introduce the following comment on it; one that you are sure to relate to as you work on your own robot designs:

Murphy was an optimist.
—Anonymous

39.1 Types of Failures

Before trying to figure out how to fix a problem, the first thing that you will have to do is determine where the problem is occurring. In the following three sections, the three primary sources of problems are listed and the types of failures that are typically ascribed to them. Once you determine where the problem lies, you can start looking at how to fix it so that it does not reoccur. A process for determining the root cause of a failure and ensuring that the corrective action will allow the robot to run until some other problem surfaces is given in the following sections of this chapter.

39.1.1 MECHANICAL FAILURE

Mechanical problems are perhaps the most common failure in robots. The typical source of the problem is that the materials or the joining methods you used were not strong enough. Avoid overbuilding your robots (that tends to make them too expensive and heavy), but at the same time strive to make them physically strong. Of course, *strong* is relative: a lightweight, scarab-sized robot needn't have the structure to support a two year old that a tricycle does. At the very least, however, your robot construction should support its own weight, including batteries.

When possible, avoid slap-together construction, such as using electrical or duct tape. These methods are acceptable for quick prototypes but are unreliable for long-term operation. When gluing parts in your robot, select an adhesive that is suitable for the materials you are using. Epoxy and hot-melt glues are among the most permanent. You may also have luck with cyanoacrylate (CA) glues, though the bond may become brittle and weak over time (a few years or more, depending on humidity and stress).

Use the pull test to determine if your robot construction methods are sound. Once you have attached something to your robot—using glue, nuts and bolts, or whatever—give it a healthy tug. If it comes off, the construction isn't good enough. Look for a better way.

39.1.2 ELECTRICAL FAILURE

Electronics can be touchy, not to mention extremely frustrating, when they don't work right. Circuits that functioned properly in a solderless breadboard may no longer work once you've soldered the components in a permanent circuit, and vice versa. There are many reasons for this, including mistakes in wiring, unexpected capacitive and inductive effects, even variations in tolerances due to heat transfer.

Certain electronic circuit construction techniques are better suited for an active, mobile robot. Wire-wrap is a fast way to build circuits, but its construction can invite problems. The long wire-wrap pins can bend and short out against one another. Loose wires can come off. Parasitic signals and stray capacitance can cause marginal circuits to work, then not work, and then work again. For an active robot it may be better to use a soldered circuit board, perhaps even a printed circuit board of your design (see Chapter 7, "Electronic Construction Techniques," for more information).

Some electrical problems may be caused by errors in programming, weak batteries, or unreliable sensors. For example, it is not uncommon for sensors to occasionally yield totally unexpected results. This can be caused by design flaws inherent in the sensor itself, spurious data (noise from a motor, for example), or corrupted or out-of-range data. Ideally, the programming of your robot should anticipate occasional bad sensor readings and basically ignore them. A perfectly acceptable approach is to throw out any sensor reading that is outside the statistical model you have decided on (e.g., a sonar ping that says an object is 1048 ft away; the average robotic sonar system has a maximum range of about 35 ft).

39.1.3 PROGRAMMING FAILURE

As more and more robots use computers and microcontrollers as their brains, programming errors are fast becoming one of the most common causes of failure. There are three basic kinds of programming bugs.

- *Compile bug or syntax error.* You can instantly recognize these because the program compiler or downloader will flag these mistakes and refuse to continue. You must fix the problem before you can transfer the program to the robot's microcontroller or computer.
- *Run-time bug, caused by a disallowed condition.* A run-time bug isn't caught by the compiler. It occurs when the microcontroller or computer attempts to run the program. An example of a common run-time bug is the use of an out-of-bounds element in an array (for instance, trying to assign a value to the thirty-first element in a 30-element array). Run-time bugs may also be caused by missing data, such as looking for data on the wrong input pin of a microcontroller.
- *Logic bug, caused by a program that simply doesn't work as anticipated.* Logic bugs may be due to simple math errors (you meant to add, not subtract) or by mistakes in coding that cause a different behavior than you anticipated.

As you become experienced in programming, you will get a lot faster at finding problems as you understand where you normally make mistakes and learn how to use the tools at your disposal for finding and fixing the problems. When you first start working on robots, you will probably feel most uncomfortable about your skills in debugging programs, but as you gain experience, you will be amazed at your ability to produce code with relatively few errors, that operates efficiently, and can be debugged easily.

39.2 The Process of Fixing Problems

When given a problem to fix, most people will try to find the easiest way to resolve it and move on; the term used for this process is *debugging*. Looking for and implementing the quick fix often yields an effective (and sometimes optimal) repair action. It does not resolve all problems and quite often masks them. With only experience in working on the most likely cause of the problem, you will become very frustrated very quickly and unable to identify the reason for the problem as well as the most effective repair. For these cases, you will need to perform a root cause failure analysis (often shortened to just failure analysis) to determine what exactly is the problem and what is the best way to fix it.

The failure analysis that is outlined in the rest of this chapter applies to all three classes of failures that you will encounter in your robots (mechanical, electrical, and programming), which may seem surprising because each class seems so different from the other. Mechanical (chassis and drivetrain design and assembly) problems do not have anything in common with electrical or programming issues. What they do have in common is the process of understanding how the structure, circuit, or program should work, characterizing the problem, developing theories regarding what is actually happening, figuring out how to repair the problem, and testing your solution before finally applying the fix to the robot. The following set of actions may seem like a lot of work to fix a problem like a nut that has fallen off due to vibration, but if you follow it faithfully, the skills you gain by fixing the simple problems will make the more difficult ones a lot easier to solve and will prevent the simple ones from happening again.

39.2.1 DOCUMENTING THE EXPECTED STATE

Do you understand exactly what should be happening in your robot? Chances are you have a good idea of what the robot, or a part of the robot, should be doing at a given time but you have probably not looked in detail at what is actually happening. For a robot that has a failed glued plastic joint, you may have done an analysis of the forces on the joint when the robot is stationary, but have you looked at what happens during acceleration and deceleration? What about forces caused by vibration or large masses (such as batteries) shifting during operation? The forces during movement as well as changes in movement must be considered when looking at a mechanical failure.

Similarly, for an electrical problem, do you understand what the actual currents flowing through the circuitry are? Starting and stopping robot drive motors when they are under load will require greater currents than on a bench being tested out. Have you calculated the temperatures of different components during operation as well as their effect on components close to them? If the robot seems to miss detecting objects in front of it, have you put in some consideration for switch bouncing or changing fields of view during operation? Electrical problems can be especially vexing when you are using third-party designs or circuitry. To predict what should be happening, you should review the basic electrical laws and make sure you fully understand the basic electrical formulas and conventions.

Finally, for software, can you trace through the source code to understand what should be happening at any one particular time? How is the operation of the robot controller documented for different situations with varying inputs? A very important tool in understanding the operation of software is the simulator and how much time has been spent at understanding how the application should work. Many robot software applications are written quickly and debugged continuously to get the robot working as desired—this makes documenting the software a difficult and confusing chore unless you are very careful to keep track of different versions of software and the changes made to them. You will often find it easier to go back over the source code and try to map out how it is supposed to work and respond to different inputs.

Documenting the expected operation of the robot at the time of failure is a time-consuming task, but one that is critical to finding and ultimately fixing the problem. In many cases when you start understanding the operation of the robot at the level of detail needed to find and fix the problem, the reason for the problem will become apparent—but you should refrain from implementing the apparent fix until you have worked through the following five steps.

39.2.2 CHARACTERIZING THE PROBLEM

After documenting and becoming very familiar with what is supposed to be happening in the robot, you will spend some time setting up experiments to observe what is actually happening. The effort required for this is not trivial and will test your ingenuity to come up with different methods of observing what is happening while having a limited budget and resources for test equipment. Spending a few minutes thinking about the problem can result in some very innovative ways of observing the different aspects of the robot in operation and help guide you to the root cause of the problem.

You will find that some failures are intermittent; that is to say they will happen at seemingly random intervals. By characterizing the operation of the robot and comparing the results to the documented expected operation, you should find situations where the operating parameters are outside the design parameters, leading to the opportunity for failure either immediately or at some later time. Once you become familiar with documenting the expected operation of your robot as well as characterizing different robot problems, you'll discover that there really is no such thing as a random failure. Each failure mode has a unique set of parameters that will cause the failure and allow you to understand exactly what is happening.

The conditions leading up to a mechanical failure can be extremely difficult to observe on the basic robot. Plastic or cardboard arrows attached to different points in the robot's structure will help illustrate flexing that is not easily observed by the naked eye. A small cup of water can also be used to show the operating angle of different components of the robot as well as the acceleration of the robot during different circumstances. A digital camera's photograph of the robot in operation, with indicators such as arrows and cups of water will help you to observe deformations of the robot's structure and allow you to measure them by printing out the picture and measuring angles using a protractor.

When searching for electrical problems during the operation of the robot, your best friend is the LM339 quad comparator along with a few LEDs and potentiometers. The potentiometers are wired as voltage dividers and used to provide different extreme values for the different electrical parameters that are going to be measured (Fig. 39-1). When the robot exceeds one of these parameters, an LED wired to the LM339 comparator output will light. This allows you to easily observe any out-of-tolerance electrical conditions during robot operation, requiring just a few minutes of setup. Depending on how the robot is powered, you may have to add a separate power supply (a 9-V radio battery works well to allow a good range on the potentiometers) to the circuit in order to test it if you suspect the robot's power supply is sagging.

If a programming failure is suspected, you will discover that the best method of characterizing what is happening is by recording the inputs followed by the outputs. Again, LEDs are your best tool for observing what are the inputs causing the bad outputs. You can also use an LCD (although this will require you to stand over the robot to see exactly what is happening) or output a different sound or message when there are specific inputs to the microcontroller. Once you have the actual inputs and output commands, you can set up a state diagram (showing the changing inputs and outputs) to help you understand exactly what the program is doing in specific cases.

When coming up with methodologies for observing what is happening in the robot when the failure is taking place, remember Heisenberg's Uncertainty Principle, which states that the apparatus used for measuring a subatomic particle parameter will affect the actual measurement. This is very possible in robotics when you are trying to characterize a failure; often the equipment used to record the failure will end up changing the behavior of the robot, hiding the true nature of the problem. For example, adding an LCD to display the

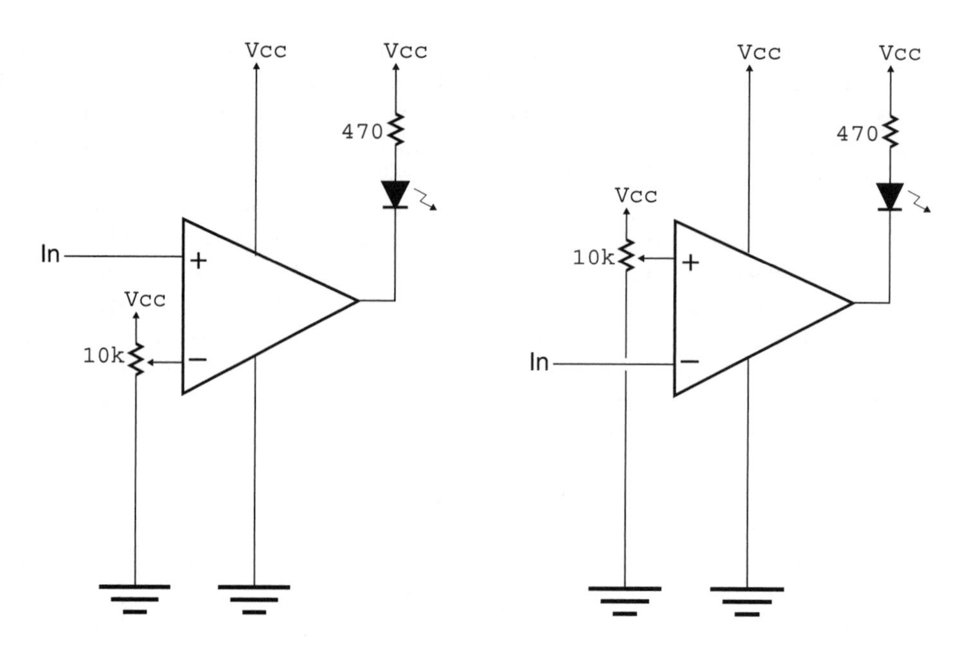

FIGURE 39-1 The LM339 comparator, along with a potentiometer, resistor, and LED are effective tools for monitoring voltage levels within a robot. Note that it may be necessary to have a separate power supply for the LM339 if the robot's power supply drops during operation.

inputs and outputs of the robot's microcontroller during operation may require you to stand over it to monitor the LCD's output, which may possibly result in you being detected by the robot and your presence causing the robot to behave differently than if you were further away from the robot.

39.2.3 HYPOTHESIZING ABOUT THE PROBLEM

With a clear understanding of how the robot should behave and how it is actually behaving, the differences should become very obvious and allow you to start making theories regarding what is the root cause of the problem. When you are hypothesizing about the problem, it is very important to (a) keep an open mind as to the cause of the problem and (b) avoid trying to come up with solutions, no matter how obvious they seem. It is easy to short-circuit this process and decide upon an obvious fix without working through the rest of the failure analysis.

Keeping an open mind is extremely difficult. To force yourself to look at different solutions, you should try to come up with at least three different possible root causes for the problem.

To illustrate this point, consider the case of a differentially driven robot with a light plastic frame that scrapes along the ground during changes in robot direction at the point where the batteries are mounted (see Fig. 39-2). As well as being scraped, the operation of the

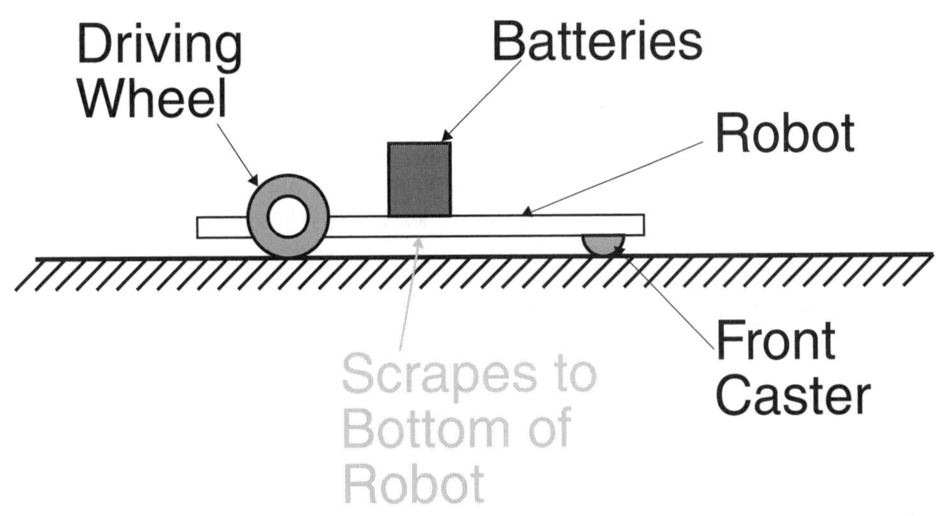

FIGURE 39-2 When the robot is sitting still, the battery pack does not touch the ground. But after operation, the chassis underneath the battery pack is scraped and the robot moves as expected.

robot seems to be erratic when the chassis comes into contact with the running surface. These observations are confirmed by photographing the robot during changes in direction.

With this information, you could make the following theories regarding the problems the robot is having:

1. The robot is accelerating too quickly, and the chassis is distorting during starting, stopping, and direction changing.
2. The inertia of the battery pack is causing the chassis to flex.
3. The motors are too powerful, and they are warping the chassis during startup or stopping.
4. The caster is digging into the running surface during operation, causing the chassis to distort.

A generic theory could be that the chassis isn't strong enough, but this will steer you toward a single solution (strengthening the chassis) while the four expanded theories give you a number of ideas to try and fix the problem.

Along with these four theories, you could probably come up with more that you can compare against the data that you collect in the first two steps and see which hypothesis best fits the data. There's a good chance that you will have to go back and look at different aspects of the robot; for example, if the caster was the problem, it should have some indications of high drag on a large part of its surface, not just the small area where it was in contact with the running surface.

39.2.4 PROPOSING CORRECTIVE ACTIONS

Once you are comfortable with understanding the different possible root causes of the failure, you can start listing out possible corrective actions. Like the multiple possible root causes listed in the previous step, you should also list out multiple possible corrective actions. While some may jump out at you, after considering different options, a much more elegant and easier to implement solution may become obvious.

In the previous section, it was mentioned that the obvious possible root cause of the robot scraping against the ground is that the chassis is simply not strong enough. The obvious solution to this problem is to strengthen the chassis.

While it may fix the problem, it is probably not the optimal solution, as strengthening the chassis could require you to effectively redesign and rebuild the robot. Before embarking on this large amount of effort, you could review a number of different solutions:

1. Strengthening the chassis.
2. Using a lighter battery pack.
3. Decreasing the robot's acceleration.
4. Relocating the battery pack.
5. Changing the front caster.
6. Using larger drive wheels.

The amount of work required for each of the different solutions varies. Along with documenting the amount of work for each solution, the potential cost and length of time needed to implement the solution can be documented in order to be able to choose the best possible fix.

39.2.5 TESTING FIXES

It is usually possible to quickly rig up a sample solution to test the effectiveness of a specific repair action before making it permanent. It is actually preferable to do this as it may become obvious that some repair actions are not going to be effective or are going to cause more problems than they solve. This is why multiple possible root cause problems are listed along with multiple repair actions for each root cause.

When testing the solution, don't spend a lot of time making it look polished; there's a good chance that it will not be as good as some other solution and you will end up having to tear down the fix and try another one. Don't be surprised if the most obvious cause of the problem and repair action aren't right; over time your ability to suggest the most effective fixes will improve, but when you are starting out, listing as many as possible and trying them all out will guide you to the most effective solution to your problem.

Finally, there's a chance that multiple corrective actions will produce the best result. For the example here, minimizing the distortion of the robot's chassis could be achieved by relocating the battery pack and using a lighter one. Testing multiple fixes together can result in even better solutions than if you were to doggedly look for a single, simple fix.

Remember to record the results of your tests (using the same testing apparatus as you used to characterize the problem). Being able to compare a difference in a robot will result in confidence that the optimal solution will be found.

39.2.6 IMPLEMENTING AND RELEASING THE SOLUTION

Once you have determined the "best" solution to the problem and implemented it, you should record it in a notebook for future use as chances are if you don't experience it again, you will experience something like it. Keeping notes listing what you did along with what you saw and expected to see will make the effort in documenting the expected state along with characterizing the actual state much easier.

Along with helping you with future problems, the notes will help you if the problem happens again (and there is a good chance that it will). In this case, you have a record of what has been tried and you can try something else.

Ideally the fix should be fairly simple; by using this process you will discover that problems you thought would require you to rebuild the robot can be resolved very easily. You will also find that after a while it takes you less time to work through the full failure analysis process than to perform a simple debug repair action. The added bonus is that your repairs will be a lot more reliable and less likely to break under extreme stress in the future.

39.3 From Here

To learn more about . . .	***Read***
Interfacing issues	Chapter 38, "Integrating the Blocks"
Mechanical structures	Chapter 3, "Structural Materials"
Electrical theory	Chapter 5, "Electrical Theory"
Programming operations	Chapter 13, "Programming Fundamentals"

CHAPTER 40

SETTING UP WORKSHOPS, DEMONSTRATIONS, AND COMPETITIONS

Robotics is an exciting and fun hobby and one that can attract a lot of interest by getting together with other hobbyists or club members and staging a public event where many people can watch and interact with the robots. The event could consist of different robots battling it out in various competitions, a chance for your club to demonstrate what its members' robots can do, or a workshop to give people a chance to experiment with robots on their own. Organizing an event can be a lot of fun and will bring some favorable publicity to you and your robotics club. Here are a few things to consider when putting together a show.

40.1 Choosing the Venue

The first order of business will be deciding where the event is going to take place. There are many different public locations to choose from, including schools, community centers, libraries, and shopping plazas. All these locations should have a room large enough to house four or five large "pit" area tables for work to be done on the robots along with a competition/demonstration area big enough for at least half of the robots to be operating at any one time. Chairs for a gallery should be arranged around the competition/demonstration area and a table for the judges and organizers should be placed in such a way that the entire room is visible. Fig. 40-1 shows a basic floor plan for just such an area.

This layout will allow the competitors/demonstrators the ability to perform last minute tweaks of their robots and give them ready access to the areas the robots will be running

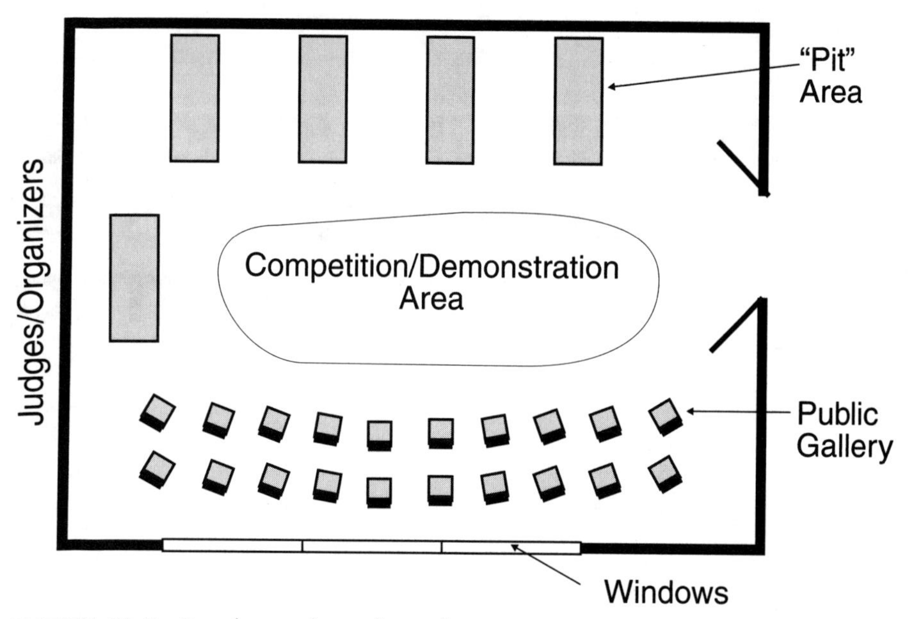

FIGURE 40-1 Sample room layout for a robot event.

in. It will also make it obvious for visitors where they should be at any given time and not get in the way of the competitors or potentially stepping on robots. Often rooms will have large glass areas to bring in the sun. These areas should be to the backs of the gallery so visitors are not squinting from the sun to see what is going on. Before arranging chairs and tables, a few minutes should be spent surveying the room and deciding where everything should go.

Along with taking time to think about where to place tables and chairs, you will also have to think about how to run power, network, public address and video wires and cables. Ideally, the cables should run around the perimeter of the room and be taped down to the edges of the floor to prevent any potential trip hazards. This is not always possible, so the cables should be located away from high-traffic areas and taped down with Gaffer's or duct tape.

If video displays are to be used in the room, then their screens should be located at a convenient location, away from where people will be traveling. The same goes for video projectors. Ideally they should be suspended from the ceiling, but if this is not possible, they should either be put on a raised platform or in a location where it is unlikely people will walk through.

40.1.1 VENUE NEEDS

Setting up the room is one thing, while a major headache will be ensuring that all the basic needs are met. These include paying rent for the room and any administrative/custodial costs of the venue, getting insurance for the event (if it is required), making sure washroom facilities are open and available, as well as providing the opportunity for everyone to get refreshments. A major concern is child management; often a parent that is interested in the event will bring along small children hoping to interest them in robotics. These concerns

can be surprisingly major and can result in having to go through several potential venues before finding the one that works best with your resources.

In today's environment of reduced funding for public facilities, you will probably find yourself having to rent the space. These costs can range from very nominal to very expensive. Along with this, a security deposit may be required to ensure that any damage will be paid for. Asking how much the room will cost should be your first question, even before how big is it and what days it is available.

Finding the money to pay for this can be difficult and may require a deposit up front and charging admission to visitors and taking a cut from selling refreshments. Club dues may be sufficient to pay for the venue, but this use should be agreed to well before the event. Fees can be minimized or avoided all together if the event is cosponsored by another group, such as a Boy Scout Troop, which already has the venue. In this case, the event could be primarily for the other group to observe and participate in.

Along with money for renting the space, you may have to arrange for liability insurance for everyone attending. The venue should have some arrangements already made that you can take advantage of in terms of getting a reduced rate as well as minimizing the amount of work required to arrange the insurance. Insurance should always be purchased if it is not provided as part of the rental of the venue; more than once somebody has broken a window moving chairs or tripped over a robot or sumo platform and injured themselves. The small headache of making sure you are protected will save you a much larger one if something happens later.

Biological needs should also be considered. Washrooms should be close by with no chance of anyone locking themselves in (or if they do, somebody with a key is available). Water, soda, doughnuts, and other snacks should also be available (proceeds can go to the robot club, the rental of the room, or prizes). The venue's management must be consulted on these issues to make sure that washrooms are available and to understand any rules they have regarding providing refreshments.

Finally, you should plan for younger children that are going to be bored with what is going on. While robots are generally fascinating for the general public, to toddlers and small children they are little more than animated toys that will become objects of frustration when they discover they can't pick them up or play with them. Some distractions that will make the event run smoothly include having a TV and a DVD player set up showing children's movies (animated movies are a good choice because they are engaging for a long period of time), and boxes of blocks or building materials should be available for the children's play and minimizing conflicts over who has a certain toy. Chances are that some of the people involved in the event will have small children, and they can help plan for the children's entertainment while the event is going on.

40.2 Competition Events

Competitions are very popular for the people taking part in the event as well as members of the public. In Table 40-1, a number of popular event types are listed along with web sites where you can get more information about them. For many types of events, the organizers will have to provide sumo rings, scales, size gauges, and so on.

In Table 40-1, you may have noticed that the FIRST ("For Inspiration and Recognition

TABLE 40-1 Different Robot Competitions

EVENT TYPE	WHERE TO FIND INFORMATION
Sumo-Bot	Robots are designed to search out and push another robot outside of a circular ring within a given amount of time. A set of rules for the different classes can be found at www.sorobotics.org/RoboMaxx/sumo-rules.html
Lego	Lego Mindstorms competitions involve teams that are given the task to design a robot that will compete in some area against other robots. Ideally, the teams are given the design constraints and are allowed to start building before visitors enter to avoid them spending time watching teams do nothing more than talk amongst themselves. Ideally, visitors should be allowed in when the robots are being tested before the competition begins.
Firefighting	Robots are given the task to find a candle in a house-like maze and extinguish it in the least amount of time. The Trinity College Fire Fighting Robot Competition is one of the most widely accepted, and their rules can be found at www.trincoll.edu/events/robot/Rules/default.asp
Line Following	Design a robot that can follow a meandering path accurately in the least amount of time. The typical track consists of a white surface with a black line about ¾ in wide. Curves should have no less than a 4 in radius.
Maze Following	Maze-following robots are very popular and instructive for competitors to build. Rules for maze-following robots are generally finding the robot that can negotiate a random maze in the least amount of time. There doesn't seem to be a standard for the size or complexity of the maze.
Combat Bots/Laser-Tag	Combat robots like the ones shown on television (e.g., *RobotWars*) are not reasonable for the open venues discussed in this book. But there are analogs to combat that can be performed that are very entertaining, such as laser-tag modified for robots. Circuitry for a type of laser-tag (called IR Tag) can be found at www.tabrobotkit.com
BEAM Robot Games	There are many different BEAM robot competitions available. Many of them are variations on the different competitions listed in this table. BEAM robots are generally inexpensive and easy to make, which makes them ideal as a way to get high school students interested in robotics. A list of BEAM robot competitions can be found at www.nis.lanl.gov/projects/robot//
Best . . .	Coming up with fun categories such as "best dressed" robot or "best robot dance" can be entertaining and fun for competitors and observers alike. This is an excellent way to introduce robotics to children and get them to start thinking about how robots work and are built.

of Science and Technology," www.usfirst.com) competition and other organized competitions are not listed. These competitions are generally large, centrally organized affairs that would be difficult to stage at a local or small level. This does not mean if you are part of a FIRST team that you cannot set up demonstrations of your robot, but you should recognize that it will be difficult for you to set up your own FIRST robot competition.

It should go without saying, especially after the discussions regarding insurance and liability, that the competitions should be safe for the audience, competitors, and the venue they are taking place in. You should watch for robots that break the rules in ways that could result in injury or property damage such as a sumo-bot that has a powerful flipper for its competition, a LEGO robot that has a spinning action which could end up throwing parts, or a laser-guided robot that can shine laser light into somebody's eyes. You should make sure to note in the competition entry form that any robot felt to be unsafe will not be allowed to run or compete.

40.2.1 SCROUNGING FOR PRIZES

Coming up with ideas for prizes for a robot competition is something that requires a great deal of imagination and perseverance. Poor prizes will be received graciously, but well thought out prizes will really excite the competitors and have them coming back for more. Really, there is no such thing as a bad prize and the winners are sure to walk away happy no matter what they get, but the perfect prize will not only be enticing for future events but be an excellent advertisement for your group in the future.

Imagination is required to come up with appropriate prizes for different people and their different skill, education, and resources. Having said this there are a number of different prizes that are always well appreciated. Tools that are appropriate for robot building such as rotary (Dremel brand) cutters, thermostatically controlled soldering irons, microcontroller development kits, and, ironically, robot kits themselves are always appreciated. Toys that mimic robots on TV or movies are good for a chuckle and will be valued for being more than they are worth. Homemade plaques and trophies, especially made from robot parts, are probably the most special type of prizes and ones that will be a source of pride for the winners for years to come.

Of course, the authors' books are considered to be the most special prizes that a competitor can receive.

Having chosen the prizes, you will then have the task of figuring out how to pay for them. Again, many methods for paying for the venue can be used for raising money for prizes. Local retailers and manufacturers can be approached for donations of tools and equipment but care must be taken to ensure that you do not appear greedy or unwilling to work with the supplier and their corporate guidelines for giving.

40.3 Alerting the Public and the Media

With the planning complete, the venue chosen and all the arrangements made, you will want to let people know that the event is going to take place. Today the obvious method for disseminating this information is either through email or updating your group's web page.

Unfortunately, these methods are probably not going to be as effective as you would like as the only people that will see these announcements are those who are actively involved in robotics. Ironically, you will have to use more traditional methods of communications.

Flyers should be left at community centers, libraries, schools, and local bulletin boards. A phone number and a web site should be available for people to find out more about the event as well as allow them to print out a page they can put up in their home to remind themselves of the event. You might want to ask people to register to come to the event to give you some idea of how many people will be coming.

Community newspapers will often mention your event free of charge and may send a reporter/photographer (they are usually one and the same) to cover the event. They will usually require an email notifying them of the event. You should first contact the newspaper to learn the procedure for submitting your event (including learning the lead time for publishing on a specific date) as well as how reporters are assigned to stories. Most newspapers' assignment editors task reporters to different events that morning. To maximize the chance that a reporter will be sent to your event, make sure that you send a media alert email to the assignment editor the day before so it is in their pile of events to cover. The same procedure should be followed for local TV stations.

Remember that it is not unusual to not have reporters come to your event, no matter how hard you work to make sure the newspapers are aware of it. Community events like yours are given to reporters based on whether they will be in the area, and if it is possible for them to make it to the event, and whether a "big story" is happening.

40.4 From Here

So far in this book the "From Here" sections of each chapter have been introspective—pointing you to different chapters of the book to get information reinforcing what is discussed in the chapter or giving you additional background information needed to work through the topics covered.

Having reached the end of the book it is really time for you to go "From Here" and start experimenting with your own robot.

Remember that the first goal is to have fun, and the second goal is to learn. Chances are your first attempt at designing and building your own robot will be long and arduous and will require you to perform a lot of redesigning, rebuilding, and reprogramming to get it to work exactly the way you want it to. When the robot starts working and running about on its own, all the hard work that took place to get you there will become totally worth it.

Try not to feel discouraged that it took longer than you expected or that you weren't as smart or as efficient as you thought you were; instead celebrate that you have built a robot. It will get easier as you build additional robots and you should feel proud that you have done something that only a small fraction of humanity has accomplished.

The authors look forward to hearing from you, seeing your creations, and we hope that if we meet in competition, your robot won't be better than ours.

APPENDIX A

Further Reading

Interested in learning more about robotics? These books are available at most better bookstores, as well as at many online bookstores, including Amazon (www.amazon.com), Barnes and Noble (www.bn.com), and Fatbrain (www.fatbrain.com).

This appendix also lists several magazines of interest to the robot experimenter. Both mailing and Internet addresses have been provided.

Contents

Interfacing to Computer Systems
Magazines
Classic Robot Fiction

A.1 Hobby Robotics

Build Your Own Robot!
Karl Lunt, A K Peters, ISBN: 1568811020

Robots, Androids, and Animatrons: 12 Incredible Projects You Can Build
John Iovine, McGraw-Hill, ISBN: 0070328048

The Personal Robot Navigator
Merl K. Miller, Nelson B. Winkless, Kent Phelps, Joseph H. Bosworth, A K Peters Ltd., ISBN: 188819300X (contains CD-ROM of robot navigation simulator)

123 Robotics Experiments for the Evil Genius
Myke Predko, McGraw-Hill, ISBN: 0071413596 (contains PCB for mounting a BS2/robot electronics)

Applied Robotics
Edwin Wise, Howard W Sams & Co, ISBN: 0790611848

Muscle Wires Project Book
Roger G. Gilbertson, Mondo-Tronics, ISBN: 1879896141

Stiquito: Advanced Experiments with a Simple and Inexpensive Robot
James M. Conrad, Jonathan W. Mills Institute of Electrical and Electronic Engineers, ISBN: 0818674083

Stiquito for Beginners: An Introduction to Robotics
James M. Conrad, Jonathan W. Mills IEEE Computer Society Press, ISBN: 0818675144

A.2 LEGO Robotics and LEGO Building

Dave Baum's Definitive Guide to LEGO Mindstorms
Dave Baum, Apress, ISBN: 1893115097

Unofficial Guide to LEGO MINDSTORMS Robots
Jonathan B. Knudsen, O'Reilly & Associates, ISBN: 1565926927

Joe Nagata's Lego Mindstorms Idea Book
No Starch Press, ISBN: 1886411409

LEGO Crazy Action Contraptions
Dan Rathjen, Klutz, Inc., ISBN: 1570541574

Extreme Mindstorms: An Advanced Guide to LEGO Mindstorms
Dave Baum et al., Apress, ISBN: 1893115097

A.3 Technical Robotics, Theory, and Design

Mobile Robots: Inspiration to Implementation
Joseph L. Jones, Anita M. Flynn, Bruce A. Seiger, A K Peters Ltd., ISBN: 1568810970

Sensors for Mobile Robots: Theory and Application
H. R. Everett, A K Peters Ltd., ISBN: 1568810482

Art Robotics: An Introduction to Engineering
Fred Martin, Prentice Hall, ISBN: 0805343369

Robot Evolution: The Development of Anthrobotics
Mark Rosheim, John Wiley & Sons, ISBN: 0471026220

Machines That Walk: The Adaptive Suspension Vehicle
Shin-Min Song, MIT Press, ISBN: 0262192748

Robot DNA, Various Titles
McGraw-Hill

Remote Control Robotics
Craig Sayers, Springer-Verlag, ISBN: 038798597

Artificial Vision for Mobile Robots: Stereo Vision and Multisensory Perception
Nicholas Ayache, Peter T. Sander, MIT Press, ISBN: 0262011247

A.4 Artificial Intelligence and Behavior-Based Robotics

An Introduction to AI Robotics
Robin Murphy, MIT Press, ISBN: 0262522632

Robot: Mere Machine to Transcendent Mind
Hans Moravec, Oxford University Press, ISBN: 0195116305

Behavior-Based Robotics: Intelligent Robots and Autonomous Agents
Ronald C. Arkin, MIT Press, ISBN: 0262011654

Artificial Intelligence and Mobile Robots: Case Studies of Successful Robot Systems
David Kortenkamp, R. Peter Bonasso, Robin Murphy, MIT Press, ISBN: 0262611376

Cambrian Intelligence: The Early History of the New AI
Rodney Allen Brooks, MIT Press, ISBN: 0262522632

Vehicles: Experiments in Synthetic Psychology
Valentino Braitenberg, MIT Press, ISBN: 0262521121

Intelligent Behavior in Animals and Robots
David McFarland, Thomas Bosser, MIT Press, ISBN: 0262132931

A.5 Mechanical Design

Five Hundred and Seven Mechanical Movements
Henry Brown, Astragal Press, ISBN: 1879335638

Mechanical Devices for the Electronics Experimenter
Britt Rorabaugh, Tab Books, ISBN: 0070535477

Mechanisms and Mechanical Devices Sourcebook, Second Edition
Nicholas P. Chironis, Neil Sclater, McGraw-Hill, ISBN: 0070113564

Home Machinist's Handbook
Doug Briney, Tab Books, ISBN: 0830615733

A.6 Electronic Components

Electronic Circuit Guidebook: Sensors
Joseph J. Carr, PROMPT Publications, ISBN: 0790610981

Electronic Circuit Guidebook (various volumes)
Joseph J. Carr, PROMPT Publications,

Volume 1: Sensors; ISBN: 0790610981
Volume 2: IC Timers; ISBN: 0790611066
Volume 3: Op Amps; ISBN: 0790611317

Build Your Own Low-Cost Data Acquisition and Display Devices
Jeffrey Hirst Johnson, Tab Books, ISBN: 0830643486

A.7 Microcontroller/Microprocessor Programming and Interfacing

123 PIC Microcontroller Experiments for the Evil Genius
Myke Predko, McGaw-Hill, ISBN: 0071451420

Programming and Customizing the BASIC Stamp Computer
Scott Edwards, McGraw-Hill, ISBN: 0079136842

Microcontroller Projects with BASIC Stamps
Al Williams, R&D Books, ISBN: 0879305878

The BASIC Stamp 2: Tutorial and Applications
Peter H. Anderson (author and publisher), ISBN: 0965335763

Programming and Customizing the Picmicro Microcontroller
Myke Predko, McGraw-Hill, ISBN: 0071361723

Design with Pic Microcontrollers
John B. Peatman, Prentice Hall, ISBN: 0137592590

Microcontroller Cookbook
Mike James, Butterworth-Heinemann, ISBN: 0750627018

Handbook of Microcontrollers
Myke Predko, McGraw-Hill, ISBN: 0079137164

Programming and Customizing the 8051 Microcontroller
Myke Predko, McGraw-Hill, ISBN: 0071341927

The 8051 Microcontroller
I. Scott MacKenzie, Prentice Hall, ISBN: 0137800088

The Microcontroller Idea Book
Jan Axelson, Lakeview Research, ISBN: 0965081907

Programming and Customizing the Hc11 Microcontroller
Thomas Fox, McGraw-Hill Professional Publishing, ISBN: 0071344063

AVR RISC Microcontroller Handbook
Claus Kuhnel Newnes ISBN: 0750699639

A.8 Electronics How-To and Theory

Teach Yourself Electricity and Electronics
Stan Gibilisco, TAB Books, ISBN: 0071377301

McGraw-Hill Benchtop Electronics Handbook
Victor Veley, McGraw-Hill, ISBN: 0070674965

The TAB Electronics Guide to Understanding Electricity and Electronics
G. Randy Slone, Tab Books, ISBN: 0070582165

Electronic Components: A Complete Reference for Project Builders
Delton T. Horn, Tab Books, ISBN: 0830633332

The Forrest Mims Engineer's Notebook
Forrest M. Mims, Harry L. Helms LLH Technology Pub, ISBN: 1878707035
Engineer's Mini-Notebook (series)
Forrest M. Mims, Radio Shack

Logicworks 4: Interactive Circuit Design Software for Windows and Macintosh
Addison-Wesley, ISBN: 0201326825 (book and CD-ROM; includes software)

Beginner's Guide to Reading Schematics
Robert J. Traister, Anna L. Lisk, Tab Books, ISBN: 0830676325

Printed Circuit Board Materials Handbook
Martin W. Jawitz, McGraw-Hill, ISBN: 0070324883

The Art of Electronics
Paul Horowitz, Winfield Hill, Cambridge University Press, ISBN: 0521370957

Student Manual for the Art of Electronics
Paul Horowitz, T. Hayes, Cambridge University Press, ISBN: 0521377099

A.9 Power Supply Design and Construction

DC Power Supplies
Joseph J. Carr, McGraw-Hill, ISBN: 007011496X

Power Supplies, Switching Regulators, Inverters, and Converters
Irving M. Gottlieb, Tab Books, ISBN: 0830644040

Motors and Motor Control: Electric Motors and Control Techniques, Second Edition
Irving M. Gottlieb, ISBN: 0070240124

A.10 Lasers and Fiber Optics

Lasers, Ray Guns, and Light Cannons: Projects from the Wizard's Workbench
Gordon McComb, McGraw-Hill, ISBN: 0070450358

Optoelectronics, Fiber Optics, and Laser Cookbook
Thomas Petruzzellis, McGraw-Hill, ISBN: 0070498407

Understanding Fiber Optics
Jeff Hecht, Prentice Hall, ISBN: 0139561455

Laser: Light of a Million Uses
Jeff Hecht, Dick Teresi, Dover, ISBN: 0486401936

A.11 Interfacing to Computer Systems

Use of a PC Printer Port for Control & Data Acquisition
Peter H. Anderson (author and publisher), ISBN: 0965335704

The Parallel Port Manual Vol. 2: Use of a PC Printer Port for Control and Data Acquisition
Peter H. Anderson (author and publisher), ISBN: 0965335755

Programming the Parallel Port
Dhananjay V. Gadre, R&D Books, ISBN: 0879305134

PC PhD: Inside PC Interfacing
Myke Predko, Tab Books, ISBN: 0071341862

PDA Robotics
Doug Williams, McGraw-Hill, ISBN: 0071417419

Real-World Interfacing with Your PC
James Barbarello, PROMPT Publications, ISBN: 0790611457

A.12 Magazines

Robot Science and Technology
3875 Taylor Road, Suite 200, Loomis, CA 95650
www.robotmag.com

Circuit Cellar Magazine
4 Park St., Vernon, Ct 06066
www.circuitcellar.com/

Servo Magazine
430 Princeland Court, Corona, CA 91719
www.servomagazine.com/

Nuts & Volts Magazine
430 Princeland Court, Corona, CA 91719
www.nutsvolts.com

Everyday Practical Electronics
Wimborne Publishing Ltd.
Allen House East Borough, Wimborne Dorset BH2 1PF United Kingdom

Elektor
www.elektor-electronics.co.uk

A.13 Classic Robot Fiction

I, Robot
Isaac Asimov

Berzerker
Fred Saberhagen

Do Androids Dream of Electric Sheep?
Philip K. Dick

Tek War
William Shatner

The Hitchhiker's Guide to the Galaxy
Douglas Adams

APPENDIX B

Sources

Contents

Note: The listing in this appendix is periodically updated at www.robotoid.com.

Internet-based companies that do not provide a mailing address on their web site are not listed. In addition, Internet-based companies hosted on a free web-hosting service (Tripod, Geocities, etc.) are also not listed because of fraud concerns.

B.1 Selected Specialty Parts and Sources

BEAM Robots
Solarbotics

Bend Sensor
Images Company

Infrared Proximity/Distance Sensors
Acroname
HVW Technologies

Infrared Passive (PIR) Sensors
Acroname
Glolab

LCD Serial Controller
Scott Edwards Electronics

Microcontroller Kits and Boards
DonTronics
microEngineering Labs
Milford Instruments
NetMedia
Parallax, Inc.
Savage Innovations
Scott Edwards Electronics, Inc.

Motor Controllers ("Set and Forget")
Solutions Cubed

Servo Motor Controller
FerretTronis
Lynxmotion
Medonis Engineering
Mister Computer
Pontech
Scott Edwards Electronics, Inc.

Shape-Memory Alloy
Mondo-Tronics

Sonar Sensors (Polaroid and Others)
Acroname

Speech Recognition
Images Company

Surplus Mechanical Parts and Electronic Components
Active Surplus
All Electronics
Alltronics
American Science & Surplus
B.G. Micro
C&H Sales
Halted Specialties Co.
Herbach & Rademan
Martin P. Jones & Assoc.

Wireless Transmitters (RF and Infrared)
Abacom Technologies
Glolab

B.2 General Robotics Kits and Parts

Acroname, P.O. Box 1894, Nederland, CO 80466 (303) 258-3161
www.acroname.com

Abacom Technologies, 32 Blair Athol Crescent, Etobicoke, Ontario M9A 1X5 Canada (416) 236-3858
www.abacom-tech.com

A.K. Peters, Ltd.
63 South Avenue, Natick, MA 01760 (508) 655-9933
www.akpeters.com

Amazon Electronics, Box 21, Columbiana, OH 44408 (888) 549-3749
www.electronics123.com

Design and Technology Index, 40 Wellington Road, Orpington, Kent, BR5 4AQ UK +44 0 1689 876880
www.technologyindex.com

Images Company, 39 Seneca Loop, Staten Island, NY 10314 (718) 698-8305
www.imagesco.com

Glolab Corp, 134 Van Voorhis, Wappingers Falls, NY 12590
www.glolab.com

HVW Technologies, Suite 473, 300-8120 Beddington Blvd., SW Calgary, Alberta T3K 2A8 Canada (403) 730-8603
www.hvwtech.com

Hyperbot, 905 South Springer Road, Los Altos, CA 94024-4833 (800) 865-7631 (415) 949-2566
www.hyperbot.com

Lynxmotion, Inc.
104 Partridge Road, Pekin, IL 61554-1403 (309) 382-1816
www.lynxmotion.com

TAB Electronics
www.tabrobotkit.com

Mekatronix, 316 Northwest 17th Street, Suite A, Gainesville, FL 32603
www.mekatronix.com

Milford Instruments, 120 High Street, South Milford, Leeds LS25 5AG UK +44 0 1977 683665
www.milinst.demon.co.uk

Mondo-Tronics, Inc., 4286 Redwood Highway, #226, San Rafael, CA 94903 (415) 491-4600
www.robotstore.com

Mr. Robot, 8822 Trevillian Road, Richmond, VA 23235 (804) 272-5752
www.mrrobot.com

Norland Research, 8475 Lisa Lane, Las Vegas, NV 89113 (702) 263-7932
www.smallrobot.com

Personal Robot Technologies, Inc.
P.O. Box 612, Pittsfield, MA 01202 (800) 769-0418
www.smartrobots.com

RobotKitsDirect, 17141 Kingview Avenue, Carson, CA 90746 (310) 515-6800 voice
www.owirobot.com

Sensory Inc., 521 East Weddell Drive, Sunnyvale, CA 94089-2164 (408) 744-9000
www.sensoryinc.com

Solarbotics, 179 Harvest Glen Way, Northeast Calgary, Alberta, T3K 3J4 Canada (403) 818-3374
www.solarbotics.com

Technology Education Index, 40 Wellington Road, Orpington, Kent, BR5 4AQ UK +44 0 1689 876880
www.technologyindex.com

Zagros Robotics, P.O. Box 460342, St. Louis, MO 63146-7342 (314) 176-1328
www.zagrosrobotics.com

B.3 Electronics/Mechanical: New, Used, and Surplus

Active Surplus
345 Queen Street West, Toronto, Ontario, Canada, M5V 2A4, (416) 593-0909
www.activesurplus.com

All Electronics, P.O. Box 567, Van Nuys, CA 91408-0567 (800) 826-5432
www.allectronics.com

Alltech Electronics, 2618 Temple Heights, Oceanside, CA 92056 (760) 724-2404
www.allelec.com

Alltronics, 2300-D Zanker Road, San Jose, CA 95101-1114 (408) 943-9773
www.alltronics.com

American Science & Surplus, 5316 North Milwaukee Avenue, Chicago, IL 60630 (847) 982-0870
www.sciplus.com

B.G. Micro, 555 North 5th Street, Suite #125, Garland, TX 75040 (800) 276-2206
www.bgmicro.com

C&H Sales, 2176 East Colorado Boulevard, Pasadena, CA 91107 (800) 325-9465
www.candhsales.com

DigiKey Corp., 701 Brooks Avenue South, Thief River Falls, MN 56701 (800) 344-4539
www.digikey.com

Edmund Scientific, 101 East Gloucester Pike, Barrington, NJ 08007-1380 (800) 728-6999
www.edsci.com

Electro Mavin, 2985 East Harcourt Street, Compton, CA 90221 (800) 421-2442
www.mavin.com

Electronic Goldmine, P.O. Box 5408, Scottsdale, AZ 85261 (480) 451-7454
www.goldmine-elec.com

Fair Radio Sales, 1016 East Eureka Street, P.O. Box 1105, Lima, OH 45802 (419) 227-6573
www.fairradio.com

Gates Rubber Company, 900 South Broadway, Denver, CO 80217-5887 (303) 744-1911
www.gates.com

Gateway Electronics, 8123 Page Boulevard, St. Louis, MO 63130 (314) 427-6116
www.gatewayelex.com

General Science & Engineering, P.O. Box 447, Rochester, NY 14603 (716) 338-7001
www.gse-science-eng.com

W. W. Grainger, Inc., 100 Grainger Parkway, Lake Forest, IL 60045-5201
www.grainger.com

Halted Specialties Co.
3500 Ryder Street, Santa Clara, CA 96051 (800) 442-5833
www.halted.com

Herbach and Rademan, 16 Roland Avenue, Mt. Laurel, NJ 08054-1012 (800) 848-8001
www.herbach.com

Hi-Tech Sales, Inc., 134R Route 1 South Newbury St., Peabody, MA 01960 (978) 536-2000
www.bnfe.com

Hosfelt Electronics, 2700 Sunset Boulevard, Steubenville, OH 43952 (888) 264-6464
www.hosfelt.com

Jameco, 1355 Shoreway Road, Belmont, CA 94002 (800) 536-4316
www.jameco.com

JDR Microdevices, 1850 South 10th Street, San Jose, CA 95112-4108 (800) 538-5000
www.jdr.com

Marlin P. Jones & Associates, Inc., P.O. Box 12685, Lake Park, FL 33403-0685 (800) 652-6733
www.mpja.com

MCM Electronics, 650 Congress Park Drive, Centerville, OH 45459 (800) 543-4330
www.mcmelectronics.com

McMaster-Carr, P.O. Box 740100, Atlanta, GA 30374-0100 (404) 346-7000
www.mcmaster.com

Mouser Electronics, 958 North Main Street, Mansfield, TX 76063 (800) 346-6873
www.mouser.com

PIC Design, 86 Benson Road, Middlebury, CT 06762 (800) 243-6125
www.pic-design.com

Scott Edwards Electronics Inc., 1939 South Frontage Road, Sierra Vista, AZ 85634 (520) 459-4802
www.seetron.com

Small Parts, Inc., 13980 Northwest 58th Court, P.O. Box 4650, Miami Lakes, FL 33014-0650 (800) 220-4242
www.smallparts.com

Supremetronic, Inc.
333 Queen Street West, Toronto, Ontario, Canada, M5V 2A4, (416) 598-9585
www.supretronic.com

Surplus Traders, P.O. Box 276, Alburg, VT 05440 (514) 739-9328
www.73.com

TimeLine, Inc., 2539 West 237 Street, Building F, Torrance, CA 90505 (310) 784-5488
www.digisys.net/timeline/

Unicorn Electronics, 1142 State Route 18, Aliquippa, PA 15001 (800) 824-3432
www.unicornelectronics.com

W.M. Berg, Inc.
499 Ocean Avenue, East Rockaway, NY 11518 (516) 599-5010
www.wmberg.com

B.4 Microcontrollers, Single-Board Computers, Programmers

Boondog Automation, 414 West 120th Street, Suite 207, New York, NY 10027
www.boondog.com/

DonTronics, P.O. Box 595, Tullamarine, 3043 Australia (check web site for phone numbers)
www.dontronics.com

Gleason Research, P.O. Box 1494, Concord, MA 01742-1494 (978) 287-4170
www.gleasonresearch.com

Kanda Systems, Ltd., Unit 17–18 Glanyrafon Enterprise Park, Aberystwyth, Credigion SY23 3JQ UK +44 0 1970 621030
www.kanda.com

microEngineering Labs, Inc., Box 7532, Colorado Springs, CO 80933 (719) 520-5323
www.melabs.com

MicroMint, Inc., 902 Waterway Place, Longwood, FL 32750 (800) 635-3355
www.micromint.com

NetMedia (BasicX), 10940 North Stallard Place, Tucson, AZ 85737 (520) 544-4567
www.basicx.com

Parallax, Inc., 3805 Atherton Road, Suite 102, Rocklin, CA 95765 (888) 512-1024
www.parallaxinc.com

Protean Logic, 11170 Flatiron Drive, Lafayette, CO 80026 (303) 828-9156
www.protean-logic.com

Savage Innovations (OOPic), 2060 Sunlake Boulevard #1308, Huntsville, AL 35824 (603) 691-7688 (fax)
www.oopic.com

Technological Arts, 26 Scollard Street, Toronto, Ontario, Canada M5R 1E9 (416) 963-8996
www.technologicalarts.com

Weeder Technologies, P.O. Box 2426, Fort Walton Beach, FL 32549 (850) 863-5723

Wilke Technology GmbH Krefelder 147 D-52070 Aachen, Germany +49 (241) 918 900
www.wilke-technology.com

Z-World, 2900 Spafford Street, Davis, CA 95616 (530) 757-3737
www.zworld.com/

B.5 Radio Control (R/C) Retailers

Tower Hobbies, P.O. Box 9078, Champaign, IL 61826-9078 (800) 637-6050 (217) 398-3636
www.towerhobbies.com

B.6 Servo and Stepper Motors, Controllers

Effective Engineering, 9932 Mesa Rim Road, Suite B, San Diego, CA 92121 (858) 450-1024
www.effecteng.com

FerretTronics, P.O. Box 89304, Tucson, AZ 85752-9304
www.FerretTronics.com

Hitec RCD Inc., 12115 Paine Street, Poway, CA 92064
www.hitecrcd.com

Medonis Engineering, P.O. Box 6521, Santa Rosa, CA 95406-0521
www.medonis.com

Mister Computer, P.O. Box 600824, San Diego, CA 92160 (619) 281-2091
www.mister-computer.com

Pontech (877) 385-9286
www.pontech.com

Solutions Cubed, 3029 Esplanade, Suite F, Chico, CA 95973 (530) 891-8045
www.solutions-cubed.com

Vantec, 460 Casa Real Plaza, Nipomo, CA 93444 (888) 929-5055
www.vantec.com

B.7 Ready-Made Personal and Educational Robots

ActiveMedia Robotics, 44–46 Concord Street, Peterborough, NH 03458 (603) 924-9100
www.activrobots.com

Advanced Design, Inc., 6052 North Oracle Road, Tucson, AZ 85704 (520) 575-0703
www.robix.com

Arrick Robotics, P.O. Box 1574, Hurst, TX 76053 (817) 571-4528
www.robotics.com

General Robotics Corporation, 1978 South Garrison Street, #6, Lakewood, CO 80227-2243 (800) 422-4265

www.edurobot.com, Newton Research Labs, Inc., 4140 Lind Avenue Southwest, Renton, WA 98055 (425) 251-9600
www.newtonlabs.com

Probotics, Inc., Suite 223, 700 River Avenue, Pittsburgh, PA 15212 (888) 550-7658
www.personalrobots.com

B.8 Construction Kits, Toys, and Parts

Valient Technologies (Inventa), Valiant House, 3 Grange Mills Weir Road, London SW12 0NE UK +44 020 8673 2233
www.valiant-technology.com

B.9 Miscellaneous

Meredith Instruments, P.O. Box 1724, 5420 West Camelback Rd., #4, Glendale, AZ 85301 (800) 722-0392
www.mi-lasers.com

Midwest Laser Products, P.O. Box 262, Frankfort, IL 60423 (815) 464-0085
www.midwewst-laser.com

Synergetics, P.O. Box 809, Thatcher, AZ 85552 (520) 428-4073
www.tinaja.com

Techniks, Inc., P.O. Box 463, Ringoes, NJ 08551 (908) 788-8249
www.techniks.com

APPENDIX C

Robot Information on the Internet

Contents

C.1 Electronics Manufacturers

Analog Devices, Inc.
www.analog.com/

Atmel Corp.
www.atmel.com/

Dallas Semiconductor
www.dalsemi.com/

Infineon (Siemens)
www.infineon.com/

Microchip Technology
www.microchip.com/

Motorola Microcontroller
www.mcu.motsps.com/

Precision Navigation
www.precisionnav.com/

Sharp Optoelectronics
www.sharp.co.jp/ecg/data.html

Xicor
www.xicor.com/

C.2 Shape-Memory Alloy

www.toki.co.jp/BioMetal/index.html

www.toki.co.jp/MicroRobot/index.html

C.3 Microcontroller Design

Peter H. Anderson—Embedded Processor Control
www.phanderson.com/

"No-Parts" PIC Programmer
www.CovingtonInnovations.com/noppp/index.html

Iguana Labs
www.proaxis.com/~iguanalabs/tools.htm

LOSA—List of Stamp Applications
www.hth.com/losa/

Myke Predko's Microcontroller Reference
www.myke.com/

PICmicro Web Ring
http://members.tripod.com/~mdileo/pmring.html

Shaun's BASIC Stamp 2 Page
www.geocities.com/SiliconValley/Orchard/6633/index.html

C.4 Robotics User Groups

Seattle Robotics Society
www.seattlerobotics.org/

Yahoo Robotics Clubs
http://clubs.yahoo.com/clubs/theroboticsclub
http://search.clubs.yahoo.com/search/clubs?p_robotics The Robot Group
www.robotgroup.org/

Robot Builders
www.robotbuilders.com/

B-9 Builder's Club
http://members.xoom.com/b9club/index.htm

San Francisco Robotics Society
www.robots.org/

Nashua Robot Club
www.tiac.net/users/bigqueue/others/robot/homepage.htm

Mobile Robots Group
www.dai.ed.ac.uk/groups/mrg/MRG.html

Dallas Personal Robotics Group
www.dprg.org/

Portland Area Robotics Society
www.rdrop.com/~marvin/

C.5 General Robotics Information

Robotics Frequently Asked Questions
www.frc.ri.cmu.edu/robotics-faq/

Legged Robot Builder
http://joinme.net/robotwise/

Tomi Engdahl's Electronics Info Page
www.hut.fi/Misc/Electronics/

Boondog Automation Tutorials
www.boondog.com\tutorials\tutorials.htm

Find Chips Search
www.findchips.com/

Robotics Resources
www.eg3.com/ee/robotics.htm

Robotics Reference
http://members.tripod.com/RoBoJRR/reference.htm

Bomb Disposal Robot Resource List
www.mae.carleton.ca/~cenglish/bomb/bomb.html

Introduction to Robot Building
www.geckosystems.com/robotics/basic.html

Robot Building Information, Hints, and Tips
www.seattlerobotics.org/guide/extra_stuff.html

Suppliers for Robotics/Control Models and Accessories
http://mag-nify.educ.monash.edu.au/measure/robotres.htm

Mobile Robot Navigation
http://rvl.www.ecn.purdue.edu/RVL/mobile-robot-nav/mobile-robot-nav.html

Robota Dolls
www-robotics.usc.edu/~billard/poupees.html

BASIC Stamp, Microchip Pic, and 8051 Microcontroller Projects
www.rentron.com/

TAB Electronics Build Your Own Robot Kit Resource Page
www.tabrobotkit.com

Hila Research QBasic
http://fox.nstn.ca/~hila/qbasic/qbasic.html

Dennis Clark's Robotics
www.verinet.com/~dic/botlinks.htm

Dissecting a Polaroid Pronto One-Step Sonar Camera
www.robotprojects.com/sonar/scd.htm

Polaroid Sonar Application Note
www.robotics.com/arobot/sonar.html

General Robot Info
www.employees.org:80/~dsavage/other/index.html

Standard Technologies of the Seattle Robotics Society
www.nwlink.com/~kevinro/guide/

Tech Wizards
www.hompro.com/techkids/

Android Workshop
www.tgn.net/~texpanda/library.htm

Hacking RAD Robot
www.netusa1.net/~carterb/radrobot.html

BEAM Robotics
http://nis-www.lanl.gov/robot/

C.6 Books, Literature, and Magazines

Robotics Book Reviews
www.weyrich.com/book_reviews/robotics_index.html

Robot Books
www.robotbooks.com/

Lindsay Publications
www.lindsaybks.com/

Circuit Cellar
www.circellar.com/

Midnight Engineering
www.midengr.com/

Robotics Bookstore
www.bectec.com/html/bookstore.html

Robohoo
www.robohoo.com/

C.7 Surplus Resources

Silicon Valley Surplus Sources
www.kce.com/junk.htm

C.8 Commercial Robots

Electrolux Vacuum Robot
www.electrolux.se/robot/meny.html

Gecko Systems Carebot
www.geckosystems.com/

iRobot
www.irobot.com

IS Robotics
www.isr.com/

Nomadic Technologies
www.robots.com/products.htm

C.9 Video Cameras

Logitech QuickCam
www.quickcam.com/

C.10 Ultrasonic Range Finders

Ultrasonic Imaging Project
http://business.netcom.co.uk/iceni/usi_project/

Interfacing Polaroid Sonar Board
www.cs.umd.edu/users/musliner/sonar/

C.11 LEGO Mindstorms Sources on the Web

LEGO Mindstorms home page
www.legomindstorms.com/

LEGO Mindstorms Internals
www.crynwr.com/lego-robotics/ RCX Software Developer's Kit (from LEGO)
www.legomindstorms.com/sdk/index.html

RCX Internals
http://graphics.stanford.edu/~kekoa/rcx/

RCX Tools
http://graphics.stanford.edu/~kekoa/rcx/tools.html

Scout Internals (from LEGO)
www.legomindstorms.com/products/rds/hackers.asp

LEGO Dacta (educational arm of LEGO)
www.lego.com/dacta/

Pitsco (educational second sourcing for LEGO)
www.pitsco-legodacta.com/

LUGNET Newsgroups (technical LEGO discussion boards; robotics group is largest)
www.lugnet.com/

NQC (Not Quite C); (popular alternative programming environment for RCX)
www.enteract.com/~dbaum/nqc/

Gordon's Brick Programmer
www.umbra.demon.co.uk/gbp.html

LEGO on My Mind
http://homepages.svc.fcj.hvu.nl/brok//LEGOmind/

Mindstorms Add-Ons
http://www-control.eng.cam.ac.uk/sc10003/addon.html

MindStorms RCX Sensor Input
www.plazaearth.com/usr/gasperi/lego.htm

C.12 Servo and Stepper Motor Information

R/C Servo Fundamentals
www.seattlerobotics.org/guide/servos.html

Modifying R/C Servos to Full Rotation
www.seattlerobotics.org/guide/servohack.html

Definitive Guide to Stepper Motors
www.cs.uiowa.edu/~jones/step/index.html

Servo-Motor 101
www.repairfaq.org/filipg/RC/F_Servo101.html

Dual Axis Stepper Motor Controller
http://members.aol.com/drowesmi/dastep.html

Using the Allegro 5804 Stepping Motor Controller/Translator
www.phanderson.com/printer/5804.html

C.13 Quick Turn Mechanical and Electronics Parts Manufacturers

Mechanical Prototypes and Parts
www.emachineshop.com/

Quick Turn PCB Prototypes
www.apcircuits.com/

Quick Turn PCB Prototypes
www.pcbexpress.com/

INDEX

ABOUT THE AUTHORS

GORDON MCCOMB is an avid electronics hobbyist who has written for TAB Books for a number of years. He wrote the best-selling *Troubleshooting and Repairing VCRs* (now in its third edition), *Gordon McComb's Gadgeteer's Goldmine,* and *Lasers, Ray Guns, and Light Cannons.*

MYKE PREDKO has 20 years of experience in the design, manufacturing, and testing of electronic circuits. An experienced author, Myke wrote McGraw-Hill's best-selling *123 Robotics Projects for the Evil Genius; 123 PIC Microcontroller Experiments for the Evil Genius; PIC Microcontroller Pocket Reference; Programming and Customizing PIC Microcontrollers, Second Edition; Programming Robot Controllers;* and others, and is the principal designer of both *TAB Electronics Build Your Own Robot Kits.*